FACULTY OF ENGINEERING

UNIVERSITY OF SHEFFIELD

Phase Diagrams of Ternary Gold Alloys by A Prince, G V Raynor and D S Evans

This book has been generously donated to the Main Library of the University of Sheffield by Professor Alan Prince to commemorate the award of the Degree of Doctor of Metallurgy (DMet)

May 1991

Phase Diagrams of Ternary Gold Alloys

The information in this publication is the result of effort by the International Programme for Alloy Phase Diagram Data, co-ordinated by The Alloy Phase Diagram International Commission, APDIC. The role of The Commission is to set overall objectives for the International Programme to meet the needs of the worldwide technical community, to establish assessment priorities in relation to the technological requirements of industry and commerce, and to set quality standards for assessments.

APDIC currently consists of representatives from the following organizations:

ASM INTERNATIONAL, USA
Deutsche Gesellschaft für Metallkunde, FRG
Groupe de Thermodynamique et Diagrammes de Phases, France
The Indian Institute of Metals, India
The Institute of Metals, UK
Japanese Committee for Alloy Phase Diagrams, Japan
Max-Planck-Institut für Metallforschung, Stuttgart, FRG
National Institute of Standards and Technology, USA
The Royal Institute of Technology, Sweden
CODATA

The phase diagram evaluations in this publication have been reviewed by noted authorities on behalf of The Institute of Metals and are presented as the most accurate and reliable information available on the subject.

Phase Diagrams of Ternary Gold Alloys

A PRINCE, G V RAYNOR AND D S EVANS

THE INSTITUTE OF METALS
1990

Book Number 294

Published in 1990 by The Institute of Metals
1 Carlton House Terrace, London SW1Y 5DB

and
The Institute of Metals
North American Publications Center
Old Post Road, Brookfield, VT 05036
U S A

Typeset by Fakenham Photosetting, Fakenham.
Illustrations supplied by the authors

Printed in Great Britain

British Library Cataloguing in Publication Data

applied for

CN/ I S B N 0-904357-50-3

Contents

Ternary Systems

Relevant Binary Systems

Appendices

Foreword

Phase diagrams are fundamental to metallurgy and materials science in terms of the extraction of metals, their processing to marketable products and their end usage, and a huge wealth of phase diagram data for a wide range of alloys is available in the literature. However, the results of individual studies are often incomplete or inconsistent, and are usually in a form that cannot be conveniently applied by the practioner. For these reasons there is a need to compile critically assessed phase diagram data for binary and multicomponent alloy systems on a systematic and planned basis, thus enabling the technological requirements of industry and commerce to be met more efficiently.

This monograph presents in one volume critical assessments of phase diagram data for gold-based ternary systems. It is an achievement by three exceptional scientists dedicated to the provision of phase diagram data for the purposes of academic and scientific research with the overall aim of application to industrial purposes, and results from research carried out meticulously and to a particularly high standard. The assessments have been made by the late Professor Geoffrey V. Raynor, the late Dr Dain S. Evans and Professor Alan Prince, and form part of the programme of The Institute of Metals' Phase Diagram Committee of which Geoffrey Raynor was first Chairman. Alan Prince is currently Chairman.

The task of critically assessing phase diagram data is immense, and requires international collaboration. The present volume forms part of a Monograph Series on Binary and Ternary Phase Diagrams resulting from the International Programme for Alloy Phase Diagram Data, now co-ordinated by The Alloy Phase Diagram International Commission (APDIC). APDIC comprises representatives of the following organizations: ASM INTERNATIONAL, The Indian Institute of Metals, The Institute of Metals, Max-Planck-Institut für Metallforschung, Stuttgart, National Institute of Standards and Technology, Deutsche Gesellschaft für Metallkunde, Groupe de Thermodynamique et Diagrammes de Phases, Japanese Committee for Alloy Phase Diagrams, The Royal Institute of Technology and CODATA.

The critical assessment of a large body of literature is a difficult and demanding task. Geoffrey Raynor was internationally renowned in the field of alloy constitution. Dain Evans excelled with brilliance as an experimentalist, and possessed a natural flair for the application of phase diagrams to practical purposes. Alan Prince, the doyen of phase diagram assessment, sets standards that are almost impossible for others to achieve. This combination of authors has provided us with a work of immense value.

Tim G. Chart
Chairman
Alloy Phase Diagram
International Commission

Introduction

This monograph forms a part of an International Programme for Alloy Phase Diagram Data, originally proposed by ASM INTERNATIONAL and the National Institute of Standards and Technology (formerly the National Bureau of Standards), and now co-ordinated by the Alloy Phase Diagram International Commission (APDIC). The programme aims to provide critically assessed phase diagram data for binary and multicomponent alloy systems on a systematic and planned basis. Currently publications from the programme include monographs on the phase diagrams of binary gold, beryllium, magnesium, titanium and vanadium alloys and monographs on the phase diagrams of ternary silver, iron (in four parts) and copper–oxygen–metal alloys. In addition ASM INTERNATIONAL have published a compilation of 'Binary Alloy Phase Diagrams' which includes current data from the programme as well as non-programme data. The well-known volumes by W. B. Pearson on the crystal structures of alloy phases have been updated and published by ASM INTERNATIONAL as 'Pearson's Handbook of Crystallographic Data for Intermetallic Phases'.

The task of critically assessing phase diagram data is onerous. Work on this monograph on the phase diagrams of ternary gold alloys started in 1981. Twenty-seven ternary systems were completed by Professor G. V. Raynor before his death in 1983. As fate would have it Professor Raynor died in the month when he had planned to begin increasing his contribution to this monograph. My other co-author, Dr D. S. Evans, was seconded by The General Electric Company to the Next European Torus team at Gartching, FRG in August 1986. He was a believer in the maxim that science is measurement and I was very fortunate to have had a long association with such a brilliant experimentalist. Dr Dain Evans died on 31 May 1989 after a long illness. Incorporated in this monograph are the results of experimental phase diagram work that we undertook at the Hirst Research Centre of GEC to clear up discrepancies in the literature or to establish new phase diagrams. All this work was founded on the incomparable expertise of my friend, the late Dr Dain Evans. As the surviving author I have had the duty of completing the monograph, updating where necessary, and assembling the whole for publication. The assessments presented are based on our best judgement of the available data. The interpretation of the equilibria are our own and these will need to be modified as new data enters the literature. Errors and omissions should be laid at my door. It is intended to update this monograph with new data and publications missed in preparing this work, at an appropriate future date. Readers are invited to submit comments and contributions for the next edition of this monograph.

Many organizations and individuals contribute to the completion of monographs such as this. In listing those to whom we are indebted we express our apologies for any inadvertent omissions. Without the financial support from the International Gold Corporation (now the World Gold Council), The General Electric Company plc, Hirst Research Centre, Johnson-Matthey Research Centre, Sonning Common, Engelhard Limited, Chessington and The Max-Planck Institut für Metallforschung, Institut für Wekstoffwissenschaften, Stuttgart, FRG this monograph would not have materialized. Their support is gratefully acknowledged. We have also had the inestimable benefit of receiving from Dr H. Okamoto and Professor T. B. Massalski their critical evaluations of gold binary phase diagrams before publication by ASM INTERNATIONAL. This lightened our task greatly and we are very appreciative of this altruistic co-operation. Experimental work on gold ternary systems initiated by us at The General Electric Company plc, Hirst Research Centre and at Brunel, The

University of West London, and by Dr B. Legendre at the University of Paris (South) and both experimental and thermodynamic calculation of gold ternary systems initiated by Dr F. Hayes at UMIST, Dr I. Ansara at the University of Grenoble and by Professor J. P. Bros and Dr M. Gaune-Escard at the University of Provence has been freely given to us prior to publication. To all these friends we extend our grateful thanks. Dr B. Gather of the University of Osnabruck collaborated with us on the evaluation of the Au–Pb–Se, Au–Se–Sn and Au–Se–Te systems whilst a visiting scientist at the Hirst Research Centre and Dr C. Gumiński of the University of Warsaw has kindly contributed evaluations of the Au–Bi–Hg and Au–Cd–Hg systems.

Many former colleagues of the Hirst Research Centre have played an essential role in our work. The majority of the text was prepared by Mrs Gill Youngman and Mrs Pam Mcaree. Drawings of the ternary phase diagrams were made by Mrs Amanda Rising in the early years and then by Mrs Jean Nunn. Pattie Dossett and Mrs Sylvie Nobes tracked down copies of the innumerable publications. All these ladies are sincerely thanked for their devoted effort. The director of the Hirst Research Centre, Dr S. L. Cundy, authorized the continuing preparation of drawings and other facilities after I retired from GEC in September 1987. This action was symptomatic of the generous support given by my former employer throughout the long gestation period of this monograph.

A great deal of help has been given by the Phase Diagram Committee of the Institute of Metals and by many friends on the staff of ASM INTERNATIONAL.

Thanks are due to the authors of the original publications. Without their dedicated labours we would not have the rich literature we now possess. The referees of the manuscript are thanked for their cogent comments which eliminated many ambiguities and errors. The staff of the Institute of Metals, in particular Mr Keith Wakelam and his colleagues, are blessed for their extreme patience in awaiting a final manuscript and then using their professional skills to bring this monograph to life.

Finally I would thank my daughter, Mrs Christine Maltby, who drew most of the binary diagrams and my wife Sheila for her support, tolerance and understanding over many years.

Alan Prince

Information for the reader

Ternary gold alloy systems form the basis of this monograph. Where sufficient ternary data exists to allow the construction of a phase diagram, especially the ternary liquidus projection, it is essential to include the relevant binary phase diagrams in order for the reader to understand the corresponding ternary equilibria. Rather than repeat the binary phase diagrams throughout the ternary text it was decided to gather all the relevant binary phase diagrams into one section. All compositions, both binary and ternary, are expressed in atomic percent. We decided against doubling the number of figures by also providing diagrams in weight (mass) percent. We also did not wish to increase the density of data on ternary diagrams by superimposing a weight percent scale along the binary edges. For those who wish to convert from atomic percent to weight percent (and vice versa), a table of atomic weights is given in Appendix II and a reminder of the necessary calculations for the conversion of compositions in Appendix III. The expression of compositions in mole fractions or mole percent is sometimes used when studying the equilibria between two binary phases such as AB_2 and AC. Appendix IV reminds the reader how to convert from mole fractions to atomic percent.

All temperatures are quoted in degrees Celsius. A table of melting points and transition points for the elements is given in Appendix I. These data have been taken from a compilation by Dr Alan Dinsdale of the National Physical Laboratory on behalf of SGTE (Scientific Group Thermodata Europe) dated 4 November 1988.

References are given in the text in the form of the last two numbers of the year of publication followed by the first three letters of the name of the (first) author. For example the text refers to [81 Pet] and the list of references at the end of each assessment is chronological with details of the publication of [81 Pet] in the usual format. Where the same author has published more than one paper in any year the text distinguishes the individual publications in terms of [81 Pet 1], [81 Pet 2] etc. For publications in the 19th century the year of publication is quoted in full in the text, i.e. [1891 Hey]. This is a system preferred to the usual sequential numbering of references.

The binary phase diagrams that are being critically evaluated under the auspices of APDIC contain information on metastable phases and, whenever possible, attempts are made to carry out a thermodynamic assessment of the phase boundaries, or to report on thermodynamically calculated equilibria contained in the literature. In considering the basis for the critical assessment of ternary phase diagrams it was recognized at an early stage that, to achieve the desired aim of encompassing all ternary data, it was necessary to exclude metastable phases. Published thermodynamic evaluations of ternary systems are included but no attempt has been made to generate ternary phase diagrams by new thermodynamic calculations.

Tie triangles in isothermal sections of ternary diagrams have been shaded. It is hoped that this will be of assistance to the reader in understanding such sections.

A standard nomenclature has been adopted for the major ternary invariant reactions involving a liquid phase. As is common practice the reaction of $L \rightleftharpoons \alpha + \beta + \gamma$ is called a ternary eutectic reaction and the liquid composition is denoted by a capital E; the reaction $L + \alpha \rightleftharpoons \beta + \gamma$ is called a ternary transition reaction and the liquid composition is denoted by a capital U; the reaction $L + \alpha + \beta \rightleftharpoons \gamma$ is called a ternary peritectic reaction and the liquid composition is denoted by a capital P.

Acknowledgements

Grateful thanks is made to the following for permission to reproduce binary phase diagrams.

McGraw-Hill Book Company.

M. Hansen and K. Anderko, CONSTITUTION OF BINARY ALLOYS (1958).

Figures 8, p. 14 (Ag–Cd); 31, p. 50 (Ag–Se); 40, p. 63 (Ag–Zn); 96, p. 165 (As–Ga); 104, p. 177 (As–S); 192, p. 333 (Bi–Sb); 195, p. 337 (Bi–Sn); 250, p. 438 (Cd–Sb); 351, p. 613 (Cu–Pd); 353, p. 617 (Cu–Pt); 527, p. 951 (Mn–Sb); 613, p. 1131 (Pd–Zn); 617, p. 1136 (Pt–Re); 618, p. 1137 (Pt–Rh); 633, p. 1173 (Sb–Se).

R. P. Elliott, CONSTITUTION OF BINARY ALLOYS, FIRST SUPPLEMENT (1965).

Figures 105, p. 201 (Bi–Te); 152, p. 292 (Cd–Tl); 366, p. 729 (Pd–Rh).

F. A. Shunk, CONSTITUTION OF BINARY ALLOYS, SECOND SUPPLEMENT (1969).

Figures 18, p. 62 (As–Se); 58, p. 215 (Cd–In); 107, p. 366 (Ga–Sn); 190, p. 615 (Pd–W).

Elsevier Science Publishing Co., Inc.

MATERIALS RESEARCH SOCIETY SYMPOSIA PROCEEDINGS, ALLOY PHASE DIAGRAMS, Volume 19 (1983). Eds: L. H. Bennett, T. B. Massalski and B. C. Giessen.

Figure 1, p. 388 (Au–Pb); Figure 3, p. 394 (In–Sn).

Elevier Sequoia.

A. Palenzona and S. Cirafici, Journal of the Less-Common Metals, *143* (1988), 167–171.

Figure 1, p. 169 (Au–U).

Pergamon Press PLC.

H. Ipser and R. Krachler, Scripta Metallurgica, *22* (1988), 1651–1654.

Figure 1, p. 1654 (Au–Zn).

The Metallurgical Society, Warrendale, PA 15086, USA.

J. L. Murray, Metallurgical Transactions A, *15* (1984), 261–268.

Figure 1, p. 264 (Ag–Cu).

Ying-Yu Chuang, Ker-Chang Hsieh and Y. Austin Chang, Metallurgical Transactions A, *17* (1986), 1373–1380.

Figure 6, p. 1377 (Fe–Ni).

Ying-Yu Chuang, Y. Austin Chang, Rainer Schmid and Jen-Chwen Lin, Metallurgical Transactions A, *17* (1986), 1361–1372.

Figure 6, p. 1369 (Fe–Ni).

Springer-Verlag, Heidelberg.

Ortrud Kubaschewski, IRON-BINARY PHASE DIAGRAMS (1982).

Figure 52, p. 89 (Fe–Pd).

International Atomic Energy Agency.

MOLYBDENUM: PHYSICO-CHEMICAL PROPERTIES OF ITS COMPOUNDS AND ALLOYS. Atomic Energy Review Special Issue No. 7 (1980). Eds: O. Kubaschewski, L. Brewer.

Figure II–35, p. 296 (Mo–Pd).

Elsevier Science Publishers, Physical Sciences and Engineering Division.

S. Bordas, M. T. Clavaguera-Mora, B. Legendre and Chhay Hancheng, Thermochimica Acta, *107* (1986), 239–265.

Figure 3, p. 242 (Sb–Te).

Ternary Systems

Ag–Al–Au

The only publication is a 500°C isothermal section, Fig. 1, determined by examination of twelve alloys using the X-ray diffraction technique [70 Fra]. No ternary phases were found. At 500°C all the binary compounds (Table 1) except Ag_3Al enter into the equilibria. There is small solubility of Ag in Au_5Al_2, Au_2Al and AuAl, or of Au in Ag_2Al. The compound $AuAl_2$ dissolves some 10 at.-% Ag and the single-phase region based on Au_4Al stretches across nearly the whole system (note the similarity in crystal structure of the compounds Au_4Al and Ag_3Al). The Au_2Al phase at 500°C consists of two low-temperature phases, αAu_2Al and βAu_2Al [74 Pus, 87 Mur]. No account was taken of this by Frank and Schubert [70 Fra]. In view of the small number of alloys studied by them [70 Fra], Fig. 1 should be regarded as indicative of the 500°C isothermal section of the Ag–Al–Au ternary system.

Table 1. Crystal structures

Solid phase	*Prototype*	*Lattice designation*
(Ag/Au)	Cu	cF4
(Al)	Cu	cF4
$AuAl_2$	CaF_2	cF12
AuAl	Monoclinically distorted MnP	mP8
γAu_2Al	$MoSi_2$	tI6
Au_5Al_2	Cu_5Zn_8	cI52
Au_4Al high-temperature	W	cI2
Au_4Al low-temperature	β–Mn	cP20
Ag_2Al	Mg	hP2
Ag_3Al high-temperature	W	cI2
Ag_3Al low-temperature	β–Mn	cP20

REFERENCES

70 Fra: K. Frank and K. Schubert: *J. Less-Common Metals*, 1970, **22**, 349–354

74 Pus: M. Puselj and K. Schubert: *J. Less-Common Metals*, 1974, **35**, 259–266

87 Mur: J. L. Murray, H. Okamoto and T. B. Massalski: *Bull. Alloy Phase Diagrams*, 1987, **8**, 20–30, 102

Ag–Au–Cd

Published work has concentrated on the ordering reactions between AgCd and Au–Cd, and no data exists that gives an overall view of the ternary equilibria. [62 Rot] established the presence of a Heusler-type phase, $AgAuCd_2$, by X-ray diffraction analysis and residual resistivity measurements. The residual resistivity data were obtained for the alloy series $Ag_{50-x}Au_xCd_{50}$ and $Ag_{52\cdot5-x}Au_xCd_{47\cdot5}$. A parabolic relation between residual resistivity and at.-% Ag was observed, except for a pronounced minimum in the region of equal Ag and Au contents. This is suggestive of ordering. Resistivity-temperature data for an alloy containing 30 Ag, 20 Au, 50 at.-% Cd indicated two order–disorder transitions. The first transition, at 201°C, is due to the disordering of the Ag and Au atoms on the Ag–Au sublattice of the Heusler-type structure. The second transition, at 551°C, is due to the disordering of the Cd sublattice. At 201°C the Heusler-type lattice transforms to an ordered CsCl-type lattice; at 551°C the latter transforms to a disordered bcc lattice. The transformation from an ordered CsCl-type lattice to a disordered bcc lattice occurs for alloys containing up to 10 at.-% Ag in the section $Ag_{50-x}Au_xCd_{50}$. With Ag >20 at.-% Ag the ordered CsCl-type lattice is stable to the liquidus, as it is reported to be in the β binary AuCd phase. X-ray diffraction analysis showed that alloys containing 20, 25 and 30 at.-% Ag on the $Ag_{50-x}Au_xCd_{50}$ section had the Heusler-type structure at room temperature; alloys containing <10 at.-% Ag and >40 at.-% Ag had ordered CsCl-type structure. Experiments were conducted at room temperature on alloy filings that had been sealed in evacuated pyrex tubes, annealed for 1 h at 350°C and cooled slowly. On the section $Ag_{52-x}Au_xCd_{47\cdot5}$ alloys containing 20, 25 and 32 at.-% Ag (scaled from original graph) had the Heusler-type structure; alloys with ≤10 at.-% Ag and ≥40 at.-% Ag had the ordered CsCl-type structure.

A paper concerned with the presence of Heusler-type structures in $AgAuZn_2$ and $AuCuZn_2$ [66 Mul 2] quotes a critical temperature of 250°C for the transition from Heusler to ordered CsCl-type structure in $AgAuCd_2$. A melting point of 650°C is also quoted (see Table II of [66 Mul 2]) for the composition $AgAuCd_2$. Muldawer [66 Mul 1] gave a lattice constant of 0·66462 nm for $AgAuCd_2$ of composition 24·82 Ag, 25·11 Au, 50·07 at.-% Cd. Cold-working $AgAuCd_2$ reduces the ternary order and produces either a fcc or hcp martensitic phase as well. Long anneals at room temperature or brief anneals at a relatively low temperature restore the ternary ordered structure and eliminate the martensitic phase.

Nakishi *et al.* [78 Nak] used differential scanning calorimetry to study the transition from the ordered CsCl to the Heusler-type structure in seven alloys on the $Ag_{52\cdot5-x}Au_xCd_{47\cdot5}$ section. Alloys were prepared by melting 99·99% pure elements in evacuated quartz tubes followed by annealing for 72 h at 720°C. If the annealing temperature were 720°C it would be anticipated that almost all the alloys would be molten. The alloys were cooled from 720°C to −100°C over a 36 h period. The ordered CsCl to Heusler-type structure transition was found to be only slightly dependent on Ag content in the alloys examined (transition temperature 235°C, 20 at.-% Ag; 237°C, 23 Ag; 233°C, 26·25 Ag; 227°C, 28 Ag; 221°C 30 Ag; 223°C, 35 Ag; 219°C, 40 Ag). This data is at variance with Rothwarf and Muldawer [62 Rot] who found that alloys on this section containing ≥40 at.-% Ag had the ordered CsCl-type structure at room temperature. The lattice parameter of the equiatomic Ag–Au alloy in the slowly cooled condition agreed with the data of Muldawer [66 Mul 1].

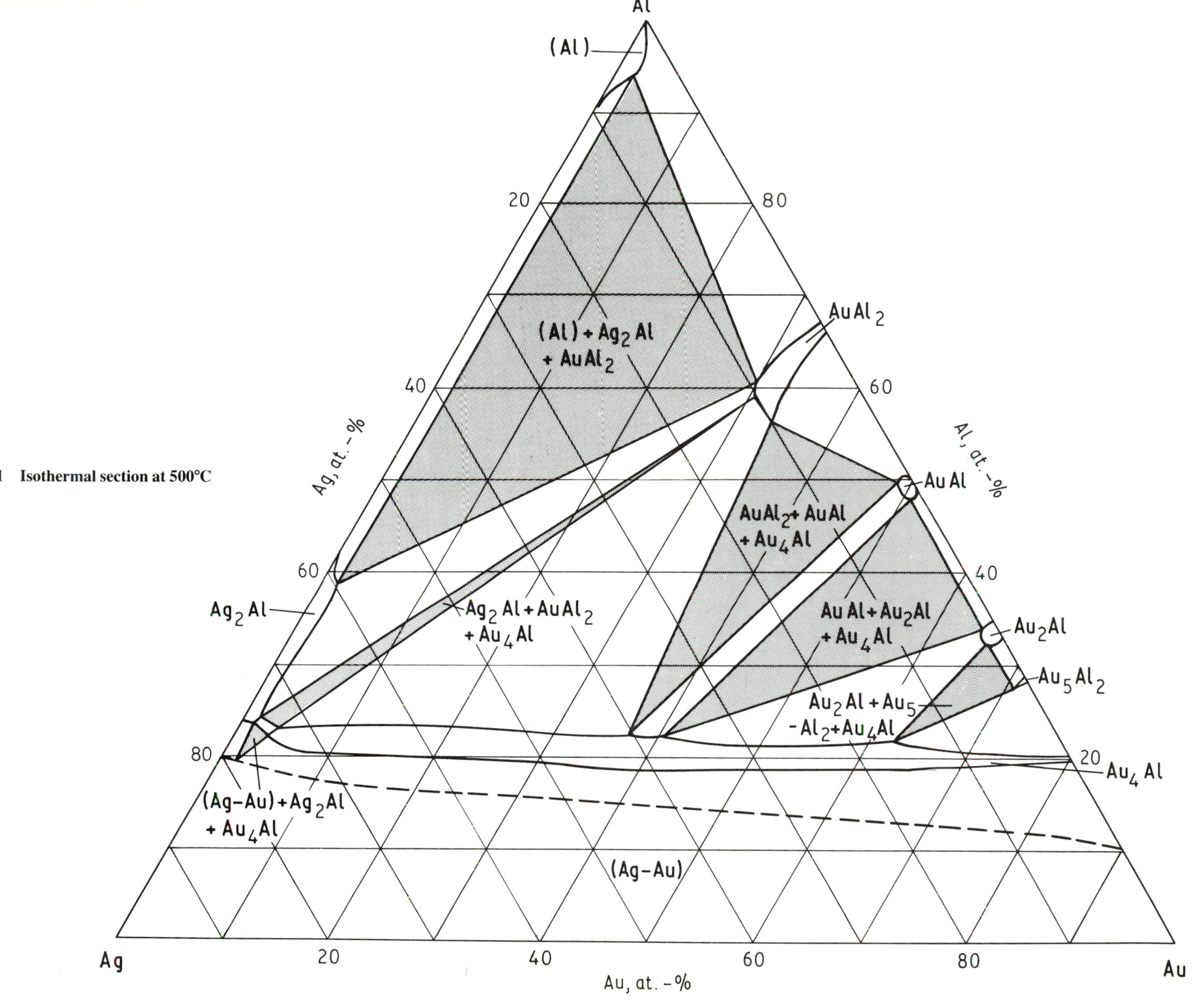

1 Isothermal section at 500°C

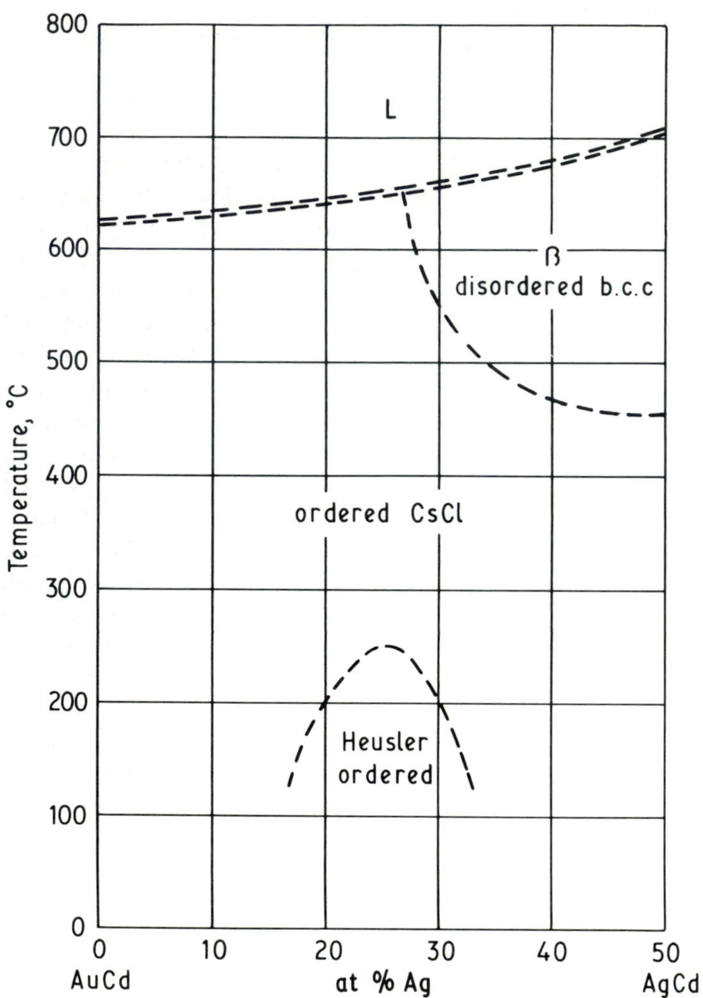

2 The AgCd–AuCd section (schematic)

Figure 2 presents a schematic sketch of the AgCd–AuCd ordered phase regions, based on the work of Rothwarf and Muldawer [62 Rot, 66 Mul 2]. It does not take into account the presence of the ζ phase in the Ag–Cd system. It may be anticipated that this phase will have limited compositional stability in the AgCd–AuCd section. The $Ag_{50-x}Au_xCd_{50}$ section is similar to the $Ag_{50-x}Au_xZn_{50}$ section [71 Mur].

Table 2. Crystal structures for the $Ag_{50-x}Au_xCd_{50}$ section

Solid phase	*Prototype*	*Lattice designation*
β′AgCd	CsCl	cP2
ζAgCd	Mg	hP2
βAgCd	W	cI2
βAuCd	CsCl	cI2
*$AgAuCd_2$	$AlCu_2Mn$	cF16

REFERENCES

62 Rot: F. Rothwarf and L. Muldawer: *J. Appl. Phys.*, 1962, **33**, 2531–2538

66 Mul 1: L. Muldawer: *Acta Cryst allogr.*, 1966, **20**, 594–595

66 Mul 2: L. Muldawer: *J. Appl. Phys.*, 1966, **37**, 2062–2066

71 Mur: Y. Murakami, S. Kachi, N, Nakanishi and H. Takehara: *Acta Metall.*, 1971, **19**, 97–105

78 Nak: N. Nakishi, M. Takano, H. Morimoto, S. Miura and F. Hori: *Scripta Metall.*, 1978, **12**, 79–83

Ag–Au–Co

INTRODUCTION

The only publication [58 Gri] on this ternary system is somewhat sketchy, but it provides sufficient data to allow a presentation of the probable equilibria. The effects of immiscibility between Ag and Co in both the liquid state and the solid state dominate the ternary phase equilibria. A liquid immiscibility gap originates at the Ag–Co binary edge and projects into the ternary system to approximately 45 at.-% Au (Fig. 3). The eutectic reaction in the Au–Co binary system is shown as running from 996·5°C (l = (Au) + (Co)), towards the Ag corner where the monovariant curve presumably ends at a degenerate eutectic reaction close to the melting point of Ag. At room temperature all alloys are essentially two-phase mixtures of (Ag/Au) and (Co) with little mutual solid solubility between the two phases.

BINARY SYSTEMS

The Ag–Co binary system shows almost complete immiscibility in the liquid and solid state (Hansen). The Ag–Au system, assessed by Okamoto and Massalski [83 Oka], is accepted. Their assessment of the Au–Co system [85 Oka] is accepted.

SOLID PHASES

No binary or ternary compounds exist. The equilibria are exclusively between the Ag–Au solid solution, denoted (Ag/Au) and (Co).

TERNARY EQUILIBRIA

Grigoriev *et al.* [58 Gri] studied 18 ternary alloys, whose compositions are plotted in Fig. 3. Alloys were prepared from the metallic components containing <0·01% impurity by melting under a layer of barium chloride. Twelve alloys were thermally analysed using cooling runs only; [58 Gri] present liquidus temperatures for 10 of the 12 alloys. Secondary arrests and the end of solidification gave weak thermal effects and no data was included for these events. The compositions of the alloys studied have been converted to atomic percent and plotted in Fig. 3 together with the liquidus temperatures. Liquidus temperatures for the binary Ag–Au and Au–Co alloys are also included in Fig. 3. [58 Gri] plot a monovariant eutectic curve as a dotted estimate of its course, but this does not conform with the vertical sections they present. Indeed, inspection of the liquidus data in Fig. 3 indicates that it is not reliable. One is probably justified in assuming that the liquid immiscibility gap in the ternary system ends in a critical tie line where l_1/l_2 = (Co), but the temperature of the critical tie line is not known. The monovariant eutectic curve e_1e_2 has associated with it a three-phase region corresponding to equilibrium of l + (Ag/Au) + (Co). This three-phase region can be expected to stretch to the degenerate reaction at about 963°C in the Ag–Co binary system. All 10 alloys for which liquidus temperatures are quoted (Fig. 3), should pass through the three-phase region on cooling, and they would therefore be expected to show thermal effects for secondary separation in the temperature range from 996°C, e_1, to 963°C, e_2. The liquidus data [58 Gri] for alloys 6 (973°C), 10 (985°C), and 11 (970°C) are considered to be more appropriate to secondary separations, rather than to represent primary separation at the liquidus temperature. The course of curve e_1e_2 (Fig. 3), is speculative.

Grigoriev *et al.* [58 Gri] also measured the Brinell hardness of alloys 1, 3, 5, 6, and 8–12 after annealing in a vacuum at 800°C for 100 h followed by slow cooling. As would be anticipated, for a constant Au content the hardness of the alloys increased with increasing Co content. This reflects the increasing proportion of (Co) phase in the (Ag/Au) + (Co) microstructure. The electrical resistivity of alloys 1, 5, 6, and 8–12 were measured after annealing samples as for the hardness study. The electrical resistivity passed through a maximum for the 90, 80 and 70 wt-% Au sections in going from the Ag–Au alloys through the ternary alloys to the Au–Co alloys.

Metallographic sections from alloys 1, 8, 10, 11, and 12 show primary (Ag/Au) present in alloy 1 and primary (Co) present in the other alloys. The significant result is the appearance of primary (Co) in alloy 8. This is justification for plotting the estimated course of curve e_1e_2 to the Ag–Au–rich side of the composition of alloy 8 (Fig. 3).

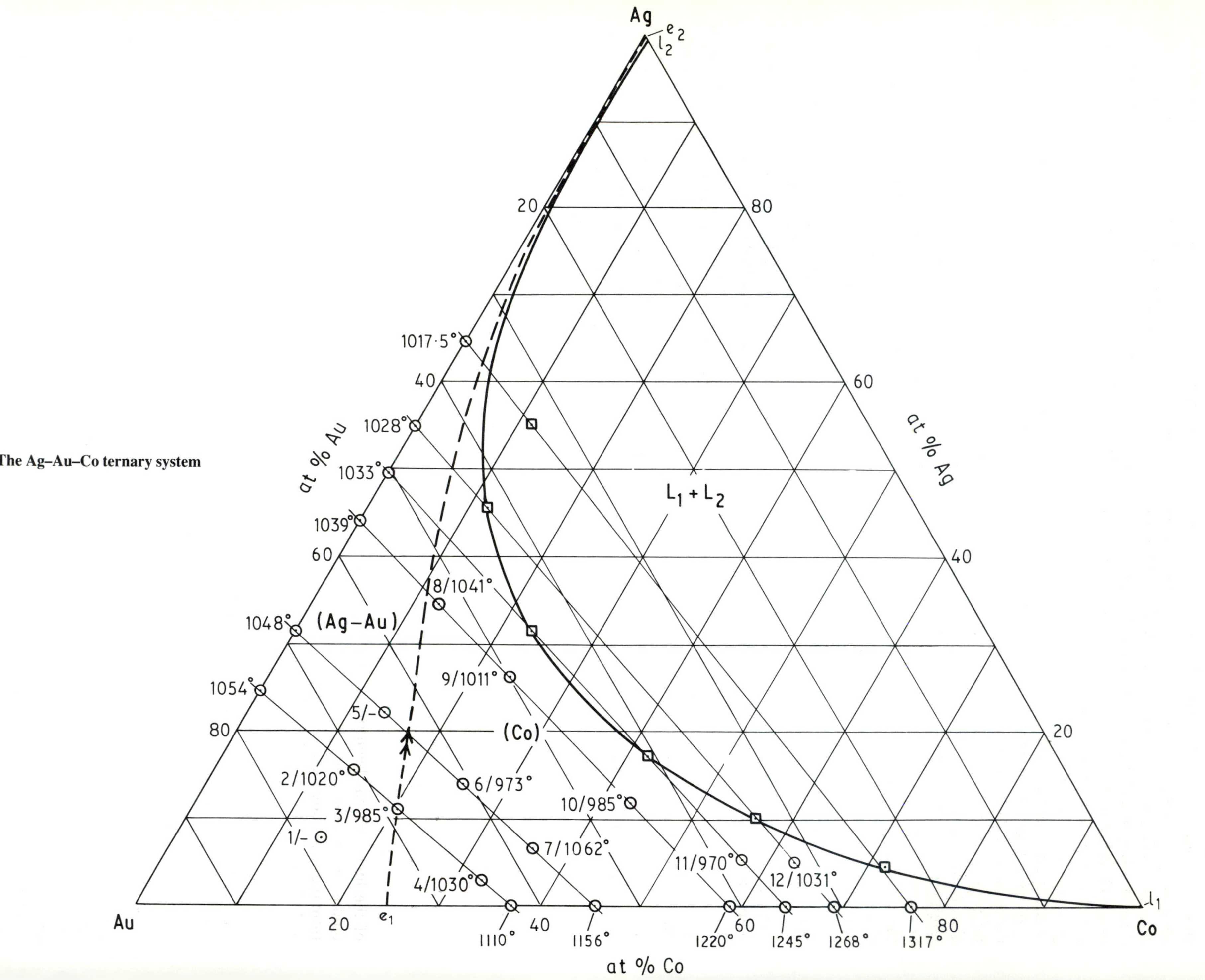

3 The Ag–Au–Co ternary system

Six alloy compositions were stated to define the limits of the liquid immiscibility gap. These have been individually plotted as small squares in Fig. 3. The alloy containing 55·2 Ag, 33·6 Au, 11·2 at.-% Co (45 Ag, 50 Au, 5 wt-% Co) lies off the curve given in Fig. 3. It is not clear how the limits for the liquid immiscibility gap were determined but reference is made to the observance of two layers in metallographic sections. To plot the limits given in Fig. 3 implies that further alloys, of unstated composition, were studied but no details are given.

At room temperature there is insignificant solubility of Co in Au or of Co in Ag. The room temperature section of the ternary system will present an extremely narrow region running parallel to the Ag–Au edge in which the (Ag/Au) solid solution exists; the rest of the section will be composed of tie lines radiating from the Co corner to the boundary running along the Ag–Au edge. Virtually all alloys can be considered as two-phase structures ((Ag/Au) + (Co)).

REFERENCES

58 Gri: A. T. Grigoriev, L. A. Panteleimonov, V. V. Kuprina and V. S. Vorobiev: *Zhur. Neorg. Khim.*, 1958, **3**, 2532–2536

83 Oka: H. Okamoto and T. B. Massalski: *Bull. Alloy Phase Diagrams*, 1983, **4**(1), 30–38

85 Oka: H. Okamoto, T. B. Massalski, M. Hasebe and T. Nishizawa: *Bull. Alloy Phase Diagrams*, 1985, **6**, 449–454

Ag–Au–Cu

INTRODUCTION

This important ternary system, acting as the basic phase diagram for an understanding of gold dental alloys, has been studied for the past 75 years. A detailed assessment of the literature has been made with respect to the liquidus and solidus surfaces, the solvus surface arising from the solid-state miscibility gap in the Ag–Cu system, and the ternary equilibria at low temperatures arising from the ordered phases in the Au–Cu system.

With the exception of a limited study of the 10 and 20 at.-% Au sections [80 Yam] and a liquidus projection together with the ternary miscibility gap isotherms at 371, 482, 593 and 704°C [73 Sis] from unpublished prime data, the data relating to the liquidus are from older literature as are those for the solidus [11 Jae, 14 Par, 26 Ste]. There is good evidence for the presence of a minimum in the monovariant eutectic curve originating at the Ag–Cu eutectic point, Fig. 4. The critical tie line, $l = \alpha_1 + \alpha_2$, is assessed at 767°C. (Throughout this assessment α_1 refers to a Ag-rich solid solution phase and α_2 to a Cu-rich solid solution phase.) The Au-rich monovariant eutectic curve originates at an estimated 800°C with a critical tie line (*see* Fig. 5*e*) joining liquid with disordered α solid solution. The three-phase, α_1+α_2+ liquid, equilibrium formed below 800°C falls to the 767°C minimum. There is a need to use modern experimental techniques to establish more reliable data for the liquidus and solidus and to confirm the minimum in the monovariant eutectic curve.

The $\alpha_1 + \alpha_2$ miscibility gap in the ternary system has been more extensively studied. The assessed isotherms at 750, 725, 700, 600, 500, 400 and 300°C are given in Fig. 6*a*. At the higher temperatures the $\alpha_1\alpha_2$ tie lines slope towards the Ag corner, at 400°C the tie lines are nearly parallel to the Ag–Cu binary edge. There is reasonable agreement between the experimental data for the $\alpha_1 + \alpha_2$ miscibility gap (Figs. 6*b*–*g*). Figure 8 is the assessed consolute point curve associated with the $\alpha_1 + \alpha_2$ miscibility gap.

At temperatures below 400°C three-phase equilibria involving the Au–Cu ordered phases will appear in the ternary system. The cluster variation method was used [80 Kik, 80 Yam, 81 de Fon] to calculate isothermal sections of the coherent phase diagram and to predict the presence of three-phase equilibria. Modern experimental work owes most to Japanese workers [80 Kog, 81 Uzu, 85 Kog]. An interpretation of the data leads to the conclusion that there are three critical tie lines along which Au–Cu ordered phases and disordered solid solution phases coexist. The critical tie line $\alpha_2 \rightleftharpoons \alpha_1 + AuCu_3$ is placed at 387°C, a few degrees below the temperature for the formation of ordered $AuCu_3$ from the disordered α solid solution. The critical tie line $\alpha_2 \rightleftharpoons \alpha_1 + AuCuII$ is placed at 374°C and the critical tie line $AuCuII \rightleftharpoons AuCuI + \alpha_1$ is placed at 358°C. A ternary eutectoid reaction, $\alpha_2 \rightleftharpoons \alpha_1 + AuCuII + AuCu_3$, is considered to occur at about 280°C, some 5°C below the $\alpha_2 \rightleftharpoons AuCuII + AuCu_3$ eutectoid in the Au–Cu binary system. A reaction scheme consistent with the suggested invariant reactions is given in Fig. 20.

Study of equilibria below 400°C requires very long annealing schedules. Nevertheless it would be very helpful in establishing the equilibria between ordered and disordered phases to study the constitution of alloys at 20°C steps, as done by [85 Kog]) (Fig. 16), on vertical sections from the Au corner to the Ag–Cu binary edge.

BINARY SYSTEMS

The Au–Cu phase diagram, recently assessed by [87 Oka], is accepted. The Ag–Au assessment [83 Oka] is also accepted, as is the Ag–Cu assessment of [80 Ell].

SOLID PHASES

Table 3 summarizes crystal structure data for the binary phases. No ternary compounds have been reported.

Table 3. Crystal structures

Solid phase	*Prototype*	*Lattice designation*
(Ag), (Au), (Cu)	Cu	cF4
α	Cu	cF4
α_1(Ag-rich)	Cu	cF4
α_2(Cu-rich)	Cu	cF4
$AuCu_3$	$AuCu_3$	cP4
AuCuI	AuCuI	tP4
AuCuII	AuCuII	Faulted tP4
Au_3Cu	$AuCu_3$	cP4

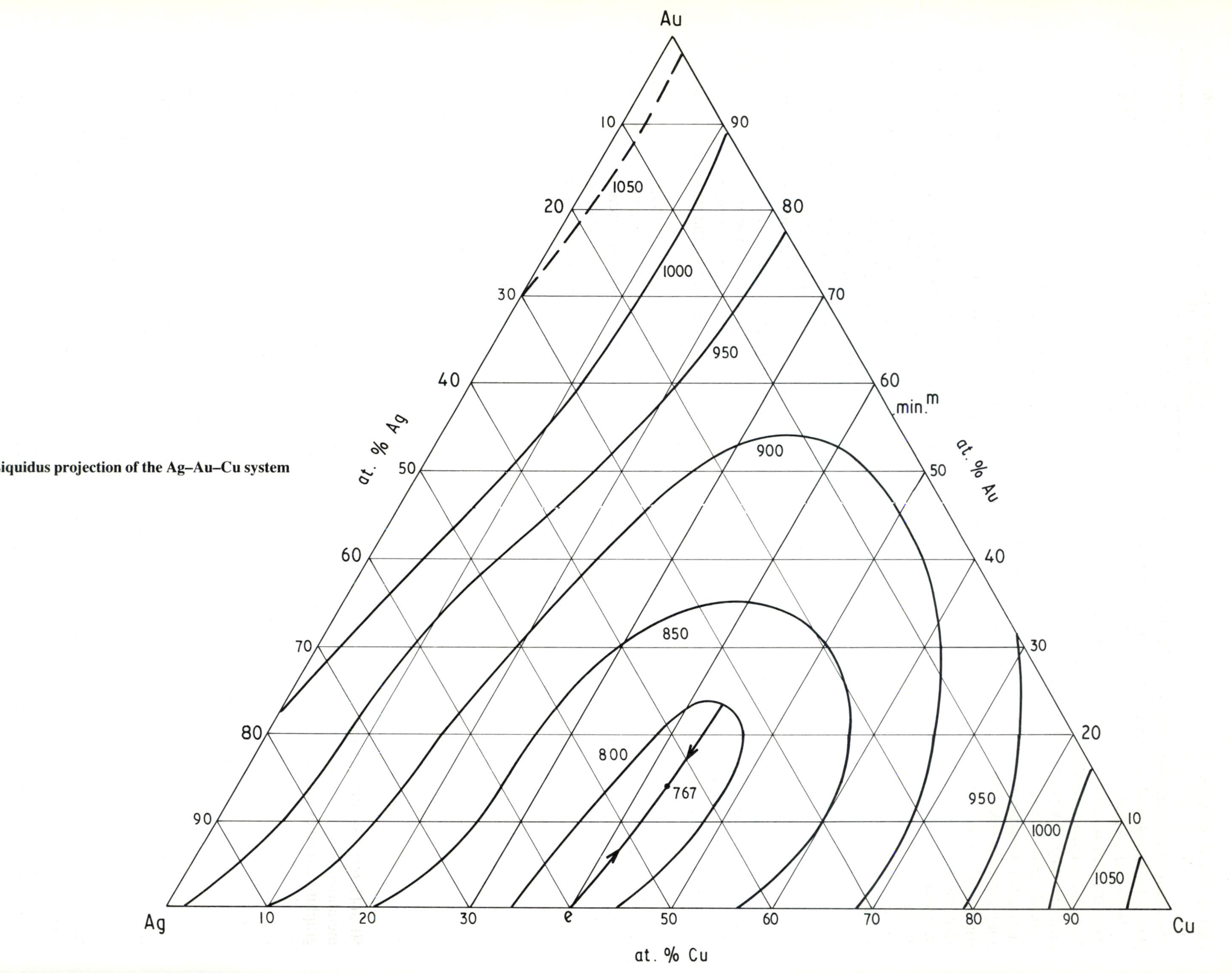

4 Liquidus projection of the Ag–Au–Cu system

LIQUIDUS SURFACE

The major studies are from the older literature. Jänecke [11 Jae] examined 62 ternary alloys whose compositions lay on eight sections from the Ag–Au edge (from 90% Ag, 10% Au to 20% Ag, 80% Au) to the Cu corner. Thermal analysis, using cooling curves only, supplemented by metallography allowed the liquidus surface to be defined to a stated accuracy of ± 5°C. The major inaccuracy would arise from the alloy preparation technique; successive amounts of Cu were added to a given Ag–Au alloy until the volume grew too large. The ternary alloy was then cut into two portions, to one half more Cu was added and to the other half Ag was added. This dilution technique would present opportunity for loss of Cu by oxidation during successive thermal analysis experiments.

The liquidus published by Jänecke [11 Jae] has the same general form as that arising from later work. The minimum in the Au–Cu liquidus, placed at 884°C rather than the currently accepted value of 911°C [87 Oka], projects into the ternary and becomes the monovariant eutectic fold ending at the binary Ag–Cu eutectic point at 779°C. Jänecke [11 Jae] interpreted the solidification equilibria correctly and located a critical tie line where liquid equilibrates with α before forming a $L + \alpha_1 + \alpha_2$ three-phase equilibrium. The compositions of the phases are given in Table 4. The monovariant eutectic separation begins with a liquid containing 21 at.-% Au at 800°C. The tabulated thermal analysis data of [11 Jae] indicate a minimum temperature for the completion of solidification, the lower boundary of the $L + \alpha_1 + \alpha_2$ phase region, at 766°C. Even with ± 5°C claimed accuracy this temperature is below that of the Ag–Cu eutectic and suggests that the monovariant eutectic curve passes through a temperature minimum before reaching the Ag–Cu eutectic point.

Parravano and co-workers [14 Par] delineated the liquidus surface by thermal analysis, using cooling curves only, of 54 ternary alloys at constant Ag concentrations of 5 wt-% (18 alloys), 10%(8), 20%(7), 30%(6), 40%(5), 50%(4), 60%(3), 70%(2) and 80%(1). Some alloys were metallographically examined. The alloys were melted in graphite crucibles and a slow current of nitrogen passed over the surface during thermal analysis. 20 g alloys were made up initially and additions made until the alloys weighed 40 g. The liquidus surface has the same form as [11 Jae], the only notable difference being the extension of the monovariant eutectic curve into the ternary. Parravano and co-workers [14 Par] place the critical liquid composition at 24% Au (Table 4), but agree with [11 Jae] on the temperature of 800°C for the beginning of three-phase formation of $L + \alpha_1 + \alpha_2$. Parravano and co-workers [14 Par] also indicate a possible minimum temperature on the monovariant eutectic curve with final solidification of the $\alpha_1 + \alpha_2$ phase at a minimum temperature of 770°C.

Sterner-Rainer [26 Ste] prepared 45 ternary alloys whose compositions were on the sections 3 carat (3 alloys), 6(5), 8(6), 12(6), 14(9), 16(5), 18(5), 20(3) and 22(3). 50 g melts were used for high carat alloys and 40 g

Table 4. Compositions of the liquid and solid phases at the temperature of the critical tie line L ⇌ α

Phase	*Composition*, at.-%			*Temperature*, °C	*Reference*
	Ag	*Au*	*Cu*		
L	36·5	21·0	42·5	800	11 Jae
α	33·5	24·0	42·5		
L	34·0	24·0	42·0	~800	14 Par
L	33·2	23·4	43·4	800	This assessment
α	33·0	24·5	42·5		

melts for lower carat alloys. Thermal analysis data on cooling were tabulated and used to produce the liquidus surface. Sterner-Rainer [26 Ste] determined the freezing points of Ag, Au, Cu, the Ag–Cu eutectic alloy and a 28·2% Ag, 71·8% Cu alloy. The values, compared with currently accepted values, were 961°C (961·9°C), 1064 (1064·4), 1084 (1083), 779 (779) and 898 (913). He accepted the Au–Cu liquidus minimum at 884°C and, as with [11 Jae] and [14 Par], drew the 900°C isotherm as intersecting the Au–Cu binary liquidus. In agreement with [11 Jae] and [14 Par], [26 Ste] noted temperatures for the end of solidification of $\alpha_1 + \alpha_2$ that were below the 779°C eutectic temperature on the Ag–Cu binary. For 6 carat alloys (25 wt-% Au) the lowest value was 767°C, again suggesting a minimum on the monovariant eutectic curve.

The more recent work of [73 Sis] suffers from a lack of data on which to base an assessment. Sistare [73 Sis] reported unpublished data in the form of a liquidus projection but without any indication of the eutectic curve. As in the earlier work, [73 Sis] took the Au–Cu liquidus minimum at a lower temperature, 889°C, than that currently accepted. In his review of the ternary equilibria [79 Cha] based the assessed liquidus projection on [26 Ste] and [73 Sis], with a correction in the liquidus isotherms for 850, 900 and 950°C to take account of the change in the Au–Cu liquidus minimum to 911°C. The currently assessed liquidus projection (Fig. 4), reflects the data of [11 Jae], [14 Par], [26 Ste] and [73 Sis] and places emphasis on the work of [26 Ste]. The isotherms should be regarded as accurate to ± 5°C within the ternary system. The monovariant eutectic curve begins at the critical liquid composition 33·2% Ag, 23·4% Au and descends to a minimum at 767°C with a liquid containing about 14 at.-% Au. Thereafter the curve ascends to the binary Ag–Cu eutectic at 779°C.

In addition to the suggestion from the data of [11 Jae], [14 Par] and [26 Ste] that there is a minimum in the monovariant eutectic curve modern data also supports this conclusion. Yamauchi *et al.* [80 Yam] determined portions of the 10 and 20 at.-% Au sections. The eutectic curve is reached at 777°C and 42 at.-% Cu in the 10 at.-% Au section; the $L + \alpha_1 + \alpha_2$ phase region extends from 777 to 770°C. In the 20 at.-% Au section the eutectic curve is met at 43 at.-% Cu at 790°C; the $L + \alpha_1 + \alpha_2$ phase region extends from 790 to 780°C. There is a remarkable consistency between these temperature values for the $L + \alpha_1 + \alpha_2$ phase region and the results of [11 Jae] and [14 Par]. For example, [11 Jae] reported 776–768°C (8·1% Au alloy), 771–769°C (8·7% Au), 772–769°C (9·1% Au) and [14 Par] 780–772°C (9·8% Au). 780–770°C (9·2% Au), 776–770°C (8·7% Au) compared with [80 Yam] 777–770°C (10% Au). For 20 at.-% Au [80 Yam] give 790–780°C whereas [11 Jae] reported 785–780°C for the two alloys containing 20·7 at.-% Au. It would appear that the $\alpha_1 + \alpha_2$ tie lines are inclined at a slight angle to the Ag-Cu binary edge with a lower Au content for the Cu-rich α_2 phase. At the critical tie line, tentatively assessed at 767°C, the liquid contains 14 at.-% Au; it is in equilibrium with α_1 of probable composition 66% Ag, 16% Au and α_2 of probable composition 8% Ag, 11% Au. This experimental data conflicts with the theoretical prediction of [78 Lup] that the eutectic temperature of the Ag–Cu binary is raised by the addition of Au.

SOLIDUS SURFACE

Solidus temperatures were tabulated for ternary alloys by [11 Jae], [14 Par] and [26 Ste] and the former two presented a projection of the solidus surface. For the ternary alloys that lie outside the semicircular-shaped $\alpha_1 + \alpha_2$ miscibility gap at 800°C, the temperature of the critical tie line between L and α_1/α_2, the measured solidus temperatures are most probably too low. An estimate of the solidus isotherms is included in Figs. *5a–f* for 1 000, 950, 900, 850, 800 and 775°C isothermal sections of the ternary. The solidus isotherms are to be regarded as indicative of the true isotherms.

SOLVUS SURFACE

The $\alpha_1 + \alpha_2$ Ag–Cu miscibility gap projects into the ternary. The ternary solvus surface has been studied by several authors [40 Mas, 49 McM, 67 Zie, 73 Sis, 75 Mur, 86 Ntu]. The assessed isotherms at 750, 725, 700, 600, 500, 400 and 300°C are shown in Fig. *6a*. It should be noted that the 300°C isotherm takes no account of the influence of ordering reactions. Figure 12 presents a suggested isothermal section at 300°C in which ordered phases are included. Masing and Kloiber [40 Mas] used 24 alloys annealed for 1 h at 750°C and water quenched followed by a 7 h anneal at 400°C to determine the 400°C isotherm by measurement of lattice spacings of the α_1 and α_2 phases. A tentative 750°C isotherm was also published. They [40 Mas] concluded that the $\alpha_1\alpha_2$ tie lines are virtually parallel to the Ag–Cu edge of the ternary system. The α_1 boundary (Ag-rich) is remarkably close to the assessed isotherm at 400°C (Fig. *6b*), but the α_2 boundary lies at lower Ag contents than are accepted.

McMullin and Norton [49 McM] determined the miscibility gap for 14 carat alloys using 7 ternary compositions (Figs. *6b–e*). A further 3 alloys were used to trace the phase boundary $\alpha/(\alpha_1 + \alpha_2)$ along a part of the section Au–37·1% Ag, 62·9% Cu. The variation of the lattice parameter of the α_1 or α_2 phase with temperature and the change in slope on crossing the solvus boundary into the α phase region was used to determine solution temperatures. Melted ingots were cold rolled, homogenized for 16 h at 700°C and filings taken for heat treatment in evacuated pyrex tubes. The heat-treatment time was unspecified other than to state that it allowed complete recrystallization but very little grain growth. The isotherms presented by [49 McM] are drawn through 3 data points in the ternary (350, 400, 500, 600°C) or 1 data point (700°C) and two points on the Ag–Cu binary edge. The isotherms can be regarded as accurate to within ± 1 at.-% Au for ternary alloys containing 14 carat or greater Au content.

Ziebold and Ogilvie [67 Zie] prepared diffusion couples between Cu-rich Au–Cu alloys and Ag-rich Ag–Au alloys and heat treated them for 48–60 h at 725 ± 5°C. Microprobe traces across the diffusion couples gave composition profiles from which the compositions of the α_1 and α_2 phases were determined. The tie lines at 725°C, (Fig. *6f*), are not parallel to the Ag–Cu binary edge as deduced by [40 Mas] but incline towards the Ag corner. Sistare [73 Sis] published isotherms at 371, 482, 593 and 704°C (Fig. *6c–e*). Since no experimental detail is given, the data originating from unpublished reports, it is not possible to assess it. The data generally agree with that of other workers, as is evident from the series of vertical sections presented in Fig. 7, and the comparison of the experimental data for the 400, 500, 600 and 700°C isotherms (Figs. *6b–e*).

Murakami *et al.* [75 Mur] used one of the experimental tie lines of [67 Zie] to calculate the ternary excess Gibbs energy and thermodynamically model the miscibility gap at 242, 350, 700 and 725°C using a subregular solution model. There is very good agreement between the 725°C isotherm of [67 Zie] and the slope of the tie lines and the calculated isotherm with tie lines.

Kogachi and Nakahigashi [80 Kog] determined the solid-state part of the 75 at.-% Ag and 75 at.-% Cu sections. The solution temperatures were measured from the variation in lattice spacing of the α_1 or α_2 phase with temperature as was done by [49 McM]. The data correspond well with the assessed isotherms for the α_2 phase but not for the α_1 phase (Figs. *6c–e*).

Uzuka *et al.* [81 Uzu] studied the solid-state part of sections from 50% Au, 50% Cu–Ag; 59·5% Au, 40·5% Cu–Ag; 41·8% Au, 58·2% Cu–Ag. Although mainly concerned with the effect of the Au–Cu ordered phases on the ternary equilibria an approximate $\alpha/(\alpha_1 + \alpha_2)$ boundary was determined for each section. The data are in agreement with other workers at 500°C but [81 Uzu] overestimate the solubility of Ag in α_2 at 600 and 700°C (Figs. *6c–e*).

Ntukogu and Cadoff [86 Ntu] determined isotherms and the corresponding $\alpha_1\alpha_2$ tie lines at 500, 600, 700 and 750°C using equiatomic Ag–Cu alloys containing 5, 10, 15, 20, 25 and 33·3 at.-% Au. Alloys were arc melted, rolled to foil, vacuum remelted in a graphite crucible and annealed for 24 h at 700°C. Each alloy was thermally cycled six times from 730°C (19 h) to 400°C (5 h) to spheroidize and coarsen the precipitating particles. The 33·3 at.-% Au alloy was cycled from 600 to 250°C. Heat-treatment times were very extended: 780 h at 500°C; 780 h at 500°C + 240 h at 600°C; 30 h at 700°C; 30 h at 700°C + 20 h at 750°C. The compositions of the α_1 and α_2 phases were determined by electron probe microanalysis. Extrapolation of the Ag-rich (α_1) boundary to the binary Ag–Cu edge gave solubilities of Cu in Ag greater than the accepted values. The ternary data were corrected for the

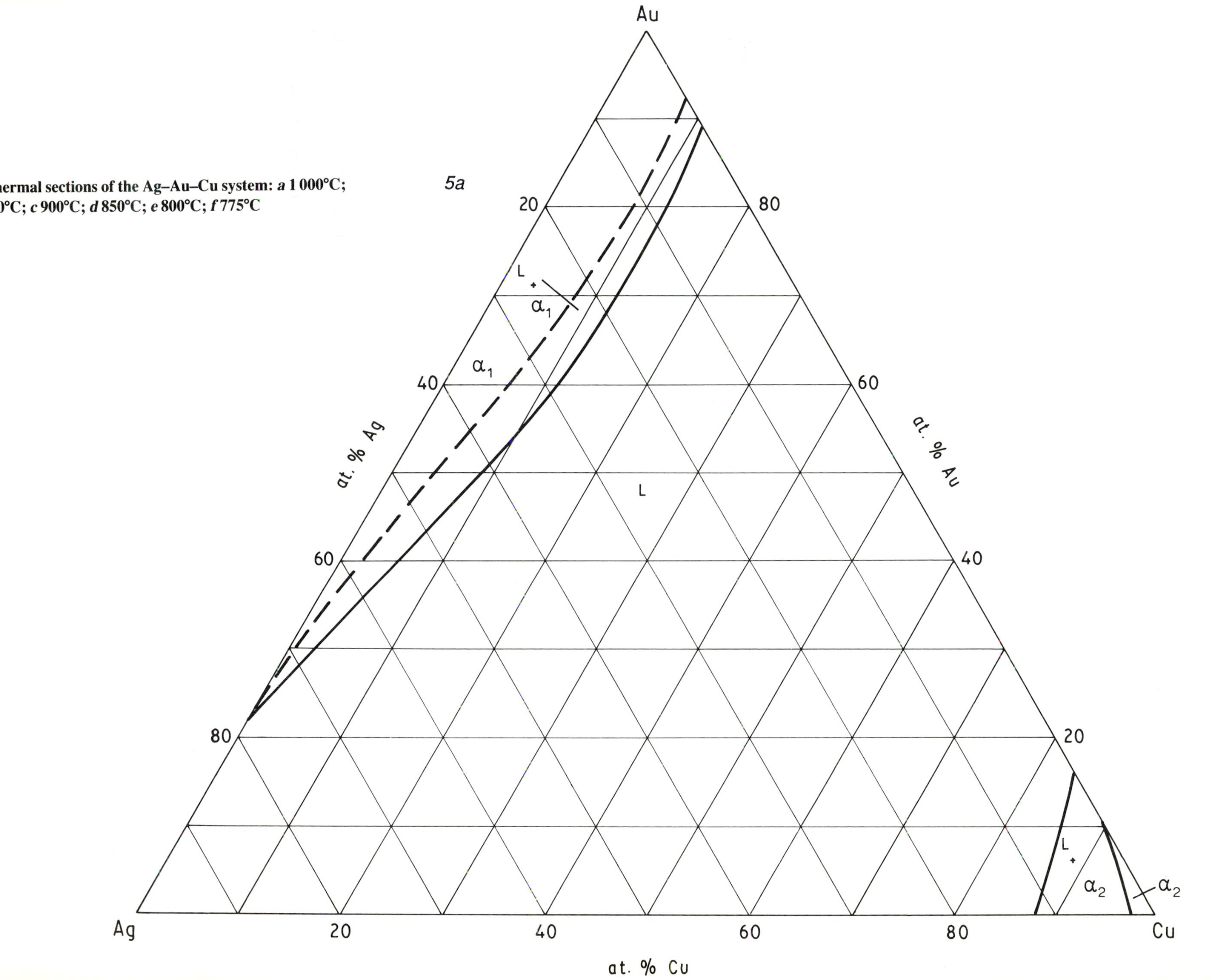

5 **Isothermal sections of the Ag–Au–Cu system: *a* 1 000°C; *b* 950°C; *c* 900°C; *d* 850°C; *e* 800°C; *f* 775°C**

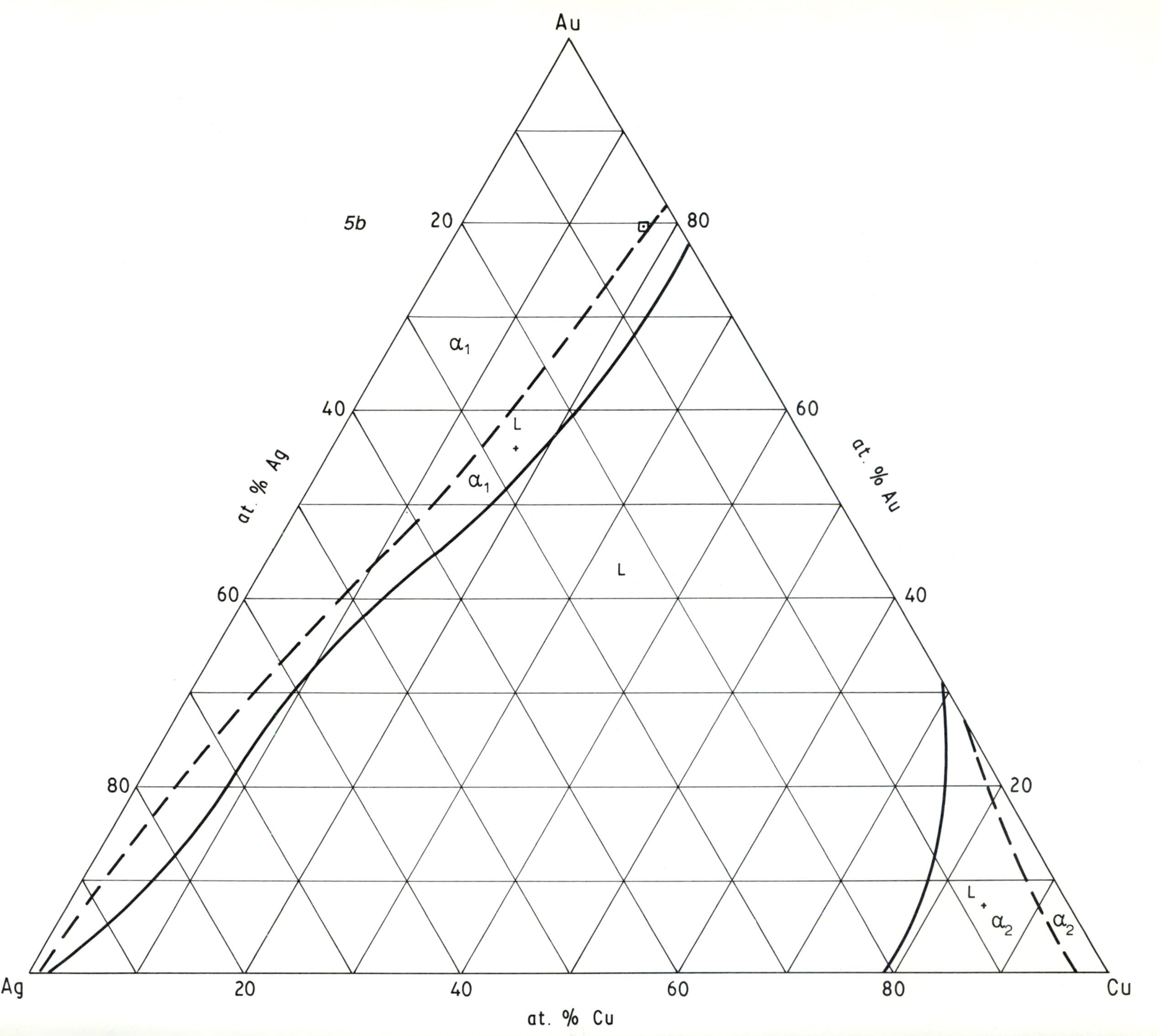

Au
5b
20
80
α_1
40
60
L + α_1
at. % Ag
at. % Au
L
60
40
80
20
L + α_2
α_2
Ag
20
40
60
80
Cu
at. % Cu

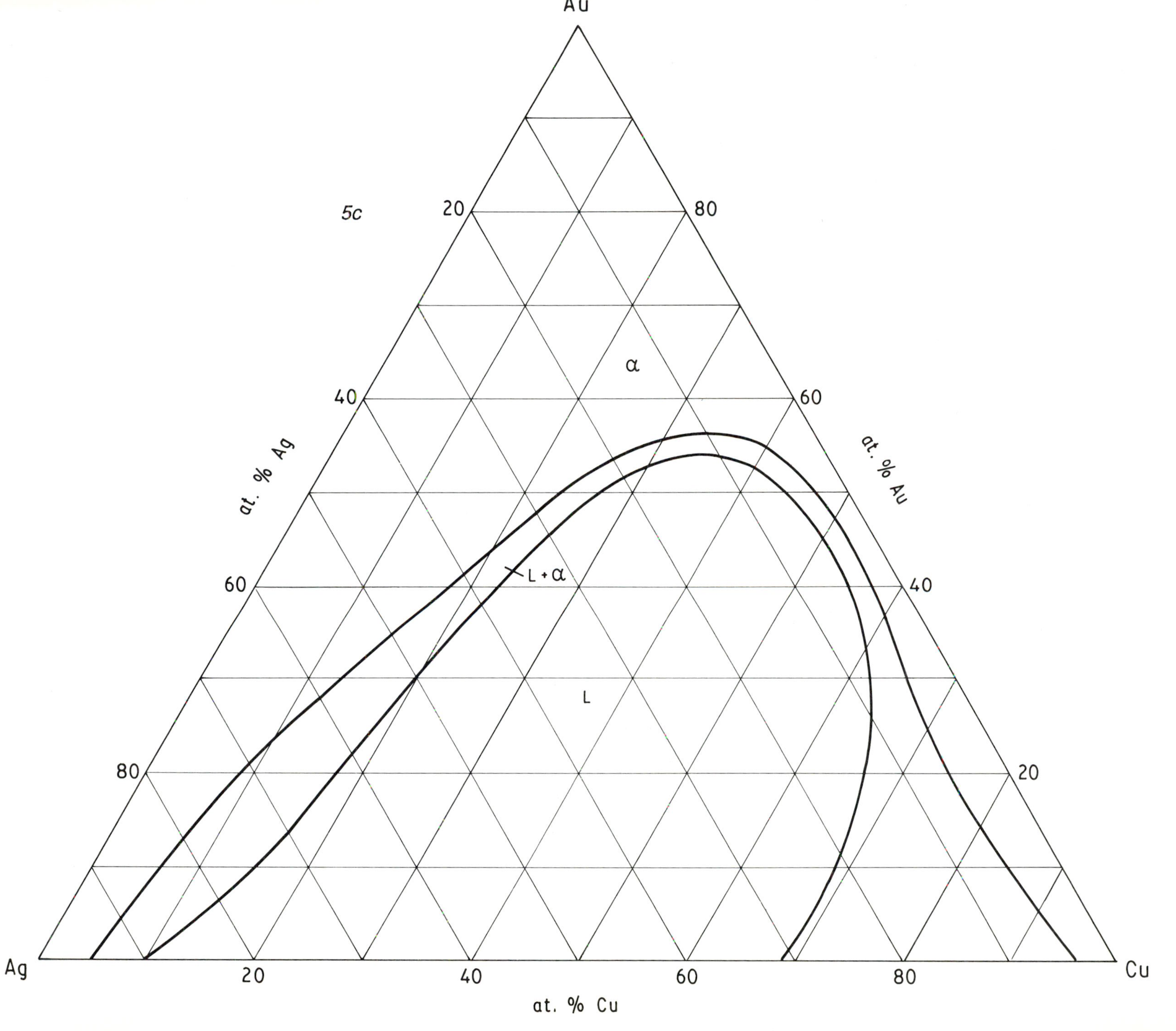
Au
5c
20
80
α
40
60
at. % Ag
at. % Au
L+α
60
40
L
80
20
Ag
20
40
60
80
Cu
at. % Cu

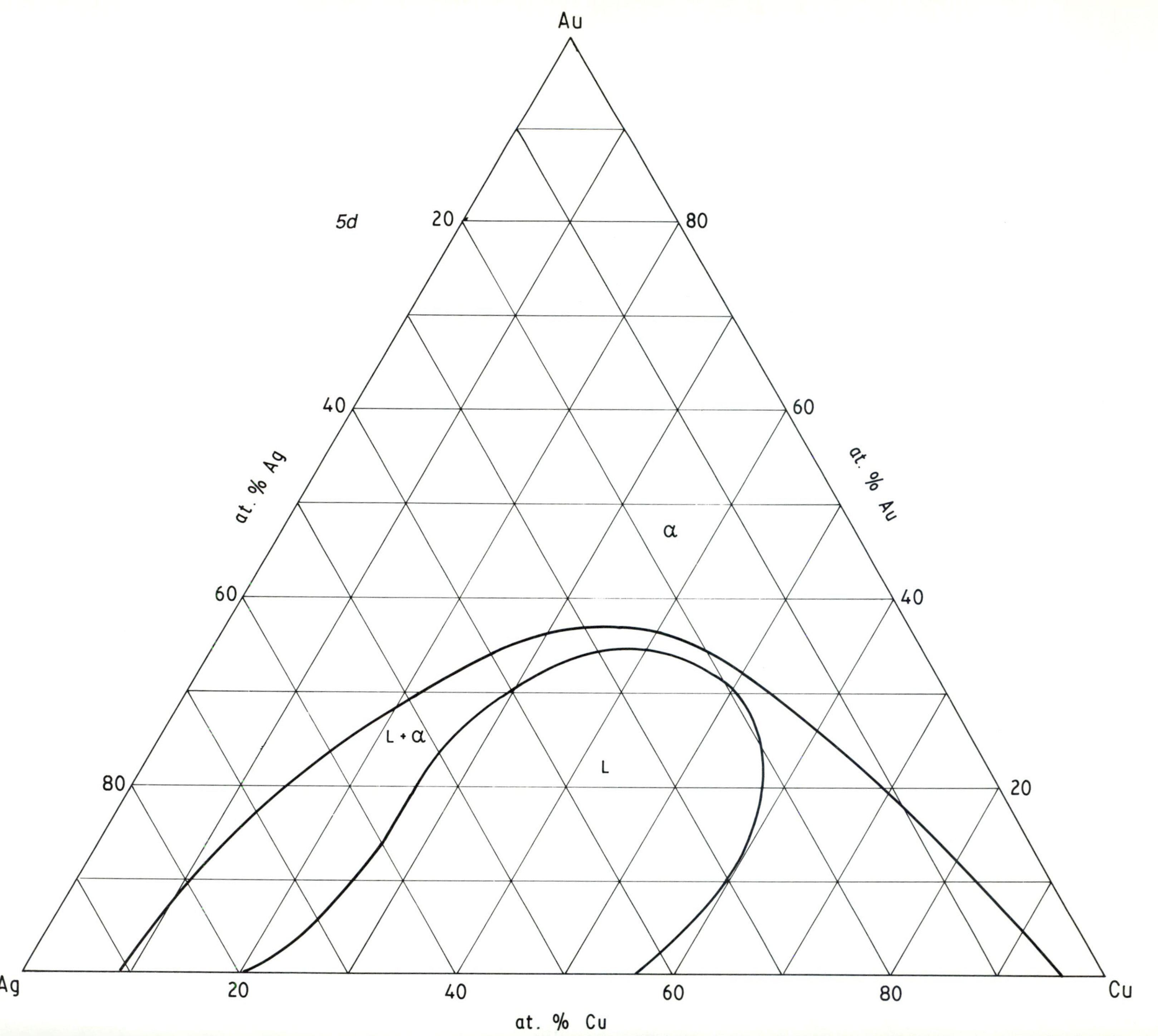

Au
5d
20
80
40
60
at. % Ag
at. % Au
α
60
40
L + α
L
80
20
Ag
20
40
60
80
Cu
at. % Cu

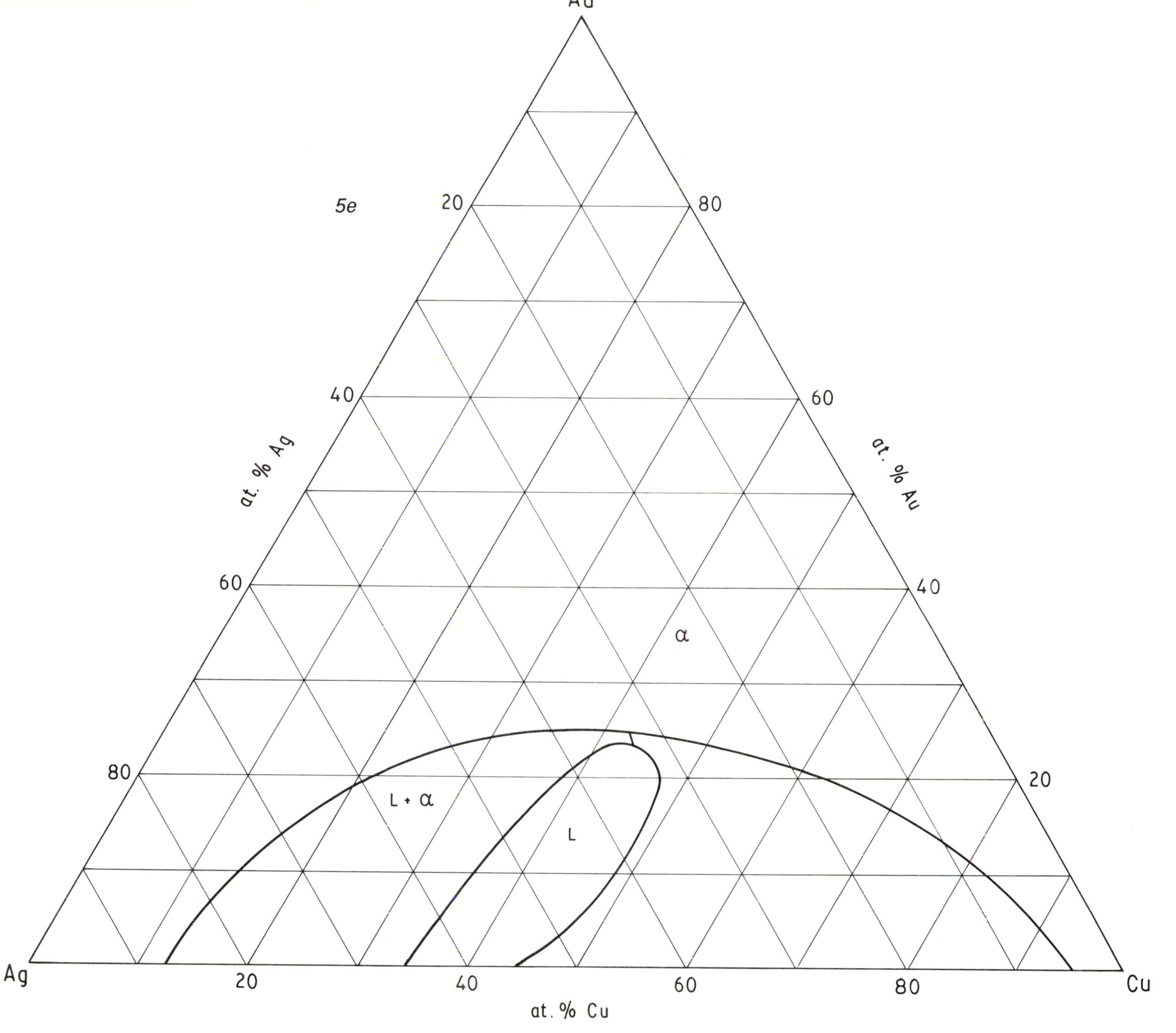
Au
5e
at. % Ag
at. % Au
α
L + α
L
Ag
Cu
at. % Cu
20
40
60
80

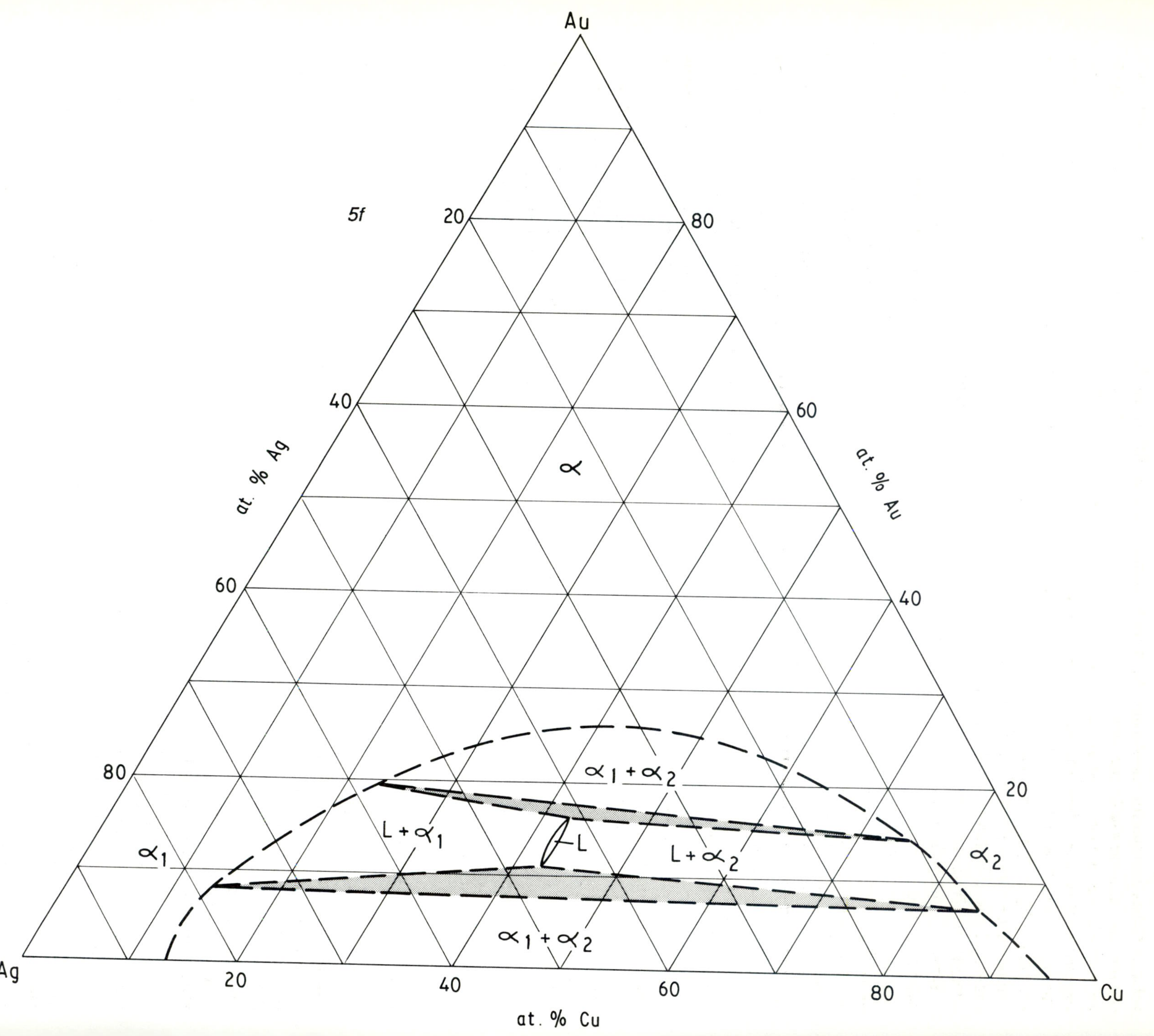

5f

α_1 phase [86 Ntu] and demonstrated good agreement with isotherms thermodynamically calculated using regular solution Gibbs energies for Ag–Au and Au–Cu alloys and semiregular solution Gibbs energies for Ag–Cu alloys to express the excess Gibbs energy of ternary alloys. The experimental tie lines slope towards the Ag corner and the magnitude of the slope is in excellent agreement with the data of [67 Zie] and the calculation of [75 Mur]. The computed tie lines of [86 Ntu] have a smaller slope and the extent of the calculated miscibility gap appears to be exaggerated (Figs. 6*c–e, g*).

Kogachi and Nakahigashi [85 Kog] determined the solid-state part of the 25 and 50 at.-% Au sections using X-ray diffraction and DTA techniques. The $\alpha/(\alpha_1 + \alpha_2)$ phase boundary was completely delineated for the 50 at.-% Au section and traced to 700°C for the 25 at.-% Au section. The data (Figs. 6*b–e*) agree with the assessed isotherms for the 25 at.-% Au section but give a small overestimate of the $\alpha_1 + \alpha_2$ miscibility gap at 400°C on the 50 at.-% Au section.

A comparison of all the experimental data is presented in Figs. 6*b–g* for isotherms at 400, 500, 600, 700, 725 and 750°C respectively.

A series of vertical sections are given in Figs. 7*a–g* for alloys containing 10 at.-% Au, 30 at.-% Au and the 6, 8, 10, 12 and 14 carat alloys respectively. Figure 5 is the assessed consolute point curve associated with the $\alpha_1 + \alpha_2$ miscibility gap. The consolute point on the isothermal section can only be established if the $\alpha_1\alpha_2$ tie line data is available; such data are relatively sparse for this ternary system. Accordingly the consolute point curve is shown as a dashed line in Fig. 8. The maximum points on the $\alpha_1 + \alpha_2$ miscibility gap are shown in terms of the Au content at temperatures from 800 to 242°C in Fig. 9.

ORDERING REACTIONS

The importance of order–disorder reactions combined with phase separation from the disordered α solid solution for the age-hardening properties of Ag–Au–Cu alloys has been recognized for a long time. Sterner-Rainer [25 Ste] defined a region in the ternary system within which ordered CuAu forms and affects the mechanical properties of ternary alloys. The lower boundary of this region was regarded as a line joining the Ag corner to 30% Au, 70% Cu and the upper boundary a line from Ag to 81% Au, 19% Cu. Similar observations were made by [35 Spa] but neither [25 Ste] nor [35 Spa] proved the presence of ordered phases in the ternary system.

The first investigations using X-ray diffraction techniques to characterize the phases was undertaken by [39 Hul] and [39 Bum]. Figure 10 summarizes the data of [39 Hul] for alloys on the 50 at.-% Cu section containing up to 5 at.-% Ag. Addition of Ag reduces the temperature of ordering of CuAuII, but the 3 at.-% Ag alloy gave anomalous results. Alloys were cold worked, homogenized *in vacuo* at 850°C for 240–408 h, and filings annealed in evacuated Pyrex tubes for 288–504 h at 400°C, 576 h at 390°C, 360–888 h at 373°C, 552 h at 360°C, 480–1 080 h at 350°C. These lengthy annealing times are exceptional for work of this era. Bumm [39 Bum] examined two series of alloys which were held for 2 h at 700°C and water quenched. They were then annealed 36 h at 350°C and quenched. Alloys on the 25 wt-% Cu section containing up to 15 wt-% Ag were homogeneous disordered α on quenching from 700°C, in agreement with Fig. 6*a*. After annealing at 350°C the alloy 5·8% Ag, 44·7% Au was disordered α whereas Fig. 11, mainly based on the work of [81 Uzu], indicates an α_1 + AuCuII structure; a 36 h anneal is insufficient time to reach equilibrium at 350°C. Alloys containing 11·4% Ag, 40·4% Au and 16·6% Ag, 36·4% Au contained $\alpha_1 + \alpha_2$ after annealing for 36 h at 350°C; this agrees with Fig. 11. The alloy series from 50 wt-% Au, 50 wt-% Cu to Ag and containing up to 20 wt-% Ag were similarly treated. After a 2 h at 700°C solution heat treatment only the alloy containing 18·2% Ag, 20% Au was heterogeneous, $\alpha_1 + \alpha_2$. According to Fig. 6*a* the alloy containing 13·6% Ag, 21·1% Au should also contain $\alpha_1 + \alpha_2$. After annealing for 36 h at 350°C no evidence of $AuCu_3$ superlattice phase was noted in the Ag-containing alloys. The alloy with 4·5% Ag, 23·3% Au contained disordered α and alloys with 90% Ag, 22·2% Au, 13·6% Ag, 21·1% Au and 18·2% Ag, 20·0% Au had an $\alpha_1 + \alpha_2$ structure. The later data, on which Fig. 11 is based, would indicate an $\alpha_1 + AuCu_3$ equilibrium structure for all four ternary alloys at 350°C.

Raub [49 Rau] prepared 13 ternary alloys which were 50% deformed prior to solution treatment. Alloys were then further deformed by 80–90% by rolling or forging, and annealed at 300, 360 and 400°C. The structures observed by X-ray diffraction analysis are shown in Fig. 12 for 300°C. As the ordered phases were referred to as tetragonal structures, and were not attributed to the AuCuI or AuCuII structure, the alloys are noted as α_1 + tetragonal (α_1t) in Fig. 12.

Modern experimental work on the occurrence of ordered phases in the Ag–Au–Cu system includes the work of [80 Kog], [81 Uzu] and [85 Kog]. Kogachi and Nakahigashi [80 Kog] examined the 75% Ag, 75% Au and 75% Cu sections by DTA and X-ray diffractometry. All alloys were melted in a graphite crucible sealed in a silica tube under an argon atmosphere. After homogenizing for 5 days at 750–800°C they were cooled to room temperature at 10°C h^{-1}. Alloys on the 75 at.-% Au section were rolled to thin plates, annealed at 500°C for 3 days in evacuated Pyrex tubes, and cooled slowly to 100°C. They were annealed for 1 year at 100°C and finally cooled to room temperature at 10°C day^{-1}. Along this section the alloys all had a disordered fcc structure with the lattice parameter changing linearly with composition from $a = 0{\cdot}4076_0$ nm for 25 at.-% Ag to $a = 0{\cdot}3990_1$ nm for 1·25 at.-% Ag. Filings from each alloy on the 75 at.-% Ag and 75 at.-% Cu sections and the remaining ingots were annealed at 400°C for 1 week in a sealed Pyrex tube filled with argon, and then cooled to room temperature at 20°C day^{-1}. Figure 13, the 75 at.-% Cu section, shows the boundaries between α_2 and $\alpha_1 + \alpha_2$ and between $\alpha_1 + \alpha_2$ and $\alpha_1 + AuCu_3$ according to [80 Kog]. There must be a three-phase region $\alpha_1 + \alpha_2 + AuCu_3$ (*see* Figs. 11 and 12) and this has been included in Fig. 13. The remainder of this figure is constructed from the assessed liquidus, solidus and solvus data. The transition from $\alpha_1 + AuCu_3$ to α_1

6a

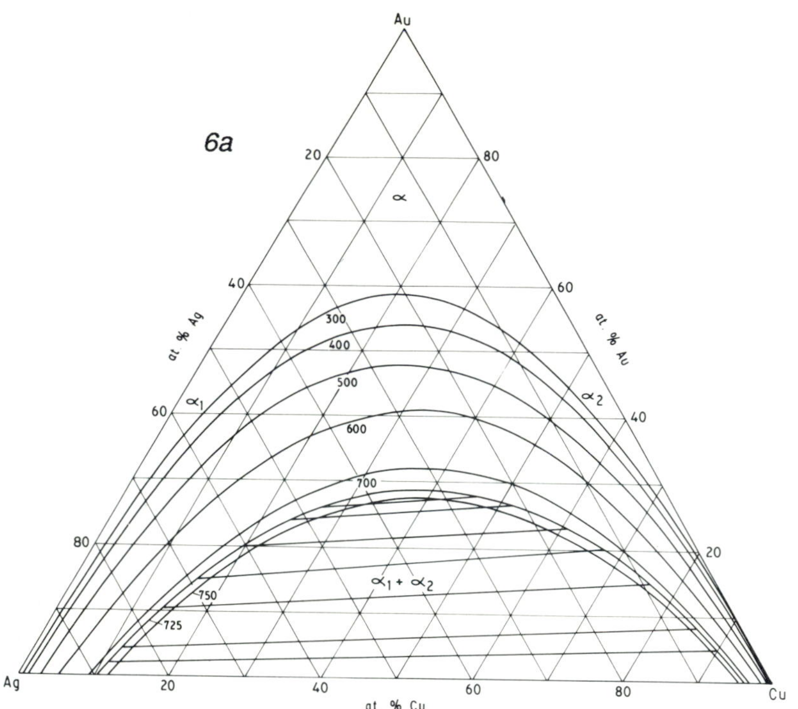

6b

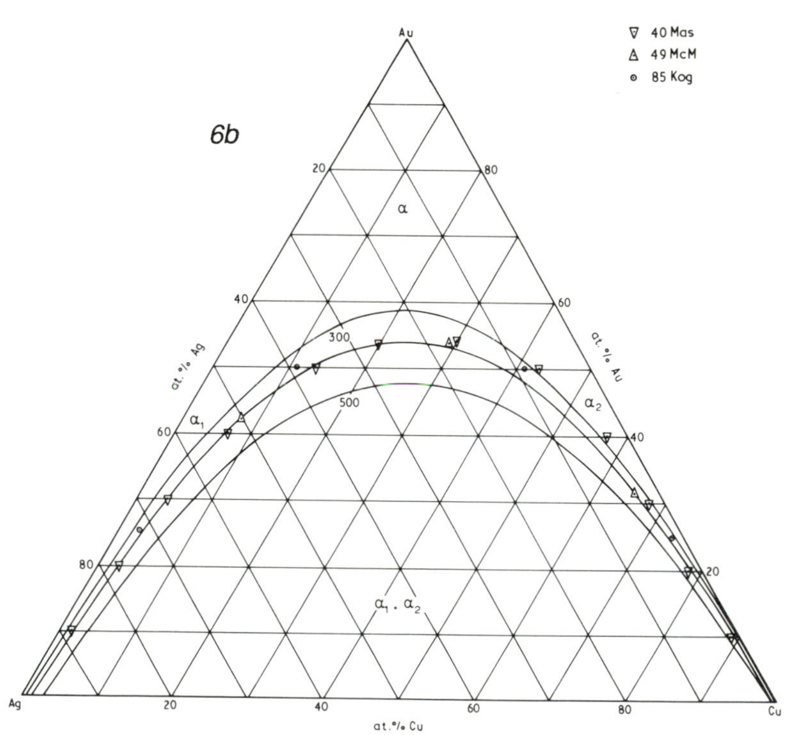

6 *a* **Assessed isothermal sections through the solvus surface. Tie lines for 725°C isotherm due to [67 Zie];** ***b*** **Comparison of data for 400°C;** ***c*** **Comparison of data for 500°C;** ***d*** **Comparison of data for 600°C;** ***e*** **Comparison of data for 700°C;** ***f*** **Data of [67 Zie] for 725°C;** ***g*** **Comparison of data for 750°C**

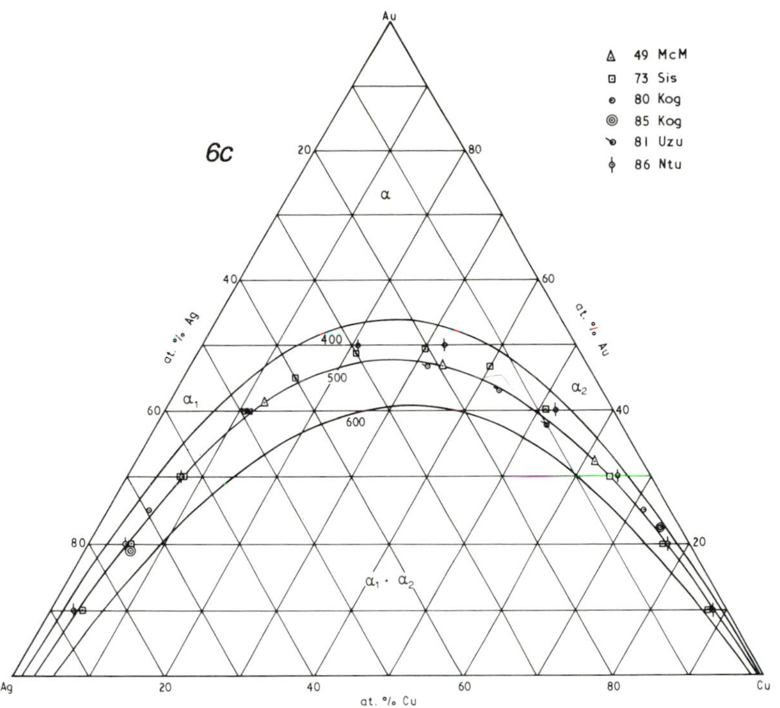
Au
6c
α
α₁
α₂
α₁ + α₂
400
500
600
at. % Ag
at. % Au
at. % Cu
Ag
Cu
49 McM
73 Sis
80 Kog
85 Kog
81 Uzu
86 Ntu

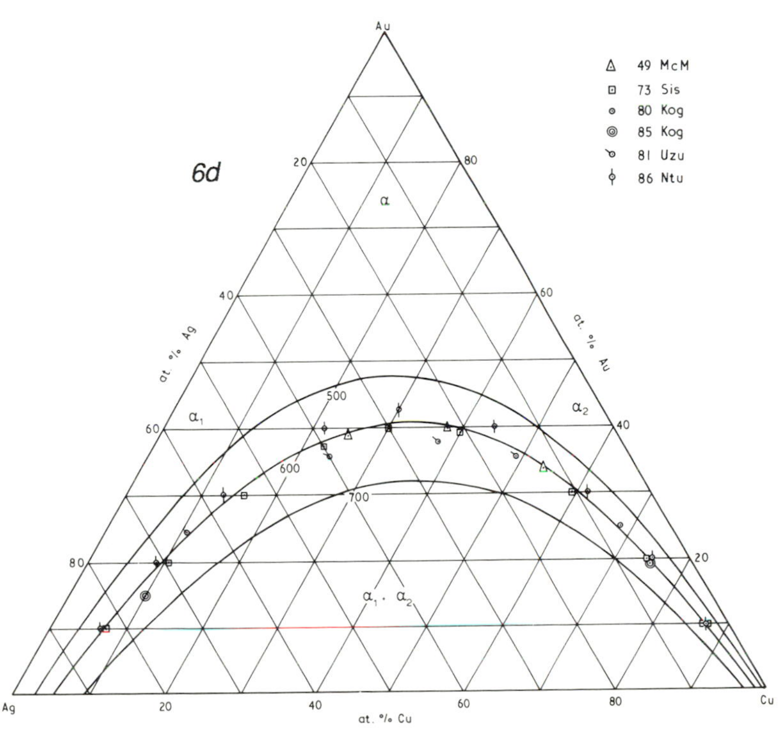
Au
6d
α
α₁
α₂
α₁ + α₂
500
600
700
at. % Ag
at. % Au
at. % Cu
Ag
Cu
49 McM
73 Sis
80 Kog
85 Kog
81 Uzu
86 Ntu

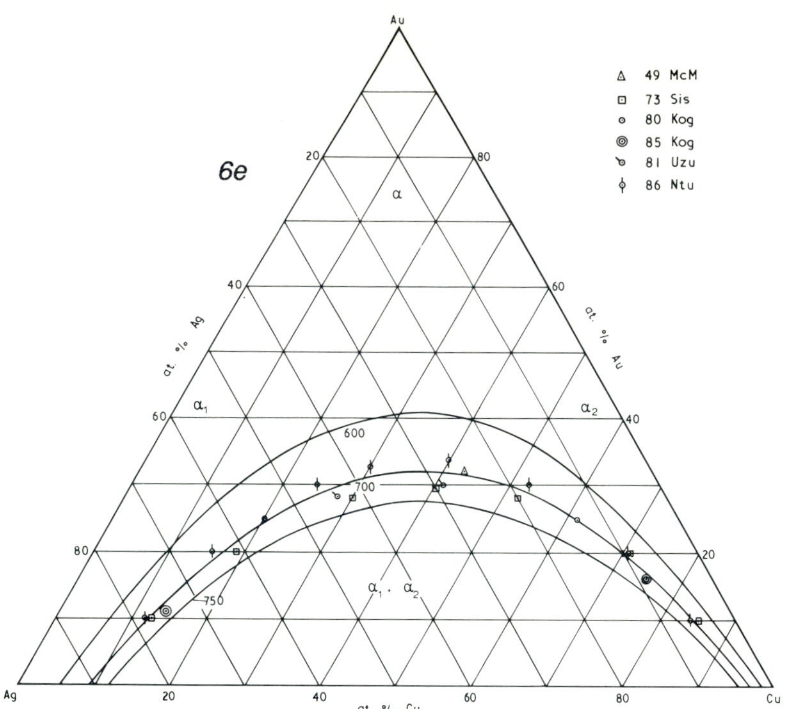
6e
Au
Ag
Cu
at. % Ag
at. % Au
at. % Cu
α
α1
α2
α1 + α2
600
700
750
Δ 49 McM
73 Sis
80 Kog
85 Kog
81 Uzu
86 Ntu
20
40
60
80

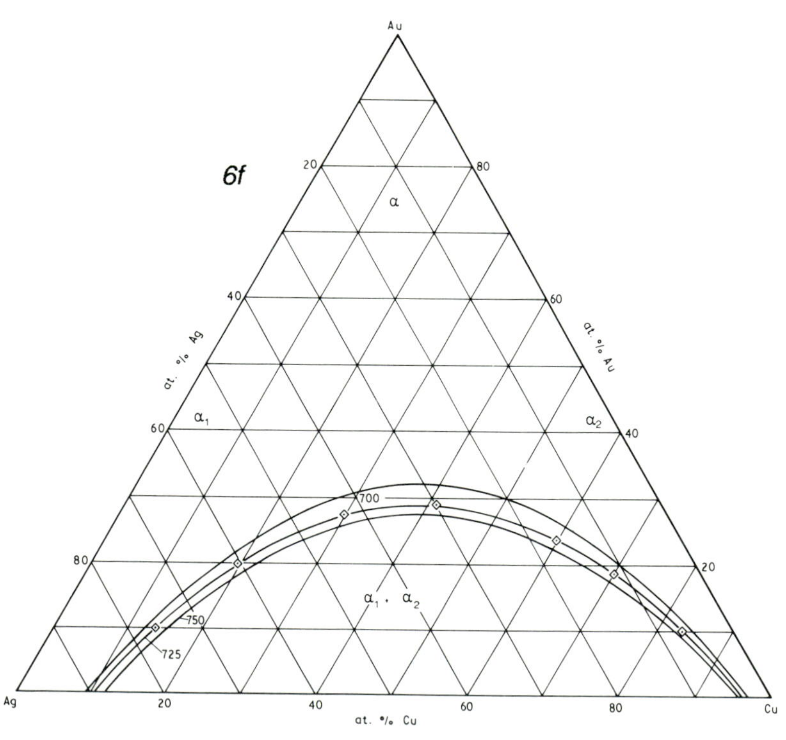
6f
Au
Ag
Cu
at. % Ag
at. % Au
at. % Cu
α
α1
α2
α1 + α2
700
750
725
20
40
60
80

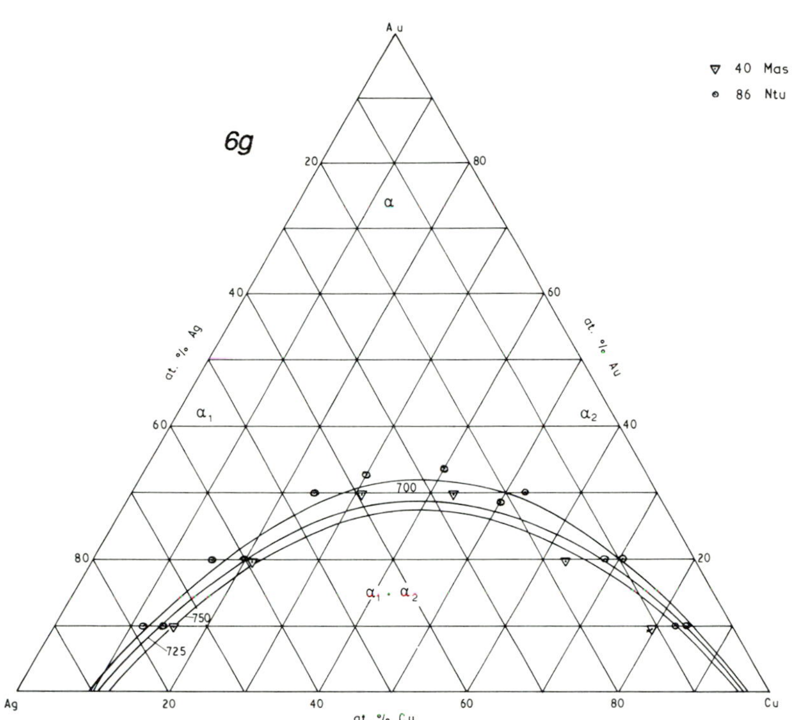
6g
Au
Ag
Cu
at. % Cu
at. % Ag
at. % Au
20
40
60
80
α
$α_1$
$α_2$
$α_1 + α_2$
700
750
725
▽ 40 Mas
⊘ 86 Ntu

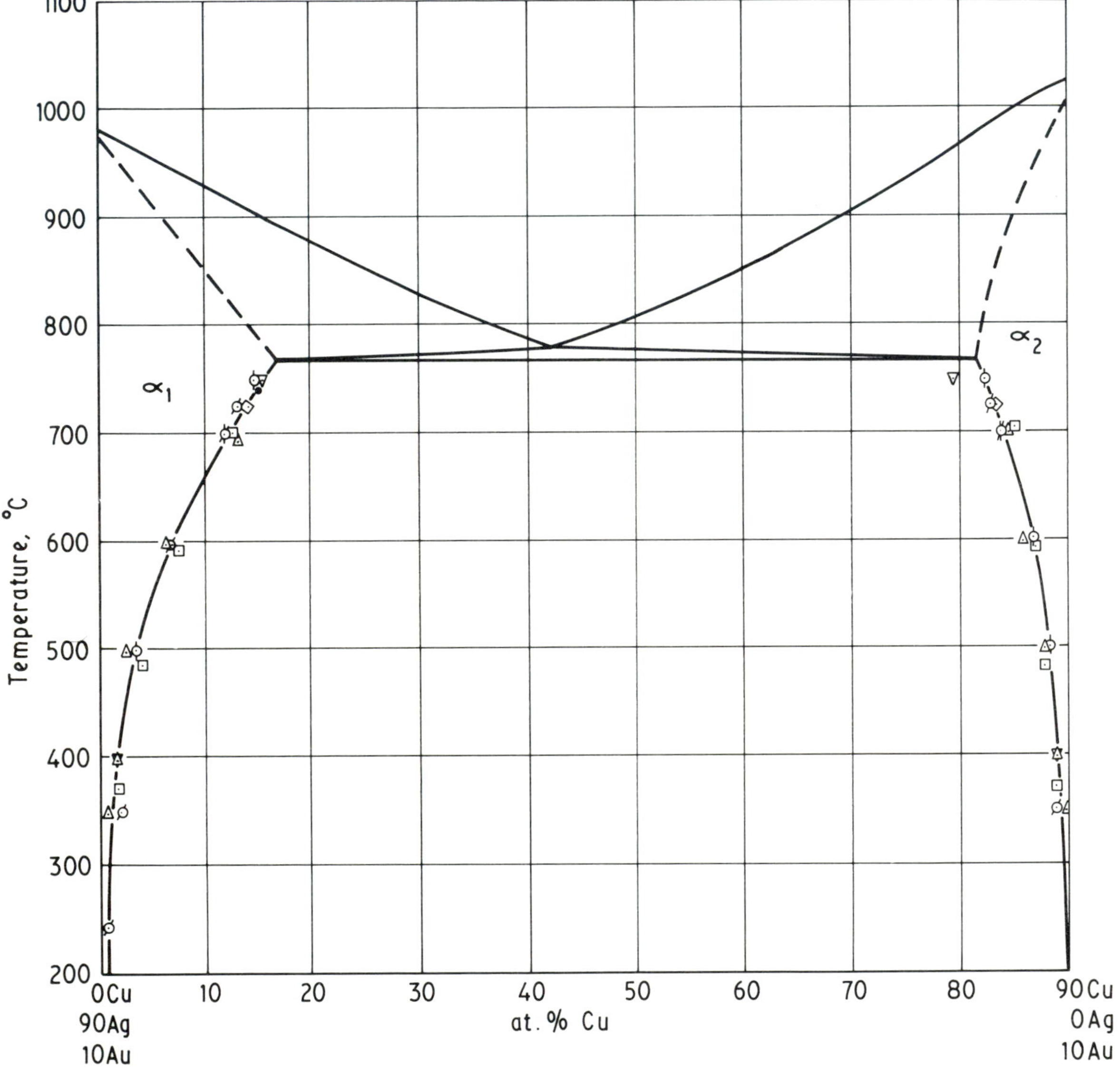

7a

7 **Vertical sections corresponding to: *a* 10 at.-% Au; *b* 30 at.-% Au; *c* 6 carat alloys; *d* 8 carat alloys; *e* 10 carat alloys; *f* 12 carat alloys; *g* 14 carat alloys**

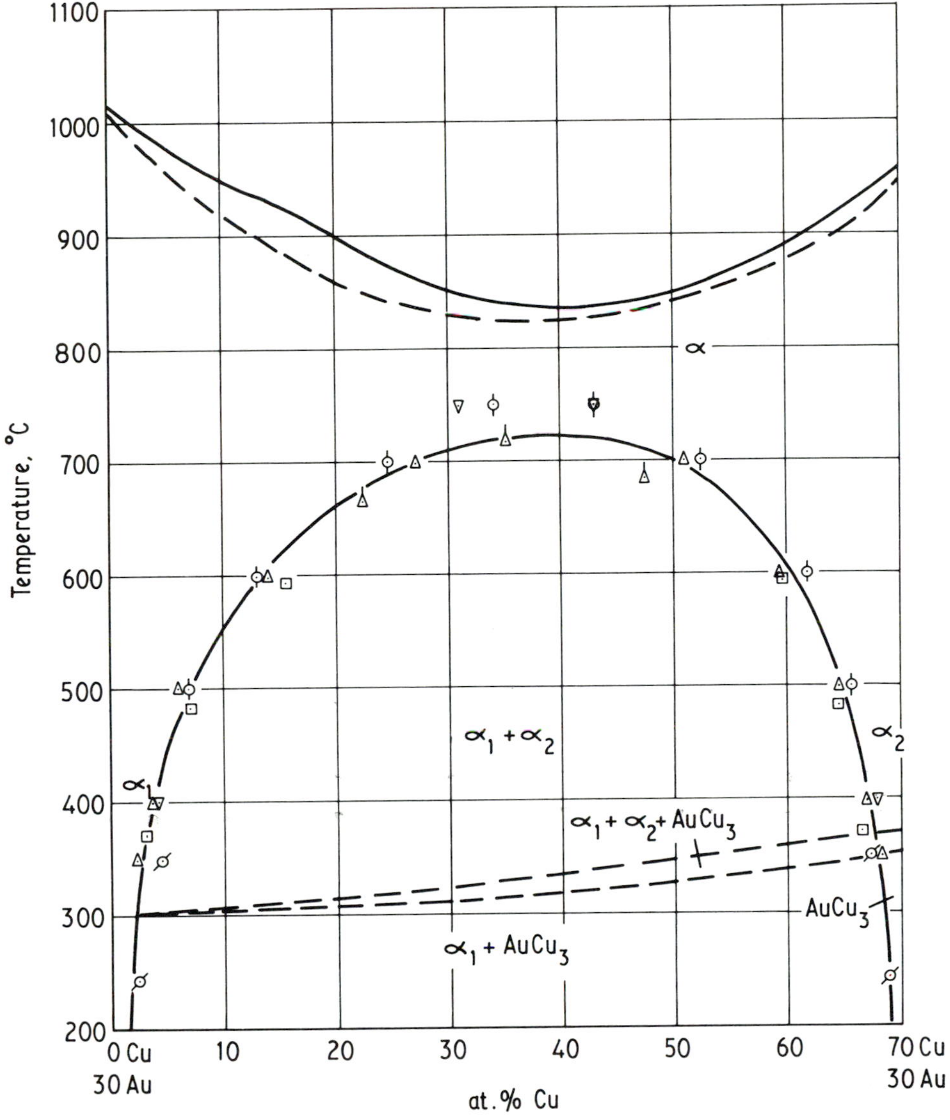
40 Mas
49 McM
73 Sis
67 Zie
75 Mur
80 Kog
80 Yam
85 Kog
86 Ntu
1100
1000
900
800
700
600
500
400
300
200
Temperature, °C
α
α₁ + α₂
α₁
α₂
α₁ + α₂ + AuCu₃
AuCu₃
α₁ + AuCu₃
0 Cu
30 Au
10
20
30
40
50
60
70 Cu
30 Au
at.% Cu

7b

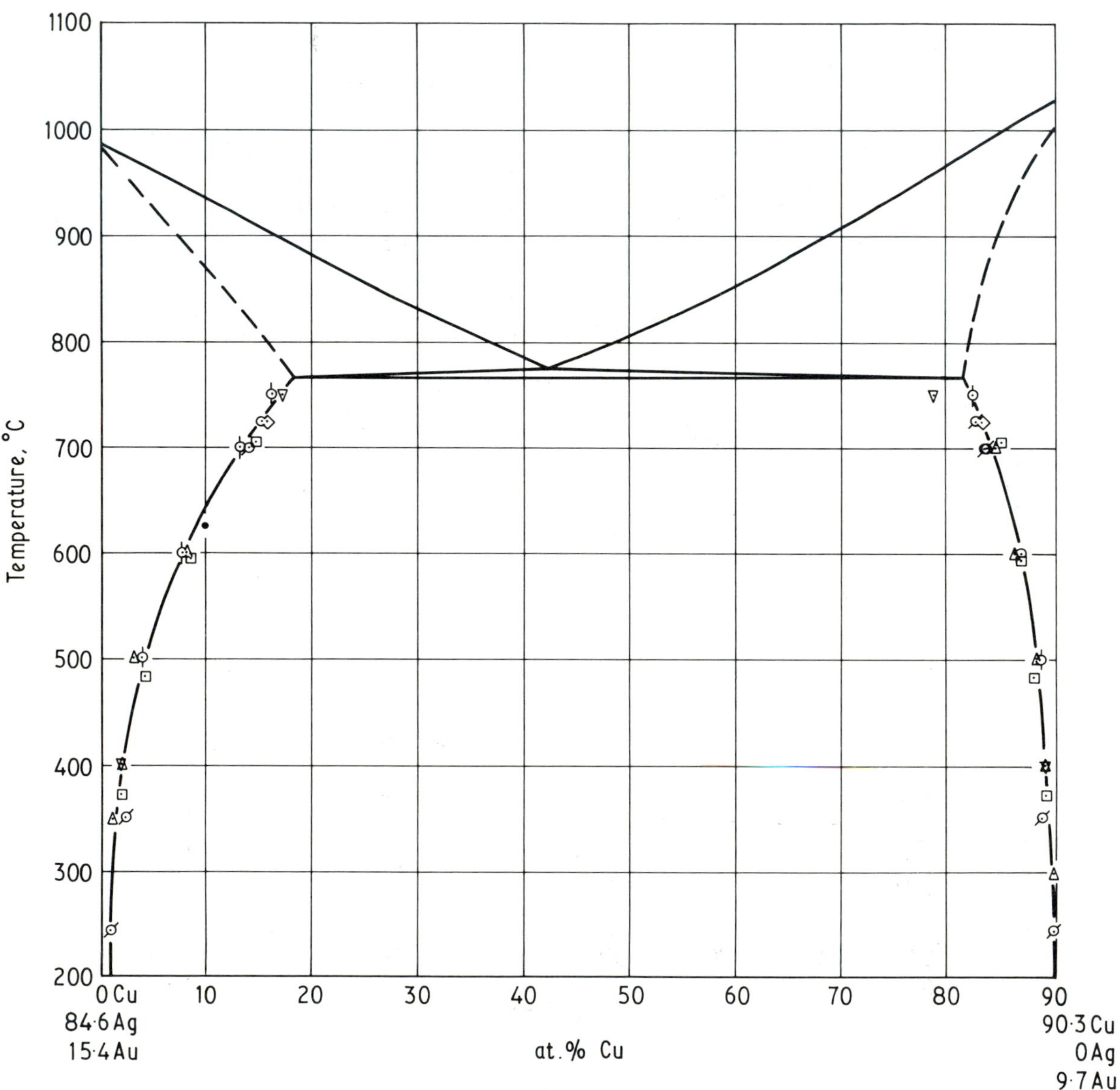

▽ 40 Mas
△ 49 McM
⊡ 73 Sis
◇ 67 Zie
⌀ 75 Mur
• 80 Kog
△ 80 Yam
⊙ 85 Kog
ϕ 86 Ntu
1100
1000
900
800
700
600
500
400
300
200
Temperature, °C
0 Cu
84·6 Ag
15·4 Au
10
20
30
40
50
60
70
80
90
90·3 Cu
0 Ag
9·7 Au
at.% Cu

7c

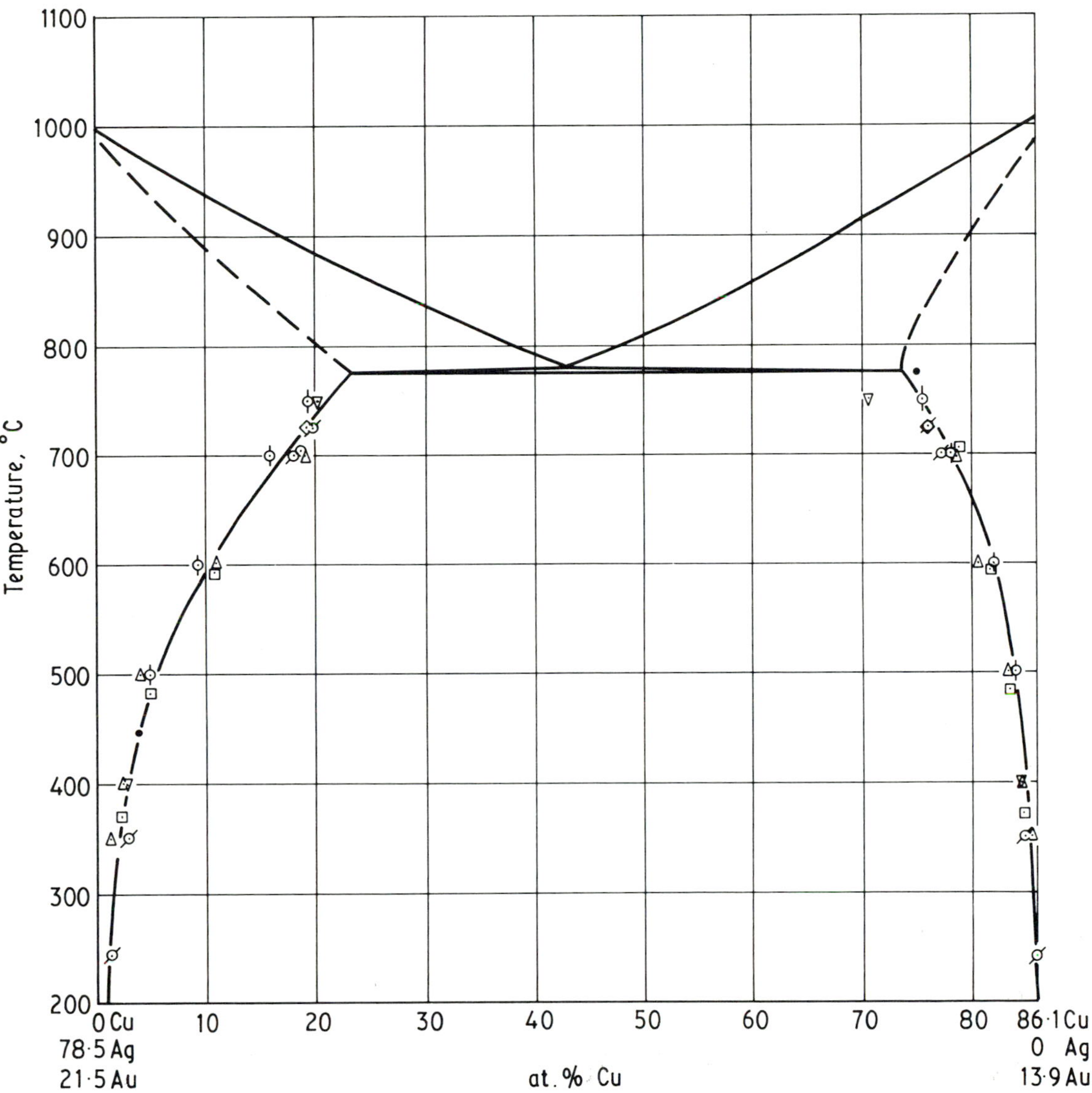

▽ 40 Mas
△ 49 McM
⊡ 73 Sis
◇ 67 Zie
⌀ 75 Mur
• 80 Kog
Å 80 Yam
⊙ 85 Kog
⌽ 86 Ntu
1100
1000
900
800
700
600
500
400
300
200
Temperature, °C
0 Cu
78·5 Ag
21·5 Au
10
20
30
40
50
60
70
80
86·1 Cu
0 Ag
13·9 Au
at.% Cu

7d

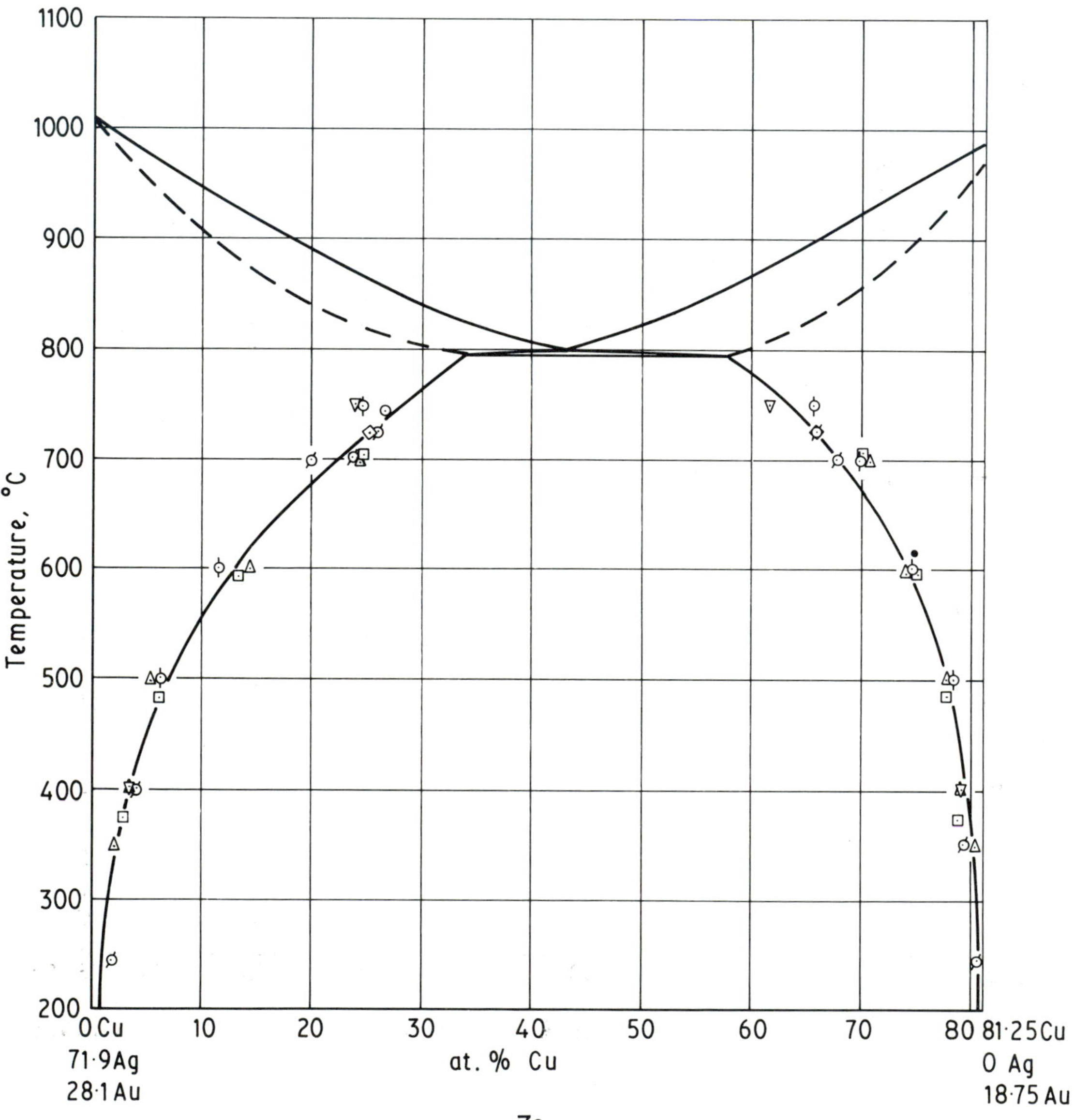

▽ 40 Mas
△ 49 McM
□ 73 Sis
◇ 67 Zie
ø 75 Mur
• 80 Kog
Å 80 Yam
○ 85 Kog
ϕ 86 Ntu
Temperature, °C
1100
1000
900
800
700
600
500
400
300
200
0 Cu
71·9Ag
28·1Au
10
20
30
40
50
60
70
80
81·25Cu
0 Ag
18·75Au
at. % Cu

7e

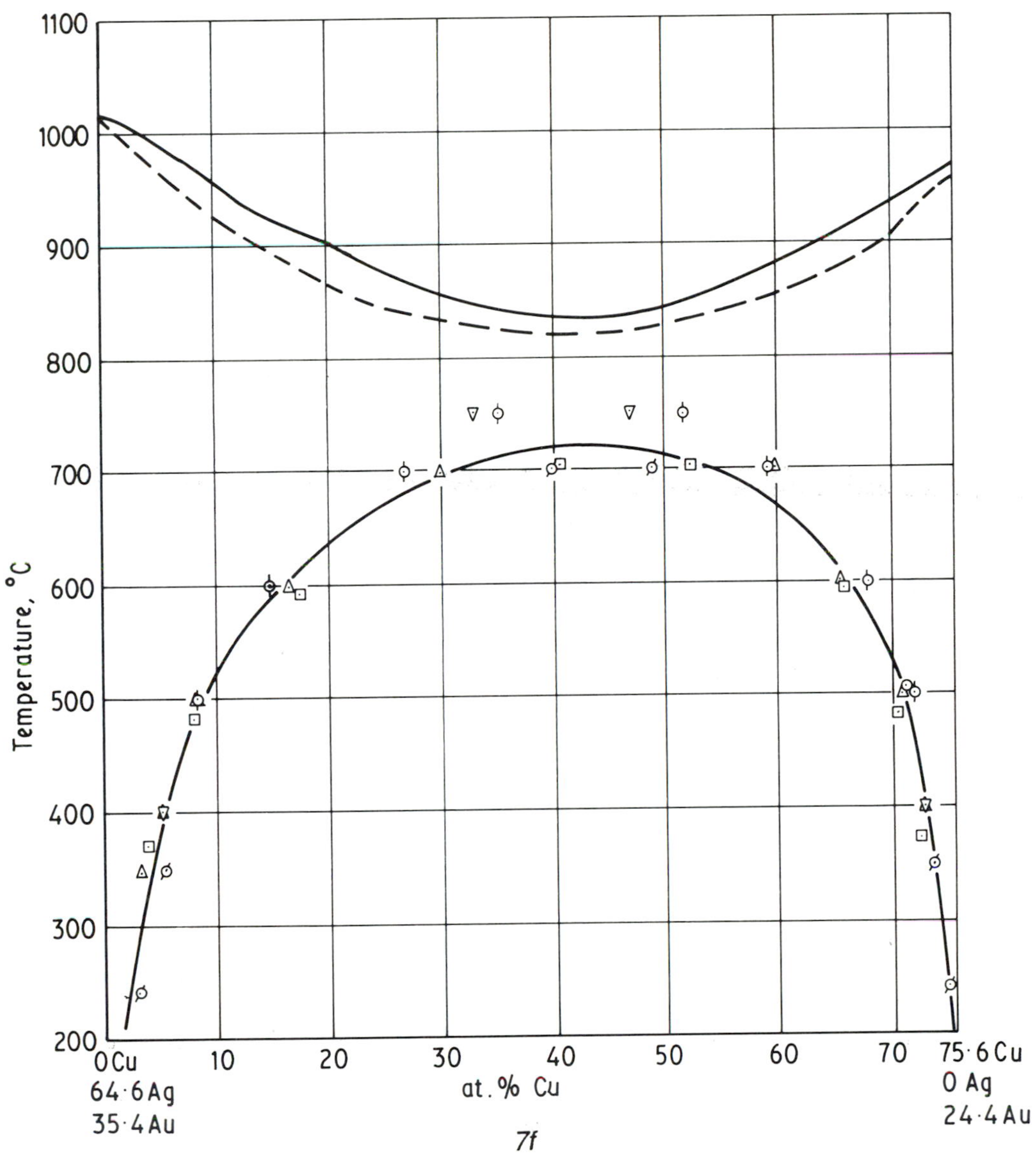

▽ 40 Mas
△ 49 McM
⊡ 73 Sis
◇ 67 Zie
ø 75 Mur
• 80 Kog
Å 80 Yam
○ 85 Kog
φ 86 Ntu
Temperature, °C
1100
1000
900
800
700
600
500
400
300
200
0 Cu
64·6 Ag
35·4 Au
10
20
30
40
50
60
70
75·6 Cu
0 Ag
24·4 Au
at.% Cu

7f

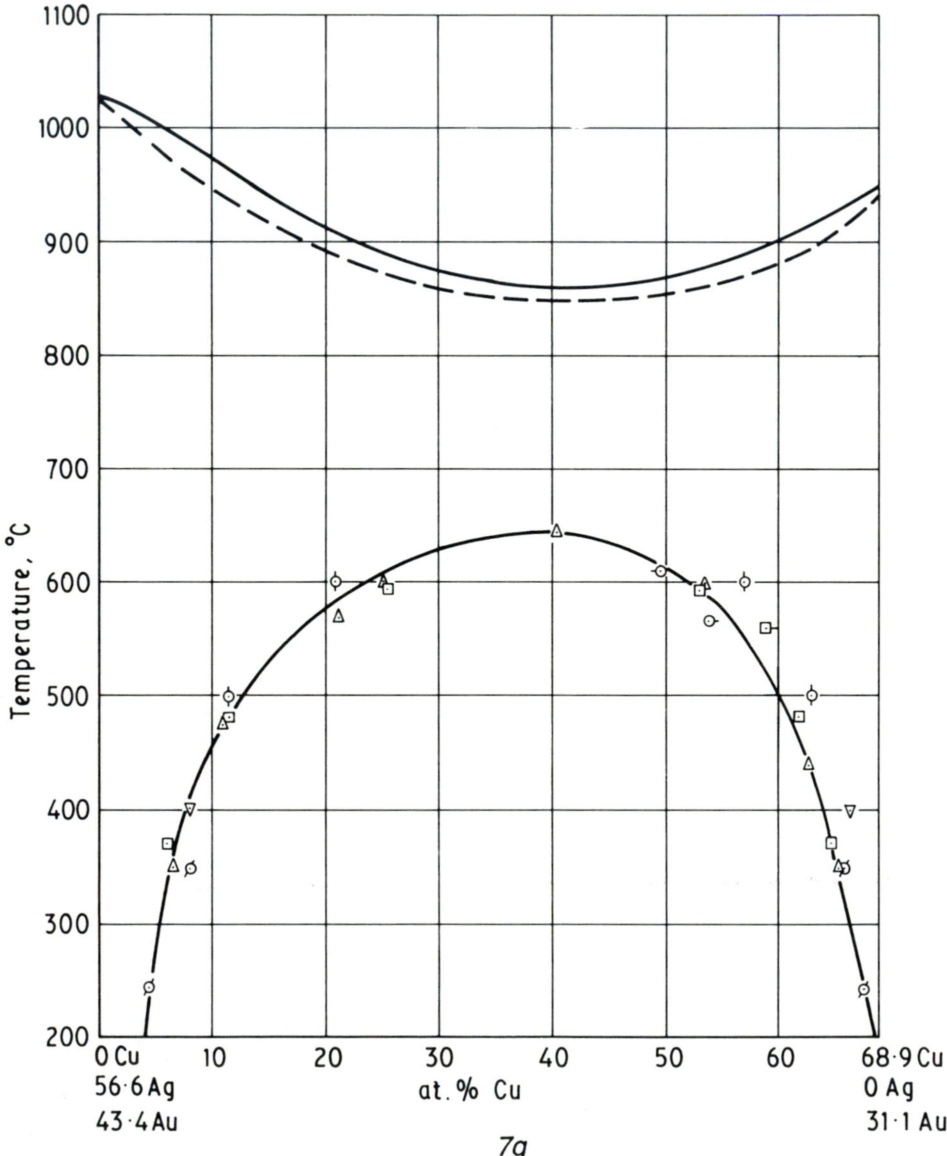

40 Mas
43 Pic
49 McM
73 Sis
74 Sha
75 Mur
79 Yas
86 Ntu
Temperature, °C
1100
1000
900
800
700
600
500
400
300
200
0 Cu
56·6 Ag
43·4 Au
10
20
30
40
50
60
68·9 Cu
0 Ag
31·1 Au
at.% Cu

7g

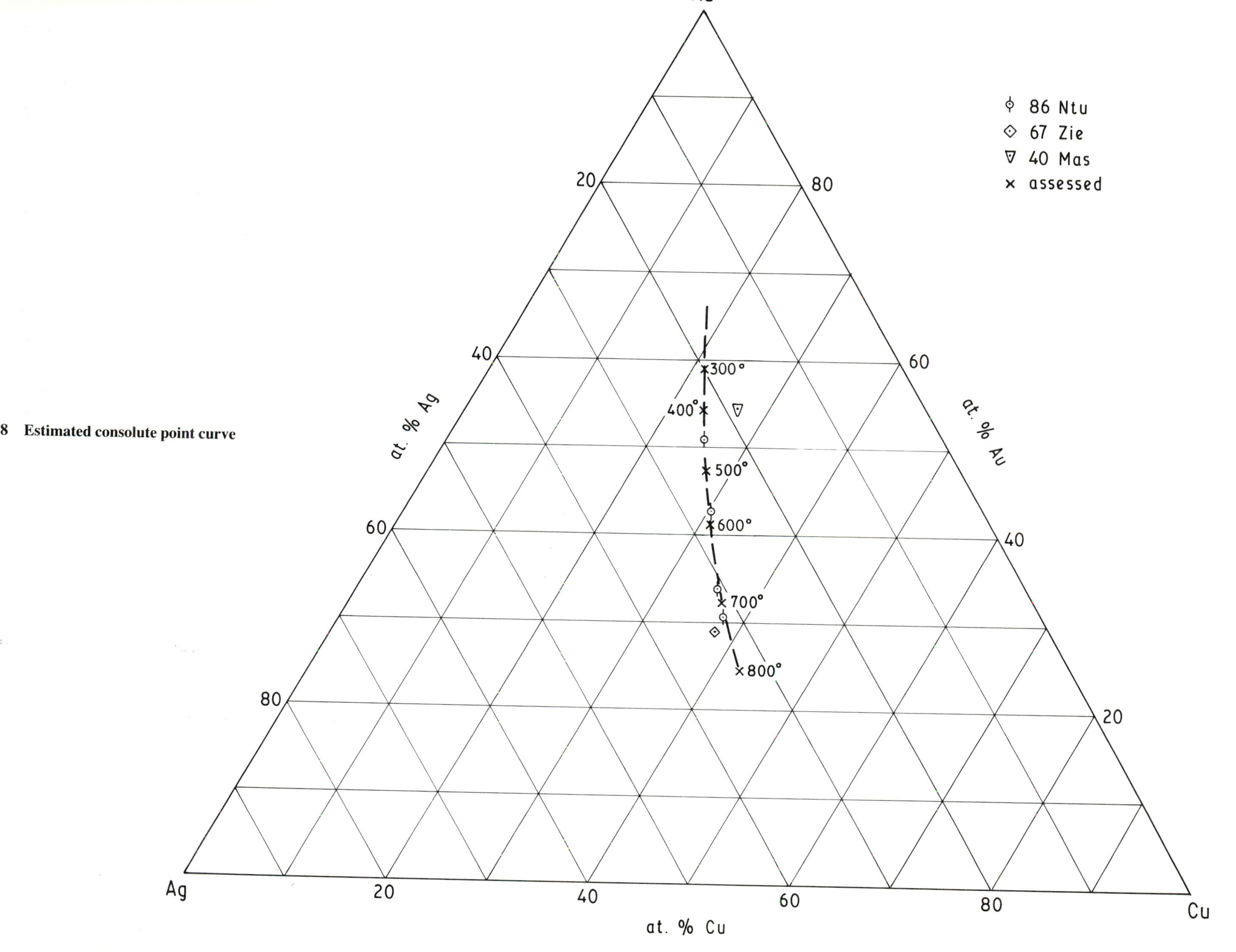

8 Estimated consolute point curve

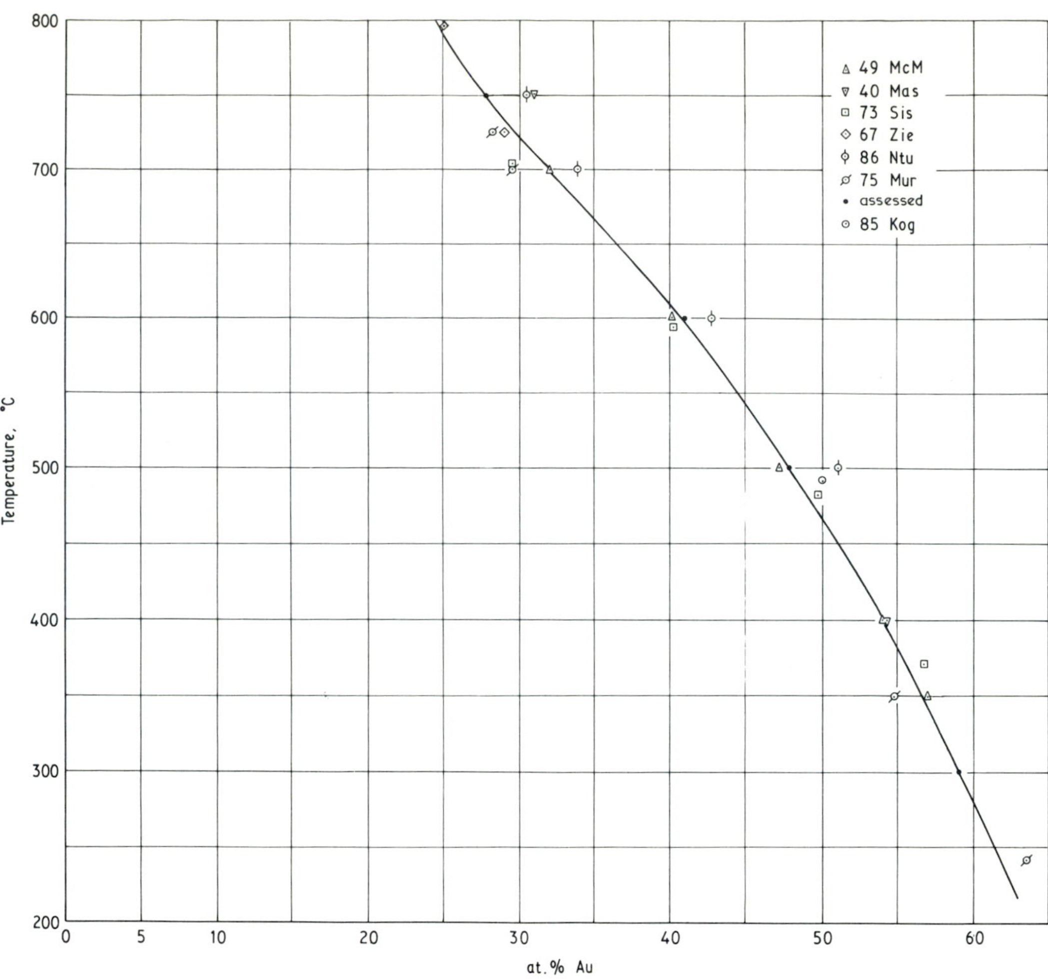

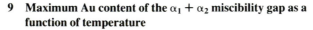

9 Maximum Au content of the $\alpha_1 + \alpha_2$ miscibility gap as a function of temperature

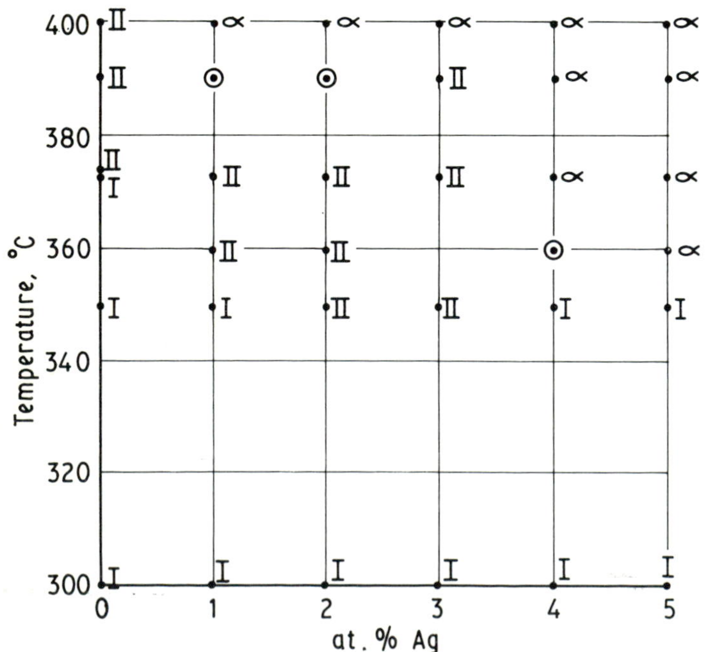

10 Constitution of 50 at.-% Cu section [39 Hul]: I = AuCuI, II = AuCuII

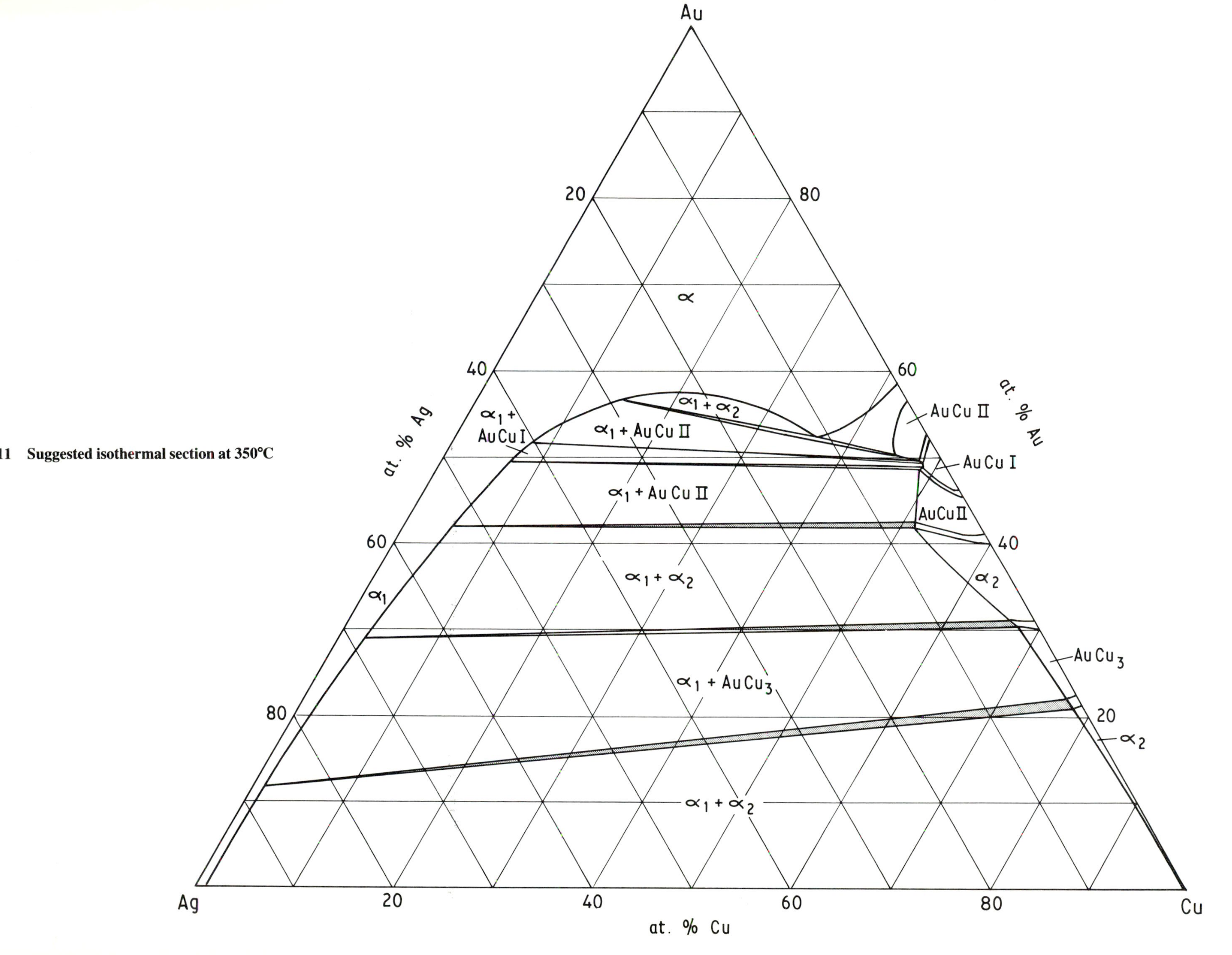

11 Suggested isothermal section at 350°C

+ α_2 was determined by both DTA on heating and from the temperature dependence of the lattice spacing of $AuCu_3$, α_2 and α solid solution (denoted as the α_2 phase region in Fig. 13). The transition from $\alpha_1 + \alpha_2$ to the α_2 phase region was detected by lattice spacing measurements as a function of temperature; DTA gave no detectable effect on crossing this phase boundary. Syutkina *et al.* [86 Syu] used the variation of specific electrical resistivity with temperature to determine the effect of Ag additions of 4, 8 and 12 at.-% on the ordering temperature of $AuCu_3$. Alloy compositions were on the Ag–$AuCu_3$ section. They found that the ordering temperature is depressed to about 360 from 390°C by the Ag additions; this is in qualitative agreement with the data of [80 Kog] for the 75 at.-% Cu section (Fig. 13). The 75 at.-% Ag section (Fig. 14) shows the boundary between α_1 + $AuCu_3$ and α_1 + α_2 as rising to a maximum of 387°C at 5 at.-% Cu. The supposition that the $\alpha_1 + \alpha_2$ + $AuCu_3$ phase region does rise to a maximum is supported by the data of [85 Kog] for the 25 at.-% Au section (Fig. 15). The alloy compositions at the maxima in Figs. 14 and 15 are 75% Ag, 5% Cu and 9% Ag, 66% Cu respectively. These compositions lie on a critical tie line that joins the α_1,α_2 and $AuCu_3$ phases at 387°C. This tie line is formed by the meeting of the dome-shaped α_2 + $AuCu_3$ phase region originating at 390°C in the Au–Cu binary system with the dome-shaped $\alpha_1 + \alpha_2$ miscibility gap, originating from the Ag–Cu binary system. Little Ag is soluble in $AuCu_3$ and this is reflected by the appearance of the critical tie line, corresponding to the reaction $\alpha_2 = \alpha_1$ + $AuCu_3$, at a temperature 3°C below the ordering temperature of $AuCu_3$. As a result of the occurrence of the critical tie line two $\alpha_1 + \alpha_2$ + $AuCu_3$ phase regions, separated by an α_1 + $AuCu_3$ region, are formed (Figs. 11 and 12). From Fig. 6*a* the α_1 phase at 387°C will contain 2·5 at.-% Cu and the α_2 phase 2 at.-% Ag. On the assumption that $AuCu_3$ takes 1 at.-% Ag into solution (Fig. 13), the compositions of α_1, α_2 and $AuCu_3$ in equilibrium at 387°C are estimated to be 77·7% Ag, 19·8% Au; 2·0% Ag, 25·5% Au; 1·0% Ag, 25·6% Au respectively. It should be noted that [80 Yam], [80 Kik] and [81 de Fon] have used the cluster variation method to calculate isothermal sections of the Ag–Au–Cu coherent phase diagram. The calculated isothermals at 350 and 310°C are very similar to Figs. 11 and 12 in showing two $\alpha_1 + \alpha_2$ + $AuCu_3$ tie triangles.

Kogachi and Nakahigashi [85 Kog] determined the low-temperature portions of the 25 and 50 at.-% Au sections, Figs. 15 and 16, respectively. Alloys were prepared in the way described above for [80 Kog]. The transition temperature from the α_1 + $AuCu_3$ phase region to the $\alpha_1 + \alpha_2$ phase region in the 25 at.-% Au section (Fig. 15), was determined by DTA on heating. The alloy ingots were annealed for 10 days at 415°C in evacuated Pyrex tubes and cooled to room temperature at 10°C/day before DTA. A three-phase region, $\alpha_1 + \alpha_2$ + $AuCu_3$, has been inserted between the α_1 + $AuCu_3$ and $\alpha_1 + \alpha_2$ phase regions in Fig. 15. The $\alpha_1 + \alpha_2$ to α transition was determined from the variation of the lattice parameter of the α_1, α_2 and α phases with composition for alloys annealed at temperatures from 700 to 450°C for times varying from 5 h to 48 days. The assessed solvus data, Fig. 6*a*, are plotted with the data of [85 Kog] in Fig. 15. There is satisfactory agreement between the data.

The 50 at.-% Au section (Fig. 16) is complex. Kogachi and Nakahigashi [85 Kog] confirmed the previously observed α, $\alpha_1 + \alpha_2$, AuCuI, AuCuII and α_2 + AuCuII phase regions. They also identified three new phase regions, α_2 + AuCuII, α_1 + AuCuI and $\alpha_1 + \alpha_2$ + AuCuII. The phase boundary between $\alpha_1 + \alpha_2$ and α was determined from the variation of the lattice parameter of the α_1, α_2 and α phases with composition for alloys annealed at temperatures from 540°C to 300°C for times varying from 6 to 181 days. The lattice parameter of the disordered α phase agrees well with previous data (Fig. 18) and shows a small positive deviation from Végard's law. For alloys containing 20, 25 and 30 at.-% Ag the lattice parameters of the α_1, α_2 and α phases were determined as a function of temperature to provide the $\alpha_1 + \alpha_2 - \alpha$ phase boundary. The phases present at temperatures from 400 to 320°C in 10°C steps and at 300°C were characterized by X-ray diffraction on samples annealed for 120 to 189 days. All the data points of [85 Kog] are plotted in Fig. 16. Certain conclusions can be derived from the data. As the lattice parameter of the α_1 and α_2 phases remain nearly constant at each annealing temperature (465, 450, 432, 400, 380°C) this implies that the tie line $\alpha_1\alpha_2$ is almost parallel to the 50 at.-% Au section over this temperature range; this agrees with the 400°C section of [40 Mas]. For alloys which contain $\alpha_1 + \alpha_2$ at temperatures of 380°C [85 Kog] observed the sequence of phase regions from 380°C to 300°C of $\alpha_1 + \alpha_2$, $\alpha_1 + \alpha_2$ + AuCuII, α_1 + AuCuII, α_1 + AuCuI. The additional phase region α_1 + AuCuI + AuCuII has been inserted in Fig. 16, as has the AuCuI + AuCuII phase region. The $\alpha_1 + \alpha_2$ + AuCuII phase region is significant in so far as it rises to a maximum for the boundary adjoining the α_1 + α_2 region; note the $\alpha_1 + \alpha_2$ structure for alloys with 9, 10, 15 and 40 at.-% Ag at 370°C. The $\alpha_1 + \alpha_2$ + AuCuII boundary with the α_1 + AuCuII phase region rises as the Ag content increases to about 30 at.-% Ag; note the three-phase structure from 4 to 12·5 at.-% Ag and the α_1 + AuCuII structure from 15 at.-% Ag. This implies the presence of a critical tie line in the ternary system at which α_1, α_2 and AuCuII are in equilibrium at a temperature near to 374°C (>370°C but <380°C, Fig. 16). The critical tie line is considered to form when the dome-shaped regions representing the α_2 surface and the AuCuII surface, emanating from the binary ordering transition, $a_2 \rightleftharpoons$ AuCuII, at 410°C meet the $\alpha_1 + \alpha_2$ dome-shaped solvus surface. Along the critical tie line $\alpha_2 \rightleftharpoons \alpha_1$ + AuCuII. Below 374°C two three-phase triangles, representing equilibrium of $\alpha_1 + \alpha_2$ + AuCuII, will appear in the isothermal sections. These are observable in the isothermal sections suggested for 350°C (Fig. 11) and 300°C (Fig. 12). In going from 350 to 300°C the middle $\alpha_1 + \alpha_2$ phase region narrows considerably as a result of the movement of the $\alpha_1 + \alpha_2$ + AuCuII phase region at about 42 at.-% Au to about 35 at.-% Au and the movement of the $\alpha_1 + \alpha_2$ + $AuCu_3$ phase region at about 29 at.-% Au to about 31 at.-% Au. It would appear that this $\alpha_1 + \alpha_2$ phase region

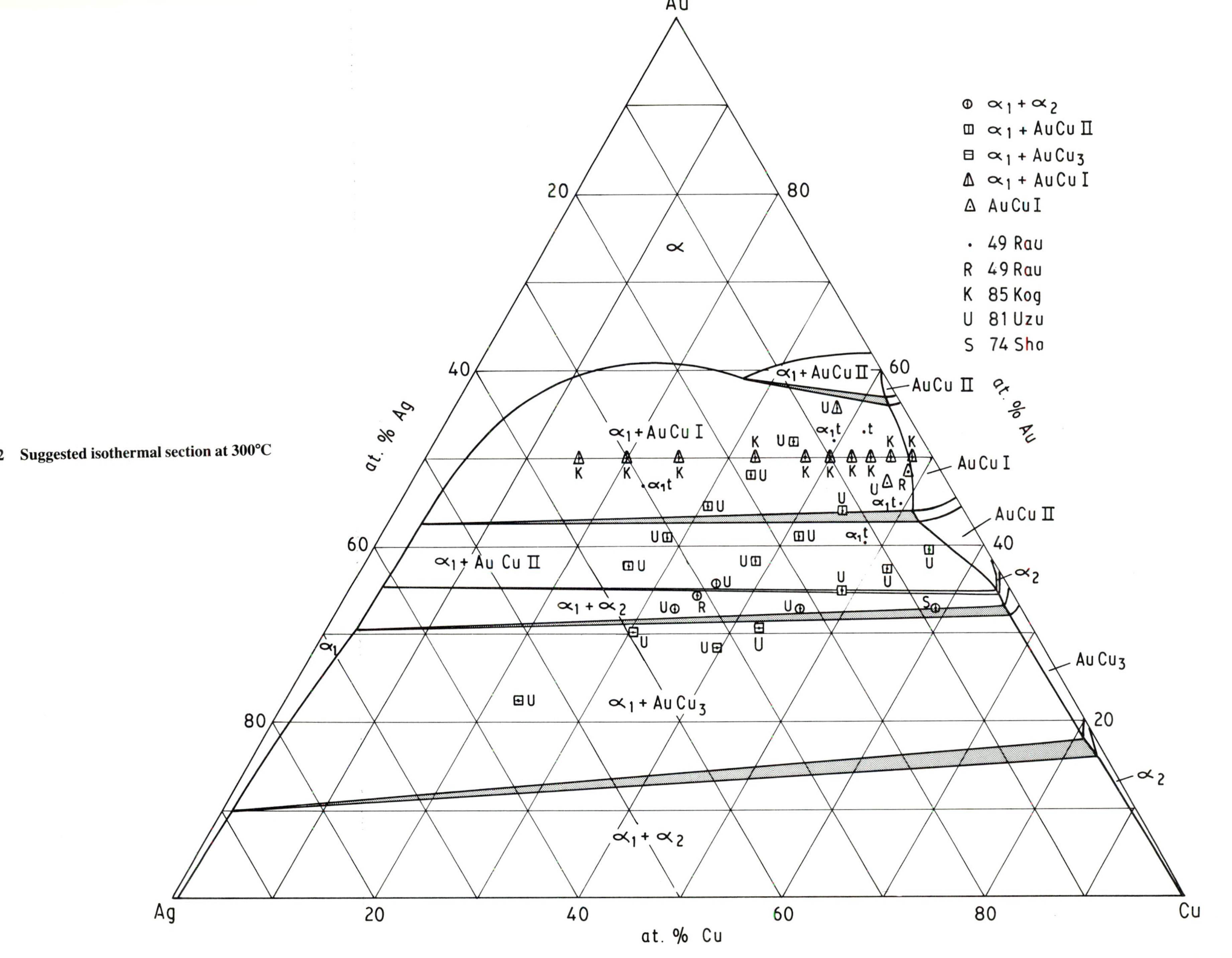

12 Suggested isothermal section at 300°C

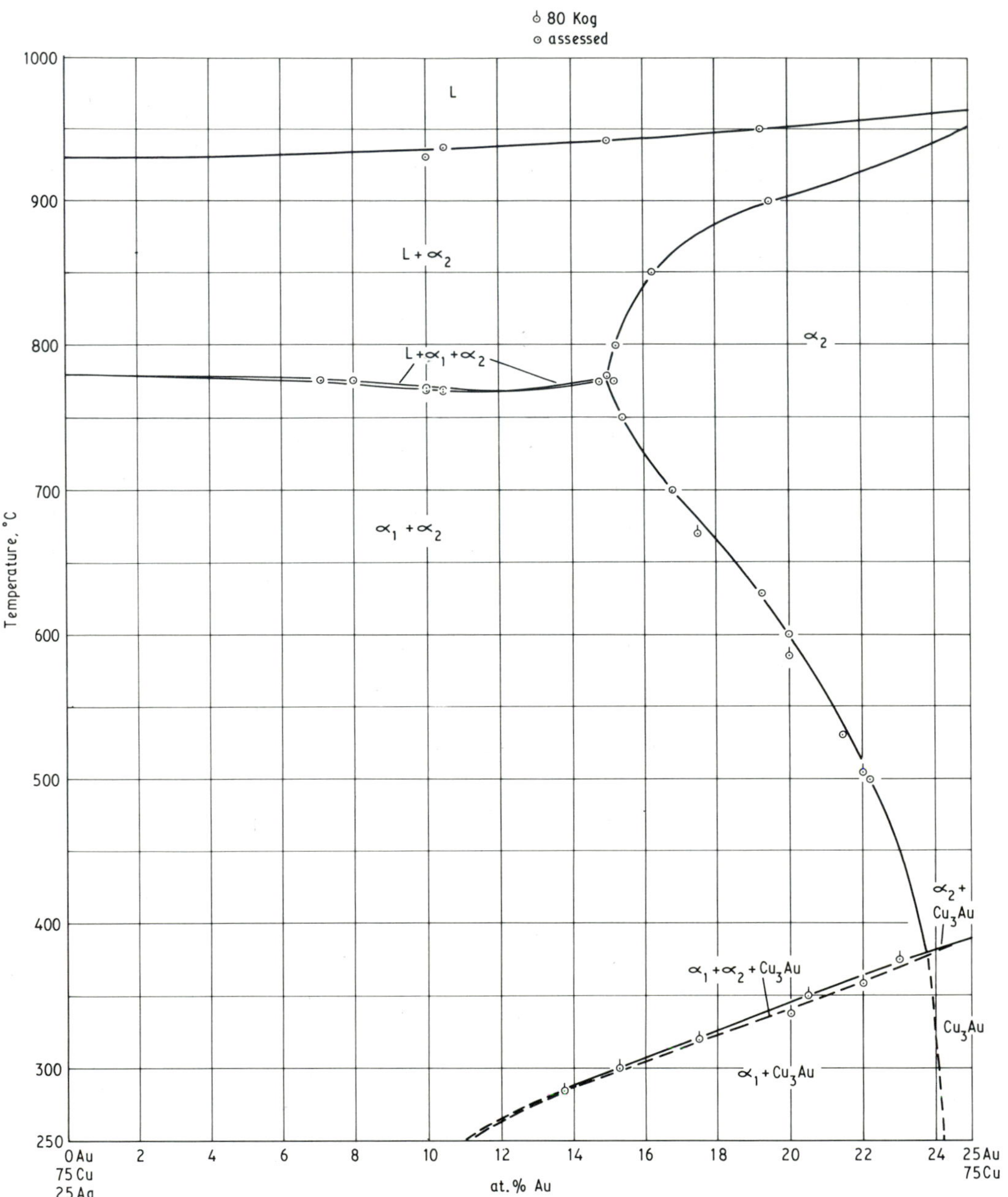

13 Vertical section at 75 at.-% Cu [80 Kog]

will disappear with the invariant reaction $\alpha_2 \rightleftharpoons \alpha_1$ + AuCuII + $AuCu_3$ at a temperature below 300°C.

The boundary between α_1 + AuCuII and α_1 + AuCuI also slopes upwards for Ag contents up to ~25 at.-%; note the α_1 + AuCuII structure for alloys with 5 and 7·5 at.-% Ag at 350°C and the α_1 + AuCuI structure for alloys with ≥10 at.-% Ag. It is suggested that another critical tie line, connecting the α_1, AuCuI and AuCuII phases, exists at a temperature of about 358°C (>350°C but <360°C, Fig. 16). Along this tie line AuCuII $\rightleftharpoons \alpha_1$ + AuCuI. The tie line is formed when the dome-shaped AuCuII and AuCuI surfaces, emanating from the AuCuII = AuCuI transition at 385°C in the Au–Cu binary system, meet the AuCuII surface of the α_1 + AuCuII phase region produced as a result of the $\alpha_2 \rightleftharpoons \alpha_1$ + AuCuII reaction at 374°C. Below 358°C the AuCuII $\rightleftharpoons \alpha_1$ + AuCuI reaction will give rise to two tie triangles representing equilibrium of α_1 + AuCuI + AuCuII. These are observed in the suggested 350° and 300°C isothermal sections (Figs. 11 and 12).

Uzuka *et al.* [81 Uzu] studied sections from 59·6% Au, 40·5% Cu to Ag; 50·0% Au, 50·0% Cu to Ag; 41·8% Au, 58·2% Cu to Ag at temperatures below 700°C. Alloys were prepared by induction melting >99·99% elements *in vacuo*, ingots were worked and then solution heat treated for 10 h at 750°C. Filings from the ingots were sealed in evacuated silica tubes, disordered by heating for several hours at 750°C and water quenched. Ageing was done in 50° steps from 300 to 750°C and X-ray diffraction analysis used to characterize the phases present. Ageing times are quoted from 20 to 130 days at 300°C and 4 days at 380°C. This should be compared with 120 to 189 days used by [85 Kog]] for the temperature range from 400 to 320°C on the 50 at.-% Au section. The results of [81 Uzu] differ from those of [85 Nog] in that the former observed an α_1 + AuCuII structure at 300°C for alloys with high Au contents on all three sections (Figs. 17*a-c*). A direct comparison of the data can be made for the 50% Au, 16% Ag alloy [85 Kog] (Fig. 16), and the same alloy at 16% Ag [81 Uzu] (Fig. 17*a*). According to [81 Uzu] this alloy transforms from $\alpha_1 + \alpha_2$ to α_1 + AuCuII at about 365°C and remains α_1 + AuCuII at 300°C. Kogachi and Nakahigashi [85 Kog] used 10°C annealing steps and identified the transitions $\alpha_1 + \alpha_2$ to $\alpha_1 + \alpha_2$ + AuCuII at 370°C; $\alpha_1 + \alpha_2$ + AuCuII to α_1 + AuCuII at 362°C; α_1 + AuCuII to α_1 + AuCuI (ignoring the narrow α_1 + AuCuI + AuCuII region) at 355°C; below 355°C the stable structure is α_1 + AuCuI. Support for [85 Kog] is the observation by [75 Kan] of α_1 + AuCuI for an alloy containing 11·2% Ag, 53·1% Au after 120 h ageing at 300°C. This alloy is very close to the 11% Ag alloy in Fig. 17*a*. Wise and Eash [33 Wis] found an α_1 + AuCuI structure for a 20·2% Ag, 50.5% Au alloy annealed for 624 h at 300°C, in agreement with [85 Kog]. The results of [81 Uzu] at 300°C for the 59·5% Au, 40·5% Cu–Ag section (Fig. 17*a*) are not accepted. The sections from 50% Au, 50% Cu–Ag and 41·8% Au, 58·2% Cu–Ag (Figs. 17*b* and 17*c*, respectively), have been used in drawing the suggested isothermal sections at 350 and 300°C (Figs. 11 and 12). The data of [85 Kog] for the 25 at.-% Au section give transition points for one alloy on Figs. 17*b* and *c*; they are in general agreement with [81 Uzu]. Age hardening studies on a 13·8% Ag, 43·0% Au alloy [78 Yas] indicated the formation of $\alpha_1 + \alpha_2$ between 510 and 380°C, of α_1 + AuCuII between 380 and 320°C and of AuCuI and AuCuII below 320°C. This data indicates a higher temperature stability of α_1 + AuCuII than [81 Uzu] (Fig. 17*b*). On the 300°C isothermal section an α_1 + AuCuI + AuCuII phase region is indicated, rather than AuCuI + AuCuII of [78 Yas]. Yasuda *et al.* [87 Yas] studied the 5·6% Ag, 46·9% Au alloy after homogenizing at 800°C for 50 h, cold rolling to 0·1 mm thick sheet and annealing at 300–350°C for 0·03–28 h. Conventional and high-resolution electron microscopy revealed the coexistence of α_1 + AuCuI at 300°C (in agreement with Fig. 12), α_1 + AuCuI + AuCuII from 310–330°C and α_1 + AuCuII at 340°C. Age-hardening of a 15·8% Ag, 34·5% Au alloy [79 Yas, 82 Yas] yielded $\alpha_1 + \alpha_2$ from 610 to 340°C and α_1 + AuCuII below 340°C in agreement with [81 Uzu] for a 300°C anneal.

The results of [81 Uzu] (Figs. 17*b* and *c*) support the conclusion that there is an invariant reaction $\alpha_2 \rightleftharpoons \alpha_1$ + AuCuII + $AuCu_3$ at a temperature below 300°C. Shaskov *et al.* [74 Sha] investigated the alloy 8·2% Ag, 32·9% Au and found an $\alpha_1 + \alpha_2$ structure at 300°C (Fig. 12). At 280°C the structure was α_1 + $AuCu_3$, suggesting that the reaction $\alpha_2 \rightleftharpoons \alpha_1$ + AuCuII + $AuCu_3$ occurs just above 280°C. As the AuCu binary eutectoid reaction, $\alpha_2 \rightleftharpoons$ AuCuII + $AuCu_3$, takes place at 285°C the proposed ternary invariant reaction occurs within a few degrees of this binary eutectoid. Kikuchi *et al.* [80 Kik] do not predict the formation of the AuCuII phase in the Au–Cu binary system but isothermal sections of the coherent ternary diagram indicate a eutectoid reaction between 310 and 280°C leading to the formation of an α_1 + AuCuI + $AuCu_3$ tie triangle.

Yamauchi *et al.* [80 Yam] provided additional low-temperature information on the 30 at.-% Au section (Fig. 7*b*). A 35% Cu alloy showed coherent precipitation of α_1 + α_2 at 330 and 350°C; a 55% Cu alloy gave coherent precipitation of α_1 + $AuCu_3$ at 320°C and coherent precipitation of $\alpha_1 + \alpha_2$ + $AuCu_3$ at 350°C. The three-phase triangle meets the α_1 solvus at a temperature slightly above 300°C, in agreement with the suggested 300°C isothermal section (Fig. 12).

Yamauchi and de Fontaine [86 Yam] have revised a plausible 300°C isothermal section given by [80 Yam] on the basis of the more recent data due to [85 Kog]. The revised isotherm contains 5 three-phase triangles and is topologically identical with the assessed 300°C isothermal section (Fig. 12). The section of [86 Yam] differs in detail from Fig. 12. Taking the tie triangle nearest to the Ag–Cu binary and going to the tie triangles with increasing Au contents, [86 Yam] place the first $\alpha_1 + \alpha_2$ + $AuCu_3$ region at higher Au contents than Fig. 12; in doing so they disregard the data of [80 Kog]. The second $\alpha_1 + \alpha_2$ + $AuCu_3$ tie triangle and the $\alpha_1 + \alpha_2$ + AuCuII tie triangles are similarly located by [86 Yam] to Fig. 12. The first α_1 + AuCuI + AuCuII tie triangle is located at higher Au contents by [86 Yam] and the second smaller α_1 + AuCuI

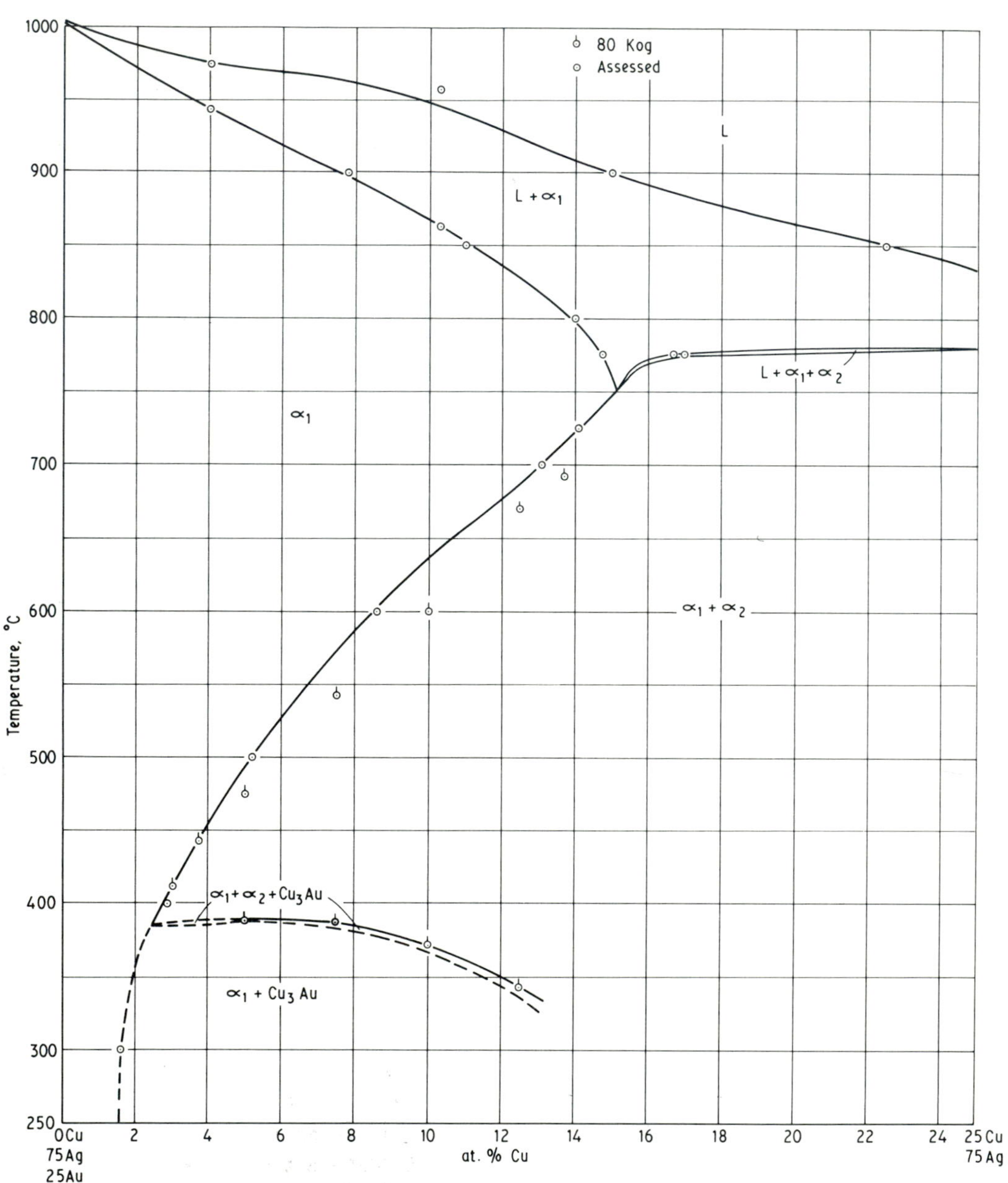

14 Vertical section at 75 at.-% Ag [80 Kog]

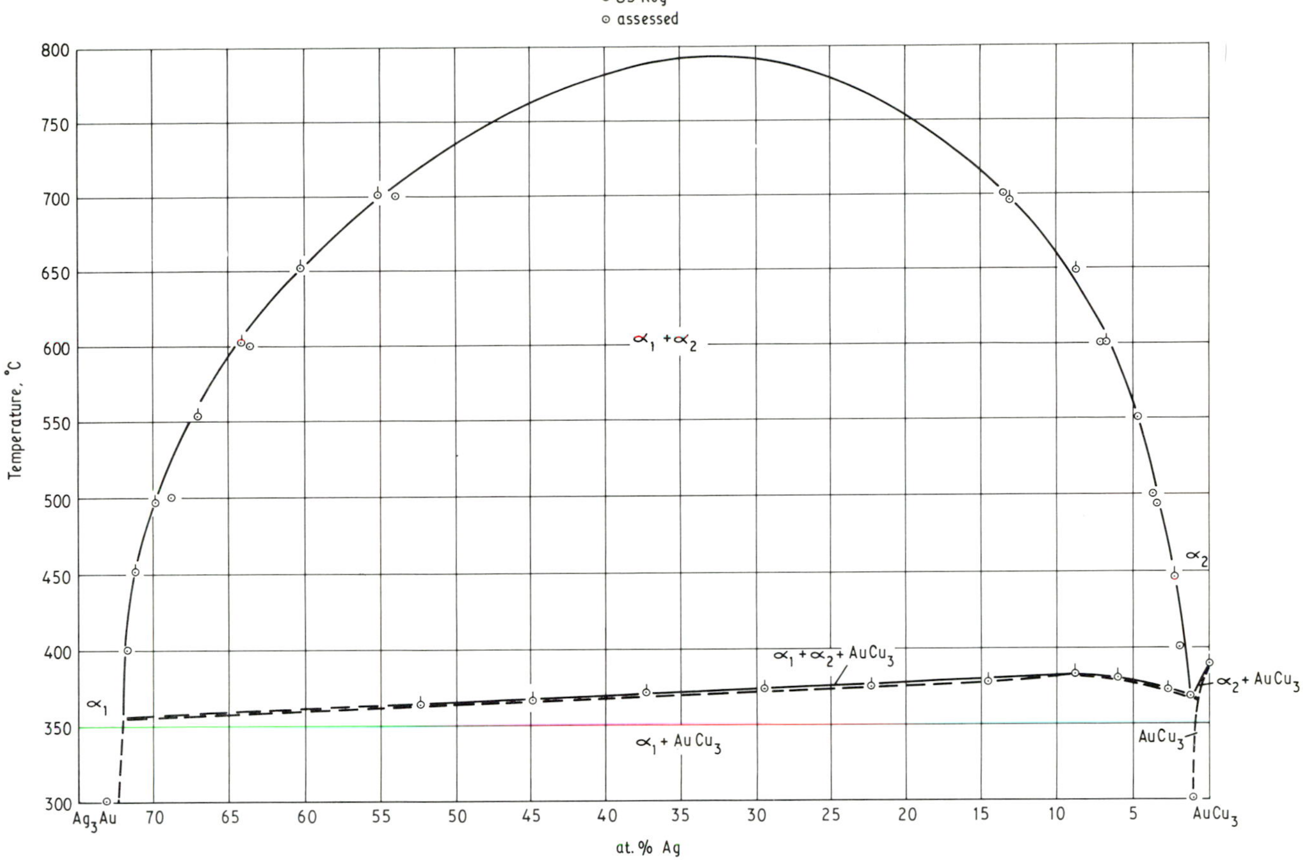

15 Vertical section at 25 at.-% Au [85 Kog]

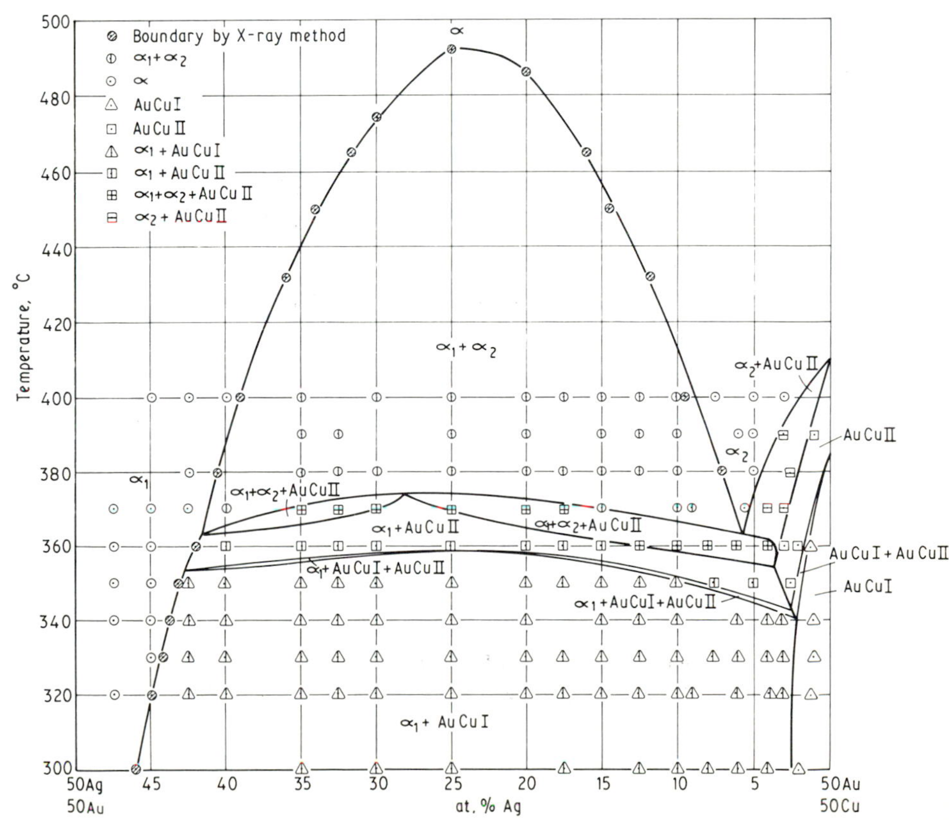

16 Vertical section at 50 at.-% [85 Kog]

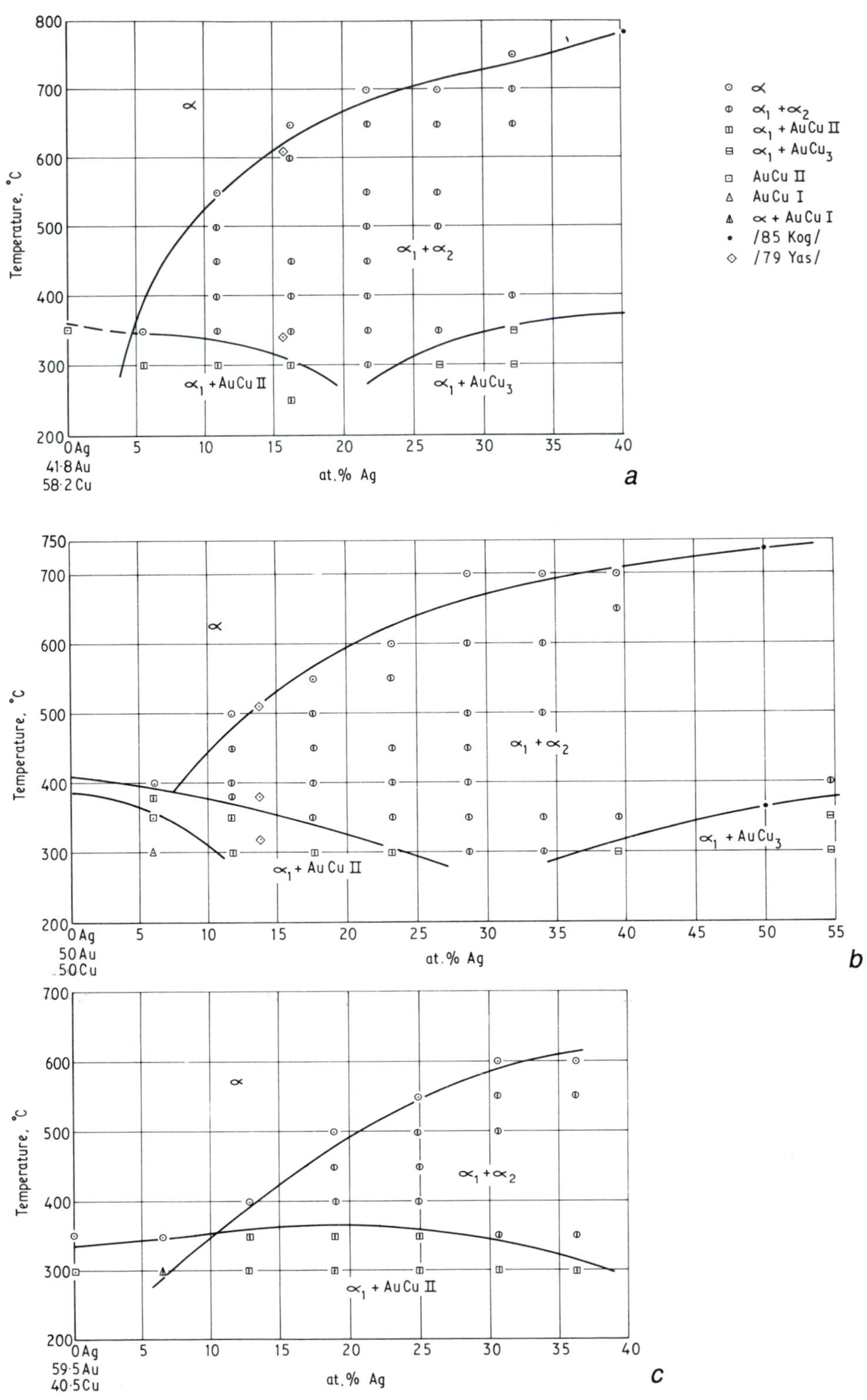

17 **Phase constitution [81 Uzu] along sections: *a* 59·5Au, 40·5Cu to Ag; *b* 50·0Au, 50·0Cu to Ag; *c* 41·8Au, 58·2Cu to Ag**

+ AuCuII tie triangle is shown with lower Au contents for AuCuII and higher Au contents for AuCuI than suggested in Fig. 12.

Study of sections from Au to compositions on the Ag–Cu binary edge in the manner of [85 Kog] would be very useful in further elucidating the complex low-temperature equilibria in the Ag–Au–Cu ternary system.

INVARIANT REACTIONS

The proposed invariant reactions are indicated on Fig. 20. Only one is a ternary four-phase reaction, $\alpha_2 \rightleftharpoons \alpha_1$ + AuCuII + $AuCu_3$. The remainder are three-phase equilibria associated with the appearance of critical tie lines at 767, 387, 374 and 358°C. A further critical tie line connects the liquidus and solidus at 800°C (Table 4).

MISCELLANEOUS

Lattice parameter values for the disordered α phase (Ag–Au–Cu) have been published by several authors [39 Hul, 40 Mas, 49 McM, 49 Rau, 72 Ber, 80 Kog, 81 Uzu and 85 Kog]. Figure 18 shows that the lines of equal lattice parameter run virtually parallel to the Ag–Au binary edge. There is excellent agreement between the data.

Kogachi and Ishibata [83 Kog] studied the effect of Ag additions up to 4 at.-% on the lattice parameter of AuCuI along the 50 at.-% Cu section for alloys homogenized for 120 h at 750°C and water quenched, annealed 144 h at 420°C followed by very slow cooling over 280 h to 300°C, annealing at 300°C for 130 days and very slow cooling over 450 h to room temperature. Figure 19 shows that Ag additions decrease *a*, slightly increase *c* and produce an overall increase in the axial ratio of AuCuI. The data of [39 Hul] indicate an AuCuI structure for alloys on the 50 at.-% Cu section up to and including 5 at.-% Ag at 300°C. Lattice parameters were given for alloys annealed at 350°C and these are included in Fig. 19 (0 and 1% Ag annealed 480 h; 4 and 5% Ag annealed 1 080 h).

REFERENCES

11 Jae: E. Jänecke: *Metallurgie*, 1911, **19**, 597–606

14 Par: N. Parravano (with P. de Cesaris, C. Mazzetti and U. Perret): *Gazz.Chim.Ital.*, 1914, **44**(ii), 321–326

25 Ste: L. Sterner-Rainer: *Z. Metallk.*, 1925, **17**(5), 162–165

26 Ste: L. Sterner-Rainer: *Z. Metallk.*, 1926, **18**(5), 143–148

33 Wis: E. M. Wise and J. T. Eash: *Trans. AIME*, 1933, **104**, 276–307

35 Spa: J. Spanner and J. Leuser: *Metallwirtschaft*, 1935, **14**(17), 319–322

39 Bum: H. Bumm: *Z. Metallk.*, 1939, **31**, 318–321

39 Hul: R. Hultgren and L. Tarnopol: *Trans. AIME*, 1939, **133**, 228–238

40 Mas: G. Masing and K. Kloiber: *Z. Metallk.*, 1940, **32**(5), 125–132

43 Pic: M. R. Pickus and I. W. Pickus: *Trans. AIME*, 1943, **152**, 94–102

49 Rau: E. Raub: *Z. Metallk.*, 1949, **40**, 46–54

49 McM: J. G. McMullin and J. T. Norton: *Trans. AMIE*, 1949, **185**(1), 46–48

67 Zie: T. O. Ziebold and R. E. Ogilvie: *Trans. AIME*, 1967, **239**(7), 942–953

72 Ber: M. Bergman, L. Holmlund and N. Ingri: *Act. Chem. Scand.*, 1972, **26**(7), 2817–2831

73 Sis: G. H. Sistare: ASM Metals Handbook, 8th Edition, Vol 8, 1973, 377–378

74 Sha: O. D. Shaskov, V. I. Syutkina and V. K. Rudenko: *Fiz. Metal. Metalloved.*, 1974, **37**(4), 782–789; *Phys. Metals Metallog.*, 1974, **37**(4), 94–100

75 Mur: M. Murakami, D. de Fontaine, J. M. Sanchez and J. Fodor: *Thin Solid Films*, 1975, **25**, 465–482

75 Kan: Y. Kanzawa, K. Yasuda and H. Metahi: *J. Less Common Metals*, 1975, **43**, 121–128

78 Lup: C. H. P. Lupis: *Metall. Trans. B*, 1978, **9B**, 231–239

78 Yas: K. Yasuda, M. Metahi and Y. Kanzawa: *J. Less Common Metals*, 1978, **60**, 65–78

79 Cha: Y. A. Chang, J. P. Neumann, A. Mikula and D. Goldberg: *INCRA Monograph VI*, 1979, 21–25

79 Yas: K. Yasuda and M. Ohta: in 'Proc. 3rd Int. Precious Metals Conf.', Chicago, IL, 1979, IPMI, 137–164

80 Yam: H. Yamauchi, H. A. Yoshimatsu, A. R. Forouhi and D. de Fontaine: in 'Proc. 4th Int. Precious Metals Conf.', 1981, Pergamon Press Canada Ltd, Ontario, 241–249

80 Kog: M. Kogachi and K. Nakahigashi: *Jpn J. Appl. Phys.*, 1980, **19**(8), 1443–1449

80 Kik: R. Kikuchi, J. M. Sanchez, D. de Fontaine and H. Yamauchi: *Acta Met.*, 1980, **28**, 651–652

80 Ell: R. P. Elliott and W. C. Giessen: *Bull. Alloy Phase Diagrams*, 1980, **1**(1), 41–45; 1980, **1**(2), 23

81 Uzu: T. Uzuka, Y. Kanzawa and K. Yasuda: *J. Dent. Res.*, 1981, **60**(5), 883–889 (*see also* T. Uzuka, *J. Jpn Soc. Dent. Appar. Mater.*, 1977, **18**(41), 67–72, in Japanese with English abstract)

81 de Fon: D. de Fontaine: *Physica*, 1981, **103B**, 57–66

82 Yas: K. Yasuda and M. Ohta: *J. Dent. Res.*, 1982, **61**(3), 473–479

83 Kog: M. Kogachi and S. Ishibata: *J. Jpn Inst. Met.*, 1983, **47**(11), 912–918 (in Japanese with English abstract)

18 Lattice parameter of the disordered α solid solution

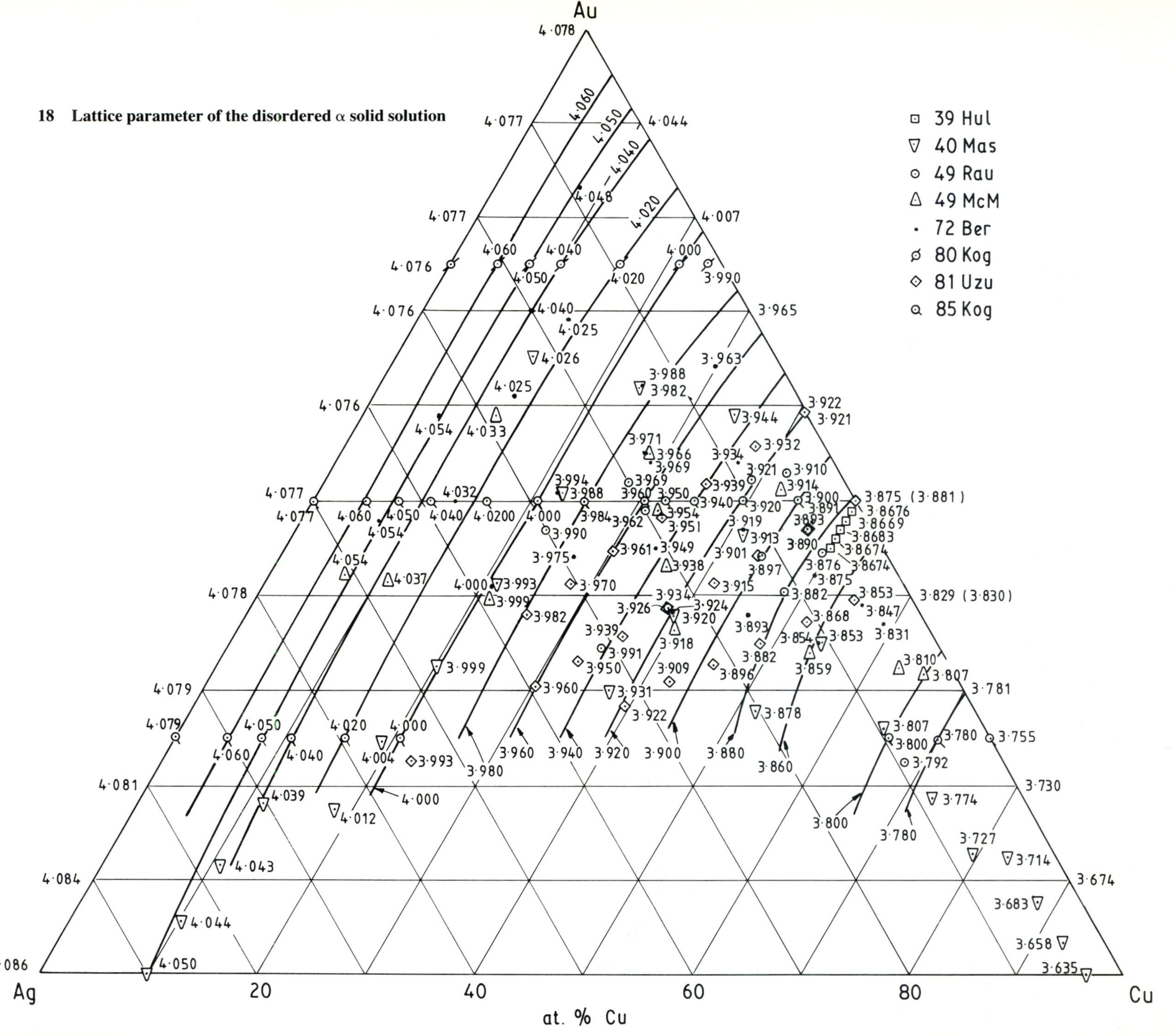

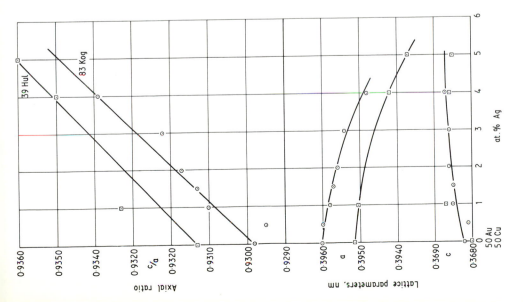

19 **Lattice parameters and axial ratio of AuCuI along the 50 at.-% Cu section**

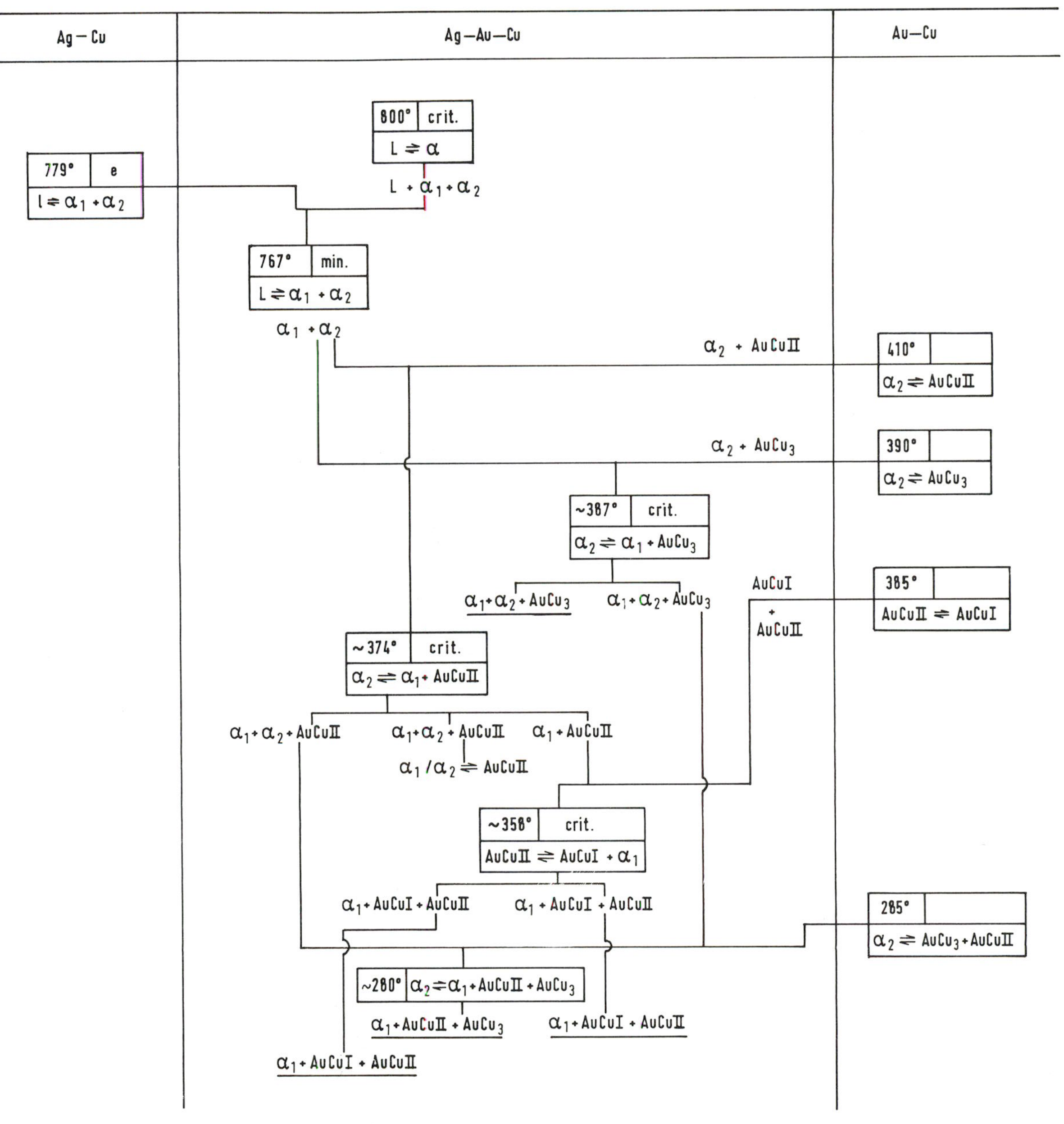

20 **Proposed reaction scheme for the Ag–Au–Cu ternary system**

83 Oka: H. Okamoto and T. B. Massalski: *Bull. Alloy Phase Diagrams*, 1983, **4**(1), 30–38

85 Kog: M. Kogachi and K. Nakahigashi: *Jpn J. Appl. Phys.*, 1985, **24**(2), 121–125

86 Ntu: T. O. Ntukogu and I. B. Cadoff: *Mater. Sci. Technol.*, 1986, **2**(6), 528–533

86 Syu: V. I. Syutkina, I. E. Kislitsyna, R. Z. Abdulov and V. K. Rudenko: *Fiz. Met. Metallov.*, 1986, **61**(3), 504–509

86 Yam: H. Yamauchi and D. de Fontaine: in 'Computer Modeling of Phase Diagrams' (ed. L. H. Bennett), 67–80 (1986), Warrendale, PA, Metallurgical Society of AIME

87 Oka: H. Okamoto, D. J. Chakrabarti, D. E. Laughlin and T. B. Massalski: *Bull. Alloy Phase Diagrams*, 1987, **8**, 454–474

87 Yas: K. Yasuda, M. Nakagawa, G. Van Tendeloo and S. Amelinckx: *J. Less Common Metals*, 1987, **135**, 169–183

Ag–Au–Gd

Lattice parameter data is available [66 Sek] on the $GdAg_{1-x}Au_x$ system for $x = 0$, 0·3, 0·5, 1·0. AgGd and AuGd melt congruently at 1100°C [72 Kie] and 1585 ± 10°C [71 McM] respectively. AgGd is bcc (CsCl type, cP2) with a lattice parameter of 0·3646 nm [66 Sek] or 0·36476 [61 Bae]. Both $GdAg_{0.7}Au_{0.3}$ and $GdAg_{0.5}Au_{0.5}$ are also bcc (CsCl type) with lattice parameters 0·3632 nm and 0·3622 nm respectively [66 Sek]. AuGd is bcc (CsCl type) at temperatures from 1475 ± 10°C [71 McM] and orthorhombic (CrB type, oC8) at temperatures below 1475 ± 10°C.

There appears to be a continuous solid solution series between AgGd and the high-temperature form of AuGd. The solid solution extends to at least 25 at.-% Au along the section AgGd–AuGd.

REFERENCES

61 Bae: N. C. Baenziger and J. L. Moriarty: *Acta Crystallogr.*, 1961, **14**, 948–950

66 Sek: K. Sekizawa and K. Yasukochi: *J. Phys. Soc. Japan*, 1966, **21**, 684–692

71 McM: O. D. McMasters, K. A. Gschneidner, G. Bruzzone and A. Palenzona: *J. Less Common Metals*, 1971, **25**, 135–160

72 Kie: G. Kiessler, E. Gebhardt and S. Steeb: *J. Less Common Metals*, 1972, **26**, 293–298

Ag–Au–Ge

INTRODUCTION

As would be anticipated from the constituent binary systems, the Ag–Au–Ge system contains a ternary monovariant eutectic curve originating at the Ag–Ge binary eutectic at 650°C and ending at the Au–Ge binary eutectic at 361°C. The ternary three-phase reaction L = (Ag–Au) + (Ge) involves the deposition of the (Ag–Au) solid solution and of (Ge) containing minute proportions of Ag and Au [64 Zwi, 85 Has].

Two ternary alloys were studied by Jaffee and Gonsor [46 Jaf] using thermal analysis and metallographic techniques. The alloys, which contained 3·65 Ag–67·09 Au–29·26 at.-% Ge and 7·13 Ag–63·22 Au–29·65 at.-% Ge, were stated to lie close to the monovariant eutectic valley. Zwingmann [64 Zwi] used 99·95% pure Ag and Au and 99·995% pure Ge to prepare an unstated number of ternary alloys. The differential thermal analysis (DTA) examination was carried out at cooling rates that were high for accurate work, 20°C min^{-1} at 700°C and 10°C min^{-1} at 350°C. Supplementary metallographic examination was used to confirm the dta results. Hassam [85 Has] studied 51 alloys by DTA at heating and cooling rates of 2°C min^{-1} and by high temperature calorimetry to produce vertical sections from Ge to 75Ag25Au, 50Ag50Au and 25Ag75Au. Figure 21 presents the liquidus surface [85 Has]. The two alloys studied by [46 Jaf] are just within the primary (Ge) phase field, but near to the monovariant curve e_1e_2.

BINARY SYSTEMS

The equilibrium diagram for the Ag–Au system given by Okamoto and Massalski [83 Oka] is accepted: the liquidus and solidus fall smoothly from the melting point of Au at 1064·43°C to that of Ag at 961·93°C. In the solid state, a continuous series of solid solutions is formed.

The Ag–Ge diagram of [75 Pre] is accepted. It is of the eutectic type with the eutectic composition at 24 at.-% Ge and temperature 650°C. The solid solubility of Ge in Ag at 651°C is 9·6 at.-%, while that of Ag in Ge reaches a maximum of $2{\cdot}3 \times 10^{-6}$ at.-% 700°C.

The Au–Ge system of [84 Oka] is accepted. The system is a simple eutectic with the eutectic composition at 28 at.-% Ge and temperature 361°C. The solubility of Ge in Au is approximately 3·1 at.-%, while that of Au in Ge is less than $1{\cdot}36 \times 10^{-3}$ at.-%.

SOLID PHASES

Only the (Ag–Au) solid solution and (Ge) exist as solid phases. The solid solubility of Ge in the (Ag–Au) phase falls from 9·6 at.-% in the Ag–Ge binary system (point a, Fig. 21) to a constant value between 2 and 3 at.-% Ge for Au contents of >50 at.-% until the Au–Ge binary is reached at 3 at.-% Ge (point b, Fig. 21).

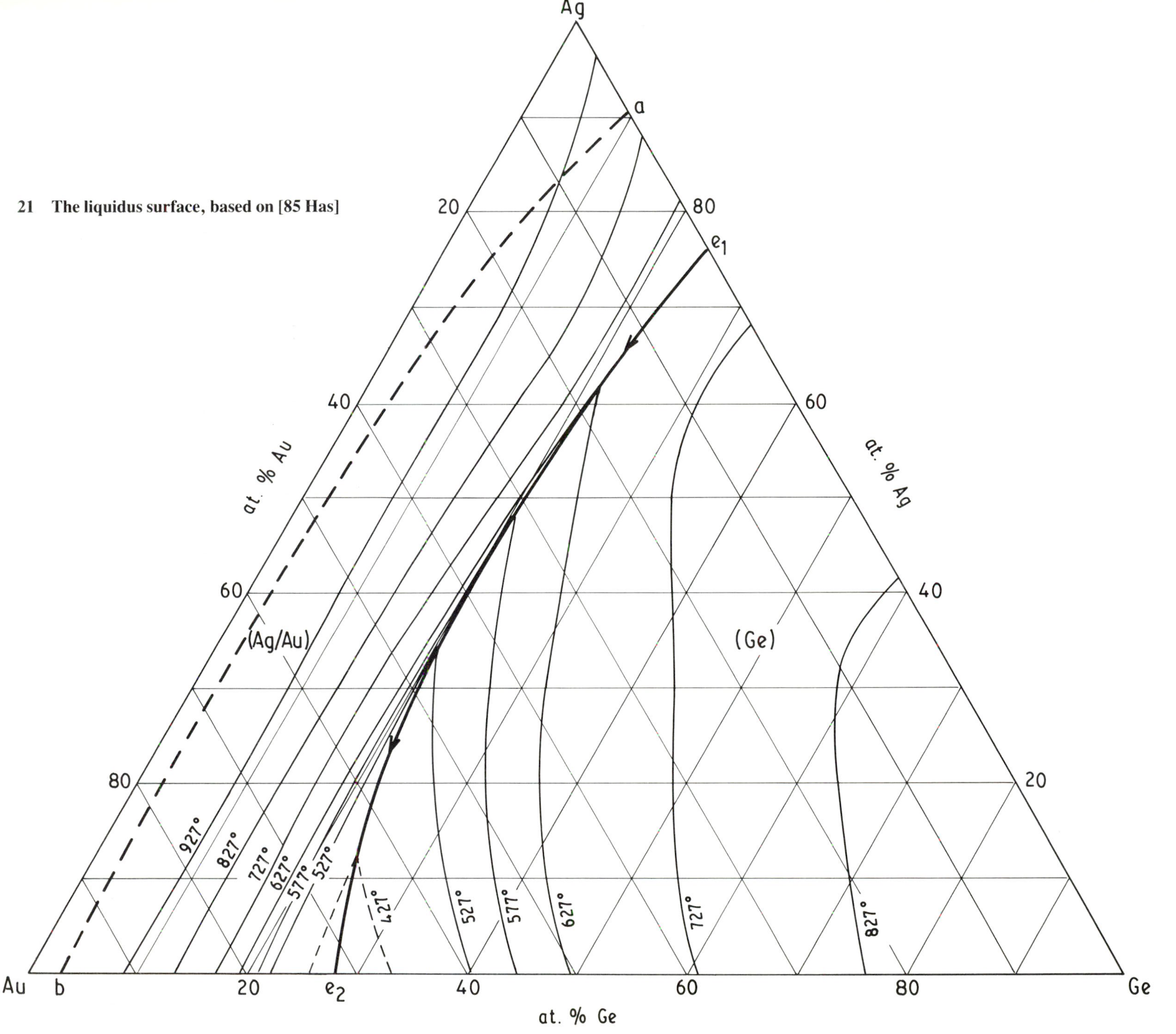

21 The liquidus surface, based on [85 Has]

LIQUIDUS SURFACE

The liquidus projection (Fig. 21), has been constructed from the three vertical sections of [85 Has] and the associated binary systems. The monovariant curve e_1e_2 is concave to the Ge corner. Zwingmann [64 Zwi] gave a liquidus projection that runs almost parallel to the Ag–Au edge at 24 at.-% Ge before bending to meet the Au–Ge binary eutectic at e_2. The data of [85 Has] is preferred on the basis of the superior experimental techniques. There is reasonable agreement between the temperature isotherms of [64 Zwi] and [85 Has] if allowance is made for a displacement of the curve e_1e_2.

VERTICAL SECTIONS

Figures 22*a–c* give the data of [85 Has]. The section Ge–25Ag75Au shows experimental points at 367°C (640 K) that have been rejected in this assessment. The phase boundary between L + (Ag/Au) + (Ge) and (Ag/Au) + (Ge) is assessed at 417°C (690 K) (Fig. 22). A value of 367°C is only 6° higher than the Au–Ge eutectic temperature and it is unlikely that the surface representing equilibrium of (Ag/Au) and (Ge) will dip to within 6° of the Au–Ge eutectic for a section from 25Ag75Au to Ge. Further work is needed on this section to clarify this discrepancy.

ISOTHERMAL SECTIONS

Figure 23 presents the three-phase triangles (L + (Ag/Au) + (Ge)) existing at 427, 477, 527 and 577°C (700, 750, 800, 850 K) constructed from the vertical sections in Fig. 22. They should be regarded as indicative of the likely isothermal sections.

MISCELLANEOUS

Hassam [85 Has] measured the enthalpy of formatiom of ternary alloys along the sections Ag — 25Au75Ge, 50Au50Ge, 75Au25Ge at 1 100°C and along the sections Au — 25Ag75Ge, 50Ag, 50Ge and 75Ag 25Ge at 1 020°C. A total of 112 alloy compositions were studied; this permitted the plotting of isoenthalpy contours for the ternary alloys.
[88 Cas] has presented similar data at 1 075°C, with reasonable agreement with the data of [85 Has].

REFERENCES

46 Jaf: R. I. Jaffee and B. W. Gonsor: *Trans. AIME*, 1946, **166**, 436–443

64 Zwi: G. Zwingmann: *Metall.*, 1964, **18**, 726–727

75 Pre: B. Predel and M. Bankstahl: *J. Less Common Metals*, 1975, **43**, 191–203

83 Oka: H. Okamoto and T. B. Massalski: *Bull. Alloy Phase Diagrams*, 1983, **4**, 30–38

84 Oka: H. Okamoto and T. B. Massalski: *Bull. Alloy Phase Diagrams*, 1985, **5**, 601–610

85 Has: S. Hassam: Thesis, Univ. Provence, France 1985

88 Cas: R. Castanet, *J. Less-Common Metals*, 1988, **136**, 287–296

Ag–Au–Ni

INTRODUCTION

Knowledge of this system is confined to the early work of de Cesaris [13 Ces 1]. A liquid miscibility gap extends from the binary Ag–Ni monotectic reaction at 1 435°C, $l_1 \rightleftharpoons l_2$ + (Ni), into the ternary system and closes at some 50–55 wt-% Au (25–35 at.-% Au) at ~1 240°C, according to de Cesaris [13 Ces 1].

Replotting the original tabulated data throws doubt on the extent and shape of the reported liquid immiscibility region. The data have been reinterpreted (Fig. 24) as showing closure of the miscibility gap at 39–40 at.-% Au and ~1 125°C. The ternary monovariant curve associated with the binary Ag–Ni eutectic reaction at 960°C was not traced with any accuracy into the ternary system. It was assumed [13 Ces 1] to lie in a flat valley ending at the Au–Ni eutectic point (now known to be a liquidus–solidus minimum). Measurements of liquidus temperatures for a limited number of Au-rich ternary alloys suggested that the ternary monovariant curve passes through a maximum of ~990°C, descending on the one hand to the Ag–Ni eutectic at 960°C and on the other hand towards the Au–Ni liquidus minimum at 955°C. The uncertain nature of the course of the ternary monovariant curve is indicated in Fig. 7 by the use of a dashed line. A redetermination of this ternary system is required, especially in the region between 50 at.-% Au and 80 at.-% Au.

BINARY SYSTEMS

The Ag–Au system evaluated in [83 Oka] is accepted. The Ag–Ni system (H, E, S) is accepted in the absence of more definitive data. The Au–Ni system evaluated in [86 Oka] is accepted.

SOLID PHASES

At temperatures below 810·3°C, the temperature at which the solid-state miscibility gap closes in the binary Au–Ni system, the ternary equilibria will be two-phase. A Ni-rich fcc solid solution is in equilibrium with an (Ag–Au)–rich fcc solid solution.

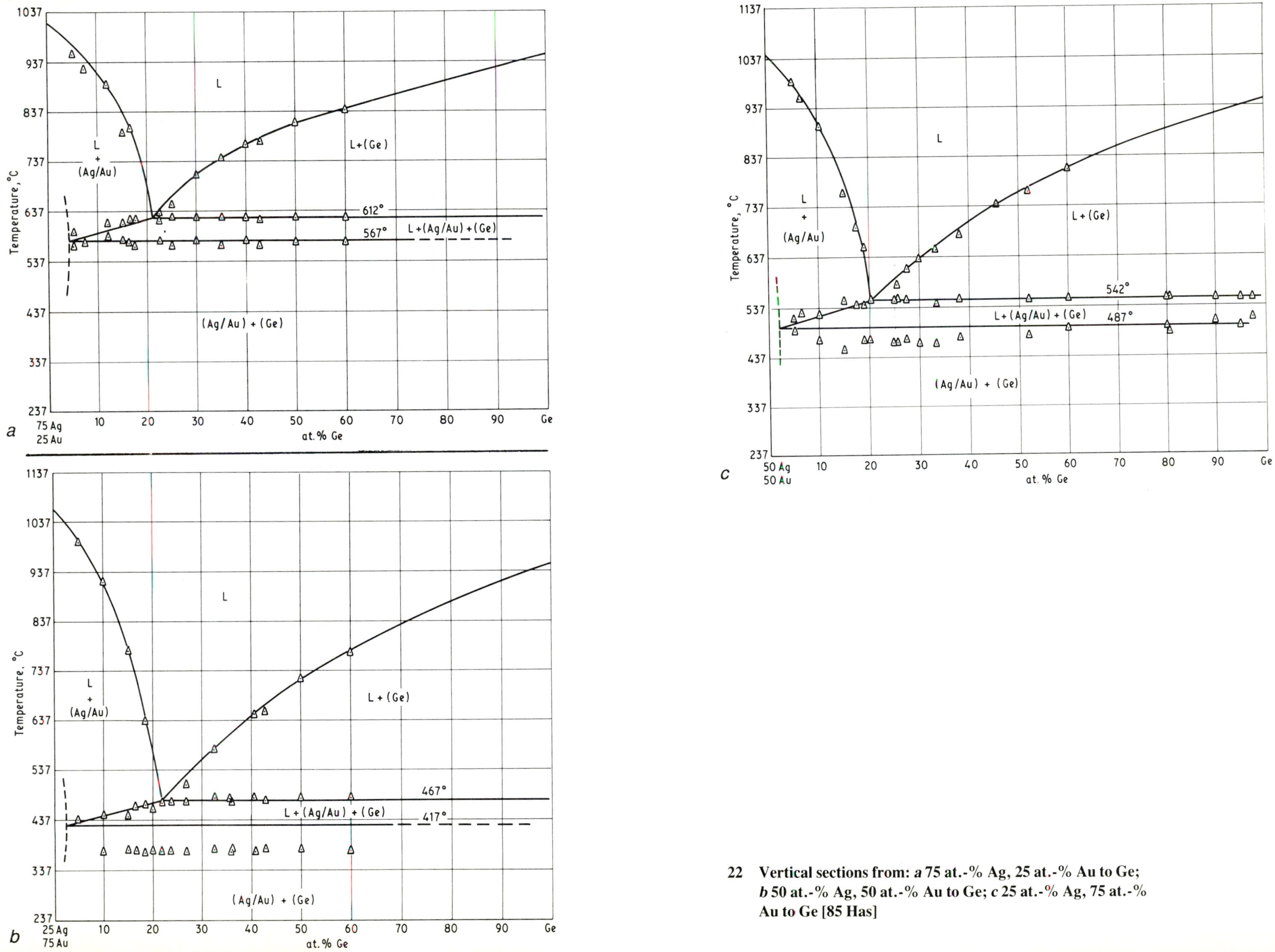

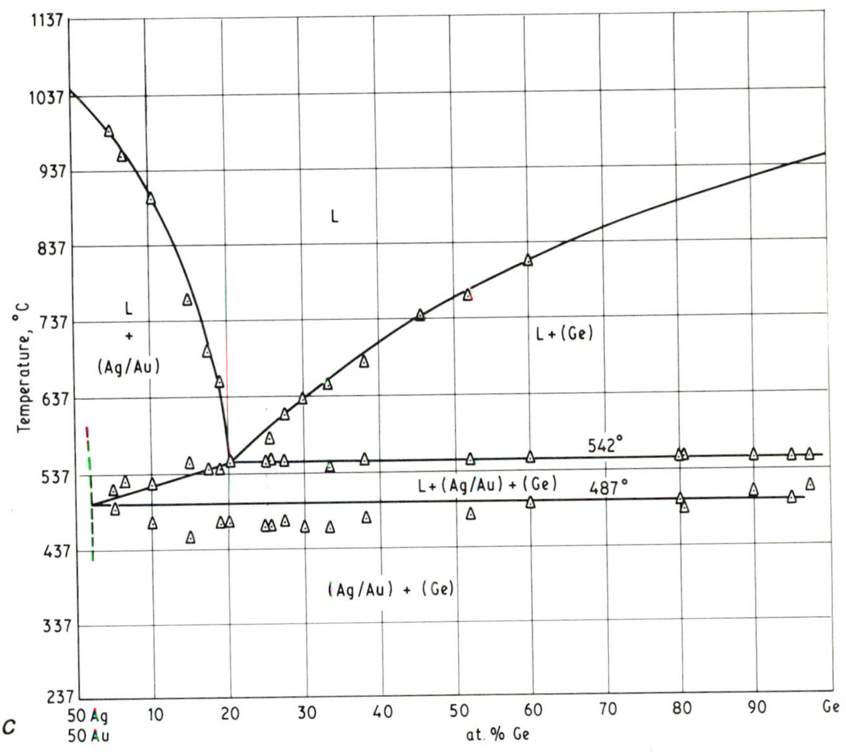

22 **Vertical sections from: *a* 75 at.-% Ag, 25 at.-% Au to Ge; *b* 50 at.-% Ag, 50 at.-% Au to Ge; *c* 25 at.-% Ag, 75 at.-% Au to Ge [85 Has]**

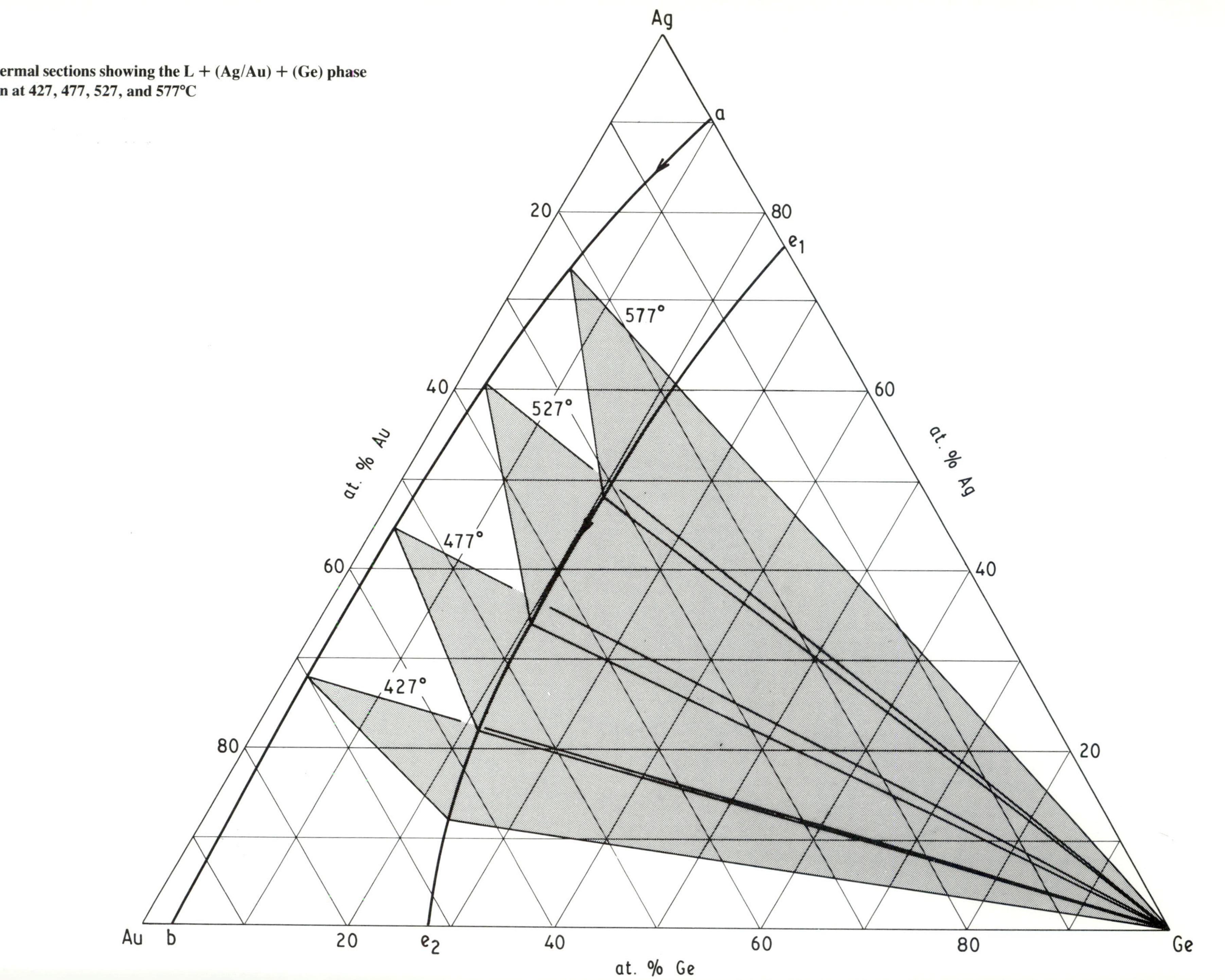

23 **Isothermal sections showing the L + (Ag/Au) + (Ge) phase region at 427, 477, 527, and 577°C**

LIQUIDUS SURFACE

The introduction described the liquidus surface which was generated by [13 Ces 1] using thermal analysis techniques on 95 alloys. Liquidus isotherms for Ag-rich alloys are uncertain, and therefore they are drawn as dashed lines (Fig. 24). The composition of l_2 at 1 435°C was plotted as 8·5 wt-% Ni (14·6 at.-% Ni) on the original liquidus surface [13 Ces 1]. A value of 1·5 wt-% Ni (2·7 at.-% Ni) was found by the same author in a concurrent paper [13 Ces 2], and this value has been adopted in Fig. 24.

ISOTHERMAL SECTIONS

de Cesaris [13 Ces 1] tabulated temperatures for the liquidus, secondary arrests and the solidus. The data quoted for secondary arrests and the solidus are not accepted. Using the same thermal analysis technique [13 Ces 1] quotes liquidus and solidus values for binary Au–Ni alloys. The large segregation obtained in cooling alloys from the melt produced spuriously low solidus values that led [13 Ces 1] to suggest the presence of a eutectic reaction in this binary system. The ternary Ag–Au–Ni melts would also be subject to similar effects. The data [13 Ces 1] do not allow any conclusions to be drawn on the form of isothermal sections. On the basis of the closure of the liquid-state immiscibility gap at ~1 125°C by a critical tie line $l_1/l_2 \rightleftharpoons$ (Ni), and the coalescence of the two-phase region l + (Ni) with the two-phase region l + (Ag–Au) at ~990°C to give two three-phase regions, l + (Ag–Au) + Ni, the experimental date [13 Ces 1] can be explained. It should be stressed that the 990°C equilibrium, implying a degenerate tie line where l $\rightleftharpoons$ (Ag–Au) + (Ni), is speculative.

REFERENCES

13 Ces 1: P. de Cesaris: *Gazz Chim. Ital.*, 1913, **43**(I), 609–620

13 Ces 2: P. de Cesaris: *Gazz Chim. Ital.*, 1913, **43**(II), 365–379

83 Oka: H. Okamoto and T. B. Massalski: *Bull. Alloy Phase Diagrams*, 1983, **4**, 30–38

86 Oka: H. Okamoto and T. B. Massalski: in 'Binary Alloy Phase Diagrams', Vol 1 (ed. T. B. Massalski), 288–290; 1986, ASM

Ag–Au–Pb

INTRODUCTION

The major work on the phase equilibria unfortunately is unpublished [70 Bha]. The authors used thermal analysis, metallography, microprobe, and X-ray techniques to define the equilibria. The liquidus surface (Fig. 25), is shown in terms of isotherms from 450 to 750°C. The system is characterized by three invariant reactions, U_1, U_2 and E, all of which occur within about 0·5 at.-% Ag from the Au–Pb binary edge (Fig. 26). The reaction scheme proposed [70 Bha] is given in Fig. 27. Recent work [87 Has], [87 Hum] throws some doubt on the interpretation of the equilibria.

Hager and Zambrano [69 Hag] studied the thermodynamic properties of the liquid Ag–Au–Pb system using the galvanic cell method. Liquidus temperatures for seven alloys were directly determined from plots of the temperature variation of the emf of the cell. This data is plotted on Fig. 25 together with an isotherm for 927°C, shown dashed, calculated from the emf and Pb activity data. There is reasonable consistency in the measured liquidus temperatures and the calculated 927°C isotherm (except for the alloy containing 45 at.-% Au, 45 at.-% Ag, 10 at.-% Pb) and the data of [69 Hag] compared with the isotherms presented in [70 Bha].

Thin film studies of the interaction between a 10 000 Å thick film of an 80 at.-% Ag, 20 at.-% Au alloy with a 5 000 Å thick Pb film [75 Tu] showed only a 200 Å thick reaction product of $AuPb_2$ after 1 week at 200°C. In bulk alloys the data of [70 Bha] would indicate equilibrium between an 80 at.-% Ag, 20 at.-% Au alloy and Pb at 200°C.

BINARY SYSTEMS

The Ag–Au system has been evaluated by [83 Oka]. The Ag–Pb system, evaluated by [86 Kar] is accepted. The Au–Pb system has been redetermined recently [83 Eva] (Fig. 28).

SOLID PHASES

Table 5 summarizes crystal structure data for the (Ag–Au) solid solution, (Pb), and the three binary Au–Pb compounds. No ternary compounds occur in the ternary system [70 Bha].

Table 5. Crystal structures

Solid phase	*Prototype*	*Lattice designation*
(Ag–Au)	Cu	cF4
(Pb)	Cu	cF4
Au_2Pb	Cu_2Mg	cF24
$AuPb_2$	$CuAl_2$	tI12
$AuPb_3$	α–V_3S	Tetragonal

INVARIANT EQUILIBRIA

As noted in Fig. 27, the ternary invariant reactions are simple. At U_1, L + $Au_2Pb \rightleftharpoons$ (Ag–Au) + $AuPb_2$. This reaction is followed by U_2, L + $AuPb_2 \rightleftharpoons$ (Ag–Au) + $AuPb_3$, and solidification ends at the ternary eutectic E, L (Ag–Au) + (Pb) + $AuPb_3$. The compositions of the (Ag–Au) solid solutions at the three invariant reaction temperatures are given in Table 6. It would appear that there is a very limited solution of Pb in the (Ag–Au) solid solutions; similarly there is very limited solubility of Ag in Au_2Pb, $AuPb_2$, and $AuPb_3$ [70 Bha]. The temperatures of the

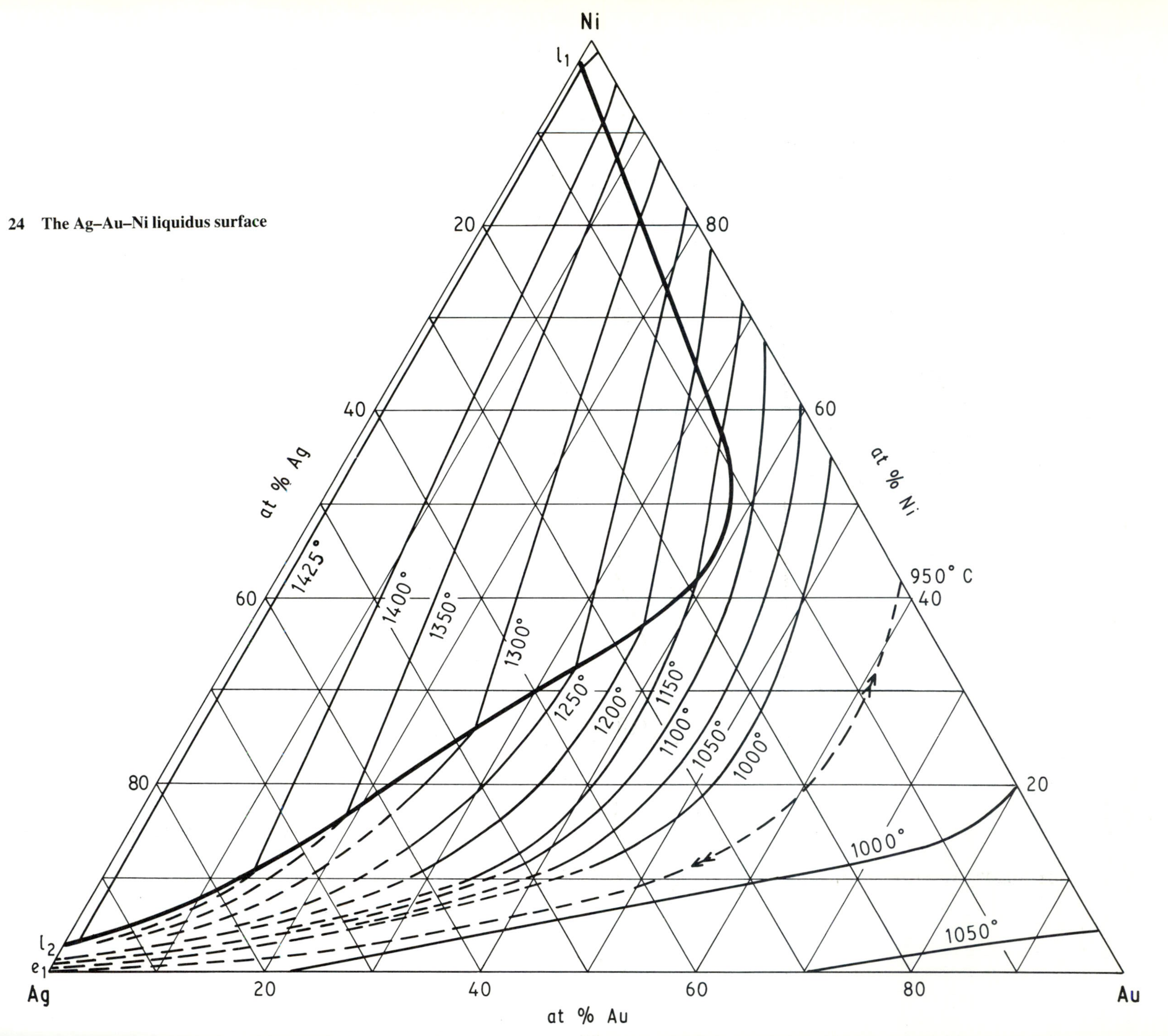

24 The Ag–Au–Ni liquidus surface

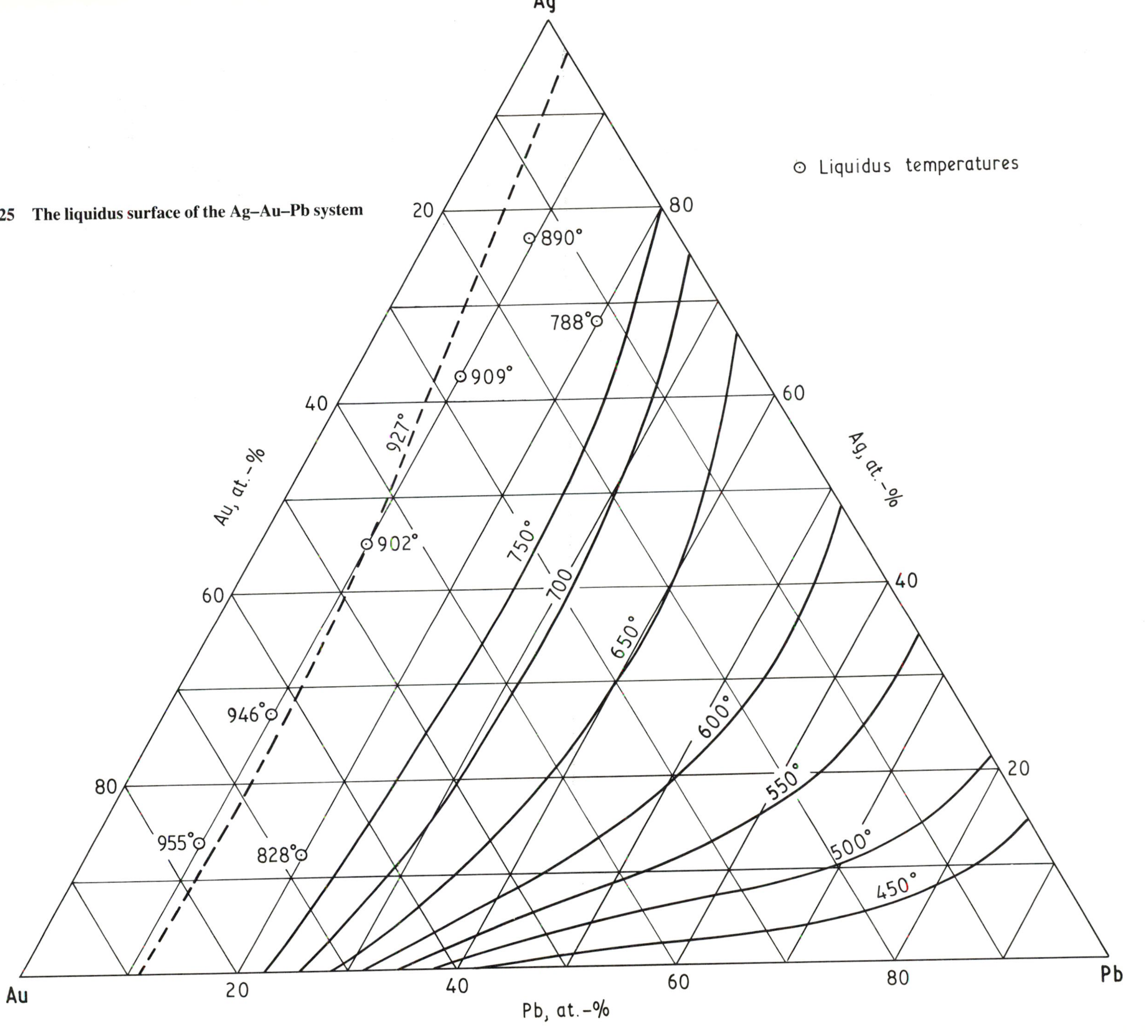

25 The liquidus surface of the Ag–Au–Pb system

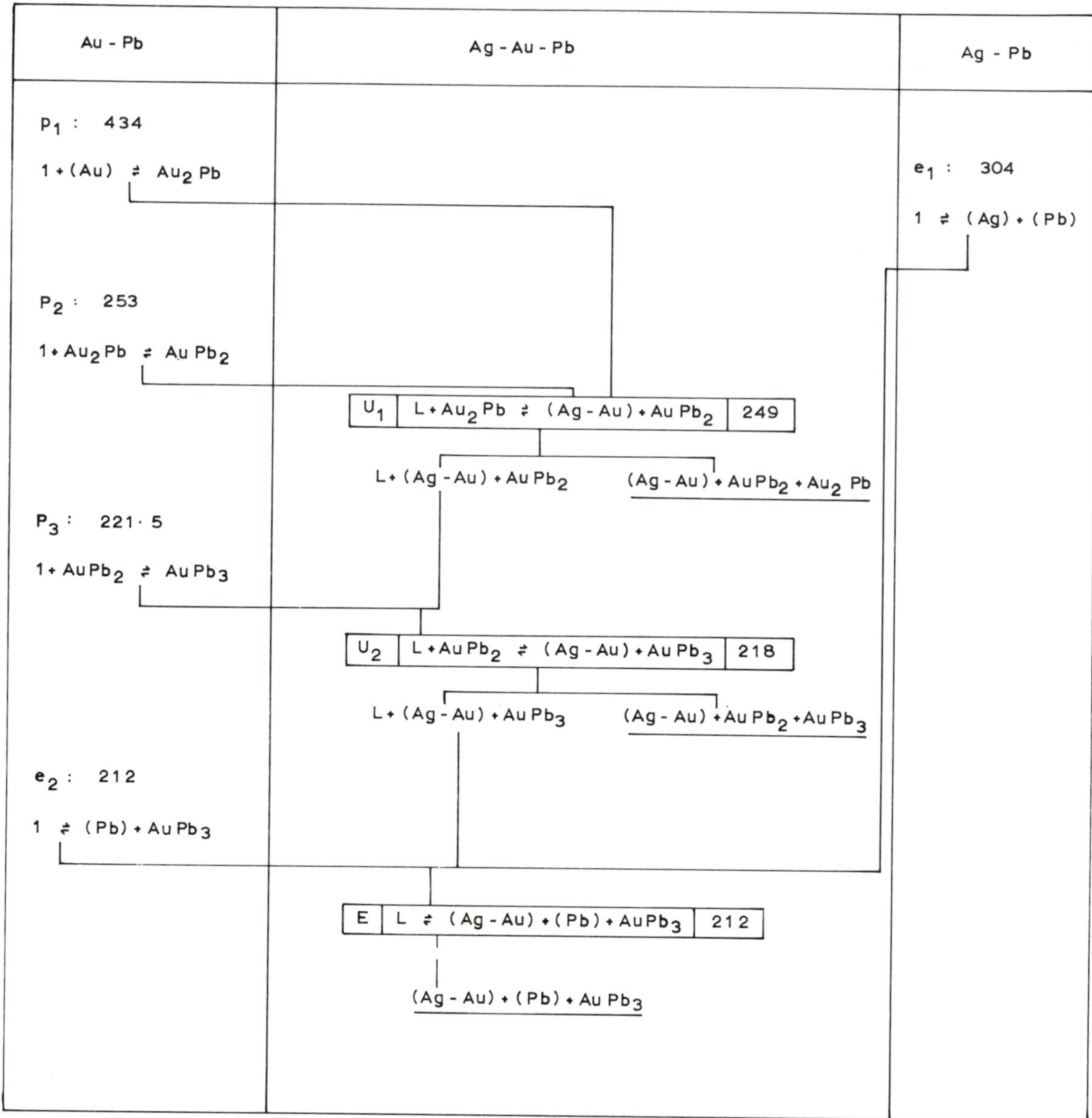

27 Reaction scheme for the Ag–Au–Pb system

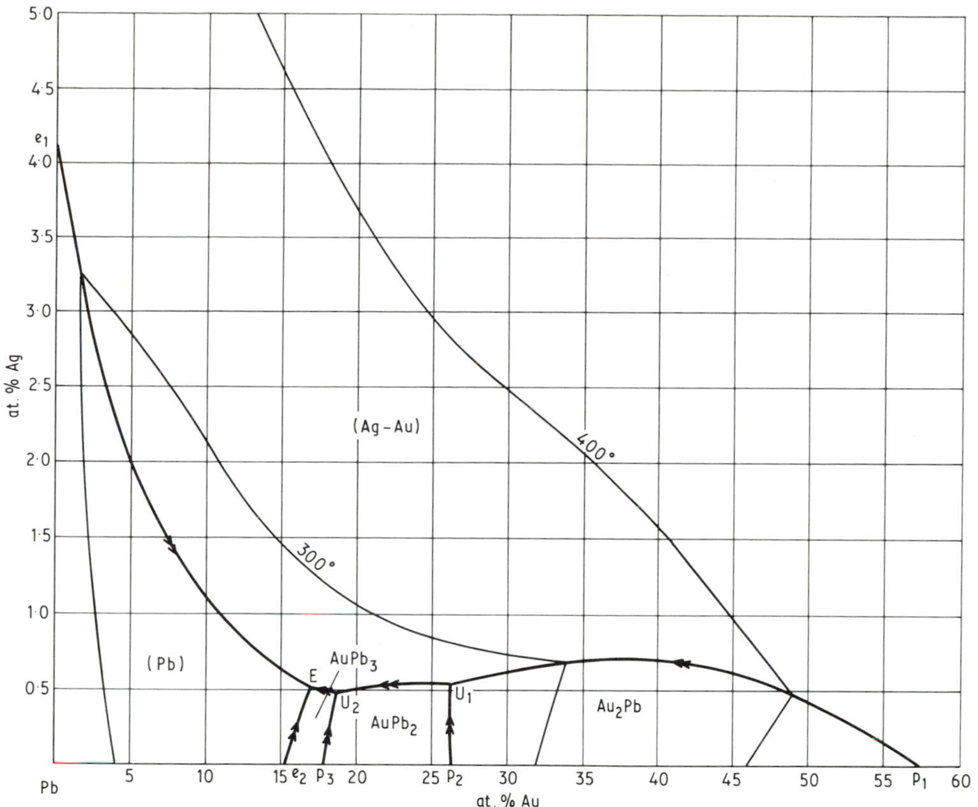

26 **The liquidus surface for Ag–Au–Pb alloys containing up to 5 at.-% Ag**

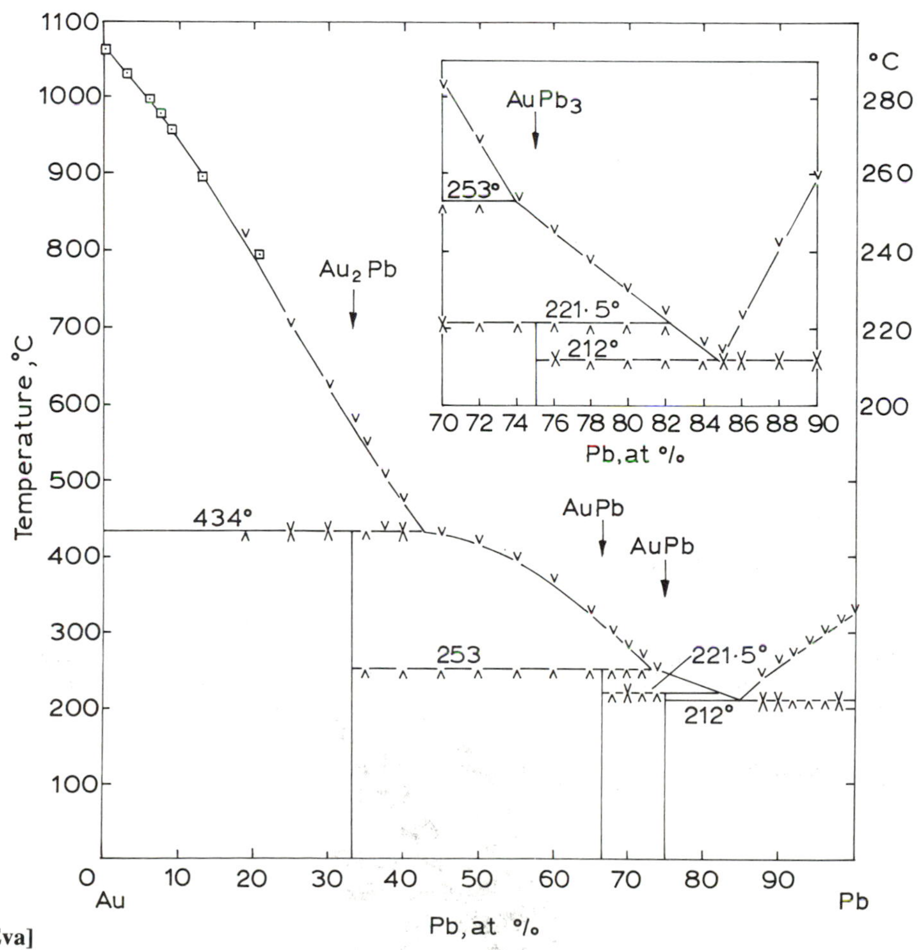

28 **The Au–Pb system [83 Eva]**

invariant equilibria are assessed as reliable to ± 1°C. The temperature of the eutectic reaction at E is probably lower than 212°C. Work by Bhattacharya and Reynolds on Au–Pb alloys [71 Bha] indicated a temperature of 213°C for the binary Au–Pb eutectic reaction, i.e. 1° above that of the ternary eutectic. The accepted value for the Au–Pb eutectic temperature [83 Eva] is 212°C, implying a lower value for the ternary eutectic temperature.

Hassam [87 Has] reported preliminary results of a DTA examination of alloys on the sections Pb–80Ag, 20Au; Ag–$AuPb_3$; and 40 at.-% Pb. The data do not confirm the findings of [70 Bha]. In particular a reaction at 210°C was found for alloy compositions well outside the (Ag–Au) + (Pb) + $AuPb_3$, tie triangle quoted in [70 Bha], i.e. 57·4Ag, 42·6Au–Pb–$AuPb_3$. Humpston and Prince [87 Hum] thermally analysed an alloy containing 8Ag, 2Au and confirmed the report of [87 Has] that it contains a low-temperature reaction at 207°C. The data of [70 Bha] should be regarded as provisional.

Table 6. Invariant equilibria

Reaction	*Temperature*, °C	*Phase*	*Composition*, at.-%		
			Ag	*Au*	*Pb*
U_1	249	L	0·53	26·10	73·37
		(Ag–Au)	19·9	80·1	—
U_2	218	L	0·47	18·61	80·91
		(Ag–Au)	43·9	56·1	—
E	212	L	0·51	16·96	82·53
		(Ag–Au)	57·4	42·6	—

REFERENCES

69 Hag: J. P. Hager and A. R. Zambrano: *Trans. Met. Soc. AIME*, 1969, **245**, 2313–2318

70 Bha: J. P. Bhattacharya and K. A. Reynolds: unpublished work, 1970, University of Aston in Birmingham.

71 Bha: J. P. Bhattacharya and K. A. Reynolds: *J. Inst. Metals*, 1971, **99**, 350–352

75 Tu: K. N. Tu and D. A. Chance: *J. Appl. Phys.*, 1975, **46**, 3229–3234

83 Oka: H. Okamoto and T. B. Massalski: *Bull. Alloy Phase Diagrams*, 1983, **4**, 30–38

83 Eva: D. S. Evans and A. Prince: *Mater. Res. Soc. Symp. Proc.*, 1983, **19**, 383–388

86 Kar: I. Karakaya and W. T. Thompson: in 'Binary Alloy Phase Diagrams', Vol 1 (ed. T. B. Massalski), 51–55; 1986, ASM

87 Has: S. Hassam, M. Gambino and J. P. Bros: *13eme Journées d'Etude des Equilibres Entre Phases*, 317–324; 1987, Lyon

87 Hum: G. Humpston and A. Prince: unpublished work, 1987

Ag–Au–Pd

INTRODUCTION

Ag, Au, and Pd are completely soluble in each other in both the molten and solid states (>900°C). The Ag–Au–Pd ternary system shows only two-phase equilibrium, l $\rightleftharpoons$ (Ag, Au, Pd), but the effect of Ag additions to the ordered compounds Au_3Pd and $AuPd_3$ has not been studied. The presence of a continuous series of solid solutions in the ternary system was established from the X-ray work of Kuznetsov [46 Kuz] and thermal analysis by Nemilov *et al.* [46 Nem] and Miane *et al.* [77 Mia].

BINARY SYSTEMS

The Ag–Au system of [83 Oka], the Au–Pd system of [85 Oka] and the Ag–Pd system of [86 Kar] were accepted.

SOLID PHASES

Table 7 summarizes the solid phases present. (Ag,Au,Pd) is the ternary solid solution phase that is formed on solidification throughout the ternary system.

Table 7. Crystal structures

Solid phase	*Prototype*	*Lattice designation*	*Lattice spacing*, nm
Ag	Cu	cF4	0·408 61 [81 Kin]
Au	Cu	cF4	0·407 84 [81 Kin]
Pd	Cu	cF4	0·389 01 [81 Kin]
(Ag,Au,Pd)	Cu	cF4	See Fig. 33
Au_3Pd	$AuCu_3$	cP4	
$AuPd_3$	?$AuCu_3$	?cP4	

LIQUIDUS SURFACE

The liquidus has been studied by Nemilov *et al.* [46 Nem] and Miane *et al.* [77 Mia]. Venudhar *et al.* [78 Ven] present a liquidus of [67 Pau], but the reference in [78 Ven] is to a 1967 doctoral thesis and it is not the thesis of [67 Pau] quoted in this assessment. Nemilov *et al.* [46 Nem] thermally analysed 47 ternary alloys prepared from >99·99% elements; they also measured hardness, strength, ductility, electrical resistance, temperature coefficient of electrical resistance, and thermoelectric force of alloys against Pt. Miane *et al.* [77 Mia] used DTA to measure the liquidus and solidus temperatures of a total of 14 ternary alloys on the sections Ag–50Au,50Pd; Au–50Ag,50Pd, and Pd–50Ag,50Au. Quoted temperatures were the mean of 4 heating + cooling cycles at rates of ~5°C h^{-1}. The liquidus isotherms presented in [78 Ven] from the work of Pauley are seriously in error in terms of the compositions at which the isotherms intersect the binary Au–Pd edge. For instance the 1 450°C isotherm should [85 Oka] meet the Au–Pd binary at 52·4 at.-% Au and not 61·3 Au from the diagram in [78 Ven]. As a result of these discrepancies the isotherms presented by [78 Ven], based on data of Pauley, have been amended so that they agree with assessed binary data. Figures 29*a–c* are sections Ag–50Au,50Pd; Au–50Ag,50Pd; Pd–

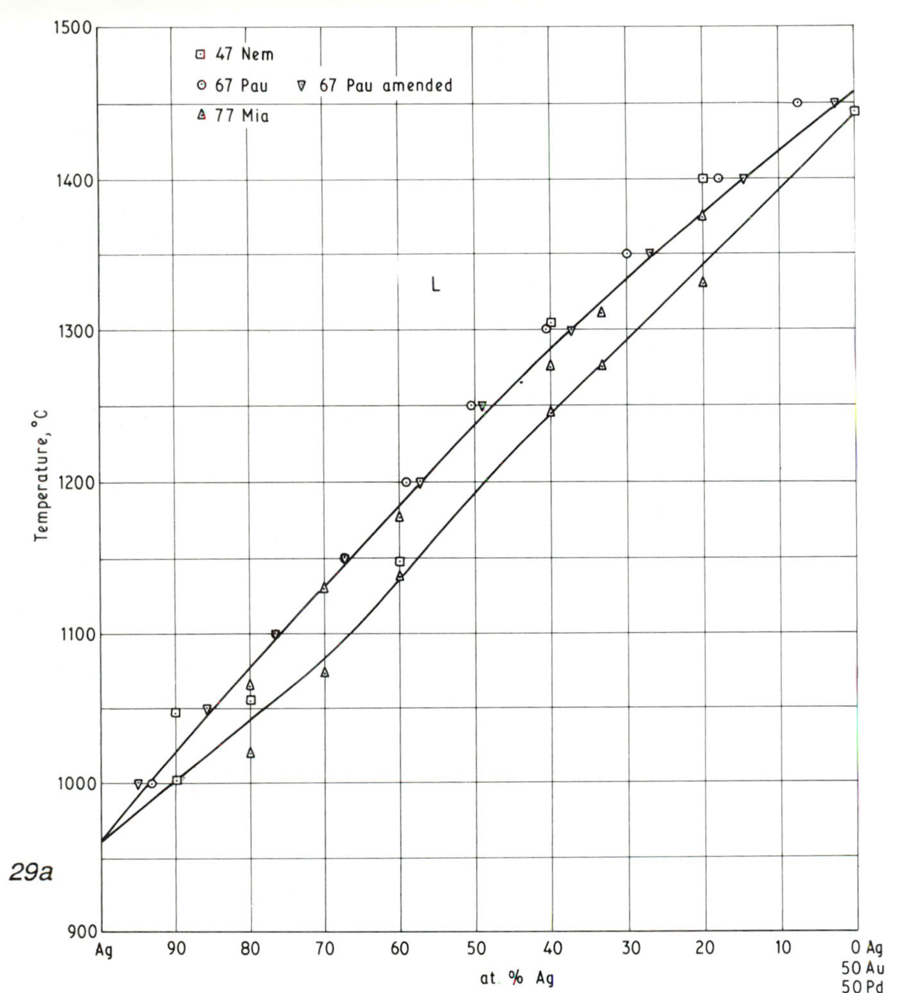

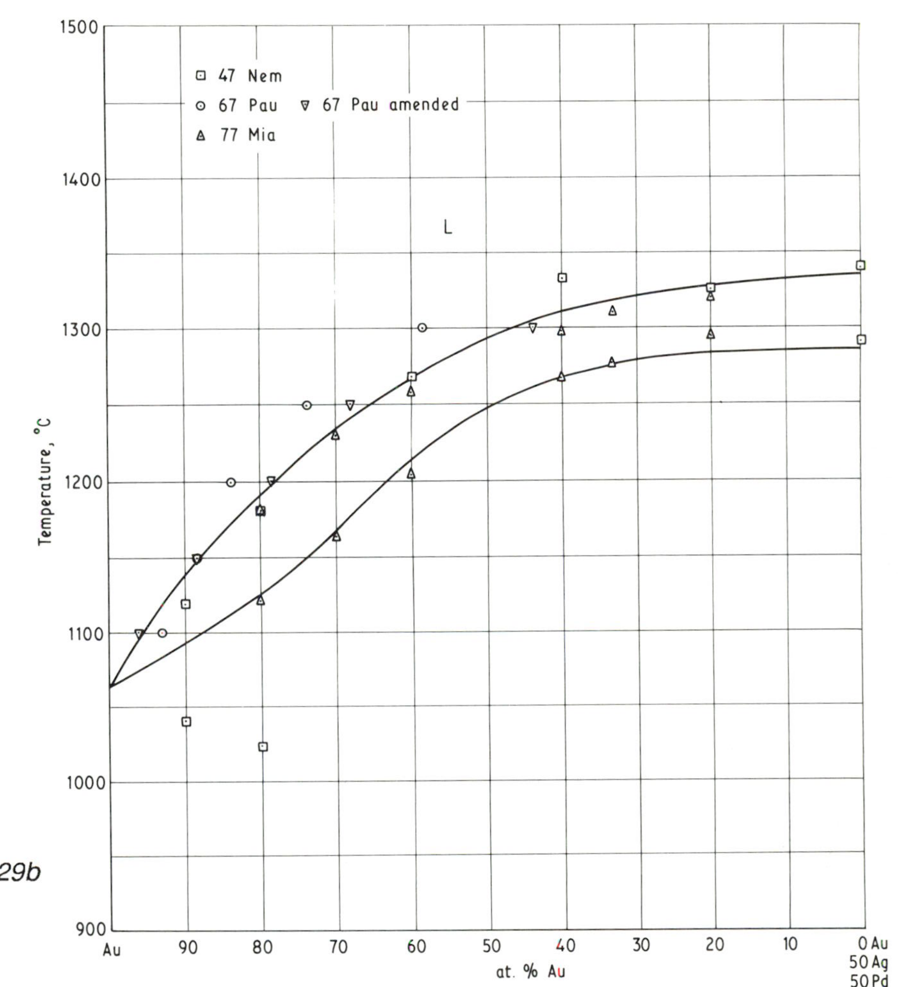

29 Sections *a* Ag–50Au, 50Pd; *b* Au–50Ag, 50Pd; *c* Pd–50Ag, 50Au

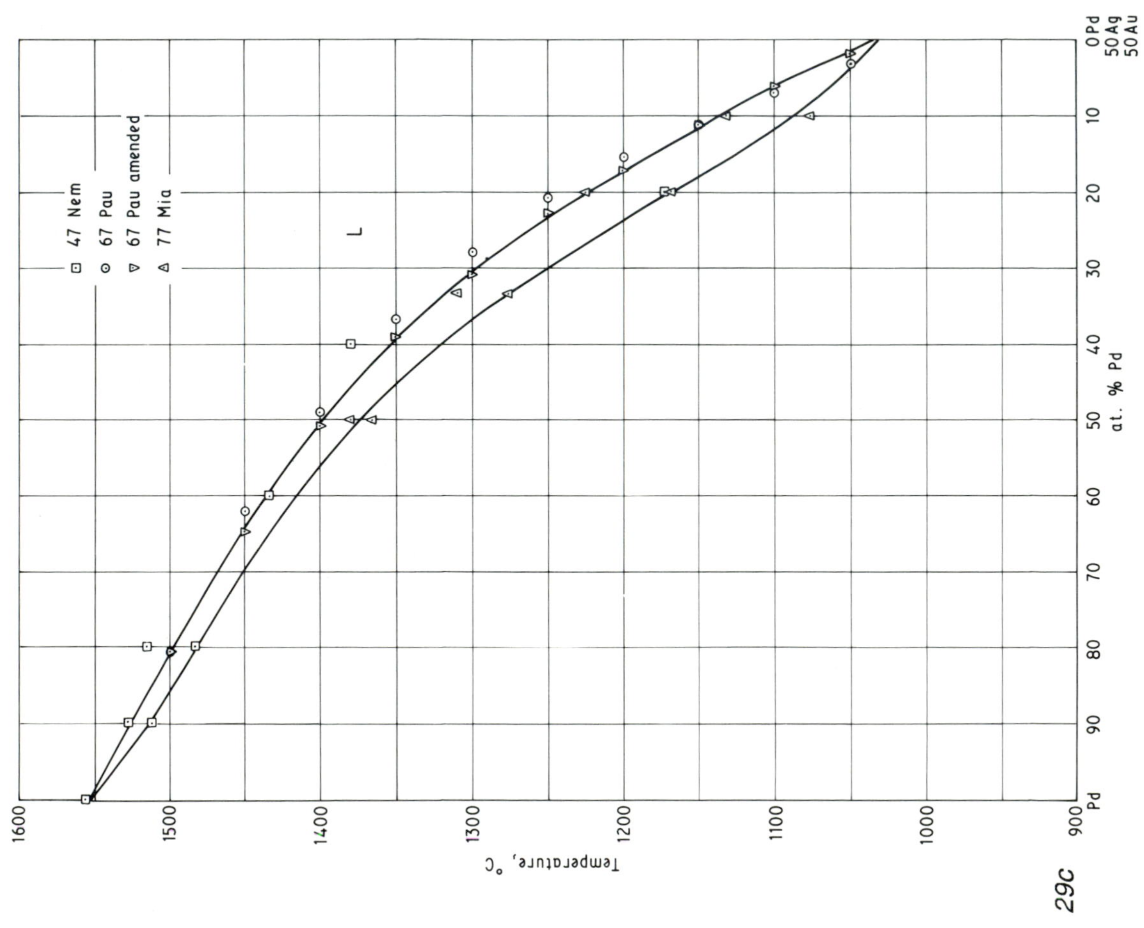

47 Nem
67 Pau
67 Pau amended
77 Mia
L
Temperature, °C
1600
1500
1400
1300
1200
1100
1000
900
Pd
90
80
70
60
50
40
30
20
10
0 Pd
50 Ag
50 Au
at. % Pd

29c

30 A liquidus projection

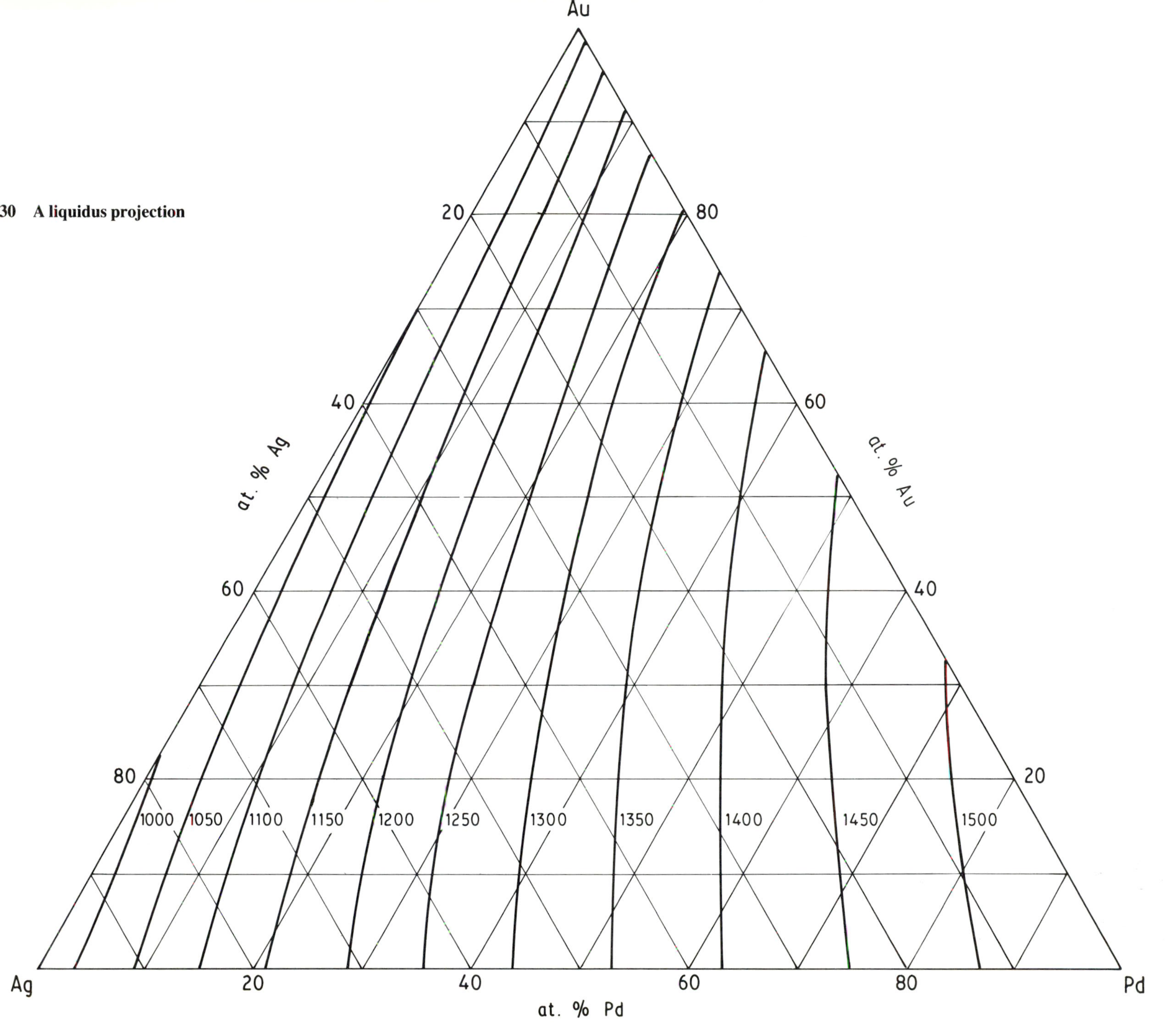

50Ag,50Au respectively. A liquidus projection, developed from the three binary systems and Figs. 29*a–c*, is given in Fig. 30. Figure 31 is an equivalent solidus projection. There is uncertainty in the solidus isotherms of ±1 at.-% Pd; at the Au-rich corner this uncertainty widens to ±2 at.-% Pd.

MISCELLANEOUS

Naidu and Houska [71 Nai] used 99·99% Ag and 99·97 Au,Pd powders to form compacts that were induction-melted under *H*. The ingots were worked and annealed at 900°C for 7 days. Filings were annealed for 2–3 h at 600°C in argon-flushed capsules and furnace cooled. X-ray diffraction data was obtained for three ternary alloys and the end members on the section Pd–50 at.-% Ag, 50 at.-% Au (Fig. 32). [78 Ven] used the same preparative technique and studied the same ternary alloys as [71 Nai]. An alloy containing 33·33 at.-% Ag, 33·33 at.-% Au, 33·33 at.-% Pd was studied by [79 Ven]. The data from [78 Ven, 79 Ven] was incorporated in [86 Ven] (Fig. 32). The agreement between [71 Nai] and [79 Ven] is very good and both sets of data indicate a small negative departure from Vegard's law on this ternary section. The temperature dependence of the lattice parameters for these ternary alloys are a linear function of temperature up to the limit of measurement (Table 8 [86 Ven]).

Kuznetsov [46 Kuz] measured the lattice spacings of 38 ternary alloys that were cold worked from 0·5 mm diameter wire to 0·2 mm diameter, annealed for 30 min at 800°C and quenched (Fig. 33). The experimental data of [46 Kuz] are in reasonably good agreement with those of [71 Nai] and [86 Ven] (Fig. 32), for alloys containing >40 at.-% Pd. At lower Pd contents [46 Kuz] reports lower values of lattice spacing. The data of [46 Kuz] has been amended to conform with that of [71 Nai] and [86 Ven] on the Pd–50Ag,50Au section and with the Au–Pd binary data [64 Mae]. Support for this amendment is provided by [67 Pau] who determined the lattice parameters of six ternary alloys on the section Ag–35 at.-% Au, 65 at.-% Pd. Figure 34 presents this data with lattice spacings derived from Fig. 33; agreement between the data is satisfactory.

Although ordering occurs in the Au–Pd system in the range of the stoichiometric compositions Au_3Pd and $AuPd_3$, [46 Kuz] found no evidence of superlattice formation in ternary alloys slowly cooled from 800°C.

Höhn and Herzig [86 Hoe] measured the activity of Au in five ternary alloys with compositions 9·3Ag,5·3Au; 9·1Ag,9·8Au; 8·6Ag,14·7Au; 8·1Ag,20·0Au and 6·6Ag,34·4Au over the temperature range 800–1 025°C.

REFERENCES

46 Kuz: V. G. Kuznetsov: *Izvest. Sekt. Platiny*: 1946, **20**, 5–20

46 Nem: V. A. Nemilov, A. A. Rudnitsky and T. A. Vidusova: *Izvest. Sekt. Platiny*, 1946, **20**, 225–239

64 Mae: A. Maeland and T. B. Flanagan: *Can. J. Phys.*, 1964, **42**, 2364–2366

67 Pau: C. L. Pauley: X-ray study of the stacking fault density near the hardness maximum of the Au–Ag–Pd system, 1967, Masters thesis, Virginia Polytechnic Institute, USA

71 Nai: S. V. Nagender Naidu and C. R. Houska: *J. Appl. Phys.*, 1971, **42**, 4971–4975

77 Mia: J.-M. Miane, M. Gaune-Escard and J.-P. Bros: *High Temp., High Pressures*, 1977, **9**, 465–469

78 Ven: Y. C. Venudhar, L. Iyengar and K. V. Krishna Rao: *J. Less-Common Metals*, 1978, **58**, P 55–P 60

79 Ven: Y. C. Venudhar, T. Ranga Prasad, L. Iyengar and K. V. Krishna Rao: *J. Less-Common Metals*, 1979, **66**, P 11–P 15

81 Kin: H. W. King: *Bull. Alloy Phase Diagrams*, 1981, **2**(3), 402

83 Oka: H. Okamoto and T. B. Massalski: *Bull. Alloy Phase Diagrams*, 1983, **4**(1), 30–38

85 Oka: H. Okamoto and T. B. Massalski: *Bull. Alloy Phase Diagrams*, 1985, **6**(3), 229–235

86 Hoe: R. Höhn and C. Herzig: *Z. Metall.*, 1986, **77**, 291–297

86 Kar: I. Karakaya and W. T. Thompson: in 'Binary Alloy Phase Diagrams' (ed. T. B. Massalski with J. L. Murray, L. H. Bennett and H. Baker), 54–55; 1986, ASM International

86 Ven: Y. C. Venudhar, L. Iyengar and K. V. Krishna Rao, *J. Less-Common Metals*, 1986, **116**, 341–350

Table 8. Analytical representation of temperature dependence of the lattice spacing for alloys on the Pd–50 at.-%Ag, 50 at.-% Au section [86 Ven]

Alloy composition, at.-%			*Analytical expression* (a = nm; T = °C)	*Temperature range,* °C
Ag	*Au*	*Pd*		
12·5	12·5	75	$a = 0{\cdot}393\,166 + (0{\cdot}549\,814)\,10^{-5}\,T$	24–899
25·0	25·0	50	$a = 0{\cdot}397\,601 + (0{\cdot}624\,546)\,10^{-5}\,T$	26–899
33·33	33·33	33·33	$a = 0{\cdot}400\,671 + (0{\cdot}663\,481)\,10^{-5}\,T$	30–710
37·5	37·5	25	$a = 0{\cdot}402\,537 + (0{\cdot}696\,193)\,10^{-5}\,T$	28–899
50·0	50·0	0	$a = 0{\cdot}407\,528 + (0{\cdot}646\,713)\,10^{-5}\,T + (0{\cdot}131\,936)\,10^{-8}\,T^2$	30–900

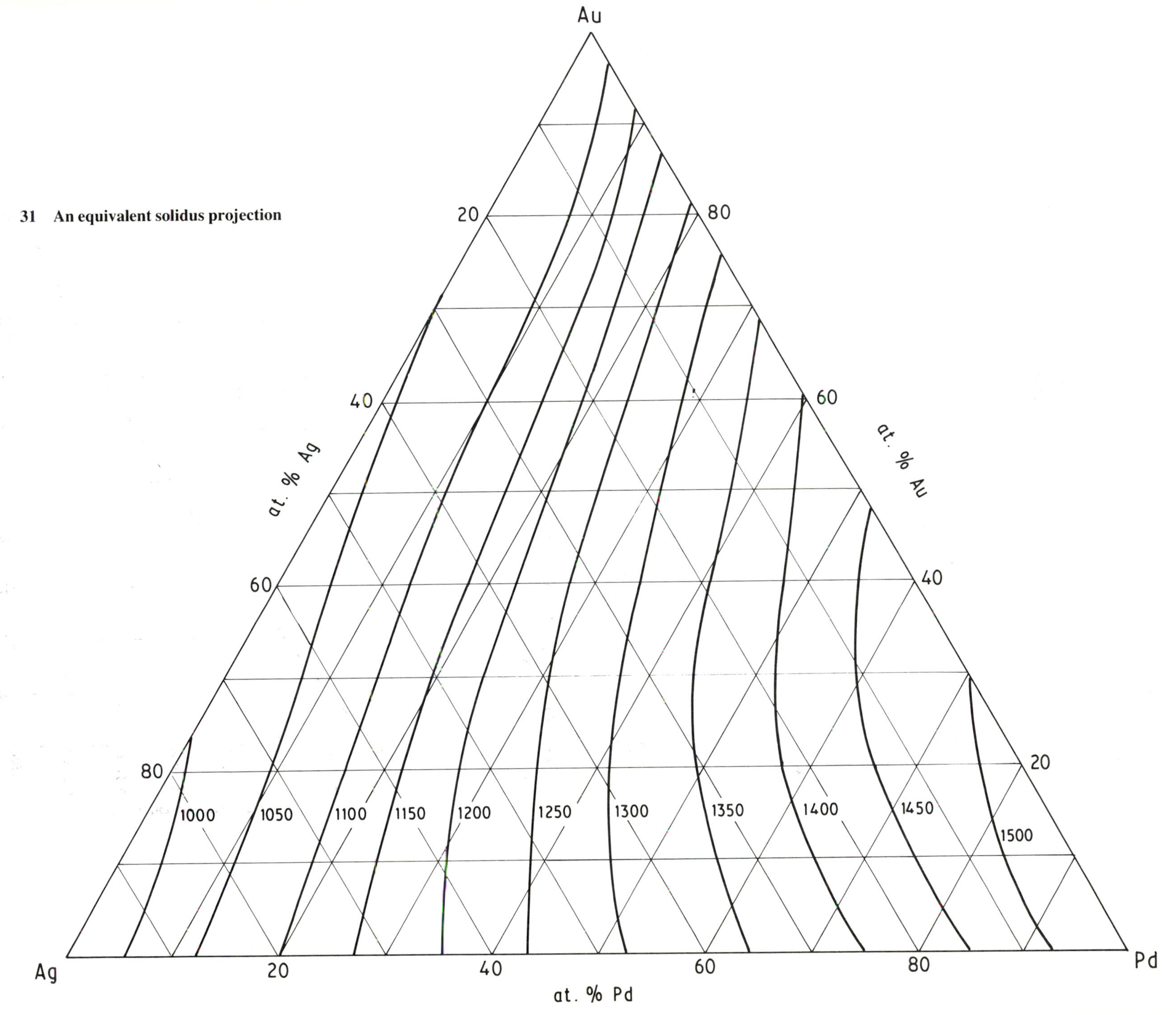

31 An equivalent solidus projection

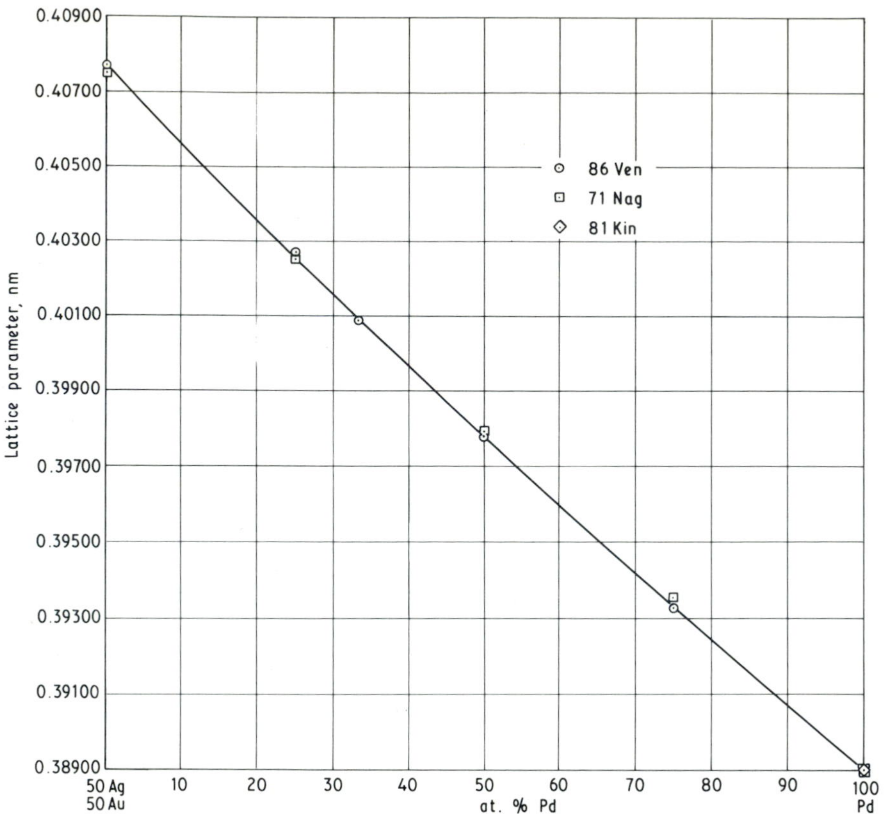

32 X-ray diffraction data for three ternary alloys and the end members on the section Pd–50 at.-% Ag, 50 at.-% Au

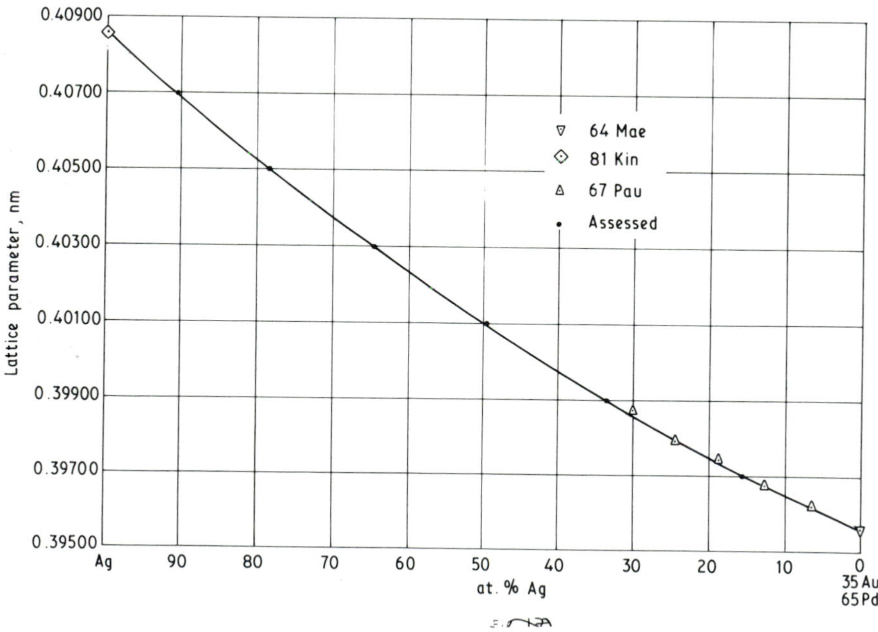

34 Lattice parameters of ternary alloys on the section Ag–35 at.-% Au, 65 at.-% Pd, with lattice spacings derived from Fig. 33

33 Lattice spacings of ternary alloys after cold-working, quenching, and annealing

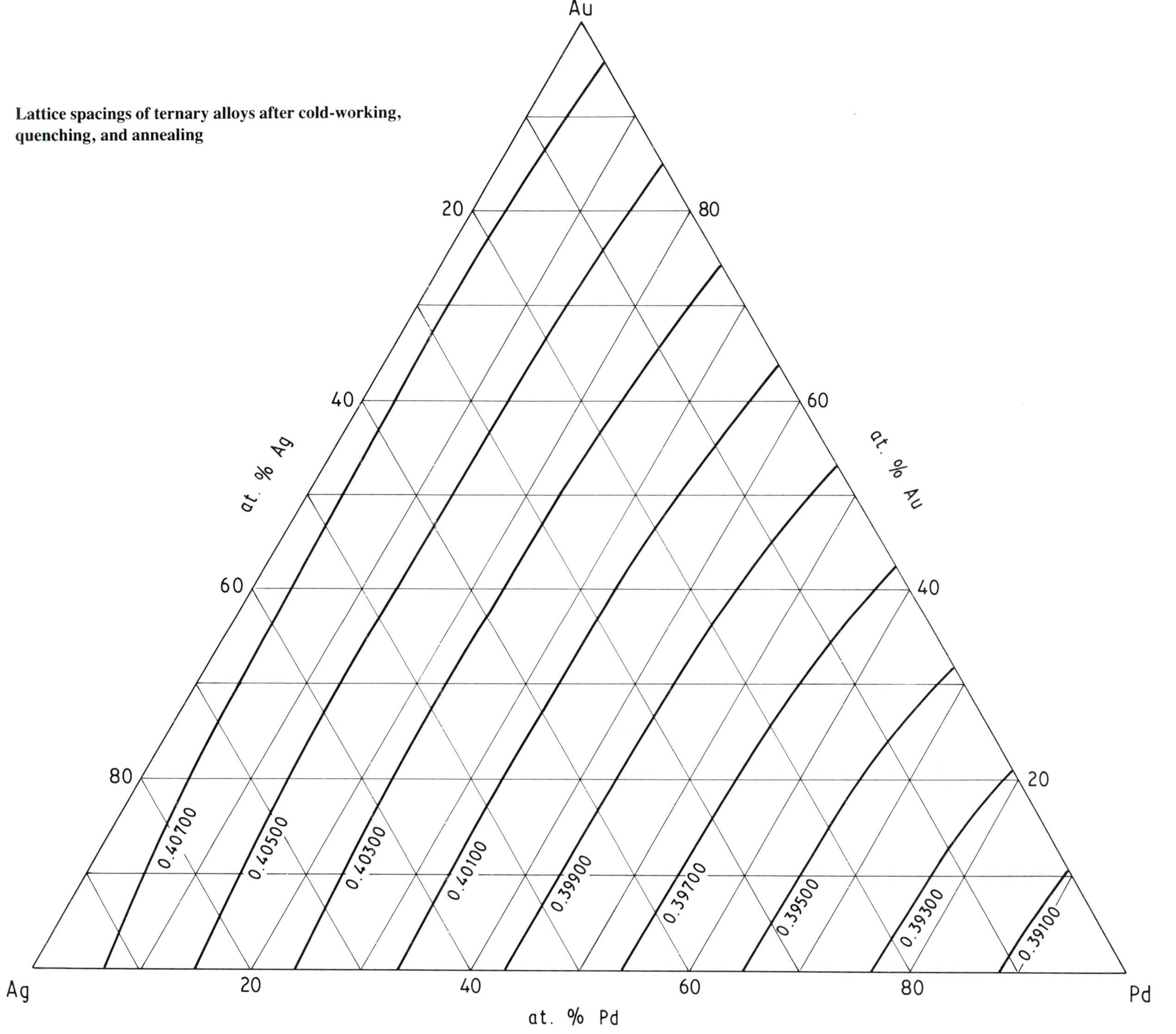

Ag–Au–Pt

INTRODUCTION

The only published work on the Ag–Au–Pt system is due to Novikova and Rudnitsky [58 Nov] who studied 33 alloys at 10 wt-% composition intervals up to and including 60 wt-% Pt. Thermal analysis on cooling was relied on to provide data on transformations involving the liquid phase. All alloys were heat-treated at 900°C for 6 h and 600°C for 18 h, water quenched, and examined metallographically, in addition to determination of their hardness, electrical resistance, temperature coefficient of electrical resistance, and thermoelectric power. All alloys were similarly studied in the annealed condition, to represent room temperature equilibria, but it is not clear what heat treatment was used to define the annealed state. Twenty-two alloys were mechanically tested, for strength and ductility, in the annealed condition.

In interpreting the data [58 Nov] accepted the presence of a peritectic reaction in the Au–Pt binary system, although it is now known that this system shows continuous solid solutions below the solidus encompassing a solid-state miscibility gap ($\alpha_1 + \alpha_2$) whose critical point lies at 61 at.-% Pt and 1 260°C [85 Oka]. According to [58 Nov] the Ag–Au–Pt ternary system possesses a simple monovariant peritectic curve, originating at the Au–Pt binary and ending at the Ag–Pt binary reaction. The presence of a monovariant ternary peritectic reaction is accepted but nothing is known of the origin of this reaction close to the Au–Pt binary edge. It is likely that a critical tie line between the liquid phase and a solid phase, α_1/α_2, is the origin of the three-phase triangle, $l + \alpha_1 + \alpha_2$, that descends to the Ag–Pt peritectic reaction at 1 185°C. The two-phase region, $\alpha_1 + \alpha_2$, enlarges in area towards the Ag–Au binary edge with fall in temperature; this is a reflection of the widening of the $\alpha_1 + \alpha_2$ miscibility gap with falling temperature in the Au–Pt system. Nothing is known of the solid state equilibria for alloys near to the Ag–Pt binary edge. [58 Nov] assumed that the Ag–Pt system was a simple peritectic system, but [58 Han] indicate complex solid state reactions that would be reflected in the ternary equilibria.

For reasons indicated above and in the section devoted to the ternary equilibria the data, [58 Nov] should be regarded as providing a guide to the equilibria in the Ag–Au–Pt system. There is a need for a more definitive study of the Ag–Pt binary system and the ternary system.

BINARY SYSTEMS

The Au–Pt and Ag–Au systems have been assessed by [85 Oka] and [83 Oka] respectively. The Ag–Pt system [58 Han] is accepted in terms of the peritectic reaction at 1 185°C, but no account has been taken of the solid state transformations in view of the lack of data on their influence on the ternary equilibria.

SOLID PHASES

No ternary compounds were detected [58 Nov]. For the alloys studied the α_1 solid solution, based on the Ag–Au series of solid solutions, and the α_2 solid solution, based on [Pt] solid solution, were the only phases identified.

TERNARY EQUILIBRIA

[58 Nov] used 99·99% Au, refined Pt of unstated purity and Ag prepared by reducing AgCl with glucose. 20 g samples were melted in alumina crucibles using high-frequency induction heating for alloys with >37·2 at.-% Pt; for alloys with lower Pt contents a tube furnace with SiC elements was used. Thermal analysis was done on cooling only, using a pyrometer for temperature measurement. Novikova and Rudnitsky admit that the rate of cooling was high and that only small thermal effects were observed [58 Nov]. Examination of the tabulated data indicates that reasonable reliance can be placed on the liquidus temperatures quoted for alloys that solidify as α_1 solid solution in a region parallel to the Ag–Au binary edge. Liquidus temperatures were not recorded for alloys separating primary α_2 (Pt solid solution) on cooling. By combining the ternary liquidus data [58 Nov] with the binary liquidus data a probable form of the ternary liquidus surface can be developed (Fig. 35). The isotherms run essentially parallel to the Ag–Au binary edge. The position of the peritectic fold on the liquidus surface is estimated from the intersection points of the experimentally determined α_1 liquidus surface and the speculative α_2 liquidus surface. The three-phase region, $l + \alpha_1 + \alpha_2$, cannot be located on vertical sections, as was done by [58 Nov], since the data tabulated for the temperatures at which the peritectic reaction begins and ends are not acceptable. Figures 36 *a–g* are vertical sections plotted in atomic percent, but representing the 10, 20, 30, 40, 50, 60, 70 wt-% Ag sections given in [58 Nov]. There is considerable scatter in the experimental data and, apart from Fig. 36*a*, all the measured temperatures are below the Ag–Pt binary peritectic temperature of 1 185°C. This is an impossibility if the ternary monovariant peritectic falls from a position near to the Au–Pt binary to the Ag–Pt binary peritectic reaction. Ternary temperatures below 1 185°C can only be valid if there is a minimum in the monovariant peritectic curve, associated with a critical tie line at the minimum temperature where liquid, α_1, and α_2 are in equilibrium. The high cooling rate used in thermal analysis has given reaction temperatures that are too low. The same remarks hold true for the solidus temperatures recorded by [58 Nov]. All the experimental points tabulated in [58 Nov] are included in Figs. 36*a–g*. The phase boundaries have been drawn in Figs. 36*a–g* to be consistent with each other and to provide isothermal sections that are also consistent with the vertical sections. It should be noted that this interpretation of the data is speculative.

The two-phase region, $\alpha_1 + \alpha_2$, widens with fall in temperature. The location of the boundary between $\alpha_1 + \alpha_2$ and α_1 is based on the metallographic examination of

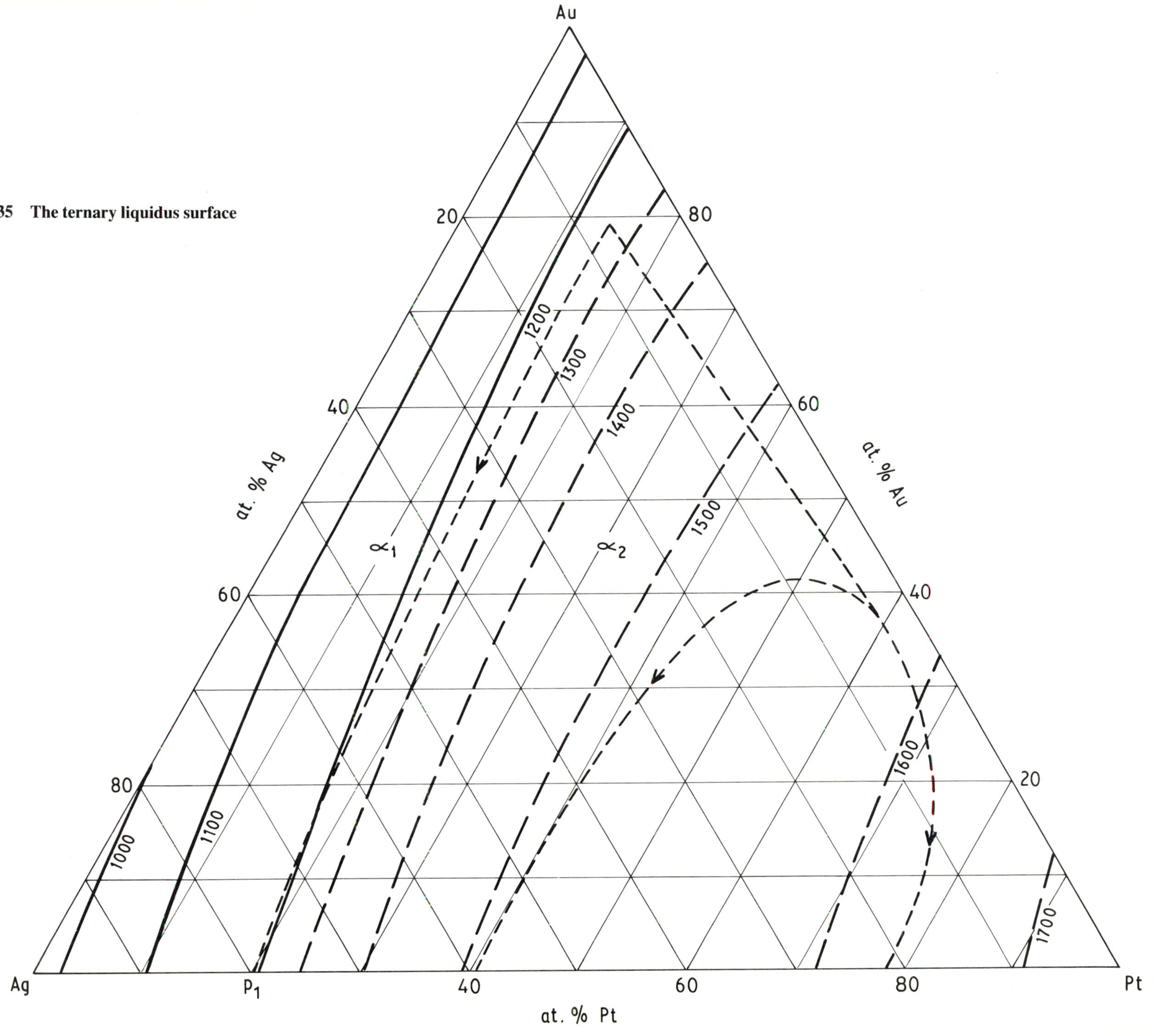

35 The ternary liquidus surface

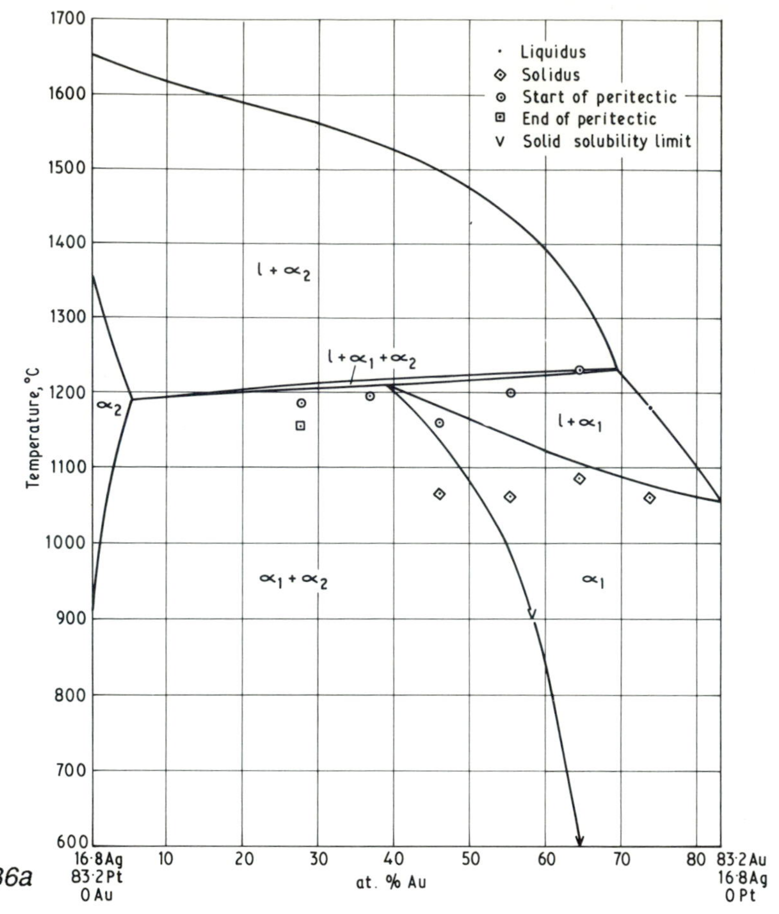

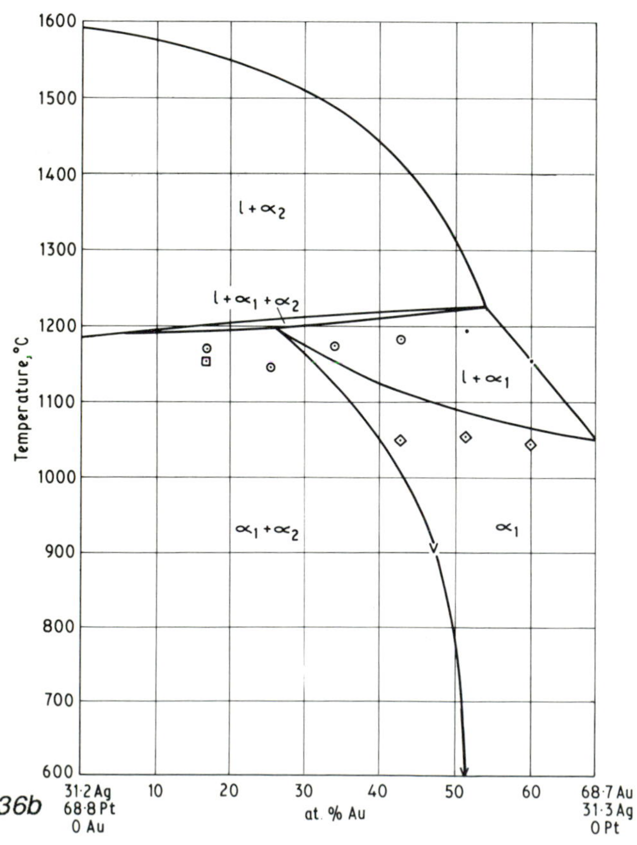

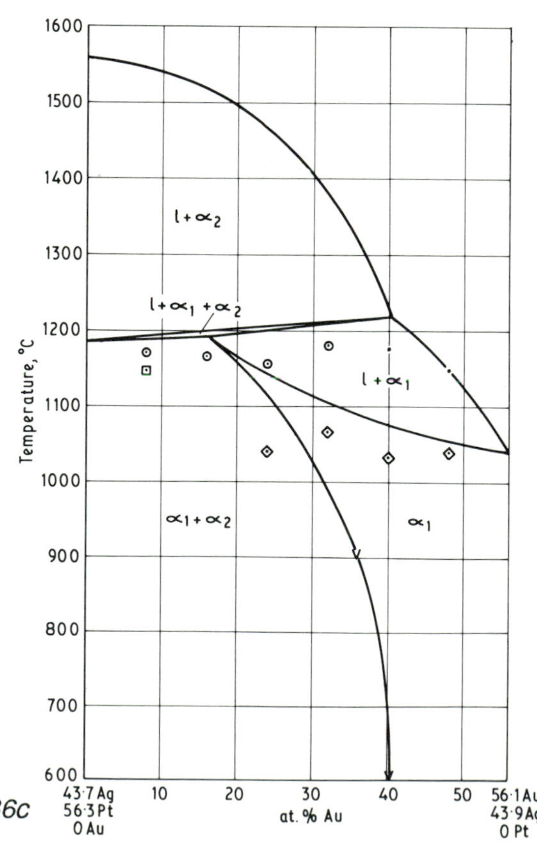

36 Vertical section at: *a* 16·8 at.-% Ag; *b* 31·3 at.-% Ag; *c* 43·8 at.-% Ag; *d* 54·8 at.-% Ag; *e* 64·5 at.-% Ag; *f* 73·2 at.-% Ag; *g* 80·9 at.-% Ag

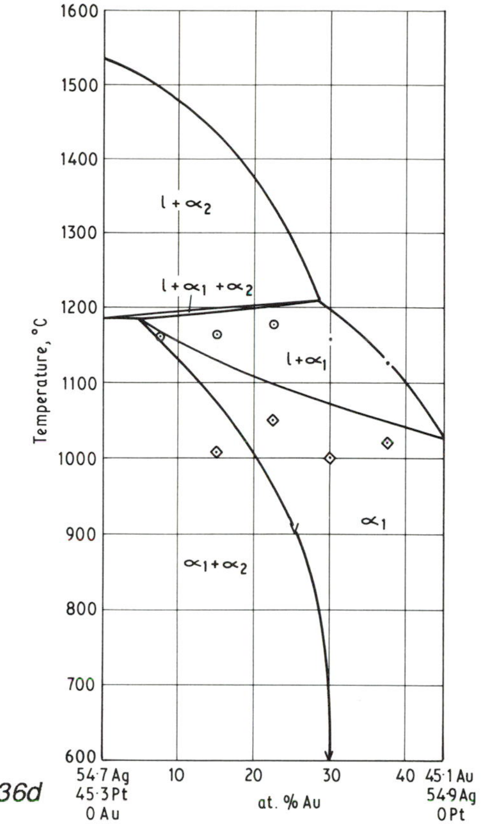

36d

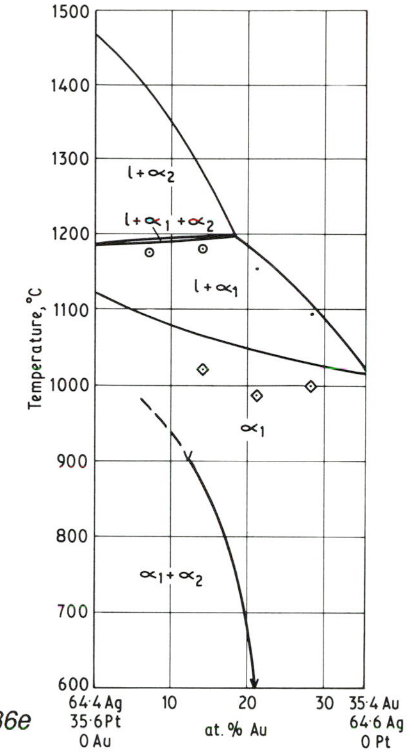

36e

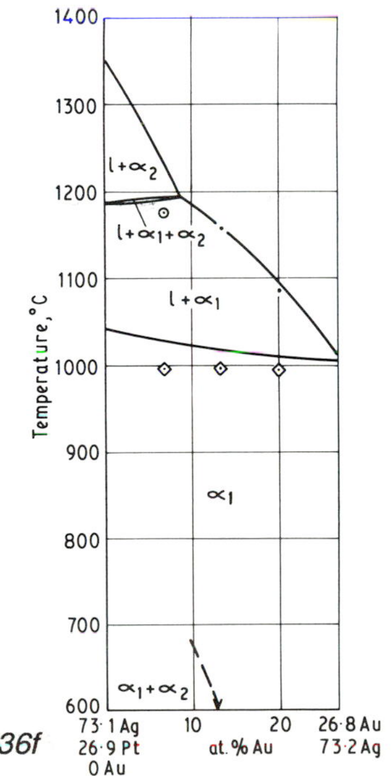

36f

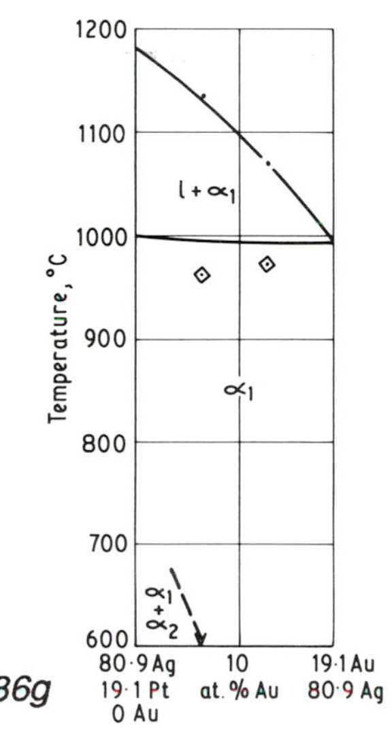

36g

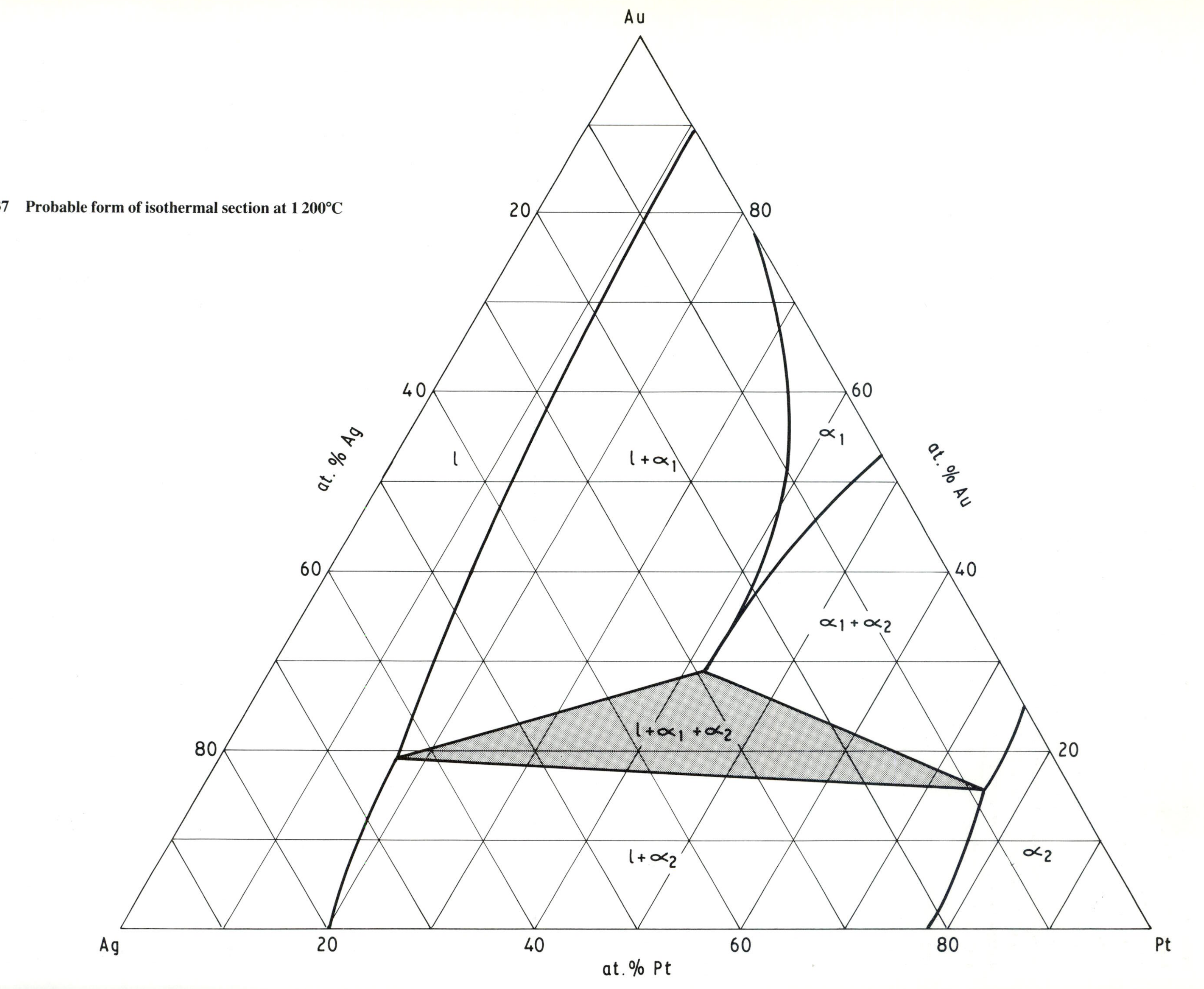

37 Probable form of isothermal section at 1 200°C

alloys annealed at 600 and 900°C and supplemented by determination of inflections in property-composition relationships. The original data has been amended to provide consistent phase boundaries for isothermal sections at 1 100, 1 000, 900 and 600°C.

Nothing is known about the ternary equilibria near to the Au–Pt binary edge. It is assumed that a critical tie line appears on the liquidus–solidus surfaces whereby liquid is in equilibrium with an α_1/α_2 solid solution. If, as appears likely, the temperature of the critical tie line is below the temperature for the closure of the solid-state miscibility gap in the Au–Pt system (1 261°C), the critical tie line $l - \alpha_1/\alpha_2$ will touch the $\alpha_1 + \alpha_2$ phase region originating from the Au–Pt binary at 1 261°C to produce a three phase region $l + \alpha_1 + \alpha_2$. It is estimated that the temperature at which this occurs is of the order of 1 235°C.

No data has been published on the ternary equilibria near to the Ag–Pt binary edge. [58 Nov] considered the Ag–Pt system to be a simple peritectic system. The presence of solid state transformations in this binary system points to a need for further work on the Ag–Au–Pt ternary system.

ISOTHERMAL SECTIONS

A series of isothermal sections can be generated from the vertical sections presented in Figs. 36*a–g*. Fig. 37 is an example of such an isothermal section, representing the probable phase boundaries at 1 200°C.

REFERENCES

58 Nov: O. A. Novikova and A. A. Rudnitsky: *Zhur. Neorg. Khim.*, 1958, **3**, 729–749

58 Han: M. Hansen and K. Anderko: 'Constitution of Binary Alloys'; 1958, New York, McGraw-Hill

83 Oka: H. Okamoto and T. B. Massalski: *Bull. Alloy Phase Diagrams*, 1983, **4**(1), 30–38

85 Oka: H. Okamoto and T. B. Massalski: *Bull. Alloy Phase Diagrams*, 1985, **6**, 46–56

Ag–Au–S

Early chemical work indicated the formation of the compounds Ag_3AuS_2 [1896 Ant] and AgAuS [1896 Mal]. Recent work has substantiated the existence of both compounds whether prepared chemically by the reaction of Ag_2S precipitates with thiogold (I) complexes or by direct reaction of the elements at 500°C in evacuated silica tubes [66 Tav, 66 Mes, 67 Tav, 68 Gra, 70 Smi, 76 Fol, 80 Bar]. The sulphide in equilibrium with Au-rich Ag–Au alloys and S vapour at 500°C contains 50 ± 10 at.-% Ag_2S [64 Bar]. Phase equilibrium data are reported for the solid state reactions in the pseudobinary system $Ag_{2-x}Au_xS$ for $x = 0$–0·5 [68 Gra] and for $x = 0$–1·0 [76 Fol, 80 Bar]. The data are in reasonably good agreement. Crystal structure determinations have been reported for Ag_3AuS_2 [66 Mes, 68 Gra, 70 Smi, 76 Fol] and AgAuS [70 Smi, 76 Fol].

BINARY SYSTEMS

The evaluations of the Ag–Au system by [83 Oka] and the Ag–S system by [86 Sha] are accepted. Little is known about the Au–S system [85 Oka], but there is a report of a metastable Au_2S compound [66 Hir].

SOLID PHASES

Both Ag_3AuS_2 and AgAuS exist in two allotropic forms (see Table 9).

PSEUDOBINARY SYSTEMS

Tavernier *et al.* [67 Tav] reported that the transition temperature of Ag_2S was reduced from 179 to 118°C on addition of up to 10 at.-% Au. [68 Gra] used 99·999% purity elements sintered in evacuated glass tubes to prepare nine alloys on the section Ag_2S–Ag_3AuS_2. Phase changes were detected by DTA (heating) and ac resistivity techniques. The section was determined up to 200°C and shown to be a pseudobinary eutectoid system. The eutectoid temperature was 113 ± 1°C and composition 7·33 at.-% Au. These results were confirmed [76 Fol, 80 Bar] and extended to give the section Ag_2S–AgAuS ($Ag_{2-x}Au_xS$) (Fig. 38). Folmer *et al.* [76 Fol] prepared alloys from 99·9% Ag, 99·99% Au, and 99·99% S by heating in evacuated silica tubes for several days at 500°C. The phase diagram was studied by DTA, DSC, and high-temperature X-ray methods. At high temperatures there is a second-order transition from bcc Ag_2S to the simple cubic $(Ag,Au)_2S$. The boundary between $Ag_2S(h_1)$ and $(Ag,Au)_2S$ meets the $Ag_2S(h_1)$ + Ag_3AuS_2 phase region at a temperature of 144°C. In addition to the eutectoid reaction $Ag_2S(h_1) = Ag_2S(l) + Ag_3AuS_2$ at 113°C there is also a eutectoid reaction $(Ag,Au)_2S(h) = Ag_3AuS_2$ + AgAuS at 181°C and ≈18·67 at.-% Au.

Barton [80 Bar] used 99·999% pure elements to synthesize (Ag,Au) alloys and AgAuS. Mixtures of (Ag,Au) + S or AgAuS + Ag were reacted for 6–7 weeks at 500 and 700°C in evacuated silica capsules and the products characterized by electron probe microanalysis and X-ray diffraction. The tie lines joining the equilibriated phases are shown in Figs. 39 and 40. DTA was used to confirm the low-temperature results of [68 Gra] and [76 Fol] for the pseudobinary section Ag_2S–AgAuS (Fig. 38), and extended to give an impression of the equilibria involving the liquid sulphide phase at high temperatures (Fig. 41). There is evidence for a minimum in the $(Ag_xAu_{1-x})_2S$ liquidus at about 720°C and 23 at.-% Au. This minimum is displaced to lower Au contents and lower temperatures with compositions that are on the S-rich side of the stoichiometric $(Ag_xAu_{1-x})_2S$ section. A similar characteristic may hold for metal-rich melts. An invariant reaction at 756° ± 10°C was proposed by [80 Bar]. This is probably the formation of $(Ag,Au)_2S$ by reaction of liquid sulphide with liquid S and a Au-rich (Ag,Au) alloy, L +

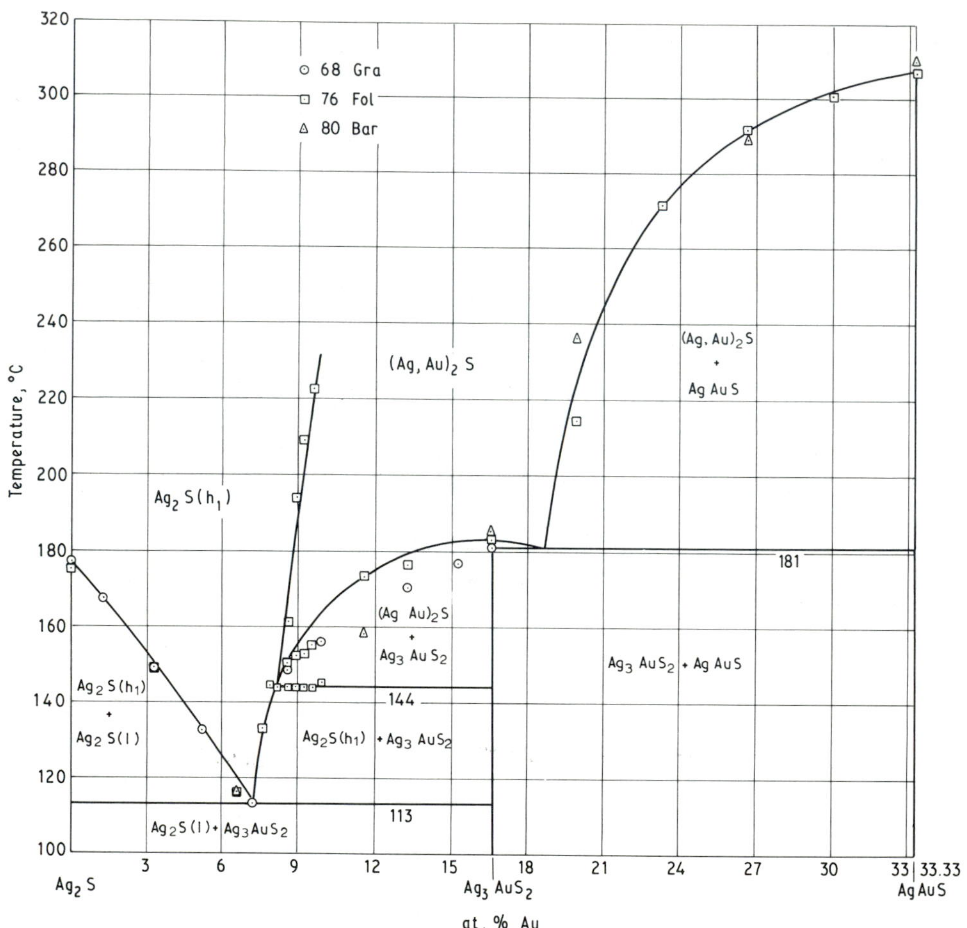

38 The pseudobinary system Ag_2S–AgAuS

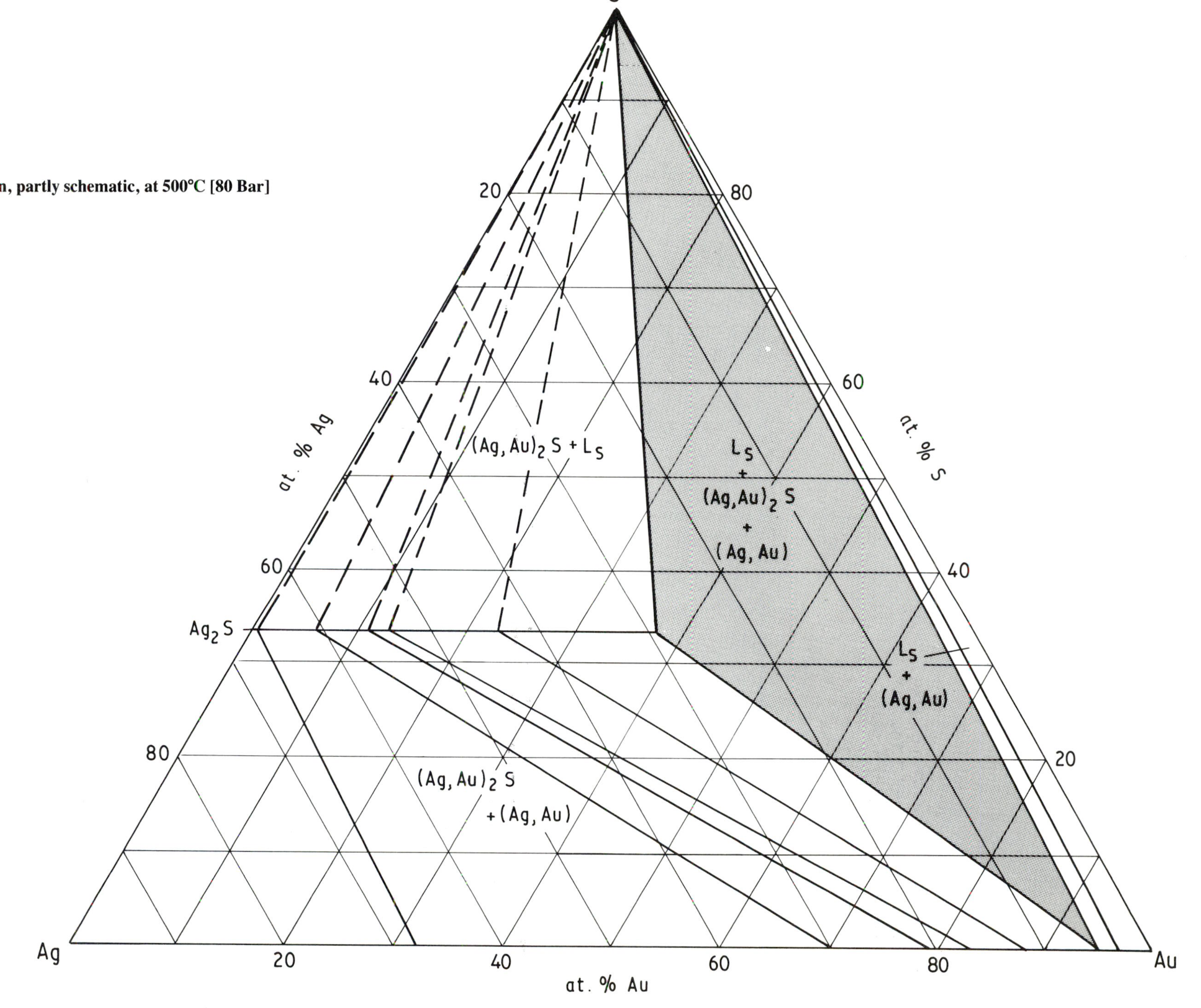

39 Isothermal section, partly schematic, at 500°C [80 Bar]

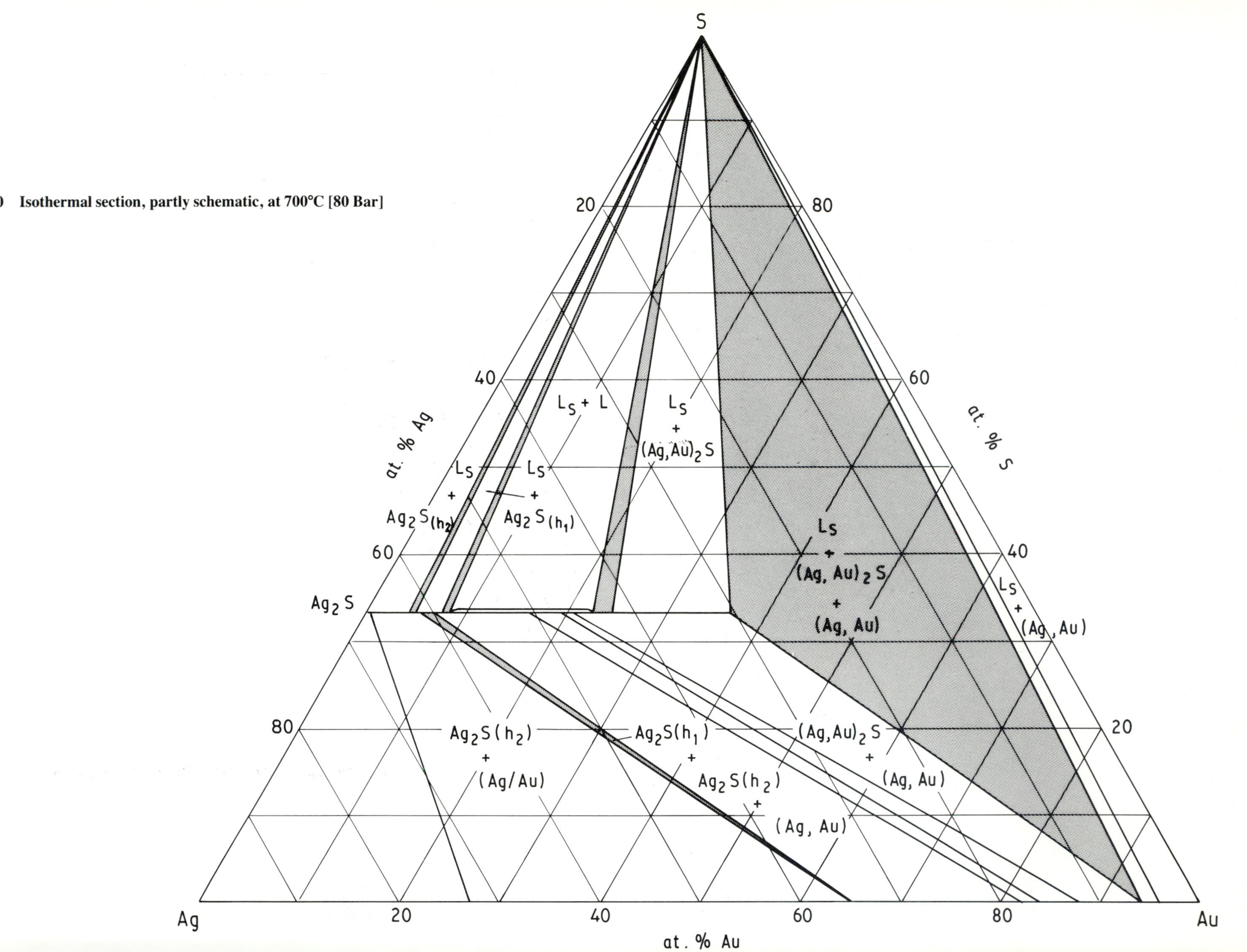

40 Isothermal section, partly schematic, at 700°C [80 Bar]

Table 9. Solid phases

Phase/temperature range, °C	*Pearson symbol/prototype*	*Lattice parameters*, pm	*Comments*
(Ag,Au)	cF4 Cu	a = 408·62 a = 407·85	pure Ag (V–C) pure Au (V–C)
S(h) 115–102	mP48 S	a = 1092 b = 1098 c = 1104 β = 83°16′	(P)
S(r) <102	oF128 S	a = 1046.46 b = 1286·60 c = 2448.60	(P)
$Ag_2S(h_2)$	cP6 $Ag_2Se(h)$	a = 634	at 650°C
$Ag_2S(h_1)$	Cubic	a = 489·0	at 300°C (P)
$Ag_2S(l)$	Monoclinic	a = 953·1 b = 692.5 c = 827·8 β = 123·85°	(P)
Au_2S	cP6 Cu_20	a = 502·0	[66 Hir]
* $Ag_3AuS_2(h)$	Cubic	a = 494 a = 497	190°C [70 Smi] 420°C [76 Fol]
* $Ag_3AuS_2(r)$	Tetragonal	a = 975 c = 985	[68 Gra]
* $(Ag,Au)_2S(h)$	Cubic	a = 501–503·3	310–490°C [70 Smi, 76 Fol]
* AgAuS	Monoclinic	a = 838 b = 1342 c = 909 β = 111·4°	[70 Smi]

L_S + (Ag,Au) = $(Ag,Au)_2S$. The sulphide equilibria (Fig. 38) are satisfactorily modelled [80 Bar] with a regular solution approach.

REFERENCES

1896 Ant: U. Antony and A. Lucchesi: *Gazz Chim. Ital.*, 1896, **26**(II), 350

1896 Mal: J. MacLaurin: *J. Chem. Soc.*, 1896, **69**, 1269–1276

64 Bar: P. B. Barton and P. Toulim: *Geochim. Cosmochim. Acta*, 1964, **28**, 619–640

66 Hir: H. Hirsch, A. de Cugnac, M. C. Gadet and J. Pouradier: *Compt. Rend. Acad. Sci. Paris*, 1966, **B263**, 1328–1330

66 Mes: P. Messien, M. Baiwir and B. Tavernier: *Bull. Soc. Roy. Sci. Liege*, 1966, **35**, 727–733

66 Tav: B. H. Tavernier: *Z. anorg. Chem.*, 1966, **343**, 323–328

67 Tav: B. H. Tavernier, J. Vervecken, P. Messien and M. Baiwir: *Z. anorg. Chem.*, 1967, **356**, 77–88

68 Gra: R. B. Graf: *Amer. Mineral.*, 1968, **53**, 496–500

70 Smi: T. J. M. Smit, E. Venema, J. Wiersma and G. A. Wiegers: *J. Solid State Chem.*, 1970, **2**, 309–312

76 Fol: J. C. W. Folmer, P. Hofmann and G. A. Wiegers, *J. Less-Common Metals*, 1976, **48**, 251–268

80 Bar: P. B. Barton: *Econ. Geol.*, 1980, **75**, 303–316

83 Oka: H. Okamoto and T. B. Massalski: *Bull. Alloy Phase Diagrams*, 1983, **4**, 30–38

85 Oka: H. Okamoto and T. B. Massalski: *Bull. Alloy Phase Diagrams*, 1985, **6**, 518–519

86 Sha: R. C. Sharma and Y. A. Chang: *Bull. Alloy Phase Diagrams*, 1986, **7**, 263–269

Ag–Au–Sb

Little is known about this system beyond an observation [59 Bur], based on X-ray diffraction studies of five ternary alloys, that Ag_3Sb can take about 10 at.-% Au into solution. Lattice parameters are given in Table 10 for the ternary phase based on Ag_3Sb which has an orthorhombic Cu_3Ti-type structure (oP8). Vyacheslavov *et al.* [70 Vya] claim that Ag–Au–Sb electrodeposits, with a maximum of 10 at.-% Sb, contain a (Ag–Au) solid solution and a ternary intermetallic compound based on gold or silver. This claim requires verification.

REFERENCES

59 Bur: W. Burkhardt and K. Schubert: *Z. Metall.*, 1959, **50**, 442–452

70 Vya: P. M. Vyacheslavov, N. P. Fedot'ev and G. A. Volyanyuk: *Zhur. Priklad. Khim.*, 1970, **43**, 79–82 (*J. Appl. Chem. USSR*, 1970, **43**, 72–75)

Table 10. Lattice parameters [59 Bur] of Ag–Au–Sb alloys

Composition, at.-%			*Heat treatment*		*Lattice parameter*, nm		
Ag	*Au*	*Sb*	*Ingot*	*Powder*	*a*	*b*	*c*
70	5	25	21 days 450°C	3 days 300°C	0·2964	0·5166	0·4796
75	5	20	10 days 400°C	6 days 400°C	0·2969	0·5160	0·4757
75	10	15	10 days 400°C	1 hour 425°C	0·2939	0·5090	0·4746
70	10	20	4 days 400°C	3 days 350°C	0·2989	0·5177	0·4813
65	20	15	4 days 400°C	3 days 350°C	0·2964	0·5133	0·4811

Ag–Au–Se

The only published data on the Ag–Au–Se system concerns the occurrence of a ternary compound Ag_3AuSe_2 and a phase diagram of the Ag_2Se–Ag_3AuSe_2 section. Tavernier [66 Tav] and Messien and Baiwir [66 Mes] prepared Ag_3AuSe_2 chemically by the reaction of Ag_2Se precipitates with thiogold (I) complexes. Smit *et al.* [70 Smi] and Wiegers [76 Wie] synthesized Ag_3AuSe_2 from the elements by heating for several days at 500°C and 450°C respectively in evacuated quartz tubes. Ag_3AuSe_2 exists in two allotropic modifications. The low-temperature form, βAg_3AuSe_2, is stable below 267°C (heating) [70 Smi, 76 Wie], 270°C (heating) and 251°C (cooling) [67 Tav]. It is cubic, cI48, with a = 0·995 nm [66 Mes], 0·9967 ± 0·0003 nm [71 Joh]. The latter data relate to a naturally occurring mineral fischesserite of the composition Ag_3AuSe_2 [71 Joh]. The high-temperature form, αAg_3AuSe_2, is bcc of structure type cP6 with a = 0·506 nm at 290°C [70 Smi].

The section Ag_2Se–Ag_3AuSe_2 (Fig. 42) was determined by [76 Wie] using dta at a heating rate of 10°C min^{-1}, electrical measurements and high-temperature X-ray diffraction analysis. It is pseudobinary. At high temperatures there is a continuous bcc solid solution from αAg_2Se to αAg_3AuSe_2. Wiegers states that this solid solution extends to Au contents higher than Ag_3AuSe_2, but no details are given [76 Wie]. βAg_2Se forms a degenerate eutectoid with βAg_3AuSe_2, the eutectoid temperature being virtually identical with the transition temperature of Ag_2Se. No (L + bcc) region is shown in Fig. 42. Wiegers states that it is probably small. The liquidus temperature at the composition Ag_3AuSe_2 is 742°C (scaled from the original figure) compared with 730°C quoted by Smit *et al.* [70 Smi].

REFERENCES

66 Tav: B. H. Tavernier: *Z. anorg. Chem.*, 1966, **343**, 323–328

66 Mes: P. Messien and M. Baiwir: *Bull. Soc. Roy. Sci. Liege*, 1966, **35**, 234–243

67 Tav: B. H. Tavernier, J. Vervecken, P. Messien and M. Baiwir: *Z. anorg. Chem.*, 1967, **356**, 77–88

70 Smi: T. J. M. Smit, E. Venema, J. Wiersma and G. A. Wiegers: *J. Solid State Chem.*, 1970, **2**, 309–312

71 Joh: Z. Johan, P. Picot, R. Pierrot and M. Kvacek: *Bull. Soc. franc. Mineral. Cristallogr.*, 1971, **94**, 381–384

76 Wie: G. A. Wiegers: *J. Less-Common Metals*, 1976, **48**, 269–283

Ag–Au–Si

INTRODUCTION

Kuprina [62 Kup] surveyed alloys at 10 wt-% intervals throughout the ternary system using thermal analysis and metallography. Alloys were synthesized from Ag and Au with ≤ 0·01% impurities, and 99·98% Si by melting under a $BaCl_2$ flux in a high frequency induction furnace. As would be anticipated from the constituent binary systems the Ag–Au–Si system contains a ternary monovariant eutectic curve originating at the Ag–Si binary eutectic and ending at the Au–Si binary eutectic. There is very little solubility of Si in the ternary (Ag/Au) solid solution or of Ag–Au in (Si). The data of [83 Cas] and [83 Has] throw doubt on the liquidus measurements reported by [62 Kup] and the latter data are not accepted. [85 Has] studied 35 alloys by differential thermal analysis and high temperature calorimetry to produce vertical sections from Si to 75Ag25Au, 50Ag50Au and 30Ag70Au.

BINARY SYSTEMS

The Ag–Au and Ag–Si binary systems are established [H,S,E]. A eutectic composition of 11·5 at.-% Si at 856°C is accepted for the Ag–Si system. The Au–Si system is of the eutectic type with a eutectic composition of 18·6 at.-% Si at 362°C.

SOLID PHASES

Only (Ag/Au) solid solution and (Si) exist as solid phases (Table 11). No ternary compounds occur in this system.

Table 11. Crystal structures

Solid phase	*Prototype*	*Lattice designation*
(Ag/Au)	Cu	cF4
(Si)	C	cF8

LIQUIDUS SURFACE

The results of [62 Kup] were presented in tabular form and as a series of vertical sections at constant wt-% Au contents. Alloy compositions have been converted into

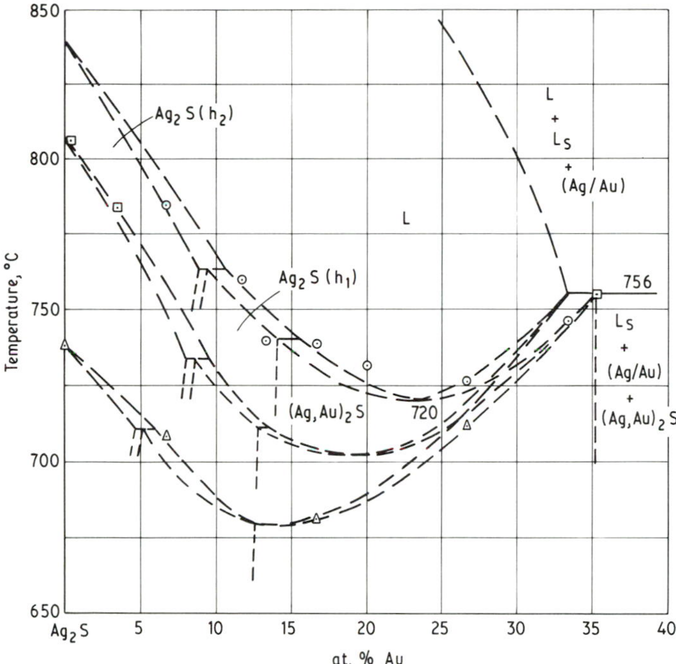

41 Equilibria on the $(Ag_xAu_{1-x})_2S$ section involving the melt [80 Bar]

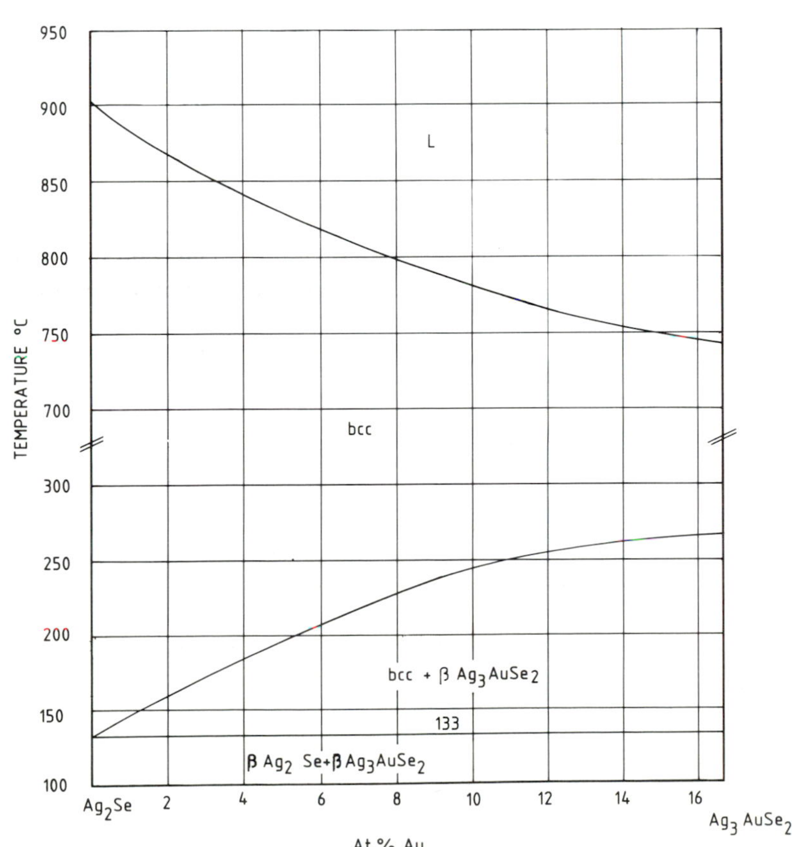

42 The Ag_2Se–Ag_3AuSe_2 section

atomic percentages (Table 12). Thermal analysis was the main experimental technique employed, with both heating and cooling curves being used to determine liquidus temperatures and the temperatures corresponding to the start and end of the secondary separation of (Ag/Au) + (Si) from the melt. Metallographic studies were made on alloys that had been annealed in vacuum for 500 h at 400°C and then cooled slowly.

The liquidus values [62 Kup] lead to a liquidus surface whose isotherms are S-shaped in the primary Si phase region. The isotherms from the Au–Si binary edge bend towards the Ag–Si binary, then away from this binary and finally bend round again to intersect the Ag–Si binary.

The unpublished data of [83 Cas] are at variance with [62 Kup]. [83 Cas] measured the enthalpy of dissolution of solid Si in molten Ag–Au alloys at 1 075°C as a function of the at.-% Si by direct reaction calorimetry using a high-temperature Calvet calorimeter. A change in slope of the enthalpy of dissolution (from positive to negative) occurs at the solubility limit of Si in the Ag–Au alloy under study. Four Ag–Au alloys were studied: 20 at.-% Au, 80 at.-% Ag; 40 Au, 40 Ag; 60 Au, 40 Ag and 80 Au, 20 Ag. Si additions were made and the solubility limit of Si determined as 27·5 at.-% Si for the 20 Au, 80 Ag binary alloy, 32% Si for the 40 Au, 60 Ag alloy, 40% Si for the 60 Au, Ag alloy and 48% Si for the 80 Au, 20 Ag alloy. This data allows the 1 075°C isotherm to be drawn (Fig. 43). It has a very different shape to the 1 100°C isotherm deduced from [62 Kup]; it covers temperatures reported by [62 Kup] ranging from 950 to 1 150°C. The binary liquidus points found by [83 Cas] and plotted in Fig. 43 are in agreement with accepted values at 1 075°C for the Au–Si system and some 30°C high for the Ag–Si system.

[83 Has] and [85 Has] used DTA on 35 ternary alloys made from 99·9999% pure elements to produce the general form of the liquidus (Fig. 43). It will be noted that the 1 075°C isotherm [83 Cas] is in good agreement with the isotherms generated from the liquidus of [83 Has] and [85 Has]. The liquidus values of [62 Kup] are not accepted.

The monovariant curve e_1e_2 runs smoothly from the Ag–Si binary eutectic to the Au–Si binary eutectic. The 400, 500, 600, 700, and 800°C isotherms are considered to intersect e_1e_2 at 6, 21, 37·5, 54, and 74 at.-% Ag. These conclusions are based on the data of [83 Has] and [85 Has], taken with a reinterpretation of the data of [62 Kup] which provides a series of consistent vertical and isothermal sections.

VERTICAL SECTIONS

The vertical sections produced by [62 Kup] do not allow the constructions of consistent isothermal sections. Figures 44*a–f* are plots of the tabulated data [62 Kup] for the start and finish of the secondary separation of (Ag/Au) + (Si) from the melt. The upper phase boundary, between the L + (Si) and the L + (Ag/Au) + (Si) phase regions in Figs. 44*a–d*, agrees reasonably well with the experimental data. The experimental results in Figs. 44*e–f* indicate failure to detect the start of the secondary separation until considerable undercooling had occurred. The lower phase boundary in Figs. 44*a–f* are at variance with the experimental data for all sections except Fig. 44*a*. Again extensive undercooling has not allowed the measurement of equilibrium temperatures.

From the vertical sections (Figs. 44*a–f*) it is possible to construct vertical sections that run from the Si corner to the Ag–Au binary edge (sections with constant Ag:Au ratio). Values for the liquidus curves were taken from Fig. 43 [83 Has, 85 Has]. Three sections studied by [85 Has] are reproduced in Figs. 45*a–c*. The data of [83 Cas] at 1 075°C is marked. Sections at constant Ag:Au ratios in a ternary system where there is extremely small solubility of the (Ag/Au) solid solution in the third component (Si) will show a nearly horizontal phase boundary for the onset of the secondary separation from the melt. In Fig. 45*a* secondary separation begins at 760°C from all alloy compositions in the primary phase field of (Si). The temperature of the monovariant curve e_1e_2 is also established as 760°C at the intersection of this vertical section

Table 12. Alloy compositions [62 Kup]. Atomic weights used were Ag 107·87, Au 196·967, Si 28·086

wt.-%			at.-%			wt.-%			at.-%		
Ag	*Au*	*Si*	*Ag*	*Au*	*Si*	*Ag*	*Au*	*Si*	*Ag*	*Au*	*Si*
80	10	10	64·6	4·4	31·0	50	40	10	45·3	19·9	34·8
70	10	20	46·0	3·6	50·4	40	40	20	28·8	15·8	55·4
60	10	30	33·2	3·0	63·8	30	40	30	18·0	13·1	68·9
50	10	40	23·9	2·6	73·5	20	40	40	10·2	11·2	78·6
40	10	50	16·8	2·3	80·9	10	40	50	4·5	9·8	85·7
30	10	60	11·3	2·1	86·6	40	50	10	37·8	25·9	36·3
20	10	70	6·8	1·9	91·3	30	50	20	22·4	20·4	57·2
10	10	80	3·1	1·7	95·2	20	50	30	12·3	16·8	70·9
70	20	10	58·6	9·2	32·2	10	50	40	5·2	14·3	80·5
60	20	20	40·6	7·4	52·0	30	60	10	29·6	32·5	37·9
50	20	30	28·4	6·2	65·4	20	60	20	15·4	25·4	59·2
40	20	40	19·6	5·3	75·1	10	60	30	6·3	20·8	72·9
30	20	50	12·9	4·7	82·4	20	70	10	20·7	39·6	39·7
20	20	60	7·6	4·2	88·2	10	70	20	8·0	30·6	61·4
10	20	70	3·4	3·8	92·8	10	80	10	10·8	47·5	41·7
60	30	10	52·3	14·3	33·4	5	90	5	6·8	67·1	26·1
50	30	20	34·9	11·5	53·6	—	—	—	—	—	—
40	30	30	23·2	9·6	67·1	—	—	—	—	—	—
30	30	40	15·0	8·2	76·8	—	—	—	—	—	—
20	30	50	8·8	7·2	84·0	—	—	—	—	—	—
10	30	60	3·9	6·4	89·7	—	—	—	—	—	—

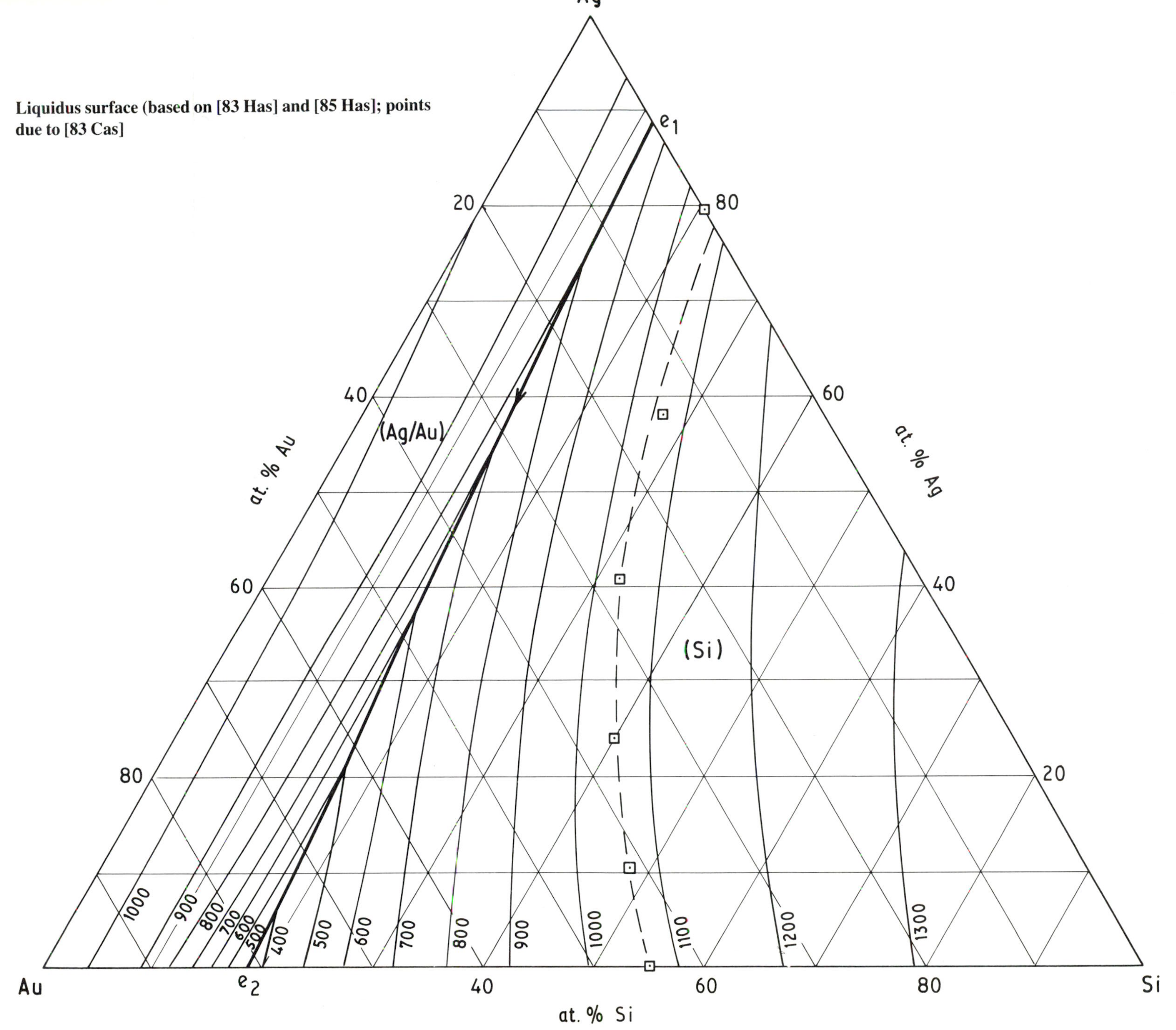

43 Liquidus surface (based on [83 Has] and [85 Has]; points due to [83 Cas]

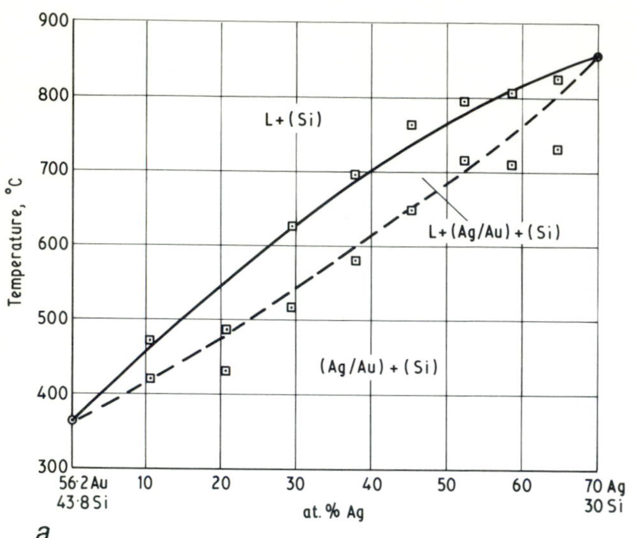

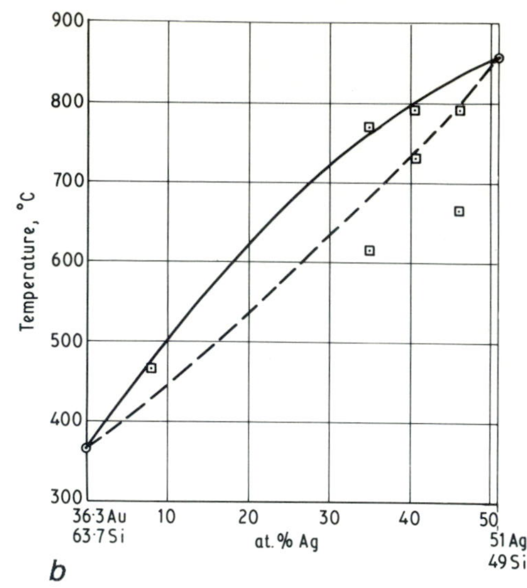

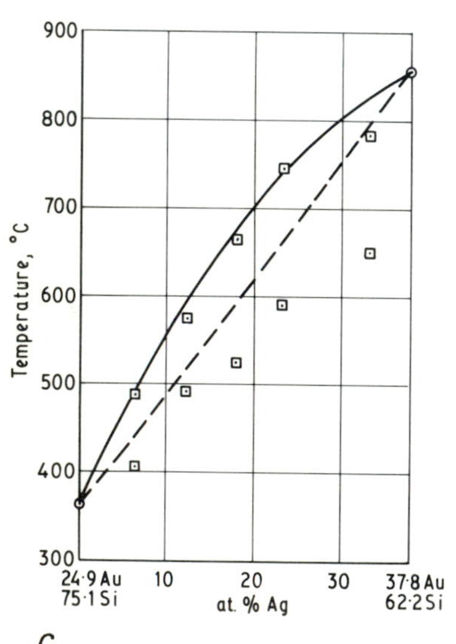

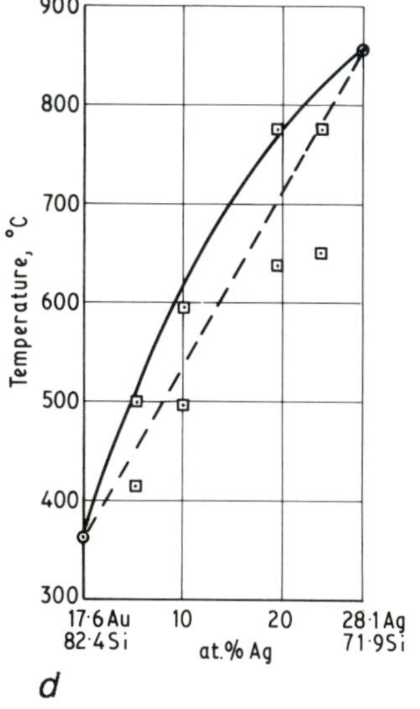

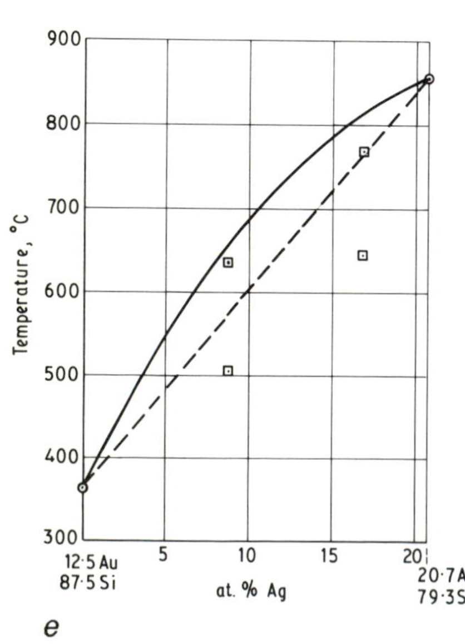

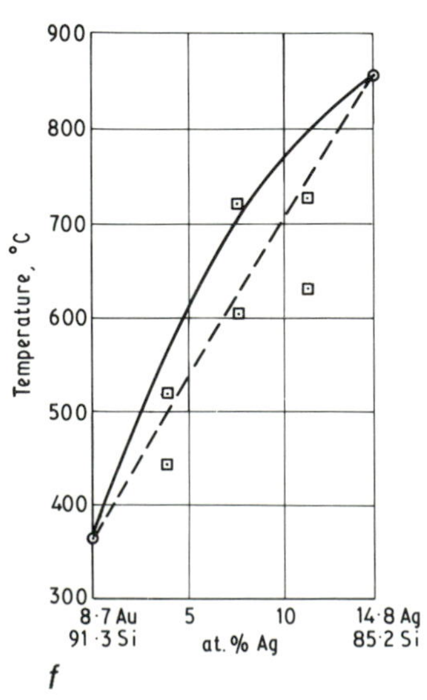

44 Vertical section at: *a* 10 wt-% Si; *b* 20 wt-% Si; *c* 30 wt-% Si; *d* 40 wt-% Si; *e* 50 wt-% Si; *f* 60 wt-% Si (based on [62 Kup])

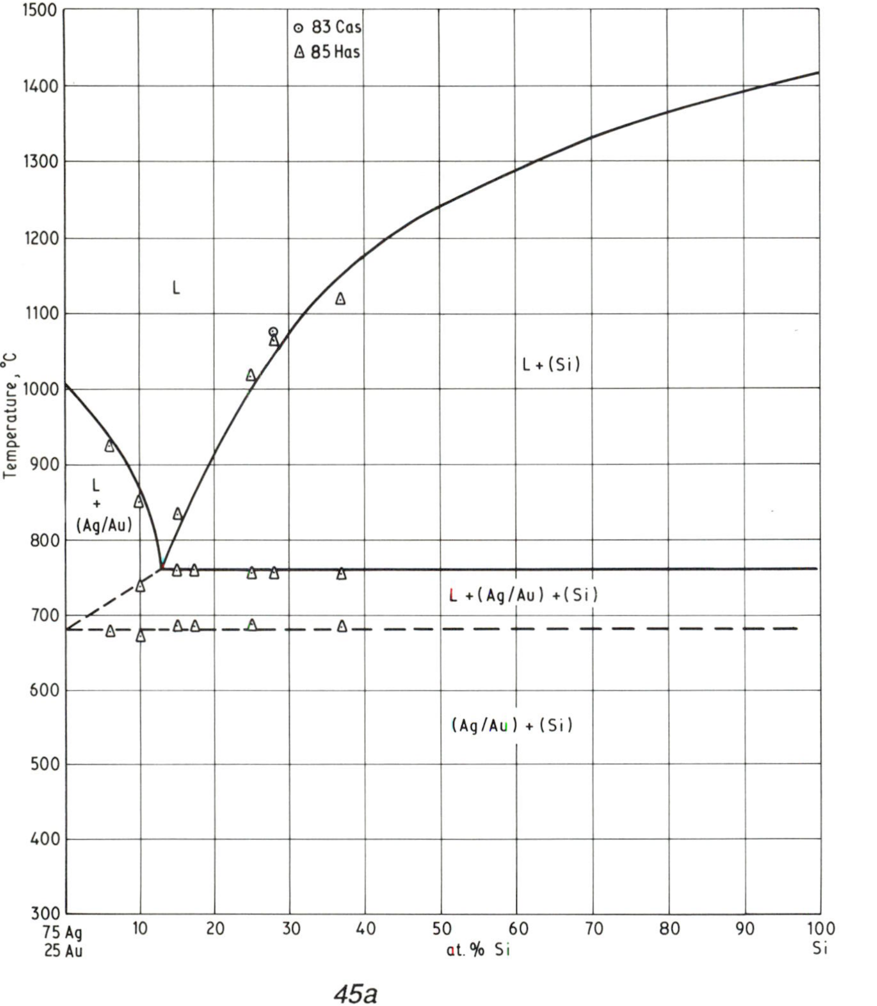

45a

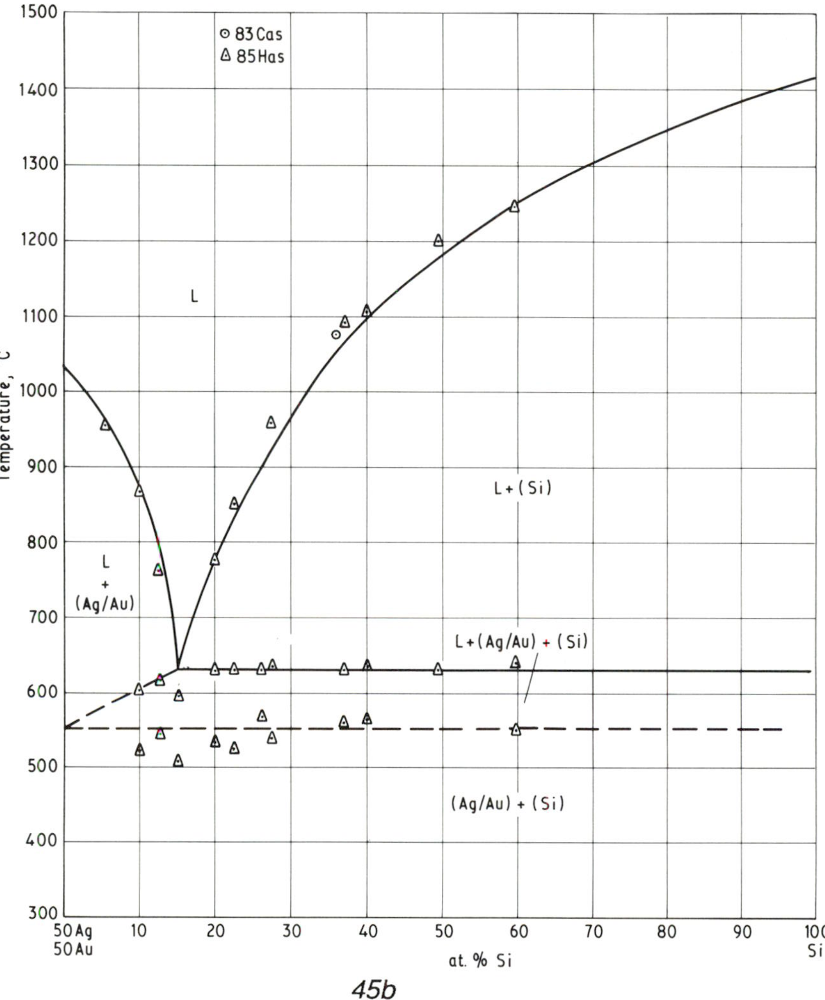

45b

45 Vertical section from: *a* 75 at.-% Ag, 25 at.-% Au to Si; *b* 50 at.-% Ag, 50 at.-% Au to Si; *c* 30 at.-% Ag, 70 at.-% Au to Si [85 Has]

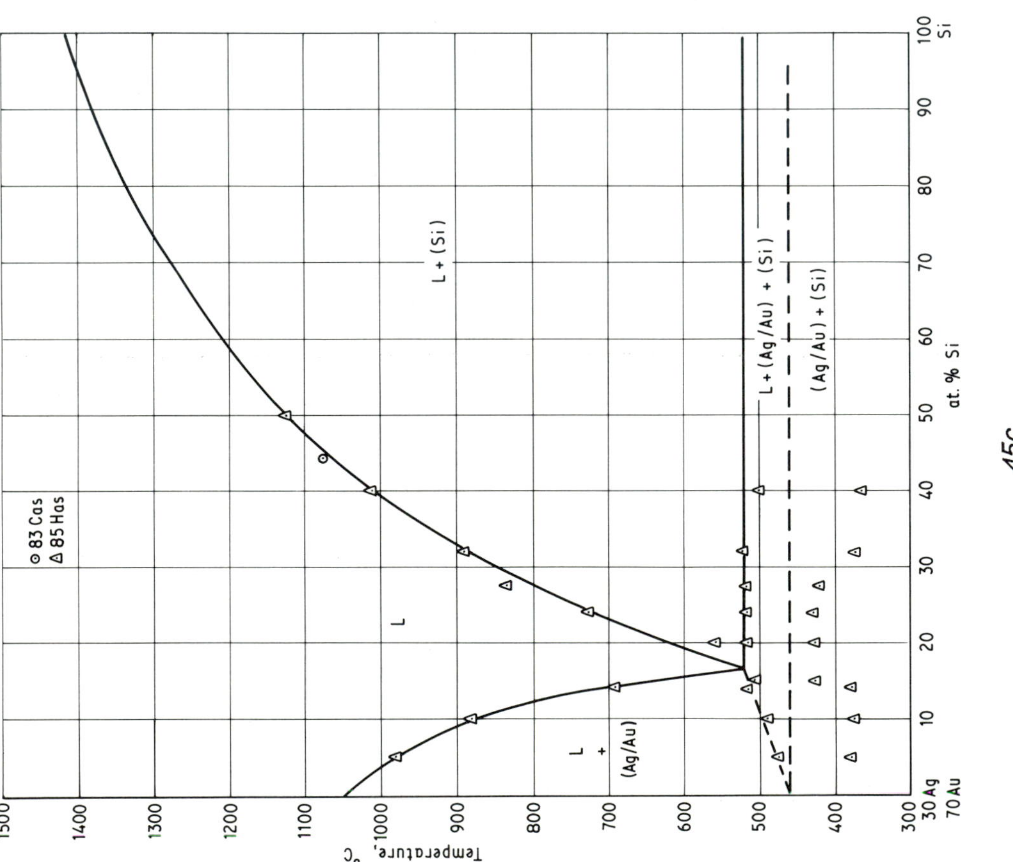
⊙ 83 Cas
Δ 85 Has
L
L + (Si)
L + (Ag/Au)
L + (Ag/Au) + (Si)
(Ag/Au) + (Si)
Temperature, °C
1500
1400
1300
1200
1100
1000
900
800
700
600
500
400
300
30 Ag
70 Au
10
20
30
40
50
60
70
80
90
100
Si
at. % Si

45c

with curve e_1e_2. Similar deductions can be made from Figs. 45*b–c*.

The interpretation of the data of [62 Kup] in terms of Figs. 44*a–f* produce vertical sections such as Figs. 45*a–c* that are consistent with the data of [83 Has] and [85 Has]. They also allow a series of consistent isothermal sections to be drawn.

Figures 46 and 47 are vertical sections from Ag to 50 Au, 50 Si, and at 20 at.-% Ag respectively. Figure 46 contains data from [85 Has] for three alloys on this section plus interpolated data points from Figs. 45*a–c*. Figure 47 contains interpolated data points from Figs. 45*a–c*.

ISOTHERMAL SECTIONS

Figure 48 gives the location of the L + (Ag/Au) + (Si) tie triangles at 400, 500, 600, 700, and 800°C, constructed from the vertical sections in Figs. 44–47 and the liquidus isotherms in Fig. 43.

MISCELLANEOUS

[83 Has] and [85 Has] measured the enthalpy of formation of ternary alloys along the sections Ag–80Au20Si, 60Au40Si, 50Au50Si; Au–90Ag10Si; Si–70Au30Ag, 50Au50Ag. A total of 200 alloy compositions were studied at 1 150°C; this permitted the plotting of isoenthalpy contours for the ternary alloys.
[88 Cas] has presented similar data at 1 075°C. More negative enthalpies of formation were found compared with the data of [85 Has].

REFERENCES

62 Kup: V. V. Kuprina: *Zhur. Neorg. Khim.*, 1962, **7**, 1611–1614; *Russian J. Inorg. Chem.*, 1962, **7**, 833–834 (English translation)

83 Cas: R. Castanet, private communication, June 1983; subsequently published as [88 Cas]

83 Has: S. Hassam, M. Gaune-Escard and J. P. Bros: *J. Calorim. d'Analyse Thermique*, 1983, **14**, 166–175

85 Has: S. Hassam: Thesis, 1985, Univ. Provence, France

88 Cas: R. Castanet, *J. Less-Common Metals*, 1988, **136**, 287–296

Ag–Au–Sn

INTRODUCTION

Massalski and Pops [70 Mas] examined a total of 42 ternary alloys along the 12·2, 13·7 and 15·2 at.-% Sn sections. Alloys were prepared from > 99·99% pure elements by melting under He in quartz tubing, homogenized for 14 days under He at temperatures ranging from 275 to 550°C, and X-ray powder diffraction analysed at 30°C after quenching from the homogenizing temperature. All the alloys had the hcp ζ-phase indicating the presence of a continuous solid solution between the ζ phases of the Ag–Sn and Au–Sn systems for temperatures above 275°C. [71 Eva] and [74 Eva] used only thermal analysis techniques to outline the ternary equilibria for alloys containing > 20 at.-% Sn. Ag and Au of purity > 99·999% and > 99·99% Sn were used to prepare alloys which were thermally analysed at heating and cooling rates of 2 to 5°C min^{-1} with continuous stirring. A pseudobinary eutectic occurs at 370°C between AuSn and ζ(Ag,Au)Sn containing 30Au, 16Sn. Four invariant ternary reactions were identified: L + ζ ⇌ ϵ + AuSn at 351°C, L + AuSn ⇌ ϵ + $AuSn_2$ at 294°C, L + $AuSn_2$ ⇌ ϵ + $AuSn_4$ at 240°C and L ⇌ ϵ + (Sn) + $AuSn_4$ at 206°C. The uncertainties in temperature are ± 1°C, in the compositions of the liquid phase (Table 13) ± 0.5 at.-%, in the compositions of the ϵ and ζ phases ± 1·5 at.-%. It was assumed that Ag had little solubility in AuSn, $AuSn_2$ or

Table 13. Composition of the phases at the ternary invariant reaction temperatures

Reaction	*Temperature*, °C*	*Phase*	*Composition*, at.-%†		
			Ag	*Au*	*Sn*
e_1: L ⇌ ζ + AuSn	370	L	16·8	43·9	39·3
		ζ	53·8	30·1	16·1
		(AuSn	0	50	50)
U_1: L + ζ ⇌ ϵ + AuSn	351	L	18·5	33·4	48·0
		ζ	53·8	30·1	16·1
		ϵ	62·4	16·1	21·5
		(AuSn	0	50	50)
U_2: L + AuSn ⇌ ϵ + $AuSn_2$	294	L	8·8	28·0	63·2
		ϵ	66·2	9·7	24·1
		(AuSn	0	50	50)
		($AuSn_2$	0	33·3	66·7)
U_3: L + $AuSn_2$ ⇌ ϵ + $AuSn_4$	240	L	5·7	12·3	82·0
		ϵ	68·0	6·5	25·5
		($AuSn_2$	0	33·3	66·7)
		($AuSn_4$	0	20·0	80·0)
E: L ⇌ ϵ + (Sn) + $AuSn_4$	206	L	4·0	2·2	93·8
		ϵ	68·9	4·1	27·0
		($AuSn_4$	0	20·0	80·0)
		(Sn	0	0	100)

* Uncertainty ± 1°C
† Uncertainty ± 0·5 at.-% for phase L, ± 1·5 at.-% for ζ and ϵ phases

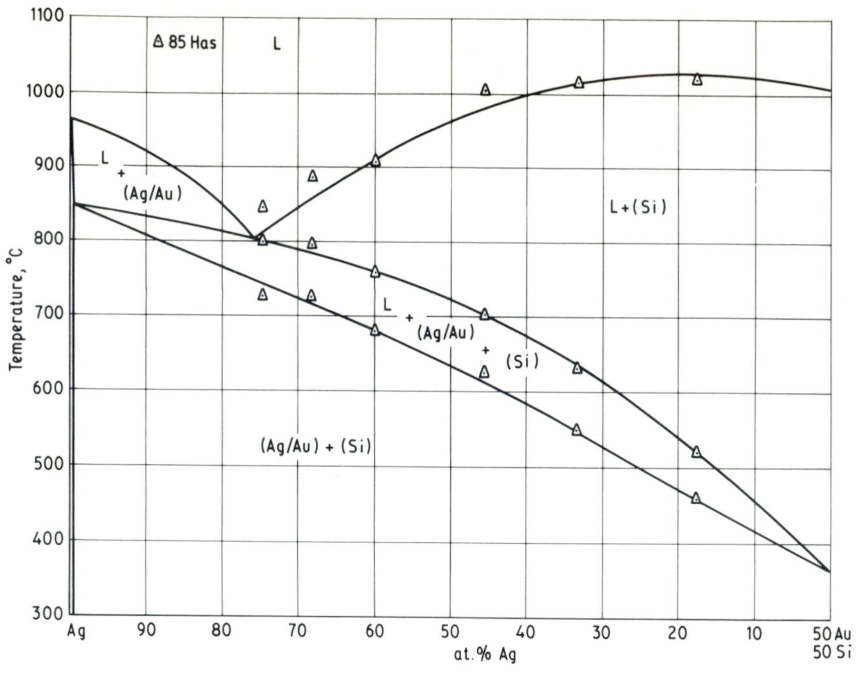

46 Vertical section from Ag to 50 at.-% Au, 50 at.-% Si [85 Has]

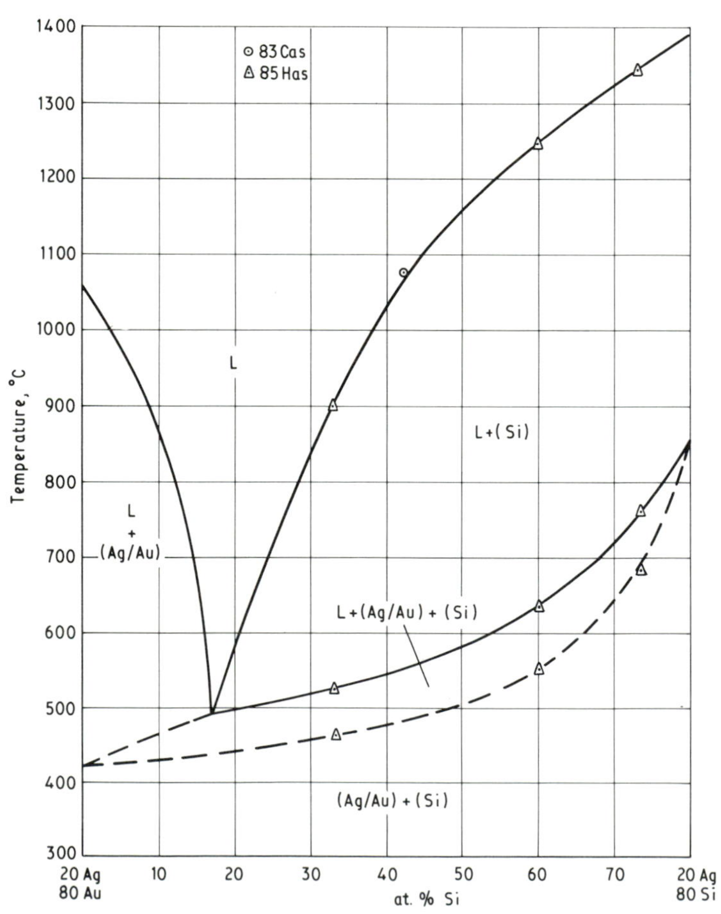

47 Vertical section at 20 at.-% Ag

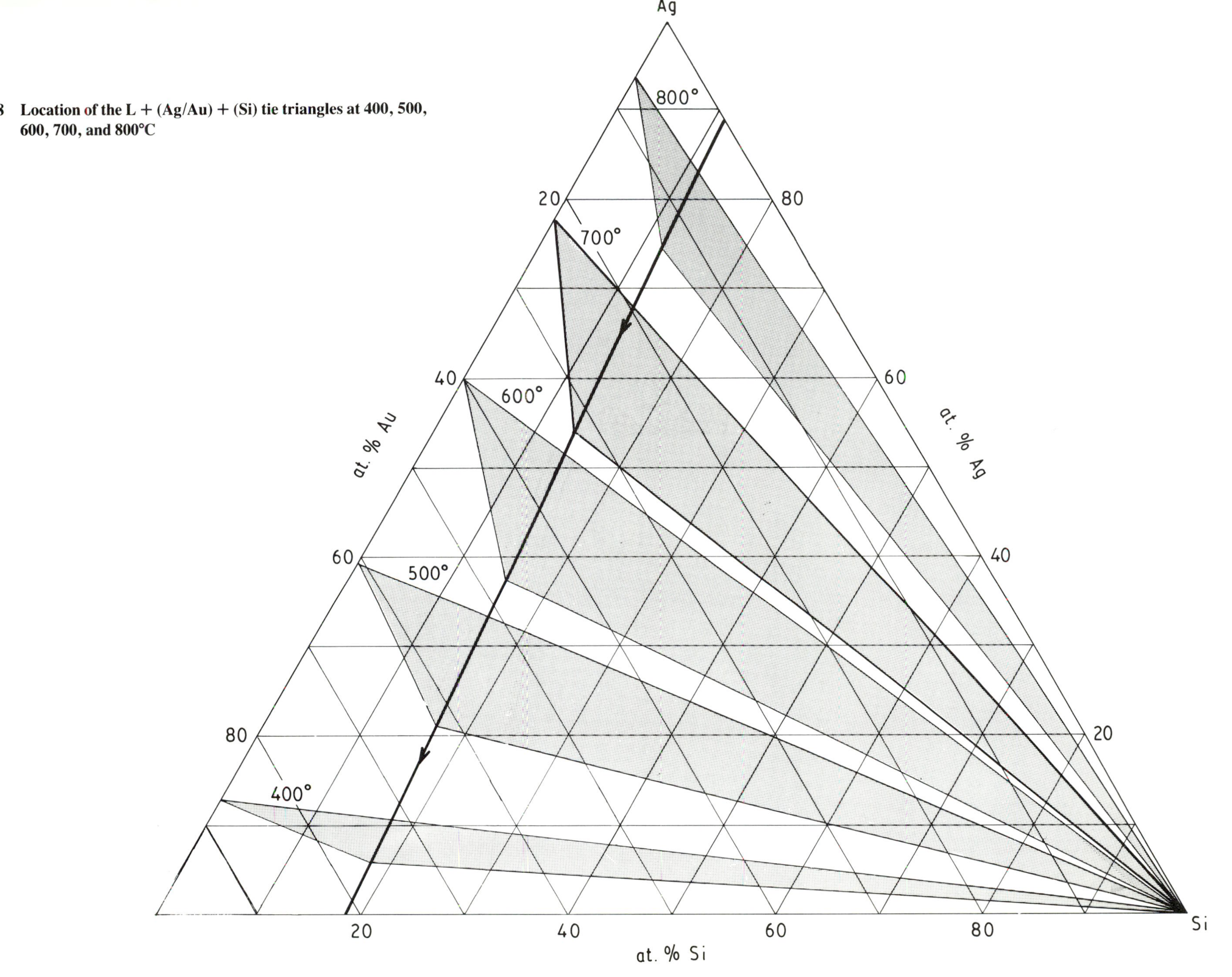

48 Location of the L + (Ag/Au) + (Si) tie triangles at 400, 500, 600, 700, and 800°C

$AuSn_4$. [59 Bur], [75 Mal] and [76 Mah] examined the structure of 3, 1, and 2 ternary alloys respectively; all contained 25 at.-% Sn and up to 10 at.-% Au. The results allow some estimate to be made of the solubility of Au in the binary ϵ phase Ag_3Sn. Thermodynamic data for the ternary system have been reported in [84 Rak] and [85 Has].

BINARY SYSTEMS

The Ag–Au system evaluated by [83 Oka] and the Ag–Sn system evaluated by [87 Kar] are accepted. The Au–Sn binary system of [84 Oka] is not accepted at the Au-rich end. A peritectic formation of β at 532°C, $l + (Au) \rightleftharpoons \beta$, and a peritectic formation of ζ at 519°C, $l + \beta \rightleftharpoons \zeta$, is more probable [87 Leg].

SOLID PHASES

Table 14 summarizes the solid phases that exist. No ternary intermetallic compound has been detected. A continuous series of solid solutions exists between the ζ phases in the Ag–Sn and Au–Sn systems at temperatures above 275°C.

PSEUDOBINARY SYSTEMS

According to [74 Eva) there is a pseudobinary eutectic system between AuSn and the ζ ternary solid solution at 370°C. The compositions of the phases at the eutectic temperature are given in Table 13.

INVARIANT EQUILIBRIA

Only the invariant equilibria beyond 20 at.-% Sn have been established [74 Eva]. On the basis of [70 Mas] it can be concluded that the peritectic reaction forming ζ in the Ag–Sn system runs through the ternary system, substantially parallel to the Ag–Au edge, as a monovariant reaction. The complex nature of the Au-rich end of the Au–Sn binary system will influence the ternary equilibria for alloys rich in Au, but this portion of the ternary system has not been studied. As noted by [87 Leg] the phases β and ζ are formed by peritectic reactions. In the Ag–Au–Sn ternary this would imply the presence of a ternary invariant reaction $L + \beta \rightleftharpoons (Au) + \zeta$, from which the monovariant separation of (Au) + ζ would fall to a minimum before rising to the Ag–Sn peritectic, $l + (Ag) \rightleftharpoons \zeta$.

Four ternary invariant reactions were detected by [74 Eva] and these are summarized in Table 13. All the reactions involve one or more of the phases AuSn, $AuSn_2$, and $AuSn_4$, and it appears that Ag has little solubility in these intermetallic compounds, but this was not verified experimentally. A reaction scheme for the Ag–Au–Sn ternary system is given in Fig. 49.

LIQUIDUS SURFACE

Figure 50 presents the liquid surface according to [74 Eva].

ISOTHERMAL SECTIONS

Isothermal sections at the temperatures of the ternary invariant reactions are given in Figs. 51*a–e* [74 Eva]. It is assumed that the compositions of the solid phases resulting from the reactions e_1, U_1, U_2, and U_3 (Figs. 50*a–d* respectively) remain constant as the temperature falls to room temperature. The constantcy of composition of the ζ phase from e_1, to U_1 requires confirmation.

[50 Bur], [75 Mal] and [76 Mah] studied the phases present in 25 at.-% Sn alloys with 3, 5–6 and 9–10 at.-% Au after heat treatment at temperatures ranging from 470 to 200°C. Table 15 summarizes the results. Referring to the isothermal sections (Figs. 51*a–e*), and recalling the ± 1·5 at.-% uncertainty in the compositions of the ϵ(Ag_3Sn) phase, the data of Table 15 can be reasonably interpreted. The 3Au, 25Sn alloy [50 Bur] would contain some liquid phase at 470°C, but further separation of ϵ would result in single phase ϵ when examined at room temperature. The 5–6Au, 25Sn alloy was partially molten at 450°C [75 Mal] and 350°C [76 Mah] since the microstructures presented indicate the presence of a liquid grain boundary film between ϵ grains, the liquid solidifying on cooling to give $AuSn_2$ and $AuSn_4$ [76 Mah]. After a 200°C anneal [50 Bur] found ϵ only, whereas [76 Mah] detected a second phase which may be $AuSn_2$ (Fig. 51*e*). The 9–10Au, 25Sn alloy was partially molten at 450°C [75 Mal], and separated $AuSn_4$ on cooling. [59 Bur] confirm this alloy to contain additional phases to ϵ after 400 and 200°C anneals.

Table 14. Crystal structures

Solid phase	*Prototype*	*Lattice designation*	*Lattice parameter*, nm			*Comment*
			a	*b*	*c*	
(Ag)	Cu	cF4	0·408 62			
(Au)	Cu	cF4	0·407 84			
(Sn)	(Sn)	tI4	0·583 16			
ζ(Ag–Sn)	Mg	hP2	0·293 86		0·478 60	(V–C) at 15 Sn
ζ(Au–Sn)	Mg	hP2	0·292 28		0·478 23	(V–C) at 14 Sn
ϵ, Ag_3Sn	Cu_3Ti	oP8	0·596 80	0·478 02	0·518 43	(V–C)
AuSn	NiAs	hP4	0·432 18		0·552 30	(V–C)
$AuSn_2$	$AuSn_2$	oP24	0·690 8	0·703 7	1·178 8	(V–C)
$AuSn_4$	$PtSn_4$	oC20	0·650 2	0·654 3	1·170 5	(V–C)
β, $Au_{10}Sn$	Ni_3Ti	hP16	0·290 2		0·951 0	(V–C)
ζ′, Au_5Sn	Au_5Sn	hR6	0·509 2		1·433 3	(V–C)
ζ(Ag,Au)Sn	Mg	hP2	See Fig. 52			[70 Mas]

49 Reaction scheme for the Ag–Au–Sn ternary system

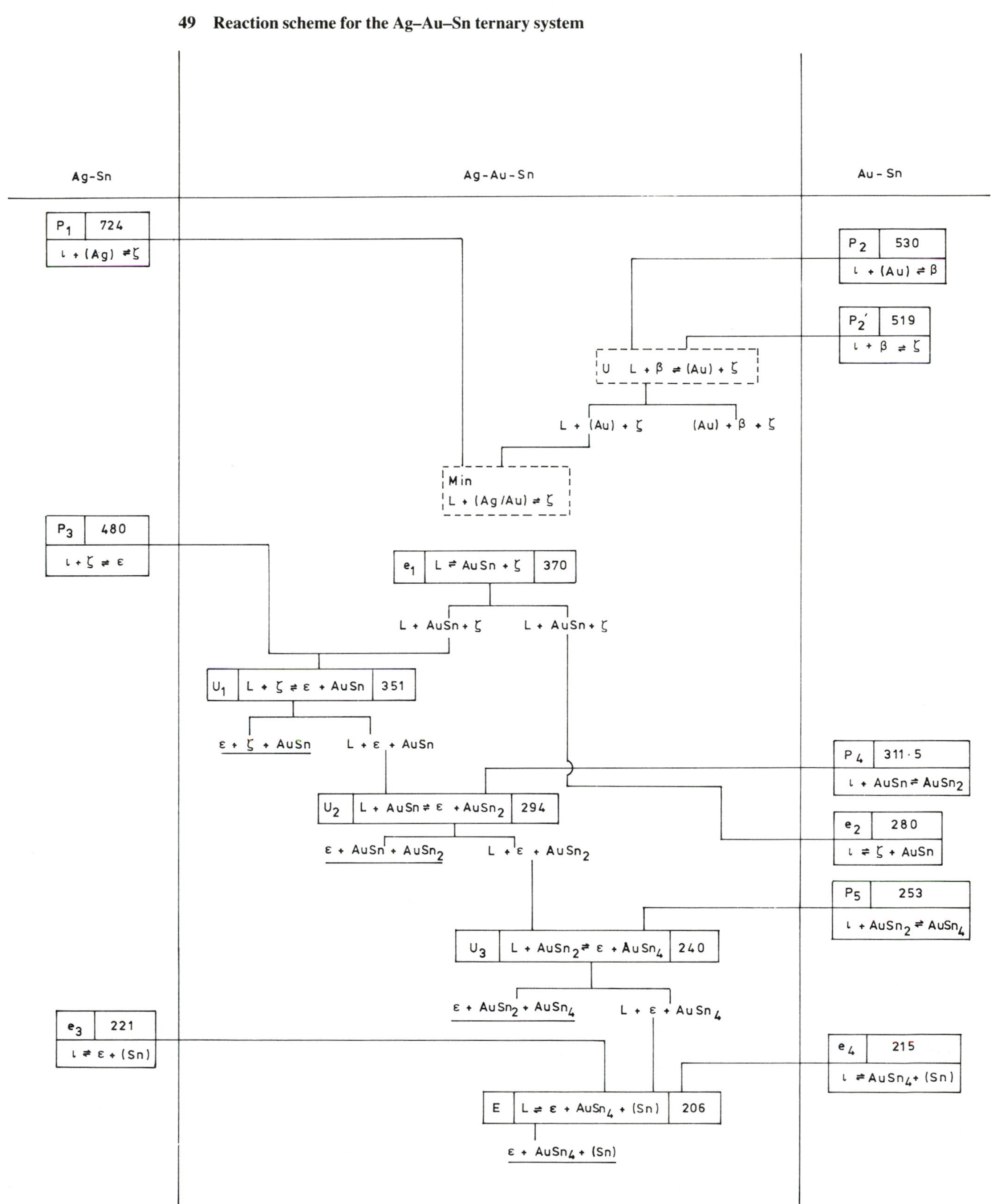

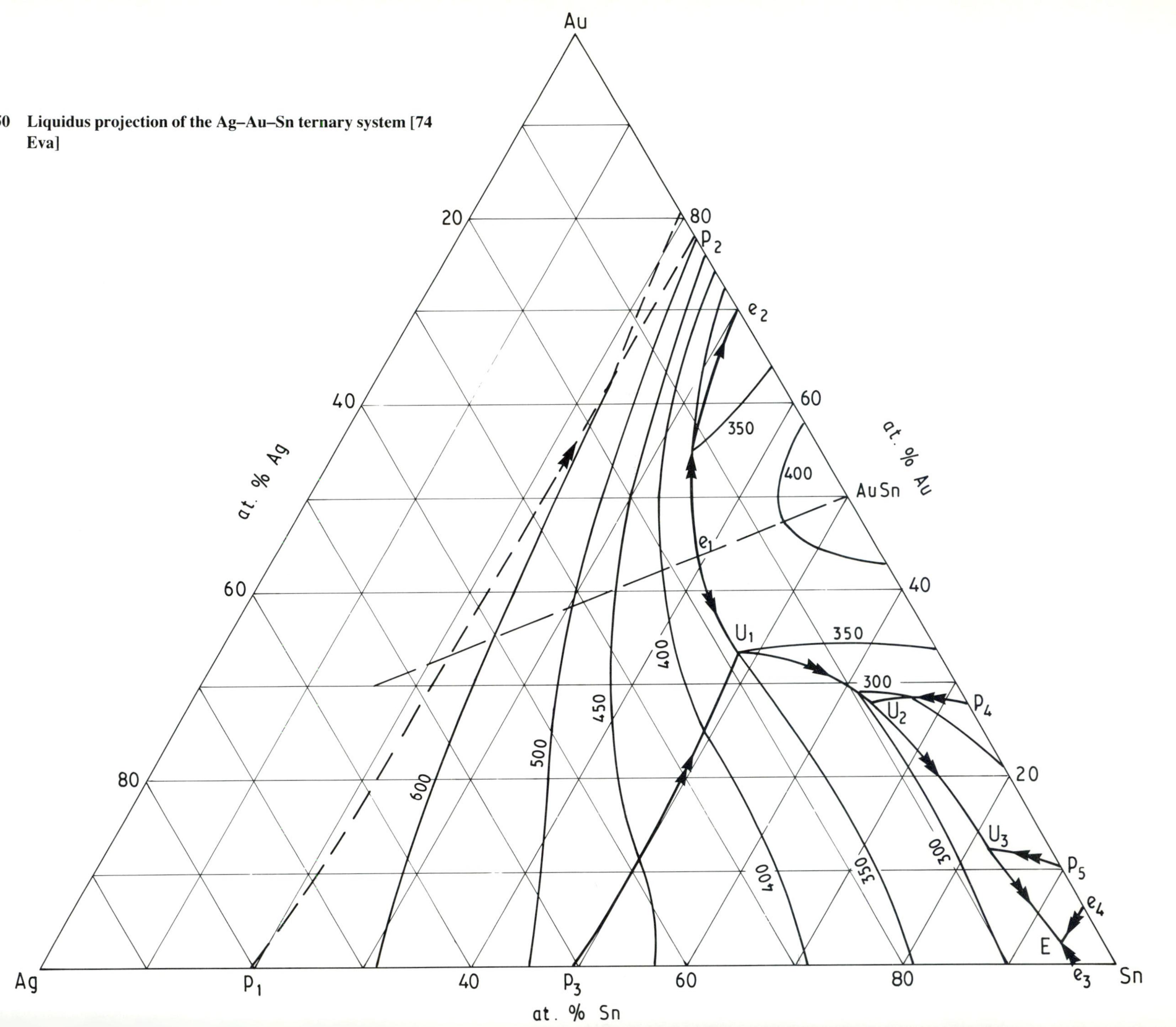

50 Liquidus projection of the Ag–Au–Sn ternary system [74 Eva]

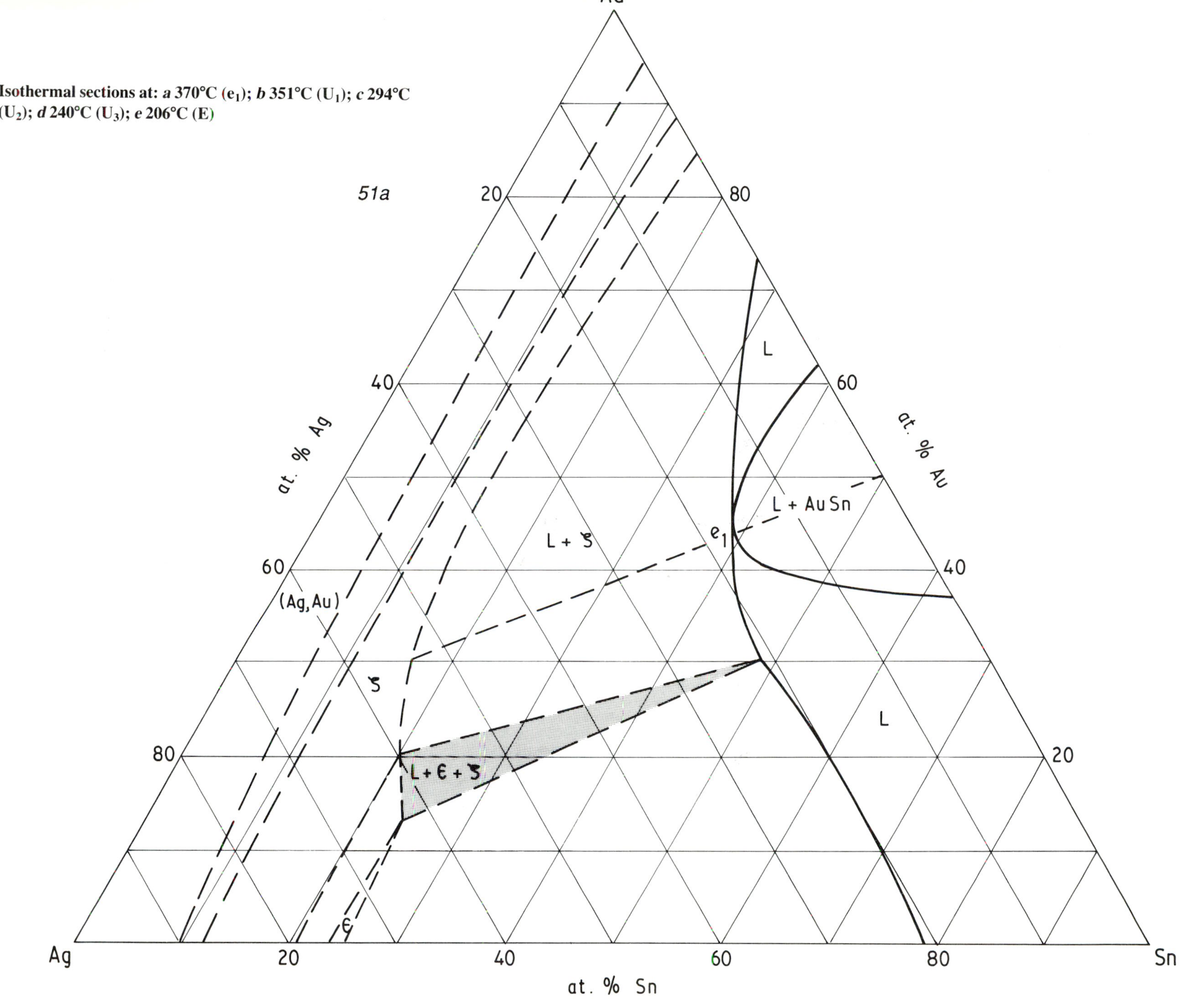

51 Isothermal sections at: *a* 370°C (e_1); *b* 351°C (U_1); *c* 294°C (U_2); *d* 240°C (U_3); *e* 206°C (E)

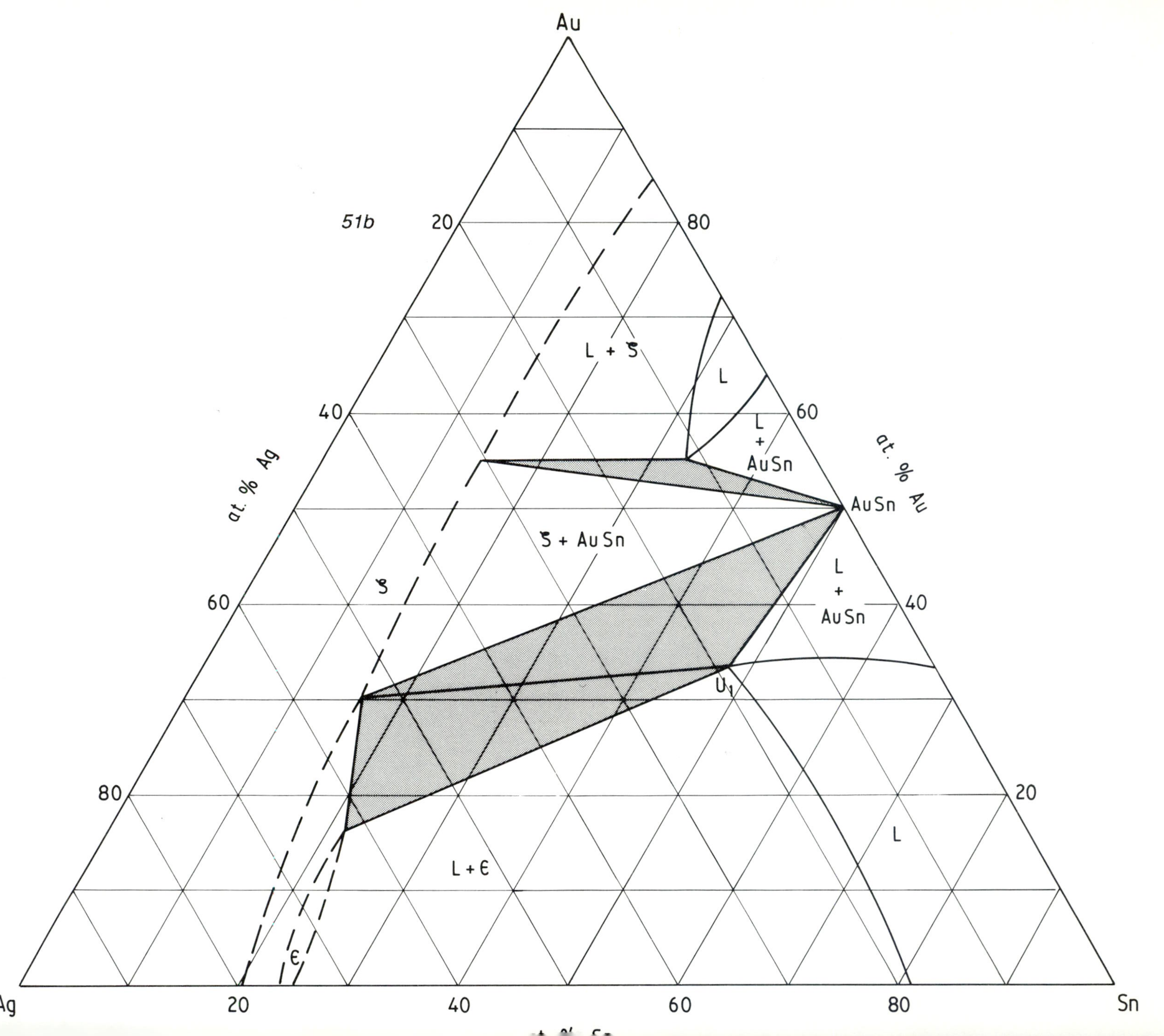

51b
Au
Ag
Sn
at. % Ag
at. % Au
AuSn
L
L + AuSn
U_1
L + ζ
ζ + AuSn
ζ
L + ε
ε
20
40
60
80

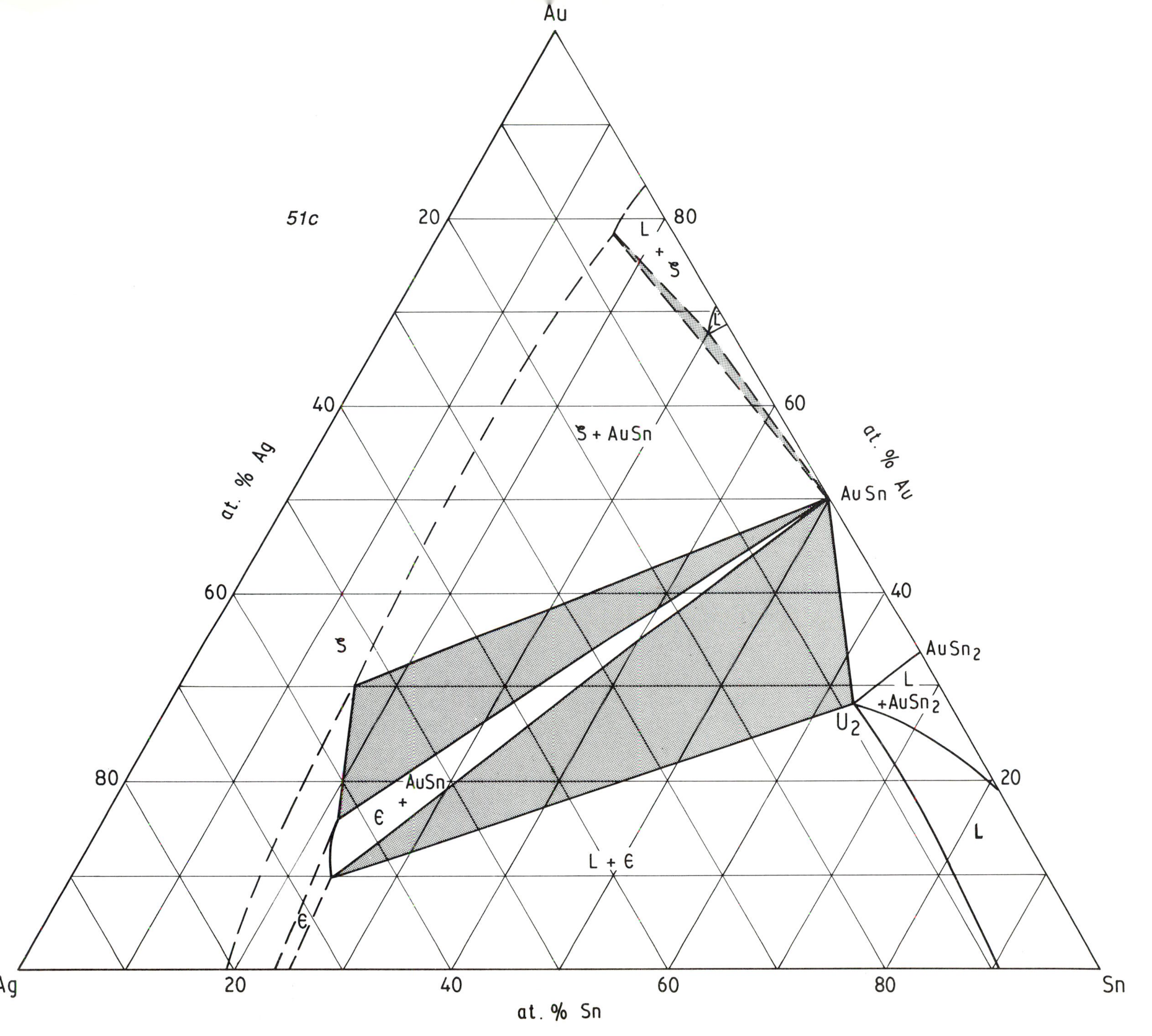
Au
51c
L + ζ
ζ + AuSn
at. % Ag
at. % Au
AuSn
ζ
AuSn$_2$
L + AuSn$_2$
U$_2$
ε + AuSn
L + ε
L
ε
Ag
Sn
at. % Sn
20
40
60
80

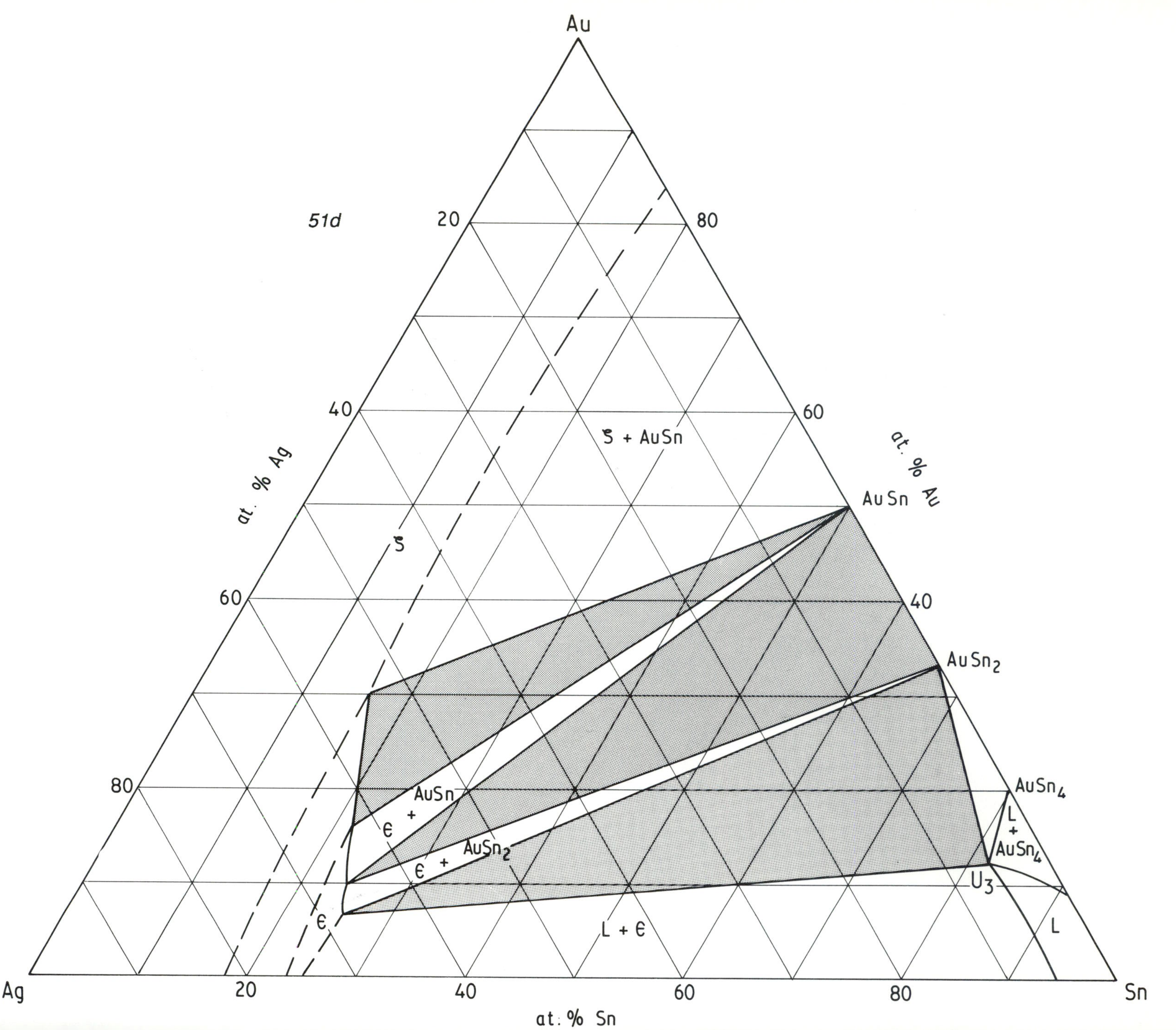
51d
Au
Ag
Sn
20
40
60
80
at. % Ag
at. % Au
at. % Sn
ζ + AuSn
ζ
AuSn
AuSn$_2$
AuSn$_4$
L + AuSn$_4$
U$_3$
L
ε + AuSn
ε + AuSn$_2$
ε
L + ε

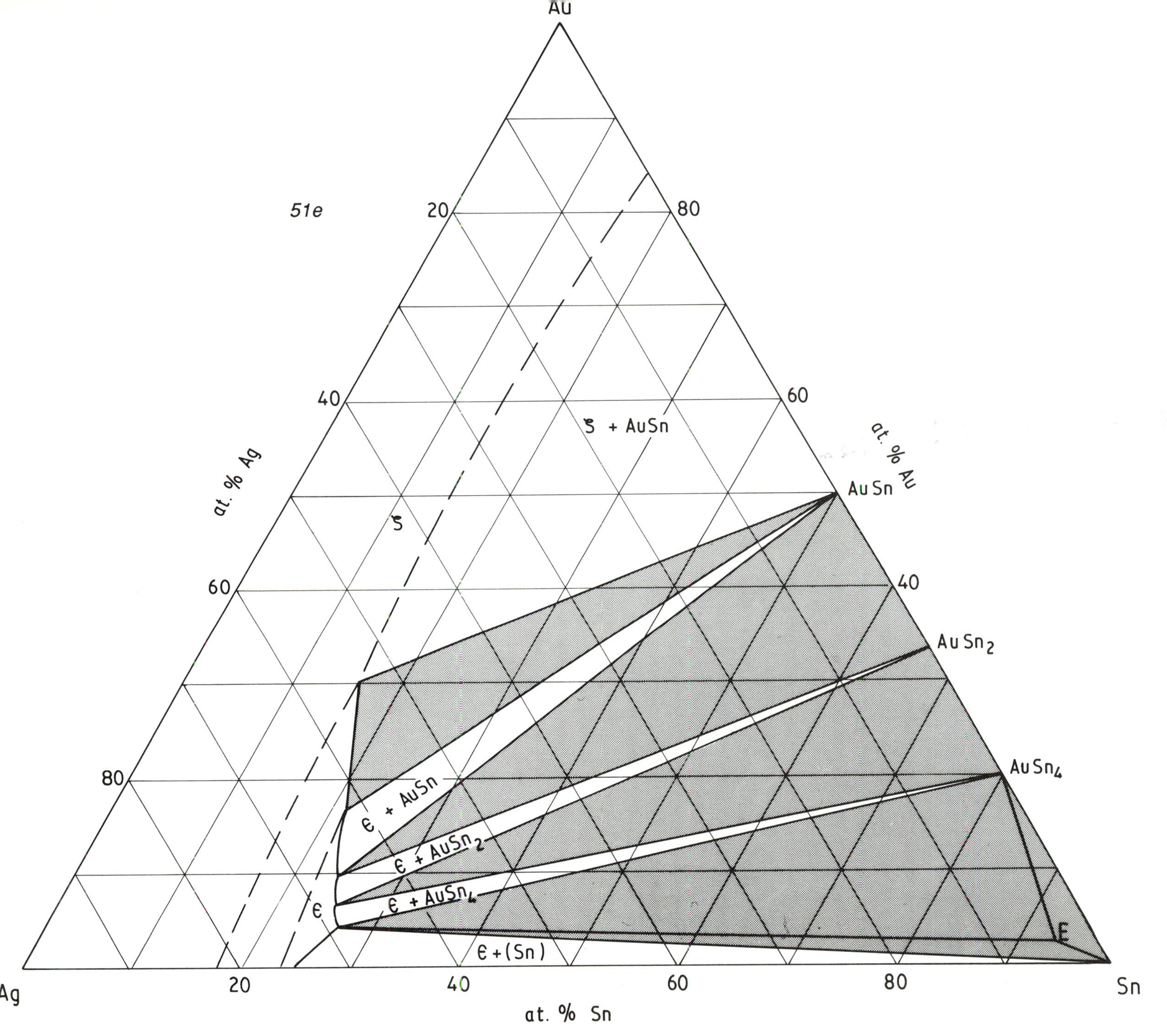
Au
Ag
Sn
51e
at. % Ag
at. % Au
at. % Sn
20
40
60
80
AuSn
AuSn$_2$
AuSn$_4$
ζ + AuSn
ζ
ε + AuSn
ε + AuSn$_2$
ε + AuSn$_4$
ε
ε + (Sn)
E

Table 15. Constitution of 25 at.-% Sn alloys containing up to 10 at.-% Au

Alloy composition, at.-%			*Annealing time and temperature*, °C	*Phases detected*	*Microstructure*	*Reference*
Ag	*Au*	*Sn*				
72	3	25	7 days, 470°C	ϵ	—	[50 Bur]
72	3	25	3 days, 400°C	ϵ	—	[50 Bur]
69·5	5·25	25·25	14 days, 450°C	$\epsilon + AuSn_4$	g.b. film of $AuSn_4$	[75 Mal]
70	5	25	3 days, 400°C	ϵ	—	[50 Bur]
68·7	5·9	25·4	7 days, 350°C	$\epsilon + AuSn_2 + AuSn_4$	Thick g.b. film of $AuSn_2 + AuSn_4$	[76 Mah]
70	5	25	16 days, 200°C	ϵ	—	[50 Bur]
68·7	5·9	25·4	10 days, 200°C	ϵ + ?$AuSn_2$	—	[76 Mah]
64·95	9·05	26	14 days, 450°C	$\epsilon + AuSn_4$	Thick g.b. film of $AuSn_4$	[75 Mal]
65	10	25	3 days, 400°C	ϵ + other phase(s)	—	[50 Bur]
65	10	25	16 days, 200°C	ϵ + other phase(s)	—	[50 Bur]

MISCELLANEOUS

[70 Mas] determined the lattice parameters and axial ratios for hcp ζ-phase alloys on the 12·2, 13·7, and 15·2 at.-% Sn sections. The data is plotted in Fig. 52 with a note of the heat treatment temperature used for each ternary alloy prior to its metallographic examination and X-ray diffraction analysis. All alloys were metallographically single phase. The variation of the axial ratio for small Ag additions to the 15·2 at.-% Sn alloy, with alloys heat treated at 300°C, could be related to transformations of the ζ-phase to ζ′ at lower temperatures.

[84 Rak] measured the enthalpies of formation of liquid Ag–Au–Sn alloys at 1 100°C. [85 Has] extended the measurements and obtained reasonable correspondence between experimental data and calculations based on Kohler's model when modified by a ternary excess term.

REFERENCES

59 Bur: W. Burkhardt and K. Schubert: *Z. Metall.*, 1959, **50**, 442–452

70 Mas: T. B. Massalski and H. Pops: *Acta Metall.*, 1970, **18**, 961–968

71 Eva: D. S. Evans, J. I. McLeod and A. Prince: 'Metallurgical Chemistry', Proc. Symp. Brunel Univ. and NPL, 1971, 459–468; 1972, London, HMSO

74 Eva: D. S. Evans and A. Prince: *Metal Sci.*, 1974, **8**, 286–290

75 Mal: M. L. Malhotra and K. R. Lawless: *J. Biomed. Mater. Res.*, 1975, **9**, 197–205

76 Mah: D. B. Mahler, J. D. Adey and J. Van Eysden: *J. Dent. Res.*, 1976, **55**, 1012–1022

83 Oka: H. Okamoto and T. B. Massalski: *Bull. Alloy Phase Diagrams*, 1983, **4**, 30–38

84 Oka: H. Okamoto and T. B. Massalski, *Bull. Alloy Phase Diagrams*, 1984, **5**, 492–503

84 Rak: J. Rakotomavo, M. Gaune-Escard, J. P. Bros and P. Gaune: *Ber. Bunsenges Phys. Chem.*, 1984, **88**, 663–670

85 Has: S. Hassam: Excess enthalpy and phase diagrams of the ternary alloys Ag–Au–Si, Ag–Au–Ge and Ag–Au–Sn, Thesis, Univ. Provence, 1985

87 Leg: B. Legendre, Chhay Hancheng, F. Hayes, C. A. Maxwell, D. S. Evans and A. Prince: *Mater. Sci. Technol.*, 1987, **3**, 875–876

87 Kar: I. Karakaya and W. T. Thompson, *Bull. Alloy Phase Diagrams*, 1987, **8**, 340–347

Ag–Au–Te

INTRODUCTION

The published data on this system are so conflicting that it cannot be said that the equilibria are known with any certainty. Greatest reliance is accorded to the study of Legendre *et al.* [80 Leg] on the basis that their alloy preparative techniques were soundly based and they used a variety of experimental methods in elucidating the proposed equilibria. A number of ternary compounds have been reported, three of which occur as natural minerals. There is no doubt that the mineral petzite, Ag_3AuTe_2, can be synthesized and is a stable phase in the ternary system. [40 Kra] were the first workers to prepare Ag_3AuTe_2. They claimed that it melts incongruently, $Ag_3AuTe_2 \rightleftharpoons (Ag/Au) + l$. [48 Tho] confirmed Ag_3AuTe_2 as a stable synthetic phase by examination of mixed 'powders' that were melted in evacuated silica capsules. [65 Cab 2] prepared an alloy of the composition Ag_3AuTe_2 by melting the mixed metals above 950°C. The product was ground and pelletized prior to a 440°C heat treatment for 10 days. [66 Tav] and [66 Mes] prepared Ag_3AuTe_2 chemically by the reaction of Ag_2Te with dithiosulphatoaurate (I) ions. [70 Smi] prepared this compound by heating the mixed metals in an evacuated quartz capsule at 500°C for 4 days followed by slow cooling to room temperature. [80 Leg] studied the ternary system in detail using alloys prepared by melting followed by slow cooling to room temperature. Again the compound Ag_3AuTe_2 was confirmed as a stable phase. There have been many studies of mineralogical samples of petzite [59 Fru, 66 Mes, 78 Cha]. Whilst there is no doubt of the existence of Ag_3AuTe_2, there is considerable dispute as to its allotropic forms and the equilibria associated with its formation. Three groups [65 Cab 2, 67 Tav, 70 Smi] claim three allotropic forms of Ag_3AuTe_2, denoted τ_α, τ_β, and τ_γ in this assessment. The transition temperature for $\tau_\alpha \rightarrow \tau_\beta$ is about 200°C [70 Smi], 210°C [59 Kra, 65 Cab 2, 80

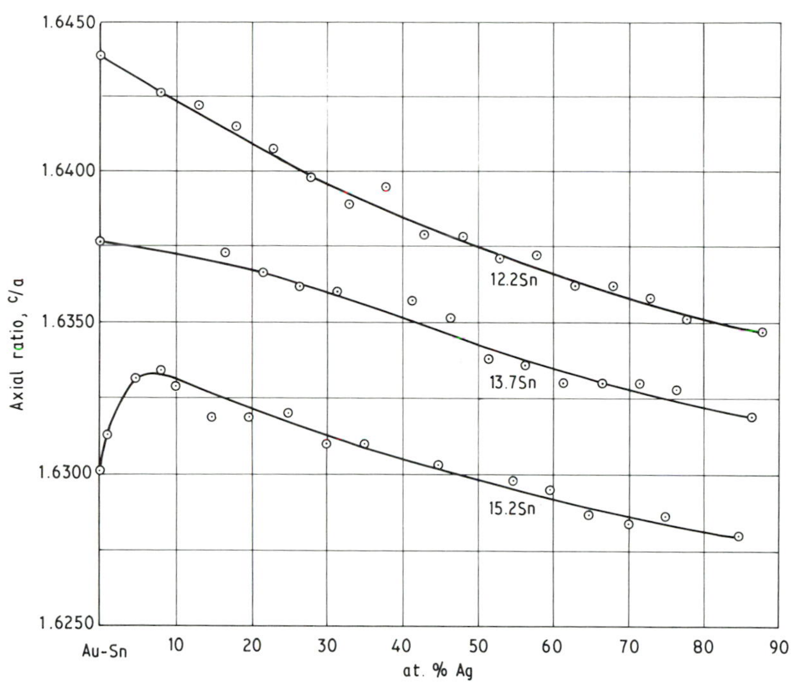

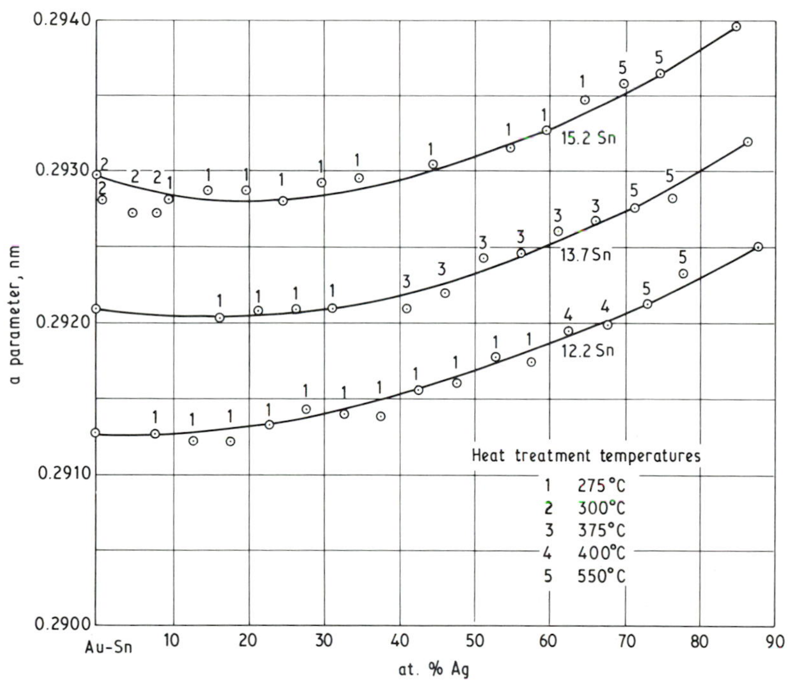

52 Lattice parameter *a* and axial ratio for ζ-phase alloys on the 12·2, 13·7, and 15·2 at.-% Sn sections

Leg] and 220°C on heating, 180°C on cooling [67 Tav]. The transition temperature for $\tau_\beta \rightarrow \tau_\gamma$ is about 320°C [70 Smi], 319°C [65 Cab 2], and 320°C on heating, 290°C on cooling [67 Tav]. In contrast to [65 Cab 2, 67 Tav, 70 Smi], who identify a high-temperature τ_γ phase, [80 Leg] finds an invariant reaction at 313°C associated with the peritectoidal breakdown of τ_β on heating ($\tau_\beta \rightleftharpoons \beta Ag_2Te + AuTe_2$ + (Ag/Au)). [40 Kra] reports incongruent melting of τ (i.e. τ_γ) to liquid + (Ag/Au); [65 Cab 2] also indicates incongruent melting of τ_γ at about 735 ± 10°C. [65 Cab 2] claims that there is a complete solid solution series between γAg_2Te and τ_γ and, if this were accepted, the alloy composition corresponding to Ag_3AuTe_2 would produce l + τ_γ/γ Ag_2Te solid solution on melting. [80 Leg] also proposes an extensive solution of Au in γAg_2Te but find this Au-containing γAg_2Te transforming by eutectoid reaction at 340°C to $\beta Ag_2Te + AuTe_2$ + (Ag/Au), which then transforms at 313°C to τ_β. It should be noted that the heat effect associated with the $\tau_\alpha \rightarrow \tau_\beta$ transition was quoted as 2·9 kcal mol^{-1} to an accuracy of ± 20% [70 Smi], but the $\tau_\beta \rightarrow \tau_\gamma$ heat effect was quoted by [70 Smi] as small. This may be anticipated if the reaction concerned is the solid state breakdown of τ_β and not its transition to τ_γ. In this assessment the data of [80 Leg] are preferred for the reasons stated in the section dealing with invariant equilibria. In summary Ag_3AuTe_2 is a stable ternary compound that undergoes transformation from τ_α to τ_β at 210°C, the τ_α being bcc and the τ_β of unknown crystal structure; τ_β decomposes at 313°C by peritectoidal reaction. There is therefore no primary Ag_3AuTe_2 phase field associated with the liquidus in the Ag–Au–Te system.

Two compounds, based on the compositions $AgAu_4Te_{10}$ (the mineral krennerite) and $AgAuTe_4$ (the mineral sylvanite), have been reported in the literature. [65 Cab 2] presents the phase relationships at 66.67 at.-% Te from $AuTe_2$ to $AgAuTe_4$. Both $AgAu_4Te_{10}$ (krennerite) and $AgAuTe_4$ (sylvanite) were synthesized. $AgAu_4Te_{10}$ melts incongruently at 382 ± 5°C to liquid + $AuTe_2$, the $AuTe_2$ containing Ag randomly substituted for Au in the $AuTe_2$ lattice; $AgAuTe_4$ melts incongruently at 354 ± 5°C to liquid + $AgAu_4Te_{10}$. [40 Kra] were not able to synthesize $AgAu_4Te_{10}$; nor could [60 Mar] in a detailed study of the 300°C isothermal section of the ternary system. [80 Leg] included the section containing 66·67 at.-% Te in their study of the ternary equilibra. They did not find the compound $AgAu_4Te_{10}$. Crystallographic data for mineralogical samples of krennerite are given in Table 16 [36 Tun, 50 Tun]. Most of the data for the section from $AuTe_2$ to $AgAuTe_4$ produced by [65 Cab 2] was obtained on alloy samples synthesized from $AuTe_2$, Ag, and Te at 320°C (i.e. in the solid state), and doubt must be thrown on the conclusions drawn from the work since the preparative conditions could lead to the formation of metastable phases. The early work of Pellini, using thermal analysis of melted alloys, shows no sign of the formation of $AgAu_4Te_{10}$ [15 Pel]. The mineral sylvanite, based on the idealized composition $AgAuTe_4$, was synthesized by [40 Kra] who stated that it melts incongruently to liquid + $AuTe_2$, by [65 Cab 2] who confirms incongruent melting but to liquid + $AgAu_4Te_{10}$ (krennerite) at 354 ± 5°C, and by [60 Mar] who found it a stable phase in the 300°C isothermal section. However [80 Leg] did not confirm these results, nor did [15 Pel]. More reliance is placed on the work of [80 Leg] for the reasons stated in the section on invariant equilibria. Mineralogical samples of sylvanite have been frequently studied (see, for example, [37 Tun, 41 Tun, 52 Tun]). A further compound, 'x', was shown in the section at 33·33 at.-% Te between Ag_2Te and Ag_3AuTe_2 [65 Cab 2]. This phase is said to be stable from 1.5 to 9.0 at.-% Au from about 50 to 415°C. The crystal structure was tentatively indexed as orthorhombic with a = 0·75, b = 0·68, c = 0·6 nm or fcc with a = 1.497 nm. The compound begins to transform to αAg_2Te after 13 h when quenched to room temperature. Slight pressure on grinding at room temperature accelerates the transformation to αAg_3AuTe_2 (petzite) and αAg_2Te. The compound 'x' has a diffraction pattern similar to petzite. It is unlikely to be a stable phase and its occurrence is regarded as requiring confirmation.

As would be anticipated from the discussion above, the ternary equilibria presented by [60 Mar], [65 Cab 2], and [80 Leg] vary considerably. The first study of the equilibria [15 Pel] concentrated on the partial ternary system $Te–AuTe_2–Ag_2Te$. Four invariant reactions were identified and, although the detailed transformations associated with these invariant reactions can be criticized, more modern work agrees in identifying four invariant reactions in this partial ternary system [60 Mar, 80 Leg]. It is not possible to reconcile the interpretations of [60 Mar] and [80 Leg] on the ternary equilibria, since [80 Leg] does not accept the formation of $AgAuTe_4$ (sylvanite) and regards Ag_3AuTe_2 (petzite) as the product of a solid state reaction. On the other hand [60 Mar] postulates $AgAuTe_4$ as a stable compound and considers Ag_3AuTe_2 as a compound which separates from the melt. [65 Cab 2] adds to the confusion by assuming stability for $AgAu_4Te_{10}$ (krennerite) and $AgAuTe_4$ (sylvanite) but does not present a view on the overall equilibria in the system. In this assessment the data presented by [80 Leg] are preferred for the reasons outlined in the section dealing with invariant equilibria. [80 Leg] propose a reaction scheme involving five invariant reactions:

1	L_{U_1} + (Ag/Au) $\rightleftharpoons AuTe_2 + \gamma Ag_2Te$	at 380°C
2	$L_{U_2} + \gamma Ag_2Te \rightleftharpoons AuTe_2 + \beta Ag_2Te$	at 360°C
3	$L_{U_3} + \beta Ag_2Te \rightleftharpoons AuTe_2 + \gamma_h$	at 358°C
4	$L_{U_4} + \gamma_h \rightleftharpoons AuTe_2 + \beta Ag_{5-x}Te_3$	at 350°C
5	$L_E \rightleftharpoons$ (Te) + $AuTe_2 + \beta Ag_{5-x}Te_3$	at 332°C

Reactions 2–5 take place within the partial ternary system $Te–AuTe_2–Ag_2Te$. It is worthy of note that all the invariant reactions associated with a liquid phase occur at very low temperatures compared with the melting points of Au and Ag and the temperature of the binary Ag–Te eutectic reaction, l $\rightleftharpoons$ (Ag) + γAg_2Te, at 869°C. It is not surprising that data produced by examination of alloys produced by the heat treatment of mixed powders at temperatures of the order of 300–400°C is misleading, since the time to produce equilibrium structures at such low temperatures will be far in excess of those used. Under the circumstances non-equilibrium effects will in-

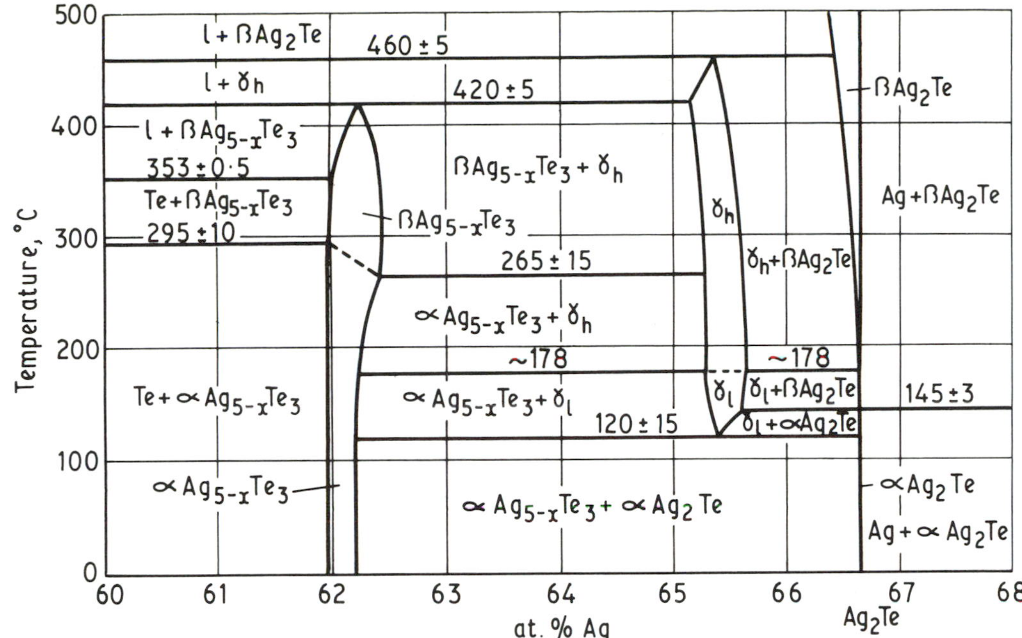

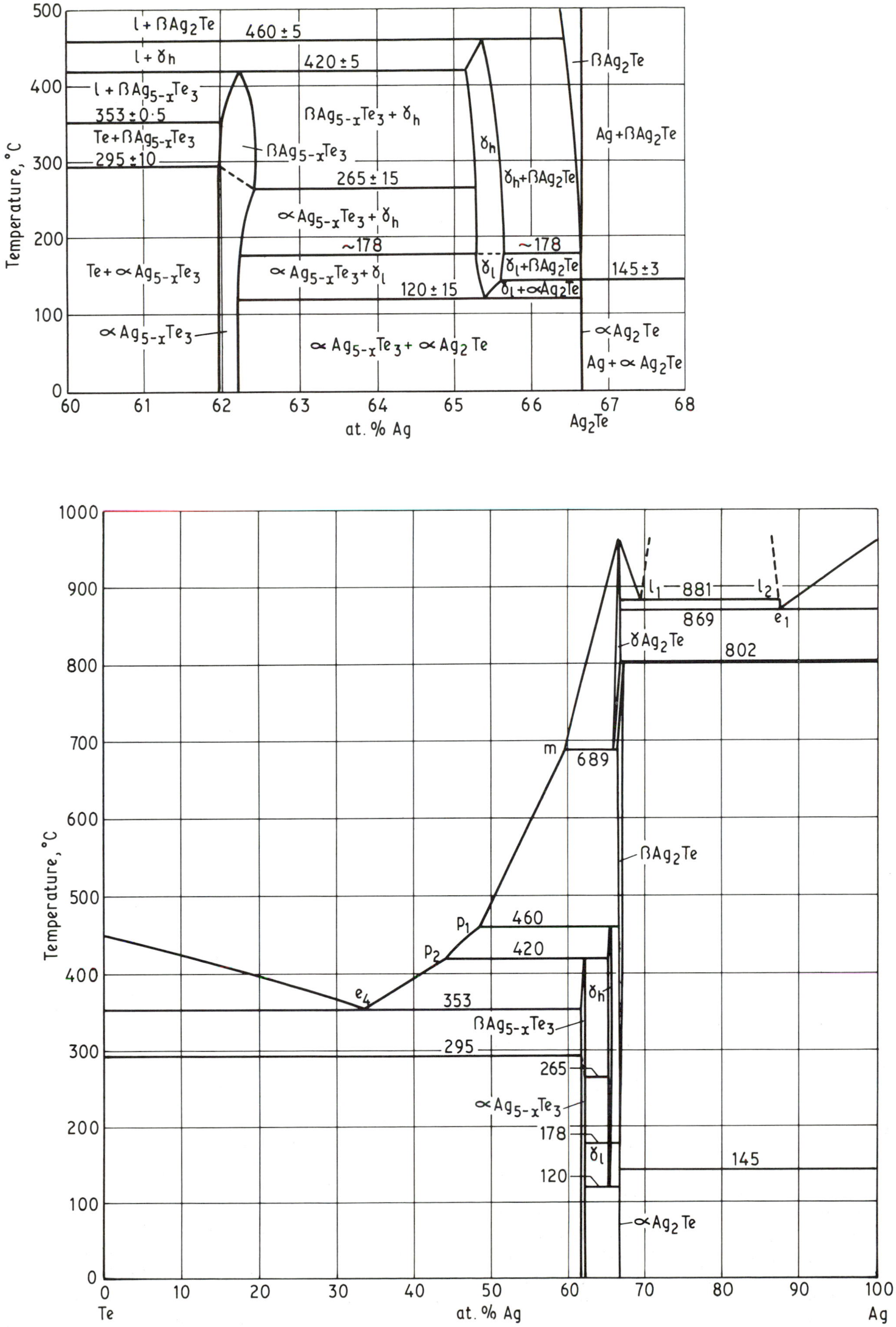

53 The Ag–Te binary system [66 Kra]

trude and metastable phases may be accepted as equilibrium phases.

Over half the ternary liquidus is associated with the primary separation of the Ag–Au solid solution, denoted (Ag / Au) (see Figure 55). There is a small region of liquid immiscibility, originating with the binary reaction $l_1 \rightleftharpoons l_2 + \gamma Ag_2Te$ at 881°C in the Ag–Te system and ending at a critical tie line $l \rightleftharpoons \gamma Ag_2Te$ in the ternary system.

BINARY SYSTEMS

The Au–Te system evaluated by [84 Oka], the Ag–Te system (Fig. 53) published by [66 Kra], and the Ag–Au system evaluated by [80 Ell] are accepted.

SOLID PHASES

One ternary compound is definitely authenticated: Ag_3AuTe_2. The ternary compounds $AgAu_4Te_{10}$ and $AuAuTe_4$ occur as the minerals krennerite and sylvanite respectively. A compound 'x' occurring between 1·5 and 9·0 at.-% Au on the 33·33 at.-% Te section [65 Cab 2] is unlikely to be a stable phase.

Table 16 summarizes the crystal structure data for the solid phases in the Ag–Au–Te system.

Table 16. Crystal structures

Solid phase	*Prototype*	*Lattice designation*
(Au)	Cu	cF4
(Ag)	Cu	cF4
(Te)	γSe	hP3
$AuTe_2$	$AuTe_2$	mC6
γAg_2Te	—	cI6
βAg_2Te	—	cF12
αAg_2Te	—	mP12
γ_h	—	—
γ_l	—	—
$\beta Ag_{5-x}Te_3$	—	—
$\alpha Ag_{5-x}Te_3$	—	hP56
γAg_3AuTe_2 (?)	—	—
βAg_3AuTe_2	—	—
αAg_3AuTe_2 (petzite)	—	cI48
$AgAu_4Te_{10}$ (krennerite)	—	oP24
$AgAuTe_4$ (sylvanite)	—	mP12

PSEUDOBINARY SYSTEMS

According to [80 Leg], the section $AuTe_2$–Ag_2Te is not a pseudobinary system. Figure 54 is a slightly amended form of the published section. The only other data is the early work of [15 Pel]. The section was regarded as pseudobinary with a eutectic at 373°C and at 35 at.-% Ag (compared with 367°C and 26·5 at.-% Ag [80 Leg]. [15 Pel] postulated the presence of a ternary compound formed by peritectic reaction at 495°C, $l + \beta Ag_2Te = Ag_{10}AuTe_7$, leading to the 373°C eutectic reaction, $l = AuTe_2 + Ag_{10}AuTe_7$. The liquidus curve [15 Pel] for alloys on the Ag_2Te side of the eutectic composition falls reasonably close to the γAg_2Te solidus curve [80 Leg]. The compound $Ag_{10}AuTe_7$ was tentatively correlated with petzite, Ag_3AuTe_2, by [15 Pel]. No trace of a 495°C reaction was found by [80 Leg] whose section is accepted.

The 66·67 at.-% Te section from $AuTe_2$ to 11 at.-% Ag and the 33·33 at.-% Te section from Ag_2Te to Ag_3AuTe_2 were regarded as pseudobinary [65 Cab 2]. This data is not accepted (see section on invariant equilibria).

INVARIANT EQUILIBRIA

There is considerable conflict between the data of [80 Leg], [40 Kra], [65 Cab 2], [60 Mar], and [15 Pel]. [15 Pel] studied the partial ternary system Te–$AuTe_2$–Ag_2Te by thermal analysis of 113 alloy compositions, 22 of which were along the $AuTe_2$–Ag_2Te section. Four invariant reactions associated with a liquid phase were identified:

$L + \gamma_h = \beta Ag_{5-x}Te_3 + \beta Ag_2Te$ (412°C)
$L + \beta Ag_2Te = \beta Ag_{5-x}Te_3 + Ag_{10}AuTe_7$ (385°C)
$L + Ag_{10}AuTe_7 = \beta Ag_{5-x}Te_3 + AuTe_2$ (340°C)
$L = \beta Ag_{5-x}Te_3 + Te + AuTe_2$ (335°C)

[15 Pel] quotes the 412°C reaction as $L + \beta Ag_2Te = \gamma_h + \beta Ag_{5-x}Te_3$; this is an error. [80 Leg] also show four invariant reactions associated with this partial ternary system, but the only agreement between [80 Leg] and [15 Pel] concerns the presence of a ternary eutectic reaction at 332°C [80 Leg] or 335°C [15 Pel], at which Te + $AuTe_2$ + $\beta Ag_{5-x}Te_3$ separate from a liquid of composition 34·2 at.-% Ag, 4·5 at.-% Au [80 Leg] or 35 at.-% Ag, 5 at.-% Au [15 Pel]. This is a remarkably good agreement. The invariant reactions preceding the ternary eutectic are necessarily in conflict since [15 Pel] found a peritectically-formed ternary compound $Ag_{10}AuTe_7$ on the $AuTe_2$–Ag_2Te section, and this enters into his proposed reaction scheme. [80 Leg] found Ag_3AuTe_2 as a ternary compound and, since this does not enter into equilibrium with the liquid phase, the reaction equilibria [80 Leg] include only $AuTe_2$, Te and the Ag–Te phases γ_h, γAg_2Te, βAg_2Te, and $\beta Ag_{5-x}Te_3$ associated with the liquid in this partial ternary system.

[40 Kra] reported a ternary eutectic reaction at 330°C in the Te–$AuTe_2$–Ag_2Te partial tenary system at a composition 35 at.-% Ag, 4 at.-% Au, in agreement with [80 Leg] and [15 Pel], but stated that the reaction was

$l \rightleftharpoons Te + Ag_{5-x}Te_3 + AgAuTe_4$ (sylvanite)

[40 Kra] published no phase diagram. [60 Mar] published the liquidus surface (Fig. 55) based on the work of [40 Kra] and [15 Pel]. [60 Mar] proposes nine invariant reactions:

1 $L_{U_1} + (Ag/Au) \rightleftharpoons \gamma Ag_2Te + Ag_3AuTe_2$
if there is a maximum on curve U_1U_5. Otherwise the reaction is
$L_{U_1} + \gamma Ag_2Te + (Ag/Au) \rightleftharpoons Ag_3AuTe_2$
2 $L_{U_2} + \gamma Ag_2Te \rightleftharpoons \beta Ag_2Te + Ag_3AuTe_2$
(additional to reaction scheme proposed by [60 Mar]
3 $L_{U_3} + \gamma_h \rightleftharpoons \beta Ag_2Te + \beta Ag_{5-x}Te_3$
4 $L_{U_4} + \beta Ag_2Te \rightleftharpoons \beta Ag_{5-x}Te_3 + Ag_3AuTe_2$
5 $L_{U_5} + (Ag/Au) \rightleftharpoons AuTe_2 + Ag_3AuTe_2$
6 $L_{U_6} + AuTe_2 \rightleftharpoons (Te) + AgAuTe_4$
7 $L_{U_7} + AuTe_2 \rightleftharpoons AgAuTe_4 + Ag_3AuTe_2$
8 $L_{U_8} + Ag_3AuTe_2 \rightleftharpoons AgAuTe_4 + \beta Ag_{5-x}Te_3$
9 $L_E \rightleftharpoons (Te) + \beta Ag_{5-x}Te_3 + AgAuTe_4$

These reactions are proposed by Markham [60 Mar] based on data of [40 Kra] and [15 Pel]. Markham studied

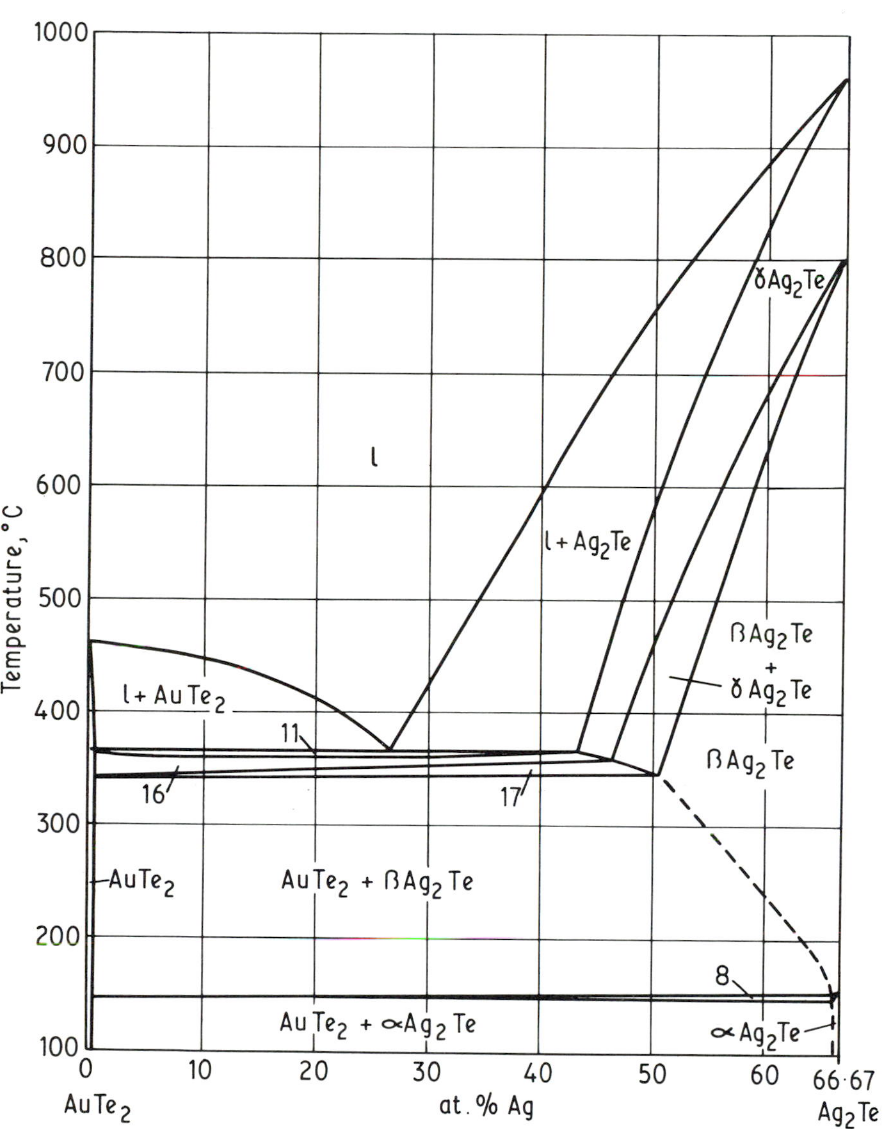

54 The section $AuTe_2$–Ag_2Te [80 Leg]

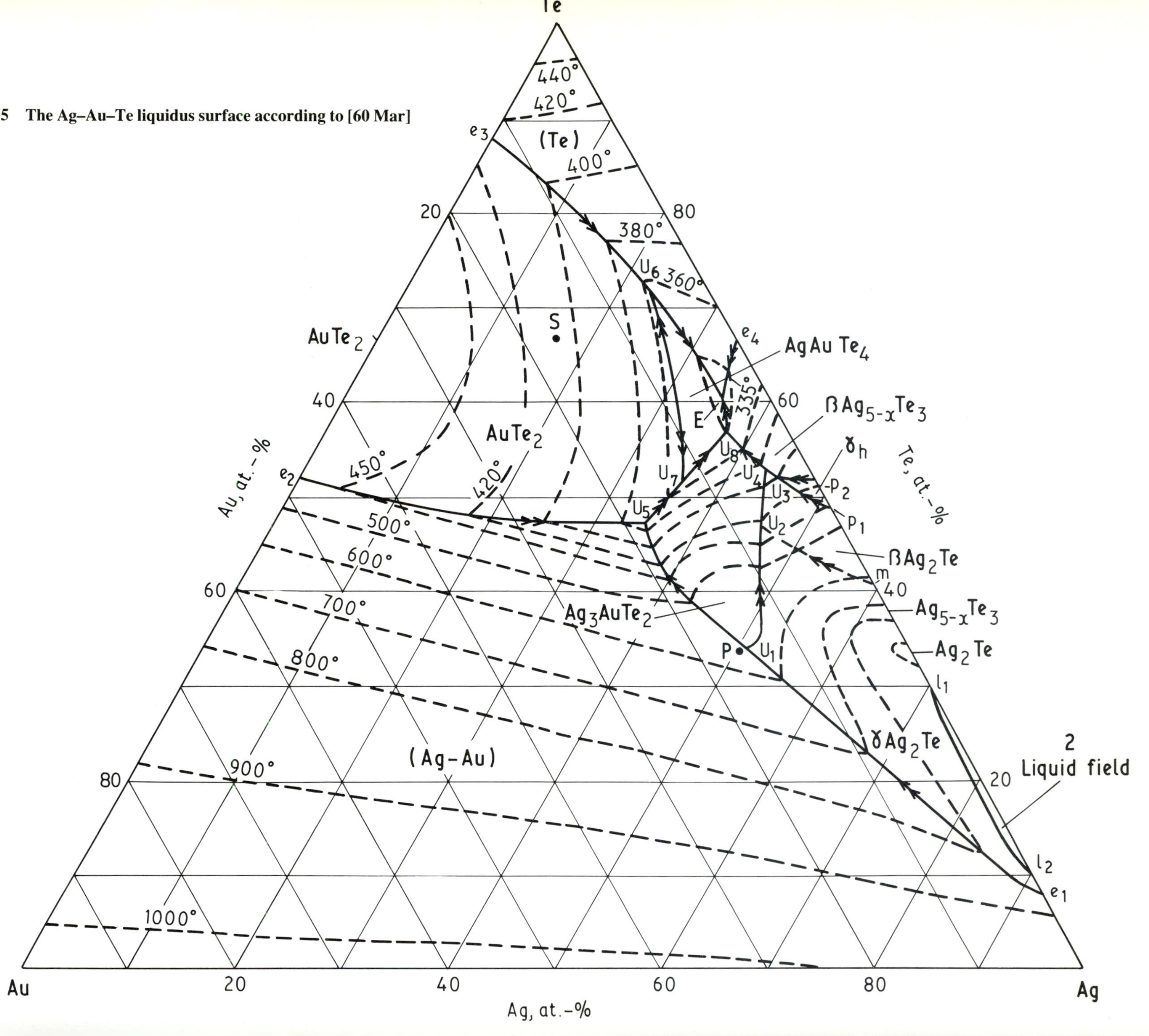

55 The Ag–Au–Te liquidus surface according to [60 Mar]

the sub-solidus phase relations as represented by the 300°C isothermal section. A total of 165 alloys prepared from 99·9% purity elemental powders were studied. Mixed powders weighing only 0·1 g were sealed in evacuated Pyrex tubes and heat treated at 300°C for times varying from 4 to 21 days. The only phase identification technique used was X-ray powder diffraction analysis; phase boundaries were deduced by the appearance or disappearance of phases between phase regions. The 300°C isothermal section [60 Mar] (Fig. 56) shows $AgAuTe_4$ (sylvanite) as a stable ternary compound in addition to Ag_3AuTe_2 (petzite). [60 Mar] did not find $AgAu_4Te_{10}$ (krennerite). It is possible to construct a reaction scheme for the ternary system based on the nine invariant reactions involving a liquid phase and three solid-state invariant reactions ($Ag_3AuTe_2 + \beta Ag_{5-x}Te_3 \rightleftharpoons AgAuTe_4 + \beta Ag_2Te$; $\beta Ag_2Te + \beta Ag_{5-x}Te_3 \rightleftharpoons AgAuTe_4 + \gamma_h$; $\gamma Ag_2Te \rightleftharpoons \beta Ag_2Te + (Ag/Au) + Ag_3AuTe_2$). Indeed [60 Mar] claims that the reaction $Ag_3AuTe_2 + \beta Ag_{5-x}Te_3 \rightleftharpoons AgAuTe_4 + \beta Ag_2Te$ occurs at 315 ± 10°C. This assertion was based on the preparation of two alloys whose compositions lay in the $AgAuTe_4$–Ag_3AuTe_2–Ag_2Te phase region at 300°C. [60 Mar] heated the two alloys at 330°C for 14 days and found the three-phase mixture $AgAuTe_4$–Ag_3AuTe_2–$\beta Ag_{5-x}Te_3$. This is understandable if one accepts the liquid reaction [60 Mar] $L_{U_8} + Ag_3AuTe_2 \rightleftharpoons AgAuTe_4 + \beta Ag_{5-x}Te_3$, since a three-phase region $AgAuTe_4 + Ag_3AuTe_2 + \beta Ag_{5-x}Te_3$ will be formed by this reaction. Although Honea [64 Hon] correlated observations on mineral phase assemblies with the 300°C isothermal section [60 Mar], a discussion in [65 Cab 1] pointed out that the two alloys prepared by [60 Mar] and heat treated for 14 days at 330°C were in the $AgAuTe_4$–Ag_3AuTe_2–Ag_2Te phase region at 300°C but, since their compositions fell on either side of the Ag_2AuTe_2–$\beta Ag_{5-x}Te_3$ join, at 330°C they should behave differently. One alloy should produce the three-phase mixture claimed by [60 Mar], $AgAuTe_4$–Ag_3AuTe_2–$\beta Ag_{5-x}Te_3$, but the other alloy should contain Ag_3AuTe_2–$\beta Ag_{5-x}Te_3$–βAg_2Te. [65 Hon] accepted the criticism of [65 Cab 1]. However when Cabri [65 Cab 2] attempted to check the claim [60 Mar] that the βAg_2Te–$AgAuTe_4$ join at 300°C changes to a $\beta Ag_{5-x}Te_3$–Ag_3AuTe_2 join at 315°C the results were inconclusive. [65 Cab 2] prepared one of the alloys of [60 Mar] that had a composition in the $AgAuTe_4$–Ag_3AuTe_2–βAg_2Te phase region at 300°C according to [60 Mar] and should convert to $AgAuTe_4$–Ag_3AuTe_2–$\beta Ag_{5-x}Te_3$ at 315°C. The alloy was prepared by heating the mixed elements at temperatures of 290, 320, 335, and 350°C for longer than 30 days. During heat treatment the mixtures were cooled and the semi-sintered samples ground on several occasions. This preparative technique would not produce equilibrium conditions and it is not surprising that [65 Cab 2] admitted a failure to reach equilibrium. Cabri reported the presence of all four phases.

The work of [65 Cab 2] was very extensive and painstakingly performed. It may well be of real value to the mineralogist interested in the sub-solidus formation of minerals in the Ag–Au–Te system but it is of little value to those concerned with the equilibrium relationships in this ternary system because the preparative techniques did not lead to the establishment of equilibrium. A variety of X-ray diffraction techniques were used by [65 Cab 2] to study alloys after heat treatment and at temperature but no other experimental technique was used. The elements used to prepare alloy compositions were all of very high purity (99·9995+% Au, 99·999+% Te, 99·999+% Ag). Weighed portions were sealed in evacuated silica tubes as loosely packed aggregates; for some of the alloys that were to be heat treated at lower temperatures the weighed charge was compacted into a 5 mm pellet using a pressure of 21 000 lb in^{-2} (145 MN m^{-2}). After heat treatment samples were quenched into water and ground before X-ray examination. [65 Cab 2] notes that compacted samples gave much sharper X-ray diffraction photographs. This would be anticipated. The experimental conditions used by [65 Cab 2] would not lead to a determination of the equilibria. At the low reaction temperatures in this system, compared to the high melting points of Au and Ag, diffusion will be very slow. The only chance of approaching equilibrium is to melt all the alloys, ensure dissolution of the components in the melt and to heat treat the solidified alloys until they show no further change in structure. Only then can an approach to equilibrium be claimed. The preparative techniques used by [65 Cab 2] were inadequate for a study of equilibrium relationships. [65 Cab 2] reported the phase relations along the $AuTe_2$–$AgAuTe_4$ (sylvanite) section and along the Ag_2Te–Ag_3AuTe_2 (petzite) section (Figs. 57 and 58). The $AuTe_2$–$AgAuTe_4$ section to 11 at.-% Ag was found to be pseudobinary (Fig. 57). As stated in the introduction [65 Cab 2] regards $AgAu_4Te_{10}$ as a phase formed at 382 ± 5°C by reaction of liquid with $AuTe_2$ containing 3 at.-% Ag. $AgAuTe_4$ forms at 354 ± 5°C by reaction of liquid with $AgAu_4Te_{10}$ containing over 8 at.-% Ag. It should be noted that the section presented [65 Cab 2] should contain an $AgAuTe_4$ + Te phase region starting at 350°C on the $AgAuTe_4$ solvus and separating the $AgAuTe_4$ region from the l + $AgAuTe_4$ + Te and $AgAuTe_4$ + Te + $Ag_{5-x}Te_3$ phase regions. [65 Cab 2] states that all samples heat treated below 382°C contained an excess of Te that varied from 0.1 to 15 wt-%. Fig. 57 is not at a constant Te content and it could not be represented as pseudobinary with a variable excess Te. Cabri's data is not accepted. The section Ag_2Te–Ag_3AuTe_2 (Fig. 58) was produced using 15 compositions prepared from Ag_2Te and Ag_3AuTe_2. It is not clear whether these two compounds were synthesized from the melt or not. The 15 alloys were produced by heating in sealed, evacuated silica tubes above 950°C (i.e. in this case the alloys should all be molten). They were then ground and compacted and heat treated at 440°C for 10 to 14 days. This preparative technique is much superior to that used by the same worker in the study of the $AuTe_2$–$AgAuTe_3$ section. After the homogenizing heat treatment at 440°C the crystal structure of the alloys was examined by high-temperature X-ray diffraction analysis using a Rigaku–Denki powder-diffraction camera. The transition temper-

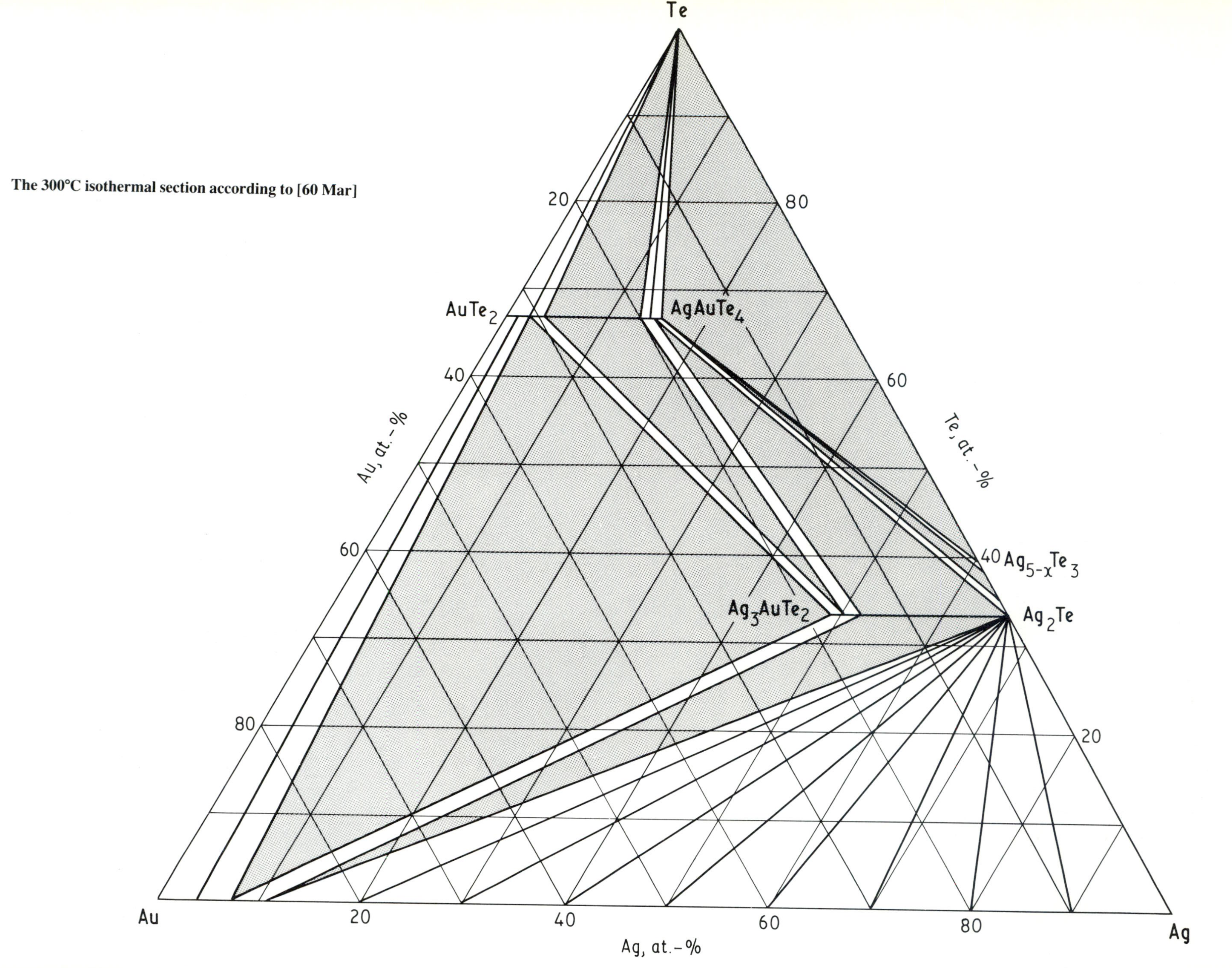

56 The 300°C isothermal section according to [60 Mar]

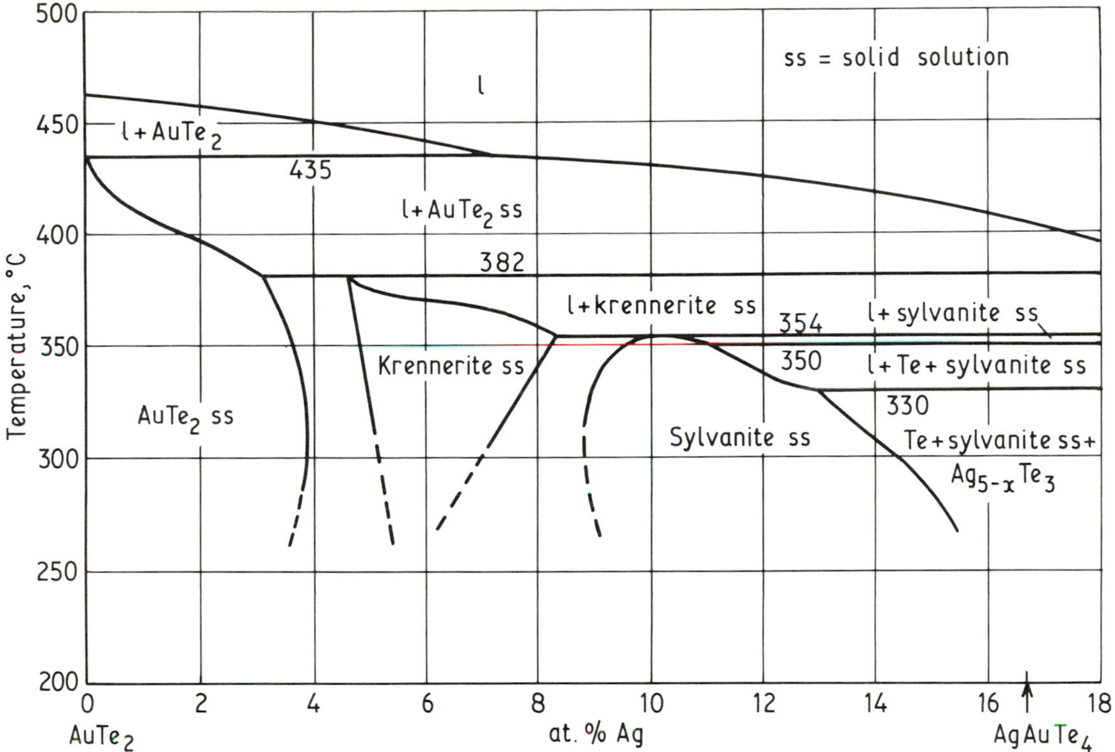

57 The section $AuTe_2$–$AgAuTe_4$ according to [65 Cab 2]

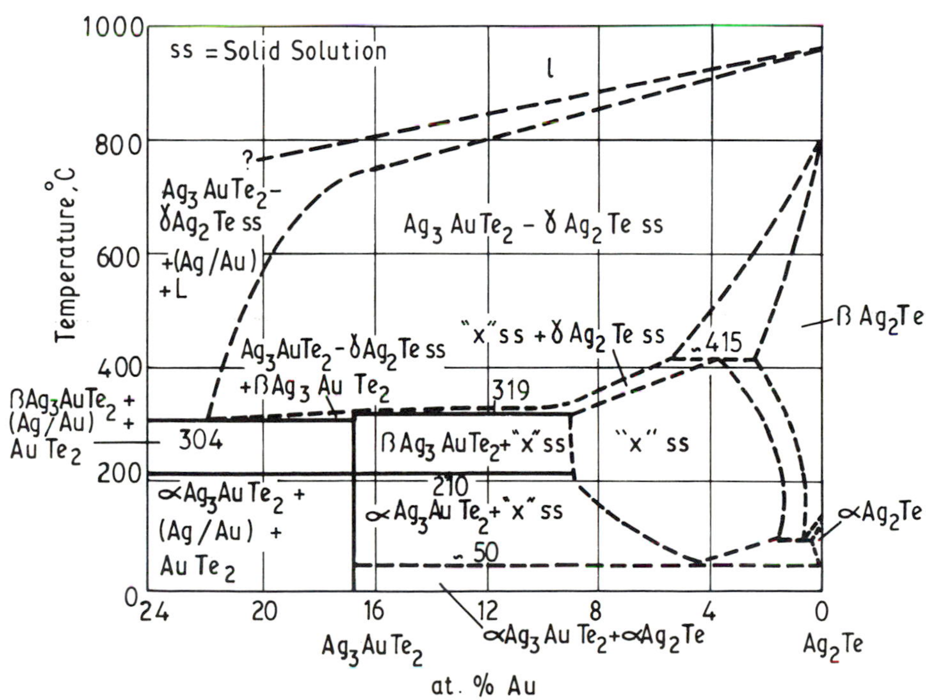

58 The section Ag_2Te–Ag_3AuTe_2 according to [65 Cab 2]

atures are quoted to ± 15°C. [65 Cab 2] regards this section as pseudobinary. It is pseudobinary from Ag_2Te to Ag_3AuTe_2. At more Au-rich compositions than Ag_3AuTe_2 the section given is not pseudobinary, and this part of the section is incorrect. As drawn [65 Cab 2] the phase region βAg_3AuTe_2 + $AuTe_2$ + (Ag/Au) exists below a 304 ± 15°C horizontal and [65 Cab 2] states that $AuTe_2$ disappears above 304°C. This is not possible. Adjoining the three-phase region βAg_3AuTe_2 + $AuTe_2$ + (Ag/Au) at either end on this section there must be a βAu_2AuTe_2 + $AuTe_2$ phase region at the Ag_3AuTe_2 end and a (Ag/Au) + $AuTe_2$ phase region at the Au-rich end. Indeed this is shown in the 290°C isothermal section [65 Cab 2]. Above the 304°C horizontal there must be phase regions also containing $AuTe_2$. [65 Cab 2] claims that γAg_2Te forms a complete solid solution series with γAg_3AuTe_2 and that a further ternary compound 'x' is formed between about 4·4 and 9 at.-% Au from about 50°C to about 415°C. This 'x' phase transforms to αAg_3AuTe_2 + αAg_2Te on holding at room temperature: slight pressure on grinding it at room temperature accelerates the transformation. Although the alloy preparation conditions were more sophisticated for alloys on this section the homogenizing anneal of 10 to 14 days at 440°C is not considered adequate to ensure equilibrium. The section at 33·33 at.-% Te published by [80 Leg] is in total disagreement with Fig. 58. Until further work is done on this section one must regard the results of [65 Cab 2] as suspect; the data of [80 Leg] are preferred.

[65 Cab 2] also postulated the presence of a 'new' ternary eutectic in the $AuTe_2$–Ag_3AuTe_2–(Ag/Au) phase field at 304 ± 10°C with the eutectic liquid having an approximate composition 21 at.-% Ag, 38 at.-% Au. [80 Leg] synthesized this composition and found a liquidus temperature of 600°C, in good agreement with the data of [60 Mar] and contradicting the claim of [65 Cab 2] to have located a ternary eutectic.

[80 Leg] is a modern study of the whole ternary system involving the examination of about 250 alloys prepared from 99·999% Ag and Te and 99·99% Au by melting the elements under a vacuum of 10^{-3} torr in sealed silica tubes. Alloys were examined by a range of experimental techniques, including differential thermal analysis, differential enthalpy analysis, metallography, and X-ray diffraction analysis at room temperature and as a function of temperature. [80 Leg] presents vertical sections at 10 at.-% Au, at 16·67, 33·33, 50, 56·67, 66·67, and 80 at.-% Te, and the sections $AuTe_2$–Ag_2Te (see section on pseudobinary systems and Fig. 54) and Au–Ag_2Te. Five invariant reactions are associated with the liquid phase:

L_{U_1} + (Ag/Au) $\rightleftharpoons$ $AuTe_2$ + γAg_2Te	at 380°C
L_{U_2} + γAg_2Te $\rightleftharpoons$ $AuTe_2$ + βAg_2Te	at 360°C
L_{U_3} + βAg_2Te $\rightleftharpoons$ $AuTe_2$ + γ_h	at 358°C
L_{U_4} + γ_h $\rightleftharpoons$ $AuTe_2$ + $\beta Ag_{5-x}Te_3$	at 350°C
L_E $\rightleftharpoons$ (Te) + $AuTe_2$ + $\beta Ag_{5-x}Te_3$	at 332°C

[80 Leg] did not publish a complete reaction scheme; Fig. 59 gives a scheme conforming with the equilibria proposed. The noteworthy fact about the equilibria is the huge fall in temperature from the Ag–Te monotectic reaction at 869°C, l_1 = γAg_2Te + l_2, to the ternary invariant reaction U_1 at 380°C. The composition of the γAg_2Te phase at the 380°C invariant plane is 16 at.-% Au, 50·7 at.-% Ag as scaled from the 380°C isothermal section. [80 Leg] did not confirm the existence of $AgAu_4Te_{10}$, $AgAuTe_4$, nor the 'x' phase of [65 Cab 2], despite heat treatment times from 1 to 3 months for alloys in the Te–$AuTe_2$–Ag_2Te partial system. Ag_3AuTe_2, petzite, was found to decompose peritectoidally on heating to 313°C. This transformation was detected by DTA, and by high temperature X-ray diffraction analysis. The compositions of the liquid phase at each of the five invariant reactions was given by [80 Leg]. Table 17 tabulates the liquid compositions and gives an estimate of the compositions of other phases. The estimated compositions were scaled from schematic isothermal sections [80 Leg] at 380°C and should not be relied upon. Of the nine vertical sections presented by [80 Leg] experimental points were plotted for the sections $AuTe_2$–Ag_2Te (Fig. 54), 66·67 at.-% Te (Fig. 60), and 80 at.-% Te (Fig. 61). Figs. 54 and 61 are amended versions of the original sections that give a consistent direction to the monovariant curve e_3E (see Fig. 68, below). The 66·67 at.-% Te section [80 Leg] shows intersection with the monovariant curve e_3E at 26·4 at.-% Ag. The 10 at.-% Te section is considered to intersect e_3E at 7 at.-% Ag. Taking the declared composition of the ternary eutectic liquid at E (Table 17) and the known eutectic composition e_3 with the two intersection points produces a curve e_3E with an unlikely kink. Moving the intersection point of the 66·67 at.-% Te section with curve e_3E from 26·4 at.-% Ag to 28 at.-% Ag produces self-consistent results, curve e_3E curving smoothly from e_3 to E. Figure 60 has been slightly amended from the original [80 Leg] in this respect. Also added to Fig. 60 are the other phase regions that must be present but that were not included by [80 Leg]. The trace of the surface of secondary separation of $AuTe_2$ + Te from the liquid is shown virtually horizontal at 370°C. This implies that the curve e_3E falls from 416°C at e_3 to 370°C at its intersection with the 66·67 at.-% Te section and then falls to 332°C in going the short distance to E. Figure 60 [80 Leg] should be compared with Fig. 57 [65 Cab 2]. The form of the section presented by [80 Leg] is similar to that obtained by study of the early data [15 Pel]. The 80 at.-% Te section does not agree with the 66·67 at.-% Te section, nor with the composition of the ternary eutectic E. As given by [80 Leg], the 80 at.-% Te section intersects the curve e_3E at 16·3 at.-% Au, 3·7 at.-% Au. Since E has a composition (Table 17) 34·2 at.-% Ag, 4·5 at.-% Au, reliance on the 80 at.-% Te section would mean that the curve e_3E bends towards the Ag–Te binary edge and then curves away to meet the point E. Taken with the previous comments on the 66·67 at.-% Te section, where its intersection with e_3E has been placed at 28 at.-% Ag, it is concluded that the 80 at.-% Te section could be redrawn, using the plotted data [80 Leg], to give intersection with e_3E at 15 at.-% Ag (Fig. 61). Curve e_3E would then bend towards the Ag–Te edge but from 80 at.-% Te to E it would be substantially parallel to the Ag–Te edge. The liquidus surface (Fig. 68) illustrates this recommendation. The remaining vertical

59 Reaction scheme for the Ag–Au–Te system

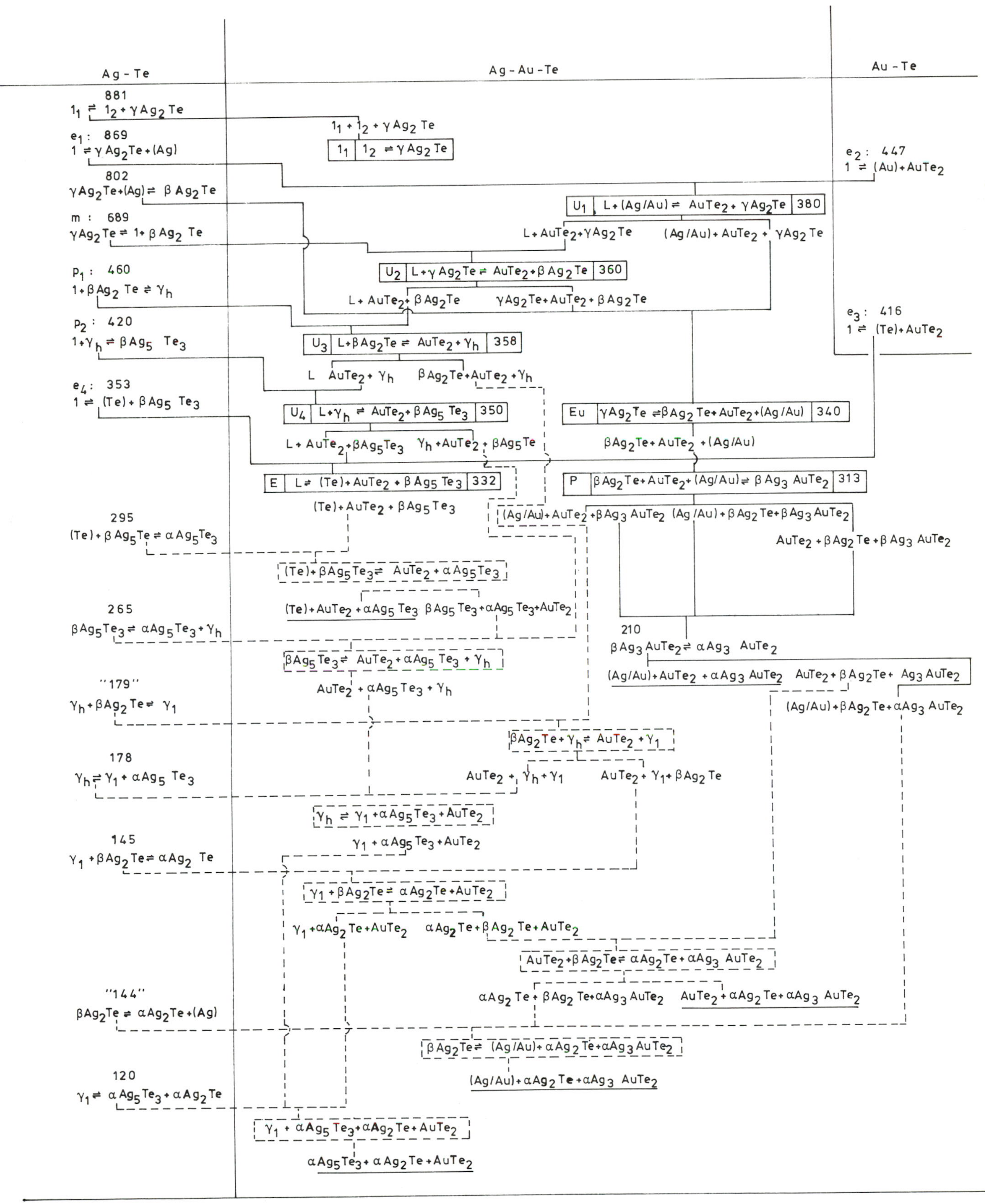

Table 17. Compositions of the co-existing phases at the invariant reaction horizontals

Reaction	*Temperature, °C*	*Liquid*		*γAg_2Te**		*βAg_2Te**		*(Ag/Au)**	
		Ag	*Au*	*Ag*	*Au*	*Ag*	*Au*	*Ag*	*Au*
U_1	380	25·3	23·3	50·7	16·0	—	—	8·0	86·2
U_2	360	33·3	12·0	53·3	6·0	58	3·0	—	—
U_3	358	35·3	8·0	—	—	—	—	—	—
U_4	350	35·8	5·6	—	—	—	—	—	—
E	332	34·2	4·5	—	—	—	—	—	—

* Scaled from isothermal sections [80 Leg] (see the section on invariant equilibria)

sections [80 Leg] contained no experimental points and are complex in their construction. Figures 62–67 are representations of the sections at 16·67, 33·33, 50, and 56·67 at.-% Te, 10 at.-% Au and the section Au–Ag_2Te respectively. They have been constructed from enlargements of the published diagrams and by correlating as far as possible the vertical sections with isothermal sections [80 Leg] to give consistent results. In view of their complexity it has been necessary to designate some phase regions by numbers. Table 18 provides a key to the numerical designation of the phase regions. The 16·67 at.-% Te section (Fig. 62) presented by [80 Leg] is incorrect in that the l + α + γAg_2Te phase region should start from the Ag–Te binary at 869°C and end at the 380°C horizontal. [80 Leg] draws two l + α + γAg_2Te regions separated by a l + γAg_2Te region. This is an impossible construction. Figure 63, the 33·33 at.-% Te section, is of interest in that [65 Cab 2] presented a portion of this section (Fig. 58). Fig. 63 is an amended version of the published diagram. [80 Leg] indicated that the phase fields βAg_2Te + γAg_2Te + $AuTe_2$ (marked 17 in Fig. 63) and βAg_2Te + $AuTe_2$ (marked 21) as originating from the Ag–Te binary. This is not so. The amended version indicates that these phase regions exist within a temperature range below 360°C. It should be stressed that the form of the vertical sections alter markedly if the composition of the γAg_2Te phase in the isothermal section at 380, 360, and 340°C is slightly changed. The stability of the γAg_2Te phase in the 33·33 at.-% Te section is worthy of note. It extends to low temperatures and is capable of taking about 16 at.-% Au into solution. The formation of βAg_3AuTe_2 (τ_β) at 313°C is clearly indicated. As τ_β lies on the 33·33 at.-% Te section the equilibria from τ_β to βAg_2Te is essentially pseudobinary below 313°C. The 50 at.-% Te section (Fig. 64) is in reasonable topological accord with the published [80 Leg] section, but there are minor variations in the extent of phase regions. The 56·67 at.-% Te section (Fig. 65) is a modified version of the published section [80 Leg]. It is based on the assumption that this section intersects the liquidus curves at U_3 (Table 17 gives 35·3 at.-% Ag, 8 at.-% Au, 56·7 at.-% Te). It differs from the published version [80 Leg] in that there is no primary phase field for the separation of βAg_2Te and all the consequences that stem from this. The 10 at.-% Au section (Fig. 66) is also an amended version of the published section. In particular the composition range for stability of γAg_2Te was overestimated by [80 Leg]. They indicate a range from 45 to 75 at.-% Ag whereas it is assessed at 52·5 to 57·5 at.-% Ag. [80 Leg] also introduce a l + γ_h phase region, but the 10 at.-% Au section passes between U_2 and U_3 with Au compositions corresponding to 12 and 8 at.-%. No l + γ_h region will be present in the 10 at.-% Au section. The Au–Ag_2Te section present difficulties. The published section [80 Leg] indicates that the section intersects the invariant reaction planes U_1 (380°C), U_2 (340°C), and the Ag_2AuTe_2 formation plane at 313°C. This is only true if the composition of the (Ag/Au) phase at 380, 340, and 313°C is moved from the composition (Table 17), scaled from the isothermal section to a composition within the triangle Au–Ag–Ag_2Te. [82 Leg] confirmed the composition of the (Ag/Au) phase on the isothermal sections to be indicative only. Accordingly the (Ag/Au) phase has been assumed to have a composition nearer to 8 at.-% Ag, 90 at.-% Au, 2 at.-% Te than that indicated in Table 17 by scaling from the published isothermal sections. The Au–Ag_2Te section published by [80 Leg] shows γAg_2Te as being capable of taking more than 50 at.-% Au into solution. This cannot be correct. Figure 67 is an attempt to present a section that is consistent with the other vertical sections. It must be stressed that Figs. 62–67 inclusive are basically schematic. Original errors in construction have been amended in deriving Figs. 62–67.

Table 18. Numerical designation of phase regions. $Ag_{5-x}Te_3$ is abbreviated to Ag_5Te_3

1	l + $AuTe_2$ + βAg_2Te	15	l + βAg_5Te_3
2	$AuTe_2$ + βAg_5Te_3 + γ_h	16	$AuTe_2$ + γAg_2Te
3	$AuTe_2$ + αAg_5Te_3 + βAg_5Te_3	17	$AuTe_2$ + βAg_2Te + γAg_2Te
4	$AuTe_2$ + αAg_5Te_3 + γ_h	18	l + $AuTe_2$ + γ_h
5	$AuTe_2$ + γ_h + γ_l	19	l + $AuTe_2$ + βAg_5Te_3
6	$AuTe_2$ + βAg_2Te + γ_l	20	l + Te + βAg_5Te_3
7	$AuTe_2$ + αAg_2Te + γ_l	21	$AuTe_2$ + βAg_2Te
8	$AuTe_2$ + αAg_2Te + βAg_2Te	22	$AuTe_2$ + βAg_2Te + γ_h
9	αAg_2Te + βAg_2Te + τ_α	23	$AuTe_2$ + αAg_5Te_3 + γ_l
10	l + α + γAg_2Te	24	$AuTe_2$ + αAg_2Te + τ_α
11	l + $AuTe_2$ + γAg_2Te	25	Te + $AuTe_2$ + βAg_5Te_3
12	l + βAg_2Te + γAg_2Te	26	Te + βAg_5Te_3
13	l + βAg_2Te + γ_h	27	Te + αAg_5Te_3
14	l + βAg_5Te_3 + γ_h	28	α + γAg_2Te

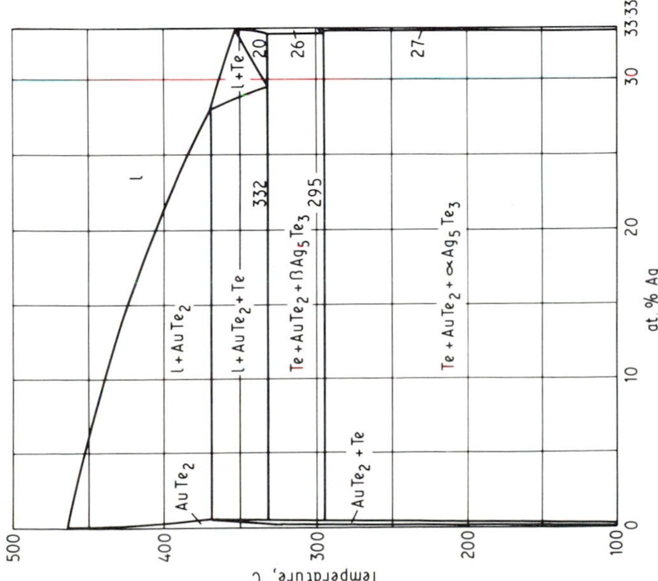

60 The 66·67 at.-% Te section [80 Leg]

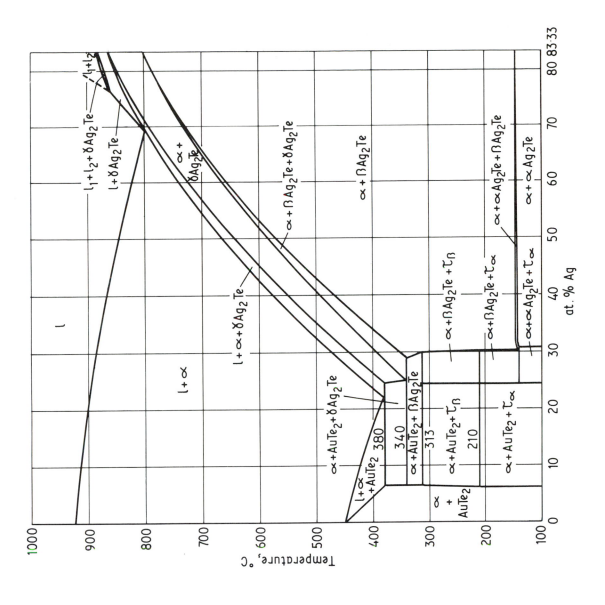

62 The 16·67 at.-% Te section [80 Leg]

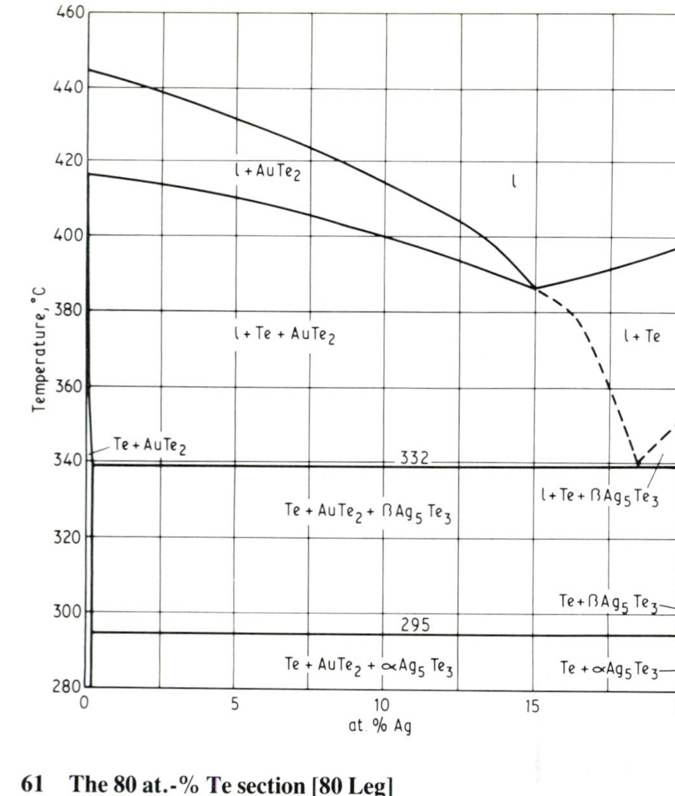

61 The 80 at.-% Te section [80 Leg]

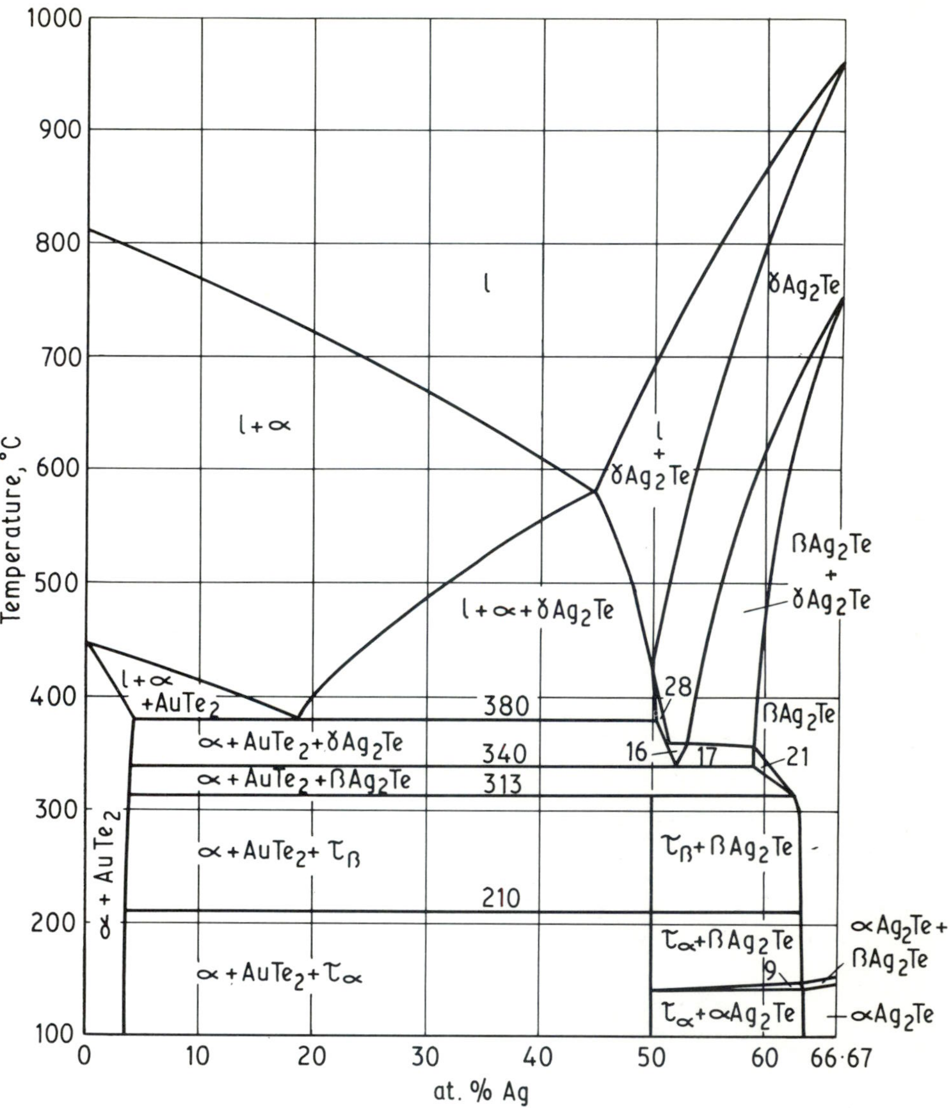

63 The 33·33 at.-% Te section [80 Leg]

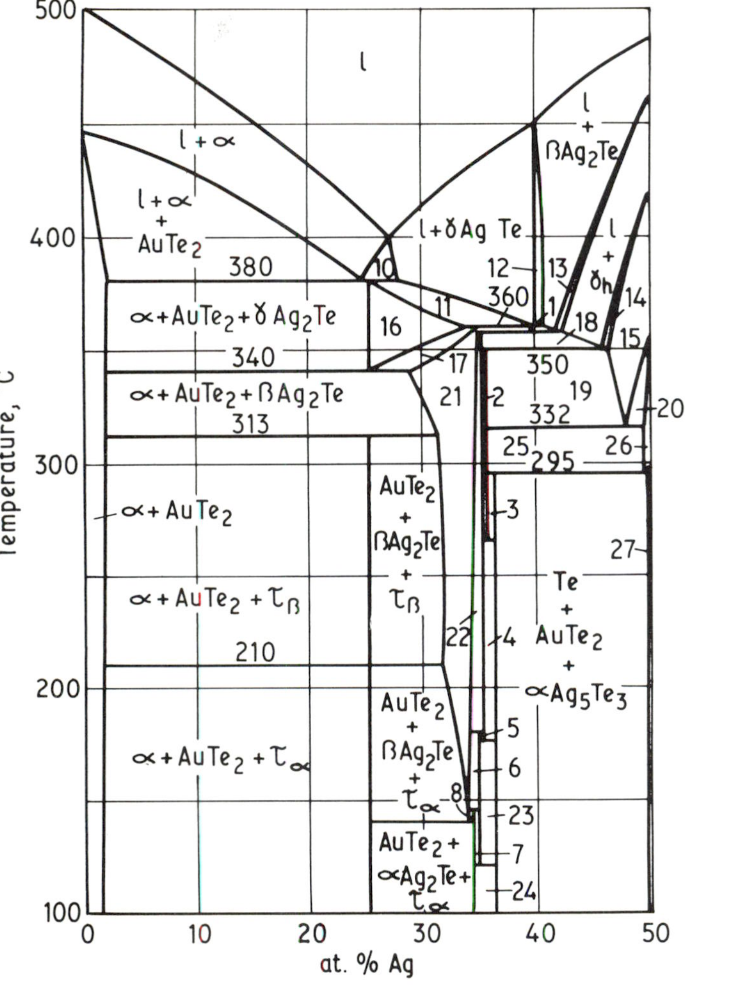

64 The 50 at.-% Te section [80 Leg]

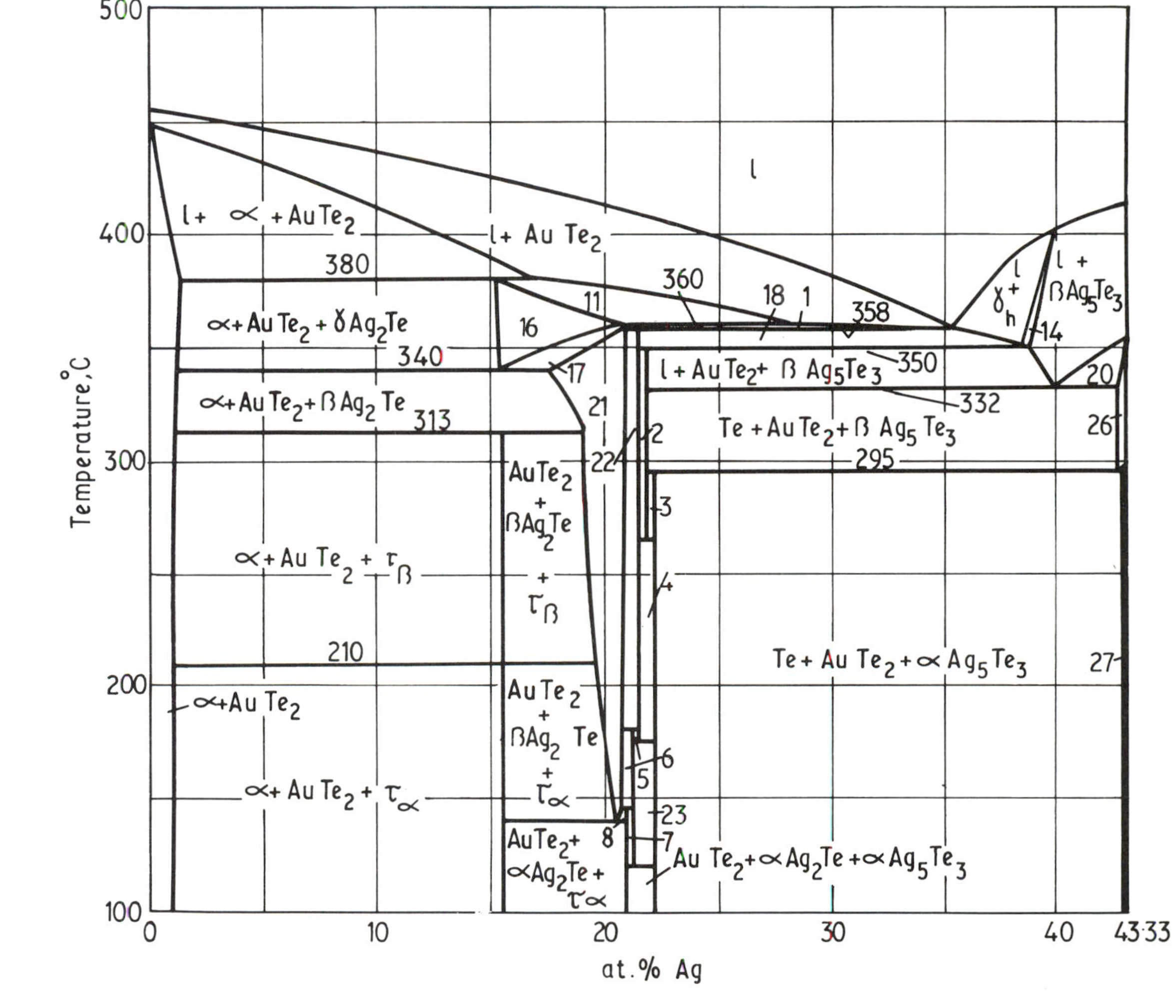

65 The 56·67 at.-% Te section [80 Leg]

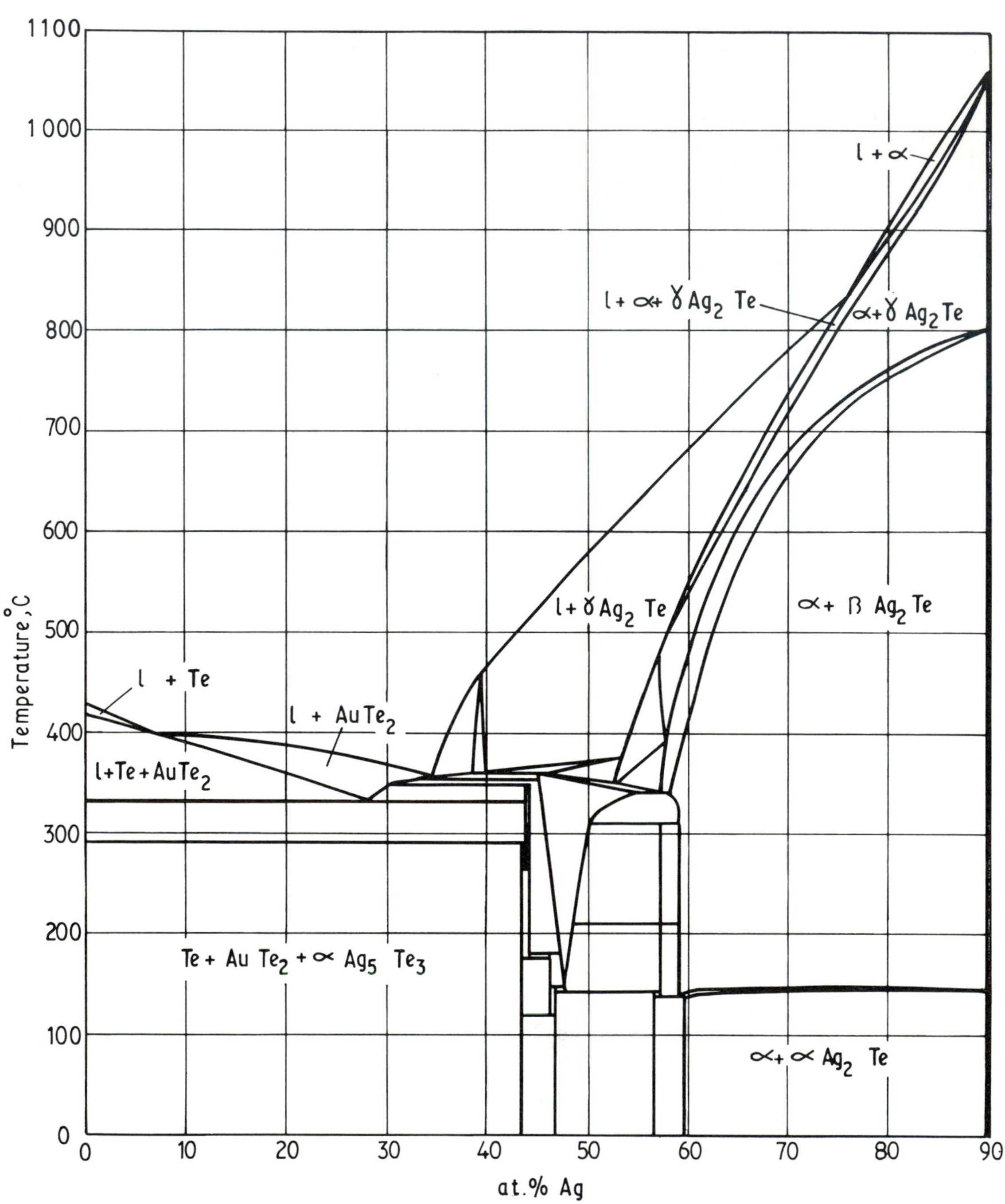

66 The 10 at.-% Au section [80 Leg]

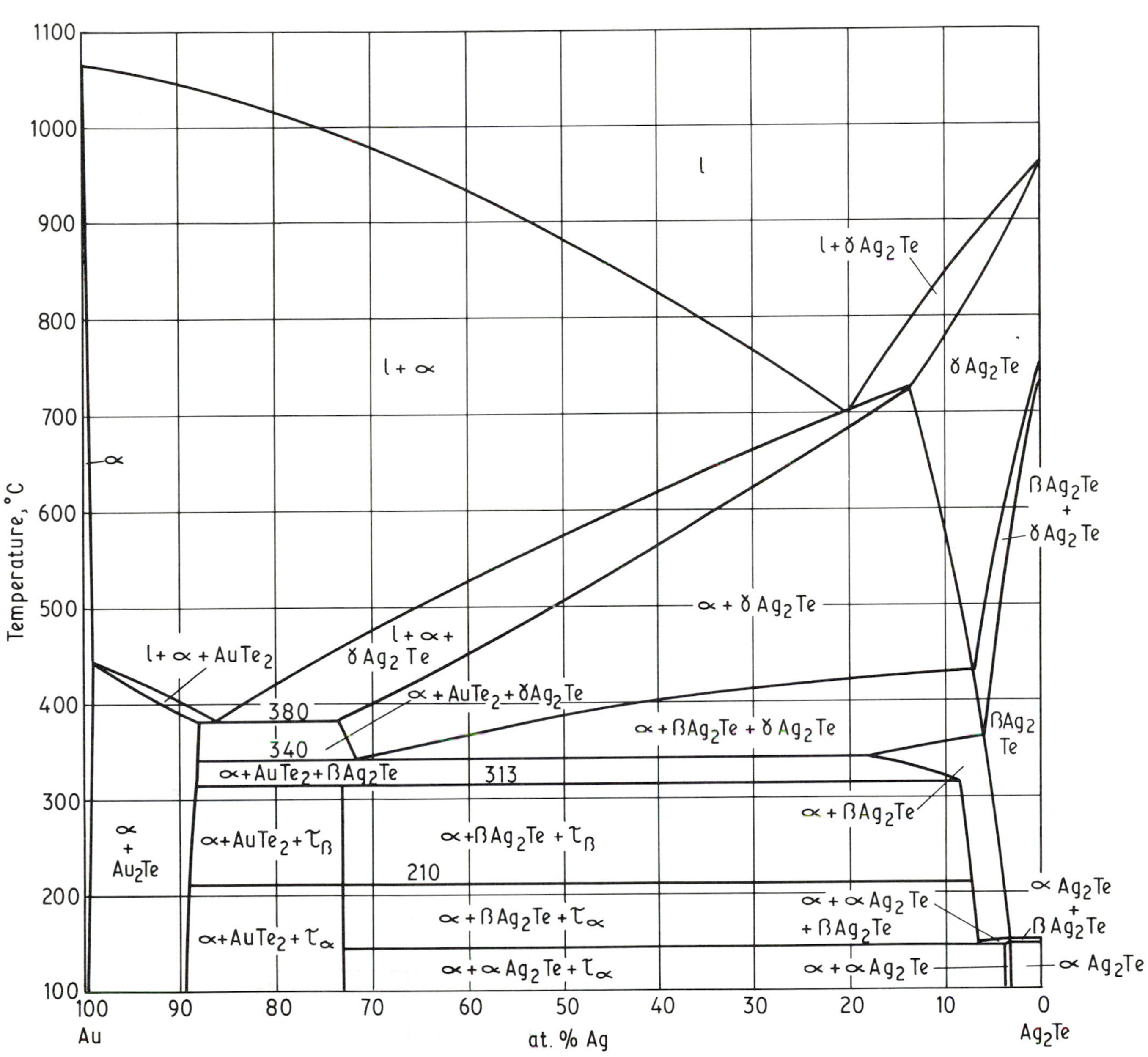

67 The section Au–Ag_2Te [80 Leg]

LIQUIDUS SURFACE

Figure 68 is the liquidus surface proposed by [80 Leg]. It is accepted as a reasonable representation of the equilibria in preference to the liquidus of [60 Mar] (Fig. 55). It should be noted that [80 Leg] did not include isotherms on their liquidus and the only guide we have are the isotherms presented by [60 Mar].

ISOTHERMAL SECTIONS

[60 Mar] determined the 300°C isothermal section (Fig. 56). [65 Cab 2] determined isothermal sections at 365, 335, 290, and 170°C. As [65 Cab 2] used inadequate heat treatment times to attain equilibrium with the alloy preparation techniques adopted, the data is suspect. The isothermal sections published by [80 Leg] relate to the temperatures of the invariant reactions U_1 and U_2 (380 and 360°C), the invariant eutectoid reaction at 340°C, and three sections at temperatures above and below the Ag_3AuTe_2 formation temperature and at temperatures below 120°C. They should be regarded as schematic sections [82 Leg]. Table 19 lists the tie-triangles at 380°C (U_1), 360°C (U_2), 340°C and $T < 120$°C.

Table 19. Tie-triangles at 380, 360, 340 and $T < 120$°C

Temperature	*Tie-triangles*
380°C	(Ag/Au) + βAg_2Te + γAg_2Te
	L + γAg_2Te + βAg_2Te
	L + γ_h + βAg_2Te
	L + γ_h + $\beta Ag_{5-x}Te_3$
	L + (Te) + $AuTe_2$
360°C	(Ag/Au) + βAg_2Te + γAg_2Te
	$AuTe_2$ + + (Ag/Au) + γAg_2Te
	L + γ_h + βAg_2Te
	L + γ_h + $\beta Ag_{5-x}Te_3$
	L + (Te) + $AuTe_2$
340°C	$AuTe_2$ + γ_h + βAg_2Te
	$AuTe_2$ + γ_h + $\beta Ag_{5-x}Te_3$
	L + (Te) + $AuTe_2$
	L + (Te) + $\beta Ag_{5-x}Te_3$
	L + $AuTe_2$ + $\beta Ag_{5-x}Te_3$
$T < 120$°C	(Te) + $AuTe_2$ + $\alpha Ag_{5-x}Te_3$
	$AuTe_2$ + $\alpha Ag_{5-x}Te_3$ + αAg_2Te
	$AuTe_2$ + αAg_2Te + τ_α
	(Ag/Au) + $AuTe_2$ + τ_α
	(Ag/Au) + αAg_2Te + τ_α

REFERENCES

15 Pel: G. Pellini: *Gazz Chim. Ital.*, 1915, **45**, 469–484

36 Tun: G. Tunell and C. J. Ksanda: *J. Washington Acad. Sci.*, 1936, **26**, 507–509

37 Tun: G. Tunell and C. J. Ksanda: *Amer. Mineral.*, 1937, **22**, 728–730

40 Kra: F. C. Kracek and C. J. Ksanda: Ann. Report Geophysics Lab., Year Book 39, 35–36; 1940, Carnegie Institute, Washington

41 Tun: G. Tunell: *Amer. Mineral.*, 1941, **26**, 457–477

48 Tho: R. M. Thompson: *Amer. Mineral.*, 1948, **33**, 209–210

50 Tun: G. Tunell and K. J. Murata: *Amer. Mineral.*, 1950, **35**, 959–984

52 Tun: G. Tunell and L. Pauling: *Acta Crystallogr.*, 1952, **5**, 375–381

59 Fru: A. J. Frueh: *Amer. Mineral.*, 1959, **44**, 693–701

59 Kra: F. C. Kracek and C. J. Ksanda: private communication reported by A. J. Frueh: *Amer. Mineral.*, 1959, **44**, 693–701 (see footnote p. 701)

60 Mar: N. L. Markham: *Econ. Geol.*, 1960, **55**, 1148–1178, 1460–1477

64 Hon: R. M. Honea: *Amer. Mineral.*, 1964, **49**, 325–338

65 Cab 1: L. J. Cabri: *Amer. Mineral.*, 1965, **50**, 795–801

65 Hon: R. M. Honea: *Amer. Mineral.*, 1965, **50**, 802–804

65 Cab 2: L. J. Cabri: *Econ. Geol.*, 1965, **60**, 1569–1606

66 Tav: B. H. Tavernier: *Z. anorg. Chem.*, 1966, **343**, 323–328

66 Mes: P. Messien and M. Baiwir: *Bull. Roy. Soc. Sci. Liege*, 1966, **35**, 234–243

66 Kra: F. C. Kracek, C. J. Ksanda and L. J. Cabri: *Amer. Mineral.*, 1966, **51**, 14–28

67 Tav: B. H. Tavernier, J. Vervecken, P. Messien and M. Baiwir: *Z. anorg. Chem.*, 1967, **356**, 77–88

70 Smi: T. J. M. Smit, E. Venema, J. Wiersma and G. A. Weigers: *J. Solid State Chem.*, 1970, **2**, 309–312

78 Cha: S. Chamid, E. A. Pobedimskaya, E. M. Spiridonov and N. V. Belov: *Kristallografiya*, 1978, **23**, 483–486 (*Soviet Phys. Crystallogr.*, 1978, **23**, 267–269)

80 Ell: R. P. Elliott and F. A. Shunk: *Bull. Alloy Phase Diagrams*, 1980, **1**(2), 45–47

80 Leg: B. Legendre, C. Souleau and Chhay-Hancheng: *Bull. Soc. Chim. France*, 1980, (5–6), I, 197–204

82 Eva: D. S. Evans and A. Prince: unpublished work, 1982

82 Leg: B. Legendre: private communication, September 1982

84 Oka: H. Okamoto and T. B. Massalski: *Bull. Alloy Phase Diagrams*, 1984, **5**(2), 172–177

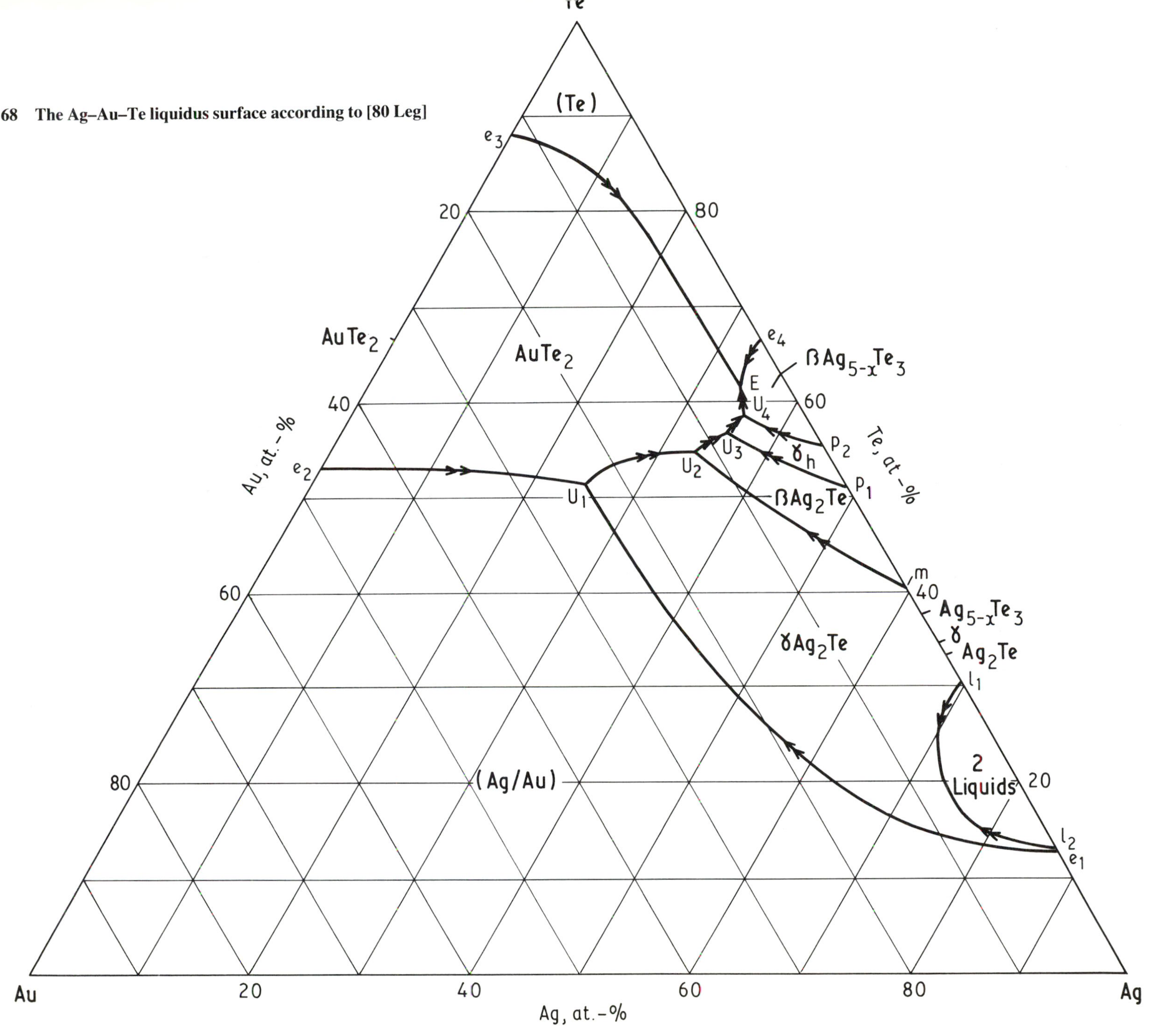

68 The Ag–Au–Te liquidus surface according to [80 Leg]

Ag–Au–U

[88 Dom] found Ag to be soluble in the compound $Au_{51}U_{14}$. An alloy containing 9 at.-% Ag, 14 at.-% U has the lattice parameters a = 1·2668, c = 0·9125 nm, with the $Ag_{51}Gd_{14}$-structure type (hP65). $Au_{51}Cu_{14}$ has lattice parameters a = 1·26521, c = 0·91381 nm.

REFERENCE

88 Dom: A. Dommann and F. Hulliger, *J. Less-Common Metals*, 1988, **141**, 261–273

Ag–Au–Zn

INTRODUCTION

Most of the work on the Ag–Au–Zn system has concentrated on the crystallographic structure of the phases found on the section AgZn–AuZn. The binary phase AgZn exists at high temperatures as a disordered bcc structure (W-type, cI2). The high-temperature phase, denoted β, transforms into a stable hexagonal structure, ζ, on slow cooling. Quenching from within the β phase region gives a metastable bcc ordered structure, denoted β′, (CsC1–type, cP2) in which Ag atoms occupy the body centre lattice sites and the Zn atoms the simple cubic sites. The binary phase AuZn has the bcc ordered β′ structure as the stable phase from room temperature to its melting point. On the section AgZn–AuZn in the ternary system, addition of Au to AgZn stabilizes the CsC1–type β′ phase (Fig. 69) and the ζAgZn phase is only stable up to 1 at.-% Au. The Ag and Au atoms in the β′ phase are randomly distributed on the body centre lattice sites but with total ordering of the Zn atoms. The transformation of β′ to β involves the disordering of the Zn sublattice to give the random atomic arrangement of Ag, Au, and Zn in the β phase. At the equiatomic composition on the AgZn–AuZn section a Heusler-type phase $AgAuZn_2$, denoted β″, is stable for temperatures up to 330°C [69 Bro] or 280 ± 20°C [66 Mul 1]. The resistivity determination of an ordering temperature of 330°C is accepted. $AgAuZn_2$ has the $L2_1$ Cu_2MnAl-type structure (cF16) with lattice parameter a = 0·62960 nm [57 Liu], 0·62956 nm [66 Mul 1], 0.62943 nm [66 Mul 2]. The parameters measured by [63 Mor] relate to samples quenched from 280°C and 600°C, a = 0·62919 and 0·62932 nm respectively, whereas the other determinations were carried out on samples slow cooled to room temperature. The variation of lattice parameter with temperature for $AgAuZn_2$ is given by [66 Mul 1]. $AgAuZn_2$ forms a super-superlattice in that the Ag and Au atoms are ordered on an NaCl-type sublattice and the Zn atoms are ordered on a simple cubic sublattice. As Ag or Au is added to $AgAuZn_2$ the ordering of the Ag–Au sublattice is decreased by the random substitution of Ag by Au, and vice versa. The transformation of the β″ phase to the β′ phase involves complete disordering of the Ag–Au atoms on their β″ sublattice to give the random arrangement of Ag–Au atoms characteristic of the CsCl-type β′ phase. The β″ phase region is shown in Fig. 69. No evidence has been found for a two-phase region between β″ and β′ or β′ and β.

The constitution of quenched alloys on the AgZn–AuZn section (Fig. 70) reflects the fact that AgZn quenches from the β phase region to produce the β′ phase. Alloys within ±3 at.-% of $AgAuZn_2$ form the β″ phase irrespective of their quenching temperature [57 Liu]. [63 Mor] also obtained the β″ phase on quenching the composition $AgAuZn_2$ from 600°C.

The remaining publications include those of [66 Kös], who outlined the (Ag–Au) solid solution region at room temperature; it is concave to the Ag–Au binary axis (Fig. 71) and has a minimum solubility of about 18 at.-% Zn. The (Ag–Au) solid solution shows short-range order except for the region next to the Au–Zn binary where the phases Au_3Zn with a faulted $L1_2$-type structure and Au_4Zn with a double-faulted $L1_2$-type structure enter the ternary system. The earlier work of [57 Wil] also defined the ternary region in which the doubly-faulted superlattice of Au_4Zn appears (Fig. 72). [49 Rau] examined an Au_3Zn alloy + 9·2 at.-% Ag. They noted little effect of Ag on the transformation temperature of the (Au) solid solution to the faulted $L1_2$ Au_3Zn phase.

BINARY SYSTEMS

The Ag–Au evaluation by [83 Oka] is accepted, as is their evaluation of the Au–Zn system [87 Oka]. The Ag–Zn system was taken from H,S,E in conjunction with data from the ternary references cited.

SOLID PHASES

Table 20 summarizes the solid phases that enter into that part of the Ag–Au–Zn system that has been studied.

Table 20. Crystal structures

Phase	*Prototype*	*Designation*
(Ag)	Cu	cF4
(Au)	Cu	cF4
(Zn)	Mg	hP2
ζAgZn		
βAgZn	W	cI2
β′AgZn*	CsCl	cP2
β′AuZn	CsCl	cP2
Au_3Zn (high temp.)	(Faulted Cu_3Au)	(Faulted cP4)
Au_4Zn	(Doubly-faulted Cu_3Au)	(Doubly-faulted cP4)
*$AgAuZn_2$	Cu_2MnAl	cF16

*Metastable; obtained by quenching βAgZn

TERNARY EQUILIBRIA

The section AgZn–AuZn

Alloys on the section AgZn–AuZn have been studied by [51 Mul], [57 Liu], [63 Mor], [65 Yee], [66 Mul 1, 2], [69 Bro], [71 Mur], [78 Bow], and [80 Mat]. The work of [80 Mat] dealt with additions of 0.5 and 1.0 at.-% Au to AgZn. In view of the small Au contents this data has been

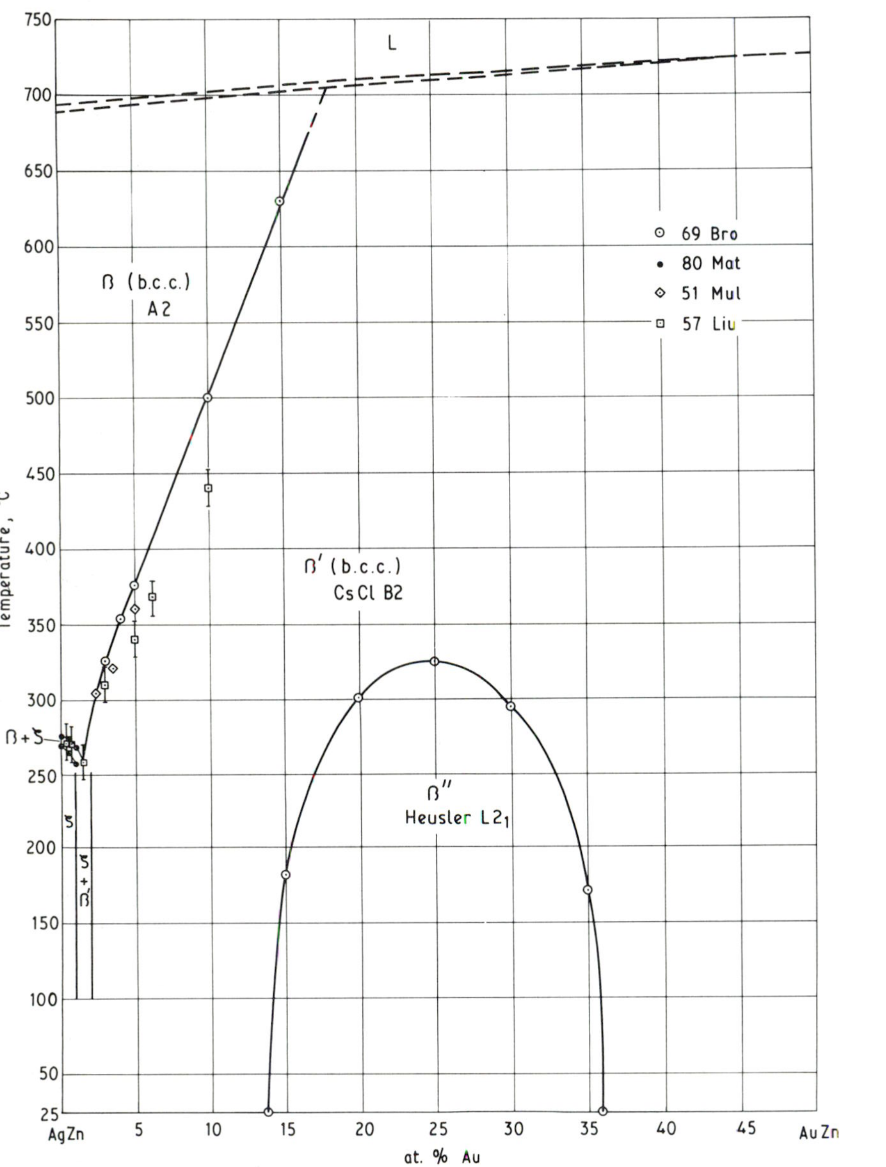

69 The section AgZn–AuZn

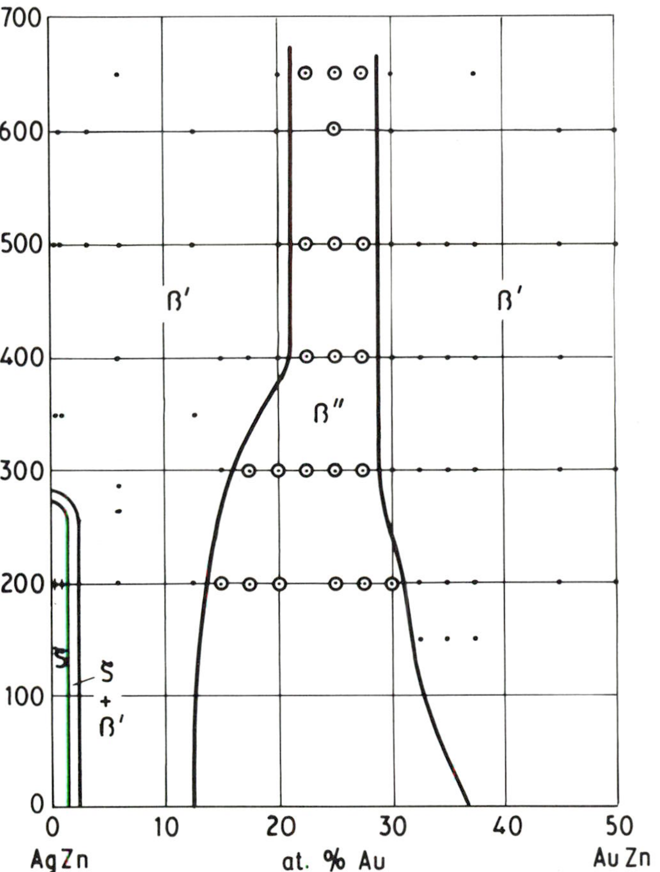

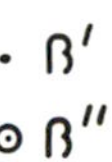

70 The section AgZn–AuZn for quenched alloys

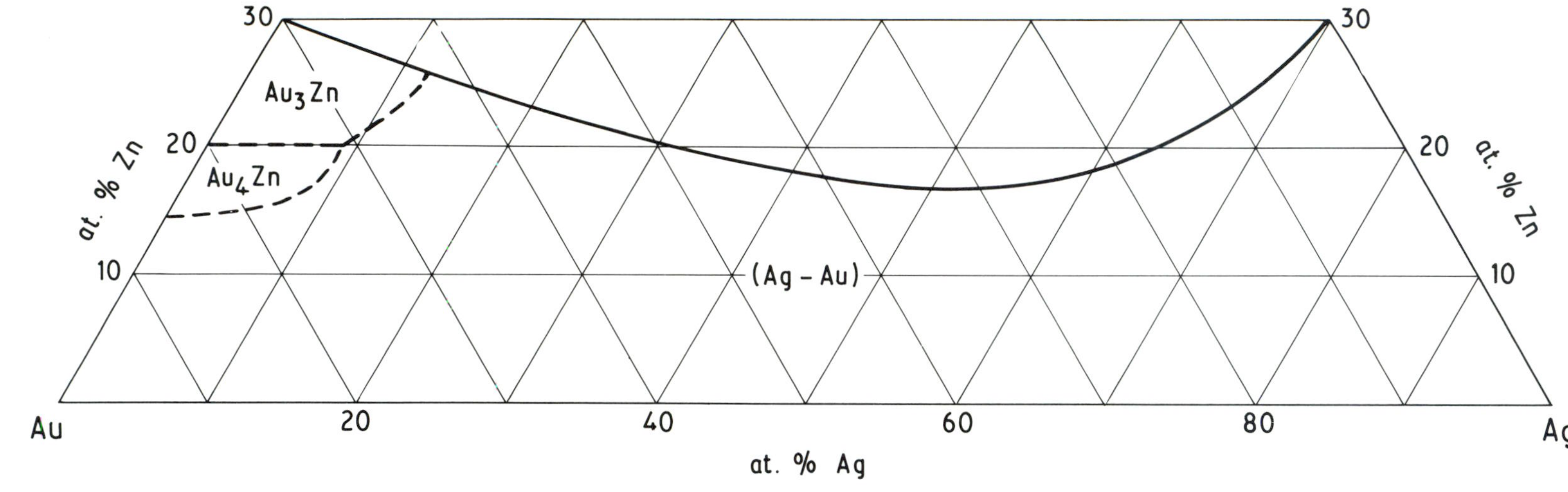

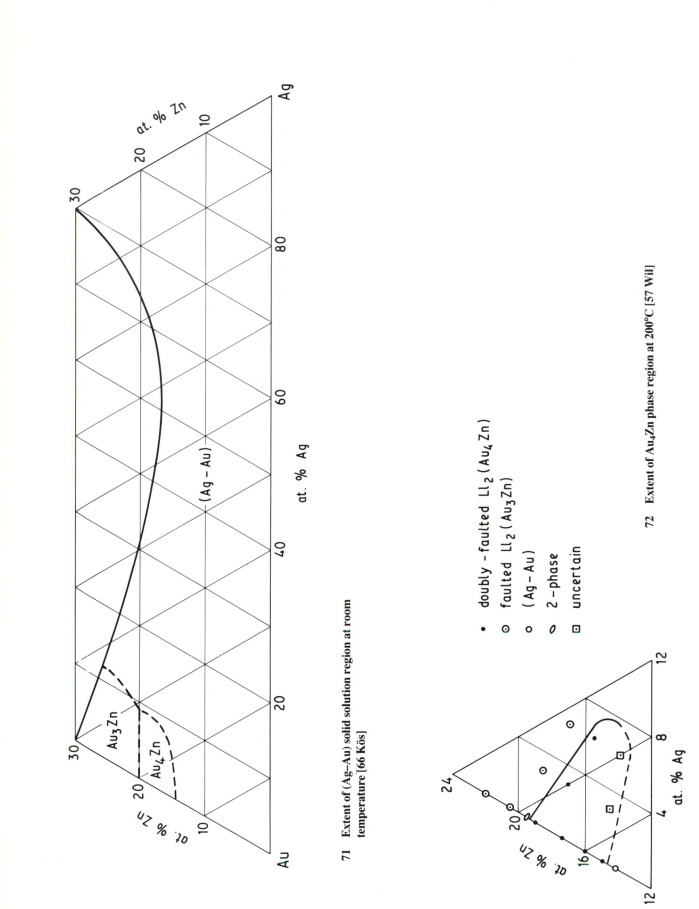

71 Extent of (Ag–Au) solid solution region at room temperature [66 Kös]

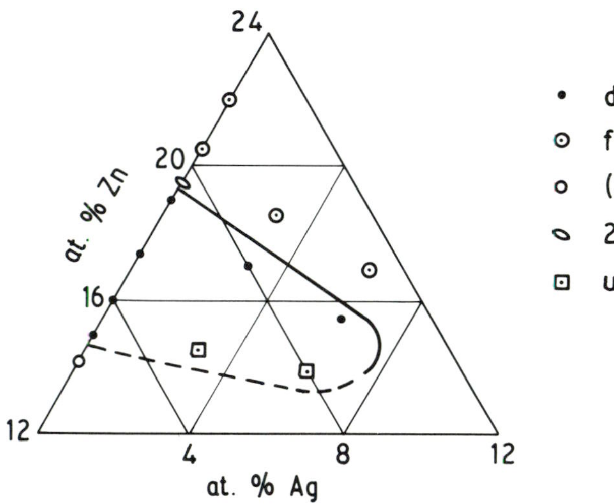

72 Extent of Au_4Zn phase region at 200°C [57 Wil]

included in that relating to the AgZn–AuZn section (Fig. 69). [51 Mul] measured the critical temperature for the order–disorder transformation (β′ → β) for three $Ag_{1-x}Au_xZn$ alloys containing 2·4, 3·5, and 5·0 at.-% Au. High-temperature X-ray powder diffraction patterns gave the temperature at which the intensity of the superstructure lines approached zero. [57 Liu] also used a high-temperature X-ray powder camera to study the β′ → β transformation. The results for 7 alloys, prepared from 99·97% Ag, 99·99% Au and spectroscopically pure Zn by melting at 1 100°C, and annealing *in vacuo* for 14 days at 500°C prior to taking filings, are given in Fig. 69. The filings were annealed for 5 h at 500°C and slowly cooled. All the high-temperature X-ray lines were sharp; temperatures were estimated to be accurate to ±12°C. The data are in good agreement with [51 Mul]. [69 Bro] used a resistivity technique to determine the disordering temperature of the β′ → β transformation for alloys with up to 15 at.-% Au. The results (Fig. 69) give higher critical temperatures than those of [51 Mul] and [57 Liu]. All three groups show that the addition of Au to AgZn, along the AgZn–AuZn section, stabilizes the ordered β′ phase. Although nothing is known of the liquidus it appears likely that β′ first appears as a stable phase up to the appearance of liquid at Au contents of about 18 at.-%. From 18 to 50 at.-% Au (AuZn) the β′ phase solidifies directly from the liquid. The ζAgZn phase has a very limited width, approximately 1 at.-% Au, on the AgZn–AuZn section. The β → ζ transformation temperature is lowered by Au additions, as is evident from [57 Liu] and [80 Mat].

At the composition $AgAuZn_2$ on the AgZn–AuZn section a Heusler phase, β″, with the $L2_1$ Cu_2MnAl-type structure (cF16) was first observed by [57 Liu] in 500°C annealed and slowly cooled samples. Table 21 summarizes the lattice parameter data. [63 Mor] confirmed the existence of $AgAuZn_2$ for samples quenched from 280 and 600°C, as did [66 Mul 1, 2] for slightly Ag-rich samples. The value of the lattice parameter obtained by [57 Liu] is accepted as the room temperature parameter for a well annealed, equilibrated alloy. The variation of lattice parameter with temperature up to 316°C was studied by [66 Mul 1], who found a large increase with temperature. Data scaled from a small graph in the original paper suggests a value of a = 0·633 8 nm at 270°C. This value should be directly comparable with the 280°C quenched value of [63 Mor], a = 0·629 19 nm. The difference may be partly due to the slightly Ag-rich alloy composition used by [63 Mor].

The region of existence of the β″ phase, which is an ordered super-superlattice at the composition $AgAuZn_2$, is plotted in Fig. 69 from the data of [69 Bro], who used resistivity measurements at high temperature to detect the β″ → β′ transformation (Cu_2MnAl-type to CsCl-type ordering). The only comparable data at high temperatures is a measurement [66 Mul 1] of the critical temperature for this transformation in the alloy 26·02 Ag, 25·21 Au, 48·77 Zn. [66 Mul 1] quotes a temperature of 280°C but notes that the accuracy of measurement was barely sufficient to locate the critical temperature; [69 Bro] give 330°C. The substitution of Ag for Au and vice versa in $AgAuZn_2$ decreases the β″ → β′ transformation temperature on the AgZn–AuZn section (Fig. 69). A more precipitate decrease may be anticipated for alloys whose composition does not lie on this section, on the reasonable assumption that the β″ phase region has an ellipsoidal shape in the ternary system with the major axis of the ellipse on the AgZn–AuZn section. If this were shown to be the case the β″ → β′ temperature of [66 Mul 1] would be below that of [69 Bro] on such alloy composition considerations. More data exists on the extent of the β″ phase region at room temperature (Table 22). The data of [57 Liu] are the most comprehensive (Fig. 73). [57 Liu] assumed Vegard's Law to apply for the β′ phase and his quoted values for the width of the β″ phase region are included in Table 22. His tabulated data has been replotted (Fig. 73) on the assumption that there is a small positive deviation from Vegard's Law (a point noted by [78 Bow] as puzzling). At most this replotting moves the β″ phase region by 0·5 at.-% Au. As can be seen from Table 22, there is reasonable agreement on the extent of the β″ phase region at room temperature. [71 Mur] applied statistical thermodynamics to a calculation of the critical temperatures for the formation of the β′ and β″ superlattices. Only first nearest neighbour potential energies were used for calculation of β′ → β temperatures; second nearest neighbour potential energies were included for calculation of the β″ → β′ temperatures. Appropriate choice of the interaction energies gave a calculated AgZn–AuZn phase diagram reproducing the diagram of [69 Bro] with the sole exception of the width of the β″ phase region at temperatures below 200°C. At room

Table 21. Lattice parameter of $AgAuZn_2$

Lattice parameter a, nm	*Alloy composition*	*Alloy condition*	*Element purity*	*Reference*
0·629 60 ± 0·000 03	$AgAuZn_2$	Ingots annealed for 14 days at 500°C Filings 5 h at 500°C; slow cooled	99·97 Ag 99·99 Au Spectroscopic Zn	[57 Liu]
0·629 19 ± 0·000 03	$AgAuZn_2$	Ingot annealed for 48 h at 600°C Sample 7 days at 280°C; quenched	Spectroscopic purity	[63 Mor]
0·629 32 ± 0·000 03	$AgAuZn_2$	Ingot annealed for 48 h at 600°C Sample 7 days at 600°C; quenched	Spectroscopic purity	[63 Mor]
0·629 56	26·02 Ag 25·21 Au 48·77 Zn	Ingot annealed; conditions not stated	>99.99	[66 Mul 1]
0·629 43 ± 0·000 01	As above	Not stated	Not stated; presumably >99·99	[66 Mul 2]

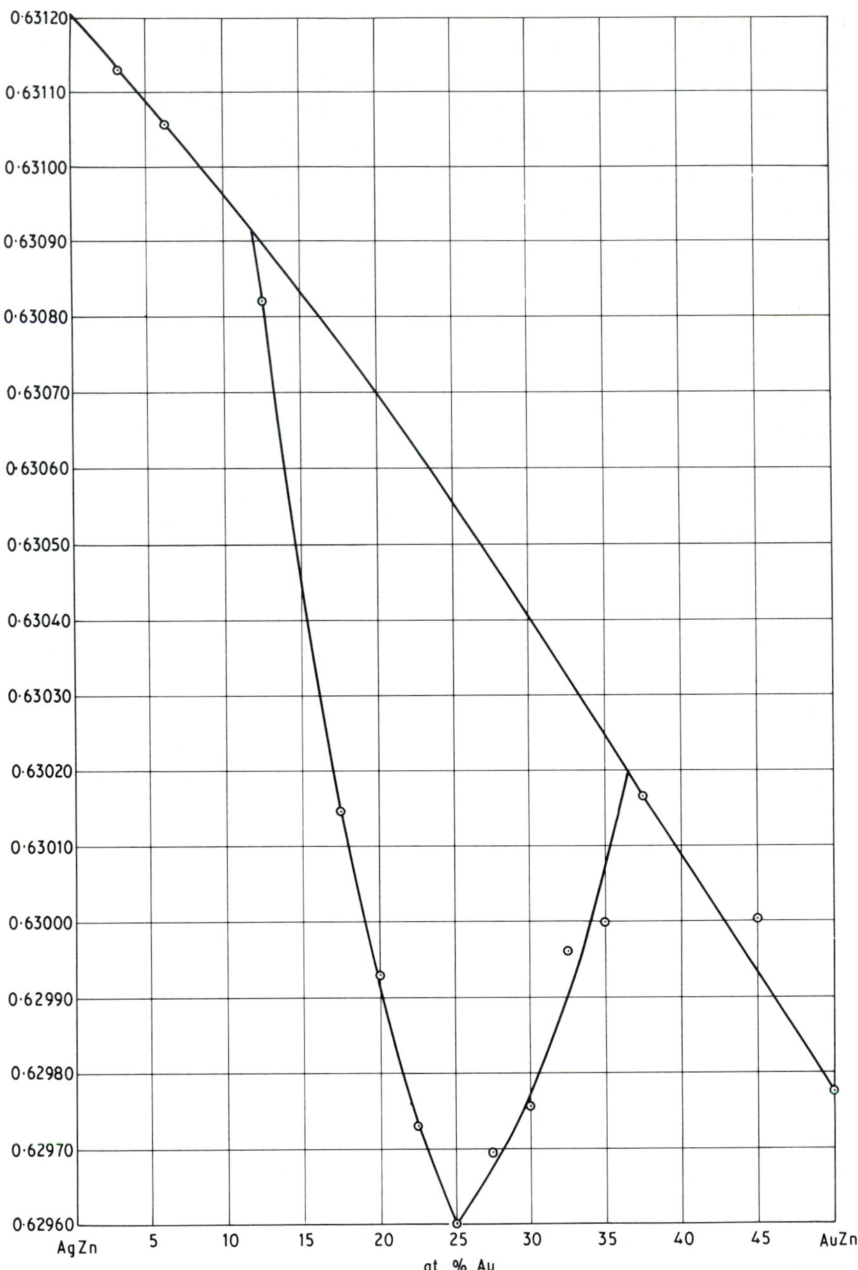

73 **Variation of lattice parameter with alloy composition for slowly cooled alloys on the AgZn–AuZn section [57 Liu]**

temperatures the calculated width is <10 at.-% Au and <40 at.-% Au (Table 22).

Yee and Liu [65 Yee] applied X-ray diffractometry to study the distribution of Ag, Au, and Zn atoms on the lattice sites in $AgAuZn_2$. Spectroscopic Zn and 99·99% Ag, Au were used to prepare an ingot which was annealed for 60 days at 500–600°C. Filings (20 μm) from the ingot were further annealed for 24 h at 600°C and then quenched in iced water. The distribution of atoms on lattice sites was determined after annealing at 100°C for times up to 240 h. It was found that the Zn atoms are completely ordered on their sub-lattice after 0.5 h at 100°C. The Ag and Au atoms are not completely ordered; 22 at.-% Ag and 3 at.-% Au are on Ag sites and 22 at.-% Au and 3 at.-% Ag on Au sites after 0.5 h at 100°C. After 40–240 h, at 100°C 24 at.-% Ag and 1 at.-% Au are on Ag sites and 24 at.-% Au, 1 at.-% Ag on Au sites. A 600°C quenched alloy of the composition $AgAuZn_2$ does not completely order on the Ag–Au sublattice when annealed at 100°C. On the other hand a slowly cooled alloy is completely ordered [57 Liu].

Figure 70 represents the phase relationships on the AgZn–AuZn section for alloys quenched from the temperatures indicated and characterized by room temperature X-ray powder diffraction analysis [57 Liu]. The binary βAgZn transforms to a CsCl-type structure, β′, when quenched from high temperatures; the addition of Au also produces the β′ structure on quenching. A two-phase β′ + ζ region, extending from 275 to 283°C, was found for the AgZn binary phase. Alloys with 0.5 at.-% Au transform to β′ on quenching, but gradually transform to the equilibrium ζ phase on annealing at 200°C. Figure 70 shows that alloys within a few atomic percent of $AgAuZn_2$ quench to the β″ structure at all temperatures from 650 to 200°C. [63 Mor] also found β″ for the $AgAuZn_2$ alloy after quenching from 600°C (Table 21). One would expect the β″ phase region from the phase diagram (Fig. 69) to be contained within the β″ phase region of quenched alloys (Fig. 70). This is so for the Ag-rich side, but the data [57 Liu] on the Au-rich side show a considerable shrinkage of the β″ phase region. The data of [57 Liu] should be regarded as requiring confirmation for quenched Au-rich alloys within 10 at.-% of the β″–β′ boundary below 300°C. No work has been reported on the lattice site occupancy of $AgAuZn_2$ when quenched to the Heusler-type structure from temperatures above the critical temperature for equilibrium conditions (330°C, Fig. 69).

Other work on the ternary equilibria

The earliest work [49 Rau] on the Ag–Au–Zn system was limited to an examination of the effect of 9·2 at.-% Ag on the transformation temperature of the disordered (Au) solid solution to the ordered, faulted $L1_2$ Au_3 Zn phase. The faulted $L1_2$ Au_3 Zn phase was detected in alloys heat treated from 250 to 400°C. By comparison with the binary Au_3Zn phase [49 Rau] concluded that Ag has little effect on the (Au) → $L1_2$ Au_3Zn transformation temperature.

[66 Kös] determined the approximate location of the (Ag–Au) solid solution region at room temperature (Fig. 71) using specific electrical resistance, Hall coefficient, and thermo-emf measurements on foil produced by melting >99·99 wt-% elements and heat treating at 500°C prior to rolling to foil.

The (Ag–Au) solid solution shows short-range order except for a narrow concentration range next to the Au–Zn binary. In this region the faulted $L1_2$ Au_3Zn phase and the doubly faulted $L1_2$ Au_4Zn phase were detected. [57 Wil] used the X-ray powder diffraction method on samples annealed for 8 days at 200°C to delineate the doubly-faulted $L1_2$ Au_4Zn phase region (Fig. 72).

[62 Pir] studied the mechanical relaxation characteristics of five ternary alloys along the 15 at.-% Zn section and of an additional four ternary alloys on the section from 50 Ag, 50 Au to Zn (up to 25 at.-% Zn). They commented that the Au_3Zn phase appears to extend a considerable distance into the ternary on the basis of damping measurements at low frequencies.

REFERENCES

49 Rau: E. Raub, P. Walter and A. Engel: *Z. Metall.*, 1949, **40**, 401–405

51 Mul: L. Muldawer: *J. Appl. Phys.*, 1951, **22**, 663–665

57 Liu: Y.-H. Liu and C.-C. Hsu: *Scienta Sinica*, 1957, **6**, 1013–1030

57 Wil: M. Wilkens and K. Schubert: *Z. Metall.*, 1957, **48**, 550–557

62 Pir: A. Pirson and C. Wert: *Acta Metall.*, 1962, **10**, 299–304

63 Mor: D. P. Morris and C. D. Price: *Nature*, 1963, **198**, 983–984

65 Yee: S.-S. Yee and Y.-H. Liu: *Chinese J. Phys.*, 1965, **21**, 1066–1078 (translation of *Acta Physica Sinica*, 1965, **21**, 839–848)

Table 22. Extent of the β″ phase region at room temperature

Ag-rich boundary, at.-% Au	*Au-rich boundary*, at.-% Au	*Measurement method*	*Reference*
12·5	36·8	X-ray lattice parameter	[57 Liu]
12·5	35·0	X-ray structure	[57 Liu]
12	38	Resistivity	[69 Bro]
14	36	Resistivity	[78 Bow]
<15	<35	X-ray structure	[78 Bow]
<10	<40	Thermodynamic calculation	[71 Mur]

66 Mul 1: L. Muldawer, *J. Appl. Phys.*, 1966, **37**, 2062–2066

66 Mul 2: L. Muldawer, *Acta Crystallogr.*, 1966, **20**, 594–595

66 Kös: W. Köster and R. Störing: *Z. Metall.*, 1966, **57**, 34–38

69 Bro: M. E. Brookes and R. W. Smith: *Scripta Metall.*, 1969, **3**, 667–669

71 Mur: Y. Murakami, S. Kachi, N. Nakanishi and H. Takehara: *Acta Metall.*, 1971, **19**, 97–105

78 Bow: R. Bowe and L. Muldawer: *Scripta Metall.*, 1978, **12**, 181–185

80 Mat: Y. Matsuo: *Trans. Japan Inst. Metals*, 1980, **21**, 174–178

83 Oka: H. Okamoto and T. B. Massalski: *Bull. Alloy Phase Diagrams*, 1983, **4**, 30–38

87 Oka: H. Okamoto and T. B. Massalski: 'Phase Diagrams of Binary Gold Alloys', 331–340; 1987, ASM International

Al–Au–Bi

[24 Tam] studied the partitioning of Au between immiscible liquid layers of Al and Bi at 715 ± 5°C. Chemical analysis of the two layers gave a Au content of 11·36 wt-% Au in the Al-rich layer and 0·36 wt-% Au in the Bi-rich layer. On the assumption that the Al–Bi liquid immiscibility gap in the ternary is parallel to the Al–Au and Au–Bi binary edges in the initial stages at 715°C, the Al-rich layer will have a composition 97·4 Al, 1·8 Au, 0·8 at.-% Bi and the Bi-rich layer will contain 19·3 Al, 0·3 Au, 80·4 at.-% Bi. The tie line joining the two liquid compositions is skewed towards the Au–Bi binary edge.

REFERENCE

24 Tam: G. Tammann and P. Schafmeister, *Z. anorg. Chem.*, 1924, **138**, 220–232

Al–Au–Cu

Raub and Walter [50 Rau] studied the effect of ternary additions on the order–disorder transformation of AuCu. The binary AuCu alloy has a transformation temperature of 410°C. Addition of 8·15 at.-% Al to AuCu lowers the ordering temperature. The disordered fcc ternary solid solution was present after 750 h at 400°C, but the AuCuI tetragonal superlattice phase was stable after 50 h at 360°C, and remained the stable phase for annealing times up to 500 h at 360°C. Chapman and Gillam [65 Cha] added 1·3 at.-% Al to the binary $AuCu_3$ alloy and found that Al also lowers the order–disorder transformation temperature of $AuCu_3$ by some 40°C.

REFERENCES

50 Rau: E. Raub and P. Walter: *Z. Metall.*, 1950, **41**, 240–243

65 Cha: N. R. Chapman and E. Gillam: *Acta Metall.*, 1965, **13**, 434–436

Al–Au–Ga

Carter *et al.* [72 Car] studied the $AuAl_2$–$AuGa_2$ section at 450°C using metallographic, X-ray diffraction, and nuclear magnetic resonance techniques. Ternary alloys were prepared by arc-melting $AuAl_2$ and $AuGa_2$; they were remelted in sealed quartz ampoules in a resistance furnace, homogenized for 10 days at 450°C and the brittle ingots crushed to −200 mesh powder. X-ray data indicated a solubility of at least 5 at.-% $AuAl_2$ in $AuGa_2$; this was confirmed by the NMR results. $AuGa_2$ is also soluble in $AuAl_2$, but the extent of the solid solution was not determined; it is probably >5 at.-% $AuGa_2$ in $AuAl_2$. Jan and Pearson [63 Jan] found a two-phase structure for an alloy containing 50 at.-% $AuAl_2$, 50 at.-% $AuGa_2$.

Both $AuAl_2$ and $AuGa_2$ have a cubic fluorite structure (cF12). Hoyt and Mota [78 Hoy] prepared three ternary alloys on the 92 at.-% Au section by arc-melting and subsequently annealing for 24 h at 500°C in argon. X-ray powder diffraction analysis showed that all alloys are in the (Au) solid solution phase region and that the lattice parameter varies linearly with composition from $Au_{92}Al_8$ to $Au_{92}Ga_8$.

REFERENCES

63 Jan: J.-P. Jan and W. B. Pearson: *Phil. Mag.*, 1963, **8**, 279–284

72 Car: G. C. Carter, I. D. Weisman, L. H. Bennett and R. E. Watson: *Phys. Rev. B*, 1972, **5**(9), 3621–3638 (X-ray data due to C. Bechtoldt)

78 Hoy: R. F. Hoyt and A. C. Mota: *J. Less-Common Metals*, 1978, **62**, 183–188

Al–Au–Gd

Brédimas *et al.* [84 Bré] found a solid solution series between the isostructural compounds Gd_2Au and Gd_2Al (orthorhombic Co_2Si-type structure, oP12, (C23) with four formula units in the unit cell). $Gd_2Au_xAl_{1-x}$ alloys were prepared from 99·99% Gd and 99·999% Al,Au and heat treated for 95 h at 600°C ($x = 0{\cdot}2, 0{\cdot}3, 0{\cdot}4, 0{\cdot}6, 0{\cdot}7$). The alloy with $x = 0{\cdot}8$ was held *in vacuo* at room temperature for 2 years. All ternary alloys were single-phase with the Co_2Si-type structure according to X-ray powder diffraction analysis.

REFERENCE

84 Bré: V. Brédimas, H. Gamari-Seale and C. Papatriantafillou: *Phys. Stat. Sol.* (b), 1984, **122**, 527–534

Al–Au–Hf

Marazza *et al.* [75 Mar] prepared the alloy Au_2AlHf by induction melting the 99·99% purity elements under argon. The alloy was annealed for 7 days at 500°C and examined by metallographic and X-ray powder diffraction techniques. The Au_2AlHf composition was two-phase; one phase was stated to be the Cu_2AlMn-type (cF16) with a = 0·655 nm. The second phase was not identified. Rieger *et al.* [65 Rie] prepared the alloys Au_2AlHf_3 (33·33 Au, 16·67 Al) and $AuAlHf_2$ (25 Au, 25 Al) by heating powders for 700 h at 850°C. Metallographically Au_2AlHf_3 was almost single phase and $AuAlHf_2$ was single phase. A Ti_2Ni-type structure (cF96) was quoted for each alloy composition with a = 1·209 nm for Au_2AlHf_3 and a = 1·210 nm for $AuAlHf_2$.

REFERENCES

65 Rie: W. Rieger, H. Nowotny and F. Benesovsky: *Montsh, Chem.*, 1965, **96**, 232–241

75 Mar: R. Marazza, R. Ferro and G. Rambaldi: *J. Less-Common Metals*, 1975, **39**, 341–345

Al–Au–In

Carter *et al.* [72 Car] studied the $AuAl_2$–$AuIn_2$ section at 450°C using metallographic, X-ray diffraction, and nuclear magnetic resonance techniques. Ternary alloys were prepared by arc-melting $AuAl_2$ and $AuIn_2$; they were remelted in sealed quartz ampoules in a resistance furnace, homogenized for 10 days at 450°C and the brittle ingots crushed to −200 mesh powder. NMR spectra indicated a solubility of 4·5 ± 0·5 at.-% $AuAl_2$ in $AuIn_2$ and 0·5 ± 0·3 $AuIn_2$ in $AuAl_2$ at 450°C. This data is consistent with metallographic and X-ray diffraction results. Jan and Pearson [63 Jan] found a two-phase structure for an alloy containing 50 at.-% $AuAl_2$, 50 at.-% $AuIn_2$ in agreement with the above results.

Both $AuAl_2$ and $AuIn_2$ have a cubic fluorite structure (cF12). Hoyt and Mota [78 Hoy] prepared three ternary alloys on the 92 at.-% Au section by arc-melting and subsequently annealing for 24 h at 500°C in argon. X-ray powder diffraction analysis showed that all alloys are in the (Au) solid solution phase region and that the lattice parameter varies linearly with composition from $Au_{92}Al_8$ to $Au_{92}In_8$.

REFERENCES

63 Jan: J.-P. Jan and W. B. Pearson: *Phil. Mag.*, 1963, **8**, 279–284

72 Car: G. C. Carter, I. D. Weisman, L. H. Bennett and R. E. Watson: *Phys. Rev. B*, 1972, **5**(9), 3621–3638

78 Hoy: R. F. Hoyt and A. C. Mota: *J. Less-Common Metals*, 1978, **62**, 183–188

Al–Au–Li

Schuster and Drews [86 Sch] synthesized the phase $Li_{1-x}AuAl_2$ and carried out X-ray diffraction studies on single crystals. It crystallizes in a filled CaF_2-type structure.

REFERENCE

86 Sch: H.-U. Schuster and J. Drews: *Z. Krist.*, 1986, **174**, 185 (abstract only)

Al–Au–Mg

An alloy containing 3·5 at.-% Al, 30 at.-% Au, 66·5 at.-% Mg, heat-treated under an argon pressure of 600 torr for 0.2 h at 600°C, had a single phase structure. X-ray analysis [69 Loo] confirmed the phase to be a ternary compound analogous to $Au_{0·9}Mg_2Si_{0·1}$ which has an orthorhombic $PbCl_2$-type structure (oP12).

REFERENCE

69 Loo: N. Van Look and K. Schubert: *Metall.*, 1969, **23**(1), 4–6

Al–Au–Mn

The only work reported on the Al–Au–Mn system concerns the alloy Au_2MnAl. Morris *et al.* [59 Mor] quenched powder samples after heat treatment of 3 h at 800°C, 40 h at 580°C and 88 h at 160°C. All showed single phase structures of the bcc CsCl-type with a lattice parameter of 0·317 9 nm at 23°C. No superlattice lines were observed. [63 Mor] used the more refined X-ray diffractometer method to examine powder samples annealed for 24 h at 700°C and cooled at different rates. Cooling to room temperature over 6 h gave odd superlattice lines indicative of a Heusler structure with lattice parameter a = 0·635 8 nm. Cooling to ambient over 3 days gave additional lines tentatively identified as the Al_2Au phase, a = 0·599 3 nm (CaF_2-type, cF12). This suggests that the composition Au_2MnAl is two-phase at low temperatures. The 700°C quenched alloy is single-phase, but [63 Mor] did not detect long-range order of the Mn and Al atoms that distinguish the Heusler structure. However Oxley *et al.* [63 Oxl] quenched Au_2MnAl from 700°C and annealed for several days at about 200°C; in this condition X-ray analysis showed the Heusler structure but long-range order of the Mn and Al atoms was incomplete.

REFERENCES

59 Mor: D. P. Morris, R. R. Preston and I. Williams: *Proc. Phys. Soc.*, 1959, **73**, 520–523

63 Mor: D. P. Morris, G. W. Davies and C. D. Price: *J. Phys. Chem. Solids*, 1963, **23**, 109–111

63 Oxl: D. P. Oxley, R. S. Tebble and K. C. Williams: *J. Appl. Phys.*, 1963, **34**, 1362–1364

Al–Au–Nb

In studying solid solution formation between Cr_3Si-type phases (cP8) von Philipsborn [70 Phi] prepared a $Nb_3Au_{0{\cdot}5}Al_{0{\cdot}5}$ alloy by arc-melting Nb_3Au and Nb_3Al on a water cooled Cu hearth under argon. Severe melt losses occurred in the preparation of this ternary alloy and a sample was annealed at 1000°C for 1 day only. X-ray powder diffraction analysis showed the presence of a Cr_3Si-type phase with a lattice parameter $a = 0{\cdot}521 \pm 0{\cdot}001$ nm together with two unidentified phases. The Cr_3Si-type phase has the same lattice parameter as Nb_3Au.

REFERENCE

70 Phi: H. von Philipsborn: *Z. Krist.*, 1970, **131**, 73–87

Al–Au–Pb

[24 Tam] studied the partitioning of Au between immiscible liquid layers of Al and Pb at 697 ± 5°C. Chemical analysis of the two layers gave a Au content of 11·0 wt-% in the Al-rich layer and 0·03 wt-% Au in the Pb-rich layer. On the assumption that the Al–Pb liquid immiscibility gap is parallel to the Al–Au and Au–Pb binary edges in its initial stages at 697°C in the ternary, the Al-rich layer will have a composition 98·0 Al, 1·7 Au, 0·3 at.-% Pb and the Pb-rich layer will contain 3·1(4) Al, 0·03 Au, 96·8(3) at.-% Pb. The tie line joining the two liquid compositions is skewed towards the Au–Pb binary edge.

REFERENCE

24 Tam: G. Tammann and P. Schafmeister, *Z. anorg. Chem.*, 1924, **138**, 220–232

Al–Au–Si

Selikson and Longo [64 Sel] reported that a black phase formed when Au wire was thermocompression-bonded to Al metallization on Si transistors. This phase was distinguished from the purple $AuAl_2$ that forms with Al–Au diffusion couples. It was suggested that Si is soluble in $AuAl_2$ and catalyses its formation. Similar observations have been reported by [62 Sch]. Although no details were given [64 Sel] states that the addition of 15 wt-% Al to the Au–Si eutectic alloy (producing an alloy with 39 at.-% Au, 9 at.-% Si, 52 at.-% Al) lowers the eutectic temperature by 10°C (i.e. to 353°C). This is difficult to reconcile with the likely equilibria in the ternary system as this alloy composition lies in the $AuAl_2$–AuAl–Si phase region where it may be anticipated that the reaction temperatures will be far higher than 353°C. [64 Sel] also added 20 wt-% Au to the Al–Si eutectic alloy (producing an alloy with 3·3 at.-% Au, 10·9 at.-% Si, 85·8 at.-% Al), and stated that this Au addition lowered the Al–Si eutectic temperature by 10°C (i.e. to 567°C). This would appear to be a reasonable result.

Philofsky [70 Phi] appears to be the first to attempt to quantify the effect of Si on phase formation in Au–Al diffusion couples. Bonds were formed between 25 thou Au wire and 25 thou Al–5 wt.-% Si wire (Al–4·8 at.-% Si) and annealed in forming gas at 250°C (21 h), 300°C (4·4 and 16 h), and 400°C (1·67 and 11·67 h). An electron microprobe trace across the 400°C/1·67 h couple indicated a phase next to the Al–Si alloy containing 23·1 at.-% Au, 62·7 at.-% Al, 14·2 at.-% Si. On the basis of this data [70 Phi] suggests that the ternary compound $AuSiAl_4$ is formed (16·7 Au, 66·7 Al, 16·7 Si). It is also noted that the Au_2Al compound contains 1 to 2 wt-% Si (up to the composition 60·5 Au, 30·2 Al, 9·3 Si). No other report of a ternary compound is to be found in the literature; the data of [70 Phi] requires confirmation. [77 Loz] determined the 1 100°C liquidus isotherm; at this temperature the liquid phase is in equilibrium with (Si). Ten binary Al–Au alloys were equilibrated with Si at 1 100°C, the liquid sampled and analysed. The isotherm (Fig. 74) is a replot in atomic percent of the original weight percent diagram. Whether the increased solubility of Si in the liquid near to the Au–Si binary edge is real or not requires verification. The data of (77 Loz] extrapolate to the 1 100°C liquidus composition adopted by [83 Oka]. The bulge in the liquidus, giving a concave curve relative to the Al–Au binary edge, suggests that the section $AuAl_2$–Si could be a pseudobinary eutectic type. The partial ternary system Al–$AuAl_2$–Si requires examination to determine the $AuAl_2$–Si equilibria and to investigate the reported occurrence [70 Phi] of a ternary compound.

Lozovsky and Kolesnichenko [81 Loz] measured the solubility of Al in Si along the sections Si–97·9 Au, 2·1 Al; 81·6 Au, 18·4 Al; 55·2 Au, 44·8 Al; 24·2 Au, 75·8 Al. Their data, scaled from the original figure, is given in Table 23. Retrograde solubility is evident for each section.

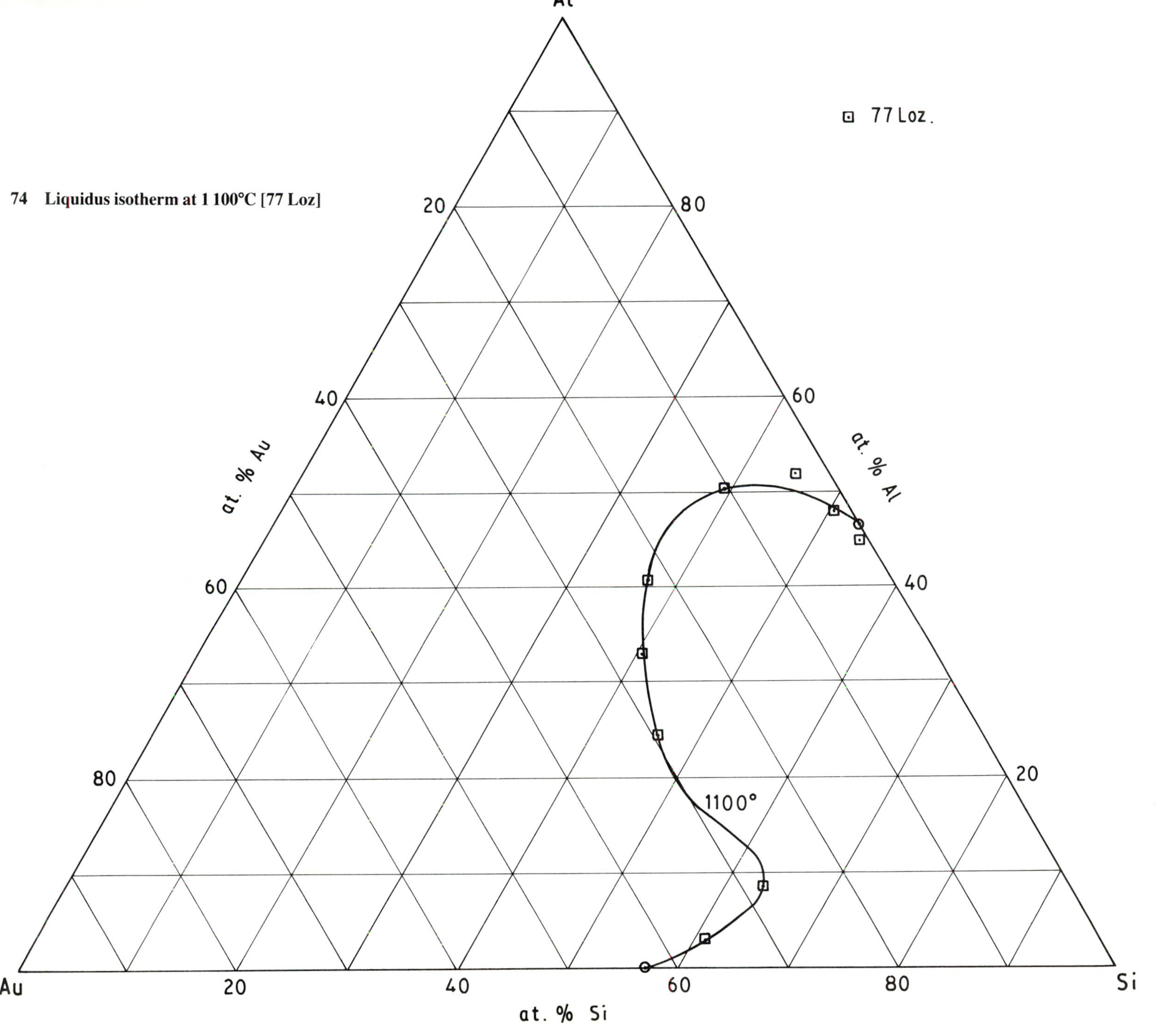

74 Liquidus isotherm at 1 100°C [77 Loz]

Table 23. Solubility of Al in Si

Section	*Temperature*, °C	at.-% Al × 10^{-4}
Si — 97·9Au, 2·1Al	1 300	0·64
	1 200	0·40
	1 100	0·16
	1 000	0·05
Si — 81·6Au, 18·4Al	1 300	10·6
	1 200	8·0
	1 100	4·0
	1 000	1·2
Si — 55·2Au, 44·8Al	1 300	86
	1 200	80
	1 100	44
Si — 24·2Au, 75·8Al	1 300	200
	1 200	280
	1 100	220
	1 000	140

REFERENCES

62 Sch: R. Schmidt: *IRE Trans. Electron. Devices*, 1962, **ED-9**, 506

64 Sel: B. Selikson and T. A. Longo: *Proc. IEEE*, 1964, **52**, 1638–1641

70 Phi: E. Philofsky: *Solid-State Electron.*, 1970, **13**, 1391–1399

77 Loz: V. N. Lozovsky and N. F. Politova: *Izvest. Akad. Nauk, SSSR, Metally.*, 1977, (3), 36–37

81 Loz: V. N. Lozovsky and A. I. Kolesnichenko, *Izvest. Akad. Nauk. SSSR, Neorg. Materialy.*, 1981, **17**, 737–738

83 Oka: H. Okamoto and T. B. Massalski: *Bull. Alloy Phase Diagrams*, 1983, **4**, 190–198

Al–Au–Sn

The only publication [1894 Hey] on the Al–Au–Sn system dealt with the depression of the freezing point of a 1·98 at.-% Al, 98·02 at.-% Sn alloy on addition of Au up to 1·55 at.-% (Fig. 75), and similar measurements for a 3·88 at.-% Al, 96·12 at.-% Sn alloy on addition of Au up to 2·36 at.-% (Fig. 76). Both figures have a maximum temperature identical with the measured melting point of the pure Sn used [1894 Hey]. In both sections the curve rises to the maximum as a linear function of the atomic percent of Au, and also falls from the maximum in a similar manner. [1894 Hey] interpreted the data to show the formation of $AuAl_2$ in these alloys, noting that the maximum point on Figs. 75 and 76 occurs at a Au : Al ratio of 1:2.

Further interpretation of the data is possible. It is considered that there is a pseudobinary system between $AuAl_2$ and (Sn). The virtual identity of temperatures found by [1894 Hey] for the melting point of Sn and the maximum temperature on the two sections indicates the presence of a degenerate eutectic in the $AuAl_2$–Sn pseudobinary system. The two lines intersecting at a maximum point in the two sections represent secondary separation of $AuAl_2$ + (Sn) from the melt. In Fig. 76 the horizontal section, at 229°C, extends to 0·91 at.-% Au. The temperature of 229°C is identical with that measured for the Al–Sn eutectic temperature [1894 Hey]. It can be concluded that the partial ternary system $AuAl_2$–Al–Sn contains a ternary eutectic melting at the same temperature as the Al–Sn eutectic and with a ternary eutectic composition virtually identical with that of the binary Al–Sn eutectic. (Plotting the binary Al–Sn data of [1894 Hey] gives a eutectic composition of 2·0 at.-% Al.) If this interpretation of the results is correct there should be a tie line joining $AuAl_2$ to the ternary eutectic composition (i.e. virtually 2·0 at.-% Al, 98 at.-%Sn), and this tie line should intersect Fig. 70 at 0·91 at.-% Au (point a). Calculation of the intersection of the tie line $AuAl_2$–2 at.-% Al, 98 at.-% Sn with the section given in Fig. 76 gives a composition of 0·93 at.-% Au (point a in Fig. 77). It appears that there is a pseudobinary system between $AuAl_2$ and (Sn), that it is of the degenerate eutectic type, and that it forms with the $AuAl_2$–Al and Al–Sn eutectic systems a ternary eutectic system. The ternary eutectic has a temperature and composition virtually identical to those of the Al–Sn system. This implies that the monovariant curve eE (Fig. 77), which is practically coincident with ee_1 due to the near identity of E and e_1, can be traced in terms of temperatures along eE. If the lines $e_1$df (Fig. 75) and abc (Fig. 76) represent the secondary separation of $AuAl_2$ + (Sn) as the melt descends from e to E(e_1), then the tie line at a fixed temperature will be defined by $AuAl_2$–b–d. This tie line extrapolates to point g on the Au–Sn binary. Point h on curve eE will almost coincide with g. The 231°C tie line extrapolates to 0·8 at.-% Al, 99·2 at.-% Sn, the 230°C tie line to 1·4 at.-% Al and the 229°C tie line to 2·0 at.-% Al. These values are within 0·1 at.-% of the binary Al–Sn liquidus values given by [1894 Hey]. These results reinforce the conclusion that the monovariant curve eE runs extremely close to the Al–Sn binary edge.

Preliminary work [85 Hum] indicates that the section AuSn–Al is not a pseudobinary. This agrees with the interpretation given of the data of [1894 Hey]. Recent work [87 Cha] indicates that alloys on the $AuAl_2$–Sn section show a characteristic eutectic behaviour with a eutectic temperature of 230 ± 0.5°C (Fig. 78).

REFERENCES

1894 Hey: C. T. Heycock and F. H. Neville: *J. Chem. Soc.*, 1894, **65**, 65–76

85 Hum: G. Humpston: The constitution of some ternary Au-based solder alloys, 1985, PhD thesis, Brunel University

87 Cha: V. Chao and F. Hayes: unpublished work, 1987, UMIST

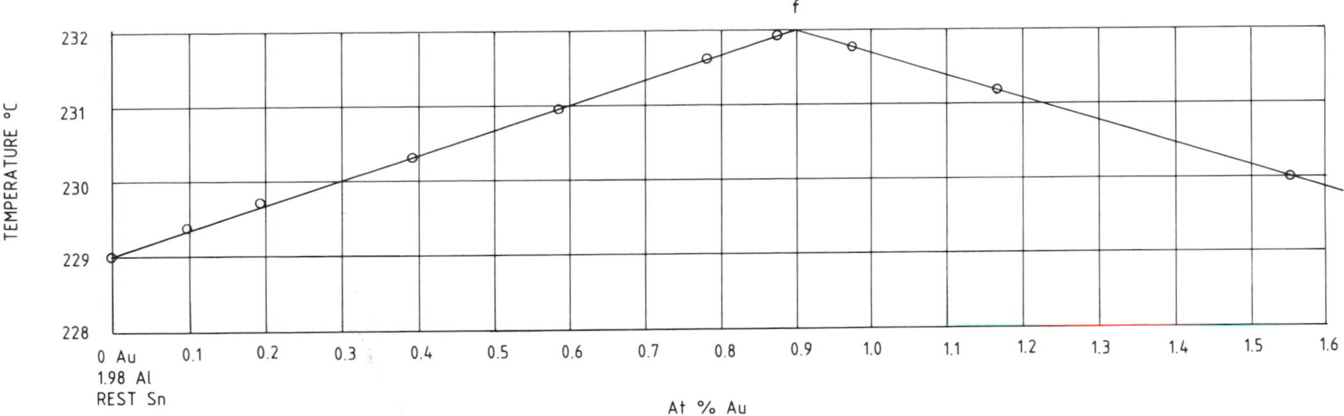

75 Experimental data [1894 Hey] relating the depression of the freezing point to the alloy composition for the section 1·98 at.-% Al, 98·02 at.-% Sn–Au

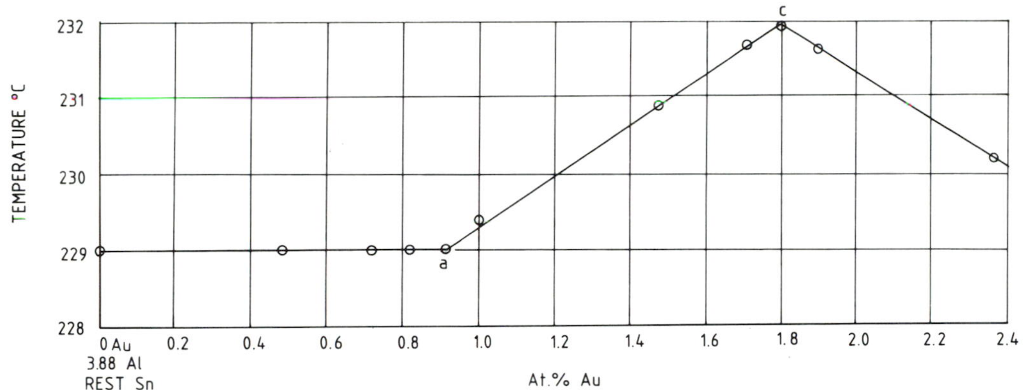

76 Experimental data [1894 Hey] relating the depression of the freezing point to the alloy composition for the section 3·88 at.-% Al, 96·12 at.-% Sn-Au

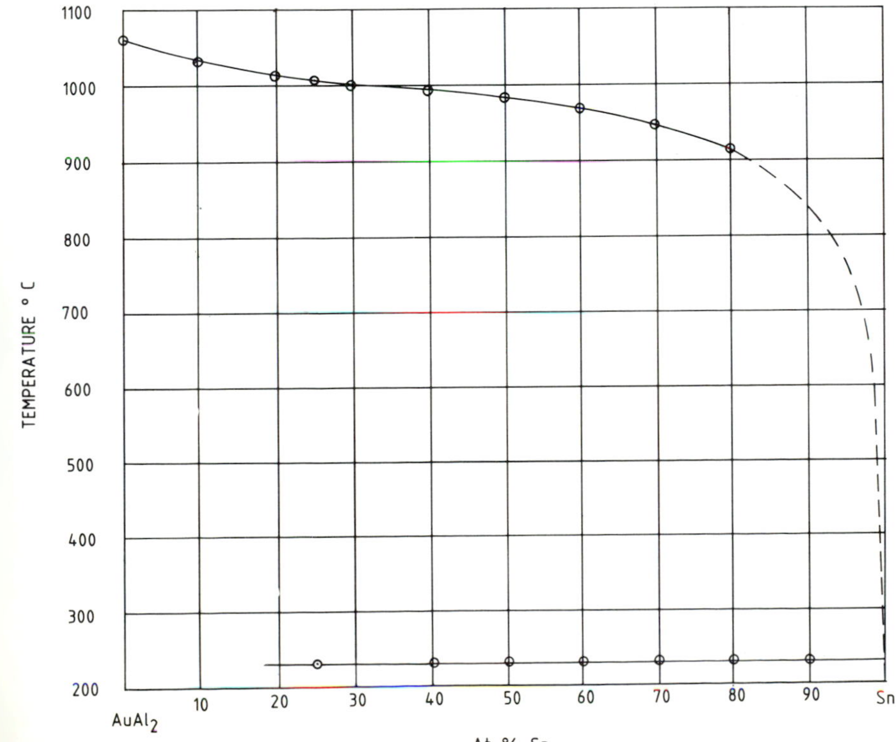

78 The $AuAl_2$–Sn section [87 Cha]

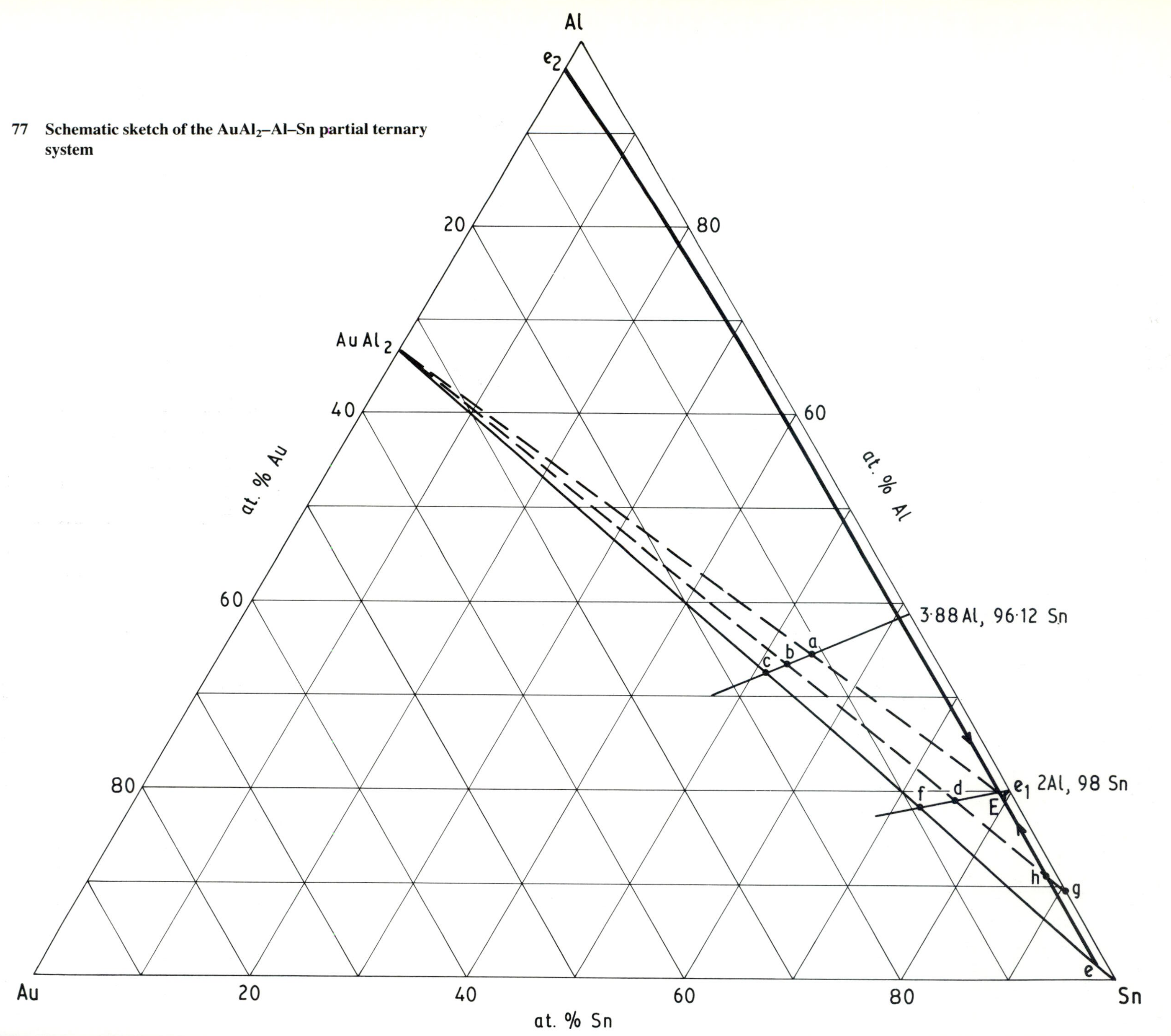

77 Schematic sketch of the $AuAl_2$–Al–Sn partial ternary system

Al–Au–Ti

Marazza *et al.* [75 Mar] prepared the alloy Au_2AlTi by induction melting the 99·99% purity elements under argon. Samples were annealed for 7 days at 500°C and examined by metallographic and X-ray powder diffraction techniques. One sample was chemically analysed as containing 50·1 Au, 24·1 Al, 25·8 Ti. Metallographically it contained one phase outlined by a solidified intergranular liquid. A CsCl-type structure (cP2) with a = 0·3200 nm was identified by X-ray analysis. Another sample of Au_2AlTi, also nearly single phase, was quoted as having a Cu_2AlMn-type structure (cF16) with double the lattice parameter. It is not clear how [75 Mar] concluded that the Cu_2AlMn-type structure was present in one preparation as no details are given of the presence of odd superlattice lines to justify the Cu_2AlMn-type structure.

Jorda *et al.* [87 Jor] prepared 49 ternary and four binary alloys from 99·99% Au, 99·999% Al and 99·9% Ti by arc-melting under argon. All alloys were annealed at 500°C for 7 days under flowing argon and then quenched. The 500°C isothermal section, derived by X-ray powder diffraction, optical metallography and microhardness techniques, is given in Fig. 79. It should be noted that the homogeneity range of some binary compounds (Ti_3Au, $TiAu_2$, $TiAu_4$, AuAl, Au_4Al) have been exaggerated and that the compound Au_5Al_2 has not been included. As only six ternary alloys contained more than 50 at.-% Au the equilibria in the Au-rich region must be regarded as tentative.

[87 Jor] confirmed the occurrence of the ternary compound $AlAu_2Ti$ and state that it forms by peritectic reaction, L + (AuTi) = $AlAu_2Ti$, at 1000°C. It has the CsCl-type structure with a = 0·3198(2) nm, in excellent agreement with one of the samples prepared by [75 Mar]. No superlattice lines were detected. An equiatomic ternary compound, AlAuTi, was also found by [87 Jor]. It melts congruently at 1180°C and has a range of homogeneity (Fig. 79) that extends to 25 at.-% Al on the 33·3 at.-% Au section. AlAuTi has a hexagonal structure, a = 0·44075(8) and c = 0·5829(1) nm, and is considered to be of the $InNi_2$-type (hP6). The two ternary compounds dominate the equilibria in the ternary system, apart from equilibria associated with the Al-rich, Au-rich and Ti-rich areas.

REFERENCES

75 Mar: R. Marazza, R. Ferro and G. Rambaldi: *J. Less-Common Metals*, 1975, **39**, 341–345

87 Jor: J. L. Jorda, J. Muller, H. F. Braun and C. Susz: *J. Less-Common Metals*, 1987, **134**, 99–107

Al–Au–Zr

Marazza *et al.* [73 Mar] prepared the alloy Au_2AlZr from elements >99·9% purity by induction-melting under argon. The alloy was annealed for 150 h at 500°C and examined by metallographic and X-ray powder diffraction techniques. Only the X-ray data are reported. The structure is either CsCl-type with a = 0·3281 nm or Cu_2AlMn-type with double the lattice parameter; it was not possible to distinguish the structure type by the powder diffraction method. Rieger *et al.* [65 Rie] prepared the compositions Au_2AlZr_3 (33·33 Au, 16·67 Al) and $AuAlZr_2$ (25 Au, 25 Al) by heating powders for 700 h at 850°C. Metallographically Au_2AlZr_3 was almost single phase with a Ti_2Ni-type structure (cF96), a = 1·227 nm. $AuAlZr_2$ was two-phase and contained two Ti_2Ni-type phases, one of which had a = 1·228 nm.

REFERENCES

65 Rie: W. Rieger, H. Nowotny and F. Benesovsky: *Monatsh. Chem.*, 1965, *96*, 232–241

73 Mar: R. Marazza, G. Rambaldi and R. Ferro: *Atti. Accad. Naz. Lincei, Rend., Classe Sci. Fis. Mat. e Nat.*, 1973, **55**, 518–521

As–Au–Eu
Au–Eu–P
Au–Eu–Sb

Tomuschat and Schuster [81 Tom] report the occurrence of ternary compounds EuBX where B = Group IB element and X = Group VB element. Gold-containing compounds correspond to EuAuP, EuAuAs and EuAuSb. The analogous copper- and silver-containing compounds also include EuAgBi and EuCuBi. The compounds were prepared by heating the elements, of unstated purity, in an alumina crucible under an argon blanket for 12–16 h at 1000°C. The product was finely ground and heat treated at 1100°C for 14 h. All compositions were brittle grey-blue powders. X-ray diffraction analysis showed that the ternary compounds are hexagonal with a Ni_2In-type structure (hP6). Lattice constants are given in Table 24. With two formula units to the unit cell the atom positions are

2 Eu	(a)	0	0	0,	0	0	½
2 Au in	(d)	⅓	⅔	¾,	⅔	⅓	¼
2 As, P, Sb in	(c)	⅓	⅔	¼,	⅔	⅓	¾

Table 24. Lattice constants

	a, nm	c, nm	c/a
AsAuEu	0·4445	0·8285	1·86
AuEuP	0·4313	0·8258	1·91
AuEuSb	0·4669	0·8486	1·82

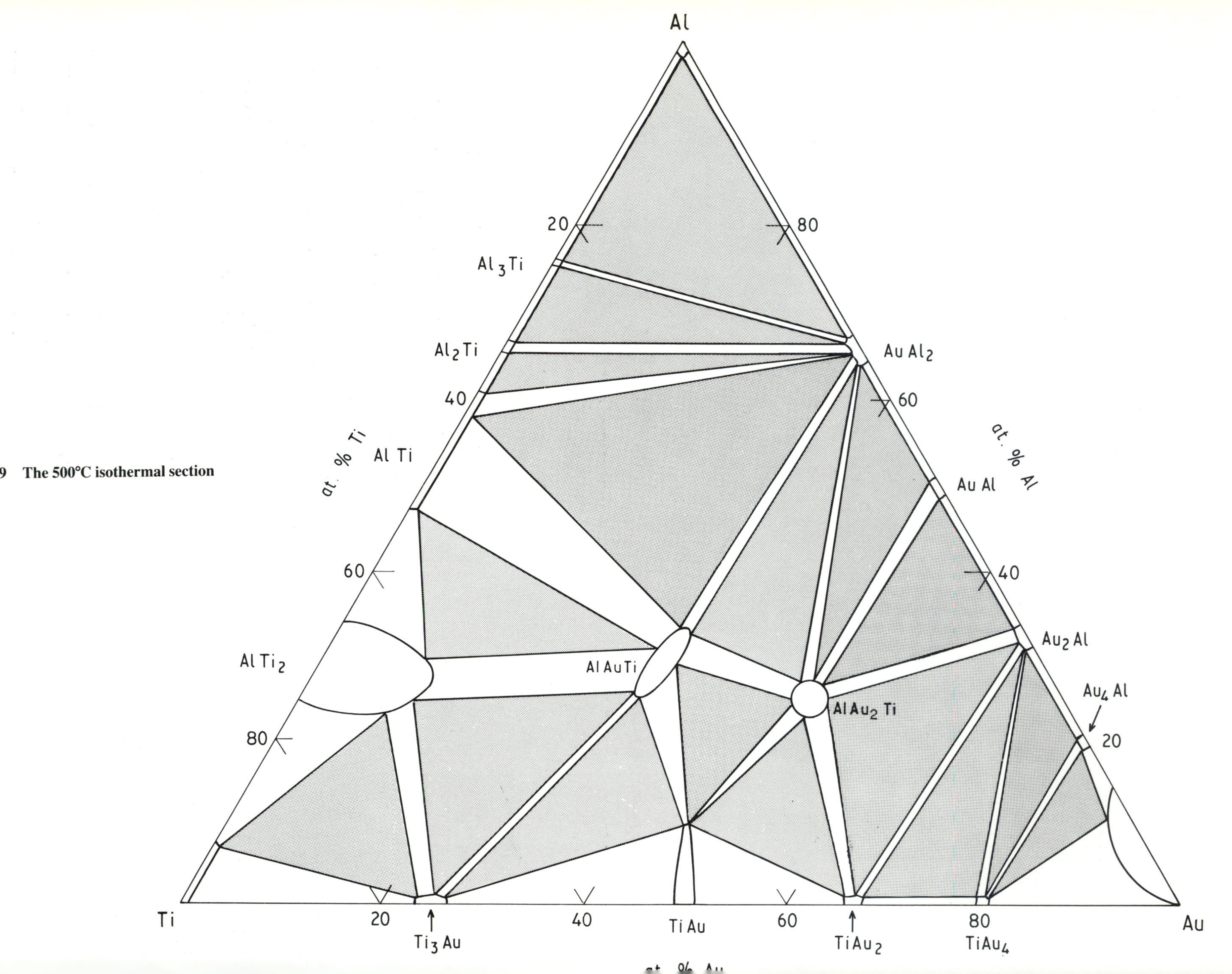

79 The 500°C isothermal section

REFERENCE

81 Tom: C. Tomuschat and H.-U. Schuster: *Z. Naturf.*, 1981, **86b**, 1193–1194

As–Au–Ga

INTRODUCTION

The only published data dealing with the equilibria of bulk samples in the As–Au–Ga system is due to [67 Pan] and [86 Tsa]. The semiconductor literature contains numerous papers relating to the interaction during heat treatment in an open system of thin gold films deposited on single crystal GaAs substrates [73 Nak, 75 Mag, 79 Kum, 80 Van, 80 Mil, 82 Ker, 83 Leu 1, 83 Leu 2, 83 Yos, 84 Yos, 84 Zen, 86 Pug]. As they do not contribute substantially to an elucidation of the ternary phase equilibria they are not included in this assessment. [67 Pan] studied 26 ternary alloys with compositions on the sections GaAs–50 at.-% Au, 50 at.-% As, GaAs–Au, GaAs–77·4 at.-% Au, 22·6 at.-% Ga and GaAs–AuGa. One alloy was studied on the GaAs–$AuGa_2$ section. It was established that the sections GaAs–$AuGa_2$ and GaAs–AuGa are pseudobinary eutectic diagrams. These findings were confirmed by [86 Tsa]. The section GaAs–Au is not pseudobinary. The ternary system is dominated by the primary phase field of GaAs. Liquidus isotherms for temperatures down to 700°C are skewed towards the Au–Ga binary edge, reflecting the absence of a pseudobinary equilibrium between GaAs and Au. [67 Pan] interpreted the data as indicating the presence of a ternary eutectic reaction in the GaAs–Au–As region of the phase diagram with a maximum temperature occurring on the monovariant curve located on the Ga-rich side of the GaAs–Au section (Fig. 80). No comment is made by [67 Pan] on the equilibria associated with the maximum on the monovariant curve, but it can be assumed to be a critical tie line, l $\rightleftharpoons$ (Au) + GaAs, where (Au) is a solid solution containing an appreciable amount of Ga and a small amount of As. It is suggested in this assessment that the data can be interpreted to show a transition reaction, L_{U_1} + (As) $\rightleftharpoons$ (Au) + GaAs, rather than a eutectic reaction (Fig. 81). In contrast with [67 Pan], [86 Tsa] conclude that the section GaAs–Au is a pseudobinary section. Their interpretation of the data can equally well agree with the suggested presence of a transition reaction. Further work is needed to clarify the equilibria.

[88 Sch] calculated an isothermal section at 25°C. It is essentially a triangulation of the ternary system and shows the same three-phase regions as those suggested in Fig. 81, although [88 Sch] regards the GaAs–Au section as a pseudobinary.

BINARY SYSTEMS

The Au–Ga diagram asssessed by [87 Oka] is accepted. The As–Ga diagram of [74 Pan] is accepted, as is the As–Au diagram due to [76 Gat]. It should be recalled that the high vapour pressure of As above As-rich alloys means that the temperature–concentration diagrams for both the As–Au and As–Ga systems are projections of *T–X–P* diagrams for variable pressure conditions. Elemental As at 1 atm pressure sublimes at 603°C; a liquid–solid–vapour triple point is located at 817°C and 37 atm.

SOLID PHASES

Table 25 summarizes crystal structure data for the binary phases. No ternary compounds were reported [67 Pan].

Table 25. Crystal structures

Solid phase	*Prototype*	*Lattice designation*
(As)	As	hR2
(Au)	Cu	cF4
(Ga)	Ga	oC8
α′(Au, 12·75–14·2 Ga)	Ni_3Ti	hP16
β(Au, 20·4–22·1 Ga)		Hexagonal
β′(Au, 21·3–24·8 Ga)		Orthorhombic
γ(Au, 29·8–31·0 Ga)		Orthorhombic
AuGa	MnP	oP8
$AuGa_2$	CaF_2	cF12
GaAs	ZnS	cF8

PSEUDOBINARY SYSTEMS

The AuGa–GaAs section is a pseudobinary eutectic system (Fig. 82). The eutectic temperature cannot be given accurately [67 Pan] due to undercooling of the melt during thermal analysis. The form of the liquidus curve indicates a eutectic composition close to AuGa. Assuming 0·5 at.-% As in the eutectic alloy the calculated eutectic temperature is only 1°C below the congruent melting point of AuGa. For a eutectic composition containing 1·0 at.-% As the eutectic temperature is 2°C below the melting point of AuGa. Figure 82 has been drawn to indicate a eutectic temperature of 460°C, compared with 461·3°C for AuGa [87 Oka]. The pseudobinary character of the AuGa–GaAs section was confirmed by X-ray powder diffraction analysis and electron beam microprobe examination of the phases present in alloys used for thermal analysis. Tsai and Williams [86 Tsa] (Table 26) prepared two alloys on this section and found GaAs + AuGa by X-ray powder diffraction analysis of alloys cooled from 1 000°C to 25°C over a period of 1 week or of one alloy cooled from 1 000°C to 600°C and water quenched. The $AuGa_2$–GaAs section is stated to be a pseudobinary system [67 Pan] on the basis of the study of one alloy. Thermal analysis gave a liquidus and a solidus effect, the solidus temperature being nearly identical to the congruent melting temperature of $AuGa_2$. X-ray powder diffraction and microprobe results showed $AuGa_2$ and GaAs as the phases present. It could be anticipated, from the melting point of $AuGa_2$ and its enthalpy of formation, that the eutectic temperature would be close to the melting point of $AuGa_2$. The eutectic composition is also assumed to be $<0·5$ at.-% As along the $AuGa_2$–GaAs section. [86 Tsa] (Table 26) prepared two alloys on this

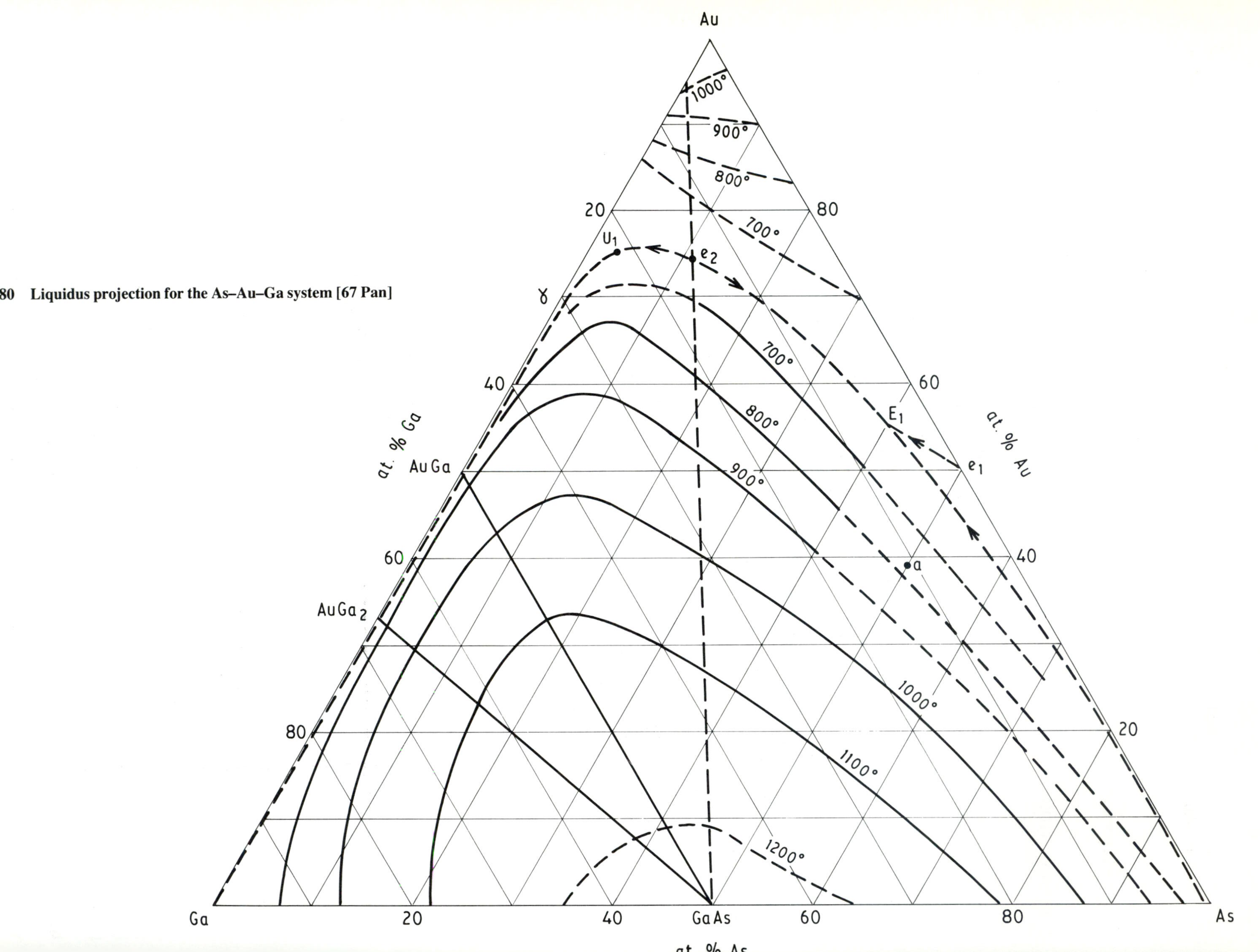

80 Liquidus projection for the As–Au–Ga system [67 Pan]

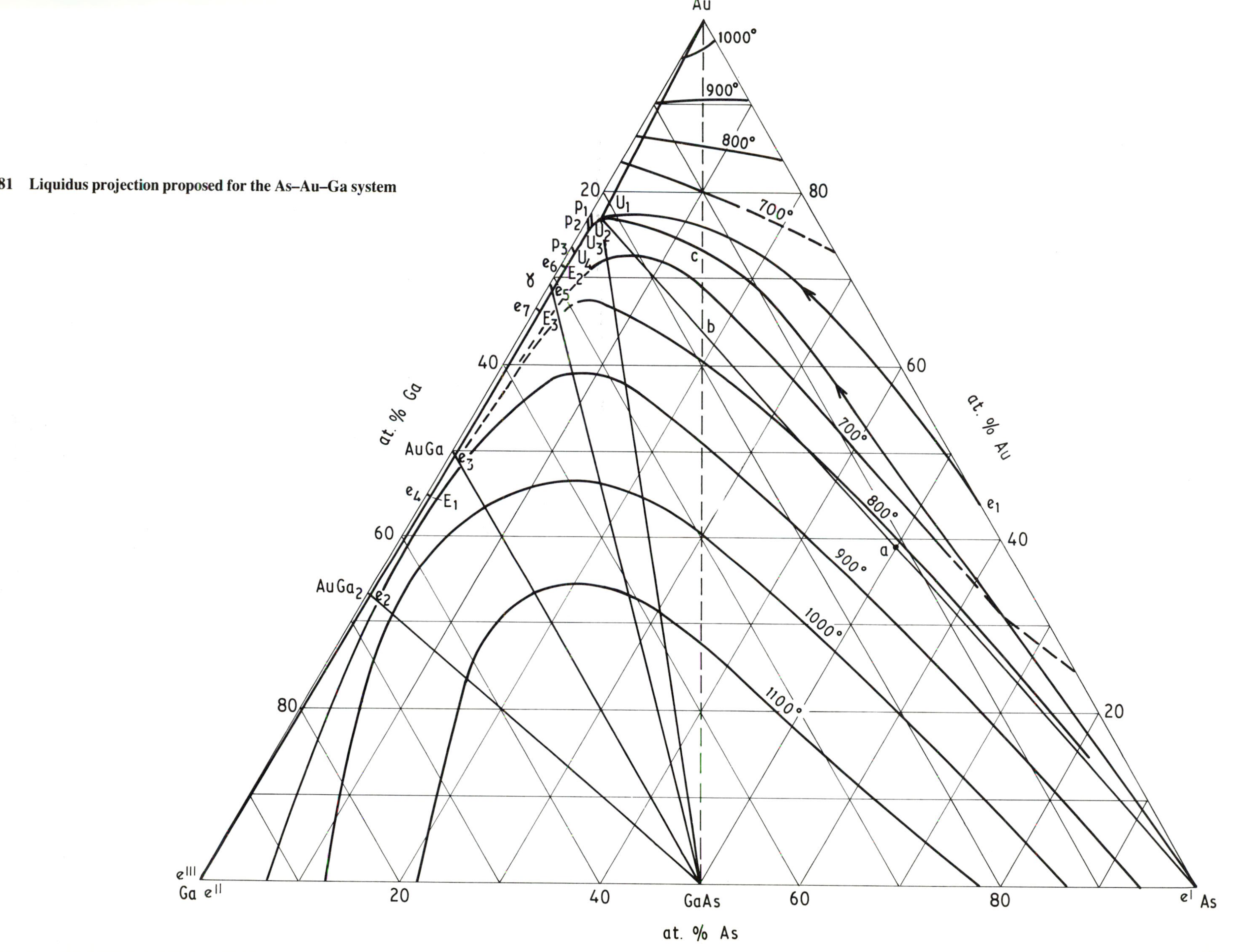

81 Liquidus projection proposed for the As–Au–Ga system

section and noted the presence of $AuGa_2$ and GaAs as the only phases after slow cooling from 1 000°C or water quenching from 600°C.

Although not studied by [67 Pan] the section γ′–GaAs is also likely to be a pseudobinary eutectic. This has been confirmed by [86 Tsa] (Table 26) on the basis of the study of three alloys. [86 Tsa] also claim that the sections GaAs–β′ and GaAs–Au are pseudobinary sections. This conclusion is not accepted; it is discussed in the following section.

INVARIANT EQUILIBRIA

[67 Pan] prepared alloys using >99·99% As, >99·99% Ga and >99·9% Au. For most of the work GaAs, prepared by passing As vapour over liquid Ga, was used instead of the elements. Alloys were melted in evacuated fused-silica capsules prior to thermal analysis, using cooling/heating rates from 1°C min^{-1} to 5°C min^{-1}. Most runs were made by cooling from the melt, but eight alloys were also examined using heating curves. X-ray and electron beam microprobe analysis of the DTA samples provided evidence of the phases present. Figure 80 reproduces the ternary equilibria proposed by [67 Pan]. A ternary eutectic reaction, $L_{E_1} \rightleftharpoons$ (Au) + (As) + GaAs, is tentatively identified as taking place at 590°C. A maximum on the monovariant curve $U_1e_2E_1$ implies the presence of a critical tie line joining GaAs to liquid e_2 and a AuGa solid solution containing a small amount of As. This tie line is formed when the (Au) liquidus surface meets the GaAs liquidus surface with falling temperature such that the tie lines e_2–(Au) and e_2–GaAs form a critical tie line (Au)–e_2–GaAs. As originally presented by [67 Pan] the (Au) liquidus and the GaAs liquidus meet on the Au–GaAs section (Fig. 83*a*) at 640°C and 73·0 at.-% Au. The temperature of the critical tie line must be somewhat higher than 640°C but, judging from the 700°C isotherms in Fig. 80, well below 700°C. [67 Pan] examined alloys on the Au–GaAs section by microprobe analysis. In all alloys GaAs and Au were detected but the As phase was only identified in alloys containing >50 at.-% Au. In terms of a ternary eutectic (Fig. 80) this is understandable since alloys rich in GaAs will solidify by primary separation of GaAs followed by secondary separation of GaAs + (Au) before the small amount of liquid that remains solidifies as the ternary eutectic. The closer the critical tie line (Au)–e_2–GaAs is to the Au–GaAs section, the less ternary eutectic liquid will be available in GaAs-rich alloys. The difficulty with adopting this interpretation of the data is that it does not accord with the DTA data for the 75 at.-% Au alloy (Fig. 83*a*). In duplicate cooling runs the liquidus was found to be 628°C and 615°C with a secondary separation at 598°C and 610°C respectively. As all other liquidus determinations fall reasonably close to the smoothed liquidus curve, with a maximum divergence for the 20 at.-% Au alloy of 22°C, it is surprising to find reported values of 628°C and 615°C against the value of 664°C required as the liquidus in Fig. 83*a*. A more fundamental query arises with the section GaAs–50 at.-% Au, 50 at.-% As (Fig. 84). Point ‘a’ on the ternary eutectic plane at 590°C must lie on the tie line joining As to E_1. The ternary eutectic E_1 was tentatively placed at about 40 at.-% As, 55 at.-% Au, 5 at.-% Ga [67 Pan]. The composition of point ‘a’ is assessed as 39 at.-% Au (Fig. 84). The tie line from As through point ‘a’ extrapolates into the ternary between the 700 and 800°C isotherms. This is an impossibility for a 590°C eutectic temperature. Thus the eutectic composition E_1 (Fig. 80) does not correspond with the section (Fig. 84). Any attempt to produce agreement between the ternary eutectic composition and Fig. 84 runs counter to the liquidus isotherms deduced from both Figs. 83*a* and 84. [67 Pan] recognized that further study was needed to clarify the ternary diagram in the region of the As–Au binary.

By interpreting the data of [67 Pan] as evidence for a transition reaction, L_{U_1} + (As) $\rightleftharpoons$ (Au) + GaAs, at 590°C some of the difficulties discussed above are removed. Figure 81 is a representation of the ternary equilibria on

Table 26. Equilibria according to [86 Tsa]

Alloy section	*Alloy*	*Alloy composition*, at.-%			*Temperature*, °C	*Dead space*, cm^3	*Phases identified*
		As	*Au*	*Ga*			
$GaAs–AuGa_2$	A	28·6	14·3	57·1	25,600	4	GaAs, $AuGa_2$
	B	12·5	25·0	62·5	25	4	GaAs, $AuGa_2$
GaAs–AuGa	C	33·33	16·67	50	25,600	4	GaAs, AuGa
	D	16·67	33·33	50	25	4	GaAs, AuGa
GaAs–γ′	E	36·8	18·4	44·7	25	4	GaAs, γ′
	K*	28·6	28·6	42·8	25	4	GaAs, γ′, AuGa(W)
	F	17·4	45·7	37·0	25,600	4	GaAs, γ′
GaAs–β′	H	37·8	18·9	43·2	25	4	GaAs, β′
	J†	25·0	37·5	37·5	25	4	GaAs, β′, γ′
	I	21·3	44·7	34·0	25,600	4	GaAs, β′
GaAs–Au	L	40·0	20·0	40·0	25,600	4	GaAs, Au, β′(W), As(W)
	Ll	40·0	20·0	40·0	25	0	GaAs, Au
	L2	40·0	20·0	40·0	350	0	GaAs, Au
	G	33·33	33·33	33·33	25	4	GaAs, Au, β′(W), As(W)
	M	18·2	63·6	18·2	25	4	GaAs, Au, β′(W), As(W)
	Ml‡	18·2	63·6	18·2	25	4	GaAs, Au, β′(W), As(W)

* The line joining GaAs to alloys E, K, F meets the Au–Ga binary at 30·3, 33·2, 29·9 at.-% Au respectively. The γ′ phase exists from 29·8 to 31 at.-% Au, i.e. AuGa would be expected to appear in alloy K

† The line joining GaAs to alloys H, J, I meets the Au–Ga binary at 22·5, 25, 22·1 at.-% Au respectively. The β′ phase exists from 22 to 23 at.-% Au at 25°C, i.e. γ′ would be expected to appear in alloy J

‡ Alloy Ml was prepared using β′ Au–Ga and As as the starting materials

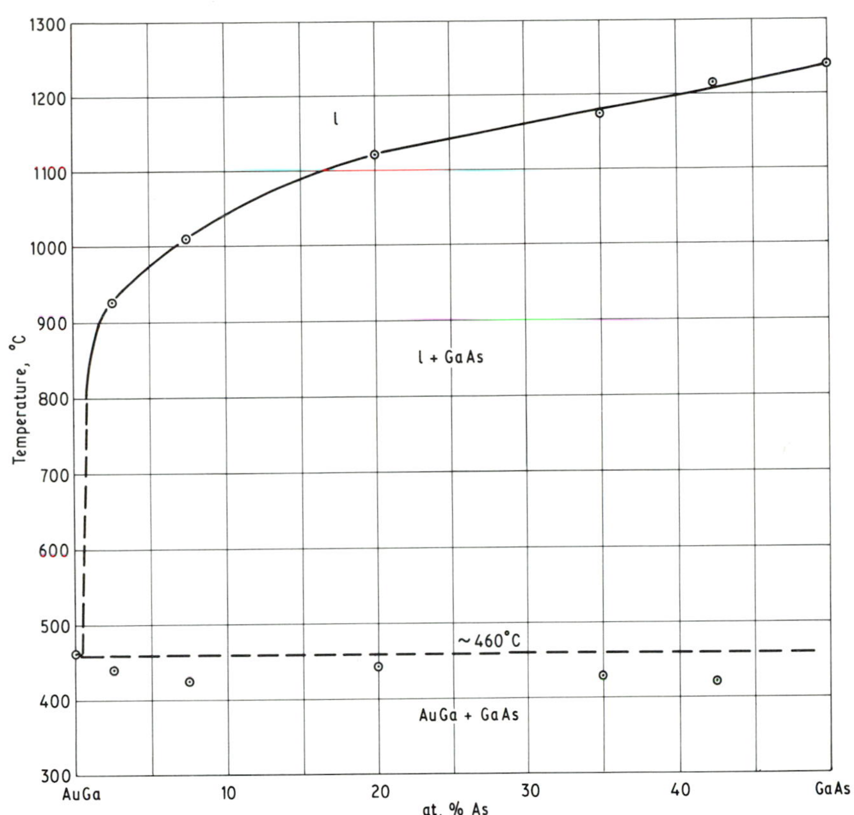

82 The AuGa–GaAs pseudobinary system

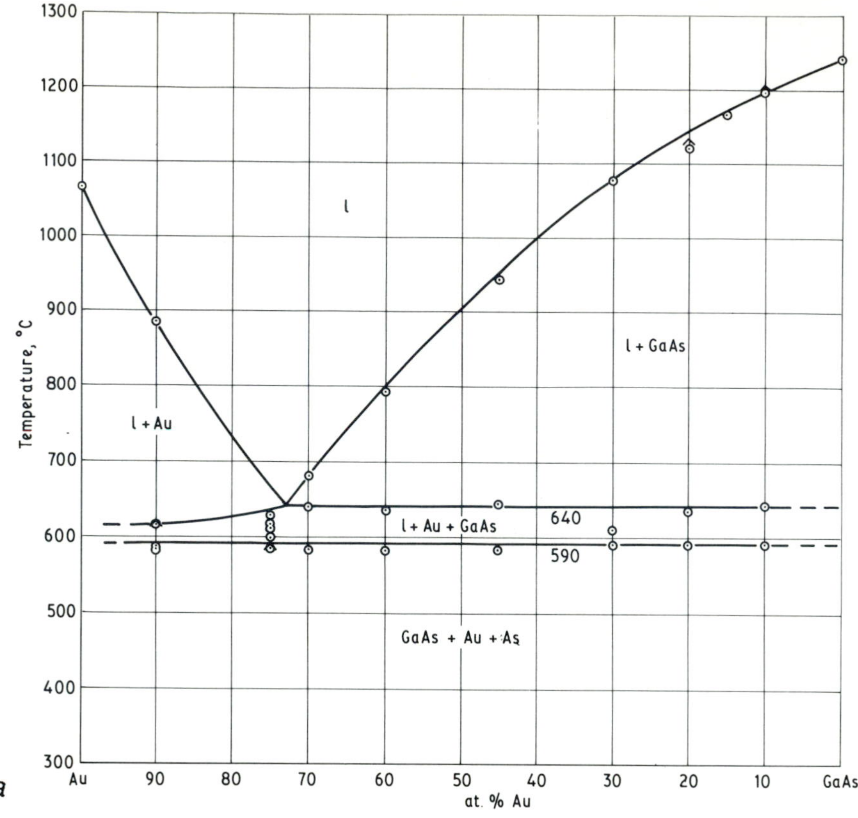

a

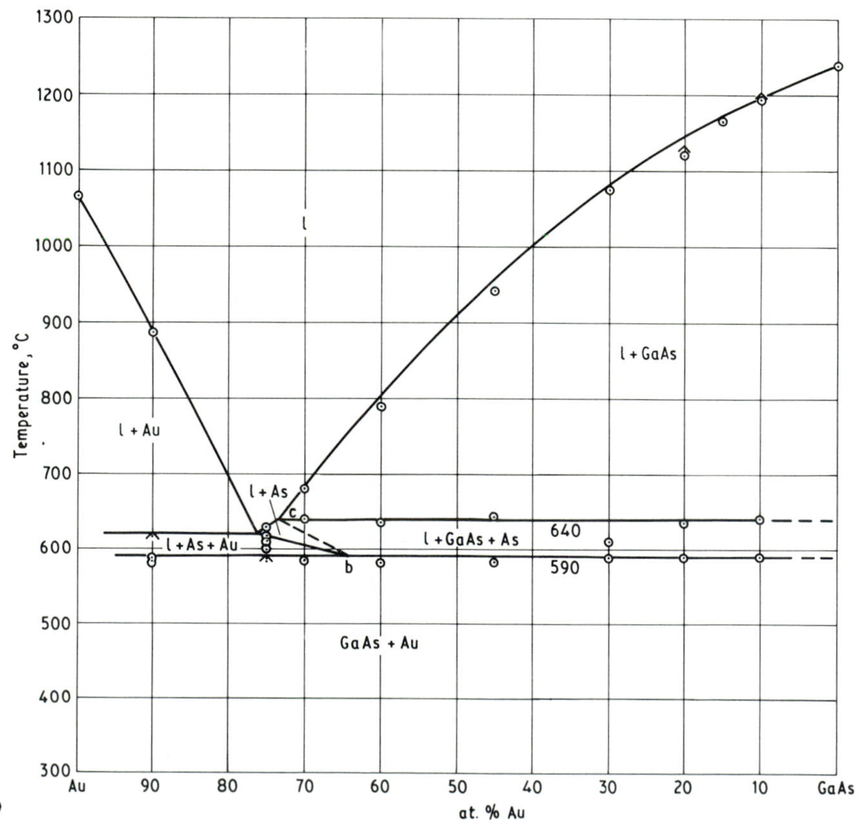

b

83 ***a*** **The Au–GaAs section [67 Pan];** ***b*** **The proposed Au–GaAs section**

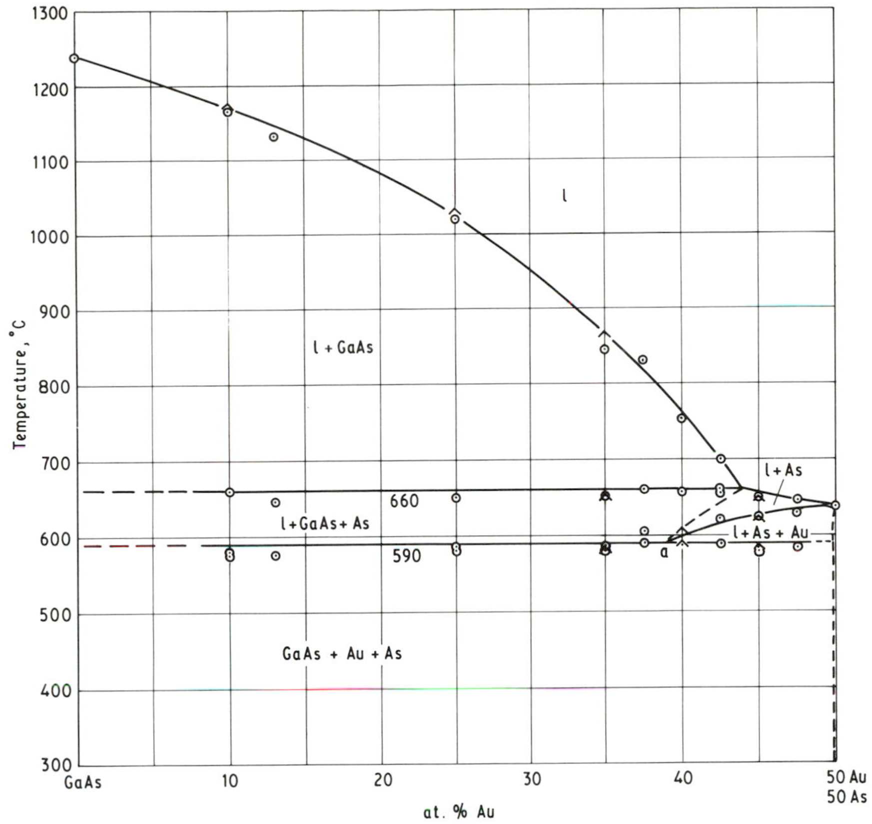

84 The GaAs–50 at.-% Au, 50 at.-% As section

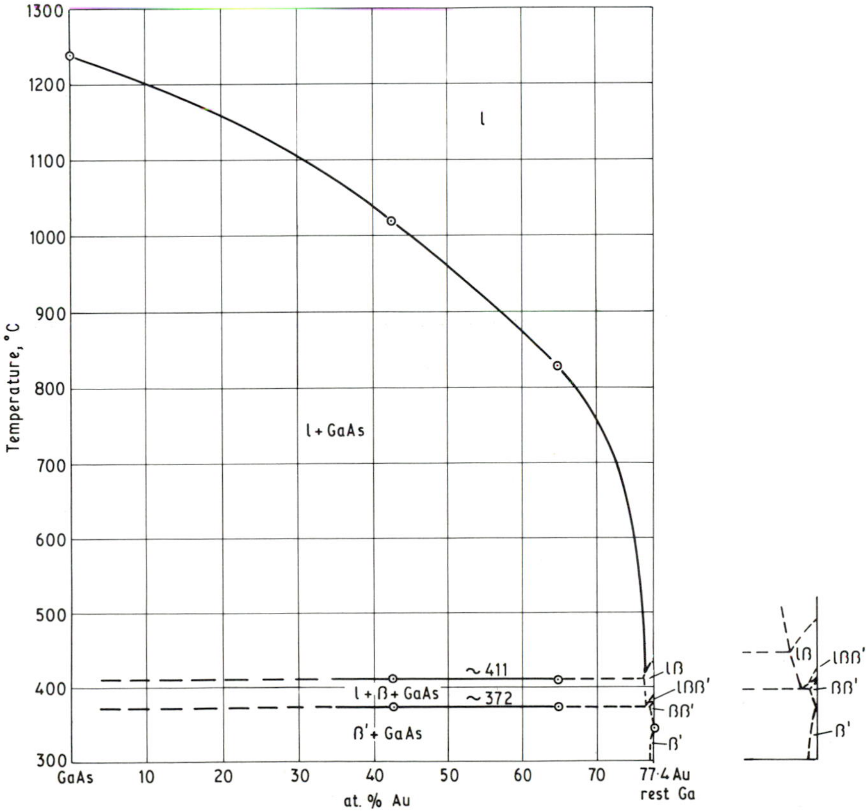

85 The GaAs–77·4 at.-% Au, 22·6 at.-% Ga section

this basis. The section GaAs–50 at.-% As, 50 at.-% Au (Fig. 84) is in accord with Fig. 81. In particular point 'a' lies on the tie line As–U_1. [67 Pan] notes that in alloys containing >35 at.-% Au (Fig. 84) it was not possible to identify GaAs as the primary phase. Figure 84 indicates that alloys containing > 39 at.-% Au will appear to have primary As. Alloys between 39 and 44 at.-% Au will separate primary GaAs until the liquid composition meets the curve As–U_1, when secondary separation of As occurs by reaction of liquid with the primary GaAs. Reaction goes to completion and the liquid enters the primary As phase field where As appears as the primary phase. Once the remaining liquid meets the monovariant curve e_1U_1 it separates (Au) in addition to As. At U_1 the residual liquid reacts with a proportion of the As to give GaAs + (Au). The solid equilibrium is a three-phase mixture of GaAs + (Au) + As.

This alternative interpretation of the data also elucidates some of the difficulties discussed with regard to the 75 at.-% Au alloy on the section Au–GaAs. This section now contains the traces of three liquidus surfaces, (Au), As and GaAs (Fig. 83*b*). The higher liquidus temperature of 628°C for the 75 at.-% Au alloy is accounted for as being a point on the liquidus surface of As. There is little evidence for the phase boundaries between the l + As and the l + GaAs + As and l + As + Au phase regions. The boundaries meet at point 'b' (Fig. 83*b*). This point has been selected on the basis that it must lie on the tie line As–U_1. Metallographically [67 Pan] found GaAs and Au in all alloy compositions but the As phase was only detected for alloys with >50 at.-% Au. According to the equilibria shown in Fig. 81 alloys with compositions from GaAs to point 'c' will separate primary GaAs, followed by secondary separation of GaAs + As along cU_1. At U_1 the residual liquid reacts with the As to form (Au) + GaAs. The detection of GaAs and Au in alloys on the Au–GaAs section is understandable. The failure to detect As for alloys with <50 at.-% Au could be associated with the small amount of As formed on cooling such alloys and the probability of the invariant reaction proceeding to completion. For alloys with >50 at.-% Au the larger amounts of As may not fully react at U_1.

[86 Tsa] also studied the GaAs–Au section. They used >99·99% As, Au, Ga, and semiconductor grade GaAs to prepare 1 g melts of ternary alloys. The components were placed in fused silica capsules that were evacuated to 10^{-6} torr and sealed. The dead volume was 4 cm^3 compared with 1 cm^3 of [67 Pan]. Alloys were heated at 1 000°C for 1 week and slowly cooled to 25 or 600°C at 5–10°C h^{-1}. Samples cooled to 600°C were then quenched into iced water. Phase identification was by X-ray powder diffractometer techniques. Table 26 shows that alloys on the GaAs–Au section contained GaAs + Au with weak reflections for As and β' when a dead space of 4 cm^3 was used. With a zero dead space only GaAs + Au were detected. [86 Tsa] speculate that the detection of As and β' was due to loss of As by partial decomposition of GaAs; this would move the alloy composition into the GaAs + Au + β' phase region, the As condensing on the walls of the capsule. On the basis of this evidence [86 Tsa] concluded that the section GaAs–Au is a pseudobinary. As noted in discussion of the results of [67 Pan] alloys on the GaAs–Au section with >50 at.-% Au contained some As, whereas [86 Tsa] detected As and β' for all alloys with a 4 cm^3 dead space. Both sets of results are explicable on the basis of the proposed transition reaction (Fig. 81) since incomplete reaction at U_1 will not only leave unreacted As but also some liquid phase which will cascade down the U_{2-4} reactions to produce some β'. In theory all alloys (Fig. 83*b*) should solidify as two-phase GaAs + Au structures, as was found by [86 Tsa] for samples with no dead volume. A redetermination of this section is needed, using a variety of investigated techniques as was done by [67 Pan].

The section from GaAs to β' (Table 26) is considered to be a pseudobinary by [86 Tsa]. Figure 85 is the equivalent section from [67 Pan]. It contains two horizontals at ~411 and ~372°C. The latter would appear to be associated with the $l + \beta \rightleftharpoons \beta'$ reaction at 375°C in the Au–Ga system, i.e. corresponding to the ternary reaction $L + \beta \rightleftharpoons GaAs + \beta'$ (U_4). The horizontal at ~411°C is considered to separate the L + GaAs phase region from the L + GaAs + β region. It should be noted that the GaAs–β' section could not be pseudobinary near the binary Au–Ga end since β' forms by a peritectic reaction.

A probable reaction scheme for the As–Au–Ga system is given in Fig. 86. In terms of the reaction of thin Au films on GaAs substrates it should be noted that the reaction products will depend on whether the system is open or closed. In the former case As will volatilize from the GaAs during heat treatment and the reaction products will include Au–Ga binary compounds [86 Tsa].

[88 Sch] calculated a 25°C isothermal section which is essentially a room temperature triangulation of the ternary system. The three-phase regions reported by [88 Sch] correspond with those given in Fig. 86. The GaAs–Au section is regarded as a pseudobinary section by [88 Sch] in contrast to the conclusions reached in this assessment. A complete thermodynamic calculation of the ternary system in terms of isothermal sections from high to low temperatures, together with an input of fresh experimental data on the redetermination of the GaAs–Au section with zero dead space, is needed.

LIQUIDUS SURFACE

The liquidus isotherms drawn in Fig. 80 are as presented by [67 Pan]. They differ slightly from the isotherms in Fig. 81 which have been produced by replotting the original tabulated data of [67 Pan]. As noted previously, the ternary system is dominated by the GaAs primary phase region.

REFERENCES

67 Pan: M. B. Panish: *J. Electrochem. Soc.*, 1967, **114**, 516–521

73 Nak: T. Nakanisi: *Japan J. Appl. Phys.*, 1973, **12**, 1818–1819

86 Proposed reaction scheme for the As–Au–Ga ternary system

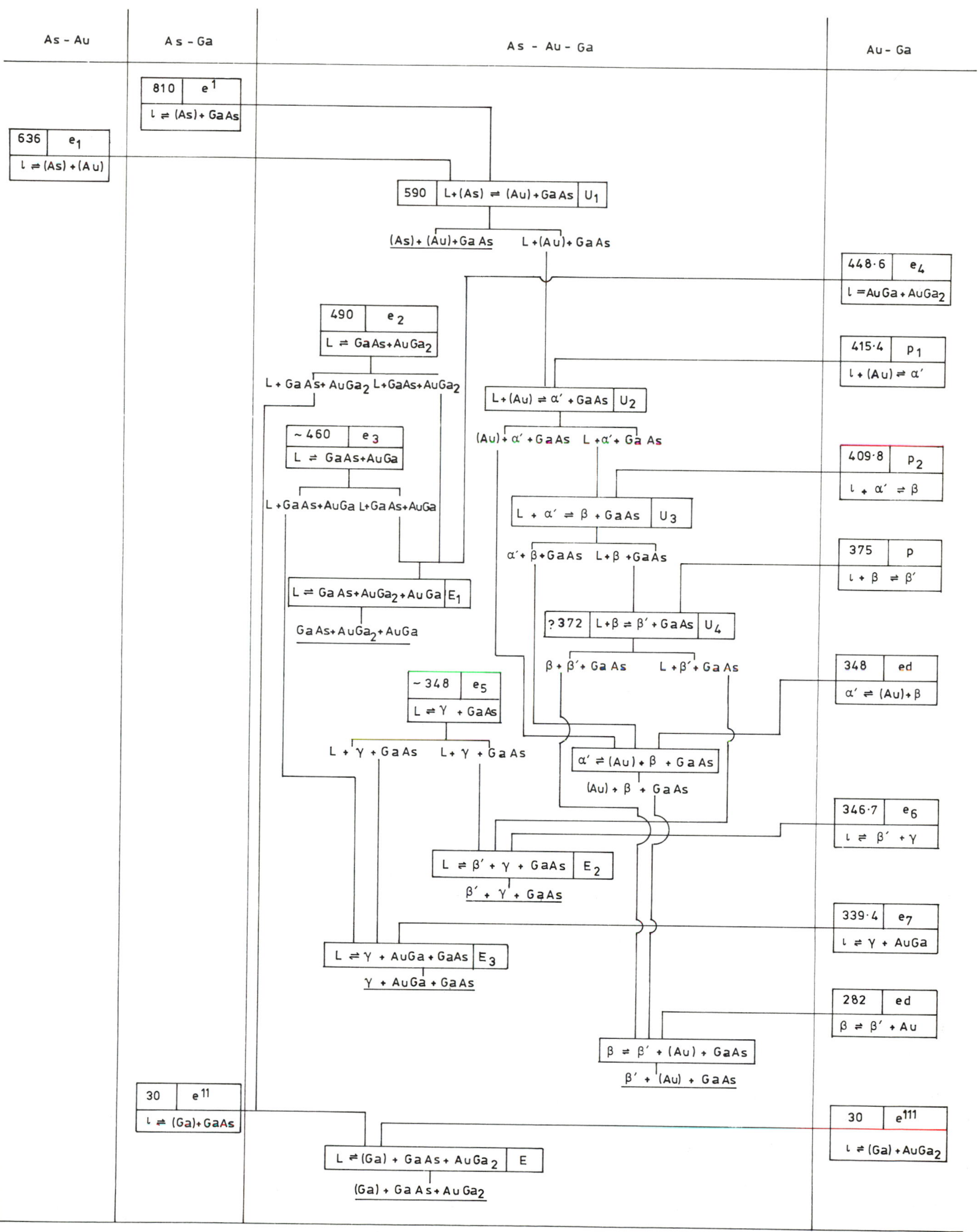

74 Pan: M. B. Panish: *J. Cryst. Growth*, 1974, **27**, 6–20

75 Mag: T. J. Magee and J. Peng: *Phys. Stat. Solidi A*, 1975, **32**, 695–700

76 Gat: B. Gather and R. Blachnik: *Z. Metall.*, 1976, **67**, 168–169

79 Kum: K. Kumar: *Japan J. Appl. Phys.*, 1979, **18**, 713–716

80 Van: J. M. Vandenberg and E. Einsbron: *Thin Solid Films*, 1980, **65**, 259–265

80 Mil: D. C. Miller: *J. Electrochem. Soc.*, 1980, **127**, 467–475

82 Ker: V. G. Keramidas: *Thin Solid Films*, 1982, **96**, 347–363

83 Leu 1: S. Leung, A. G. Milnes and D. D. L. Chung: *Thin Solid Films*, 1983, **104**, 109–131

83 Leu 2: S. Leung, T. Yoshiie, C. L. Bauer and A. G. Milnes: *J. Electrochem. Soc.*, 1983, **130**, 462–468

83 Yos: T. Yoshiie and C. L. Bauer: *J. Vac. Sci. Technol.*, 1983, **A1**(2), 554–557

84 Yos: T. Yoshiie, C. L. Bauer and A. G. Milnes: *Thin Solid Films*, 1984, **111**, 149–166

84 Zen: X. Zeng and D. D. L. Chung: *Solid-State Electron.*, 1984, **27**, 339–345

86 Pug: J. H. Pugh and R. S. Williams: *J. Mater. Res.*, 1986, **1**, 343–351

86 Tsa: C. T. Tsai and R. S. Williams: *J. Mater. Res.*, 1986, **1**, 352–360

87 Oka: H. Okamoto and T. B. Massalski: 'Phase Diagrams of Binary Gold Alloys', 111–118; 1987, ASM International

88 Sch: R. Schmid-Fetzer, *J. Electron. Mater.*, 1988, **17**, 193–200

As–Au–Ge

A search for ternary compounds of the composition $A^{I}B_2^{IV}C_3^{V}$ (A = Cu, Ag, Au; B = Si, Ge, Sn, Pb; C = P, As, Sb, Bi) [64 Sok] was confined to one experiment on the Au-containing alloys. They report that the compound $AuGe_2As_2$ was not formed and that Au only partially reacts with Ge and As. No further data was given.

A ternary compound AsAuGe was identified [81 Ste] in the interface microstructure of Au–Ge–In contacts to GaAs substrates. AsAuGe was identified by scanning electron microscopy with energy dispersive X-ray analysis of the interface after amalgamation to remove the top contact metals, and by transmission electron microscopy of the interface region after thinning by ion bombardment. Bulk samples prepared by melting the elements and annealing for 10 weeks at 500°C contained AsAuGe. The ternary compound is monclinic, space group C 2/c with lattice parameters a = 0·71 nm, b = 0·63 nm, c = 0·68 nm, β = 126°. Convergent beam electron diffraction was used [84 Vin1] to confirm the crystal structure of AsAuGe. The ternary compound is isostructural with NiP_2. It has 12 atoms in the unit cell and lattice parameters $a = 0{\cdot}711 \pm 0{\cdot}003$ nm, $b = 0{\cdot}634 \pm 0{\cdot}003$ nm, $c = 0{\cdot}684 \pm 0{\cdot}003$ nm, $\beta = 126 \pm 1°$. The Au atoms occupy 4(a) sites and the Ge and As atoms are randomly distributed on 8(f) sites. These workers [84 Vin2] refined the location of the Ge and As atoms at x = 0·105 (standard deviation 0·001), y = 0·370 (sd 0·001), z = 0·161 (sd 0·002).

Four points on the liquidus were determined [69 Gub] along a section from Au–2·33As to Ge. Foil of Au–2·33As of known thickness was placed on a single crystal Ge substrate and heated to a temperature of 600, 700, 750 and 800°C. Measurement of the depth of Ge dissolved by the Au–2·33As gives the solubility of Ge in the alloy at each temperature. Liquidus compositions were found at 51·8, 62·4, 67·4, and 74·1 Ge at temperatures of 600, 700, 750, and 800°C respectively. Judging from the corresponding binary liquidus compositions for the Au–Ge and As–Ge systems, the ternary isotherms between 600 and 800°C are convex when viewed from the Au–As binary edge or concave when viewed from the Ge corner.

REFERENCES

64 Sok: V. I. Sokolova and E. V. Tsvetkova: *Issled. po Poluprov., Akad. Nauk. Moldavsk SSR, Inst. Fiz. i Mat.*, 1964, 168–172

69 Gub: A. Ya Gubenko and M. B. Miller: *Izvest. Akad. Nauk. SSSR, Neorg. Materialy.*, 1969, **5**, 235–238

81 Ste: J. W. Steeds, G. M. Rackham and D. Merton-Lyn, 'Microscopy of Semiconducting Materials, 1981', Inst. Physics Conference Series No 60 (ed. A. G. Cullis and D. C. Joy), 387–396; 1981, London, Institute of Physics

84 Vin 1: R. Vincent, D. M. Bird and J. W. Steeds: *Phil. Mag.*, 1984, **50A**, 745–763

84 Vin 2: R. Vincent, D. M. Bird and J. W. Steeds: *Phil. Mag.*, 1984, **50A**, 765–786

As–Au–In

The only published data [86 Tsa] on the As–Au–In system is a study of 16 alloy compositions and the characterization of the phases present by X-ray diffraction analysis. Alloys were prepared in 1 g quantities from elements of >99·99% purity. Each alloy was contained in an evacuated, 10^{-6} torr, and sealed fused-silica capsule with a 4 cm^3 dead space. Alloying was effected by heating at 1 000°C for 1 week and samples were slowly cooled to either 25°C or 600°C at a rate of 5–10°C h^{-1}. The samples cooled to 600°C were then quenched into iced water.

Table 27 lists the alloys studied and their phase constitution. [86 Tsa] concluded that the sections from InAs

87 Triangulation of the As–Au–In ternary system [86 Tsa]

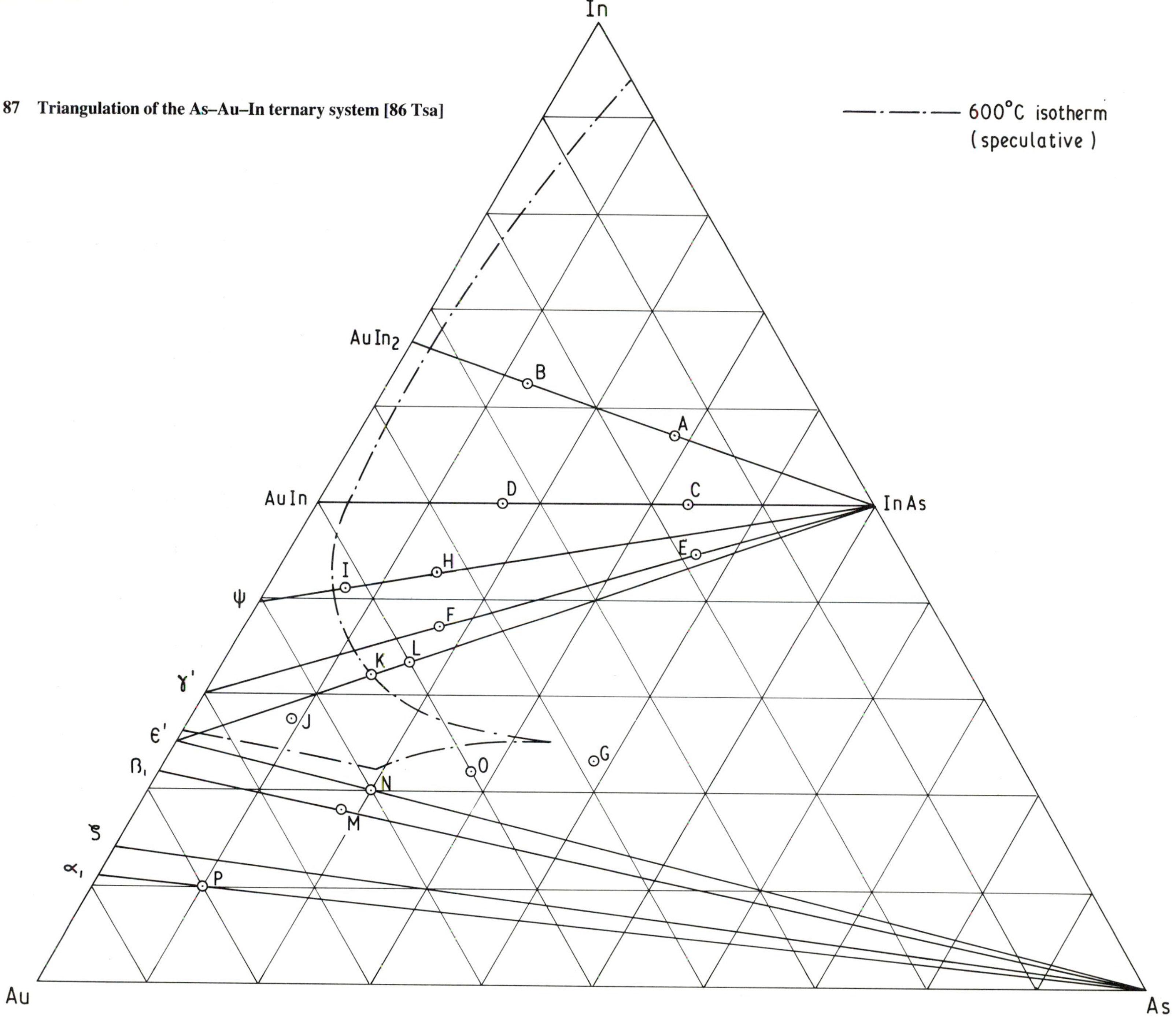

to $AuIn_2$, AuIn, ψ, γ′, and ϵ′ are pseudobinary in nature. The sections from As to ϵ′, β_1, ζ, and α_1 were also considered to be pseudobinary (Fig. 87). [86 Tsa] appear to assume that all the ternary boundaries of three-phase regions are pseudobinary sections. Figure 87 is actually a triangulation of the ternary system into its constituent three-phase regions for alloys cooled slowly to room temperature. The sections $AuIn_2$–InAs, AuIn–InAs, and ψ–InAs are most probably pseudobinary sections; the remaining sections cannot be so in view of the peritectic mode for the formation of γ′, ϵ′, ζ, and α_1 and the solid state peritectoid formation of β_1. (The assessment of [87 Oka] is accepted for the Au–In binary system.)

[86 Tsa] found the binary ψ phase (Au_3In_2) to be a stable phase on the section ψ–InAs for alloys slowly cooled to room temperature, although ψ decomposes eutectoidally in binary Au–In alloys at 224°C. Either the ψ phase is a metastable phase in the ternary system, i.e. equilibrium was not established despite the slow cooling, or As additions stabilize the ψ phase. A speculative 600°C isotherm has been shown in Fig. 87. There is a need for further work on this ternary system, using a range of experimental techniques, to determine the liquidus surface and the nature of the invariant reactions.

REFERENCES

86 Tsa: C. T. Tsai and R. S. Williams: *J. Mater. Res.*, 1986, **1**(2), 352–360

87 Oka: H. Okamoto and T. B. Massalski, 'Phase Diagrams of Binary Gold Alloys', 142–153; 1987, ASM International

As–Au–Na

Mues and Schuster [80 Mue] prepared the ternary compound Na_2AuAs by heating the elements in a quartz ampoule contained in a Ta crucible for 24 h at 700°C under argon. The alloy was annealed for 24 h at 600°C and its crystal structure characterized as orthorhombic (oC16), space group Cmcm, with a = 0·887 1(2), b = 0·712 9(2), c = 0·576 0(2) nm. Na_2AuAs is isotypic to Na_2AuSb and K_2AuSb.

REFERENCE

80 Mue: C. Mues and H.-U. Schuster: *Z. Naturf.*, 1980, **35B**, 1055–1058

As–Au–S

Gather [76 Gat] has speculated on the equilibria present in the As–Au–S system, assuming condensed phases only are present. Figure 88, adapted from [76 Gat], presents a schematic view of the liquidus projection. Four invariant reactions are envisaged, $L_1 = L_2$ + (As) + (Au); L_{E1} = (As) + (Au) + AsS; L_{E2} = (Au) + AsS + As_2S_3 and L_{E3} = (Au) + (S) + As_2S_3.

REFERENCE

76 Gat: B. Gather, Thesis, 1976, Technischen Universität Clausthal

As–Au–Se

The only reference to this system is speculative [76 Gat]. Figure 89 is a representation of the liquidus surface and Fig. 90 the associated reaction scheme.

Table 27. Alloys studied by [86 Tsa]

Alloy section	*Alloy*	*Alloy composition*, at.-%			*Temperature*, °C	*Phases identified*
		As	*Au*	*In*		
$AuIn_2$–InAs	A	28·6	14·3	57·1	25,600	$AuIn_2$, InAs
	B	12·5	25·0	62·5	25	$AuIn_2$, InAs
AuIn–InAs	C	33·33	16·67	50·0	25,600	AuIn, InAs
	D	16·67	33·33	50·0	25	AuIn, InAs
ψ–InAs	H	14·3	42·9	42·9	25,600	ψ, InAs
	I	7·0	52·0	41·0	25	ψ, InAs
γ′–InAs	E	36·8	18·4	44·7	25	γ′, InAs
	F	17·4	45·7	37·0	25,600	γ′, InAs
ϵ′–InAs	L	16·67	50·0	33·33	25,600	ϵ′, InAs, As(W)
	K	14·0	54·0	32·0	25	ϵ′, InAs
ϵ′–As	N	20·0	60·0	20·0	25,600	ϵ′, As
β_1–As	M	18·2	63·6	18·2	25,600	β_1, As
α_1–As	P	10·0	80·0	10·0	25	α_1, As
	G	33·33	33·33	33·33	25	ϵ′, InAs, As
	J	9·1	63.6	27·3	25	ϵ′, InAs, As
	O	28·0	50·0	22·0	25	ϵ′, InAs, As

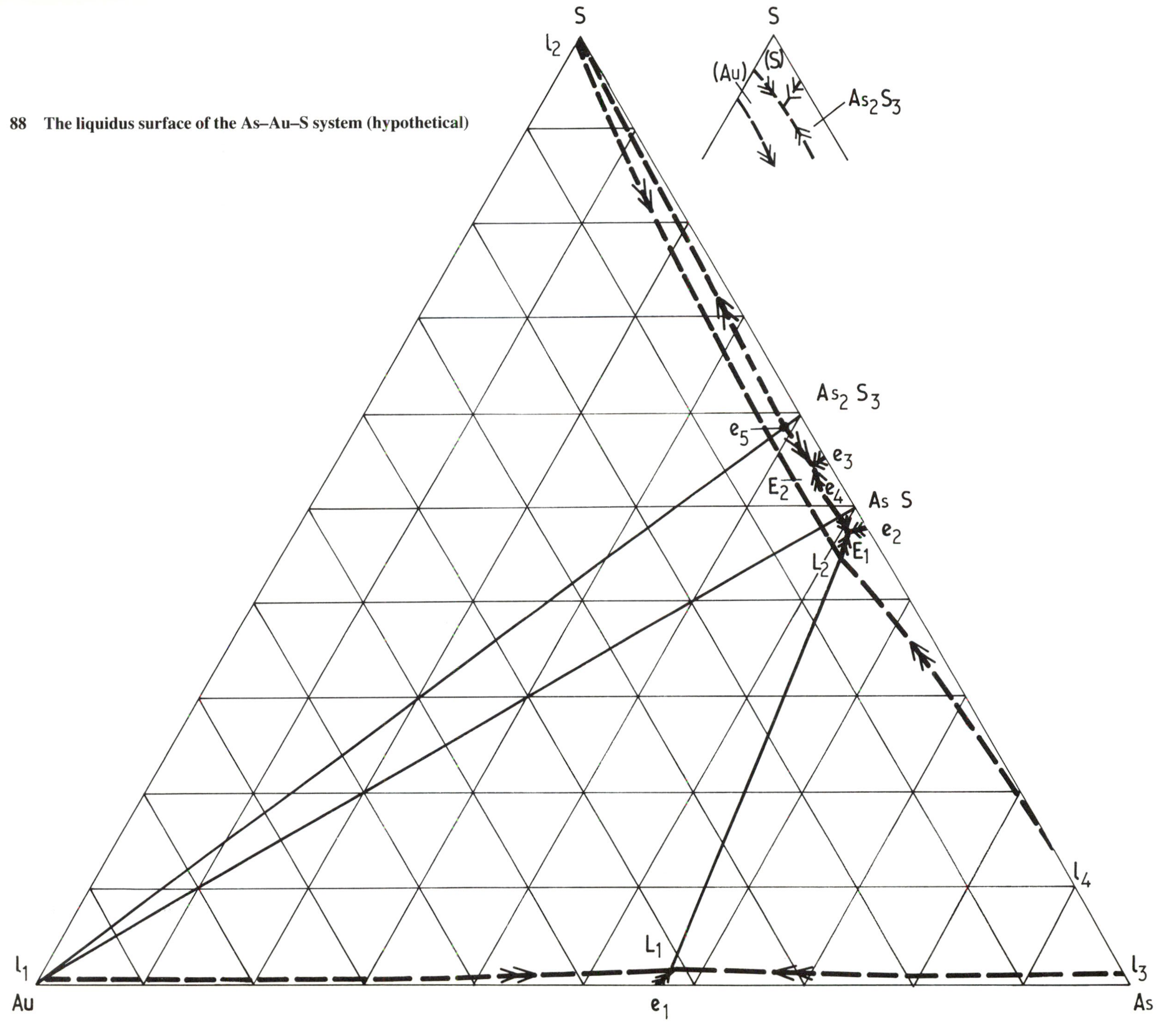

88 The liquidus surface of the As–Au–S system (hypothetical)

89 The liquidus surface of the As–Au–Se system (speculative)

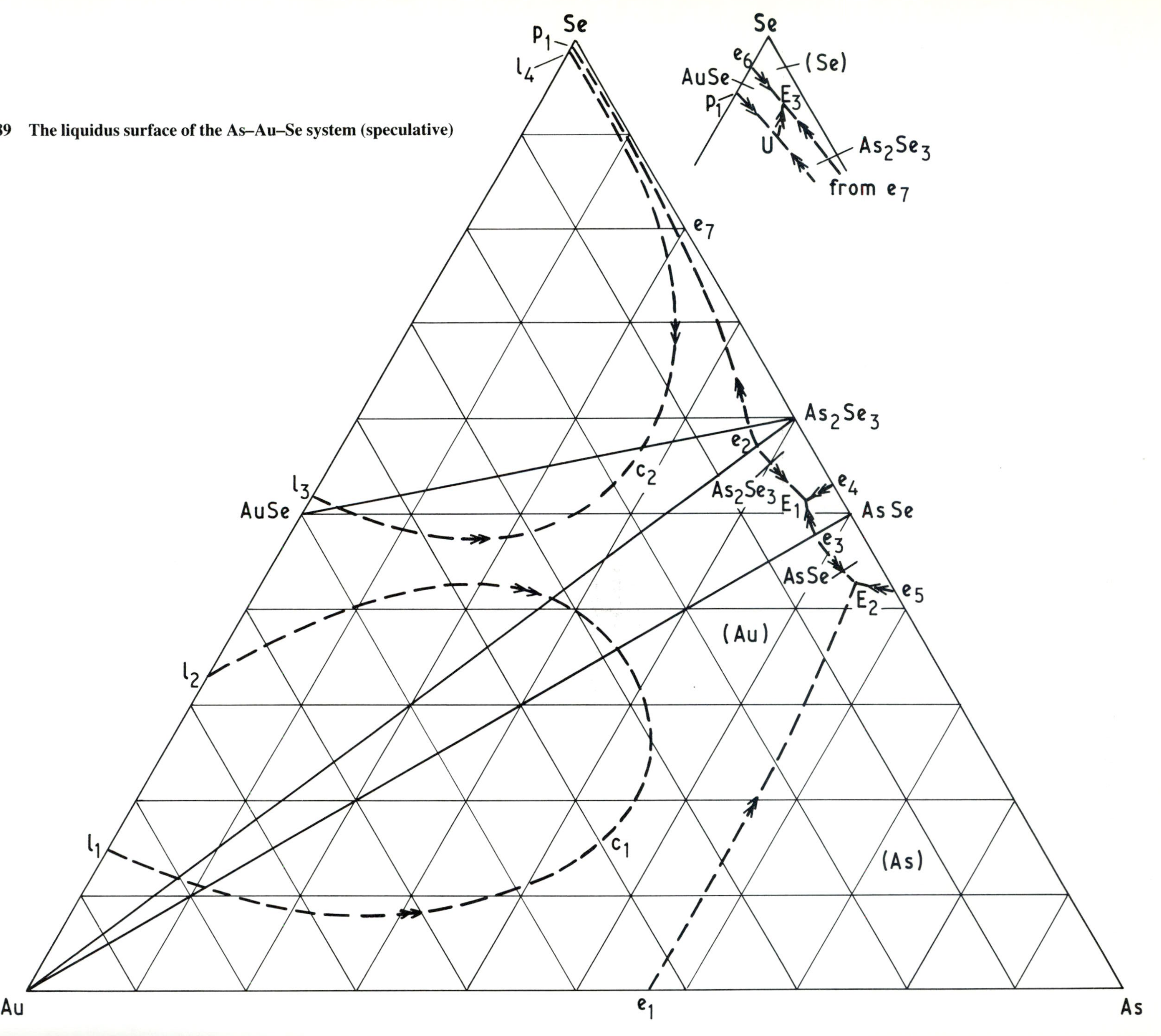

90 Reaction scheme for the As–Au–Se system (speculative)

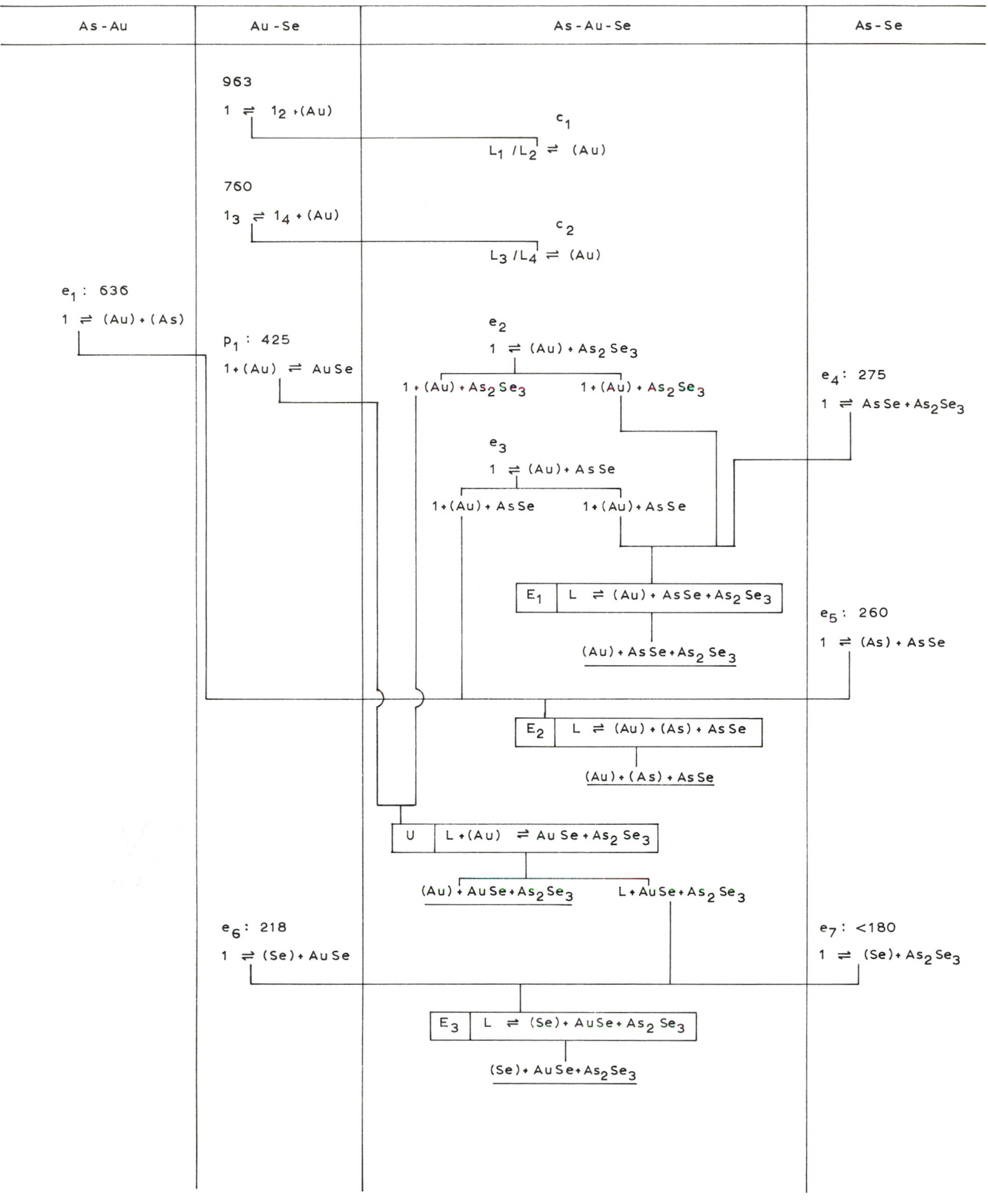

REFERENCE

76 Gat: B. Gather: Thesis, 1976, Technischen Universität Clausthal

As–Au–Si

Cagnini [69 Cag] used radiotracer methods to study the solubility of Au in As-doped Si at 1 100 and 1 000°C. Little effect of As on the solubility of Au in Si was observed at 1 000°C until the As concentration reached 10^{19} atoms cm^{-3} (202 atomic ppm As). This observation agrees with data of [71 Gub] who used similar techniques and showed only a small solubility increase of Au in Si with 4×10^{18} atoms cm^{-3} As (81 atomic ppm As). With increased As doping at 7×10^{19} atoms cm^{-3} (1 417 atomic ppm As) [69 Cag] found greater solubility of Au in Si, 0·88 atomic ppm Au at 1 000°C, and 1·68 atomic ppm Au at 1 100°C. The data are compared with the binary Au–Si evaluation [83 Oka] in Fig. 91.

REFERENCES

69 Cag: S. F. Cagnina: *J. Electrochem. Soc.*, 1969, **116**, 498–502

71 Gub: A. Ya Gubenko and Yu. I. Shmelev: *Izvest. Akad. Nauk. SSSR, Neorg. Mater.*, 1971, **7**, 731–733

83 Oka: H. Okamoto and T. B. Massalski: *Bull. Alloy Phase Diagrams*, 1983, **4**, 190–198

As–Au–Te

INTRODUCTION

The complete ternary system has been characterized [76 Gat 2] by preparing alloys, from 99·995% pure elements, melted in evacuated silica capsules, and subsequently heat treated for 30 days at 300°C. The experimental techniques used to examine alloys were differential thermal analysis, metallography, and X-ray diffraction analysis. In addition to the primary phase fields associated with Au, As and Te, $AuTe_2$ and As_2Te_3, a primary phase field associated with the ternary compound $Au_2As_2Te_3$ was found. The primary Au and As fields are large and characterized by liquid surfaces that fall steeply towards the Te corner. $AuTe_2$ is the primary phase over a broad flat region and the ternary compound Au_2As_2Te has a very small region of primary crystallization. Pseudobinary eutectic reactions occur in the sections Au–$Au_2As_2Te_3$, As–$Au_2As_2Te_3$, $AuTe_2$–$Au_2As_2Te_3$ and $AuTe_2$–As_2Te_3. The ternary system is therefore divided into four partial ternary systems. Simple ternary eutectics were found in the partial ternary system Au–$Au_2As_2Te_3$–$AuTe_2$, Au–As–$Au_2As_2Te_3$, and Te–$AuTe_2$–As_2Te_3. The fourth partial ternary system, As–$Au_2As_2Te_3$–$AuTe_2$–As_2Te_3 contains a transition reaction, $L_U + Au_2As_2Te_3 \rightleftharpoons (As) + AuTe_2$ followed by a ternary eutectic reaction, $L_{E_3} \rightleftharpoons (As) + AuTe_2 + As_2Te_3$. The ternary compound forms congruently at 417°C and its primary phase field, being very small, spans only 10 to 407°C (U). [76 Bla] give further information on $Au_2As_2Te_3$ but did not index the powder diffraction data. Their X-ray data is presented as relative intensities as a function of 2θ.

BINARY SYSTEMS

The Au–Te system evaluated by [84 Oka] is accepted. The Au–As was studied [76 Gat 1] using closed containers. It is a simple eutectic system with a eutectic temperature of 636 ± 2°C and a eutectic composition of 56·5 ± 1·5 at.-% As. The As–Te system was also studied [75 Bla] in closed containers. As_2Te_3 melts congruently at 381°C, just 1°C above the eutectic temperature for As–As_2Te_3 alloys. The eutectic composition is 45 at.-% As. The As_2Te_3–Te alloys exhibit a eutectic at 363°C and 28 at.-% As. In both the Au–As and As–Te systems the usual assumption of constant pressure does not apply. The high vapour pressure of As above As-rich alloys means that the temperature–concentration diagrams for both the Au–As and Au–Te systems are projections of *T–X–P* diagrams for variable pressure conditions.

SOLID PHASES

Table 28 summarizes the crystal structures of all the phases entering into the ternary equilibria.

Table 28. Crystal structures

Phase	*Prototype*	*Lattice designation*
(Au)	Cu	cF4
(As)	As	hR2
(Te)	γSe	hP3
$AuTe_2$	$AuTe_2$	mC6
As_2Te_3	$AuTe_2$	mC6
$Au_2As_2Te_3$	—	—

PSEUDOBINARY SYSTEMS

[76 Gat 2] found four pseudobinary eutectic sections. Phase diagrams are given for the $AuTe_2$–As_2Te_3 and the Au–$Au_2As_2Te_3$ pseudobinary systems. The As–$Au_2As_2Te_3$ and $AuTe_2$–$Au_2As_2Te_3$ sections are also pseudobinary eutectic systems. Table 29 summarizes the compositions and temperatures for these invariant reactions, e_4–e_6 and e_8. Compositions are stated to be within ± 1·5 at.-% and temperatures to within ± 2°C. The composition for the pseudobinary eutectic e_5 was given as 28 at.-% Au, 29·5 at.-% As, 42·5 at.-% Te [76 Gat 2]. This composition does not quite lie on the As–$Au_2As_2Te_3$ section. On the assumption that the stated As content is correct the Te content would be 42·3 at.-% and the Au content 28·2 at.-%. This is a slight change and well within the quoted error limits of ± 1·5 at.-% for all the invariant liquid compositions. Figures 92 and 93 illustrate the $AuTe_2$–As_2Te_3 and Au–As_2Te_3 (including the Au–$Au_sAs_2Te_3$ portion) systems.

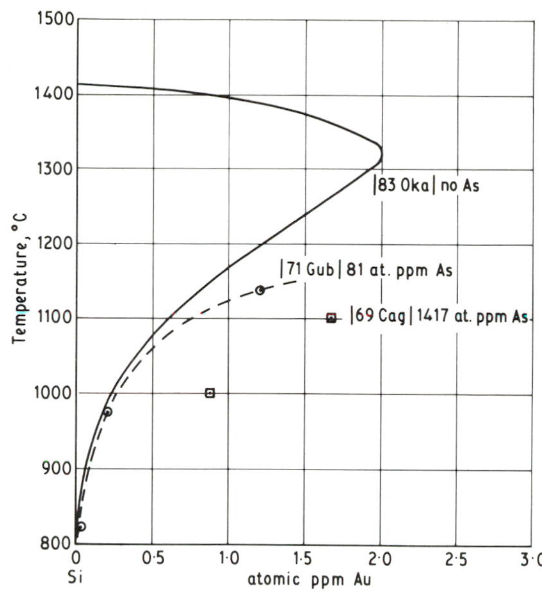

91 The effect of As on the solubility of Au in Si

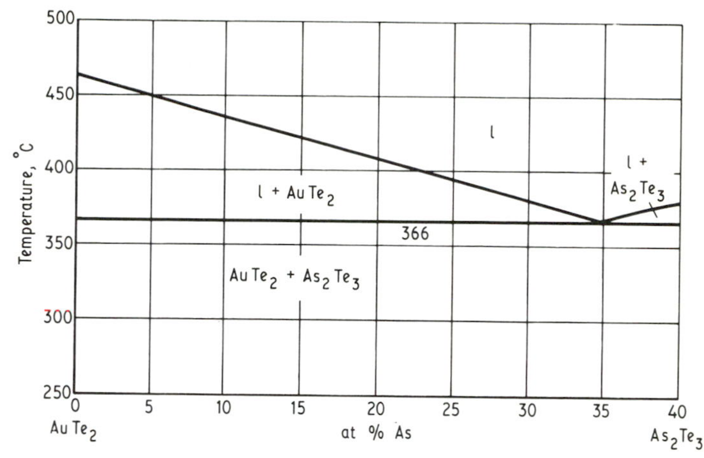

92 The $AuTe_2$–As_2Te_3 section

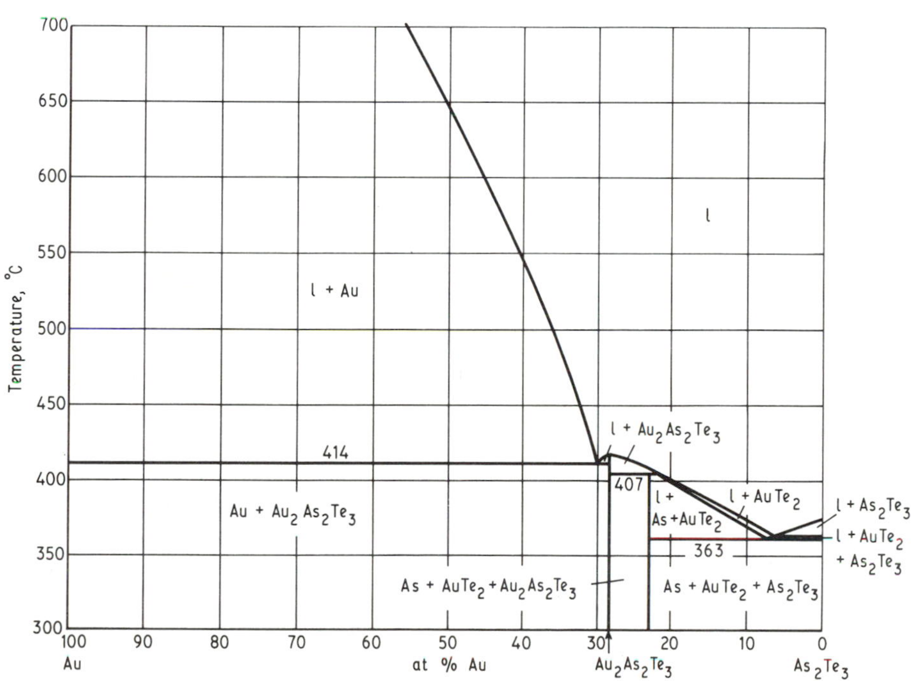

93 The Au–As_2Te_3 section

INVARIANT EQUILIBRIA

Table 29 and Figure 94 give the details of the invariant equilibria. The ternary system is divided into four partial ternary systems by the four pseudobinary eutectic systems mentioned in the previous section. The ternary compound $Au_2As_2Te_3$ melts congruently at 417°C and forms pseudobinary eutectics with Au at 414°C, with As at 413°C, and with $AuTe_2$ at 411°C. The small temperature difference between these eutectics and the melting point of $Au_2As_2Te_3$, places the pseudobinary eutectic compositions, e_4, e_5, and e_6, close to $Au_2As_2Te_3$. In turn this imposes a very small field of primary crystallization for $Au_2As_2Te_3$. As would be anticipated the Au–$AuTe_2$–$Au_2As_2Te_3$ and Au–As–$Au_2As_2Te_3$ partial ternary systems are simple eutectics with eutectic compositions, E_2 and E_1 respectively, near to the composition of $Au_2As_2Te_3$. The fourth pseudobinary eutectic system, $AuTe_2$–As_2Te_3, takes part in the ternary eutectic system Te–$AuTe_2$–As_2Te_3 in which the ternary eutectic E_4 has a composition within 0·5 at.-% of the Te–As_2Te_3 eutectic composition. The pseudobinary eutectic e_8, like e_4 to e_6, is a saddle point on the liquidus surface. The partial ternary system As–$Au_2As_2Te_3$–$AuTe_2$–As_2Te_3 contains a invariant transition reaction, $L_U + Au_2As_2Te_3 \rightleftharpoons$ (As) + $AuTe_2$, followed by a ternary eutectic reaction, $L_{E_3} \rightleftharpoons$ (As) + $AuTe_2$ + As_2Te_3. The invariant reaction at U produces (As) + $AuTe_2$. At temperatures below 407°C, the section As–$AuTe_2$ (Fig. 95) should show just the (As) + $AuTe_2$ phase region. The detection of the eutectic reaction at 363°C in the As–$AuTe_2$ section is evidence of incomplete reaction of the liquid at U.

Table 29. Invariant equilibria

Reaction	*Temperature*, °C	*Phase*	*Composition*, at.-%	
			As	*Te*
e_4	414	L	28·0	42·0
e_5	413	L	29·5	42·3
e_6	411	L	26·0	45·0
e_8	366	L	34·8	60·9
E_1	412	L	29·0	41·5
E_2	410	L	25·0	45·0
E_3	363	L	40·5	55·5
E_4	355	L	28·0	71·5
U	407	L	31·0	47·0

The composition of E_3 was quoted as 4 at.-% Au, 60·5 at.-% As, 55·5 at.-% Te (Table 1 of [76 Gat 2]). It is amended to 4 at.-% Au, 40·5 at.-% As, 55·5 at.-% Te in Table 29. [76 Gat 2] reproduce ternary vertical sections at 35 at.-% Au (Fig. 96), 15 at.-% As (Fig. 97), and 30 at.-% Te (Fig. 98), in addition to the As–$AuTe_2$, the $AuTe_2$–As_2Te_3 and the Au–As_2Te_3 sections. Figure 95 has been amended insofar as the l + As + $AuTe_2$ phase region descending to the 363°C invariant reaction was incorrectly designated l + $AuTe_2$ + As_2Te_3. Similarly the phase region As + $AuTe_2$ + $Au_2As_2Te_3$ (Fig. 98) was incorrectly designated As + Au + $Au_2As_2Te_3$, and the phase region, l + As_2Te_3 + $AuTe_2$ (Fig. 93) was incorrectly designated as l + As + As_2Te_3.

LIQUIDUS SURFACE

The liquidus surface is illustrated in Fig. 99. The inset is a larger scale representation of the reactions associated with $Au_2As_2Te_3$.

ISOTHERMAL SECTIONS

[76 Gat 2] gave an isothermal section at 355°C, E_4, on the assumption that Au, As and $AuTe_2$ were soluble in $Au_2As_2Te_3$ to an extent of 3 at.-%. Isothermal sections may be constructed from the liquidus surface and the vertical sections (Figs. 92, 93, 95–98).

REFERENCES

75 Bla: R. Blachnik, A. Jäger and G. Enninga: *Z. Naturf.*, 1975, **30B**, 191–197

76 Bla: R. Blachnik and B. Gather: *Z. Naturf.*, 1976, **31B**, 526–527

76 Gat 1: B. Gather and R. Blachnik: *Z. Metall.*, 1976, **67**, 168–169

76 Gat 2: B. Gather and R. Blachnik: *Z. Metall.*, 1976, **67**, 223–227

84 Oka: H. Okamoto and T. B. Massalski: *Bull. Alloy Phase Diagrams*, 1984, **5**, 172–177

Au–B–Co
Au–B–Cr
Au–B–Fe
Au–B–Mn
Au–B–Ni
Au–B–Ti
Au–B–V

[84 Mat] reported on the surface hardening by boronizing of binary Au alloys containing Co, Cr, Fe, Mn, Ni, Ti and V. A range of Au–Ni alloy compositions were boronized at 900°C for 6 h; boronizing at 950°C for 6 h was used for Au–Cr, Au–Fe and Au–Ti alloys. The surface hardened alloys were examined by metallography, X-ray diffraction analysis and on occasion by electron probe microanalysis. Boronizing Au–Cr alloys gave a surface layer containing predominantly CrB, with minor amounts of CrB_2. The Au–Fe alloys contained FeB as the major phase in the surface layer, with minor amounts of Fe_2B. Au–Ni alloys contained Ni_2B as the major phase with minor amounts of Ni_4B_3 and Ni_3B depending on the initial Ni content of the alloy. The higher the Ni content the higher the Ni content of the boride formed as the minor phase. The Au–Ti alloys showed a very thin surface layer of TiB_2. It was also stated that Au–Co, Au–Mn and Au–V alloys could be surface hardened by boronizing.

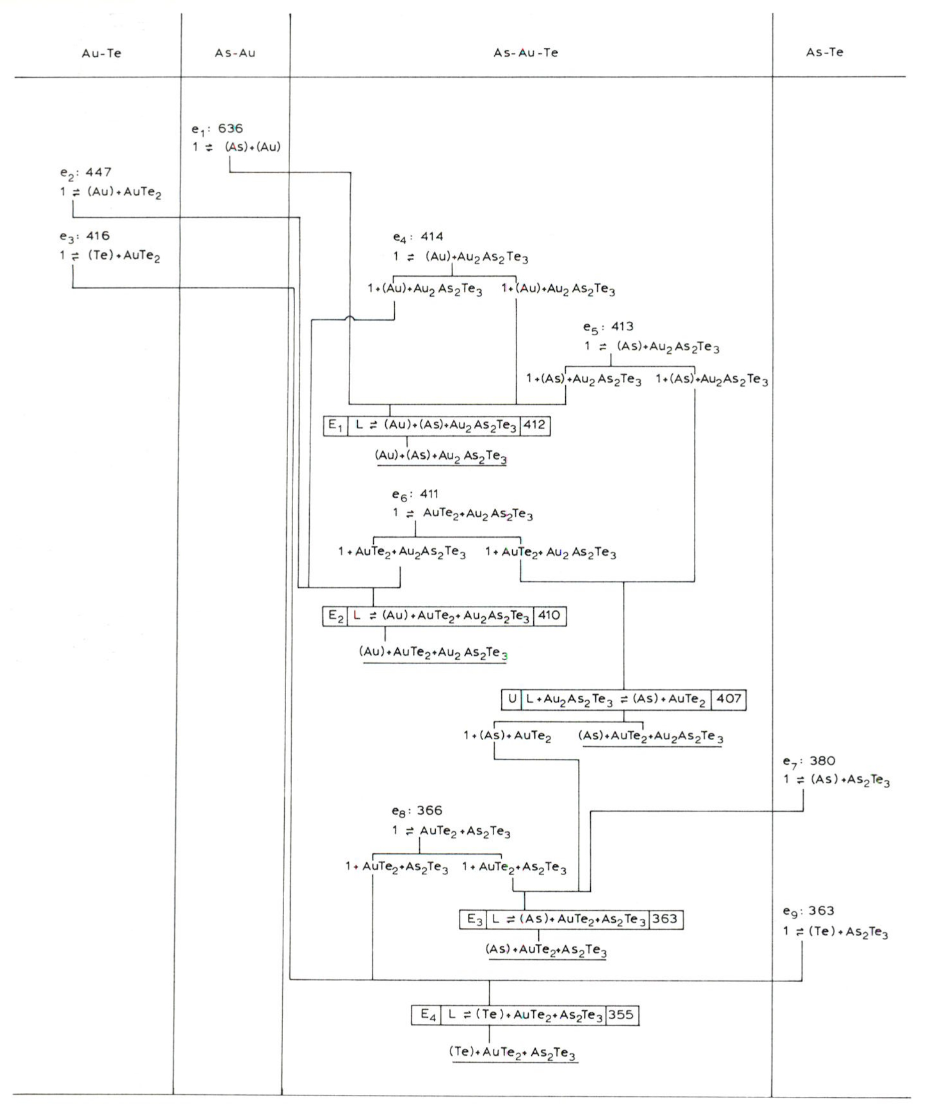

94 **Reaction scheme for the As–Au–Te system**

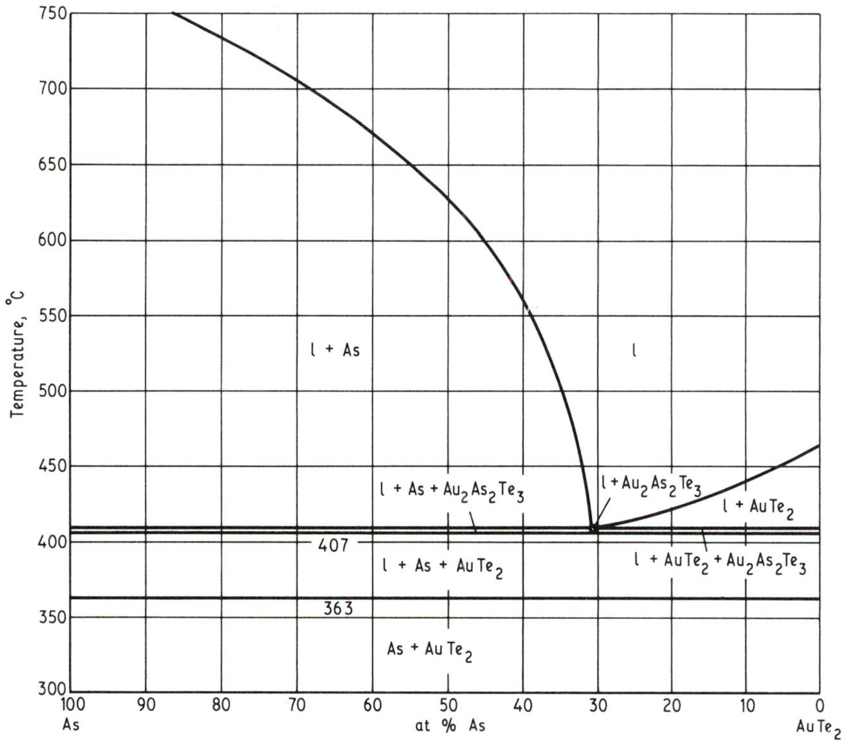

95 **The As–$AuTe_2$ section**

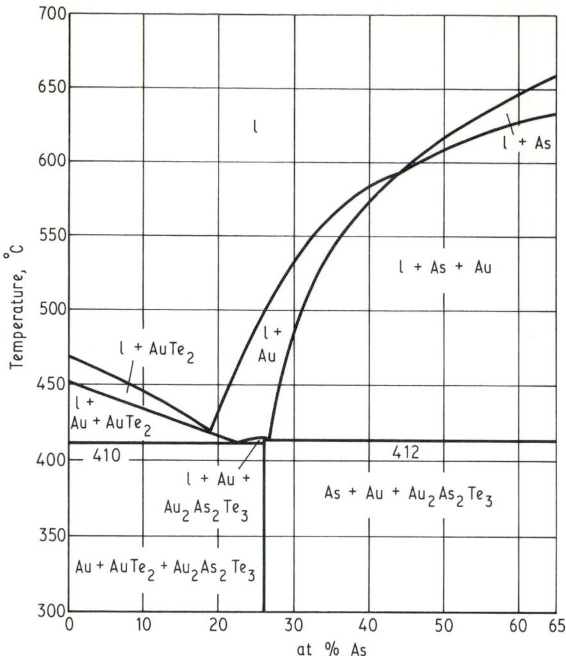

96 The 35 at.-% Au section

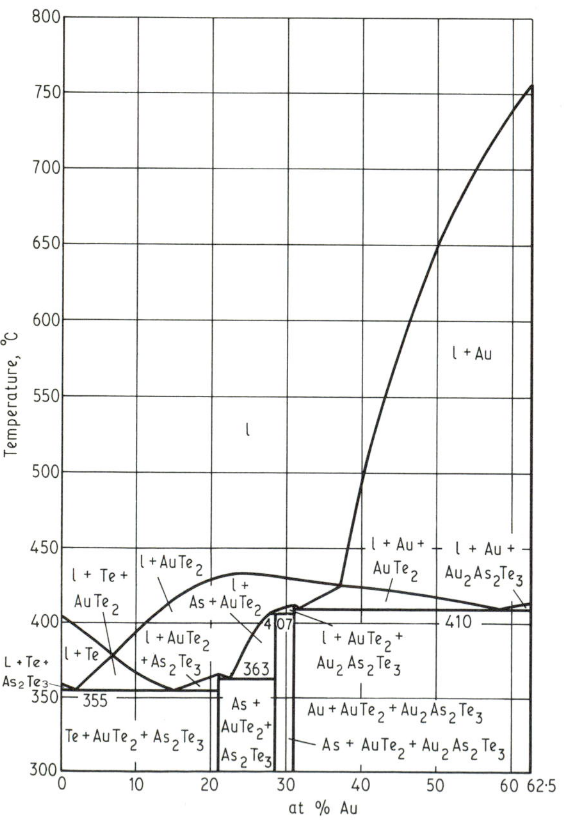

97 The 15 at.-% As section

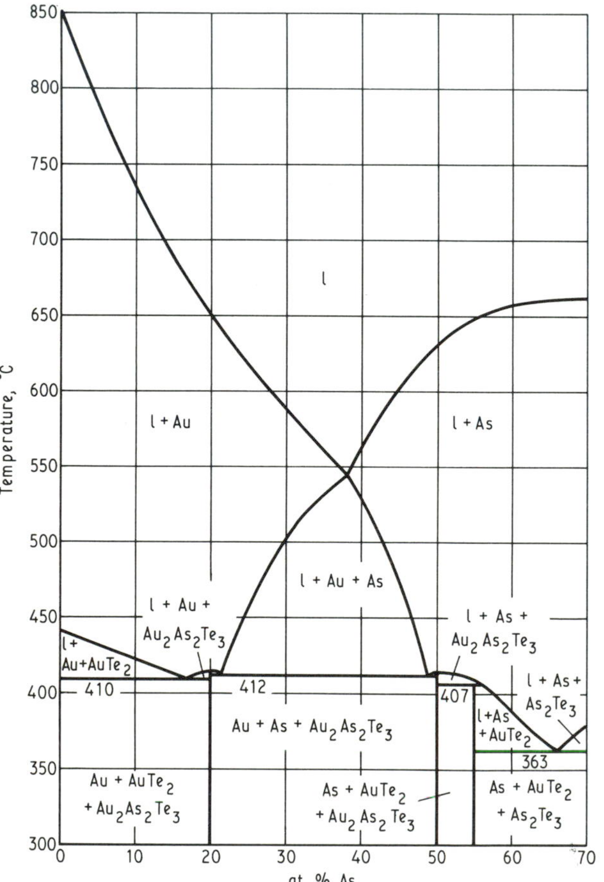

98 The 30 at.-% Te section

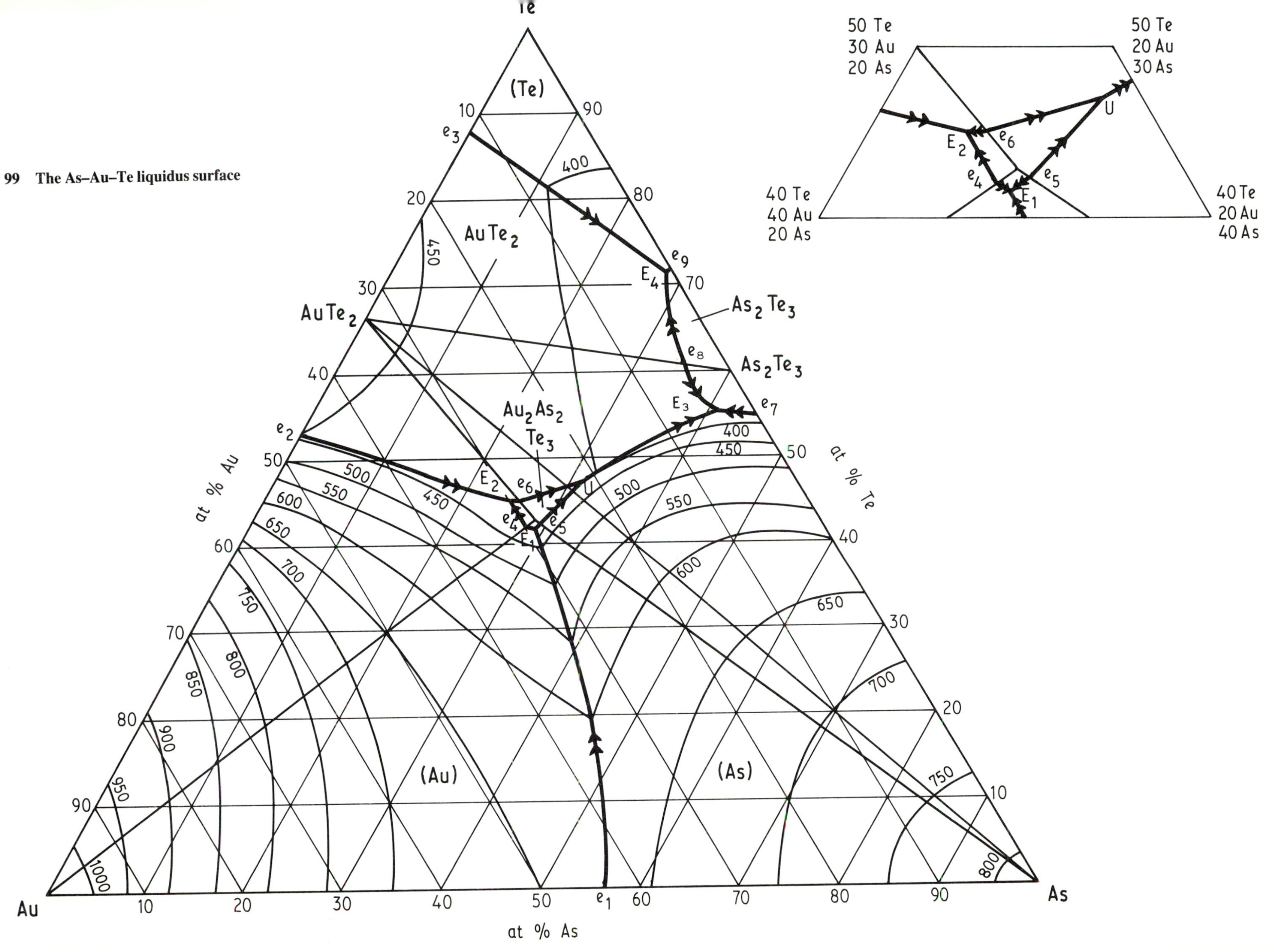

99 The As–Au–Te liquidus surface

REFERENCE

84 Mat: F. Matsuda and K. Nakata, *Gold Bull.*, 1984, **17**, 55–61

Au–Be–Si

[86 Rei] state that they have determined the Au–Be–Si phase diagram. The published data indicates the existence of a ternary eutectic E_1 at 591 ± 5°C, Fig. 100, whose composition scales from the published diagram as 54·1 Au, 37·7 Be, 8·2 at.-% Si. A second ternary eutectic E_2 at 357 ± 5°C occurs at 80 Au, 3 Be, 17 at.-% Si. An alloy containing 70 Au, 15 Be, 15 at.-% Si has a liquidus temperature of 610°C; it solidifies by separating primary Si, meets the monovariant curve e_1U_2 at 553°C with the secondary separation of Si + Au_2Be, undergoes the transition reaction L + Au_2Be = Au_3Be + (Si) at U_2, separates Si + Au_3Be from U_2 to E_2 where solidification ends with the eutectic reaction L = (Au) + (Si) + Au_3Be. It is probable that the section Au_2Be − Si is a pseudobinary eutectic. The ternary eutectic E_1 would be associated with the reaction L = Au_2Be + Au_4Be_3 + (Si). The transition reaction at U_1, L + AuBe = Au_4Be_3 + (Si), is speculative. The phase Au_4Be_3 transforms at 555°C in the Au–Be binary system into Au_2Be + AuBe. It is probable that a ternary eutectoid reaction, Au_4Be_3 = Au_2Be + AuBe + (Si), also occurs.

REFERENCE

86 Rei: D. F. Reich, D. J. Fray, A. F. Evason, J. R. A. Cleaver and H. Ahmed, *Microelectron. Eng.*, 1986, **5**, 171–178

Au–Bi–Cd

The only publication [1894 Hey] on the Au–Bi–Cd system dealt with the examination of the depression of the freezing point of a 4·79 at.-% Au, 95·21 at.-% Bi alloy on addition of Cd up to 12·38 at.-%. The behaviour is similar to that found in the more extensive work on the Au–Cd–Sn system [1891 Hey]. There is almost certainly a pseudobinary system between a phase containing a Au : Cd ratio of 1:1 and (Bi) with a reaction temperature of 266·4°C referred to the melting point of Bi measured at 267·54°C (compared with the currently accepted value of 271·44°C). The Bi used [1894 Hey] must have contained substantial impurities. By analogy with the Au–Cd–Sn system there is a possibility of the presence of a ternary compound AuBiCd.

The section examined (Fig. 101) indicates an increase in liquidus temperature on addition of Cd. This is followed at 1·67 at.-% Cd by the appearance of an invariant reaction horizontal at 264°C. It is not known what reaction is associated with this horizontal beyond the melt, (Bi), $AuBi_xCd$ (x = 0 or 1), and a fourth phase. Beyond the horizontal this curve rises to a maximum at 266·4°C and thereafter falls. The parabolic-shaped curve represents secondary separation of (Bi) + $AuBi_xCd$. Beyond approximately 10·8 at.-% Cd the curve changes slope.

Given that a pseudobinary system exists, $L_e \rightleftharpoons AuBi_xCd$ + (Bi), with a reaction temperature only 1·1°C below the melting point of Bi, it would appear that the eutectic point e lies very close to pure Bi. The path of the monovariant curves descending from e cannot be derived from the data available.

REFERENCES

1891 Hey: C. T. Heycock and F. H. Neville: *J. Chem. Soc.*, 1891, **59**, 936–966

1894 Hey: C. T. Heycock and F. H. Neville: *J. Chem. Soc.*, 1894, **65**, 65–76

Au–Bi–Hg
(C. Gumiński)

Differential thermal analysis of the Au–Bi–Hg system was performed by [77 Tre] using components of purity greater than 99·99%. The liquidus isotherms (Fig. 102) were constructed using the data of [77 Tre] and data from Hansen for the binary systems. Data points from modern evaluations of the Au–Bi system [83 Oka] and the Au–Hg system [87 Oka] are shown circled in Fig. 102 with their corresponding temperatures. Although the uncertainty with regard to the ternary liquidus projection is signified by using dashed lines for the isotherms in Fig. 102, the non-availability of the thesis of [77 Tre] precluded the correction of the isotherms.* No formation of a ternary compound was noted [76 Koz, 77 Tre].

The Hg-rich alloys were intensively investigated using X-rays [76 Koz], hydrostatic separation [76 Koz], calorimetry [24 Tam], potentiometry [65 Nig, 76 Koz, 77 Gum], amalgam polarography [65 Nig] and voltammetry [77 Gum]. There is complete agreement from all the experiments that no interaction leading to the formation of intermetallic compounds between Au and Bi in liquid Hg is observed. Even in the mutually saturated amalgam at 25°C the solubilities determined are practically the same as in the binary amalgams [76 Koz]; however at higher temperatures some influence might be observable.

* The general shape of the isotherms in Fig. 102, taken with the known binary invariant reactions, suggests that a eutectic monovariant curve descends from e_1 and ends near the Hg corner. Similarly a monovariant peritectic curve could descend from p_3 to meet the Au–Hg monovariant peritectic curve descending from p_1 in the area enclosed by the 400° and 300°C isotherms. A series of ternary transition reactions would then precede a degenerate ternary eutectic reaction. L + (Au) $\rightleftharpoons \alpha_1$ + Au_2Bi, L + $\alpha_1 \rightleftharpoons \zeta$ + Au_2Bi, L + $\zeta \rightleftharpoons Au_2Hg$ + Au_2Bi, L + $Au_2Bi \rightleftharpoons Au_2Hg$ + (Bi) and L $\rightleftharpoons Au_2Hg$ + (Bi) + (Hg) are suggested reactions. *A. Prince*

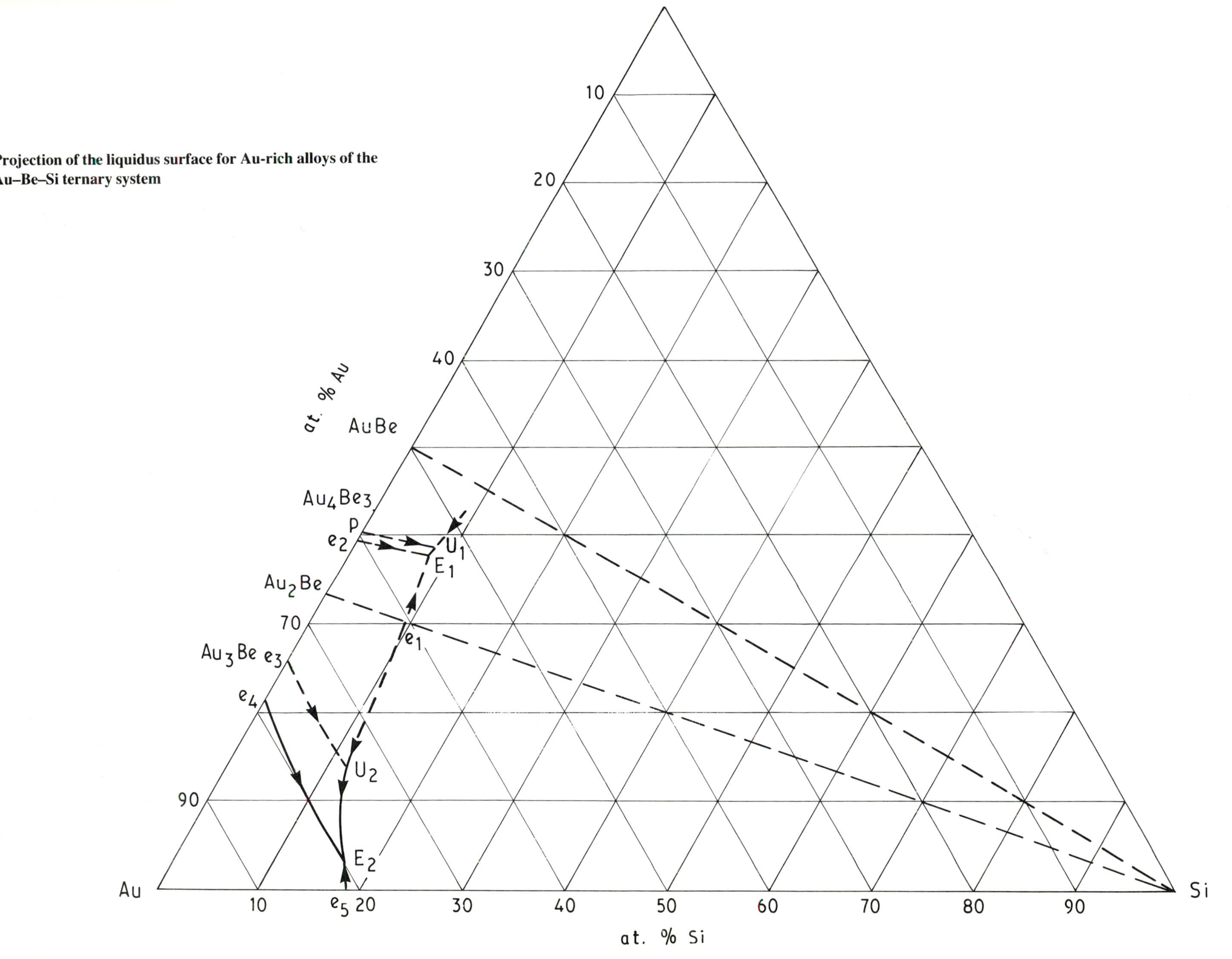

100 **Projection of the liquidus surface for Au-rich alloys of the Au–Be–Si ternary system**

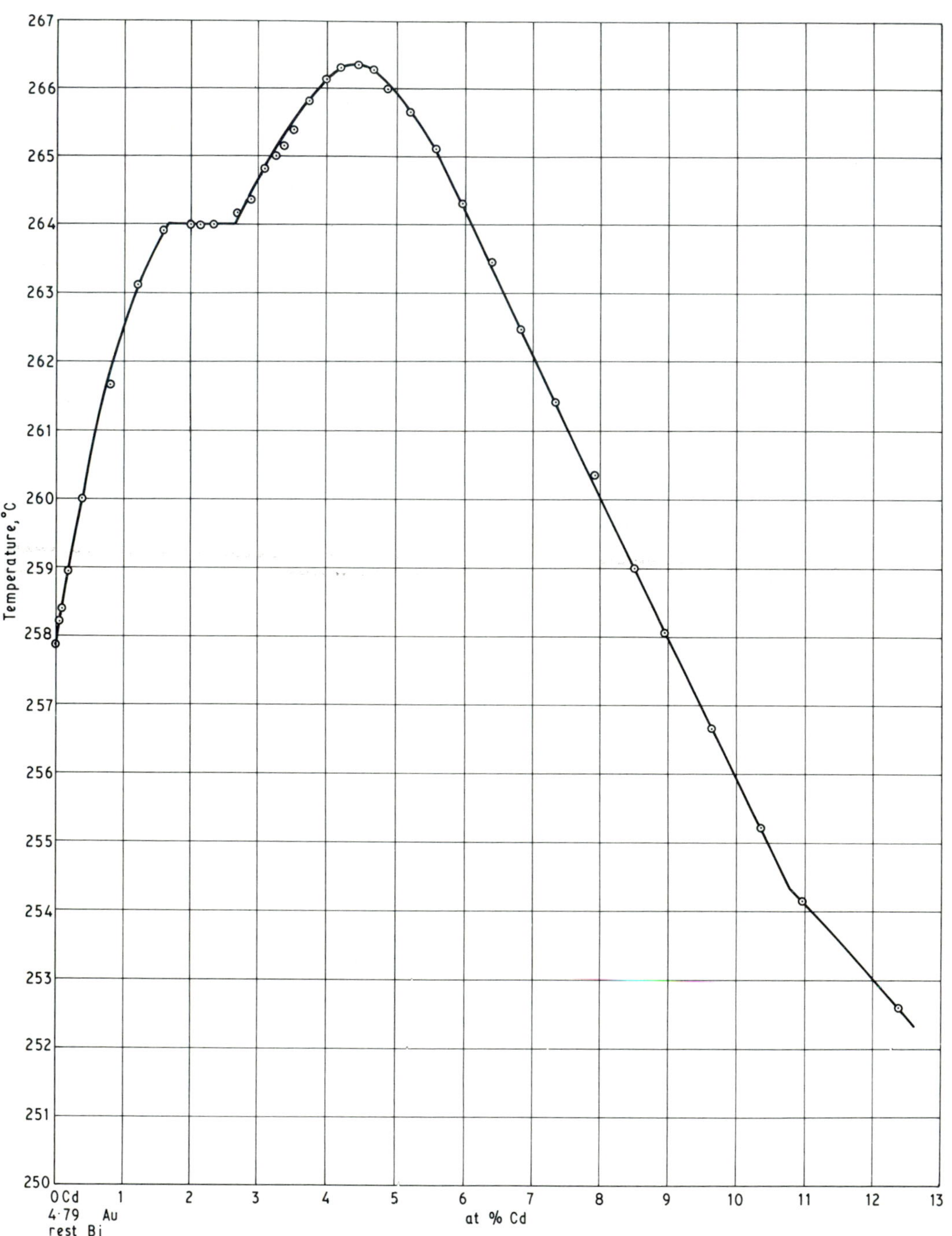

101 Experimental data [1894 Hey] relating the depression of the freezing point to alloy composition for the 4·79 at.-% Au, 95·21 at.-% Bi–Cd section

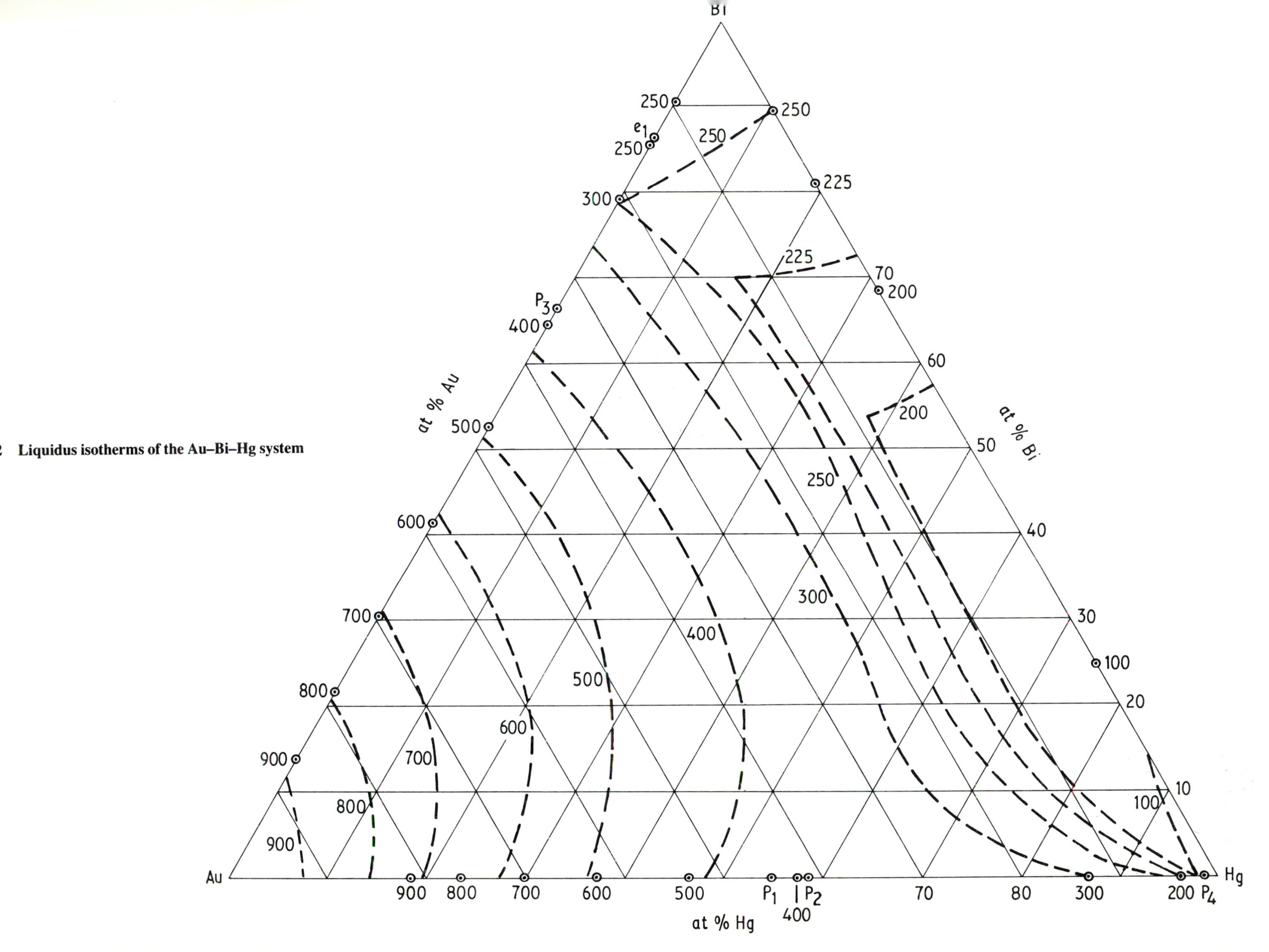

102 Liquidus isotherms of the Au–Bi–Hg system

REFERENCES

24 Tam: G. Tammann and E. Ohler, *Z. anorg. Chem.*, 1924, **135**, 118–126

65 Nig: A. A. Nigmatulina, *Sbor. Stat. Aspir. Kaz. Univ., Khim. Khim. Tekhnol*, 1965 (3–4), 164–167

76 Koz: L. F. Kozin, M. B. Dergacheva and N. L. Nikushkina, *Trudy Inst. Khi. Nauk., Akad. Nauk. Kaz. SSR*, 1976, **42**, 82–87

77 Tre: A. A. Trebukhov, Ph.D. Thesis, *Akad. Nauk. Kaz. SSR, Alma-Ata*, 1977; quoted in M. V. Nosek and N. M. Atamanova, 'Amalgamnye Sistemy', *Nauka, Alma-Ata*, 1980, pp. 113–115

77 Gum: C. Gumiński and Z. Galus, *J. Electroanal. Chem.*, 1977, **83**, 139–150

83 Oka: H. Okamoto and T. B. Massalski, *Bull. Alloy Phase Diagrams*, 1983, **4**, 401–407

87 Oka: H. Okamoto and T. B. Massalski, 'Phase Diagrams of Binary Gold Alloys', ASM, Metals Park, Ohio, 1987, pp. 132–139

Au–Bi–Li
Au–Ga–Li
Au–Ge–Li
Au–In–Li
Au–Li–Mg
Au–Li–Pb
Au–Li–Sb
Au–Li–Si
Au–Li–Sn
Au–Li–Tl
Au–Mg–Sn
Au–Sc–Sn

Intermetallic compounds of the composition $Li_2M^1M^{3-5}$ occur in many of the ternary systems of Li with Group 1B elements (M^1 = Cu, Ag, Au) and Group 3B to 5B elements (M^{3-5} = Al, Ga, In, Tl, Si, Ge, Sn, Pb, Sb, Bi). The ternary compounds Li_2AuM^{3-5}, where M^{3-5} = Ga, In, Tl, Si, Ge, Sn, Pb, Sb, Bi, were reported by [68 Pau 1]. The compound Li_2AuSn had been reported earlier by [66 Sch] with subsequent studies by [69 Sch] and [80 Ebe]. A ternary compound Li_2AuMg was identified by [68 Pau 2]. There is agreement between various investigators that these ternary compounds crystallize with a fcc structure containing 16 atoms in the unit cell. Lattice parameter values are summarized in Table 30. There is less agreement on the occupancy of atom sites in the crystal structure adopted by Li_2AuM^{3-5} compounds. According to [68 Pau 1] the compounds Li_2AuBi, Li_2AuGa, Li_2AuIn, Li_2AuSn, Li_2AuSb, and Li_2AuPb have ordered arrangements of the three atom species on lattice sites with four Li atoms on 4a sites 000, four Li atoms on 4b sites ¼¼¼, four Au atoms on 4c sites ½½½ and four Group 3B–5B atoms on 4d sites ¾¾¾. The structure was regarded as an ordered ternary Zintl phase. For Li_2AuBi and Li_2AuTl there is an equal probability of the lattice containing a random distribution of Au and Bi or Tl atoms on atom sites (B32A type structure). Li_2AuGe is 50% ordered arrangement of Au and Ge atoms and 50% random occupancy of atomic sites. Li_2AuSi was stated to be B32A type. [68 Pau 1] noted that the samples of Li_2AuBi and Li_2AuTl contained a considerable proportion of another unspecified phase. [69 Sch] claim Li_2AuSn to have a modified Li_3Bi — type structure ($L2_1C$ type). They further noted that Li_2AuSi yielded a diffraction pattern with additional reflections. In a later publication [75 Sch 2] confirmed that the Li_2AuSi composition contained extraneous diffraction lines and claimed that the ternary fcc compound has the composition $Li_8Au_3Si_5$ with lattice parameter a = 0·603(7) nm. [80 Ebe] are critical of the alloy preparation previously adopted by [68 Pau 1] and [69 Sch] and state that X-ray examination of single crystals isolated from the alloys show that the structure is a 'filled' zinc blende type for Li_2AuM^{3-5} alloys. There is a pressing need for confirmatory X-ray analysis of single crystals of the compounds Li_2AuM^{3-5} and for their study by additional techniques to that of X-ray diffraction. In none of the work reported to date has metallography been used to check whether alloys are homogeneous single phase or contain other phases. Alloys have not been proved to be in an equilibrium condition, and little attempt has been made to anneal isothermally until equilibrium was established.

Table 30. Lattice constants of Li_2AuM^{3-5} compounds

Compound	*Lattice parameter*, nm				*Colour*
	[68 Pau 1]	[68 Pau 2]	[66 Sch] [69 Sch] [80 Ebe]	[68 Pau 3]	[68 Pau 1]
Li_2AuBi	0·6650 ± 0·0010				grey
Li_2AuGa	0·6188 ± 0·0003				greenish-yellow
Li_2AuGe	0·6170 ± 0·0001				yellowish-orange
Li_2AuIn	0·6466 ± 0·0003			0·6466 ± 0·0003	greenish-yellow
Li_2AuMg		0·6442 ± 0·0001			
Li_2AuPb	0·6601 ± 0·0001				violet
Li_2AuSb	0·6442 ± 0·0003			0·6441 ± 0·0005	grey
Li_2AuSi	0·6078 ± 0·0003				yellowish
Li_2AuSn	0·6438 ± 0·0003		0·641(7)		reddish-orange
Li_2AuTl	0·6549 ± 0·0005				greenish-yellow

[68 Pau 3] present data for the AuLi–InLi and AuLi–'LiSb' sections. Only X-ray powder diffraction data are quoted (Table 31). The B32 type congruently melting compound InLi forms a complete series of solid solutins with Li_2AuIn. Alloys with more Au than Li_2AuIn are not single phase. In the AuLi–'LiSb' section only the composition Li_2AuSb was single phase; all other alloys were heterogeneous. A second ternary compound, AuLiSb, has been reported in the Au–Li–Sb system [75 Sch 1]. It has a fcc structure of the ordered ternary Zintl phase type (space group T_d^2–F43m) and is reddish violet in colour. Phase analysis along the section AuLiSb–Li_2AuSb indicated that AuLiSb is homogeneous to 37·5 at.-% Li, 31·25 at.-% Au, 31·25 at.-% Sb. With increased Li content a distorted tetragonal structure was observed, with a decreasing *c/a* ratio as the Li content increased, but this phase disappeared before the composition Li_2AuSb was reached. The lattice parameter of AuLiSb is 0·6328 nm. [75 Sch 1] note that AuLiSb is a stable phase whereas Li_2AuSb decomposes within a few weeks at room temperature to $AuLi_3$ and $AuSb_2$. This is the only comment on the instabiity of the Li_2AuM^{3-5} ternary compounds. The observation requires confirmation with respect to all the reported Li_2AuM^{3-5} compounds.

[69 Sch] stated that Li_2AuSn has only a small region of solid solution in the direction towards Li_2AuSn (along the 25% Sn section). At the composition 30 at.-% Au, 45 at.-% Li, 25 at.-% Sn another phase was observed, which increases as the Li content decreases. The colour of the alloys varies from bright orange for Li_2AuSn, a distinctly weaker colour at 30 at.-% Au, 45 at.-% Li and a metallic grey at 35 at.-% Au, 40 at.-% Li.

[80 Ebe] report the occurrence of the ternary compounds AuMgSn and AuScSn with lattice parameters of 0·6410 and 0·6422 nm respectively. They crystallize in a 'filled' zinc blende type structure and are considered to be ordered ternary Zintl-phases. The atom positions are given as four Mg (Sc) in 4b ½½½, four Au in 4c ¼¼¼ and four Sn in 4a 000. AuMgSn has a reddish violet colour whereas AuScSn is grey. [76 Dwi] summarized data for Au–Rare Earth–Sn compounds. In contrast to [80 Ebe], AuScSn is assigned the AgAsMg-type structure with four Au atoms in 4a 000, four Sc atoms in 4c ¼¼¼, and four Sn atoms in 4d ¾¾¾. The atom positions found by [80 Ebe] and [76 Dwi] are equivalent. [76 Dwi] finds an identical lattice parameter to [80 Ebe] for AuScSn.

REFERENCES

66 Sch: H.-U. Schuster: *Naturwiss.*, 1966, **53**, 360–361

68 Pau 1: H. Pauly, A. Weiss and H. Witte: *Z. Metall.*, 1968, **59**, 47–58

68 Pau 2: H. Pauly, A. Weiss and H. Witte: *Z. Metall.*, 1968, **59**, 414–418

68 Pau 3: H. Pauly, A. Weiss and H. Witte: *Z. Metall.*, 1968, **59**, 554–558

69 Sch: H.-U. Schuster, D. Thiedemann and H. Schönemann: *Z. anorg. Chem.*, 1969, **370**, 160–170

75 Sch 1: H.-U. Schuster and W. Dietsch: *Z. Naturf.*, 1975, **30B**, 133

75 Sch 2: H.-U. Schuster and W. Seelentag: *Z. Naturf.*, 1975, **30B**, 804

76 Dwi: A. E. Dwight, 12th Rare Earth Res. Conf., 1976, Vol. II, 480–489

80 Ebe: U. Eberz, W. Seelentag and H.-U. Schuster: *Z. Naturf.*, 1980, **35B**, 1341–1343

Au–Bi–S

[76 Gat] has speculated on the equilibria present in the Au–Bi–S system, assuming condensed phases only are present. Figure 103, adapted from [76 Gat] presents a schematic view of the liquidus projection. Four invariant reactions are envisaged, $L_1 \rightleftharpoons L_2 + (Au) + Bi_2S_3$; $L_U + (Au) \rightleftharpoons Au_2Bi + Bi_2S_3$; $L_{E_1} \rightleftharpoons (Bi) + Au_2Bi + Bi_2S_3$ and $L_{E_2} \rightleftharpoons (Au) + (S) + Bi_2S_3$.

REFERENCE

76 Gat: B. Gather: Thesis, 1976, Technischen Universität Clausthal

Table 31. Lattice constants of the ordered ternary Zintl phase in the AuLi–InLi and AuLi–'LiSb' sections [68 Pau 3]

Alloy composition, at.-%			*Lattice parameter*, nm	*Colour*
Au	*In*	*Li*		
0	50	50	0·6792 ± 0·0001	grey
6·25	43·75	50	0·6704 ± 0·0003	grey
12·5	37·5	50	0·6637 ± 0·0002	grey (violet tinge)
18·75	31·25	50	0·6549 ± 0·0003	violet
21·25	28·75	50	0·6517 ± 0·0002	golden yellow
25	25	50	0·6466 ± 0·0003	greenish yellow
31·25	18·75	50	0·6468 ± 0·0004	—
37·5	12·5	50	0·6470 ± 0·0005	—
Au	*Sb*	*Li*		
17·5	32·5	50	0·6428 ± 0·0002	
22·5	27·5	50	0·6434 ± 0·0002	
25	25	50	0·6441 ± 0·0005	
32·5	17·5	50	0·6453 ± 0·0005	

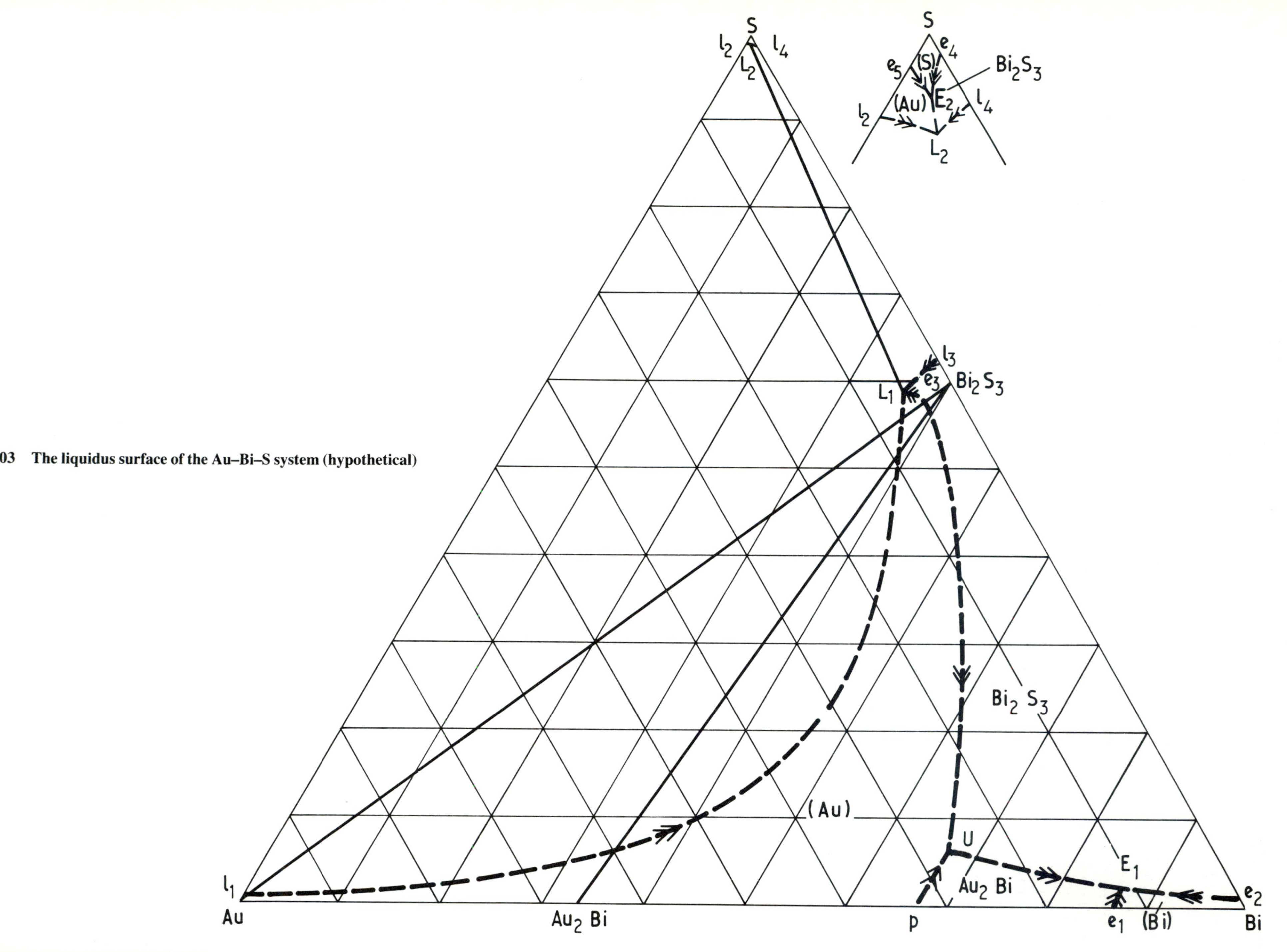

103 The liquidus surface of the Au–Bi–S system (hypothetical)

Au–Bi–Sb

INTRODUCTION

The only publication [81 Gat] is a study of alloys prepared from 99·99% Au and 99·995% Bi and Sb in 1 g quantities by melting in evacuated silica capsules. All alloys were subsequently annealed at 220°C for 12 months. After 6 and 8 months the alloys had not attained equilibrium. Differential thermal analysis, differential scanning calorimetry, X-ray diffraction, and electron beam microprobe techniques were used to establish vertical and isothermal sections. The Au–Bi–Sb system contains two invariant reactions, a transition reaction at 288 ± 2°C (L_U + (Au) $\rightleftharpoons$ Au_2Bi + $AuSb_2$) and a eutectic reaction at 238·6 ± 1°C ($L_E \rightleftharpoons$ (Bi/Sb) + Au_2Bi + $AuSb_2$). The compositions of the liquid phase associated with the two reactions are given in Table 32. There is uncertainty in the liquid compositions [81 Gat] and examination of the published data suggests that the point U (Fig. 105) could be at 29 at.-% Bi, 22 at.-% Sb. This composition is still within the range quoted by [81 Gat]. The composition of the Bi/Sb solid solution at 238·6°C is quoted as 96·0 ± 0·5 at.-% Bi, 4·0 ± 0·5 at.-% Sb with <0·1 at.-% Au. There must be some measurable solubility of Au in the Bi–Sb solid solution series but it appears to be very slight. The compound $AuSb_2$ takes a maximum of ~4·5 at.-% Bi into solution at 220°C. The binary compound has a negligible phase width; the ternary solution has a width of ~1 at.-% and occurs at a constant Au:Sb ratio.

The liquidus falls steeply from the Au corner to the monovariant curves e_1U and Up_2. The central region of the ternary system from e_1 to the Bi corner is a relatively flat region.

Table 32. Invariant equilibria

Reaction	*Temperature*, °C	*Phase*	*Composition*, at.-%	
			Bi	*Sb*
U	288	L	28·0 ± 1·5	23·0 ± 1·5
E	238·6	L	73·5 ± 2·0	8·2 ± 1·7
		Bi/Sb	96·0 ± 0·5	4·0 ± 0·5
		$AuSb_2$	4·5	~62·2

BINARY SYSTEMS

The Au–Bi system [83 Oka], with a peritectic temperature of 377·5°C [83 Eva], is accepted. The Au–Sb system evaluated by [84 Oka] indicates a peritectic liquid containing 66·6 at.-% Sb, only slightly differing from the composition of the compound $AuSb_2$. [84 Eva] report the peritectic liquid at 65·6 at.-% Sb at 458 ± 1°C. The eutectic temperature is 356 ± 1°C at 36·1 at.-% Sb. The Bi–Sb system evaluated by [58 Han] is accepted.

SOLID PHASES

Table 33 summarizes crystal structure data for the phases present in the Au–Bi–Sb system. No ternary compounds exist.

Table 33. Crystal structures

Solid phase	*Prototype*	*Lattice designation*
(Au)	Cu	cF4
(Bi)	As	hR2
(Sb)	As	hR2
Au_2Bi	Cu_2Mg	cF24
$AuSb_2$	FeS_2	cP12

INVARIANT EQUILIBRIA

Table 32 and Fig. 104 give details of the invariant equilibria. The phase relationships at 220°C (Fig. 105) were established after very long term annealing (12 months). The monovariant eutectic curve e_1U meets the monovariant peritectic curve p_2U at the point U whose composition is given [81 Gat] as 49·0 ± 1·5 at.-% Au, 28·0 ± 1·5 at.-% Bi, 23·0 ± 1·5 at.-% Sb. U is the corner of the four-phase plane at 288 ± 2°C corresponding to the liquid composition in the invariant reaction L_U + (Au) $\rightleftharpoons$ Au_2Bi + $AuSb_2$. Au dissolves <0·1 at.-% Bi and Sb, the phase Au_2Bi exists as a line compound with negligible homogeneity range, whereas $AuSb_2$ dissolves a maximum of ~4·5 at.-% Bi whilst maintaining the Au:Sb ratio constant. The location of the curves e_1U and p_2U were established with reference to the vertical sections at 20 at.-% Bi (Fig. 106), 60 at.-% Bi (Fig. 107), 20 at.-% Au (Fig. 108), and 20 at.-% Sb (Fig. 109). The intersection of the 20 at.-% Sb section with e_1U and p_2U appears to occur at slightly lower Bi contents (Fig. 109) than shown by [81 Gat]. The data can be interpreted as giving a composition for U of 49 at.-% Au, 29 at.-% Bi, 22 at.-% Sb. However this composition is within the limits of composition given [81 Gat] for U (Table 32).

From U the monovariant eutectic curve UE descends to a ternary eutectic point E of composition 18·3 ± 1·2 at.-% Au, 73·5 ± 2·0 at.-% Bi, 8·2 ± 1·7 at.-% Sb (Table 32). At the ternary eutectic temperature of 238·6 ± 1°C the eutectic liquid separates Au_2Bi, $AuSb_2$ containing 4·5 at.-% Bi in solution and a Bi/Sb solid solution whose composition is given as <0·1 at.-% Au, 96·0 ± 0·5 at.-% Bi, 4·0 ± 0·5 at.-% Sb. This composition implies a very slight solubility of Au in the Bi–Sb solid solution series. In Figs. 106, 107, 109, and 110 the dashed lines outline a schematic representation of the presence of Bi/Sb solid solution containing a small amount of Au. The monovariant curve p_1E will be peritectic in character from p_1, but it will change to a eutectic curve before it has progressed very far into the ternary system.

LIQUIDUS SURFACE

A projection of the liquidus surface is illustrated in Fig. 112. The isotherms are consistent with data derived from the vertical sections to within 1 at.-%.

ISOTHERMAL SECTIONS

Figure 105 represents the 220°C isothermal with the addition of the monovariant curves and the invariant points U and E. A 300°C isothermal section is presented in Fig. 113. The tie lines in the Bi/Sb–$AuSb_2$ phase region were

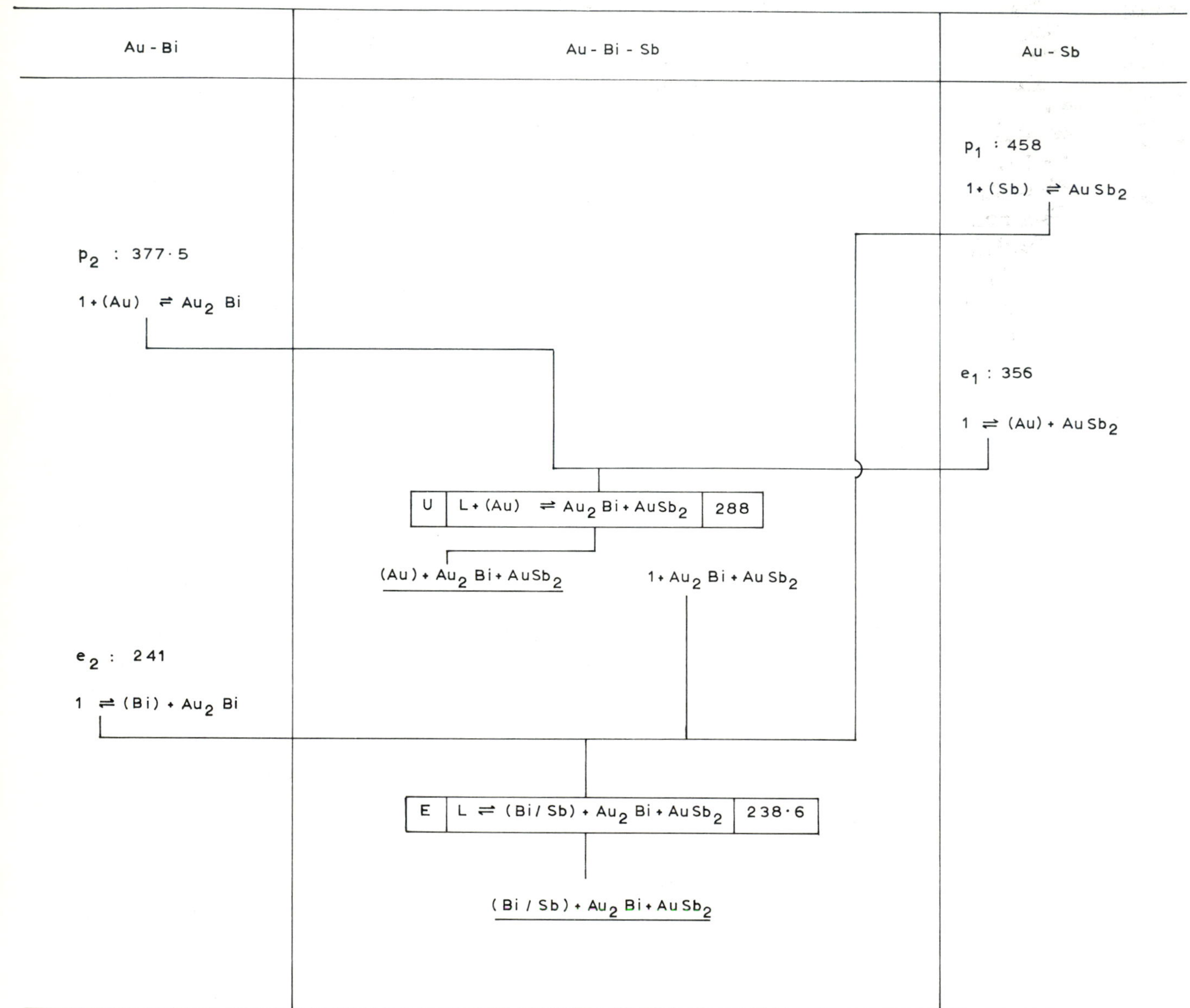

104 Reaction scheme for the Au–Bi–Sb system

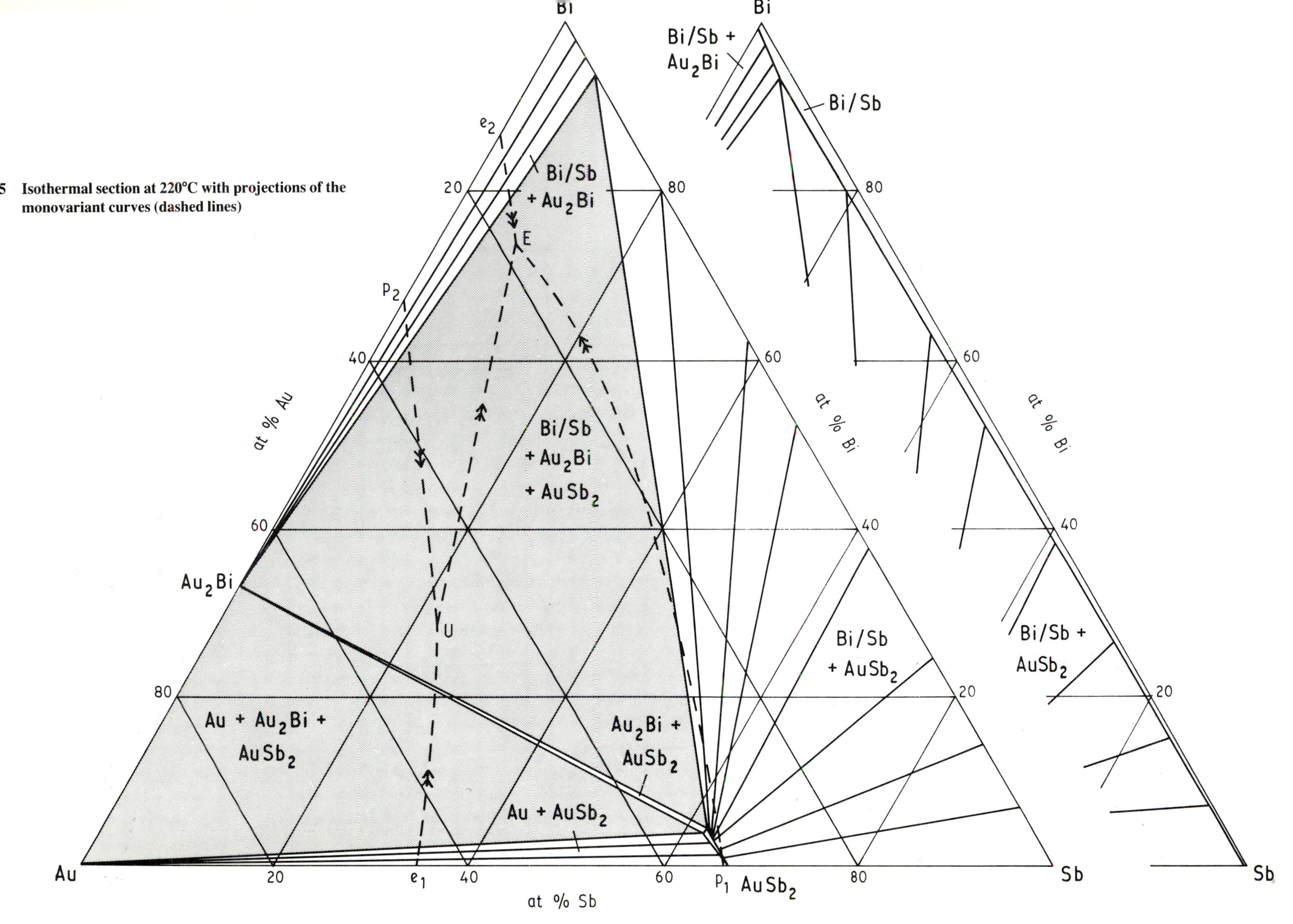

105 Isothermal section at 220°C with projections of the monovariant curves (dashed lines)

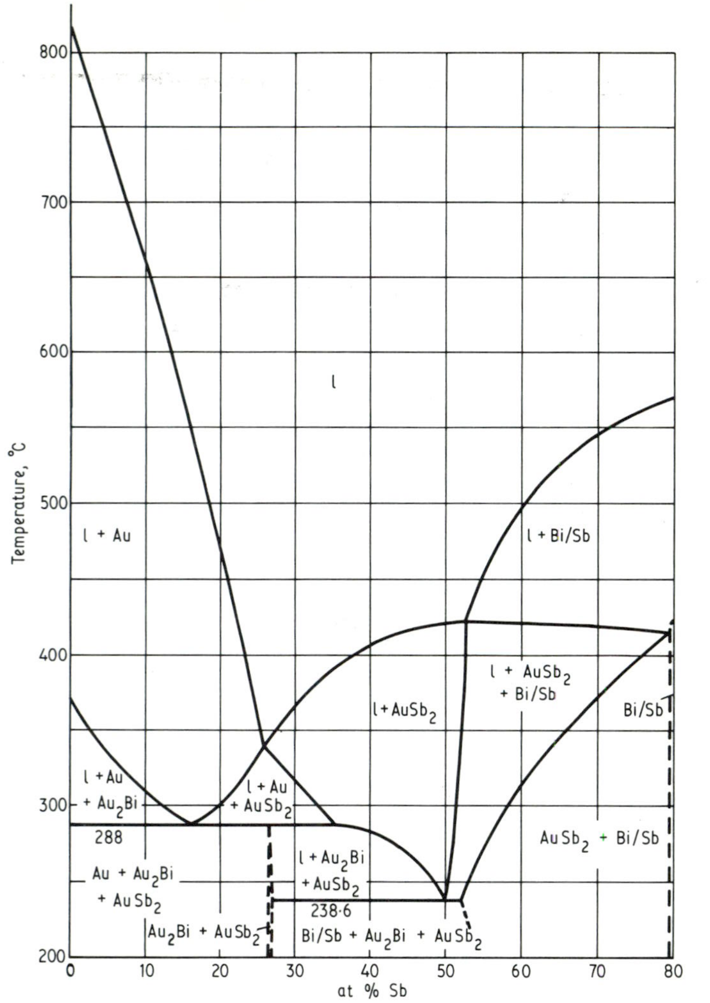

106 The 20 at.-% Bi section

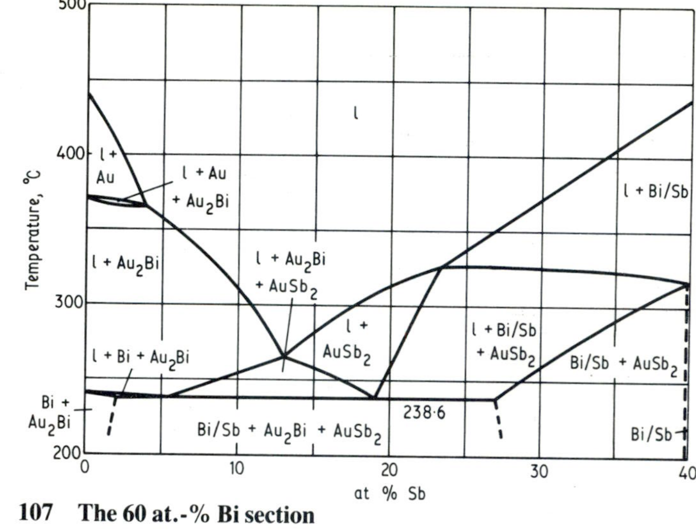

107 The 60 at.-% Bi section

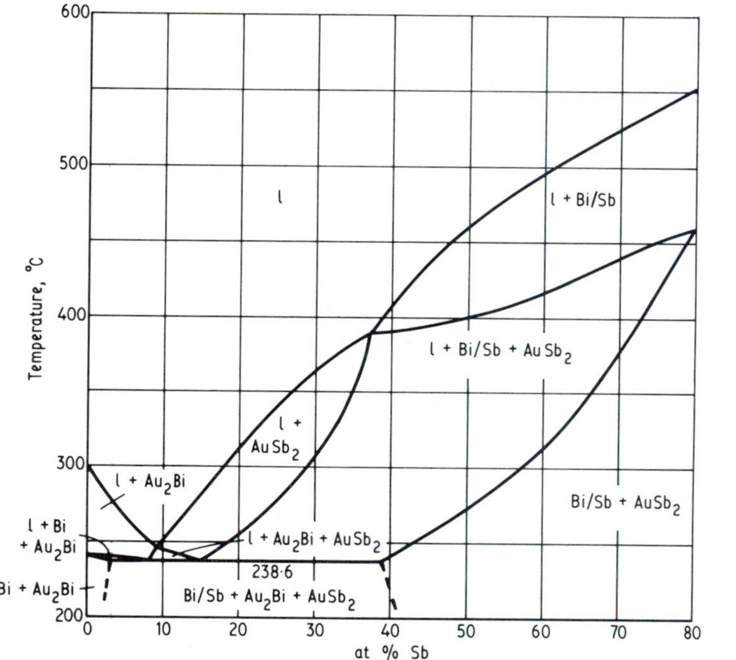

108 The 20 at.-% Au section

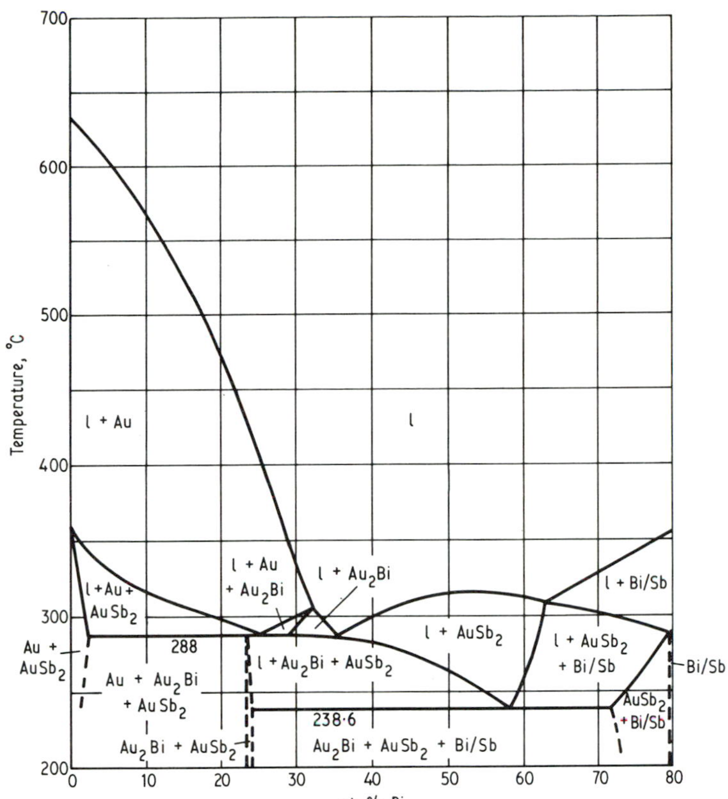

109 The 20 at.-% Sb section

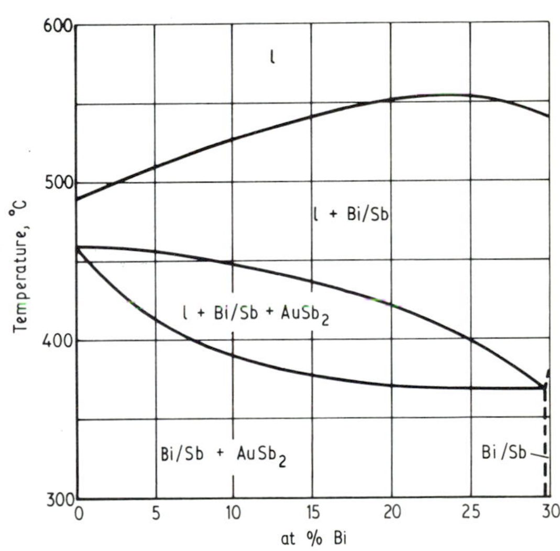

110 The 70 at.-% Sb section

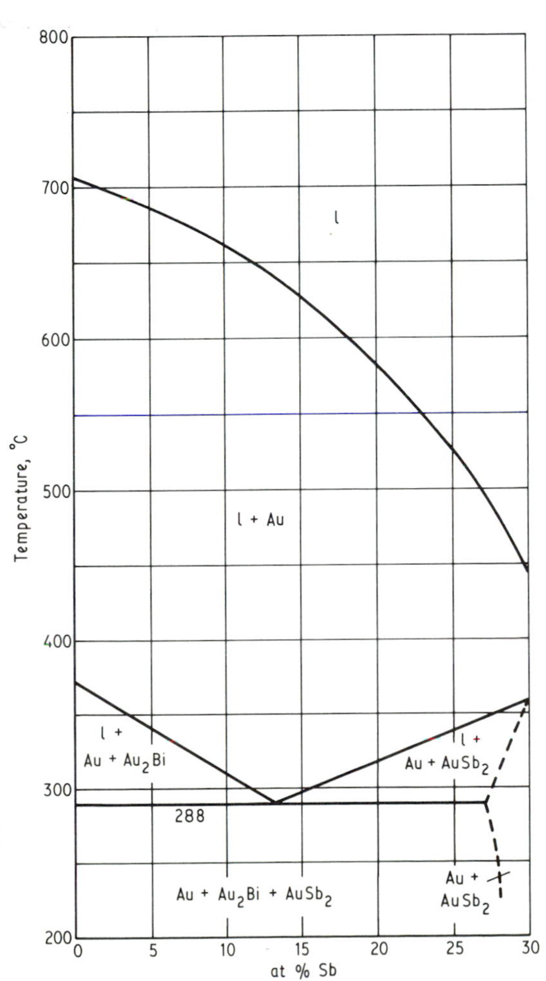

111 The 70 at.-% Au section

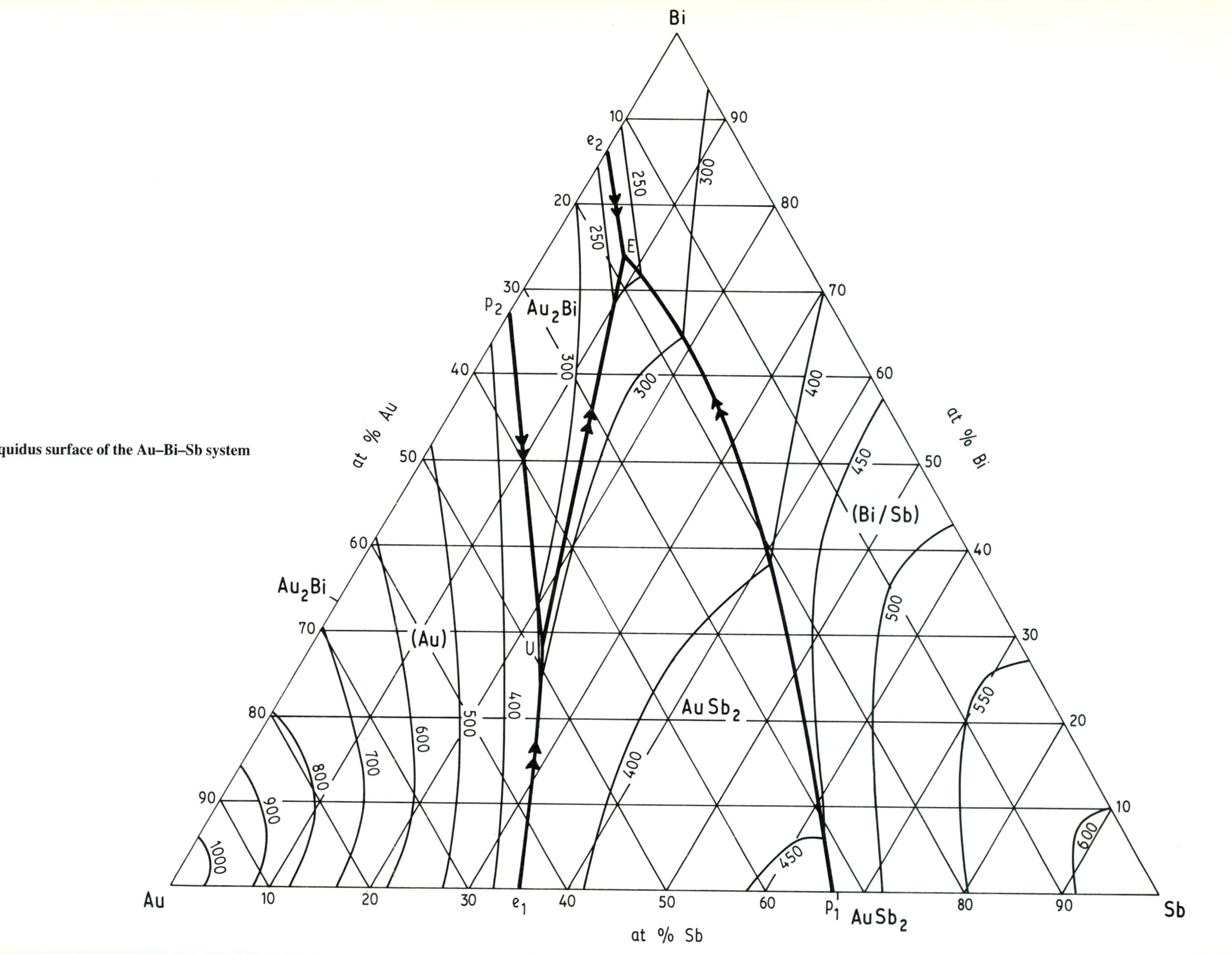

112 Liquidus surface of the Au–Bi–Sb system

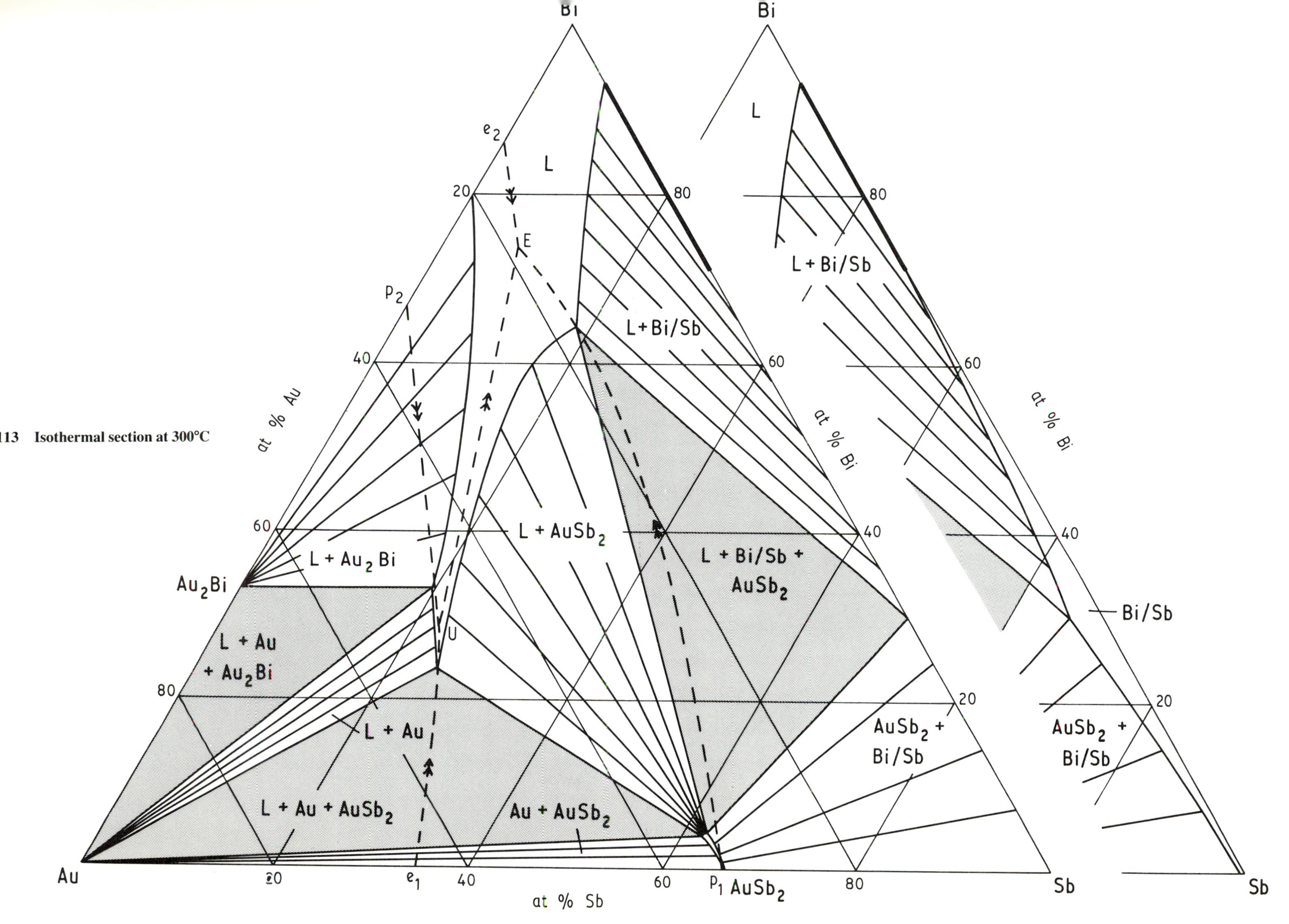

113 Isothermal section at 300°C

determined by electron microprobe analysis of the compositions of the equilibrated phases. The tie lines in the Bi/Sb–L phase region were constructed with the help of the binary tie line at 300°C in the Bi–Sb system. An enlarged schematic representation of the equilibria close to the Bi–Sb edge is given to the right of the Bi–Sb edge in Fig. 113. A similar schematic representation of the phase relationships close to the Bi–Sb edge has been drawn in Fig. 105 for the 220°C isothermal.

REFERENCES

58 Han: M. Hansen and K. Anderko: 'Constitution of Binary Alloys'; 1958, New York, McGraw-Hill

62 Nat: M. W. Nathans and M. Leider: *J. phys. Chem.*, 1962, **66**, 2012–2015

81 Gat: B. Gather: *Z. Metall.*, 1981, **72**, 507–511

83 Oka: H. Okamoto and T. B. Massalski: *Bull. Alloy Phase Diagrams*, 1983, **4**, 401–407

83 Eva: D. S. Evans and A. Prince: CALPHAD XII, Liège, 1983

84 Oka: H. Okamoto and T. B. Massalski: *Bull. Alloy Phase Diagrams*, 1984, **5**, 166–171

84 Eva: D. S. Evans and A. Prince: CALPHAD XIII, Villard de Lans, 1984

Au–Bi–Se

INTRODUCTION

The only published data is a survey [75 Gat] of an unspecified number of alloys made by melting 99·995% purity elements in evacuated quartz capsules. Alloys were subsequently homogenized for 90 days at 300°C and examined by DTA, X-ray, and metallographic methods. Sections were examined at constant Au content in steps of 10 at.-% Au, but only the 50 at.-% Au section was published. The Au–Bi_2Se_3 section has pseudobinary characteristics and this section was also published. Isothermal sections for six temperatures were reproduced with a projection of the monovariant curves associated with the liquid phase. There is good correspondence between the two vertical sections and the relevant isothermal sections, but the absence of published data does not allow the construction of liquidus isotherms for the ternary system.

The ternary system is dominated by a very large field of primary solidification of Au, inside which is a large phase region of liquid demixing. The liquid immiscibility gap associated with the monotectic reaction $l_1 \rightleftharpoons l_2 + (Au)$ at 963°C in the Au–Se system extends into the ternary system and ends in the critical tie line K–Au at 545°C (Fig. 114). The Au–Bi_2Se_3 section divides the ternary system into two. From the eutectic saddle point e_1 at 635°C the partial ternary system Au–Bi–Bi_2Se_3 contains a series of three transition reactions, U_1, U_3, and U_4, with final solidification at the ternary eutectic point E_2. The partial ternary system Au–Se–Bi_2Se_3 contains a ternary monotectic reaction $L_1 \rightleftharpoons L_2 + (Au) + Bi_2Se_3$ at 516°C followed by a transition reaction U_2 and a ternary eutectic reaction E_1, both of which are located very close to the Se corner of the concentration triangle.

BINARY SYSTEMS

The Au–Bi diagram [83 Oka], with a peritectic temperature of 377·5°C [83 Eva], is accepted. The Au–Se diagram [71 Rab] was substantiated by [75 Gat], with the exception of the extent of the liquid miscibility gap at 760°C. According to [71 Rab] this miscibility gap extends to 100 at.-% Se, but the interpretation placed upon the ternary reactions by [75 Gat] is based on the 760°C miscibility gap ending at 99 at.-% Se (l_4, Fig. 115) followed by a peritectic reaction $l_{p_3} + (Au) \rightleftharpoons AuSe$ at 425°C where p_3 is located at 99·5 at.-% Se. Solidification ends at 218°C with the eutectic reaction $l_{e_4} \rightleftharpoons (Se) + AuSe$ where e_4 has the composition 99·9 at.-% Se. Considerable doubt exists on the allotropy of AuSe. [75 Gat] note that both αAuSe and βAuSe were detected together in ternary alloys and, in view of the uncertainty concerning the stability and equilibria between αAuSe and βAuSe, the AuSe compound is referred to as AuSe. The Bi–Se system (Fig. 116) determined by [75 Gat] is accepted.

SOLID PHASES

Table 34 summarizes crystal structure data for the binary compounds. No ternary compounds were found [75 Gat].

Table 34. Crystal structures

Solid phase	*Prototype*	*Lattice designation*
(Au)	Cu	cF4
(Bi)	As	hR2
(Se)	γSe	hP3
Au_2Bi	$MgCu_2$	cF24
AuSe	—	mC24
Bi_3Se_2	—	—
BiSe	—	—
Bi_2Se_3	Bi_2STe_2	hR5

PSEUDOBINARY SYSTEMS

Strictly speaking there are no pseudobinary systems in the Au–Bi–Se ternary system. However the Au–Bi_2Se_3 section (Fig. 117) can be regarded as pseudobinary below 725°. A eutectic occurs at e_1 with the composition 7 at.-% Au, 37·2 at.-% Bi, 55·8 at.-% and temperature 635 ± 3°C.

INVARIANT EQUILIBRIA

Table 35 and Fig. 118 give details of the invariant equilibria. The ternary system is dominated by a large area of liquid immiscibility originating from the monotectic reaction $l_1 \rightleftharpoons l_2 + (Au)$ at 963°C in the Au–Se system. The monovariant equilibrium involving $l_1 + l_2 + (Au)$ descends from 963°C and ends at a critical tie line, K–(Au),

114 The liquidus surface of the Au–Bi–Se system

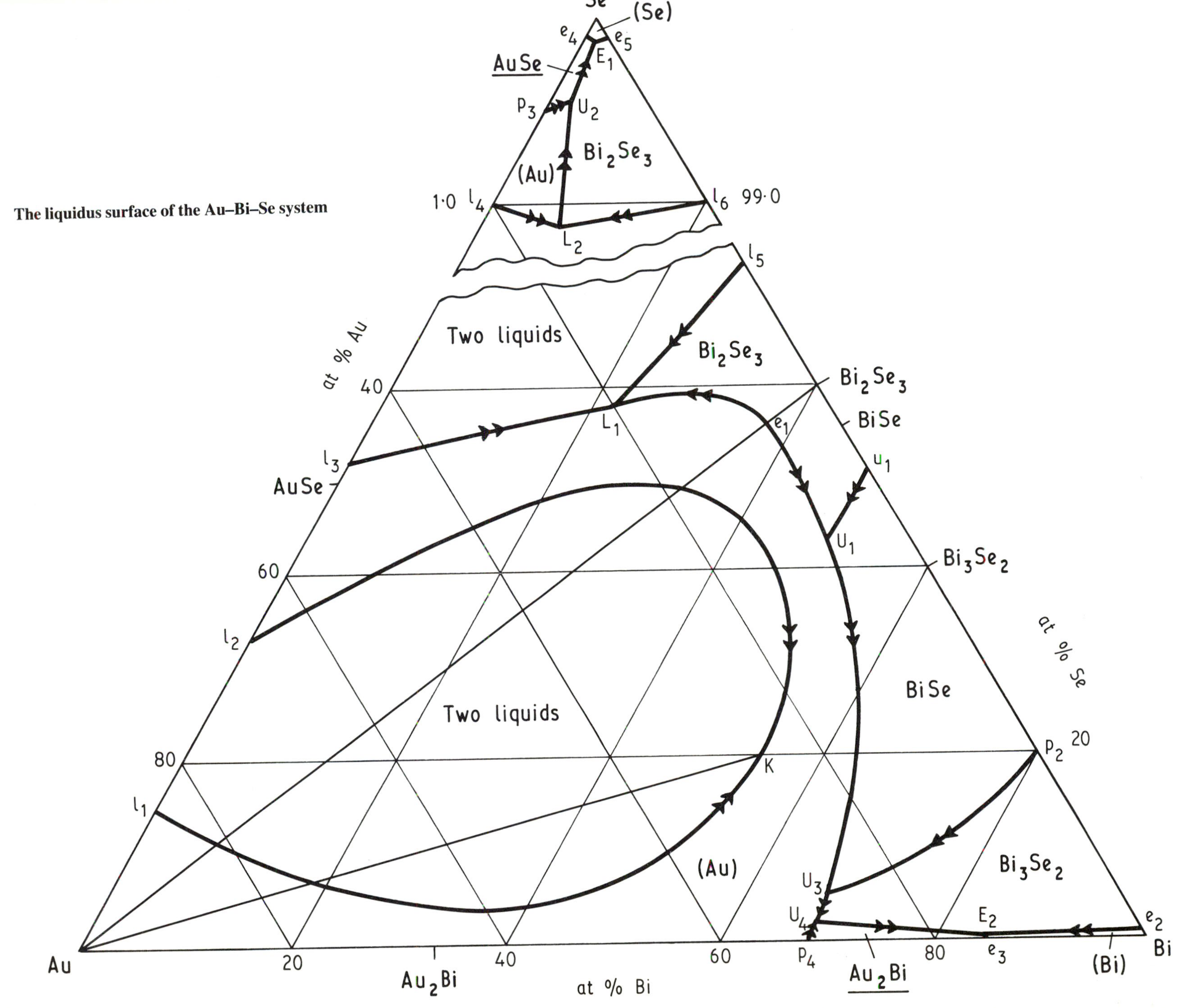

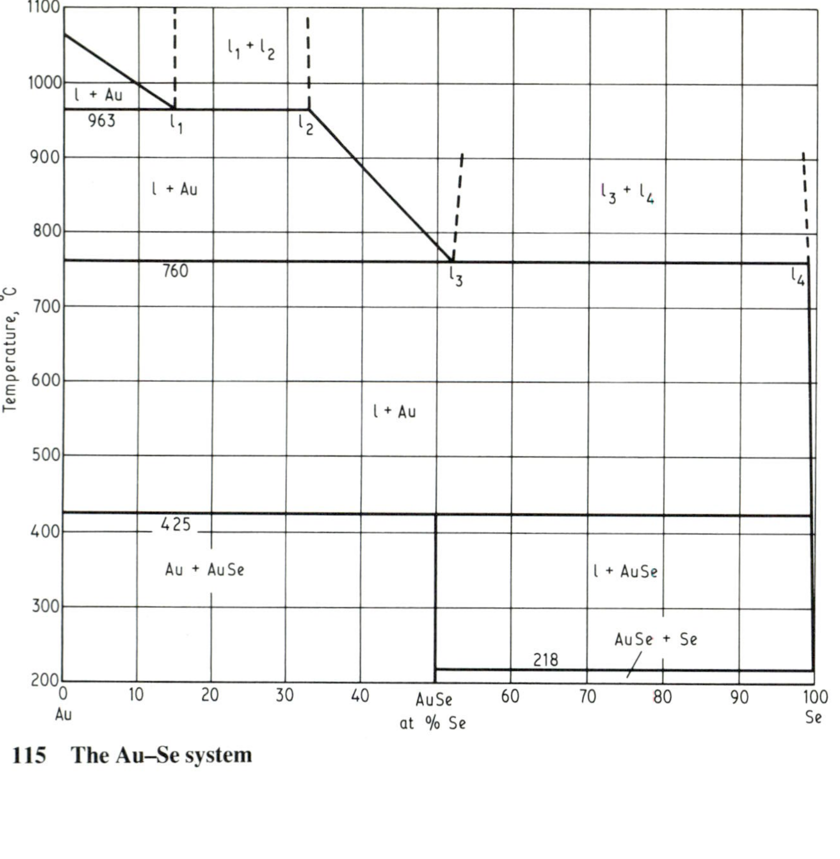

115 The Au–Se system

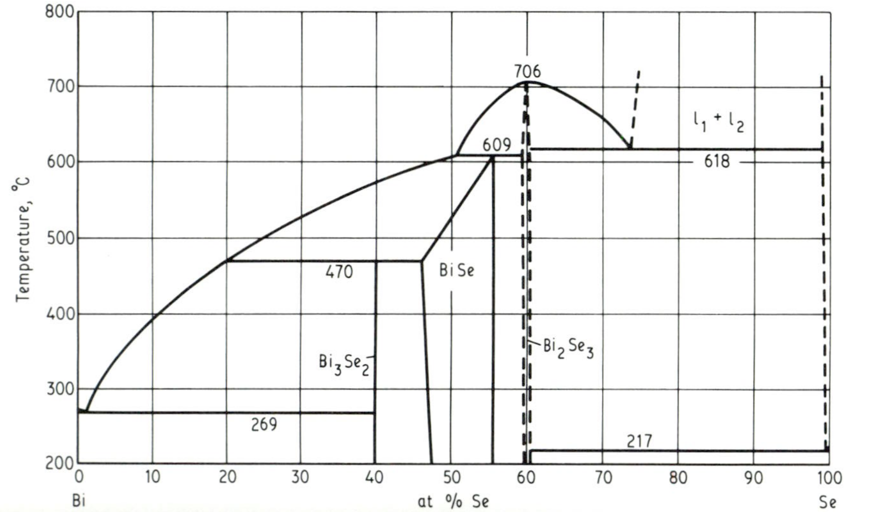

116 The Bi–Se system

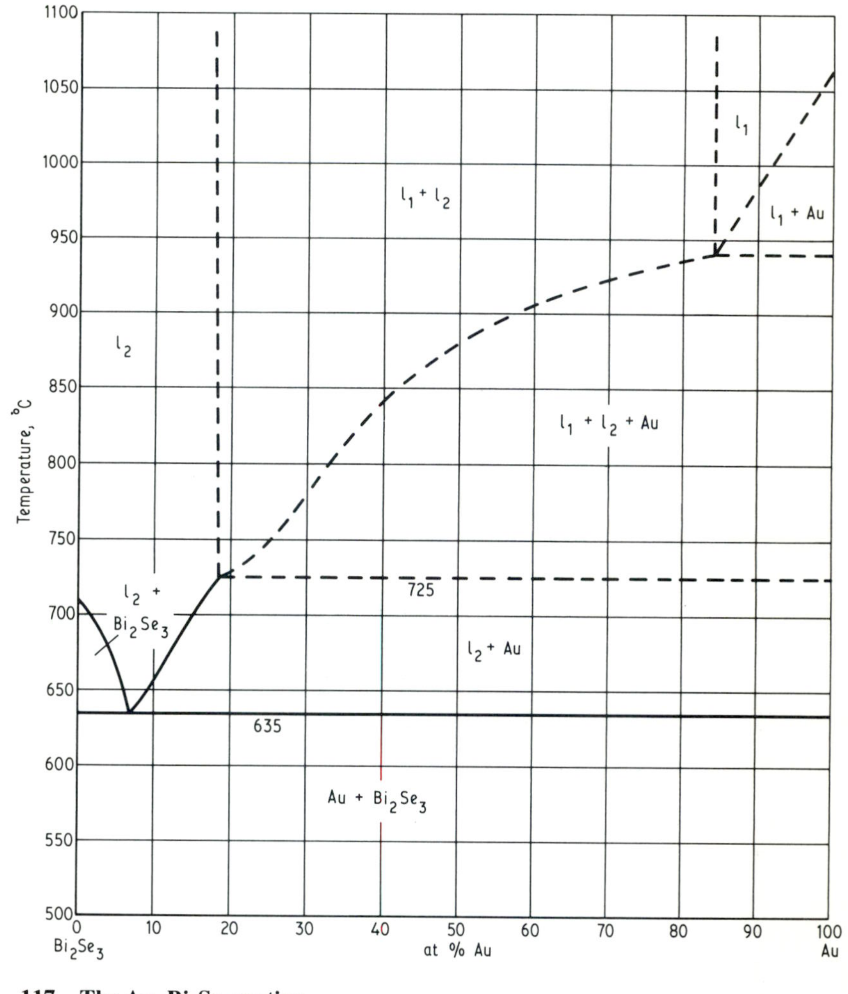

117 The Au–Bi_2Se_3 section

118 Reaction scheme for the Au–Bi–Se system

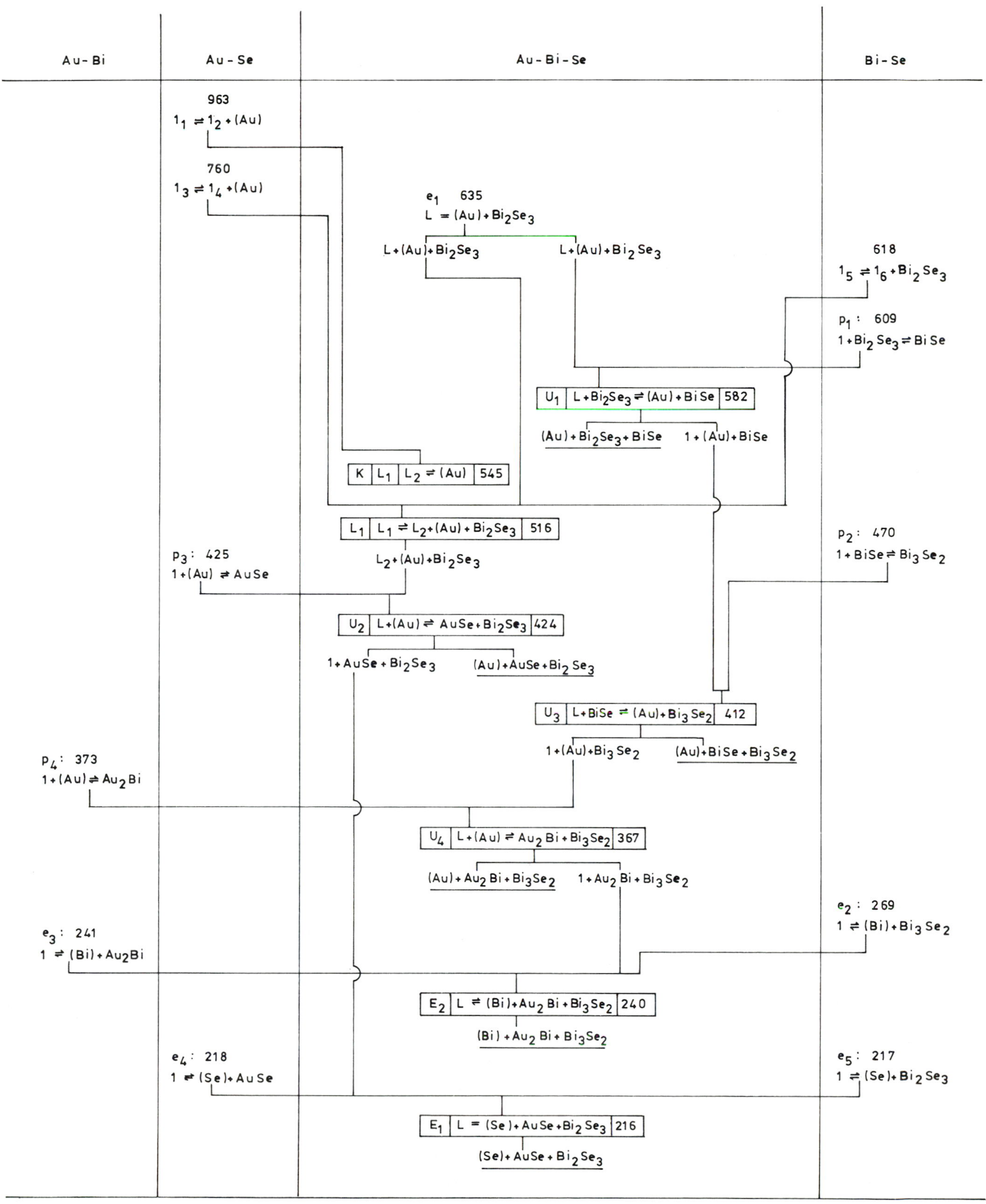

at 545°C. The point K has the approximate composition [75 Gat] 26 at.-% Au, 54 at.-% Bi, 20 at.-% Se. At K the liquids descending from the 963°C monotectic have the same composition, that of the point K. [75 Gat] do not indicate the tie lines for l_1–l_2 in the miscibility gap l_1Kl_2, but the isothermal sections at 800°C and 582°C (Figs. 119 and 120) give two l_1–l_2 tie lines. The 800°C tie line is skewed with respect to the Au–Se binary in that l_1 lies at a higher Bi content than l_2. The reverse skew is apparent at 582°C. The region of liquid immiscibility defined by the curve l_1Kl_2 was determined [75 Gat] by studying sections through the ternary system parallel to the Bi–Se binary at 10 at.-% Au intervals. Only one section, for 50 at.-% Au, was reproduced by [75 Gat] (Fig. 121). Additionally wet chemical analysis was used on the separated liquid layers.

Table 35. Invariant equilibria

Reaction	*Temperature*, °C	*Phase*	*Composition*, at.-%	
			Bi	*Se*
e_1	635	L	37·2	55·8
K–Au	545	L	~54	~20
U_1	582	L	49	43
U_2	424	L	~ 0·1	~99·55
U_3	412	L	67·5	5
U_4	367	L	68	2
L_1	516	L	22	58
L_2	516	L	~ 0·38	~99·87
E_1	216	L	~ 0·05	~99·88
E_2	240*	L	84·5	0·5

* Quoted as 210°C by [75 Gat]

Another wide area of liquid immiscibility stretches from the monotectic reaction $l_3 \rightleftharpoons l_4$ + (Au) at 760°C in the Au–Se system to the $l_5 \rightleftharpoons l_6 + Bi_2Se_3$ monotectic reaction at 618°C in the Bi–Se system. This region of liquid immiscibility is contained within the area $l_3L_1l_5l_6L_2l_4$. The section Au–Bi_2Se_3 contains a pseudobinary eutectic, e_1, at 635°C with a composition 7 at.-% Au, 37·2 at.-% Bi, 55·8 at.-% Se. Au and Bi_2Se_3 separate from the liquid along the monovariant curve e_1L_1. Along the curve l_5L_1 liquid l_6 and Bi_2Se_3 separate. At L_1 a ternary monotectic reaction occurs, $L_1 \rightleftharpoons L_2$ + (Au) + Bi_2Se_3. The composition of L_1 is given [75 Gat] as 20 at.-% Au, 22 at.-% Bi, 58 at.-% Se. L_2 lies close to the Se corner and [75 Gat] provide a diagram of the Se corner at a magnification of twenty (Fig. 114). They do not quote the compositions of the invariant reactions. Measurement from the original [75 Gat] give a composition for L_2 of 0·75 at.-% Au, 0·375 at.-% Bi, 98·875 at.-% Se. From L_2 liquid separates Au + Bi_2Se_3 along the monovariant curve L_2U_2 (Fig. 114, insert). At U_2 a transition reaction L_{U_2} + (Au) $\rightleftharpoons$ AuSe + Bi_2Se_3 occurs. The reaction temperature is only 1°C below the peritectic reaction l_{p_3} + (Au) $\rightleftharpoons$ AuSe in the AuSe binary system. The composition of U_2 is estimated to be 0·35 at.-% Au, 0·1 at.-% Bi, 99·55 at.-% Se. Solidification ends at the ternary eutectic E_1 at 216°C with the reaction $L_{E_1} \rightleftharpoons$ (Se) + AuSe + Bi_2Se_3. The composition of E_1 is estimated to be 0·075 at.-% Au, 0·05 at.-% Bi, 99·875 at.-% Se.

The section Au–Bi_2Se_3 (Fig. 117) has pseudobinary characteristics below 725°C. [75 Gat] note that alloys containing up to 27 at.-% Au showed a weak thermal effect at 582°C. This is associated with the presence of some liquid phase that remains after the eutectic solidification of Au + Bi_2Se_3 at 635°C. The remnant liquid transforms to solid U_1 by the reaction L_{U_1} + $Bi_2Se_3 \rightleftharpoons$ (Au) + BiSe. The presence of a 582°C thermal effect for alloys at the Bi_2Se_3 end of the Au–Bi_2Se_3 section suggests that such alloys cut through the four-phase plane associated with the invariant reaction at 582°C (U_1–BiSe–Bi_2Se_3–Au). One possible reason for this behaviour is that the composition corresponding to the congruent melting of Bi_2Se_3 is slightly Se-rich compared to stoichiometric Bi_2Se_3.

The Au–Bi_2Se_3–Bi region can be regarded as a partial ternary system. Invariant reactions occur at 582°C as noted above, at 412°C with L_{U_3} + BiSe $\rightleftharpoons$ (Au) + Bi_3Se_2, at 367°C with L_{U_4} + (Au) $\rightleftharpoons$ Au_2Bi + Bi_3Se_2, and at 240°C with the final ternary eutectic solidification $L_{E_2} \rightleftharpoons$ (Bi) + Au_2Bi + Bi_3Se_2. The compositions of U_1, U_3, U_4, and E_2 are given in Table 35 [75 Gat]. The eutectic temperature for the reaction at E_2 was quoted as 210°C [75 Gat] but this is unlikely when it is recalled that the ternary eutectic composition contains 0·5 at.-% Se. It would be anticipated that the ternary eutectic temperature would lie close to the binary eutectic temperature, e_3, of 241°C. The only independent check in the original publication [75 Gat] is from two measurements of the eutectic temperature on the 50 at.-% Au section (Fig. 121). Scaling of the data points from the published diagram gives values of 227 and 232°C. Measurement of the ternary eutectic temperature [82 Eva] gave a value of 240°C. This value is accepted in preference to the 210°C quoted by [75 Gat].

The phase regions below 367 and 240°C have been labelled l + Au_2Bi + Bi_3Se_2 and Bi + Au_2Bi + Bi_3Se_2 in Fig. 121. In the original [75 Gat] they are incorrectly designated. A schematic drawing of the section close to AuSe is shown as an enlargement to the right of Fig. 121.

LIQUIDUS SURFACE

Figure 114 is a representation of the liquidus surface. Isotherms were not included [75 Gat].

ISOTHERMAL SECTIONS

[75 Gat] published isothermal sections at 800°C (Fig. 119) and at temperatures corresponding to the invariant reactions U_1 (Fig. 120), L_1 (Fig. 122), U_3 (Fig. 123), U_4 (Fig. 124), and E_2 (Fig. 125). The Se corner in Figs. 120, 122 and 125 is enlarged twentyfold to allow representation of the phase regions. Figure 125 is modified from the original [75 Gat] to account for the acceptance of 240°C as the temperature of the ternary eutectic E_2. The ternary eutectic E_1 at 216°C is below the temperature of the isothermal section and the phase regions in the Se corner will reflect this fact (Fig. 125).

In Figs. 119 and 120 the two regions of liquid immiscibility are designated l_1 + l_2 (originating from the $l_1 \rightleftharpoons l_2$ + (Au) monotectic (Fig. 114) and l_3 + l_4 (originating from the $l_3 \rightleftharpoons l_4$ + (Au) monotectic (Fig. 114)).

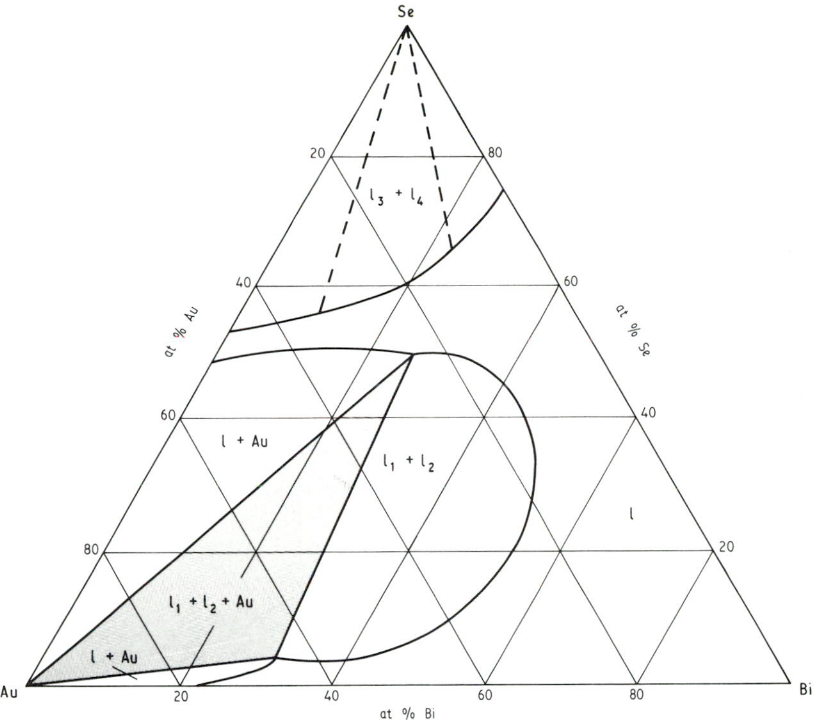

119 The 800°C isothermal section

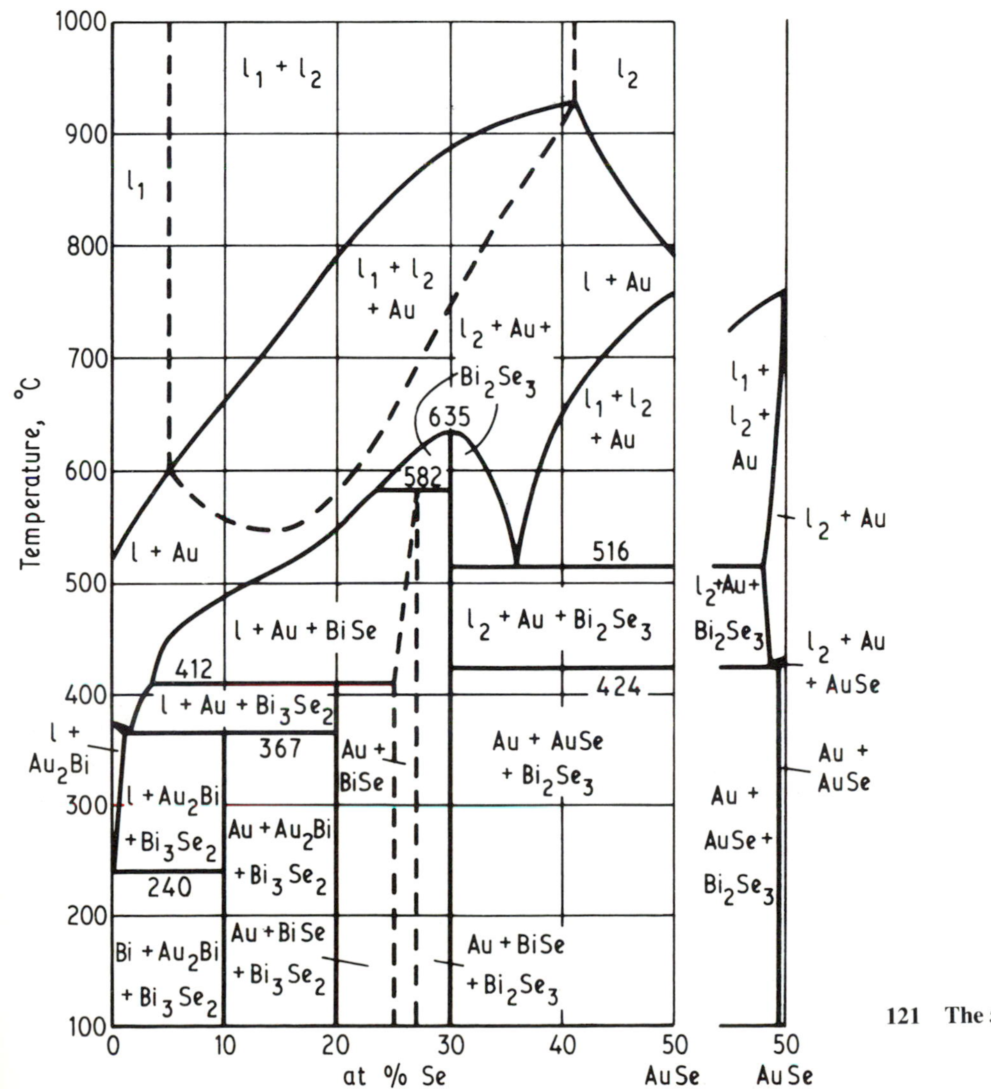

121 The 50 at.-% Au section

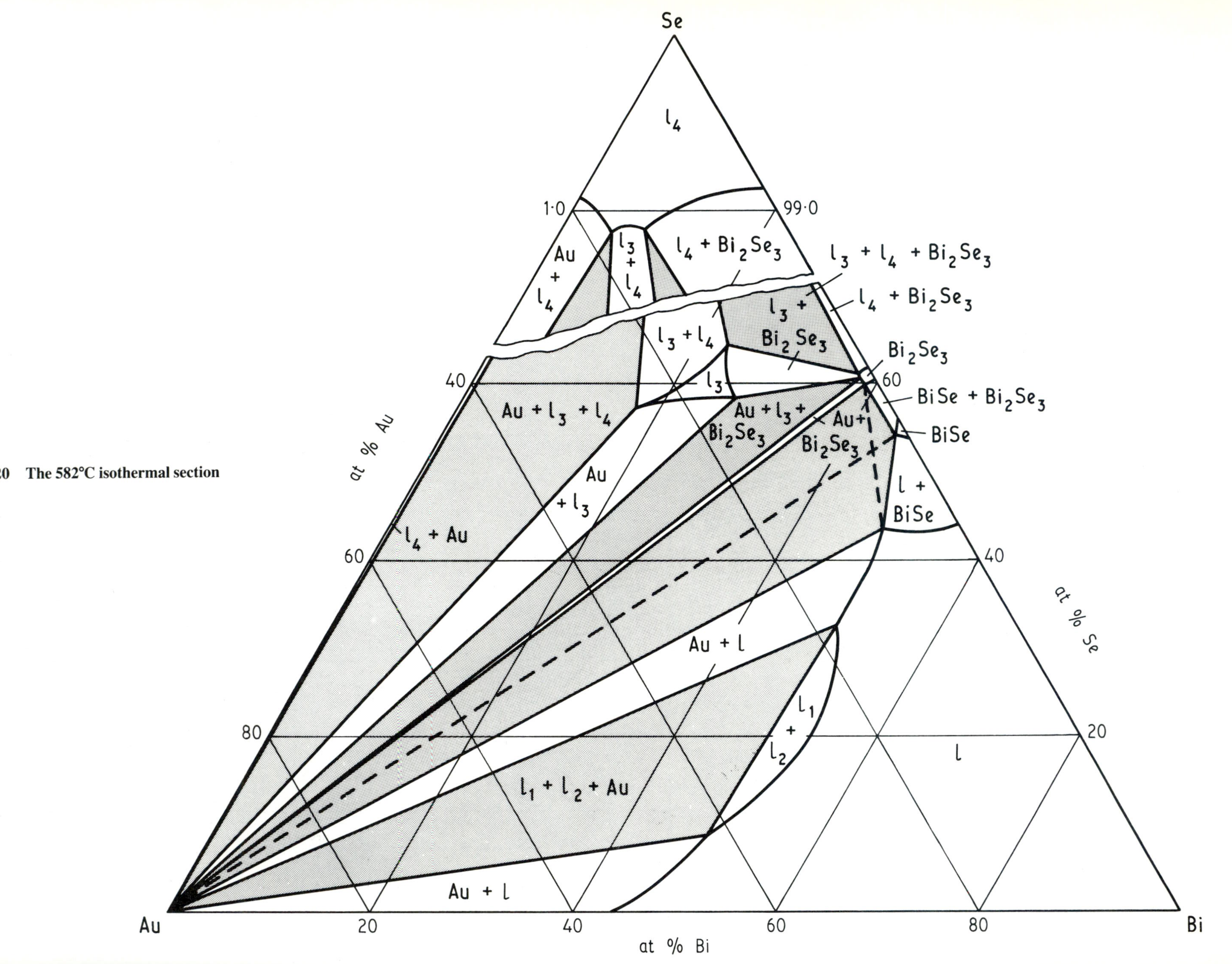

120 The 582°C isothermal section

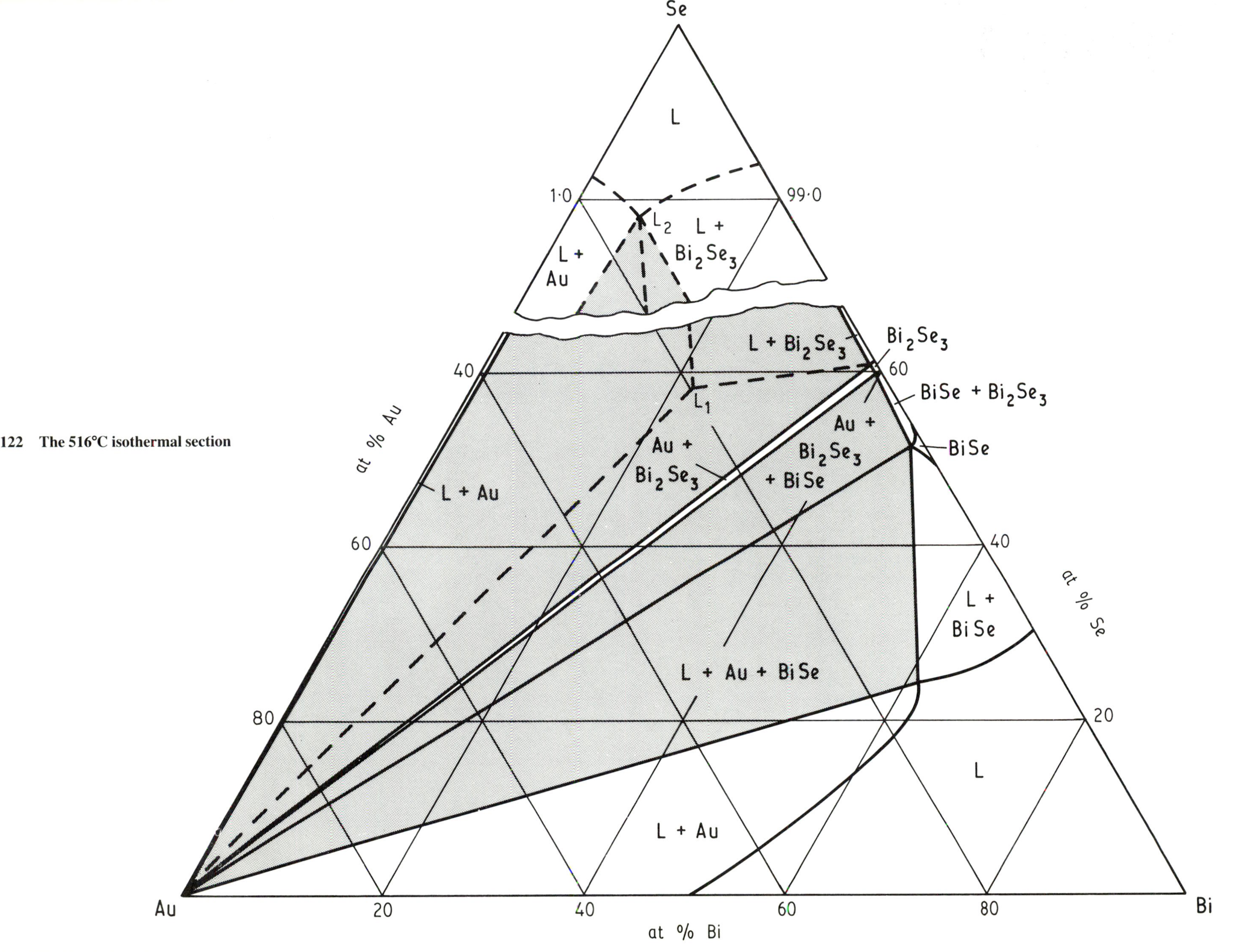

122 The 516°C isothermal section

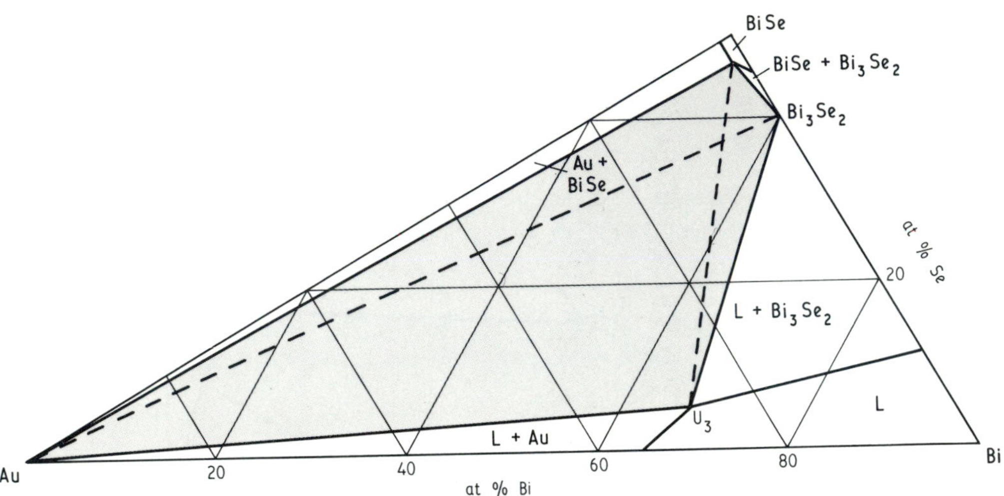

123 The 412°C isothermal section

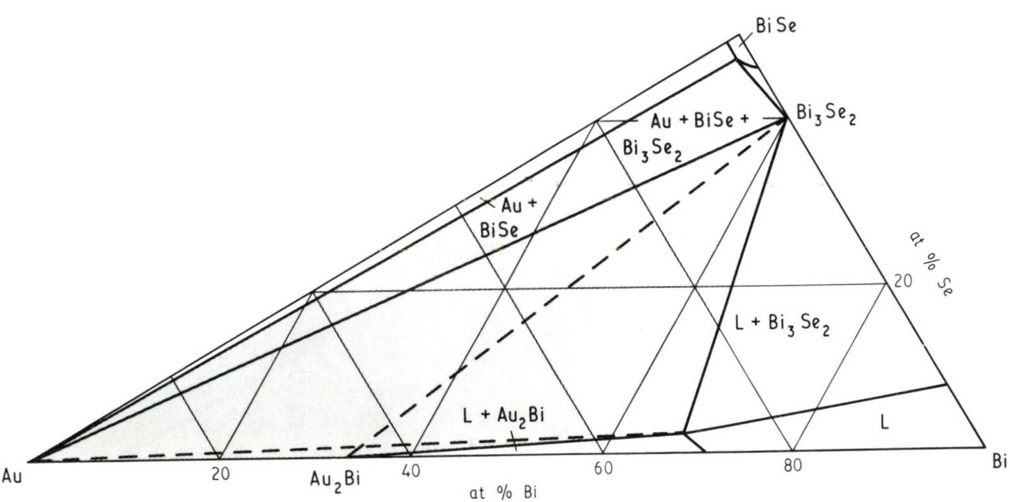

124 The 367°C isothermal section

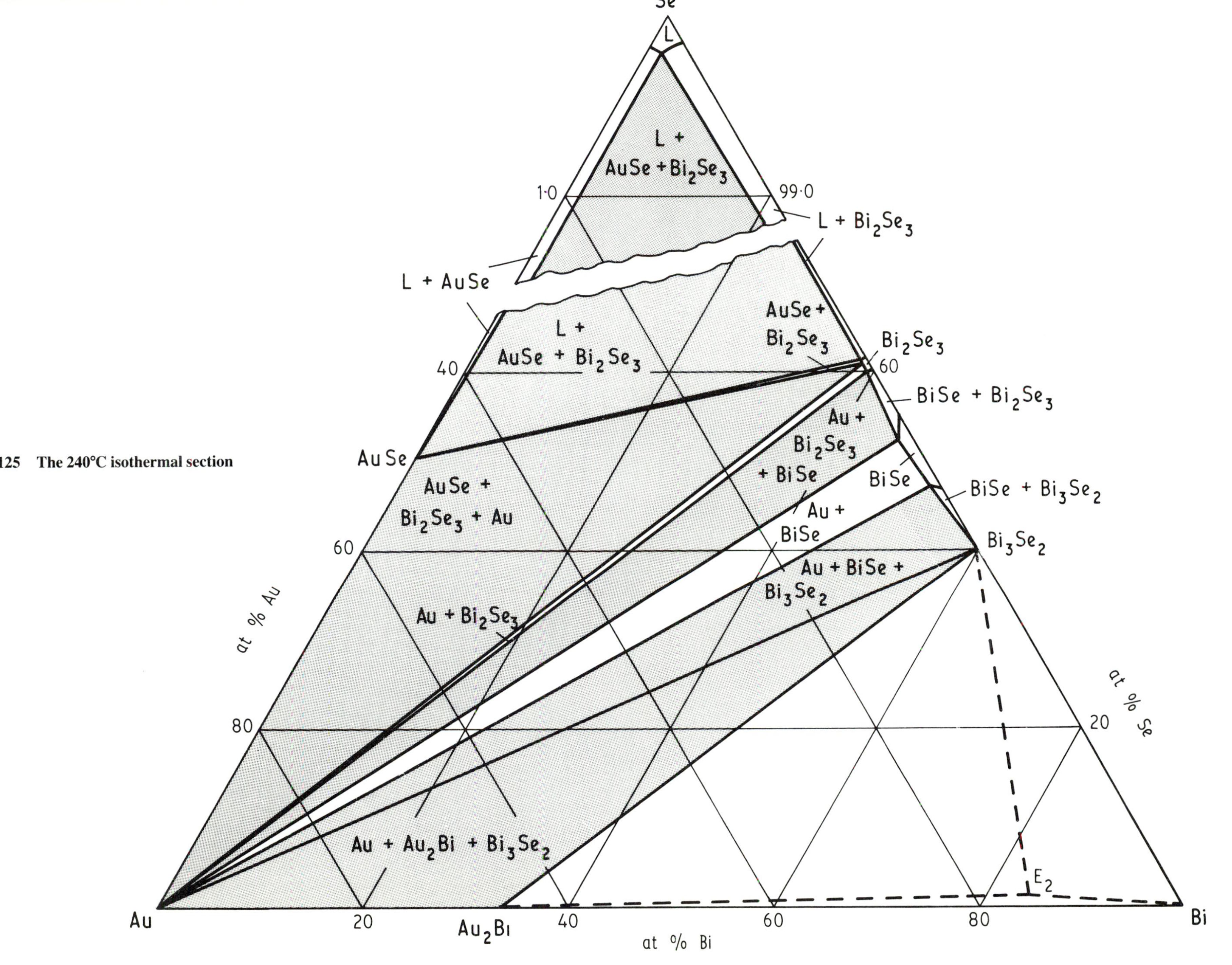

125 The 240°C isothermal section

REFERENCES

62 Nat: M. W. Nathans and M. Leider: *J. phys. Chem.*, 1962, **66**, 2012–2015

71 Rab: A. Rabenau, H. Rau and G. Rosenstein: *J. Less-Common Metals*, 1971, **24**, 291–299

75 Gat: B. Gather and R. Blachnik: *Z. Metall.*, 1975, **66**, 356–359

82 Eva: D. S. Evans and A. Prince: unpublished work, 1982

83 Eva: D. S. Evans and A. Prince: CALPHAD XII, Liege, 1983

83 Oka: H. Okamoto and T. B. Massalski: *Bull. Alloy Phase Diagrams*, 1983, **4**, 401–407

Au–Bi–Si

Legendre and Chhay Hancheng [89 Leg] used 99·99% Au and 99·999 9% Bi and Si to prepare 80 alloy compositions on sections through the ternary system corresponding to 6·66 at.-% Si, 10 Si, 20 Si, 50 Bi and Au–33·33 Bi, 66·67 Si. Alloys, contained in evacuated silica capsules, were melted, quenched and heat treated for 3–7 days at temperatures appropriate to their compositions. The ternary equilibria were studied by dta at heating rates of 2 or 5°C min^{-1} and by dsc at heating rates of 2 and 0·1°C min^{-1}. The liquidus projection (Fig. 126), indicates that the liquid immiscibility gap, originating at the 1 400°C monotectic reaction on the Bi–Si binary edge, extends to about 38 at.-% Au. The liquid immiscibility gap terminates at a critical point whose temperature is estimated to be about 500°C at a critical liquid composition of approximately 35 Au, 52 Bi, 13 at.-% Si. There are two invariant reactions; a transition reaction, L + (Au) = Au_2Bi + (Si), occurs at 351°C with liquid U of composition 32 Au, 65·7 Bi, 2·3 at.-% Si and a degenerate ternary eutectic reaction occurs at the temperature and composition of the binary Au–Bi eutectic reaction, 1 = Au_2Bi + (Bi), at 241°C. The degenerate ternary eutectic reaction is denoted in Fig. 126 by the symbol D and in Fig. 127a by the reaction L = (Bi) + Au_2Bi, (Si) [86 Luk]. The drop in temperature along the monovariant curve e_1U is only 12°C and this is reflected in the suggested course of the 400°C isotherms. The section Au–33·33 Bi, 66·67 Si contains the alloy composition corresponding to the ternary compound Au_2BiSi_2 reported by [74 Laz]. [89 Leg] found no evidence for the formation of this compound as a result of their thermal analysis data (Fig. 127b). Further study, using metallography and X-ray diffraction analysis, is indicated. The reaction scheme [88 Leg], is relatively simple.

REFERENCES

74 Laz: V. B. Lazarev and I. S. Shaplygin, 'Tezisy Dokl-Vses Konf Kristallokhim Intermet Soedin', 2nd, 1974, pp. 19–20. Ed: R. M. Rykhal, L'vov Gos Univ: Lvov, USSR

86 Luk: H. L. Lukas, E.-T. Henig and G. Petzow, *Z. Metallk.*, 1986, **77**, 360–367

89 Leg: B. Legendre and Chhay Hancheng, *Bull. Soc. Chim. France* 1989, **1**, 53–57, for publication.

Au–Bi–Sn

INTRODUCTION

Humpston [85 Hum] studied the AuSn–Bi section and found it to be a pseudobinary eutectic with a virtually horizontal liquidus over the composition range 35–65 at.-% Bi. Recent work [87 Max] (Fig. 128) has disclosed a liquid immiscibility region associated with a monotectic reaction at 386°C. The eutectic reaction is placed at 262·5°C and 95·5 at.-% Bi. There appears to be a considerable solubility of Bi in AuSn but limited solubility of AuSn in Bi. [85 Hum] also examined the section 50 at.-% Au, 50 at.-% Bi–Sn from 36 to 80 at.-% Sn. Three invariant reactions occur in the partial ternary system AuSn–Bi–Sn, L + AuSn $\rightleftharpoons$ (Bi) + $AuSn_2$ at 223°C, L + $AuSn_2$ $\rightleftharpoons$ (Bi) + $AuSn_4$ at 144°C, and L $\rightleftharpoons$ (Bi) + (Sn) + $AuSn_4$ at 134°C. Within the Au–Bi–AuSn partial ternary system [85 Hum] states that the equilibria is complex and includes a region of liquid immiscibility. [86 Max] extended the work of [85 Hum] on the partial ternary system AuSn–Bi–Sn and confirmed the presence of three invariant reactions (Table 36). The data of [86 Max] and [87 Max] are accepted.

Table 36. Composition of the liquid phase entering into the invariant reactions [86 Max]

Reaction	*Temperature*, °C	*Composition of liquid phase*, at.-%		
		Au	*Bi*	*Sn*
U_1	224	1·9	76·8	21·3
U_2	146	2·7	44·7	52·6
E	137	0·8	40·0	59·2

BINARY SYSTEMS

The Au–Sn system has been evaluated by [84 Oka], but considerable uncertainties exist concerning equilibrium reactions for Au-rich alloys. [87 Leg] have shown the existence of two peritectic reactions, l + (Au) $\rightleftharpoons$ β at 532°C and l + β $\rightleftharpoons$ ξ at 519°C. The solid-state transformations require further study. The Au–Bi diagram [83 Oka] with a peritectic temperature of 377·5°C [83 Eva] is accepted. The Bi–Sn diagram presented by [58 Han] is accepted.

SOLID PHASES

Table 37 summarizes crystal structure data for the binary phases relevant to the AuSn–Bi–Sn partial ternary system. No ternary compounds occur [85 Hum, 86 Max].

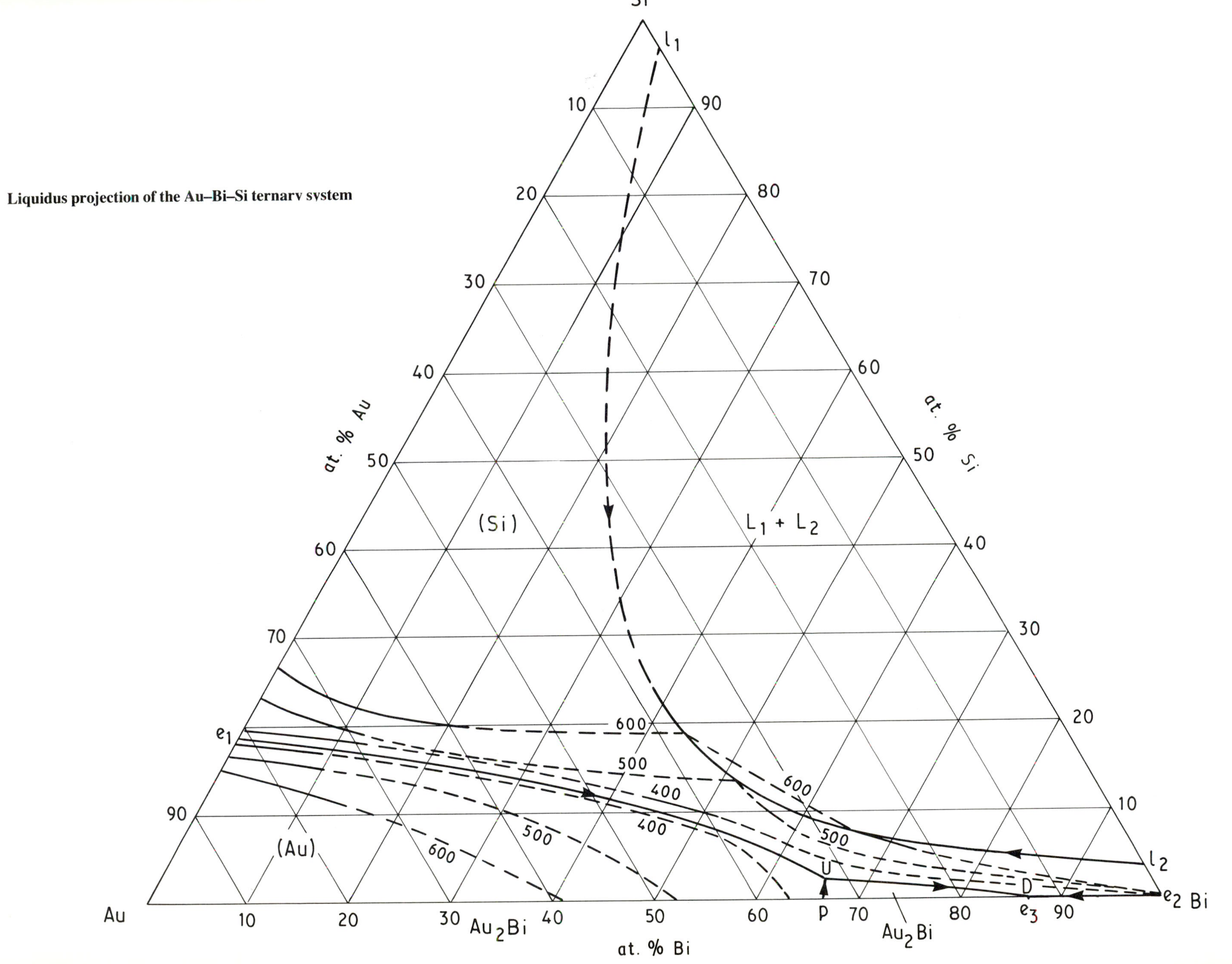

126 Liquidus projection of the Au–Bi–Si ternary system

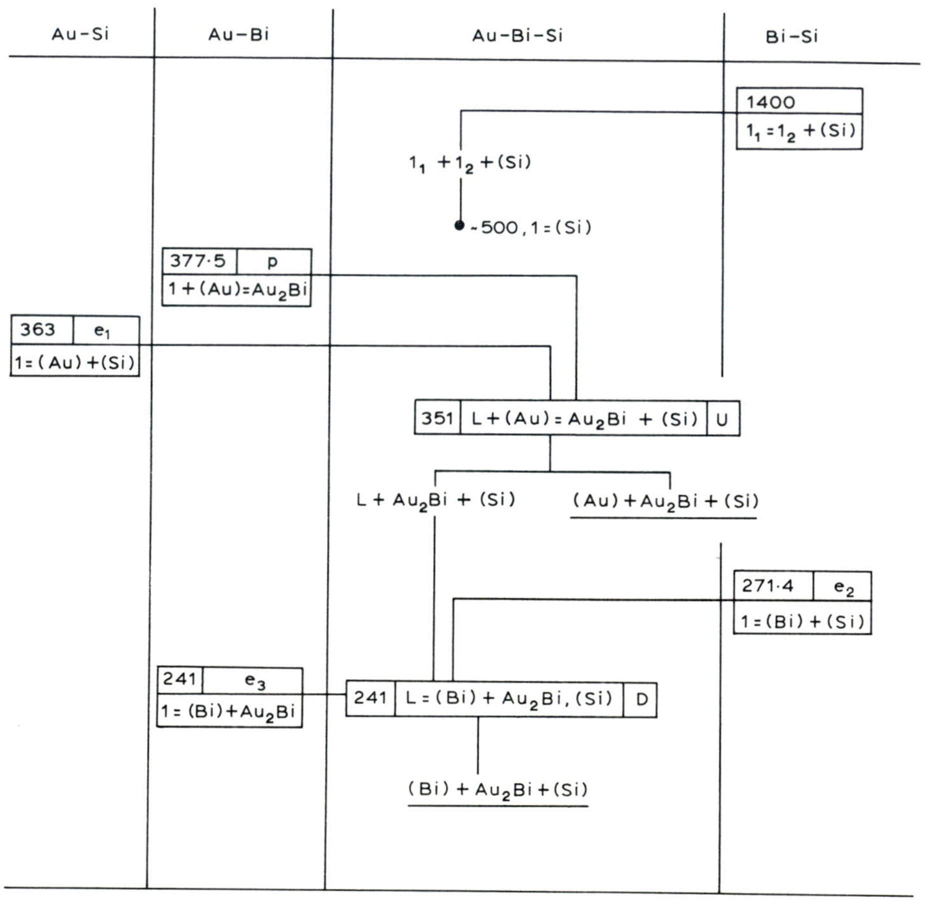

127a Reaction scheme for the Au–Bi–Si ternary system

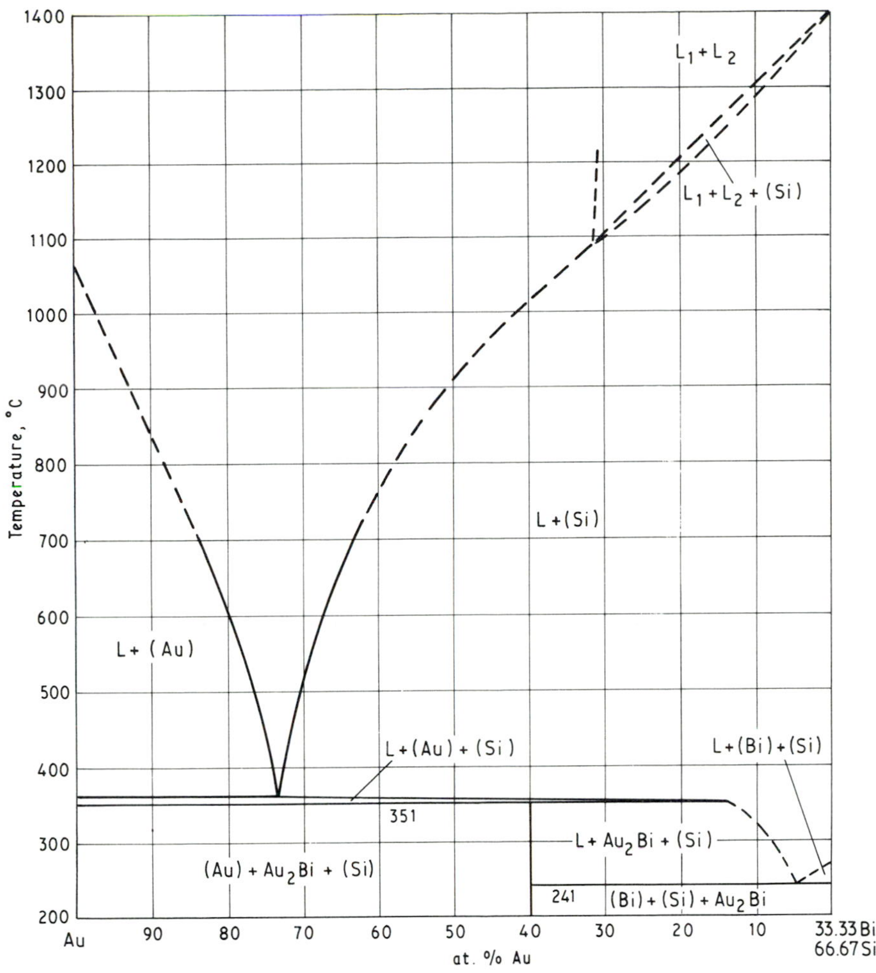

127b The section Au–33·33Bi, 66·7Si

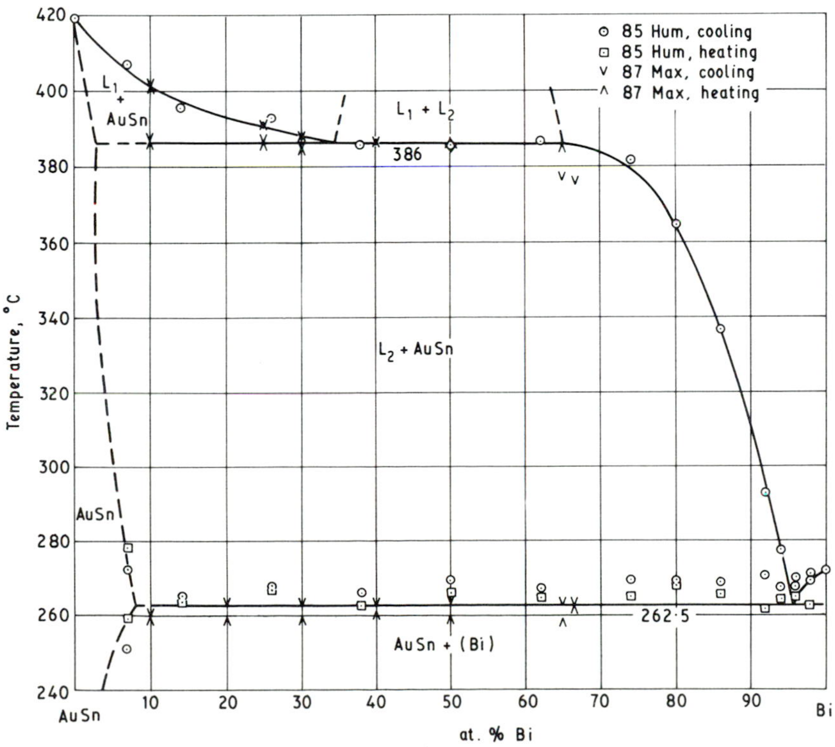

128 The AuSn–Bi section

Table 37. Crystal structures

Solid phase	*Prototype*	*Lattice designation*
(Bi)	αAs	hR2
(Sn)	βSn	tI4
AuSn	NiAs	hP4
$AuSn_2$	—	oP24
$AuSn_4$	$PtSn_4$	oC20

PSEUDOBINARY SYSTEMS

The AuSn–Bi section is a pseudobinary (Fig. 128). A monotectic reaction occurs at 386°C, $L_1 \rightleftharpoons L_2 + AuSn$. The composition of L_1 is 34·5 at.-% Bi, that of L_2 about 65 at.-% Bi. Bi may be soluble in AuSn at 386°C but the solubility relations have not been determined. A eutectic reaction occurs at 262·5°C and 95·5 at.-% Bi. The data of [85 Hum] suggests a solubility of about 8 at.-% Bi in AuSn at the eutectic temperature.

INVARIANT EQUILIBRIA

[86 Max] determined vertical sections from 80Bi, 20Sn–Au (to 15·5 at.-% Au), 60Bi, 40Sn–Au (to 28 at.-% Au), 30Bi, 70Sn–Au (to 30 at.-% Au), and from 50Au, 50Bi–Sn (from 80–96 at.-% Sn). Figures 129–132 present the results of thermal analysis. Figure 132 includes the data of [85 Hum] from 36–80 at.-% Sn. Alloys were prepared from >99·99% Au, 99·9995% Bi, and 99·99% Sn by melting in fused quartz crucibles under a protective H_2 atmosphere. Thermal analysis was done under a N_2 atmosphere. Phase identification included optical metallography, microprobe analysis by the SEM–EDAX technique, and X-ray diffraction analysis. [86 Max] accepted the pseudobinary section AuSn–Bi of [85 Hum] and did not discover any region of liquid immiscibility in the partial ternary system AuSn–Bi–Sn. The later work of [87 Max] shows that a region of liquid immiscibility will appear in this partial ternary system and will be reflected in the vertical sections (Figs. 130–132). No account has been taken of the presence of a $L_1 + L_2 + AuSn$ phase region in the sections represented by Figs. 130–132.

Three invariant reactions were found in the AuSn–Bi–Sn partial ternary system (see the introduction and Table 36). A reaction scheme is given in Fig. 133.

LIQUIDUS SURFACE

Figure 134 illustrates the projection of the liquidus surface [86 Max]. The region of liquid immiscibility projecting from the AuSn–Bi section is not known. It is shown as a dashed curve disappearing at a critical tie line $L \rightleftharpoons AuSn$ at a temperature below 386°C. Most of the liquidus is taken up by the primary surfaces of crystallization of AuSn, $AuSn_2$, and $AuSn_4$. The regions of primary crystallization of (Bi) and (Sn) are confined to narrow regions adjoining the Bi–Sn binary edge.

ISOTHERMAL SECTIONS

Figures 135 – 138 are isothermal sections at 240, 200, 160, and 139°C. They have been slightly amended from [86 Max] to produce consistency with the vertical sections.

MISCELLANEOUS

Jena *et al.* [70 Jen] reported calorimetric measurements at 350°C of the partial molar enthalpy of Au in dilute solutions of Au (up to 3·2 at.-%) in two Bi–Sn alloys.

REFERENCES

58 Han: M. Hansen and K. Anderko: 'Constitution of Binary Alloys'; 1958, New York, McGraw-Hill

70 Jen: A. K. Jena, J. H. Smith and M. B. Bever: *Metall. Trans.*, 1970, **1**, 1257–1261

83 Oka: H. Okamoto and T. B. Massalski: *Bull. Alloy Phase Diagrams*, 1983, **4**, 401–407

83 Eva: D. S. Evans and A. Prince: CALPHAD XII, Liege, 1983.

84 Oka: H. Okamoto and T. B. Massalski: *Bull. Alloy Phase Diagrams*, 1984, **5**, 492–503

85 Hum: G. Humpston: PhD thesis, Brunel Univ., 1985

86 Max: C. A. Maxwell: MSc Thesis, Univ. Manchester, Faculty of Technology, 1986

87 Max: C. A. Maxwell and F. Hayes: private communication, 1987

87 Leg: B. Legendre, Chhay Hancheng, F. Hayes, C. A. Maxwell, D. S. Evans and A. Prince: *Mater. Sci. Technol.*, 1987, **3**, 875–876

Au–Bi–Te

INTRODUCTION

The $AuTe_2$–Bi_2Te_3 and Au–Bi_2Te_3 sections are both pseudobinary eutectic systems. The results of [64 Win] and [74 Gat] are in good agreement in terms of eutectic compositions and temperatures. [74 Gat] studied the complete ternary system. It is divided into three partial ternary systems, Te–$AuTe_2$–Bi_2Te_3, Au–$AuTe_2$–Bi_2Te_3, and Au–Bi_2Te_3–Sn. The Te–$AuTe_2$–Bi_2Te_3 partial ternary system is a simple ternary eutectic system, as is the Au–$AuTe_2$–Bi_2Te_3 partial ternary system. The Au–Bi_2Te_3–Sn partial ternary system contains three transition reactions, $l_{U_1} + Bi_2Te_3 \rightleftharpoons (Au) + BiTe$ at 456°C, l_{U_2} + $BiTe \rightleftharpoons (Au) + Bi_5Te_3$ at 374°C, and $l_{U_3} + (Au) \rightleftharpoons Au_2Bi + Bi_5Te_3$ at 346°, followed by a ternary eutectic reaction $l_{E_1} \rightleftharpoons (Bi) + Au_2Bi + Bi_5Te_3$ at 235°C. [74 Gat] prepared alloys by melting 99·995% pure elements in evacuated quartz tubes and annealing the alloys for 80 days at 300°C. They state that these preparation conditions did not lead to the formation of the Bi_7Te_3 phase, and consequently it was not detected in the Bi-rich region of the ternary system. Under equilibrium conditions there should appear a fourth transition reaction, $l + Bi_5Te_3 \rightleftharpoons Au_2Bi + Bi_7Te_3$, and the ternary eutectic reaction should be $l \rightleftharpoons (Bi) + Au_2Bi + Bi_7Te_3$. [74 Gat] quote compositions for the liquid compositions associated with each invariant reaction and produce isothermal sections for each reac-

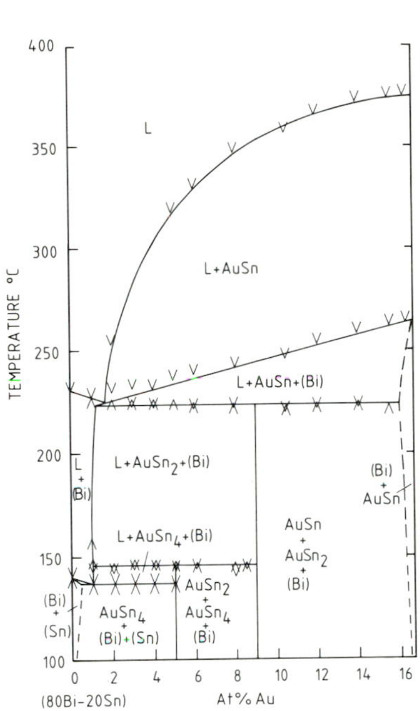

129 The section 80Bi, 20Sn–Au

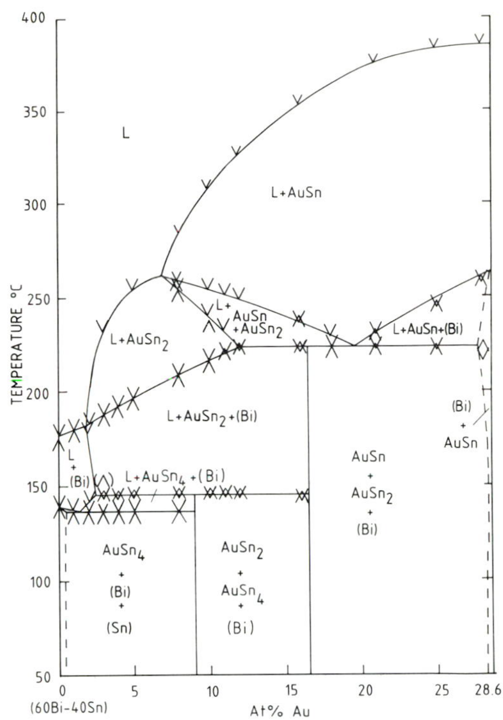

130 The section 60Bi, 40Sn–Au

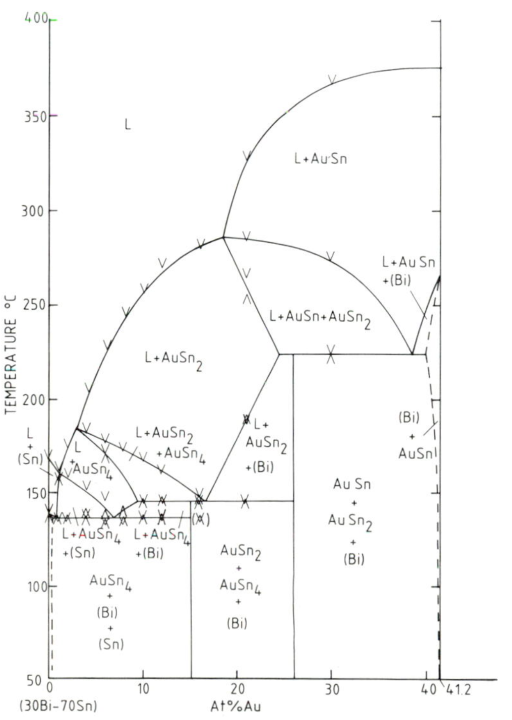

131 The section 30Bi, 70Sn–Au

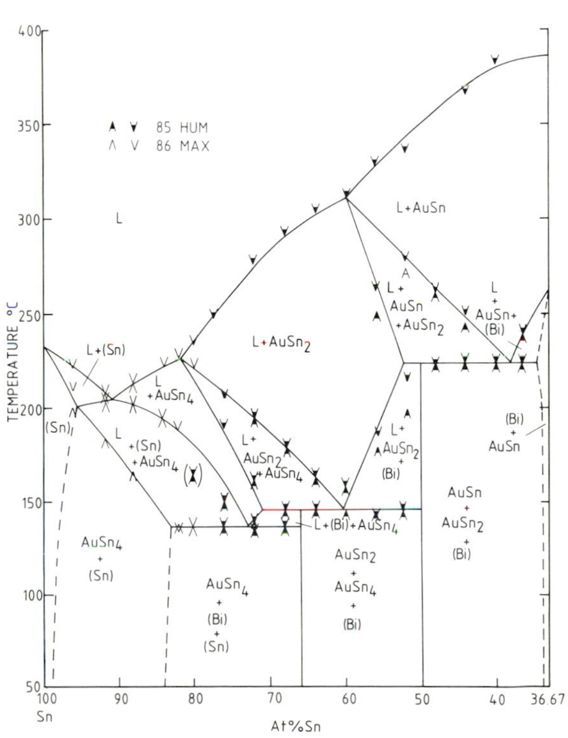

132 The section 50Au, 50Bi–Sn

133 Reaction scheme for the AuSn–Bi–Sn partial ternary system

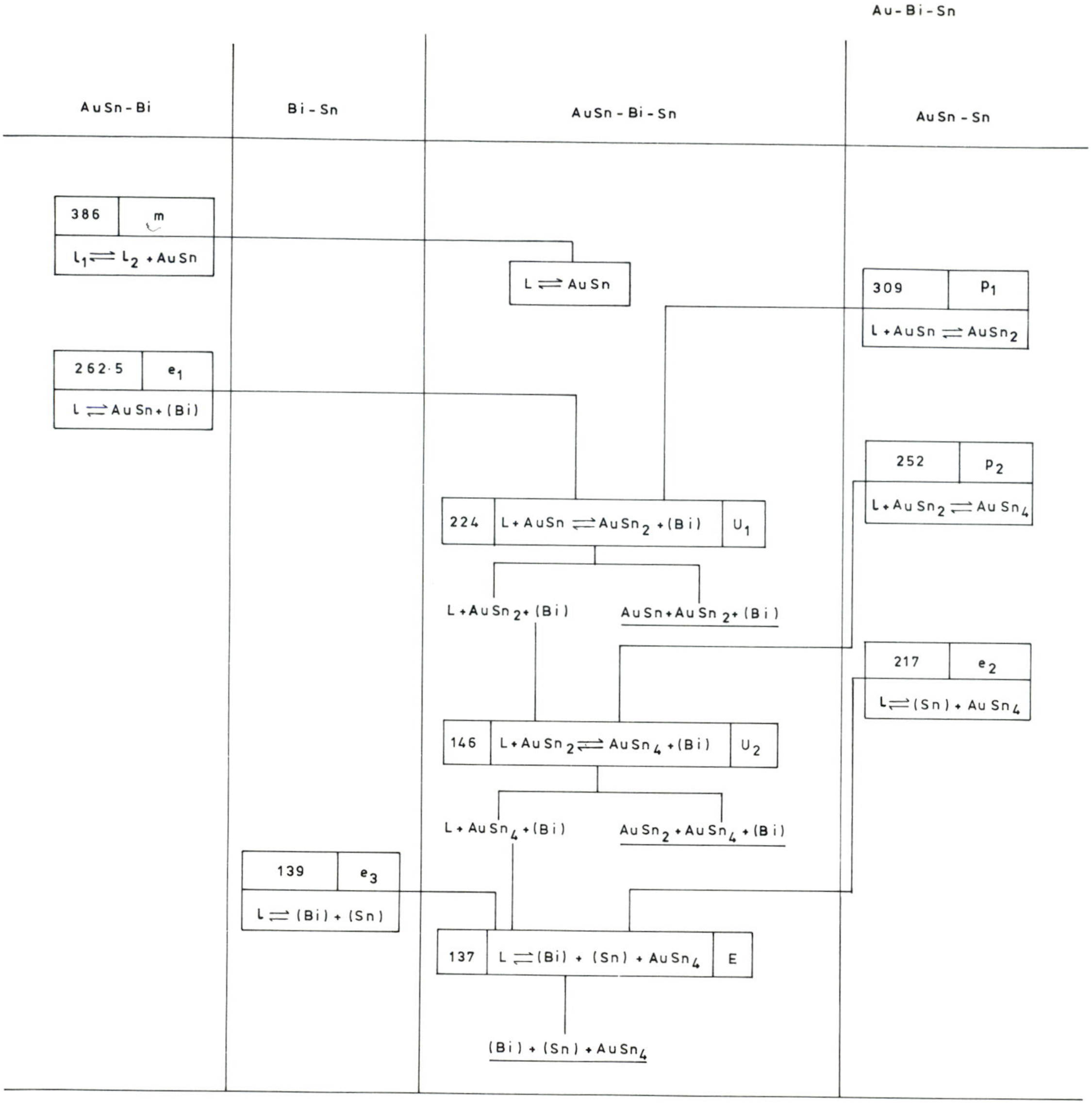

Au-Bi-Sn
AuSn-Bi
Bi-Sn
AuSn-Bi-Sn
AuSn-Sn
386
m
L1 ⇌ L2 + AuSn
L ⇌ AuSn
309
P1
L + AuSn ⇌ AuSn2
262·5
e1
L ⇌ AuSn + (Bi)
252
P2
L + AuSn2 ⇌ AuSn4
224
L + AuSn ⇌ AuSn2 + (Bi)
U1
L + AuSn2 + (Bi)
AuSn + AuSn2 + (Bi)
217
e2
L ⇌ (Sn) + AuSn4
146
L + AuSn2 ⇌ AuSn4 + (Bi)
U2
L + AuSn4 + (Bi)
AuSn2 + AuSn4 + (Bi)
139
e3
L ⇌ (Bi) + (Sn)
137
L ⇌ (Bi) + (Sn) + AuSn4
E
(Bi) + (Sn) + AuSn4

Au Sn
60
Au Sn2
70
At%Sn
P1
L1
(Au Sn)
Au Sn4
360
L2
280
P2
90
(Au Sn2)
e1
e2
(Au Sn4)
240
200
(Sn)
E
U2
(Bi)
U1
Sn
10
20
30
40
e3
50
60
70
80
90
Bi
At% Bi

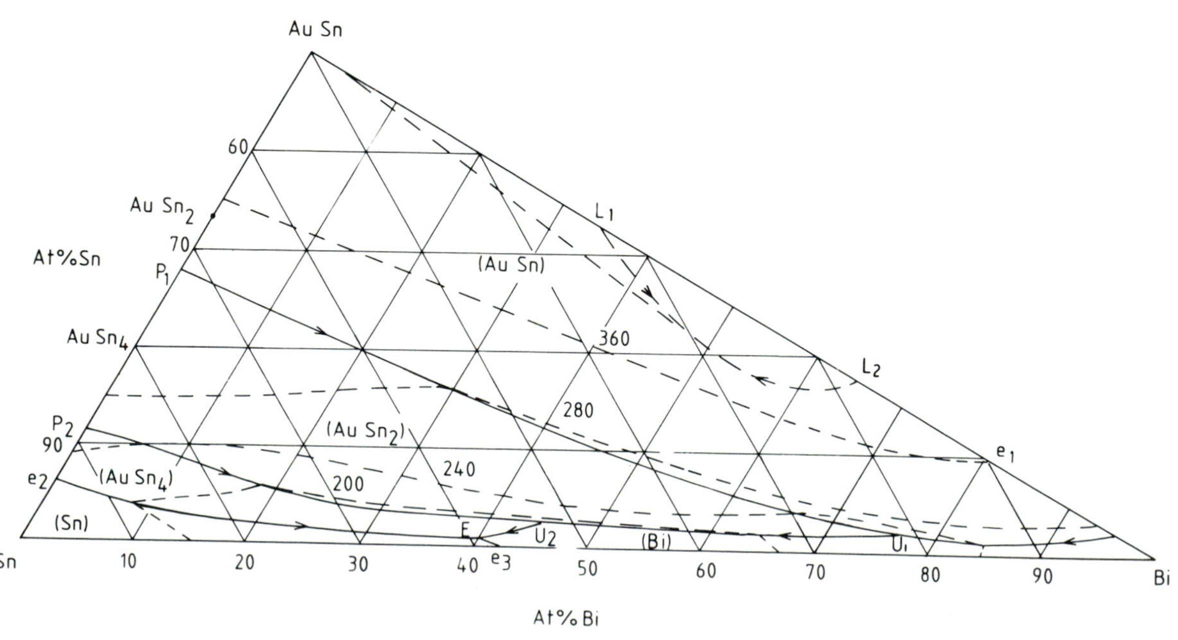

134 Projection of the liquidus surface (AuSn–Bi–Sn)

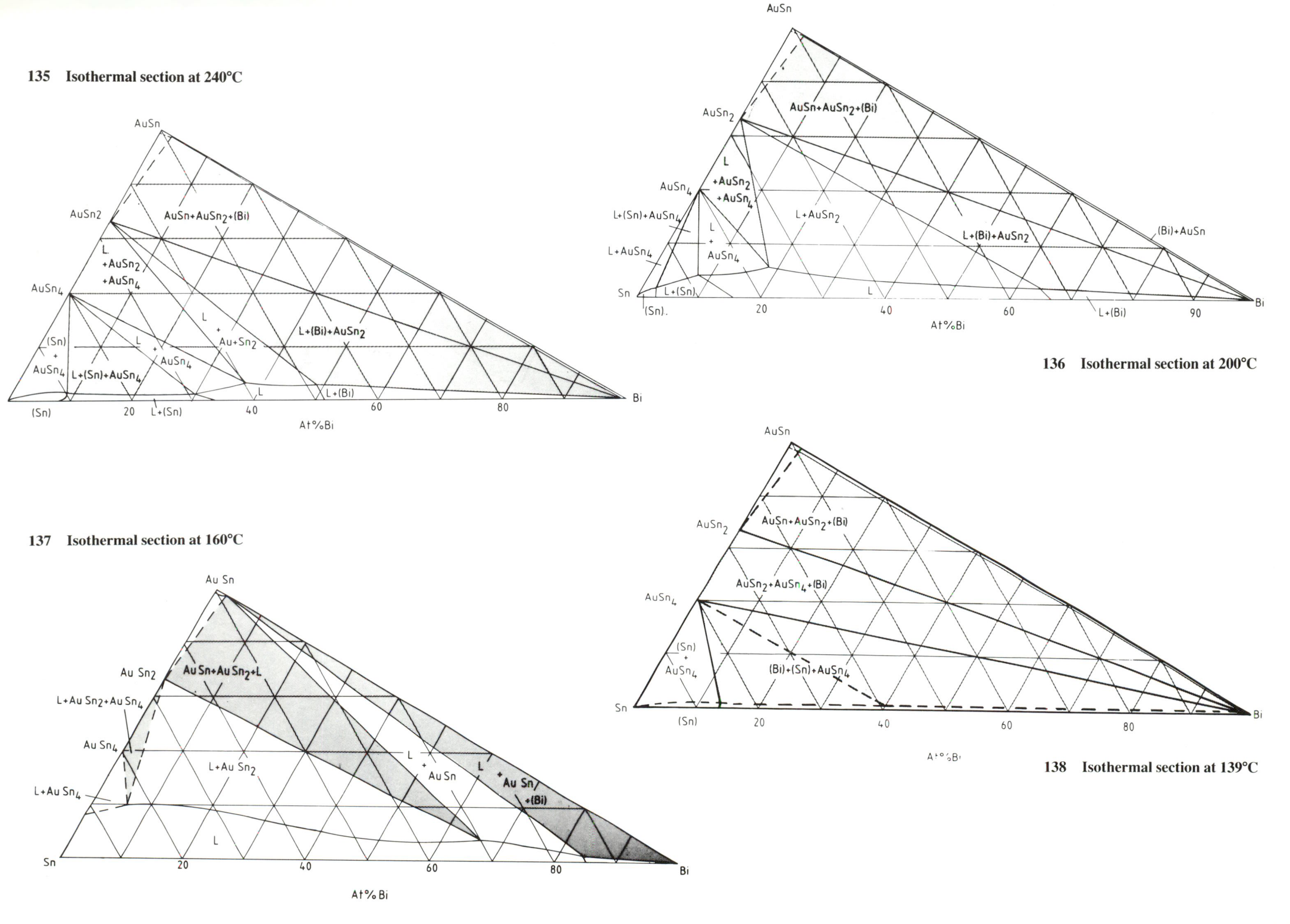

135 Isothermal section at 240°C

136 Isothermal section at 200°C

137 Isothermal section at 160°C

138 Isothermal section at 139°C

tion temperature. They do not provide any experimental evidence to justify the isothermal sections which must therefore be accepted as presented. The sections are consistent with the proposed reaction equilibria. No ternary compounds were found by [74 Gat]. In summarizing work intended for later publication [64 Win] found no evidence for ternary compound formation, but quote the presence of a ternary eutectic reaction at 210°C, compared with 235°C [74 Gat], at a composition close to that found by [74 Gat]. Since the ternary eutectic composition lies close to the Au–Bi binary edge it is to be anticipated that the ternary eutectic temperature will lie near to the Au_2Bi–Bi eutectic temperature of 241°C. On this basis the data of [74 Gat] are preferred. [63 Key] determined the solubility of Au in Bi_2Te_3. It varies from 0·000 2 at.-% Au at 200°C to 0·02 at.-% Au at 400°C. [61 Kin] reported the formation of a compound $AuBiTe_2$ on quenching from the melt. This compound is not a stable phase in the Au–Bi–Te system, nor was the formation of $AuBiTe_2$ by rapid cooling of the melt confirmed by [74 Gat].

BINARY SYSTEMS

The Au–Te system evaluated by [84 Oka] is accepted. The Au–Bi system [83 Oka], with a peritectic temperature of 377·5°C [83 Eva] is accepted. The Bi–Te system presented by [65 Ell] is accepted, except that the phase Bi_2Te is replaced by Bi_5Te_3 according to [74 Gat].

SOLID PHASES

No stable ternary compounds have been identified. Table 38 summarizes crystal structure data for the binary compounds.

Table 38. Crystal structures

Solid phase	*Prototype*	*Lattice designation*
(Au)	Cu	cF4
(Bi)	As	hR2
(Te)	γSe	hP3
$AuTe_2$	$AuTe_2$	mC6
Au_2Bi	Cu_2Mg	cF24
Bi_2Te_3	Bi_2STe_2	hR5
BiTe	—	?hR5
Bi_5Te_3	Bi_2STe_2	hR5
Bi_7Te_3	Bi_2STe_2	hR5

PSEUDOBINARY SYSTEMS

The $AuTe_2$–Bi_2Te_3 section is a pseudobinary eutectic system [64 Win, 74 Gat]. Alloys were prepared by melting 99·999% [64 Win] or 99·995% [74 Gat] pure elements in evacuated silica tubes. [74 Gat] subsequently heat treated the alloys at 300°C for 80 days. Examination [64 Win, 74 Gat] was by DTA, X-ray, and metallographic methods. [64 Win] defined the system with 23 alloys and quoted a eutectic composition of 22·7 at.-% Au, 12·8 at.-% Bi, 64·5 at.-% Te at 408°C. [74 Gat] used 18 alloys and gave a eutectic composition of 23·1 at.-% Au, 12·3 at.-% Bi, 64·6 at.-% Te at 417°C. The data of [74 Gat] shows less scatter than that of [64 Win]. The recommended value (Fig. 139) of the eutectic composition is 22·9 at.-% Au, 12·5 at.-% Bi, 64·6 at.-% Te at 415 ± 4°C.

The Au–Bi_2Te_3 section is also a pseudobinary eutectic system [64 Win, 74 Gat]. 13 alloys were used by [64 Win] and 19 by [74 Gat] to characterize the equilibria (Fig. 140). The eutectic composition and temperature quoted are 29·5 at.-% Au, 28·2 at.-% Bi, 42·3 at.-% Te at 472°C [64 Win], and 31·8 at.-% Au, 27·3 at.-% Bi, 40·9 at.-% Te at 476°C [74 Gat]. The recommended values are 30 at.-% Au, 28 at.-% Bi, 42 at.-% Te at 475 ± 2°C. The data agree well for the Bi_2Te_3 liquidus, but the Au liquidus values of [64 Win] are preferred to those of [74 Gat] on the basis that the data of [74 Gat] are less consistent and show evidence of undercooling in Au-rich alloys.

INVARIANT EQUILIBRIA

Table 39 and Fig. 141 give details of the invariant equilibria. As stated in the introduction, [74 Gat] did not observe the formation of the Bi_7Te_3 phase in the ternary system and attribute this to annealing the alloys for 80 days at 300°C. In contrast [74 Gat] confirmed the existence of Bi_7Te_3 in binary Bi–Te alloys after an annealing schedule of three months at 250°C. It is concluded that Bi_7Te_3 is a phase that enters into both the binary Bi–Te and the partial ternary system Au–Bi_2Te_3–Bi equilibria. The ternary eutectic reaction at E_3 should be considered at $L_{E_3} \rightleftharpoons Au_2Bi + (Bi) + Bi_7Te_3$. This implies the existence of a transition reaction $L_{U_4} + Bi_5Te_3 \rightleftharpoons Au_2Bi + Bi_7Te_3$ at a temperature between E_3 and p_4. In Fig. 141 the reaction scheme has been amended to include the equilibrium reactions in the Bi corner (in dashed lines) but no attempt has been made to indicate reaction temperatures for U_4 or E_3 for reasons discussed immediately below. [64 Win] mentioned the presence of a ternary eutectic point at 12 at.-% Au, 86 at.-% Bi, 2 at.-% Te at 210°C. If this data is accepted the eutectic reaction proposed by [74 Gat] could be interpreted as the transition reaction U_3 at 235°C followed by the eutectic reaction E_3 at 210°C, the compositions of U_4 and E_3 being close together. Alternatively the 210°C eutectic temperature [64 Win] must be discounted and the ternary eutectic temperature at E_3 placed at 235°C [74 Gat] which is reasonably close to the binary Au–Bi eutectic temperature of 241°C. This interpretation is accepted pending further experimental work to elucidate the equilibria in the Bi corner. It implies that the transition reaction at U_4 is at a temperature between 235°C (E_3) and 312°C (p_4).

Table 39. Invariant equilibria

Reaction	*Temperature*, °C	*Phase*	*Composition*, at.-%	
			Bi	*Te*
U_1	456	L	46	33
U_2	374	L	62·5	9·5
U_3	346	L	66	6·5
E_1	402	L	10	53
E_2	383	L	7·5	82
E_3	235	L	84·5	1·5

The compositions of the liquid phase associated with the invariant reactions U_1–U_3 and E_1–E_3 (Table 39) are as

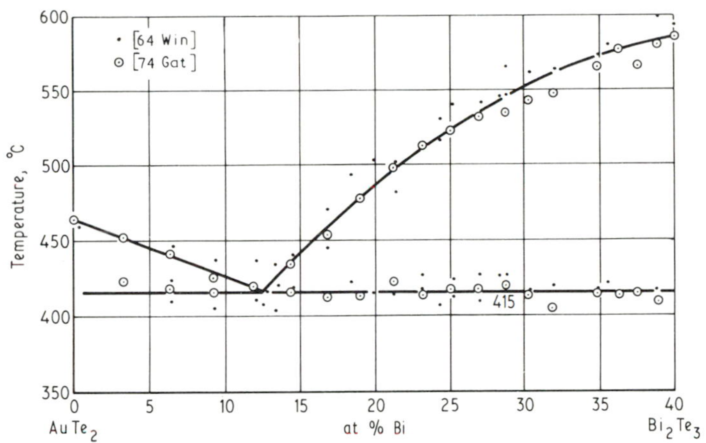

139 The $AuTe_2$–Bi_2Te_3 section

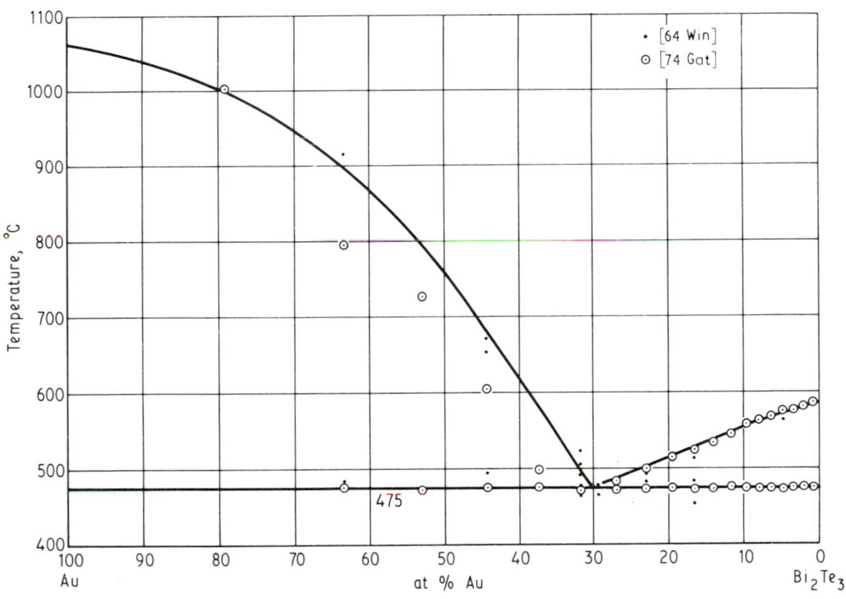

140 The Au–Bi_2Te_3 section

141 Reaction scheme for the Au–Bi–Te system

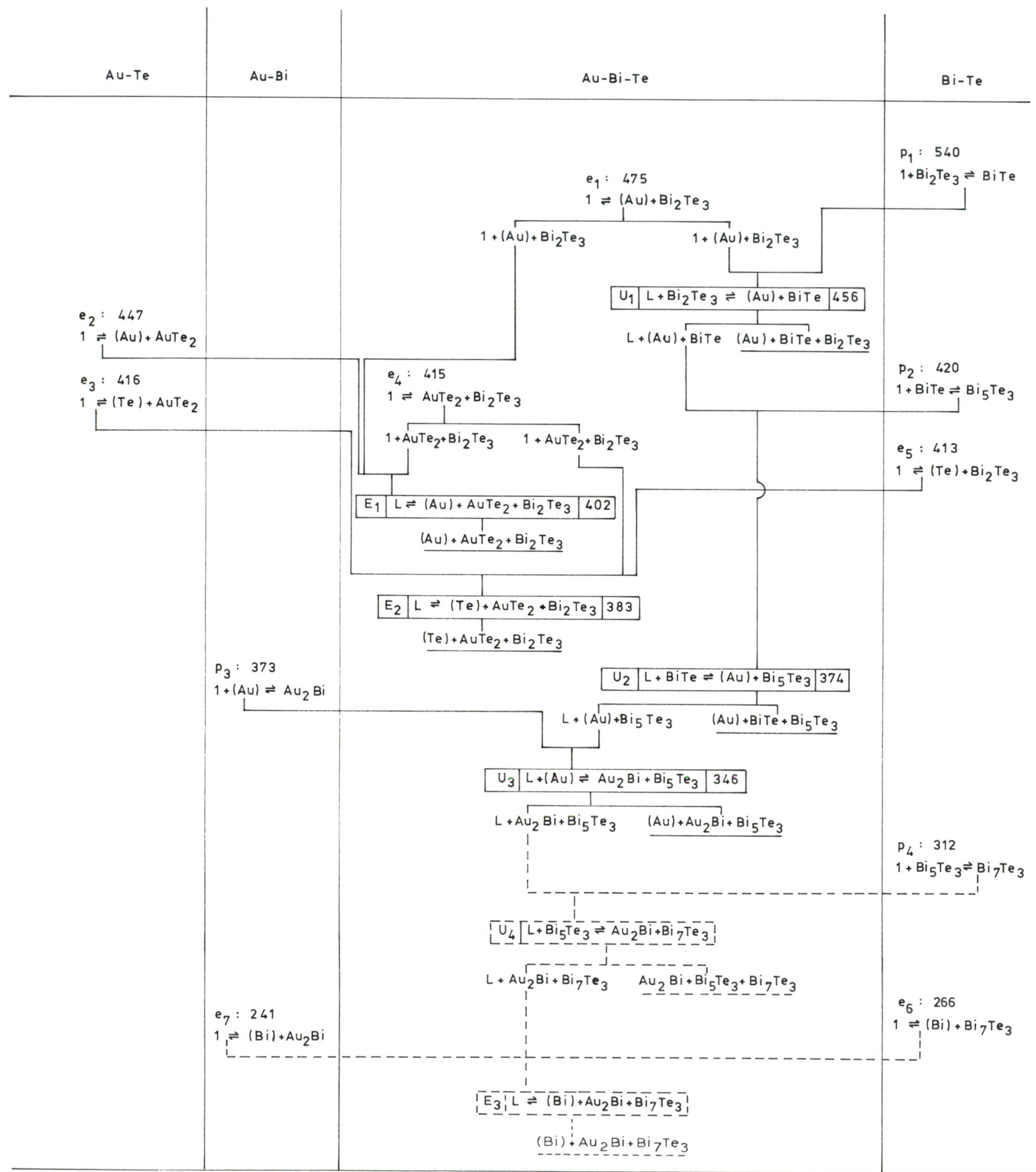

given [74 Gat]. Apart from an error in transcription for U_1 (quoted by [74 Gat] as 21 at.-% Au, 46 at.-% Bi, 38 at.-% Te and amended to 21 at.-% Au, 46 at.-% Bi, 33 at.-% Te) the compositions have been accepted. [74 Gat] provide isothermal sections for temperatures corresponding to U_1–U_3 and E_1–E_3 but no data points are included in the sections. Their validity cannot therefore be checked by an independent assessment of the experimental data.

LIQUIDUS SURFACE

The liquidus surface [74 Gat] is given in Fig. 142 with the inclusion of a dotted monovariant curve originating at p_4 and terminating at U_4 to reflect the necessity for the presence of the transition reaction at U_4. The use of a dotted line indicates lack of experimental verification for the composition of the liquid phase at U_4.

ISOTHERMAL SECTIONS

Figures 143–148 reproduce isothermal sections [74 Gat] at temperatures corresponding to the invariant reactions U_1–U_3 and E_1–E_3. The sections in Figs. 143–148 are consistent with the reaction scheme (Fig. 142) and the liquidus surface (Fig. 143). The section at 235°C (Fig. 148) has been amended to recognize the entry of Bi_7Te_3 into the ternary equilibria. The sections (Figs. 143–148) are drawn on the assumption [74 Gat] of 1% solubility of Au in the Bi–Te intermetallic compounds and of Au_2Bi in Bi_5Te_3, and Bi_7Te_3, and of $AuTe_2$ in Bi_2Te_3.

REFERENCES

61 Kin: V. J. King: Lockheed Missiles and Space Division, Report LMSD–325511, 1961

63 Key: J. D. Keys and H. M. Dutton: *J. Appl. Phys.*, 1963, **34**, 1830–1831

64 Win: E. W. Winkler and N. F. H. Bright: *Solid State Commun.*, 1964, **2**, 293–295

65 Ell: R. P. Elliott: 'Constitution of Binary Alloys', First Supplement; 1965, New York, McGraw-Hill

74 Gat: B. Gather and R. Blachnik: *Z. Metall.*, 1974, **65**, 653–656

83 Eva: D. S. Evans and A. Prince: CALPHAD XII, Liege, 1983

83 Oka: H. Okamoto and T. B. Massalski: *Bull. Alloy Phase Diagrams*, 1983, **4**, 401–407

84 Oka: H. Okamoto and T. B. Massalski: *Bull. Alloy Phase Diagrams*, 1984, **5**, 172–177

Au–Bi–Zn

[24 Tam] studied the partitioning of Au between immiscible liquid layers of Bi and Zn at 475 ± 5°C. Chemical analysis of the two layers gave a Au content of 0·16 wt-% in the Bi-rich layer and 8·97 wt-% in the Zn-rich layer. On the assumption that the Bi–Zn liquid immiscibility gap in the ternary is parallel to the Au–Bi and Au–Zn binary edges in the initial stages, the Bi-rich layer has a composition 0·1 Au, 52·3 Bi, 47·6 at.-% Zn. The Zn-rich layer contains 3·3 Au, 2·1 Bi, 94·6 at.-% Zn. The tie line joining the two liquid compositions is skewed towards the Au–Bi binary edge.

[66 Glu] reported measurement of the emf–temperature relation in galvanic cells for three series of alloys with constant Zn and variable Au contents (1·5 at.-% Zn with 0·5, 1·0, 1·5, 3·75, 5·0 Au; 2·5 at.-% Zn with 0·5, 1·0, 1·5, 2·0 Au; 5·0 at.-% Zn with 1·0, 1·5, 2·5 Au). Changes in slope of the emf–temperature data at specific temperatures are directly related to the liquid composition at such liquidus temperatures. Independent confirmation of liquidus temperatures was attempted by differential thermal analysis of three ternary alloys at a heating rate of 16°C min^{-1} and unspecified cooling rates. Figure 149 gives isotherms at 650, 600, and 550°C in the Bi-rich corner of the ternary system. The liquidus temperatures determined by DTA are included. There is good agreement with emf data for the ternary alloy containing 5·0 Au, 1·5 Zn.

The noteworthy feature of the partial liquidus surface (Fig. 149) is the large increase in liquidus temperatures of Bi–Au alloys on additions of small quantities of Zn.

REFERENCES

24 Tam: G. Tammann and P. Schafmeister, *Z. anorg. Chem.*, 1924, **138**, 220–232

66 Glu: J. V. Gluck and R. D. Pehlke: *Trans. Met. Soc. AIME*, 1966, **236**, 1238–1240

Au–C–Th

Krikorian *et al.* [67 Kri] reacted mixtures of ThC and Au powder in a graphite crucible at 850°C for 114 h under a vacuum of 10^{-6} torr. X-ray analysis indicated the formation of a Au–Th compound, probably Au_3Th by analogy with the Au–C–U system. At this temperature the stable join in the Au–C–Th system appears to be between (Au_3Th) and C, indicating that Au + C + (Au_3Th) form a stable phase mixture. [73 Bre] considers that it is more likely that the ThC disproportionated to ThC_2 in the experiment of [67 Kri], implying the existence of a stable join between Au_3Th and ThC_2.

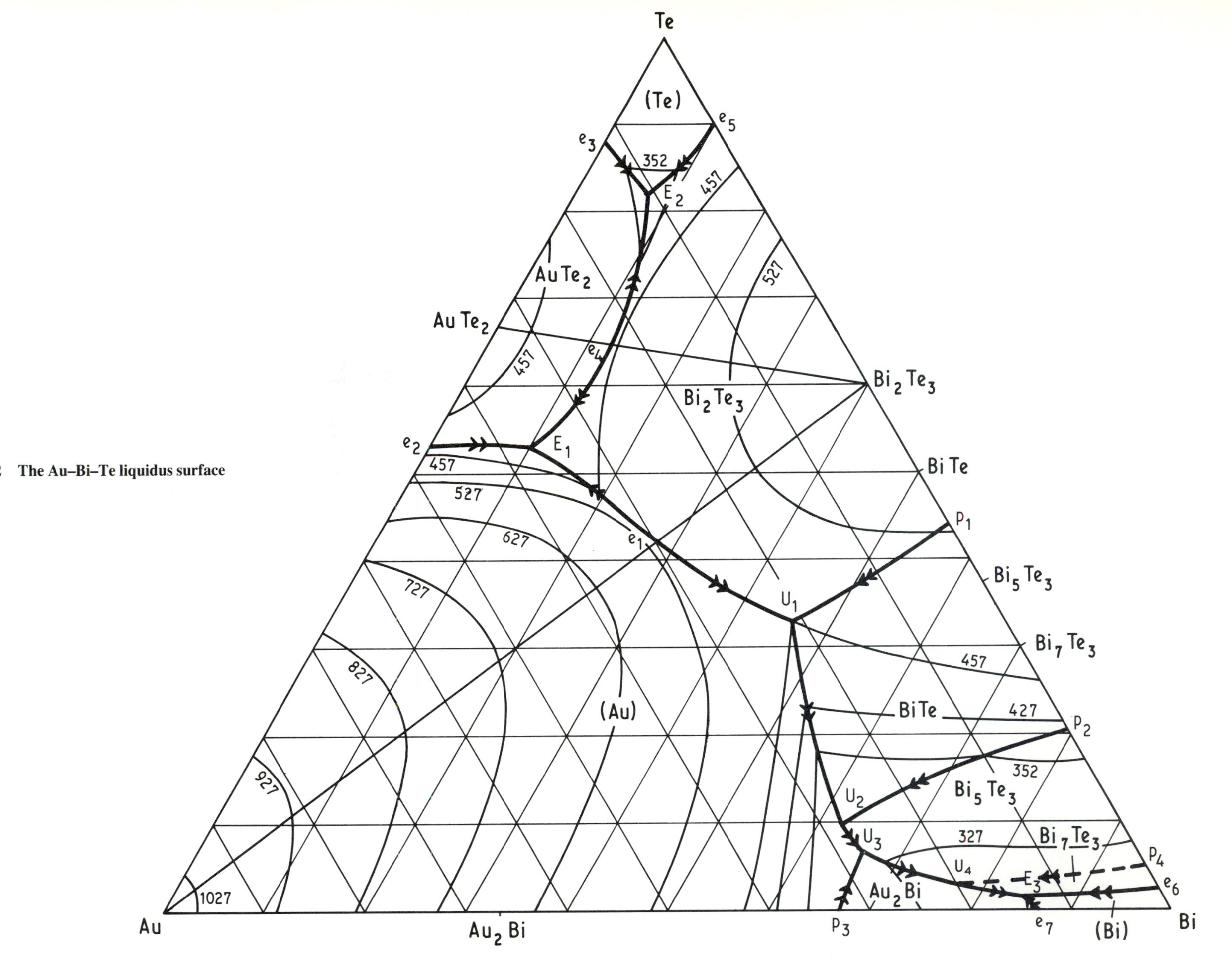

142 The Au–Bi–Te liquidus surface

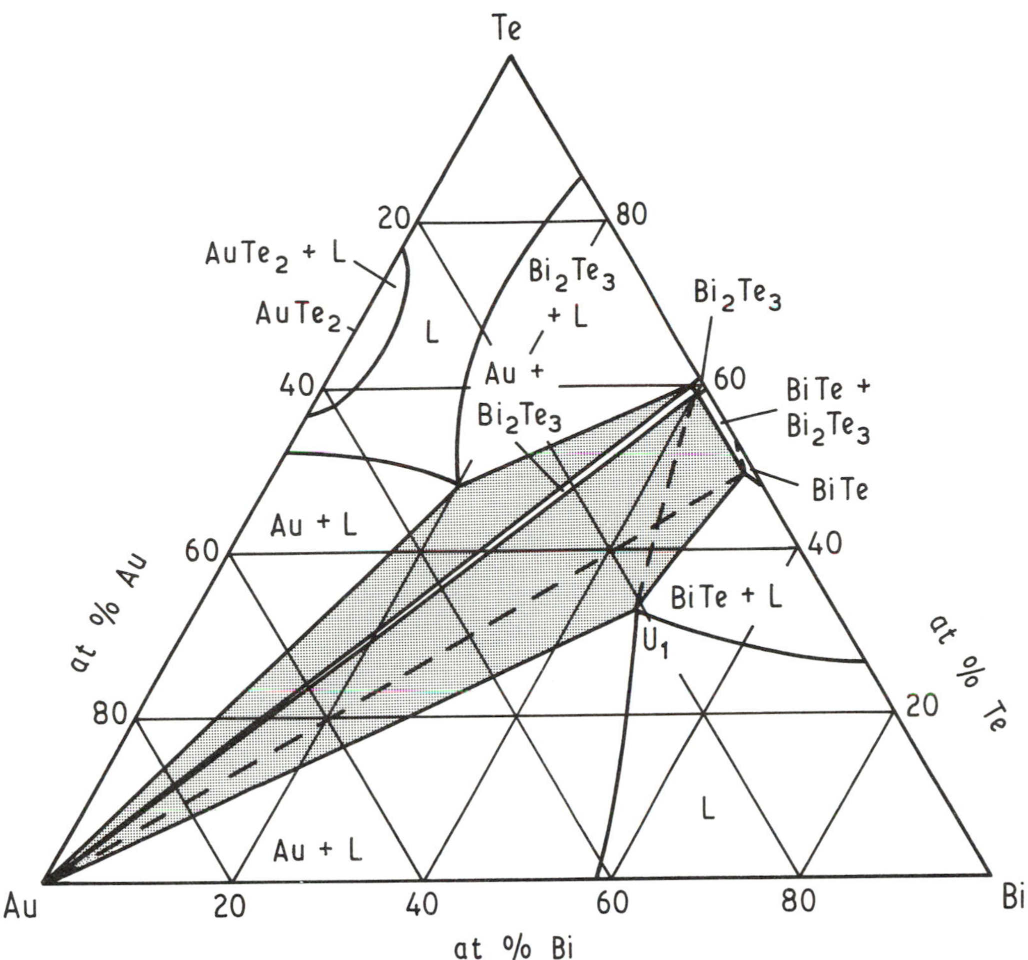

143 **Isothermal section at 456°C (U_1)**

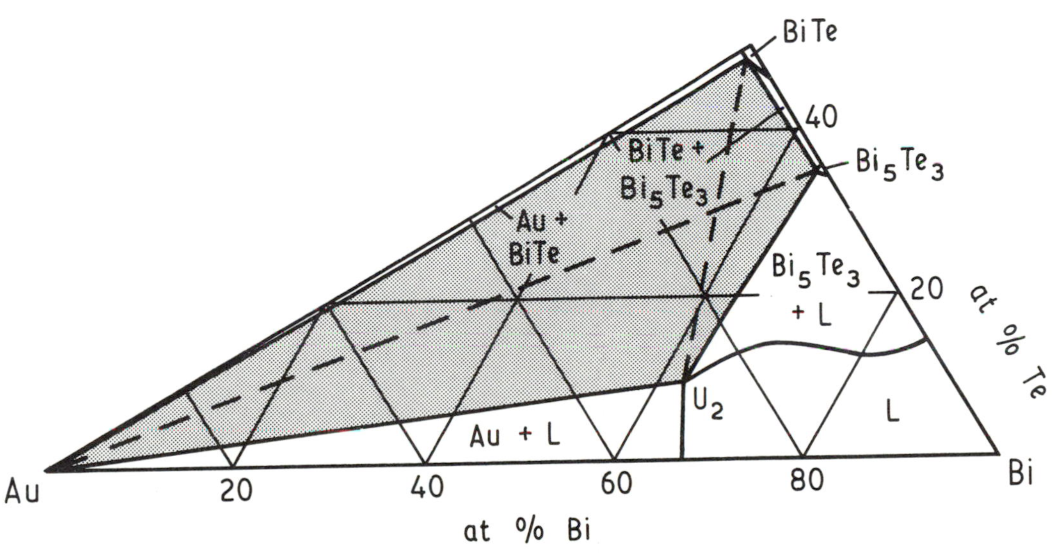

144 **Isothermal section at 374°C (U_2)**

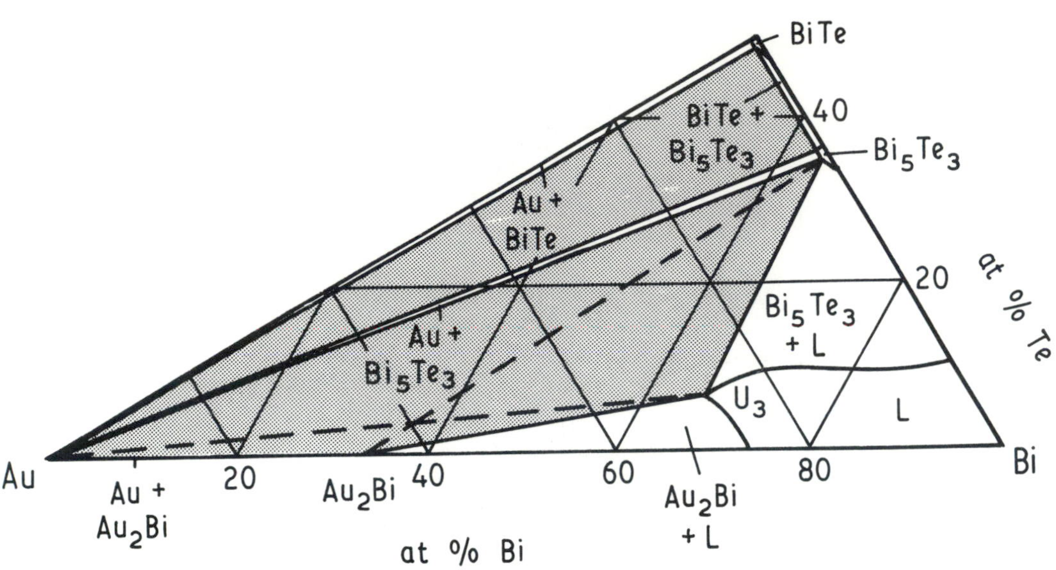

145 Isothermal section at 346°C (U_3)

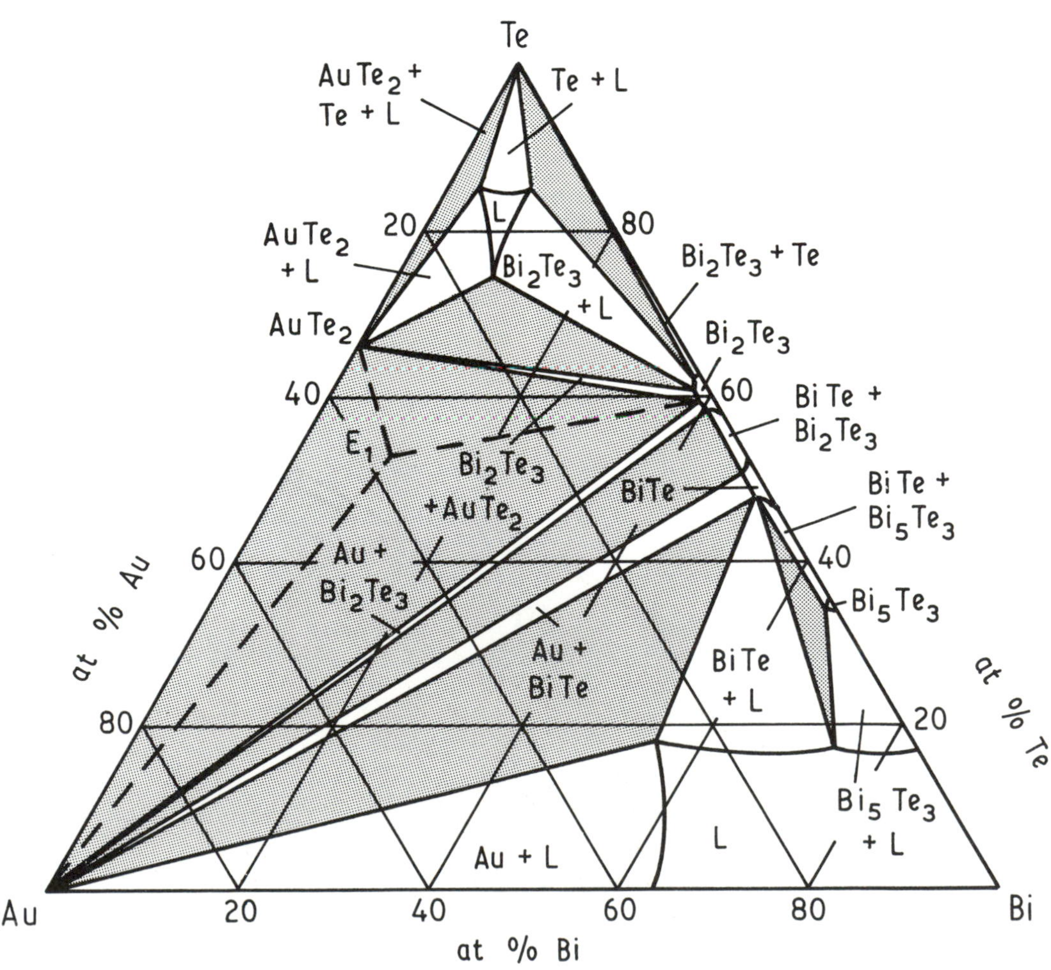

146 Isothermal section at 402°C (E_1)

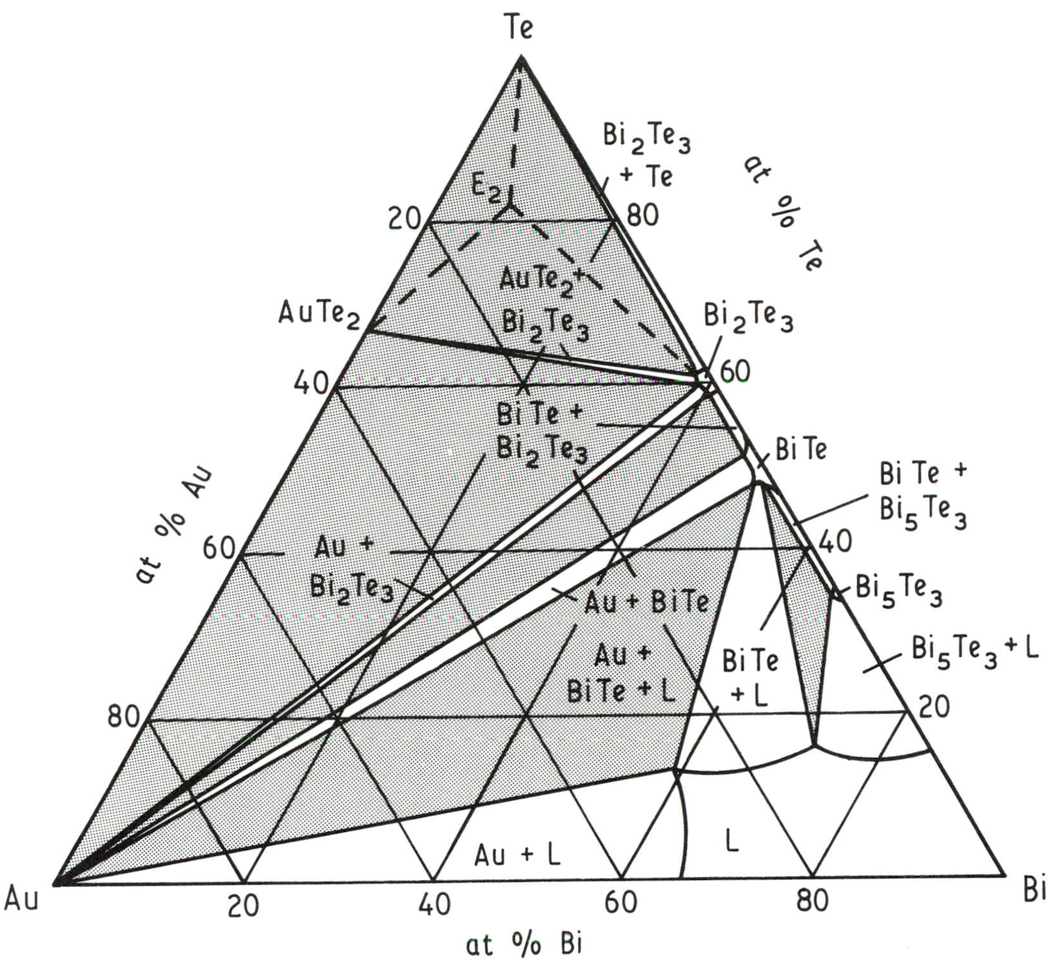

147 Isothermal section at 383°C (E_2)

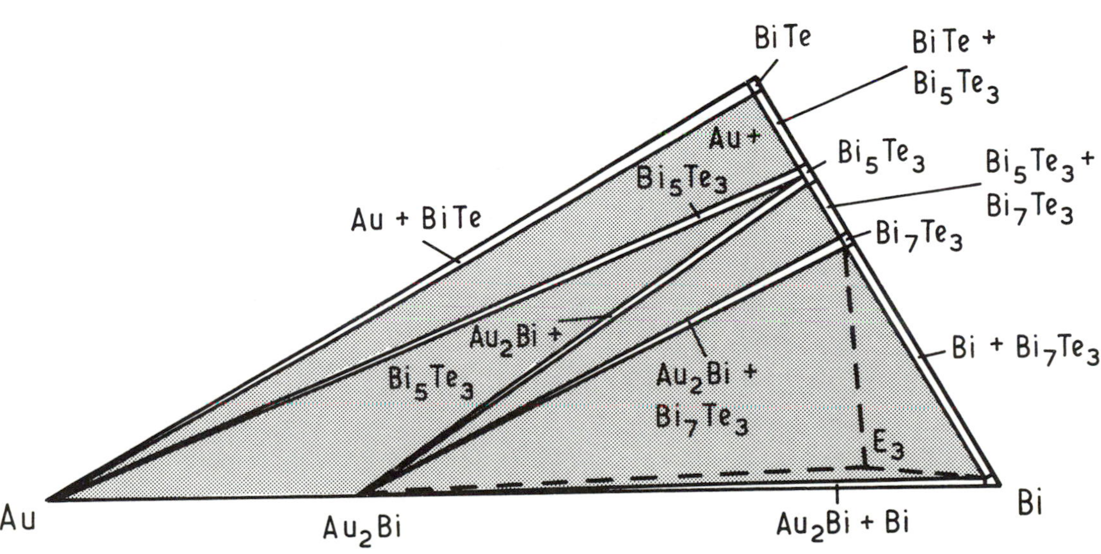

148 Isothermal section at 235°C (E_3)

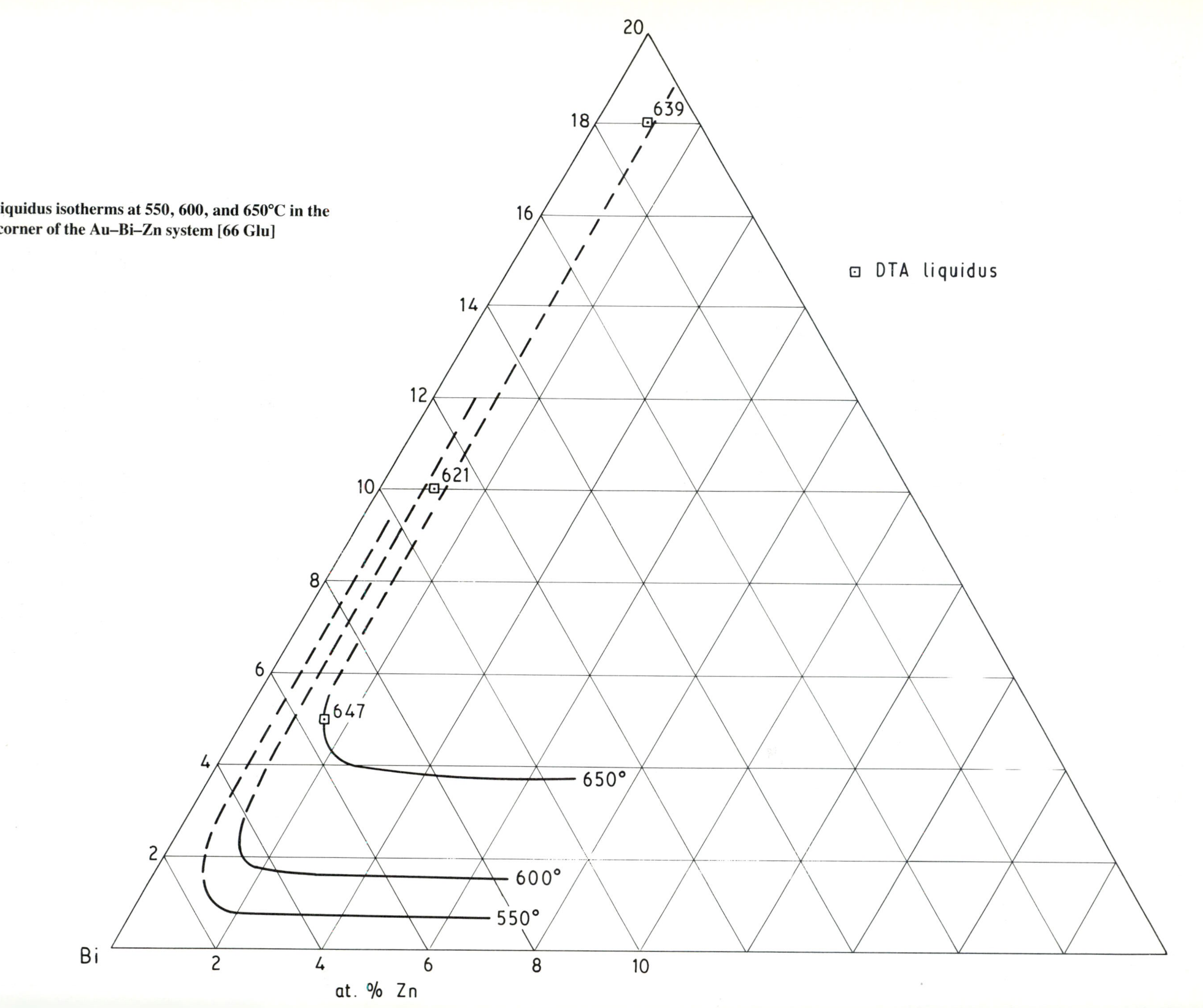

149 Partial liquidus isotherms at 550, 600, and 650°C in the Bi-rich corner of the Au–Bi–Zn system [66 Glu]

REFERENCES

67 Kri: N. H. Krikorian, T. C. Wallace, M. C. Krupka and C. L. Radosevich: *J. Nucl. Mater.*, 1967, **21**, 236–238

73 Bre: L. Brewer and P. R. Wengert: *Metall. Trans.*, 1973, **4**, 83–104

Au–C–Ti

Von Philipsborn and Laves [64 Phi] found that $AuTi_3$ with the Cr_3Si-type structure (cP8) forms a two-phase structure when it contains small amounts of the interstitial elements O, N, and C. The second phase stabilized by the interstitial elements has the ordered Cu_3Au (Ll_2)-type structure (cP4). Additions of C to $AuTi_3$ have a similar influence to oxygen additions; in the latter case about 4·5 at.-% O gives some 40% of the Cu_3Au phase and 16 at.-% O gives 100% of the Cu_3Au phase.

REFERENCE

64 Phi: H. von Philipsborn and F. Laves: *Acta Crystallogr.*, 1964, **17**, 213–214

Au–C–U

Krikorian *et al.* [67 Kri] reacted pressed plugs of $UC_{0·95}$ and Au for 18 h above 1 100°C and for 17 h below 950°C in graphite crucibles under a vacuum of 10^{-6} torr. In both cases the observed reaction was UC + 3Au = $Au_{51}U_{14}$ + C. X-ray analysis identified $Au_{51}U_{14}$ + C as reaction products. At these temperatures the stable join in the Au–C–U system is between $Au_{51}U_{14}$ and C, indicating that Au + C + $Au_{51}U_{14}$ form a stable phase mixture.

REFERENCE

67 Kri: N. H. Krikorian, T. C. Wallace, M. C. Krupka and C. L. Radosevich: *J. Nucl.*, *Mater.*, 1967, **21**, 236–238

Au–C–Y

Krupka *et al.* [69 Kru] reported the high-temperature high-pressure synthesis of the Y–Au carbide $(Y_{0·9}Au_{0·1})C_{1·3}$ with a Pu_2C_3-type structure (cI40) and lattice parameter a = 0·825 03 ± 0·000 06 nm. $YC_{1·3}$ prepared similarly (15–25 kbar, 1 200–1 450°C) had a = 0·823 86 ± 0·000 04 nm, suggesting that Au replaces Y to form a ternary solid solution phase based on $YC_{1·3}$.

REFERENCE

69 Kru: M. C. Krupka, A. L. Giorgi, N. H. Krikorian and E. G. Szklarz: *J. Less-Common Metals*, 1969, **19**, 113–119

Au–C–Zr

Brewer and Wengert [73 Bre] studied the reaction of mixtures of Au + ZrC and Au_4Zr + Au_3Zr + C when hot pressed at 925–950°C for 90–110 h under 0·8 atm argon. No reaction occurred. When liquid state reactions were studied at 1 500°C with a mixture of Au_4Zr + Au_3Zr + C under 1 atm helium, the reaction products were identified as Au + C + ZrC the Au phase containing 9·5 ± 3·5 at.-% Zr. A mixture of Au + C + Zr heated at 1 500°C remained Au–0·87 at.-% Zr + C + Zr, the liquid Au forming a non-wetting bead on top of the C + Zr powders. These results indicate that the Au–ZrC join is stable at 1 500°C and that therefore Au + C + ZrC are equilibrium phases at 1 500°C.

REFERENCE

73 Bre: L. Brewer and P. R. Wengert: *Metall. Trans.*, 1973, **4**, 83–104

Au–Ca–Fe

Sponseller and Flinn [64 Spo] determined the effect of Au on the solubility of Ca in liquid Fe at 1 607 ± 6°C. The Ca–Fe system contains two immiscible liquids at 1 607°C, an Fe-rich liquid containing 0·045 at.-% Ca, and a Ca-rich liquid. Additions of Au were made at nominal 2, 5, 10, and 20 at.-%. The addition of Au drastically reduces the solubility of Ca in liquid Fe at 1 607°C (Table 40). The Au distributes itself preferentially in the Ca-rich liquid. Indeed [64 Spo] state that the Ca-rich liquid had a greater density, due to its Au content, than the Fe-rich liquid for the 80 Fe, 20 at.-% Au alloy that was used. It is apparent that the tie lines between the Fe-rich liquid and the Ca-rich liquid at 1 607°C fan out from the Fe corner of the ternary system to give Ca-rich liquids with considerable concentrations of Au.

Table 40. Effect of Au on the solubility of Ca in liquid Fe at 1 607°C

Ca content of Fe-rich liquid layer, at.-%	*Au content of Fe-rich liquid layer*, at.-%
0·045	0
0·034	0·06
0·017	0·10
0·006	0·26
0·003	2·57

REFERENCE

64 Spo: D. L. Sponseller and R. A. Flinn: *Trans. Metall. Soc. AIME*, 1964, **230**, 876–888

Au–Ba–Ge
Au–Ca–Ge
Au–Ge–Sr

May and Schäfer [72 May] synthesized $CaAu_2Ge_2$, $SrAu_2Ge_2$, and $BaAu_2(Au,Ge)_2$ by melting the elements (purity unspecified) in Al_2O_3 crucibles under an argon atmosphere at temperatures of 1 250–1 300°C. No chemical analysis was done to check for the presence of Al (or oxygen) in the compounds, as would arise if the alkaline-earth metals reacted with the Al_2O_3 crucibles under the preparation conditions. The compounds were in the form of thin platelets with a metallic appearance. X-ray diffraction analysis of single crystal showed that the compounds have the $ThCr_2Si_2$ structure type, tI10, with unit cell constants given in Table 41. The compound $BaAu_2(Au,Ge)_2$ does not occur at the 1:2:2 stoichiometry, the Ge atom sites being substituted by Au to give a stoichiometry of $BaAu_2Au_{1\cdot16}Ge_{0\cdot84}$. [72 May] regard this compound as falling between the $ThCr_2Si_2$ and the $BaAl_4$ structure types.

Table 41. Unit cell constants

Compound	*a* ± 0·002, nm	*c* ± 0·002, nm
$BaAu_2(Au,Ge)_2$	0·465	1·056
$CaAu_2Ge_2$	0·443	1·023
$SrAu_2Ge_2$	0·451	1·035

REFERENCE

72 May: N. May and H. Schäfer: *Z. Naturf.*, 1972, **27B**, 864–865

Au–Ca–Si
Au–Si–Sr

Dörrscheidt [76 Doe] synthesized $CaAu_2Si_2$ and $SrAu_2Si_2$ by melting the elements (purity unspecified) in Al_2O_3 crucibles under an argon atmosphere. No chemical analysis was done to check for the presence of Al (or oxygen) in the compounds, as would arise if the alkaline-earth metals reacted with the Al_2O_3 crucibles under the preparation conditions. The compounds were in the form of thin platelets that hydrolysed when exposed to a humid atmosphere. X-ray diffraction analysis of single crystals showed that $CaAu_2Si_2$ and $SrAu_2Si_2$ have the $ThCr_2Si_2$ structure type, tI10, with unit cell constants given in Table 42.

Table 42. Unit cell constants

Compound	*a* ± 0·001, nm	*c* ± 0·002, nm
$CaAu_2Si_2$	0·432	1·002
$SrAu_2Si_2$	0·437	1·014

REFERENCE

76 Doe: W. Dörrscheidt, N. Niess and H. Schäfer: *Z. Naturf.*, 1976, **31B**, 890–891

Au–Cd–Cu

Raub and Walter [50 Rau] studied the effect of ternary additions on the order–disorder transformation of AuCu. The binary AuCu alloy has a transformation temperature of 410°C. Cd additions strongly depress the transformation temperature. With 5 at.-% Cd the disordered fcc ternary solid solution is stable to 400°C and the CuAuI tetragonal superlattice is stable at 350°C and lower; with 7·8 at.-% Cd the disordered fcc ternary solid solution is stable to 300°C and the CuAuI phase from 250°C to lower temperatures; with 11 at.-% Cd the corresponding temperatures are 350 and 300°C. Little Cd appears to be contained in the ordered CuAuI phase and, in all cases, a Cd-rich second phase was detected.

[87 Dem] tabulated the solidus temperatures of 15 Au-rich alloys containing from 70 to 86 at.-% Au, without giving details of the measurement technique. The data indicate that the solidus isotherms are concave to the Au corner and generally agree with the binary Au–Cd and Au–Cu solidus temperatures.

REFERENCES

50 Rau: E. Raub and P. Walter: *Z. Metall.*, 1950, **41**, 240–243

87 Dem: G. Demortier, D. Decroupet and S. Mathot, 'Precious Metals 1987', Proc. 11th International Precious Metals Institute Conf., Brussels, Belgium, June 1987. IPMI, Allentown, PA, USA, 1987, pp. 335–350

Au–Cd–Ge

The section AuCd–Ge was studied by [89 But] using DTA and metallographic techniques. A pseudobinary eutectic system was observed, with the eutectic at 555°C and 10·0 at.-% Ge. The AuCd phase dissolves 1·3 at.-% Ge at the eutectic temperature.

REFERENCE

89 But: M. T. Z. Butt, C. Bodsworth and A. Prince, unpublished work, Brunel University, 1989

Au–Cd–Hg
(C. Gumiński)

Only the Hg-rich corner of this ternary system has been investigated at ambient or moderately elevated temperatures using the following methods: calorimetry [24 Tam, 56 Har, 70 Jak], potentiometry [22 Tam, 56 Har, 69 Koz, 72 Gum, 75 Pal, 75 Der, 77 Ost, 78 Mik], electroanalytical oxidation [58 Kem, 64 Ste, 75 Mes, 75 Naz, 81 Rod] and chemical analysis of the complex amalgam after hydrostatic segregation [69 Koz].

In the first potentiometric experiments of [22 Tam] no interaction between Au and Cd in the amalgam was found, but in the next calorimetric study [24 Tam] a distinct exothermic effect accompanying the mixing of the simple amalgams was found. None of the subsequent investigators of the system denied formation of intermetallics in the complex amalgam; however the form of the compound formed in Hg have been the subject of many discussions.

Originally, [56 Har] reported formation of AuCd being soluble in Hg, although its calculated dissociation constant was dependent on both component concentrations. [58 Zeb, 72 Gum, 75 Mar, 78 Mik] recalculated the data of [56 Har] to show that the solubility product $K_s = [Au][Cd]$ had a constant value of $(2{\cdot}6 \pm 0{\cdot}2) \times 10^{-5}$ (mol %)^2 at 20°C, that suggested the existence of AuCd in crystalline form in Hg medium. The precipitation of solid AuCd was found in the course of chemical analysis [69 Koz] and confirmed in several electrochemical investigations [58 Kem, 64 Ste, 69 Koz, 72 Gum, 75 Der, 75 Pal, 77 Ost, 81 Rod]. Contrary claims for the formation of soluble AuCd in Hg were again more recently presented by [75 Mes, 75 Naz, 78 Mik] who supported their statement by briefly described electrochemical experiments.

In order to find a compromise, [78 Mik] proposed the following scheme of equilibria in the system:

$$Au + Cd \rightleftharpoons AuCd \rightleftharpoons AuCd \downarrow$$

The existence of soluble AuCd molecules was not sufficiently documented because several investigators [22 Tam, 56 Har, 69 Koz, 77 Ost, 81 Rod] did not observe any potential difference, which should be observed in such a case, for the same input concentration of Cd in Hg as well as in the Au diluted amalgam when the product of Au and Cd concentrations was lower than K_s.

Quantitative results related to the formation of AuCd in Hg are collected in Table 45. One may observe that the K_D values in Table 43, as well as those formally calculated from [69 Koz, 72 Gum, 75 Pal, 77 Ost], are unstable and concentration dependent on both metals. The recommended values of K_s are those from [69 Koz, 72 Gum, 75 Der, 81 Rod]. [69 Koz] were the first to observe that when more than a twofold excess of Au with regard to Cd was present in the amalgam then the Au_3Cd compound was also formed. The formation of crystalline Au_3Cd was examined by chemical analysis [69 Koz] and later confirmed in potentiometric experiments of [75 Pal, 77 Ost]. Other investigators found no evidence for Au_3Cd formation and [78 Mik] suggested the formation of compounds with different formulae. However, the conclusions of [78 Mik] were based on a purely mathematical treatment of potentiometric results whereas the chemical analysis of [69 Koz] is a much more convincing argument for the existence of Au_3Cd in Hg. Apart from Au_3Cd and AuCd, no other Au–Cd compounds are formed in diluted amalgams. Values of $K_s = [Au]^3[Cd]$ are summarized in Table 44; the most tentative results were obtained by [69 Koz]. Experimental results of the thermodynamic functions of Au_3Cd and AuCd formation in Hg were recently collected by [86 Gum]. The selected values of the enthalpies

Table 43. Constants characterizing equilibrium of AuCd in diluted amalgam $K_s = [Au][Cd]$; $K_D = [Au][Cd]/[AuCd]$

Type of constant	*Value*	*Temperature, °C*	*Method*	*Reference*
K_D, mol%	$1{\cdot}8 \times 10^{-3} - 9{\cdot}9 \times 10^{-3}$	20	Potentiometry	(56 Har)
K_s, (mol%)^2	$(1{\cdot}86 \pm 0{\cdot}30) \times 10^{-5}$	20	Potentiometry	(69 Koz)
	$(3{\cdot}83 \pm 1{\cdot}0) \times 10^{-5}$	25		
	$6{\cdot}57 \times 10^{-5}$	35		
	$(1{\cdot}52 \pm 0{\cdot}4) \times 10^{-4}$	45		
	$2{\cdot}38 \times 10^{-4}$	60		
	$(3{\cdot}54 \pm 0{\cdot}4) \times 10^{-4}$	70		
K_s, (mol%)^2	$(3{\cdot}5 \pm 1{\cdot}1) \times 10^{-5}$	25	Potentiometry	(72 Gum)
K_s, (mol%)^2	$(2{\cdot}6 \pm 0{\cdot}4) \times 10^{-5}$	5	Potentiometry	(75 Pal)
	$(5{\cdot}1 \pm 0{\cdot}8) \times 10^{-5}$	15		
	$(8{\cdot}8 \pm 2{\cdot}0) \times 10^{-5}$	25		
	$(1{\cdot}34 \pm 0{\cdot}22) \times 10^{-4}$	35		
	$(2{\cdot}4 \pm 0{\cdot}4) \times 10^{-4}$	45		
	$(4{\cdot}0 \pm 0{\cdot}4) \times 10^{-4}$	55		
	$(5{\cdot}3 \pm 0{\cdot}4) \times 10^{-4}$	65		
	$(8{\cdot}6 \pm 0{\cdot}2) \times 10^{-4}$	75		
K_D, mol%	$1{\cdot}5 \times 10^{-5}$	20	Electrochem. oxid.	(75 Mes)
K_D, mol%	$(1{\cdot}2 - 3{\cdot}0) \times 10^{-4}$	20	Electrochem. oxid.	(75 Naz)
K_s, (mol%)^2	$1{\cdot}9 \times 10^{-5}$	20	Potentiometry	(75 Der)
	$1{\cdot}7 \times 10^{-4}$	50		
	$4{\cdot}1 \times 10^{-4}$	75		
	$7{\cdot}9 \times 10^{-4}$	90		
K_s, (mol%)^2	$(5{\cdot}7 \pm 0{\cdot}8) \times 10^{-5}$	25	Potentiometry	(77 Ost)
K_D, mol%	$1{\cdot}8 \times 10^{-5}$	16	Potentiometry	(78 Mik)
K_s, (mol%)^2	$(3{\cdot}76 \pm 0{\cdot}31) \times 10^{-5}$	25	Electrochem. oxid.	(81 Rod)

of formation are 89·5 ± 4·0 kJ/mol Au_3Cd and 41·2 ± 0·4 kJ/mol AuCd. The Gibbs energies may be easily calculated from ${}^{Au_3Cd}K_s$ and ${}^{AuCd}K_s$ but no entropy values could be recommended at present.

REFERENCES

22 Tam: G. Tammann and W. Jander, *Z. Anorg. Chem.*, 1922, **124**, 105–122

24 Tam: G. Tammann and E. Ohler, *Z. Anorg. Chem.*, 1924, **135**, 118–126

56 Har: H. Hartmann and K. Schölzel, *Z. Phys. Chem.*, N.F., 1956, **9**, 106–126

58 Kem: W. Kemula, Z. Galus and Z. Kublik, *Bull. Acad. Polon. Sci., Ser. Sci. Chim. Geol. Geogr.*, 1958, **6**, 661–668

58 Zeb: A. I. Zebreva, *Vestn. Akad. Nauk Kaz. SSR*, 1958 (11), 88–91; *Zh. Fiz. Khim.*, 1961, **35**, 948–949; *Tr. Inst. Khim. Nauk Akad. Nauk Kaz. SSR*, 1962, **9**, 55–70

64 Ste: O. S. Stepanova, M. S. Zakharov, L. F. Trushinina and V. I. Aparina, *Izv. Vyssh. Ucheb. Zaved., Khim. Khim. Tekhnol.*, 1964, **7**, 184–188

69 Koz: L. F. Kozin, G. F. Cherkasova and M. I. Erdenbaeva, *Izv. Akad. Nauk Kaz. SSR, Ser. Khim.*, 1969 (3), 42–49

70 Jak: B. Jakuszewski, J. Grabowski and S. Partyka, *Roczniki Chem.*, 1970, **44**, 1025–1030

72 Gum: C. Gumiński and Z. Galus, *Bull. Acad. Polon Sci., Ser. Sci. Chim.*, 1972, **20**, 1037–1044

75 Mar: B. M. Maryanov, *Elektrokhimiya*, 1975, **11**, 1808–1812

75 Naz: B. F. Nazarov, *Sovrem. Probl. Polarografii s Nakopleniem, Tomsk*, 1975, 47–56

75 Der: M. B. Dergacheva, *Tr. Inst. Org. Katal. Elektrokhim. Akad. Nauk Kaz. SSR*, 1975, **11**, 36–42

75 Mes: N. A. Mesyats and N. I. Mikheeva, *Izv. Tomsk. Politekhn. Inst.*, 1975, **195**, 43–45

75 Pal: D. Pallyska, M.Sc. thesis, University of Warsaw, 1975

77 Ost: P. Ostapczuk and Z. Kublik, *J. Electroanal. Chem.*, 1977, **83**, 1–17

78 Mik: N. P. Mikheeva and A. G. Stromberg, *Zh. Anal. Khim.*, 1978, **33**, 1726–1731

81 Rod: R. S. Rodgers and L. Meites, *J. Electroanal. Chem.*, 1981, **125**, 167–176

86 Gum: C. Gumiński, *Z. Metallk.*, 1986, **77**, 87–96

Au–Cd–In

The Au-rich part of the Au–Cd–In system has been studied by X-ray powder diffraction techniques by [57 Sch], [58 Weg], and [60 Bha]. No other work has been reported. The 350°C isothermal section (Fig. 150 [58 Weg]) indicates a narrow homogeneity region for the α_1 phase based on the Au–In binary α_1 phase (Nd-type structure, hP4). The ternary α_1 phase extends up to 17.5 at.-% Cd along an approximately constant valency electron concentration of 1·25. The ζ phase based on the Au–In binary ζ phase (Mg-type, hP2) extends to 11 at.-% Cd, and its In-rich boundary is almost constant at 79 at.-% Au. [58 Weg] differentiate five ternary phases, τ_1 to τ_5, which will have close-packed long-period superlattices based on hexagonal or orthorhombic structures. The atom stacking sequence is quoted as ABAB (τ_1), ABABACAC (τ_2), ABABCBCAC (τ_3), ABABCBCAC (τ_4), and ABABAC (τ_5). The phase τ_5 does not occur at 425°C. In view

Table 44. Solubility product of Au_3Cd ($K_s = [Au]^3\,[Cd]$) in the diluted amalgam

K_s *value* (mol%)^4	*Temperature*, °C	*Method*	*Reference*
$(2{\cdot}16 \pm 0{\cdot}41) \times 10^{-9}$	20	Potentiometry	(69 Koz)
$(4{\cdot}25 \pm 0{\cdot}40) \times 10^{-9}$	25		
$7{\cdot}63 \times 10^{-9}$	35		
$(4{\cdot}29 \pm 0{\cdot}81) \times 10^{-8}$	45		
$1{\cdot}94 \times 10^{-7}$	60		
$(3{\cdot}66 \pm 0{\cdot}49) \times 10^{-7}$	70		
$(8{\cdot}6 \pm 1{\cdot}9) \times 10^{-10}$	5	Potentiometry	(75 Pal)
$(2{\cdot}4 \pm 0{\cdot}2) \times 10^{-9}$	15		
$(6{\cdot}2 \pm 1{\cdot}5) \times 10^{-9}$	25		
$(1{\cdot}34 \pm 0{\cdot}53) \times 10^{-8}$	35		
$(3{\cdot}7 \pm 1{\cdot}1) \times 10^{-8}$	45		
$(2{\cdot}9 \pm 0{\cdot}7) \times 10^{-7}$	55		
$(7{\cdot}7 \pm 2{\cdot}5) \times 10^{-7}$	65		
$(2{\cdot}9 \pm 0{\cdot}4) \times 10^{-6}$	75		
$(2{\cdot}0 \pm 0{\cdot}3) \times 10^{-8}$	25	Potentiometry	(77 Ost)

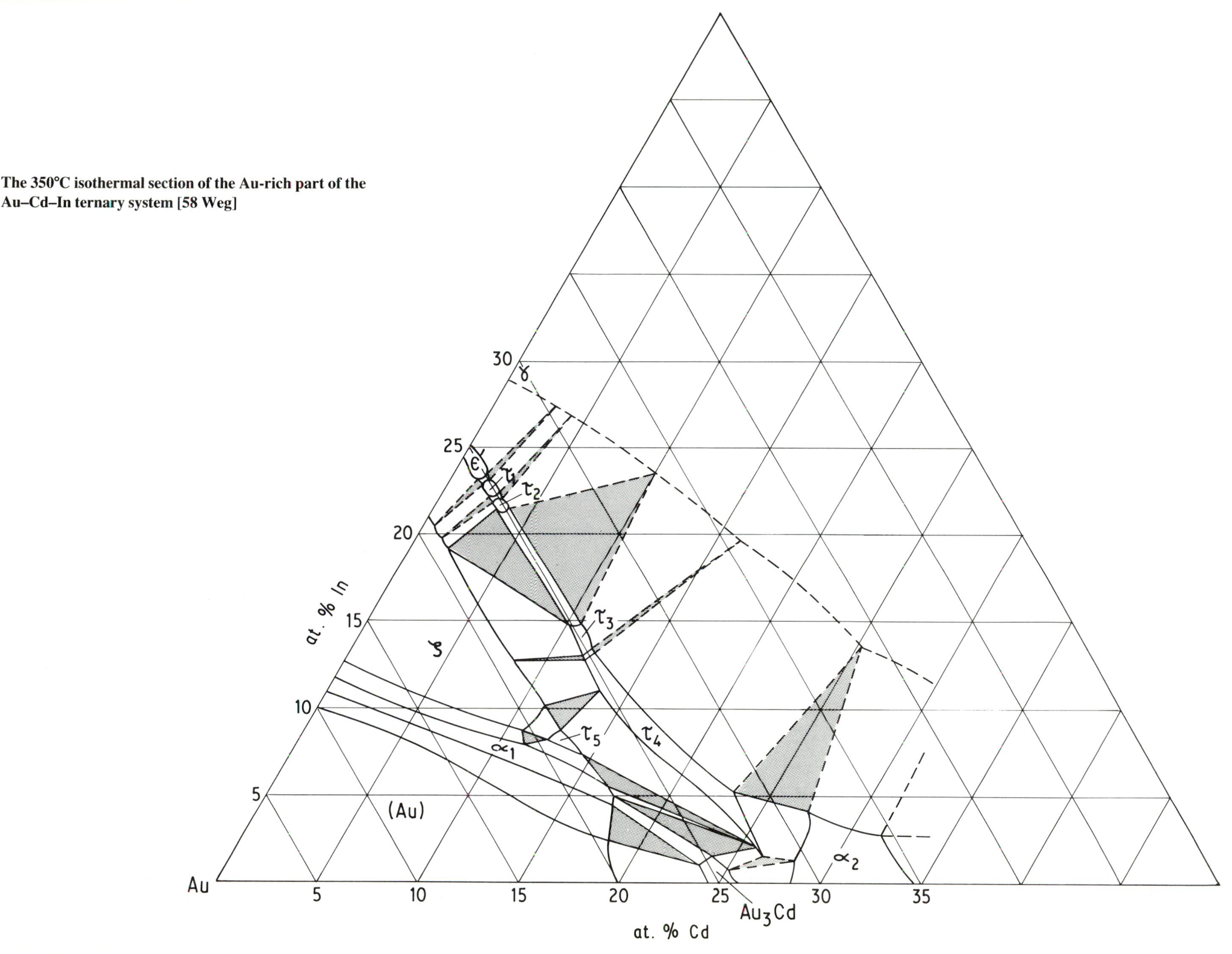

150 The 350°C isothermal section of the Au-rich part of the Au–Cd–In ternary system [58 Weg]

of the uncertainties surrounding the crystal structures of the Au-rich phases in the Au–In and Au–Cd binary systems [87 Oka] the isothermal section (Fig. 150) should be regarded as provisional. The γ phase of the Au–In binary, with a γ-brass structure, appears to be shown in Fig. 150 as extending towards the corresponding γ-brass phase in the Au–Cd system as a continuous solid solution series at 350°C.

Schubert *et al.* [57 Sch] studied four alloys in the Au-rich part of the ternary system; their data is superseded by that of [58 Weg]. [60 Bha] report that an alloy 70Au, 25Cd, 5 at.-% In after heat treatment for 1 day at 425°C had the same structure as the β phase in the Au–In binary system, i.e. hexagonal $Cu_{4\cdot5}Sb$-type (see [85 Oka]). At 350°C [58 Weg] place this alloy in the same three-phase field $\tau_4 + \gamma + \alpha_2$ (Fig. 150).

REFERENCES

57 Sch: K. Schubert, H. Breimer, W. Burkhardt, E. Günzel, R. Haufler, H. L. Lukas, H. Vetter, J. Wegst and M. Wilkens: *Naturwiss.*, 1957, **44**, 229–230

58 Weg: J. Wegst and K. Schubert: *Z. Metall.*, 1958, **49**, 533–544

60 Bha: S. Bhan and K. Schubert: *Z. Metall.*, 1960, **51**, 327–339

87 Oka: H. Okamoto and T. B. Massalski: 'Phase Diagrams of Binary Gold Alloys', 42–57 (Au–Cd), 142–153 (Au–In); 1987, ASM International

Au–Cd–Mg

Matsuo *et al.* [81 Mat] studied 13 alloys on the 50 at.-% Au section. Both binary phases AuCd and AuMg crystallize in the CsCl-type ordered structure, cP2, and both these β′ phases are stable to their melting points. The β′ AuCd phase transforms martensitically to β″ at 30° ± 2 on cooling. Additions of Mg up to 8 at.-% depress the β′ = β″ transformation temperature to −147°C. The β′ phases AuCd and AuMg form a continuous series of solid solutions with lattice parameters varying linearly from β′ AuCd to β′AuMg. No Heusler-type phase was detected at the Au_2CdMg stoichiometry.

REFERENCE

81 Mat: Y. Matsuo, S. Minamigawa and K. Katada: *Trans. Japan Inst. Metals*, 1981, **22**, 367–368

Au–Cd–Mn

The only work on the Au–Cd–Mn ternary system relates to the replacement of some of the Au atoms in AuMn by Cd atoms. Ido *et al.* [70 Ido] found by X-ray diffraction analysis at room temperature that alloys on the section $Au_{1-x}Cd_xMn$ had a tetragonally distorted CsCl structure up to a composition 42Au, 8Cd, 50Mn. A 40Au, 10Cd, 50Mn alloy was bcc CsCl, implying that at least 10 at.-% Cd is soluble in AuMn at room temperature on the 50 at.-% Mn section.

REFERENCE

70 Ido: T. Ido, H. Teramoto, T. Kasai, K. Sato and K. Adachi: *J. Phys. Soc. Japan*, 1970, **28**, 1589.

Au–Cd–Pb

The only publication [1894 Hey] on the Au–Cd–Pb system dealt with the examination of the depression of the freezing point of an 4·66 at.-% Cd, 95·34 at.-% Pb alloy on addition of Au up to 6·09 at.-% Au, and similar measurement for a 3·85 at.-% Au, 96·15 at.-% Pb alloy on addition of Cd up to 8·68 at.-% Cd. The behaviour is similar to that found in the more extensive work on the Au–Cd–Sn system [1891 Hey]. There is almost certainly a pseudobinary system between a phase containing a Au : Cd ratio of 1 : 1 and (Pb) with a reaction temperature of 318·8°C, referred to the melting point of Pb measured at 327·64°C (compared with the currently accepted value of 327·50°C). By analogy with the Au–Cd–Sn system there is a possibility of the presence of the ternary compound AuCdPb. The data of [1894 Hey] do not allow any judgement to be made of the phase in equilibrium with (Pb) at 318·8°C; it could be AuCd or AuCdPb.

The two sections studied by [1894 Hey] are presented in Figs. 151 and 152. The 4·66 at.-% Cd, 95·34 at.-% Pb–Au section (Fig. 151) indicates an increase in liquidus temperature on addition of Au to just below 1 at.-%. This is followed by a distinct break in the data points at a temperature of about 314°C; the experimental data for further Au additions are temperatures for secondary separation of (Pb) + $AuCdPb_x$ where x could be 0 or 1. The maximum on the curve coincides with a Au : Cd ratio of 1 : 1 at 318·8°C. The 3·85 at.-% Au, 96·15 at.-% Pd–Cd section (Fig. 152) indicates a small reduction in liquidus temperature with 0·2 at.-% Cd additions followed by a curve that rises to a maximum at a Au : Cd ratio of 1 : 1. There is a change in slope when Cd additions exceed approximately 5·8 at.-%. The parabolic-shaped curve represent secondary separation of (Pb) + $AuCdPb_x$ as in Fig. 151.

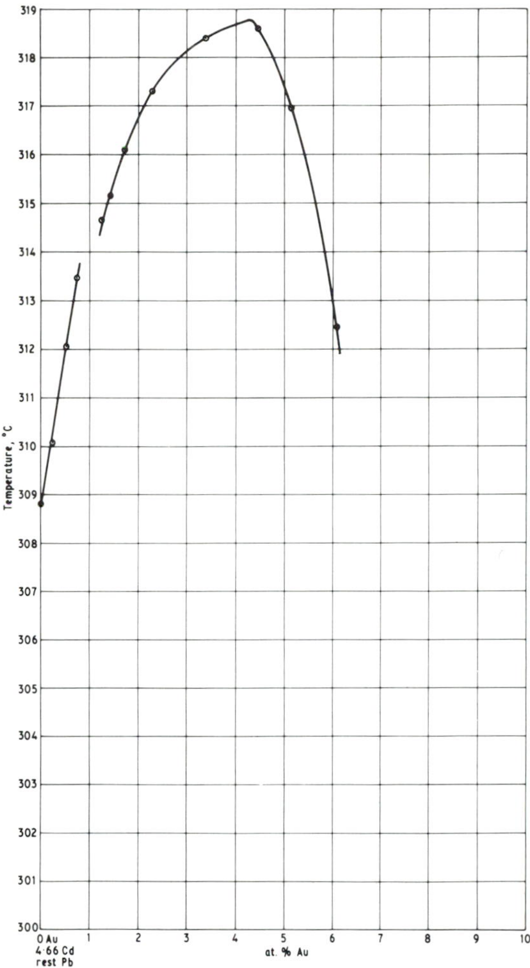

151 **Experimental data [1894 Hey] relating the depression of the freezing point to alloy composition for the 4·66 at.-% Cd, 95·34 at.-% Pb–Au section**

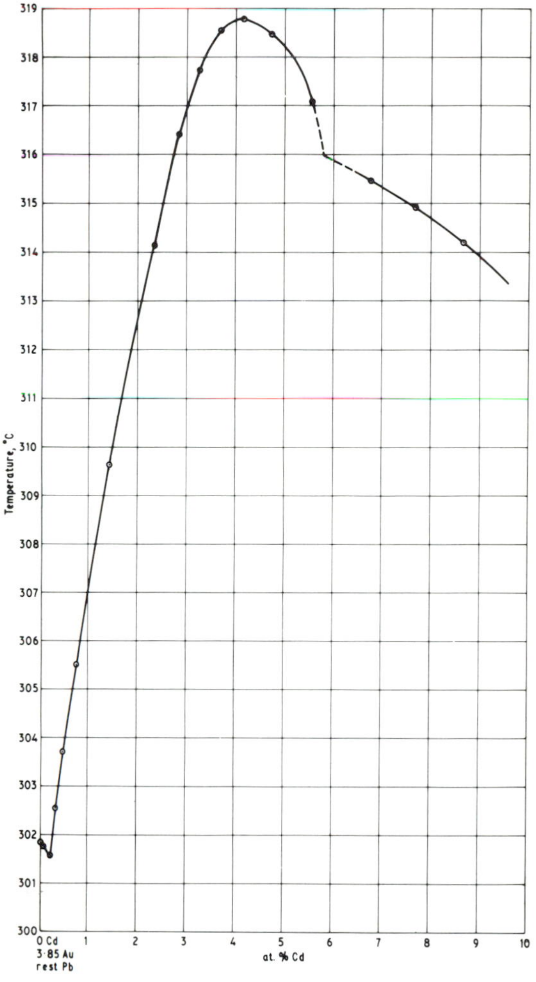

152 **Experimental data [1894 Hey] relating the depression of the freezing point to alloy composition for the 3·85 at.-% Au, 96·15 at.-% Pb–Cd section**

Given that a pseudobinary system exists, L_e = $AuCdPb_x$ + (Pb), with a reaction temperature some 8·8°C below the melting point of Pb, it would appear that the eutectic point e must lie within the region 97–98 at.-% Pb. Monovariant curves, representing the separation of $AuCdPb_x$ + (Pb) from the melt, will descend towards both the Au–Pb and CdPb binary axes. The break in Fig. 151 at about 1 at.-% Au and 314°C can be construed as the composition at which the monovariant curve from e cuts this section. The sharp change in slope at 0·2 at.-% Cd (Fig. 152) can be interpreted as the composition at which the monovariant curve from e cuts this section.

REFERENCES

1891 Hey: C. T. Heycock and F. H. Neville: *J. Chem. Soc.*, 1891, **59**, 936–966

1894 Hey: C. T. Heycock and F. H. Neville: *J. Chem. Soc.*, 1894, **65**, 65–76

Au–Cd–S

[68 Neb] measured the solubility of Au in single crystal CdS using a radioactive tracer technique. At a pressure of S_2 vapour of 500 Torr the solubility obeys an Arrhenius-type relation with the log of solubility inversely proportional to temperature (Kelvin). The solubility increases with temperature from 0·0010 at.-% Au at 500°C to 0·0025 at.-% Au at 600°C and 0·0058 at.-% Au at 725°C. The solubility of Au in CdS decreases with decrease of the S_2 vapour pressure from 500 to 10^{-3} Torr.

REFERENCE

68 Neb: E. Nebauer, *Phys. Status Solidi*, 1968, **29**, 269–281

Au–Cd–Sb

The three publications [70 Bel, 73 Bel, 75 Bel] on this system originate from the same research group. All concern the pseudobinary section of Au–CdSb in which a congruently melting ternary compound AuCdSb was identified [70 Bel, 73 Bel]. Alloys were prepared from single crystal CdSb and high purity Au, of unstated purity. After sealing in quartz tubes evacuated to 10^{-4} mm the alloys were heated for 10 h at 1 100°C, slowly cooled at 25°C h^{-1} to 800, 600, and 470°C, soaking for 10 h at each temperature. Finally the alloys were given a 170 h anneal at 250°C. AuCdSb melts at 513°C [70 Bel] or 515°C [73 Bel]. It has a cubic structure with a = 0·62532 nm [70 Bel, 73 Bel]. AuCdSb forms a eutectic with CdSb located at 20 at.-% Au and 430°C [70 Bel, 73 Bel]; it also forms a eutectic with Au located at 70 at.-% Au and 460°C [70 Bel, 73 Bel]. Fig. 153 gives the Au–CdSb phase diagram as scaled from the original diagram [73 Bel]. There is a break in the temperature scale used by [73 Bel] for Au-rich alloys, but no indication is given of the temperature scale for the Au-rich liquidus. The liquidus rises steeply from the AuCdSb–Au eutectic point and may tentatively be assigned coordinates of 71·5, 79, and 88 at.-% Au at 700, 900, and 1 000°C respectively. The solubility of CdSb in Au is about 5 mol.-% at 250°C, as determined by X-ray analysis and confirmed by measurements of the composition dependence of microhardness and electrical conductivity in addition to metallographic examination. Alloys were annealed at 250°C and examined at room temperature. The solubility of Au in CdSb [75 Bel] was determined by metallographic analysis of alloys equilibrated at 25, 150, 250, 350, and 400°C, and by measurement of the composition dependence of microhardness on alloys equilibrated at these temperatures. Alloys were heated to temperature and held for 1 600, 1 500, 1 000, and 900 h respectively at temperatures of 150, 250, 350, and 400°C respectively. To measure the room temperature solubility alloys were heat treated for 1 200 h at 100°C and then slowly cooled to room temperature over 700 h. The results are presented in Table 45. Extrapolation of the solubility curve to the eutectic temperature of 430°C indicates a maximum solubility of Au in CdSb of 0·76 at.-% Au.

Table 45. Solid solubility of Au in CdSb [75 Bel]

Temperature, °C	*Solubility*, at.-% Au
400	0·70
350	0·60
250	0·45
150	0·30
25	0·16

REFERENCES

70 Bel: D. P. Belotskii, M. K. Makhova, V. G. Galichanskii and M. P. Kotsyumakha: *Izvest. Akad, Nauk. SSSR, Neorg. Materialy.*, 1970, **6**, 1593–1597

73 Bel: D. P. Belotskii and V. G. Galichanskii: *Ukrain. Khim, Zhur.*, 1973, **39**, 906–909

75 Bel: D. P. Belotskii and M. K. Makhova: *Izvest. Akad. Nauk. SSSR, Neorg. Materialy.*, 1975, **11**, 1882–1883

Au–Cd–Te

A short conference abstract, [74 Gri], states that the solubility of AuCd in CdTe was studied by the reaction of a radioactive Au layer on CdTe with Cd vapour to form AuCd and carrying out diffusion experiments under unstated conditions. The solubility of AuCd in CdTe was not quantified. A statement is made that the solubility is lower than that of $AuTe_2$ in CdTe. [74 Aku 1] report preliminary data on the solubility of Au in CdTe but a more definitive account was published by [74 Aku 2]. Single crystal CdTe was grown by the Bridgman technique and doped with Au in the melt or by diffusion of Au into undoped crystals. Samples were annealed at 800 and 900°C in a two-zone furnace, the colder end containing either Cd or Te to establish an overpressure. After quenching the Au concentration was determined by ion channelling techniques using 2MeV $^4He^+$ ions. The solubility of Au in CdTe, expressed as the total Au concentration given by the random backscattering spectrum, decreases with increasing partial pressure of Cd and increases with temperature (Table 46). At 800°C the solubility increases from 0·008 at.-% Au near the Cd solidus (high p_{Cd}) to 0·075 at.-% Au near the Te solidus. At 900°C the solubility increases from 0·08 at.-% Au near the Cd solidus to 0·34 at.-% Au near the Te solidus. [78 Gri] investigated the Au–CdTe phase diagram by DTA, metallography and X-ray diffraction methods. The section (Fig. 154), is a pseudobinary eutectic with a eutectic at 75 ± 5 at.-% Au and 810 ± 10°C. The authors comment that the 30 at.-% Au alloy showed a tendency to demixing, according to metallographic evidence, but this observation does not correspond with the CdTe liquidus found by DTA. Examination of the $AuTe_2$–CdTe and AuCd–CdTe sections would give a further guide to the ternary equilibria.

REFERENCES

74 Gri: V. I. Grytsiv, O. Z. Panchuk and D. P. Belotskii, 'Tezisy, Dokl.-Vses. Konf. Kristallokhim. Intermet. Soedin', 2nd, 1974, p. 173. Ed: R. M. Rykhal, L'vov. Gos. Univ.: Lvov, USSR

74 Aku 1: W. Akutagawa, D. Turnbull, W. K. Chu and J. W. Mayer, *Solid State Commun.*, 1974, **15**, 1919–1922

74 Aku 2: W. Akutagawa, D. Turnbull, W. K. Chu and J. W. Mayer, *J. Phys. Chem. Solids*, 1974, **36**, 521–528

78 Gri: V. I. Grytsiv, O. Z. Panchuk and D. P. Belotskii, *Izvest. Akad. Nauk. USSR, Neorg. Mater*, 1978, **14**, 1348–1349

Table 46. The solubility of Au in CdTe [74 Aku 2]

Preparation conditions	*Temperature* °C	p_{Cd} *atm.*	*Total Au concentration* *atoms.* cm^{-3}	at.-%
melt doped	800	0·85	$2{\cdot}3 \times 10^{18}$	0·008
melt doped	800	5×10^{-2}	$5{\cdot}3 \times 10^{18}$	0·018
melt doped	800	7×10^{-4}	$2{\cdot}2 \times 10^{19}$	0·075
melt doped	900	2·5	$2{\cdot}3 \times 10^{19}$	0·078
diffused	900	2·5	$3{\cdot}2 \times 10^{19}$	0·108
melt doped	900	$1{\cdot}5 \times 10^{-1}$	$5{\cdot}2 \times 10^{19}$	0·176
diffused	900	$1{\cdot}5 \times 10^{-1}$	$6{\cdot}6 \times 10^{19}$	0·224
melt doped	900	9×10^{-3}	$9{\cdot}2 \times 10^{19}$	0·312
diffused	900	9×10^{-3}	$1{\cdot}0 \times 10^{20}$	0·339

Au–Cd–Tl

The only publication [1894 Hey] on the Au–Cd–Tl system dealt with the examination of the depression of the freezing point of a 3·85 at.-% Cd, 96·15 at.-% Tl alloy on addition of Au up to 5·66 at.-%. Cadmium was then added to the ternary alloy containing 5·66 at.-% Au, 3·63 at.-% Cd up to a Cd content of 6·65 at.-%. Fig. 155 gives the temperatures measured by [1894 Hey] on the initial addition of Au to the binary Cd–Tl alloy. Fig. 156 relates to the addition of Cd to the alloy in Fig. 155 containing 5·66 at.-% Cd.

The behaviour is similar to that found in the more extensive work on the Au–Cd–Sn system [1891 Hey]. There is almost certainly a pseudobinary system between a phase containing a Au : Cd ratio of 1 : 1 and (βTl) with a reaction temperature of 294·8°C, referred to the melting point of Tl measured at 301·67°C (compared with the currently accepted value of 303·85°C). By analogy with the Au–Cd–Sn system there is the possibility of the presence of the ternary compound AuCdTl, but the binary phase AuCd may be the phase in equilibrium with (βTl) at 294·8°C.

The initial fall in the liquidus temperature on adding 0·25 at.-% Au to the 3·85 at.-% Cd, 96·15 at.-% Tl alloy (Fig. 155) is followed by a curve that rises to a maximum at a Au : Cd ratio of 1 : 1 and falls thereafter. This parabolic-shaped curve represents the secondary separation of (βTl) + $AuCdTl_x$ from the melt ($x = 0$ or 1). On adding Cd (Fig. 156) to the 5·66 at.-% Au in Fig. 155 a curve with a maximum is produced, again representing the same secondary separation. Both the sections given in Figs. 155 and 156 intersect a line drawn with a Au : Cd ratio of 1 : 1 to the Tl corner of the ternary system.

Given that a pseudobinary eutectic reaction is present, $L_e \rightleftharpoons$ (βTl) + $AuCdTl_x$, monovariant curves will descend from e towards the Cd–Tl and Au–Tl axes. The data of [1894 Hey] indicate that the monovariant curve from e cuts Fig. 155 at 0·25 at.-% Au and 285°C on the Cd-rich side of the pseudobinary section.

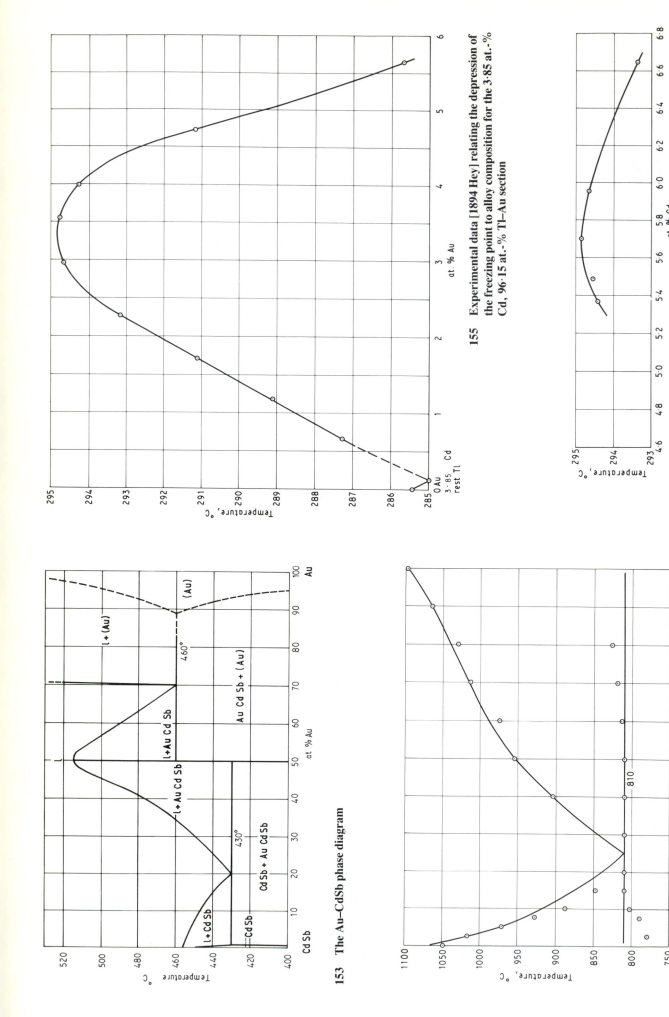

153 The Au–CdSb phase diagram

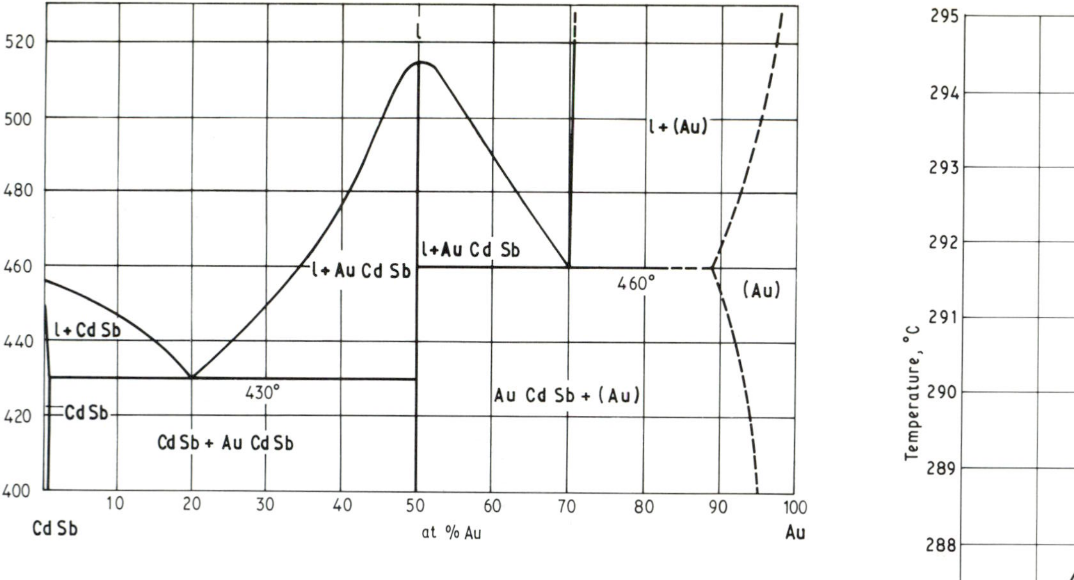

154 The Au–CdTe section [78 Gri]

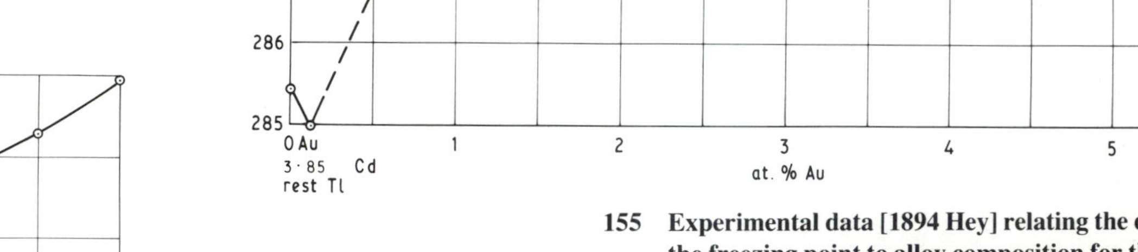

155 Experimental data [1894 Hey] relating the depression of the freezing point to alloy composition for the 3·85 at.-% Cd, 96·15 at.-% Tl–Au section

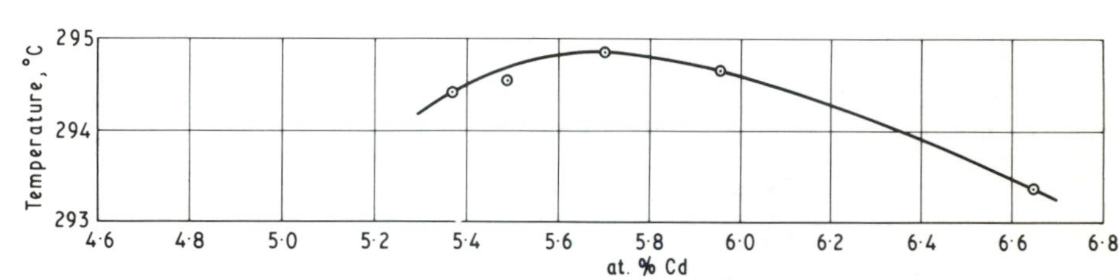

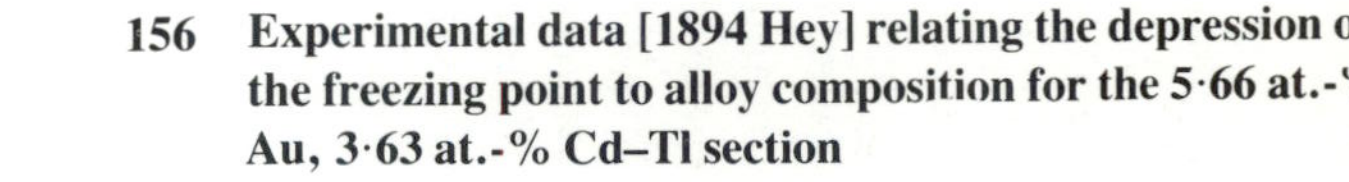
156 Experimental data [1894 Hey] relating the depression of the freezing point to alloy composition for the 5·66 at.-% Au, 3·63 at.-% Cd–Tl section

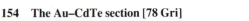

REFERENCES

1891 Hey: C. T. Heycock and F. H. Neville: *J. Chem. Soc.*, 1891, **59**, 936–966

1894 Hey: C. T. Heycock and F. H. Neville: *J. Chem. Soc.*, 1894, **65**, 65–76

Au–Cd–Zn

No overall picture is available for the ternary equilibria in the Au–Cd–Zn system. The initial work of [49 Rau, 54 Sch, 55 Sch] concentrated on the study by X-ray powder diffraction analysis of alloys between Au_3Zn and Au_3Cd. [54 Sch] reported the presence of the faulted $L1_2$-type Au_3Zn phase α_1 from 5 at.-% Cd to 33·3 at.-% Cd. The axial ratio of the Au_3Zn (α_1) phase was found to be proportional to the Cd content. Transformation of the Au_3Zn (α_1) phase to the Au_3Zn (α_2) phase was observed for alloys containing up to 5 at.-% Cd. [55 Sch] heat treated alloys on the Au_3Zn–Au_3Cd section for times up to 12 h at 300°C. All alloys were single phase with the Au_3Zn (α_1) structure (Table 48). [49 Rau] studied one alloy containing 7·2 Cd, 23·2 Zn on the Au_3Zn–Cd section. The faulted $L1_2$-type Au_3Zn (α_1) phase was present for heat treatments of 100 h at 300°C, 750 h at 360°C and 170 h at 400°C. Addition of Cd to Au_3Zn appears to have little effect on the transformation temperature of disordered (Au) to Au_3Zn (α_1) since binary Au–Zn alloys undergo this transformation at 420°C [87 Oka].

Table 47. Crystal structures

Solid phase	*Prototype*	*Lattice designation*
(Au)	Cu	cF4
(Cd)	Mg	hP2
(Zn)	Mg	hP2
Au_4Zn	Doubly-faulted Cu_3Au	Doubly-faulted cP4
Au_3Zn (α_1)	Faulted Cu_3Zu	Faulted cP4
Au_3Zn (α_2)	Orthorhombic	oC32
γAu–Zn	γbrass	—
αAu–Cd	cph	—
Au_3Cd	Faulted Cu_3Au	Faulted cP4
β′AuCd	CsCl	cP2
β′AuZn	CsCl	cP2

[57 Wil] delineated the region of existence of the doubly-faulted $L1_2$-type Au_4Zn phase by X-ray powder diffraction analysis of alloys heat treated for 8 days at 200°C (Fig. 157). [67 Mac] studied the effect of Cd additions up to 5·18 at.-% Cd on the structure of γ Au–Zn alloys on the section $AuZn_2$–Cd. All alloys retained the γ-brass structure for heat treatments of 12 days at 550°C followed by 3 days at 250°C and water quenching. The single phase structure was confirmed by metallographic examination.

[81 Mat, 83 Ips] reported on the AuZn–AuCd section. β′ AuCd undergoes a martensitic transformation at 30° ± 2°C on cooling. The addition of Zn strongly depresses the martensitic transformation temperature in a linear manner from 30°C to −145°C at 3·5 at.-% Zn. The phases β′ AuCd and β′ AuZn form a continuous series of solid solutions with the lattice parameter of the CsCl-type ordered solid solution falling linearly from β′ AuCd to β′ AuZn (Fig. 158). No Heusler-type phase was found at the Au_2CdZn stoichiometry [81 Mat]. The work of [831 Ips] combined X-ray powder diffraction analysis and DTA at heating rates of 0·1 to 2°C min^{-1} to delineate the homogeneity range of the β′ AuCd(Zn) phase region at 500°C (Fig. 159), and the ternary liquidus projection in the range from 40 to 60 at.-% Au (Fig. 160). They confirmed the occurrence of a continuous single-phase region between β′ AuCd and β′ AuZn. The lattice parameters at Au contents of 46·9, 50·0, 53·4, and 57 at.-% Au fall linearly in all cases with increasing substitution of Cd by Zn. The liquidus surface falls from the congruent melting point of β′ AuZn (758°C at 48·5 at.-% Zn) to the congruent melting point of β′ AuCd (629°C at 47·0 at.-% Cd). The liquidus surface is very shallow near to β′ AuCd; with 20 at.-% Zn the liquidus is only raised from 629°C for β′ AuCd to 640°C. [831 Ips] speculated on the existence of a ternary eutectic reaction, L $\rightleftharpoons$ (Au) + α(Au–Cd) + β′ AuCd(Zn), originating from the binary Au–Cd reactions l + (Au) $\rightleftharpoons$ α(Au–Cd) at 627°C and l $\rightleftharpoons$ α(Au–Cd) + β′ AuCd at 626·6°C and the binary Au–Zn reaction l + (Au) $\rightleftharpoons$ β′ AuZn at 684°C. No firm evidence is given to support the speculation.

Table 48. Lattice parameters of the Au_3Zn (α_1) phase at 75 at.-% Au

Alloy composition, at.-%			*Lattice parameters*, nm (± 0·0003)	
Au	*Cd*	*Zn*	*a*	*c*
74	0	26	0·4019	0·4094
75	4	21	0·4019	0·4095
75	5	20	0·4023	0·4092
75	10	15	0·4047	0·4103
75	15	10	0·4071	0·4113
75	20	5	0·4093	0·4122
75	25	0	0·4117	0·4131

REFERENCES

49 Rau: E. Raub, P. Walter and A. Engel, *Z Metallk.*, 1949, **40**, 401–405

54 Sch: K. Schubert, B. Kiefer and M. Wilkens, *Z Natur.*, 1954, **9A**, 987–988

55 Sch: K. Schubert, B. Kiefer, M. Wilkens and R. Haufler, *Z Metallk.*, 1955, **46**, 692–715

57 Wil: M. Wilkens and K. Schubert, *Z Metallk.*, 1957, **48**, 550–557

67 Mac: R. B. Maciolek, J. A. Mullendore and R. A. Dodd, *Acta Met.*, 1967, **15**, 259–264

81 Mat: Y. Matsuo, S. Minamigawa and K. Katada, *Trans. Japan Inst. Metals*, 1981, **22**, 367–368

83 Ips: H. Ipser, A. Mikula and P. Terzieff, *Monatsh. Chem.*, 1983, **114**, 1177–1184

87 Oka: H. Okamoto and T. B. Massalski, 'Phase Diagrams of Binary Gold Alloys', ASM, Metals Park, Ohio, 1987, pp. 331–340

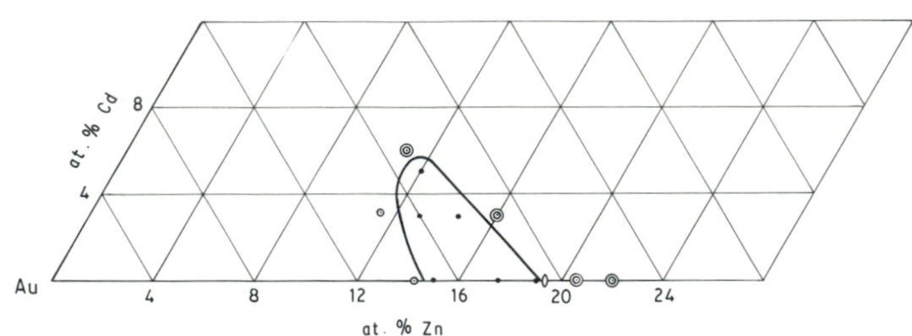

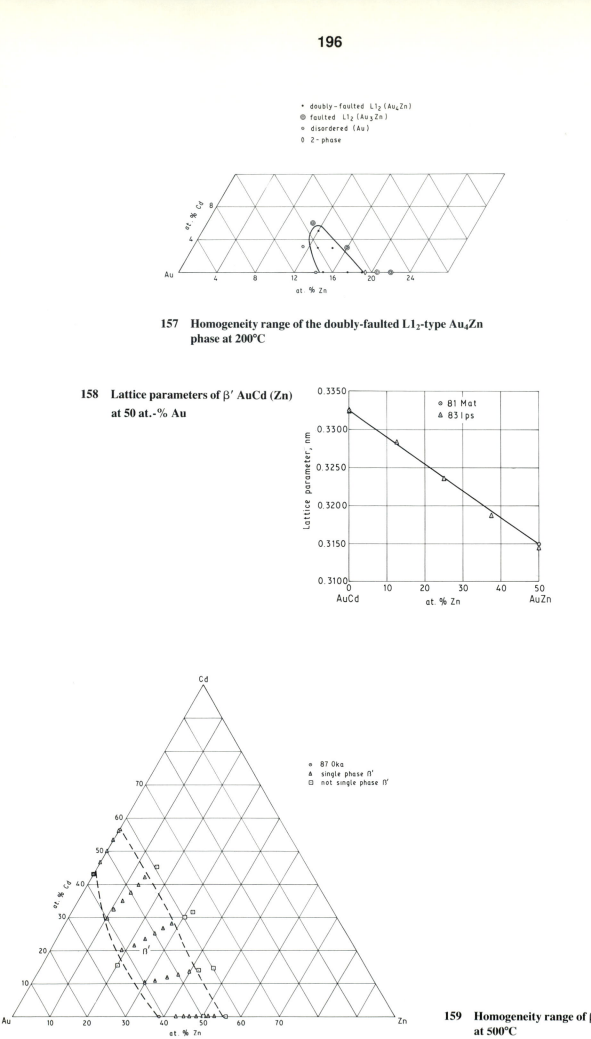

157 Homogeneity range of the doubly-faulted $L1_2$-type Au_4Zn phase at 200°C

158 Lattice parameters of β′ AuCd (Zn) at 50 at.-% Au

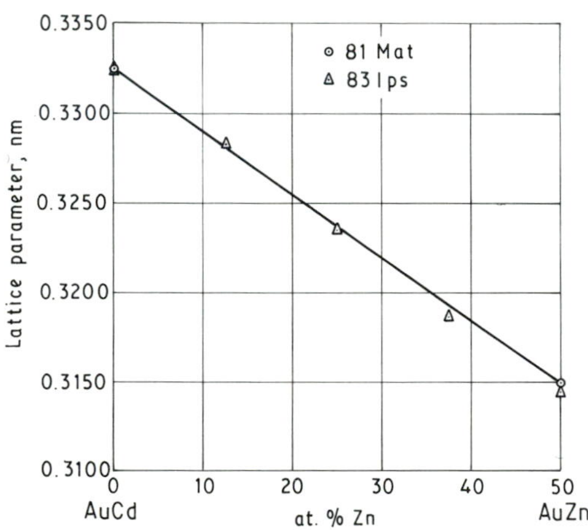

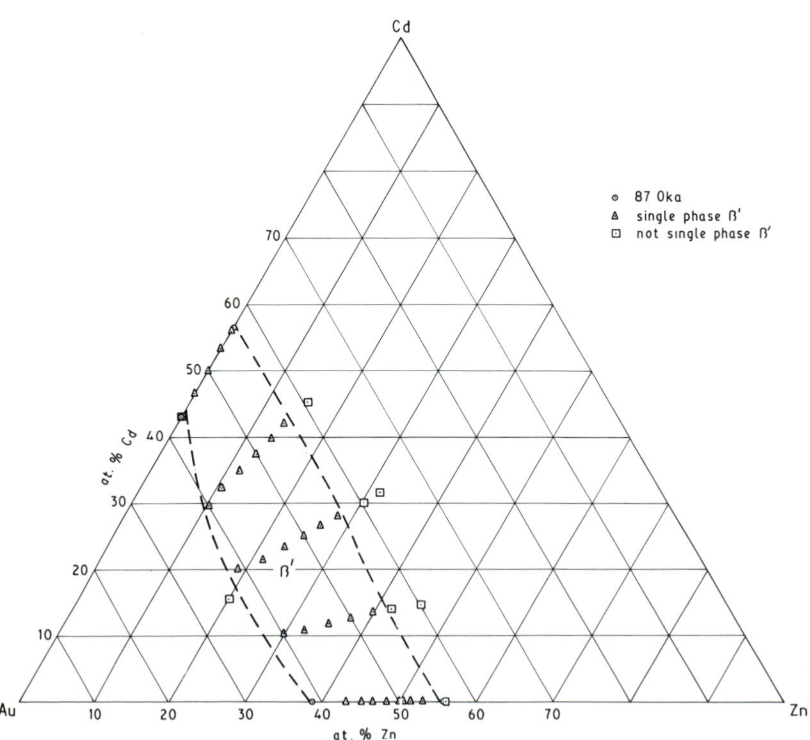

159 Homogeneity range of β′ AuCd (Zn) at 500°C

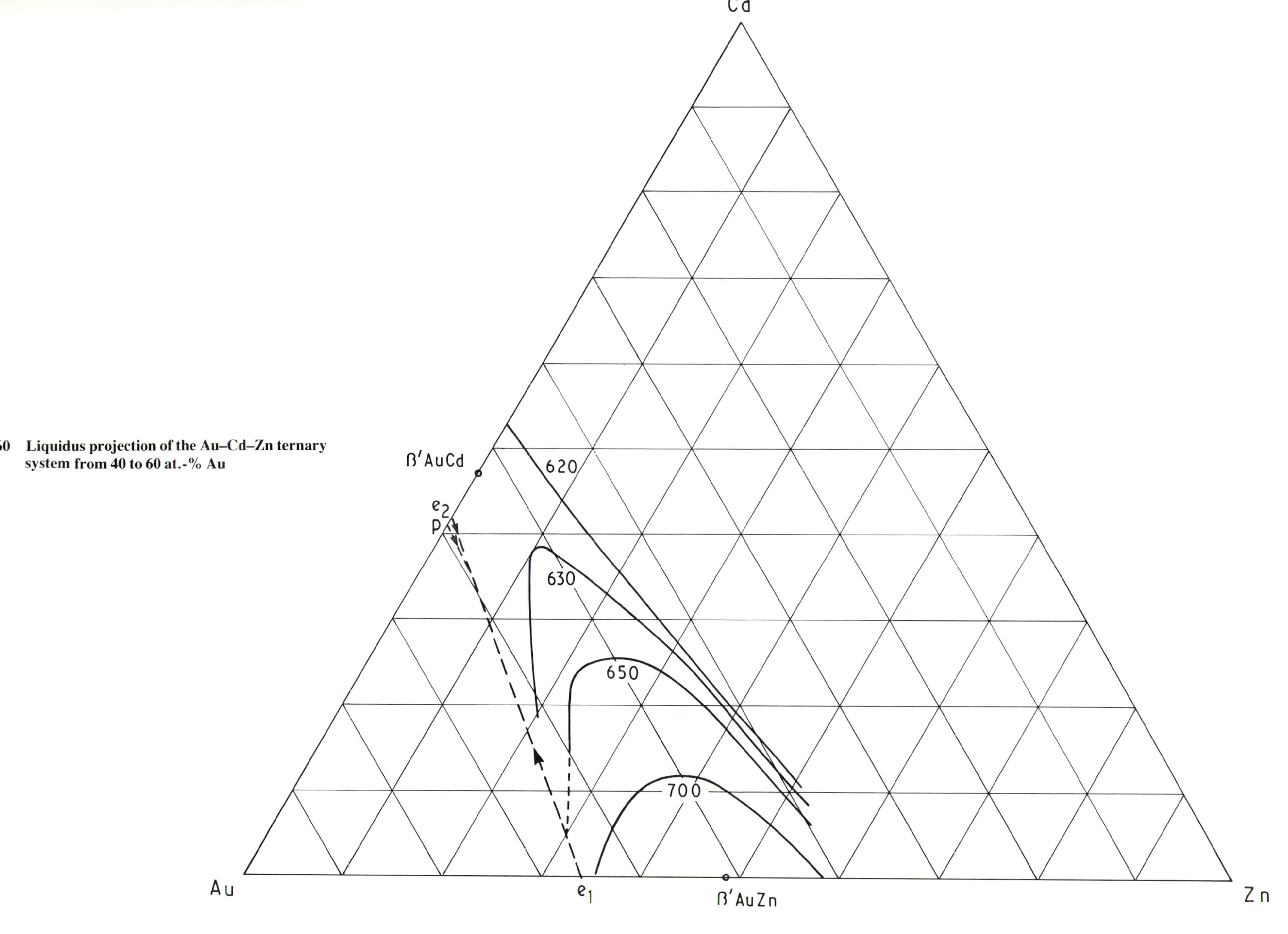

160 **Liquidus projection of the Au–Cd–Zn ternary system from 40 to 60 at.-% Au**

Au–Ce–Ga
Au–Ga–La
Au–Ga–Nd
Au–Ga–Pr
Au–Ga–Sm

[87 Gri] prepared $REAu_xGa_{4-x}$ alloys containing up to 30 at.-% Au by arc melting 1 g samples of the high-purity elements in a gettered argon atmosphere. Weight losses after melting were less than 1 wt-%. Each alloy was homogenized for 170 h at 600°C in evacuated silica capsules, quenched and examined by X-ray diffraction techniques. A ternary compound with the Al_4Ba structure-type (tI10) was identified for alloys containing 20 at.-% Ce, La, Nd, Pr and Sm over the homogeneity range 5·5–26 at.-% Au ($CeAu_xGa_{4-x}$), 6·0–18 at.-% Au ($LaAu_xGa_{4-x}$), 8·5–25·5 at.-% Au ($NdAu_xGa_{4-x}$), 7·5–25 at.-% Au ($PrAu_xGa_{4-x}$) and 9·0–26·5 at.-% Au ($SmAu_xGa_{4-x}$). The Al_4Ba-type compound is in equilibrium at 600°C with a second ternary compound of the $CaBe_2Ge_2$-type (tP10) which has a narrow range of homogeneity centred on 30 at.-% Au (see Table 49).

No formation of $REAu_xGa_{4-x}$ compounds of the Al_4Ba-type was observed with the heavier rare-earths Gd, Tb and Dy.

REFERENCE

87 Gri: Y. N. Grin, P. Rogl, K. Hiebl, F. E. Wagner and H. Noël, *J. Solid State Chem.*, 1987, **70**, 168–177

Table 49

Alloy composition, at.-%			*Constitution*	*Compound Structure-type*	*Lattice parameters*, nm	
					a	*c*
Au	*Ce*	*Ga*				
4	20	76	multiphase	Al_4Ba	0·436 4(1)	1·013 4(3)
7	20	73	single phase	Al_4Ba	0·436 4(1)	1·028 2(4)
10	20	70	single phase	Al_4Ba	0·436 1(2)	1·045 9(5)
12	20	68	single phase	Al_4Ba	0·435 3(2)	1·058 6(9)
15	20	65	single phase	Al_4Ba	0·434 3(1)	1·064 1(3)
20	20	60	single phase	Al_4Ba	0·434 0(2)	1·066 3(4)
22	20	58	single phase	Al_4Ba	0·433 5(2)	1·069 9(3)
27	20	53	multiphase	Al_4Ba	0·433 0(2)	1·072 6(6)
30	20	50	single phase	$CaBe_2Ge_2$	0·434 9(2)	1·067 1(6)
Au	*Ga*	*La*				
4	76	20	multiphase	Al_4Ba	0·440 6(1)	1·023 1(3)
7	73	20	single phase	Al_4Ba	0·440 5(1)	1·038 4(4)
9	71	20	single phase	Al_4Ba	0·440 3(1)	1·053 1(4)
12	68	20	single phase	Al_4Ba	0·439 6(1)	1·056 8(5)
13·5	64·5	20	single phase	Al_4Ba	0·439 2(1)	1·057 9(3)
17	63	20	single phase	Al_4Ba	0·437 9(1)	1·063 3(3)
23	57	20	multiphase	Al_4Ba	0·437 9(1)	1·063 2(2)
30	50	20	single phase	$CaBe_2Ge_2$	0·437 9(1)	1·063 2(2)
Au	*Ga*	*Nd*				
7	73	20	multiphase	Al_4Ba	0·432 4(1)	1·003 1(5)
12	68	20	single phase	Al_4Ba	0·431 2(1)	1·030 1(5)
15	65	20	single phase	Al_4Ba	0·430 0(1)	1·052 9(6)
17	63	20	single phase	Al_4Ba	0·429 0(2)	1·070 0(8)
20	60	20	single phase	Al_4Ba	0·428 9(1)	1·072 6(3)
24	56	20	single phase	Al_4Ba	0·429 0(1)	1·073 2(3)
27	53	20	multiphase	Al_4Ba	0·428 7(1)	1·074 8(3)
30	50	20	single phase	$CaBe_2Ge_2$	0·430 3(1)	1·070 5(3)
Au	*Ga*	*Pr*				
6	74	20	multiphase	Al_4Ba	0·433 4(1)	1·009 6(3)
10	70	20	single phase	Al_4Ba	0·433 7(1)	1·037 6(8)
12	68	20	single phase	Al_4Ba	0·432 5(1)	1·051 6(4)
15	65	20	single phase	Al_4Ba	0·432 0(1)	1·065 3(3)
19	61	20	single phase	Al_4Ba	0·431 4(1)	1·068 0(4)
24	56	20	single phase	Al_4Ba	0·431 3(1)	1·069 9(2)
27	53	20	multiphase	Al_4Ba	0·430 9(1)	1·073 9(4)
30	50	20	single phase	$CaBe_2Ge_2$	0·432 4(1)	1·068 2(4)
Au	*Ga*	*Sm*				
7	73	20	multiphase	Al_4Ba	0·428 0(1)	1·000 9(2)
11	69	20	single phase	Al_4Ba	0·428 3(1)	1·017 3(2)
15	65	20	single phase	Al_4Ba	0·425 9(1)	1·059 8(4)
19	61	20	single phase	Al_4Ba	0·425 1(1)	1·080 9(1)
23	57	20	single phase	Al_4Ba	0·423 8(1)	1·084 5(3)
27	53	20	multiphase	Al_4Ba	0·424 7(1)	1·081 8(3)
32	48	20	single phase	$CaBe_2Ge_2$	0·425 5(1)	1·079 1(5)

Au–Ce–In
Au–Dy–In
Au–Er–In
Au–Gd–In
Au–Ho–In
Au–In–La
Au–In–Lu
Au–In–Nd
Au–In–Pr
Au–In–Sm
Au–In–Tb
Au–In–Tm
Au–In–Y
Au–In–Yb

Marazza *et al.* [75 Mar] synthesized intermetallic compounds of nominal composition $REInAu_2$ where the rare earth RE = Y, La, Ce, Pr, Nd, Sm, Gd, Tb, Dy, Ho, Er, Yb. Rare earths of 99·9% purity were induction melted under argon with Au and In of 99·99% purity. The compounds produced were all very brittle and their synthesis was accompanied by a 'remarkable exothermic reaction'. Prior to X-ray and metallographic analysis all compounds were annealed for 150 h at 500°C. They were all either single phase or close to being single phase. The powder diffraction patterns were indexed on the basis of the W-type structure, cI2, with the lattice parameters given in Table 50. [86 Bes] confirmed the occurrence of $REInAu_2$ compounds and extended the series to include $LuInAu_2$ and $TmInAu_2$. At room temperature all the compounds except $LaInAu_2$ have the cubic W-type structure. Weak superstructure lines suggest a Heusler-type ordering and the lattice parameters quoted (Table 50) are based on a cubic cell twice that of the CsCl cell. $LaInAu_2$ has a tetragonal structure at room temperature, in contrast to the findings of [75 Mar]. It undergoes structural transition to the cubic structure at 82 ± 5°C. The light rare-earth compounds $CeInAu_2$, $PrInAu_2$ and $NdInAu_2$ transform from the cubic to a tetragonal symmetry at −42 ± 5, −123 ± 10, and −203 ± 20°C respectively. [75 Mar] commented on the possible existence of compounds with the nominal composition REInAu. [77 Ros] reported the synthesis of REInAu compounds with similar preparation conditions to those used to prepare the $REInAu_2$ compounds. All compounds were rather brittle. Prior to X-ray and metallographic analysis they were annealed for 1 week at 500°C. The powder diffraction patterns were indexed on the basis of the Fe_2P-type structure, hP9, with the lattice parameters given in Table 51. [76 Dwi] also found REInAu compounds to have the Fe_2P-type structure. Alloys were arc-melted, using AuIn master alloys to which the lanthanide was added, on a water-cooled Cu hearth under argon. Samples were subsequently homogenized at 800°C in evacuated silica capsules for an unspecified time. Powders were examined by X-ray diffraction to yield the unit cell constants quoted in Table 51. There is good agreement between the data of [77 Ros] and [76 Dwi].

Table 50. Unit cell constants of $REInAu_2$ Compounds

Compound	*Unit cell constants* *a*, nm [75 Mar]	*a*, nm [86 Bes]	*c*, nm [86 Bes]
$YInAu_2$	0·3443	0·6893	
$LaInAu_2$	0·3560	0·7048	0·7318
*$Ce_{0·245}In_{0·26}Au_{0·495}$	0·3544	0·7090	($CeInAu_2$)
$PrInAu_2$	0·3524	0·7055	
$NdInAu_2$	0·3516	0·7030	
$SmInAu_2$	0·3488	0·6973	
*$Gd_{0·27}In_{0·24}Au_{0·49}$	0·3469	0·6940	($GdInAu_2$)
$TbInAu_2$	0·3455	0·6913	
$DyInAu_2$	0·3449	0·6885	
$HoInAu_2$	0·3437	0·6870	
*$Er_{0·22}In_{0·24}Au_{0·54}$	0·3430	0·6864	($ErInAu_2$)
$TmInAu_2$		0·6856	
*$Yb_{0·25}In_{0·22}Au_{0·53}$	0·3428	0·6860	($YbInAu_2$)
$LuInAu_2$		0·6822	

* Analysed compositions

Table 51. Unit cell constants of REInAu compounds

Compound	*Unit cell constants* [77 Ros] *a* ± 0·0002, nm	*c* ± 0·0001, nm	*Metallographic structure* [77 Ros]	*Unit cell constants* [76 Dwi] *a*, nm	*c*, nm
YInAu	0·7691	0·3907	Nearly single phase	0·769	0·392
YInAu	0·7684	0·3909	Fair quantity of a second phase		
LaInAu	0·7727	0·4315	Nearly single phase	—	—
CeInAu	0·7698	0·4256	Nearly single phase	0·771	0·425
CeInAu	0·7691	0·4260	Nearly single phase		
PrInAu	0·7700	0·4195	Single phase	0·769	0·422
PrInAu	0·7688	0·4211	Nearly single phase		
$Pr_3In_2Au_3$	0·7735	0·4137	Nearly single phase	—	—
NdInAu	0·7719	0·4124	Fair quantity of a second phase	0·772	0·413
SmInAu	0·7716	0·4043	Single phase	0·771	0·404
GdInAu	0·7693	0·3991	Single phase	0·770	0·398
TbInAu	0·7691	0·3951	Single phase	0·769	0·394
DyInAu	0·7686	0·3902	Nearly single phase	0·767	0·392
HoInAu	0·7683	0·3877	Nearly single phase	0·768	0·387
ErInAu	0·7674	0·3847	Fair quantity of a second phase	0·768	0·383
TmInAu	—	—	—	0·766	0·383
YbInAu	0·7708	0·4027	Nearly single phase	—	—
LuInAu	—	—	—	0·7635	0·382

[75 Mar] used the Pr–In–Au system to study alloys other than the nominal $PrInAu_2$ composition in order to provide some idea of the phase equilibria in RE–In–Au systems. $PrInAu_2$ [75 Mar] and PrInAu [77 Ros] are ternary compounds, the latter probably having a range of homogeneity including compositions close to $Pr_3In_2Au_3$ (Table 51). Compositions Pr_3In_2Au and Pr_3InAu_2 on the section AuPr–InPr are heterogeneous. Pr_2InAu_3 and Pr_3InAu_2 are also heterogeneous although [75 Mar] state that they show practically the same diffraction lines as $PrInAu_2$. This data does not allow any conclusions to be drawn on the phase equilibria in the RE–In–Au ternary systems.

REFERENCES

75 Mar: R. Marrazza, R. Ferro and D. Rossi: *Z. Metall.*, 1975, **66**, 110–111

76 Dwi: A. E. Dwight: 'Proc. 12th Rare Earth Research Conference', Vail, Colorado, Vol. I, 480–489; 1976, Denver Research Institute (preprints)

77 Ros: D. Rossi, R. Ferro, V. Contardi and R. Marazza: *Z. Metall.*, 1977, **68**, 493–494

86 Bes: M. J. Besnus, J. P. Kappler, M. F. Ravet, A. Meyer, R. Lahiouel, J. Pierre, E. Siaud, G. Nieva and J. Sereni: *J. Less-Common Metals*, 1986, **120**, 101–112

Au–Ce–Ni
Au–Dy–Ni
Au–Er–Ni
Au–Gd–Ni
Au–Ho–Ni
Au–Lu–Ni
Au–Ni–Sc
Au–Ni–Tb
Au–Ni–Tm
Au–Ni–Y
Au–Ni–Yb

Dwight [75 Dwi] synthesized intermetallic compounds of nominal composition $RENi_4Au$ where the rare earth RE = Sc, Y, Gd, Tb, Dy, Ho, Er, Tm, Yb, Lu. All alloys were prepared from 99·9% purity elements by arc-melting on a water-cooled Cu hearth under argon. They were subsequently homogenized for an unspecified time at 800°C and examined by X-ray powder diffraction analysis. All $RENi_4Au$ compounds were found to be fcc of the $MgSnCu_4$-type, cF24. Unit cell dimensions are reproduced in Table 52. [77 Fel] confirmed the existence of $RENi_4Au$ compounds for RE = Gd, Tb, Dy, Ho, Er, Tm, Yb. Alloys were synthesized by melting elements of unspecified purity in an induction furnace under argon. X-ray diffraction analysis gave the unit cell constants quoted in Table 52. [77 Fel] provide no information on homogenizing treatments and it must be assumed that they examined the cast samples. In work on other systems alloys were held for 30 min to 1 h at melting temperature in an attempt at homogenization [73 May]. There is reasonable agreement between the structural data of [75 Dwi] and [77 Fel]. Additions of 0·1–0·7 at.-% Gd to a 10 at.-% Au, 90 at.-% Ni alloy are reported [74 Res] to form GdNi and $GdNi_2$ as grain boundary phases. [77 Olc] found no detectable solid solubility of Au in $CeNi_2$, all alloys being two-phase.

Table 52. Unit cell constants of $RENi_4Au$ compounds

Compound	*Unit cell constant*, nm	
	[75 Dwi]	[77 Fel]
$ScNi_4Au$	0·682 8	—
YNi_4Au	0·705 9	—
$GdNi_4Au$	0·712 8	0·713 6
$TbNi_4Au$	0·707 4	0·709 4
$DyNi_4Au$	0·703 6	0·705 2
$HoNi_4Au$	0·699 9	0·703 4
$ErNi_4Au$	0·699 3	0·701 1
$TmNi_4Au$	0·696 7	0·698 2
$YbNi_4Au$	0·696 9	0·695 2
$LuNi_4Au$	0·694 8	—

REFERENCES

73 May: I. Mayer, J. Cohen and I. Felner: *J. Less-Common Metals*, 1973, **30**, 181–184

74 Res: Research Group of Gold Alloys, Chin Shu Hsueh Pao, 1974, **10**, 48–53

75 Dwi: A. E. Dwight: *J. Less-Common Metals*, 1975, **43**, 117–120

77 Fel: I. Felner: *Solid State Commun.*, 1977, **21**, 267–268

77 Olc: G. L. Olcese: *J. Phys. Chem. Solids*, 1977, **38**, 1239–1241

Au–Ce–Pb
Au–Dy–Pb
Au–Er–Pb
Au–Gd–Pb
Au–Ho–Pb
Au–La–Pb
Au–Nd–Pb
Au–Pb–Pr
Au–Pb–Sm
Au–Pb–Tb
Au–Pb–Y

[88 Mar] synthesized intermetallic compounds of composition REAuPb where the rare-earth, RE, = La, Ce, Pr, Nd, Sm, Gd, Tb, Dy, Ho, Er and Y. Alloys were induction melted from 99·99 wt-% Au, Pb and 99·9 wt-% RE in a Ta crucible under argon. The cast alloys were subsequently annealed for 1 week at 500°C and examined by optical metallography and X-ray powder diffraction. The lattice parameters are given in the Table below. The light rare-earths (La to Sm) form 1:1:1 compounds with the hexagonal $CaIn_2$-type structure, hP6. The heavy rare-earths and Y form 1:1:1 compounds with the MgAgAs-type structure, cF12.

Table 53. Lattice parameters of REAuPb compounds

Compound	*Prototype*	*Lattice designation*	*Lattice parameter* *a* (nm)	*c* (nm)
LaAuPb	$CaIn_2$	hP6	0·4820	0·7842
CeAuPb	$CaIn_2$	hP6	0·4802	0·7739
PrAuPb	$CaIn_2$	hP6	0·4785	0·7681
NdAuPb	$CaIn_2$	hP6	0·4764	0·7628
SmAuPb	$CaIn_2$	hP6	0·4742	0·7575
GdAuPb	MgAgAs	cF12	0·6775	
TbAuPb	MgAgAs	cF12	0·6747	
DyAuPb	MgAgAs	cF12	0·6728	
HoAuPb	MgAgAs	cF12	0·6718	
ErAuPb	MgAgAs	cF12	0·6694	
YAuPb	MgAgAs	cF12	0·6729	

REFERENCE

88 Mar: R. Marazza, D. Rossi and R. Ferro, *J. Less-Common Metals*, 1988, **138**, 189–193

Au–Ce–Si
Au–Dy–Si
Au–Er–Si
Au–Eu–Si
Au–Gd–Si
Au–Ho–Si
Au–La–Si
Au–Lu–Si
Au–Nd–Si
Au–Pr–Si
Au–Si–Sm
Au–Si–Tb
Au–Si–Tm
Au–Si–Y
Au–Si–Yb

Mayer *et al.* [73 May 1] synthesized intermetallic compounds of the composition $REAu_2Si_2$ where the rare earth RE = La, Ce, Pr, Nd, Sm, Eu, Dy, Er. The elements of 99·9% purity were induction-melted under argon. In an unspecified number of preparations the alloys were kept for 30 min to 1 h at the furnace temperature of 1600–1800°C before cooling. The compounds were examined by X-ray diffraction analysis and the powder diffraction patterns indexed on the basis of a body-centred tetragonal cell of the ordered $ThCr_2Si_2$-type, tI10. Unit cell constants are listed in Table 54. [73 May 1] noted a failure to prepare $YbAu_2Si_2$ and $LuAu_2Si_2$. All compounds quoted in Table 54 were considered to be single phase from their powder patterns.

[75 Fel] prepared the compounds shown in Table 54 and determined their unit cell constants as part of a study of magnetic characteristics. [75 Fel] state that $TmAu_2Si_2$, $YbAu_2Si_2$, and $LuAu_2Si_2$ could not be prepared. The identity of unit cell constants for $EuAu_2Si_2$ reported by [73 May 1] and [75 Fel] suggests that [75 Fel] quoted the data of [73 May 1] for $EuAu_2Si_2$. [75 Fel] quotes the unit cell constants of [73 May 1] for compounds containing Ce, Pr, Sm, and Dy.

Rossi *et al.* [79 Ros] used rare earth metals with a purity of >99·9%, Au and Si with a purity >99·99%. Alloys were prepared by induction-melting under argon but they were subsequently annealed at 500°C for 1 week. They were hard and brittle. X-ray powder diffraction analysis and metallographic techniques were used to confirm the existence of $GdAu_2Si_2$ and establish the occurrence of $YbAu_2Si_2$. The latter compound was nearly single phase according to metallographic observation. Unit cell constants are quoted in Table 54. The data of

Table 54. Unit cell constants of $REAu_2Si_2$ compounds

Compound	*Unit cell constants* [73 May 1] ± 0·000 5		[75 Fel]		[79 Ros]		[86 Bus]	
	a, nm	*c*, nm	*a*, nm	*c*, nm	*a*, nm	*c*, nm	*a*, nm	*c*, nm
YAu_2Si_2	—	—	0·423 0	1·01 8	—	—	—	—
$LaAu_2Si_2$	0·433 7	1·018	—	—	—	—	—	—
$CeAu_2Si_2$	0·432 6	1·024	—	—	—	—	—	—
$PrAu_2Si_2$	0·430 3	1·019	—	—	—	—	—	—
$NdAu_2Si_2$	0·428 0	1·017	0·427 6	1·019	—	—	—	—
$SmAu_2Si_2$	0·426 0	1·017	—	—	—	—	—	—
$EuAu_2Si_2$	0·435 2	1·011	0·435 2	1·011	—	—	—	—
$GdAu_2Si_2$	—	—	0·423 6	1·010	0·424 4	1·015 9	0·424 30	1·016 47
$TbAu_2Si_2$	—	—	0·423 0	1·016	—	—	—	—
$DyAu_2Si_2$	0·421 8	1·015	—	—	—	—	—	—
$HoAu_2Si_2$	—	—	0·422 4	1·012	—	—	—	—
$ErAu_2Si_2$	0·421 4*	1·017	0·421 4	1·017	—	—	—	—
$YbAu_2Si_2$	—	—	—	—	0·428 7	1·005 0	—	—

* [73 May 1] give 0·424 4 nm but refer to a regular decrease of the lattice constant attributable to the lanthanide contraction of the trivalent rare-earth metals on proceeding from La to Er

[73 May 1] and [75 Fel] can only be compared for $ErAu_2Si_2$ and $NdAu_2Si_2$ and there is some discrepancy between them. In terms of a plot of the volume of the unit cell as a function of the ionic radius of the trivalent rare earth the data of [75 Fel] fall closer to a linear plot exhibiting the lanthanide contraction with increasing atomic number. Similarly the data of [79 Ros] for $GdAu_2Si_2$ are preferred to the data of [75 Fel].

Buschow and De Mooij [86 Bus] prepared $GdAu_2Si_2$ from >99·9% pure elements by arc-melting. The alloy was subsequently wrapped in a Ta foil and vacuum annealed in a silica tube for 3 weeks at 800°C. X-ray powder diffraction study confirmed the bc tetragonal structure of the $ThCr_2Si_2$-type with lattice parameters a = 0·424 30, c = 1·016 47 nm and c/a = 2·395 6.

[73 May 2] studied the composition $EuAu_{0·5}Si_{1·5}$ prepared from 99·9% purity elements by induction-melting under argon. X-ray powder diffraction analysis showed that this composition is not single phase. The major phase has the AlB_2-type structure, hP3. The unit cell constants are a = 0·415 0 nm, c = 0·451 5 nm.

REFERENCES

73 May 1: I. Mayer, J. Cohen and I. Felner: *J. Less-Common Metals*, 1973, **30**, 181–184

73 May 2: I. Mayer and I. Felner, *J. Solid State Chem.*, 1973, **8**, 355–356

75 Fel: I. Felner: *J. Phys. Chem. Solids*, 1975, **36**, 1063–1066

79 Ros: D. Rossi, R. Marazza and R. Ferro: *J. Less-Common Metals*, 1979, **66**, P17–P25

86 Bus: K. H. J. Buschow and D. B. De Mooij: *Philips J. Res.*, 1986, **41**, 55–76

Au–Ce–Sn
Au–Dy–Sn
Au–Er–Sn
Au–Gd–Sn
Au–Ho–Sn
Au–Lu–Sn
Au–Nd–Sn
Au–Pr–Sn
Au–Sc–Sn
Au–Sm–Sn
Au–Sn–Tb
Au–Sn–Tm
Au–Sn–Y

Dwight [76 Dwi] synthesized intermetallic compounds of composition ReSnAu where the rare earth RE = Sc, Y, Ce, Pr, Nd, Sm, Gd, Tb, Dy, Ho, Er, Tm, Lu. Alloys were arc-melted, using AuSn master alloys to which the lanthanide was added, on a water-cooled Cu hearth under argon. Samples were subsequently homogenized at 800°C in evacuated silica capsules for an unspecified time. Powders were examined by X-ray diffraction to yield structure-types and unit cell constants quoted in Table 55. Y and the lighter lanthanides form the $CaIn_2$-type structure, hP6, whereas Sc and the heavier lanthanides form the MgAgAs-type structure, cF12. The transition in structure occurs at Ho in that with HoSnAu the $CaIn_2$-type structure coexists with the MgAgAs-type structure.

Table 55. Unit cell constants of RESnAu compounds

Compound	*Prototype*	*Lattice designation*	*Unit cell constant*	
			a, nm	*c*, nm
YAuSn	$CaIn_2$	hP6	0·463	0·738
CeAuSn	$CaIn_2$	hP6	0·473	0·771
PrAuSn	$CaIn_2$	hP6	0·471	0·764
NdAuSn	$CaIn_2$	hP6	0·470	0·759
SmAuSn	$CaIn_2$	hP6	0·467	0·748
GdAuSn	$CaIn_2$	hP6	0·466	0·743
TbAuSn	$CaIn_2$	hP6	0·464	0·739
DyAuSn	$CaIn_2$	hP6	0·463	0·736
HoAuSn	$CaIn_2$	hP6	0·462	0·735
HoAuSn	MgAgAs	cF12	0·662 4	
ErAuSn	MgAgAs	cF12	0·660 6	
TmAuSn	MgAgAs	cF12	0·659 1	
LuAuSn	MgAgAs	cF12	0·656 3	
ScAuSn	MgAgAs	cF12	0·642 2	

REFERENCE

76 Dwi: A. E. Dwight: 'Proc. 12th Rare Earth Research Conference', Vail, Colorado, Vol. I, 480–489; 1976, Denver Research Institute (preprints)

Au–Co–Cu
(G. V. Raynor)

INTRODUCTION

The gold–cobalt–copper system has been examined by [67 Kup], using thermal analysis and conventional metallography. Thirty-six alloys lying at intervals of 10 wt-% Au along composition lines corresponding to constant weight percentages of Cu were investigated, and the thermal arrests, on the interpretation of which the proposed constitution is based, were given in tabular form. Also included was an illustrative diagram showing the derived partial polythermal section at 10 wt-% Cu. In this assessment the results given in [67 Kup] have been plotted, in terms of atomic percentages, to give a series of polythermal sections, which have been used to derive a diagram showing, in atomic percentage, the projection of the surfaces of primary crystallization, together with proposed liquidus isothermal contours. Possible isothermal sections for temperatures of 1 200 and 900°C have also been derived.

BINARY SYSTEMS

The system Au–Co has been assessed by [85 Oka]. The Au–Cu phase diagram, recently assessed by [87 Oka] is accepted. The assessment by [84 Nis] of the Co–Cu system is accepted.

THE SOLID PHASES

No discrete ternary intermediate phases have been reported. The solid phases concerned in the equilibria are therefore the fcc (Au–Cu) solid solution, and the restricted solid solutions in fcc and cph Co. At low temperatures the ordered phases $AuCu_3$ and AuCu will also be involved. The crystal structures of the solid phases are summarized in Table 56.

Table 56. Crystal structures

Solid phase	*Prototype*	*Lattice designation*
(Au–Cu) solid solution	Cu	cF4
$AuCu_3$	$AuCu_3$	cP4
AuCuI	AuCuI	tP4
AuCuII	AuCuII	tP4 (faulted)
(γCo)	Cu	cF4
(εCo)	Mg	hP2

THE TERNARY SYSTEM

The 36 alloys studied [67 Kup] had compositions lying along the lines of constant Cu content spaced at intervals of 10 wt-% from 10 to 80 wt-% Cu. Along each composition line the alloys lay at intervals of 10 wt-% Au. The specimens were prepared by induction melting in carborundum crucibles under a protective layer of molten $BaCl_2$. The constitution was derived from thermal analysis; the hardness of alloys annealed for 300 h at 800°C was also measured, and confirmatory metallography was also carried out. Arrest points were tabulated, and an illustrative polythermal section at 10 wt-% Cu was presented. In this assessment the thermal arrest values have been plotted in terms of atomic percentages for the eight constant wt-% Cu sections. In atomic percentages, the terminal compositions of these sections are as follows:

at.-% Cu	at.-% Au	to	at.-% Cu	at.-% Au
9·3	0		25·6	74·4
18·8	0		43·7	56·3
28·4	0		57·1	42·9
38·2	0		67·4	32·6
48·1	0		75·6	24·4
58·2	0		82·3	17·7
68·4	0		87·9	12·1
78·8	0		92·5	7·5

The resulting polythermal section using the results reported in [67 Kup] for the first of these sections is shown, in atomic percentages, in Fig. 161. From this section, and the additional seven diagrams of the same form, the projection of primary surfaces may be derived, together with isothermal liquidus contours. The derived projection is shown in Fig. 162. The compositions of the monovariant line, at which L $\rightleftharpoons$ (Co) + (Au–Cu) for the various polythermal sections, have been read from the diagrams and plotted in Fig. 162 to give the curve ab, which joins the liquid compositions at the Au–Co eutectic (996·5°C) and at the Co–Cu peritectic reaction (1 112°C). The results obtained in [67 Kup] indicate that the projection of the monovariant curve ab is convex towards the Co corner of the composition diagram; the Co content of the liquid along the curve increases to a maximum of ~36 at.-% Co at ~20 at.-% Cu. At some stage the nature of the reaction changes from the peritectic type to the eutectic type; the data do not allow the point at which this occurs to be located.

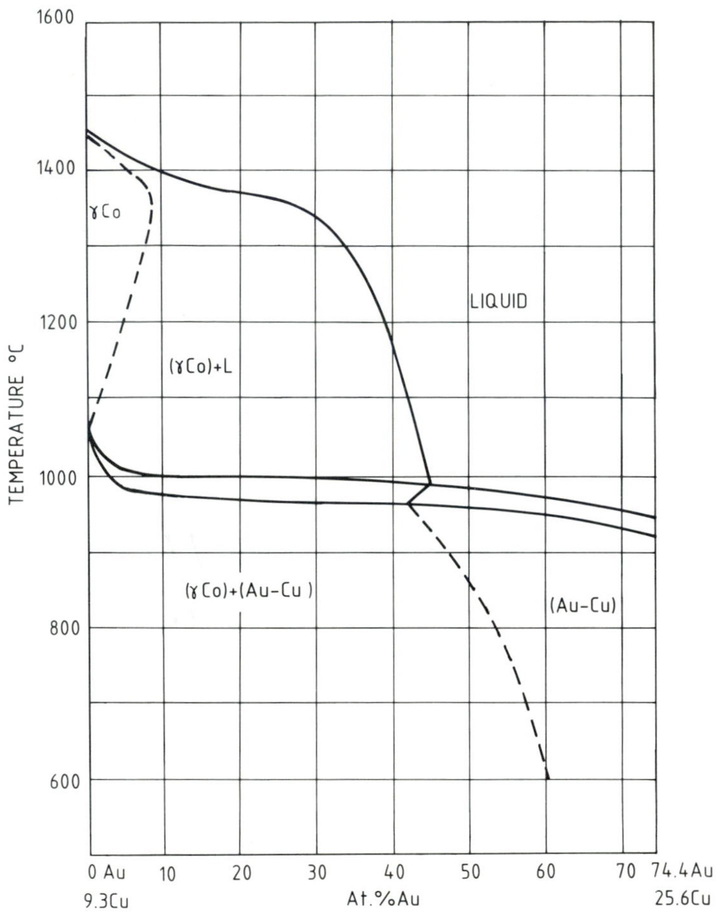

161 System Au–Co–Cu: polythermal section along composition line 9·3 at.-% Cu, 0 at.-% Au to 25·6 at.-% Cu, 74·4 at.-% Au

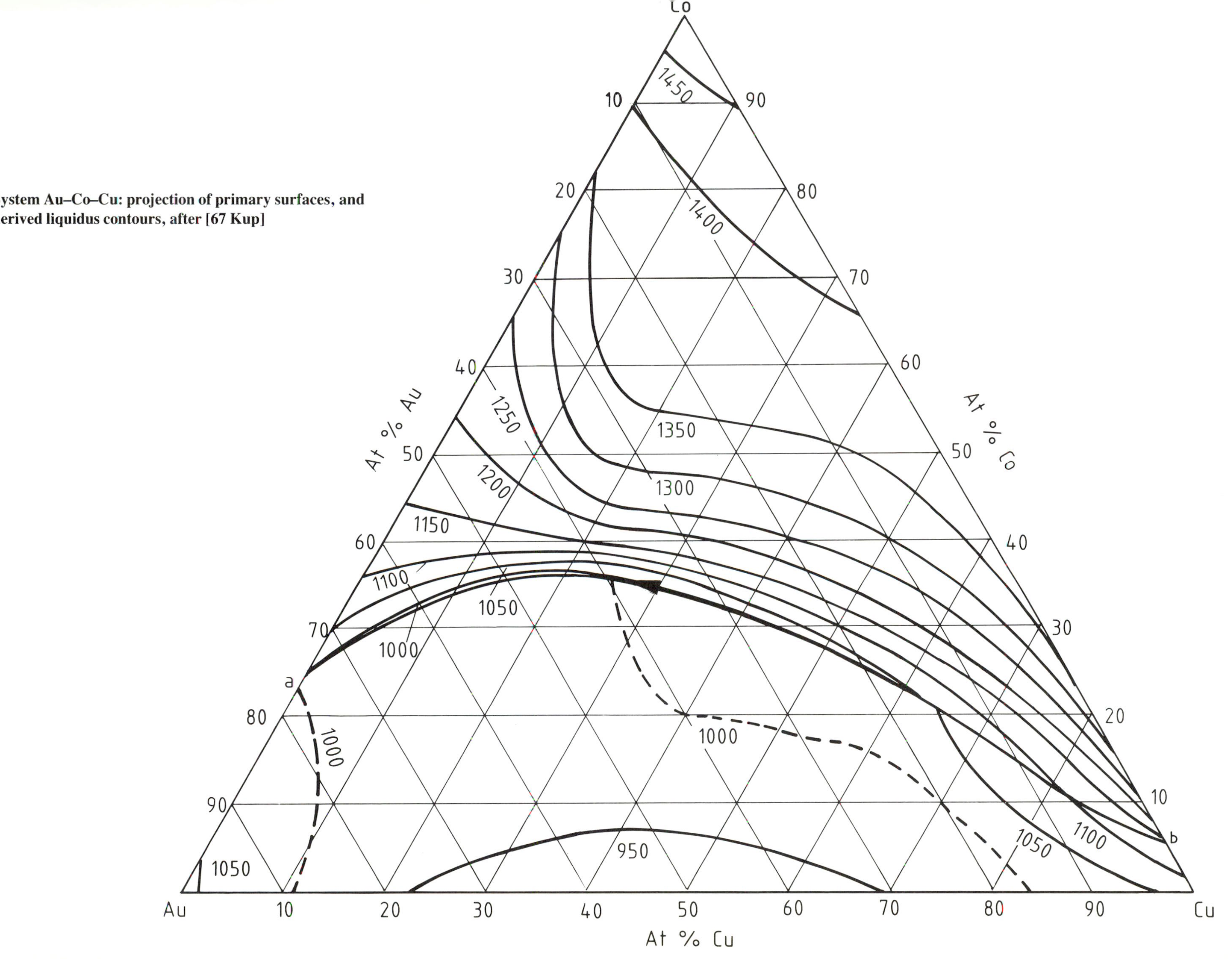

162 System Au–Co–Cu: projection of primary surfaces, and derived liquidus contours, after [67 Kup]

From the polythermal diagrams, compositions for various liquidus temperatures have been read off, and are also plotted in Fig. 162 to give isothermal liquidus contour lines. If the results of [67 Kup] are accepted, the contour lines take a somewhat unusual form. The shape of the contours for the primary (Co) field appears to be related to the flat portion of the Co–Cu liquidus curve. The monovariant curve ab is closely parallel to a 1 000°C contour, as it penetrates the ternary model from the Au–Co axis; the 1 000°C contour, however, crosses the monovariant curve at ~24 at.-% Cu and ~36 at.-% Co. The projection of the monovariant curve is also crossed successively by the 1 050 and 1 100°C contours, as the monovariant curve itself rises to the temperature (1 112°C) of the Co–Cu eutectic reaction. The shape of the 1 000°C isothermal contour after crossing the projection of the monovariant curve is, however, uncertain, and is shown in Fig. 162 as a broken curve. The 1 000°C isothermal joining the Au–Co and Au–Cu axes is also drawn as a broken curve. The shape of the primary (Au–Cu) surface thus appears complex, and is in urgent need of verification. Figure 162, however, represents the indications arising from the only experimental data available.

In spite of the complications introduced by the uncertainties in the form of the liquidus surfaces, the Au–Co–Cu system is fundamentally simple. Thus at a temperature such as 1 200°C (Fig. 163) equilibrium is between (γCo) and the liquid; the (γCo) + L/L boundary follows the course of the 1 200°C isothermal. In the solid state (e.g. 900°C, Fig. 164) (γCo) is in equilibrium with solid (Au–Cu); the (γCo) + (Au–Cu)/(Au–Cu) boundary is consistent with the results in [67 Kup], but is shown as a broken curve to emphasize its uncertainty. In Figs. 163 and 164 the boundaries of the (γCo) solid solution are not known, but they are anticipated to be similar to those shown.

At temperatures below ~410°C, phases based on the binary ordered structures $AuCu_3$ and AuCu are expected to take part in the equilibria. No constitutional details are available, but a limited number of $AuCu_3$ alloys containing Co have been examined by [77 Mar]. Five alloys of the series $(Cu_3Au)_{1-x}Co_x$ containing up to 2 at.-% Co were prepared from pure components and homogenized at 925°C, and water-quenched. Filings made from these alloys were solution treated for 15 min at 925°C and again quenched in water. After this treatment these were nonmagnetic, and a linear decrease in lattice spacing up to 2 at.-% Co indicated that the Co was in solid solution. Subsequent experiments showed that Co up to this concentration was without influence on the critical temperature for order (T_c), or on the degree of long range order at temperatures below Tc. The experiments did not permit determination of the disposition of the Co atoms on the Au and Cu crystallographic sites. No information is available on the effect of Co on the AuCuI and AuCuII structures. It is probable that ordered structures based on $AuCu_3$ and AuCu persist in the ternary model, without significant change in ordering temperature, up to the solubility limit of Co in (Au–Cu), and, within the appropriate composition and temperature ranges, enter into equilibrium with (γCo) or (εCo). No experimental evidence, however, is yet available.

REFERENCES

67 Kup: V. V. Kuprina and V. B. Bernard: *Vestn. Moskov Univ., Khim.*, 1967, (3), 41–43

77 Mar: N. Mardesich, C. N. T. Wagner and A. J. Ardell: *J. Appl. Cryst.*, 1977, **10**, (6), 468–472

84 Nis: T. Nishizawa and K. Ishida: *Bull. Alloy Phase Diagrams*, 1984, **5**, 161–165

85 Oka: H. Okamoto, T. B. Massalski, M. Hasebe and T. Nishizawa: *Bull. Alloy Phase Diagrams*, 1985, **6**. 449–454

87 Oka: H. Okamoto, D. J. Chakrabarti, D. E. Laughlin and T. B. Massalski, *Bull. Alloy Phase Diagrams*, 1987, **8**, 454–474

Au–Co–Fe

(G. V. Raynor)

INTRODUCTION

The gold–cobalt–iron system is of potential importance in the development of magnetic materials. Alloys of approximately 90 wt-% Co and 10 wt-% Fe are nonmagnetostrictive, but the addition of gold increases the coercive force significantly, following solution treatment, cold work, and aging. The desirable magnetic properties are considered to be due to an incoherent precipitate of gold in a matrix of fcc cobalt–iron solid solution. The control of properties demands a knowledge of the constitution of the ternary system, and a limited experimental X-ray study of Au–Co–Fe alloys having a constant proportion of Co atoms to Fe atoms of 6·633 was accordingly undertaken by [70 Lyn]. More recently, diffusion studies carried out by [78 Ass] have provided data from which a partial isothermal section at 900°C may be derived.

THE BINARY SYSTEMS

The system Au–Co has been assessed by [85 Oka]. The Au–Fe system, as assessed by [84 Oka], is accepted. The constitution of the Co–Fe alloys as assessed by [84 Nis] is accepted.

THE SOLID PHASES

No ternary intermediate phases have been reported, and the solid phases between which equilibrium is established include fcc (Au), (γCo), and (γFe). In addition, within appropriate temperature ranges, the bcc solid solutions in δFe and αFe are involved, and there is also the cph (εCo) solid solution. The crystal structures are summarized in Table 57.

Table 57. Crystal structures

Solid phase	*Prototype*	*Lattice designation*
(Au)	Cu	cF4
(γCo)	Cu	cF4
(γFe)	Cu	cF4
(δFe)	W	cI2
(αFe)	W	cI2
a_1 (system Co–Fe)	CsCl	cP2
(εCo)	Mg	hP2

THE TERNARY SYSTEM

In the work of [70 Lyn], nine alloys with a Co : Fe atomic ratio of 6·633 : 1 and Au contents from 0·24 to 6·12 at.-% Au were cast in an inert atmosphere, cold swaged, and homogenized at 1 155°C for 40 h. Fine wire specimens were drawn, heated at temperatures of 750, 900, 1 000, 1 050, and 1 100°C and water quenched to retain the gold in solid solution. Lattice spacings of the (γFe–Co) phase were determined by the Debye–Scherrer method. The discontinuity in the plot of spacings against gold content was taken as the limit of solubility at the temperature concerned. The derived solubility curve is shown in Fig. 165, which represents the partial vertical section of the ternary model for a constant Co : Fe atomic ratio of 6·633 : 1. The broken curves refer to temperatures and composition regions which were not experimentally examined. The lower limit of the (γFe–Co) + (Au) + L region for this series of alloys was established by differential thermal analysis; a phase change was observed, on heating alloys containing 3·19, 4·61, and 6·12 at.-% Au, at 1 015 ± 6°C. Extrapolation of this boundary indicates that the limit of the three-phase region occurs at 1·47 at.-% Au, and the (γFe–Co)/(γFe–Co) + L boundary shows a retrograde solubility of Au in the (γFe–Co) solid solution. This solubility reaches a maximum of 2·32 at.-% Au at 1 155°C. It is particularly to be noted that the upper boundary of the (γFe–Co) + (Au) + L region for this section is not accurately determined. The form of Fig. 165 was confirmed by metallographic experiments, and shows that the width of the (γFe–Co) region in the ternary model for the Co : Fe atomic ratio of 6·633 : 1 cannot exceed 2·32 at.-% Au.

In the work of [78 Ass], diffusion couples were prepared such that cylindrical specimens of different compositions, made from metals of 99·99% purity, were pressure bonded at 750°C across a circular face. The couples studied were chosen as follows:

(i) Fe–AuCo
(ii) AuCo–AuFe
(iii) AuCo–CoFe

The two-symbol formulae denote equiatomic compositions, which for the systems Au–Fe and Co–Fe represent single phase alloys at the temperature of the experiments. The composite samples, wrapped in tantalum foil and sealed into quartz tubes in an atmosphere of helium, were annealed for a week at 900°C. They were then cooled in air to ~400°C and quenched in water. Specimens were section longitudinally (parallel to the cylindrical axis) and the exposed surfaces were polished. During the annealing process, diffusion took place, creating concentration gradients, and the concentration profiles were measured by electron probe microanalysis. In some zones of the diffused samples the microstructures were two-phase, either (Au) + fcc Co–Fe solid solution, or (Au) + bcc Co–Fe solid solution; in these cases the compositions of the phases in equilibrium with each other at 900°C were determined by probe measurements on both sides, and close to, the interface between matrix and second phase. Since a concentration gradient parallel to the cylindrical axis existed in the specimens, measurements were possible for a range of matrix compositions. The work concentrated on the boundaries of the two-phase regions derived from the Au–Co miscibility gap, and gave results only up to ~50 at.-% Fe. The accuracy of the method is higher for the boundary on the Co-rich side of the two-phase region than for that on the Au-rich side, which may therefore be subject to some uncertainty. The results of the work are shown in Fig. 166. No attempt was made to investigate in detail the position of the three-phase fields, which are shown bounded by broken lines, though evidence from the diffusion experiments exists for their approximate position.

REFERENCES

70 Lyn: L. Lynch, G. Krauss and P. S. Venkatesan: *Metall. Trans.*, 1970, **1**, 1471–1472

78 Ass: F. C. R. Assuncão: PhD Thesis, University of Florida, 1978; published 1983, Ann Arbor, Michigan, USA, University Microfilms International

84 Nis: T. Nishizawa and K. Ishida: *Bull. Alloy Phase Diagrams*, 1984, **5**, 250–259

84 Oka: H. Okamoto, T. B. Massalski, L. J. Swartzendruber and P. A. Beck: *Bull. Alloy Phase Diagrams*, 1984, **5**, 592–601

85 Oka: H. Okamoto and T. B. Massalski: *Bull. Alloy Phase Diagrams*, 1985, **6**, 449–454

Au–Co–Ni
(G. V. Raynor)

INTRODUCTION

In spite of the possible development of alloys with desirable magnetic properties, the constitution of the gold–cobalt–nickel system has received very little attention. The only relevant work in the metallurgical literature is that by [78 Ass, 80 Ass] which record data for the solid alloys arising from a study of diffusion in certain ternary alloys.

THE BINARY SYSTEMS

The system Au–Co has been assessed by [85 Oka]. The Au–Ni system has been assessed by [86 Oka] and the Co–Ni system by [83 Nis].

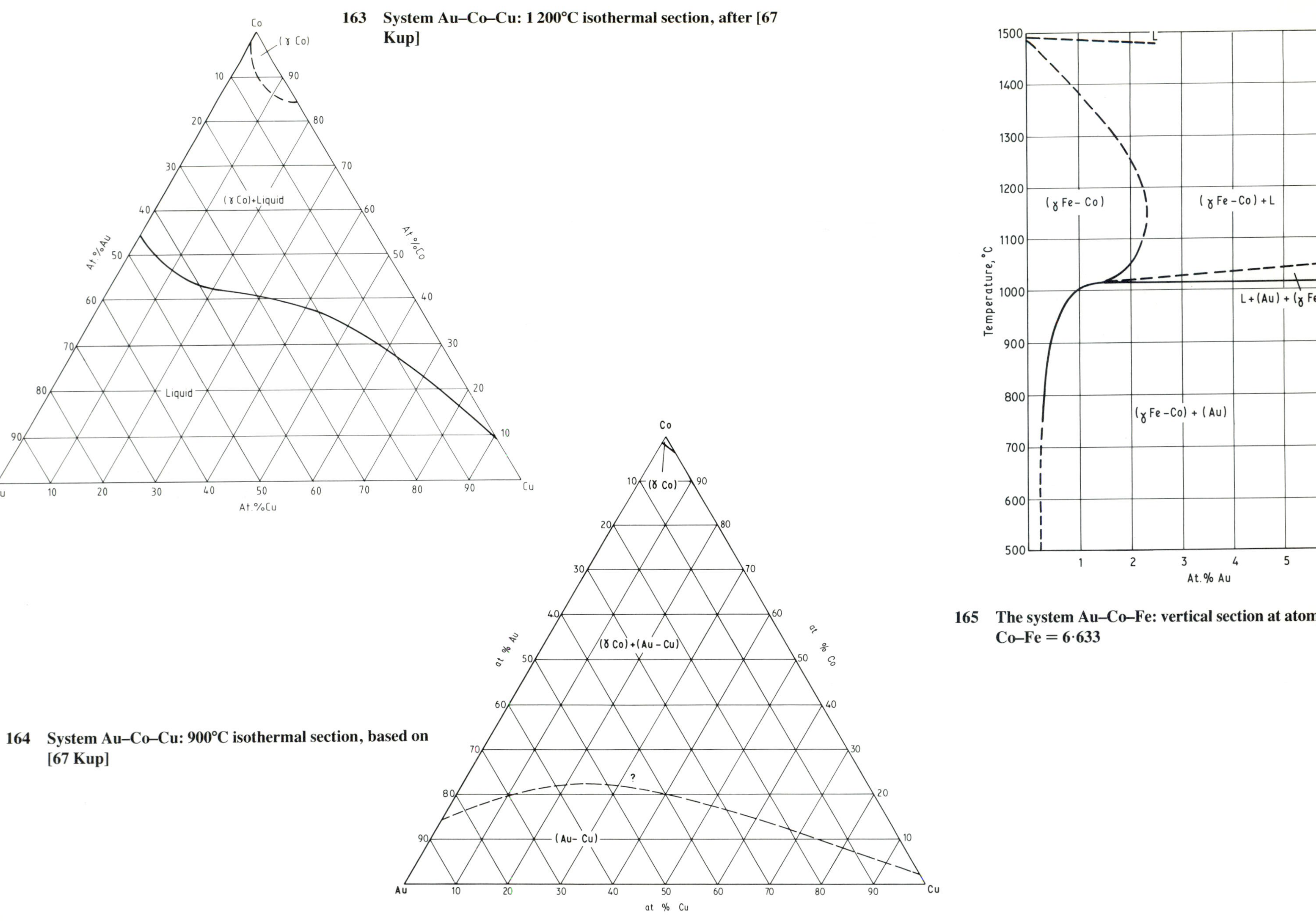

163 System Au–Co–Cu: 1 200°C isothermal section, after [67 Kup]

164 System Au–Co–Cu: 900°C isothermal section, based on [67 Kup]

165 The system Au–Co–Fe: vertical section at atomic ratio Co–Fe = 6·633

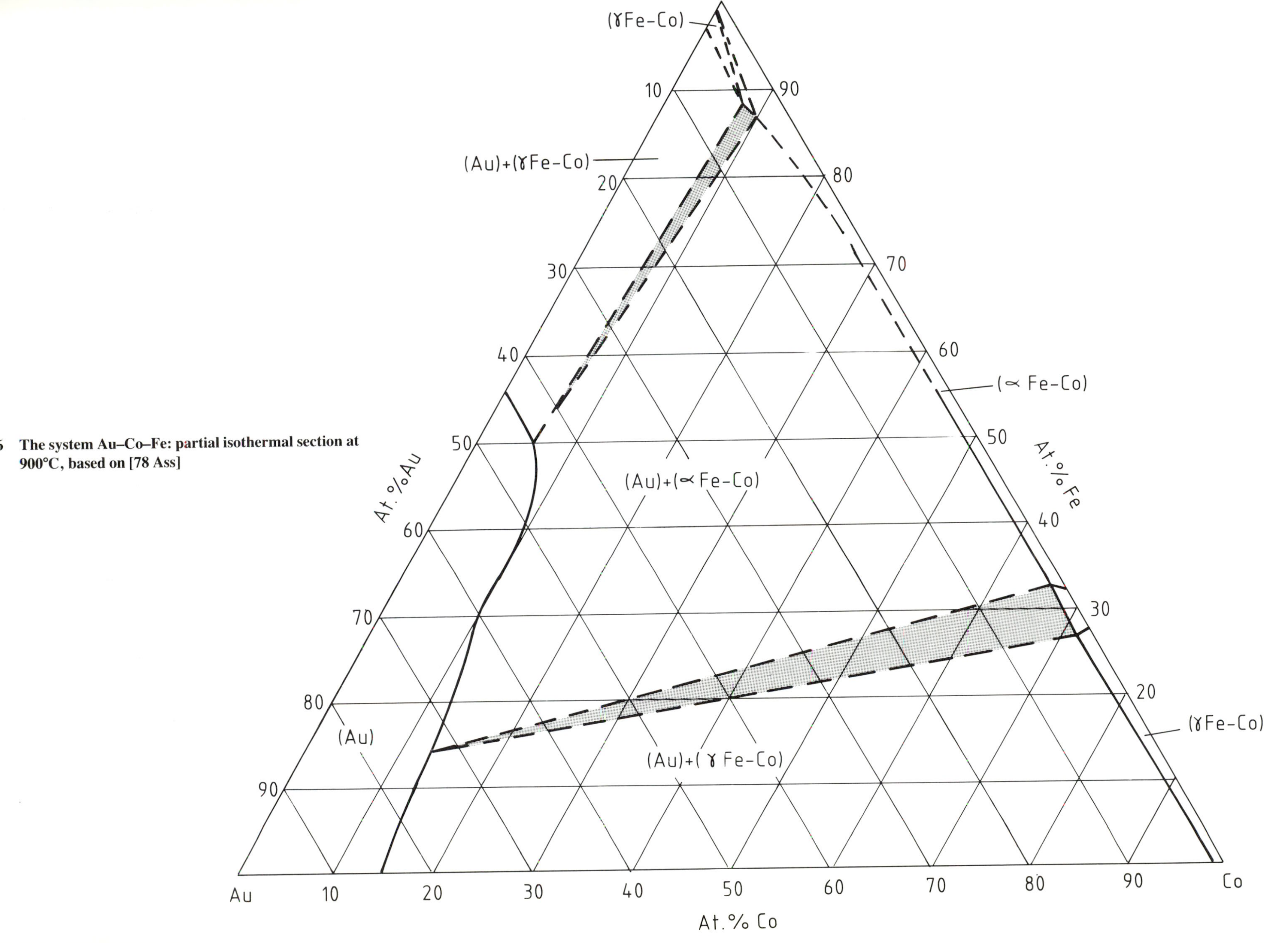

166 The system Au–Co–Fe: partial isothermal section at 900°C, based on [78 Ass]

THE SOLID PHASES

The solid phases encountered in the ternary system are the complete solid solution formed between Ni and γCo, (γNi–Co), the complete solid solution formed at high temperatures between Au and Ni, (Au–Ni), and at certain temperatures the Au-rich (Au–Ni) limited solid solution, and the solid solution of Ni in ϵCo. The crystal structures in the solid state are summarized in Table 58. No discrete ternary phases have been reported.

Table 58. Crystal structures

Solid phase	*Prototype*	*Lattice designation*
(γNi–Co)	Cu	cF4
(Au–Ni)	Cu	cF4
(ϵCo)	Mg	hP2

THE TERNARY SYSTEM

In the course of a comprehensive study of diffusion in certain ternary alloys, [78 Ass, 80 Ass] obtained data from which a partial isothermal diagram at 900°C may be derived. Using components of purity 99·99%, cylindrical specimens of compositions corresponding to AuCo, AuNi, and CoNi were prepared. At the temperature of the experiments, the composition AuCo represented a two-phase alloy, while the other two compositions represented single-phase alloys. The cylindrical specimens of different compositions were pressure-bonded together at 750°C across circular faces to form the following diffusion couples:

(i) AuCo–CoNi
(ii) AuCo–AuNi
(iii) Ni–AuCo

The composite samples were wrapped in tantalum foil, sealed into quartz tubes in an atmosphere of hydrogen, and annealed for a week at 900°C; they were then cooled in air to ~400°C and quenched in water. Specimens were section longitudinally (parallel to the cylindrical axis) and the exposed surfaces were polished. The concentration profiles for Au, Co, and Ni were measured, again parallel to the cylindrical axis, using electron-probe microanalysis techniques.

In each case the two phases (Au–Ni) and (γNi–Co) were present in the portion of the couple richer in Au and Co. The compositions of these phases in equilibrium with each other at 900°C could then be determined by electron probe microanalysis measurements close to, and on both sides of, the interface between the matrix and the second phase. Since a concentration gradient parallel to the cylindrical axis existed in the specimens as a result of diffusion at 900°C, it was possible to make measurements for an extensive range of matrix compositions using the same diffusion couple. By this means the (Au–Ni) + (γNi–Co)/(γNi–Co) phase boundary was determined with good reproducibility up to ~53 at.-% Ni. Similarly, the (Au–Ni)/(Au–Ni) + (γNi–Co) phase boundary was determined up to ~33 at.-% Ni, but with rather poorer reproducibility, for reasons arising out of experimental conditions. The results are shown in Fig. 167, in which the Ni-rich portion of the two-phase boundary, though taken from the phase diagram given in [78 Ass, 80 Ass] and substantially correct, must be regarded as less well established than the remainder of the boundary. The Au–Co miscibility gap penetrates deeply into the ternary system, and according to the diagram in [78 Ass, 80 Ass] the gap closes, at 900°C, at the composition 11 at.-% Au, 17 at.-% Co and 72 at.-% Ni. The Ni-rich corner of the isothermal diagram corresponds with a Ni-rich solid solution containing both Au and Co.

REFERENCES

78 Ass: F. C. R. Assuncão: PhD Thesis, University of Florida, 1978; published 1983, Ann Arbor, Michigan, USA, University Microfilms International

80 Ass: F. C. R. Assuncão: *Metall. ABM (São Paulo)*, 1980, **36**(267), 101–104

83 Nis: T. Nishizawa and K. Ishida: *Bull. Alloy Phase Diagrams*, 1983, **4**, 390–395

85 Oka: H. Okamoto and T. B. Massalski: *Bull. Alloy Phase Diagrams*, 1985, **6**, 449–454

86 Oka: H. Okamoto and T. B. Massalski: 'Binary Alloy Phase Diagrams' (ed. T. B. Massalski), 288–290; 1986, ASM

Au–Co–Pd
(G. V. Raynor)

INTRODUCTION

The metallurgical literature appears to contain only one reference to work on the constitution of the gold–cobalt–palladium system [56 Gri]. In this work, alloys were examined by the methods of thermal analysis, microscopic examination of the microstructures of annealed and quenched alloys, and by techniques involving measurement of hardness, electrical resistivity, and the temperature coefficient of the latter. The general nature of the equilibrium relationship was established.

BINARY SYSTEMS

The equilibrium diagram for the Au–Pd system evaluated by [85 Oka 1] is accepted, as is the evaluation by [85 Oka 2] for the Au–Co system.

According to [58 Han], the form of the Co–Pd equilibrium diagram is simple, with complete miscibility in both solid and liquid states. The liquidus and solidus curves fall from the melting points of the pure metals to a minimum of 1 219°C at the equiatomic composition. There is no evidence of the development of long-range order in the solid state, and there appears to be no transformations apart from that arising from the polymorphic change in pure Co.

SOLID PHASES

From the description of the binary systems, it is clear that the only structures to be encountered in the solid state are the fcc structure typical of the three components above the temperature of the polymorphic change point of Co, and the hexagonal solid solution in the low-temperature form of Co. The equilibrium relationships involving the hexagonal phase in the ternary system have not been investigated.

THE TERNARY SYSTEM

In the comprehensive work of [56 Gri], a selection of experimental methods was used, as noted above. The thermal analysis and metallography (using some 60 ternary compositions) adequately define the nature of the system; the remaining results were largely confirmatory of the conclusions reached. Alloys were prepared from spongy Au and Pd, with not more than 0·01 wt-% impurity, and Co containing 0·01% C. The metals were melted together in a resistance furnace in corundum-lined crucibles under molten $BaCl_2$. In most cases the resulting specimens were chemically analysed, and satisfactory agreement with intended compositions was demonstrated. Thermal analysis was carried out using a Pt/Pt–Rh thermocouple. In annealing experiments samples were heated *in vacuo* for 100 to 150 h at a temperature close to that of the solidus for the relevant compositions, and then slowly cooled to room temperature. In addition, alloys were heat-treated at 900, 1 000, and 1 100°C for periods of 4 h and quenched in iced water. The thermal analysis, hardness, and resistivity results are given in tabular form for the alloys examined, the compositions of which lay upon composition lines at constant weight percentages varying in steps of 10 wt-% Pd from 10 to 90 wt-% Pd; in addition, experiments were carried out for alloys containing 25, 45, and 55 wt-% Pd. The thermal analysis and micrographic results are also presented in the form of eleven vertical sections at constant weight percentages of Pd corresponding with those quoted above. Throughout, the compositions of samples are quoted in weight percentages of the components. For the present assessment, all compositions have been converted into atomic percentages, and the vertical sections have been reconstructed on this basis. The results of [56 Gri] were used for the ternary alloys; for the binary alloys at the extremes of the sections, liquidus and solidus values were taken from the binary assessments. An example is given in Fig. 168, which represents the section at 10 wt-% Pd, and therefore extends from the composition 17 at.-% Pd, 83 at.-% Au to the composition 5·8 at.-% Pd, 94 at.-% Co. The temperature of the eutectic reaction in the Au–Co system (996·5°C) is raised by the addition of Pd, and the two-phase region between the Au-rich solid solution (Au–Pd) and the Co-rich solid solution (Co–Pd) occupies a large proportion of the vertical section. A further example representing the results for the section at 30 wt-% Pd is given in Fig. 169. From the eleven replotted vertical sections a liquidus surface may be constructed. This has been done in Fig. 170, which shows the most probable form of surface, together with isothermal contours from 1 000 to 1 500°C. Minor uncertainties exists, as follows:

(a) For vertical section at 70 wt-% Pd (81·2 at.-% Pd, 18·8 at.-% Au to 56·3 at.-% Pd, 43·7 at.-% Co) the compositions of the points at 1 300 and 1 350°C lie at too high a Co content in relation to results for sections less rich in Pd and to the binary Co–Pd liquidus curve, but this section contains only two experimentally determined ternary liquidus values. Similarly the vertical sections at 80 and 90 wt-% Pd give rise to points at 1 350, 1 400, and 1 450°C which are inconsistent with other sections and with binary Co–Pd values, but in each of these sections only one experimental point is available. In the case of these three sections, therefore, contours have been drawn in conformity with the results for sections less rich in Pd and with the binary Au–Pd liquidus.

(b) In the section at 55 wt-% Pd (69·3 at.-% Pd, 30·7 at.-% Au to 40·3 at.-% Pd, 59·7 at.-% Co) the composition of the point for the 1 250°C contour lies at too high a Co content in relation to results for neighbouring sections and has been rejected.

Though the ternary liquidus contours in the Pd corner of Fig. 170 are therefore subject to some uncertainty, the remainder of the diagram appears well established. The effect of the minimum in the liquidus curve of the Co–Pd alloys is marked. The temperature of the Au–Co eutectic (996·5°C) is raised by Pd, and the results given in [56 Gri] show clearly that in the ternary alloys the Co content of the liquid taking part in the monovariant reaction L $\rightleftarrows$ (Au–Pd) + (Co–Pd) is also raised by the addition of Pd. The monovariant reaction line terminates at the approximate composition Au 37 at.-%, Co 40 at.-%, Pd 23 at.-%. At Au contents less than 37 at.-% and temperatures exceeding 1 150°C, the liquid deposits on cooling only the ternary solid solution denoted (Au–Co–Pd). At this point, the three-phase triangle L + (Au–Pd) + (Co–Pd) has degenerated into a critical tie line relating to the simple equilibrium L + (Au–Co–Pd). The solidus surface is not presented, as inconsistency of results introduces unsatisfactory uncertainties.

The vertical section at 20 wt-% Pd (31·6 at.-% Pd, 68·4 at.-% Au to 12·1 at.-% Pd, 87·9 at.-% Co) is of generally similar form to Fig. 168. For the 30 wt-% Pd section (Fig. 169), however, the (Au–Pd) + (Co–Pd) two-phase region does not make contact with the solidus, and between temperatures of approximately 1 030 and 1 190°C the solid alloys consist of the ternary solid solution (Au–Co–Pd). This confirms that the monovariant eutectic line terminates at a lower Pd content than corresponds to this section; it is also stated [56 Gri] that the eutectic is absent from the section at 25 wt-% Pd (38·1 at.-% Pd, 61·9 at.-%

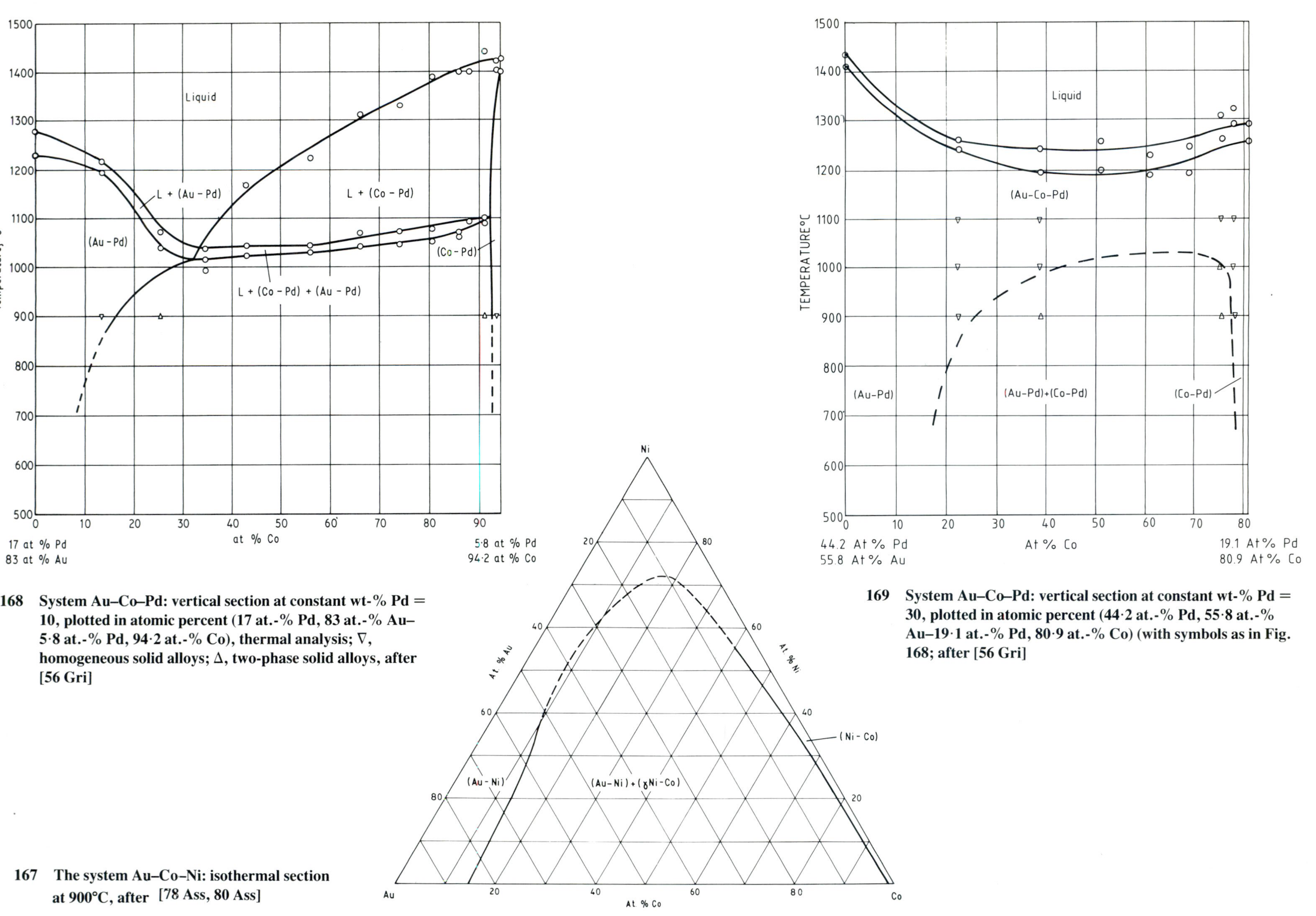

168 System Au–Co–Pd: vertical section at constant wt-% Pd = 10, plotted in atomic percent (17 at.-% Pd, 83 at.-% Au–5·8 at.-% Pd, 94·2 at.-% Co), thermal analysis; ∇, homogeneous solid alloys; Δ, two-phase solid alloys, after [56 Gri]

169 System Au–Co–Pd: vertical section at constant wt-% Pd = 30, plotted in atomic percent (44·2 at.-% Pd, 55·8 at.-% Au–19·1 at.-% Pd, 80·9 at.-% Co) (with symbols as in Fig. 168; after [56 Gri]

167 The system Au–Co–Ni: isothermal section at 900°C, after [78 Ass, 80 Ass]

Au to 15·5 at.-% Pd, 84·5 at.-% Co). The two-phase (Au–Pd) + (Co–Pd) region persists in the 40 wt-% Pd section (55·2 at.-% Pd, 44·8 at.-% Au to 26·9 at.-% Pd, 73·1 at.-% Co), and in the 45 wt-% Pd section (60·2 at.-% Pd, 39·8 at.-% Au to 31·1 at.-% Pd, 68·9 at.-% Co), but is not observed in alloys of the 50 wt-% Pd section (64·9 at.-% Pd, 35·1 at.-% Au to 35·6 at.-% Pd, 64·4 at.-% Co) or in more Pd-rich alloys. At these compositions, the solid alloys consist only of the fcc ternary (Au–Co–Pd) solid solution.

From the metallographic results, contours defining the limits of the two-phase field at various temperatures may be derived. This has been done in Fig. 171 for 1 000 and 900°C; the limit at room temperature as given by [56 Gri] is also included, though it is possible that the area enclosed by this boundary is too small and corresponds to some unspecified temperature at which diffusion during slow cooling of the specimens became negligible. At 900°C the two-phase area in the binary Au–Co alloys projects into the ternary diagram; its width in terms of Co content decreases with the introduction of Pd and the maximum Pd content on the 900°C boundary appears to be about 40 at.-% at a Co content of 40 at.-%. As shown in Fig. 170 the temperature of the monovariant reaction L $\rightleftharpoons$ (Au–Pd) + (Co–Pd) rises with the addition of Pd, so that (Au–Pd) + (Co–Pd) alloys exist in the ternary system at temperatures exceeding that of the binary eutectic (996·5°C) in the Au–Co system. This is clearly shown by the 1 000°C contour in Fig. 170. This boundary cannot, however, intersect directly the binary Au–Co axis, and the inclusion of two-phase regions involving equilibrium between the liquid and solid alloys rich in Au and Co respectively must be included, together with the three-phase (L + (Au–Pd) + (Co–Pd)) triangle. It is this triangle which, as noted above, degenerates into a critical tie line at approximately 1 150°C. The probable positions of the boundaries of the necessary phase fields at 1 000°C are shown as broken lines in Fig. 171. According to Fig. 170 a region of coexistence of two solid phases must persist, continually decreasing in area as the temperature rises, up to a limiting temperature of approximately 1 150°C, at which temperature its area has decreased to vanishing point. Little accurate experimental evidence exists, however, for the positions of the boundaries of the two phase region between 1 150 and the 1 000°C contour plotted in Fig. 171; from the limited results of [56 Gri] it may be inferred that the very small area occupied by two-phase alloys at a temperature just below 1 150°C surrounds a very approximate composition of 40 at.-% Co, 26 at.-% Pd.

The equilibrium relationships involving Co-rich ternary alloys based upon the low-temperature cp hexagonal form of Co have not been investigated.

REFERENCES

56 Gri: A. T. Grigor'ev, E. M. Sokolovskaya, L. D. Buddennaya, I. A. Iyutina and M. V. Maksimova: *Zhur. Neorg. Khim.*, 1956, **1**, 1052–1063; *Russ. J. Inorg. Chem.*, 1956, **1**(5), 181–193

58 Han: M. Hansen and K. Anderko: 'Constitution of Binary Alloys'; 1958, New York, McGraw-Hill

85 Oka 1: H. Okamoto and T. B. Massalski: *Bull. Alloy Phase Diagrams*, 1985, **6**. 229–235

85 Oka 2: H. Okamoto and T. B. Massalski: *Bull. Alloy Phase Diagrams*, 1985, **6**, 449–454

Au–Co–V

In studying solid solution formation between Cr_3Si-type phases (cP8), von Philipsborn [70 Phi] prepared a $V_3Au_{0.5}Co_{0.5}$ alloy by arc-melting V_3Au and V_3Co on a water-cooled Cu hearth under argon. Samples were annealed for 45 days at 550°C, 4 days at 850°C, and 20 days at 1 050°C in evacuated quartz ampoules. The phases present were characterized by the X-ray powder diffraction technique.

The as-cast, 550 and 850°C annealed samples contained a Cr_3Si-type phase with lattice parameter $a = 0{\cdot}478 \pm 0{\cdot}001$ nm and a V_3AuO phase with a Cu_3Au-type structure. The 1 050°C annealed sample contained a Cr_3Si-type phase with $a = 0{\cdot}476 \pm 0{\cdot}001$ nm and more of the V_3AuO phase. The absence of binary compounds in the Au–Co system means that V_3Au and V_3Co should form a stable tie line in the ternary system. It is worth noting that the lattice parameter for the ternary Cr_3Si-type phase is the mean of the lattice parameters of V_3Au (0·488 nm) and V_3Co (0·468 13 nm).

REFERENCE

70 Phi: H. von Philipsborn: *Z. Kristallogr.*, 1970, **131**, 73–87

Au–Cr–Fe

The phase Au_4Cr has the ordered Ni_4Mo-type structure, tI10. No corresponding phase occurs in the Au–Fe phase diagram. Toth *et al.* [69 Tot] determined the temperature dependence of resistivity for the ternary alloy $Au_4Cr_{0.5}Fe_{0.5}$ (80Au, 10Cr, 10 at.-% Fe) with a heating rate of ~1°C min^{-1}. The ordered Ni_4Mo-type structure disordered at 240°C. This temperature, scaled from the published figure, should be compared with 380°C for the disordering temperature of Au_4Cr [85 Oka] and the absence of an Au_4Fe compound.

REFERENCES

69 Tot: R. S. Toth, A. Arrott, S. S. Shinozaki, S. A. Werner and H. Sato: *J. Appl. Phys.*, 1969, **40**, 1373–1375

85 Oka: H. Okamoto and T. B. Massalski: *Bull. Alloy Phase Diagrams*, 1985, **6**, 224–228

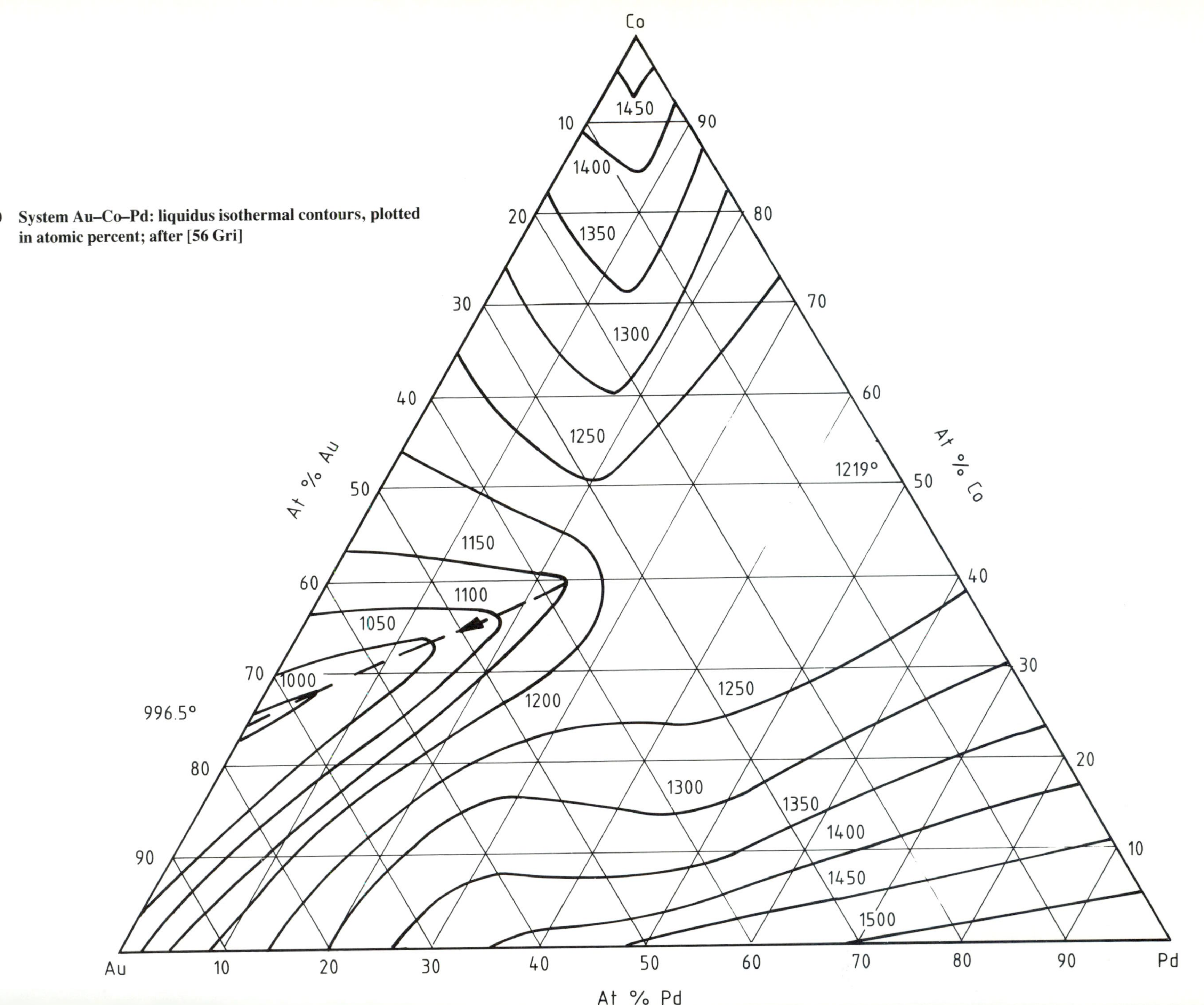

170 System Au–Co–Pd: liquidus isothermal contours, plotted in atomic percent; after [56 Gri]

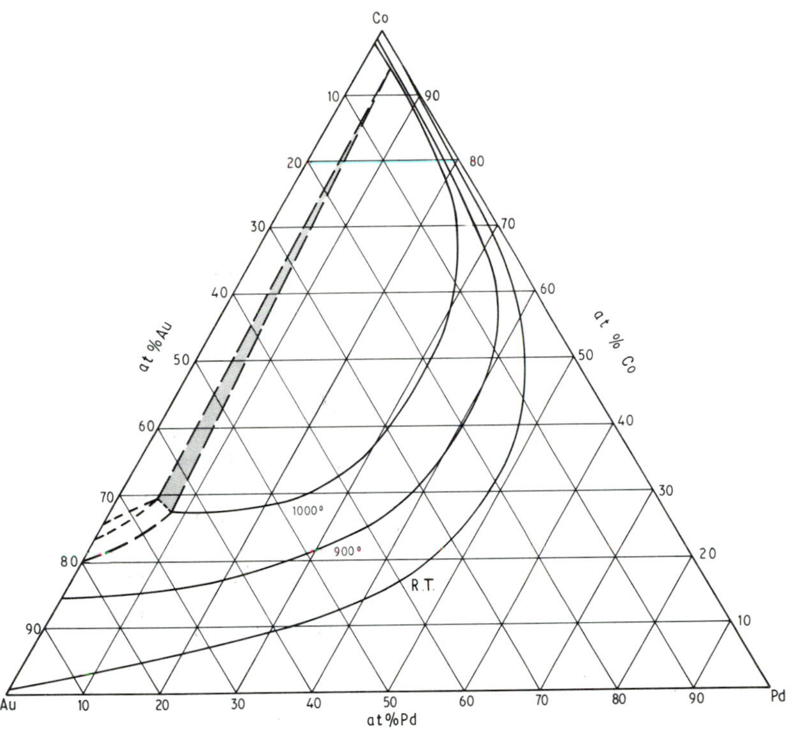

171 System Au–Co–Pd: regions of immiscibility at 1 000 and 900°C, and at room temperature; after [56 Gri]

Au–Cr–Mn

Toth *et al.* [69 Tot] state, without elaboration, that the ordered Au_4X phase is stable on the section $Au_4(Cr_{1-x}Mn_x)$ with $0 \leq x \leq 1$. This would be expected since Au_4Cr and Au_4Mn both have the ordered Ni_4Mo-type structure, tI10, with disordering temperatures of 380 and 450°C respectively.

REFERENCE

69 Tot: R. S. Toth, A. Arrott, S. S. Shinozaki, S. A. Werner and H. Sato: *J. Appl. Phys.*, 1969, **40**, 1373–1375

Au–Cr–Ti

Both Au_4Cr and Au_4Ti have the ordered Ni_4Mo-type structure, tI10. The disordering temperatures are 380°C [87 Oka] and 1 172°C [83 Mur] respectively. [69 Tot] determined the temperature dependence of resistivity for the ternary alloy $Au_4Cr_{0\cdot5}Ti_{0\cdot5}$ (80 Au, 10Cr, 10 at.-% Ti) at a heating rate of ~1°C min^{-1}. No disordering was detected up to the maximum temperature of 900°C. This suggests a higher disordering temperature than 900°C which is possible in view of the stability of ordered Au_4Ti to 1 172°C.

REFERENCES

69 Tot: R. S. Toth, A. Arrott, S. S. Shinozaki, S. A. Werner and H. Sato: *J. Appl. Phys.*, 1969, **40**, 1373–1375

83 Mur: J. L. Murray: *Bull. Alloy Phase Diagrams*, 1983, **4**, 278–283

87 Oka: H. Okamoto and T. B. Massalski: 'Phase Diagrams of Binary Gold Alloys', 68–72; 1987, ASM International

Au–Cr–V

Toth *et al.* [69 Tot] state, without elaboration, that the ordered Au_4X phase is stable on the section $Au_4(V_{1-x}Cr_x)$ with $0 \leq x \leq 1$. This would be expected since Au_4V and Au_4Cr both have the ordered Ni_4Mo-type structure, tI10, with disordering temperatures of 565 and 380°C respectively.

REFERENCE

69 Tot: R. S. Toth, A. Arrott, S. S. Shinozaki, S. A. Werner and H. Sato: *J. Appl. Phys.*, 1969, **40**, 1373–1375

Au–Cs–O

Wasel-Nielen and Hoppe [68 Was] prepared AuCsO by heating AuCs in oxygen in a closed system at 400°C for 8 h. It was produced as a brittle pale-golden powder which is very sensitive to humidity. AuCsO has a tetragonal structure, $a = 1\cdot0160 \pm 0\cdot0005$, $c = 0\cdot6170 \pm 0\cdot0003$ nm. It is isotypic to AgKO with the space group S_4^2-I$\bar{4}$ and eight formula units in the elementary cell.

REFERENCE

68 Was: H.-D. Wasel-Nielen and R. Hoppe: *Z. anorg. Chem.*, 1968, **359**, 36–40

Au–Cu–Dy
Au–Cu–Er
Au–Cu–Gd
Au–Cu–Ho
Au–Cu–Tb

The compound $TbAuCu_4$ was synthesized and shown to be cubic with a = 0·7175 nm [83 Kan] and to have the $AuBe_5$ structure-type [84 Kan]. The intermetallic compounds $REAuCu_4$ (RE = Dy, Er, Gd, Ho, Tb) were prepared by arc-melting 99·9% RE, 99·99% Au and 99·999% Cu under a purified argon atmosphere [86 Kan]. Weight losses were only 0·1% after melting. Room-temperature X-ray powder diffraction analysis of powder specimens previously annealed for 2 days at 500°C indicated a $AuBe_5$ (cF24) structure type with no additional phases. Lattice parameters, scaled from the published figure, are presented below. The lattice parameter given by [83 Kan] for $TbAuCu_4$ disagrees with that of [86 Kan]. The latter value is preferred on the basis of the trend shown by both $REAgCu_4$ and $REAuCu_4$ compounds to have decreasing lattice parameters with increasing atomic number in the series from Gd to Er.

	Lattice parameter (nm)
$GdAuCu_4$	0·7138
$TbAuCu_4$	0·7134
$DyAuCu_4$	0·7098
$HoAuCu_4$	0·7080
$ErAuCu_4$	0·7079

REFERENCES

83 Kan: T. Kaneko, S. Abe, K. Kamigaki and M. Ohashi: *J. Magn. Magn. Mater.*, 1983, **31–34**, 253–254

84 Kan: T. Kaneko, S. Abe, K. Kamigaki, M. Ohashi and S. Arai: *J. Appl. Phys.*, 1984, **55**, 2028–2030

86 Kan: T. Kaneko, S. Arai, S. Abe and K. Kamigaki: *J. Phys. Soc. Japan*, 1986, **55**, 4441–4447

Au–Cu–Fe
(G. V. Raynor)

INTRODUCTION

The gold–copper–iron system was investigated [67 Kup] using the methods of thermal analysis, together with limited metallographic work. The report includes a table of thermal arrest points for 36 alloys with compositions lying at intervals of 10 wt-% Au on lines of constant Cu percentages. The results are illustrated by a single vertical section at 10 wt-% Cu. In this assessment the thermal arrests have been plotted in terms of atomic percentages, and eight vertical sections were constructed. From these it is possible to derive a projection of the surfaces of primary crystallization and to establish a reasonably accurate series of liquidus contours.

THE BINARY SYSTEMS

The Au–Cu phase diagram, assessed by [87 Oka] is accepted. The equilibrium diagram of the Au–Fe system has been assessed by [84Oka]. The Cu–Fe phase diagram assessed by [82 Kub] with amendments to the Fe-rich region by [79 Cha] is accepted.

THE SOLID PHASE

No discrete ternary intermediate phases have been reported in the Au–Cu–Fe system. The only solid phases encountered are therefore solid solutions based upon the three components, and the ordered structures of the Au–Cu system. The crystal structures are summarized in Table 59.

Table 59. Crystal structures

Solid phase	*Prototype*	*Lattice designation*
(Au)	Cu	cF4
(Cu)	Cu	cF4
$AuCu_3$	$AuCu_3$	cP4
AuCuI	AuCuI	tP4
AuCuII	AuCuII	tP4 (faulted)
(γFe)	Cu	cF4
(δFe)	W	c12
(αFe)	W	c12

THE TERNARY SYSTEM

In the work of [67 Kup], 36 ternary alloys were prepared by induction melting in carborundum crucibles under a protective layer of molten $BaCl_2$. All were subjected to thermal analysis using conventional methods. Specimens were also prepared, and annealed for 300 h at 800°C in evacuated quartz capsules, for metallographic examination and the determination of hardness. The thermal analysis results were classified into those arising from primary crystallization and from the beginning and end of the ternary monovariant reaction involving liquid, (γFe), and the (Au–Cu) solid solution. Some arrests corresponding to the separation of (γFe) from the (Au–Cu) solid solution were also noted. In the case of the ternary alloys, no examination was made of the reaction resulting from the γFe ⇌ αFe transformation in the binary Au–Fe and Cu–Fe alloys. Though metallographic results confirmed the constitution proposed, the thermal arrests constitute the quantitative evidence. From these results, after the conversion of compositions to atomic percentages, it has been possible to construct vertical sections for sections corresponding in atomic percentages to the constant Cu weight percentages examined by [67 Kup]. The composition sections were as follows:

at.-% Cu	at.-% Au	to	at.-% Cu	at.-% Au
8·9	0		25·6	74·4
18·0	0		43·7	56·3
27·4	0		57·1	42·9
36·9	0		67·4	32·6
46·8	0		75·6	24·4
56·9	0		82·3	17·7
67·2	0		87·9	12·1
77·8	0		92·5	7·5

The first of these vertical sections is shown in Fig. 172, which illustrates the nature of the results. A very small region of primary separation of (δFe) has been ignored. The ternary reactions arising from the L + (γFe) ⇌ (Cu) reaction at 1 094°C in the Cu–Fe system persists to approximately 41 at.-% Au in this section and the liquidus surface is clearly separated into two distinct regions. The same form is preserved in all the vertical sections constructed, so that it is possible to plot the course of the monovariant reaction involving liquid, (γFe) and (Au–Cu) on a projection of the surfaces of primary crystallization. This is shown in Fig. 173 as the curve ab, which joins the compositions of the liquids at the reactions L + (γFe) ⇌ (Au) (1 173°C) and L + (γFe) ⇌ (Cu) (1 094°C). A similar curve will join the liquid compositions of the L + (δFe) ⇌ (γFe) reactions in the systems Au–Fe and Cu–Fe (curve cd in Fig. 173). There is no experimental evidence for the exact position of this curve, which is therefore drawn as a broken curve.

From the same eight vertical sections the compositions corresponding to various liquidus temperatures may be derived, allowing constant temperature contours to be drawn on the composition diagram, as shown in Fig. 173. The derived contours are consistent with each other and with the liquidus compositions in the binary systems, and very little adjustment or 'smoothing' was necessary. Consequently, although it is difficult to estimate the overall accuracy, it is felt that Fig. 173 gives a very reasonable picture of the constitution. The two major surfaces are those across which the solid solutions (γFe) and (Au–Cu) crystallize on cooling, and are separated by the monovariant line ab. The secondary thermal arrests do not appear to be sufficiently consistent for the construction of reliable secondary or solidus surfaces.

In the Au–Fe system the saturated solid solution of Fe in Au is formed peritectically at 1 173°C; at 1 100°C, therefore, an isothermal section takes the form of Fig. 174. Although the boundaries of the (γFe) and solid solution (Au) areas are hypothetical, the positions of the boundaries of the liquid region and of the three-phase triangle

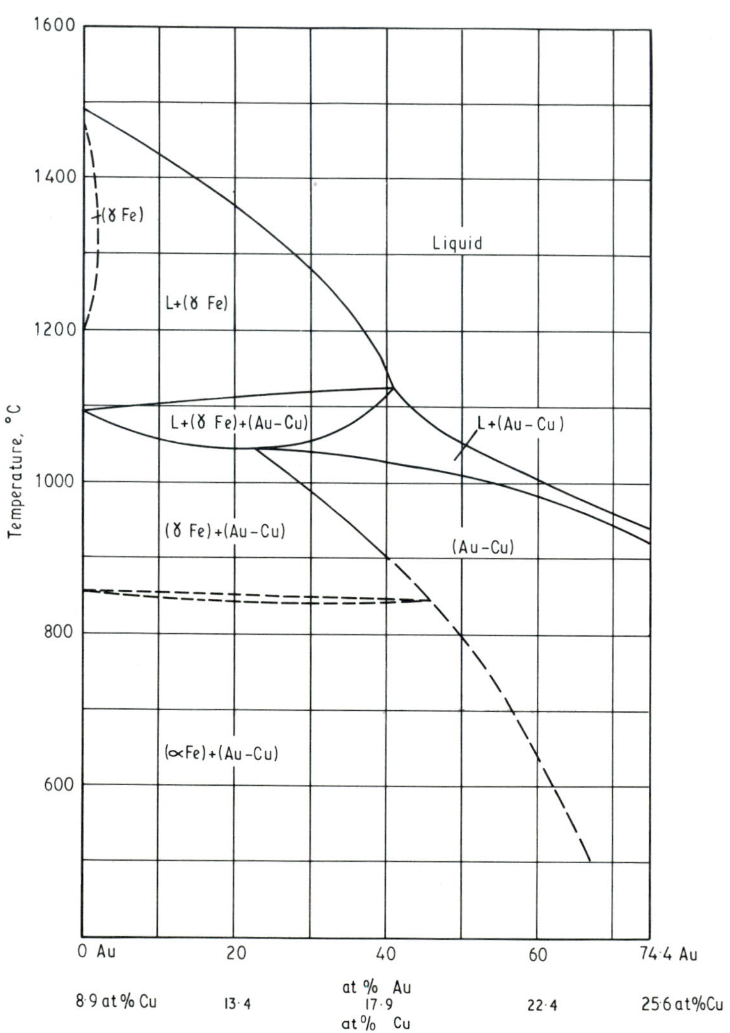

172 System Au–Cu–Fe: Vertical section between compositions 8·9 at.-% Cu, 91·1 at.-% Fe to 25·6 at.-% Cu, 74·4 at.-% Au, after [67 Kup]

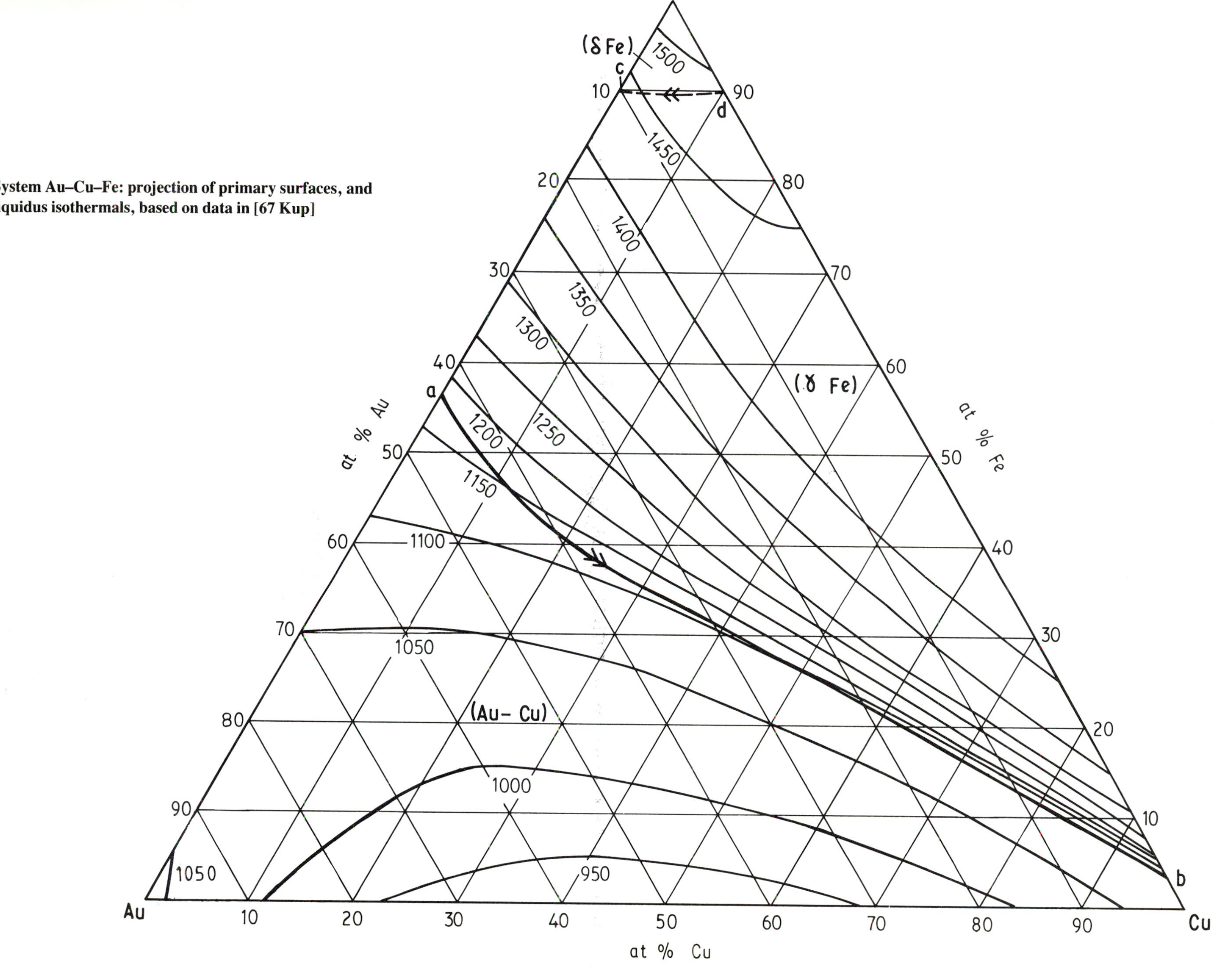

173 System Au–Cu–Fe: projection of primary surfaces, and liquidus isothermals, based on data in [67 Kup]

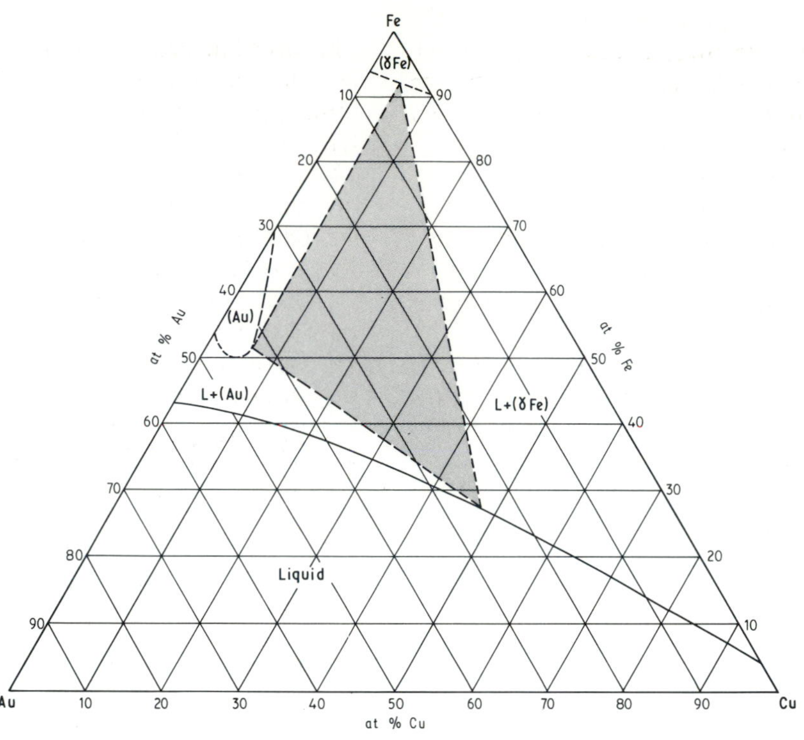

174 System Au–Cu–Fe: isothermal section at 1 100°C. The broken curves are hypothetical, after [67 Kup]

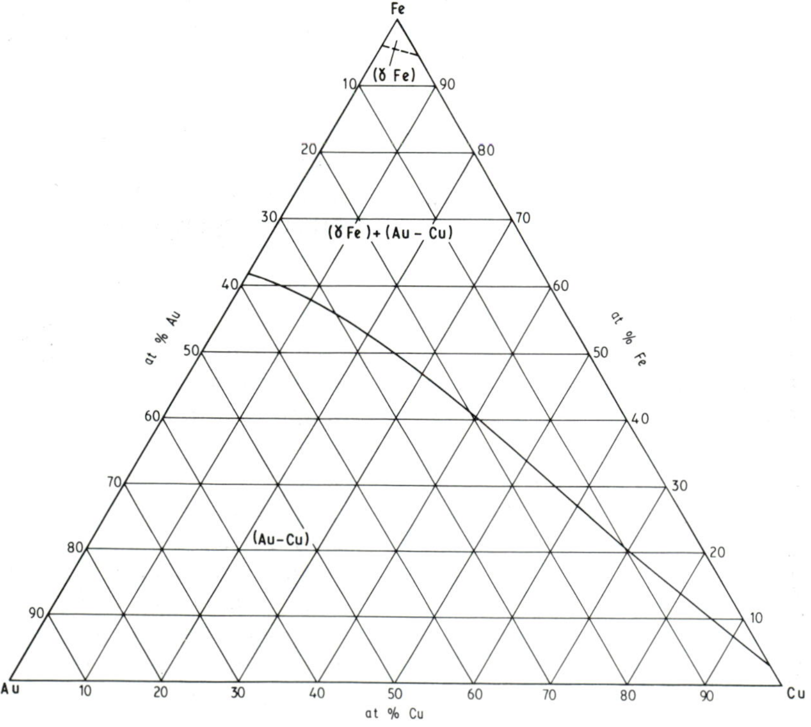

175 System Au–Cu–Fe: isothermal section at 1 000°C. The broken curve is hypothetical, after [67 Kup]

are consistent with Fig. 172 and with the other relevant polythermal diagrams. It is perhaps surprising that the (γFe)–liquid tie line of the three-phase triangle is not nearer to the Cu–Fe binary edge since the temperature of the isothermal section is only 6°C above that of the binary peritectic reaction, liquid + (γFe) ⇌ (Cu).

At 1 000°C the alloys are completely solid, and equilibrium involves (γFe) and (Au–Cu) as shown in Fig. 175, where again the boundary of the (Au–Cu) region is established by the vertical sections derived from the results in [67 Kup], while that of the limited (γFe) solid solution is hypothetical.

There is no information with regard to the solubility of Fe in the ordered phases $AuCu_3$ and AuCu, nor to the effect of Fe on their respective critical temperatures for order or their degrees of long range order. It is however expected that at temperatures below their critical ordering temperatures both $AuCu_3$ and AuCu will be involved in the equilibria, though it is not certain that they will enter into equilibrium with (αFe).

REFERENCES

67 Kup: V. V. Kuprina and V. B. Bernard: *Vest. Moscov Univ., Khim.*, 1967, (2), 63–66

79 Cha: Y. A. Chang, J. P. Neumann, A. Mikula and D. Goldberg: 'Phase Diagrams and Thermodynamic Properties of Ternary Cu-Metal Systems', INCRA Monograph VI, 1979, 35–39

82 Kub: O. Kubaschewski: 'Iron-Binary Phase Diagrams', 35–37; 1982, Berlin/Heidelberg, Springer-Verlag

84 Oka: H. Okamoto, T. B. Massalski, L. J. Swartzendruber and P. A. Beck: *Bull. Alloy Phase Diagrams*, 1984, **5**, 592–601

87 Oka: H. Okamoto, D. J. Chakrabarti, D. E. Laughlin and T. B. Massalski: *Bull. Alloy Phase Diagrams*, 1987, **8**, 454–474

Au–Cu–Ga

(G. V. Raynor)

INTRODUCTION

Although little work has been carried out on the system Au–Cu–Ga the effect of Ga on the ordered equiatomic structures in the Au–Cu system has been studied [68 Fai, 72 Dir, 75 Dir, 77 Dir, 77 Gad], while a limited examination of isothermal sections at 300, 250 and 200°C has been carried out [77 Gad].

BINARY SYSTEMS

The evaluations of the Au–Cu system by [87 Oka 1] and of the Au–Ga system by [87 Oka 2] are accepted. The Cu–Ga system presented by [86 Mas] is accepted.

SOLID PHASES

The three binary systems contain numerous phases whose crystal structures are summarized in Table 60. Certain ternary phases have also been reported [68 Fai, 72 Dir, 75 Dir, 77 Dir, 77 Gad]. A ternary phase at 5 at.-% Ga on the 50 at.-% Au section has an orthorhombic structure based on a distorted AuCd-type structure: it was denoted AuCu III [68 Fai, 72 Dir, 75 Dir, 77 Dir] and K phase by [77 Gad]. A further five ternary phases have been reported [77 Gad]. It is considered that verification of these structures is required and reference is made to them in the discussion of the ternary equilibria. [77 Dir] also refer to the ternary phases Au_2CuGa and $AuCu_2Ga$.

Table 60. Crystal structures

Solid phase	*Prototype*	*Lattice designation*
(Au)	Cu	cF4
(Cu)	Cu	cF4
(Ga)	Ga	oC8
Au_3Cu	$AuCu_3$	cP4
AuCuI	AuCuI	tP4
AuCuII	AuCuII	tP4 faulted
$AuCu_3$	$AuCu_3$	cP4
β	W	cI2
γ	Cu_9Al_4	cP52
ζ_1 (Cu_3Ga)	Mg	hP2
γ_1	Cu_9Al_4	cP52
γ_2	Cu_5Zn_8	cI52
γ_3	Cu_5Zn_8	cI52
ϵ ($CuGa_2$)	$FeSi_2$	tP3
α′	Ni_3Ti	hP16
β (Au_7Ga_2.h)	—	hP27
β′ (Au_7Ga_2.l)	—	Orthorhombic
γ (Au_7Ga_3.h)	$AsPd_2$	oC24
γ′ (Au_7Ga_3.l)	—	Orthorhombic
AuGa	MnP	oP8
$AuGa_2$	CaF_2	cF12

TERNARY EQUILIBRIA

The initial interest in work on this system [68 Fai] lay in the equilibria involving ordered and disordered phases on the 50 at.-% Au section, $Au_{50}Cu_{50-x}Ga_x$. The work refers to a partial diagram due to J. Hertz (Theses Sciences, Nancy, 1967); this work was not available for assessment. Though no details of experimental methods are given, in similar work on the section $Au_{50}Cu_{50-x}Ni_x$ the same workers employed alloys melted in Vycor tube, homogenized and powdered. Part of the work was carried out on quenched samples which had been annealed in hydrogen at various temperatures; some experiments were made using a high-temperature X-ray camera. The diagram, Fig. 176, appears to have been established by observing the appearance or disappearance of the prominent diffractions of the phases concerned. According to this work AuCu I can dissolve about 2 at.-% Ga, and at ca. 4·5 at.-% Ga it has been eliminated from the equilibria. The AuCu II phase can dissolve up to ca.4·4 at.-% Ga; above this limit other phases are also involved. The major feature is the appearance of a phase AuCu III whose crystal structure differs from those of the binary Au–Cu ordered phases. It may be noted that AuCu II is shown as stable both above 300 and below 167°C. The associated two-phase and three-phase fields are as proposed [68 Fai]. The

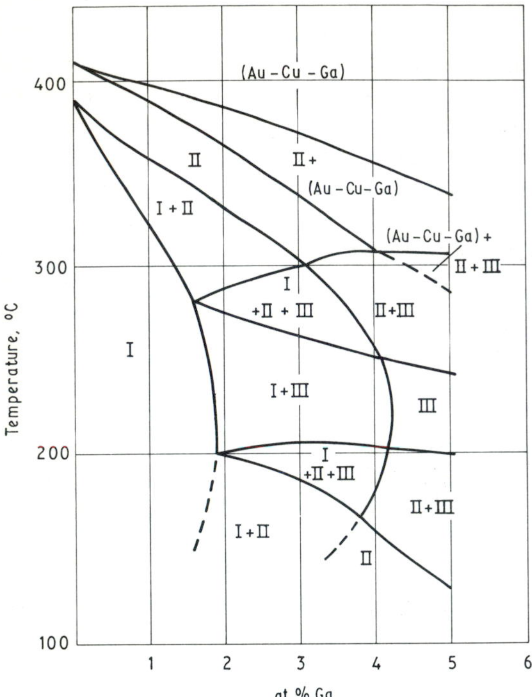

176 The 50 at.-% Au section, up to 6 at.-% Ga

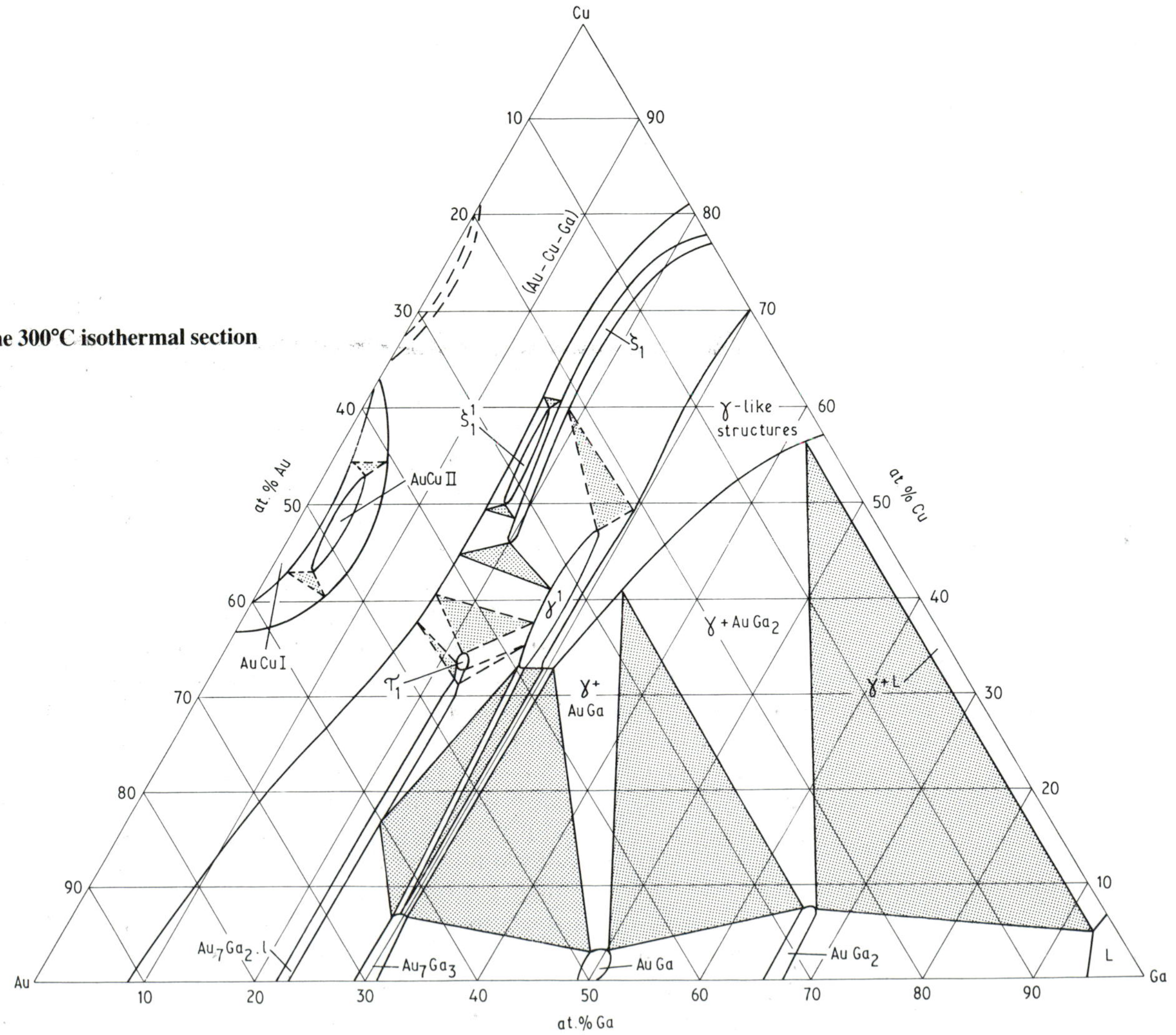

177 The 300°C isothermal section

crystal structure of AuCu III was studied [72 Dir, 75 Dir], using a single-phase alloy of composition 50 Au, 5 at.-% Ga annealed for 15 days at 250°C. The structure is orthorhombic with $a = 0{\cdot}892$, $b = 0{\cdot}456$, $c = 0{\cdot}283$ nm with 8 atoms in the unit cell. The phase is isomorphous with alloys of composition close to that of Au_2CuZn in the Au–Cu–Zn system. The structure is based on a distorted AuCd-type of structure in which one type of atom site is occupied by Au atoms and the other randomly by Cu and Ga atoms.

The binary AuCu II phase may be regarded as based upon 10 cells of AuCu I placed side by side in a row, the long period arising from a fault (a shift of one interatomic distance in the direction [001] after traversing the initial five cells. Using alloys made from spectrographically pure Au and Cu, and 99·999% Ga, fused together in a sealed tube *in vacuo*, furnace cooled, and annealed for 15 days at 720°C and finally quenched. [72 Dir] examined the structure of alloys on the 50 at.-% Au section ordered at various temperatures for 10 to 20 days. The high temperature structures are retained by quenching. For the alloy containing 50 Au, 3 at.-% Ga, quenched from 323°C after 10 days, the nature of the faulting was found to have changed in such a way that on average the number of basic cells side by side in the structure was about 7·4. The frequency of faulting was increased by Ga.

Further crystallographic information was provided [77 Dir]. An investigation of the composition Au_2CuGa indicated that this alloy, initially disordered fcc, undergoes ordering at ca. 600°C to a CsCl-type structure and subsequently at 400°C to the Cu_2AlMn-type structure. The alloy $AuCu_2Ga$ had the β–Mn structure after annealing for 15 days at 400°C; these findings are difficult to reconcile [77 Gad].

The only other significant contribution to the Au–Cu–Ga phase equilibria is that of [77 Gad] who presented the results of an X-ray study of isothermal sections at 300, 250 and 200°C. The materials used were of a purity exceeding 99·5% and alloys were prepared in argon in a high frequency furnace. After homogenizing for 1 day at 300°C the quenched alloys were powdered and the powders were heat treated *in vacuo* at the appropriate temperature and finally quenched. The phases present were identified by their diffraction patterns. The system presented by [77 Gad] is complex. At 300°C the constitution is as shown in Fig. 177. The solid solubilities of Au in the Cu–Ga ζ_1 phase and the γ-type phases are very extensive, as is that of Cu in Au_7Ga_2.r. The solubilities in the remaining binary phases are relatively small. It should be noted that [77 Gad] shows $CuGa_2$ as taking part in the equilibria at 300°C. Since this phase is not stable above 254°C in the binary system, it and the associated equilibria have been omitted from Fig. 176. It may also be noted that AuCu I and AuCu II are shown as coexisting at 300°C, in reasonable conformity with Fig. 177. The chief interest lies in the discovery of ternary phases related to the extensive solid solutions in the binary phases. Thus a γ^1 phase occurs, with a deformed structure of the general γ-brass type. A ζ_1^1 phase is postulated close to the composition 55 Cu, 18 at.-% Ga; the structure is reported to be $TiNi_3$-type (hP16) with $a = 0{\cdot}5375$ and $c = 0{\cdot}8786$ nm at the composition 25 Au, 15 at.-% Ga (quenched from 235°C). A further phase, denoted τ_1 in this assessment, is stated to occur at about 34 Cu, 22 at.-% Ga. This is also said to have the $TiNi_3$-type structure.

It is, however, not clearly shown in the 300°C isothermal section in [77 Gad], but it has been tentatively inserted in Fig. 177. Certain phase boundaries in the diagram published by [77 Gad] were confusingly drawn. Figure 177 represents an attempt to clarify the constitution in accordance with the textual description. Experiments were also made at 250 and 200°C. New features appear. The AuCu III phase [72 Dir, 75 Dir] and the AuCu II phase are both stable. Alloys containing 25 at.-% Au and from 7 to 12·5 at.-% Ga are stated to have structures of the faulted $AuCu_3$ type and occur at both 250 and 200°C. The ζ_1^1 phase, Fig. 177, is stable at both 250 and 200°C but its homogeneity range shrinks with decreasing temperature from 300 to 200°C. The τ_1 phase is stable at 250°C but not at 200°C. A new phase, based on the composition 50 Au, 16 at.-% Ga, appears in the 200°C section. The equilibria at 250 and 200°C for alloys containing less than 30 at.-% Ga are not clear from the sections given by [77 Gad] and have not been reproduced in this assessment.

REFERENCES

68 Fai: R. Faivre, J. Hertz and M. Gantois, *Mater. Res. Bull.*, 1968, **3**, 661–670

72 Dir: M. Dirand, L. Rimlinger and J. Hertz, *Mem. Sci., Rev. Mét.*, 1972, **69**, 903–917

75 Dir: M. Dirand, A. Courtois, J. Hertz and J. Protas, *Compt. rend.*, 1975, **280C**, 559–561

77 Dir: M. Dirand, A. Courtois, H. Lelaurain and J. Hertz, *J. Chim. Phys.*, 1977, **74**, 971–983

77 Gad: A. A. Gadalla, Proc. 1st Egypt Conf. Mining and Met. Technol. *Assint.*, 1977, **2**, pp. 153–164. Ed: A. Soliman, Univ. Assint, Egypt.

86 Mas: T. B. Massalski (ed), *Binary Alloy Phase Diagrams*, ASM International, Metals Park, Ohio, Vol. 1, p. 917, 1986

87 Oka 1: H. Okamoto, D. J. Chakrabarti, D. E. Laughlin and T. B. Massalski, *Bull. Alloy Phase Diagrams*, 1987, **8**, 454–474

87 Oka 2: H. Okamoto and T. B. Massalski, 'Phase Diagrams of Binary Gold Alloys', ASM International, Metals Park, Ohio, pp. 111–118, 1987

Au–Cu–Ge

Jaffee and Gonser [46 Jaf] studied the effect of Cu additions on the Au–Ge eutectic alloy (Table 61). A monovariant eutectic curve (Fig. 178) was traced to nearly 50 at.-% Cu. As might be anticipated from the constituent

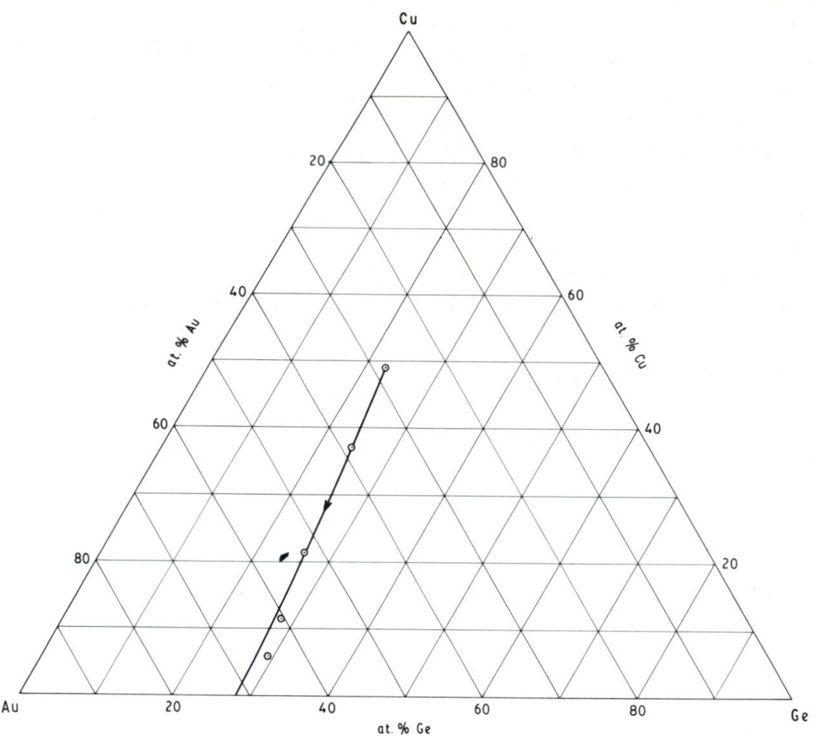

178 The monovariant eutectic curve [46 Jaf]

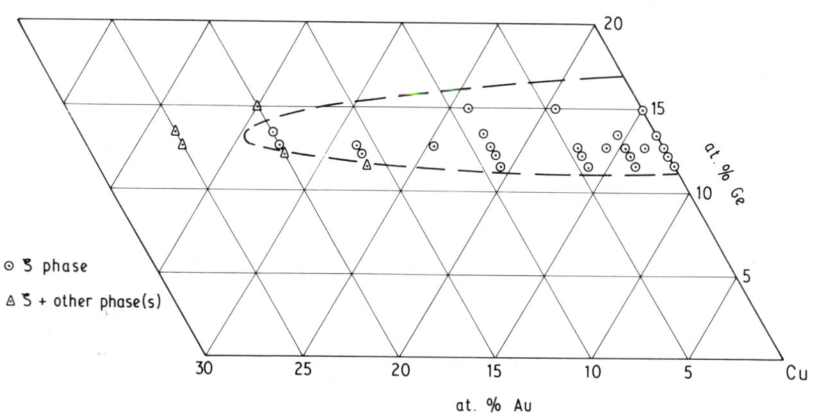

179 The ζ phase region at 400°C [63 Kin]

binary systems, Cu additions lead to increasing temperatures for separation of the (Au–Cu) + (Ge) binary eutectic complex in the ternary compared with binary eutectic separation of (Au) + (Ge). X-ray examination of 1 g melts solidified by cooling from 1 050°C in graphite crucibles indicated (Au–Cu) solid solution (disordered fcc with $a = 0{\cdot}400$ nm) + Ge for alloys up to and including 21·4 at.-% Cu. The 37 at.-% and 48·8 at.-% Cu alloys contained Ge with a second phase that was not identified; this second phase was present whether the alloy was slow cooled from the melt or the melt quenched into water. Metallographic evidence showed a (Au–Cu) solid solution + Ge for all five ternary alloys, with coreing of the (Au–Cu) solid solution becoming more pronounced at the higher Cu contents.

Thurmond and Logan [56 Thu] found that at 700°C a molten Au–Cu–Ge alloy containing 0·4 at.-% Cu was in equilibrium with solid Ge containing $8{\cdot}5 \times 10^{-8}$ at.-% Cu.

King *et al.* [63 Kin] outlined the extent of the hcp ζ phase field at 400°C (Fig. 179). Alloys were prepared from spectroscopically pure elements by melting under He in transparent quartz tubing. They were annealed for 14 days at 400°C and quenched. Metallography was used to characterize single phase ζ compositions and these 20 ternary alloys, with five binary Cu–Ge ζ phase alloys, were examined by X-ray powder diffraction to determine the lattice parameters and axial ratios (Table 61). Powder samples were requenched from 400°C and lattice spacings measured at 30°C. Alloy compositions that contained phases additional to ζ were not analysed by X-ray powder diffraction but were used to locate the approximate limit of the ternary ζ phase region at 400°C.

Table 61. Arrest temperatures along the monovariant eutectic curve [46 Jaf]

Alloy composition, at.-%			*Eutectic–line arrest*, °C
Au	*Cu*	*Ge*	
73	—	27	356
64·6	6·0	29·4	360
60·2	11·5	28·3	368
52·4	21·4	26·2	385
38·8	37·0	24·2	430
28·4	48·8	22·8	477

[88 Cas] measured the enthalpy of formation of ternary alloys along the sections 20 Au, 80 Cu–Ge; 40 Au, 60 Cu–Ge; 60 Au, 40 Cu–Ge and 80 Au, 20 Cu–Ge. A plot of isoenthalpy contours for a temperature of 1 075°C was presented.

REFERENCES

46 Jaf: R. I. Jaffee and B. W. Gonser: *Trans. Amer. Inst. Min. Met. Eng.*, 1946, **166**, 436–443

56 Thu: C. D. Thurmond and R. A. Logan: *J. Phys. Chem.*, 1956, **60**, 591–595

63 Kin: H. W. King, T. B. Massalski and L. L. Isaacs: *Acta Metall.*, 1963, **11**(12), 1355–1361

88 Cas: R. Castanet, *J. Less-Common Metals*, 1988, **136**, 287–296

Table 62. Lattice parameters and axial ratios of hcp ζ phase alloys in the Au–Cu–Ge ternary system [63 Kin]

Alloy composition, at.-%			Lattice spacings, nm		*Axial ratio*	*Microstructure*
Au	*Cu*	*Ge*	*a*	*c*	*c/a*	
0	88·2	11·8	0·258 24	0·422 32	1·635 4	ζ
2·0	86·2	11·8	0·259 04	0·423 64	1·635 4	ζ
4·5	83·7	11·8	0·260 05	0·425 27	1·635 3	ζ
9·0	79·2	11·8	0·261 86	0·428 18	1·635 2	ζ
16·0	72·2	11·8	—	—	—	ζ + other phase(s)
0	87·8	12·2	0·258 34	0·422 44	1·635 2	ζ
2·0	85·8	12·2	0·259 11	0·423 70	1·635 2	ζ
4·5	83·3	12·2	0·260 15	0·425 34	1·635 0	ζ
9·0	78·8	12·2	0·261 94	0·428 22	1·634 8	ζ
16·0	71·8	12·2	0·264 73	0·432 65	1·634 3	ζ
20·0	67·8	12·2	—	—	—	ζ + other phase(s)
0	87·3	12·7	0·258 47	0·422 47	1·634 5	ζ
1·0	86·3	12·7	0·258 88	0·423 13	1·634 5	ζ
2·0	85·3	12·7	0·259 29	0·423 80	1·634 4	ζ
3·0	84·3	12·7	0·259 69	0·424 44	1·634 4	ζ
4·5	82·8	12·7	0·260 30	0·425 42	1·634 3	ζ
9·0	78·3	12·7	0·262 12	0·428 30	1·634 0	ζ
12·0	75·3	12·7	0·263 30	0·430 18	1·633 8	ζ
16·0	71·3	12·7	0·264 90	0·432 74	1·633 6	ζ
20·0	67·3	12·7	0·266 48	0·435 24	1·633 3	ζ
25·0	62·3	12·7	—	—	—	ζ + other phase(s)
0	86·5	13·5	0·258 72	0·422 49	1·633 0	ζ
2·0	84·5	13·5	0·259 59	0·423 90	1·633 0	ζ
9·0	77·5	13·5	0·262 43	0·428 47	1·632 7	ζ
20·0	66·5	13·5	0·266 78	0·435 38	1·632 0	ζ
25·0	61·5	13·5	—	—	—	ζ + other phase(s)
0	85·0	15·0	0·259 23	0·422 47	1·629 7	ζ
4·5	80·5	15·0	0·261 17	0·425 55	1·629 4	ζ
9·0	76·0	15·0	0·262 99	0·428 40	1·629 0	ζ
20·0	65·0	15·0	—	—	—	ζ + other phase(s)

Au–Cu–H

The solubility of H in Cu_3Au drops sharply as the alloy goes through the disorder to order transformation at 390°C on cooling [70 Gol]. The solubility data is reproducible on heating the ordered alloy through its order–disorder transformation. Solubilities were calculated from measured hydrogen permeabilities and diffusion coefficients during step cooling from 625 to 275°C at an average rate of 5°C h^{-1}, and also during step heating. Solubilities found by isothermal anneals at 350°C for 25–60 h were little different than those found by step cooling and step heating. Fig. 180 shows that the solubility of H in Cu_3Au is exponentially dependent on temperature above and below 390°C. Vykhodets *et al.* [70 Vyk] give a theoretical treatment for H solubility changes in the presence of order–disorder transformations.

REFERENCES

70 Gol: V. A. Gol'tsov, V. B. Vykhodets, P. V. Gel'd and T. A. Krylova: *Fiz. Metallov Metalloved*, 1970, **30**, 657–659

70 Vyk: V. B. Vykhodets, V. A. Gol'tsov and P. V. Gel'd: *Ukrain. Fiz. Zhur.*, 1970, **15**, 107–110

Au–Cu–In

King *et al.* [63 Kin] outlined the extent of the hcp ζ phase field at 475°C (Fig. 181). Alloys were prepared from spectroscopically pure elements by melting under He in transparent quartz tubing. They were annealed for 14 days at 475°C and quenched. Metallography was used to characterize single phase ζ compositions and these 12 ternary alloys, with four binary Au–In ζ phase alloys, were examined by X-ray powder diffraction to determine the lattice parameters and axial ratios (Table 64). Powder samples were requenched from 475°C and lattice spacings measured at 30°C. Alloy compositions that contained phases additional to ζ were not analysed by X-ray powder diffraction but were used to locate the approximate limit of the ternary ζ phase region at 475°C.

REFERENCE

63 Kin: H. W. King, T. B. Massalski and L. L. Isaacs: *Acta Metall.*, 1963, **11**(12), 1355–1361

Au–Cu–Mg

Raub and Walter [50 Rau] studied the effect of ternary additions on the order–disorder transformation of AuCu. Addition of 6·6 at.-% Mg has little effect on the transformation temperature of binary AuCu, 410°C. Heat treatment of the ternary alloy at 400°C showed a disordered fcc lattice after 25 h but this transformed to the CuAuI tetragonal superlattice after 50 h at 400°C.

REFERENCE

50 Rau: E. Raub and P. Walter: *Z. Metall.*, 1950, **41**, 240–243

Au–Cu–Mn

Raub and Walter [50 Rau] studied the effect of ternary additions on the order-disorder transformation of AuCu. Addition of 12·9 at.-% Mn strongly depressed the transformation temperature of AuCu, 410°C. After 300 h heat treatment at 360°C the ternary alloy had a disordered fcc structure, and after 500 h at 300°C the ternary alloy had transformed to the CuAuI tetragonal superlattice.

REFERENCE

50 Rau: E. Raub and P. Walter: *Z. Metall.*, 1950, **41**, 240–243

Au–Cu–Ni
(G. V. Raynor)

INTRODUCTION

The system Au–Cu–Ni is of special importance in the jewellery industry, in which, as 'white gold', it is used as a substitute for platinum in setting precious stones. In this context the system has been briefly reviewed [78 Mac] with special reference to alloys containing 75 wt-% Au (18 carat), 58·3 wt-% Au (14 carat) and 41·7 wt-% Au (10 carat). Experimental work exists for the liquidus surface of the system [14 Ces] and the solid state has been systematically examined [47 Rau 1, 2).

BINARY SYSTEMS

The Au–Cu phase diagram, as recently assessed [87 Oka 1], is accepted.

The assessment of the Au–Ni phase diagram [87 Oka 2] is accepted. The published Cu–Ni diagram [73 Met] is accepted.

THE SOLID PHASES

The solid phases involved in the equilibria are the ternary solid solution (Au–Cu–Ni) and phases based upon the ordered structures at the compositions $AuCu_3$, AuCu, and Au_3Cu. The crystal structures in the solid state are summarized in Table 63.

Table 63. Crystal structures

Solid phase	*Prototype*	*Lattice designation*
(Au–Cu–Ni)	Cu	cF4
$AuCu_3$	$AuCu_3$	cP4
AuCuI	AuCuI	tP4
AuCuII	AuCuII	tP4 faulted

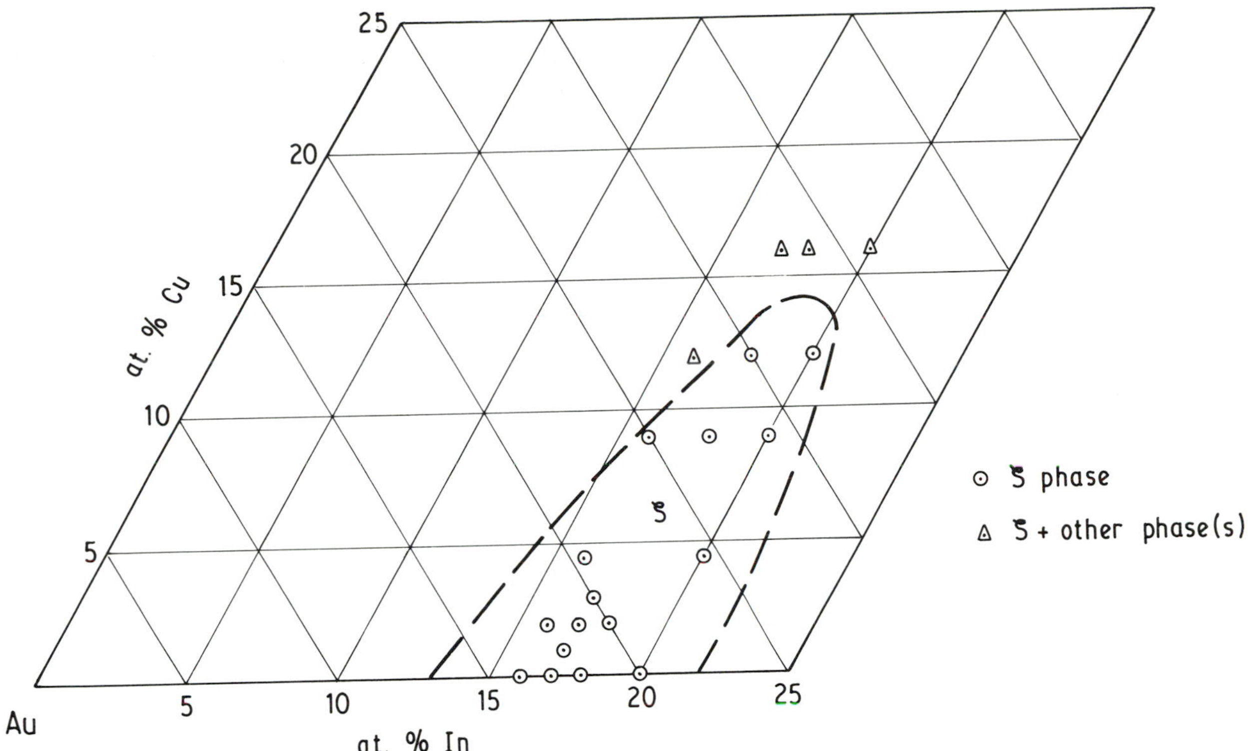

181 The ζ phase region at 475°C

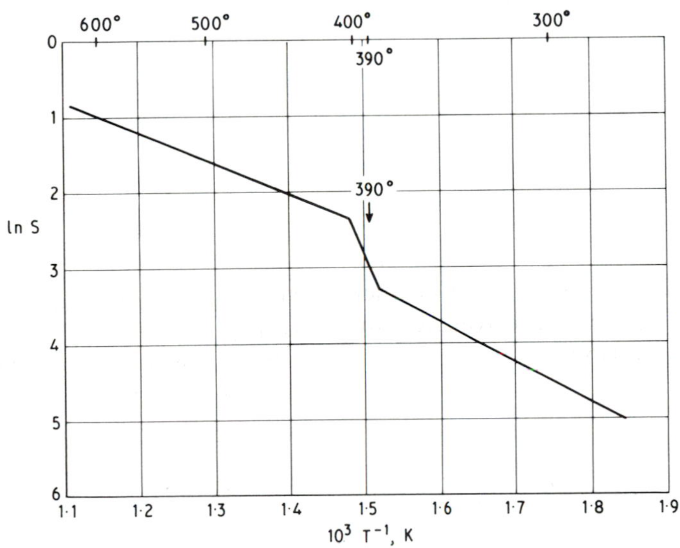

180 Temperature dependence of the solubility of H in Cu_3Au [70 Gol]; (S = $cm^3(H_2)/cm^3$ alloy $atm^{\frac{1}{2}}$)

Table 64. Lattice parameters and axial ratios of hcp ζ phase alloys in the Au–Cu–In ternary system

Alloy composition, at.-%			*Lattice spacing*, nm		*Axial ratio*	*Microstructure*
Au	*Cu*	*In*	*a*	*c*	*c/a*	
84	0	16	0·290 53	0·478 82	1·648 1	ζ
82	2	16	0·290 06	0·478 05	1·648 1	ζ
79·5	4·5	16	0·289 45	0·477 04	1·648 1	ζ
75	9	16	0·288 36	0·475 25	1·648 1	ζ
72	12	16	—	—	—	ζ + other phase(s)
83	0	17	0·290 74	0·479 14	1·648 0	ζ
82	1	17	0·290 50	0·478 74	1·648 0	ζ
81	2	17	0·290 24	0·478 33	1·648 0	ζ
80	3	17	0·290 00	0·477 92	1·648 0	ζ
67	16	17	—	—	—	ζ + other phase(s)
82	0	18	0·290 96	0·479 27	1·647 2	ζ
80	2	18	0·290 48	0·478 45	1·647 1	ζ
73	9	18	0·288 79	0·475 52	1·646 6	ζ
70	12	18	0·288 07	0·474 22	1·646 2	ζ
66	16	18	—	—	—	ζ + other phase(s)
80	0	20	0·291 59	0·479 23	1·643 5	ζ
75·5	4·5	20	0·290 51	0·477 34	1·643 1	ζ
71	9	20	0·289 44	0·475 43	1·642 6	ζ
68	12	20	0·288 70	0·474 10	1·642 2	ζ
64	16	20	—	—	—	ζ + other phase(s)

THE TERNARY SYSTEM

The only systematic investigation of the liquidus surface of the Au–Cu–Ni system was carried out as long ago as 1914 [14 Ces]. Although at this date the liquid–solid relationships in the Au–Cu and Cu–Ni systems were generally understood, the Au–Ni phase diagram was thought to exhibit a eutectic relationship between the Au- and Ni-rich solid solutions, whereas the currently accepted diagram shows that the miscibility gap closes in the solid state at a sub-solidus temperature. Thus the liquidus surface for the ternary alloys can contain no monovariant reaction lines, and it is necessary to reinterpret the results of [14 Ces]. Fifty-six ternary alloys were investigated by thermal analysis, and though no experimental details were given, the liquidus temperatures were recorded in tabular and graphical form. Figure 182 shows liquidus temperature contours (using atomic percentages) which are based on the results in [14 Ces] for the ternary alloys, but drawn to intersect the binary axes at compositions consistent with the currently accepted binary diagrams. The experimental contours are satisfactorily consistent with the liquidus curves of the Au–Ni and Cu–Ni systems, but not with that of the Au–Cu system, in which a minimum liquidus temperature of 886°C was assumed, instead of the 910°C minimum currently accepted. The isotherm at 900°C in the region of the Au–Cu axis has therefore been omitted in Fig. 182. The contours for 1 100°C and above extend smoothly from the Au–Ni axis to the Cu–Ni axis. Below this temperature some curvature develops, and a pronounced valley links the liquidus minima in the Au–Ni and Au–Cu systems. The solid which separates over the whole composition range is the ternary (Au–Cu–Ni) fcc solid solution. The results in [14 Ces] are insufficiently detailed to permit any representation of the solidus.

The solid state was investigated by Raub and Engel [47 Rau 1] with special reference to the two-phase region arising from the miscibility gap in the Au–Ni system. High purity components (Au 99·99% pure, electrolytic Cu, and carbonyl nickel) were melted together in hydrogen or argon. After homogenization, alloys were re-annealed at various temperatures and examined by X-ray powder diffraction methods. Lattice spacings were first determined for the quenched homogeneous ternary solid solution; spacings were then determined for the phases present in two-phase alloys at lower temperatures. Difficulty in attaining equilibrium was noted, and long annealing, up to ¾–1 year at 300°C, was necessary. From the measured spacings the boundary of the two-phase region was deduced, together with tie lines in certain cases. The boundary is shown in Fig. 183 for 100°C temperature steps from 300 to 900°C, and for 950°C. Two features are of interest. The addition of Cu to the Au–Ni system causes an increase in the width of the miscibility gap characteristics of the latter, and the boundary approaches very close to the Cu–Ni axis, though without touching it. Also the critical temperature at which the gap closes in the binary system (810·3°C) is raised by Cu, leading to a domed volume in the ternary model. It is estimated in [47 Rau 1] that the highest critical temperature attained is ~970°C at Au 20 at.-%, Cu 20 at.-%, and Ni 60 at.-%. Direct determination of tie lines indicates that they tend to fan out from the nickel-rich region of the isothermal diagram towards the Au–Cu axis. Thus Figs. 184, 185, and 186 represent the conditions at 400, 700, and 900°C respectively. The approximate critical compositions on the various isothermal boundaries at which the miscibility gap closes, and therefore at which the respective tie lines vanish, are shown in Fig. 183 as open circles.

The results of less detailed work [69 Pop] on nine alloys containing 70, 80, and 90 at.-% Ni is broadly in agreement with Fig. 183.

ORDERED STRUCTURES

The influence of Ni on the ordered structures based on the compositions AuCu and $AuCu_3$ was studied [47 Rau 2] at temperatures of 360 and 400°C, both with and without the

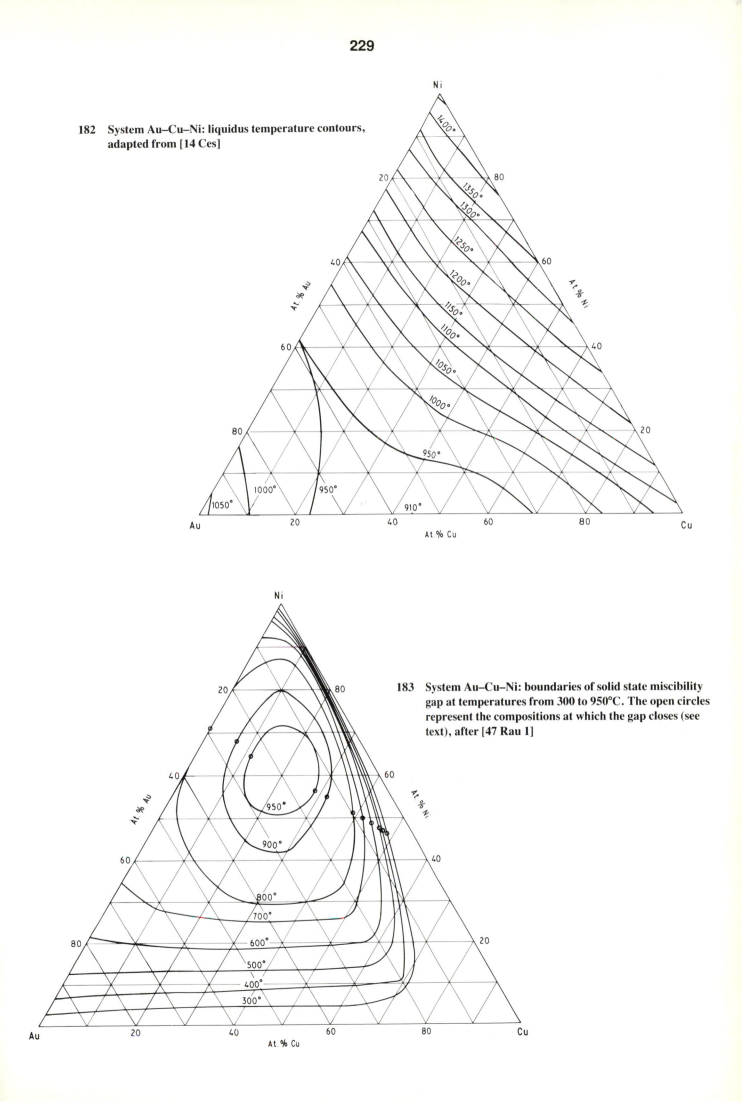

182 **System Au–Cu–Ni: liquidus temperature contours, adapted from [14 Ces]**

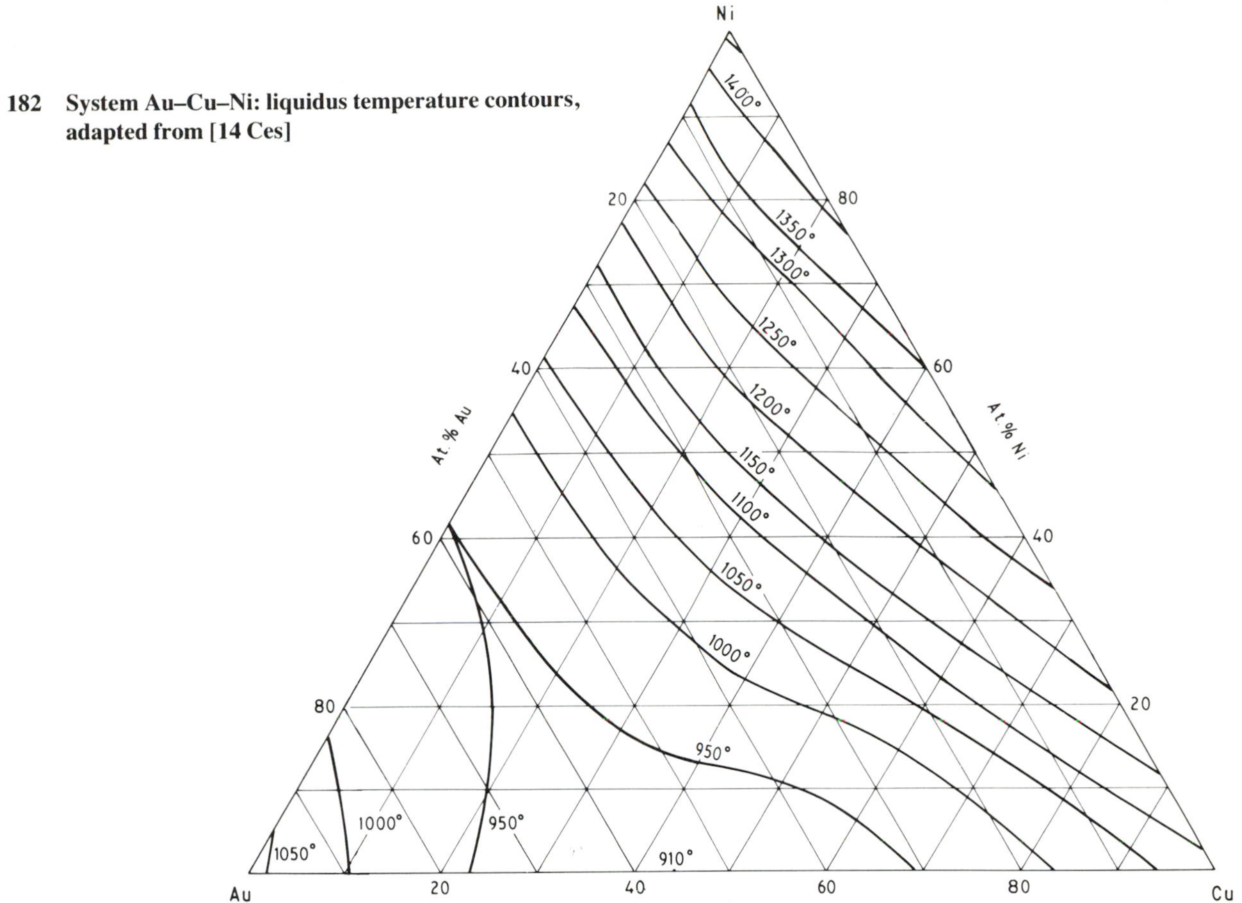

183 **System Au–Cu–Ni: boundaries of solid state miscibility gap at temperatures from 300 to 950°C. The open circles represent the compositions at which the gap closes (see text), after [47 Rau 1]**

184 System Au–Cu–Ni: boundary of miscibility gap at 400°C, with tie lines, after [47 Rau 1]

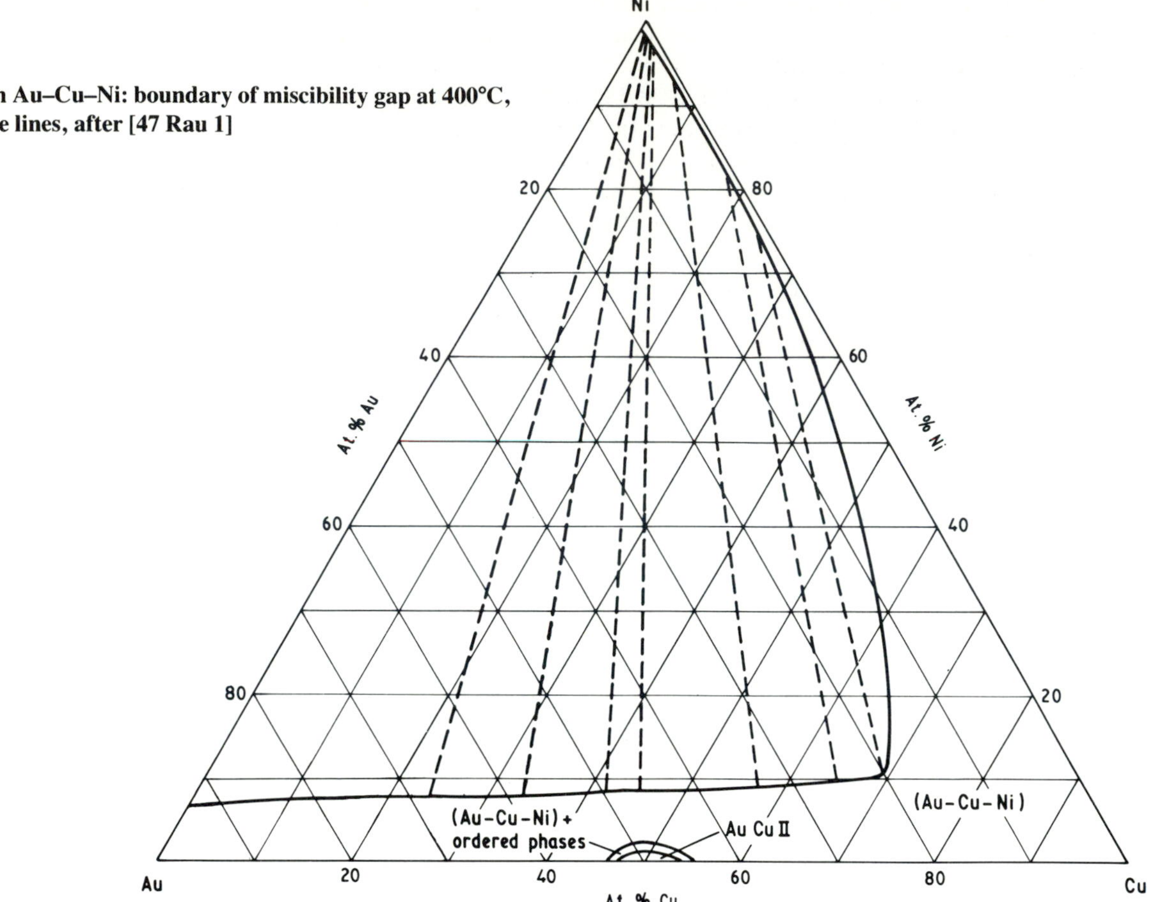

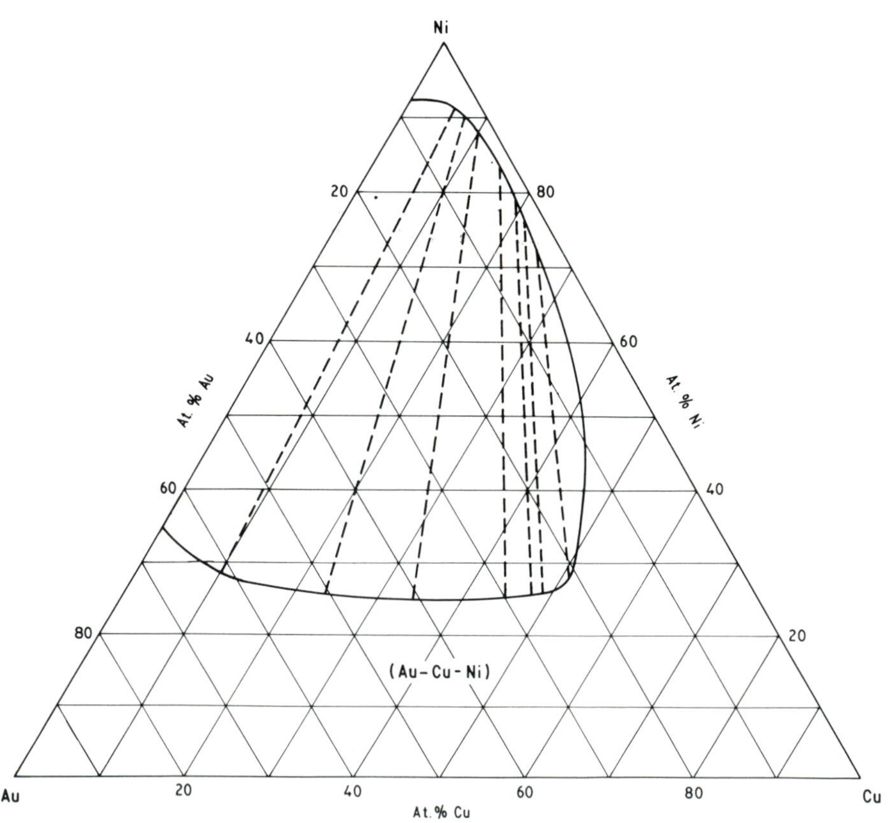

185 System Au–Cu–Ni: boundary of miscibility gap at 700°C, with tie lines, after [47 Rau 1]

application of cold work after the homogenization treatment but before final annealing. The results at a given temperature varied considerably with time of annealing and the use or neglect of cold-working, which was found to accelerate ordering. The degree of tetragonality of the ordered AuCuI phase observed was always greater after cold work than without it. In the case of AuCuI the experiments indicated that order persisted up to the limit of the homogeneous solid solution area in the region close to 50 at.-% Cu. Later work to be considered below, however, indicates strongly that this is unlikely, and in Fig. 184 the ordered structure stable at this temperature (AuCuII) is shown as restricted to a small composition region, separated from the (Au–Cu–Ni) solid solution phase field by two-phase regions in which ordered and disordered phases may coexist, and which is considered in more detail later. The $AuCu_3$ phase is not stable at 400°C; at 360°C however it was observed by [47 Rau 2], who again suggested that the ordering persisted to the solubility boundary in the region of 75 at.-% Cu. Though no later evidence exists on this point, it is considered unlikely. In view of the variability of the results in [47 Rau 2] it would appear possible that equilibrium may not in all cases have been achieved. The results obtained for two-phase alloys, however, largely confirmed the miscibility gap boundaries previously established (Fig. 183).

The influence of Ni on the binary Au–Cu ordered structures in the region of the equiatomic composition has been examined in a series of researches by Gantois and his colleagues, the results of which may now be summarized.

Working with four alloys of the series $Au_{50}Cu_{50-x}Ni_x$, [63 Gan 1] showed that in specimens homogenized for 10 days at 800°C and subsequently annealed at 150°C for as long as 3000 h, the degree of long range order remained at unity for Ni contents up to 12·5 at.-%, indicating complete order. Fig. 183, however, suggests that at this temperature an alloy of the composition quoted should contain the Ni-rich phase, and it may be that although the equilibrium degree of order was reached, final constitutional equilibrium was not attained. In a subsequent X-ray diffraction study of the variation of the degree of order in quenched samples with quenching temperature [63 Gan 2] it was pointed out that the substitution for copper of 3 at.-% Ni was sufficient to eliminate the orthorhombic AuCuII ordered structure from the equilibria, so that above this limit, the ordered structure formed on cooling was the tetragonal ordered phase AuCuI. By comparing the intensities of superlattice and fundamental diffraction lines present together, it was shown that a two-phase region existed between the disordered tenary solid solution and AuCuI at Ni contents exceeding ~3 at.-%, again referring to the 50 at.-% Au section. Experiments on the influence of quenching rate on the results indicated that a sufficiently rapid quench was attained, while simultaneous high temperature X-ray diffraction and differential thermal analysis for an alloy containing 4 at.-% Ni were consistent with the results on quenched alloys [65 Gan 1, 2]. In more detailed work [68 Gan 1, 2; 68 Fai] using materials of purity not less than 99·99%, alloys of the series $Au_{50}Cu_{50-x}Ni_x$ were melted *in vacuo*, homogenized and powdered, and examined under high temperature X-ray diffraction techniques or room temperature X-ray work on samples quenched from various temperatures at a rate of ~300°C s^{-1}. An equilibrium diagram was established for the 50 at.-% Au section by observing the appearance or disappearance of the prominent diffractions of the phases concerned. Enthalpy of formation measurements, using liquid tin calorimetry, was also carried out on alloys equilibrated at various temperatures and quenched. The resulting phase diagram is shown in Fig. 187, and indicates that the equilibria are complex below ~3 at.-% Ni, but involve only the disordered phase and AuCuI above this limit. The critical temperature for the onset of order is decreased by the substitution of Ni for Cu; at 400°C the alloys containing more than 2·4 at.-% Ni are disordered, and this fact has been used in Fig. 184. The nature of the polythermal diagram at 50 at.-% Au suggests that the equilibria in the region of the AuCu-type ordered phases are as illustrated schematically on a larger scale in Fig. 188.

It may be noted that according to [68 Gan 1, 2; 68 Fai] a two-phase region exists between the disordered and ordered phases at both the stoichiometric compositions AuCu and $AuCu_3$ in the binary alloys; the thermodynamics of this type of transformation have been considered by [66 Her]. This is, however, in conflict with the accepted Au–Cu phase diagram, and in Fig. 187 it has been assumed that no two-phase regions exist between the disordered solid solution and the ordered AuCuII phase, or between AuCuII and AuCuI, at the exact stoichiometric compositions.

THERMODYNAMIC REVIEWS

The system has been examined thermodynamically by [78 Tar] and by [77 Don] using somewhat different methods. Both claim reasonable agreement with [47 Rau 1]. The agreement in [77 Don] is best for the Ni-rich corner, and appreciable divergences occur as the Au and Cu contents are increased. The miscibility gap contours deduced by [78 Tar] are, however, in very good agreement with the experimental results summarized in Fig. 183, except that at certain temperatures they actually intersect the Cu–Ni axis, whereas experimentally the miscibility gap does not exist in the binary Cu–Ni alloys. Both analyses confirm the tendency towards a widening of the Au–Ni miscibility gap with the addition of Cu, which is a prominent feature of Fig. 183.

REFERENCES

14 Ces: P. de Cesaris: *Gazz Chim. Ital.*, 1914, **44**(i), 27–35

47 Rau 1: E. Raub and A. Engel: *Z. Metall.*, 1947, **38**, 11–16

47 Rau 2: E. Raub and A. Engel: *Z. Metall.*, 1947, **38**, 147–158

63 Gan 1: M. Gantois: *Compt. rend.*, 1963, **256**, 3629–3631

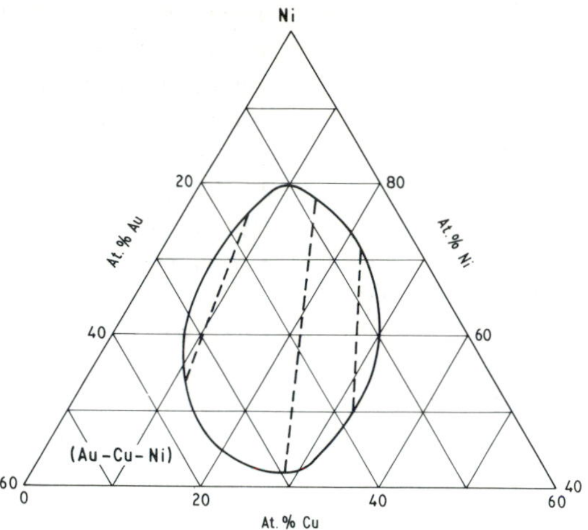

186 **System Au–Cu–Ni: boundary of miscibility gap at 900°C, with tie lines, after [47 Rau 1]**

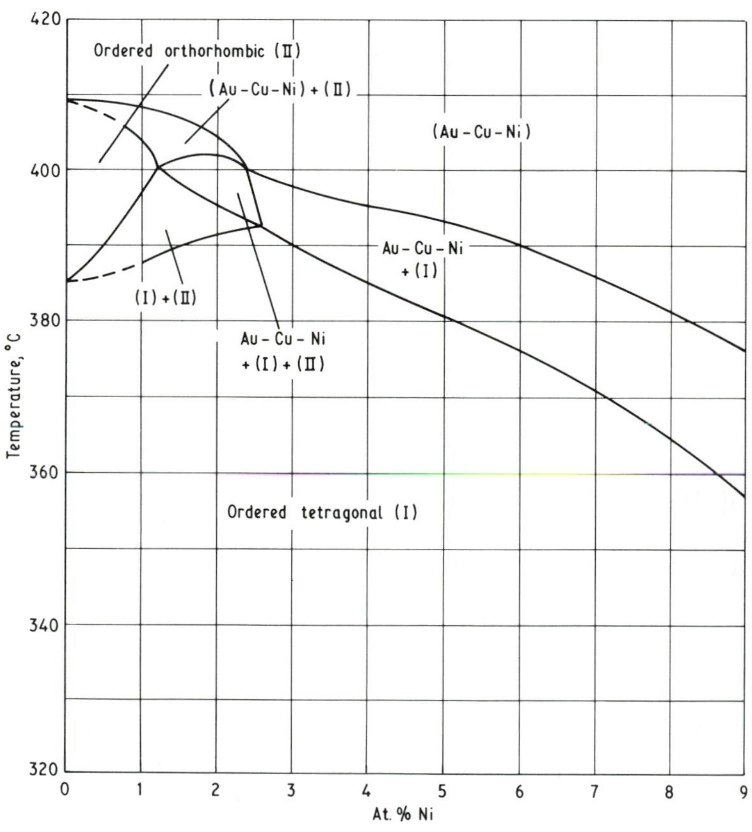

187 **System Au–Cu–Ni: polythermal diagram for section at 50 at.-%, after [68 Gan 1, 2] [68 Fai]**

63 Gan 2: M. Gantois: *Compt. rend.*, 1963, **257**, 2104–2107

65 Gan 1: M. Gantois, A. Pianelli and R. Faivre: *Compt. rend.*, 1965, **260**, 3643–3646

65 Gan 2: M. Gantois: *Compt. rend.*, 1965, **261**, 1543–1546

66 Her: J. Hertz: *Compt. rend.*, 1966, **263C**, 363–366

68 Gan 1: M. Gantois: *Mem. Sci. Rev. Met.*, 1968, **65**, 129–139

68 Fai: R. Faivre, J. Hertz and M. Gantois: *Mater. Res. Bull.*, 1968, **3**, 661–670

68 Gan 2: M. Gantois: *J. Appl. Cryst.*, 1968, **1**, 263–271

69 Pop: I. Pop, L. Maxim and I. Maxim: *Stud. Univ. Babes-Bolyai, Ser. Math.-Phys.*, 1969, **14**, 113–118

73 Met: Metals Handbook: 'Metallography, Structures and Phase Diagrams', Vol. 8, 294, 8th edn.; 1973, ASM

77 Don: Dong Nyung Lee: *J. Korean Inst. Metals*, 1977, **15**, 564–573

78 Tar: S. K. Tarby, C. J. Van Tyne and M. L. Buyle: *NBS Special Publ.* (WS), No. 496, Vol. 2, 726–743; 1977 (published 1978)

78 Mac: A. S. Macdonald and G. H. Sistare: *Gold Bull.*, 1978, **11**(4), 128–131

87 Oka 1: H. Okamoto, D. J. Chakrabarti, D. E. Laughlin and T. B. Massalski: *Bull. Alloy Phase Diagrams*, 1987, **8**, 454–474

87 Oka 2: H. Okamoto and T. B. Massalski: 'Phase Diagrams of Binary Gold Alloys', 193–208; 1987, ASM International

Au–Cu–Pb

The only phase diagram work on the Au–Cu–Pb ternary system was reported by [55 Ura]. The original publication was not available and reliance has had to be placed on a short description of the result in [55 Age] showing the liquidus projection and an evaluation of the ternary system by [79 Cha]. The system was constructed from an examination by thermal analysis, using cooling runs, of three sections from the Au–Pb binary system to Cu. Supplementary metallographic study was undertaken. [79 Cha] noted that the vertical sections appear to be incorrect in some regions but do not elaborate on this statement. The liquidus surface (Fig. 189), shows that the liquid immiscibility gap in the Cu–Pb binary system extends to 9·4 at.-% Au in the ternary system and closes at a lower critical point c at 9 Au, 21 at.-% Pb and 883°C. The Au–Pb invariant reactions at p_1 . p_2 p_3 and e_2 [83 Eva] do not appear to extend into the ternary system so that the ternary invariant reactions U_1 (L + Au_2Pb = $AuPb_2$ + (Au–Cu)), U_2 (L + $AuPb_2$ = (Au–Cu) + $AuPb_3$) and E (L = $AuPb_3$ + (Pb) + (Au–Cu)) occur at essentially the same temperatures as the binary reactions p_2, p_3 and e_2 respectively. The (Au–Cu) phase is the primary phase to separate over virtually the entire ternary composition range. It should be noted that the $AuPb_3$ phase was not recognized by [55 Ura] and therefore not included in the ternary equilibria. The order-disorder temperature for the formation of AuCu II is slightly lowered by the addition of Pb. No further information was reported on the solid-state equilibria.

REFERENCES

55 Ura: G. G. Urazov and A. V. Vanyukov, *Sborn Nauch Trudov, Moskov Inst Tsvet Met i Zolota*, 1955 (25), 112–124

55 Age: N. V. Ageev, ed: 'Diagrammy Sostoyaniya Metallicheskikh Sistem', 1955, Publ. 1959, **I**, 111

79 Cha: Y. A. Chang, J. P. Neumann, A. Mikula and D. Goldberg, 'Phase Diagrams and Thermodynamic Properties of Ternary Copper–Metal Systems', INCRA Monograph VI, 1979, pp. 276–279. Int. Copper Res. Assoc., Inc., New York.

83 Eva: D. S. Evans and A. Prince, *Mater. Res. Soc. Symp. Proc.*, 1983, **19**, 383–388

Au–Cu–Pd
(G. V. Raynor)

INTRODUCTION

The major interest with regard to the gold–copper–palladium system lies in the understanding of the effect of copper and palladium on the ordered structures which exist in the Cu–Pd and Au–Cu binary systems respectively. The system has been investigated [49 Nem, 55 Rau] by methods described below, while the regions in which ordered structures are found have also been examined [55 Sch, 65 Nag, 66 Nag, 79 Nak]. [86 Udo] has calculated a 310°C isothermal section of the coherent phase diagram (in which all structures are derived from an fcc lattice) using the cluster variation method. Other contributions are noted below.

BINARY SYSTEMS

The Au–Cu phase diagram as assessed [87 Oka] is accepted. The assessment of the Au–Pd system by [85 Oka] is accepted, and also that of the Cu–Pd system by [58 Han].

THE SOLID PHASES

No distinct ternary phases have been reported in the system Au–Cu–Pd. The structures of the phases discussed above are summarized in Table 65. The coherent phase diagram [86 Udo] exhibits ternary ordered phases $AuCu_2Pd$ and Au_2CuPd, as detected in thin film samples by [66 Nag].

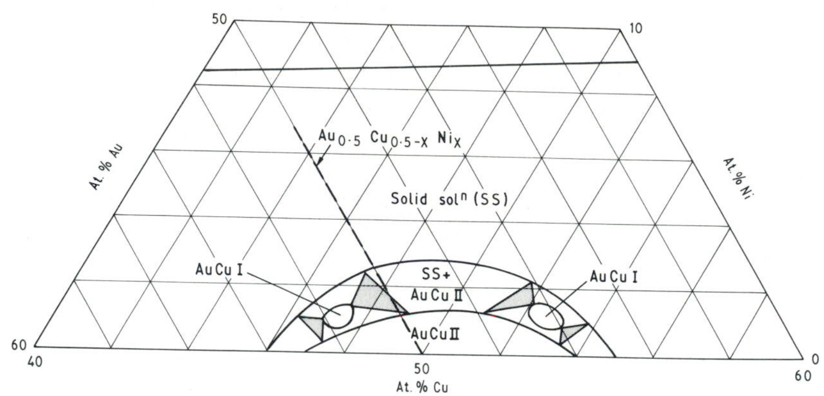

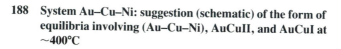

188 System Au–Cu–Ni: suggestion (schematic) of the form of equilibria involving (Au–Cu–Ni), AuCuII, and AuCuI at ~400°C

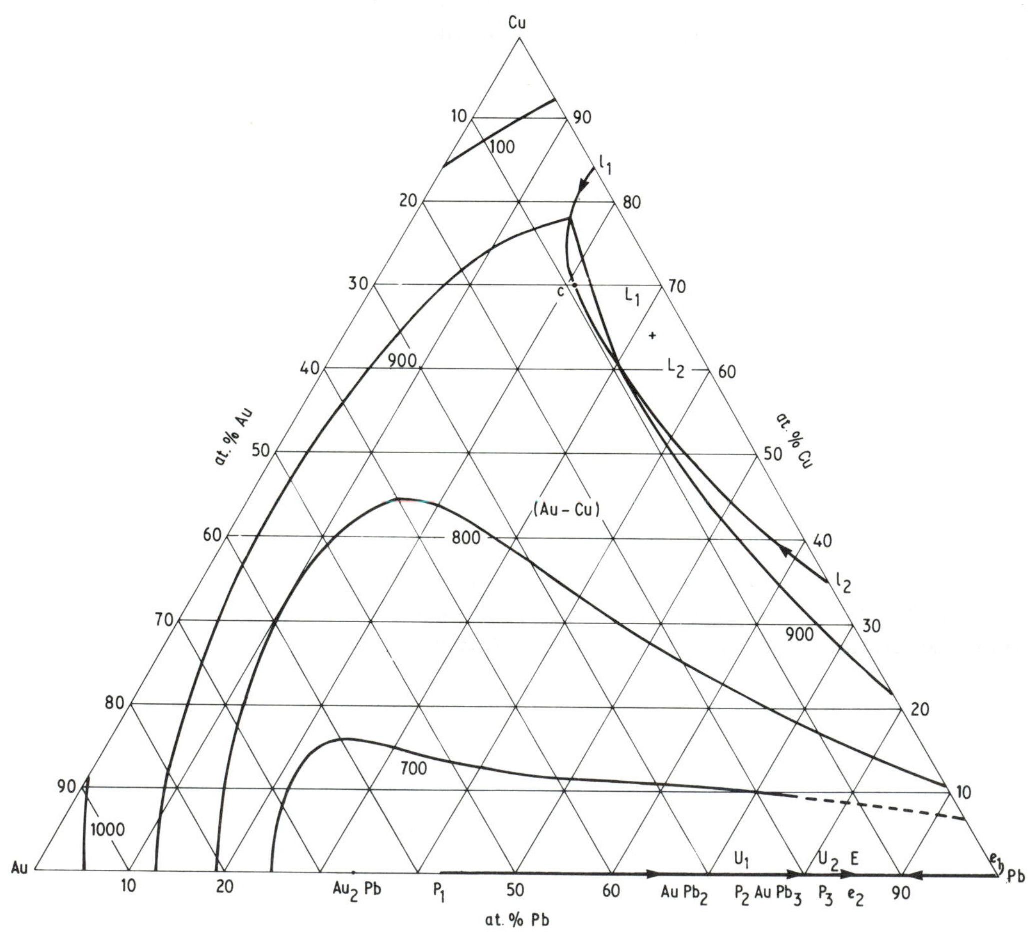

189 Liquidus surface of the Au–Cu–Pb ternary system

Table 65. Crystal structures

Solid phase	*Prototype*	*Lattice designation*
Au–Cu–Pd	Cu	cF4
$AuCu_3$	$AuCu_3$	cP4
AuCuI	AuCuI	tP4
AuCuII	AuCuII	tP4 faulted
Cu_3Pd (α')	$AuCu_3$	cP4
Cu_3Pd (α'')	—	cP4 faulted
CuPd	CsCl	cP2

THE TERNARY SYSTEM

A comprehensive survey of the ternary system was carried out by [49 Nem] using the techniques of thermal analysis, electrical resistance, microhardness, and microscopic observation on alloys prepared from pure materials. The thermal analysis results are the most significant, those for the application of the other techniques serving mainly to confirm the interpretation of the thermal analysis. Forty-five alloy compositions were studied, lying on composition sections at 10, 20, 30, 40, 50, 60, 70, 80, and 90 at.-% Cu.

Liquidus

In the thermal work at the higher temperatures the liquidus arrests only were recorded; no solidus results are available from [49 Nem]. From these liquidus arrests a contour diagram may be constructed, and is shown in Fig. 190. In conformity with the relevant binary diagrams the liquidus surface indicates the existence of a complete series of solid solutions between Au, Cu, and Pd. Figure 190 has been drawn to be consistent with the liquidus temperatures in the binary systems. In general the results of [49 Nem] are in good agreement with the binary diagrams adopted. Only in one case (1 000°C contour) is there any discrepancy, and in this case the contour has been slightly modified to cut the Au–Cu axis at a somewhat smaller percentage of Cu than would be indicated by the work of [49 Nem]. Very limited work on the ternary system by [62 Rhy] has demonstrated that the liquidus temperatures for two alloys (44·6 at.-% Au, 49·5 at.-% Cu, and 27·7 at.-% Au, 57·2 at.-% Cu) are consistent with the contours of Fig. 190, but more importantly show only small liquidus–solidus separations of 27 and 33°C respectively. It is therefore probable that the undetermined solidus surface lies close to the liquidus surface.

The solid state

The thermal analysis of alloys lying on the composition sections at 70, 75, and 80 at.-% Cu showed arrests in the solid state [49 Nem]. These were interpreted as due to an ordering reaction corresponding to the formation of the binary ordered structures $AuCu_3$ and Cu_3Pd. Since these phases have either the same or very closely related crystal structures, complete solid solution formation would be expected and the results are consistent with this. In each section the transformation curves rise from those characteristic of the binary ordering reactions to maxima of 596°C for the 80 at.-% Cu section, and 630°C for the 75 and 70 at.-% Cu sections, within the ternary system. These maxima occur for approximately equal atomic proportions of Au and Pd. The ordering temperatures in the Au–Cu and Cu–Pd systems are therefore raised by the addition of Pd and Au respectively. In the 60 at.-% Cu section the ordering temperature for the AuCu structure is very sharply raised by the addition of Pd, and in the 50 at.-% Cu section two maxima are clearly revealed, corresponding to the presence of ordered structures derived from AuCu and CuPd respectively, which, since their crystal structures differ, are not expected to form mutual solid solutions. At 40 at.-% Cu only the formation of the ordered phase derived from AuCu was observed; at 30, 20, and 10 at.-% Cu no transformations in the solid state were recorded.

For the other experiments in the solid state alloys were annealed *in vacuo* at 900–1 000°C for 10 days, and then either quenched or annealed for 5 days at 600°C, 3 days at 500°C, and 2 days at 300–400°C before furnace cooling. The results from quenched alloys were typical of uninterrupted solid solutions, while the furnace-cooled alloys showed results in conformity with the presence of the ordered structures.

A detailed re-examination of the relationships between AuCu and CuPd was carried out by [55 Rau], using X-ray diffraction, conventional metallography, electrical resistance measurements, and thermal expansion experiments. Alloys were melted from chemically pure metals under hydrogen and homogenized *in vacuo*. Homogenized samples were filed for X-ray experiments and the samples brought to equilibrium by annealing at the desired temperatures for 10–12 days in evacuated quartz tubes. Alloys for other experiments were appropriately annealed. The constitutional work concentrated on isothermals from 350 to 550°C at 50°C intervals, and diagrams were given, in weight percentages, for 550, 450 and 350°C. These are reproduced in atomic percentages in Figs. 191, 192, and 193. At 550°C (Fig. 191) the ordered structure based on CuPd projects into the ternary diagram. The two-phase region between disordered α and ordered CuPd is confirmed. There is also an area within which the alloys correspond with the ordered AuCu structure, surrounded by an area within which the disordered and ordered alloys coexist, although the ordered structure is not stable at this temperature in the binary Au–Cu alloys. This agrees with the observation of [49 Nem] that the ordering temperature for AuCu is raised by Pd. According to [55 Rau] the maximum ordering temperature (about 600°C) occurs close to the composition $Au_3Pd_2Cu_5$.

At 450°C AuCu is still not stable in the binary alloys, and consequently the ordered AuCu area in the ternary diagram, though considerably enlarged, does not extend to the binary Au–Cu axis. The homogeneity range of CuPd is moderately enlarged. The important feature is the establishment of a two-phase (AuCu + CuPd) equilibrium, with the associated three-phase triangles. The form of the 350°C isothermal is very similar, except that, since AuCu is stable in the binary Au–Cu system at this temperature, the corresponding ternary area makes contact with the binary Au–Cu axis. The results shown in Figs. 191, 192, and 193 allow the construction of a vertical

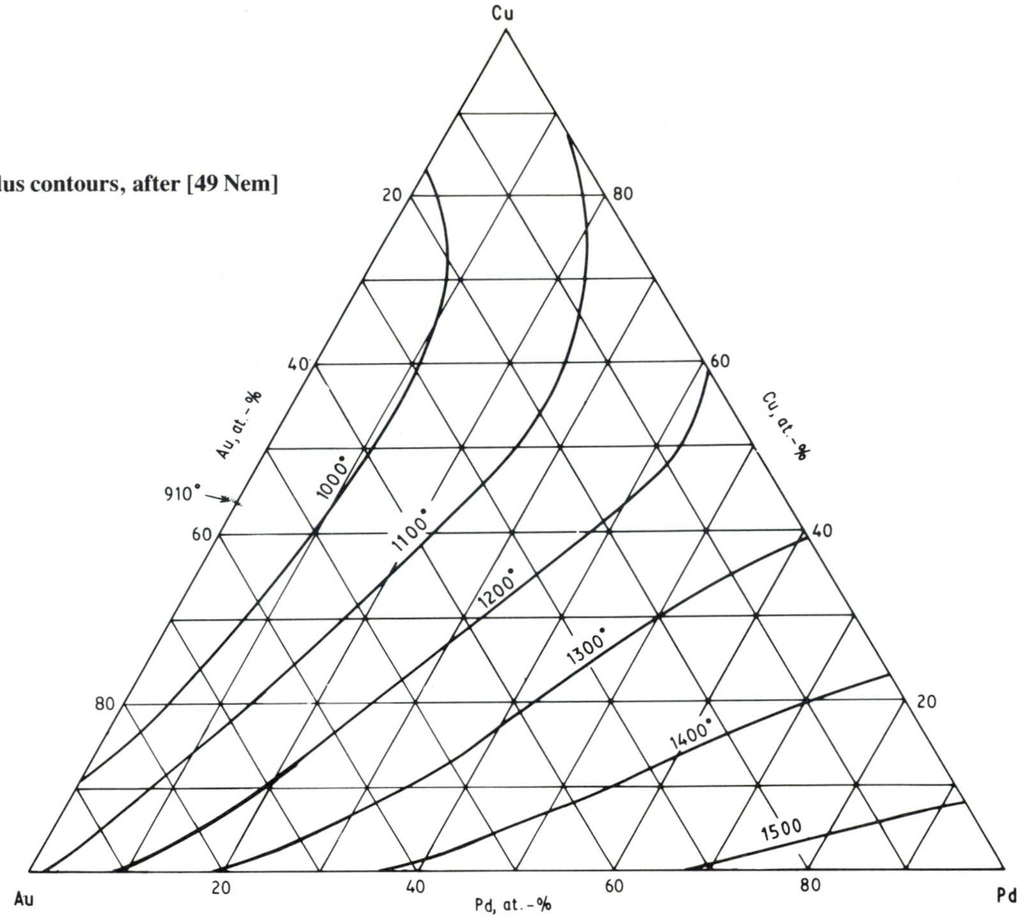

190 System Au–Cu–Pd: liquidus contours, after [49 Nem]

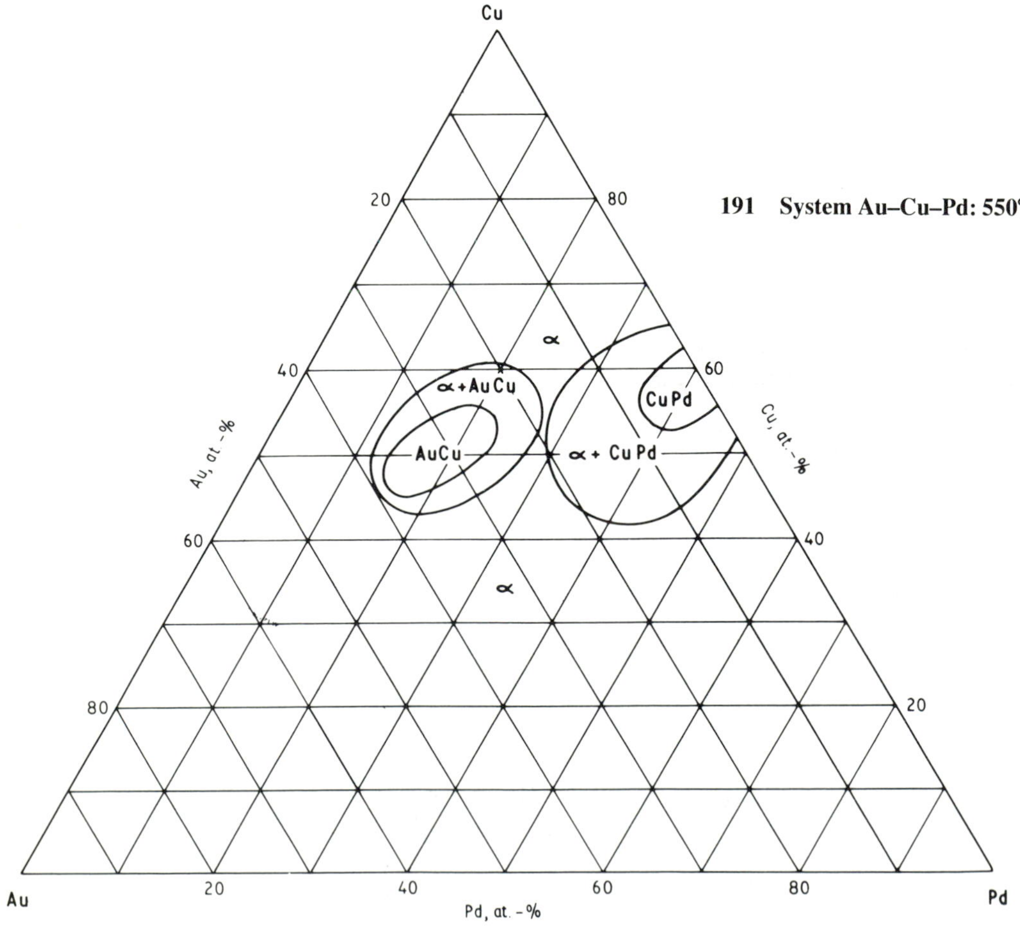

191 System Au–Cu–Pd: 550°C isothermal, after [55 Rau]

192 System Au–Cu–Pd: 450°C isothermal, after [55 Rau]

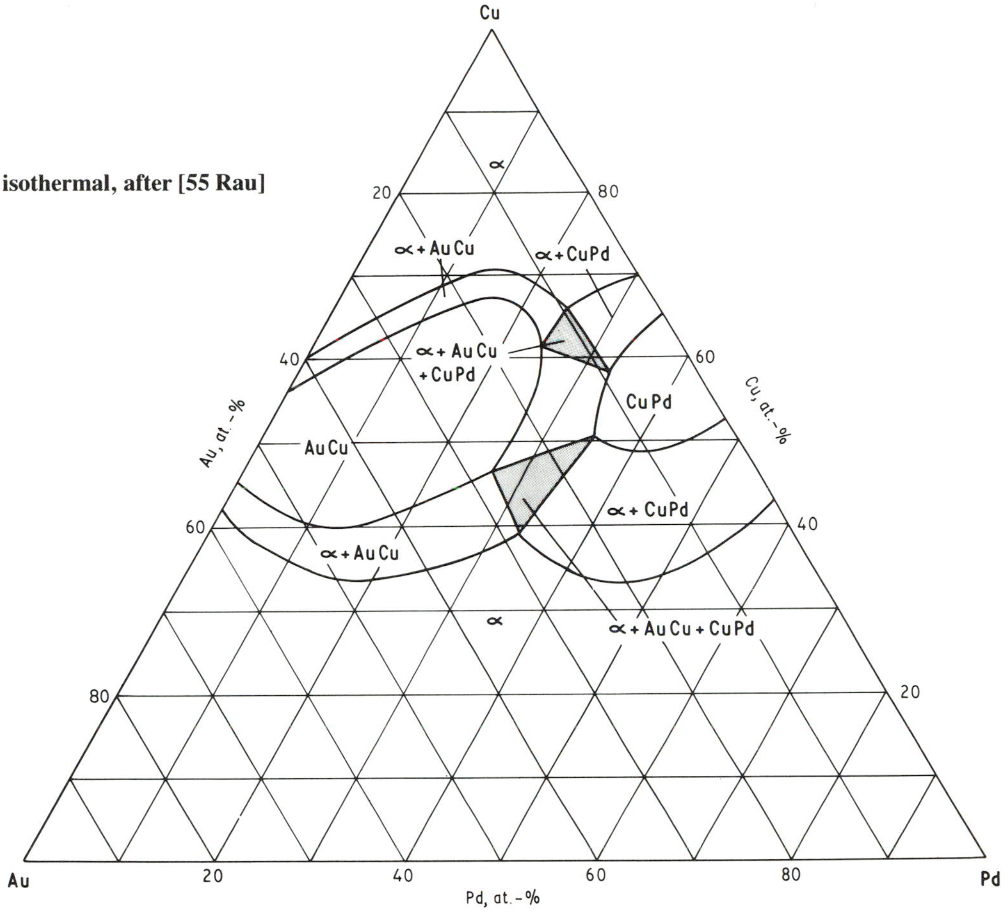

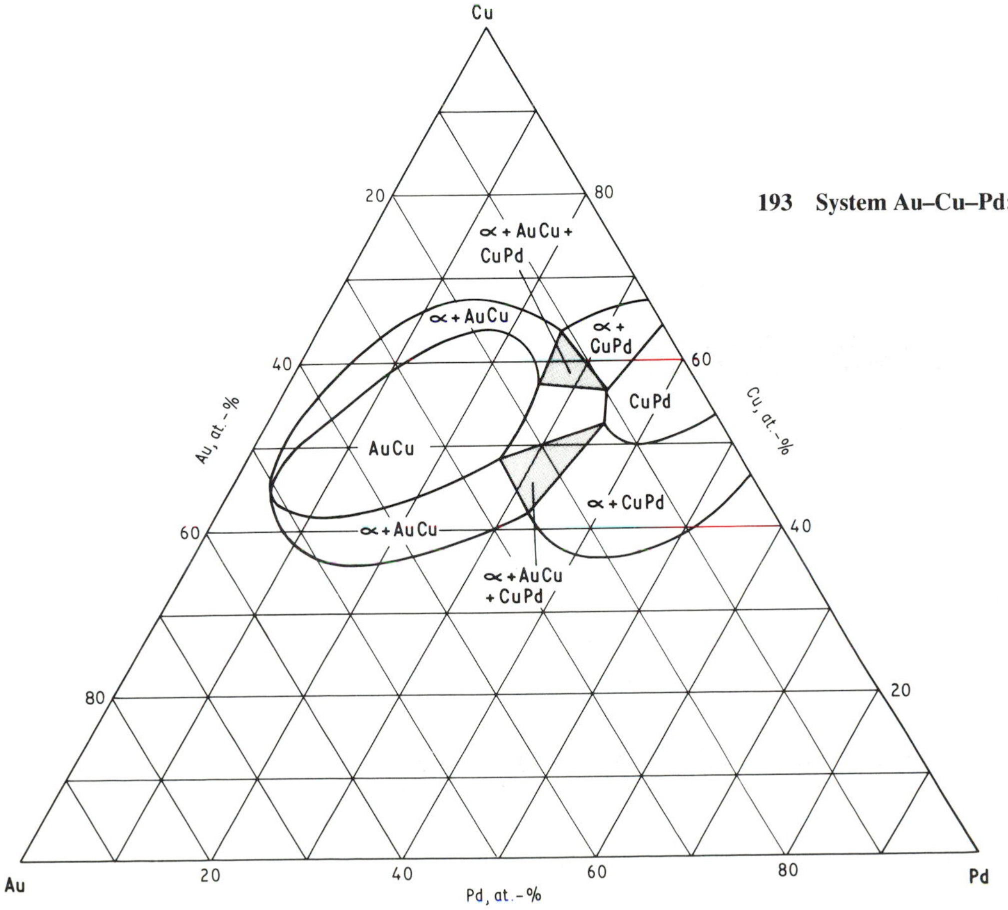

193 System Au–Cu–Pd: 350°C isothermal, after [55 Rau]

section along the composition line connecting the maximum ordering temperatures for CuPd and AuCu in the relevant binary systems. According to [55 Rau] this line runs from the composition 41·1 at.-% Pd, 58·9 at.-% Cu (in good agreement with independent work on the binary system) to 50 at.-% Cu, 50 at.-% Au. The vertical section is shown in atomic percentages in Fig. 194, and it is confirmed that the binary ordering temperatures are raised in the ternary system. Though the composition section of Fig. 194 is not identical with the 50 at.-% Cu section in the work of [49 Nem], the double maximum is confirmed; the ordering temperature for CuPd and with Au in solid solution rises to 642°C, while that of AuCu with Pd in solid solution is raised to 600°C. The (CuPd + AuCu) two-phase area was clearly observed. It will be noted, however, that no distinction is made between the two ordered structures for AuCu, and it is stated in [55 Rau] that the diffraction pattern for AuCuII was not observed in the ternary alloys. The observation is confirmed by Sato and Toth [61 Sat], who also studied the effect of Pd on AuCu, with particular reference to the AuCuII structure. This was maintained on ordering in the region of 410 to 436°C only up to about 1 at.-% Pd. At higher Pd contents (up to 12 at.-%) the ordering temperature continued to rise, but the ordered phase had the AuCuI structure.

In the work of [55 Rau] no evidence was obtained for the existence of a phase or structure based on Au_3Cu. It will be noted that the isothermals proposed by [55 Rau] are broadly consistent with the solid state thermal analysis results of [49 Nem].

The study of order in alloys containing less than 50 at.-% Cu was carried further by [65 Nag], who examined superlattice structures in thin evaporated Au–Cu–Pd films by electron diffraction. The method was to prepare by evaporation single crystal films of a binary alloy, for example the composition Au_2Pd, which were quenched after heating to a high temperature. To establish composition the lattice spacings were compared with existing X-ray data. The third element was then evaporated onto the binary film and the ternary alloy thus formed again quenched from a high temperature. The estimate of composition again relied on lattice spacing measurements, assuming that the spacing of the ternary alloy would vary linearly from that of the original binary alloy to that of the third element. The possible error was assessed as $<\pm 5\%$. A film of composition Au25 Pd23 (approximately Au_2CuPd) proved to have the AuCuI structure when quenched from 325 and 350°C, but to be fcc on quenching from 450°C. By similar methods, the AuCuI structure was also identified at the approximate composition $AuCuPd_2$. If results on evaporated thin films can be taken as typical of the bulk material, the ordered region of the ternary system must be extended considerably in comparison with Figs. 191, 192 and 193.

In development of this work the structures of thin evaporated single crystal films of alloys of compositions $AuCu_2Pd$ and Au_2CuPd were examined by similar methods [66 Nag]. The first composition lies within the AuCuI area proposed by [55 Rau] at 350 and 450°C, but not at 550°C. According to [66 Nag] the transformation temperature is about 550°C; although the ordered structure is very closely related to that of AuCuI, it differs slightly in that the Pd atoms are themselves in ordered positions at the centres of the basal planes, rather than sharing with Au atoms the sites not occupied by Cu. Similarly the thin film of composition Au_2CuPd was found to have the same type of structure below about 430°C with the Cu and Au atom sites transposed. The exact details of the states of order in the ternary system have therefore yet to be clarified.

Limited work has also been carried out on a composition section running from that of $AuCu_3$ towards Pd [55 Rau]. This showed that the ordering temperature of $AuCu_3$ is raised by Pd, in this section, to a maximum of 532°C at 9·1 at.-% Pd; it then falls to 518°C at 13·9 at.-% Pd. At this composition $AuCu_3$ comes into equilibrium with AuCu containing Pd. These relationships are summarized in Fig. 195. No two-phase region between $AuCu_3$ and the disordered α was observed in the work of [55 Rau], who also found that along the section $AuCu_3$–Cu_3Pd only the ordered fcc phase was observed (with change in lattice spacing), thus confirming solid solution formation between the binary phases. Although no close study was made of the crystal structure of the Cu_3Pd phase containing Au, it was suggested that the effect of Au is to discourage the formation of the faulted structure discussed above. [86 Syu] used the variation of specific electrical resistivity with temperature to determine the effect of Pd additions of 4, 8, and 12 at.-% on the ordering temperature of $AuCu_3$. Alloy compositions were on the $AuCu_3$–Pd section. They confirmed [55 Rau] in that Pd increases the ordering temperature of $AuCu_3$.

More recently the quasi-binary section $AuCu_3$–Cu_3Pd was studied by [79 Nak], using alloys which had been prepared in a plasma jet arc furnace, homogenized at 997°C for 5 days, and cooled to room temperature at 0·25°C min^{-1}. Specimens for differential thermal analysis using heating and cooling rates of 3·5°C min^{-1}, and powders for high-temperature X-ray diffraction experiments, were then annealed for 5 days at 500°C and cooled to room temperature at 0·1°C min^{-1}. In both thermal analysis and high-temperature X-ray lattice spacing determinations, the critical temperature for order showed hysteresis; the difference between the temperatures observed on heating and cooling was 5–13°C. The variation of critical temperature with composition as defined by the results of heating curves is shown in Fig. 196. Complete solid solution formation was confirmed. In conformity with the work of [55 Rau] the critical temperature for ordering of $AuCu_3$ is raised by Pd; that of Cu_3Pd is also raised by Au, so that in this quasi-binary section the ordering temperature reaches a maximum of 586°C close to the composition $Cu_3Au_{0.5}Pd_{0.5}$. It may be noted that the tetragonal distortion typical of the composition Cu_3Pd was recognized only in alloys $Cu_3Au_{1-x}Pd_x$ for which x was equal to or greater than 0·8, that is, at 5 at.-% Au or less, confirming the conclusion of [55 Rau] and [55 Sch] that Au discourages the formation of the faulted ordered structures. In the work of [79 Nak] no evidence was found

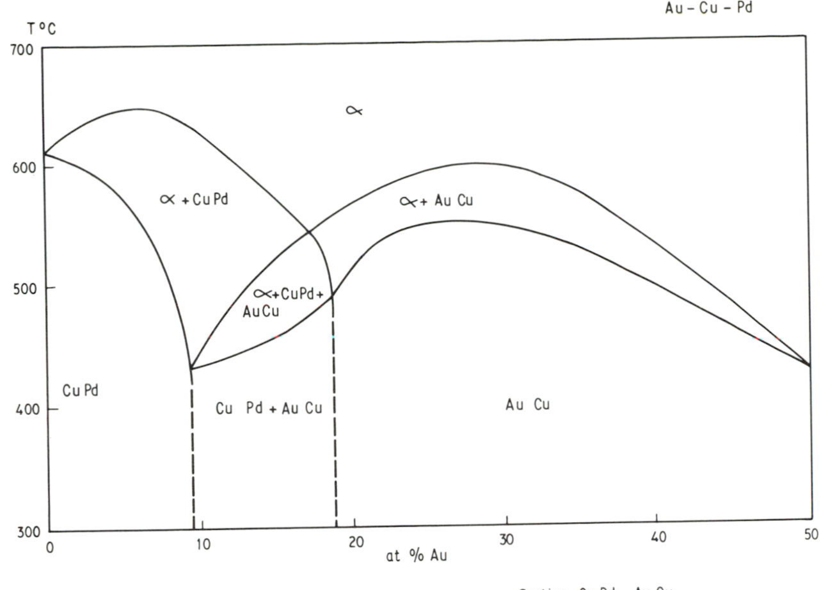

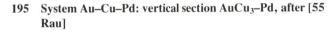

194 System Au–Cu–Pd: vertical section CuPd–AuCu (58·9 at.-% Cu, 41·1 at.-% Pd to 50 at.-% Au, 50 at.-% Cu), after [55 Rau]

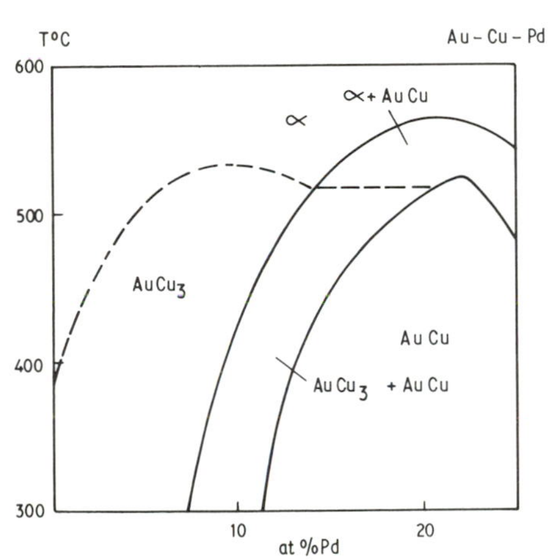

195 System Au–Cu–Pd: vertical section $AuCu_3$–Pd, after [55 Rau]

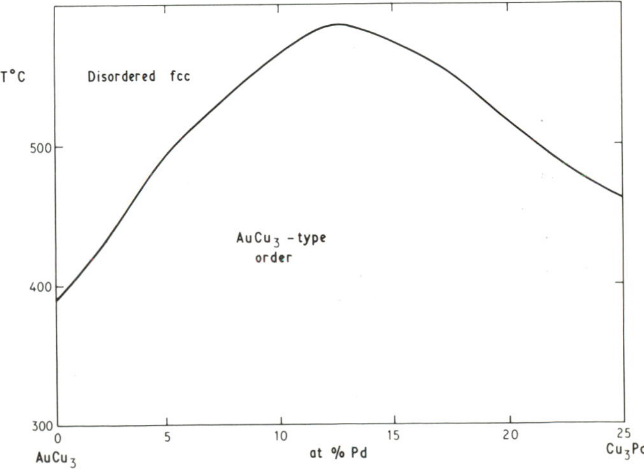

196 Critical temperature for order in quasi-binary system $AuCu_3$–Cu_3Pd [79 Nak]

for the formation of a ternary superlattice in which the Cu and Au atoms were ordered with respect to each other.

The solid solution of Au in Cu_3Pd has been examined in detail by [55 Sch]. For a constant copper content of 75 at.-%, the structure of the alloy of composition Cu_3Pd is, in conformity with work on the binary system, the complex form of the cP4 structure. Substitution of up to 5 at.-% Au for Pd, however, leads to the simpler faulted structure, with the fault density decreasing as the Au content rises. At 6 at.-% the structure has become of the unfaulted cubic cP4 type, and thus, apart from lattice spacing changes, identical with $AuCu_3$, so that complete solid solution formation may be taken as confirmed. In alloys for which the (Cu + Au) content is maintained constant at 75 at.-% the complex structure is preserved up to 4 at.-% Au, but the structure with 6 at.-% Au is disordered fcc, so that the boundary between ordered and disordered phases occurs between 4 and 6 at.-% Au for this section. No further detail, however, is available.

[86 Udo] reported a calculated 310°C isothermal (Fig. 197) using the tetrahedron approximation in the cluster variation method. The coherent phase diagram thus calculated is based on the presence of only phases with lattice structures derived from the fcc lattice. The bcc phase CuPd cannot appear as an equilibrium phase in the calculated coherent phase diagram and [86 Udo] replaced it with a hypothetical CuPd phase of the $L1_0$ (tP4) structure-type. The section should be taken as a guide for further experimental work. The ternary phases $AuCu_2Pd$ and Au_2CuPd reported in thin film samples by [66 Nag] appear in the coherent phase diagram when an appropriate choice of ternary interaction–energy parameters is made. The transition between the $AuCu_2Pd$ phase and AuCu appears to be higher-order. The 310°C isothermal section includes the binary ordered phases Cu_3Au and Au_3Cu — not observed by [55 Rau] — and the Cu_3Pd ordered phase. An alloy containing 62·5 Au, 12·5 at.-% Pd after annealing for 7 days at 300°C was found, [86 Udo], to contain an $L1_2$-type phase (Au_3Cu) in agreement with the observed increase of the critical temperatures of the Au–Cu ordered phases on the addition of Pd. Contrary to the findings of [55 Rau, 55 Sch, 79 Nak] the section Cu_3Au–Cu_3Pd does not show a series of solid solutions in the calculated section at 310°C. It should be noted that the coherent phase relations illustrated in Fig. 197 are not necessarily identical or similar to the normal incoherent phase relations.

DISCUSSION

There is no doubt that at the higher temperatures in the solid state the three components are mutually completely soluble, and the liquidus surface is simple in form. The relationships between the ordered structures based on AuCu and CuPd are well understood in principle, though electron diffraction work on thin films indicates that further ordered compositions may be involved. Complete solid solution formation between the ordered structures based on $AuCu_3$ and Cu_3Pd appears to be established, but further investigation is needed to establish the equilibrium relationships between this solid solution and the other ordered structures.

REFERENCES

49 Nem: V. A. Nemilov, A. A. Rudnitsky and R. S. Polyakova: *Izvest. Sekt. Platiny*, 1949, **24**, 35–51

55 Rau: E. Raub and G. Wörwag: *Z. Metall.*, 1955, **46**, 119–128

55 Sch: K. Schubert, R. Kieffer, M. Wilkens and R. Heuffer: *Z. Metall.* 1955 , **46**, 692–715

58 Han: M. Hansen and K. Anderko: 'Constitution of binary alloys'; 1958, New York, McGraw-Hill

61 Sat: H. Sato and R. S. Toth: *Phys. Rev.*, 1961, **124**, 1833–1847

62 Rhy: D. W. Rhys and R. D. Berry: *Metallurgia*, 1962, **66**, 255–263

65 Nag: A. Nagasawa: *J. Phys. Soc. Japan*, 1965, **20**, 1520

65 Ell: R. P. Elliott: 'Constitution of binary alloys' (1st Suppl.); 1965, New York, McGraw-Hill

66 Nag: A. Nagasawa: *J. Phys. Soc. Japan*, 1966, **21**, 955–960

79 Nak: K. Nakahigashi and M. Kogachi: *Jap. J. Appl. Phys.*, 1979, **18**, 1915–1922

85 Oka: H. Okamoto and T. B. Massalski: *Bull. Alloy Phase Diagrams*, 1985, **6**, 229–235

86 Syu: V. I. Syutkina, I. E. Kislitsyna, R. Z. Abdulov and V. V. Rudenko: *Fiz. Metallov Metalloved*, 1986, **61**(3), 504–509

86 Udo: K. Udoh, K. Yasuda and H. Yamauchi, *Met. Soc. AIME*, TMS Technical Paper No. A86–37, pp. 1–10, 1986

87 Oka: H. Okamoto, D. J. Chakrabarti, D. E. Laughlin and T. B. Massalski: *Bull. Alloy Phase Diagrams*, 1987, **8**, 454–474

Au–Cu–Pt

(G. V. Raynor)

INTRODUCTION

The binary systems Au–Cu and Cu–Pt both contain ordered phases at low temperatures, and the interactions between these are of potential practical and scientific interest. Very little experimental work has been carried out on the Au–Cu–Pt system, however, and the only significant contributions are an investigation of the section $AuCu_3$–Cu_3Pt [79 Nak] and a study of the effect of Pt additions on the ordering temperature of $AuCu_3$ [86 Syu].

BINARY SYSTEMS

The Au–Cu phase diagram as assessed [87 Oka] is accepted. The AuPt assessment by [86 Oka] is also accepted.

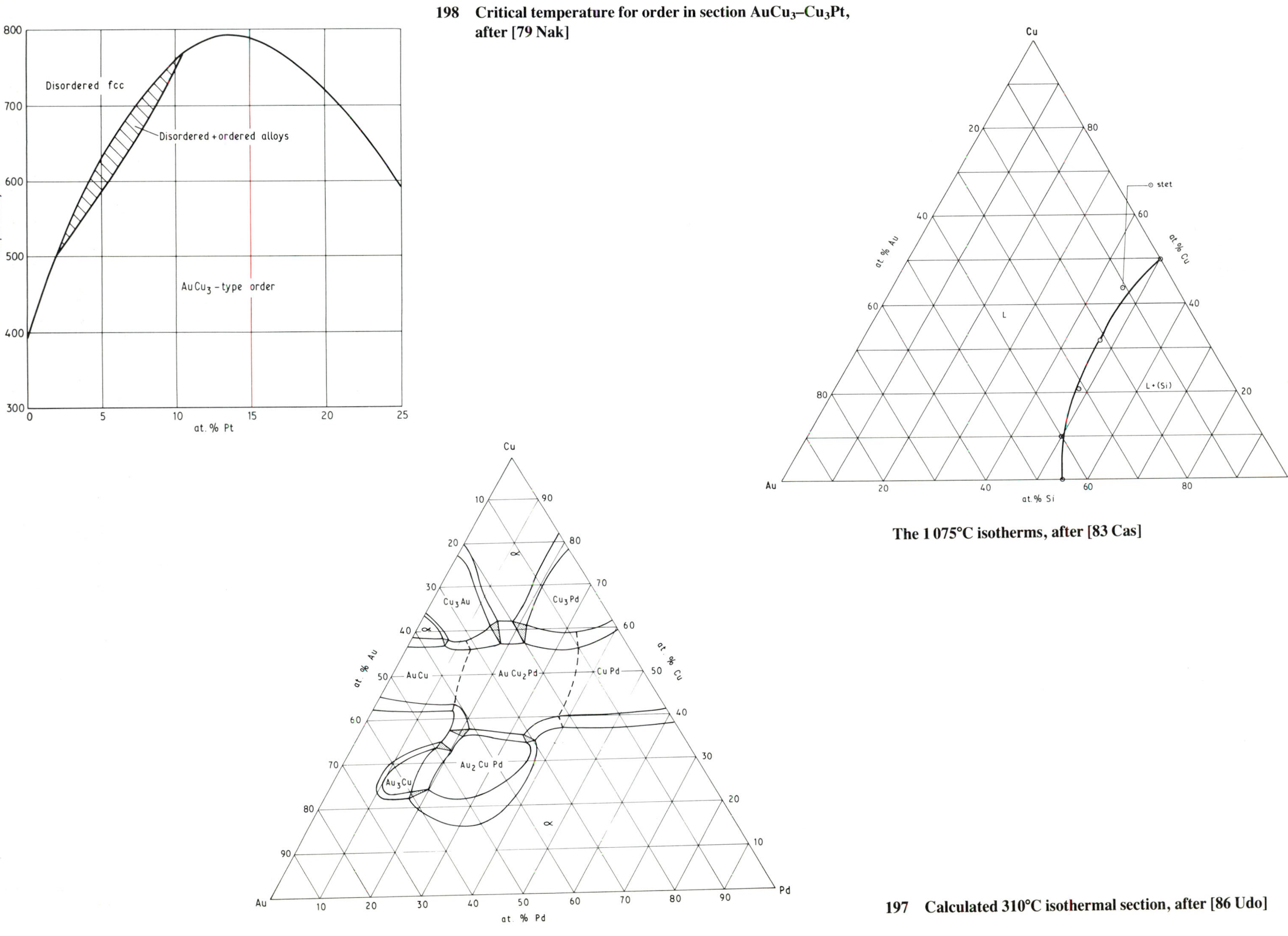

The 1 075°C isotherms, after [83 Cas]

197 Calculated 310°C isothermal section, after [86 Udo]

198 Critical temperature for order in section $AuCu_3$–Cu_3Pt, after [79 Nak]

The Cu–Pt equilibrium given in [58 Han] is accepted as a basis for interpreting the ternary system.

THE SOLID PHASES

No distinct ternary phases have been reported in the system Au–Cu–Pt. The structures of the phases referred to above are summarized in Table 66.

Table 66. Crystal structures

Solid phase	*Prototype*	*Lattice designation*
Au–Cu–Pt solid solution	Cu	cF4
$AuCu_3$	$AuCu_3$	cP4
AuCuI	AuCuI	tP4
AuCuII	AuCuII	tP4 (faulted)
Cu_3Pt	—	cP4 (faulted)
CuPt	—	Rhombohedral

THE TERNARY SYSTEM

[79 Nak] examined eleven alloys lying in the section $AuCu_3$–Cu_3Pt, prepared from pure materials in a plasma jet arc furnace, homogenized at 997°C for 5 days, and cooled to room temperature at 0·25°C min^{-1}. The critical temperature for order was measured by differential thermal analysis and by determining the temperature dependence of lattice spacings. Specimens for thermal analysis were annealed in suitable sealed containers for 5 days at 450°C and cooled to room temperature at 0·1°C min^{-1}. Powders prepared for high-temperature X-ray diffraction experiments were annealed for 5 days at 500°C and also cooled as above. Differential thermal analysis was carried out at a heating or cooling rate of 3·5°C min^{-1}, and the experiments revealed that the observed critical temperature for ordering was subject to hysteresis; the difference between temperatures observed on heating and cooling was 4–15°C. The variation of critical temperature with composition as derived from heating curves is shown in Fig. 198, and the critical temperature rises to a maximum of 795°C at 13·75% Pt, that is, at the composition $Cu_3(Au_{0{\cdot}45}Pt_{0{\cdot}55})$. Complete solid solution formation was confirmed, and the critical temperatures for order of both $AuCu_3$ and Cu_3Pt are raised by the inclusion of the third element. The variation of lattice spacing with temperature for alloys containing between approximately 2 and 10·5% Pt gave clear evidence for the presence of two structures of different lattice spacings within a certain temperature range: the spacings indicated that the two structures were the disordered fcc and the partially ordered $Cu_3(Au,Pt)$ structure. The shaded portion of Fig. 198 defines the area within which the disordered and ordered structures can coexist under the conditions of the experiments. It may be noted that although a slight tetragonal distortion was observed in the alloy of composition Cu_3Pt, no evidence was found in this work for the existence of the antiphase domain structure proposed in [55 Sch]. [86 Syu] used the variation of specific electrical resistivity with temperature to determine the effect of Pt additions of 4, 8, and 12 at.-% on the ordering temperature of $AuCu_3$. Alloy compositions were on the $AuCu_3$–Pt section. They found a marked increase in ordering temperature with additions of Pt. The ordering temperature increases from 390°C at 0 at.-% Pt to 490°C at 4 at.-% Pt, 690°C at 8 at.-% Pt and 740°C at 12 at.-% Pt.

The liquidus surface of the ternary system has not been determined, and the relationships between the phases AuCu and CuPt have not been examined.

REFERENCES

55 Sch: K. Schubert, R. Kieffer, M. Wilkens and R. Heuffer: *Z. Metall.*, 1955, **46**, 692–715

58 Han: M. Hansen and K. Anderko: 'Constitution of binary alloys'; 1958, New York, McGraw-Hill

65 Ell: R. P. Elliott: 'Constitution of binary alloys' (1st Suppl.); 1965, New York, McGraw-Hill

79 Nak: K. Nakahigashi and M. Kogachi: *Jap. J. Appl. Phys.* 1979, **18**, 1915–1922

86 Oka: H. Okamoto and T. B. Massalski: *Bull. Alloy Phase Diagrams*, 1986, **6**, 46–56

86 Syu: V. I. Syutkina, I. E. Kislitsyna, R. Z. Abdulov and V. V. Rudenko: *Fiz. Metallov Metalloved*, 1986, **61**(3), 504–509

87 Oka H. Okamoto, D. J. Chakrabarti, D. E. Laughlin and T. B. Massalski: *Bull. Alloy Phase Diagrams*, 1987, **8**, 454–474

Au–Cu–S

The work reported on the Au–Cu–S ternary system relates to the distribution of Au between molten Cu and a molten Cu_2S phase (containing 32·9 at.-% S [83 Cha]). [71 Asa] reported values of the partition coefficient, defined as the wt-% Au in the sulphide divided by the wt-% Au in the Cu, at 1 200°C of $5{\cdot}8 \times 10^{-3}$. [74 Bur 1] quoted a value of $5{\cdot}0 \times 10^{-3}$ at 1 300°C using data of [72 Rom]. Experimental determinations of the partition coefficient at 1 150 and 1 300°C [74 Bur 2], gave values of $5{\cdot}1 \times 10^{-3}$ and $5{\cdot}0 \times 10^{-3}$ respectively. A value of 8×10^{-3} at 1 200°C was found by [75 Sch] but the authors worked with low concentrations of Au and their analysis of the small quantity of Au in the molten sulphide phase, 0·003 to 0·007 wt-% Au, is subject to high uncertainty limits. The value of 8×10^{-3} is clearly too high in relation to the good agreement between the data of [71 Asa, 72 Rom and 74 Bur 2]. [74 Bur 2] presented a thermodynamic analysis of melts in the Au–Cu–S system and calculated a rapid reduction in the solubility of S in Cu at 1 150 and 1 300°C with additions of Au up to 10 wt.-% (see table following).

REFERENCES

71 Asa: N. Asano, K. Wase and M. Kobayashi, *Nippon Kogyo Kaishi*, 1971, **87**, 347–352

72 Rom: V. D. Romanov, V. V. Mechev, B. P. Burylev, L. Sh. Tsemekhman and S. E. Vaisburd, 'The

Theory of Regular Solutions, its Development and Application to Melts', Abstracts, All-Union Conf., Krasnodar, 1972, page 48 (in Russian)

74 Bur 1: B. P. Burylev, V. D. Romanov, L. Sh. Tsemekhman, V. V. Mechev and S. E. Vaisburd, *Zhur. Fiz. Khim.*, 1974, **48**, 154–155

74 Bur 2: B. P. Burylev, V. V. Mechev, L. Sh. Tsemekhman, V. D. Romanov and S. E. Vaisburd, *Izvest. Vyssh. Ucheb. Zaved. Tsvet. Met.*, 1974 (4), 27–33

75 Sch: W. J. Schlitt and K. J. Richards, *Metall. Trans.*, 1975, **6B**, 237–243

83 Cha: D. J. Chakrabarti and D. E. Laughlin, *Bull. Alloy Phase Diagrams*, 1983, **4**, 254–271

The effect of Au on the solubility of S in molten Cu

Temperature, °C	at.-% Au	at.-% S
1 150	0	2·04
	2·5	1·58
	5·0	1·19
	10·0	0·72
1 300	0	2·70
	2·5	2·10
	5·0	1·58
	10·0	0·92

Au–Cu–Si

Little data exists on the Au–Cu–Si ternary system. Sankur [75 San] prepared 118 alloys, whose compositions were close to the Au–Cu binary edge, to determine the solid solubility of Si in the (Au/Cu) solid solution. Alloys were prepared from high-purity elements (99·999 9% Au, 99·999% Cu,l 10 Ωcm resistivity semiconductor grade Si) by RF furnace melting in a quartz tube continuously purged with dry argon. The molten alloys were quenched in their quartz tubes into water. X-ray powder diffraction and metallographic examination indicated alloy homogeneity. The alloys were heat treated for 8 days at 349 ± 1°C in a quartz tube under vacuum. The solubility limit of Si in the (Au/Cu) solid solution was determined by metallography and X-ray powder diffraction analysis. Table 67 gives the experimental results. No trace of ordered Au–Cu phases was found. The (Au/Cu) solid solution is stated to be in equilibrium at 349°C with the cubic γ phase of the Cu–Si binary system.

Castanet [83 Cas] determined the 1 075°C isotherm by measuring the enthalpy of dissolution of solid Si in molten Au–Cu alloys (of compositions 80 at.-% Au, 20 at.-% Cu; 60 Au, 40 Cu; 40 Au, 60 Cu and 20 Au, 80 Cu) as a function of the Si addition using direct reaction calorimetry. The slope of the enthalpy–at.-% Si curve changes, from positive to negative, when the limit of solubility of Si in the Au–Cu melt is reached. Table 68 gives the experimental data on which the isotherm on page 241 is based.

[88 Cas] measured the enthalpy of formation of ternary alloys along the sections 20 Au, 80 Cu–Si; 40 Au, 60 Cu–Si; 60 Au, 40 Cu–Si and 80 Au, 20 Cu–Si. A plot of isoenthalpy contours for a temperature of 1 075°C was presented.

Table 67. Composition of alloys in atomic percent at the (Au–Cu) solid solution boundary [75 San]

Au	*Cu*	*Si*	*Au*	*Cu*	*Si*
0	92·6	7·4	37·5	61·6	0·9
3·9	91·2	4·9	41·2	57·5	1·1
10·1	86·5	3·4	43·8	54·9	1·3
14·2	83·1	2·8	44·9	53·3	1·8
19·8	78·3	1·9	53·2	44·8	2·0
22·6	75·7	1·7	53·6	43·8	2·6
26·2	72·1	1·7	57·5	39·4	3·1
32·1	66·4	1·5	61·6	36·2	2·2
34·4	64·5	1·1	71·8	27·6	0·4
36·4	62·7	0·9	75·3	24·5	0·2

Table 68. Solubility limit of Si in molten (Au–Cu) alloys at 1 075°C [83 Cas], compositions in atomic percent

Au	*Cu*	*Si*
10·9	43·6	45·5
21·2	31·8	47·0
31·2	20·8	48·0
40·0	10·0	50·0
45·0	0·0	55·0

REFERENCES

75 San: H. Sankur: Thesis, 1975, California Institute of Technology.

83 Cas: R. Castanet: private communication, June 1983

88 Cas: R. Castanet, *J. Less-Common Metals*, 1988, **136**, 287–296

Au–Cu–Sn

[84 Kim] studied the microstructures of four ternary alloys which had been homogenized at 700°C followed by an anneal at 250°C for 30 min. The results are given in the table below.

Alloy composition, at.-%			*Phases detected*
Au	*Cu*	*Sn*	
0·3	99·2	0·5	(Cu)
0·4	97·2	2·4	(Cu) matrix + Cu_3Sn (< 1%)
0·3	94·2	5·5	(Cu) matrix + Cu_3Sn (~ 9%)
0·4	84·6	15·0	Cu_3Sn (60%) + (Cu)

No Au–Sn compounds were detected by optical metallography, transmission electron microscopy or X-ray diffractometry. The phases detected and their volume fractions agree with those anticipated from the binary Cu–Sn phase diagram. At the very low Au contents used, 0·3 to 0·4 at.-%, the 0·5 at.-% Sn alloy lies in the (Cu) phase region at 250°C and the other three ternary alloys lie in the (Cu) + Cu_3Sn phase region. At 250°C one would anticipate the Cu-rich corner to show a narrow three-phase region of (Cu) + Cu_3Sn + $AuCu_3$. Thin film specimens showed similar phases after a 250°C anneal as bulk alloys.

REFERENCE

84 Kim: J. Kim, *J. Electron. Mater.*, 1984, **13**, 191–209

Au–Cu–Tb

[83 Kan] reported that an alloy with the stoichiometry $AuTbCu_4$ had the $AuBe_5$-type structure, cF24, with lattice parameter a = 0·717 5 nm. It is uncertain whether this phase is a ternary compound, as referred to by the authors, or a solution phase based on $TbCu_5$. The binary phase $TbCu_5$ also has the $AuBe_5$-type structure below 895°C and a lattice parameter a = 0·704 1 nm [88 Sub]. Substitution of the larger Au atom for Cu in $TbCu_5$ would be expected to increase the lattice parameter, as observed by [83 Kan].

REFERENCES

83 Kan: T. Kaneko, S. Abe, K. Kamigaki and M. Ohashi, *J. Magn. Magn. Mater.*, 1983, **31–34**, 253–254

88 Sub: P. R. Subramanian and D. E. Laughlin, *Bull. Alloy Phase Diagrams*, 1988, **9**, 390–394

Au–Cu–Ti

Raub and Walter [50 Rau] studied the effect of ternary additions on the order–disorder transformation of AuCu. The binary AuCu alloy has a transformation temperature of 410°C. A ternary alloy containing 45·36 Au, 47·44 Cu, 7·2 at.-% Ti, on the Cu-rich side of the AuCu–Ti section, was heat treated at 700, 400, 360, 300 and 250°C. The disordered fcc ternary solid solution with a second phase (assumed to be a Au–Ti compound phase) was detected after 2 h at 700°C and 700 h at 400°C. Separation of the second phase leaves only a small amount of Ti in solution in the fcc ternary solid solution and this has only a small effect in depressing the transformation temperature of AuCu. A CuAuI tetragonal superlattice was detected at 360°C and below.

REFERENCE

50 Rau: E. Raub and P. Walter: *Z. Metall.*, 1950, **41**, 240–243

Au–Cu–U

Umarji *et al.* [87 Uma] synthesized the ternary phase UCu_4Au by arc-melting 99·9% U, 99·999% Cu, and 99·99% Au under argon. The sample was subsequently annealed in an evacuated quartz ampoule for 8 days at 727°C. X-ray diffraction analysis confirmed that UCu_4Au has the fcc $AuBe_5$-type structure (cF24) with $a = 0{\cdot}7079 \pm 0{\cdot}0005$ nm. Preliminary nuclear magnetic resonance studies indicated that the Au atoms have a strong preference to order on the (4c) sites. It should be noted that UCu_5 also has the $AuBe_5$-type structure with $a = 0{\cdot}7035 \pm 0{\cdot}0005$ nm. It would appear that Au is soluble to at least 16·7 at.-% in the phase UCu_5. Zolnierek *et al.* [87 Zol] also report the $AuBe_5$-type structure for the composition UCu_4Au but the lattice parameter, $a = 0{\cdot}7427$ nm, is far greater than that found by [87 Uma].

[88 Dom] reported considerable solubility of Cu in the compound $Au_{51}U_{14}$. An alloy containing 9 at.-% Cu, 14 at.-% U has lattice parameters a = 1·248 3, c = 0·906 7 nm with the $Ag_{51}Gd_{14}$ structure-type (hP65). An alloy containing 27 at.-% Cu had not the $Ag_{51}Gd_{14}$-structure type.

REFERENCES

87 Uma: A. M. Umarji, J. V. Yakhmi, N. Nambudripad, R. M. Iyer, L. C. Gupta and R. Vijayaraghaven: *J. Phys. F: Met. Phys.*, 1987, **17**, L25–L28

87 Zol: Z. Zolnierek, R. Troc and D. Kaczorowski: *J. Magn. Magn, Mater.*, 1987, **63–64**, 184–186

88 Dom: A. Dommann and F. Hulliger, *J. Less-Common Metals*, 1988, **141**, 261–273

Au–Cu–Yb

The ternary compound $AuCu_4Yb$ has been prepared by Rossel *et al.* [87 Ros]. It was produced by arc-melting the elements under argon on a Cu hearth, by heating in a BeO crucible, and by heating in a sealed Ta Tube. X-ray powder diffraction and metallographic analysis indicated a single-phase which has the $AuBe_5$ structure, cF24, with a = 0.705 19.

REFERENCE

87 Ros: C. Rossel, K. N. Yang, M. B. Maple, Z. Fisk, E. Zirngiebl and J. D. Thompson: *Phys. Rev. B. Condensed Matter*, 1987, **35**, 1914–1918

Au–Cu–Zn

INTRODUCTION

Most of the published work on the Au–Cu–Zn system has concentrated on the crystallographic structure of the phases found between β′ AuZn and the β/β′ Cu–Zn phases and on the complex series of superlattice structures found in the (Au–Cu) solid solution phase region at low temperatures. On sections near 50 at.-% Zn there appears to be a continuous series of solid solutions between the low-temperature β′ phase in the Cu–Zn system and the β′ AuZn phase (Fig. 199). The β′ phase has an ordered b.c.c. structure in which the Au and Cu atoms are randomly distributed on the body centre lattice sites but with complete ordering of the Zn atoms. The transformation from β′ to β involves the disordering of the Zn sublattice to give a random arrangement of Au,Cu and Zn

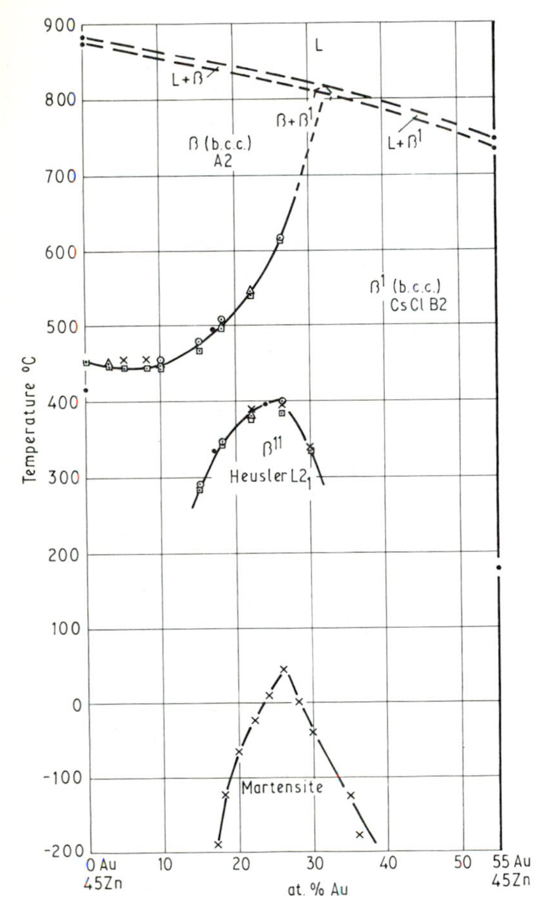

199 The 45 at.-% Zn section

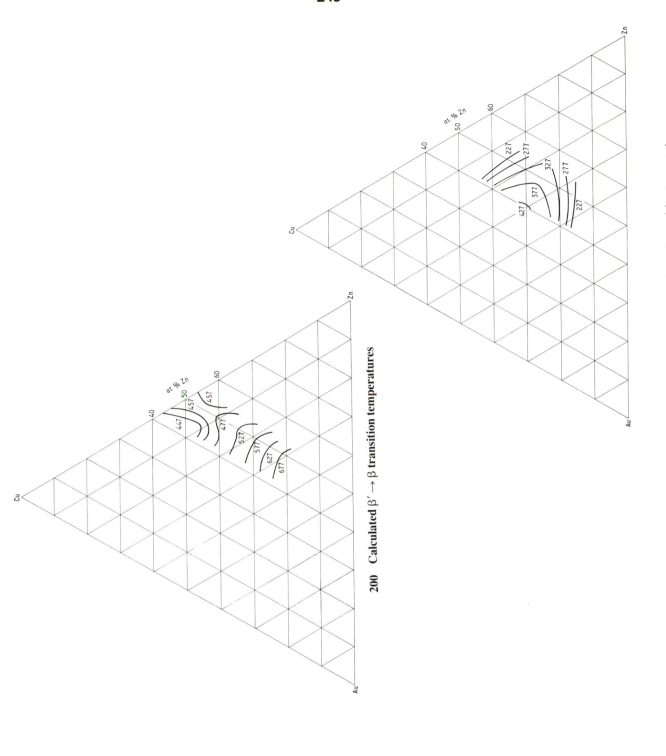

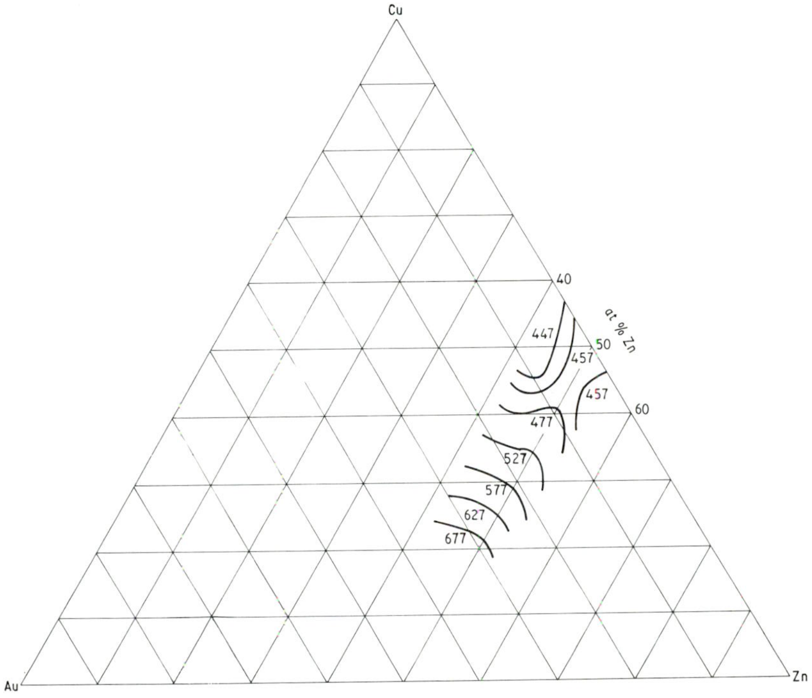

200 Calculated $\beta' \rightarrow \beta$ transition temperatures

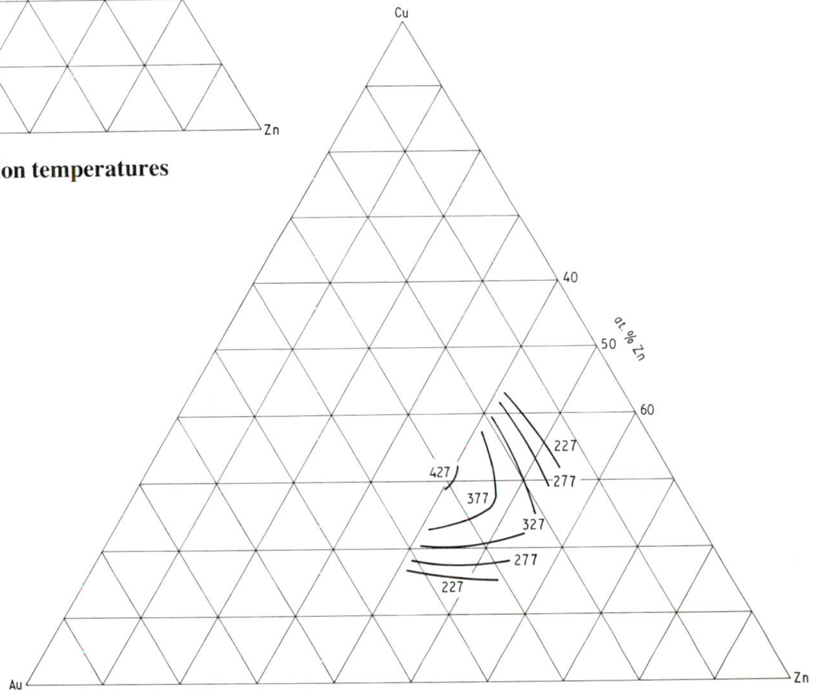

201 Calculated $\beta'' \rightarrow \beta'$ transition temperatures

in the β phase. At the equiatomic composition a Heusler-type phase $AuCuZn_2$, denoted β″, is stable at temperatures below 400°C. This phase forms a super-superlattice in that the Au and Cu atoms are ordered on an NaCl-type sublattice and the Zn atoms are ordered on a simple cubic sublattice. As Au or Cu is added to the $AuCuZn_2$ the ordering of the Au–Cu sublattice is decreased by the random substitution of Au by Cu and vice versa. The transformation of the β″ phase to the β′ phase involves complete disordering of the Au–Cu atoms on their β″ sublattice. There is evidence for the presence of a two-phase region between β″ and β′ but not for a two-phase region between β′ and β. Long term annealing precipitates the disordered (Au–Cu) phase in β″ [67 Mur, 71 Mur 1]. In view of this statement and the discrepancy between the isothermal section at 600°C [87 Dob] (Fig. 204), and the findings in Fig. 199, further work is needed to study sections such as that at 50 at.-% Zn using long term anneals.

The low-temperature superlattice structures that occur in the region of the (Au–Cu) phase have been exhaustively studied by German workers [49 Rau, 50 Rau, 54 Sch, 57 Wil, 58 Wil, 58 Koe, 66 Koe]. The picture that emerges is represented by a 200°C section (Fig. 202), in which three ternary phases have been detected. The phase Au_2CuZn was first reported by [50 Rau] and confirmed by [55 Sch, 57 Wil]. It exists from 10 to 30 at.-% Zn on the 50 at.-% Au section [58 Wil]. The ternary phases Au_5CuZn_2 [55 Sch, 57 Wil] and $Au_4Cu_5Zn_5$ [50 Rau, 57 Wil] have also been reported.

No work has been published on the liquidus of the Au–Cu–Zn system. [64 Lee] gave some data on the solidus temperatures for Cu–Au alloys containing 5 and 10 at.-% Zn (Fig. 203).

BINARY SYSTEMS

Evaluations of the Au–Cu system [87 Oka], and the Au–Zn system, [89 Oka] are accepted. The Cu–Zn system reproduced by [86 Mas] is accepted.

SOLID PHASES

Table 69 summarizes the solid phases that enter into that part of the Au–Cu–Zn system that has been studied.

Table 69. Crystal structures

Solid phase	*Prototype*	*Lattice designation*
(Au)	Cu	cF4
(Cu)	Cu	cF4
(Zn)	Mg	hP2
$AuCu_3$	$AuCu_3$	cP4
AuCuII	AuCuII	oI40
AuCuI	AuCu	tP4
Au_3Cu	$AuCu_3$	cP4
β(Cu–Zn)	W	cI2
β′(Cu–Zn)	CsCl	cP2
γ_c(Cu–Zn)	Cu_5Zn_8	cI52
ϵ_c(Cu–Zn)	Mg	hP2
$\alpha_3(Au_4Zn)$	Doubly-faulted $AuCu_3$	Doubly-faulted cP4
$\alpha_2(Au_3Zn)$	—	Orthorhombic
$\alpha_1(Au_3Zn)$	Faulted $AuCu_3$	Faulted cP4
Au_5Zn_3	—	Orthorhombic
β′AuZn	CsCl	cP2
γ(Au–Zn)	γ brass	Cubic
γ_2(Au–Zn)	H_3U	cP32
γ_3(Au–Zn)	—	Hexagonal
ε(Au–Zn)	Mg	hP2
$AuCuZn_2$ (β″)	Cu_2MnAl	cF16
Au_2CuZn	Stacking variant AuCd	oP4
Au_5CuZn_2	—	Orthorhombic
$Au_4Cu_5Zn_5$	—	Ordered tetragonal

TERNARY EQUILIBRIA

1. The section AuZn — 'CuZn'

Interest in sections near to 50 at.-% Zn stems from the occurrence of an Heusler-type ternary ordered phase $AuCuZn_2$, denoted β″, in the ternary system. [56 Sch] noted the formation of this phase between 200 and 400°C with a lattice parameter a = 0·611 nm. [64 Dug] prepared an alloy containing 23·5 Au, 31·2 Cu, 45·3 at.-% Zn and confirmed the presence of $AuCuZn_2$ with lattice parameter a = 0·6098 ± 0·0007 nm. [66 Mul 1] reported a lattice parameter a = 0·61273 nm for an alloy close to the stoichiometric composition, 25·22 Au, 24·72 Cu, 50·06 at.-% Zn. In further work [66 Mul 2] used the same alloy composition to study the ordering of Au and Cu atoms in $AuCuZn_2$. After melting 99·99% purity metals in evacuated Vycor tubes the melt was shaken and quenched. An unspecified annealing treatment was continued until good resolution was obtained for high angle lines in elevated temperature X-ray diffraction studies. The transition from the fully ordered Heusler-type phase to the ordered CsCl-type phase, denoted β′, occurred at 350°C. A major contribution from [67 Mur] was concerned with the phase relations along the 45 at.-% Zn section (see Fig. 199). Metals of 99·999% purity were melted in evacuated silica tubes, held at 600°C for 15–20h and then quenched into brine. Transition temperatures were measured by DTA, dilatometry and by recording the variation of electrical resistivity and specific heat with temperature. The Heusler-type phase was found in the range 15–30 at.-% Au and the maximum disordering temperature for the transition β″ to β′ was 400°C. The transition from the ordered CsCl-type β′ phase to the disordered β Cu–Zn phase varied little with Au additions up to 10 at.-%. With increased Au additions the β′ to β transition temperature rose rapidly. [67 Mur] commented that long term annealing of the 17 at.-% Au alloy at 270°C, within the β″ phase region, led to precipitation of the (Au–Cu) disordered solid solution phase. The enthalpy of β″ to β′ disordering is 1·25 kJ/g.atom for a 17 at.-% Au alloy; the enthalpy of β′ to β disordering is 0·24 kJ/g.atom for a 24 at.-% Au alloy.

[71 Mur 1] represented the 45 at.-% Zn section with essentially the features reported by [67 Mur]. The β′ to β transition was found to have a minimum at 445°C at about 8 at.-% Au and the β″ to β′ transition has a maximum temperature near the $AuCuZn_2$ composition. There is hysteresis in the transition of β″ to β′ between heating and cooling experiments. The authors claim this as evidence that this transition is first order. No temperature hysteresis was detected for the β′ to β transition. [71 Mur 1]

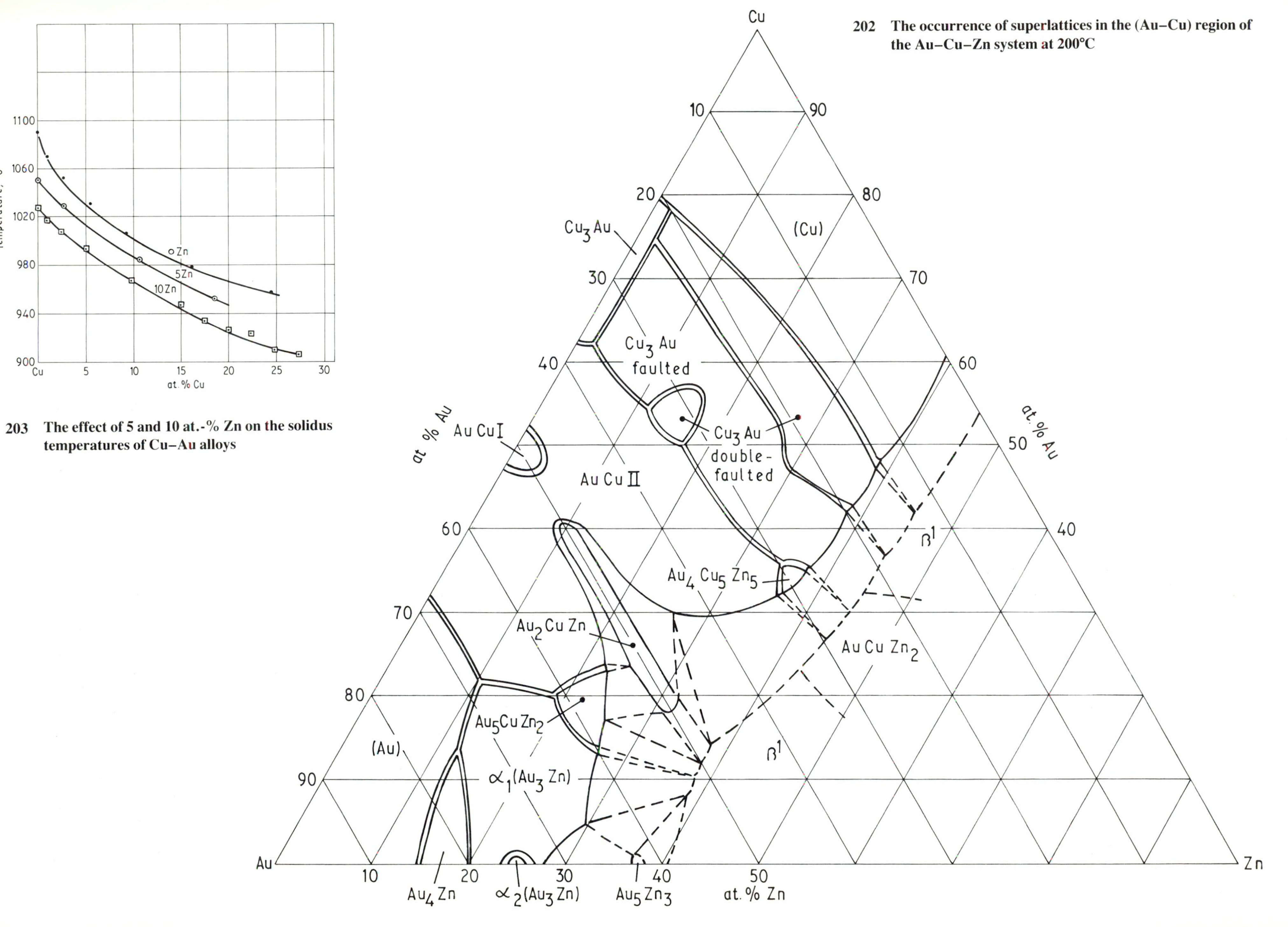

202 The occurrence of superlattices in the (Au–Cu) region of the Au–Cu–Zn system at 200°C

203 The effect of 5 and 10 at.-% Zn on the solidus temperatures of Cu–Au alloys

also state that long term annealing precipitates the (Au–Cu) solid solution phase with the Heusler-type phase. The diagram given in Fig. 199 is to be regarded as a provisional phase diagram and not necessarily a representation of the equilibrium conditions on the 45 at.-% Zn section. As noted below the 600°C isothermal section reported by [87 Dob] conflicts with the data in Fig. 199. [71 Mur 2] calculated the critical temperatures for the formation of the β′ and β″ superlattice phases using statistical thermodynamics with interactions between first and second nearest neighbours taken into account. The calculated 45 at.-% Zn section was obtained by appropriate choice for the interaction energies of atom pairs. A good agreement between calculated and experimental [71 Mur 1] transition temperatures was found for the β′ to β transition. The calculated β″ to β′ transition is displaced some 3 to 5 at.-% towards the Cu–Zn axis compared with the experimental data. Temperature contours for the β′ to β and β″ to β′ transitions were calculated for alloys with compositions ranging from 42 to 54 at.-% Zn. Figure 200 shows the calculated β′ to β transition temperatures and Fig. 201 the calculated β″ to β′ transition temperatures.

[72 Bow] studied the β′ to β transition for the 48 at.-% Zn section. As with the 45 at.-% Zn section, the transition temperature decreased on initial addition of Au, attaining a minimum at about 10 at.-% Au and then increasing with further Au additions. Transition temperatures were detected by measuring the electrical resistivity as a function of temperature for alloys that had been annealed at 600°C for 20 h and slow cooled to room temperature (Table 70).

Table 70.

Alloy composition, at.-%			*Transition temperature*, °C
Au	*Cu*	*Zn*	
0	52	48	470
1	51	48	465·5
5	47	48	454
10	42	48	445·5
15	37	48	452

It should be noted that the calculated β′ to β transition temperatures for a 48 at.-% Zn section (Fig. 200), agree with the experimental values of [72 Bow] up to 5 at.-% Au. At higher Au contents there is an increasing disagreement; at 10 at.-% Au the calculated temperature is 20°C higher than the experimental and at 15 at.-% Au some 55°C higher. In this respect the calculated transition temperatures on the 48 at.-% Zn section by [74 Mos] are noteworthy. They used a Monte Carlo simulation of the β′ to β transition with nearest neighbour interactions only to calculate the effect of Au additions. Good agreement was obtained with the experimental data of [72 Bow] over the range of Au contents studied (0–15 at.-% Au).

An orthorhombic martensitic phase forms at low temperatures on sections approximating to AuZn — ‘CuZn’ (Fig. 199). [68 Nak] reported a similar behaviour for the 49 at.-% Zn section, with the additional comment that the formation of the martensitic phase appears to occur at a minimum temperature at about 45 at.-% Au and to rise in temperature on approaching the Au–Zn binary. The elastic anistropy of the alloys follows closely the martensitic transformation behaviour. The enthalpy of the martensitic transformation is 0·71 kJ/g.atom at 26 at.-% Au on the 47 at.-% Zn section [71 Nak].

2. Ordering transformations in the (Au–Cu) phase region.

Solid-state transformations occurring in the phase region occupied by the disordered (Au–Cu) solid solution at high temperatures have been studied extensively by [49 Rau, 50 Rau, 54 Sch, 55 Sch, 57 Wil, 58 Wil, 58 Koe, 66 Koe]. Although X-ray diffraction techniques have been used by most workers, [58 Koe, 66 Koe] associated the variation of properties with the appearance of ordered phases. They measured specific electrical resistance, Hall constant, elastic modulus and thermal EMF as functions of composition. The transformation temperature from the disordered (Au) phase to AuCu II is gradually depressed by the addition of Zn. According to [58 Koe] the transformation temperature falls by 2°C per at.-% Zn up to 8 at.-% Zn whereas [50 Rau] detected the AuCu II phase at 400°C in alloys with up to 32·6 at.-% Zn. [55 Sch] reported the presence of orthorhombic AuCu II for alloys from 5 to 35 at.-% Zn after annealing for 3 days at 300°C. There is reasonable agreement between the lattice parameters given by [50 Rau, 54 Sch, 55 Sch] for the AuCu II phase with Zn in solution. The addition of Zn has a greater effect on the AuCu II ⇌ AuCu I transformation temperature. [50 Rau] detected AuCu II in a 3·3 at.-% Zn alloy after annealing for 70 h at 300°C. Annealing at 250°C for 250 h caused a transformation to f.c. tetragonal AuCu I. [58 Koe] also noted a more rapid decrease in the AuCu II ⇌ AuCu I transformation temperature but they found a 300°C transition temperature for an alloy with 6.5 at.-% Zn. The data of [50 Rau] are supported by [55 Sch] who reported AuCu II as stable at 300 and 150°C in an alloy containing 5 at.-% Zn. The AuCu I phase region only extends to about 4 at.-% Zn in the ternary system. At higher Zn contents it is replaced by the AuCu II phase region.

The disordered (Au) ⇌ $AuCu_3$ ordering reaction is depressed along the 25 at.-% Au section by Zn additions. [58 Koe] reported ordering temperatures falling from 390°C at 0% Zn to 365°C at 2% Zn, 360°C at 4% Zn, 320°C at 10% Zn and 340°C at 25 at.-% Zn. The $AuCu_3$ structure is faulted for alloys with ⩾4 at.-% Zn [55 Sch]. These data are supported by the work of [50 Rau, 54 Sch, 57 Wil]. The effect of Cu on the disordered (Au) ⇌ ordered $\alpha_1(Au_3Zn)$ transformation temperature was studied by [49 Rau, 50 Rau, 54 Sch, 55 Sch, 57 Wil, 58 Wil]. Zn stabilises the ordered $\alpha_1(Au_3Zn)$ phase and sharply depresses the $\alpha_1(Au_3Zn)$ ⇌ $\alpha_2(Au_3Zn)$ transformation. A ternary phase Au_2CuZn, detected by [50 Rau] and confirmed by [55 Sch, 57 Wil], exists from 10 to 30 at.-% Zn on the 50 at.-% Au section [58 Wil]. It has lattice parameters $a = 0.9007$, $b = 0.2915$, $c = 0.4547$ nm at the stoichiometric composition. According to [58 Wil] the crystal structure is a stacking variant of the AuCd/type (B19).

The $\alpha_3(Au_4Zn)$ phase was studied by [57 Wil]. It occurs for Au contents >75 at.-% and its structure is isotypic to

Cu_3Pd_{1+} (doubly-faulted LI_2 structure). [57 Wil] gave a representation of the complex regions for the existence of superlattice phases at 200°C in the (Au–Cu) region of the ternary system (Fig. 202). In addition to the elongated region of existence of Au_2CuZn, ternary phases were observed at Au_5CuZn_2 [55 Sch] with an orthorhombic structure ($a = 0.3873$, $b = 0.3954$, $c = 0.4064$ nm [55 Sch, 57 Wil]) and a phase denoted by $Au_{1+}Cu_{1.5}Zn_{1.5}$ by [57 Wil]. The latter phase is homogeneous at the composition 28 Au, 35 Cu, 37 at.-% Zn and is possibly the ordered tetragonal phase previously reported by [50 Rau] as having the AuCuZn stoichiometry. It is denoted in this assessment as $Au_4Cu_5Zn_5$.

3. Solidus
The only reported solidus data are due to [64 Lee] (Fig. 203). The solidus values for binary Cu–Au alloys are taken from [87 Oka]. Additions of 5 and 10 at.-% Zn depress the solidus temperatures as would be anticipated from the binary Cu–Zn solidus curve. Solidus temperatures were determined on alloys prepared from high-purity metals, 99·99 Au, pp.994 Cu and Zn, homogenized at 750°C for ≮7 days. The solidus was determined metallographically to ±5°C.

ISOTHERMAL SECTIONS

[87 Dob] published an isothermal section, shown in Fig. 204, based on the examination of 51 alloys by X-ray powder diffraction and metallographic techniques. Alloys containing <70 at.-% Zn were annealed at 600°C for 480 h for metallographic study and powders at 600°C for 120–230 h for X-ray analysis. Alloys containing >70 at.-% Zn were annealed at 400°C for 330–850 h (metallography) and powders at 400°C for 120–260 h and 300°C for 160–300 h. There is disagreement with the findings of [67 Mur] (Fig. 199), for the 45 at.-% Zn section. Neither did [87 Dob] detect the γ_2 and γ_3 Au–Zn binary phases [89 Oka] in the ternary system. In view of the large amount of work reported on superlattice structures it is surprising that more attention has not been paid to elucidating the higher temperature reactions in the Au–Cu–Zn ternary system.

REFERENCES

49 Rau: E. Raub, P. Walter and A. Engel, *Z. Metallkunde*, 1949, **40**, 401–405

50 Rau: E. Raub and P. Walter, *Z. Metallkunde*, 1950, **41**, 425–433

54 Sch: K. Schubert, B. Kiefer and M. Wilkens, *Z. Naturforschung*, 1954, **9A**, 987–988

55 Sch: K. Schubert, B. Kiefer, M. Wilkens and R. Haufler, *Z. Metallkunde*, 1955, **46**, 692–715

56 Sch: K. Schubert, W. Burkhardt, P. Esslinger, E. Günzel, H. G. Meissner, W. Schütt, J. Wegst and M. Wilkens, *Naturwiss*, 1956, **43**, 248–249

57 Wil: M. Wilkens and K. Schubert, *Z. Metallkunde*, 1957, **48**, 550–557

58 Koe: W. Köster and W. Lang, *Z. Metallkunde*, 1958, **49**, 443–449

58 Wil: M. Wilkens and K. Schubert, *Z. Metallkunde*, 1958, **49**, 633–646

64 Lee: K. T. Lee and B. B. Argent, *J. Inst. Metals*, 1964–65, **93**, 167

64 Dug: M. J. Duggin and W. A. Rachinger, *Acta Met.*, 1964, **12**, 1015–1024

66 Koe: W. Köster and R. Störing, *Z. Metallkunde*, 1966, **57**, 156–161

66 Mul 1: L. Muldawer, *Acta Cryst.*, 1966, **20**, 594–595

66 Mul 2: L. Muldawer, *J. Appl. Phys.*, 1966, **37**, 2062–2066

67 Mur: Y. Murakami, H. Asano, N. Nakanishi and S. Kachi, *Japan J. Appl. Phys.*, 1967, **6**, 1265–1271

68 Nak: N. Nakanishi, Y. Murakami and S. Kachi, *Scripta Met.*, 1968, **2**, 673–676

71 Nak: N. Nakanishi, Y. Murakami and S. Kachi, *Scripta Met.*, 1971, **5**, 433–440

71 Mur 1: Y. Murakami, N. Nakanishi and S. Kachi, *Acta Met.*, 1971, **19**, 93–96

71 Mur 2: Y. Murakami, S. Kachi, N. Nakanishi and H. Takehara, *Acta Met.*, 1971, **19**, 97–105

72 Bow: R. C. Bowie and L. Muldawer, *J Appl. Phys.*, 1972, **43**, 4495–4499

74 Mos: J. Mośiski and Z. Rycerz, *J Phys. F: Metal Phys.*, 1974, **4**, 1853–1858

86 Mas: T. B. Massalski, 'Binary Alloy Phase Diagrams', ASM, Metals Park, Ohio, *1986*, Vol. 1, page 981

87 Oka: H. Okamoto, D. J. Chakrabarti, D. E. Laughlin and T. B. Massalski, *Bull. Alloy Phase Diagrams*, 1987, **8**, 454–474

87 Dob: M. Dobersek and I. Kosovinc, 'Precious Metals 1987', Proc. 11th. Int. Conf., Brussels, Belgium, June 1987. Int. Precious Metals Inst., Allentown, Pa, *1987*, pp. 325–331

89 Oka: H. Okamoto and T. B. Massalski, *Bull. Alloy Phase Diagrams*, 1989, **10**, 59–69

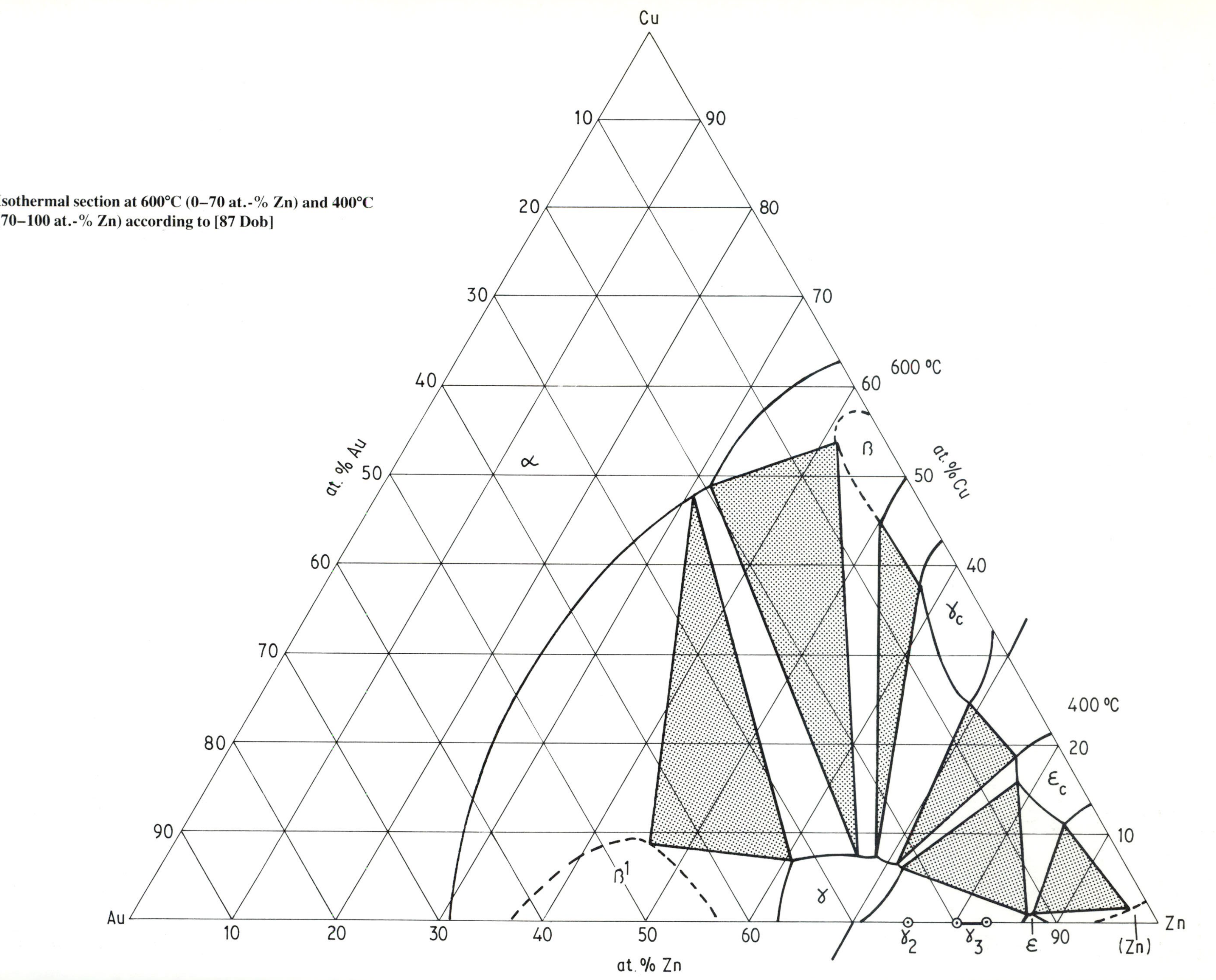

204 Isothermal section at 600°C (0–70 at.-% Zn) and 400°C (70–100 at.-% Zn) according to [87 Dob]

Au–Dy–Ga
Au–Er–Ga
Au–Ga–Gd
Au–Ga–Ho
Au–Ga–Lu
Au–Ga–Tb
Au–Ga–Tm
Au–Ga–Y

Dwight [82 Dwi] reported on the occurrence of ternary equiatomic compounds, AuGaRE, where RE = Dy, Er, Gd, Ho, Lu, Tb, Tm, Y. Alloys were melted under argon and homogenized for an unspecified time at 700°C. X-ray diffraction analysis indicated orthorhombic structures which are considered to be of the $CeCu_2$-type (oI12). Unit cell constants are given in Table 71.

REFERENCE

82 Dwi: A. E. Dwight: 'Rare Earths in Modern Science Technology', Vol 3, 359–360; 1982

Table 71. Lattice constants of AuGaRE compounds (nm ± 0·000 1)

Compound	*a*	*b*	*c*
AuGaDy	0·446 9	0·708 3	0·769 3
AuGaEr	0·445 5	0·699 5	0·767 7
AuGaGd	0·4502	0·7161	0·773 3
AuGaHo	0·446 6	0·703 6	0·768 3
AuGaLu	0·444 1	0·688 5	0·765 6
AuGaTb	0·449 3	0·710 4	0·771 9
AuGaTm	0·444 0	0·695 4	0·767 8
AuGaY	0·447 4	0·707 5	0·772 1

Au–Dy–Sb
Au–Er–Sb
Au–Gd–Sb
Au–Ho–Sb
Au–Lu–Sb
Au–Nd–Sb
Au–Sb–Sc
Au–Sb–Sm
Au–Sb–Tb
Au–Sb–Tm
Au–Sb–Y

The only data located are in a report [77 Dwi] on the occurrence of $Au_3RE_3Sb_4$ compounds, where the rare earth element is Nd, Sm, Gd, Tb, Dy, Ho, Er, Tm, Lu, isotypic with the compound $Au_3Y_3Sb_4$. A single phase composition was identified at $Au_3Y_3Sb_4$, the compound having an invariant unit cell constant implying no substitution of Au for Sb. The unit cell contains four formula weights with 12 Y atoms on (a) sites, 12 Au atoms on (b) sites, and 16 Sb atoms on (c) sites. The structure is cubic, cI40. Y has four close Au neighbours at 0·3006 ± 0·000 1 nm and four close Sb neighbours at 0·3349 ± 0·0021 nm. Au has four close Sb neighbours at 0·276 6 ± 0·001 0 nm.

The unit cell constants for the $Au_3RE_3Sb_4$ compounds are given in Table 72. [77 Dwi] noted that Sc failed to form this structure.

REFERENCE

77 Dwi: A. E. Dwight: *Acta Crystallogr.*, 1977, **33B**, 1579–1581

Table 72. Unit cell constants, nm (+ 0·000 2)

$Au_3Nd_3Sb_4$	0·996 1	$Au_3Dy_3Sb_4$	0·981 1
$Au_3Sm_3Sb_4$	0·990 9	$Au_3Ho_3Sb_4$	0·978 8
$Au_3Gd_3Sb_4$	0·986 4	$Au_3Er_3Sb_4$	0·976 8
$Au_3Tb_3Sb_4$	0·983 4	$Au_3Tm_3Sb_4$	0·975 2
$Au_3Y_3Sb_4$	0·982 1	$Au_3Lu_3Sb_4$	0·972 3

Au–Er–Pd

[69 Hav] reported the occurrence of a compound $AuErPd_2$ with the Ni_3Ti-type structure, hP16, and lattice parameters a = 0·576, c = 0·952 nm.

REFERENCE

69 Hav: E. E. Havinga, J. H. N. van Vucht and K. H. J. Buschow, *Philips Res. Repts*, 1969, **24**, 407–426

Au–Fe–Mn

Toth *et al.* [69 Tot] state, without elaboration, that the ordered Au_4X phase is stable on the section $Au_4(Mn_{1-x}Fe_x)$ with $0 \leq x \leq 1$. This would be expected for low values of x since Au_4Mn has the ordered Ni_4Mo-type structure, tI10, with a disordering temperature of 450°C, but no Au_4Fe phase exists.

REFERENCE

69 Tot: R. S. Toth, A. Arrott, S. S. Shinozaki, S. A. Werner and H. Sato, *J. Appl. Phys.*, 1969, **40**, 1373–1375

Au–Fe–Ni
(G. V. Raynor)

INTRODUCTION

The ternary system Au–Fe–Ni was examined in detail [61 Koe] as part of a general programme to develop ferromagnetic alloys which were hardenable by appropriate heat treatments. The possibility of enhanced mechanical properties lay in the circumstance that in two of the relevant binary systems (Au–Fe, Au–Ni) miscibility gaps existed in the solid state, giving rise to the possibility of precipitation on cooling of certain ternary compositions from a high-temperature homogeneous condition. In the work reported [61 Koe] a representative series of alloys was examined by the methods of thermal analysis, metallography, and X-ray powder diffraction, supplemented by dilatometry. In addition, the composition dependence of hardness, electrical resistance and its temperature coefficient, and magnetic properties was examined.

THE BINARY SYSTEMS

The Au–Fe system has been assessed by [84 Oka] and the Au–Ni system by [87 Oka]. The Fe–Ni system has been assessed by [86 Chu 1,2].

THE SOLID PHASES

All solid phases with the exception of $FeNi_3$ take the form of solid solutions based on one of the components but containing all three. Solid solutions in δ- or α-Fe are body-centred cubic (cI2) in structure; those in γ-Fe, Ni, and Au have the face-centred cubic structure (cF4). At low temperatures where the solid solutions are relatively limited it will be convenient to refer to them in terms of the major components, for example, (Au), (γFe) or (Ni). At higher temperatures, at which solid solutions based on specific components merge and become continuous, descriptions such as (γFe–Ni) or (γFe–Ni–Au) become more appropriate. The structure of $FeNi_3$ is of the $AuCu_3$ (cP4) type. The structures in the solid state are summarized in Table 73. No discrete ternary phases have been reported.

Table 73. Crystal structures

Solid phase	*Prototype*	*Lattice designation*
(δFe)	W	cI2
(αFe)	W	cI2
(γFe)	Cu	cF4
(Au)	Cu	cF4
(γFe–Au)	Cu	cF4
(γFe–Ni)	Cu	cF4
(γFe–Au–Ni)	Cu	cF4
$FeNi_3$	$AuCu_3$	cP4

THE TERNARY SYSTEM

The interpretation of the work described in [61 Koe] was based upon phase diagrams for the Au–Fe and Au–Ni systems which were closely similar to those now accepted. The diagram assumed for the Fe–Ni system was, however, incomplete in that the existence of $FeNi_3$ below 516°C and its equilibrium relationships with the (γFe–Ni) solid solution (γ_2) were not recognized. Also, a narrower two-phase region between (αFe) and (γFe–Ni) was assumed than is now established. Consequently the constitution below 516°C as proposed by [61 Koe] is incomplete, while the restricted portions of the proposed ternary model involving equilibrium between (αFe) and (γFe–Ni] need modification. The essential character of the ternary constitution is, however, adequately established by [61 Koe].

Some 80 alloys were prepared from Armco iron, Mond nickel, and 'fine' gold, of unspecified purity. Alloys were melted in an arc-furnace under argon; the resulting ingots were cold-rolled and subsequently homogenized at 900 to 1 150°C for 12 h. The compositions were verified by chemical analysis and corresponded very closely with those intended. Compositions lay along six ternary sections radiating from the Au corner of the composition diagram to points on the Fe–Ni axis at 10, 30, 48, 65, 85, and 90 wt-% Fe. The corresponding approximate compositions in atomic percentages are similar because of the small atomic weight difference between Fe and Ni. The alloy compositions also lay along certain sections of constant Au percentages by weight, giving additional information. For the metallographic work, alloys were heat-treated for 40 h at 900°C (or 1 000°C for the section for which Ni : Fe = 15 : 85 by weight), and quenched. Equilibrium was then established at temperatures from 800 to 400°C by annealing treatments of 15 h duration, terminated by quenching. 31 alloys were thermally analysed; surfaces of primary crystallization were established by cooling curves, and some solidus points were determined by heating curves. In addition, dilatometric techniques were used to investigate the αFe/γFe transformation in the section containing Ni and Fe in the ratio by weight of 15:85. For phase recognition X-ray powder diffraction methods were used.

Fig. 205 shows the projection of the surfaces of primary separation according to [61 Koe], together with the determined isothermal contours. The latter are taken from [61 Koe] but plotted in terms of atomic percentages, and are drawn to cut the axes at compositions which conform to the accepted binary diagrams. The main characteristic of Fig. 205 is the large area over which the fcc ternary solid solution of Au, Fe, and Ni with each other crystallizes. In the Fe-rich corner the monovariant curve p_1–p_2 divides the limited surface over which (δFe) crystallizes from the remainder of the surface. In the system Au–Fe, liquid of composition 57 at.-% Fe reacts with (γFe) to form (Au) (p_3: 1 173°C). This reaction gives rise to a monovariant curve which projects into the ternary model, but over a short composition range only, terminating in the critical point K_s (44.4 at.-% Au; 48.9 at.-% Fe). At compositions less rich in Ni, the reaction which takes place along the line p_3–K_s is liquid + (γFe) $\rightleftharpoons$ (Au–Fe), each of the solid phases containing some Ni. As the point K_s is approached with increasing Ni content, the compositions of (γFe) and (Au–Fe) approach each other more and more closely, and coincide when the liquid composition reaches K_s. This condition corresponds with a critical tie line, and at

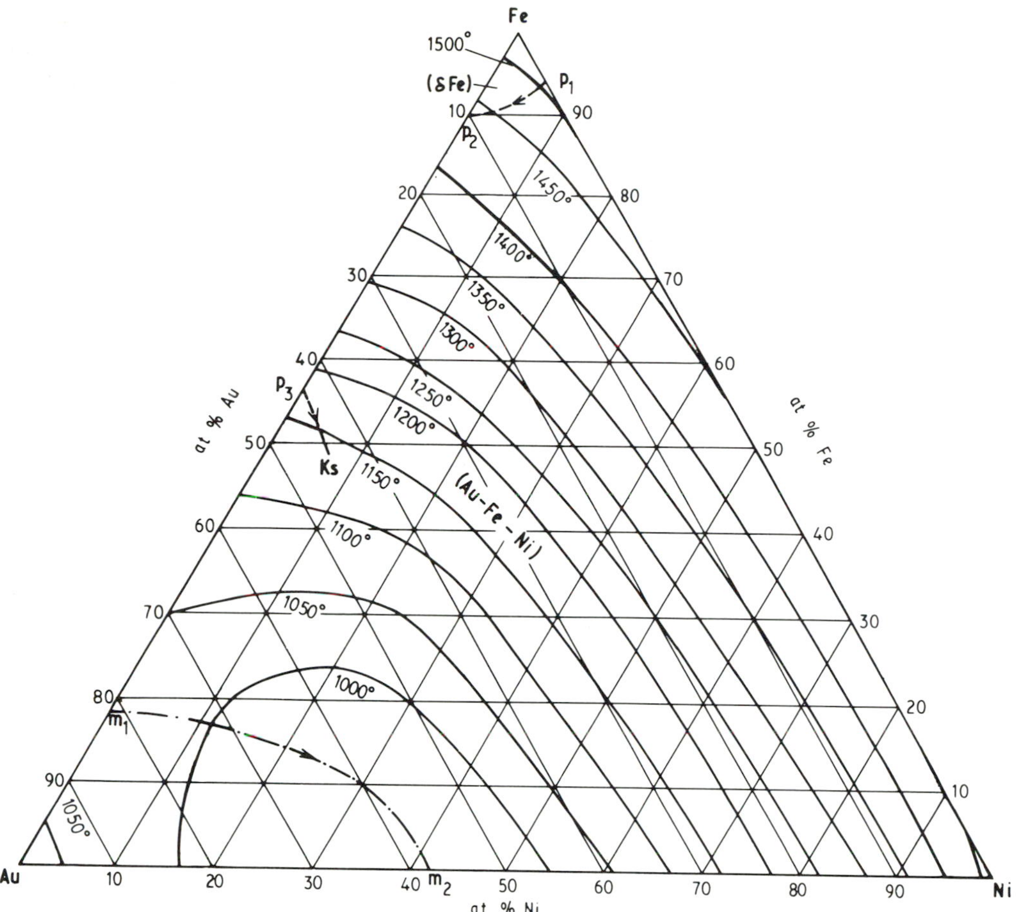

205 **System Au–Fe–Ni: projection of liquidus surfaces, and liquidus isothermal contours, after [61 Koe]. The curve p_1–p_2 is the projection of the monovariant ternary peritectic; the curve m_1–m_2 is the projection of the liquidus minimum. For explanation of curve p_3–K_s, see the text**

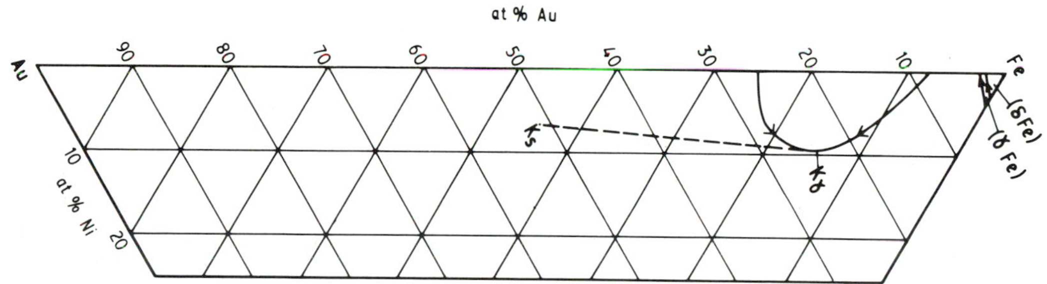

206 **System Au–Fe–Ni: projection of solid compositions arising from ternary monovariant reactions, after [61 Koe]**

compositions still richer in Ni, the liquid deposits one solid solution phase (Au–Fe–Ni) only. In Fig. 205 the projection of the minimum arising from those in the systems Au–Fe and Au–Ni is also shown (m_1–m_2). The projections of the solid compositions arising from the two ternary monovariant lines are shown in Fig. 206. The (δFe) and (γFe) compositions are confined to a region very close to the Fe corner. The compositions of the (γFe) and (Au–Fe) solid solutions arising from the p_3–K_s reaction are also included, as determined by [61 Koe]. The critical tie line K_s–K_γ is shown as a broken line, and the approximate composition of the point K_γ is 14·9 at.-% Au; 75·9 at.-% Fe. The results for solidus compositions are insufficiently critical for the construction of a general solidus projection.

The constitution in the solid state was reported in [61 Koe] in the form of polythermal sections corresponding with the composition sections described above, and with sections with constant Au compositions of 10, 20, 40, 60, and 70 wt-% Au. From these polythermal sections it is possible to construct isothermal diagrams showing the details of the equilibrium at various temperatures in terms of atomic percentages. Thus at 1 000°C the equilibria take the form of Fig. 207. The liquid region in the Au–Ni system projects into the ternary diagram as indicated, enclosed by the 1 000°C isotherm of Fig. 205. The corresponding solidus compositions are not well established, and the probable position of the boundary is given as a broken curve. The (γFe) + (Au–Fe) region of the Au–Fe system projects into the ternary diagram, and the maximum Ni content corresponding to immiscibility is ~12·8 at.-% at this temperature. The remainder of the isotherm is occupied by a ternary fcc solid solution denoted in Fig. 208 as (Au–Fe–Ni). At 900°C all compositions are solid and the constitution is represented by Fig. 208. The (γFe) + (Au–Fe) miscibility gap has expanded and now extends to a maximum of ~20·4 at.-% Ni. The major region of Fig. 208 is again that corresponding to the fcc ternary solid solution. A similar form of diagram, with some enlargement of the region of immiscibility, persists down to 810·3°C; this is the critical temperature for the onset of immiscibility in the Au–Ni system. At 800°C, therefore, the miscibility gap in the system Au–Ni has been established and the two-phase area, now to be considered as (Au) + (γFe–Ni), extends from near the Au–Fe to the Au–Ni axis, as shown in Fig. 209. The forms of the boundaries of this region are derived from [61 Koe]. The width of the (αFe) + (γFe–Ni) two-phase region has also increased, and the position of the three-phase triangle has again been fixed approximately by reference to the dilatometric results in [61 Koe]. The change in constitution between 1 000 and 800°C is considerable. At temperatures above 516°C the general form of equilibrium remains similar to that in Fig. 209, with further extension of the two-phase and three-phase areas. Thus at 600°C the results of [61 Koe] indicate the isothermal section shown in Fig. 210, which is a development of that shown in Fig. 209. The data provided by [61 Koe] would be consistent with the persistence of the form of equilibrium in Fig. 210 to low temperatures. The currently accepted Fe–Ni phase diagram, however, shows that the phase $FeNi_3$ is formed congruently from (γFe–Ni) at 516°C, and needs to be included in the equilibria below this temperature.

The Fe–Ni equilibria at low temperature [86 Chu 1] show the existence of paramagnetic (γFe–Ni), γ_1, ferromagnetic (γFe–Ni), γ_2, and (αFe) at 389°C. At 345°C, γ_2 transforms to (αFe) + $FeNi_3$. It is likely that the low-temperature equilibria in the Au–Fe–Ni system will be similar to those proposed for the Cr–Fe–Ni system [87 Chu], leading to room-temperature equilibrium between (Au) + (αFe) + $FeNi_3$ and (Au) + (γFe–Ni) + $FeNi_3$.

An examination of the lattice spacings of a layer of Au deposited on an Fe–Ni substrate and subjected to diffusion at temperatures up to 1 000°C [68 Oka] indicates that diffusion >~700°C is rapid, but gives no additional constitutional data.

REFERENCES

61 Koe: W. Köster and W. Ulrich: *Z. Metall.*, 1961, **52**, 383–391

68 Oka: H. Okamoto ; *Jap. J. Appl. Phys.*, 1968, **7**, 685–686

84 Oka: H. Okamoto, T. B. Massalski, L. J. Swartzendruber and P. A. Beck: *Bull. Alloy Phase Diagrams*, 1984, **5**, 592–601

86 Chu 1: Ying-Yu Chuang, Y. A. Chang, R. Schmid and Jen-Chwen Lin: *Metall. Trans. A*, 1986, **17A**, 1361–1372

86 Chu 2: Ying-Yu Chuang, Ker-Chang Hsieh and Y. A. Chang: *Metall. Trans. A*, 1986, **17A**, 1373–1380

87 Chu: Ying-Yu Chuang and Y. A. Chang: *Metall. Trans. A*, 1987, **18A**, 733–745

87 Oka: H. Okamoto and T. B. Massalski: 'Phase Diagrams of Binary Gold Alloys', 193–208; 1987, ASM International

Au–Fe–Pd
(G. V. Raynor)

INTRODUCTION

Alloys of the gold–iron–palladium system within certain composition ranges have a relatively high electrical resistivity and are promising as resistance materials, and their electrical properties are sensitive to mechanical and heat treatments. In spite of this comparatively little work has been carried out on the system. In the work of Kuranov *et al.* [71 Kur] a single alloy containing 46 at.-% Pd and 31·4 at.-% Au was investigated by electrical resistivity and microhardness methods, together with a survey of mechanical properties. The work gives little information on constitutional features, except that after quenching from 900°C and aging at various temperatures between 400 and 600°C, X-ray investigation revealed superlattice lines in

207 System Au–Fe–Ni: 1 000°C isothermal diagram, after [61 Koe]

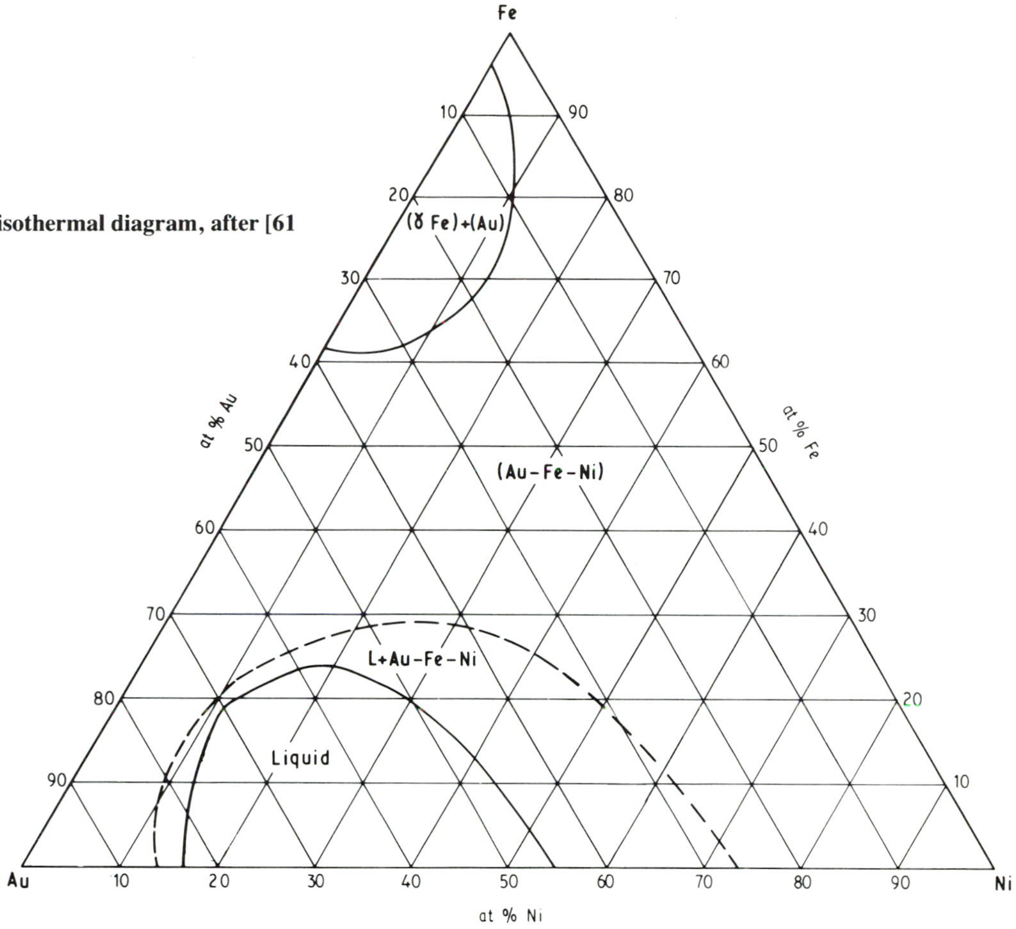

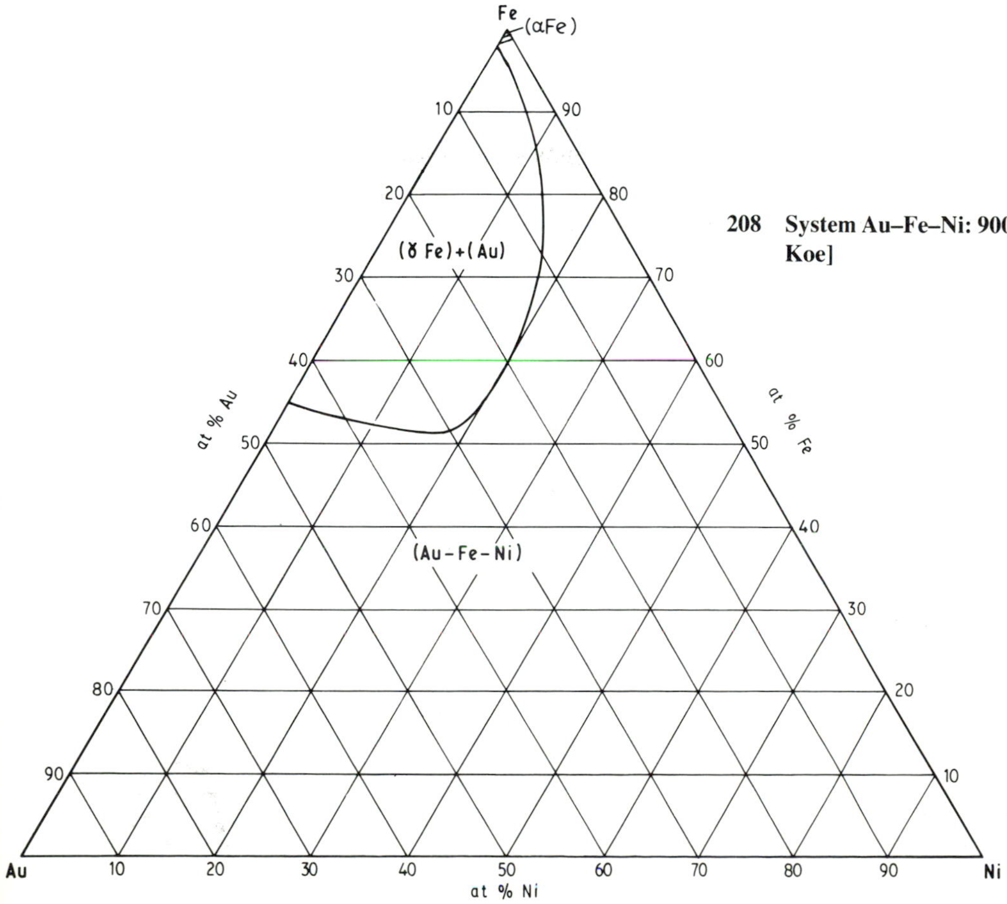

208 System Au–Fe–Ni: 900°C isothermal diagram, after [61 Koe]

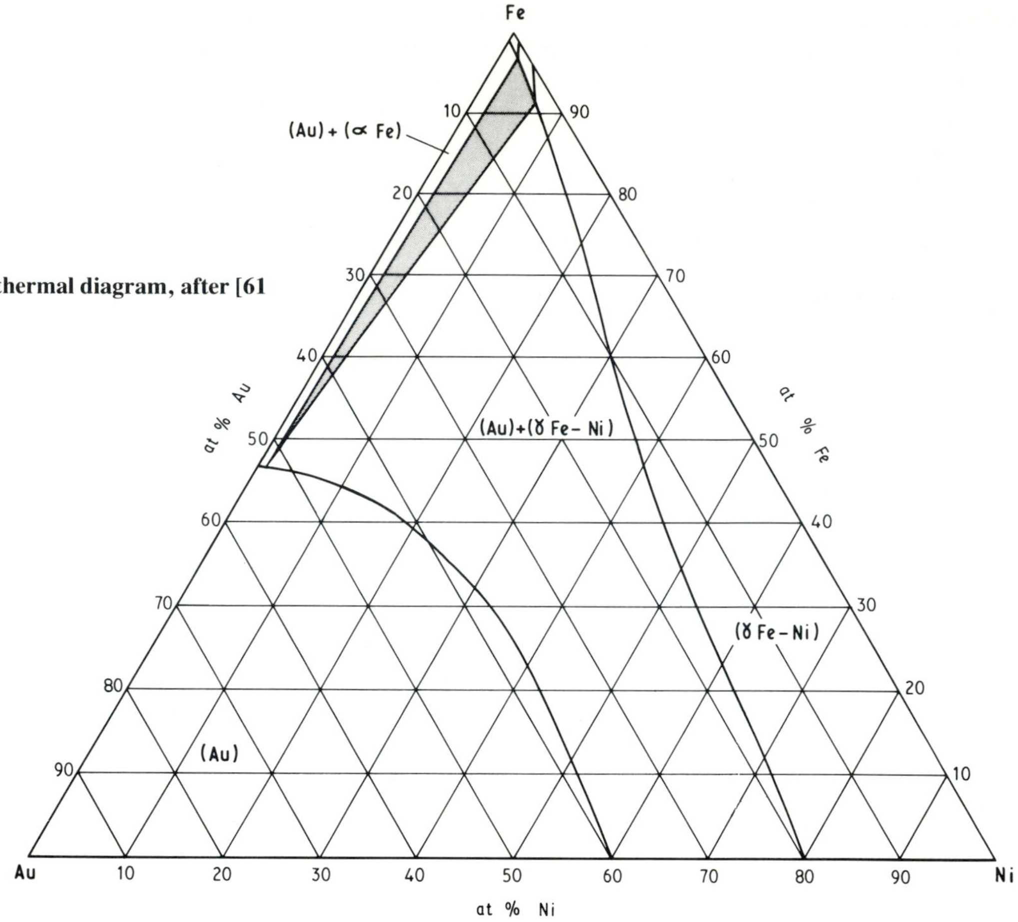

209 System Au–Fe–Ni: 800°C isothermal diagram, after [61 Koe]

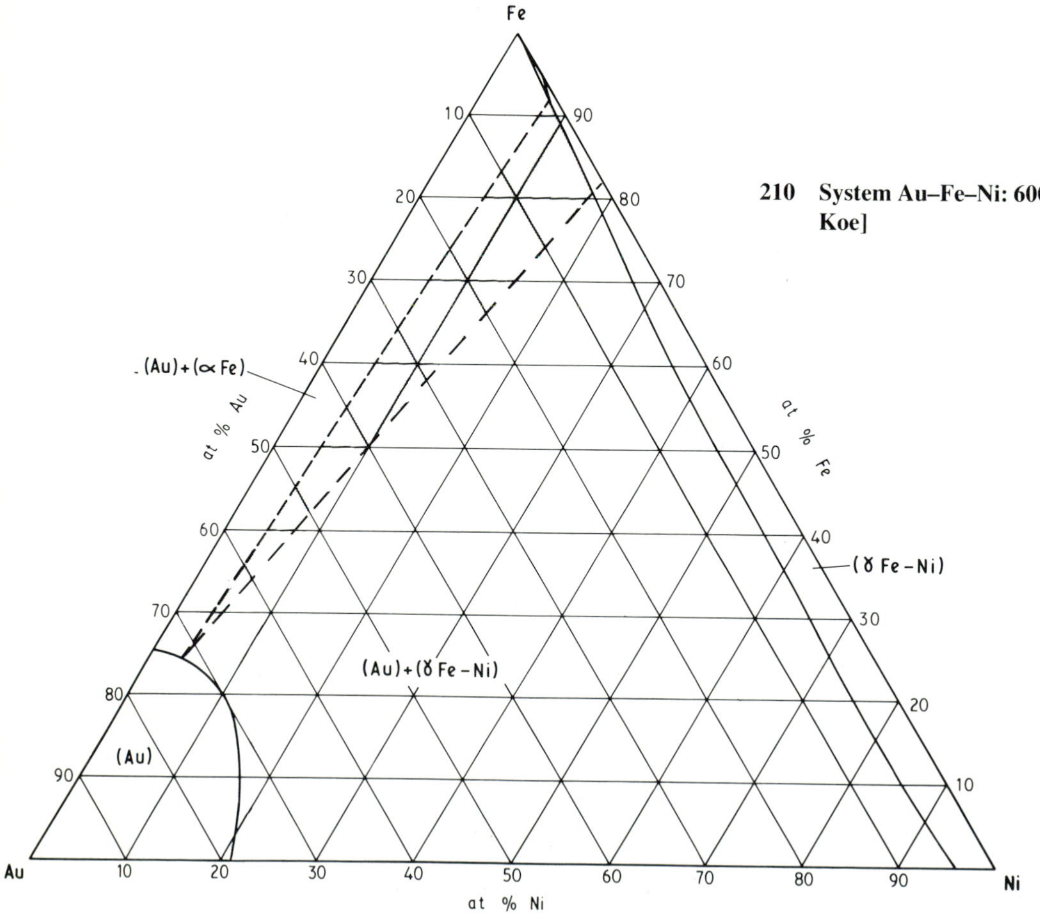

210 System Au–Fe–Ni: 600°C isothermal diagram, after [61 Koe]

diffraction photographs. The quenched alloy consisted of homogeneous solid solution. Later a more complete examination was carried out by [76 Pal]. Thirty alloys in the composition range 33–75 at.-% Pd and 0–50 at.-% Au were examined by X-ray diffraction methods; other measurements made included electrical resistivity, magnetization, and Curie points derived from the temperature dependence of magnetization in the range 300–900 K.

BINARY ALLOYS

The Au–Pd system has been assessed by [85 Oka] and the Au–Fe system by [84 Oka]. The Fe–Pd assessment by [82 Kub] is accepted.

THE TERNARY SYSTEM

Following the earlier work of [71Kur], Pal'guev [76 Pal] prepared 30 alloys, within the composition ranges quoted above, from Pd (99·5 wt-%), Fe (99·98 wt-%) and Au (99·99 wt-%). The components were melted together in appropriate quantities *in vacuo* in an induction furnace, using corundum crucibles. Alloys were annealed for 1 h at 947°C and either quenched or slowly cooled at 50°C h^{-1}. The constitutions of the alloys were then examined by X-ray diffraction methods. In addition electrical resistivity, its temperature coefficient, and Curie points were measured. All quenched alloys within the composition range studied possessed the disordered fcc structure. Slow cooling produced long range order of the AuCu type only in the binary Fe–Pd alloys containing 50 and 60 at.-% Pd, and in the ternary alloy 5 at.-% Au, 47.5 at.-% Pd, and 42.5 at.-% Fe. All other compositions studied were of the ordered $AuCu_3$ type. Apart from values for the parameters listed above, this is the whole of the constitutional information available, but when combined with the binary diagram data a probable low-temperature constitution for the ternary system may be proposed. It is not known to what temperature the constitution produced by slow cooling at the rate quoted approximates. It may be assumed to be approximately 400°C. With this assumption a diagram for this approximate temperature may be attempted, as in Fig. 211. The form and extent of the $FePd_3$ region is drawn to include all alloys shown by [76 Pal] to have the $AuCu_3$ structure. Probable positions for the narrow two-phase regions between $FePd_3$ and FePd, and between $FePd_3$ and the Au–Pd solid solution are included, necessitating the inclusion of a three-phase triangle (γ + FePd + $FePd_3$), the position of which is hypothetical. It is thought that the low temperature constitution can differ little from that illustrated. Atomic ordering in an alloy containing 25 at.-% Au, 25 at.-% Fe ($FePd_2Au$) has been studied [82 Ryz] by Mössbauer spectroscopy.

DISCUSSION

The work of [71 Kur] and [76 Pal] indicates that the production of a state of high electrical resistance in Au–Fe–Pd alloys is connected with the ordering reactions which occur on cooling. The maximum resistivity values correspond with $AuCu_3$-type order at compositions in the region of 25 at.-% Au, 25 at.-% Fe, and 40 at.-% Pd ($FePd_2Au$).

REFERENCES

71 Kur: A. A. Kuranov, I. B. Klyuyeva, A. F. Laptevskiy and R. A. Sasinova: *Trudy Inst. Fiz. Metal. Akad. Nauk, SSSR, Ural Filial*, 1971, **28**, 165–169

76 Pal: E. V. Pal'guev, A. A. Kuranov, P. N. Syutkin and F. A. Sidorenko: *Fiz. Metallov Metalloved*, 1976, **41**, 1208–1211; *Phys. Metals Metallogr.*, 1976, **41**(3), 84–88

82 Ryz: B. V. Ryzhenko, B. Yu. Goloborodskii, F. A. Sidorenko and P. V. Gel'd: *Soviet Phys. — Solid St.*, 1983, **24**, 18–22

82 Kub: O. Kubaschewski: 'Iron-Binary Phase Diagrams', 88–91; 1982, Berlin/Heidelberg, Springer-Verlag

84 Oka: H. Okamoto, T. B. Massalski, L. J. Swartzendruber and P. A. Beck: *Bull. Alloy Phase Diagrams*, 1984, **5**, 592–601

85 Oka: H. Okamoto and T. B. Massalski: *Bull. Alloy Phase Diagrams*, 1985, **6**, 229–235

Au–Fe–S

[44 Mas] melted Au and Fe with excess S. On cooling, two layers were observed; Au was the lower layer and pyrrhotite ($Fe_{1-x}S$) the upper layer. The pyrrhotite contained 1·97 wt-% Au but this was distributed as Au globules within the pyrrhotite grains and as films round the grains. It is evident that Au and $Fe_{1-x}S$ form immiscible liquids but no further data is available.

REFERENCE

44 Mas: I. Maslenitsky, *Compt. rend. (Doklady) Acad. Sci. URSS*, 1944, **45**, 385–388

Au–Ga–Ge

The AuGa–Ge and $AuGa_2$–Ge sections are both pseudobinary eutectic systems. The eutectic point is at 5·5 at.-% Ge and 446°C on the AuGa–Ge section, [89 But] and at 5·0 at.-% Ge and 476°C on the $AuGa_2$–Ge section [90 But].

REFERENCES

89 But: M. T. Z. Butt, C. Bodsworth and A. Prince, *J. Less-Common Metals*, 1989, **154**, 229–231

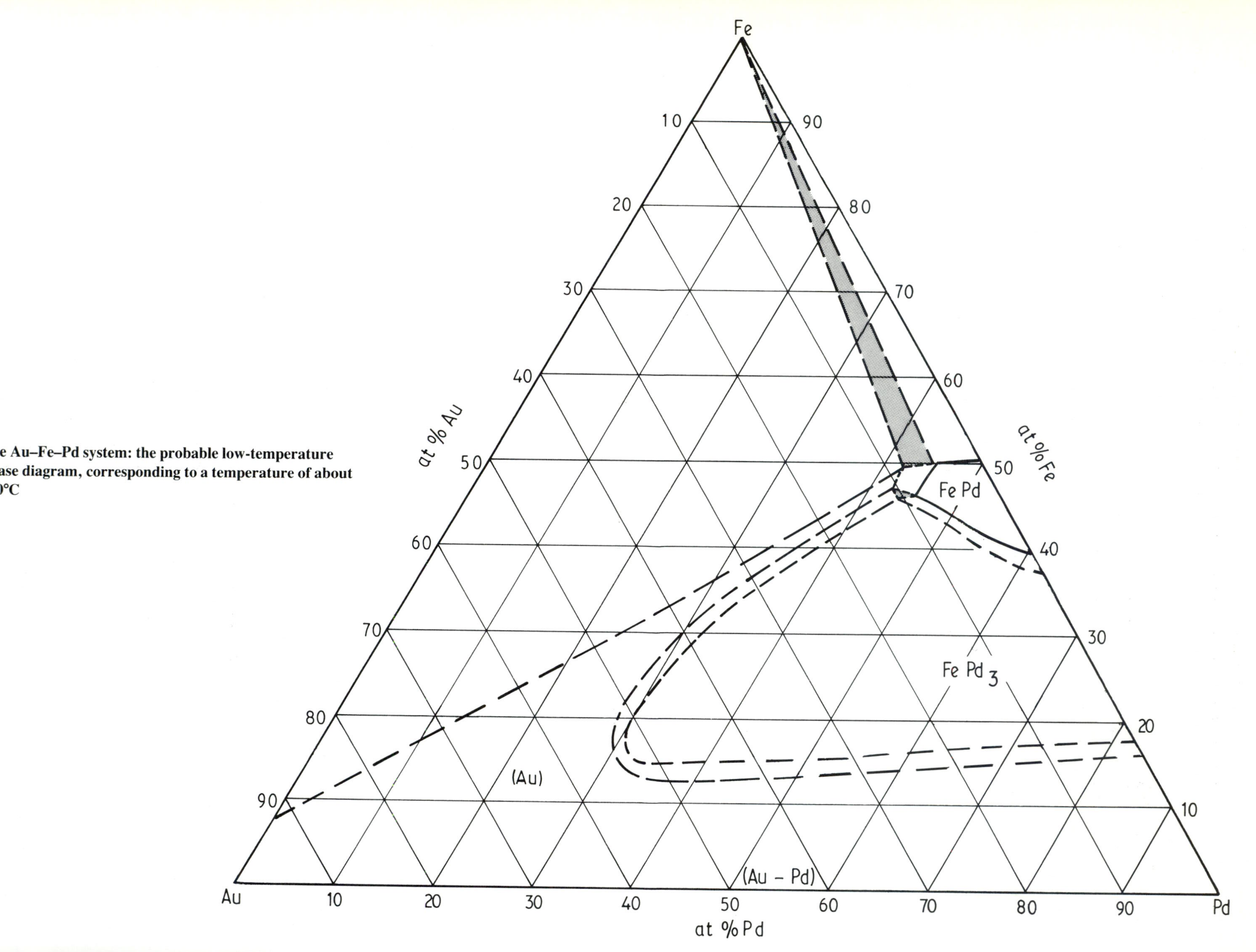

211 The Au–Fe–Pd system: the probable low-temperature phase diagram, corresponding to a temperature of about 400°C

90 But: M. T. Z. Butt, C. Bodsworth and A. Prince, *Scripta Met. Mater.*, 1990, **24**, 481–484

Au–Ga–In

Carter *et al.* [72 Car] studied the $AuGa_2$–$AuIn_2$ section at 450°C using metallographic, X-ray diffraction, and nuclear magnetic resonance techniques. Ternary alloys were prepared by melting $AuGa_2$ and $AuIn_2$ in sealed quartz ampoules and subsequently homogenizing for 10 days at 450°C. The brittle ingots were crushed to −200 mesh powder. X-ray data indicates a considerable solubility of $AuIn_2$ in $AuGa_2$ at 450°C; $AuGa_2$ is single phase up to at least 13·3 at.-% In. Limited data indicate a solubility of at least 3·3 at.-% Ga in $AuIn_2$ at 450°C. Jan and Pearson [63 Jan] found a two-phase structure for an alloy containing 33·33Au, 33·33Ga, 33·33 at.-% In.

Both $AuGa_2$ and $AuIn_2$ have a cubic fluorite structure (cF12). [78 Hoy] prepared three ternary alloys on the 91 at.-% Au section by arc-melting and subsequently annealing for 24 h at 500°C in argon. X-ray powder diffraction analysis showed that all alloys are in the (Au) solid solution phase region and that the lattice parameter varies linearly with composition from $Au_{92}Ga_8$ to $Au_{92}In_8$.

REFERENCES

63 Jan: J.-P. Jan and W. B. Pearson: *Phil. Mag.* 1963, **8**, 279–284

72 Car: G. C. Carter, I. D. Weisman, L. H. Bennett and R. E. Watson: *Phys. Rev. B.*, 1972, **5**(9), 3621–3638 (X-ray data due to C. Bechtoldt)

78 Hoy: R. F. Hoyt and A. C. Mota: *J. Less-Common Metals*, 1978, **62**, 183–88

Au–Ga–Mg

An alloy containing 30 at.-% Au, 3·5 at.-% Ga, 66·5 at.-% Mg, heat-treated under an argon pressure of 600 torr for 0·2 h at 600°C followed by water quenching, was not single phase but it contained the X-ray pattern of a ternary phase identified in the Au–Mg–X systems (X = Al, Ge, Si) [69 Loo] as $Au_{0 \cdot 9}Mg_2X_{0 \cdot 1}$. This phase is orthorhombic with a $PbCl_2$-type structure (oP12).

REFERENCE

69 Loo: N. Van Look and K. Schubart: *Metall.*, 1969, **23**(1), 4–6

Au–Ga–P

[78 Dzh] reported relatively low solubilities of Au in (111)-oriented single crystal GaP over the temperature range 1 050 to 1 300°C. Values ranged from 0·01 ppm Au (5×10^{14} atoms Au. cm^{-3}) at 1 050°C to 0·20 ppm Au (1×10^{16} atoms Au. cm^{-3}) at 1 300°C.

REFERENCE

78 Dzh: T. D. Dzhafarov, A. A. Litvin and S. V. Khudyakov, *Fiz. Tverd. Tela*, 1978, **20**, 267–269

Au–Ga–Pd
(G. V. Raynor)

The only published work on the Au–Ga–Pd ternary system [69 Wer] relates to the superconductivity, and other physical properties, of solid solutions of Pd in $AuGa_2$. The alloys used for the physical property measurements were of the series $Au_{1-x}Pd_xGa_2$, where x took the values 0·02, 0·05, and 0·1. Stoichiometric amounts of the three metals, of purity exceeding 99·999%, were melted together in recrystallized alumina crucibles by induction heating in argon. Ingots were subsequently zone-melted. Specimens were analysed by atomic absorption analysis, and the intended compositions were confirmed. Room temperature lattice spacing determinations were made on alloys, presumably similarly prepared, covering a wider composition range up to $x = 0 \cdot 5$. The lattice spacing of $AuGa_2$ is contracted by the solution of Pd, and a graph included in [69 Wer] shows that the plotted lattice spacing curve intersects the constant spacing line correspondingly to saturation of $AuGa_2$, with Pd at $x = 0 \cdot 19$ (27·00Au–66·67Ga–6·33% Pd). However, the alloy $Au_{0 \cdot 85}Pd_{0 \cdot 15}Ga_2$ (28·33Au–66·67Ga–5·00% Pd), although its lattice spacing was plotted on the decreasing spacing curve for the solid solution, was stated to contain some second phase. The maximum solubility at room temperature was therefore stated to be below this value. Since the thermal history of the samples is not given in details, and no details are given of measures taken to approach equilibrium, the solid solubility of Pd in $AuGa_2$ cannot be taken as accurately defined. It must, however, be close to 5% Pd at 66·67% Ga. No information is available with regard to the liquidus surfaces, or to the general constitution in the solid state, and further work is required.

REFERENCE

69 Wer: J. H. Wernick, A. Menth, T. H. Geballe, G. Hull and J. P. Maita: *J. Phys. Chem. Solids*, 1969, **30**, 1949–1956

Au–Ga–Sb

The only published data [86 Tsa] on the Au–Ga–Sb system is a study of 15 alloy compositions and the characterization of the phases present by X-ray diffraction analysis. Alloys were prepared in 1 g quantities from elements of >99·99% purity and semiconductor grade GaSb. Each alloy was contained in an evacuated, 10^{-6} torr, and sealed fused-silica capsule with 4 cm^3 dead space. Alloying was effected by heating at 1 000°C for 1 week and samples were slowly cooled to either 25 or 600°C at a rate of 5–10°C h^{-1}. The samples cooled to 600°C were then quenched into iced water.

Table 74 lists the alloys studied and their phase constitution. Tsai and Williams [86 Tsa] concluded that the sections GaSb–$AuGa_2$, GaSb–AuGa, γ'–$AuSb_2$, and β'–$AuSb_2$ are pseudobinary sections on the basis that the alloys studied on these sections contained the appropriate two phases only. Whereas this conclusion is probably correct for the GaSb–$AuGa_2$, GaSb–AuGa, and the γ'–$AuSb_2$ sections (γ' contains 70Au, 30Ga), the section between β' (containing 78Au, 22Ga) and $AuSb_2$ cannot be a pseudobinary near the Au–Ga binary since β' is formed by peritectic reaction. The alloys in the lower half of Table 74 all contained four or more phases and were not in equilibrium. Alloys E, F, H, I, G, J, and O are most probably situated within the primary phase field of GaSb (Fig. 212). Alloys A, B, C, and D are also most probably within the GaSb primary phase region. [86 Tsa] did not define the triangulation of the ternary bounded by γ'–AuGa–GaSb–Sb–$AuSb_2$, but speculate on the possibility of an invariant reaction near to room temperature. One can anticipate three invariant reactions however this region of the ternary is triangulated; none would be expected to involve a liquid phase at temperatures lower than about 300°C.

For the alloys quenched from 600°C it is worth noting that alloys A, C, and O would contain primary GaSb + liquid at 600°C, whereas alloys F, L, and M would most probably be entirely liquid at 600°C (Fig. 212). Quenching alloys on a pseudobinary from the liquid phase or a liquid + solid phase region would produce the two-phase structure identified by [86 Tsa]. Alloys such as F and O would produce non-equilibrium structures if they did not lie on pseudobinary sections. For alloy F the γ'–GaSb and AuGa–$AuSb_2$ sections are thus eliminated as pseudobinary sections; for alloy O the GaSb–$AuSb_2$ and AuGa–Sb sections are eliminated as pseudobinary sections. [86 Tsa] refer to unpublished work of [86 Lin] on the reaction of Au films deposited on single crystal GaSb upon heating to an unspecified temperature. Strong Sb, medium AuGa, and weak $AuSb_2$ peaks were identified. [86 Tsa] tentatively assigned the sections AuGa–Sb and AuGa–$AuSb_2$ as pseudobinary sections at room temperature. This is not accepted. [86 Tsa] seem to assume that all the ternary boundaries of three-phase regions are pseudobinary sections.

There is a need for further work on this ternary system, using a range of experimental techniques, to determine the liquidus surface and the nature of the invariant reactions.

REFERENCES

86 Tsa: C. T. Tsai and R. S. Williams: *J. Mater. Res.*, 1986, **1**(2), 352–360

86 Lin: J. R. Lince and R. S. Williams: *Thin Solid Films* (to be published)

Au–Ga–Si

Gubenko and Shmelev [71 Gub] used the radioactive tracer method to study the effect of Ga doping on the solubility of Au in Si. Figure 213*a* shows that doping with Ga at 2×10^{17} atoms cm^{-3} (4 atomic ppm Ga) and 1×10^{18} atoms cm^{-3} (20 atomic ppm Ga) increases the solubility of Au in Si by a relatively small amount. The binary L + (Si)/(Si) solidus is due to [83 Oka]. Figure 213*b*

Table 74. Alloys studied by [86 Tsa]

Alloy section	*Alloy*	*Alloy composition*, at.-%			Temperature, °C	*Phases identified*
		Au	*Ga*	*Sb*		
GaSb–$AuGa_2$	A	14·3	57·1	28·6	25,600	GaSb, $AuGa_2$
	B	25·0	62·5	12·5	25	GaSb, $AuGa_2$
GaSb–AuGa	C	16·67	50	33·33	25,600	GaSb, AuGa
	D	33·33	50	16·67	25	GaSb, AuGa
γ'–$AuSb_2$	M	51·2	14·6	34·2	25,600	$AuSb_2$, γ'
β'–$AuSb_2$	L	66·67	16·67	16·67	25,600	$AuSb_2$, β'
	E	18·4	44·7	36·9	25	GaSb, $AuSb_2$(m), AuGa(m), γ'(w), Sb(w)
	F	45·7	37·0	17·3	25	GaSb, $AuSb_2$(m), AuGa, γ'(m)
	F	45·7	37·0	17·3	600	GaSb, $AuSb_2\gamma'$, Sb(m)
	H	18·9	43·2	37·9	25	GaSb, $AuSb_2$(m), AuGa(m), Sb(m)
	I	44·7	34·0	21·3	25	GaSb, $AuSb_2$(m), AuGa, γ'(m), Sb(w)
	G	33·33	33·33	33·33	25	GaSb, $AuSb_2$, AuGa, Sb
	J	42·9	28·6	28·6	25	GaSb, $AuSb_2$(m), AuGa(m), γ'(w), Sb
	K	38·9	16·7	44·4	25	$AuSb_2$ AuGa(w), γ', Sb
	N	53·8	23·1	23·1	25	$AuSb_2$, AuGa(w), γ', Sb(w)
	O	20·0	20·0	60·0	25	GaSb, $AuSb_2$(m), AuGa(m), Sb
	O	20·0	20·0	60·0	600	GaSb, AuGa(m), Sb, Au(w)

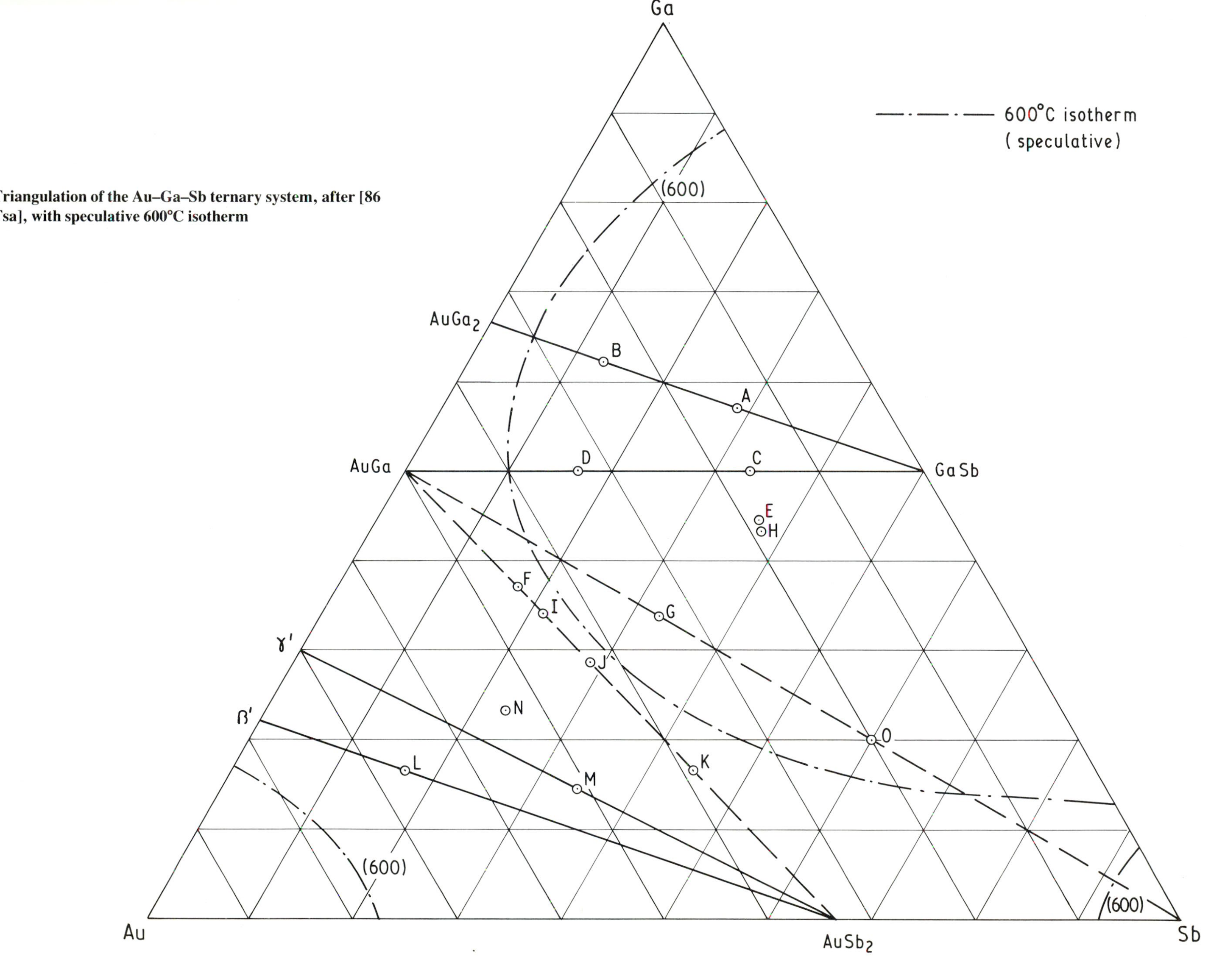

212 Triangulation of the Au–Ga–Sb ternary system, after [86 Tsa], with speculative 600°C isotherm

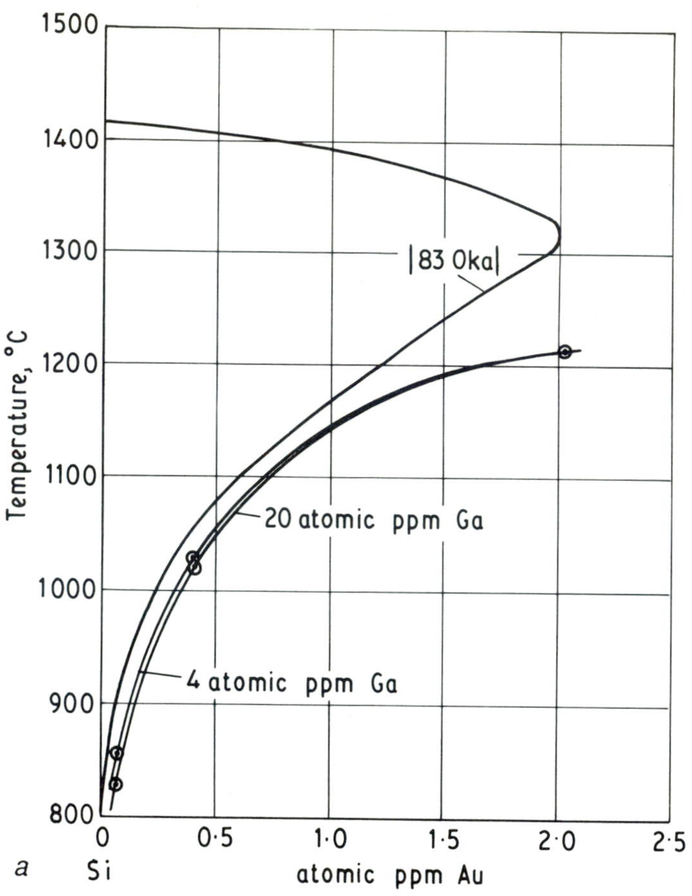

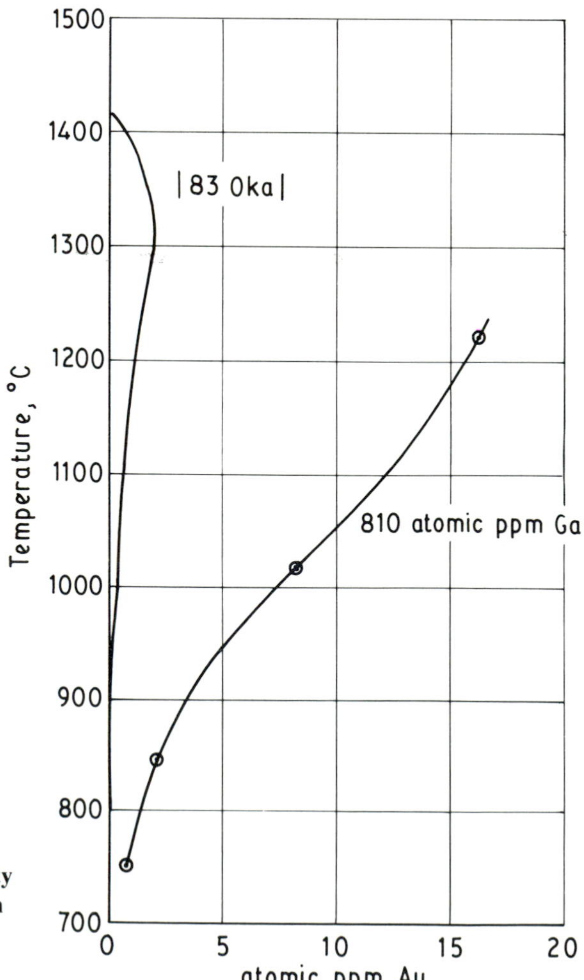

213 ***a*** **The effect of Ga (4 and 20 atomic ppm) on the solubility of Au in Si [71 Gub];** ***b*** **Effect of Ga (810 atomic ppm) on the solubility of Au in Si [71 Gub]**

demonstrates the marked increase in the solubility of Au in Si when the level of Ga doping is increased to 4×10^{19} atoms cm^{-3} (810 atomic ppm Ga).

Kuznetsov and Shmelev [74 Kuz] studied the sections Au, 2·8 at.-% Ga–Si and Au, 9·3 at.-% Ga–Si. Thermal analysis was used to detect the liquidus for the three ternary alloys with low Si contents (6·5, 12·3, and 26·6 at.-% Si for the former section and 6·2, 11·9, and 25·8 at.-% Si for the latter section). Metallography on quenched samples was used to detect the liquidus for Si-rich alloys (43·3, 54·8 and 63·3 at.-% Si for the former section and 42·3, 53·8, 62·2 at.-% Si for the latter section). Figure 214 shows that the Au, 2·8 at.-% Ga–Si section is of the anticipated type with a monovariant eutectic separation of (Au) + (Si) over a narrow temperature range. At low temperatures one would expect to see a phase boundary separating the (Au) + (Si) phase region from a (Au) + (Si) + β′ phase region since the binary Au, 2·8 at.-% Ga alloy begins to separate β′ from the (Au) solid solution at about 75°C [87 Oka]. The section Au, 9·3 at.-% Ga–Si (Fig. 215) is more complex, and [74 Kuz] noted that they observed four phases in the solidified alloys. They assumed that this was due to the invariant reactions L + (Au) = α′ + (Si) and L + α′ = β + (Si). This cannot be so in this section, since both [74 Kuz] and [87 Oka] show that the binary Au, 9·3 at.-% Ga alloy does not undergo any invariant reaction associated with a liquid phase. This binary alloy separates β from the (Au) solid solution and the β transforms eutectoidally at 282°C to produce (Au) + β′. The proposed disposition of phase regions is in agreement with the Au–Ga binary constitution, and it agrees with [74 Kuz] in that it includes the phase regions L + (Au) + (Si), (Au) + β + (Si) and (Au) + β′ + (Si). The four-phase invariant reaction, shown by the dashed line, corresponds to the ternary eutectoid transformation of β to (Au) + β′ + (Si). Incomplete reaction would lead to the detection of four phases as noted by [74 Kuz]. It is possible to suggest a reaction scheme for the Au-rich part of the Au–Ga–Si ternary system (Fig. 216) consistent with the ternary sections determined by [74 Kuz].

REFERENCES

71 Gub: A. Ya. Gubenko and Yu. I. Shmelev: *Izvest. Akad. Nauk. SSSR, Neorg. Materialy.*, 1971, **7**, 731–733

74 Kuz: G. M. Kuznetsov and Yu. I. Shmelev: *Izvest. Akad. Nauk. SSSR, Neorg. Materialy.*, 1974, **10**, 1257–1262

83 Oka: H. Okamoto and T. B. Massalski: *Bull. Alloy Phase Diagrams*, 1983, **4**, 190–198

87 Oka: H. Okamoto and T. B. Massalski: 'Phase Diagrams of Binary Gold Alloys', 111–118; 1987, ASM International

Au–Ga–Sn

Humpston [85 Hum] studied a number of AuSn–X systems. The AuSn–Ga section is not a pseudobinary. [85 Eva] studied the section 70·7 Au 29·3 at.-% Sn–Ga up to its intersection with the AuGa–AuSn join (Fig. 217) by DTA techniques. A projection of the probable liquidus surface is given in Fig. 218. The AuGa–AuSn section is a pseudobinary eutectic system with a eutectic temperature of 357·1 ± 1°C. The composition of the eutectic, e_1 in Fig. 218 was estimated to be at 22 at.-% Ga. The γ–AuSn section is also a pseudobinary eutectic system with a eutectic temperature of 296·7 ± 1°C and an estimated eutectic composition, e_4, of 21·5 at.-% Ga (γ is the congruently melting Au–Ga compound existing from 29·8 to 31 at.-% Ga). Three ternary invariant reactions were identified. A ternary eutectic reaction E_1 takes place at 293·6°C. The partial system γ–AuGa–AuSn is a simple ternary eutectic system. The composition of E_1 is estimated to be 61·1 Au, 22·5 Ga, 16·4 at.-% Sn. A transition reaction U occurs at 269·7°C. Liquid of composition U lies on the extension of the line AuSn–b (Fig. 218). It has been arbitrarily located at 71·3 Au, 14·7 Ga–14 at.-% Sn on this line. The reaction at U is $L + \gamma \rightleftharpoons \beta' + AuSn$. A ternary eutectic reaction E_2 occurs at 255·2°C, $L \rightleftharpoons \beta' + \zeta + AuSn$. E_2 lies on the extension of the line AuSn–a (Fig. 218). It has been arbitrarily located at 72·4 Au–8·8 Ga–18·8 at.-% Sn on this line.

REFERENCES

85 Hum: G. Humpston: PhD Thesis, 1985, Brunel University

85 Eva: D. S. Evans and A. Prince: llèmes Journées d'Etudes des Equilibres entre Phases, Marseille, March 1985

Au–Ga–V

In studying solid solution formation between Cr_3Si-type phases (cP8) von Philipsborn [70 Phi] prepared a $V_3Au_{0.5}Ga_{0.5}$ alloy by arc-melting V_3Au and V_3Ga on a water-cooled Cu hearth under argon. Samples were annealed 45 days at 550°C, 4 days at 850°C, and 20 days at 1 050°C in evacuated quartz ampoules. The phases present were characterized by the X-ray powder diffraction technique.

The as-cast sample contained a Cr_3Si-type phase with lattice parameter a = 0·487 ± 0·001 nm and a W-type phase with a = 0·307 ± 0·001 nm. The samples annealed at 550 and 850°C contained a Cr_3Si-type phase with a = 0·484 ± 0·001 nm and a V_3AuO phase with a Cu_3Au-type structure. The sample annealed at 1 050°C contained two Cr_3Si-type phases with a = 0·4886 and 0·4855 ± 0·0003 nm and a V_3AuO phase. Without further data it is not possible to interpret these results.

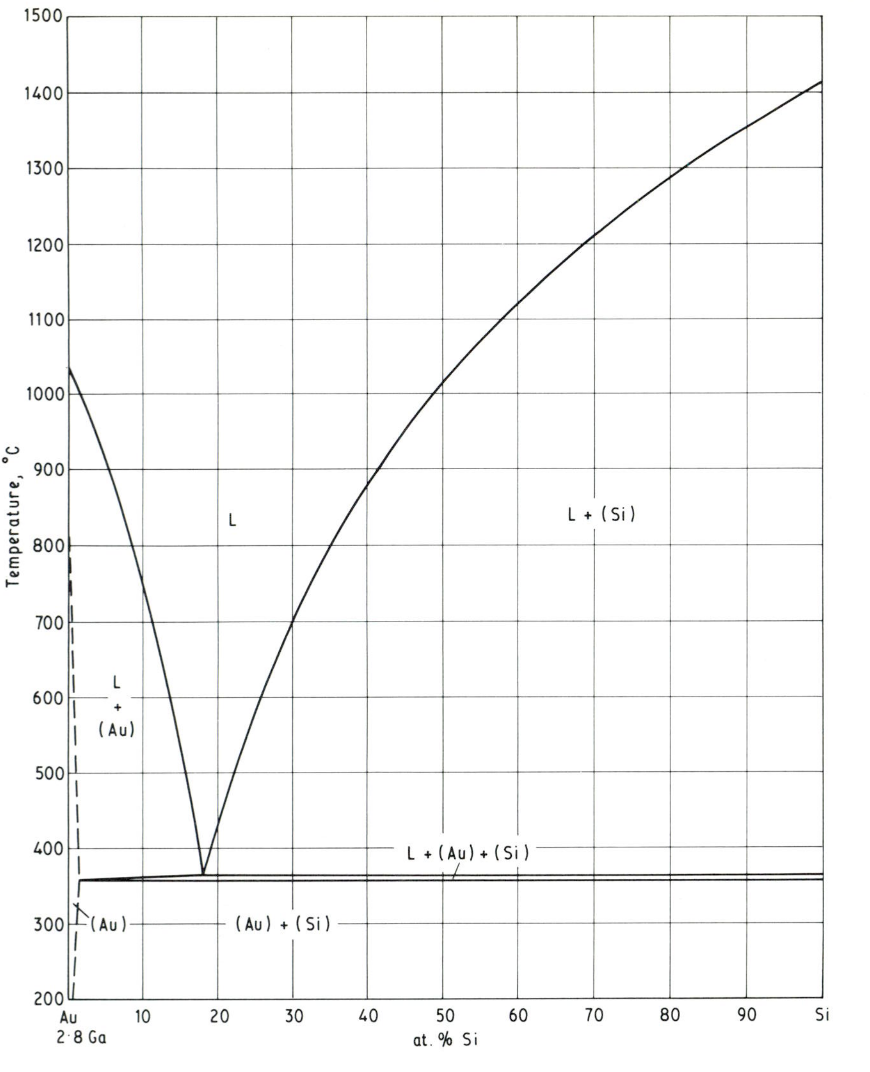

214 The section Au, 2·8 at.-% Ga–Si [74 Kuz]

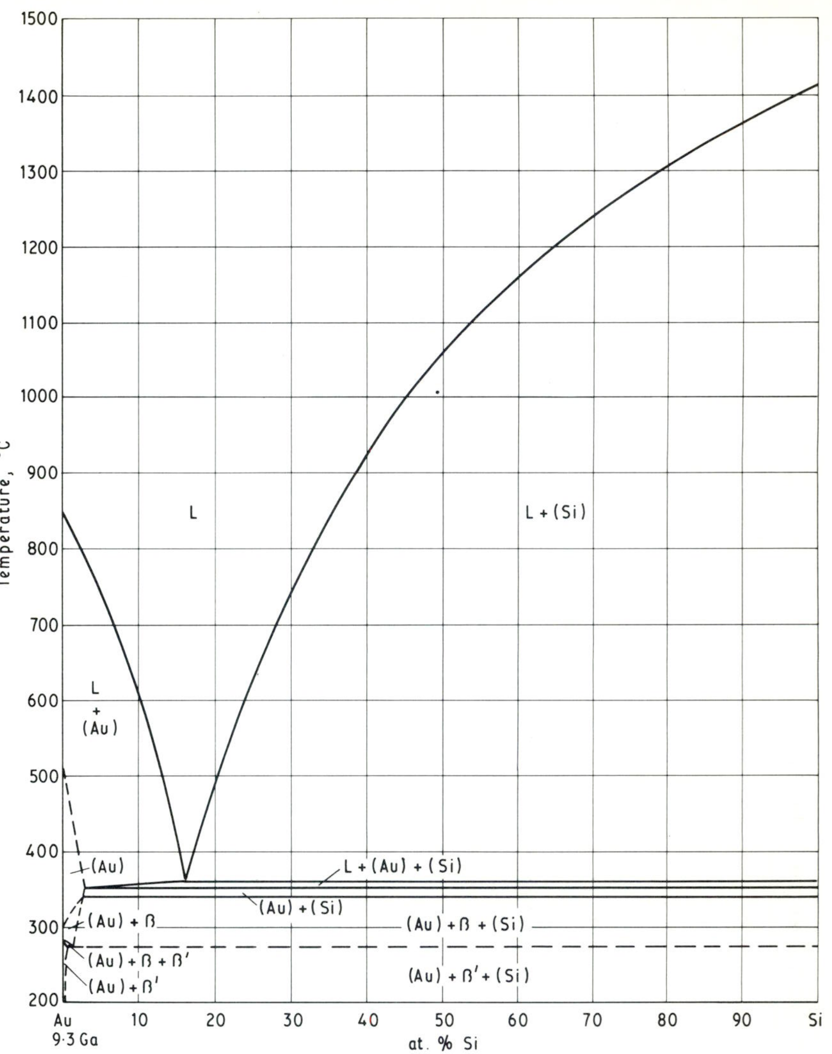

215 The section Au, 9·3 at.-% Ga–Si [74 Kuz]

216 Suggested reaction scheme for the Au-rich part of the Au–Ga–Si ternary system

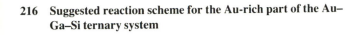

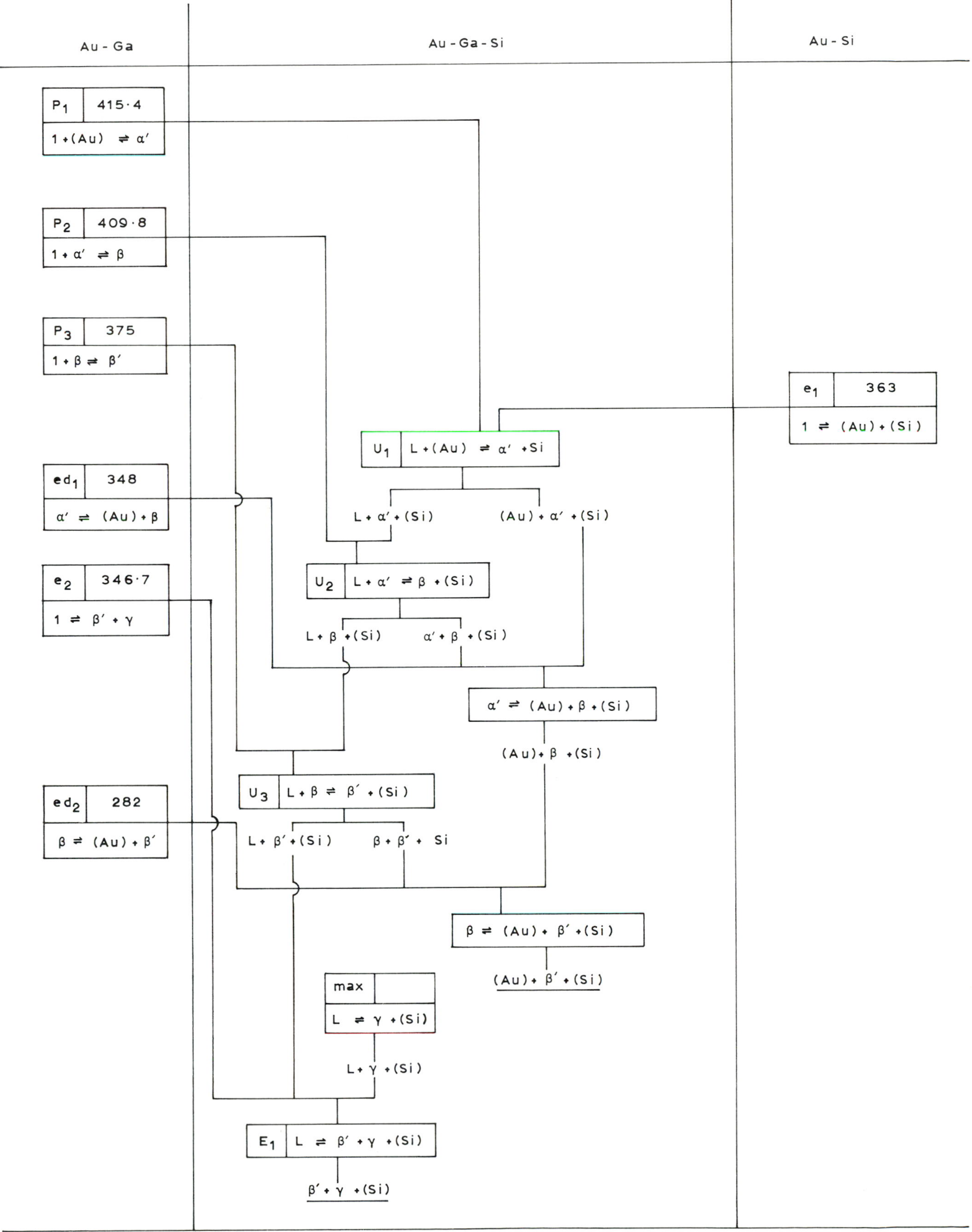

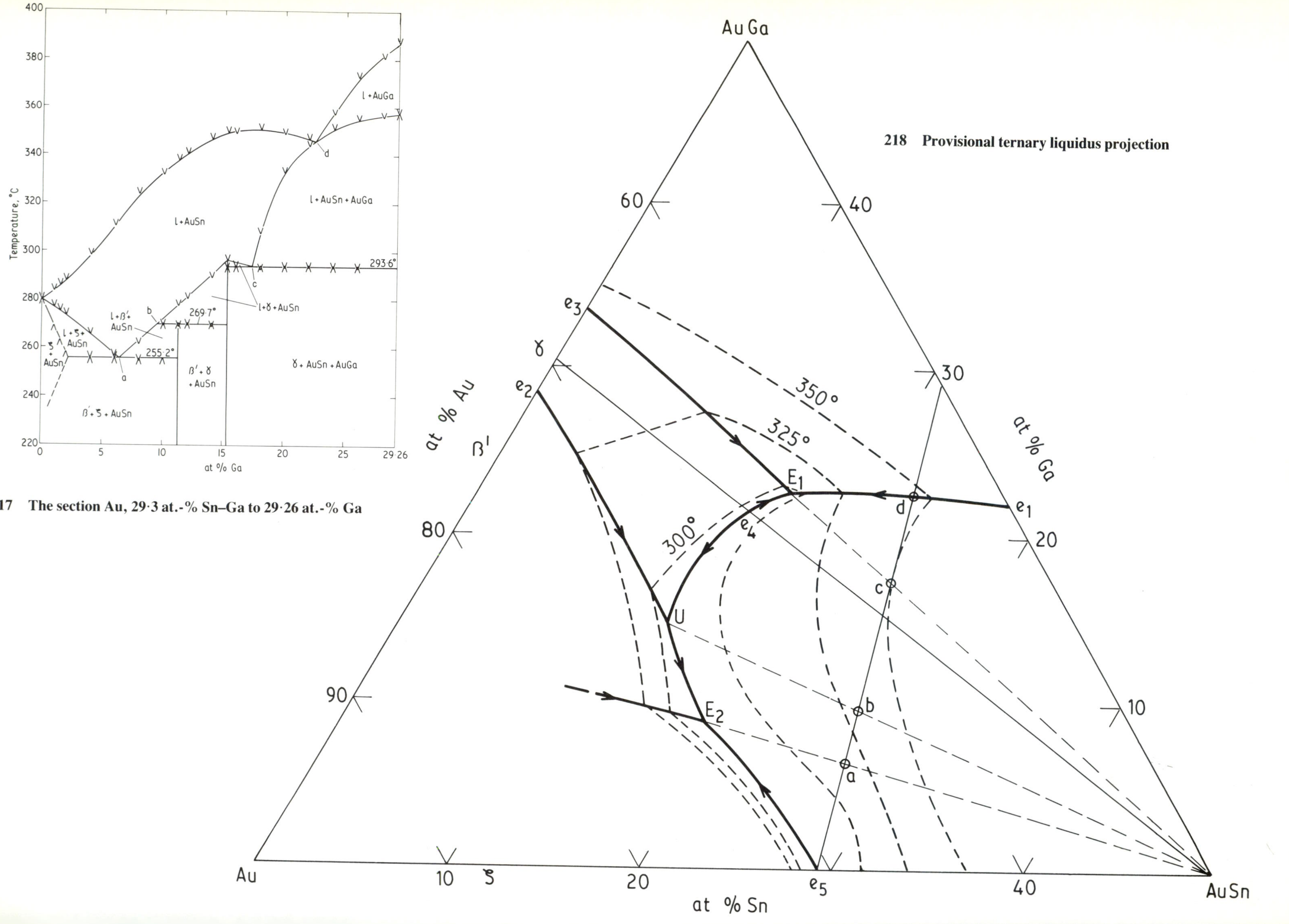

217 The section Au, 29·3 at.-% Sn–Ga to 29·26 at.-% Ga

218 Provisional ternary liquidus projection

REFERENCE

70 Phi: H. von Philipsborn: *Z. Kristallogr.*, 1970, **131**, 73–87

Au–Ga–Zn

Wilkens and Schubert [57 Wil] used the X-ray powder diffraction method on nine ternary alloys, annealed for 8 days at 200°C, to delineate the phase region for the doubly-faulted LI_2-type structure based on Au_4Zn (Fig. 219).

REFERENCE

57 Wil: M. Wilkens and K. Schubert: *Z. Metall.*, 1957, **48**, 550–557

Au–Gd–Tb

Gamari-Seale and Bredimas [85 Gam] found a solid solution series between the isostructural compounds Gd_2Au and Tb_2Au (orthorhombic Co_2Si-type structure, oP12 (C23) with four formula units in the unit cell). $Gd_xTb_{2-x}Au$ alloys were induction-melted under argon using 99·99% Gd, Tb and 99·999% Au. The alloys were heat treated for 48 h at 550°C (x = 0·2, 0·8, 1·0, 1·2, 1·8), crushed to powder and X-ray analysed. All were single-phase with the Co_2Si-type structure.

REFERENCE

85 Gam: H. Gamari-Seale and V. Bredimas: *J. Magn. Magn. Mater.*, 1985, **49**, 155–160

Au–Ge–In

The $AuIn_2$–Ge section is a pseudobinary eutectic system with a eutectic at 4·1 at.-% Ge and 522°C [88 But 1]. The AuIn–Ge section is also a pseudobinary eutectic system with a eutectic at 2·0 at.-% Ge and 488°C [88 But 2]. The $AuIn_2$–AuIn–Ge partial ternary system is a simple ternary eutectic system with a ternary eutectic temperature of 471°C. The ternary eutectic composition was not established [88 But 1].

REFERENCES

88 But 1: M. T. Z. Butt and C. Bodsworth, submitted for publication, 1988

88 But 2: M. T. Z. Butt and C. Bodsworth, submitted for publication, 1988

Au–Ge–Mg

An alloy containing 30 at.-% Au, 3·5 at.-% Ge, 66·5 at.-% Mg, heat-treated under an argon pressure of 600 torr for 0·2 h at 600°C followed by water quenching, had a single phase structure. X-ray analysis [69 Loo] confirmed the phase to be a ternary compound analogous to $Au_{0\cdot9}Mg_2Si_{0\cdot1}$, which has an orthorhombic $PbCl_2$-type structure (oP12).

REFERENCE

69 Loo: N. Van Look and K. Schubert: *Metall.*, 1969, **23**(1), 4–6

Au–Ge–Na

Wrobel and Schuster [75 Wro] mixed equimolar fractions of AuNa and GeNa, melted them for a short time in a Ta crucible at 1 000°C under argon, and subsequently annealed the alloy at 600°C, followed by slow furnace cooling. X-ray analysis indicated that a ternary compound, chemically analysed as Na_2AuGe, was orthorhombic. These investigators [77 Wro] also carried out a single crystal structure determination and quote the lattice parameters for Na_2AuGe as a = 0·722(7), b = 0·752(9), c = 0·441(7) nm. It belongs to space group Immm (oI16). It was noted, without elaboration, that several ternary phases were found in the Au–Ge–Na system. Döring *et al.* [79 Dör] heated the elements, with an excess of Na, in fused quartz ampoules in a Ta crucible. An initial mixture, corresponding to a composition $Na_{3\cdot3}Au_{2\cdot5}Ge$, after preheating for 12 h at 500°C, 20 min melting at 1 000°C, and 5 days annealing at 500°C, produced a chemically analysed ternary phase $Na_{1\cdot75}Au_{2\cdot5}Ge$. Its crystal structure was characterized from single crystals as bcc with a = 1·462 nm and related to the $Mg_{32}(Zn,Al)_{49}$ structure. The composition corresponding to the structural analysis was $Na_{52}Au_{80}Ge_{30}$, a structure with 162 atoms in the unit cell (cI162).

The ternary compounds $Na_{1\cdot75}Au_{2\cdot5}Ge$, Na_2Au_3Sn and $Na_{1\cdot5}Au_2Si$ [79 Dör] all occur at 33·33 at.-% Na with Au contents ranging from 44·4 to 50 at.-%; they all have the same crystal structure.

Döring and Schuster [80 Dör] reported the synthesis of a ternary compound $NaAu_3Ge$ from a Na-rich mixture of the composition $Na_{1\cdot5}Au_3Ge$ using the same preparative technique as [79 Dör]. The crystal structure, determined by single crystal X-ray diffraction, is cubic, space group $Pa\bar{3}$ (cP40) with unit cell parameter a = 0·902 1 nm. There are eight formula units in the elementary cell. The compound $NaAu_3Ge$ tends to decompose at unspecified high temperatures with the appearance of Au reflections in the X-ray spectra. The analogous compound $NaAu_3Si$ has the same crystal structure as $NaAu_3Ge$; the compound $NaAu_3Sn$ was not formed.

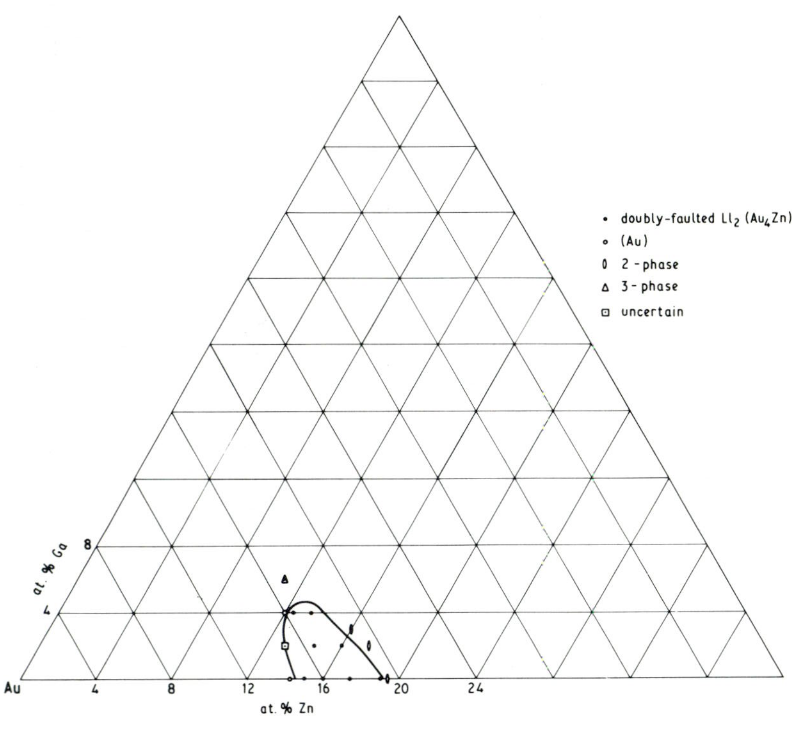

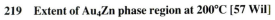

219 Extent of Au_4Zn phase region at 200°C [57 Wil]

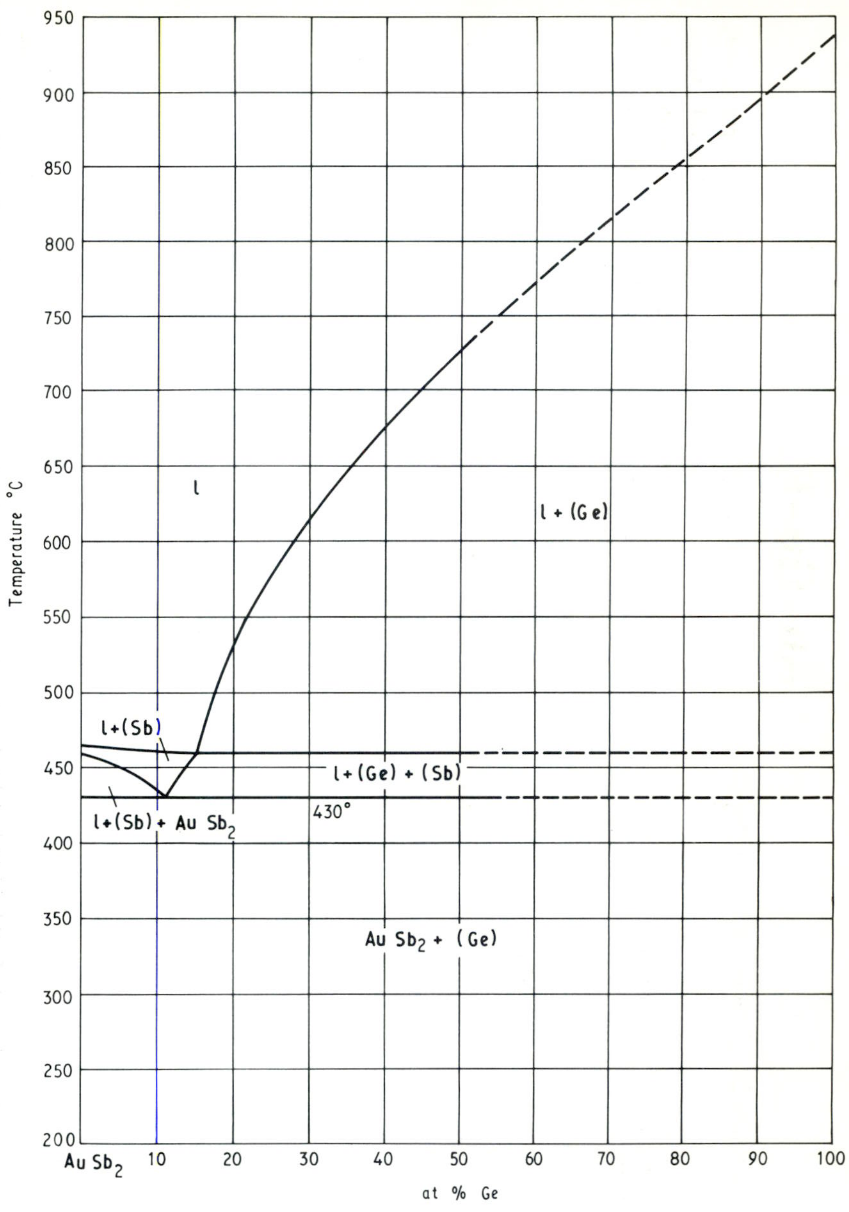

220 The $AuSb_2$–Ge section

REFERENCES

75 Wro: G. Wrobel and H.-U. Schuster: *Z. Naturf.*, 1975, **30B**, 806

77 Wro: G. Wrobel and H.-U. Schuster: *Z. anorg. Chem.*, 1977, **432**, 95–100

79 Dör: W. Döring, W. Seelentag, W. Buchholz and H.-U. Schuster: *Z. Naturf.*, 1979, **34B**, 1715–1718

80 Dör: W. Döring and H.-U. Schuster: *Z. Naturf.*, 1980, **35B**, 1482–1483

Au–Ge–Ni

The only published data [46 Jaf] tabulates the effect of small Ni additions on the Au–Ge eutectic temperature. The course of the monovariant eutectic curve, originating at the Au–Ge eutectic, was estimated by metallographic examination of alloy samples. Thermal analysis of alloys intended to lie on the monovariant eutectic curve showed that Ni has little effect on the Au–Ge eutectic temperature (Table 75).

Table 75. The effect of Ni on the Au–Ge eutectic temperature

Alloy composition, at.-%			*Eutectic line arrest*, °C
Au	*Ge*	*Ni*	
73	27	—	356
65·9	27·5	6·6	357
58·2	29·6	12·2	357

Both ternary alloys undercooled on meeting the monovariant eutectic curve. The microstructure of the higher Ni alloy indicated a complex structure; primary Ge and a fine eutectic structure were identifiable but other phases were also present. X-ray examination confirmed the presence of Au and Ge; a third phase was not identified. This phase could be GeNi.

Christou [79 Chr] quotes a ternary eutectic temperature of 425°C and states that it was determined by a resistivity technique. No other details were given. It is unlikely that a reaction at 425°C is a ternary eutectic reaction. It is more probably associated with a ternary transition reaction of the type L + GeNi $\rightleftharpoons$ (Au–Ni) + (Ge).

REFERENCES

46 Jaf: R. I. Jaffee and B. W. Gonser: *Trans. Amer. Inst. Min. Met. Eng.*, 1946, **166**, 436–443

79 Chr: A. Christou: *Solid St. Electron.*, 1979 , **22**, 141–149

Au–Ge–P

Graham and Steeds [83 Gra] reported the occurrence of the ternary compound AuGeP, isostructural with AsAuGe, at the interface of Au–Ge–Ni contacts to an InP substrate. Convergent beam electron diffraction of isolated single crystals shows a monoclinic structure (C2/c) with a = 0·645 nm, b = 0·584 nm, c = 0·635 nm, β = 127°.

REFERENCE

83 Gra: R. J. Graham and J. W. Steeds: ‘Microscopy of Semiconducting Materials, 1983’, *Inst. Phys. Conf. Ser.* No. 67. (ed. A. G. Cullis, S. M. Davidson and G. R. Booker), 507–512; 1983, London, Institute of Physics

Au–Ge–Pt

The only published data [46 Jaf] tabulates the effect of small Pt additions on the Au–Ge eutectic temperature. The course of the monovariant eutectic curve, originating at the Au–Ge eutectic, was estimated by metallographic examination of alloy samples. Thermal analysis of alloys intended to lie on the monovariant eutectic curve showed that Pt slightly lowers the Au–Ge eutectic temperature (Table 76).

Both ternary alloys undercooled by about 12°C on meeting the monovariant eutectic curve. X-ray examination of the 4·1 at.-% Pt alloy demonstrated the presence of a nearly pure Au phase (a = 4·075 Å), Ge and an unidentified third phase. This phase could be Ge_2Pt.

REFERENCE

46 Jaf: R. I. Jaffee and B. W. Gonser: *Trans. Amer. Inst. Min. Met. Eng.*, 1946, **166**, 436–443

Table 76. The effect of Pt on the Au–Ge eutectic temperature

Alloy composition, at.-%			*Eutectic line arrest*, °C
Au	*Ge*	*Pt*	
73	27	—	356
68·33	29·67	2·0	354
65·3	30·6	4·1	352

Au–Ge–Sb

INTRODUCTION

Zwingmann [64 Zwi] examined 39 alloys on the sections Au–90 at.-% Sb, 10 at.-% Ge, $AuSb_2$–Ge and 15 at.-% Ge. Vertical sections are reproduced, with minor amendments, in Figs 220–222. Differential thermal analysis at cooling rates of about 19°C min^{-1} at 600°C and about 8°C min^{-1} at 300°C was the primary experimental method,

221 The Au–90% Sb, 10% Ge section

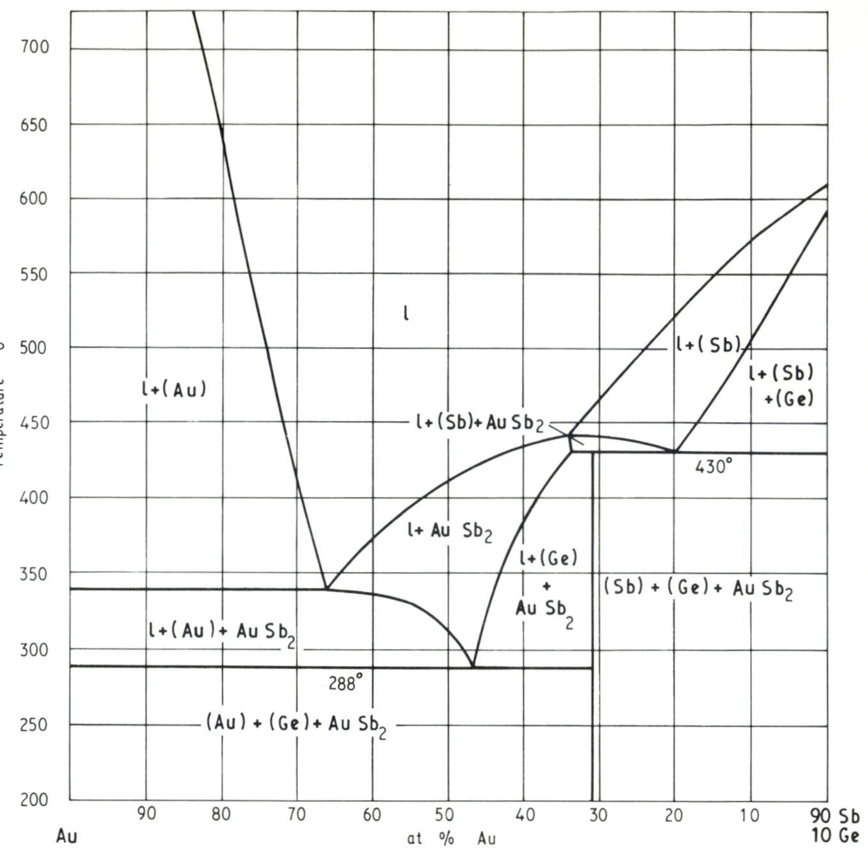

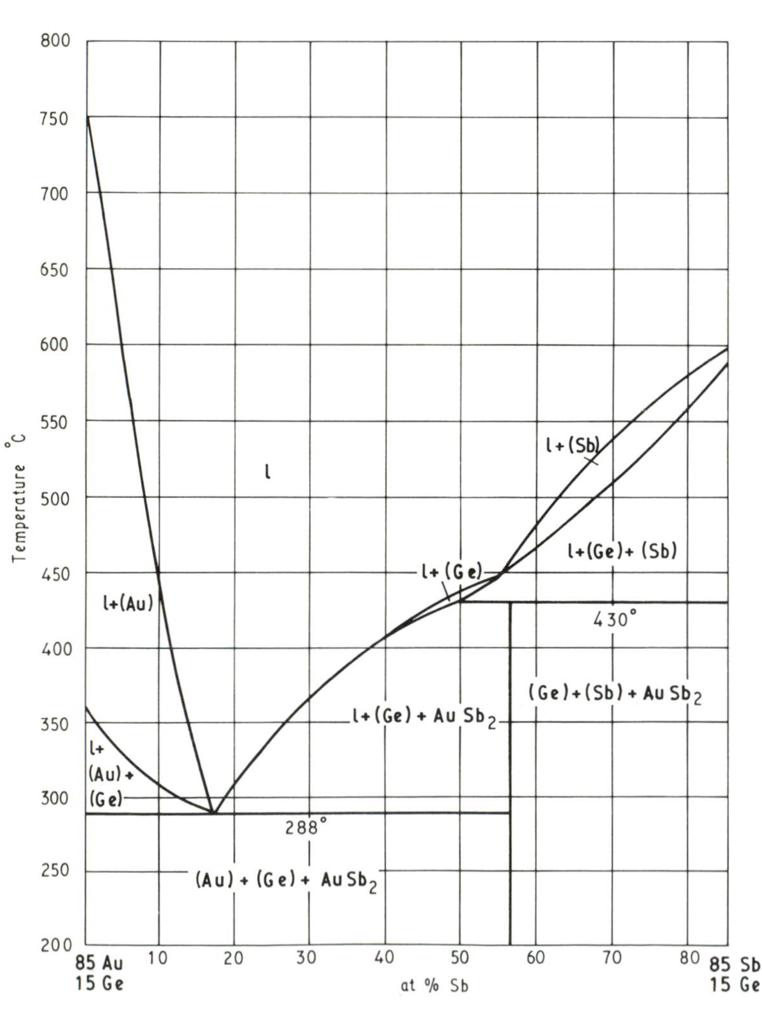

222 The 15% Ge section

supplemented by metallographic examination of selected alloys. All alloys were prepared from 99·95 wt-% Au, 99·95 wt-% Sb and 99·9 wt-% Ge by electrically melting in an alumina crucible under a hydrogen atmosphere. The Au–Ge–Sb ternary system contains two invariant reactions. The reaction L_U + (Sb) ⇌ (Ge) + $AuSb_2$ occurs at 430 ± 3°C, and is followed by a ternary eutectic reaction, L_E ⇌ (Au) + (Ge) + $AuSb_2$ at 288 ± 5°C. The compositions of the liquid phase associated with the invariant reactions is given in Table 77.

Table 77. Invariant equilibria

Reaction	*Temperature*, °C	*Phase*	*Composition*, at.-%		
			Au	*Ge*	*Sb*
U	430 ± 3	L	35	14	51
E	288 ± 5	L	68	15	17

Gubenko and Kononykhina [68 Gub] determined the liquidus curve for four alloys containing ⩾50 at.-% Ge on the section 98·4 at.-% Au, 1·6 at.-% Sb–Ge. Figure 223 compares this data with that for the binary Au–Ge liquidus as determined [68 Gub] and as assessed [84 Oka 1,2]. According to [68 Gub] the addition of very small amounts of Sb to Au–Ge alloys depresses the liquidus temperature significantly. This does not agree with the 600 and 700°C isotherms given by [64 Zwi] (Fig. 224) on the basis of liquidus values for the $AuSb_2$–Ge section and the Au–Ge and Ge–Sb binary systems. Although the isotherms due to [64 Zwi] (Fig. 224) should be regarded as tentative, they are accepted in preference to the data of [68 Gub]. The work of [71 Gub] on the solid solubility of Sb in Ge for the sections from Ge to a concentration of 98·4 at.-% Au, 1·6 at.-% Sb are not reported since [71 Gub] state that they did not determine the equilibrium values.

BINARY SYSTEMS

The Ge–Sb system presented by [58 Han] is accepted. The Au–Ge system assessed by [84 Oka 2] is accepted; the Au–Sb system evaluated by [84 Oka 1] indicates a peritectic liquid containing 66·6 at.-% Sb, only slightly differing from the composition of the compound $AuSb_2$. [84 Eva] report the peritectic liquid at 65·6 at.-% Sb at 458 ± 1°C. The eutectic temperature is 356 ± 1°C at 36·1 at.-% Sb.

SOLID PHASES

Table 78 summarizes the crystal structure data for the binary phases. No ternary compounds were observed.

INVARIANT REACTIONS AND LIQUIDUS SURFACE

In establishing the equilibria three vertical sections were studied [64 Zwi]. The $AuSb_2$–Ge section (Fig. 220) was determined up to 50 at.-% Ge. The Au–90% Sb, 10% Ge section (Fig. 221) is presented on the basis of very small solubility of Ge and Sb in Au. [64 Zwi] did not study alloys with >90 at.-% Au. The 15% Ge section (Fig. 222) was an unfortunate choice since the ternary eutectic composition lies on this section and the monovariant curve UE (Fig. 224) begins at 14% Ge and runs along the 15% Ge composition from about 40% Sb to the eutectic point at E. From the eutectic at 17% Sb to 40% Sb there will be no primary separation of either Ge or $AuSb_2$. Ge and $AuSb_2$ will separate simultaneously from the liquid over this composition range. The trace of the surface of secondary separation of Ge and Sb from 55% Sb to 85% Sb in Fig. 222, i.e. the boundary between the l + (Sb) and the l + (Ge) + (Sb) phase regions, has been adjusted to make it consistent with the liquidus isotherms for temperatures of 450, 500, and 550°C.

Table 78. Crystal structures

Solid phase	*Prototype*	*Lattice designation*
(Au)	Cu	cF4
(Ge)	C	cF8
(Sb)	As	hR2
$AuSb_2$	FeS_2	cP12

From the sections presented in Figs 220–222 the liquidus surface (Fig. 224) has been derived. The binary compound $AuSb_2$ was regarded as an example of a limiting case between congruent formation from the liquid and incongruent formation. The work of [84 Eva] has established that $AuSb_2$ forms by peritectic reaction. The monovariant curve p_1U, Fig. 224, has been amended to indicate this fact. In general the isotherms drawn by Zwingmann [64 Zwi] have been accepted. Changes have been made to the isotherms at 450, 500, and 550°C in the region of primary separation of Sb. They are not as markedly convex to the Sb corner as originally drawn [64 Zwi]. The isotherms from 450 to 700°C in the primary Ge region rely in the main on data points for the Ge–Sb and Au–Ge binary systems and the $AuSb_2$–Ge section. The 600 and 700°C isotherms are particularly tentative. The 800 and 900°C isotherms in the primary Ge field are more so since they depend on an extrapolation of the liquidus on the $AuSb_2$–Ge section (Fig. 220). The dashed curves originating at the 600, 700, 800, and 900°C liquidus points on the Au–Ge binary edge are an estimation of the liquidus isotherms from the data of [68 Gub]. They do not confirm the trend shown by [74 Zwi] and are not accepted.

The compositions of invariant points U and E are given in Table 77. They are considered to be within ±1% for each element in terms of U and ±0·5% for each element in terms of E.

ISOTHERMAL SECTION

Isothermal sections can be generated from the liquidus isotherms of Fig. 224. In this simple system there is no need to present them.

REFERENCES

58 Han: M. Hansen and K. Anderko: 'Constitution of binary alloys'; 1958, New York, McGraw-Hill

64 Zwi: G. Zwingmann: *Z. Metall.*, 1964, **55**, 192–194

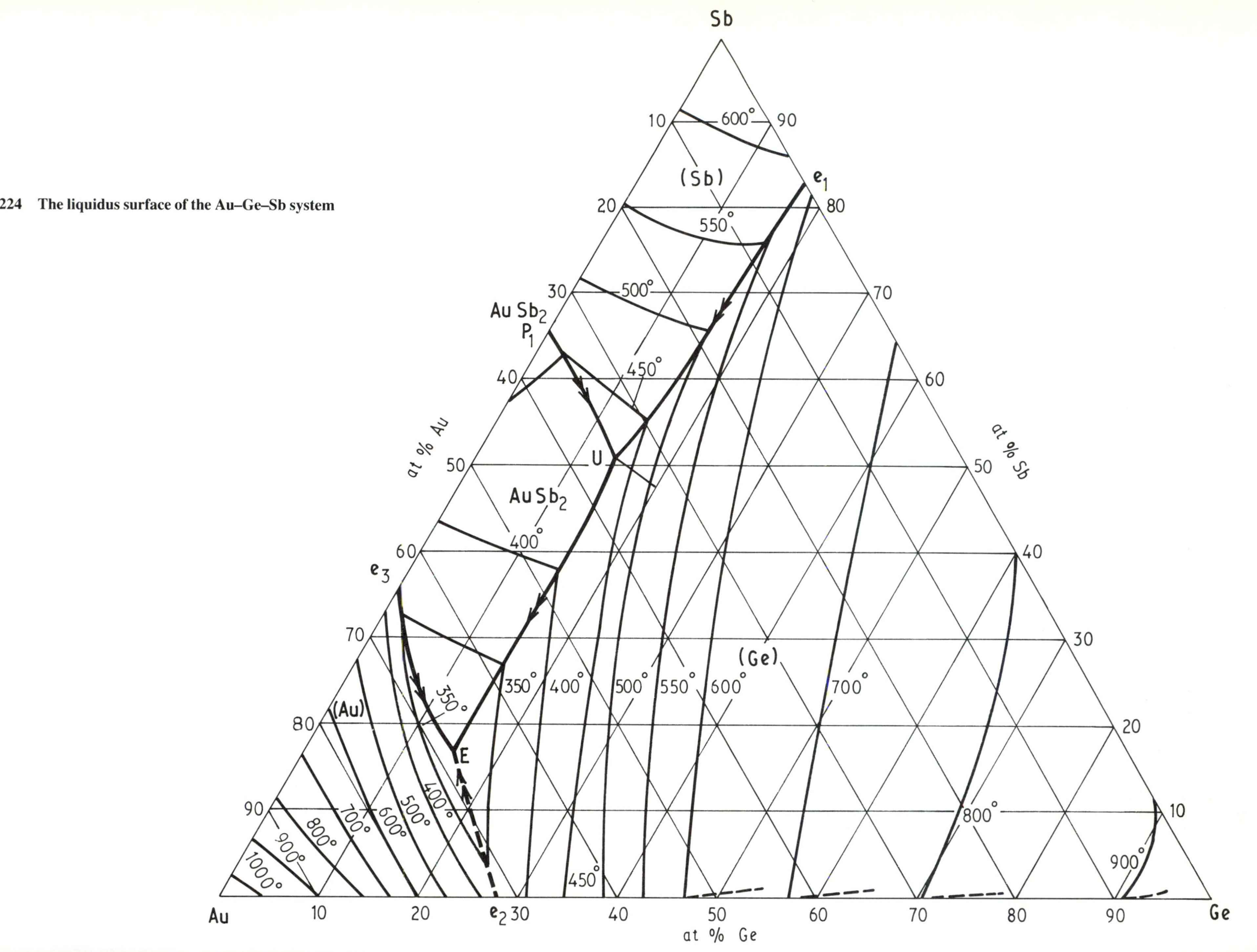

224 The liquidus surface of the Au–Ge–Sb system

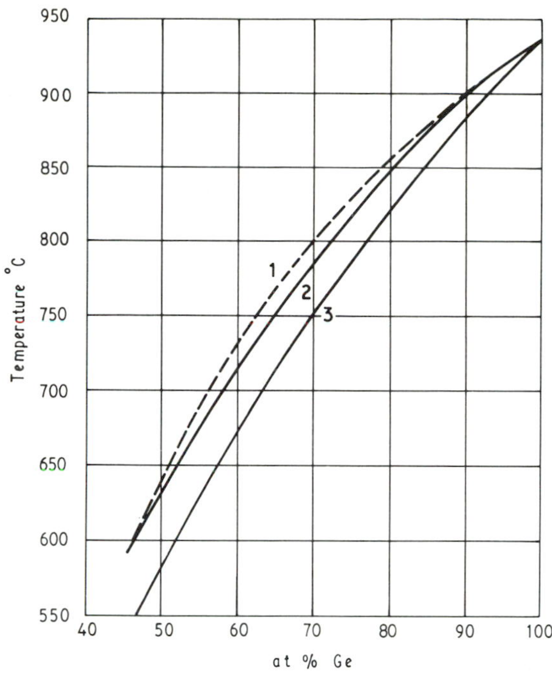

223 **The section 98·4% Au, 1·6% Sb–Ge; *a* Au–Ge binary liquidus [84 Oka 2]; *b* Au–Ge binary liquidus [68 Gub]; *c* 98·4% Au, 1·6% Sb–Ge liquidus [68 Gub]**

Au - Sb	Au - Ge - Sb	Ge - Sb	Au - Ge
		590° 1 ⇌ (Ge) + (Sb)	
458° 1 + (Sb) ⇌ AuSb$_2$			
	U | L + (Sb) ⇌ (Ge) + AuSb$_2$ | 430°		
	(Ge) + (Sb) + AuSb$_2$ L + (Ge) + AuSb$_2$		
356° 1 ⇌ (Au) + AuSb$_2$			361° 1 ⇌ (Au) + (Ge)
	E | L ⇌ (Au) + (Ge) + AuSb$_2$ | 288°		
	(Au) + (Ge) + AuSb$_2$		

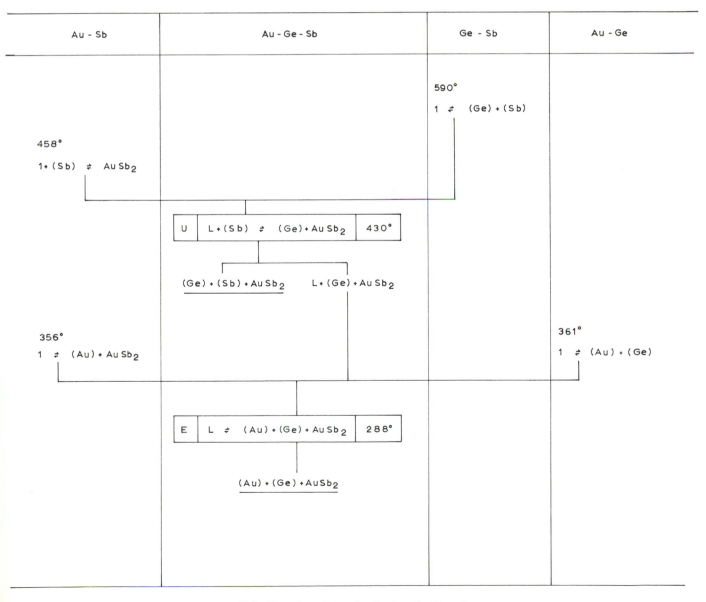

225 **Reaction scheme for the Au–Ge–Sb system**

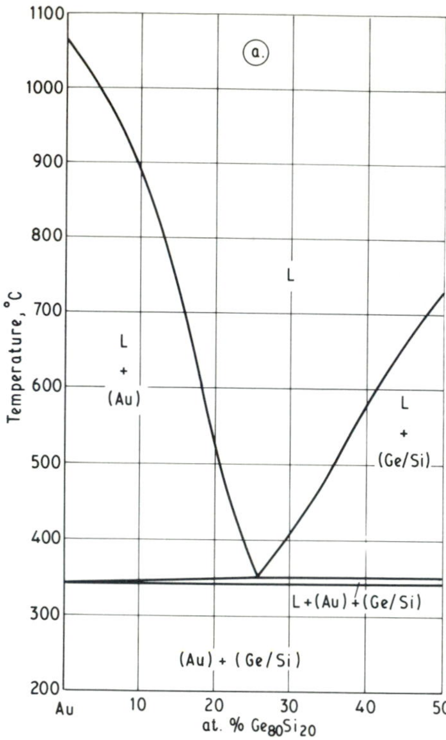

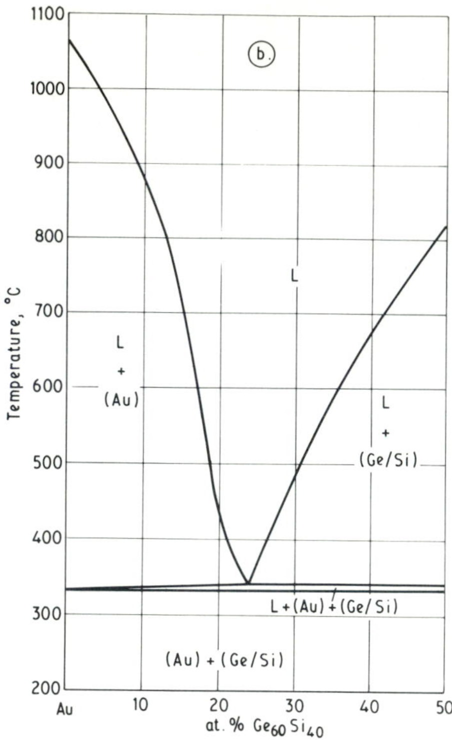

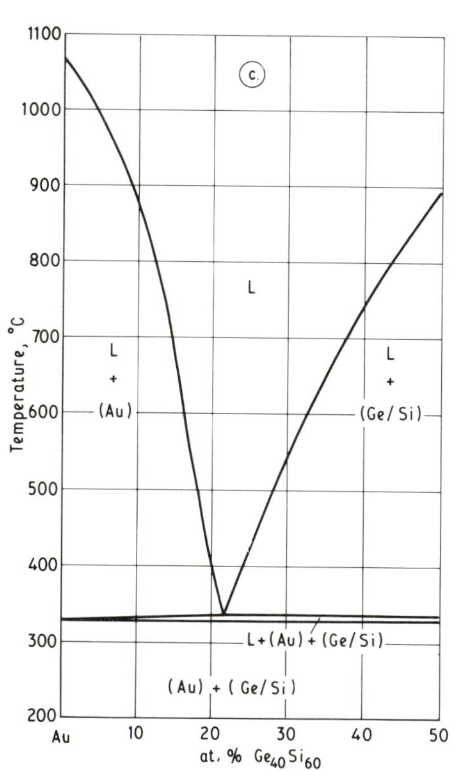

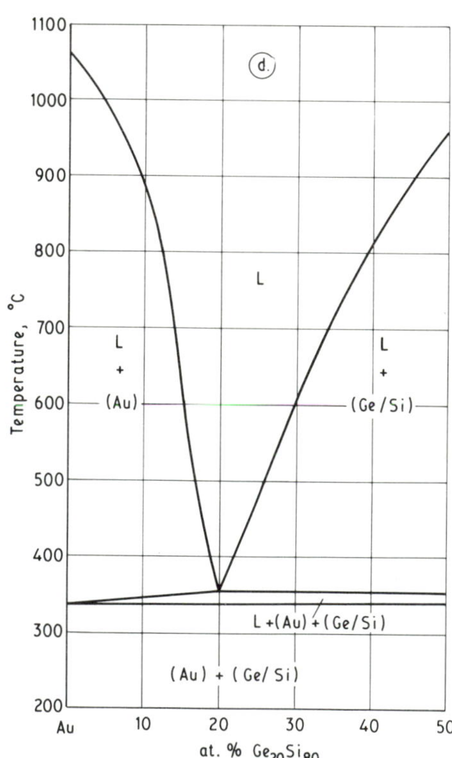

226 Vertical sections from Au to: *a* $Ge_{80}Si_{20}$; *b* $Ge_{60}Si_{40}$; *c* $Ge_{40}Si_{60}$; *d* $Ge_{20}Si_{80}$

68 Gub: A. Ya. Gubenko and N. A. Kononykhina: *Izvest. Akad. Nauk. SSSR, Neorg. Materialy.*, 1968, **4**, 1787–1788

71 Gub: A. Ya. Gubenko, M. B. Miller and N. A. Kononykhina: *Izvest. Akad. Nauk. SSSR, Neorg. Materialy.*, 1971, **7**, 1153–1156

84 Oka 1: H. Okamoto and T. B. Massalski: *Bull. Alloy Phase Diagrams*, 1984, **5**, 166–171

84 Oka 2: H. Okamoto and T. B. Massalski: *Bull. Alloy Phase Diagrams*, 1984, **5**, 601–610

84 Eva: D. S. Evans and A. Prince: CALPHAD XIII, Villard de Lans, 1984

Au–Ge–Si

Predel and Bankstahl [76 Pre] studied 55 ternary Au–Ge–Si alloys in delineating the equilibria for Au contents >50 at.-%. Alloys were prepared from very pure elements, 99·995% Au, 30 Ωcm Ge, 1 500 Ωcm Si, by melting under argon in corundum crucibles. Sections from Au to 50 Au, 40 Ge, 10 at.-% Si (on the section Au–80 Ge, 20 Si), Au to 50 Au, 30 Ge, 20 at.-% Si (Au–60 Ge, 40 Si), Au to 50 Au, 20 Ge, 30 at.-% Si (Au–40 Ge, 60 Si) and Au to 50 Au, 10 Ge, 40 at.-% Si (Au–20 Ge, 80 Si) were constructed by thermal analysis using a cooling rate of 2°C min^{-1}. The vertical sections are reproduced in Fig. 226*a–d*. Metallographic examination of selected solidified alloys confirmed the thermal analysis data. The liquidus surface, derived from the vertical sections and the binary Au–Ge and Au–Si phase diagrams, is given in Fig. 227. Both Au–Ge and Au–Si phase diagrams are simple eutectic types. The investigators [76 Pre] show that the monovariant eutectic curve e_1e_2 in the Au–Ge–Si system passes through a minimum temperature of 326°C at a composition 79 Au, 7·5 at.-% Ge. The critical tie line connecting the (Au), melt, and (Ge/Si) phases at 326°C is shown by the broken line in Fig. 227. Vertical sections at 70, 80, 90 at.-% Au, derived from the vertical sections in Fig. 227*a–d*, are presented in Fig. 228*a–c*. They illustrate the presence of a minimum in the monovariant curve e_1e_2 and the associated critical tie line. The liquidus isotherms at 400°C and at 500–700°C in the region of primary (Au) solidification have been slightly changed from the original to conform with the assessed liquidus given by [83 Oka]. It should be noted that [76 Pre] use the [75 Pre] value for the Au–Si eutectic temperature, 345°C. This temperature is well below the assessed value of 363 ± 3°C. As only cooling curves were used in the construction of the ternary equilibria the temperature values given in Fig. 226*a–d* may be on the low side.

The only other work reported is concerned with the solubility of Au in the (Ge/Si) solid solution. Belokurova *et al.* [72 Bel] showed that the solidus on the Ge–4·5 at.-% Si to Au section has a retrograde nature. The maximum solubility of Au in Ge, 4·5 at.-% Si is 0·65 × 10^{-4} at.-% Au at 850°C. This value is higher than the maximum solubility of Au in Ge, 0·5 × 10^{-4} at.-% Au, according to [72 Bel], but the assessed value [84 Oka] for the maximum solubility of Au in Ge is 0·8 × 10^{-4} at.-% Au. [84 Mav] also reported the retrograde nature of the (Ge/Si) solidus in the Au–Ge–Si system and obtained values consistent with [72 Bel] and a thermodynamic model of the temperature dependence of the solubility of Au in (Ge–Si) solid solution alloys (Table 79). The thermodynamic model predicts that the temperature for maximum solubility increases with increasing Si content of the Ge/Si solid solution, in accord with maximum solubility of Au in (Si) at 1 325°C [83 Oka].

Table 79. Solubility of Au in (Ge/Si) solid solution

Temperature, °C	*Ge–Si alloy composition*, at.-%	*Au solubility* × 10^4 at.-% ± 10%	*Reference*
950°C	Ge, 4·5Si	0·18	[72 Bel]
927°C	Ge, 4·0Si	0·19	[84 Mav]
850°C	Ge, 4·5Si	0·65	[72 Bel]
827°C	Ge, 5·0Si	0·37	[84 Mav]
927°C	Ge, 24Si	0·27	[84 Mav]
	Ge, 85Si	0·11	[84 Mav]
	Ge, 90Si	0·10	[84 Mav]
827°C	Ge, 82Si	0·05	[84 Mav]

REFERENCES

75 Pre: B. Predel and H. Bankstahl: *J. Less-Common Metals*, 1975, **43**, 191–203

76 Pre: B. Predel, H. Bankstahl and T. Gödecke: *J. Less-Common Metals*, 1976, **44**, 39–49

72 Bel: N. N. Belokurova, V. S. Zemskov, V. E. Il'inykh, A. P. Rusin, N. I. Shadeev and V. S. Shibanov: *Izvest. Akad. Nauk. SSSR, Neorg. Materialy.*, 1972, **8**, 1726–1729

83 Oka: H. Okamoto and T. B. Massalski, *Bull. Alloy Phase Diagrams*, 1983, **4**, 190–198

84 Mav: O. I Mavrin and A. A. Novikov: *Izvest. Akad. Nauk. SSSR, Neorg. Materialy.*, 1984, **14**, 1204–1207

84 Oka: H. Okamoto and T. B. Massalski: *Bull. Alloy Phase Diagrams*, 1984, **5**, 601–610

Au–Ge–Sn

Using differential thermal analysis, metallography, and X-ray diffraction techniques, Humpston [85 Hum] determined the section AuSn–Ge to be a pseudobinary eutectic with a eutectic temperature of 410 ± 2·5°C. The experimental results are given in Table 80. No reliable measurements of the liquidus were obtained. The eutectic composition, determined metallography, was placed at 7 ± 1 at.-% Ge. AuSn is soluble in Ge to a level just beyond 5 at.-% at the eutectic temperature.

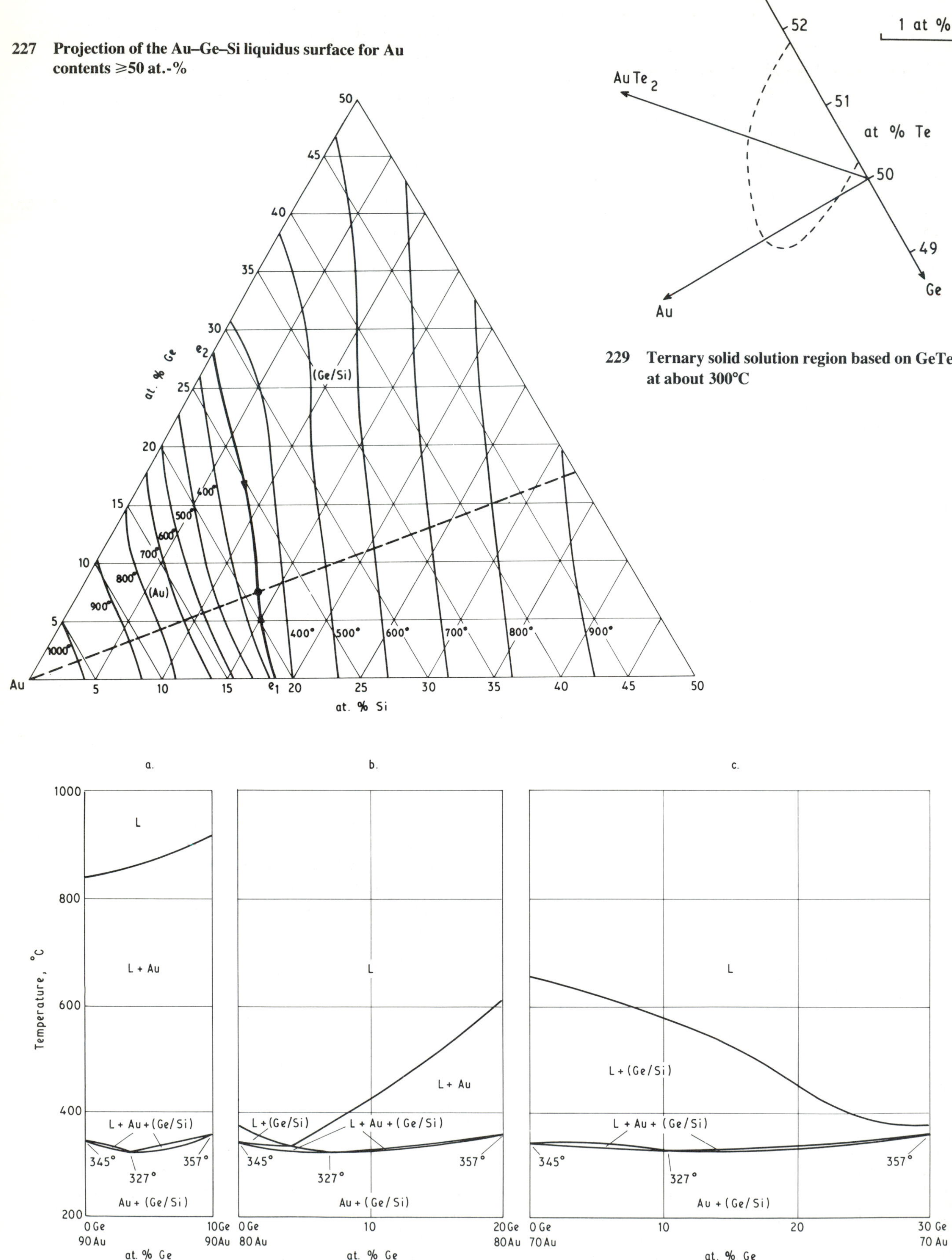

227 Projection of the Au–Ge–Si liquidus surface for Au contents ≥50 at.-%

229 Ternary solid solution region based on GeTe at about 300°C

228 Vertical sections at: *a* 90 at.-% Au; *b* 80 at.-% Au; *c* 70 at.-% Au

REFERENCE

85 Hum: G. Humpston: PhD Thesis, 1985, Brunel University

Table 80. Thermal analysis data for the AuSn–Ge section

Alloy composition, at.-% Ge	*Heating arrest,* °C	*Cooling arrest,* °C
0	419·8	419·8
4	409·5	412·9
8	409·6	411·1
11	411·4	414·0
14	408·9	411·1
26	408·1	411·2
38	405·4	407·9
52	405·4	407·4
62	407·8	412·6
76	410·6	408·6
90	409·1	406·7
95	355·5, 431·0	358·6, 436·9
100	—	(938·3)

Au–Ge–Te

INTRODUCTION

Legendre *et al.* [74 Leg] showed that the sections $AuTe_2$–GeTe and Au–GeTe are both pseudobinary eutectic systems, except for compositions within a few atomic percent of GeTe. The ternary solid solution region based on αGeTe was traced at about 300°C from microhardness measurements (Fig. 229). Legendre and Souleau [77 Leg] include the [74 Leg] data in an extensive study of the Ge–Te binary and the Au–Ge–Te ternary system. Some 180 alloy compositions were studied using thermal analysis techniques, supplemented by metallography and X-ray diffraction analysis of powder samples. The two pseudobinary sections divide the ternary system into three partial ternary systems. [77 Leg] present vertical sections for 30, 40, 56·67, 66·67, and 80 at.-% Te; 60 at.-% Ge; 23·33 at.-% Au. They claim that each of the three partial ternary systems are simple ternary eutectic systems.

The partial ternary system Te–$AuTe_2$–GeTe cannot be a simple ternary eutectic system since a metatectic reaction occurs at 400°C in the GeTe–Te system, βGeTe ⇌ αGeTe + l [77 Leg]. A transition reaction , L_U + βGeTe ⇌ $AuTe_2$ + αGeTe, is postulated as a necessary precursor to the ternary eutectic reaction, L_{E_2} ⇌ $AuTe_2$ + (Te) + αGeTe at 364°C.

The partial ternary system Au–$AuTe_2$–GeTe is a simple ternary eutectic system, L_{E_1} ⇌ (Au) + $AuTe_2$ + βGeTe at 382°C [77 Leg]. Although this is accepted, the experimental data on which this conclusion is based (the vertical sections at 40 and 56·67 at.-% Te in particular) are not consistent with each other and must be in error. A check on one alloy composition [87 Hum] reveals large errors in some of the data of [77 Leg]; further discussion is to be found in the section on invariant reactions.

The partial ternary system Au–Ge–GeTe solidifies by eutectic reaction, L_{E_3} ⇌ (Au) + (Ge) + βGeTe. [77 Leg] regard this eutectic as producing αGeTe, but this cannot be correct. This partial ternary system was not well defined and it is not possible to trace the monovariant curves e_1E_3 and e_2E_3 with any degree of certainty.

Although the general form of the liquidus surface and the associated invariant reactions have been given [74 Leg, 77 Leg] there is a need for further experimental work before the ternary equilibria can be regarded as established.

BINARY SYSTEMS

The Au–Ge system evaluated by [84 Oka 2] and the Au–Te system evaluated by [84 Oka 1] are accepted. The Ge–Te system was studied by [60 McH] in terms of the eutectic equilibrium between Ge and βGeTe. The eutectic contains 49·85 at.-% Te at 723°C; βGeTe melts congruently at 50·6 at.-% Te and 724°C. [77 Leg] is accepted for the equilibria below 450°C and [85 Kor, 85 Rog] for the range of homogeneity of βGeTe.

SOLID PHASES

No ternary compounds have been reported [77 Leg]. Table 81 summarizes crystal structure data for the solid phases.

Table 81. Crystal structures

Solid phase	*Prototype*	*Lattice designation*
(Au)	Cu	cF4
(Ge)	C	cF8
(Te)	γSe	hP3
$AuTe_2$	$AuTe_2$	mC6
βGeTe	NaCl	cF8
αGeTe	Distorted NaCl	cF8
γGeTe	GeS	oP8

PSEUDOBINARY SYSTEMS

The 50 at.-% Ge composition in the Ge–Te system solidifies as (Ge) + βGeTe and transforms to (Ge) + αGeTe by peritectoid reaction at 430°C [77 Leg]. The $AuTe_2$–GeTe and Au–GeTe pseudobinary systems (Figs 230 and 231 respectively) will depart from their pseudobinary character for compositions close to GeTe. The $AuTe_2$–GeTe systems contains a eutectic, e_5, at 20·5 ± 0·5 at.-% Ge. [77 Leg] quote 20 at.-% Ge. The eutectic temperature is 400°C. The Au–GeTe system contains a eutectic, e_2, at 26·67 ± 0·5 at.-% Ge and 480°C. All alloys were studied by DTA using a heating rate of 5°C min^{-1}. Nineteen alloys were analysed in each section.

INVARIANT EQUILIBRIA

The ternary invariant reactions [77 Leg] are summarized in Table 82 with respect to the composition of the liquid phase involved. The partial ternary system Te–$AuTe_2$–GeTe must contain two ternary invariant reactions. Final solidification occurs at a ternary eutectic E_2 at 364°C with the separation of (Te) + $AuTe_2$ + αGeTe. In the 56·67 and 66·67 at.-% Te sections [77 Leg] showed a thermal effect at 400°C for alloys originally separating primary

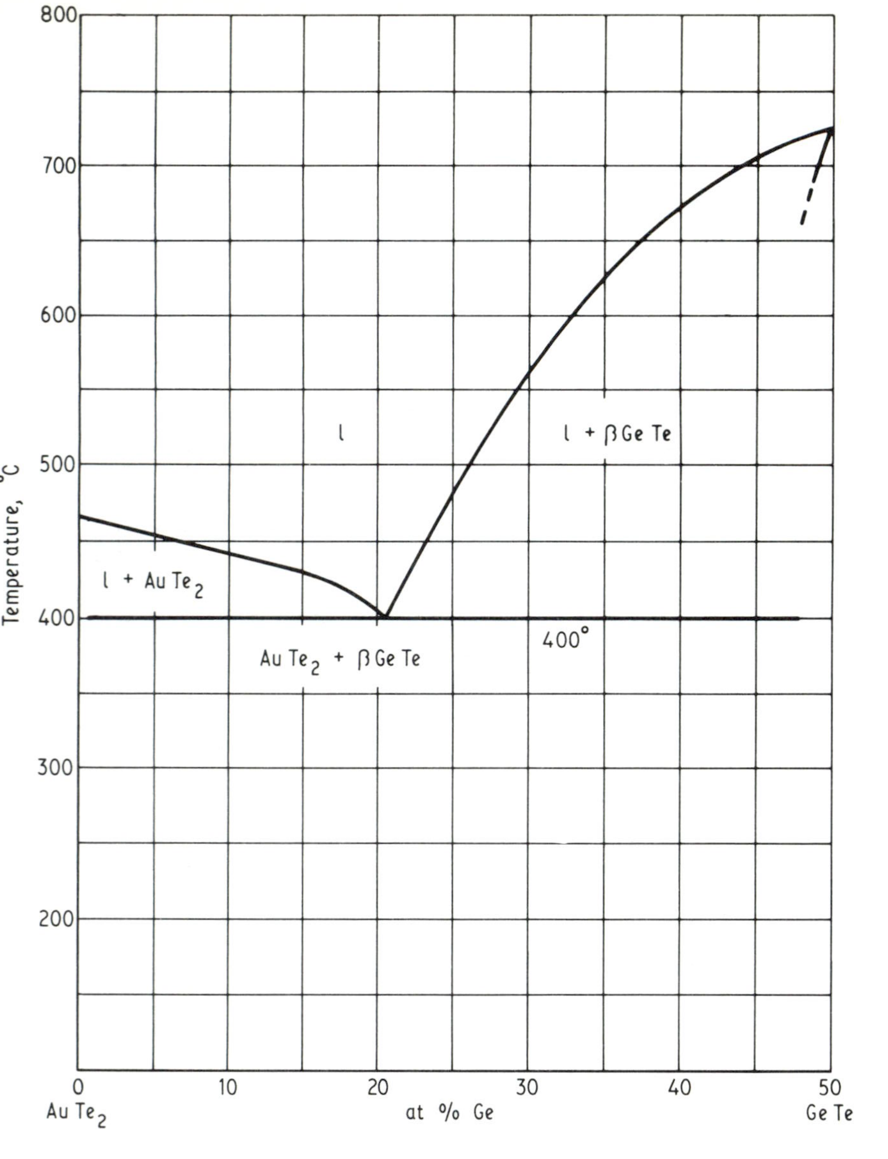

230 The AuTe$_2$–GeTe section

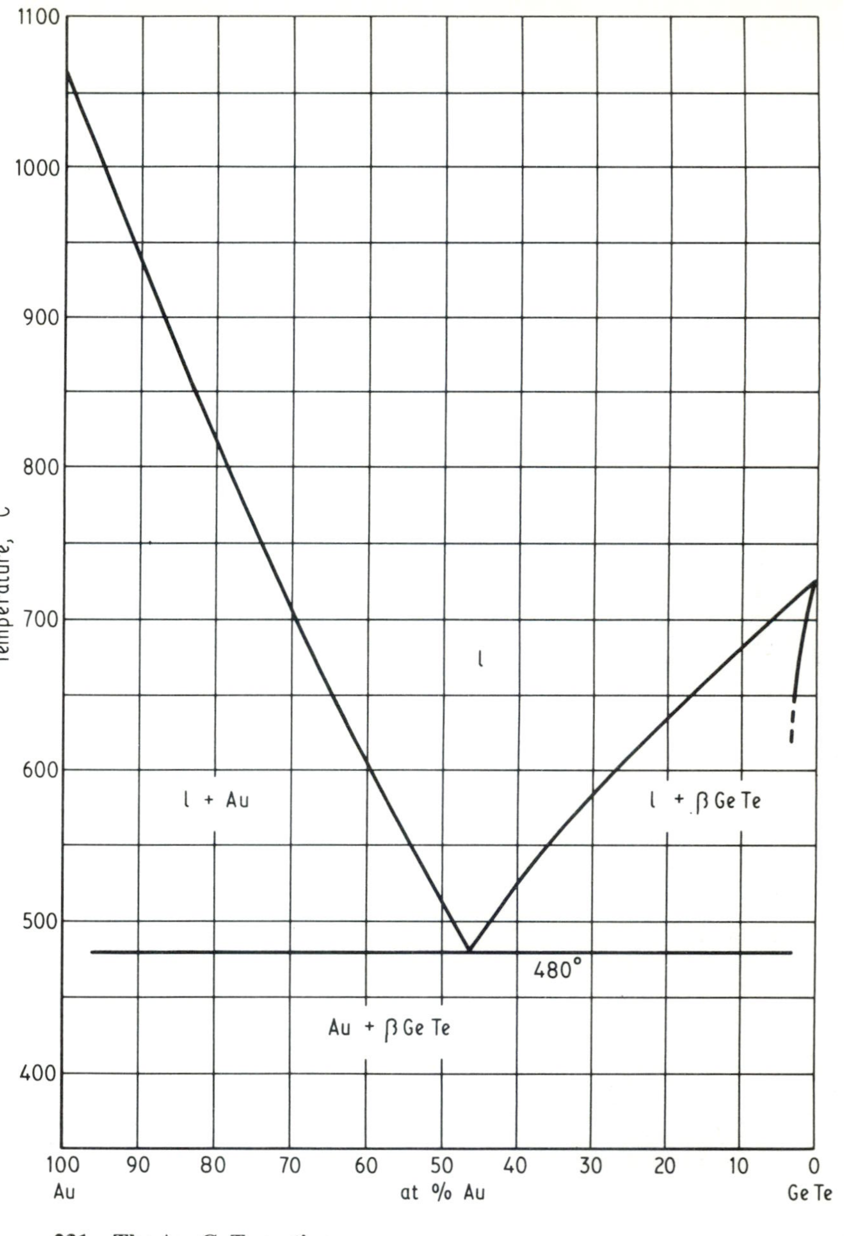

231 The Au–GeTe section

Table 82. Invariant equilibria

Reaction	*Temperature*, °C	*Liquid composition*, at.-%	
		Au	*Ge*
e_5	400	19·66	20·5
e_2	480	46·66	26·66
E_1	382	30	17·6
E_2	364	5	15
E_3	361	71	27·6

βGeTe (in the region mUe_5GeTe of Fig. 232). This thermal effect is a reflection of the binary metatectic reaction βGeTe ⇌ αGeTe + l at 400°C. It implies the presence of a transition reaction U, close to the pseudobinary eutectic e_5, at which the liquid reacts with βGeTe to give $AuTe_2$ + αGeTe (Fig. 233). The pseudobinary eutectic e_5 is associated with the separation of $AuTe_2$ + βGeTe.

The partial ternary system Au–$AuTe_2$–GeTe appears to be a simple ternary eutectic system. [77 Leg] defined it using sections at 30, 40, 56·67 at.-% Te and 23·33 at.-% Au. Unfortunately the 40 at.-% Te section is indicated as intersecting the monovariant curve e_2E_1 (Fig. 232) at 508°C and 22·3 at.-% Ge. As e_2 lies at 480°C and E_1 at 382°C and e_2E_1 is shown descending from e_2 to E_1, the thermal data of [77 Leg] for the 40 at.-% Te section are not acceptable. The trace of the surface of secondary separation of (Au) + βGeTe for alloys on the 40 at.-% Te section with 22.3 to 40 at.-% Te is shown with an impossible negative slope by [77 Leg], the L + βGeTe/L + (Au) + βGeTe boundary falling from 508°C at 22·3 at.-% Ge to 480°C at 40 at.-% Te (e_2). A recent thermal analysis of an alloy containing 30Ge, 40 at.-% Te [87 Hum] gave agreement with [77 Leg] on the ternary eutectic temperature of 382°C, but the L + βGeTe/L + (Au) + βGeTe boundary was located at 471·5°C (502°C, [77 Leg]) giving the correct slope to this boundary. The liquidus temperature of 525°C is 75°C lower than that given by [77 Leg]. A thermal effect was observed [87 Hum] at 359°C on heating. This may be associated with the proposed solid-state transition reaction, $AuTe_2$ + βGeTe ⇌ (Au) + αGeTe, which generates the equilibrium (Au) + $AuTe_2$ + αGeTe constitution (Fig. 233). The 56·67 at.-% Te section has the same error as the 40 at.-% Te section. Although the plotted thermal data points in both sections appear consistent they are incorrect in terms of the liquidus values and the L + βGeTe/L + (Au) + βGeTe boundaries. The 56·67 at.-% Te section is given as intersecting the monovariant curve e_5E_1 at 18·6 at.-% Ge and 412°C [77 Leg], although e_5E_1 descends from 400°C at e_5 to 382°C at E_1.

The partial ternary system Au–Ge–GeTe is stated to be a simple ternary eutectic with a eutectic E_3 at 361°C, virtually identical with that of the binary Au–Ge eutectic temperature. [77 Leg] attempted to define the equilibrium from sections at 30, 40 at.-% Te, 23·33 at.-% Au and 60 at.-% Ge. There is no consistency between these sections; one cannot trace the course of the monovariant curve from e_1 to the 30 at.-% Te section. Naturally it is not possible to trace e_1E_3 from the 30 at.-% Te section to E_3 for lack of data. Fig. 233 is a reaction scheme based on [77 Leg] and including the amendments to the ternary equilibria discussed above.

LIQUIDUS SURFACE

Fig. 232 reproduces the compositions of the liquid phases associated with invariant reactions in the binary systems and the ternary system. The paths of the monovariant curves are not well established; they are given as dashed lines. Isotherms were drawn by [77 Leg] but they are not included in this assessment.

ISOTHERMAL SECTIONS

It was not possible to construct isothermal sections from the vertical sections of [77 Leg]. It is concluded that further experimental work is required to accurately establish the equilibria in this ternary system.

REFERENCES

60 McH: J. P. McHugh and W. Tiller: *Trans. AIME*, 1960, **218**, 187–188

74 Leg: B. Legendre, J.-C. Rouland and C. Souleau: *Compt. rend.*, 1974, **284C**, 451–454

77 Leg: B. Legendre and C. Souleau: *J. Chem. Res. (S)*, 1977, (12), 3701–3745

84 Oka 1: H. Okamoto and T. B. Massalski: *Bull. Alloy Phase Diagrams*, 1984, **5**, 172–177

84 Oka 2: H. Okamoto and T. B. Massalski: *Bull. Alloy Phase Diagrams*, 1984, **5**, 601–610

85 Rog: E. I. Rogacheva, A. N. Melikhova, S. A. Laptev, N. V. Rusinov and A. G. Ob'edkov: *Izvest. Akad. Nauk. SSSR, Neorg. Materialy.*, 1985, **21**, 397–401

85 Kor: N. N. Koren', V. I. Levchenko, V. V. Dikareva and A. V. Kazushchik: *Izvest. Akad. Nauk. SSSR, Neorg. Materialy.*, 1985, **21**, 578–580

87 Hum: G. Humpston and A. Prince: unpublished work 1987

Au–Ge–Th

The ternary compound $ThAu_2Ge_2$ was prepared [77 Mar] by induction-melting >99·99% Ge and >99·9% Au,Th under argon. The alloy was subsequently annealed in a silica capsule under an inert atmosphere for 1 week at 500°C. Metallographically the alloy was observed to be nearly homogeneous. X-ray powder diffraction analysis established that $ThAu_2Ge_2$ has a tetragonal $ThCr_2Si_2$-type structure, tI10, with lattice parameters a = 0·437 8 ± 0·000 2, c = 1·009 2 ± 0·000 5 nm and c/a = 2·305. The analogous compounds $ThAu_2Si_2$ and UAu_2Si_2 have the same crystal structure, but homogeneous samples of UAu_2Ge_2 were not successfully prepared by the techniques quoted [77 Mar].

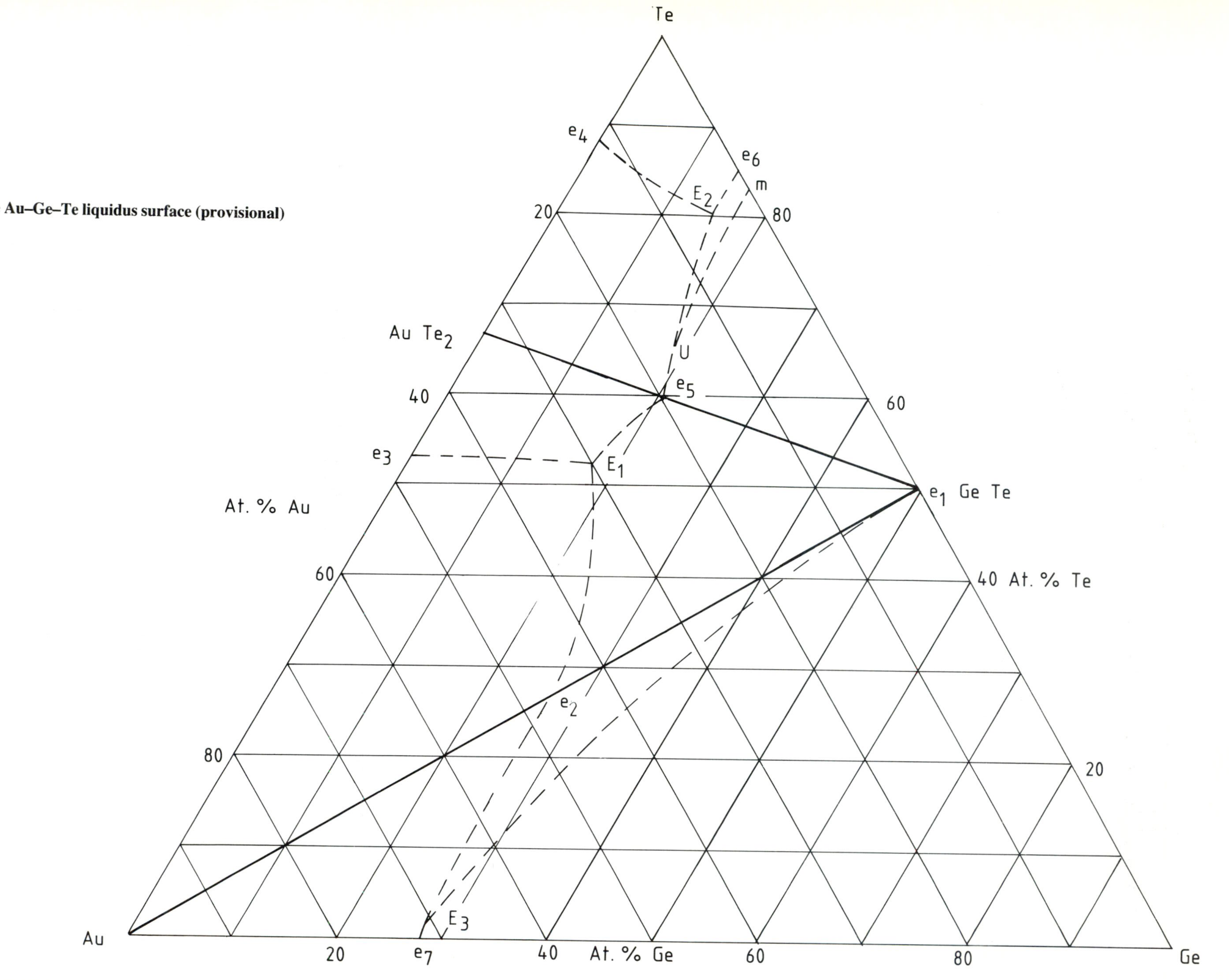

232 The Au–Ge–Te liquidus surface (provisional)

233 Reaction scheme for the Au–Ge–Te system

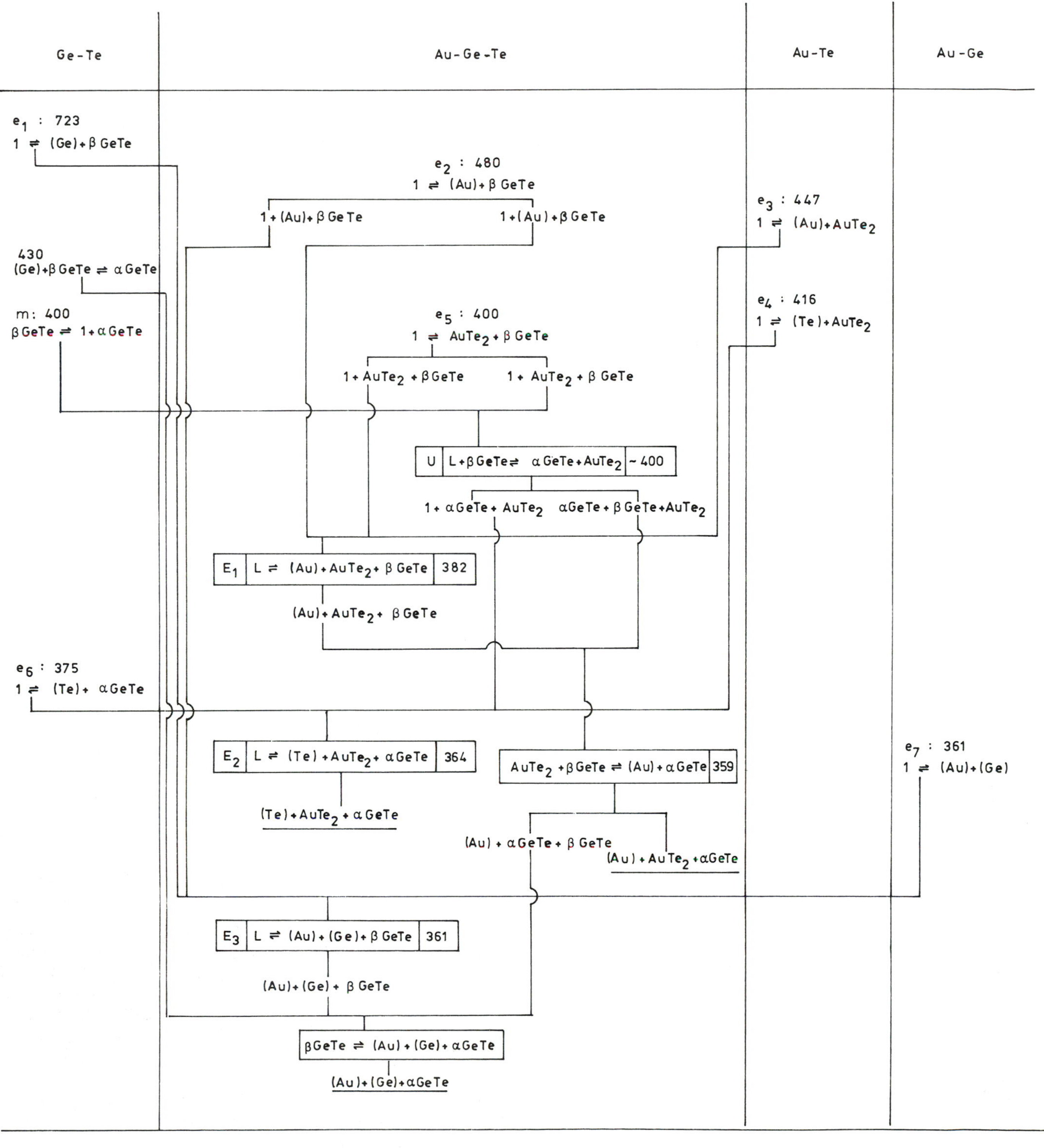

REFERENCE

77 Mar: R. Marazza, R. Ferro, G. Rambaldi and G. Zanicchi: *J. Less-Common Metals*, 1977, **53**, 193–197

Au–Ge–U

Although [77 Mar] prepared and characterized the ternary compounds $ThAu_2Si_2$, UAu_2Si_2 and $ThAu_2Ge_2$, they reported that it was not possible to obtain homogeneous samples of UAu_2Ge_2 by induction melting the components and annealing for 1 week at 500°C.

REFERENCE

77 Mar: R. Marazza, R. Ferro, G. Ramaldi and G. Zanicchi: *J. Less-Common Metals*, 1977, **53**, 193–197

Au–Ge–V

In studying solid solutions formed between Cr_3Si-type phases (cP8) von Philipsborn [70 Phi] prepared a $V_3Au_{0\cdot5}Ge_{0\cdot5}$ alloy by arc-melting V_3Au and V_3Ge (both Cr_3Si-type) on a water cooled Cu hearth under argon. Samples were annealed 45 days at 550°C, 4 days at 850°C and 20 days at 1050°C in evacuated quartz ampoules. The phases present were characterized by the X-ray powder diffraction technique. V_3Au and V_3Ge have lattice parameters $a = 0{\cdot}488$ nm and $a = 0{\cdot}4783$ nm respectively.

The as-cast and 850°C annealed samples contained two Cr_3Si-type phases with lattice parameters $a = 0{\cdot}4825$ and $0{\cdot}4813 \pm 0.0003$ nm. In addition a V_3AuO phase with a Cu_3Au-type structure was found. The sample annealed at 550°C contained two Cr_3Si-type phases with lattice parameters $a = 0{\cdot}4846$ and $0{\cdot}4822 \pm 0{\cdot}003$ nm. The sample annealed at 1050°C contained more V_3AuO phase than was present in the 850°C annealed samples, a Cr_3Si type phase with lattice parameter $a = 0{\cdot}479 \pm 0{\cdot}001$ nm and an unidentified phase. Without further data it is not possible to interpret these results [70 Phi]. The absence of binary compounds in the Au–Ge system means that V_3Au and V_3Ge should form a stable tie line unless a ternary silicide phase prevents this.

REFERENCE

70 Phi: H. von Philipsborn: *Z. Kristallogr.*, 1970, **131**, 73–87

Au–Ge–Zn

[46 Jaf] examined the effect of ternary additions on the Au–Ge eutectic. The course of the monovariant curve, originating at the Au–Ge eutectic, was estimated by metallographic examination of alloy samples. Thermal analysis of alloys intended to lie on the monovariant eutectic curve gave a guide to the effect of the ternary addition on the binary eutectic temperature. Only one Au–Ge–Zn alloy was thermally analysed. Its composition was 60·8 at.-% Au, 31·1 at.-% Ge, 8·1 at.-% Zn. The eutectic line arrest of 448°C compared with a measured value for the Au–Ge eutectic of 356°C. It is doubtful whether the single alloy studied fell on a ternary monovariant curve. A photomicrograph of the alloy could be interpreted as showing primary Ge.

[76 Kro] stated that the alloy AuGeZn was not single phase, but no further details were given. [89 But] studied the Au Zn–Ge section by DTA and metallographic techniques. The section is a pseudobinary eutectic system with a eutectic point at 12·2 ± 0·2 at.-% Ge and 673°C. Ge is soluble in AuZn up to 1·3 at.-% Ge at the eutectic temperature. The observation of [76 Kro] is in agreement with the data of [89 But] since the alloy AuGeZn would contain AuZn + Ge after eutectic solidification.

REFERENCES

46 Jaf: R. I. Jaffee and B. W. Gonser: *Trans. Amer. Inst. Min. Met. Eng.*, 1946, **166**, 436–443

76 Kro: K. Krompholz and A. Weiss: *Z. Metall.*, 1976, **67**, 400–403

89 But: M. T. Z. Butt, C. Bodsworth and A. Prince, *Scripta Met.*, 1989, **23**, 1105–1108

Au–H–Ni

[69 Lan] determined the effect of Au additions on the solubility of H in Ni over the temperature range 1050–1600°C using a Sieverts apparatus. All measurements were at a partial H pressure of 1 atm. Au reduces the solubility of H in molten Ni (see Fig. 234). It also reduces the solubility of H in solid Ni for alloys containing up to 7 at.-% Au (a value near the limit of the (Ni) solid solution at 1200°C). The interaction parameter, $\epsilon^{Au}_{(H)Ni} = 3{\cdot}5$ at 1600°C. Polynomial equations were derived for the heat and entropy of solution of H in Au–Ni alloys. [71 Pet] presented a thermodynamic calculation of the solubility of H in molten Au–Ni alloys which gave excellent agreement with the data of [69 Lan]. [69 Lip] cathodically charged Au–Ni alloys with H at 20°C and presented desorption isotherms. These can be interpreted as showing two solid solution phases, an α solution phase of H in Ni and a β solution phase at H contents of approximately 40 at.-% for H–Ni alloys. The addition of Au sharply reduces the width of the two-phase region for Au contents up to 10 at.-%; from 10 to 35 at.-% Au the two-phase region has a constant width and it closes at a composition below 40 at.-% Au. This parallels the behaviour in the Au–H–Pd system [65 Mae].

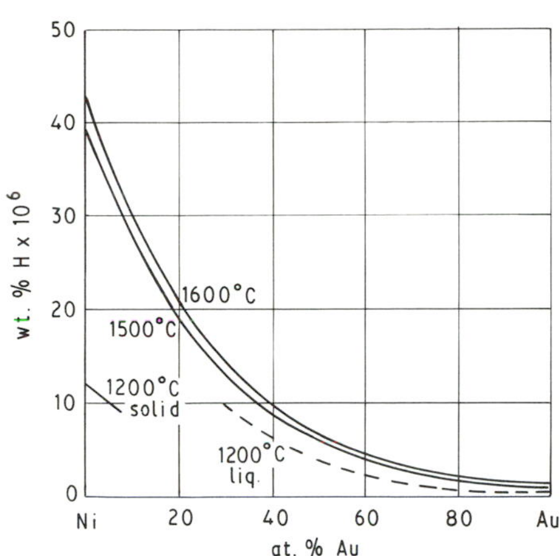

234 The effect of Au on the solubility of H in Ni.

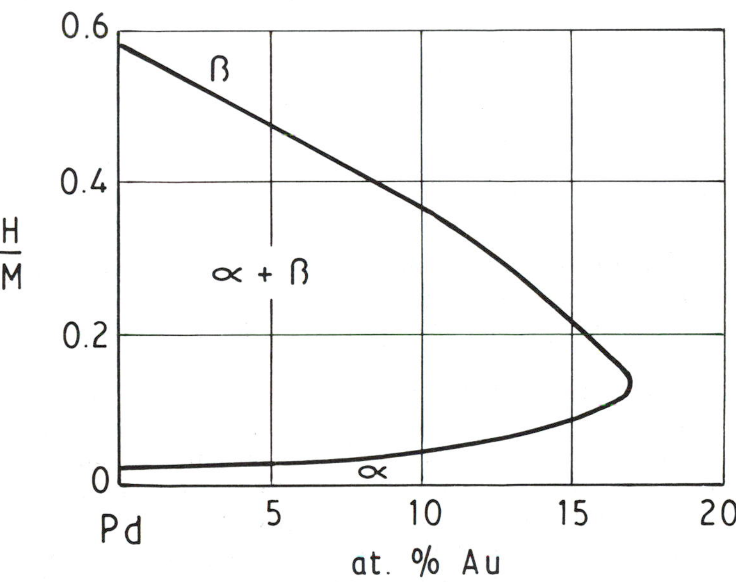

235 Phase boundaries of the hydride phases at 25°C in the Au–Pd–H system.

REFERENCES

65 Mae: A. Maeland and T. B. Flanagan: *J. Phys. Chem.*, 1965, **69**, 3575–3581

69 Lan: K. W. Lange and H. Schenck: *Z. Metallkde.*, 1969, **60**, 638–642

69 Lip: T. V. Lipets, Zh. L. Vert and I. P. Tverdovsky: *Zhur. Fiz. Khim.*, 1969, **43**, 1841–1844

71 Pet: M. S. Petrushevsky and P. V. Gel'd: *Izvest. Vyssh. Ucheb. Zaved. Chern. Met.*, *1971*, (2), 9–13

Au–H–Pd

The Pd–H binary system contains a two-phase region, $\alpha + \beta$, where the α phase is saturated with H to a H : Pd ratio of 0·02 (2 at.-% H) and the β phase is saturated with H at a H : Pd ratio of 0·58 (36·7 at.-% H) at 1 atm H_2 pressure at r.t. Both the α and β phases have a f.c.c. lattice with H atoms incorporated into the large octahedral interstices. The effect of Au additions to Pd is to reduce the width of the phase separation region until it finally closes (Fig. 235). [34 Mun] reported the two-phase region extending to 47 at.-% Au at room temperature. Using a 15·3 at.-% Au–Pd alloy [46 Ben] found the lattice parameters of electrolytically charged samples to vary from a = 0·391 nm at zero H content to a = 0·401 nm at H saturation. This implies a single phase region and contradicts the data of [34 Mun]. Figure 235 is taken from [65 Mae] who charged Au–Pd alloys containing 2·77, 5·66, 8·70, 11·90 and 15·26 at.-% Au with H and measured the changes in electrode potential and resistivity of the samples as a function of time. The H contents were mainly determined by vacuum degassing. The $\alpha + \beta$ phase region closes at about 17 at.-% Au at room temperature under 1 atm H_2 pressure. The lattice parameters of the α and β phase are given in Table 83 together with the enthalpy of absorption of H [65 Mae].

The equilibrium solubility of H in Pd at 25°C and 1 atm H_2 pressure is reduced by alloying with Au [65 Mae] (Fig. 236). [11 Ber] reported a similar behaviour although he noted somewhat higher solubilities than [65 Mae]. [68 Mae] used a neutron diffraction technique to study the location of H atoms in the α phase of an alloy containing 15·1 at.-% Au–Pd. The H occupies octahedral sites in the α phase. This alloy has a narrow $\alpha + \beta$ phase region according to [64 Mak, 65 Bro], in agreement with the data of [65 Mae]. At high temperatures with H at atmospheric pressure the solubility of H in Au–Pd alloys attains a maximum before falling to very low values with increasing Au contents. The original data of [15 Sie] show the same characteristics as the data plotted in Fig. 237 from [85 Yos]. [77 Tim] also confirmed the presence of a maximum solubility at about 25 at.-% Au at 500°C although only data for two alloy compositions were quoted for comparison with pure Pd. Absorption isotherms were determined for H in Au–Pd alloys containing up to 68 at.-% Au at 25°C and 40 to 10^4 atmospheres (81 Sza). The H content is linearly dependent on the log H fugacity.

Table 83.

Alloy composition, at.-% Au	*Lattice parameter*, nm		*Enthalpy of absorption of H* ($kcal/mole\ H_2$)
	α	β	
2·77	0·390 0_0	0·402 1_8	−9·33
5·66	0·390 6_4	0·401 2	−9·50
8·70	0·391 4_6	0·400 4_0	−9·63
11·90	0·392 4_0	0·399 1_8	−9·83
15·26	0·394 1_8	0·396 9_6	−9·83

REFERENCES

11 Ber: A. J. Berry: *J. Chem. Soc.*, 1911, **99**, 463

15 Sie: A. Sieverts, E. Jurisch and A. Metz: *Z. anorg. Chem.*, 1915, **92**, 329

34 Mun: H. Mundt: *Ann. Physik*, 1934, **19**, 721

46 Ben: J. Bénard and J. Talbot: *Compt. rend.*, 1946, **222**, 493–495

64 Mak: A. C. Makrides: *J. Phys. Chem.*, 1964, **68**, 2160

65 Mae: A. Maeland and T. B. Flanagan: *J. Phys. Chem.*, 1965, **69**, 3575–3581

65 Bro: H. Brodowsky and E. Poeschel: *Z. Phys. Chem., Neue Folge*, 1965, **44**, 143

68 Mae: A. Maeland: *Canad. J. Phys.*, 1968, **46**, 121–124

77 Tim: N. I. Timofeev, V. A. Gol'tsov and G. E. Kagan: 'Splavy Blagorod Met', *1977*, pp. 73–75. Publ. 'Nauka', Moscow

81 Sza: A. W. Szafranski: *Pol. J. Chem.*, 1981, **55**, 2137–2141

85 Yos: M. Yoshihara and R. B. McLellan: *J. Phys. Chem. Solids*, 1985, **46**, 357–362

Au–H–Ti

Ti_3Au absorbs H to form the Ti_3AuH stoichiometry with the H atoms assumed to be in interstitial solution in Ti_3Au. The lattice parameter increases from a = 0·509 6 nm for Ti_3Au to a = 0·515 nm for Ti_3AuH. The composition $Ti_3AuH_{2·8}$, with a = 0·529 0 nm also has the Cr_3Si-type structure, cP8, of Ti_3Au and the lower hydride [67 Vet].

REFERENCE

67 Vet: J. B. Vetrano, G. L. Guthrie and H. E. Kissinger, *Phys. Lett.*, 1967, **26A**, 45–46

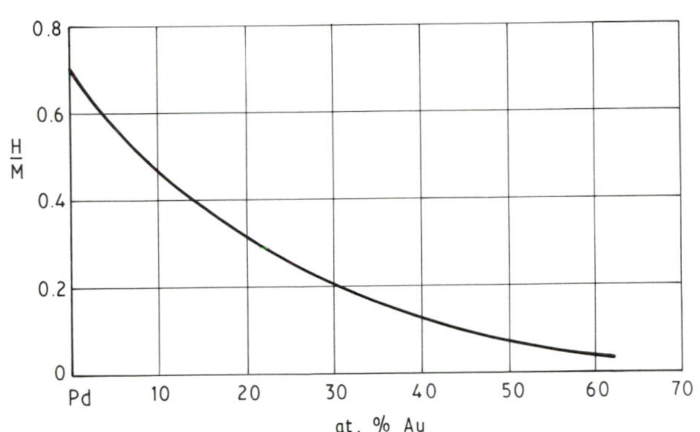

236 **The equilibrium solubility of H at 25°C and 1 atm H_2 pressure in Au–Pd alloys.**

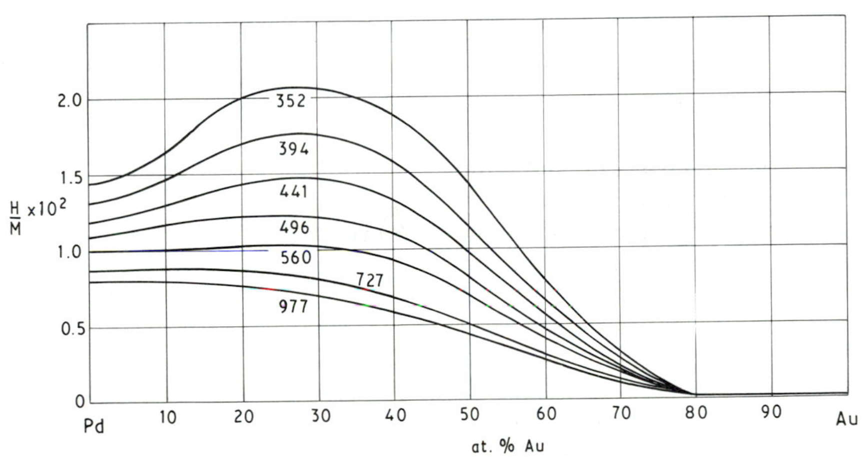

237 **Hydrogen solubility isobars in the Au–Pd–H ternary solid solution at 1 atm H_2 pressure.**

Au–Hf–In

The alloy Au_2HfIn was prepared [75 Mar] by induction-melting the 99·99% purity elements under argon. The alloy was annealed for 7 days at 500°C and examined by metallographic and X-ray powder diffraction techniques. The alloy was two-phase, no indication being given of the second phase. The major phase had a Cu_2AlMn-type structure (cF16) with a = 0.651 4 nm. It is not clear how [75 Mar] distinguished the Heusler-type structure from a CsCl structure by X-ray powder diffraction techniques. No details are given of the presence of odd superlattice lines to justify the Cu_2AlMn-type structure.

REFERENCE

75 Mar: R. Marazza, R. Ferro and G. Rambaldi: *J. Less-Common Metals*, 1975, **39**, 341–345

Au–Hg–Mg

The only reported work [84 Daa] used X-ray diffraction analysis to determine the phase equilibria for ternary alloys containing ⩾50 at.-% Mg. A total of 13 binary alloys and 8 ternary alloys were studied (Fig. 238). Very pure metals were used: Au with 74 ppm impurities, Mg with 133 ppm impurities and double-distilled Hg. The elements were reacted at 800°C in a closed Mo crucible with an internal argon atmosphere. Subsequent annealing was done at temperatures from 500 to 800°C. No trace of Mo was detected in the reacted products. The schematic isothermal (Fig. 238), represents the phase constitution at a temperature of 600–800°C. Although both AuMg and HgMg have the CsCl-type structure (cP2) they do not appear to form a continuous series of solid solutions. The phase $HgMg_2$ has a large solubility for Au. Similarly Hg replaces more than ⅓Au in phase $AuMg_3$.

REFERENCE

84 Daa: J. L. C. Daams and J. H. N. Van Vucht: *Philips J. Res.*, 1984, **39**, 275–292

Au–Hg–Mn

Hori and Nakagawa [86 Hor] studied the crystal structure of an unspecified number of alloys on the AuMn–HgMn section. Alloys were prepared from liquid Hg and powdered Au and Mn sealed in evacuated quartz tubes and heated at 300°C for >4 days. X-ray diffraction analysis at room temperature showed that the ordered CsCl structure (cP2) of HgMn extends to 12·5 at.-% Au on the 50% Mn section of the ternary. Addition of Au to HgMn reduces the lattice spacing from 0·331 4 nm for HgMn [62 Nak] to 0·330 nm for 2·5 at.-% Au, 47·5 at.-% Hg and to 0·328 nm for 12·5 at.-% Au, 37·5 at.-% Hg. At −93°C (180 K) the 12·5 at.-% Au, 37·5 at.-% Hg alloy changes from an ordered cubic structure to a transitional tetragonal structure which transforms at about −123°C (150 K) to an orthorhombic structure. At −185°C (88 K) the lattice spacing is a = 0·471, b = 0·308, c = 0·481 nm (oP4, AuCd type). The transformation from ordered cubic to orthorhombic structure was observed for all alloys with Au contents between 2·5 and 12·5 at.-%.

REFERENCES

62 Nak: Y. Nakagawa and T. Hori: *J. Phys. Soc. Japan*, 1962, **17**, 1313–1314

86 Hor: T. Hori and Y. Nakagawa: *J. Phys. Soc. Japan*, 1986, **55**, 4025–4029

Au–In–Mn

The published work on the Au–In–Mn ternary system relates to the effect of replacing some of the Au atoms in AuMn with In. The Au–Mn binary diagram evaluated by [85 Mas] is accepted. It shows the congruent formation of AuMn with the bcc CsCl structure at 1260°C. At lower temperatures this phase transforms into tetragonally distorted CsCl-type structures β_1 with $c:a < 1$ for alloys with >50 at.-% Au and β_2 with $c:a > 1$ for alloys with <50 at.-% Au. The homogeneity range of AuMn is a maximum of 33·4 to 69·2 at.-% Mn. At room temperature the range is probably from 40 to 60 at.-% Mn.

Morris *et al.* [59 Mor] reported that the alloy Au_2InMn was CsCl-type at room temperature. The Au atoms lie on one simple cubic lattice and the In and Mn atoms are randomly distributed on the second simple cubic lattice that interpenetrates with the Au cubic lattice. The alloy was not entirely single phase, possibly due to slight departure from stoichiometry. A cubic to tetragonal transformation was noted at −37°C. [62 Mor] state, without further detail, that In is soluble in AuMn up to approximately the composition $Au_2In_{0·75}Mn_{1·25}$. All four alloys studied on the section $Au_2Mn_{2-x}In_x$ (x = 0·25, 0·5, 0·75, 1·00) showed a cubic to tetragonal transformation. The transformation temperature is depressed by In additions and occurs at ~130°C in $Au_2In_{0·25}Mn_{1·75}$, ~50°C in $Au_2In_{0·5}Mn_{1·5}$ and ~−40°C in $Au_2In_{0·75}Mn_{1·25}$ and Au_2InMn. In all cases the tetragonal structures had $c:a >$ 1.

Ido *et al.* [70 Ido] prepared three ternary alloys by unspecified methods and studied their structure by X-ray diffraction at room temperature. The alloys were on the section $Au_{1-x}In_xMn$ and contained 2·5 at.-% In, 50 at.-% Mn; 5 at.-% In, 50 at.-% Mn; and 7.5 at.-% In, 50 at.-% Mn. The crystal structure at room temperature became cubic at 5 at.-% In; the 7·5 at.-% In alloy was also cubic, indicating at least this solubility of In in AuMn along the 50% Mn section at room temperature. Both sets of limited data [62 Mor], and [70 Ido], demonstrate a considerable ternary solid solution region based on bcc AuMn in the Au–In–Mn system.

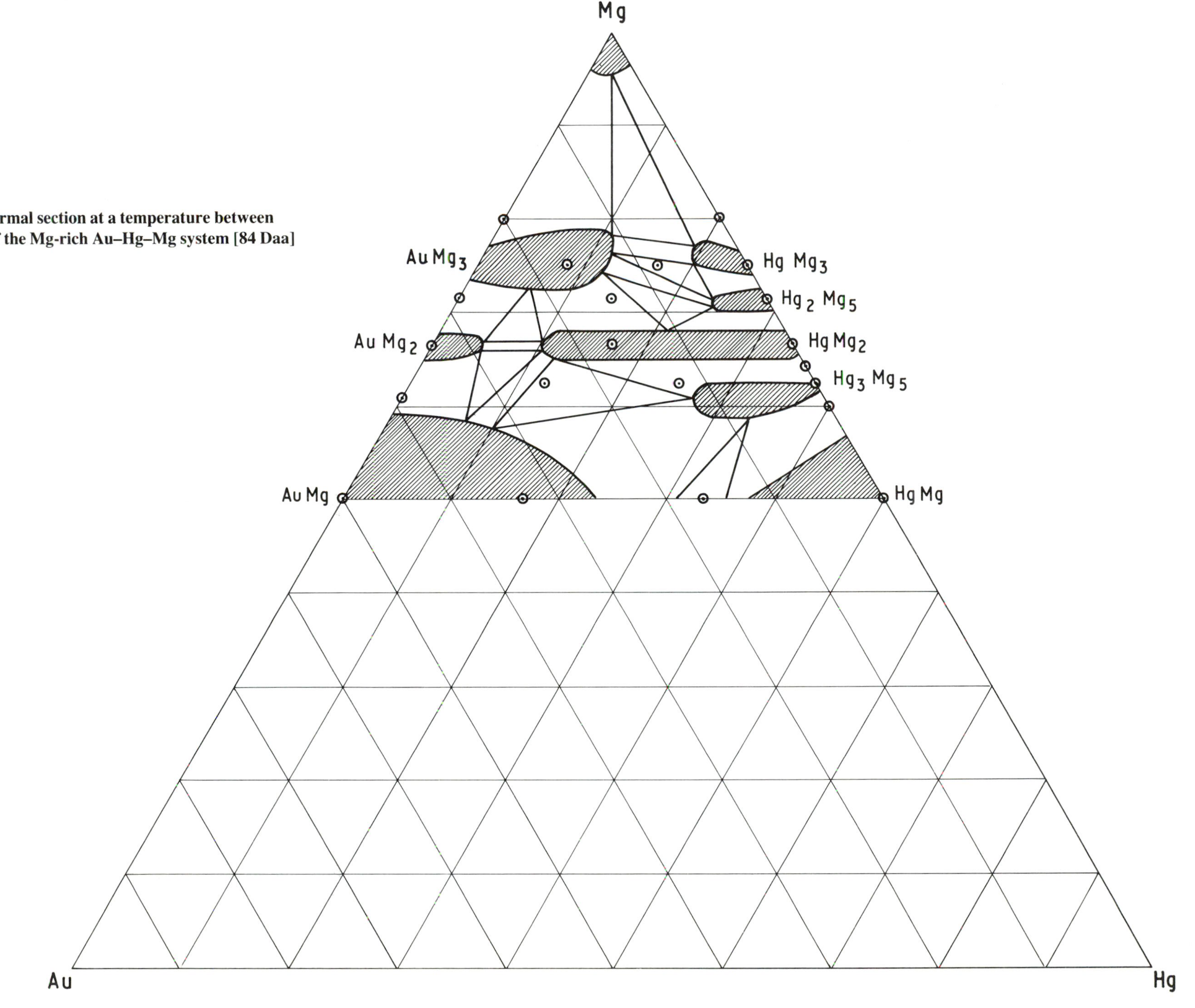

238 Schematic isothermal section at a temperature between 600 and 800°C of the Mg-rich Au–Hg–Mg system [84 Daa]

REFERENCES

59 Mor: D. P. Morris, R. R. Preston and I. Williams: *Proc. Phys. Soc.*, 1959, **73**, 520–523

62 Mor: D. P. Morris, G. W. Davies and C. D. Price: *J. Phys. Chem. Solids*, 1962, **23**, 109–111

70 Ido: T. Ido, H. Teramoto, T. Kasai, K. Sato and K. Adachi: *J. Phys. Soc. Japan*, 1970, **28**, 1589

85 Mas: T. B. Massalski and H. Okamoto: *Bull. Alloy Phase Diagrams*, 1985, **6**, 454–467

Au–In–P

Arseni [68 Ars] determined the solubility of Au in InP between 600 and 900°C (Fig. 239). It has a retrograde character with a maximum Au solubility near 850°C. An increase in the vapour pressure of P above InP increases the solubility of Au in InP. Samples of InP maintained at 720°C in contact with P heated to 350, 450, and 570°C gave Au solubilities of 0·008, 0·014, and 0·021 atomic ppm Au. The remaining literature is concerned with the reactions between thin Au films and InP substrates and the reported results are conflicting. Piotrowska *et al.* [81 Pio] observed a strong reaction between Au films and InP in the temperature range 320–360°C. The InP dissociates, the In diffusing into the Au metallization and the Au into the semiconductor to form Au_2P_3 and Au_3In. The phase Au_2P_3 is not regarded as a stable phase in the Au–P system [84 Oka] but it undoubtedly exists in thin film reaction products. [82 Van] did not substantiate the results of [81 Pio]; on continuous heating of Au films on InP substrates they found the presence of Au by glancing angle X-ray diffraction analysis up to 320°C, Au_4In beginning to form at 340°C, Au_9In_4 at 390°C, and $AuIn_2$ at 450°C. No Au_3P_2 phase was detected although [82 Van] comment that the X-ray technique of [81 Pio] would give a greater penetration depth and detect Au–P interactions deeper in the substrate. Camlibel *et al.* [82 Cam] reported on the reaction between Au films and InP substrates heated for 10 min at 420°C. Using X-ray microanalysis on a scanning electron microscope they showed that the InP dissociates and In enters the Au metallization film to form an alloy containing 15 at.-% In. The Au reacts with the InP to form a Au–P compound which [82 Cam] refer to as Au_2P_3. No Au–In compounds were detected as interaction layers at the metal–semiconductor interface.

REFERENCES

68 Ars: K. A. Arseni: *Fiz. Tekhn. Poluprovod.*, 1968, **2**, 1758–1761 (*Soviet Phys. — Semicond.*, 1969, **2**, 1464–1467)

81 Pio: A. Piotrowska, P. Auvray, A. Guivarc'h, G. Pelous and P. Henoc: *J. Appl. Phys.*, 1981, **52**, 5112–5117

82 Van: J. M. Vandenberg, H. Temkin, R. A. Hamm and M. A. DiGiuseppe: *J. Appl. Phys.*, 1982, **53**, 7385–7389

82 Cam: I. Camlibel, A. K. Chin, F. Ermanis, M. A. DiGiuseppe, J. A. Lourenco and W. A. Bonner: *J. Electrochem. Soc.*, 1982, **129**, 2585–2590

84 Oka: H. Okamoto and T. B. Massalski: *Bull. Alloy Phase Diagrams*, 1984, **5**, 490–491

Au–In–Pb

INTRODUCTION

Ternary liquid immiscibility was reported by [76 Kar] and confirmed by [77 Eva 1, 2] and [87 Nab 1]. The AuIn–Pb section (Fig. 240), is pseudobinary with a monotectic reaction at 487 ± 2°C and a eutectic reaction at 318 ± 2°C [77 Eva 1]. The $AuIn_2$–Pb section (Fig. 241) shows a tendency to incipient liquid immiscibility but the ternary liquid immiscibility gap does not extend to this section. There is disagreement as to whether the section $AuIn_2$–Pb is also a pseudobinary. [76 Kar] reported it as such but [77 Eva 2] noted invariant reactions at 321 and 317°C and interpreted them as evidence for a transition reaction U_8, L + AuIn = $AuIn_2$ + (Pb), where the liquid composition lies on the In side of the $AuIn_2$–Pb section (see Fig. 242). A reaction scheme for the AuIn–In–Pb partial ternary system is incorporated in Fig. 243. Fig. 244 shows a vertical section from 64·3 In, 37·5 at.-% Pb towards the Au corner (50 In, 50 wt-% Pb to Au). The invariant reactions $L_1'' = L_2''$ + AuIn + $AuIn_2$ at 472°C, reactions at U_9 and U_{10} are considered as experimentally established, but there is uncertainty as to the compositions of L_1'' and L_2'' (see Fig. 242). [87 Nab 1] measured the activities of In and Pb in ternary alloys and presented a thermodynamic calculation of the ternary equilibria. The partial ternary system AuIn–Au–Pb (Figs 242 and 243) is based on the calculated equilibria and supported by some experimental observations of invariant reactions by [77 Eva 2] and [87 Nab 1]. The numerous transition reactions that take place close to the Au–Pb binary edge (Fig. 242), require experimental verification.

BINARY SYSTEMS

The Au–Pb diagram determined by [83 Eva] is accepted. The Au–In phase diagram evaluated by [87 Oka] and the In–Pb phase diagram evaluated by [87 Nab 2] are accepted.

SOLID PHASES

Table 85 summarizes the crystal structure data for the binary phases. No ternary compounds have been reported.

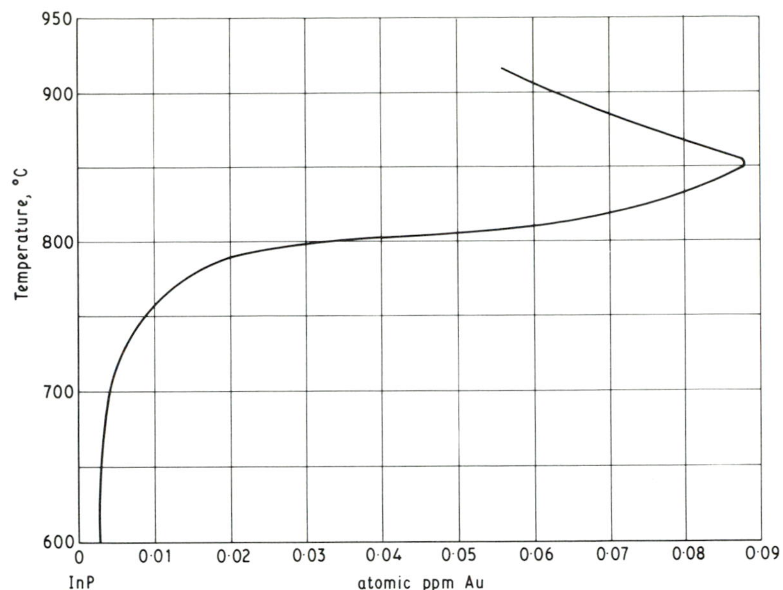

239 Solubility of Au in InP [68 Ars]

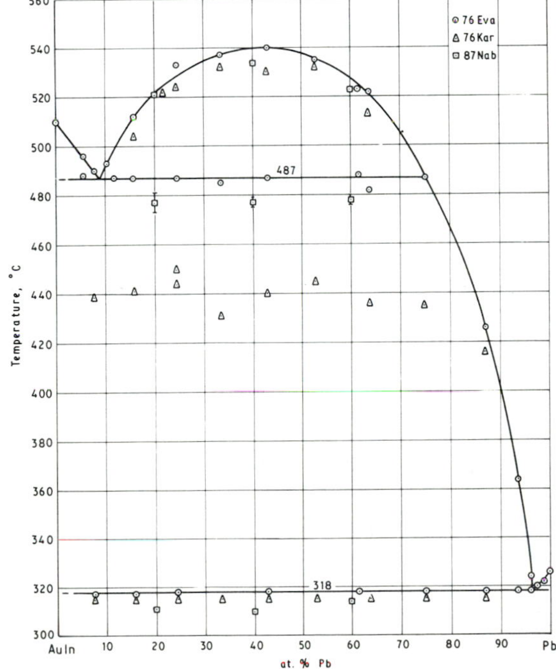

240 The AuIn–Pb section

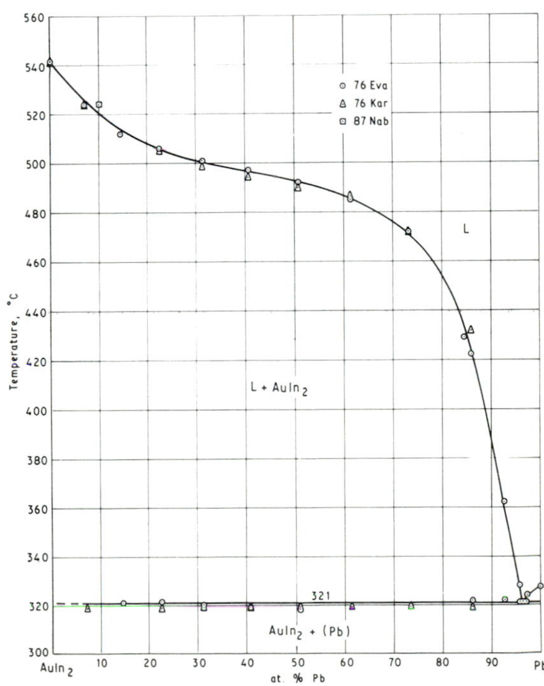

241 The $AuIn_2$–Pb section

Table 84. Crystal structures

Solid phase	*Prototype*	*Lattice designation*
(Au)	Cu	cF4
(In)	In	tI2
(Pb)	Cu	cF4
(αIn)	In	tI2
α_1(Au–In)	Ni_3Ti/Nd	hP16/hP4
ζ	Mg	hP2
ϵ(Au_3In)	orthorhombic	—
ϵ′(Au_3In.r)	βCu_3Ti	oP8
γ(Au_7In_3)	Cu_9Al_4	cP52
γ′(Au_7In_3.r)	Au_7In_3	hP60
ψ($Au_{62}In_{38}$)	Ni_2Al_3	hP5
AuIn	triclinic	—
$AuIn_2$	CaF_2	cF12
Au_2Pb	Cu_2Mg	cF24
$AuPb_2$	Al_2Cu	tI12
$AuPb_3$	αV_3S	tI2

PSEUDOBINARY SYSTEMS

The AuIn–Pb section is pseudobinary. The original work of [76 Kar] showed a monotectic reaction at 441°C with a ±10°C scatter in the DTA data (Fig. 240), followed by a eutectic reaction at 315 ± 2°C. [77 Eva 1] confirmed the pseudobinary characteristics but placed the monotectic temperature at 487 ± 2°C and the eutectic temperature at 318 ± 2°C. They noted gross undercooling of up to 40°C at the 487°C reaction, although all alloy compositions reheated to within a degree or so of 487°C when the melts were stirred during thermal analysis. [76 Kar] did not stir the melt during thermal analysis and the low value for the monotectic temperature and the scatter of results are considered to be due to undercooling. [87 Nab 1] thermally analysed three alloys and found a monotectic temperature of 477 ± 3°C with a eutectic temperature of 312 ± 2°C. He also used mass spectrometry to measure the activities of In and Pb in the ternary Au–In–Pb system, thereby allowing an optimised calculation of the ternary equilibria. The calculated AuIn–Pb section shows a monotectic reaction at 467°C but below this temperature the calculated section deviates from a pseudobinary. The data of [77 Eva 1] have been preferred on the basis of a more careful experimental technique, with stirring of the melt providing a closer approach to equilibrium conditions.

The $AuIn_2$–Pb section was reported as a pseudobinary eutectic by [76 Kar] with a eutectic temperature of 319°C. However, unpublished work [77 Eva 2], included in [87 Nab 1], indicated two arrests, the first at 321°C and the second at 317°C. The liquidus curve shows excellent agreement with that reported by [76 Kar] (see Fig. 241). The calculated section [87 Nab 1] indicated liquid immiscibility extending from 20·5 to 60 at.-% Pb with the presence of a three-phase region of L_1 + L_2 + $AuIn_2$. Experimental work confirmed the absence of liquid immiscibility on this section [87 Nab 1]. The liquidus projection, based on the experimental work of [76 Kar], [77 Eva 2] and [87 Nab 1], for the AuIn–Pb–In partial ternary system indicates that the ternary liquid immiscibility region approaches very close to the $AuIn_2$–Pb section (Fig. 242), but does not intersect this section.

INVARIANT EQUILIBRIA

No satisfactory experimental determination of the ternary invariant equilibria has been reported for the AuIn–Au–Pb partial ternary system. [76 Kar] identified the invariant reactions in the AuIn–Pb–In partial ternary system. A reaction $L_1'' = L_2''$ + AuIn + $AuIn_2$ was found at 433°C compared with a temperature of 472°C by [77 Eva 2] and a comparable temperature of 469 ± 10°C by [87 Nab 1]. A critical tie line between liquid of estimated composition c_1 (Fig. 242), and $AuIn_2$ occurs at about 495°C [77 Eva 2]. A ternary eutectic reaction, L = (Pb) + AuIn + $AuIn_2$, at 316°C was indicated by [76 Kar]. This reaction cannot occur at 316°C if the AuIn–Pb eutectic is placed at 315°C [76 Kar]. In contrast to the interpretation of [76 Kar] a transition reaction, L + AuIn = $AuIn_2$ + (Pb), was proposed by [77 Eva 2]. Such a reaction accords with the observance of two thermal arrests at 321 and 317°C on the $AuIn_2$–Pb section, corresponding to the presence of a three-phase region L + AuIn + $AuIn_2$ between 321 and 317°C on this section. The transition reaction is denoted U_8 in Figs 242 and 243. From U_8 the liquid traces a path close to the In–Pb binary edge. Transition reactions U_9, L + (Pb) = $AuIn_2$ + (αIn), were detected at 172°C [76 Kar] and 173°C [77 Eva 2] and U_{10}, L + (αIn) = $AuIn_2$ + (In), at 159°C [76 Kar, 77 Eva 2].

The Au–AuIn–Pb partial ternary system contains a number of invariant reactions that were not characterized by [77 Eva 2]. Valuable insight into the ternary equilibria has emerged from the thermodynamically calculated equilibria [87 Nab 1]. Figs 242 and 243 are based on the calculated equilibria, taking into account experimental phase diagram data from [77 Eva 2] and [87 Nab 1]. Although they should not be regarded as definitive the liquidus projection (Fig. 242), and the proposed reaction scheme (Fig. 243), are considered a good guide to the probable equilibria. There is agreement between [77 Eva 2] and [87 Nab 1] on a temperature of 432 ± 3°C for the invariant reaction $L_1' = \gamma + \psi + L_2'$. The calculated temperature is some 40°C lower. The transition reactions U_1 and U_2 were calculated at 392 and 312°C respectively. The transition reactions U_3 and U_4 were similarly calculated at 252 and 250°C respectively; [77 Eva 2] found evidence for invariant reactions at 253 and 250°C but could not characterize these reactions. The reactions U_5, U_6, U_7 and E_1 were not characterized by [77 Eva 2] but evidence was found for the occurrence of invariant reactions at temperatures of about 245, 240, 215 and 207·5°C respectively. The steep slope of the liquidus surface and the clustering of the invariant reactions U_1 to U_7 close to the Au–Pb binary edge make experimental study difficult. Further study of alloy compositions at less than 1 at.-% In intervals for alloys containing up to 10 at.-% In is required.

LIQUIDUS SURFACE

Figure 242 is a presentation of the probable liquidus surface. Isotherms are given for the AuIn–Pb–In partial ternary system, based on the work of [76 Kar, 77 Eva 2, 81 Mar].

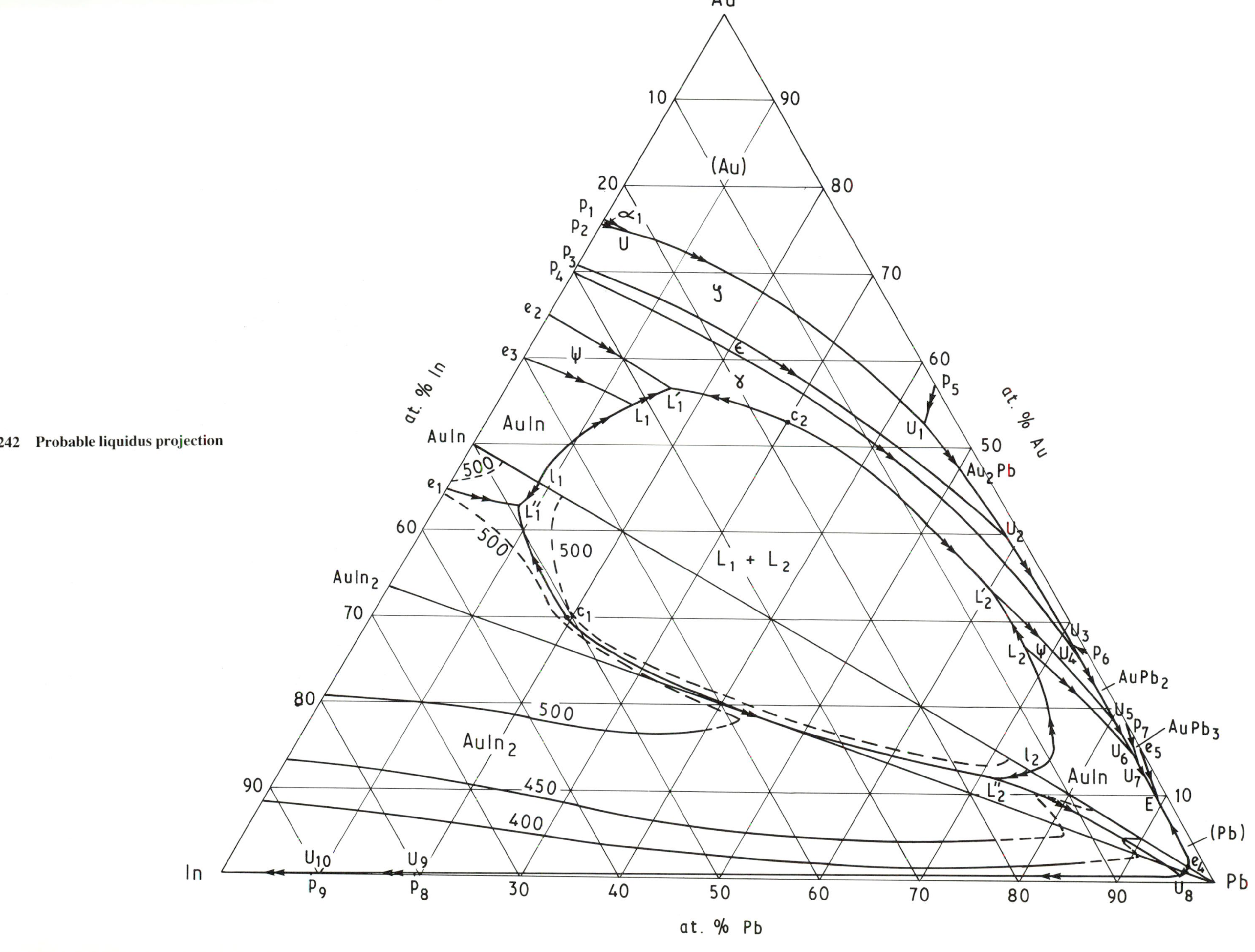

242 Probable liquidus projection

243 Probable reaction scheme

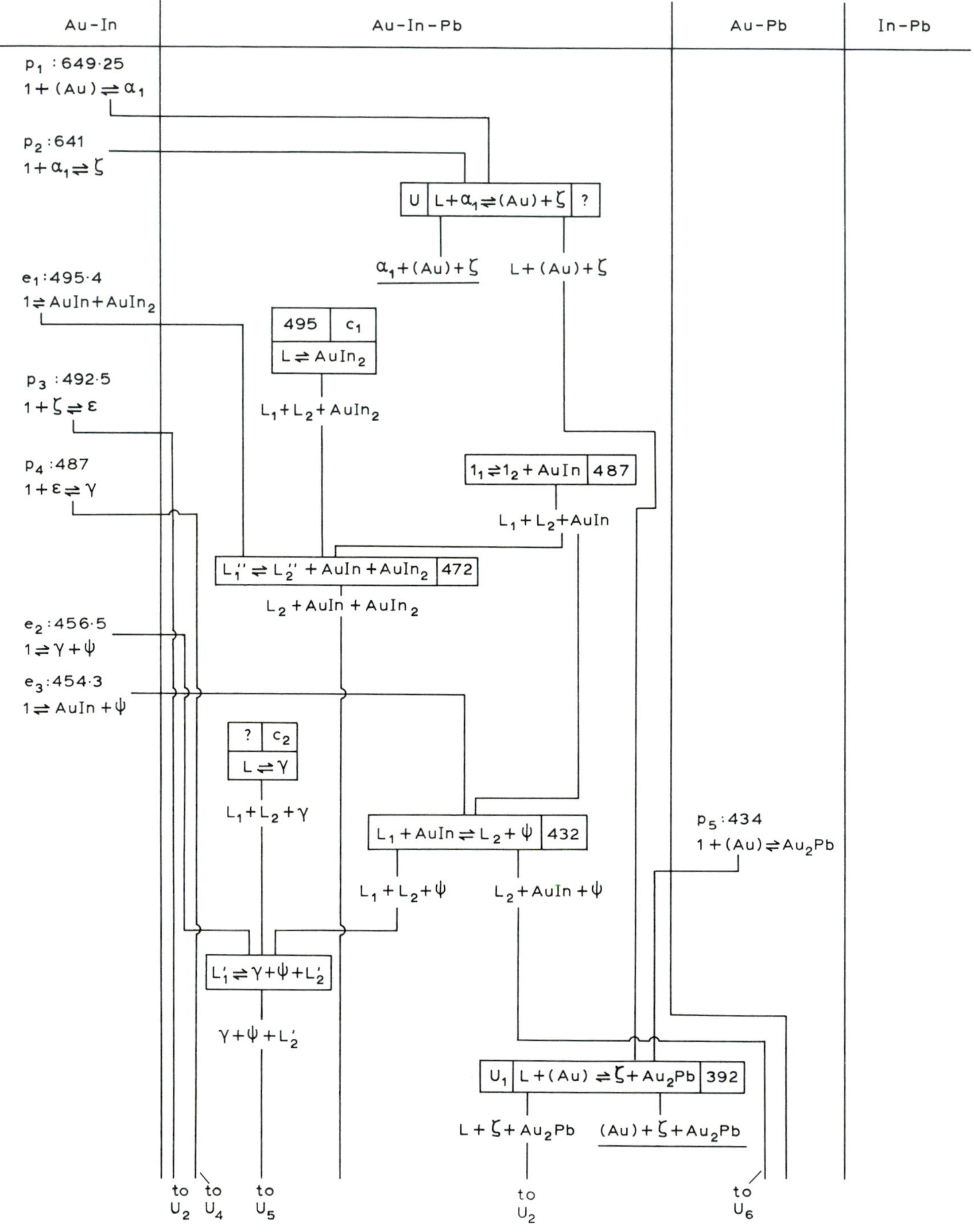

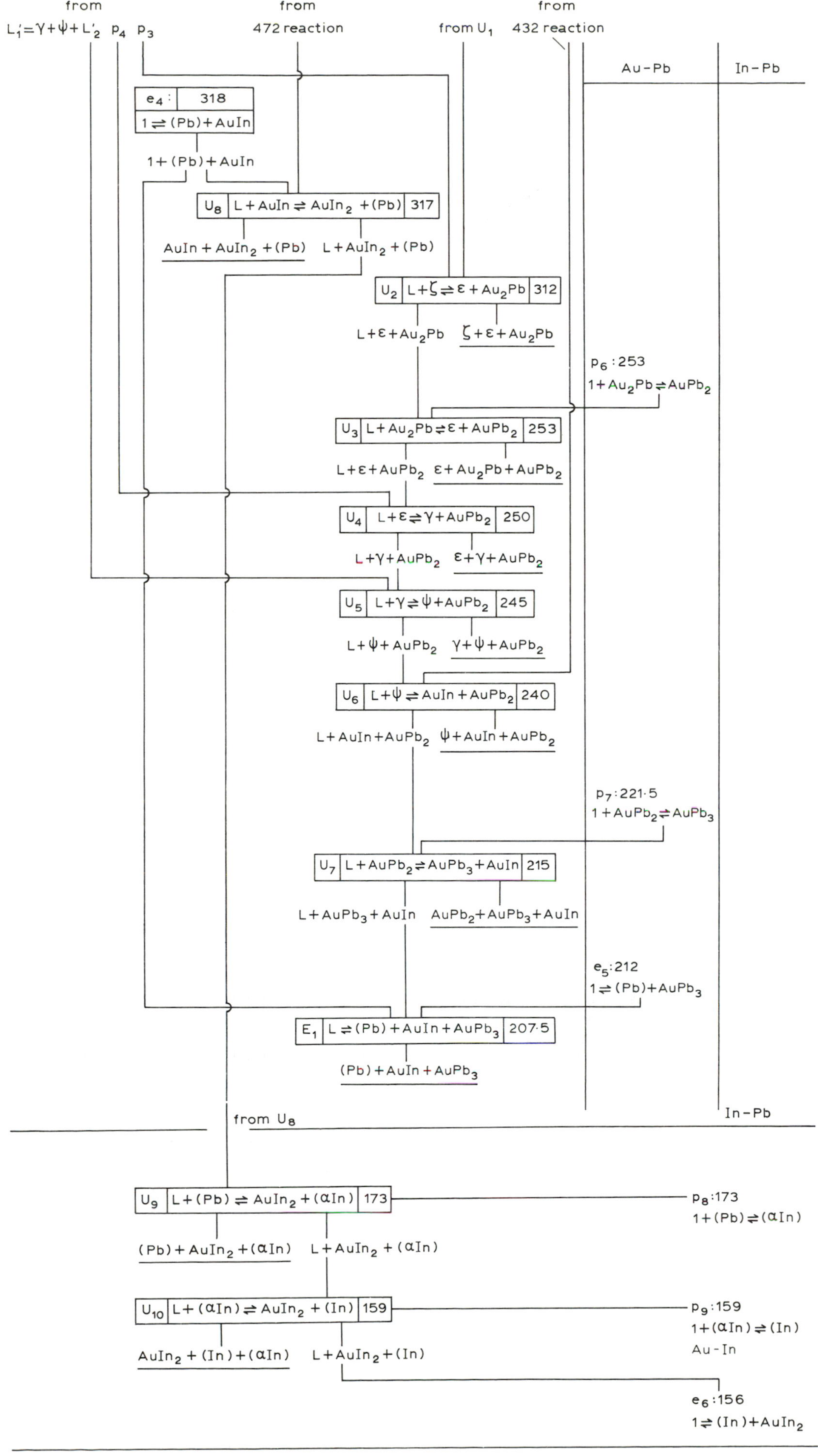
from
L′1=γ+ψ+L′2 p4 p3
from
472 reaction
from U1
from
432 reaction
Au-Pb
In-Pb
e4: 318
1⇌(Pb)+AuIn
1+(Pb)+AuIn
U8 L+AuIn⇌AuIn2+(Pb) 317
AuIn+AuIn2+(Pb)
L+AuIn2+(Pb)
U2 L+ζ⇌ε+Au2Pb 312
L+ε+Au2Pb
ζ+ε+Au2Pb
p6:253
1+Au2Pb⇌AuPb2
U3 L+Au2Pb⇌ε+AuPb2 253
L+ε+AuPb2
ε+Au2Pb+AuPb2
U4 L+ε⇌γ+AuPb2 250
L+γ+AuPb2
ε+γ+AuPb2
U5 L+γ⇌ψ+AuPb2 245
L+ψ+AuPb2
γ+ψ+AuPb2
U6 L+ψ⇌AuIn+AuPb2 240
L+AuIn+AuPb2
ψ+AuIn+AuPb2
p7:221·5
1+AuPb2⇌AuPb3
U7 L+AuPb2⇌AuPb3+AuIn 215
L+AuPb3+AuIn
AuPb2+AuPb3+AuIn
e5:212
1⇌(Pb)+AuPb3
E1 L⇌(Pb)+AuIn+AuPb3 207·5
(Pb)+AuIn+AuPb3
from U8
In-Pb
U9 L+(Pb)⇌AuIn2+(αIn) 173
p8:173
1+(Pb)⇌(αIn)
(Pb)+AuIn2+(αIn)
L+AuIn2+(αIn)
U10 L+(αIn)⇌AuIn2+(In) 159
p9:159
1+(αIn)⇌(In)
Au-In
AuIn2+(In)+(αIn)
L+AuIn2+(In)
e6:156
1⇌(In)+AuIn2

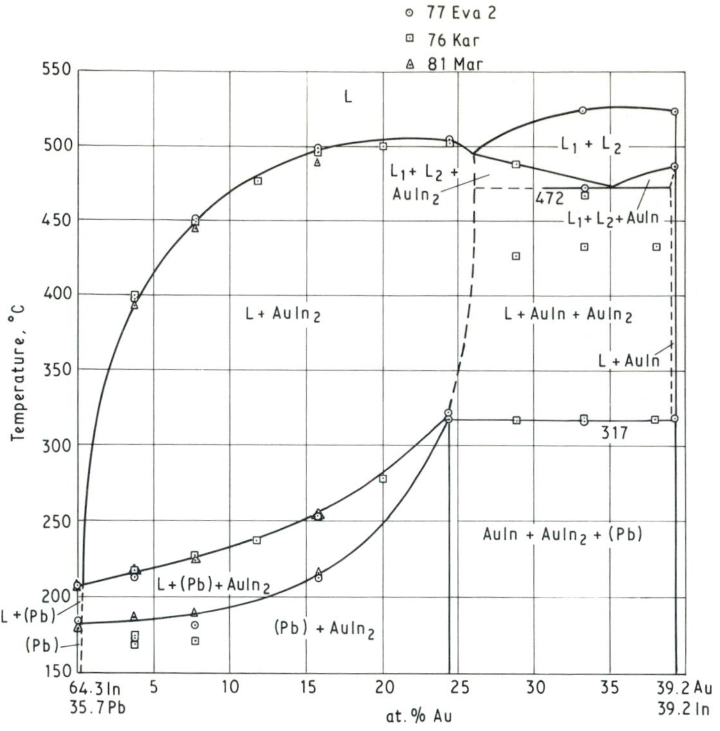

244 The 64·3 In, 35·7 at.-% Pb–Au section up to its intersection with the AuIn–Pb section

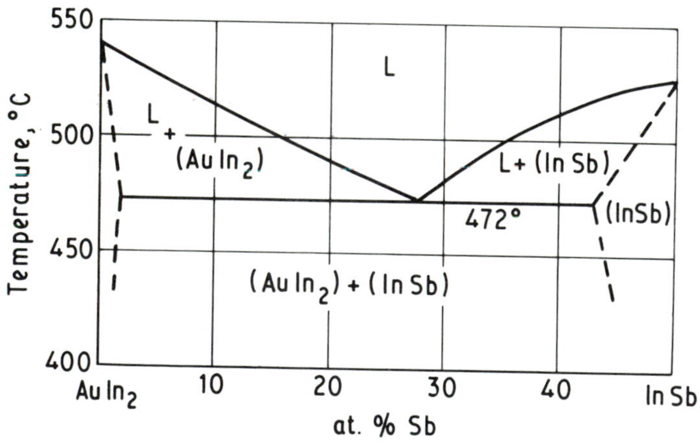

245 The $AuIn_2$–InSb section [67 Nik]

MISCELLANEOUS

A partial vertical section from 64·3 In, 35·7 at.-% Pb to Au, extending to the AuIn–Pb pseudobinary section, was originally presented by [76 Kar]. [81 Mar] showed that a homogenizing anneal of 500 h at 170°C, a temperature just below that of the ternary invariant reaction U_9, eliminated the (αIn) phase shown by [76 Kar, 77 Eva 2] for alloy compositions containing less than 10 at.-% Au. The section, Fig. 244, shows good agreement for the liquidus and for the phase boundary between L + $AuIn_2$ and L + (Pb) + $AuIn_2$. The data of [81 Mar] are accepted in terms of the absence of the (αIn) phase in this section. This implies that the composition of the (Pb) phase associated with the transition reaction at U_9 contains less than 35·7 at.-% Pb. It should be noted that the composition of the (Pb) phase in the binary peritectic reaction l + (Pb) = (αIn) is 30 at.-% Pb [87 Nab 2]. The schematic sketch for low Au contents [76 Kar] is incorrect, as are the constructions of isothermal sections corresponding to the invariant reactions U_9 and U_{10}. The reaction $L_1'' = L_2''$ + AuIn + $AuIn_2$, located at 472°C by [77 Eva 2] was given as 433°C by [76 Kar], as noted in the section on invariant equilibria. The low temperature reported by [76 Kar] is probably caused by undercooling arising from a lack of stirring of melts on cooling.

[76 Yos] examined the interaction between a 64·3 In, 35·7 at.-% Pb solder alloy and a vapour deposited Au film at 250 ± 5°C. The molten solder was reacted for 5 sec with the Au and the assembly subsequently aged at temperatures varying from 70 to 170°C. Bulk samples of the In–Pb solder to Au were also prepared and aged at 150°C. As would be anticipated from Fig. 244 a mixture of $AuIn_2$ + (Pb) is the reaction product with a layer of $\alpha(Au_7In_3)$ at the Au interface.

REFERENCES

76 Yos: F. G. Yost, F. P. Ganyard and M. M. Karnowsky: *Metall. Trans. A*, 1976, **7A**, 1141–1148

76 Kar: M. M. Karnowsky and F. G. Yost: *Metall. Trans. A*, 1976, **7A**, 1149–1156

77 Eva 1: D. S. Evans and A. Prince: *Metal. Sci.*, 1977, **11**, 597

77 Eva 2: D. S. Evans and A. Prince, unpublished data, The General Electric Company plc, Hirst Research Centre, Wembley, 1977

81 Mar: V. C. Marcotte and M. W. Ricker: *Metall. Trans.* A., 1981, **12A**, 2136–2138

83 Eva: D. S. Evans and A. Prince: *Mater. Res. Soc. Symp. Proc.*, 1983, **19**, 383–388

87 Oka: H. Okamoto and T. B. Massalski: 'Phase Diagrams of Binary Gold Alloys', *ASM International*, Metals Park, Ohio, 1987, pp. 142–153

87 Nab 1: J. P. Nabot, 'Thermodynamic Analysis of the Au–In–Pb System' Thesis, Institut National Polytechnique de Grenoble, April 1987

87 Nab 2: J. P. Nabot and I. Ansara: *Bull. Alloy Phase Diagrams*, 1987, **8**, 246–255

Au–In–Pt

According to [59 Bur] the addition of 0·5 at.-% Pt to Au_3In hardly alters the lattice parameters of orthorhombic Au_3In (Table 84). With 1·0 at.-% Pt additions a second phase, Au_7In_3, was detected. The solubility of Pt in Au_3In is somewhat less than 1 at.-%.

REFERENCE

59 Bur: W. Burkhardt and K. Schubert: *Z. Metall.*, 1959, **50**, 442–452

Table 85. Lattice parameters of Au–In–Pt alloys

Composition, at.-%			*Heat treatment, powder samples*	*Lattice parameter*, nm		
Au	*In*	*Pt*		*a*	*b*	*c*
75	25	0	—	0·586	0·517	0·475
74·5	25	0·5	4 days at 400°C	0·584	0·514	0·472
74	25	1	4 days at 400°C	0·586	0·510	0·475

Au–In–Sb

INTRODUCTION

The $AuIn_2$–InSb section has been found to be a pseudobinary eutectic type with a eutectic temperature of 472°C and composition 28 at.-% Sb [67 Nik] (Fig. 245). Nikitina and Lobanova [74 Nik] found that the section AuIn–Sb was also a pseudobinary eutectic with a eutectic temperture of 420°C and composition 28 at.-% Sb (Fig. 246). Babitsyna and Luzhnaya [72 Bab] studied the AuIn–InSb section (Fig. 247) and found it not to be a pseudobinary system. In contrast to this work [86 Tsa] reported that two alloys on the AuIn–InSb section, slowly cooled to room temperature, contained only AuIn + InSb. Kubiak and Schubert [80 Kub] prepared 16 ternary alloys and annealed the Au-rich compositions for 10 days at 300 and 200°C; low-Au compositions were annealed for 10 days at 190°C. They presented an 'isothermal' section (Fig. 248) covering the temperature range 250–300°C. This work confirms the two-phase nature of the sections previously determined by [67 Nik] and [74 Nik]. A ternary intermediate phase $AuIn_{0.9}Sb_{0.1}$ was stable at 200–300°C; its crystal structure is orthorhombic (TlI-type, Amma, space group No. 63). A second ternary phase, $AuIn_{0.7}Sb_{0.3}$, was found to exist above 380°C. It has the hexagonal NiAs-type structure (hP4). [86 Tsa] proposed a triangulation of the ternary system, based on the X-ray diffraction analysis of phases found in alloys slowly cooled to room temperature from 1 000°C (Fig. 249).

Boltaks and Sokolov [64 Bol] found the solubility of Au in InSb to vary from 0·03 ppm Au to 0·14 ppm Au over the temperature range 300 to 500°C. Further work on this system should be directed at the AuIn–InSb section, clarification of the occurrence of ternary phases, and a survey of the equilibria for alloys containing <50 at.-% In.

BINARY SYSTEMS

The Au–In diagram evaluated by [87 Oka] is accepted. The Au–Sb system [84 Eva] is also accepted, as is the In–Sb diagram of H,S,E.

SOLID PHASES

The solid phases are summarized in Table 86.

Table 86. Crystal structures

Phase	*Prototype*	*Lattice designation*
(Au)	Cu	cF4
(In)	In	tI2
(Sb)	As	hR2
InSb	ZnS	cF8
$AuSb_2$	FeS_2	cP12
α_1	Ni_3Ti/Nd	hP16/hP4
ζ	Mg	hP2
$\beta_1(Au_{10}In_3.r)$	$Cu_{10}Sb_3$	Hexagonal
$\epsilon'(Au_3In)$	βCu_3Ti	oP8
$\gamma'(Au_7In_3)$	Au_7In_3	Hexagonal
ψ	Ni_2Al_3	hP5
AuIn	—	Triclinic
$AuIn_2$	CaF_2	cF12
$AuIn_{0.9}Sb_{0.1}$	TlI	Orthorhombic
$AuIn_{0.7}Sb_{0.3}$	NiAs	hP4

PSEUDOBINARY SYSTEMS

The $AuIn_2$–InSb section is a pseudobinary eutectic according to [67 Nik] who used differential thermal analysis (only heating data tabulated; cooling data showed large undercooling), metallography, and X-ray analysis on 13 alloys prepared from 99·99 wt-% pure elements. The eutectic temperature quoted (Fig. 245) is probably accurate to ±5°C and the eutectic composition to ±5 at.-% Sb. At the eutectic temperature there is considerable solubility of $AuIn_2$ in InSb (4·67 Au, 43 at.-% Sb) but little solubility of InSb in $AuIn_2$ (32·2 Au, 1·7 at.-% Sb). Tsai and Williams [86 Tsa] found $AuIn_2$ + InSb as the phases in two alloys (Fig. 249 and Table 87) cooled slowly to room temperature.

The AuIn–Sb section is also a pseudobinary eutectic [74 Nik]. In addition to the techniques used by [6, Nik] the variation of overall hardness and the microhardness of the AuIn and Sb phases with composition was determined. The eutectic temperature of 420°C (Fig. 246) is probably accurate to ±3°C and the eutectic composition of 28 at.-% Sb to ±2 at.-% Sb. At the eutectic temperature 12 at.-% Sb is soluble in AuIn and the (Sb) solid solution extends to about 94 at.-% Sb. All the techniques gave consistent data; the microhardness of AuIn increases from 190 kg mm^{-2} for AuIn to 250 kg mm^{-2} at 12 at.-% Sb and remains constant for higher Sb contents. The microhardness of Sb increases from 70 kg mm^{-2} for Sb to 90 kg mm^{-2} for alloys in the two-phase region. [86 Tsa] found AuIn + Sb in one alloy cooled slowly to room temperature (Fig. 249 and Table 87).

Tsai and Williams [86 Tsa] prepared alloys (Fig. 249 and Table 87) from >99·99% pure elements and semiconductor grade InSb. They were contained in fused-silica capsules, evacuated to 10^{-6} torr and sealed with a 4 cm^3 dead space. After heating at 1 000°C for 1 week the alloys were slowly cooled at 5–10°C h^{-1} to room temperature or to 600°C. The 600°C samples were quenched into iced water. Phases were characterized by X-ray powder diffraction. Two-phase structures were found on the sections $AuIn_2$–InSb, AuIn–Sb, AuIn–InSb, ψ–$AuSb_2$, γ'–$AuSb_2$, ϵ'–$AuSb_2$, β'–$AuSb_2$ and α_1–$AuSb_2$. Not all these sections will be pseudobinary sections. As α_1, β', ϵ', and γ' are not congruently formed compounds the sections containing these phases will not be pseudobinary.

[72 Bab] studied the AuIn–InSb section (Fig. 247) and found that it was not a pseudobinary. There is undoubtedly an invariant reaction at 414°C and evidence for the separation of three primary phases from the melt; accord-

Table 87. Alloys studied by [86 Tsa]

Alloy section	*Alloy*	*Alloy composition*, at.-%			*Temperature*, °C	*Phases identified*
		Au	*In*	*Sb*		
$AuIn_2$–InSb	A	14·3	57·1	28·6	25,600	$AuIn_2$, InSb
	B	25·0	62·5	12·5	25	$AuIn_2$, InSb
AuIn–InSb	C	16·67	50·0	33·33	25,600	AuIn, InSb
	D	33·33	50·0	16·67	25	AuIn, InSb
AuIn–Sb	G	33·33	33·33	33·33	25,600	AuIn, Sb
ψ–$AuSb_2$	K	50·0	25·0	25·0	25,600	ψ, $AuSb_2$
	N	46·67	20·0	33·33	25	ψ, $AuSb_2$
γ'–$AuSb_2$	H	56·3	18·7	25·0	25,600	γ', $AuSb_2$
ϵ'–$AuSb_2$	M	63·6	18·2	18·2	25,600	ϵ', $AuSb_2$(m)
	O	57·1	14·3	28·6	25	ϵ', $AuSb_2$
β'–$AuSb_2$	L	66·67	16·67	16·67	25,600	β', $AuSb_2$
α_1–$AuSb_2$	S	81·0	9·5	9·5	25	α_1, $AuSb_2$
	E	18·4	44·7	36·8	25	InSb, AuIn(m), Sb(w)
	Q	52·9	35·3	11·8	25	ψ, AuIn(m), $AuSb_2$(m)
	I	43·0	29·0	28·0	25	ψ, Sb, AuIn, $AuSb_2$
	J	42·9	28·6	28·6	25	ψ, Sb, AuIn, $AuSb_2$
	J	42·9	28·6	28·6	600	ψ, Sb, AuIn, $AuSb_2$(m)
	P	24·0	16·0	60·0	25	ψ(m), Sb
	T	50·0	45·0	5·0	25	AuIn, ψ, Sb(w)
	T	50·0	45·0	5·0	325	AuIn, ψ(w), Sb(w)
	F	45·7	37·0	17·4	25	AuIn, $AuSb_2$(m), Sb(w)
	U	40·0	20·0	40·0	25	$AuSb_2$, ψ(m), Sb(m)
	U	40·0	20·0	40·0	325	$AuSb_2$, ψ(m), Sb(m)
	R	30·0	13·0	58·0	25	Sb, $AuSb_2$(m), ψ(m)

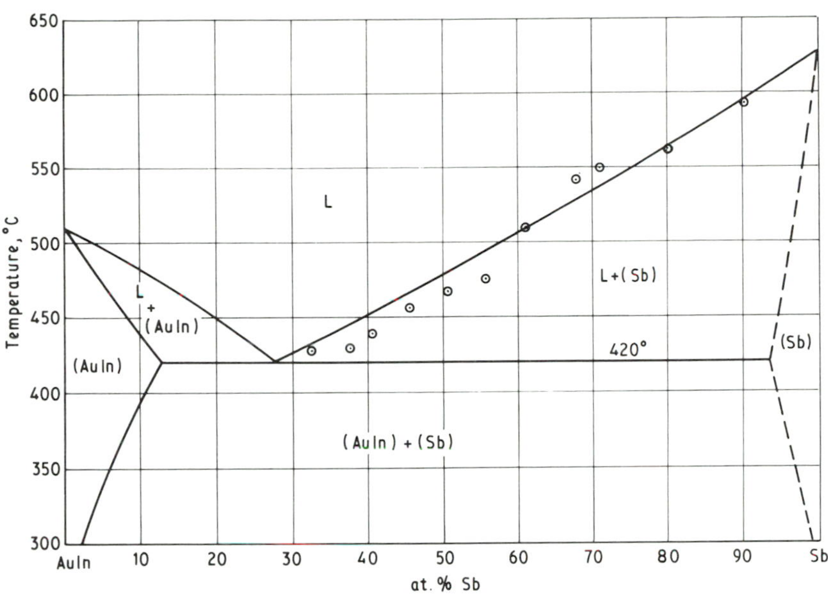

246 The AuIn–Sb section [74 Nik]

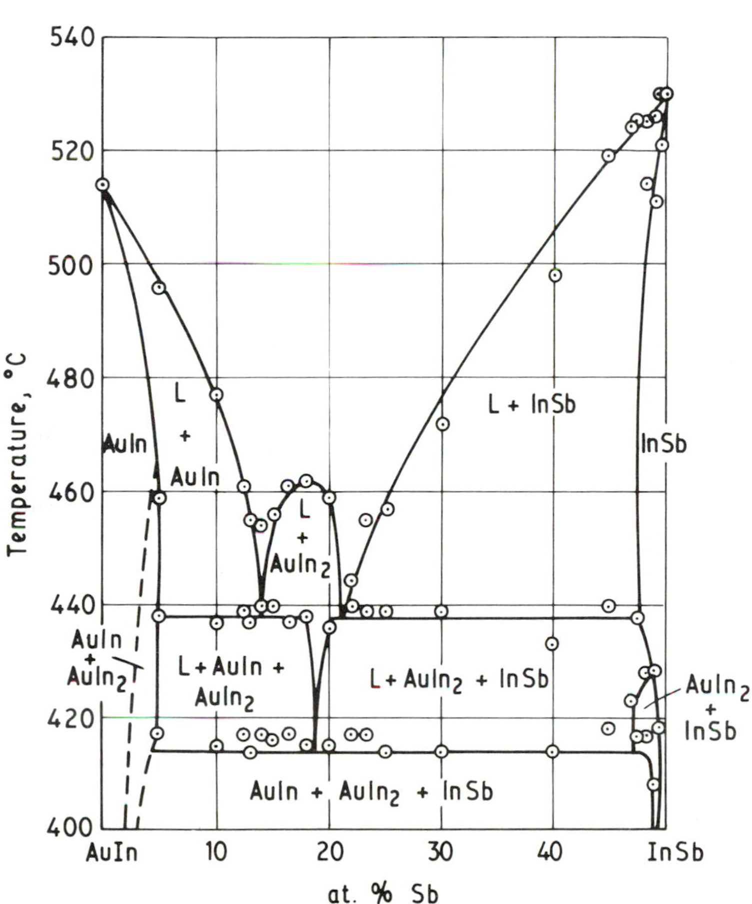

247 The AuIn–InSb section [72 Bab]

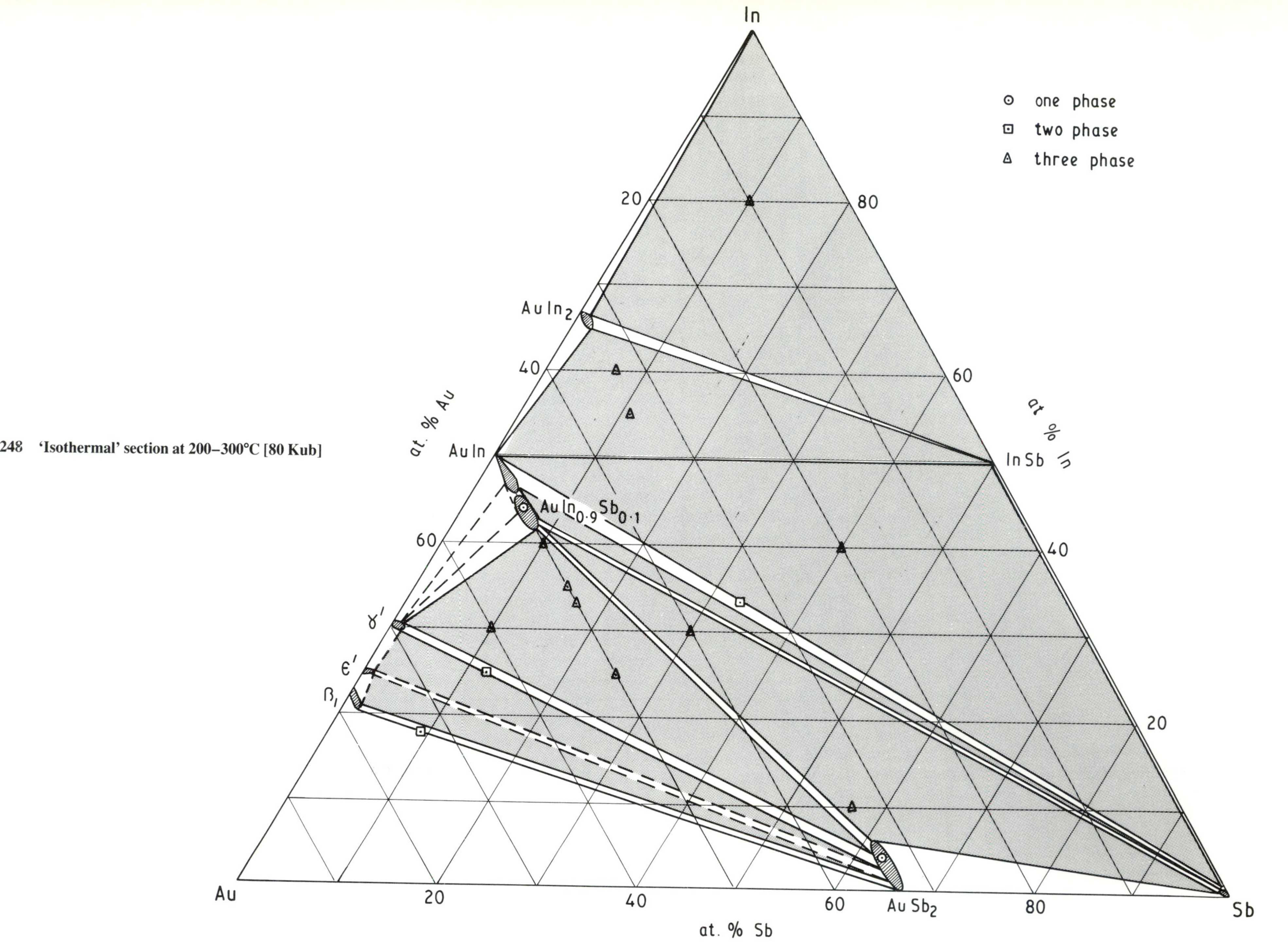

248 'Isothermal' section at 200–300°C [80 Kub]

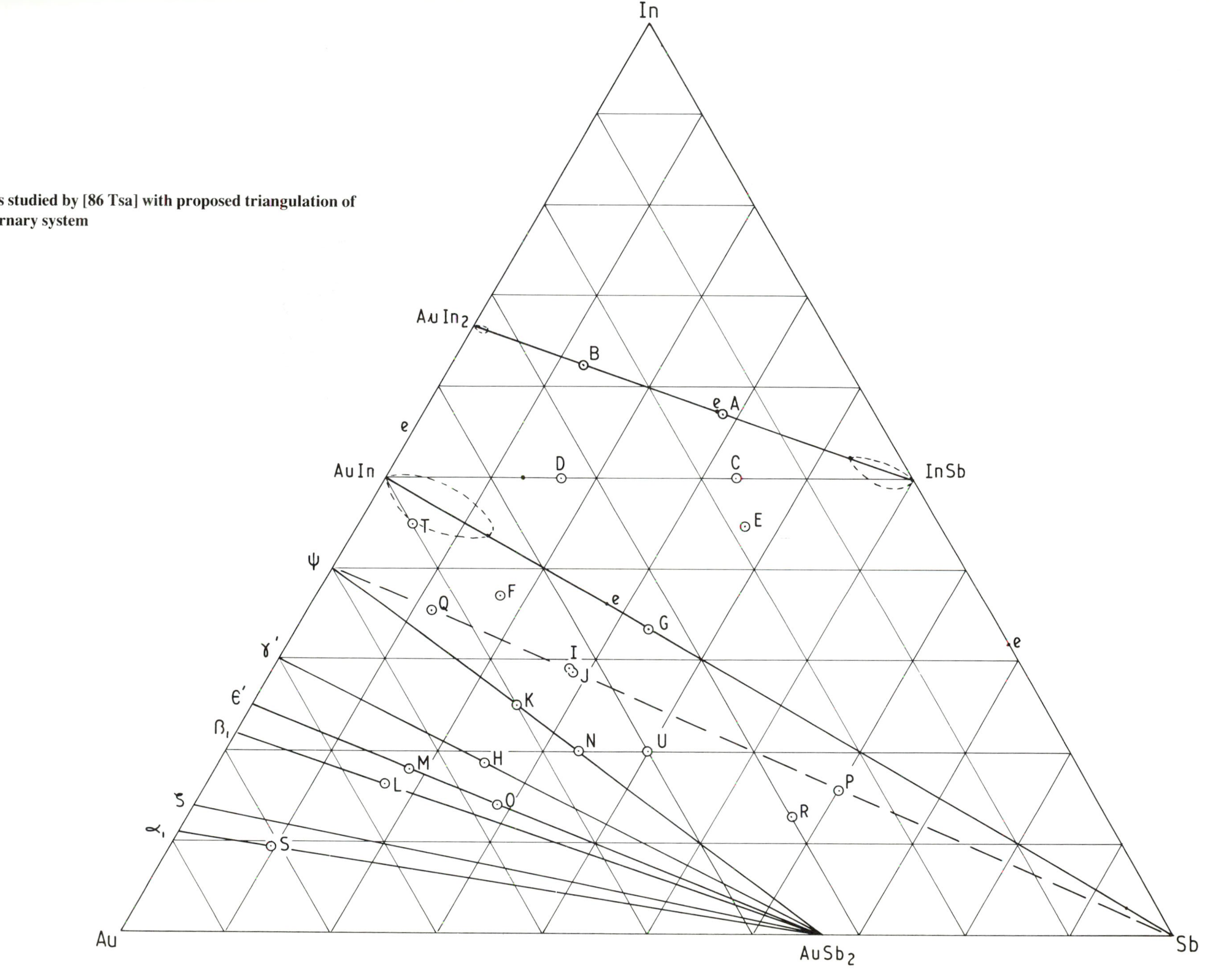

249 **Alloys studied by [86 Tsa] with proposed triangulation of the ternary system**

ing to [72 Bab] there are primary AuIn, $AuIn_2$, and InSb phase regions. However to produce a thermodynamically correct diagram [72 Bab] were constrained to construct an unlikely sharp curvature in the phase boundaries between the l + $AuIn_2$ phase region and the l + $AuIn_2$ + InSb and l + AuIn + $AuIn_2$ phase region. The diagram is reproduced in Fig. 247, but it would seem that there could be a 438°C invariant reaction. The implication of the work of [72 Bab] is that there is a transition reaction of the type L + $AuIn_2$ = AuIn + InSb with the AuIn–InSb tie line at In contents somewhat below 50 at.-% In. This contradicts [86 Tsa] and requires further study.

ISOTHERMAL SECTIONS

Kubiak and Schubert [80 Kub] determined the 'isothermal' section (Fig. 248) representative of a temperature between 200 and 300°C. It is not strictly accurate for the In-rich corner. At these temperatures there will be a liquid phase region at the In corner and a three-phase tie triangle corresponding to the equilibrium of the In-rich liquid with $AuIn_2$ and InSb. Figure 248 shows that AuIn enters into equilibrium with $AuIn_2$ + InSb, Sb + InSb, Sb + $AuIn_{0\cdot9}Sb_{0\cdot1}$, γ' + $AuIn_{0\cdot9}Sb_{0\cdot1}$. The presence of the ternary phase $AuIn_{0\cdot9}Sb_{0\cdot1}$ prevents AuIn equilibrating with $AuSb_2$. The ternary phase $AuIn_{0\cdot9}Sb_{0\cdot1}$ is in equilibrium with AuIn + Sb, $AuSb_2$ + Sb, $AuSb_2$ + γ', AuIn + γ'. The three-phase tie triangles that are drawn with dashed lines (Fig. 248) were not present in [80 Kub]. [86 Tsa] prepared the alloy $AuIn_{0\cdot9}Sb_{0\cdot1}$ (T in Fig. 249 and Table 87) and found it contained ψ, AuIn, and Sb rather than the ternary phase.

MISCELLANEOUS

The ternary phase $AuIn_{0\cdot9}Sb_{0\cdot1}$ with an orthorhombic structure has lattice parameters a = 0·4300(2), b = 1·0582(4), c = 0·3580(3) nm [80 Kub]. A second ternary phase, $AuIn_{0\cdot7}Sb_{0\cdot3}$, was found when this alloy composition was quenched from 800 to 400°C. If quenched from 380°C and below the ternary phase was not present and in its place the three-phase structure ($AuIn_{0\cdot9}Sb_{0\cdot1}$ + γ' + $AuSb_2$) was present (Fig. 248). Thermal analysis of the 50Au, 35In, 15Sb alloy showed thermal effects at 430 and 380°C; these were interpreted as the temperatures for solidification and eutectoidal decomposition of the ternary phase $AuIn_{0\cdot7}Sb_{0\cdot3}$. This phase has a hexagonal structure with lattice parameters a = 0·43014(4), c = 0·55347(6) nm. Further work is needed on the occurrence of ternary phases and the equilibria associated with any such compounds.

Tsai and Williams [86 Tsa] produced evidence for the triangulation of the ternary system (Fig. 249 and Table 87). The region ψ–AuIn–Sb–$AuSb_2$ most likely contains the section ψ–Sb as the stable join and this section is probably a pseudobinary section. It is shown dashed in Fig. 249. The ψ phase is not stable below 224°C in the Au–In binary system. The observation of ψ in ternary alloys slowly cooled to room temperature either implies that Sb stabilizes the ψ phase in this ternary system or that the transformation of ψ by a ternary solid-state reaction does not go to completion during the slow cooling used by [86 Tsa].

It should be noted that only alloys in the Au corner and the Sb corner will contain some solid at 600°C; all other ternary alloys would be expected to be molten at 600°C. All the alloys studied by [86 Tsa] would be molten upon quenching from 600°C.

REFERENCES

64 Bol: B. I. Boltaks and V. I. Sokolov: *Soviet Phys. — Solid St.*: 1964, **6**, 600–603 (*Fiz. Tverd. Tela*, 1964, **6**, 771–775)

67 Nik: V. K. Nikitina, A. A. Babitsyna and Yu. K. Lobanova: *Izvest. Akad. Nauk. SSSR, Neorg. Mater.*, 1967, **3**, 311–314

72 Bab: A. A. Babitsyna and N. P. Luzhnaya: *Zhur. Neorg. Khim.*, 1972, **17**, 1741–1742 (*Russian J. Inorg. Chem.*, 1972, **17**, 902–903)

74 Nik: V. K. Nikitina and Yu. K. Lobanova: *Izvest. Akad. Nauk. SSSR, Neorg. Mater.*, 1974, **10**, 1596–1599

80 Kub: R. Kubiak and K. Schubert: *Z. Metall.*, 1980, **71**, 635–637

84 Eva: D. S. Evans and A. Prince: CALPHAD XIII, Villard de Lans, France, 1984

86 Tsa: C. T. Tsai and R. S. Williams: *J. Mater. Res.*, 1986, **1**(2), 352–360

87 Oka: H. Okamoto and T. B. Massalski: 'Phase Diagrams of Binary Gold Alloys', 142–153; 1987, *ASM International*

Au–In–Sn

The presence of a ternary compound, $Au_4In_3Sn_3$, isotypic with Pt_2Sn_3 was announced [58 Sch]. The lattice constants were quoted as a = 0·450 nm, c/a = 0·290 nm (after conversion from kX units). In a later paper [59 Sch] published a section of the ternary system representing alloy equilibria after heat treatment of alloys at various temperatures for times ranging from 0 to 340 days. Alloys were melted in evacuated silica tubes: both solidified alloys and powder samples were vacuum heat-treated before examination by metallographic and X-ray diffraction techniques respectively. Of the 20 alloy compositions for which data is tabulated one half were heat treated at 250°C. The other half were examined as cast or after heat treatment at temperatures ranging from 100 to 350°C (one cast alloy was heat treated at 600°C). The 'isothermal' section developed (Fig. 250) can be taken to approximate to a temperature of 250°C. At 250°C In–Sn binary alloys are all molten, Au–In alloys have a L + $AuIn_2$ region stretching to 98·8% In, and Au–Sn alloys have an L + $AuSn_4$ region from 80 to 89 at.-% Sn followed by a liquid region to Sn. Fig. 250 is therefore a guide to the 250°C

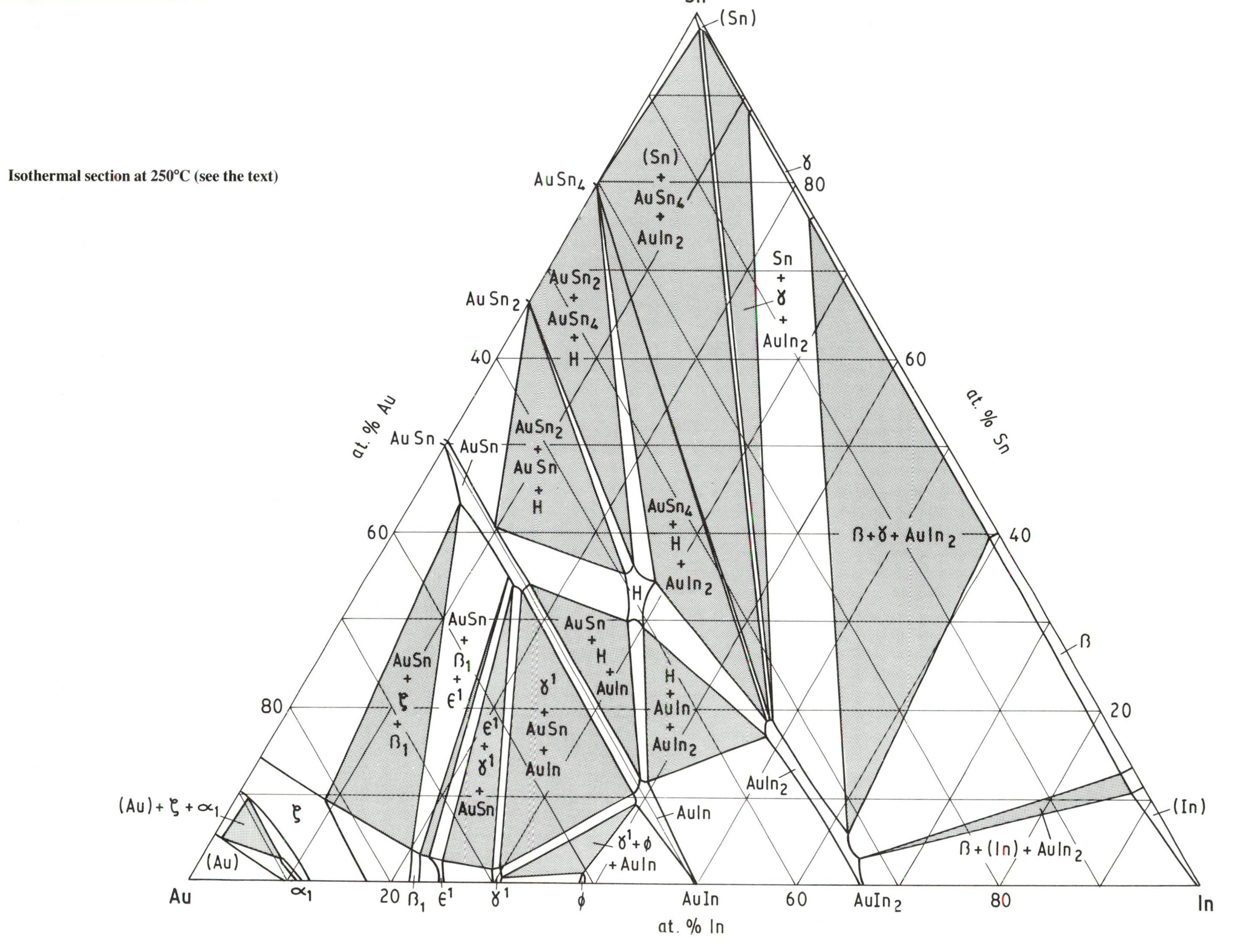

250 Isothermal section at 250°C (see the text)

isothermal for alloys whose compositions are on the Au-rich side of a line joining $AuIn_2$ to $AuSn_4$. For alloys on the Au-poor side of this line the equilibria is to be interpreted as equivalent to a 100°C section. The section has been amended to take into account the composition limits of the binary phases at 250°C. It should be noted that the constitution of 42 alloys were included in the section [59 Sch] but the tabulated experimental data refers to 20 alloy compositions only.

The α_1 phase ($Au_{88}In_{12}$) forms a continuous series of solid solutions with the β phase ($Au_{10}Sn$) over the range of temperature corresponding to stability of β [59 Sch]. As the β phase was not detected in the isothermal section (Fig. 250) it is probable that β transforms to (Au) + ζ at a temperature above 250°C. Similarly the ζ phase in both the Au–In and Au–Sn systems form continuous series of solid solutions. The disposition of the three-phase regions (Fig. 250) implies that the ternary system is dominated by regions of primary crystallization of AuIn, $AuIn_2$, $Au_4In_3Sn_3$ (H), and AuSn.

A brief reference is made to the crystal structure of an alloy containing 50 at.-% Au, 37 at.-% In, 13 at.-% Sn with lattice parameters a = 0·360 nm, b = 1·049 nm, c = 0·428 nm [68 Sch]. An alloy slightly richer in In (50 at.-% Au, 40 at.-% In, 10 at.-% Sn) was quoted as having lattice constants a = 0·428 nm, b = 1·050 nm, c = 0·360 nm [59 Sch]. The a and c parameters should be interchanged [84 Sch] in the report of [59 Sch]. The lattice parameters of an alloy 75 at.-% Au, 22 at.-% In, 3 at.-% Sn were given by [59 Bur]. The alloy was substantially ϵ' (Au_3In) with some ζ. It should be noted that the same alloy is quoted by [68 Sch] as being 100% ϵ' by X-ray examination of powders, and as containing 95% ϵ' with 5% AuSn in the grain boundaries by metallography on a solidified alloy heat treated for 48 days at 250°C. Figure 250 [68 Sch] has been modified to indicate three-phase regions for $\epsilon' + \beta_1$ + AuSn and $\beta_1 + \zeta$ + AuSn in place of one three-phase region $\epsilon' + \beta_1/\zeta$ + AuSn in the original, the phase β_1/ζ encompassing the structurally related hexagonal β β_1 and ζ phases.

The section AuIn–AuSn shows considerable solubility of Sn in AuIn (10 at.-% at ~250°C) and extensive solubility of In in AuSn (17 at.-% at ~250°C). Recent work [85 Hum] has confirmed that there is a two-phase region below ~325°C but the AuIn–AuSn section is not pseudobinary. The ternary compound melts congruently at 431°C [85 Hum]. The section AuSn–In is not a pseudobinary [59 Sch, 85 Hum].

REFERENCES

58 Sch: K. Schubert, H. Breimer, R. Gohle, H. L. Lukas, H. G. Meissner and E. Stolz: *Naturwiss.*, 1958, **45**, 360–361

59 Sch: K. Schubert, H. Breimer and R. Gohle: *Z. Metall.*, 1959, **50**, 146–153

59 Bur: W. Burkhardt and K. Schubert: *Z. Metall.*, 1959, **50**, 442–452

68 Sch: K. Schubert, S. Bhan, T. K. Biswas, K. Frank and P. K. Panday: *Naturwiss.*, 1968, **55**, 542–543

84 Sch: K. Schubert: private communication, July 1984

85 Hum: G. Humpston: PhD Thesis, Brunel University, 1985

Au–In–Te
(B. Gather and A. Prince)

The only publication [62 Zal] on the Au–In–Te system implies the occurrence of the ternary compound $AuInTe_2$ formed by a peritectic reaction. Cooling curves, with an average cooling rate of 7°C per minute measured from the cooling curve, indicate the beginning of solidification at about 617°C and a lengthy arrest at about 420°C (Fig. 2 of [62 Zal]). It is stated that shifts and elimination of arrests could be achieved by different cooling rates and by different heat treatments in the partially molten region. Metallographically the composition $AuInTe_2$ appears to be three-phase in slowly cooled ingots. X-ray powder diffraction analysis did not indicate the presence of a chalcopyrite structure, as is observed for the congruent ternary compounds $AgInTe_2$ and $CuInTe_2$. All that can be stated with certainty is that no compound occurs at the composition $AuInTe_2$. The alloys of this composition prepared by [62 Zal] were not in equilibrium after cooling from the melt. As the compositions of the phases observed were not determined, it is fruitless to speculate on the equilibria involved.

REFERENCE

62 Zal: S. M. Zalar and I. B. Cadoff: *Trans. Met. Soc. AIME*, 1962, **224**, 436–447

Au–In–Th
Au–In–U

[88 Bes] synthesized $ThInAu_2$ and $UInAu_2$ by high-frequency induction melting stoichiometric quantities of the metals in a cold Ag boat under purified argon. Purities used were 99·9% Th, 99·95% U and 99·99% Au, In. X-ray diffraction analysis indicated single phase alloys; no other checks were made. At 27°C the ternary compounds have the cubic CsCl-type structure, cP2, with weak superstructure lines suggesting Heusler-type ordering. The lattice parameters are 0·696 9 nm for $UInAu_2$ and 0·707 0 nm for $ThInAu_2$. The compound $ThInAu_2$ undergoes a structural transformation from cubic to tetragonal symmetry at −83°C with c/a of approximately 1·03 for the tetragonal phase.

REFERENCE

88 Bes: M. J. Besnus, M. Benakki, J. P. Kappler, P. Lehmann, A. Meyer and P. Panissod, *J. Less-Common Metals*, 1988, **141**, 121–131

Au–In–Ti

The alloy Au_2InTi was prepared by induction-melting the 99·99% purity elements under argon [75 Mar]. The alloy was annealed for 7 days at 500°C and examined by metallographic and X-ray powder diffraction techniques. The alloy was two-phase, no indication being given of the second phase. The major phase had a Cu_2AlMn-type structure (cF16) with a = 0·652 7 nm. It is not clear how [75 Mar] distinguished the Heusler-type structure from a CsCl structure by X-ray powder diffraction techniques. No details are given of the presence of odd superlattice lines to justify the Cu_2AlMn-type structure.

REFERENCE

75 Mar: R. Marazza, R. Ferro and G. Rambaldi: *J. Less-Common Metals*, 1975, **39**, 341–345

Au–In–Zn

According to [59 Bur] the addition of 2 at.-% Zn to Au_3In hardly alters the lattice parameters of orthorhombic Au_3In (Table 88). With 3 at.-% Zn additions a second phase with a Mg-type structure (hP2) was detected. This is likely to be the ζAu–In phase [87 Oka]. It was noted that the equilibrium relationships were dependent on the annealing temperature but no further details were given. Along the 75 at.-% Au section the solubility of Zn in Au_3In is about 1 at.-% Zn.

Wilkens and Schubert [57 Wil] used the X-ray powder diffraction method on 10 ternary alloys, annealed for 8 days at 200°C, to delineate the phase region for the doubly-faulted $L1_2$-type structure based on Au_4Zn (Fig. 251).

REFERENCES

59 Bur: W. Burkhardt and K. Schubert: *Z. Metall.*, 1959, **50**, 442–452

57 Wil: M. Wilkens and K. Schubert: *Z. Metall.*, 1957, **48**, 550–557

87 Oka: H. Okamoto and T. B. Massalski: 'Phase Diagrams of Binary Gold Alloys', 142–153; 1987, ASM International

Table 88. Lattice parameters of Au–In–Zn alloys

Composition, at.-%			*Heat treatment, powder samples*	*Lattice parameter*, nm		
Au	*In*	*Zn*		*a*	*b*	*c*
75	25	0	—	0·586	0·517	0·475
75	23	2	8 days at 300°C	0·585	0·516	0·475
75	22	3	4 days at 400°C	0·586	0·511	0·475

Au–In–Zr

Marazza, Ferro and Rambaldi [75 Mar] prepared the alloy Au_2InZr by induction-melting the 99·99% purity elements under argon. The alloy was annealed for 7 days at 500°C and examined by metallographic and X-ray powder diffraction techniques. The alloy was single phase but the crystal structure was reported as CsCl-type (cP2) with a = 0·336 5 nm and Cu_2AlMn-type (cF16) with double the lattice parameter. It is not clear how [75 Mar] concluded that the Cu_2AlMn-type structure was present in one preparation as no details are given of the presence of odd superlattice lines to justify the Cu_2AlMn-type structure.

REFERENCE

75 Mar: R. Marazza, R. Ferro and G. Rambaldi: *J. Less-Common Metals*, 1975, **39**, 341–345

Au–Ir–Pd
(G. V. Raynor)

INTRODUCTION

Very little work has been carried out on the system Au–Ir–Pd, apart from that described in [72 Jon], in which isothermal sections at 1 200 and 1 450°C were examined. The ternary alloys were prepared by melting components of unspecified purity, either by induction-melting under argon or, for alloys rich in iridium, in an argon-arc furnace. Boundaries between single- and two-phase regions were examined by metallographic work on alloys which had been homogenized for not less than 65 h at the relevant temperature and water-quenched before examination. In addition three tie lines were established at 1 200°C, and two at 1 450°C. The compositions of the phases present in quenched two-phase specimens were obtained either by X-ray determinations of lattice spacings or by electron-probe microanalysis. The former method was considered the more reliable. There is need for further work to extend the range of temperatures over which information is available.

BINARY SYSTEMS

For the Au–Ir system no phase diagram is available. A broad liquid-state immiscibility region exists at and below the melting point of Ir (2 447°C) [84 Oka].

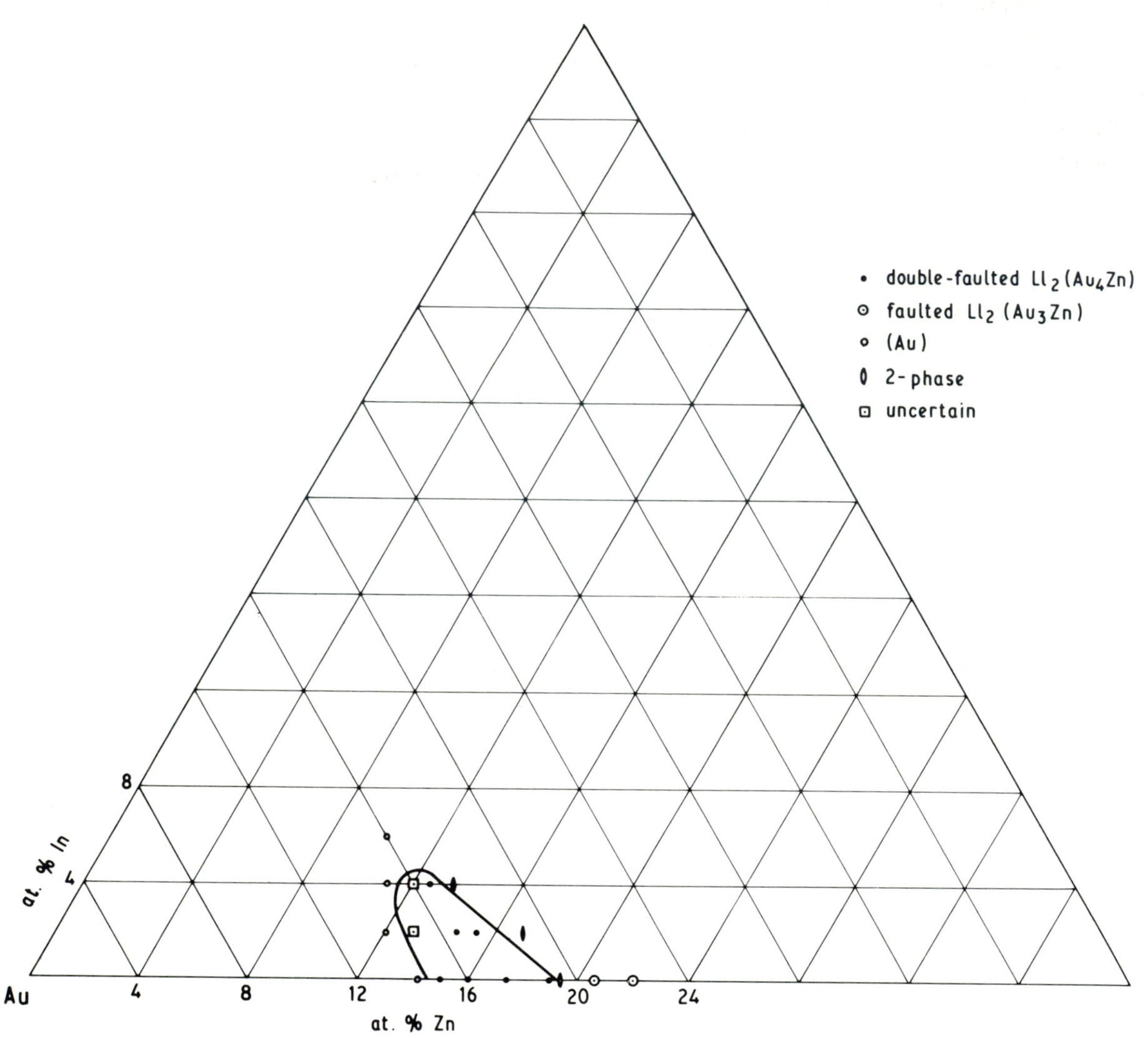

251 Extent of the Au_4Zn phase region at 200°C [57 Wil]

The Au–Pd system has been assessed by Okamoto and Massalski [85 Oka]. Ir and Pd are completely miscible in the solid state above 1 480°C [65 Ell], and the liquidus and solidus curves rise from the melting point of Pd; only the solidus curve has been experimentally established (up to 1 700°C). Below the critical point of 1 480°C the solid solution splits into two phases, respectively rich in Ir and Pd. The miscibility gap between solid phases of the same fcc crystal structure is extensive.

SOLID PHASES

No ternary compounds were reported in the work of [72 Jon]. (Au), (Ir) and (Pd) have crystal structures of the Cu-type, cF4. Solid solution formation between Ir and Pd, and between Pd and Au, is clearly to be expected within the appropriate temperature ranges.

THE TERNARY SYSTEM

Figure 252 shows the (Ir) + (Au–Pd) phase boundary at 1 200°C, where (Au–Pd) denotes the wide solution between Au and Pd, as suggested, but not firmly established, by [72 Jon], and drawn to conform to the solubility limits at this temperature as described by [85 Oka]. Since at this temperature the Au-rich alloys (>90 at.-% Au) are in the liquid state, a two-phase liquid + (Au–Pd) region, and a corresponding three-phase liquid + (Au–Pd) + (Ir) region have been added to complete the indication of the equilibria involved. Also included in Fig. 252 are the determinations [72 Jon] of the terminations of tie lines for two-phase alloys. These terminal compositions do not agree exactly with the suggested boundary, nor should tie lines intersect on an isothermal section, but the results serve to show the general direction of the tie lines in the two-phase region. The termination of one tie line at the Ir corner indicates that the Ir-rich corner of the hypothetical three-phase triangle cannot vary greatly from the position shown. The influence of the phase separation in the Ir–Pd system is clearly illustrated.

The constitution at 1 450°C is shown in Fig. 253. Again the suggested (Ir) + (Au–Pd) boundary is drawn near the Ir–Pd axis to conform to the Ir–Pd phase diagram. The L + (Au–Pd) region is now at higher percentages of Pd. Also plotted are the tie lines determined by [72 Jon], and it is seen that again the terminal compositions do not lie on the suggested Ir + (Au–Pd)/(Au–Pd) boundary; the general tie line direction is however established.

This limited information is sufficient to suggest strongly the forms of constitution at 1 200 and 1 450°C.

REFERENCES

72 Jon: B. Jones, M. W. Jones and D. W. Rhys: *J. Inst. Metals*, 1972, **100**, 136–141

65 Ell: R. P. Elliott: 'Constitution of Binary Alloys, First Supplement'; 1965, New York, McGraw-Hill

84 Oka: H. Okamoto and T. B. Massalski: *Bull. Alloy Phase Diagrams*, 1984, **5**, 381

85 Oka: H. Okamoto and T. B. Massalski: *Bull. Alloy Phase Diagrams*, 1985, **6**, 229–235

Au–Ir–Pt
(G. V. Raynor)

INTRODUCTION

The Au–Ir–Pt system received limited attention in the work of Jones *et al.* [72 Jon], who considered the suitability of Au–Pt alloys as coatings to protect from oxidation refractory metals such a Ir. Although it is stated [72 Jon] that an isothermal section was determined at 1 100°C, no details are given and no isothermal diagram presented. The only firm data available relate to the determination of one tie line at 1 100°C by an X-ray method involving the determination of lattice spacings.

THE BINARY SYSTEMS

The Au–Pt equilibrium diagram has been assessed [85 Oka].

For the Au–Ir system no phase diagram is available. A broad liquid-state immiscibility region exists at and below the melting point of Ir (2 447°C) [84 Oka].

According to [58 Han], Ir and Pt are completely miscible in the solid state above 975°C. The liquidus curve falls smoothly from the melting point of Ir to that of Pt, but the solidus curve is not accurately known. Below ~975°C the fcc solid solution (Ir–Pt) breaks down into two solid solutions of different compositions. The consequent two-phase region widens rapidly as the temperature falls, and at 800°C stretches from 10 at.-% Ir to ~96 at.-% Ir.

THE SOLID ALLOYS

The only crystal structure to occur in the ternary alloys is the fcc structure typical of the three components, though miscibility gaps occur in the systems Au–Pt and Ir–Pt, and the mutual solid solubility of Au and Ir is extremely limited. It is convenient therefore to distinguish the various fcc phases which might occur as follows:

(a) single symbols in brackets (e.g. (Au)) denote solid solutions containing little of the other two components;
(b) symbols such as (Au–Pt) and (Ir–Pt) denote solid solutions containing little of the third component.

THE TERNARY SYSTEM

In the work of [72 Jon] alloys were prepared by melting the components together in an argon-arc furnace, if rich in refractory metal, or by induction-melting under argon. The boundary between single and two-phase regions in the phase diagram was determined by optical microscopy after homogenizing for not less than 65 h at the relevant temperature and water quenching. The results, however,

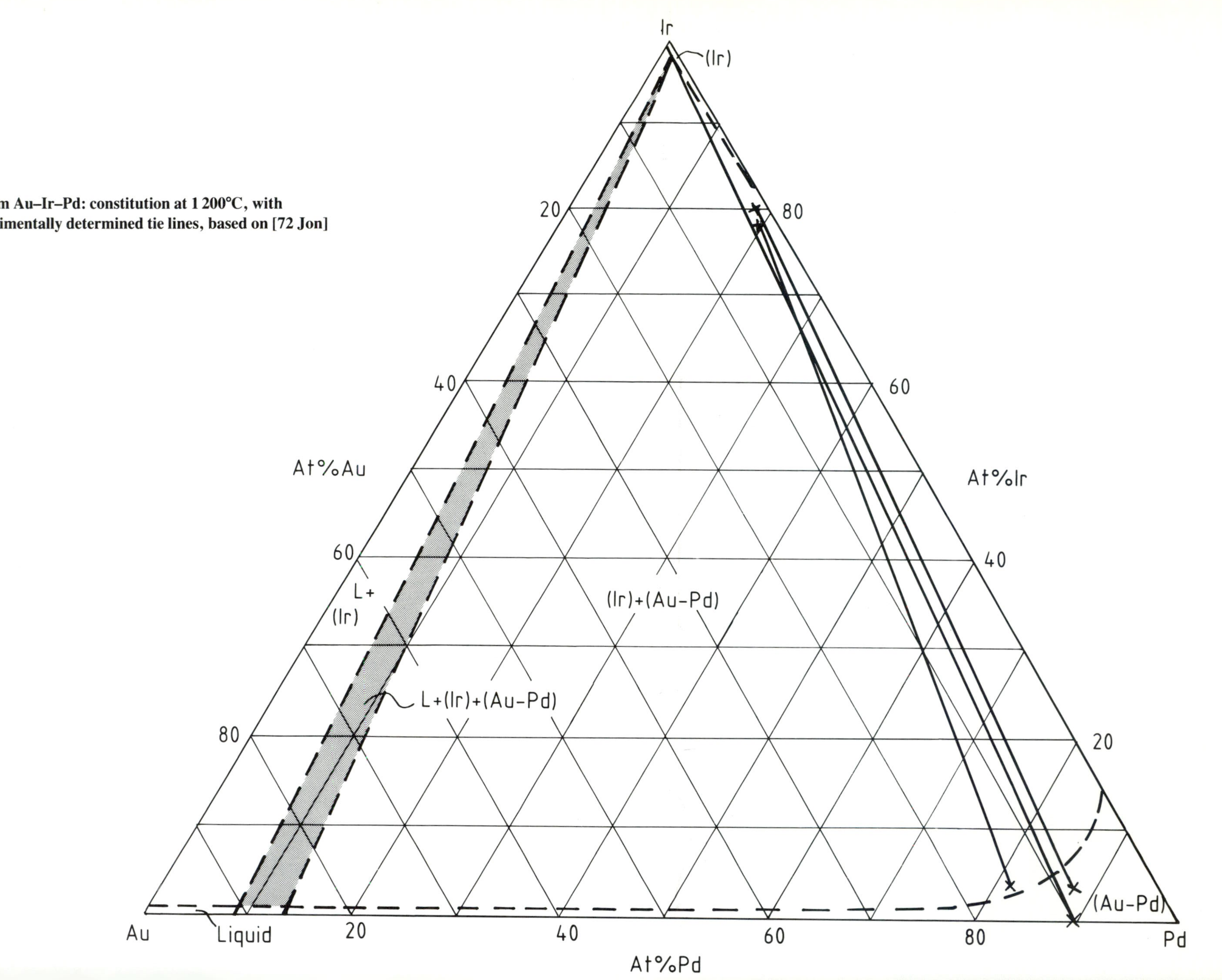

252 System Au–Ir–Pd: constitution at 1 200°C, with experimentally determined tie lines, based on [72 Jon]

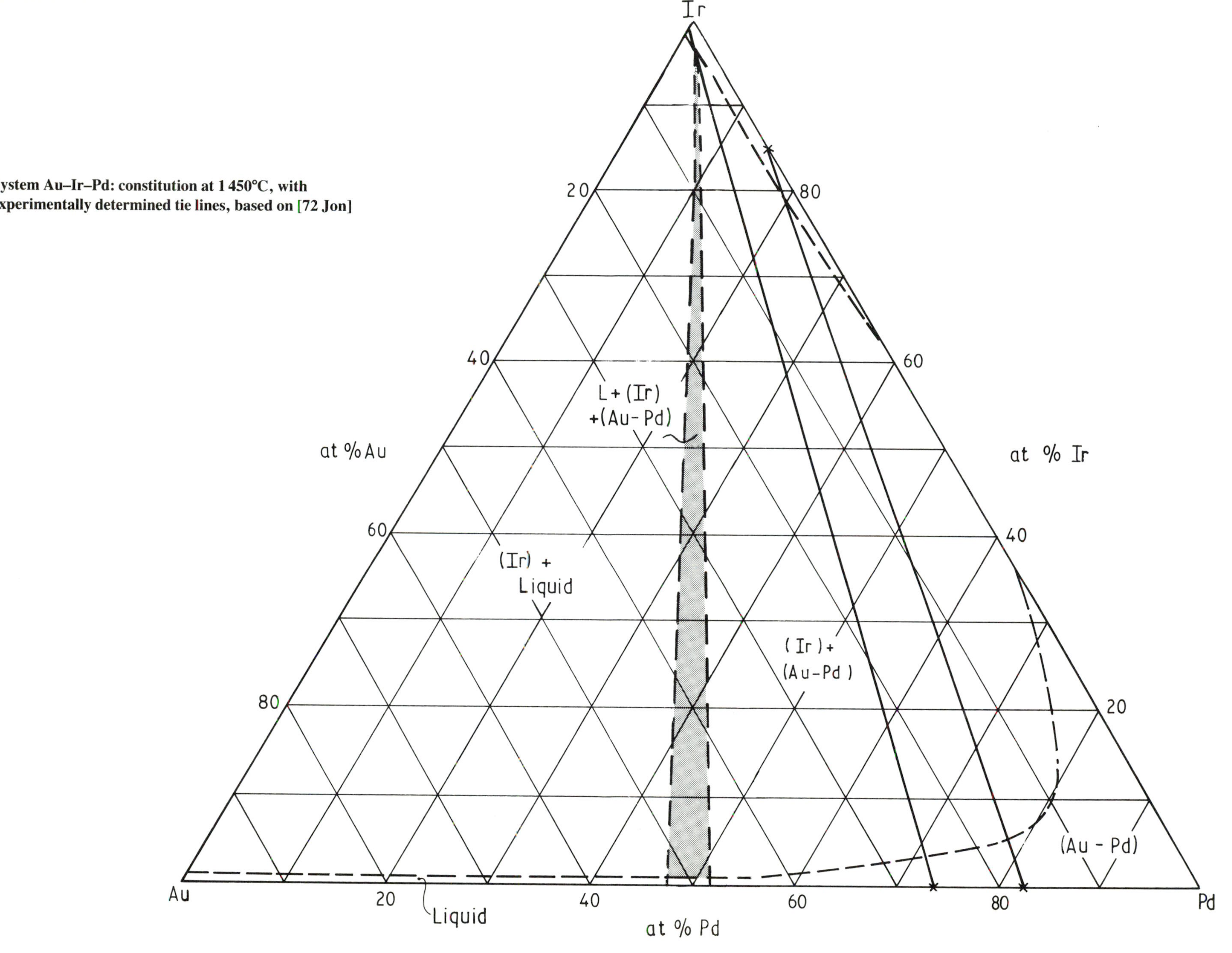

253 System Au–Ir–Pd: constitution at 1 450°C, with experimentally determined tie lines, based on [72 Jon]

are not quoted in the account, so that no experimental phase boundary can be derived. A tie line at 1 100°C was examined by determining the lattice spacings of the two phases present in two-phase alloys annealed at, and quenched from, this temperature. The measured lattice spacings were then related to composition from the known variation of spacings with composition in the Au–Pt and Ir–Pt systems. Since no allowance was apparently made for the effect of the third component on these lattice spacings the tie line compositions are subject to error, which is, however, small owing to the low solubility of Ir in the Au–Pt alloys. The general direction of the tie line is well defined. Using a suitable two-phase ternary alloy, the compositions for the tie line at 1 100°C was given as follows:

	at.-% Pt	at.-% Au	at.-% Ir
Au-rich phase	29·2	70·8	–
Pt-rich phase	94·9	–	5·1

This tie line is plotted in Fig. 254, and suggests that the solid solubility of Au in the Pt-rich solid solution is indeed small; that of Ir in (Au) is known to be very limited. The phase boundary enclosing the two-phase region in Fig. 254 has therefore been drawn as a broken curve which approaches closely to the Pt corner of the diagram. This boundary is likely to be qualitatively correct but is not accurately established. At 1 100°C the miscibility gap in the Au–Pt system extends from 38 at.-% Pt to 83 at.-% Pt, as shown in Fig. 254; also the liquid phase is present at the extreme Au corner of the diagram. This necessitates two phase regions involving liquid and both (Ir) and (Au), together with the three-phase triangle L + (Au) + (Ir). These regions have been included in Fig. 254 but are hypothetical, in that there is no experimental evidence for their exact positions. At a temperature below the melting point of Au but exceeding ~975°C the constitution will clearly involve simple equilibrium between the Au-rich solid solution and the (Ir–Pt) solid solution containing little Au.

REFERENCES

58 Han: M. Hansen and K. Anderko: 'Constitution of Binary Alloys'; 1958, New York, McGraw Hill

72 Jon: B. Jones, M. W. Jones and D. W. Rhys: *J. Inst. Metals*, 1972, **100**, 136–141

84 Oka: H. Okamoto and T. B. Massalski: *Bull. Alloy Phase Diagrams*, 1984, **5**, 381

85 Oka: H. Okamoto and T. B. Massalski: *Bull. Alloy Phase Diagrams*, 1985, **6**, 46–56

Au–K–S

Klepp and Bronger [87 Kle] prepared K_2S from 99·9% K and 99·99% S by reaction in aqueous ammonia. Admixture of 99·99% Au and S to produce the 1:1:1 stoichiometry were melted in sealed quartz tubes, heated to 750°C and maintained at 750°C for 1 day and finally cooled to 350°C over a 2 week period. Metallographically the structure indicated a peritectic formation of AuKS whose crystal structure is orthorhombic with a = 0·630 4(3), b = 0·785 0(4), c = 0·652 03 nm (Z = 4, where Z is the number of moles of AuKS in the unit cell).

[88 Kle] mixed K_2S, Au and S to produce the 4:6:5 stoichiometry. The cold-pressed powder mixture was melted at 600°C in a corundum crucible enclosed in a sealed quartz tube. The melt was maintained at 600°C for several days before cooling to room temperature at 2°C h^{-1}. Hexagonal crystals of $K_4Au_6S_5$ were extracted for crystal structure determination by X-ray diffractometry. The greenish-yellow compound has the space group P62c with lattice parameters a = 0·963 1, c = 0·985 6 nm, with two formula units of $K_4Au_6S_5$ in the unit cell.

REFERENCE

87 Kle: K. O. Klepp and W. Bronger: *J. Less-Common Metals*, 1987, **127**, 65–71

88 Kle: K. O. Klepp and W. Bronger, *J. Less-Common Metals*, 1988, **137**, 13–20

Au–K–Sb

The ternary compound K_2AuSb was prepared by heating the elements in a Ta crucible under argon for 24 h at 700°C and subsequently cooling slowly in the furnace to room temperature [80 Mue]. X-ray powder diffraction showed K_2AuSb to be isotypic with Na_2AuAs and Na_2AuSb. It is orthorhombic with a = 1·045 9(5), b = 0·786 1(3), c = 0·650 0(5) nm; (oC16) space group Cmcm.

REFERENCE

80 Mue: C. Mues and H.-U. Schuster: *Z. Naturf.*, 1980, **35B**, 1055–1058

Au–K–Se

Klepp and Bronger [87 Kle] prepared K_2Se from 99·9% K and 99·999% Se by reaction in aqueous ammonia. Admixture of 99·99% Au and Se to produce the 1:1:1 stoichiometry were melted in sealed quartz tubes, heated to 750°C and maintained at 750°C for 1 day and finally cooled to 350°C over a 2 week period. Metallographically the structure indicated a peritectic formation of AuKSe whose crystal structure is orthorhombic with a = 0·649 4(2), b = 0·810 7(3), c = 0·667 2(1) nm (Z = 4, where Z is the number of moles of AuKSe in the unit cell).

REFERENCE

87 Kle: K. O. Klepp and W. Bronger: *J. Less-Common Metals*, 1987, **127**, 65–71

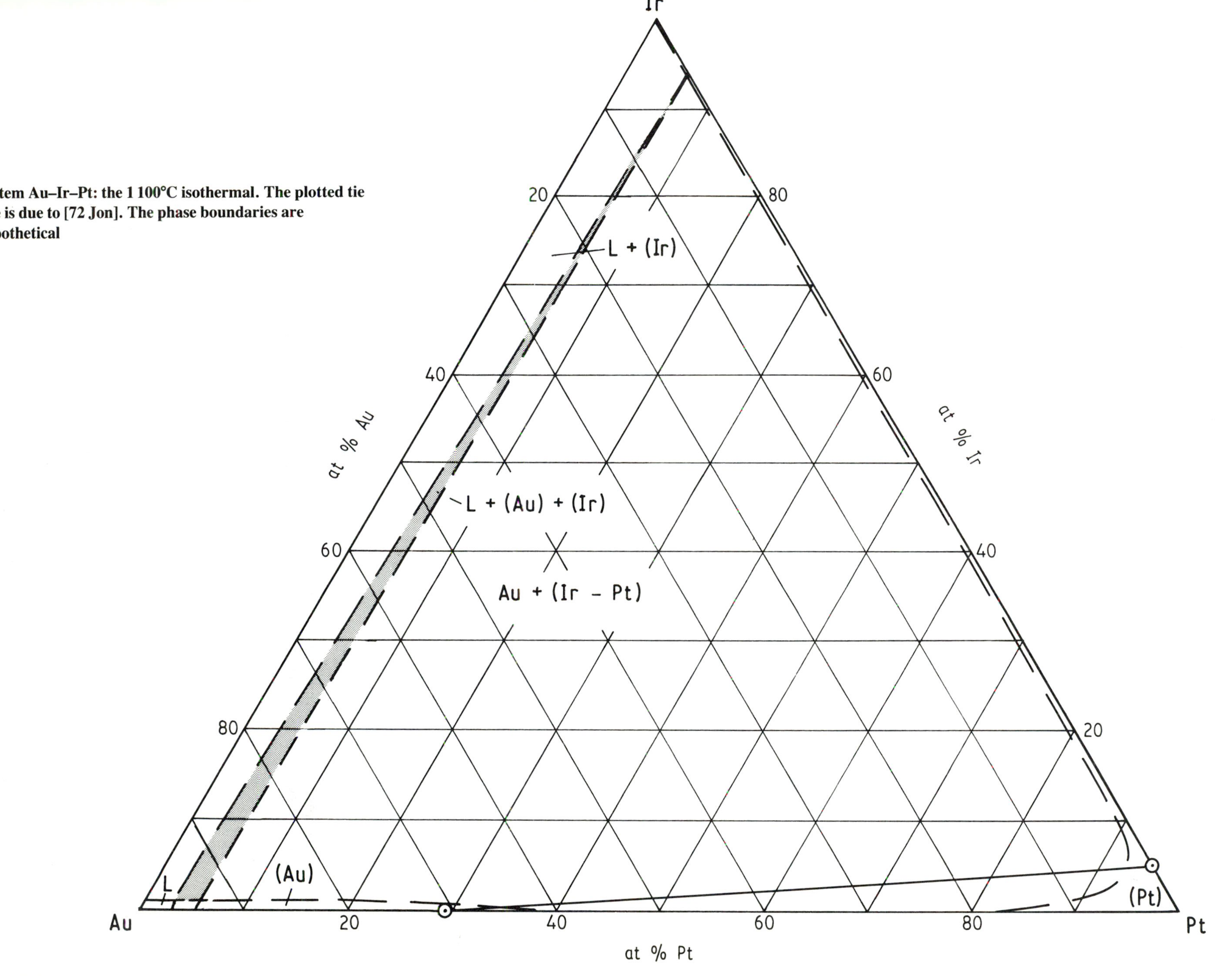

254 System Au–Ir–Pt: the 1 100°C isothermal. The plotted tie line is due to [72 Jon]. The phase boundaries are hypothetical

Au–K–Sn

Sinnen and Schuster [78 Sin] prepared the composition '$K_2Au_4Sn_2$' from the elements in a Ta crucible under argon. After melting, the alloy was annealed for 3–4 days at 400–600°C. Chemical analysis showed the presence of the ternary compound KAu_4Sn_2. It crystallizes in the tetragonal structure (tI28), space group I$\bar{4}$c2, with unit-cell parameters a = 0·8847 ± 0·0004, c = 0·8178 ± 0·0004 nm, $c:a$ = 0·92. The similarity of the structure of KAu_4Sn_2 to Tl_2Se is noted; the lattice positions of the Au atoms are occupied by Tl in Tl_2Se and those of K and Sn by the Se atoms.

REFERENCE

78 Sin: H.-D. Sinnen and H.-U. Schuster: *Z. Naturf.*, 1978, **33B**, 1077–1079

Au–Mg–Si

Van Look and Schubert [69 Loo] studied 14 alloys containing ≧50 at.-% Mg. The alloys were prepared in 2 g amounts from 99·9% purity elements by melting at 1 080°C in an Al_2O_3 crucible under argon. The loss of Mg by vaporization was stated to be <2%. Apart from metallographic examination, the major technique used was the X-ray analysis of powders, water quenched after annealing for 0·5 h at 770°C in quartz ampoules filled with pressurized argon. A 770°C isothermal section was presented (Fig. 255). The Mg corner has been amended from the original publication to reflect the fact that there must be a liquid phase at 770°C. The liquid phase region and the L + $AuMg_3$ + Mg_2Si tie triangle are shown with dashed lines in Fig. 255 to indicate that this region was not experimentally verified. The sole alloy examined, containing 10 at.-% Au, 10 at.-% Si, consisted of (Mg) + $AuMg_3$ + Mg_2Si after quenching to room temperature from 770°C [69 Loo]. This structure would be expected, given that $AuMg_3$ and Mg_2Si are in equilibrium with each other at 770°C.

The 770°C isothermal section does not include the binary phase Au_2Mg_5 as a stable phase. [69 Loo] did not observe Au_2Mg_5 on X-ray analysis of powdered binary alloys quenched from 770°C. The addition of Si to an alloy containing 70 at.-% Mg apparently stabilizes a ternary phase, τ, whose composition can be represented as $Au_{0·9}Mg_2Si_{0·1}$. This phase enters into equilibrium with AuMg + Mg_2Si, AuMg + $AuMg_2$ and $AuMg_3$ + Mg_2Si. An alloy containing 27·5 at.-% Au, 2·5 at.-% Si contained τ + $AuMg_3$. No indication was given by [69 Loo] of any equilibrium of τ with $AuMg_2$ + $AuMg_3$ and the low Si boundary of the τ single-phase region must be regarded as requiring confirmation. X-ray examination of single crystals fragments of the τ phase (30 at.-% Au, 3·5 at.-% Si), annealed for 0·2 h at 790°C and water quenched, indicate an orthorhombic primitive structure of the $PbCl_2$-type (oP12) with lattice constants a = 0·8450 0·0001 nm, b = 0·6110 ± 0·0001 nm, c = 0·4435 ± 0·0001 nm.

It is not clear whether τ is a ternary phase or is the binary Au_2Mg_5 phase containing Si. [69 Loo] did not observe Au_2Mg_5 as a stable binary phase and they state that an alloy with 27·5 at.-% Au, 71·5 at.-% Mg, 1·0 at.-% Si (printed in error as 27·5 at.-% Au, 31·5 at.-% Mg, 1·0 at.-% Si) showed only τ + $AuMg_3$ after 1 h at 770°C followed by water quenching. [69 Loo] claim τ to be a high-temperature phase on the basis that annealing for 1 h at 600°C or 2 h at 500°C produces a structure containing $AuMg_2$ + AuMg in an alloy of initial composition 30 at.-% Au, 3·5 at.-% Si. However it is stated that this alloy apparently lost a lot of Mg during annealing. If this were the case it would not be necessary to postulate that τ was only stable at high temperatures. [76 Kro] noted that an alloy with a composition corresponding to AuMgSi was not single phase; no further details were given. Such an alloy would be expected to contain AuMg + (Si) or AuMg + (Si) + Mg_2Si from the work of Van Look and Schubert [69 Loo].

The precise nature of the phase relations between $AuMg_2$, τ, and $AuMg_3$ cannot be regarded as established and further work is desirable. The Au–Si–AuMg portion of the ternary system has not been studied, although it would be anticipated that alloys within this region would solidify with (Au) + (Si) + AuMg structures in the absence of ternary compounds.

REFERENCES

69 Loo: N. Van Look and K. Schubert: *Metall.*, 1969, **23**(1), 4–6

76 Kro: K. Krompholz and A. Weiss: *Z. Metall.*, 1976, **67**, 400–403

Au–Mn–N

The occurrence of the nitride Mn_3AuN with a perovskite structure (cP5) and lattice parameter a = 0·4023(5) nm has been reported [67 Mad].

REFERENCE

67 Mad: R. Madar, L. Gilles, A. Rouault, J.-P. Bouchard, E. Fruchart, G. Lorthioir and R. Fruchart: *Compt. rend.*, 1967, **264C**, 308–311

Au–Mn–Pd
(G. V. Raynor)

INTRODUCTION

The very limited work which has been carried out on this system relates to the effect on replacing, with Pd, some of

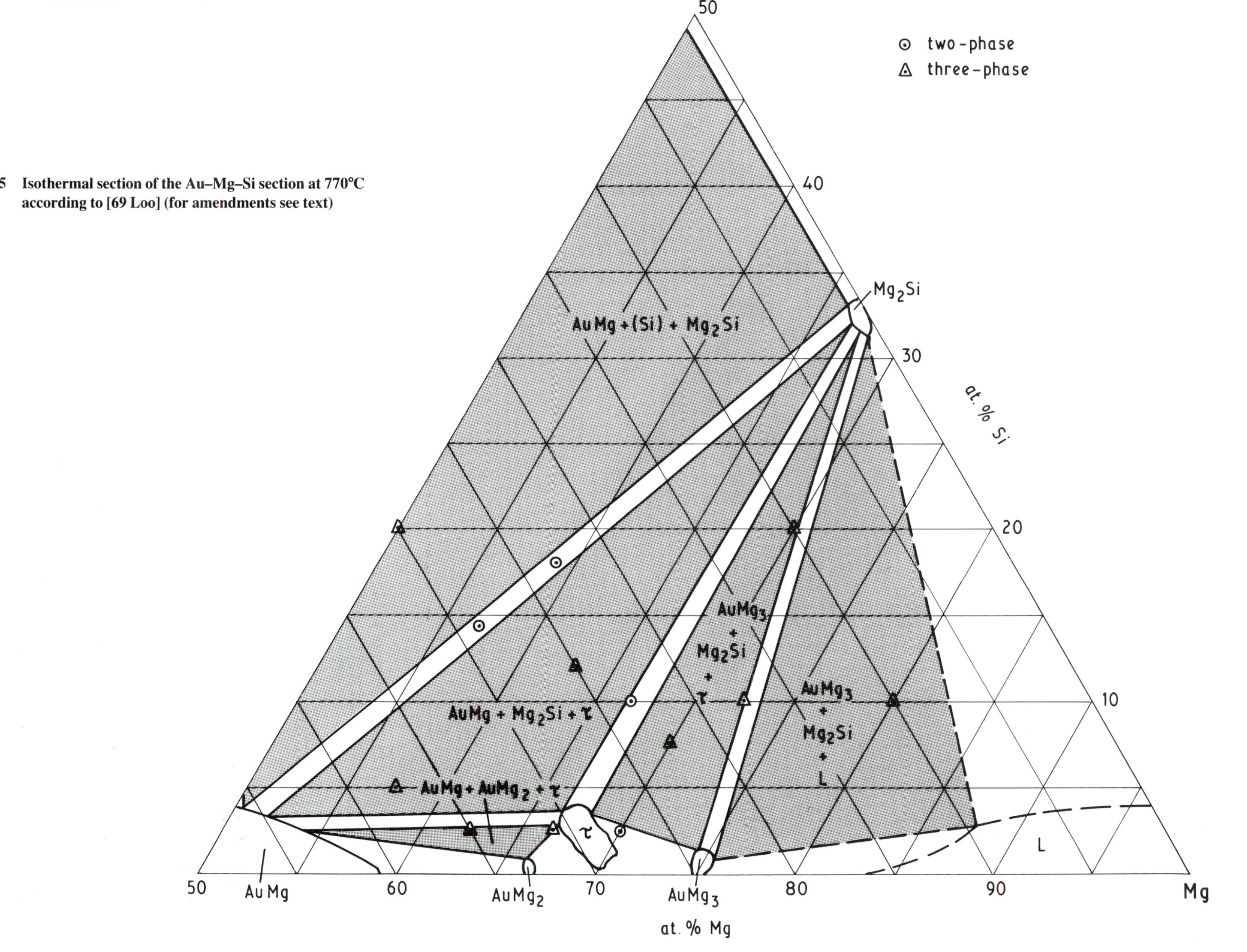

255 Isothermal section of the Au–Mg–Si section at 770°C according to [69 Loo] (for amendments see text)

the Au atoms in the phase AuMn, which is antiferromagnetic with a Neel point of 480 K (207°C).

BINARY SYSTEMS

In relation to the only available reference to work on the ternary system, only the nature of the Au–Mn binary system is of significance, and there is still doubt about the precise nature of the equilibria involved in this system [85 Oka]. A phase with the CsCl structure, and a considerable range of homogeneity (maximum 30·8–66.6 at.-% Au) is formed at a maximum on the liquidus of 1 260°C and the composition AuMn. This phase undergoes transformations at a lower temperature to tetragonally distorted forms.

THE TERNARY SYSTEM

A limited number of ternary alloys was prepared by [70 Ido], using unspecified methods, and examined by X-ray diffraction and magnetic susceptibility methods. The alloys were of compositions $(Au_{1-x}Pd_x)$ Mn, and the authors describe the phase AuMn at the temperatures involved as body-centred tetragonal of distorted CsCl type, with c/a = 1·04–0·95, according to composition. The Neel point is decreased by the solution of Pd, but the most significant constitutional observation is that a tetragonal structure is maintained at room temperature from an electron concentration of 1 to 0·85, but that the structure becomes cubic at electron concentrations of 0·8 and 0·75. The authors use valency values of unity for Au and Mn, and it is assumed that a value of zero is assigned to Pd. With this assumption the crystal structure becomes cubic when x = 0·4, that is, at the composition Au 30 at.-%, Pd 20 at.-%, Mn 50 at.-%. The solid solution of Pd in AuMn must extend at least to the composition $Au_{25}Pd_{25}Mn_{50}$, and the homogeneity range is thus considerable.

REFERENCES

70 Ido: T. Ido, H. Teramoto, T. Kasai, K. Sato and K. Adachi: *J. Phys. Soc. Japan*, 1970, **28**, 1589

85 Oka: H. Okamoto and T. B. Massalski: *Bull. Alloy Phase Diagrams*, 1985, **6**, 454–467

Au–Mn–Sb

The presence of the ternary compound AuMnSb has been established [71 Elb, 71 Mas]. [71 Elb] found the compound to be a CaF_2-type structure (cF12) with lattice parameter a = $0{\cdot}63_{44}$ nm (alloy heat treated for 1 day at 350°C. In a detailed study of the equilibria associated with Au–Mn–Sb [71 Mas] examined 77 alloys by X-ray diffraction analysis, supplemented by metallography (Fig. 256). Au, Mn, and Sb powders, of purity >99·9%, 99·999%, and 99·999% respectively, were heated in evacuated silica ampoules to between 850 and 1 350°C. The melts were held at temperature for 1 to 20 h and the ampoule quenched in iced water. To homogenize the alloys specimens were ground to powder, remelted in evacuated silica ampoules and quenched. This procedure was repeated several times. The powder or bulk samples were further homogenized by heat treatment at 400–850°C for 5–300 h, followed by slow cooling to room temperature.

The stoichiometric alloy AuMnSb melts at 525°C; it is not clear whether congruent or incongruent melting was found. The crystal structure was determined as the AgAsMg prototype (cF12) with ordered arrangement of Mn and Sb atoms on lattice sites. [71 Elb] suggested the CaF_2-type structure with mixed occupancy of the Mn and Sb atoms on the 4c and 4d sites. The data of [71 Mas] is accepted, with Au atoms at 4a sites (000, 0½½, ½0½, ½½0), Mn atoms at 4c sites (¼¼¼, ¼¾¾, ¾¼¾, ¾¾¼) and Sb atoms at 4d sites (¾¾¾, ¾¼¼, ¼¾¼, ¼¼¾). The lattice parameter of stoichiometric AuMnSb, heat-treated for 100 h at 500°C followed by slow cooling, was found to be a = 0·637 3 nm. Metallographic examination confirmed the single phase structure.

Alloys within the dashed line in Fig. 256 gave relatively strong diffraction lines for the AuMnSb structure. This suggests that AuMnSb enters into equilibrium with MnSb, AuMn, Au_2Mn, and $AuSb_2$. The range of homogeneity of AuMnSb was determined for alloys along the section Au–MnSb, Sb–AuMn, and Mn–'AuSb' (Fig. 257). Alloy compositions marked by an open circle or a double circle were found to be single phase by X-ray analysis; only alloys with an open circle were single phase when metallographically examined. Weak diffraction lines of MnSb or Au_2Mn appear beyond this composition range; such alloys are distinguished by filled circles in Fig. 257. [71 Mas] also presented the variation of lattice parameter for alloys along the sections Au–MnSb (full line in Fig. 258), and Mn–'AuSb' (dotted line in Fig. 258). In the former case the lattice parameter decreases from $0{\cdot}637\,7_5$ nm to 0·636 8 nm as the Au increases from 32·33 to 34·5 at.-%; in the latter case the lattice parameter increases from 0·637 1 to 0·637 7 nm as the Mn increases from 32·33 to 34·5 at.-%. The range of homogeneity of AuMnSb is of the order of 1·2 at.-% along each section (Fig. 258). As metallographic examination indicates a more restricted range of single phase AuMnSb the composition limits for AuMnSb derived from X-ray analysis should be regarded as overestimates of the region of solid solution based on AuMnSb.

REFERENCES

71 Elb: M. El-Boragy and K. Schubert: *Z. Metall.*, 1971, **62**, 667–675

71 Mas: H. Masumoto and K. Watanabe: *Trans. Japan Inst. Metals*, 1971, **12**, 256–260

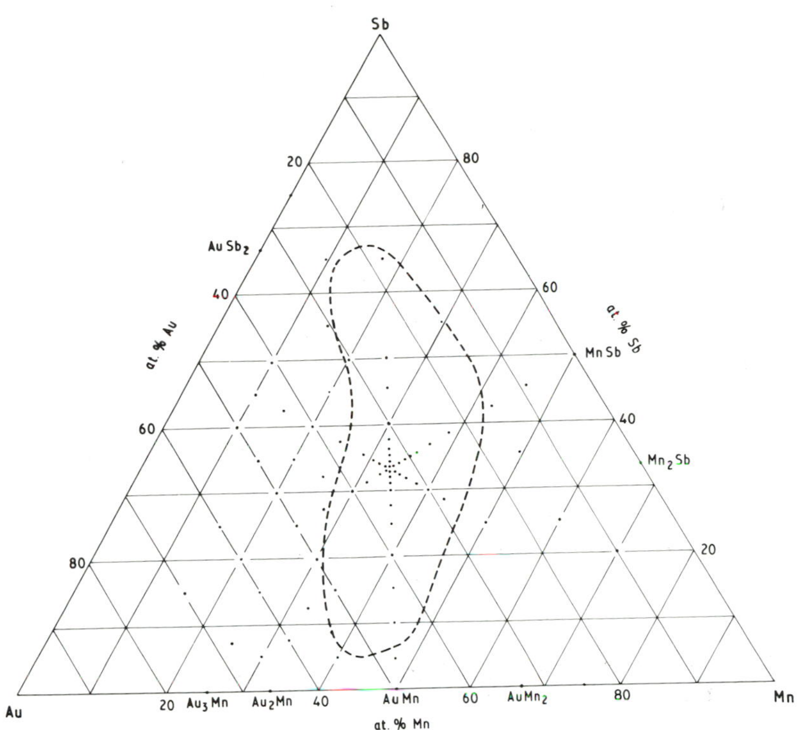

256 Composition of alloys studied [71 Mas]

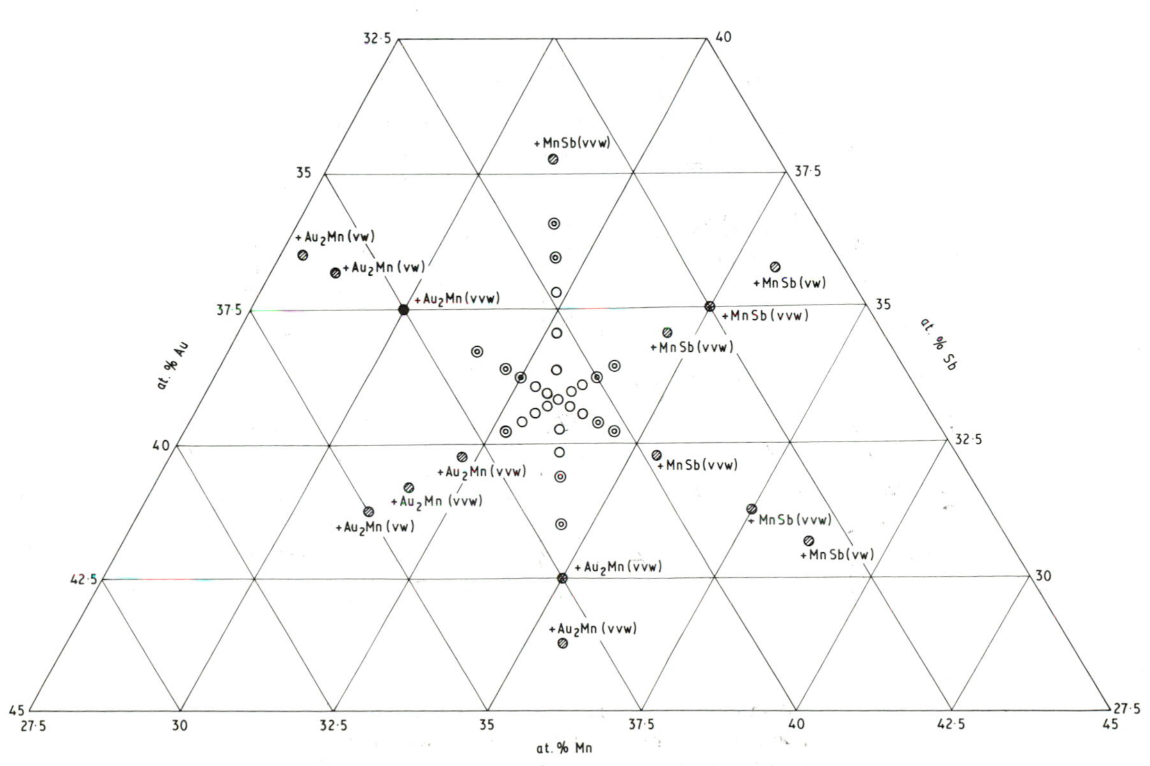

257 Crystal structures of alloys close to AuMnSb [71 Mas]

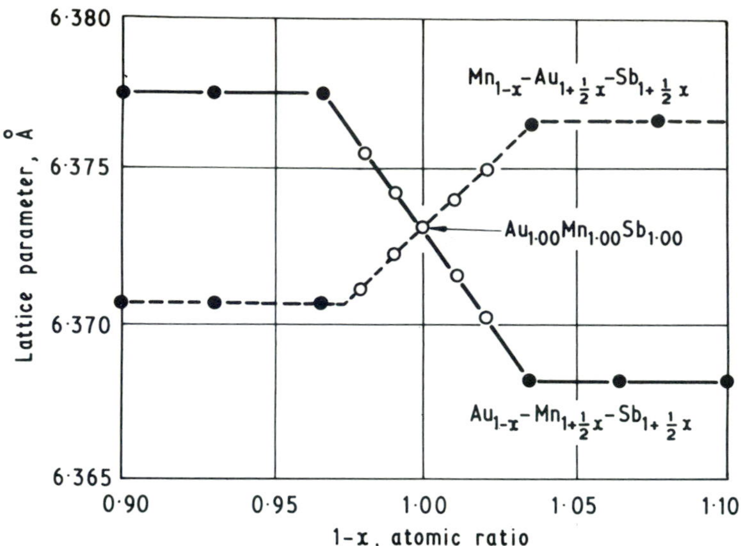

258 **Lattice parameters of alloys along the sections Au–MnSb and Mn–'AuSb' [71 Mas]**

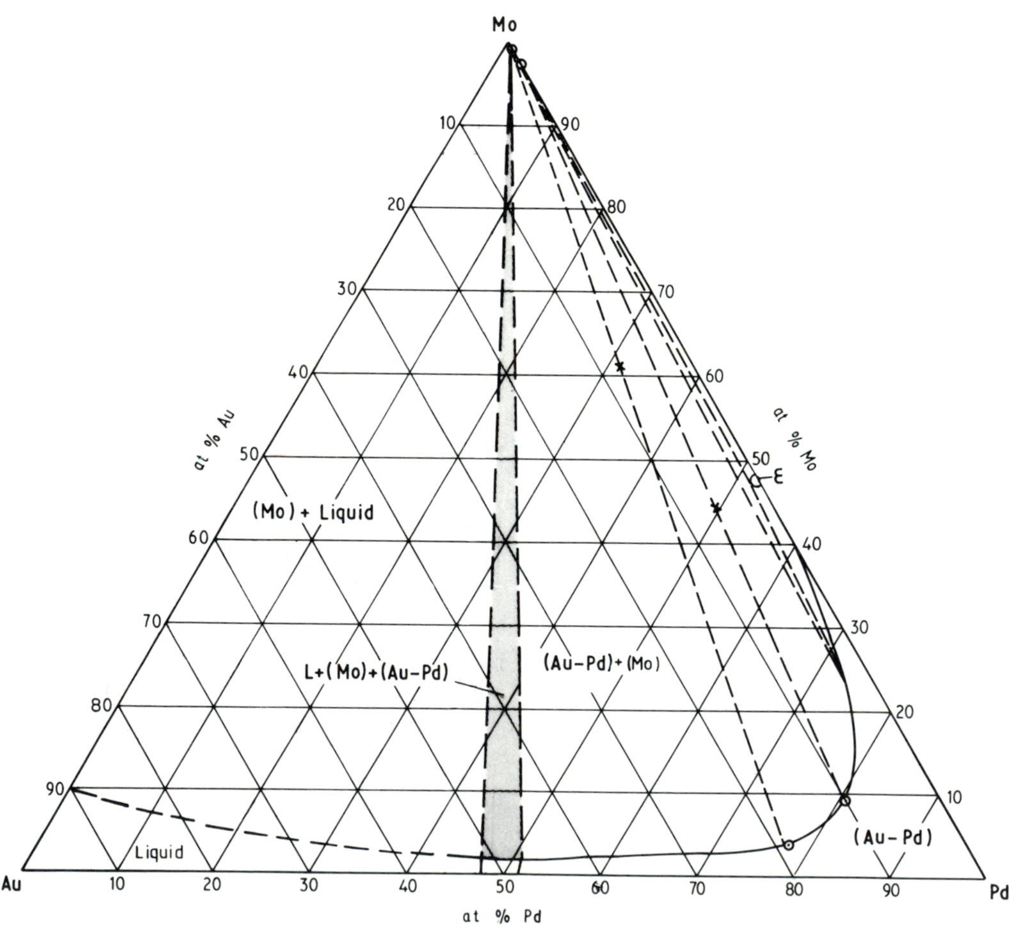

259 **System Au–Mo–Pd: isothermal section at 1 450°C; X denotes compositions of alloys used to determine coexistent phase compositions, marked O, according to [72 Jon]. The L/L + (Mo) boundary, and the (Mo) + ε + (Au–Pd) triangle are hypothetical**

Au–Mn–Sn

The alloy 47·35 Au, 50 Mn, 2·65 at.-% Sn has been prepared [70 Ido]. At room temperature it had the body-centred tetragonal CsCl-type structure of low-temperature AuMn. The bcc CsCl-type structure of AuMn [85 Mas] must occur at higher Sn concentrations.

REFERENCES

70 Ido: T. Ido, H. Teramoto, T. Kasai, K. Sato and K. Adachi: *J. Phys. Soc. Japan*, 1970, **28**, 1589

85 Mas: T. B. Massalski and H. Okamoto: *Bull. Alloy Phase Diagrams*, 1985, **6**, 454–467

Au–Mn–V

Both Au_4Mn and Au_4V have the ordered Ni_4Mo-type structure, tI10. [69 Tot] determined the temperature dependence of resistivity for the ternary alloy $Au_4Mn_{0·5}V_{0·5}$ (80 Au, 10Mn, 10 at.-% V) with a heating rate of ~1°C min^{-1}. The ordered Ni_4Mo-type structure disordered at 570°C. The temperature, scaled from the published figure, is comparable with the transformation temperature of Au_4V at 565 ± 5°C [81 Smi] and above that of Au_4Mn at 450°C [87 Oka].

REFERENCES

69 Tot: R. S. Toth, A. Arrott, S. S. Shinozaki, S. A. Werner and H. Sato: *J. Appl. Phys.*, 1969, **40**, 1373–1375

81 Smi: J. F. Smith: *Bull. Alloy Phase Diagrams*, 1981, **2**, 344–347

87 Oka: H. Okamoto and T. B. Massalski: 'Phase Diagrams of Binary Gold Alloys', 169–182; 1987, *ASM International*

Au–Mn–Zn

The composition Au_2MnZn has a CsCl-type structure similar to that of Au_2MnAl and Au_2MnIn [59 Mor]. No further details were given. Ido *et al.* [70 Ido] studied the $(Au_{1-x}Zn_x)Mn$ section at room temperature. The body-centred tetragonal CsCl-type structure of low-temperature AuMn was still evident with $x = 0·2$ (40Au, 10Zn, 50 at.-% Mn). The bcc CsCl-type structure of AuMn [85 Mas] must occur at higher Zn concentrations.

REFERENCES

59 Mor: D. P. Morris, R. R. Preston and I. Williams: *Proc. Phys. Soc.*, 1959, **73**, 520–523

70 Ido: T. Ido, H. Teramoto, T. Kasai, K. Sato and K. Adachi: *J. Phys. Soc. Japan*, 1970, **28**, 1589

85 Mas: T. B. Massalski and H. Okamoto: *Bull. Alloy Phase Diagrams*, 1985, **6**, 454–467

Au–Mo–Pd
(G. V. Raynor)

INTRODUCTION

In the course of experiments to examine the possibility of protecting refractory metals from oxidation by the use of gold–palladium alloys [72 Jon], some information on the constitution of the ternary system Au–Mo–Pd was obtained by the application of X-ray methods.

THE BINARY SYSTEM

The Au–Mo phase diagram has been assessed by [86 Mas], the Au–Pd by [85 Oka] and the Mo–Pd by [80 Bre].

THE SOLID PHASES

No study of the system sufficiently comprehensive to reveal the presence of discrete ternary phases has been made. The solid phases in the system are the Mo-rich solid solution (bcc), the continuous solid solution (Au–Pd) formed between Au and Pd (fcc), and the ϵ-phase (cph). The structures are summarized in Table 89.

Table 89. Crystal structures

Solid phase	*Prototype*	*Lattice designation*
(Au–Pd) solid solution	Cu	cF4
(Mo)	W	cI2
ϵ in Mo–Pd system	Mg	hP2

THE TERNARY SYSTEM

The ternary alloys for the determination of equilibria were prepared by melting. Those rich in Mo were melted in an argon-arc furnace; otherwise induction-melting was used. The solid alloys were homogenized by annealing for not less than 65 h at 1 450 and 1 200°C, and water-quenched. The phase compositions were obtained using powders of the quenched samples, stress-relieved at the relevant annealing temperatures, and again quenched, by the X-ray determination of lattice spacings. In order to arrive at the phase compositions a knowledge of the variation of lattice spacings with composition is required. This variation is not explicitly described in [72 Jon], and it is possible that the effect of Mo on the spacings of the (Au–Pd) solid solution was calculated assuming an additive relationship between the atomic diameter of Mo and the mean atomic diameter of the solid solution. The results are tabulated [72 Jon] and no corresponding phase diagram is given. It is however possible to present probable isothermal sections for 1 450 and 1 200°C using the tie line results quoted in the report, together with a knowledge of the relevant binary diagrams. Thus Fig. 259 shows the two tie lines determined for 1 450°C. It is clear that the Mo-rich phase contains very little Pd or Au, in conformity with the Au–Mo and Mo–Pd phase diagrams. The solubility of Mo in the (Au–Pd) phase varies from that characteristic of the solid solubility of Mo in Pd to about 2 at.-% at 50 at.-% Pd. The tie line results are consistent with both the Au–Pd and Mo–Pd binary diagrams. At 1 450°C compositions rich in Au are in the liquid state. Since the liquidus of the Au–Mo system is unknown, the L + (Mo) region in Fig. 259 is bounded by a purely hypothetical Au-rich boundary, shown as a broken curve. The L + (Au–Pd) region conforms to the Au–Pd binary diagram, and the location of the required three-phase region L + (Mo) + (Au–Pd) cannot be much different from that shown. The ϵ-phase of the Mo–Pd system remains stable down to 1 370 ± 30°C, and therefore takes part in the equilibria at 1 450°C, most probably as shown in Fig. 259. The position of the three-phase triangle (Mo) + ϵ + (Au–Pd) is hypothetical. The results of tie line determinations at 1 200°C are shown in Fig. 260. At this temperature the L + (Au–Pd) region has moved much closer to the Au corner of the diagram, and its composition range defines the position of the L + (Mo) + (Au–Pd) three-phase triangle. 1 200°C is below the temperature of the eutectoidal decomposition of ϵ, which therefore disappears from the phase diagram of Fig. 260.

REFERENCES

72 Jon: B. Jones, M. W. Jones and D. W. Rhys: *J. Inst. Metals*, 1972, **100**, 136–141

80 Bre: L. Brewer and R. H. Lamoreaux: 'Molybdenum: Physicochemical properties of its compounds and alloys', *Atomic Energy Review*, Special Issue No. 7, 1980, pp. 295–297. International Atomic Energy Agency, Vienna

85 Oka: H. Okamoto and T. B. Massalski: *Bull. Alloy Phase Diagrams*, 1985, **6**, 229–235

86 Mas: T. B. Massalski, H. Okamoto and L. Brewer: 'Binary Alloy Phase Diagrams', ed.: T. B. Massalski, ASM, 1986, pp. 283–284

Au–N–Nb

Rieger *et al.* [65 Rie] consider that Au_2Nb_3 is only a stable phase when interstitial impurities such as O, C, and N are present. This is contrary to the later work of [73 Ros] who found Au_2Nb_3 from 1 220°C to room temperature. [65 Rie] report a N-containing phase, $Au_2Nb_3N_x$, to have the β-Mn structure, cP20, with lattice parameter $a = 0{\cdot}708\,5$ nm.

REFERENCES

65 Rie: W. Rieger, H. Nowotny and F. Benesovsky: *Monatsh. Chem.*, 1965, **96**, 232–241

73 Ros: E. Roschel, O. Loebich and C. J. Raub: *Z. Metall.*, 1973, **64**, 359–361

Au–N–Ti

Von Philipsborn and Laves [64 Phi] found that $AuTi_3$ with the Cr_3Si-type structure (cP8) forms a two-phase structure when it contains small amounts of the interstitial elements, O, N, and C. The second phase stabilized by the interstitial elements has the ordered Cu_3Au ($L1_2$)-type structure (cP4). Additions of N to $AuTi_3$ have a similar influence to oxygen additions; in the latter case about 4·5 at.-% O gives some 40% of the Cu_3Au phase and 16 at.-% O gives 100% of the Cu_3Au phase.

REFERENCE

64 Phi: H. von Philipsborn and F. Laves: *Acta Crystallogr.*, 1964, **17**, 213–214

Au–N–U

Krikorian *et al.* [67 Kri] reacted pressed plugs of UN with Au for 47 h at 860°C under a vacuum of 4×10^{-6} torr. X-ray analysis identified the solid reaction product as Au_3U according to the reaction $2UN + 6Au = 2Au_3U + N_2$. When UN and Au were heated for 93 h at 860°C at 0·75 atm N_2 the UN was converted to U_2N_3 and no reaction occurred with the Au. At 860°C and low N pressures the stable join in the Au–N–U system is between Au_3U and N_2, indicating that Au + N_2 + Au_3U form a stable phase mixture.

REFERENCE

67 Kri: N. H. Krikorian, T. C. Wallace, M. C. Krupka and C. L. Radosevich: *J. Nucl. Mater.*, 1967, **21**, 236–238

Au–N–V

Von Philipsborn and Laves [64 Phi] found that AuV_3 with the Cr_3Si-type structure (cP8) forms a two-phase structure when it contains small amounts of the interstitial elements O, N, and C. The second phase stabilized by the interstitial elements has the ordered Cu_3Au ($L1_2$)-type structure (cP4). Addition of N to AuV_3 stabilizes the Cu_3Au-type phase; no quantification of the N content required to produce 100% of the Cu_3Au-type phase was given. Rieger *et al.* [65 Rie] reported a nitrogen-containing phase AuV_3N_x with a perovskite structure and almost identical lattice parameter to the Cu_3Au-type phase described by [64 Phi].

REFERENCES

64 Phi: H. von Philipsborn and F. Laves: *Acta Crystallogr.*, 1964, **17**, 213–214

65 Rie: W. Rieger, H. Nowotny and F. Benesovsky: *Monatsh. Chem.*, 1965, **96**, 232–241

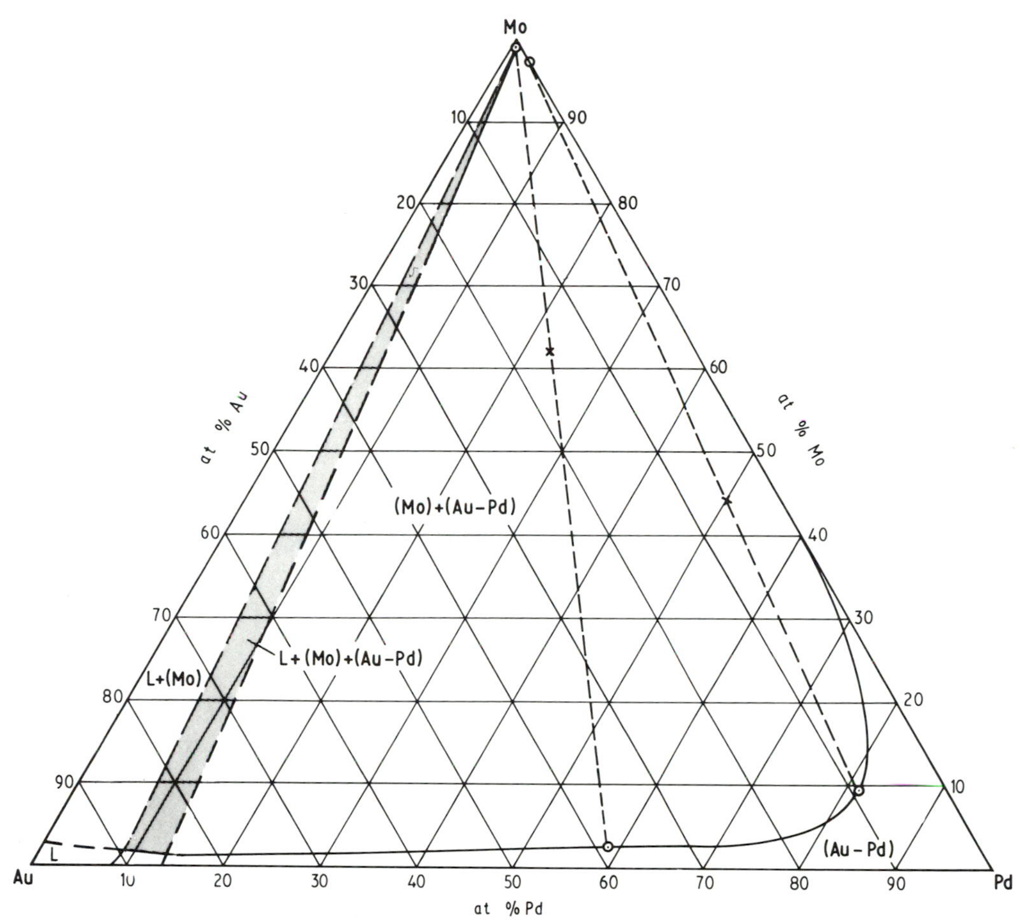

260 System Au–Mo–Pd: isothermal section at 1 200°C; symbols X and O have the same meaning as in Fig. 259. The L/L + (Mo) boundary is hypothetical

Au–Na–S

The early chemical literature [1896 Ant] contains details of the preparation of the compound Na_3AuS_2. Klepp and Bronger [87 Kle] prepared Na_3AuS_2 by mixing stoichiometric amounts of purified Na_2S, 99·999% Au powder, and 99·99% S, cold compacting and heating to 600°C in a corundum crucible enclosed in an evacuated quartz ampoule. After several days at 600°C the reaction product was cooled to room temperature at 2°C h^{-1}. The crystal structure was determined from four-circle diffractometer data. Na_3AuS_2 crystallizes in the space group R$\bar{3}$c with a = 0·762 3(2), c = 1·667 2(5) nm and Z = 6, where Z is the number of moles of Na_3AuS_2 in the unit cell.

REFERENCES

1896 Ant: U. Antony and A. Lucchesi: *Gazz. Chim. Ital.*, 1896, **26**, 350

87 Kle: K. O. Klepp and W. Bronger: *J. Less-Common Metals*, 1987, **132**, 173–179

Au–Na–Sb

Mues and Schuster [80 Mue] prepared the ternary compound Na_2AuSb by heating the elements in a Ta crucible for 24 h at 700°C. Its crystal structure was characterized as orthorhombic (oC16), space group Cmcm, with a = 0·927 9(3), b = 0·756 2(2), c = 0·584 1(5) nm. Na_2AuSb is isotypic with Na_2AuAs and K_2AuSb.

REFERENCE

80 Mue: C. Mues and H.-U. Schuster: *Z. Naturf.*, 1980, **35B**, 1055–1058

Au–Na–Si

Döring *et al.* [79 Dör] heated the elements, with an excess of Na, in fused quartz ampoules under argon. An initial mixture corresponding to the composition Na_2Au_2Si, after a 12 h preheat at 500°C, 20 min melting at 1 000°C and 5 days annealing at 500°C, produced a chemically analysed ternary phase $Na_{1.5}Au_2Si$. Its crystal structure was characterized from single crystals as bcc with a = 1.443 8 nm, related to the $Mg_{32}(Zn,Al)_{49}$ structure. The composition of the ternary phase according to structural analysis was $Na_{52}Au_{81}Si_{29}$, a structure with 162 atoms in the unit cell (cI162).

The ternary compounds $Na_{1.75}Au_{2.5}Ge$, Na_2Au_3Sn, and $Na_{1.5}Au_2Si$ [79 Dör] all occur at 33·33 at.-% Na with Au contents ranging fom 44·4 to 50 at.-%; they all have the same crystal structure.

[80 Dör] reported the synthesis of a ternary compound $NaAu_3Si$ from a Na-rich mixture of the composition Na_2Au_3Si using the same preparative technique as [79 Dör]. The crystal structure, determined by single crystal X-ray diffraction, is cubic, space group Pa$\bar{3}$ (cP40) with unit cell parameter a = 0·891 6 nm. There are eight formula units in the elementary cell. The compound $NaAu_3Si$ tends to decompose at unspecified high temperatures with the appearance of Au reflections in the X-ray spectra. The analogous compound $NaAu_3Ge$ has the same crystal structure as $NaAu_3Si$; the compound $NaAu_3Sn$ was not formed.

REFERENCES

79 Dör: W. Döring, W. Seelentag, W. Buchholz and H.-U. Schuster: *Z. Naturf.*, 1979, **34B**, 1715–1718

80 Dör: W. Döring and H.-U. Schuster: *Z. Naturf.*, 1980, **35B**, 1482–1483

Au–Na–Sn

Wrobel and Schuster [75 Wro] prepared the ternary compound NaAuSn from the elements in a Ta crucible under argon. The material was preheated for 20 h at 400°C, melted at 1 000°C for 20 min and subsequently annealed for 20 h at 550°C before cooling to room temperature. X-ray diffraction analysis indicated that the compound, chemically analysed as NaAuSn, was orthorhombic. [77 Wro] carried out a single crystal structure determination and quote the lattice parameters for NaAuSn as a = 0·747(6), b = 0·808(8), c = 0·453(0) nm. It belongs to the space group $Pna2_1$(oP12). There are four formula units in the unit cell. Dietsch [77 Die] used differential thermal analysis to detect a polymorphic transformation in NaAuSn at 542 ± 3°C. The crystal structure of the high-temperature polymorph of NaAuSn is unknown. Döring *et al.* [79 Dör] heated the elements, with an excess of Na, in fused quartz ampoules under argon. An initial mixture corresponding to Na_3Au_3Sn, after 1 h preheat at 400°C and a 24 h anneal at 700°C, produced a chemically analysed ternary phase Na_2Au_3Sn. Its crystal structure was characterized as a bcc lattice with a = 1·498 9 nm, related to the $Mg_{32}(Zn,Al)_{49}$ structure. The composition corresponding to the structural analysis was $Na_{60}Au_{78}Sn_{24}$, a structure with 162 atoms in the unit cell (cI162). [79 Dör] state that the bcc structure was detected over a range of compositions, from '$NaAu_2Sn$' to 'NaAuSn', with a variation of a = 1·489 0 to 1·515 7 nm. They make no comment on their previous preparation of the compound NaAuSn. It is concluded that NaAuSn and Na_2Au_3Sn exist as separate phases and that they are in equilibrium with each other as two- or three-phase mixtures in the ternary system. The ternary compounds $Na_{1.75}Au_{2.5}Ge$, Na_2Au_3Sn, and $Na_{1.5}Au_2Si$ [79 Dör] all occur atg 33·33 at.-% Na with Au contents ranging from 44·4 to 50 at.-%; they all have the same crystal structure. Döring and Schuster [80 Dör] prepared and characterized the compounds $NaAu_3Ge$ and $NaAu_3Si$, but found that the compound $NaAu_3Sn$ did not form.

REFERENCES

75 Wro: G. Wrobel and H.-U. Schuster: *Z. Naturf.*, 1975, **30B**, 806

77 Wro: G. Wrobel and H.-U. Schuster: *Z. anorg. Chem.*, 1977, **432**, 95–100

77 Die: W. Dietsch: private communication, quoted by [77 Wro]

79 Dör: W. Döring, W. Seelentag, W., Buchholz and H.-U. Schuster: *Z. Naturf.*, 1979, **34B**, 1715–1718

80 Dör: W. Döring and H.-U. Schuster: *Z. Naturf.*, 1980, **35B**, 1482–1483

Au–Nb–Pd

Wang *et al.* [67 Wan] studied alloys on the 25 at.-% Nb section by X-ray diffraction techniques. They report the presence of three ordered close-packed phases on this section, but no further details were given.

REFERENCE

67 Wan: R. Wang, B. C. Giessen and N. J. Grant: *Abst. Bull. Inst. Metals Div., Met. Soc. AIME*, 1967, **2**(2), 120

Au–Nb–Pt

Wang *et al.* [67 Wan] studied alloys on the 25 at.-% Nb section by X-ray diffraction techniques. They report the presence of three ordered close-packed phases on this section but no further details were given. Flükiger *et al.* [69 Flü] prepared three ternary alloys on the 75 at.-% Nb section and found the W_3O-type structure (cP8) for $AuNb_3$, Nb_3Pt, and the ternary alloys (Table 90). Weight losses on argon-arc-melting were <1% and as-cast alloys were vacuum heat treated at temperatures up to 1 400°C for 2 h. The lattice constants were independent of heat treatment up to 1 400°C to within the accuracy of measurement (±0·000 2 nm;. Khan *et al.* [73 Kha] studied the constitution of alloys on the 70, 75, and 80 at.-% Nb section by X-ray powder diffraction and metallographic techniques. Alloys were prepared by arc-melting the Au–Nb and Nb–Pt binary compositions, using 99·9% pure elements, prior to arc-melting mixtures of the binary alloys to give the required ternary alloy compositions. The as-cast alloys were annealed for 14 days at 1 200°C in a vacuum of 10^{-5} torr (Table 90). The major phase in all the sections was the W_3O-type structure, cP8. On the 70 at.-% Nb section the W_3O-type structure was accompanied by a second unidentified phase that accounted for about 15% of the X-ray lines. Metallographically the 22·5Au, 70Nb, 7·5Pt alloy showed a dispersion of a second phase in the annealed state; the higher Pt content alloy with 15Au, 70Nb, 15Pt exhibited a dendritic structure after annealing indicating that it was not equilibrated. The alloys on the 75 at.-% Nb section were all homogeneous W_3O-type phase according to X-ray analysis. Metallographically the annealed alloy containing 18·75Au, 75Nb, 6·25Pt showed a dispersed second phase associated with the primary W_3O-type phase; the higher Pt content alloy with 8·75Au, 75Nb, 16·25Pt again exhibited a dendritic structure in the annealed state indicating that it was not equilibrated. The lattice spacing of the W_3O-type phase shows a discontinuity for the alloy 6·25Au, 75Nb, 18·75Pt [73 Kha], but this is not confirmed by the data of [69 Flü]. The alloys on the 80 at.-% Nb section were all homogeneous W_3O-type phase according to X-ray analysis. Metallographically the annealed alloy with 5Au, 80Nb, 15Pt again exhibited a dendritic structure indicating a lack of equilibrium. The 20Au, 80Nb binary alloy was two-phase after annealing, whereas the tentative diagram [87 Oka] indicates a single phase $AuNb_3$ structure. It would appear that Nb_3Pt and Nb_3Au form a solid solution series at 1 200°C when in an equilibrium condition.

REFERENCES

67 Wan: R. Wang, B. C. Giessen and N. J. Grant: *Abst. Bull. Inst. Metals Div., Met. Soc. AIME*, 1967, **2**(2), 120

69 Flü: R. Flükiger, P. Spitzli, F. Heiniger and J. Muller: *Phys. Lett.*, 1969, **29A**, 407–408

73 Kha: H. R. Khan, E. Röschel and Ch. J. Raub: *Z. Phys.*, 1973, **262**, 279–293

87 Oka: H. Okamoto and T. B. Massalski: 'Phase Diagrams of Binary Gold Alloys', 188–190; 1987, ASM International

Table 90. Lattice parameters and metallographic structures of alloys on the 70, 75 and 80 at.-% Nb sections of the Au–Nb–Pt system

Composition, at.-%			*Lattice parameter*, nm, *of W_3O-type phase*		*Metallographic structure*
Au	*Nb*	*Pt*	[73 Kha]	[69 Flü]	
30	70	0	0·5196		
22·5	70	7·5	0·5185		2-phase
15	70	15	0·5170		2-phase
7·5	70	22·5	0·5152		
0	70	30	0·5144		
25	75	0	—	0·5211	
18·75	75	6·25	0·5196	0·5197	2-phase
12·5	75	12·5	0·5181	0·5185	
8·75	75	16·25	0·5174	—	2-phase
6·25	75	18·75	0·5178	0·5174	
3·75	75	21·25	0·5170	—	
0	75	25	0·5154	0·5159	
20	80	0	0·5220		2-phase
15	80	5	0·5209		
10	80	10	0·5195		
5	80	15	0·5186		2-phase
0	80	20	0·5174		

Au–Nb–Rh

Wang *et al.* [67 Wan] studied alloys on the 25 at.-% Nb section by X-ray diffraction techniques. They report the presence of four ordered close-packed phases on this section, but no further details were given. Zegler [65 Zeg] prepared alloys on the 75 at.-% Nb section by arc-melting 99·95% Nb with >99·9% Au,Rh in an argon–helium atmosphere on a water-cooled hearth. The as-solidified alloys were annealed 1 week at 1 400°C in a vacuum of 2×10^{-5} mm Hg, furnace-cooled to room temperature, encapsulated in evacuated quartz capsules, and re-annealed for 3 to 6 weeks at 900°C and water quenched. Phase identification was by metallographic and X-ray techniques. Table 91 gives the lattice parameter for the W_3O-type phase (cP8) as a function of composition. Minor amounts of a second phase, observed metallographically, were attributed to small composition variations. The W_3O-type phases Nb_3Au and Nb_3Rh form a solid solution series at 900°C in the ternary system.

REFERENCES

65 Zeg: S. T. Zegler: *Phys. Rev.*, 1965, **137**(5A), A1438–A1440

67 Wan: R. Wang, B. C. Giessen and N. J. Grant: *Abst. Bull. Inst. Metals Div., Met. Soc. AIME*, 1967, **2**(2), 120

Table 91. Lattice parameters of the W_3O-type phase on the 75 at.-% Nb section of the Au–Nb–Rh system [65 Zeg]

Composition, at.-%			*Lattice parameter*, nm
Au	*Nb*	*Rh*	
25	75	0	0·520 27
24·5	75	0·5	0·520 3
23·75	75	1·25	0·520 0
22·5	75	2·5	0·519 60
17·5	75	7·5	0·518 27
12·5	75	12·5	0·516 88
7·5	75	17·5	0·515 73
2·5	75	22·5	0·514 12
1·25	75	23·75	0·513 7
0·5	75	24·5	0·513 3
0	75	25	0·513 17

Au–Nb–Sn

In studying solid solution formation between Cr_3Si-type phases (cP8) von Philipsborn [70 Phi] prepared a $Nb_3Au_{0\cdot5}Sn_{0\cdot5}$ alloy by arc-melting Nb_3Au and Nb_3Sn on a water cooled Cu hearth under argon. Samples were annealed for 45 days at 550°C, 30 days at 800°C, and 9 days at 1 050°C. The phases present were characterized by the X-ray powder diffraction technique.

The as-cast and 550°C annealed samples contained a Cr_3Si-type phase with lattice parameter $a = 0{\cdot}525 \pm 0.001$ nm, and a W-type phase with $a = 0{\cdot}332 \pm 0{\cdot}001$ nm. The 800°C annealed sample contained the Cr_3Si-type phase with $a = 0{\cdot}525 \pm 0{\cdot}001$ nm and the 1 050°C annealed sample also contained this phase together with an unidentified phase. It is worth noting that the lattice parameter of the Cr_3Si-type phase is the mean of the lattice parameters of Nb_3Au (0·520 nm) and Nb_3Sn (0·528 0 nm on the Nb-rich side).

REFERENCE

70 Phi: H. von Philipsborn: *Z. Kristallogr.*, 1970, **131**, 73–87

Au–Ni–Pd
(G. V. Raynor)

INTRODUCTION

The Au–Ni–Pd system has been investigated by [62 Gri], using 77 ternary alloys lying on composition lines at each interval of 10 wt-% Pd from 10 to 90 wt-% Pd. The component metals employed were gold containing less than 0·01% impurity, electrolytic nickel, and spongy refined palladium (99·96% pure). Alloys were prepared by melting in a high-frequency induction furnace under a layer of $BaCl_2$ and borax. Many of the alloys were chemically analysed, with sufficiently close agreement with synthetic compositions for the latter to be accepted. Specimens were investigated by thermal analysis for the liquidus and solidus temperatures, by differential thermal analysis for solid state reactions, and by conventional metallography. In addition, hardness, electrical resistivity and the temperature coefficient of resistivity were examined. Specimens for solid state experiment were first vacuum-annealed for 500 h at 900°C and then slowly cooled. The conditions under which solid alloys were examined do not therefore correspond with equilibrium conditions at any given temperature. The observed constitution must correspond with the temperatures at which diffusion for alloys in various composition ranges becomes negligible on cooling. These temperatures are unlikely to remain similar at all compositions. Nevertheless the results of the work allow the general nature of the ternary system to be understood; they confirm, extend, and largely supplant the very early work of Fraenkel and Stern [27 Fra], in which the approximate position of the liquidus surface for part of the ternary system was investigated by thermal analysis.

BINARY SYSTEMS

The Au–Pd phase diagram has been assessed by [85 Oka], the Au–Ni by [86 Oka], and the Ni–Pd by [84 Nas].

SOLID PHASES

Between the solidus surface of the ternary system and the critical temperature of the region of immiscibility in the ternary model, the ternary system consists of a complete series of solid solutions involving Au, Ni, and Pd. At lower temperatures a volume of immiscibility exists, involving Au-rich and Ni-rich regions of the ternary solid solution. No ternary intermediate phases have been reported, and no crystal structure for the equilibrium state has been observed other than the face-centred cubic structure of the components.

THE TERNARY SYSTEM

In the work of [62 Gri] thermal analysis was carried out on alloys with compositions lying along lines of constant weight percentages of Pd. The composition sections were arranged at intervals of 10 wt-% Pd from 10 wt-% to 90 wt-% inclusive. An example of the results obtained is shown in Fig. 261, which shows liquidus and solidus curves, plotted in *atomic* percentages, for the section investigated at a constant Pd content of 10 wt-%. The two curves show minima corresponding to the minimum in the liquidus and solidus curves of the Au–Ni system. The other sections investigated by [62 Gri], which are not reproduced here, also show minima. The results allow the construction of liquidus contours, though this was not done in the original publication. Figure 262 shows the contour diagram constructed from the liquidus results of [62 Gri], and clearly reflects the influence of the minima in the Au–Ni and Ni–Pd liquidus curves. Doubt attaches to the positions of the 1 150 and 1 100°C contours, owing to the fact that the results for the 20 wt-% Pd section are not consistent with those for the other sections, but their general forms are clear. Also there is no direct experimental evidence for the position of the 1 000°C contour within the ternary model, except that it does not extend to 10 at.-% Pd, but its position cannot differ greatly from that shown in Fig. 262.

The differential thermal analysis results of [62 Gri] for the solid alloys of sections at 10 wt-% and 20 wt-% Pd contained thermal effects due to the separation of the homogeneous ternary solid solution into Au-rich and Ni-rich constituents, corresponding with the miscibility gap in the binary Au–Ni system. Whereas the critical temperature below which the miscibility gap exists in the binary alloys is 810°C, a critical temperature for the 10 wt-% Pd section of 908°C was recorded. This evidence appears to be sound and if accepted means that immiscibility in the solid state extends to higher temperatures than in the Au–Ni system itself. From the thermal analysis results recorded in [62 Gri] contours may be constructed for the surface which encloses the volume in the ternary model correponding to the coexistence of two solid phases. This has been done, in atomic percentage, in Fig. 263 (down to 300°C). The surface contains a peak, towards the Ni-rich corner, for which the temperature exceeds 900°C. The maximum recorded temperature of 908°C was obtained for the alloy containing 9·5 at.-% Au and 83·5 at.-% Ni. From Fig. 263 the variation of critical temperature with Pd content may be derived. This variation for the section at a constant Au content of 9·5 at.-% (which contains the recorded maximum of 908°C) is shown in Fig. 264, curve a, and the form of this curve suggests that the maximum critical temperature for this section of the ternary system is approximately 922°C at a composition close to 9·5 at.-% Au and 5·7 at.-% Pd. Figure 264, curve b, shows a section at 29 at.-% Au, corresponding to the composition of the maximum critical temperature in the binary Au–Ni system. For this section the maximum critical temperature for the ternary alloys lies at approximately 870°C. By making use of these sections and of sections at 15 and 20 at.-% Au it can be shown that maxima occur at approximately 5·7 at.-% Pd in each case. Construction of a section at a constant Pd content of 5·7 at.-% then suggests that the maximum critical temperature in the ternary system as a whole is approximately 927°C at a composition of 15 at.-% Au and 5·7 at.-% Pd.

The thermal analysis results, particularly with reference to the region of immiscibility, were supplemented by microstructural observations, hardness measurements, and measurement of electrical resistivity and its temperature coefficient. The results were consistent with those described above.

REFERENCES

27 Fra: W. Fraenkel and A. Stern: *Z. anorg. Chem.*, 1927, **166**, 161–170

62 Gri: A. T. Grigoriev, L. A. Panteleimonov, V. V. Kuprina, G. V. Goldobina and M. A. Rudnitskii: *Zhur. Neorg. Khim.*, 1962, **7**, 1110–1113 (*Russian J. Inorg. Chem.*, 1962, **7**, 570–573)

84 Nas: A. Nash and P. Nash: *Bull. Alloy Phase Diagrams*, 1984, **5**, 446–450

85 Oka: H. Okamoto and T. B. Massalski: *Bull. Alloy Phase Diagrams*, 1985, **6**, 229–235

86 Oka: H. Okamoto and T. B. Massalski: 'Binary Alloy Phase Diagrams' (ed. T. B. Massalski), 288–290; 1986, *ASM International*

Au–Ni–Pt
(G. V. Raynor)

INTRODUCTION

The only comprehensive examination of the constitution of the Au–Ni–Pt alloys which has been traced in the metallurgical literature was carried out by [73 Car]. The work involved a study of isothermal sections ranging in temperature from 812 to 1 260°C by thermodynamic calculation and by experiment, using microprobe analysis. For the computations the investigators used the subregular approximation, deriving the parameters to be used in their equations from characteristic points in the binary boundary systems, such as the critical points in the Au–Ni and Au–Pt systems, the minimum in the liquidus of the Au–Ni system, and a tie line in the Au–Pt system. From the equations derived, the calculated versions of the Au–Pt and Au–Ni phase diagrams were computed,

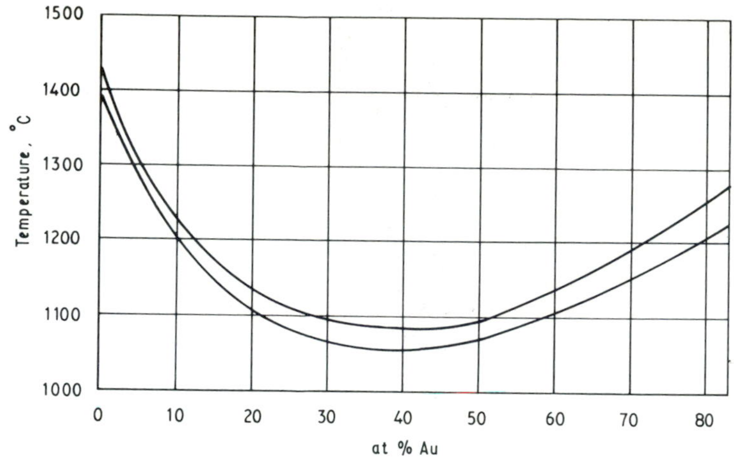

261 System Au–Ni–Pd: liquidus and solidus curves for section at constant Pd content of 10 wt-% Pd, plotted in atomic percentages (after [62 Gri])

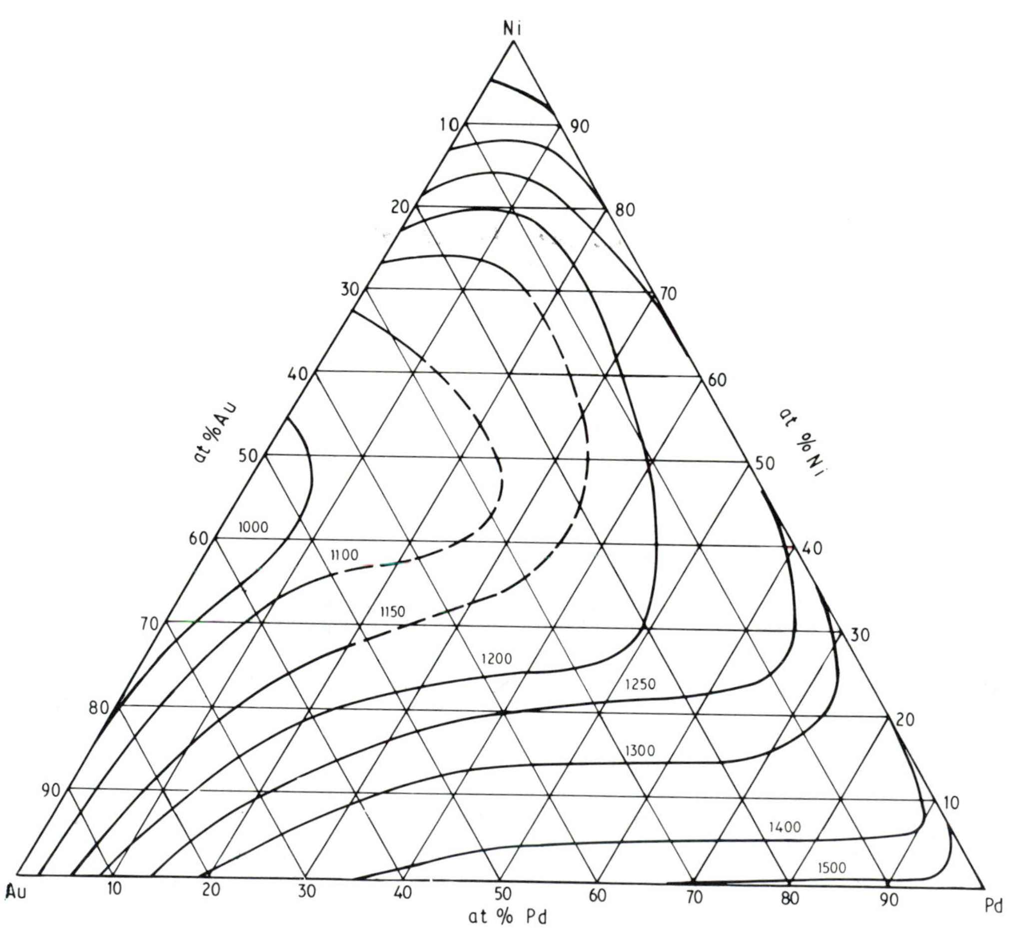

262 System Au–Ni–Pd: liquidus contours, constructed from [62 Gri]

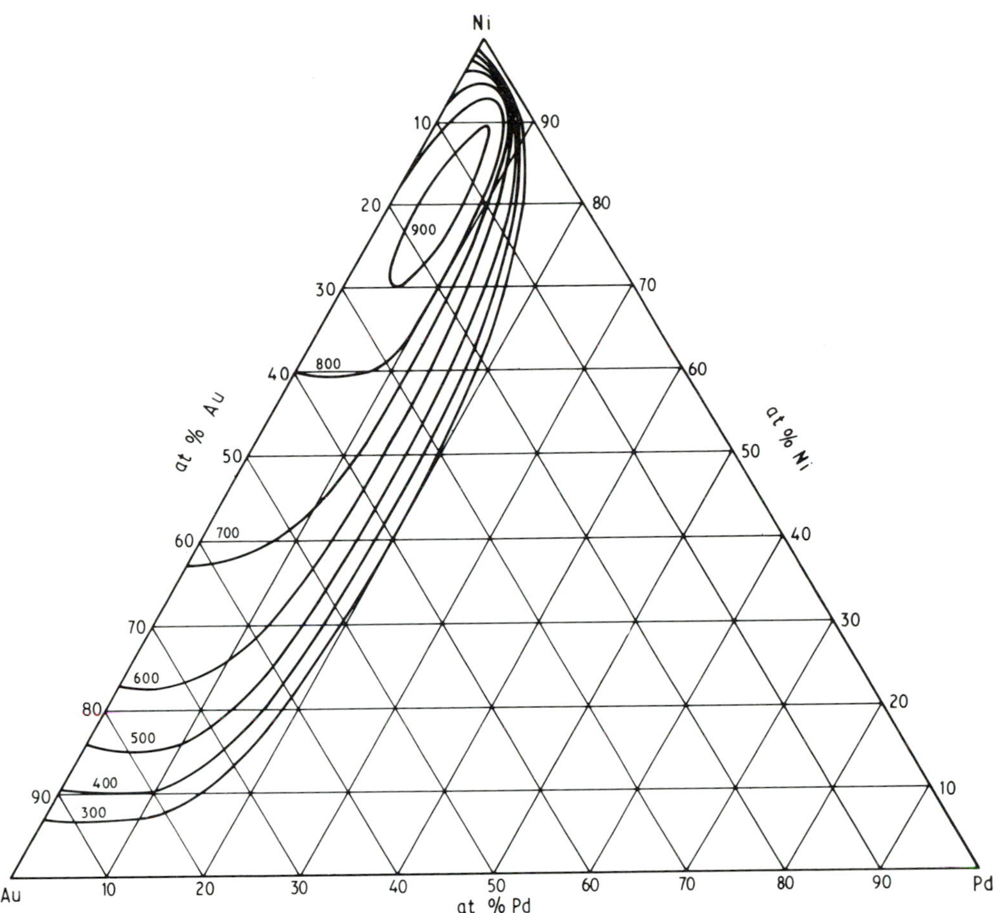

263 System Au–Ni–Pd: contours for the region of immiscibility in the solid state, constructed from [62 Gri]

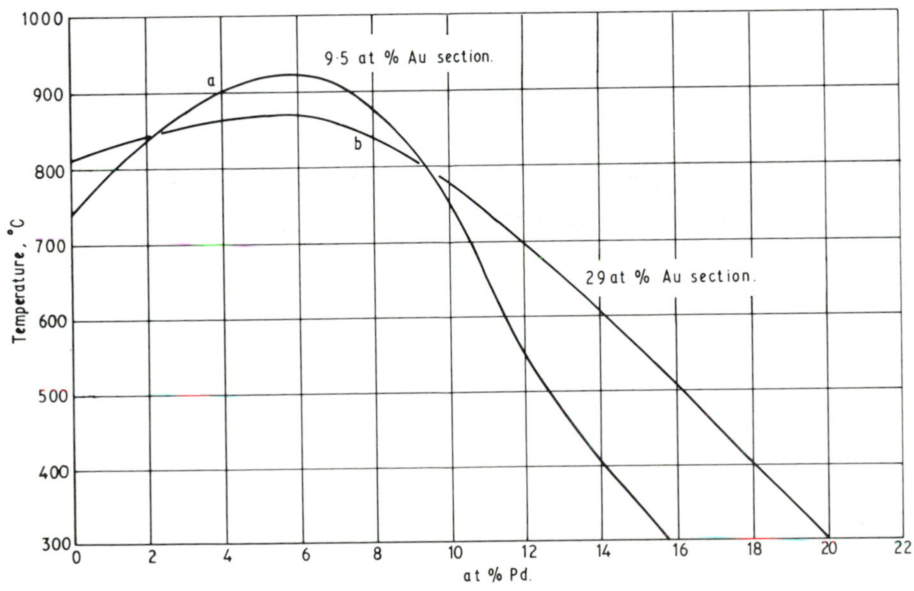

264 System Au–Ni–Pd: regions of immiscibility for sections at *a* 9·5 at.-% Au; *b* 29 at.-% Au

together with isothermal section of the Au–Ni–Pt ternary system at 812, 950, 975, 1 000, 1 025, 1 150, 1 200, and 1 260°C. The results were compared with the established forms of the binary diagrams, and with experimentally determined ternary isothermals. Inspection indicates that the general nature of the ternary equilibria were correctly interpreted; the calculated and experimentally determined diagrams, however, differ quantitatively. It has been felt better to refer in this assessment only to the experimentally determined diagrams.

BINARY SYSTEMS

The Au–Pt phase diagram has been assessed by [86 Oka 1], the Au–Ni by [86 Oka 2], and the Ni–Pt by [86 Nas].

SOLID PHASES

No ternary intermediate phases have been observed in the ternary system. The only crystal structure to be encountered is therefore the fcc structure of the components. Miscibility gaps in the solid state necessitate the recognition of separate composition ranges, and it is convenient to refer to Au-rich alloys containing Pt and Ni as (Au), while the solid solution formed between Ni and Pt, and containing Au, may be denoted (Ni–Pt). Where the solid solution contains appreciable quantities of all components, the symbol (Au–Ni–Pt) may be appropriated. NiPt and Ni_3Pt do not occur at the temperatures of the experimental survey, and their involvement in the ternary alloys is not known.

THE TERNARY SYSTEM
Experimental Studies

Alloys for the experimental study of the ternary system were prepared from pure materials (Au: 99·99 at.-%; Ni: 99·998 at.-%; Pt: 99·9 at.-%). Weighed quantities were induction-melted in Al_2O_3 crucibles using a protective atmosphere, and held in the liquid state for 15 min before cooling. Specimens weighing approximately 4 g were annealed for 300 to 800 h at the required temperatures in Al_2O_3 crucibles in an atmosphere of N_2 + 10 wt-% H_2. After quenching from the annealing temperature the specimens were examined metallographically and the compositions of the phases present were determined by electron microprobe methods.

The results may be presented in the form of isothermal diagrams. At 812°C the solid-state miscibility gap in the Au–Pt system is wide; the components Ni and Pt, however, are mutually completely soluble, while at this temperature the (Au–Ni) solid solution begins to decompose into Au-rich and Ni-rich constituents. The isothermal section thus takes the form of Fig. 265, where the curved boundary between the single phase and two-phase areas approaches closely to the Ni–Pt axis and just touches the Au–Ni axis. At lower temperatures the two-phase area would include a section of the Au–Ni axis. In Fig. 265 the plotted points represent electron probe microanalysis results, and the lines joining them are the determined tie lines. The portions of the boundary drawn as broken lines are not accurately established, and the intersections of the boundary with the Au–Pt axis have been drawn to be consistent with the established Au–Pt phase diagram. Similar remarks apply to the diagrams relevant to the other temperatures involved. The isothermal section for 950°C is shown in Fig. 266; the boundary of the two-phase region has now withdrawn from the Au–Ni axis and the solid solution region surrounding the two-phase area stretches continuously from the Au-rich corner of the diagram to the Pt-rich corner. At 1 000°C the liquid phase makes its appearance owing to the minimum in the liquidus of the Au–Ni system; a long narrow three-phase region, L + (Au) + (Ni–Pt) is thereby introduced into the equilibria (Fig. 267). The temperature of 1 150°C exceeds the melting point of Au, so that at this temperature the liquid region extends to the Au-rich corner (Fig. 268), and the three-phase triangle has shifted its position considerably. This shift is continued at 1 260°C (Fig. 269); also, because of the widening of the liquidus–solidus separation in the Au–Pt system, the corner of the three-phase triangle corresponding to (Au) has moved to a composition much richer in Pt. Further, 1 260°C is the temperature below which two solid phases occur in the Au–Pt system. The boundary of the region corresponding to two solid phases in the ternary system thus just touches the Au–Pt axis at this temperature.

It is stated in [73 Car] that isothermal sections were also investigated at 975, 1 025, and 1 200°C. No results were given for these temperatures as no essentially new information was provided, additional to that summarized in Figs 265–269. The temperatures at which the three-phase equilibrium is introduced and disappears were not located experimentally, but, as expected from the binary Au–Pt diagram, it is present at 975°C. Equilibrium relationships involving the Ni–Pt ordered phases were not examined. No information exists with regard to the liquidus and solidus surfaces.

Development of the ternary equilibria associated with the liquidus minimum in the Au–Ni system [73 Car]

The diagrams given in Figs 265–269 refer to the results of experimental investigations, with which thermodynamically calculated phase diagrams are in reasonable agreement. By making use of such calculations it is possible to indicate the manner in which the constitution develops as the temperature rises. Thus the temperature of 950°C relevant to Fig. 266 is 5°C below that of the minimum in the Au–Ni liquidus and solidus curves. A little above this temperature (e.g. 960°C) the constitution will take the form shown in Fig. 270*a*, where the boundaries are schematic. The calculations in [73 Car] indicate that the temperature of the isothermal section in which the solidus curve makes contact with the curve enclosing the ((Au) + (Ni–Pt)) field is ~969°C (Fig. 270*b*). This contact establishes a critical tie line lab between the liquid and two solid alloy compositions. The calculated compositions for phases 1, a, and b are 55·4 Au, 43·9 Ni; 41·4 Au, 54·9 Ni; and 21·2 Au, 70·9 Ni respectively [73 Car].

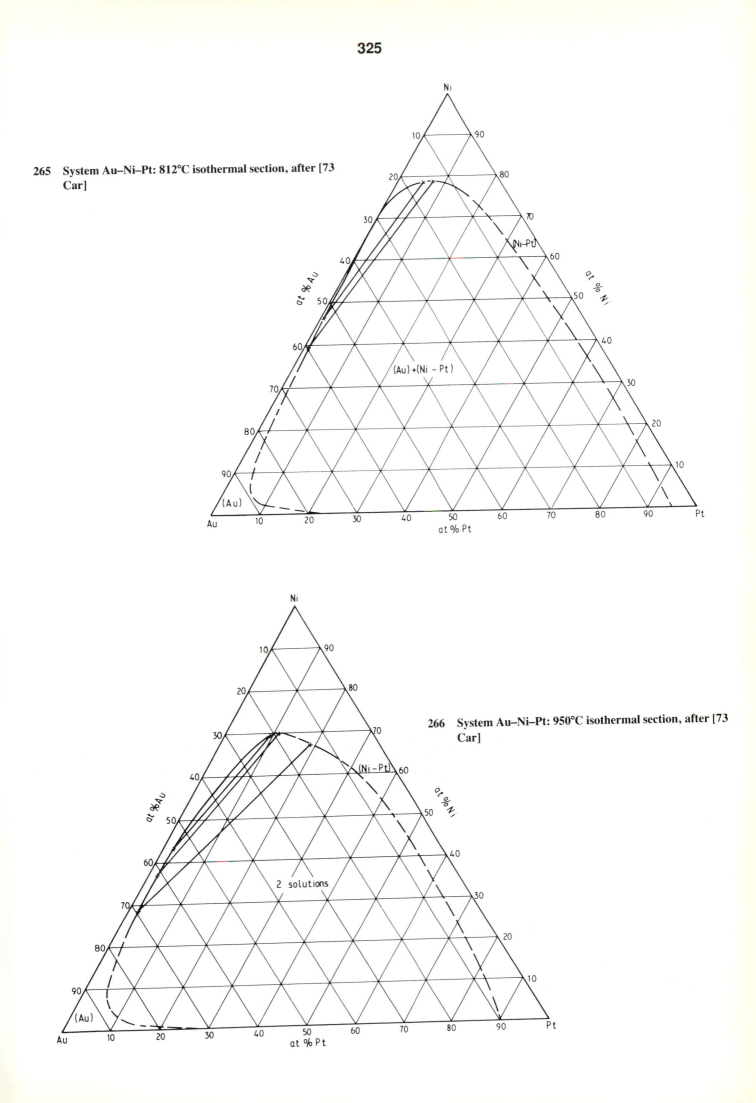

265 **System Au–Ni–Pt: 812°C isothermal section, after [73 Car]**

266 **System Au–Ni–Pt: 950°C isothermal section, after [73 Car]**

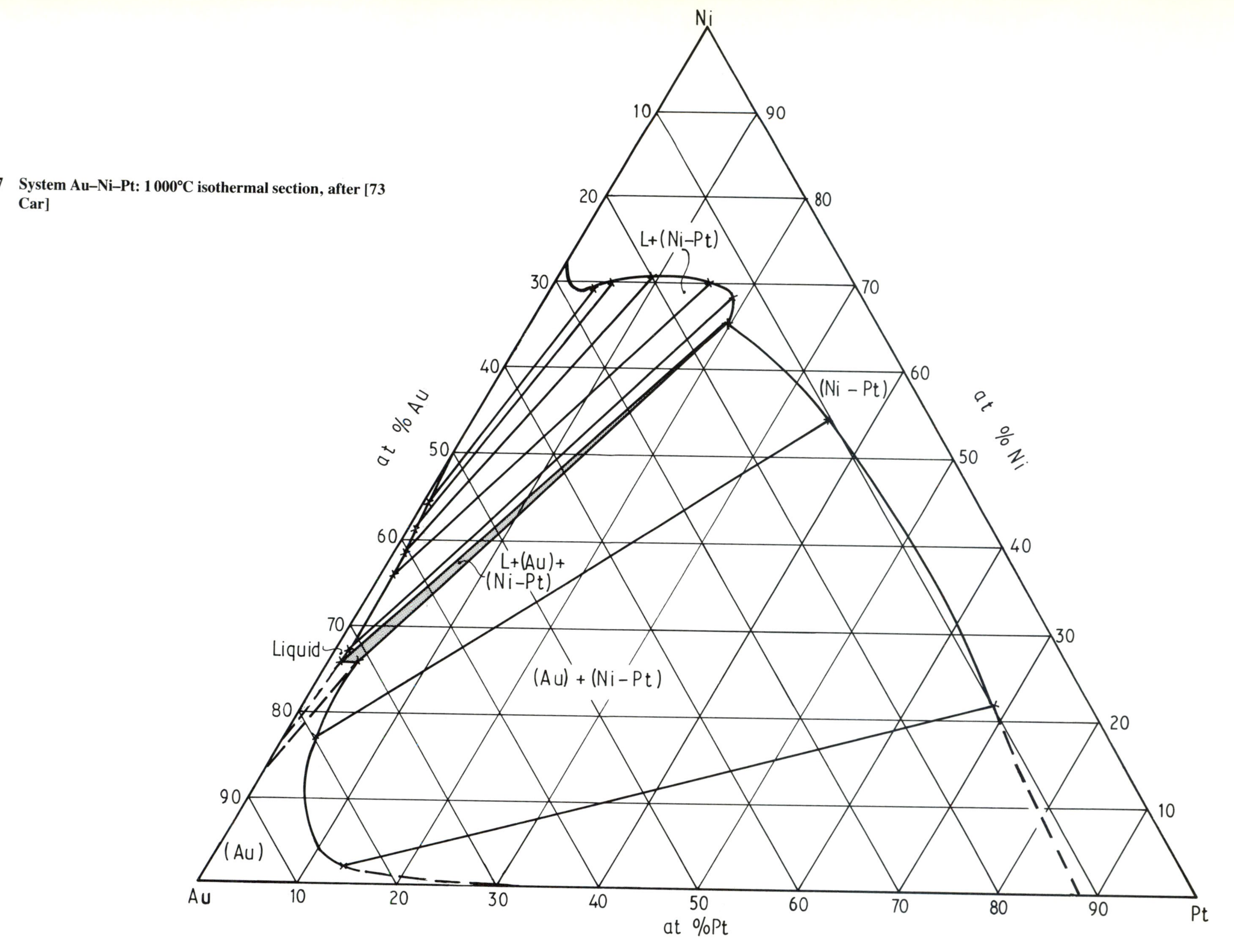

267 System Au–Ni–Pt: 1 000°C isothermal section, after [73 Car]

268 System Au–Ni–Pt: 1 150°C isothermal, after [73 Car]

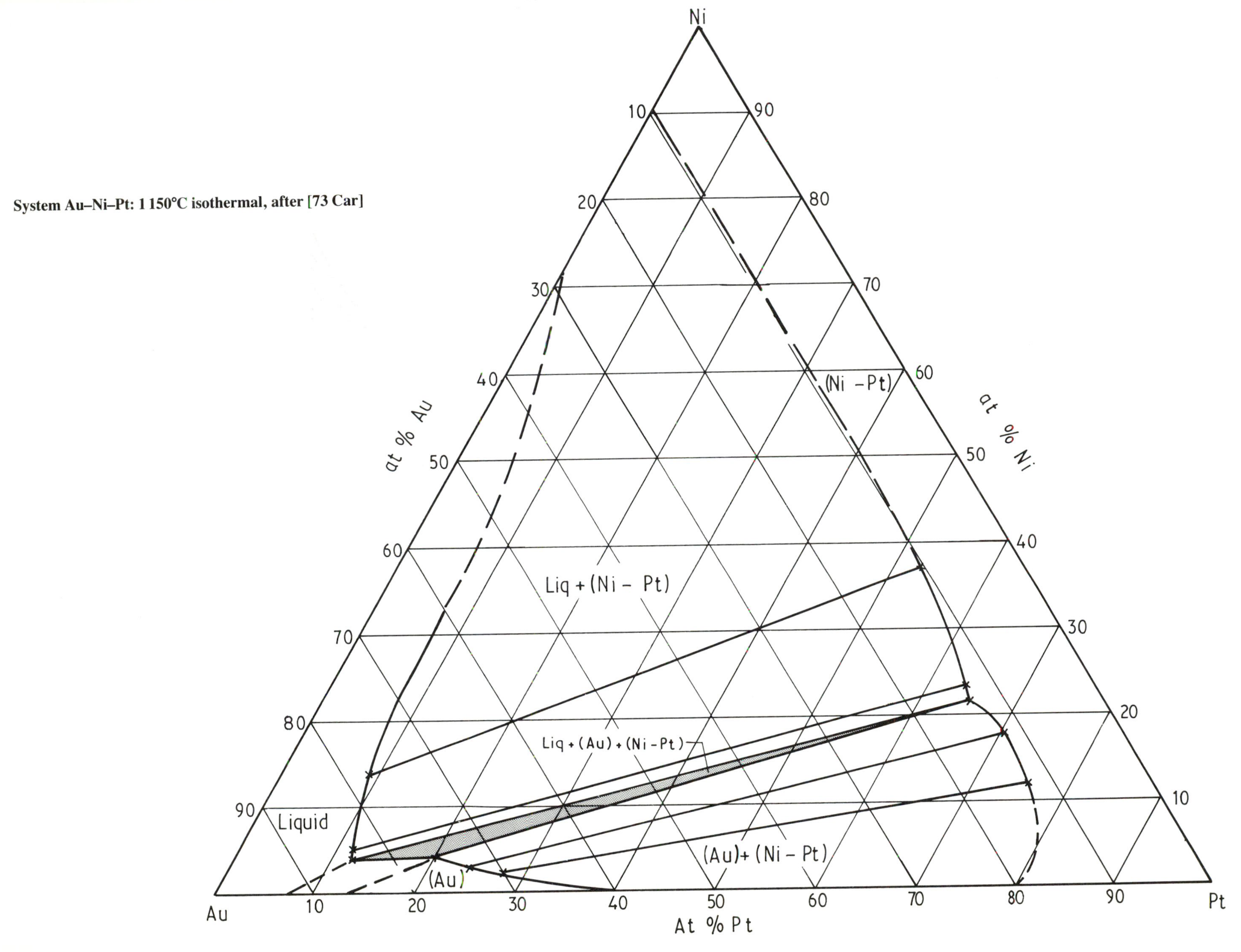

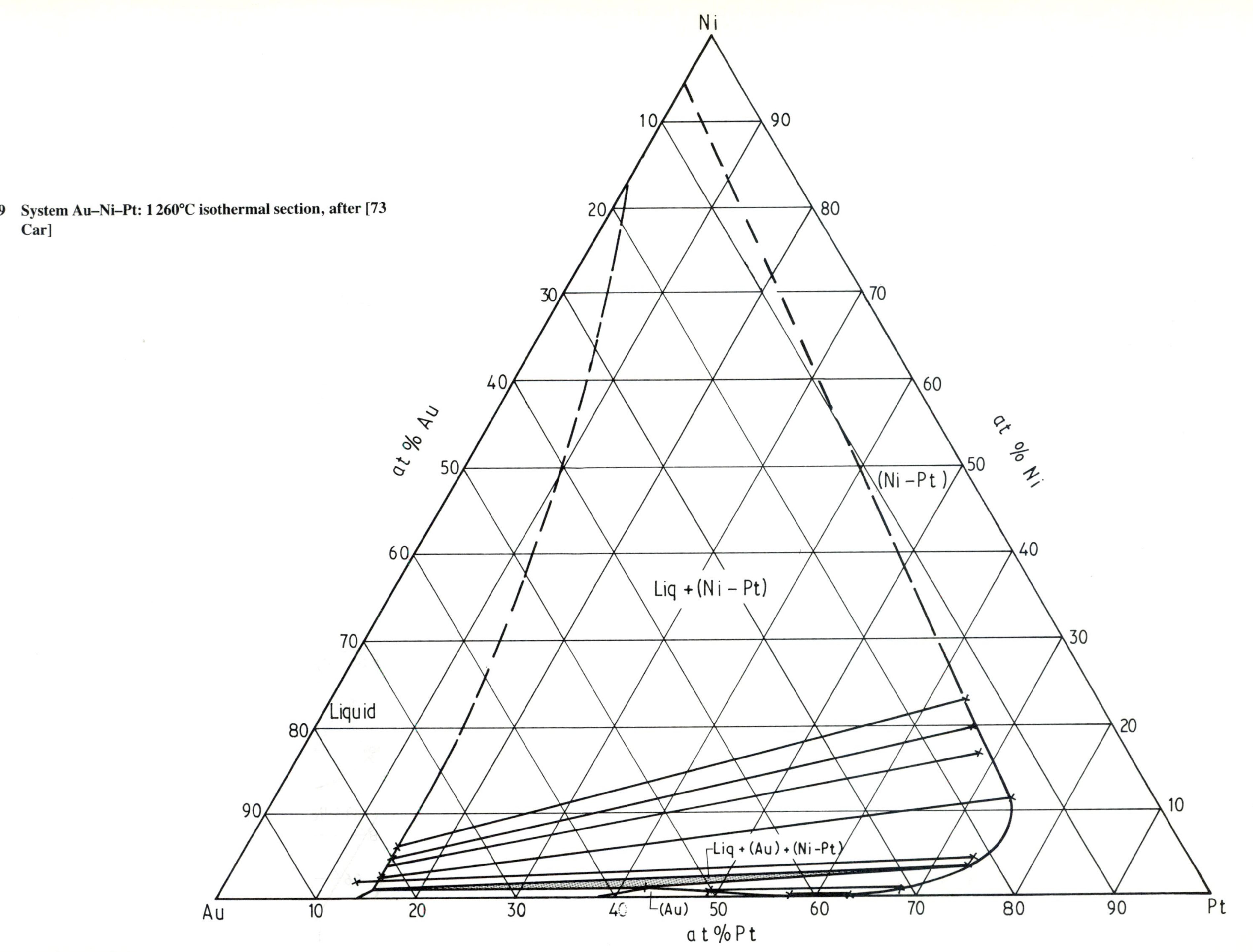

269 System Au–Ni–Pt: 1 260°C isothermal section, after [73 Car]

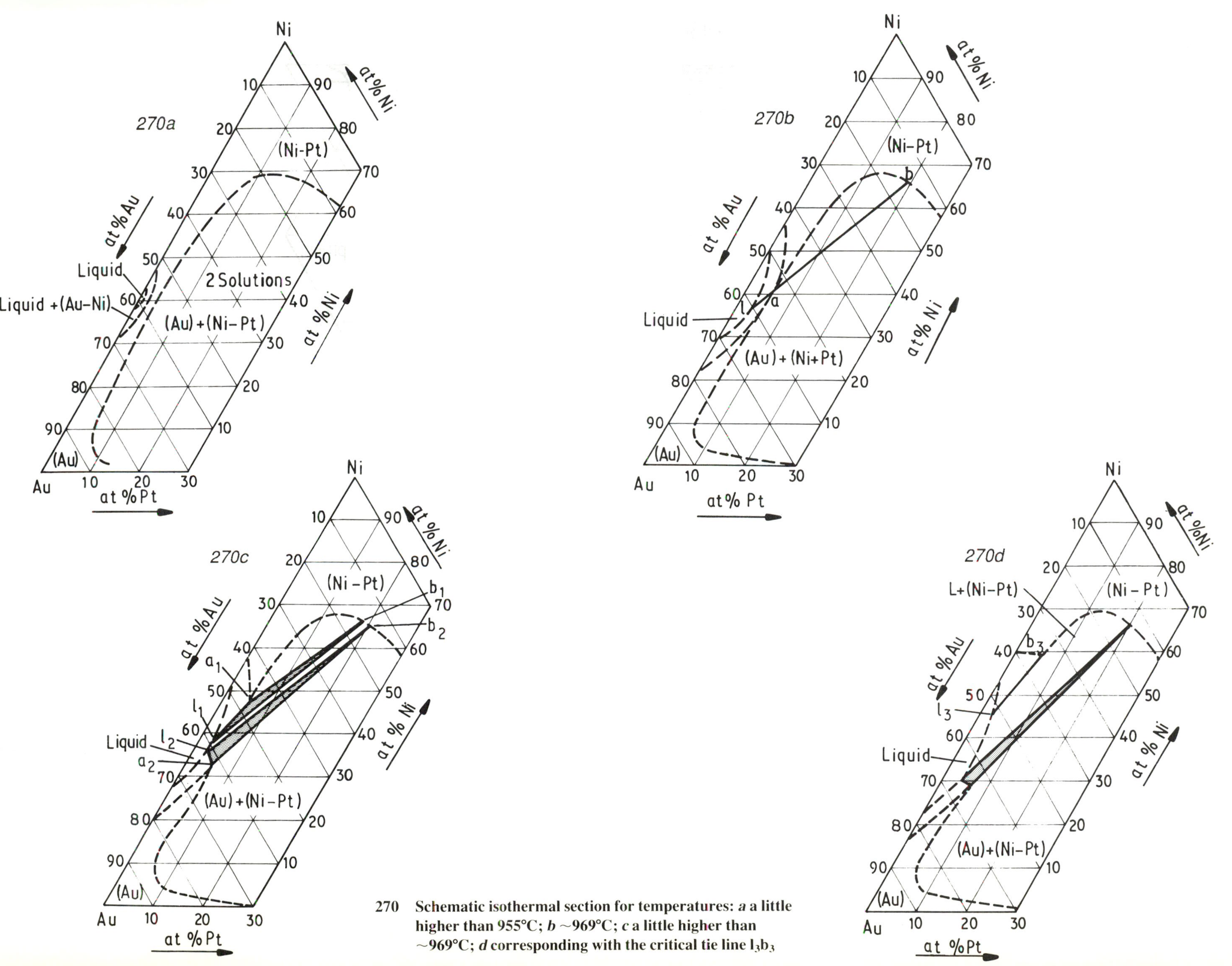

270 Schematic isothermal section for temperatures: *a* a little higher than 955°C; *b* ~969°C; *c* a little higher than ~969°C; *d* corresponding with the critical tie line l_3b_3

With further rise in temperature two points of intersection of the solidus and immiscibility curves develop and move apart, leading to the situation in Fig. 270*c*. In this figure, two three-phase triangles $1_1a_1b_1$ and $1_2a_2b_2$, involving liquid and two solid phases, flank a two-phase (liquid + (Ni–Pt)) field. With further rise in temperature the triangle $1_1a_1b_1$ degenerates into a critical tie line 1_3b_3 establishing the (liquid + (Ni–Pt)) equilibrium (Fig. 270*d*). According to the calculations of [73 Car] the critical tie line 1_3b_3 is formed at a temperature only slightly above that for the formation of the critical tie line lab. When this temperature is exceeded the constitution is essentially that already presented in the experimentally established Fig. 267 (1 000°C), which is maintained, apart from the changing position of the remaining (liquid + (Au) + (Ni–Pt)) triangle, to 1 260°C. At this temperature the ((Au) + (Ni–Pt)) immiscibility region just touches the Au–Pt axis. On further increase of temperature, this two-phase region withdraws from the Au–Pt axis and progressively diminishes in area until it becomes a point on a liquid + (Ni–Pt) tie line. The calculations in [73 Car] indicate that the liquid + (Au) + (Ni–Pt) triangle degenerates to a liquid + (Ni–Pt) tie line at ~1 283°C. The calculated composition of the (Ni–Pt) phase at 1 283°C is 40·2Au, 2·2Ni. It may be noted that equilibrium between two solid phases persists in the ternary system at higher temperatures than in either of the Au–Ni and Au–Pt binary systems.

At temperatures exceeding ~1 283°C but below the melting point of Ni the system will exhibit the simple equilibrium between liquid and the (Ni–Pt) solid solution. Above the melting point of Ni the liquid region will include the Ni-rich corner, and at successively higher temperatures the (Ni–Pt) solid solution area will continue to shrink towards the Pt-rich corner of the diagram.

The equilibria are summarized in Fig. 271, using the designation of phase compositions given in Fig. 270*a–d*.

REFERENCES

73 Car: S. M. Carmio and J. L. Meijering: *Z. Metall.*, 1973, **64**, 170–175

86 Oka 1: H. Okamoto and T. B. Massalski: *Bull. Alloy Phase Diagrams*, 1986, **6**, 46–56

86 Oka 2: H. Okamoto and T. B. Massalski: 'Binary Alloy Phase Diagrams' (ed. T. B. Massalski), 288–290; 1986, ASM International

86 Nas: P. Nash and M. F. Singleton: 'Binary Alloy Phase Diagrams' (ed. T. B. Massalski), 1741–1744; 1986, ASM International

Au–Ni–Si

A limited amount of evidence from [83 Irv] indicates that the addition of Si to Au–Ni alloys raises the temperature of the consolute point and widens the solid-state miscibility gap. In the Au–Ni–Ni_3Si part of the ternary system it can be speculated that a ternary eutectic reaction occurs as shown in Fig. 272. The $\alpha_1 + \alpha_2$ phase region associated with the L + α_1 + α_2 tie triangle approaches the Au–Ni edge with fall in temperature, touching it at the temperature of the consolute point in the Au–Ni binary system and thereby isolating the α_1 and α_2 phase regions at lower temperatures. The L + α_1 + Ni_3Si three-phase region is assumed to emanate from the reaction $L + Ni_5Si_2 \rightleftharpoons \alpha_1 + Ni_3Si$.

REFERENCE

83 Irv: D. Irving: PhD Thesis, University of Surrey, UK, 1983

Au–Ni–Zn
(G. V. Raynor)

INTRODUCTION

Very little work has been carried out on the system Au–Ni–Zn; attention has been confined to alloys of composition close to that of Au_2NiZn [76 Dir], to limited experiments on the quasibinary system AuZn–NiZn [76 Kro], and to the delineation of the Au_4Zn phase region at 200°C [57 Wil].

BINARY SYSTEMS

The liquidus and solidus curves for the system Au–Ni show a minimum at 950°C and approximately 42 at.-% Ni. There is some doubt about the position of the solidus on the Ni-rich side of the minimum, and the equilibrium diagram given by [58 Han] presents alternative solidus curves. This difference is without significance in the assessment of the limited work on the ternary system. At temperatures immediately below the solidus curve Au and Ni are miscible in all proportions. At lower temperatures, however, this solid solution splits up into two separate solid solutions, respectively Au- and Ni-rich. The miscibility gap in the solid state is roughly symmetrical in the phase diagram and widens with decreasing temperature. The maximum on the curve which encloses the area corresponding to the mixture of two solid phases, denoted (Au) and (Ni), lies at 812°C and 71 at.-% Ni. At 300°C the solid solution (Au) contains 6·5 at.-% Ni; the Au content of (Ni) at the same temperature is 1·5 at.-% Au. According to [58 Han] short range order may exist at 900°C, though this is disputed in work summarized by [69 Shu], which suggests instead that clustering of like atoms occurs. In discussion of the system [65 Ell] adds that during the separation of the solid solution which exists above the miscibility gap into two phase, transitional structures may occur. Since these are not equilibrium phases they are not considered further in this assessment.

The Au–Zn system is characterized by the existence of at least nine intermediate phases [87 Oka]. There is con-

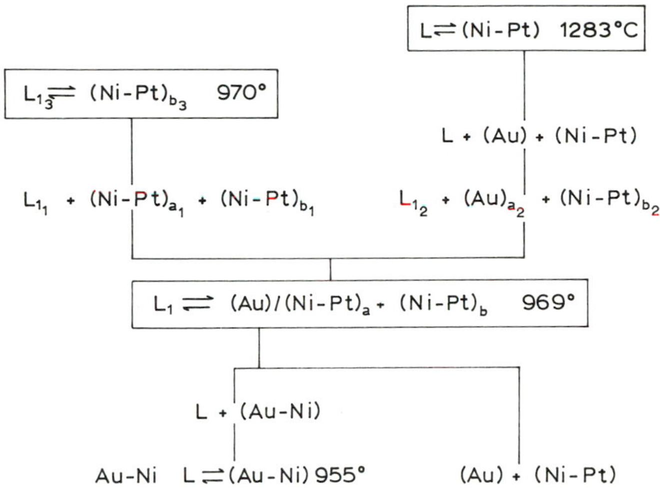

271 Development of the (Au) + (Ni–Pt) low-temperature equilibrium

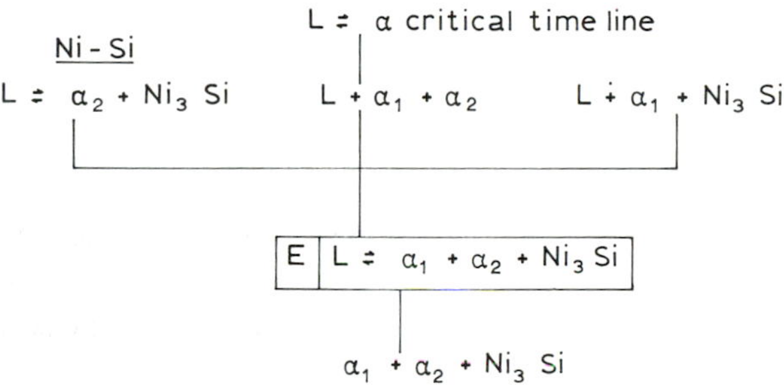

α_1 ⇌ Au-rich solid solution, α_2 ⇌ Ni-rich solid solution

272 The ternary eutectic reaction in the Au–Ni–Si system

siderable solid solubility of Zn in Au reaching a maximum of 31 at.-% Zn. The liquidus and solidus curves for this solid solution fall from the melting point of Au to a eutectic at 683°C (34 at.-% Zn), at which L $\rightleftharpoons$ (Au) + β, where (Au) represents the Au-rich solid solution. The β-phase melts at a maximum on the liquidus of 751°C, which also corresponds with the composition AuZn. The β-phase has a wide range of homogeneity and the crystal structure is of the CsCl type. At 654°C (63 at.-% Zn), L $\rightleftharpoons$ β + γ; the γ-phase has the typical γ-brass crystal structure. With further increase in zinc content the phases γ_3 and ϵ are formed peritectically, and finally the Zn-rich solid solution is formed peritectically from liquid and ϵ. A phase γ_2 is formed from γ and γ_3 in the solid state (~520°C). The phases γ_2 and γ_3, neither of which appears to have the ideal γ-brass structure, and the ϵ phase are not particularly relevant to the interpretation of the limited amount of work on the Au–Ni–Zn system. A series of gold-rich phases also exists below ~420°C and 30 at.-% Zn, which are of importance. At 420°C the phase Au_3Zn h forms congruently from (Au) and at 403°C comes into equilibrium with β as a result of the reaction (Au) $\rightleftharpoons$ Au_3Zn h + β. At the exact stoichiometric composition, Au_3Zn h transforms into Au_3Zn r, but at slightly lower zinc contents the two phases can coexist to room temperature. A further phase, Au_4Zn, is formed by reaction of (Au) and Au_3Zn h at 300°C. All the phases mentioned have a small homogeneity range. The complex phase relationship below 30 at.-% Zn are established between phases of basically similar crystal structures, which are given, where known, in Table 92.

Table 92. Crystal structures of relevant phases

Solid phase	*Prototype*	*Lattice designation*
(Au–Ni–Zn) solid solution	Cu	cF4
Au_4Zn	Doubly-faulted Cu_3Au	Doubly-faulted cP4
Au_3Zn h	Faulted Cu_3Au	Faulted cP4
Au_3Zn r	—	Orthorhombic; faulted cP4
Au_5Zn_3	—	Orthorhombic
β(Au–Zn)	CsCl	cP2
β(Ni–Zn)	CsCl	cP2
β_1(Ni–Zn)	AuCuI	tP4

The equilibrium diagram for the Ni–Zn system is given in [58 Han]. The solid solubility of Zn in Ni reaches a maximum of 39·5 at.-% Zn at 1040°C, where the fcc Ni-rich solid solution reacts with the liquid to form the β phase (48·5–58·5 at.-% Zn) which has the CsCl structure. At 875°C, L $\rightleftharpoons$ β + γ (70–85 at.-% Zn); the γ-phase has the typical γ-brass structure. A further phase $NiZn_8$ is formed from γ and liquid at 490°C. A eutectic reaction L $\rightleftharpoons$ δ + (Zn), where (Zn) represents the Zn-rich solid solution containing only very small amounts of Ni, occurs at 418·5°C and 99·7 at.-% Zn. Both the β and γ phases undergo transformations as the temperature is lowered. Between 810 and 675°C, according to composition, β $\rightleftharpoons$ β_1 which has a structure of the AuCuI type (tP4) and remains stable to room temperature. The γ phase and β_1 react together at ~500°C to form γ_1 which has a slightly distorted γ-brass structure, and coexists with γ down to low temperatures. The system thus, at temperatures below 418·5°C, contains four intermediate phases.

THE SOLID PHASES

The number of solid phases which may be involved in the ternary equilibria is considerable, but since experiments in the ternary system are confined to Zn-contents of less than approximately 50 at.-% Zn, this assessment is concerned only with the more Au-rich or Ni-rich intermediate phases in the Au–Zn and Ni–Zn systems. The phases of major importance are listed in Table 92. No ternary intermediate phases have been reported.

THE TERNARY SYSTEMS

The experiments of [76 Dir] concentrated on the composition Au_2NiZn, which might have been expected to be homogeneous (Au–Ni–Zn) solid solution at temperatures in the region of 700°C. Differential thermal analysis, however, indicates the occurrence of a transformation between 660 and 720°C. The nature of this transformation was investigated by X-ray diffraction methods using powdered samples of Au_2NiZn annealed in evacuated ampoules at various temperatures and quenched. After quenching from 800 and 380°C (temperatures respectively above and below the tranformation range) the observed crystal was in each case of the $AuCu_3$ (cP4) type. The lattice spacings of the high- and low-temperature phases were respectively a = 0·3895 and a = 0·3883 nm; the lattice spacing difference was attributed to a difference in the degree of order. Superlattice diffractions, and therefore order, persisted above the temperature of the observed transformation. After annealing at 714°C, the diffraction patterns showed the presence of two isomorphous phases, a = 0·3917 and a = 0·3883 nm. The transformation was therefore interpreted as the separation into two phases as a result of encountering a miscibility gap on traversing the temperature range 660–720°C. The presence of two phases at 710°C was confirmed micrographically. The energy of separation was measured, and compared with the results of calculations of the relevant energy based on known bond energies, for various assumed partitions of the composition Au_2NiZn. The best correspondence between theory and experiment was attained by the separation $Au_{50}Ni_{25}Zn_{25} \rightarrow \frac{1}{3}$ $Au_{24}Ni_{51}Zn_{25} + \frac{2}{3}\, Au_{63}Ni_{12}Zn_{25}$. The structures of the alloys of the two new compositions were shown to be both of the cP4 type.

A schematic phase diagram was proposed by [76 Dir], for a temperature of 700°C. Though this may be accepted as summarizing the experimental results, certain other features are doubtful. Thus the diagram indicates the presence of the Au_3Zn phase which does not exist >420°C in the binary system. Also the existence of a similar phase in the Ni–Zn system is implied, which is in conflict with the accepted binary diagram. The suggested diagram has therefore been replaced in this assessment by Fig. 273, in which the separation of the Au_2NiZn composition pro-

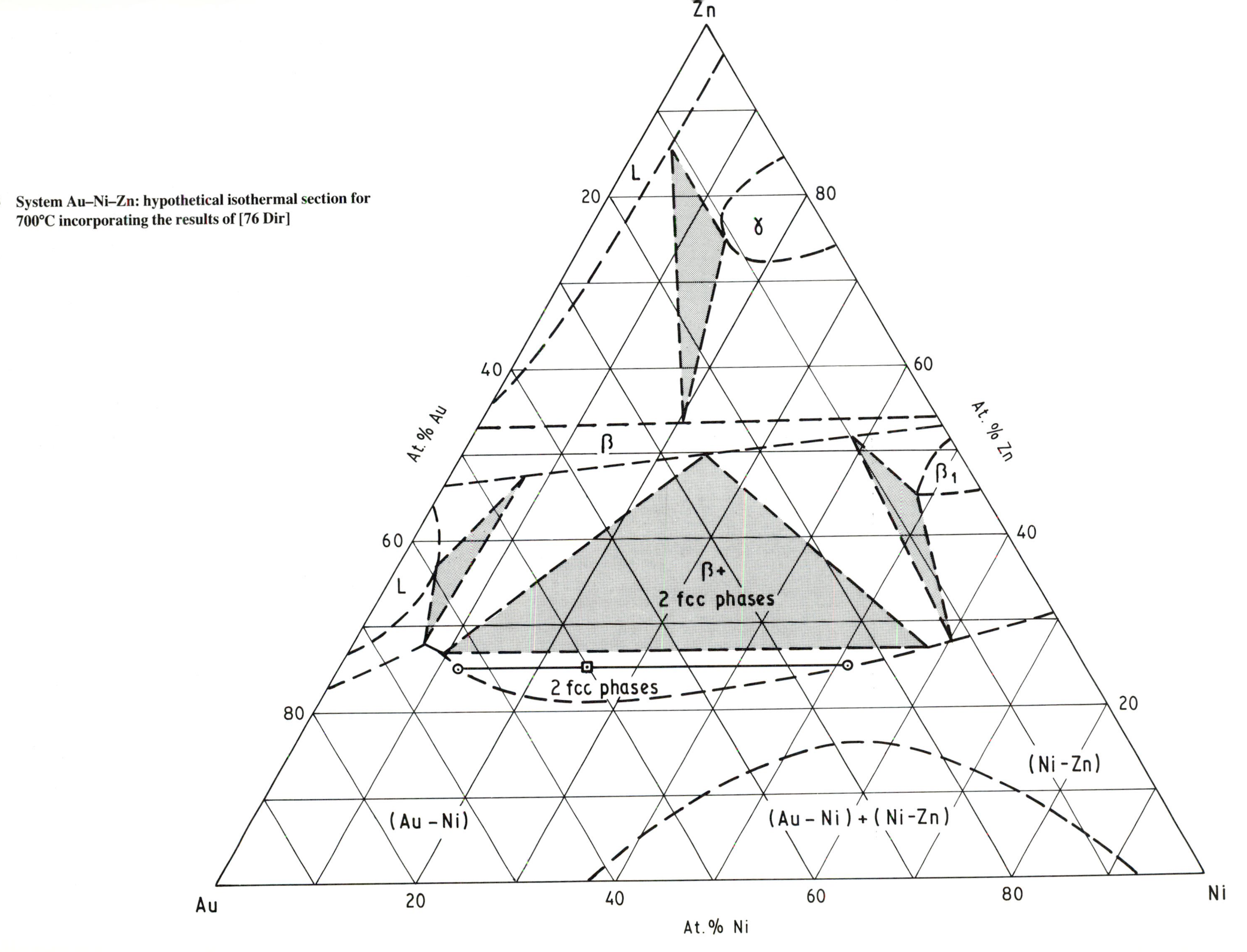

273 System Au–Ni–Zn: hypothetical isothermal section for 700°C incorporating the results of [76 Dir]

posed by [76 Dir] is included leading to a region corresponding to the coexistence of two phases. The remainder of the diagram is hypothetical but conforms to the accepted binary diagram. It is assumed that the Au–Ni miscibility gap closes at 700°C within the ternary system, and that the Au–Zn and Ni–Zb β phases (CsCl structure) form a continuous series of solid solutions.

The general nature of the equilibria for temperatures at which the Au_4Zn, Au_3Zn, and Au_5Zn_3 phases are stable is difficult to discuss in view of the paucity of information on the relative extents to which the corresponding phase fields project into the ternary model. One possible form at ~350°C (above the stability limits of Au_4Zn and Au_5Zn_3) is illustrated in Fig. 274, which is consistent with the binary diagrams and also interprets the finding in [76 Dir] that the Au_2NiZn alloy is single phase at temperatures in the region of 320°C. The structure of Au_3Zn and that reported for the low temperature phase at the Au_2NiZn composition are closely related and a continuous transition between the two is possible. It is emphasized that Fig. 274 is mainly hypothetical. Wilkens and Schubert [57 Wil] used the X-ray powder diffraction method on six ternary alloys, annealed for 8 days at 200°C, to delineate the phase region for the doubly-faulted $L1_2$-type structure based on Au_4Zn, Fig. 275.

In a very limited study of the section AuZn–NiZn [76 Kro], alloys were melted in quartz capsules in argon and shaken before cooling. Prior to X-ray diffraction experiments samples were annealed at 530°C for more than 100 h. Results showed that the alloys $Au_{0\cdot8}Ni_{0\cdot2}Zn$ and $Au_{0\cdot5}Ni_{0\cdot5}Zn$ were respectively single-phase and two-phase. This is consistent with Fig. 274, in which it is assumed that the Au–Zn β phase (CsCl structure) is in equilibrium with the Ni–Zn β_1 phase (AuCuI-type structure).

Owing to the many uncertainties over the manner in which the low-temperature phases of the Au–Zn system enter into the equilibria, speculation with regard to the nature of the low-temperaure phase diagrams, other than that shown in Fig. 274, is not justified. Similarly the interactions of the several γ-brass type phases in the ternary system is unknown but likely to be complex. It may be expected, however, that the two CsCl structures will form a continuous series of solid solutions with each other (see Fig. 273), so that the compositions of the liquid at the L = (Au) + β and L + (Ni) = β reactions will be joined by a monovariant line ab in the projection of liquidus surfaces, as shown in Fig. 276. Within the area of the projection bounded by this line and the Au and Ni corners of the projection, hypothetical liquidus contours may be sketched in as shown in Fig. 276.

CONCLUSIONS

The limited amount of available evidence suggests that the constitution of the Au–Ni–Zn alloys is complex, and that between 660 and 720°C a region of immiscibility exists, such that an alloy of composition Au_2NiZn separates into two isomorphous phases within this temperature range. Further experimental work is urgently needed.

REFERENCES

57 Wil: M. Wilkens and K. Schubert: *Z. Metall.*, 1957, **48**, 550–557

58 Han: M. Hansen and K. Anderko: 'Constitution of Binary Alloys'; 1958, New York, McGraw-Hill

65 Ell: R. P. Elliott: 'Constitution of Binary Alloys' (1st Suppl.); 1965, New York, McGraw-Hill

69 Shu: F. A. Shunk: 'Constitution of Binary Alloys' (2nd Suppl.); 1969, New York, McGraw-Hill

76 Dir: M. Dirand and J. Hertz: Journées de Calorimetre et D'Analyse Thermique (C.R.); 1976, (3–10), 1–11

76 Kro: K. Krompholz and A. Weiss: *J. Less-Common Metals*, 1976, **50**, 213

87 Oka: H. Okamoto and T. B. Massalski: 'Phase Diagrams of Binary Gold Alloys', 331–340; 1987, ASM International

Au–O–Ti

Von Philipsborn and Laves [64 Phi] found that $AuTi_3$ with the Cr_3Si-type structure (cP8) forms a two-phase structure when it contains small amounts of the interstitial elements O, N, and C. The second phase stabilized by the interstitial elements has the ordered Cu_3Au ($L1_2$)-type structure (cP4). With about 4·5 at.-% O some 40% of the Cu_3Au phase forms. At 16 at.-% O no $AuTi_3$ is stable; the alloy is 100% Cu_3Au phase.

REFERENCE

64 Phi: H. von Philipsborn and F. Laves: *Acta Crystallogr.*, 1964, **17**, 213–214

Au–O–V

Von Philipsborn and Laves [64 Phi] found that AuV_3 with the Cr_3Si-type structure (cP8) forms a two-phase structure when it contains small amounts of the interstitial elements O, N, and C. The second phase stabilized by the interstitial elements has the ordered Cu_3Au ($L1_2$)-type structure (cP4). With about 3 at.-% O the Cr_3Si-type phase is completely transformed to the Cu_3Au-type phase.

REFERENCE

64 Phi: H. von Philipsborn and F. Laves: *Acta Crystallogr.*, 1964, **17**, 213–214

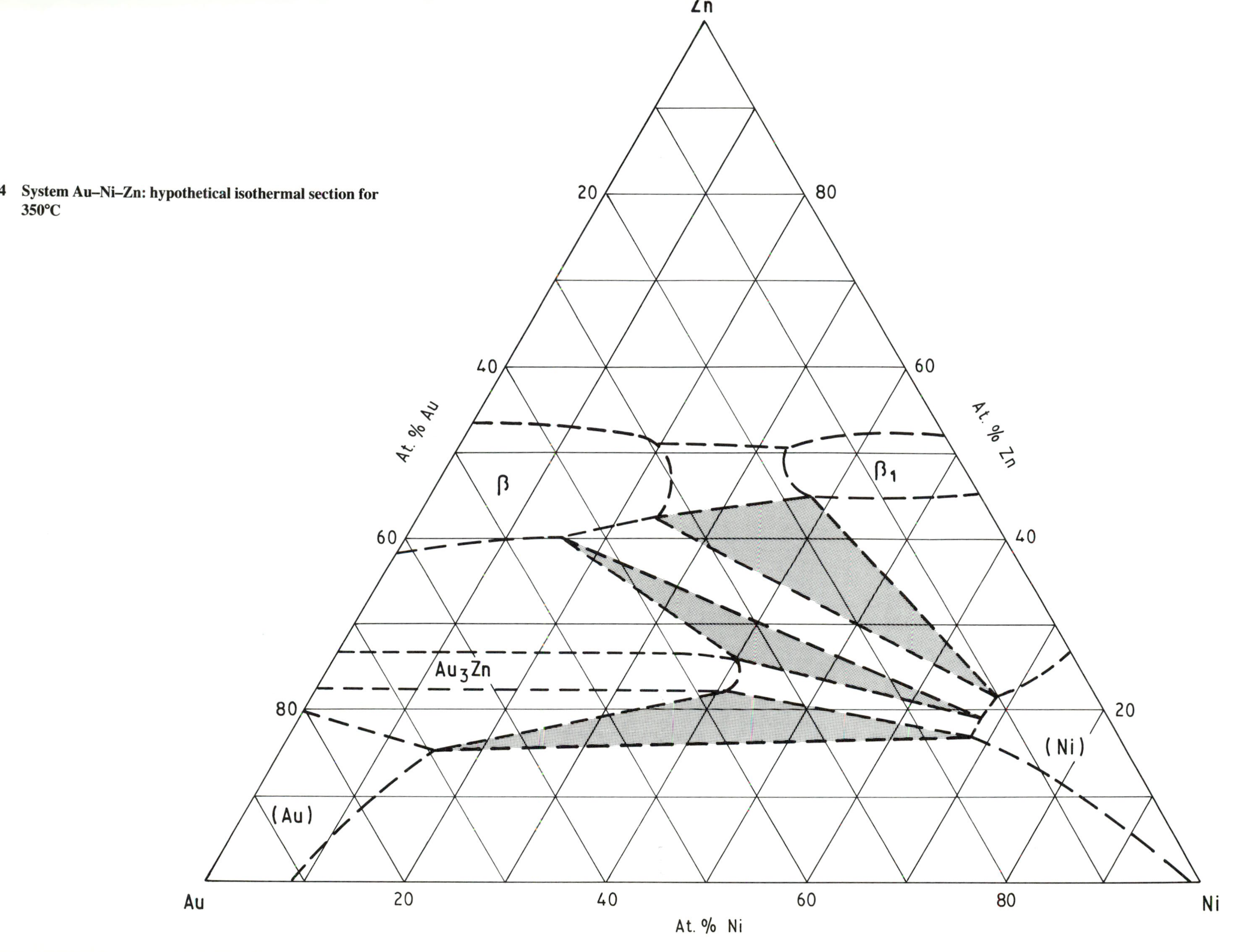

274 System Au–Ni–Zn: hypothetical isothermal section for 350°C

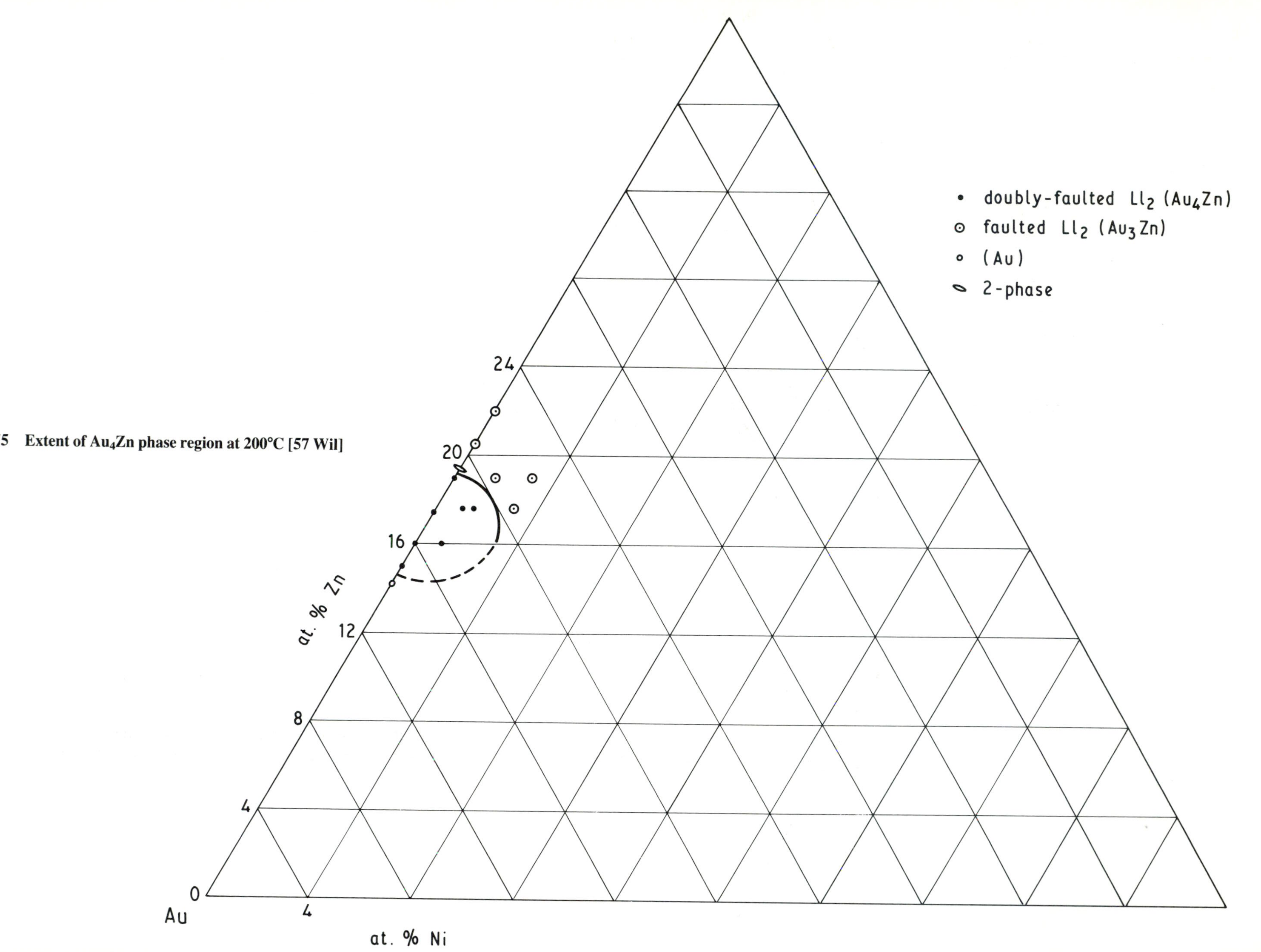

275 Extent of Au_4Zn phase region at 200°C [57 Wil]

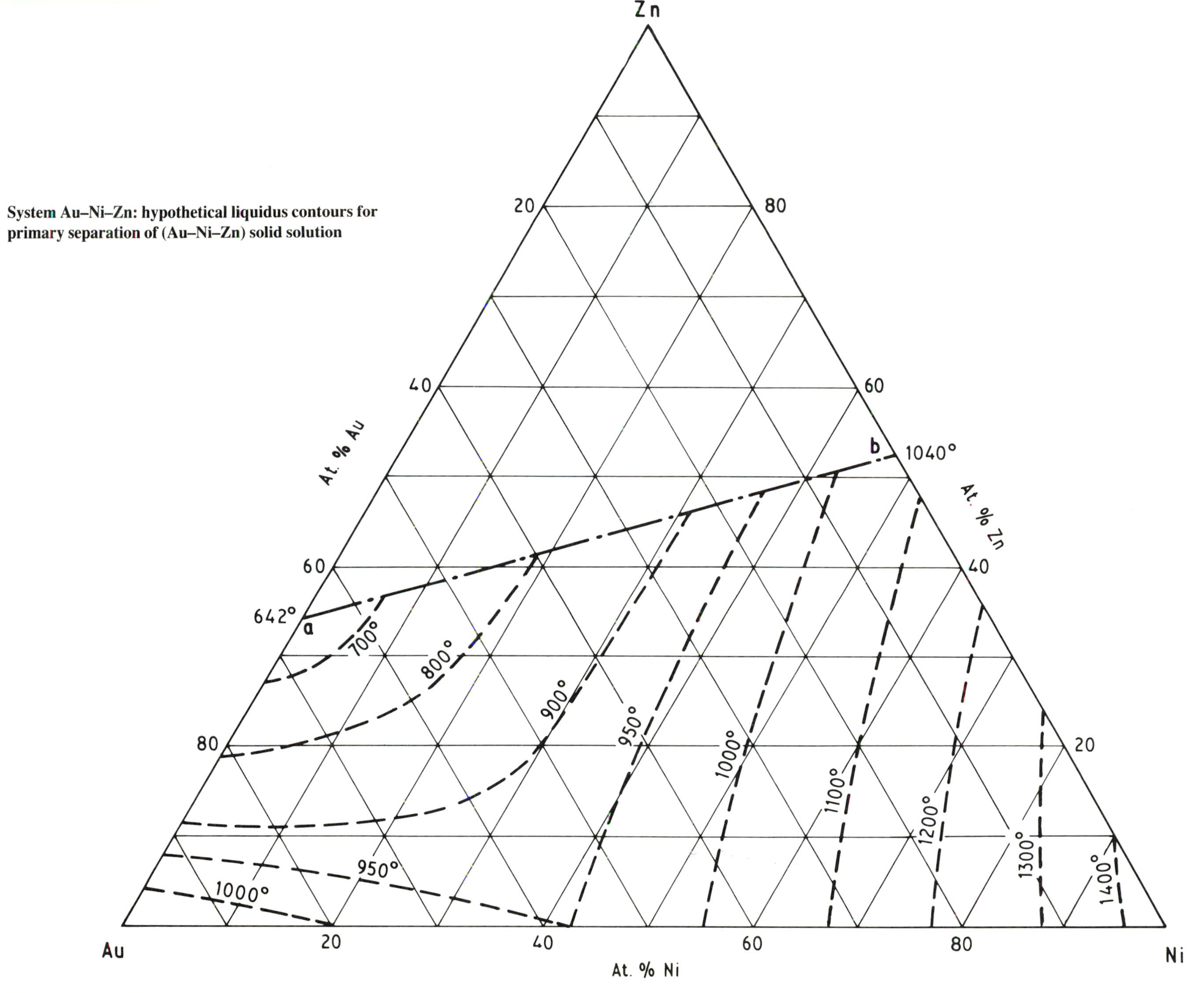

276 System Au–Ni–Zn: hypothetical liquidus contours for primary separation of (Au–Ni–Zn) solid solution

Au–P–Si

All the published data shows that doping of Si with P increases the solubility of Au in Si (Fig. 277). To convert the concentrations of Au and P from atoms cm^{-3} to atomic ppm in Fig. 277 it is necessary to divide by $4{\cdot}94 \times 10^{16}$, Wilkens *et al.* [64 Wil] and Joshi and Dash [66 Jos] were early workers to note an enhanced Au solubility in Si containing a high P concentration. Cagnina [69 Cag] used a radioactive tracer method to measure the Au solubility as a function of P doping at 1 000 and 1 100°C. O'Shaughnessy *et al.* [74 O'Sh] similarly determined the solubility of Au in Si doped with 4×10^{19} atom cm^{-3} P (810 atomic ppm P) at 1 000, 1 050, 1 100, and 1 200°C, Meek and Seidel [75 Mee] included one solubility measurement at 900 and 1 000°C for high P concentrations. Chou and Gibbons [75 Cho] gave two data points at 1 100°C again for high P concentrations and developed a model for enhanced Au solubility taking into account electronic interactions and ion pairing of Au and P atoms on adjacent substitutional sites in the Si lattice. The curves presented in Fig. 277 are based on this model. It will be noted that the data of [69 Cag] agrees with the theoretical model, [74 O'Sh] solubility data falls below [69 Cag], and the high P concentration solubility data of [75 Mee] and [75 Cho] gives considerably higher Au solubilities than predicted by the theoretical model.

REFERENCES

64 Wil: W. R. Wilcox, T. J. LaChappelle and D. H. Forbes: *J. Electrochem. Soc.*, 1964, **111**, 1377–1380

66 Jos: M. L. Joshi and S. Dash: *J. Appl. Phys.*, 1966, **37**, 2453–2457

69 Cag: S. F. Cagnina: *J. Electrochem. Soc.*, 1969, **116**, 498–502

74 O'Sh: T. A. O'Shaughnessy, H. D. Barber, D. A. Thompson and E. L. Heasell: *J. Electrochem. Soc.*, 1974, **121**, 1350–1354

75 Mee: R. L. Meek and T. E. Seidel: *J. Phys. Chem. Solids*, 1975, **36**, 731–740

75 Cho: S. L. Chou and J. F. Gibbons: *J. Appl. Phys.*, 1975, **46**, 1197–1203

Au–Pb–Pd

Cabri and Traill [66 Cab] examined a Pd_3Pb mineral by X-ray powder diffraction analysis and electron-probe microanalysis of two grains, both of which were shown to be homogeneous. The mineral contained 3 to 4 at.-% Au substituting for Pd in Pd_3Pb. X-ray analysis confirmed the Cu_3AuI-type structure, cP4, with $a = 0{\cdot}4025$ nm compared to $a = 0{\cdot}4024$ nm for a synthetic binary alloy Pd_3Pb prepared by [66 Cab]. The $(Pd,Au)_3$ Pb mineral was associated with a (Ag–Au) alloy whose composition was not determined. It would appear that there is some solubility of Au in Pd_3Pb and that a two-phase region (Ag–Au) + Pd_3Pb is formed in the quarternary system.

REFERENCE

66 Cab: L. J. Cabri and R. J. Traill: *Canad. Mineral.*, 1966, **8**, 541–550

Au–Pb–Se
(B. Gather and A. Prince)

INTRODUCTION

The only published data is a comprehensive survey of 103 alloys prepared from 99·999% pure elements [77 Rou] and a determination of the solubility of Au in p-type PbSe single crystals of very high purity (65 Kha]. Vertical sections were examined by [77 Rou] using differential thermal analysis with supporting X-ray diffraction and metallographic assessment. The ternary system is divided into two partial systems, Au–PbSe–Se and Au–PbSe–Pb, by an invariant reaction at 955°C, $L_1 + L_2 \rightleftharpoons$ (Au) + PbSe. In effect the tie line Au–PbSe is two-phase below 955°C, although it is not a pseudobinary section above 955°C. The Au–PbSe–Se partial system exhibits three invariant reactions. Monovariant monotectic reactions, originating from the binary monotectic reactions in the Au–Se system ($l_3 \rightleftharpoons l_4$ + (Au)) and the Pb–Se system ($l_5 \rightleftharpoons l_6$ + PbSe), meet the L_2 + (Au) + PbSe monovariant equilibrium originating from the 955°C invariant reaction plane to produce the reaction $L_3 \rightleftharpoons L_4$ + (Au) + PbSe at 530°C. The composition of L_4 is very close to the Se corner and evidence is provided [77 Rou] that the transition reaction, L_{U_2} + (Au) $\rightleftharpoons$ AuSe + PbSe ocurs at about 395°C. Solidification ends in the Au–PbSe–Se partial system with the ternary eutectic reaction, $L_{E_1} \rightleftharpoons$ AuSe + PbSe + (Se) at 217°C.

The partial system Au–PbSe–Pb includes four invariant reactions, all based on the binary reactions occurring in the Au–Pb system. The ternary reactions were all stated to take place at temperatures identical with the corresponding binary reactions [77 Rou], implying that the compositions of liquids U_1, U_3, U_4, and E_2 (Fig. 278) are very close to the Au–Pb binary edge. In the reaction scheme (Fig. 279) for the ternary system the data of [83 Eva] are accepted for the Au–Pb binary invariant reaction temperatures. It is likely that the ternary reaction temperatures quoted by [77 Rou] for reactions U_1, U_3, U_4, and E_2 should be increased to near the Au–Pb binary reaction temperatures found by [83 Eva]. The reactions in the Au–PbSe–Pb partial system are L_{U_1} + (Au) $\rightleftharpoons Au_2Pb$ + PbSe at 429°C, $L_{U_3} + Au_2Pb \rightleftharpoons AuPb_2$ + PbSe at 250°C, $L_{U_4} + AuPb_2 \rightleftharpoons AuPb_3$ + PbSe at 217°C, and $L_{E_2} \rightleftharpoons$ (Pb) + $AuPb_3$ + PbSe at 210°C (temperatures from [77 Rou]).

A large proportion of the Au–Pb–Se ternary system is covered at high temperatures by liquid miscibility gaps,

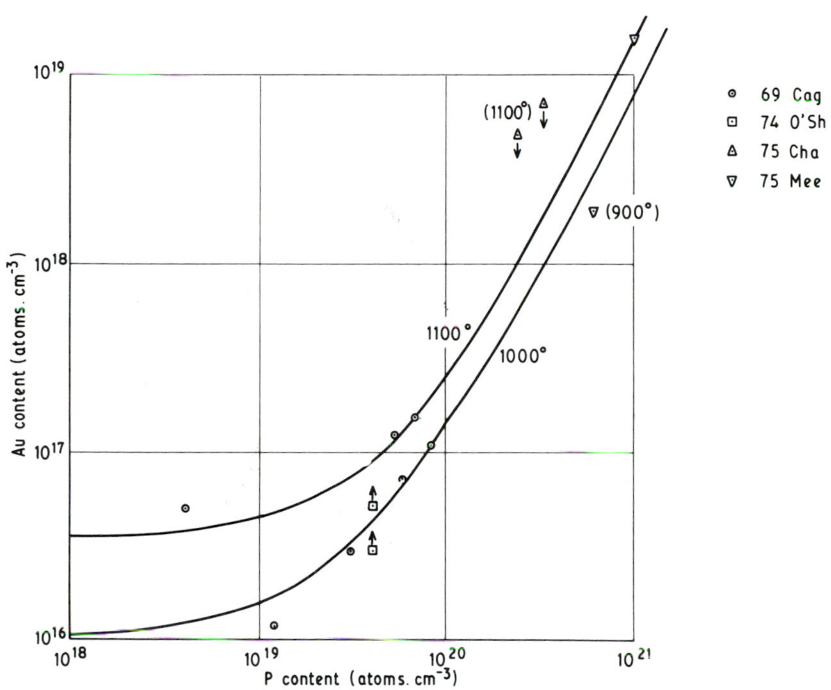

277 Effect of P on the solubility of Au in Si

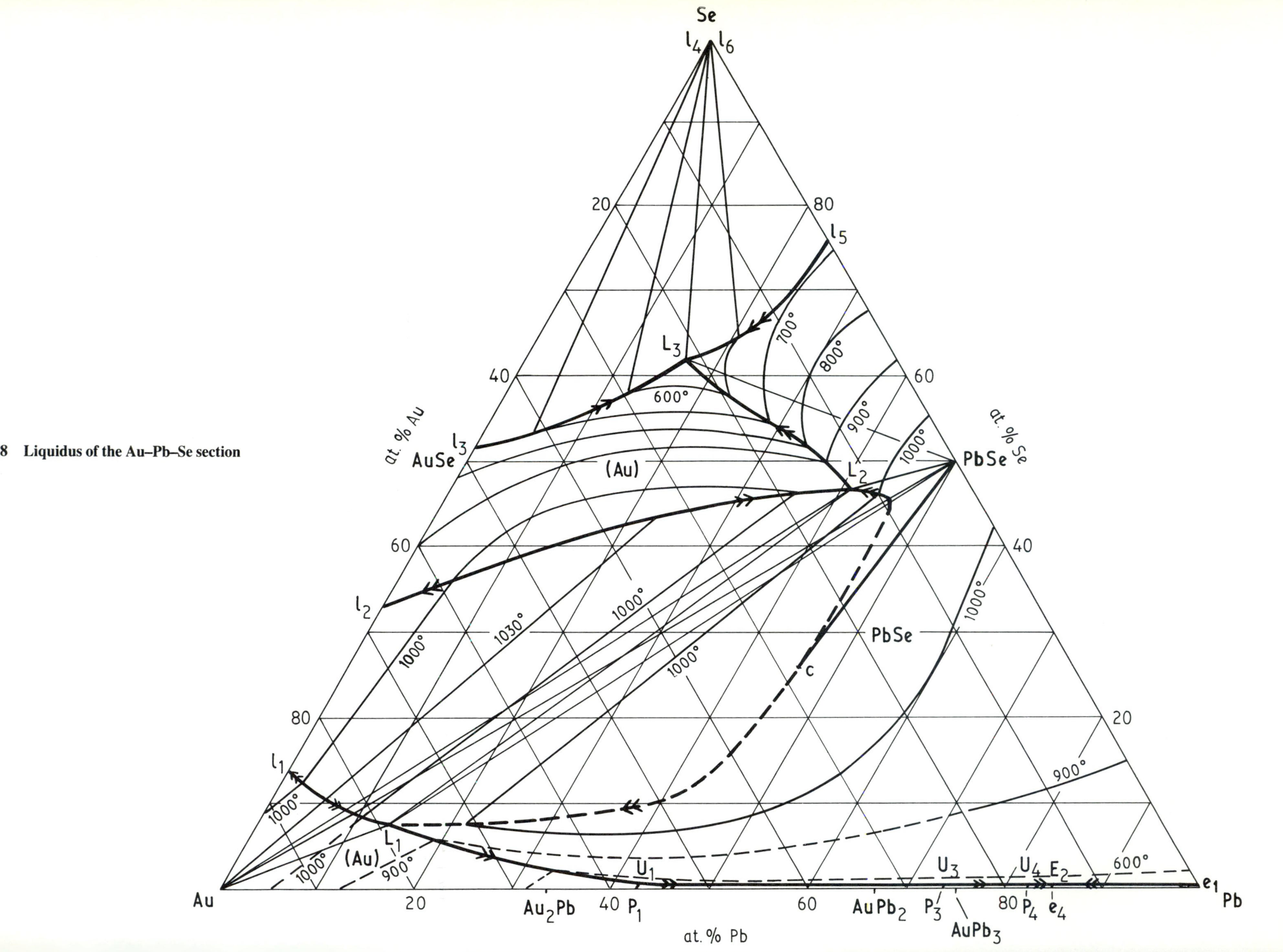

278 Liquidus of the Au–Pb–Se section

279 Reaction scheme for the Au–Pb–Se system

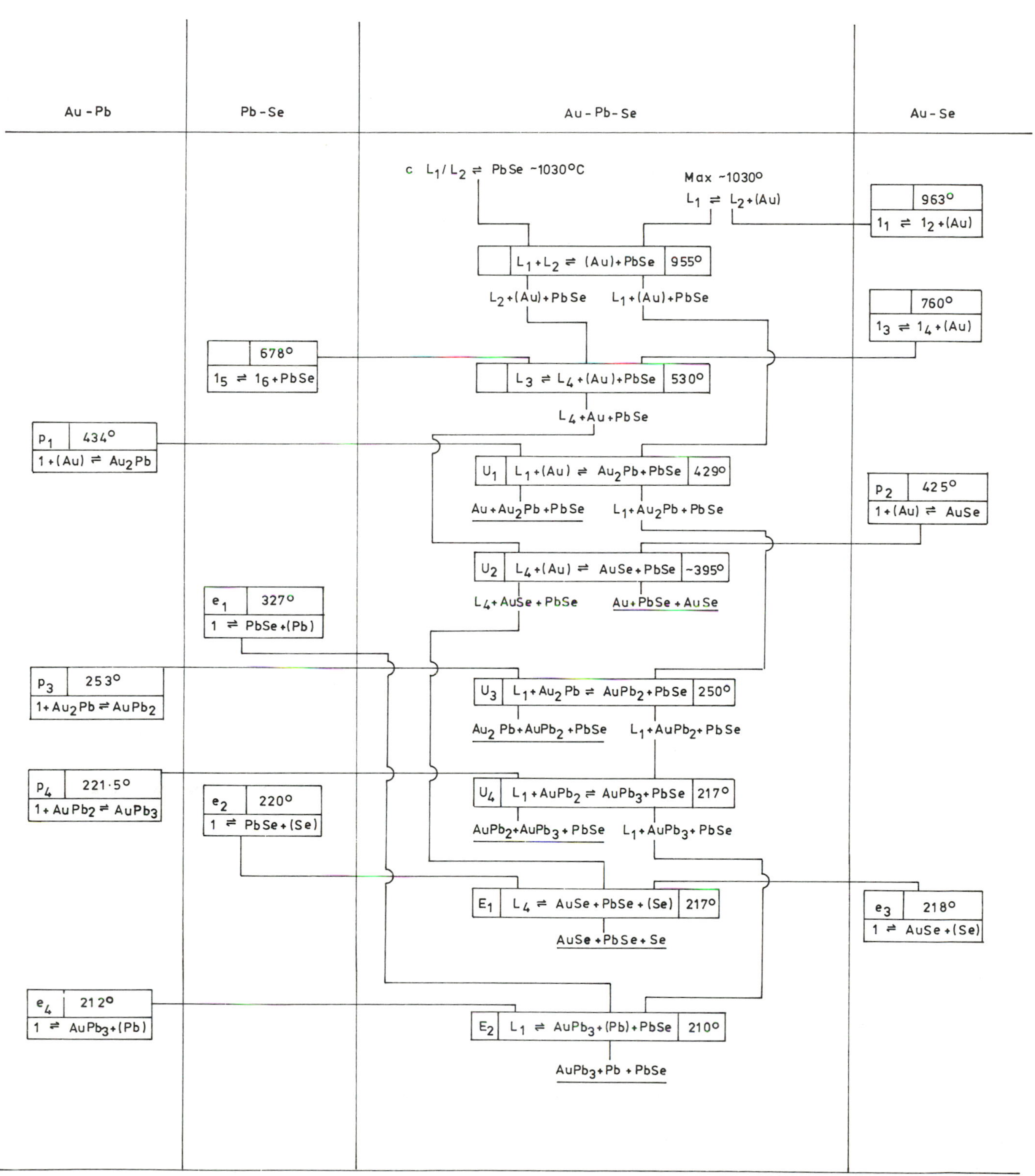

one originating from the 963°C Au–Se monotectic reaction and the second spanning from the 760°C Au–Se monotectic reaction to the 678°C Pb–Se monotectic reaction. The dominant region of primary crystallization is centred on the congruent compound PbSe. The monovariant equilibrium of L_1 + L_2 + PbSe closes at a critical temperature near to 1 030°C with the merging of the compositions L_1 and L_2 to give a critical tie line L_1/L_2–PbSe. The vertical sections presented by [77 Rou] imply the existence of a second critical tie line, also close to 1 030°C, where L_1 + L_2 + (Au) are in equilibrium. From this temperature maximum the monovariant curves representing equilibrium between L_1 + L_2 + (Au) descend to the 963°C monotectic reaction on the Au–Se binary system and to the 955°C invariant reaction, $L_1 + L_2 \rightleftharpoons$ (Au) + PbSe. [77 Rou] do not comment on the presence of this second critical tie line. Amendments have been made to the data presented by [77 Rou] to produce a series of self-consistent sections. The major difference between this assessment and [77 Rou] is the relocation of the composition L_1 at a higher Se content.

BINARY SYSTEM

The Au–Pb diagram determined by [83 Eva] is accepted. The compounds Au_2Pb, $AuPb_2$, and $AuPb_3$ all form peritectically at temperatures of 434, 253, and 221·5°C respectively. The compositions of the liquid phase at the peritectic horizontals are 42·6, 73·8, and 82·2% Pb respectively. $AuPb_3$ and (Pb) form a eutectic at 212°C and 84·8% Pb. The Au–Se system evaluated by [86 Oka] and the Pb–Se system evaluated by [86 Lin] are accepted.

SOLID PHASES

Table 93 summarizes crystal structure data for the binary phases. No ternary compounds were found [77 Rou].

Table 93. Crystal structures

Solid phase	*Prototype*	*Lattice designation*
(Au)	Cu	cF4
(Pb)	Cu	cF4
(Se)	γSe	hP3
Au_2Pb	Cu_2Mg	cF24
$AuPb_2$	$CuAl_2$	tI12
$AuPb_3$	α–V_3S	tI32
PbSe	NaCl	cF8
AuSe	—	mC24

PSEUDOBINARY SYSTEMS

No true pseudobinary system exists in the Au–Pb–Se system. As noted in the introduction the section Au–PbSe (Fig. 280) is pseudobinary below 955°C but not at temperatures above 955°C. The two-phase nature of the Au–PbSe section below 955°C was verified by heat treating an alloy containing 46·67 Au, 26·67 Pb, 26·67 Se for 1 week at 400°C. X-ray examination revealed (Au) and PbSe only. [65 Kha] determined the solid solubility of Au in PbSe as 0·01% at 750°C and 0·001 4% at 650°C.

INVARIANT EQUILIBRIA

The major invariant reaction is $L_1 + L_2 \rightleftharpoons$ (Au) + PbSe at 955°C. The direction of the L_1–L_2 tie line is determined from the vertical sections at 10%, 26·67%, 43·33% Se, 13·33% Au and the Au–PbSe section (Figs 281–284 and 280 respectively). Amendments have been made to the sections [77 Rou] at 10% Se (Fig. 281), 13·33% Au (Fig. 284) and Au–PbSe (Fig. 280) to give consistent results in terms of the compositions at which the L_1 + L_2 + (Au) phase region meets the L_1 + L_2 + PbSe phase region along the line L_1L_2. The compositions of L_1 and L_2 are obtained by noting the point of intersection of the tie lines L_1–L_2 and L_1–PbSe (for L_1) and L_1–L_2 with Au–L_2 (for L_2). The original data [77 Rou] does not allow a precise calculation of the intersection points but an estimate of L_1 at 78·5 Au, 14 Pb, 7·5% Se is considered valid. This places L_1 at a much higher Se content than suggested by [77 Rou]. The primary Au region meets the L_1 + L_2 phase region on the Au–PbSe section at 82% Au, in contrast to 84% Au plotted by [77 Rou]. Intersection at 82% Au lends support to the higher Se content suggested for the L_1 composition. The composition of L_2 is also subject to uncertainty. An estimate of 12·3 Au, 41 Pb, 46·7% is considered valid, in reasonable agreement with the composition suggested by [77 Rou], table 94. The section Au–PbSe (Fig. 280) intersects the tie line L_1–L_2 at 43·5% Au, rather than the 37% Au given [77 Rou]. The critical tie line between L_1/L_2 and PbSe is located at approximately 26% Se on the curve L_1cL_2 (Fig. 278). This composition is indicated by the skewing of the tie lines from L_1–L_2 at 955°C to the point c at which compositions L_1 and L_2 coalesce. The critical tie line c–PbSe is estimated to be situated at 1 030°C, some 50°C below the congruent MP of PbSe. The sections at 10, 26·67, and 43·33% Se, Figs 281–3, show a maximum in the L_1 + L_2 + (Au) phase region. The maximum occurs at about 1 030°C. It is not discussed by [77 Rou], who show an incorrect construction for the L_1 + L_2 + (Au) phase regions. The presence of a maximum implies that the three-phase equilibrium between L_1 + L_2 + (Au), originating at the monotectic horizontal (Au)–1_1–1_2 at 963°C reduces to a tie line connecting the three phases at the temperature maximum and then is converted into a tie-triangle which descends to meet the 955°C ternary invariant reaction as the tie-triangle L_1–L_2–(Au). The compositions of the two liquid phases on the tie line maximum are estimated to be 33.5 Au, 23 Pb, 43·5 Se, and 84 Au, 5·5 Pb, 10·5 Se.

The partial system Au–PbSe–Pb contains four invariant ternary reactions. All occur at temperatures close to those quoted by [77 Rou] for the corresponding Au–Pb binary reactions and the compositions of the invariant liquid compositions U_1, U_3, U_4, and E_2 will lie close to the Au–Pb binary edge. In Table 94 the liquid compositions are quoted as binary compositions with 0% Se. This is intended to express a very low, but undetermined Se content in the liquid phases. The reactions are specified as L_{U_1} + (Au) $\rightleftharpoons Au_2Pb$ + PbSe (429°C), $L_{U_3} + Au_2Pb \rightleftharpoons AuPb_2$ + PbSe (250°C), $L_{U_4} + AuPb_2 \rightleftharpoons AuPb_3$ + PbSe (217°C) and $L_{E_2} \rightleftharpoons$ (Pb) + $AuPb_3$ + PbSe (210°C).

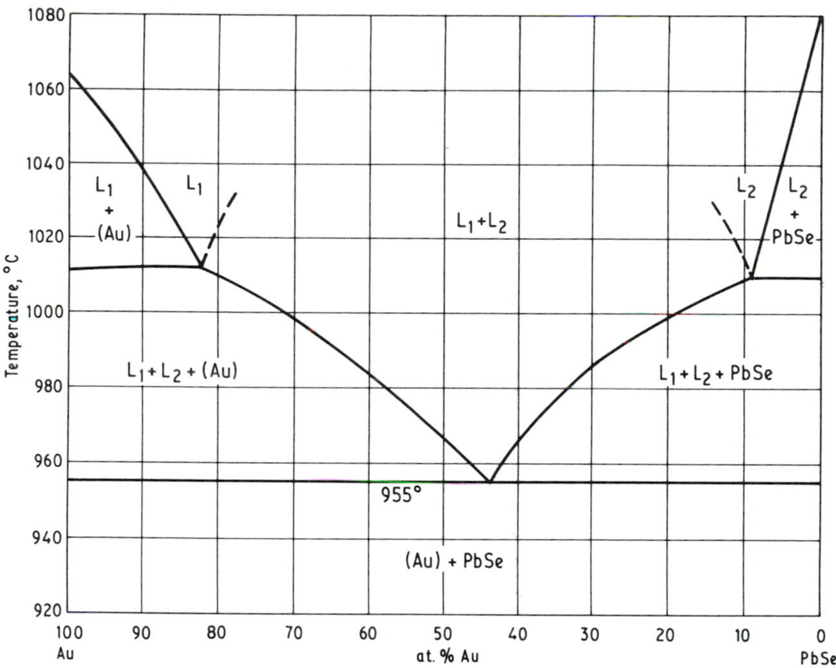

280 The Au–PbSe section

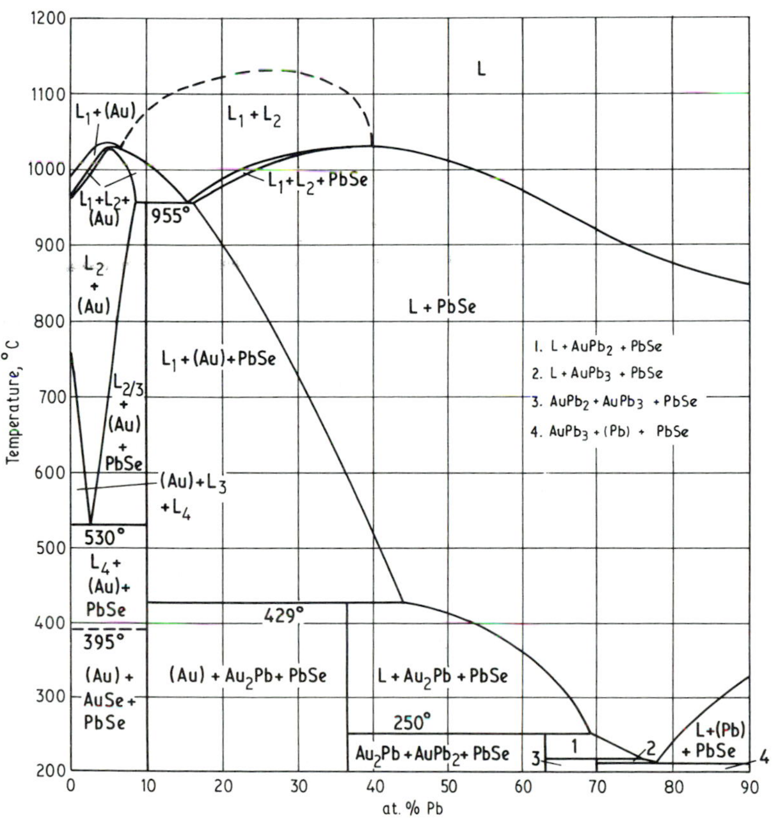

281 The 10% Se section

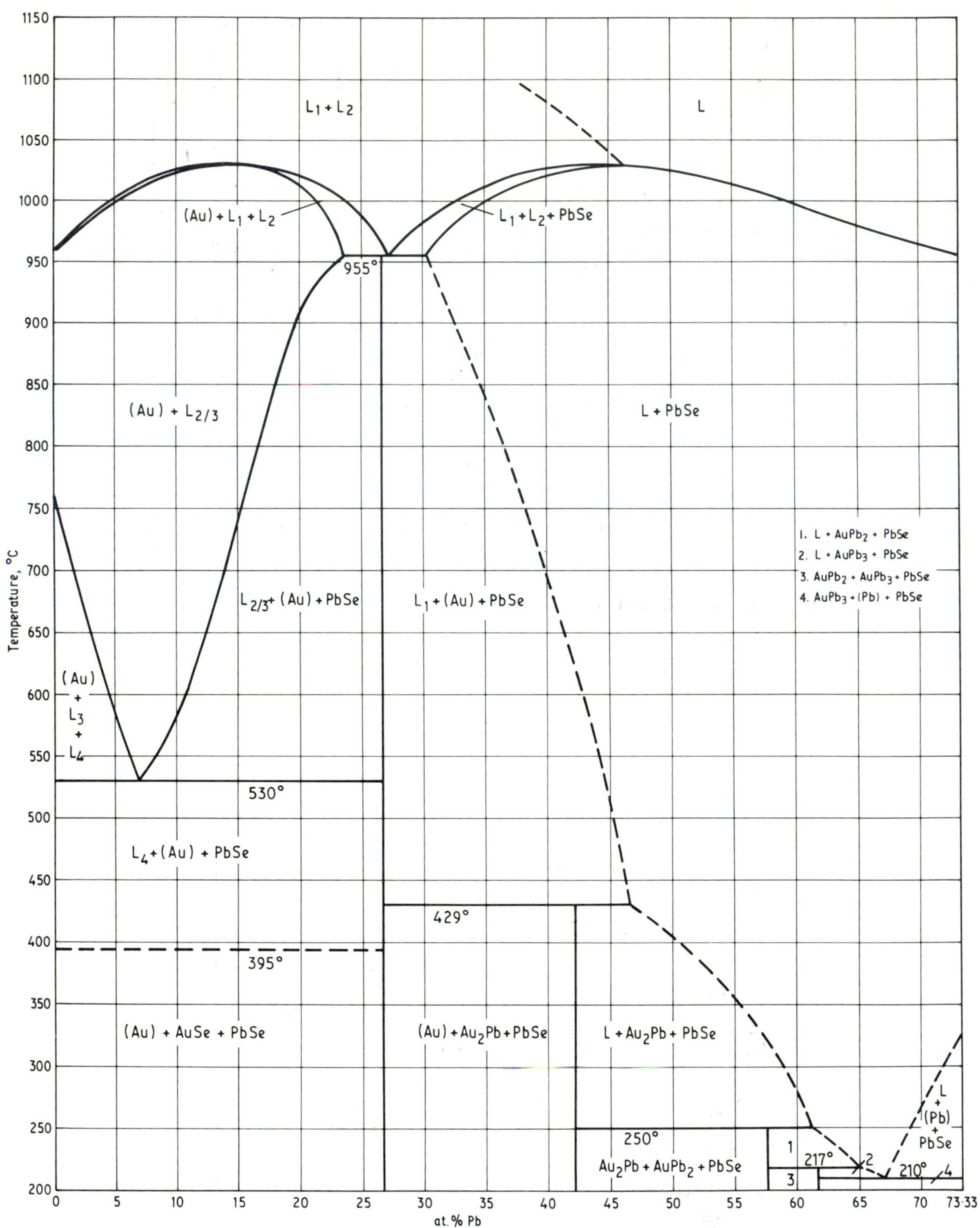

L1 + L2
L
(Au) + L1 + L2
L1 + L2 + PbSe
955°
(Au) + L2/3
L + PbSe
1. L + AuPb2 + PbSe
2. L + AuPb3 + PbSe
3. AuPb2 + AuPb3 + PbSe
4. AuPb3 + (Pb) + PbSe
L2/3 + (Au) + PbSe
L1 + (Au) + PbSe
(Au) + L3 + L4
530°
L4 + (Au) + PbSe
429°
395°
(Au) + AuSe + PbSe
(Au) + Au2Pb + PbSe
L + Au2Pb + PbSe
L + (Pb) + PbSe
250°
217°
210°
Au2Pb + AuPb2 + PbSe
Temperature, °C
at.% Pb

282 The 26·67% Se section

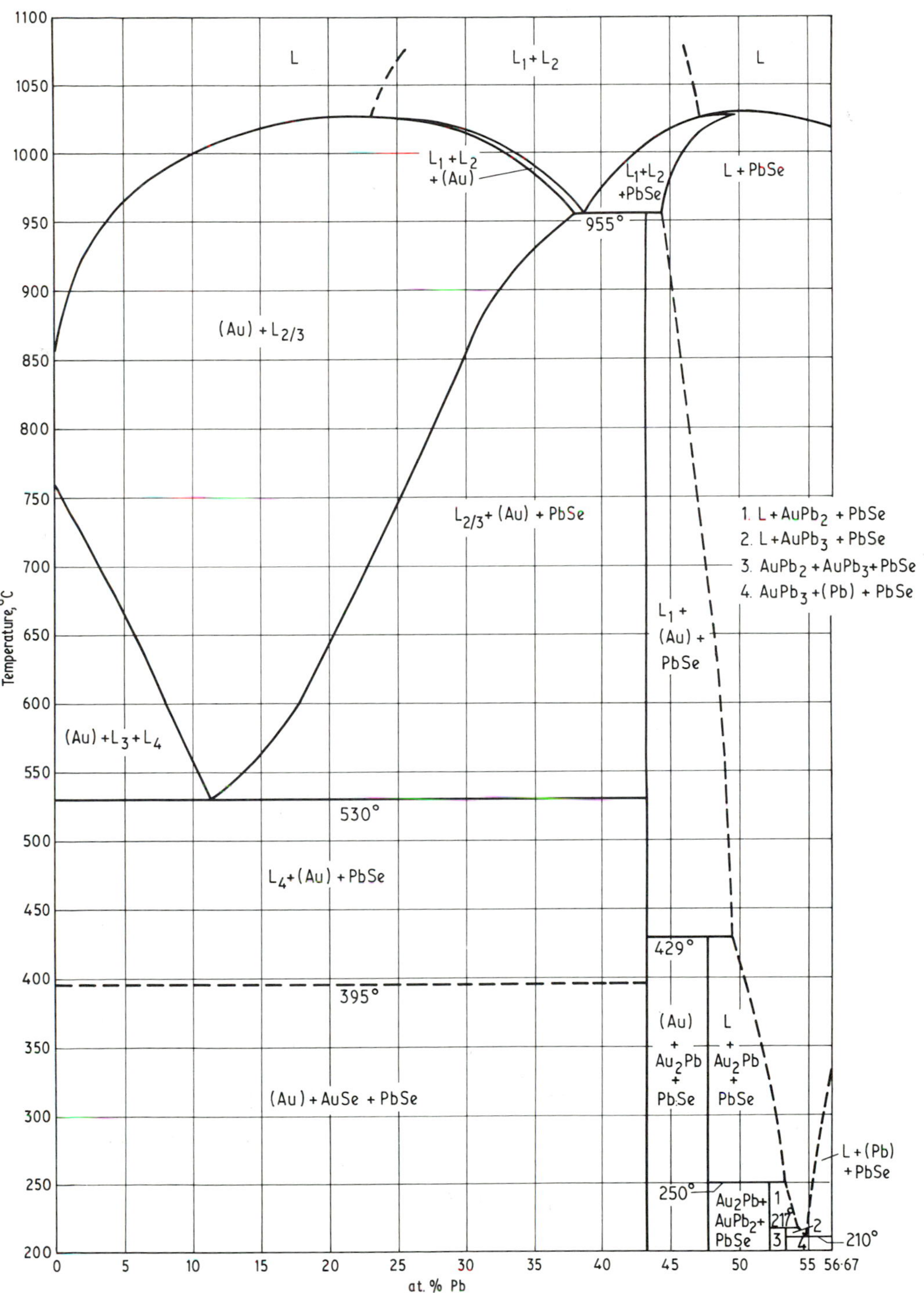
L
L1+L2
L
L1+L2
+(Au)
L1+L2
+PbSe
L+PbSe
955°
(Au)+L2/3
L2/3+(Au)+PbSe
1. L+AuPb2 +PbSe
2. L+AuPb3 +PbSe
3. AuPb2+AuPb3+PbSe
4. AuPb3+(Pb)+PbSe
L1+
(Au)+
PbSe
(Au)+L3+L4
530°
L4+(Au)+PbSe
429°
395°
(Au)
+
Au2Pb
+
PbSe
L
+
Au2Pb
+
PbSe
(Au)+AuSe+PbSe
L+(Pb)
+PbSe
250°
Au2Pb+
AuPb2+
PbSe
210°
Temperature, °C
at. % Pb
56·67

283 **The 43·33% Se section**

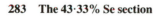

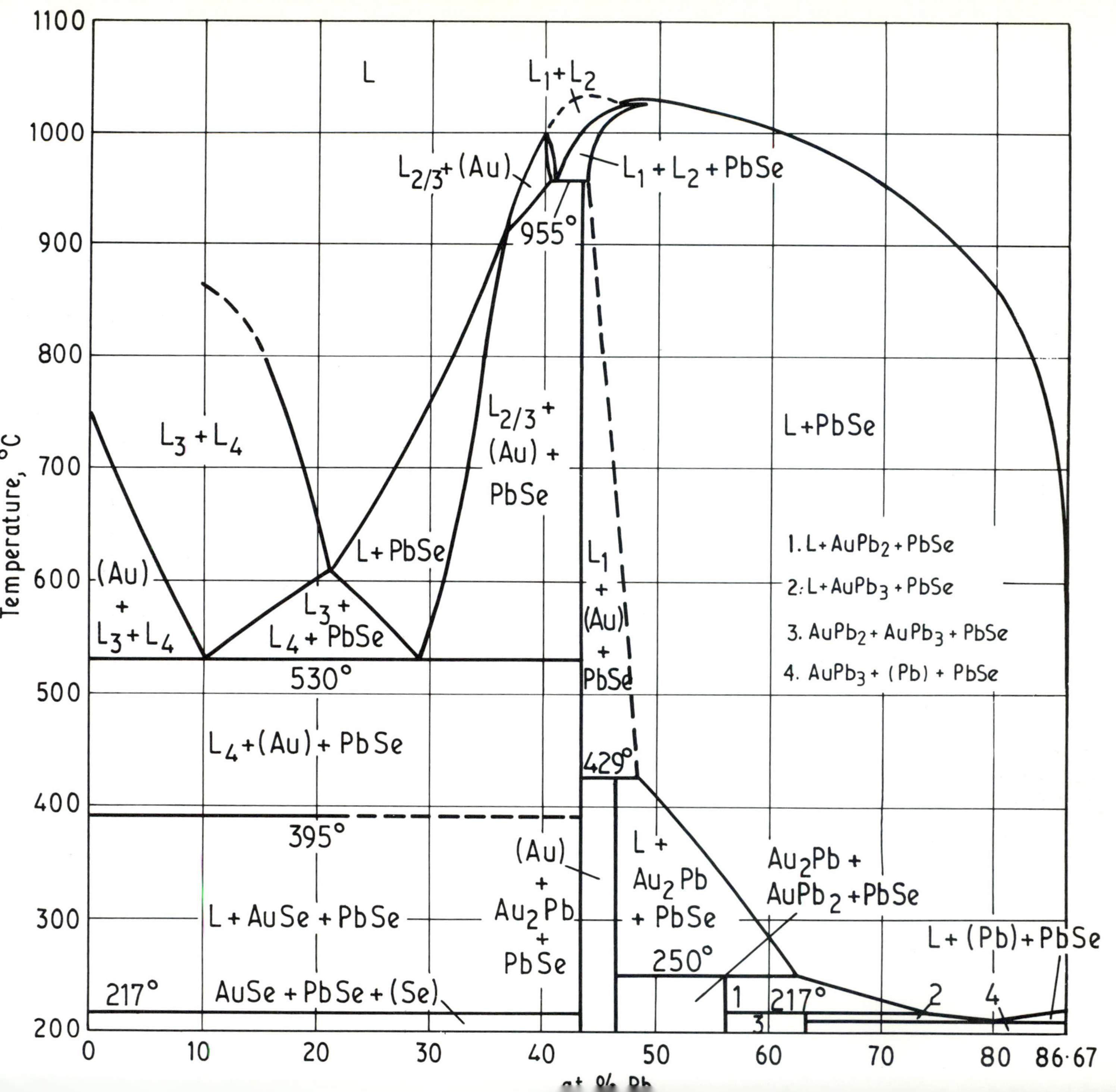

284 The 13·33% Au section

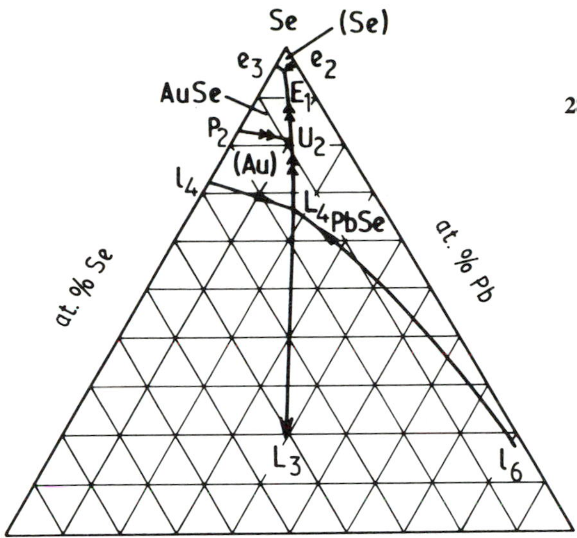

285 Schematic representation of equilibria in the Se-rich region

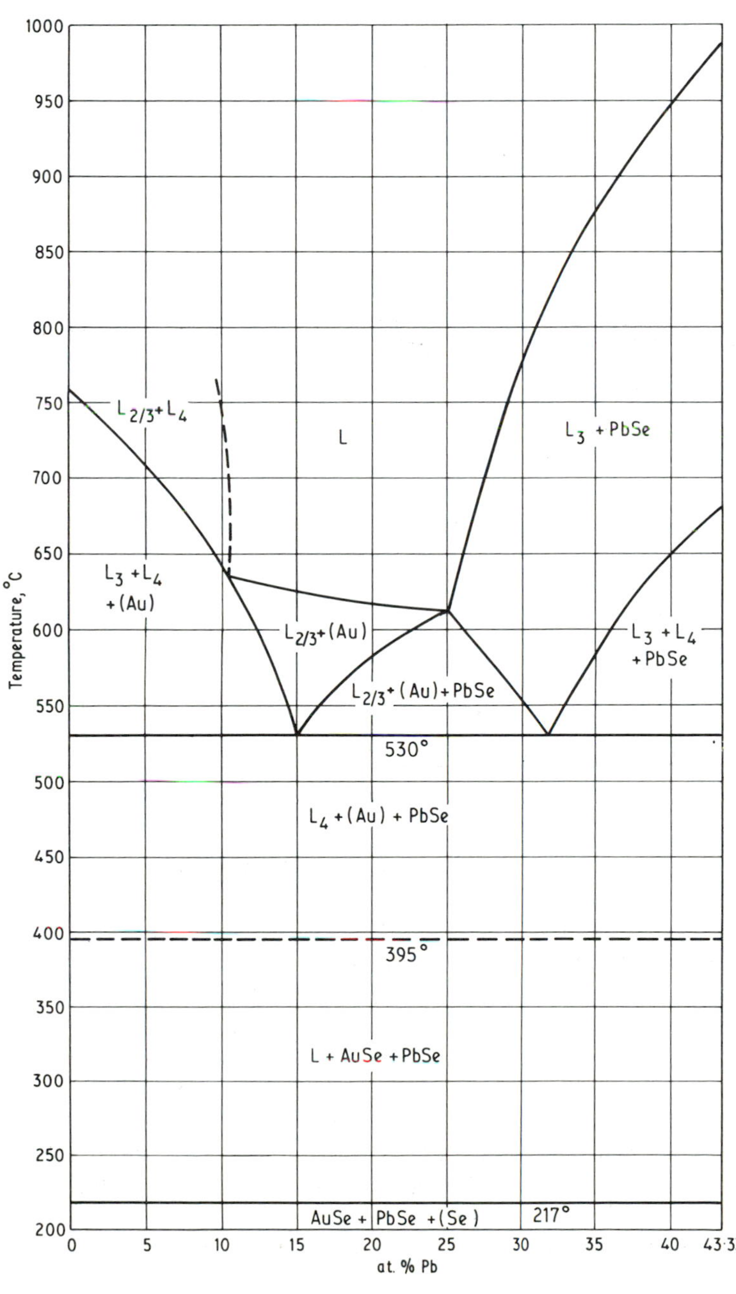

286 The 56·67% Se section

Table 94. Invariant equilibria

Reaction	*Temperature*, °C	*Phase*	*Composition*, at.-%		
			Au	*Pb*	*Se*
E_1	217	L			>99·9
E_2	210	L	15	85	0
U_1	429	L	57	42·5	0·5
U_2	395	L			>99·9
U_3	250	L	26	74	0
U_4	217	L	17·5	82·5	0
L_1/L_2	955	L_1	81	15·33	3·67
			78·5 ± 2*	14 ± 2*	7·5 ± 2*
	955	L_2	11	41	48
			12·3 ± 2*	41 ± 2*	46·7 ± 2*
L_3L_4	530	L_3	21·33	16·33	62·33
	530	L_4			>99·0

* Suggested amendments to the compositions: see the discussion of invariant equilibria

The partial ternary system Au–PbSe–Se contains three invariant reactions. Tie-triangles descend from the monotectic reactions $l_3 \rightleftharpoons l_4$ + (Au) in the Au–Se system and $l_5 \rightleftharpoons l_6$ + PbSe in the Pb–Se system to meet on the 530°C ternary reaction plane along the tie line L_3–L_4. A third tie-triangle descends from the 955°C reaction plane and represents the equilibrium of L_2 + (Au) + PbSe. At 530°C the L_2 and L_3 phases are the same phase; they are denoted by $L_{2/3}$ in the vertical sections. The reaction at 530°C is $L_3 \rightleftharpoons L_4$ + (Au) + PbSe. The composition of L_3 given by [77 Rou] (Table 94) is accepted. L_4 lies very close to the Se corner and Fig. 285 is a schematic representation of the ternary equilibria in the Se-rich corner. The direction of the tie line L_3–L_4 and the composition of L_3 can be deduced from the sections at 13·33% Au, 56·67, 70, and 80% Se (Figs 284, 286, 287, 288). The surface of secondary separation of L_3 + L_4 + (Au), represented by the boundary between the L_3 + L_4 and L_3 + L_4 + (Au) phase regions, has been amended in the 70 and 80% Se sections (Figs 287 and 288) to show the separation of Au at higher temperatures than drawn by [77 Rou]. The data in Figs 286, 287, and 288 are now self-consistent. From L_4 the monovariant equilibrium L_4 + (Au) + PbSe descend to the transition reaction at U_2: L_{U_2} + (Au) $\rightleftharpoons$ AuSe + PbSe (Fig. 285). AuSe forms slowly, and the ternary invariant reaction at U_2 was only detected in alloys containing less than 10% Pb in the Au–PbSe–Se partial system. Thermal effects associated with the reaction at U_2 were very weak and the invariant reaction temperature of 395°C [77 Rou] must not be considered accurate to less than ±10°C. Solidification ends at a degenerate ternary eutectic with the reaction $L_{E_1} \rightleftharpoons$ (Se) + AuSe + PbSe at a temperature of 317°C.

The composition of the liquid phase associated with the ternary invariant reactions in the Au–Pb–Se system is summarized in Table 94. Suggested amendments to the original data [77 Rou] are noted.

LIQUIDUS SURFACE

The liquidus surface (Fig. 278) differs from [77 Rou] with respect to the location of the invariant points L_1 and L_2. Isotherms have been included in Fig. 278 from 600 to 1 000°C at 100° intervals. Tie lines between L_1 and L_2 and between L_3 and L_4 have also been incorporated in Fig. 278.

ISOTHERMAL SECTIONS

The vertical sections have been amended so as to be self-consistent. Isothermal sections have been generated from the vertical sections for temperatures of 1 000, 800, and 600°C (Figs 289, 290, and 291 respectively).

REFERENCES

65 Kha: F. F. Kharakhorin, D. A. Gambarova and V. V. Aksenov: *Izvest. Akad. Nauk. SSSR, Neorg. Materialy.*, 1965, **1**, 1502–1505

77 Rou: J. C. Rouland, B. Legendre and C. Souleau: *J. Chem. Res. (S)*, 1977, (10) 2745–2784

83 Eva: D. S. Evans and A. Prince: *Mat. Res. Soc. Symp. Proc.*, 1983, **19**, 383–388

86 Oka: H. Okamoto and T. B. Massalski: 'Binary Alloy Phase Diagrams' (ed. T. B. Massalski), Vol. 1, 309–312; 1986, ASM International

86 Lin: J. C. Lin, R. C. Sharma and Y. A. Chang: 'Binary Alloy Phase Diagrams' (ed. T. B. Massalski), Vol. 2, 1842–1845; 1986, ASM International

Au–Pb–Si

The only published work on the Au–Pb–Si ternary phase diagram is due to [88 Leg]. High-purity elements, 99·99% Au and 99·9999% Pb, Si, were melted in evacuated silica ampoules, quenched in water and then heat treated for 3 to 7 days at a temperature 'chosen as a function of alloy composition'. A total of 75 alloy compositions were studied on five sections at 50 at.-% Au, 20 at.-% Pb, 10 at.-% Si, Au_2Pb–80Au,20Si and Au_2Pb–66·67Au, 33·33Si. Reliance was placed on the use of DTA with heating rates of 2 or 5°C min^{-1} and differential scanning calorimetry with heating rates of 0·1 or 2°C min^{-1}. The

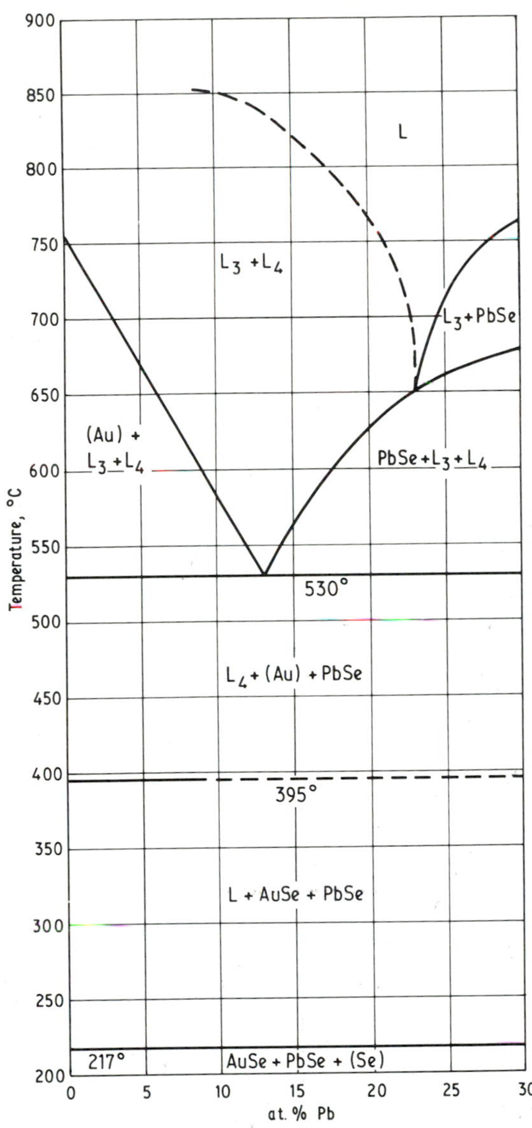

287 The 70% Se section

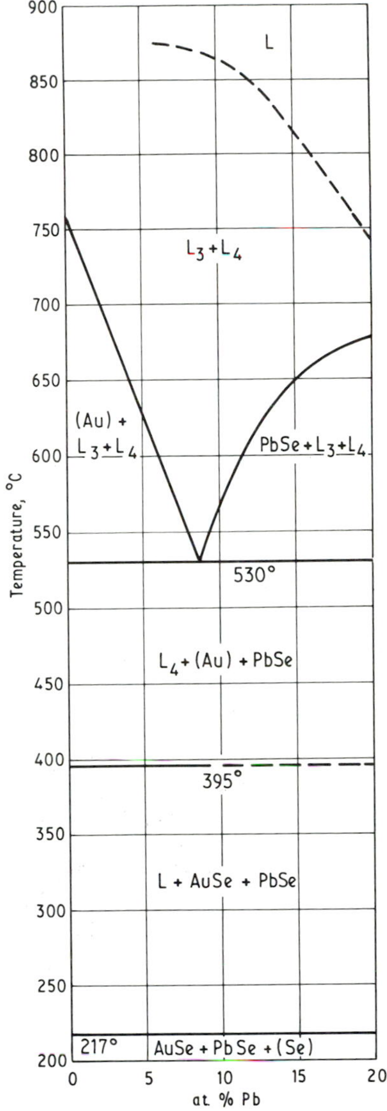

288 The 80% Se section

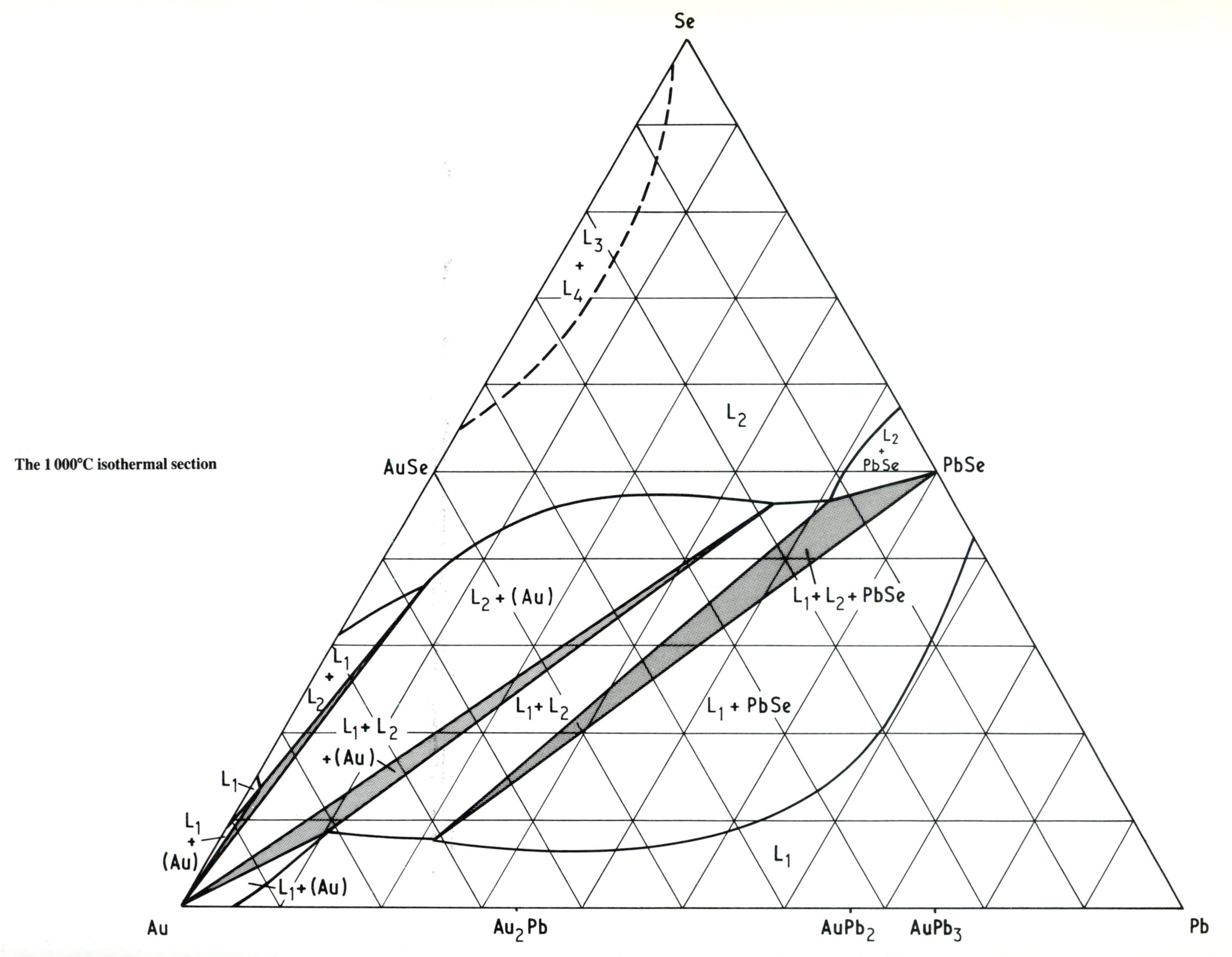

289 The 1 000°C isothermal section

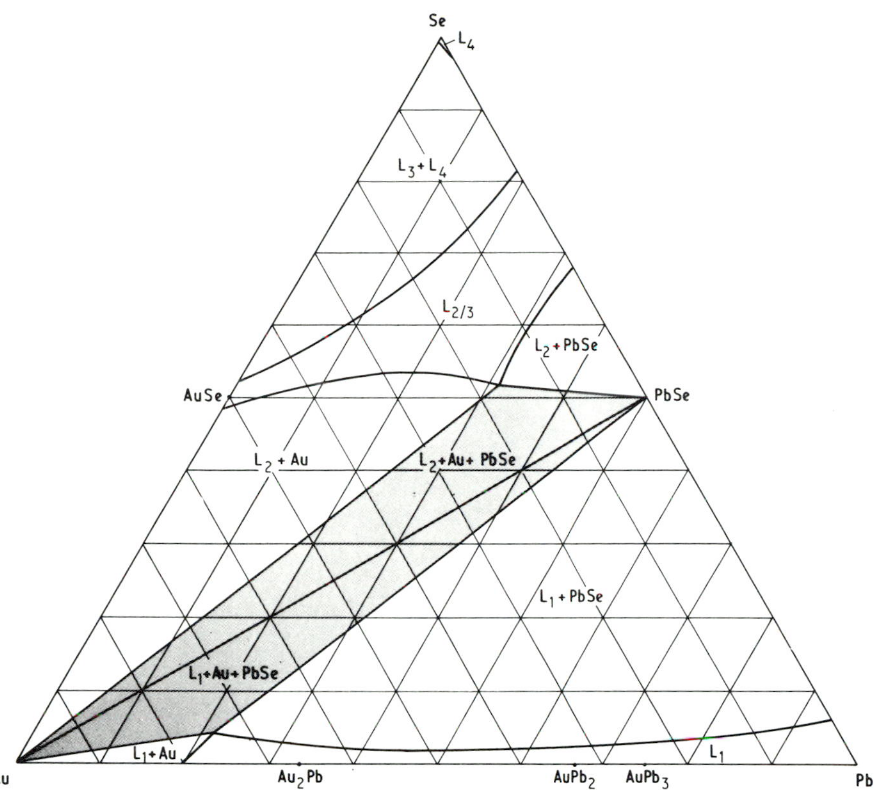

290 The 800°C isothermal section

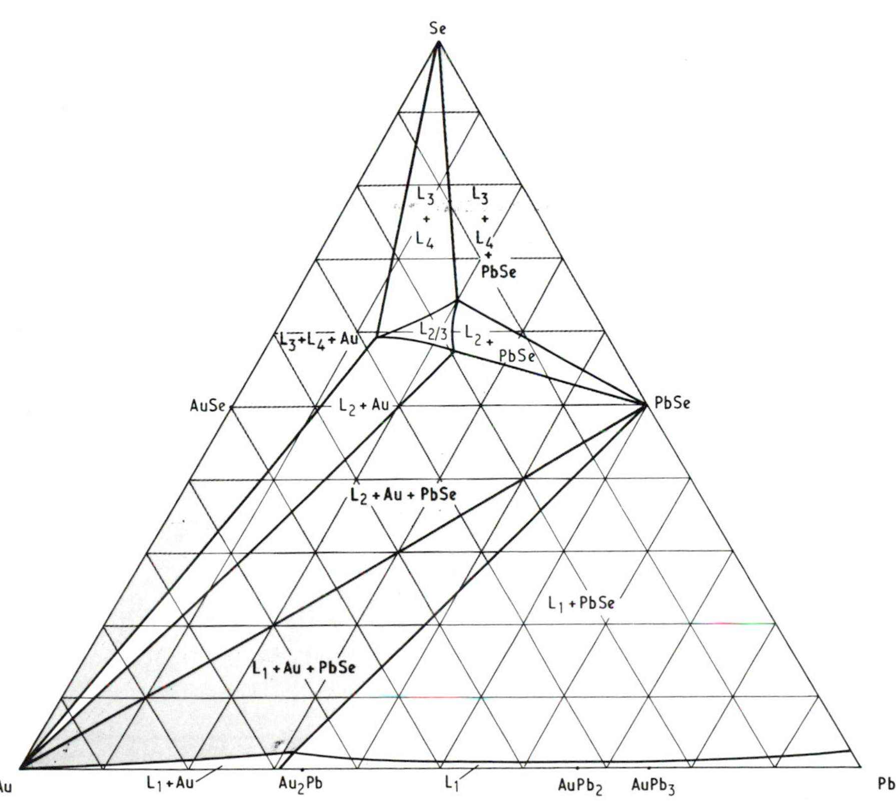

291 The 600°C isothermal section

ternary system, Fig. 292, is characterized by a very extensive region of liquid immiscibility extending to 68 at.-% Au. This liquid immiscibility region originates at the Pb–Si monotectic reaction, $l_1 = l_2 + (Si)$, at 1 402°C [84 Ole] and ends at a ternary invariant monotectic reaction, $L_{M_1} + L_{m_2} \rightleftharpoons Au_2Pb + (Si)$, at 341 ± 1°C. The compositions of M_1 and M_2 are 67·5Au, 19·0Pb, 13·5 at.-% Si and 37·5Au, 55·5Pb, 7·0 at.-% Si respectively. A critical tie line joining liquid of composition c and solid Au_2Pb is estimated to appear at about 415°C at a composition for liquid c of about 30Pb, 5 at.-% Si judged from the maximum temperatures of the $L_1 + L_2 + Au_2Pb$ phase region in the vertical sections presented by [88 Leg]. Solidification ends in the Au–Si–Au_2Pb region of the ternary with the ternary eutectic reaction E_1, $L \rightleftharpoons (Au) + (Si) + Au_2Pb$, at 324·5°C. The composition of E_1 was reported to be 74·5 ± 1Au, 9·5 ± 1Pb, 16·0 ± 1 at.-% Si.

The ternary transition reactions U_1 and U_2 and a ternary eutectic reaction E_2 are associated with the binary Au–Pb invariant reactions p_2, p_3 and e_3 respectively. The temperatures of these ternary reactions are virtually identical with those of the binary invariant reactions and the compositions of U_1, U_2 and E_2 must therefore lie close to the Au–Pb binary edge (Fig. 292). A reaction scheme is presented in Fig. 293. An isothermal section at 350°C, deduced from the vertical sections, is given in Fig. 294.

REFERENCES

84 Ole: R. W. Olesinski and G. J. Abbaschian: *Bull. Alloy Phase Diagrams*, 1984, **5**, 271–273

88 Leg: B. Legendre and Chhay Hancheng: *Bull. Soc. Chim. France*, *1988*, (1), 32–38

Au–Pb–Sn

INTRODUCTION

This system is of industrial importance in relation to the soft soldering of Au surfaces. The AuSn–Pb section is a pseudobinary eutectic (Fig. 295 and Table 96), according to [66 Pri, 68 Kar, 84 Hum]. The data are in reasonable agreement but those of [84 Hum] are preferred. The partial ternary system AuSn–Pb–Sn was studied extensively by [67 Pri, 68 Kar, 85 Hum]. Three invariant reactions were idenified by [67 Pri] and confirmed by [68 Kar, 85 Hum]. There is reasonable agreement between the data of [67 Pri] and [85 Hum], Fig. 296 and Table 97, but substantial disagreement with the work of [68 Kar]. The more refined thermal analysis techniques used by [85 Hum] allow reliance to be placed on their quoted results. The invariant reactions are transition reactions U_2 ($L + AuSn \rightleftharpoons (Pb) + AuSn_2$), U_6 ($L + AuSn_2 \rightleftharpoons (Pb) + AuSn_4$) and a ternary eutectic reaction E_2 ($L \rightleftharpoons (Pb) + (Sn) + AuSn_4$).

The partial ternary system AuSn–Au–Pb was investigated by [68 Kar, 87 Hum]. The later work is considered to supersede that of [68 Kar]. Apart from a suggested transition reaction U ($L + \beta \rightleftharpoons (Au) + \zeta$) arising from the two peritectic reactions in the Au–Sn binary system [87 Leg], [87 Hum] detected six invariant reactions (Fig. 300 and Table 98). An extensive region of ternary liquid immiscibility, associated with the reaction $L_1 \rightleftharpoons L_2 + \zeta + AuSn$, was detected but its whole boundary was not determined. The temperatures of the critical tie lines at c_1 and c_2 were not established but they are estimated to be at about 340 and 350°C respectively. A reaction scheme for the complete ternary system is given in Fig. 306.

Table 95. Crystal structures

Solid phase	*Prototype*	*Lattice designation*
(Au)	Cu	cF4
(Pb)	Cu	cF4
(Sn)	βSn	tI4
$\beta(Au_{10}Sn)$	Ni_3Ti	hP16
ζ	Mg	hP2
AuSn	NiAs	hP4
$AuSn_2$	orthorhombic	oP24
$AuSn_4$	$PtSn_4$	oC20
$AuPb_3$	αV_3S	tI32
$AuPb_2$	Al_2Cu	tI12
Au_2Pb	Cu_2Mg	cF24

BINARY SYSTEMS

The Au–Pb phase diagram determined by [83 Eva] is accepted, as is the evaluation of the Pb–Sn system by [88 Kar]. The Au–Sn diagram evaluated by [84 Oka] is not accepted for the Au-rich side. The work of [87 Leg] has shown the presence of two peritectic reactions corresponding to the formation of the β phase at 532°C and the ζ phase at 519°C.

SOLID PHASES

Table 95 summarizes crystal structure information for the phases in the Au–Pb–Sn ternary system. No ternary compounds have been reported.

PSEUDOBINARY SYSTEMS

The AuSn–Pb section is a pseudobinary eutectic system (Fig. 295). The data of [84 Hum] are accepted in preference to the earlier data of [66 Pri, 68 Kar] (Table 96). The

Table 96. The AuSn–Pb pseudobinary system

Reaction	*Composition of the phases*, at.-%			*Temperature*, °C	*Reference*
	liquid	*AuSn*	*(Pb)*		
$l \rightleftharpoons AuSn + (Pb)$	89·8	—	>98·7	292	66Pri
	91·0	—	—	288	68Kar
	90·0	3·0 ± 0·5	94·0 ± 0·5	294·5 ± 0·5	84Hum

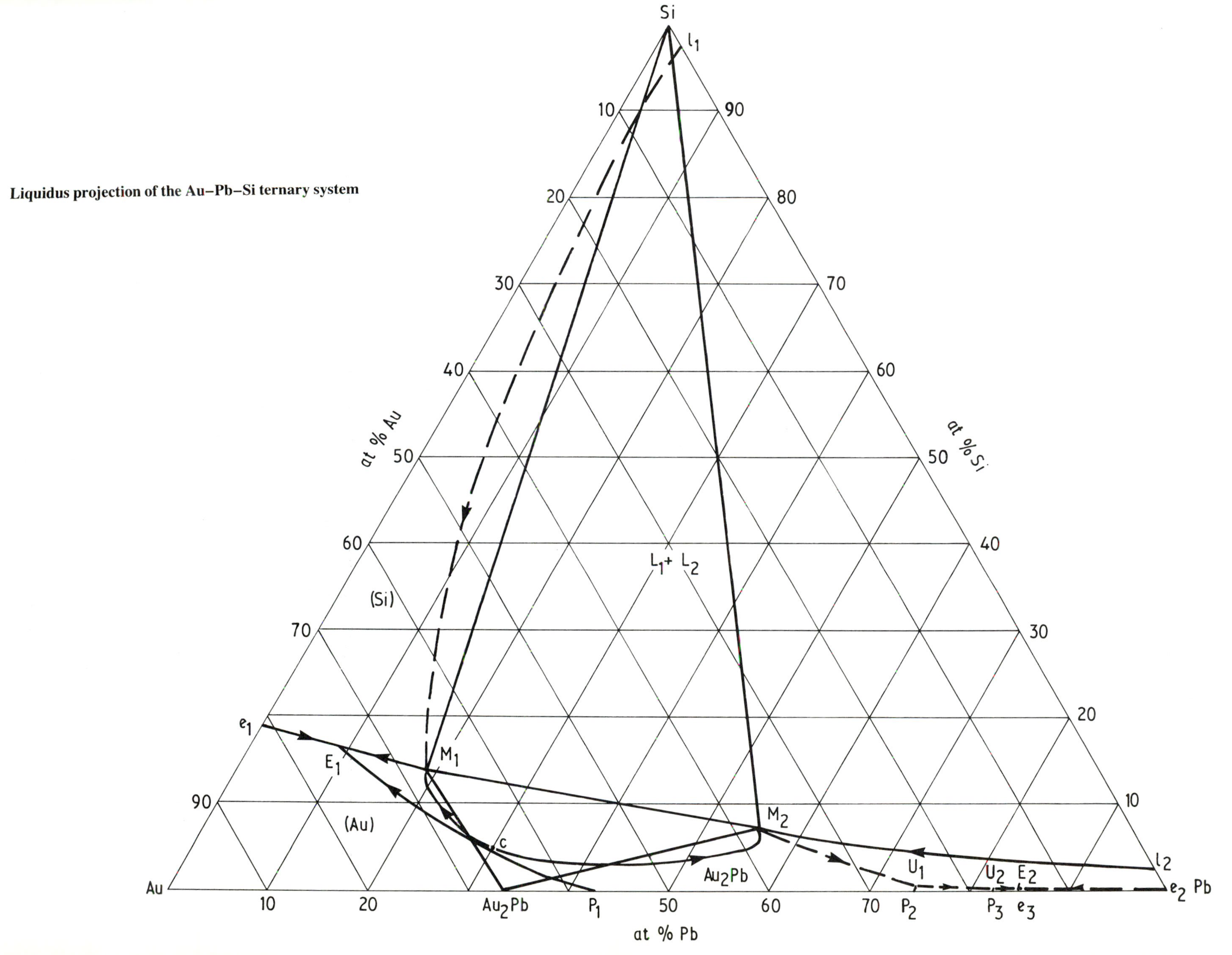

292 Liquidus projection of the Au–Pb–Si ternary system

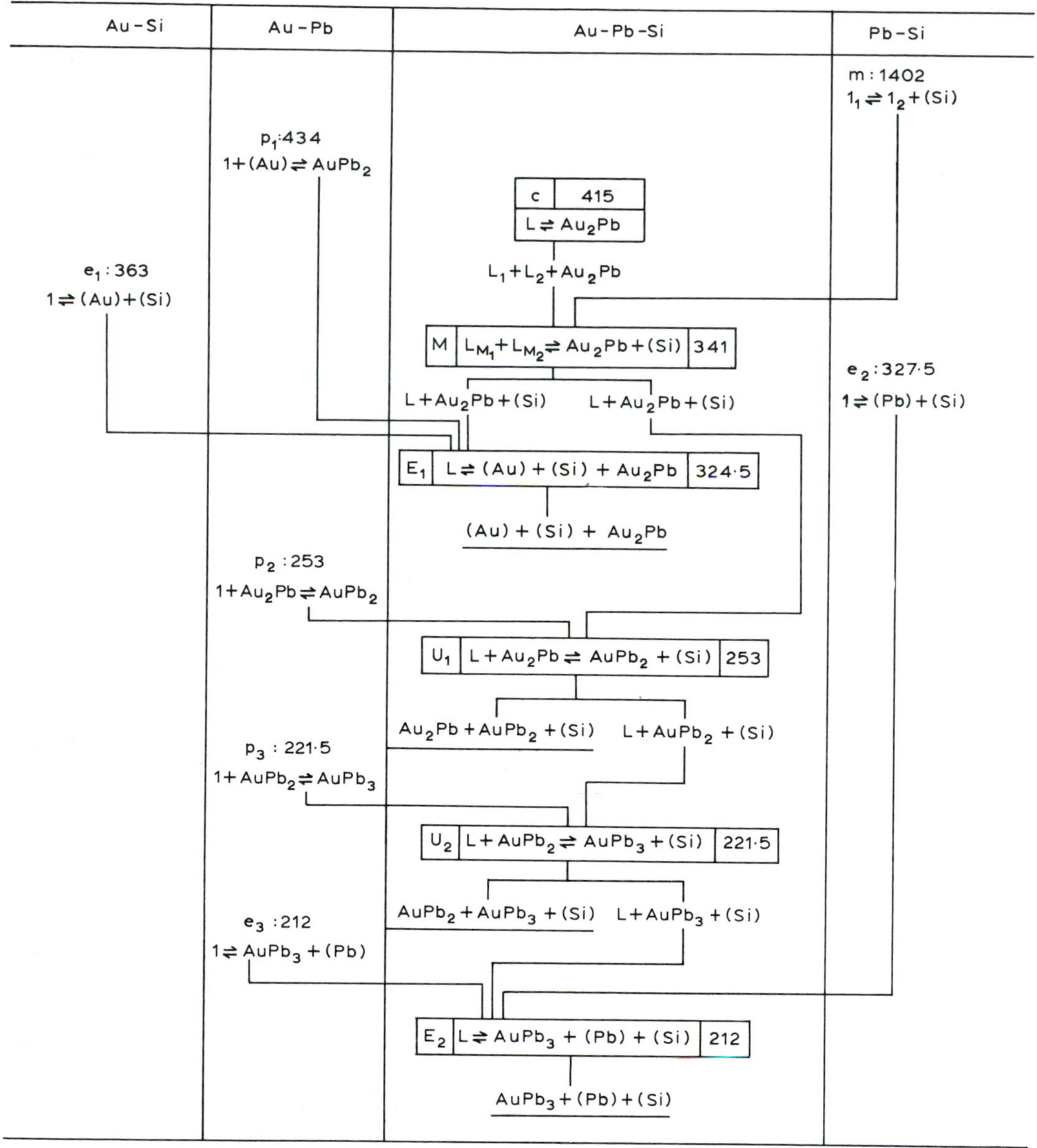

293 Reaction scheme for the Au–Pb–Si ternary system

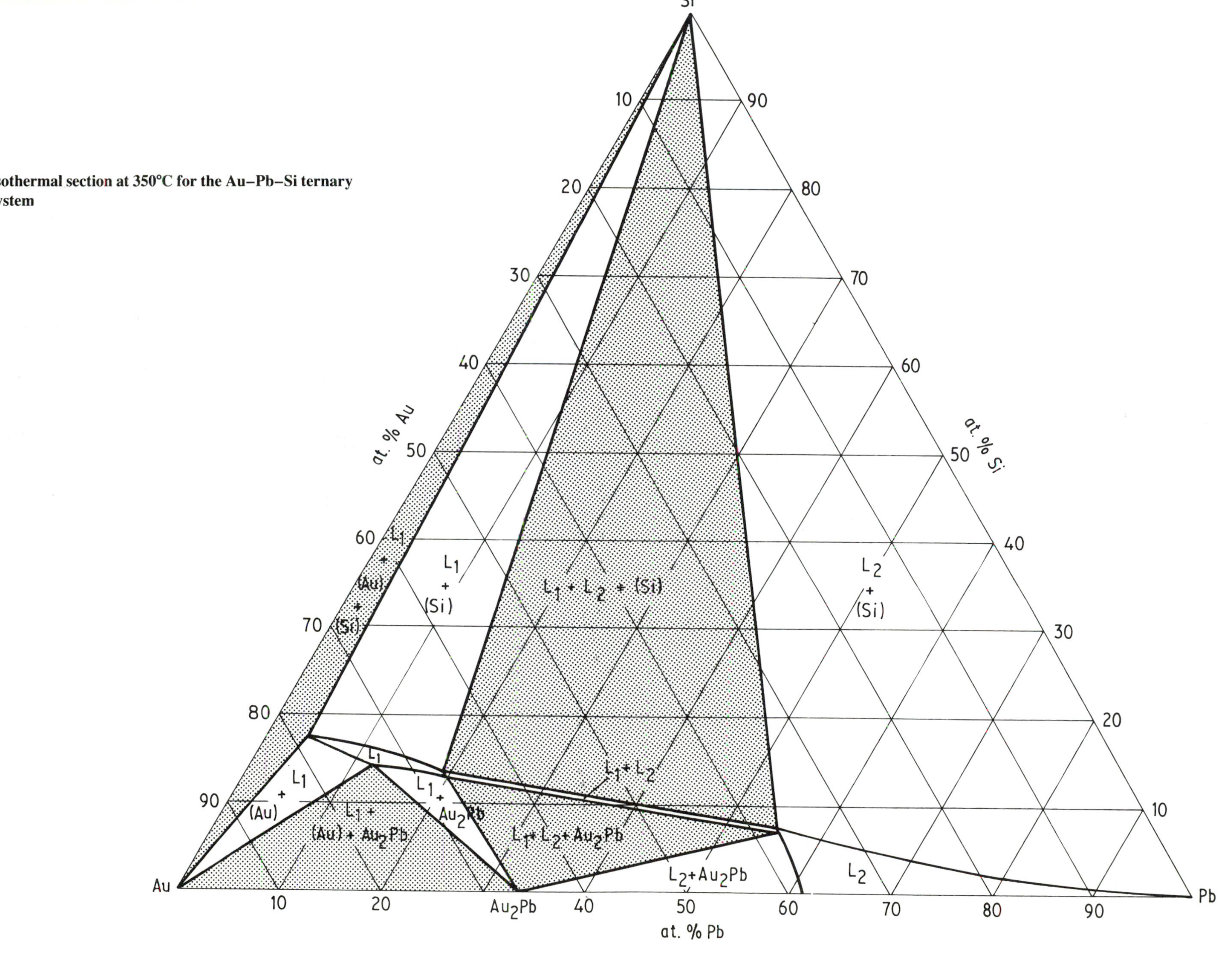

294 Isothermal section at 350°C for the Au–Pb–Si ternary system

liquidus curve shows a tendency towards demixing. The liquidus data agree very well up to 60 at.-% Pb, but beyond this limit the liquidus values reported by [66 Pri, 68 Kar] are considerably below those found by [84 Hum] in the primary AuSn region. This can be attributed to a strong undercooling tendency which was combated by great diligence in stirring as the liquidus was crossed [84 Hum]. The solubility of AuSn in Pb was based on metallographic examination of air-cooled alloys from DTA experiments [66 Pri] and on the metallographic examination of alloys containing 91·5, 92·5, 93·5, 94·5 and 95·5 at.-% Pb heat treated at 292 ± 2°C for 200 h [84 Hum]. Greater reliance can be placed on the latter data.

INVARIANT EQUILIBRIA

The AuSn–Pb pseudobinary system divides the Au–Pb–Sn ternary system into two partial ternary systems. The AuSn–Pb–Sn partial ternary system has been studied by [64 Ber, 67 Pri, 68 Kar, 85 Hum]. A ternary eutectic was located at 175°C [64 Ber] for alloys in the section 25·2 Pb, 74·8 at.-% Sn–Au containing up to 11 at.-% Au. [67 Pri] identified three ternary invariant reactions; L + AuSn ⇌ (Pb) + $AuSn_2$, L + $AuSn_2$ ⇌ (Pb) + $AuSn_4$ and L ⇌ (Pb) + (Sn) + $AuSn_4$ (see Fig. 296 and Table 97). These reactions were confirmed by [68 Kar, 85 Hum]. There is good agreement on the temperatures of the transition reactions U_2 and U_6 and of the ternary eutectic reaction E_2. The temperatures quoted by [85 Hum] are accepted since they were based on a superior experimental technique than those used by [67 Pri, 68 Kar]. The latter workers relied on thermal analysis of stirred melts during cooling only [67 Pri] or of unstirred melts [68 Kar]. In contrast [85 Hum] used both cooling and heating runs with conventional DTA and with an adaptation of the Smith technique to provide refined data. The compositions of the liquid phases involved in the ternary invariant reactions are tabulated in Table 97. There is reasonable agreement between [67 Pri] and [85 Hum] but the data of [68 Kar] differ greatly. Unfortunately [68 Kar] did not quote liquid compositions for U_2, U_6 and E_2. The values given in Table 97 are those scaled from a liquidus projection and from isothermal sections at the temperatures of the three invariant reactions. As can be seen the compositions deduced from the liquidus projection bear no correspondence with those from the isothermal sections for the transition reactions U_2 and U_6. The compositions quoted by [85 Hum] for the liquid phases are accepted. The AuSn phase dissolves 3·0 ± 0·5 at.-% Pb at the pseudobinary eutectic temperature [84 Hum]. The compositions of the AuSn, (Pb) and (Sn) phases entering into the ternary invariant reactions are uncertain. They are quoted in Table 97 but it should be noted that [85 Hum] placed wide uncertainty limits on these compositions. For AuSn the composition is 47·5 ± 1·0 at.-% Au, 3·5 ± 2·0 Pb, 49 ± 1·0 at.-% Sn at 275°C; for the (Pb) phase 4·5 ± 3·0 Au, 69 ± 3 Pb, 26·5 ± 3 at.-% Sn between 275 and 177°C and for the (Sn) phase 1 ± 1 Au, 1 ± 1 Pb, 98 ± 1 at.-% Sn at 177°C.

Vertical sections studied by [85 Hum] in elucidating the ternary equilibria were the AuSn–20 at.-% Pb to Sn section (Fig. 297), the AuSn–50 at.-% Pb to Sn section (Fig. 298), and the 50 at.-% Pb section from AuSn–50 at.-% Pb to 50Pb, 50 at.-% Sn (Fig. 299). They are consistent with the liquidus projection (Fig. 296).

The AuSn–Au–Pb partial ternary system has been investigated by [68 Kar, 87 Hum]. Both identified the transition reactions U_1 (L + (Au) ⇌ Au_2Pb + ζ), U_3 (L + ζ ⇌

Table 97. Invariant reactions in the AuSn–Pb–Sn partial ternary system

Reaction	*Phase*	*Phase composition*, at.-%			*Temperature*, °C	*Reference*
		Au	*Pb*	*Sn*		
U_2						
L + AuSn ⇌ (Pb) + $AuSn_2$	L	13·2	50·2	36·6	275	67Pri
	L	13·0	48·0	39·0	275	85Hum
	L	3·4	71·1	25·5	277	68Kar†
	L	7·0	46·0	47·0	277	68Kar††
	AuSn	47·5	3·5	49·0	275	85Hum
	(Pb)	4·5	69·0	26·5	275	85Hum
	(Pb)	5·5	83·0	11·5	277	68Kar††
	$AuSn_2$	33·33	—	66·67	275	85Hum
U_6						
L + $AuSn_2$ ⇌ (Pb) + $AuSn_4$	L	5·2	27·7	67·1	208	67Pri
	L	8·5	28·0	63·5	210	85Hum
	L	3·5	50·8	45·7	205	68Kar†
	L	1·0	35·0	64·0	205	68Kar††
	(Pb)	4·5	69·0	26·5	210	85Hum
	(Pb)	5·0	79·0	16·0	205	68Kar††
	$AuSn_2$	33·33	—	66·67	210	85Hum
	$AuSn_4$	20·0	—	80·0	210	85Hum
E_2						
1 ⇌ (Pb) + (Sn) + $AuSn_4$	L	2·1	21·9	76·0	177	67Pri
	L	3·5	20·0	76·5	177	85Hum
	L	3·0	25·8	71·2	176	68Kar†
	L	4·0	24·5	71·5	176	68Kar††
	(Pb)	4·5	69·0	26·5	177	85Hum
	(Pb)	5·0	66·5	28·5	176	68Kar††
	(Sn)	1·0	1·0	98·0	177	85Hum
	$AuSn_4$	20·0	—	80·0	177	85Hum

† Liquidus projection
†† Isothermal section

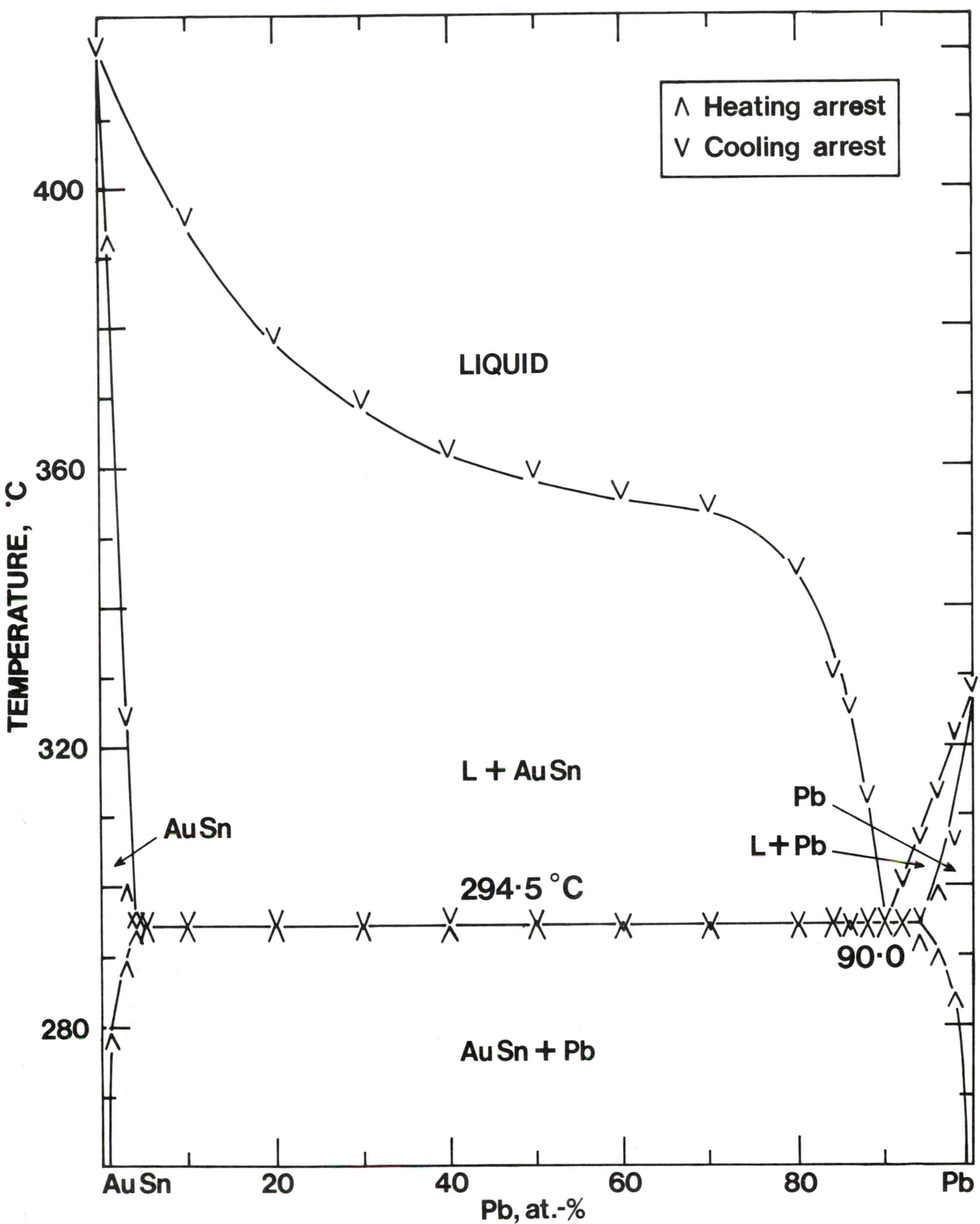

295 The AuSn–Pb section

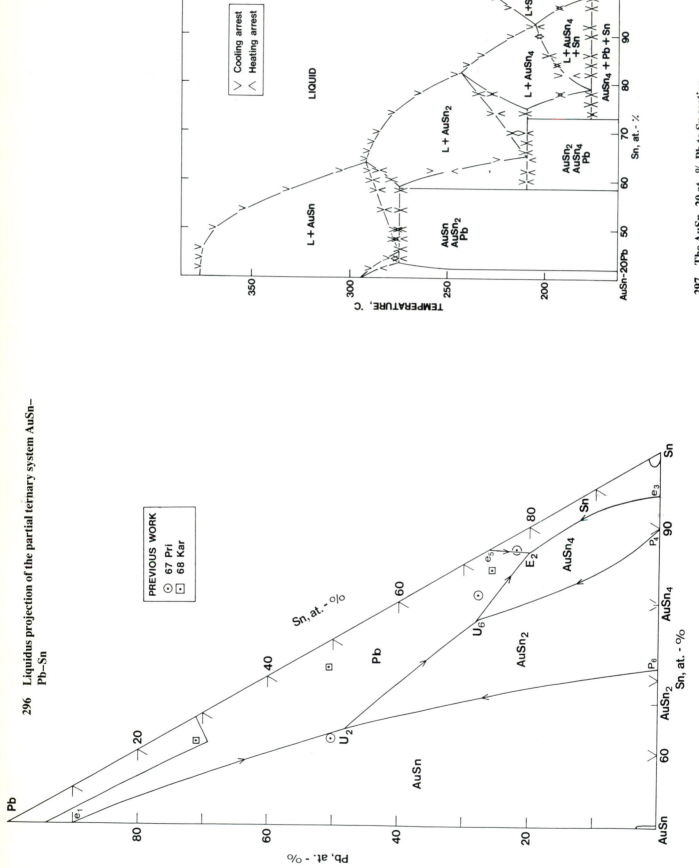

296 **Liquidus projection of the partial ternary system AuSn–Pb–Sn**

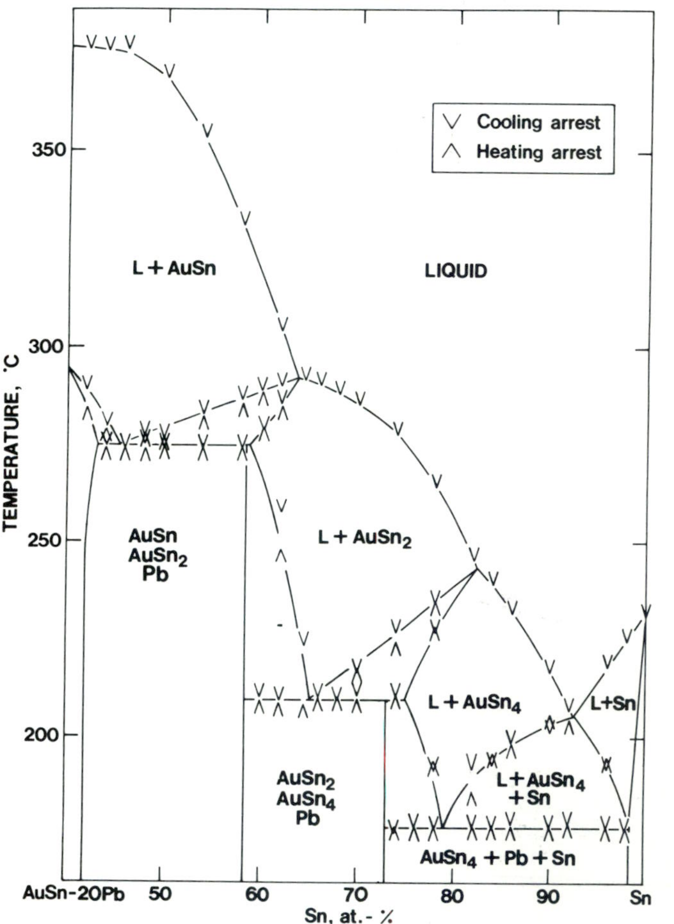

297 **The AuSn–20 at.-% Pb to Sn section**

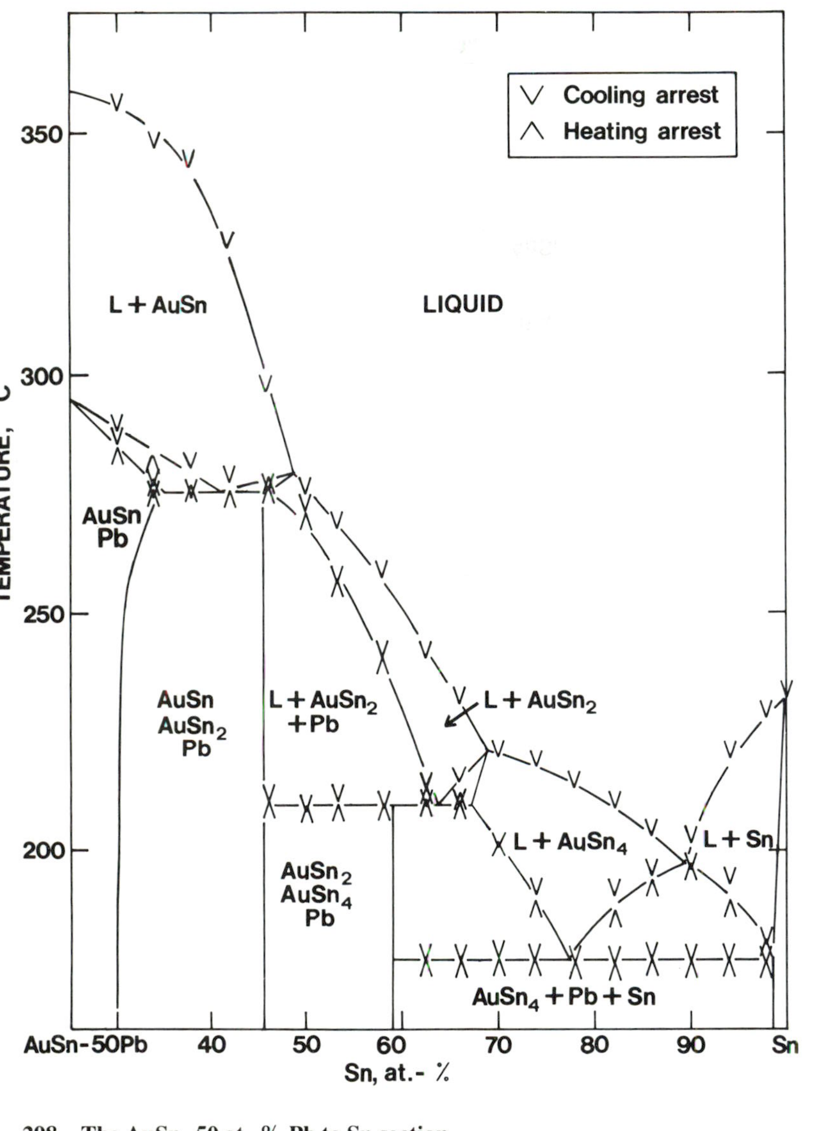

298 The AuSn–50 at.-% Pb to Sn section

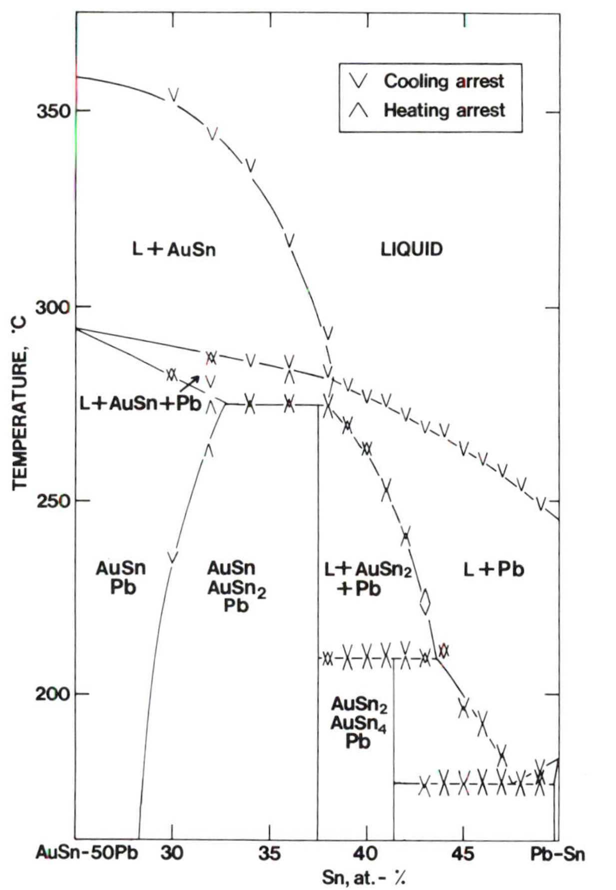

299 The AuSn–50 at.-% Pb to 50Pb, 50 at.-% Sn section

Au_2Pb + AuSn) and U_4 (L + $Au_2Pb \rightleftharpoons AuPb_2$ + AuSn) (Table 98). [68 Kar] were unaware of the formation of the compound $AuPb_3$ in the Au–Pb binary system and therefore quoted solidification as ending with the ternary eutectic reaction L $\rightleftharpoons$ (Pb) + AuSn + $AuPb_2$, whereas [87 Hum] took into account the $AuPb_3$ phase and identified a transition reaction U_5 (L + $AuPb_2 \rightleftharpoons AuPb_3$ + AuSn) preceding the ternary eutectic reaction E_1 (L $\rightleftharpoons$ (Pb) + AuSn + $AuPb_3$). It is possible to compare the data of [68 Kar, 87 Hum] for the reactions U_1, U_3, U_4 and E_1. Table 98 lists the compositions of the liquid phases associated with these reactions. [68 Kar] gave a liquidus projection and isothermal sections at the four invariant reaction temperatures. For reactions U_1 and U_3 there is no correspondence between the liquid compositions scaled from the liquidus projection and the appropriate isothermal section. Better agreement is evident for reactions U_4 and E_1. Comparing the compositions scaled from the liquidus projection of [68 Kar] with the tabulated values of [87 Hum] reasonable agreement is found for reactions U_1, U_4 and E_1. The values for U_3 differ considerably. The temperatures for the invariant reactions are in good agreement for reaction U_4; otherwise [68 Kar] report lower temperatures than [87 Hum]. This is a consequence of the use of an experimental technique by [68 Kar] that precluded stirring of the melt during thermal analysis. The data of [87 Hum) are accepted.

In contrast to [68 Kar] the work of [87 Hum] revealed an extensive elongated area of ternary liquid immiscibility (Fig. 300), associated with the invariant reaction $L_1 \rightleftharpoons L_2$ + AuSn + ζ. The liquidus projection was constructed from five vertical sections; at 85 at.-% Pb (Fig. 301), from $AuPb_2$ to AuSn–50 at.-% Pb (Fig. 302), from Au_2Pb to AuSn–20 at.-% Pb (Fig. 303), from 80Au, 20Pb('Au_4Pb') to 48Au, 40Sn (Fig. 304) and from 66·67Au, 33·33Sn ('Au_2Sn') towards Pb (Fig. 305). The compositions quoted by [87 Hum] for the invariant liquid compositions associated with reactions U_1 to U_5 and M are consistent, within the error limits, with the vertical sections given in Figs 301–305. The composition of the ternary eutectic E_1 is considered to lie at 84Pb, 0·5 at.-% Sn; this value gives better consistency with the 85 at.-% Pb section (Fig. 301).

Two critical tie lines, representing equilibria between liquid c_1 and ζ and liquid c_2 and AuSn, must be present in the ternary system. Each gives rise to a three-phase equilibrium, $L_1 + L_2 + \zeta$ and $L_1 + L_2$ + AuSn, that fall to the monotectic invariant reaction $L_1 \rightleftharpoons L_2 + \zeta$ + AuSn at 257·5°C. [87 Hum] indicated the tie lines by dashed lines (Fig. 300), but did not give the temperatures associated with them. Points c_1 and c_2 scale from the liquidus projection at 57Au, 28Pb, 15 at.-% Sn and 39Au, 35Pb, 26 at.-% Sn. It is estimated that the tie line c_1–ζ lies at a temperature of 340°C and that the tie line c_2–AuSn lies at a temperature of about 350°C. [87 Hum] did not study alloy compositions close to the Au–Sn binary edge. From the work of [87 Leg] it can be suggested that a transition reaction U, L + $\beta \rightleftharpoons$ (Au) + ζ, occurs near to the Au–Sn edge followed by the observed transition reaction at U_1. [87 Hum] noted a thermal effect at about 154°C in Pb-rich ternary alloys, suggesting the presence of solid-state

Table 98. Invariant reactions in the AuSn–Au–Pb partial ternary system

Reaction	*Phase*	*Phase composition*, at.-% *Au*	*Pb*	*Sn*	*Temperature*, °C	*Reference*
U						
L + $\beta \rightleftharpoons$ (Au) + ζ	L	—	—	—		
U_1						
L + (Au) $\rightleftharpoons Au_2Pb + \zeta$	L	63	27	10	363	68Kar†
	L	53	32	15	363	68Kar††
	L	63·5 ± 2	26·0 ± 2	10·5 ± 2	382·5 ± 1·1	87Hum
M						
$L_1 \rightleftharpoons L_2$ + AuSn + ζ	L_1	64·0	9·0	27·0	257·5 ± 0·7	87Hum
	L_2	45·0 ± 1	40·0 ± 1	15·0 ± 1	257·5 ± 0·7	87Hum
U_3						
L + $\zeta \rightleftharpoons Au_2Pb$ + AuSn	L	57·0	23·5	19·5	246	68Kar†
	L	49·5	25·5	25·0	246	68Kar††
	L	42·5 ± 1	47·5 ± 1	10·0 ± 1	254 ± 1·1	87Hum
U_4						
L + $Au_2Pb \rightleftharpoons AuPb_2$ + AuSn	L	29·0	64·0	7·0	225	68Kar†
	L	25·5	65·5	9·0	225	68Kar††
	L	30·5 ± 1	60·5 ± 1	9·0 ± 1	224 ± 0·7	87Hum
U_5						
L + $AuPb_2 \rightleftharpoons AuPb_3$ + AuSn	L	20·5 ± 2	75·5 ± 2	4·0 ± 2	214 ± 0·5	87Hum
E_1						
L $\rightleftharpoons$ (Pb) + AuSn + $AuPb_3$	L	16·3	80·0	3·7	204	68Kar†
	L	14·3	81·4	4·3	204	68Kar††
	L	15·0 ± 1	84·0 ± 1	1·0 ± 1	211 ± 0·5	87Hum

† Liquidus projection
†† Isothermal section

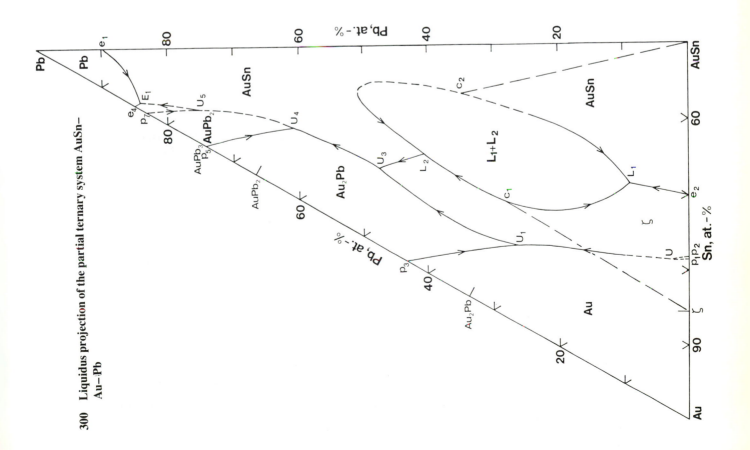

300 Liquidus projection of the partial ternary system AuSn–Au–Pb

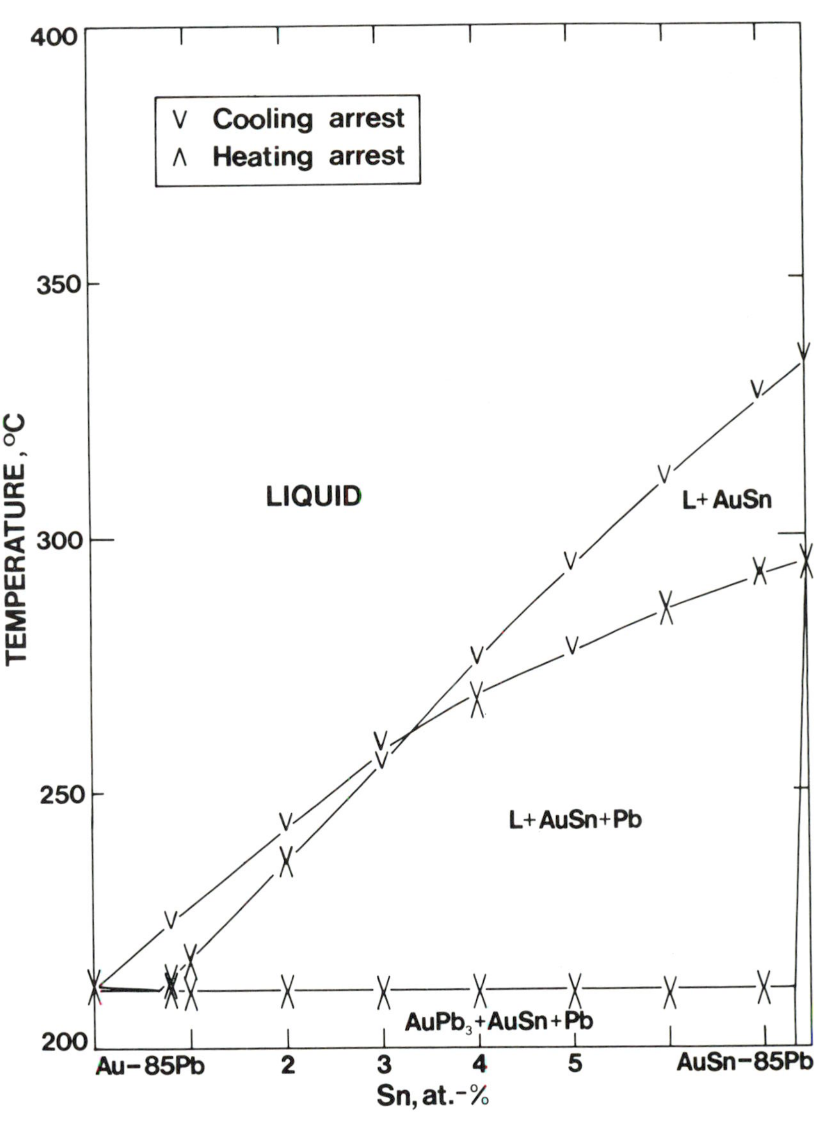

301 The 85 at.-% Pb section (partial)

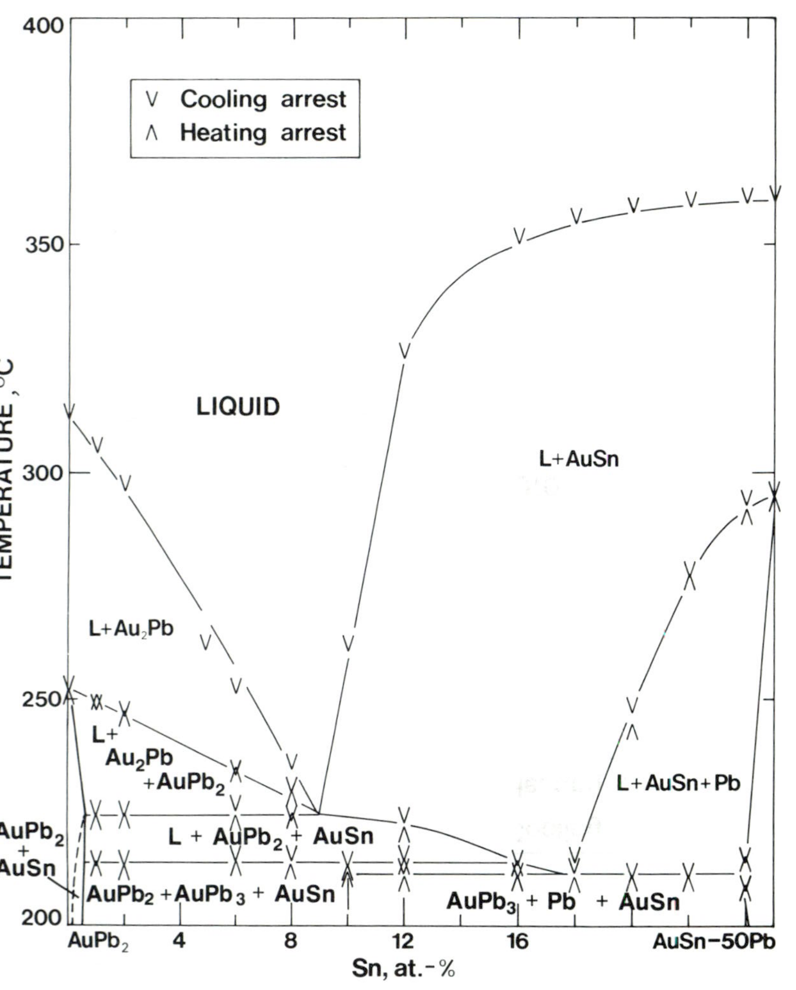

302 The $AuPb_2$ to AuSn–50 at.-% Pb section

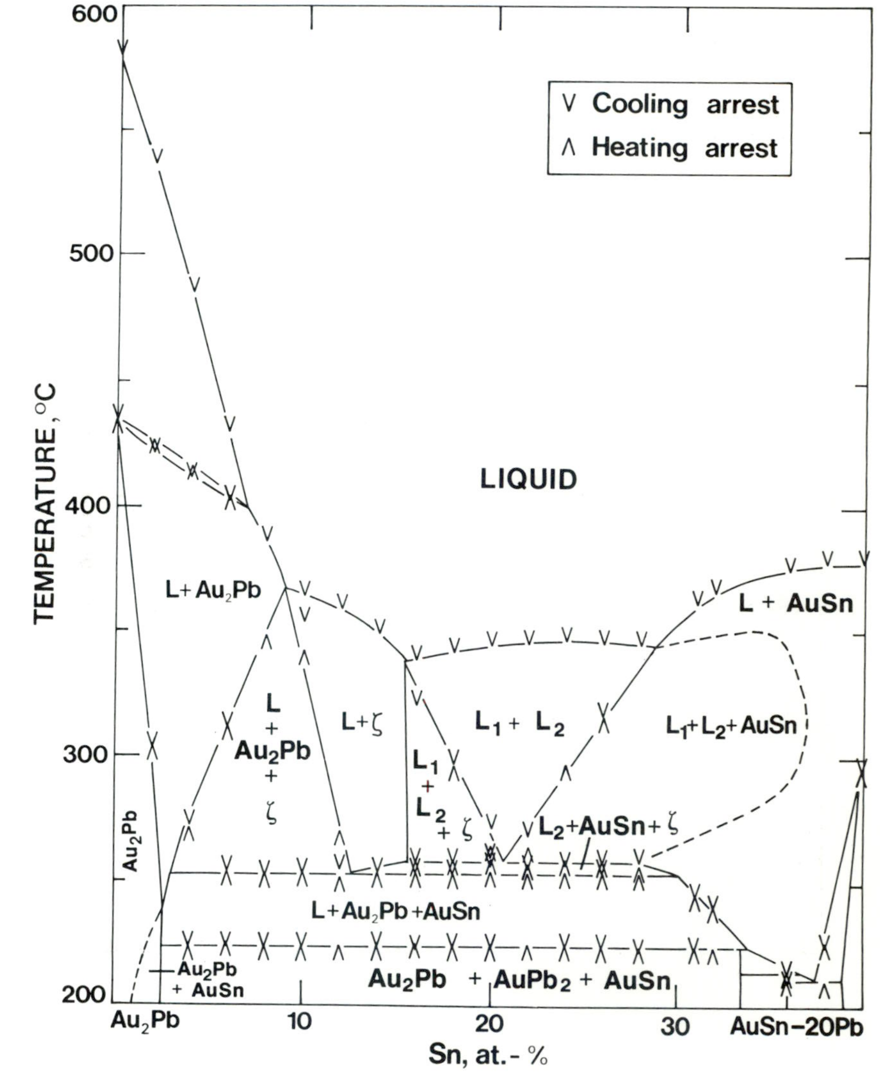

303 The Au_2Pb to AuSn–20 at.-% Pb section

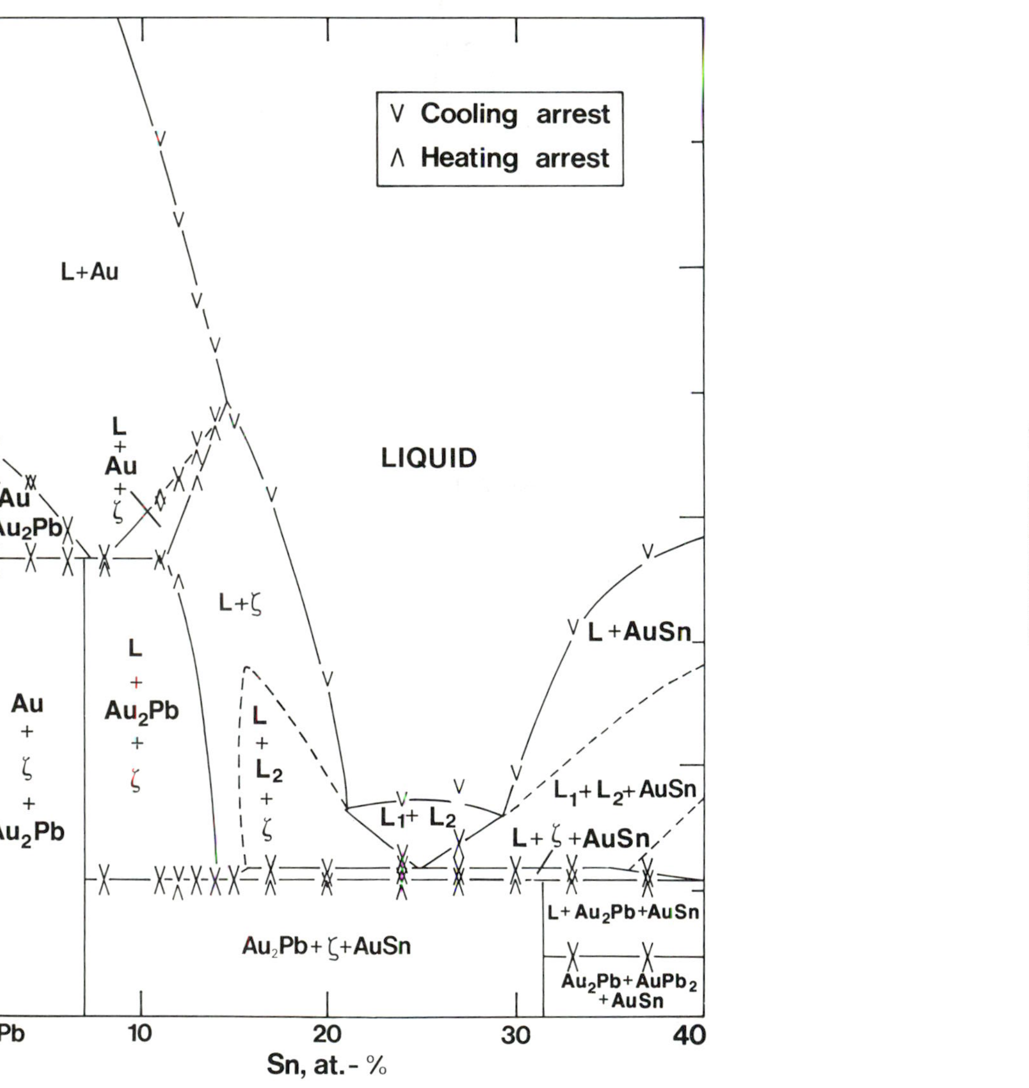

304 The 80Au, 20Pb ('Au$_4$Pb') to 48Au, 40Sn section

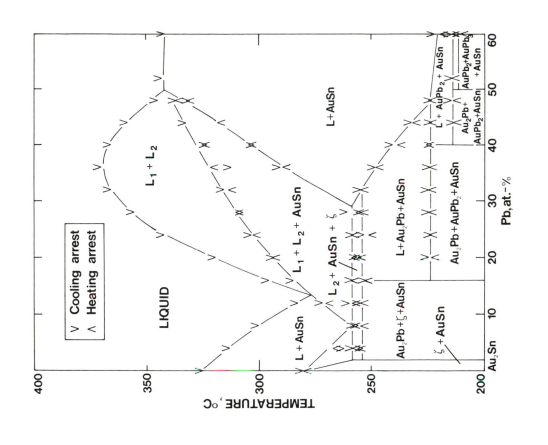

305 The 66·67Au, 33·33Sn ('Au$_2$Sn') to Pb section (partial)

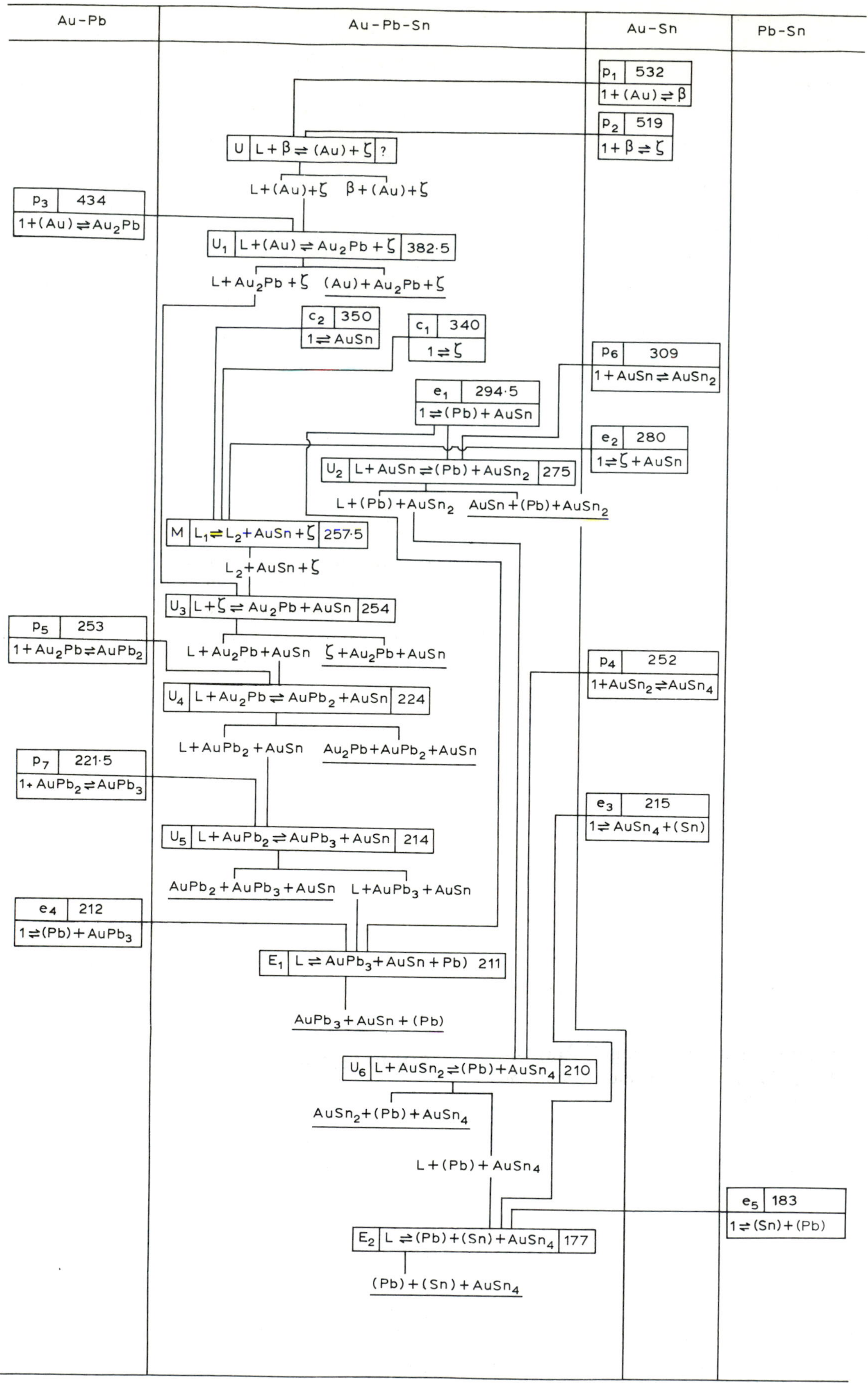

306 **Reaction scheme for the Au–Pb–Sn ternary system**

transformations. A reaction scheme for the complete ternary system is given in Fig. 306.

LIQUIDUS SURFACE

Figure 296 is the liquidus surface projection for the AuSn–Pb–Sn partial ternary system and Fig. 300 is the projection of the liquidus surface for the partial ternary system AuSn–Au–Pb. The region of primary separation of AuSn tends to dominate the ternary system.

ISOTHERMAL SECTIONS

Figure 307 is a representation of the 300°C isothermal section, deduced from the vertical sections.

REFERENCES

64 Ber: R. D. Berry and R. W. Johnson: *Inst. Welding Conf. on Brazing and Soldering*, Nov. 1964, Paper 9

66 Pri: A. Prince: *J. Less-Common Metals*, 1966, **10**, 365–368

67 Pri: A. Prince: *J. Less-Common Metals*, 1967, **12**, 107–116

68 Kar: M. M. Karnowsky and A. Rosenzweig: *Trans. Met. Soc. AIME*, 1968, **242**, 2257–2261

83 Eva: D. S. Evans and A. Prince: *Mater. Res. Soc. Symp. Proc.*, 1983, **19**, 383–388

84 Hum: G. Humpston and B. L. Davies: *Metal Sci.*, 1984, **18**, 329–331

84 Oka: H. Okamoto and T. B. Massalski: *Bull. Alloy Phase Diagrams*, 1984, **5**, 492–503 (correction, ibid., 1986, **7**, 522)

85 Hum: G. Humpston and B. L. Davies: *Mater. Sci. Technol.*, 1985, **1**, 433–441

87 Hum: G. Humpston and D. S. Evans: *Mater. Sci. Technol.*, 1987, **3**, 621–627

87 Leg: B. Legendre, Chhay Hancheng, F. Hayes, C. A. Maxwell, D. S. Evans and A. Prince: *Mater. Sci. Technol.*, 1987, **3**, 875–876

88 Kar: I. Karakaya and W. T. Thompson, *Bull. Alloy Phase Diagrams*, 1988, **9**, 144–152

Au–Pb–Te

INTRODUCTION

Mottern and Wald [68 Mot] reported results of a study of the Au–PbTe section indicating its quasibinary eutectic nature. Legendre *et al.* [72 Leg] made a more extensive study of the ternary system, confirming *inter alia* the conclusions of [68 Mot] on the Au–PbTe section. The ternary reactions proposed by [72 Leg] are consistent with the known binary equilibria and are accepted. However, a detailed examination of the data presented by [72 Leg] reveals inconsistencies in the interpretation of the experimental results. As a consequence the liquidus surface presented in Fig. 313 differs markedly from [72 Leg]; also the composition of the ternary eutectic in the partial system $AuTe_2$–Te–PbTe is not in accord with the conclusions reached by [72 Leg].

Both [68 Mot] and [72 Leg] used thermal analysis techniques, supplemented by metallography and X-ray diffraction analysis of powder samples to elucidate the equilibria. Neither incorporated any facility for nucleating the separation of the primary phase either by stirring or vibration. As a consequence the reported liquidus values are likely to be too low in temperature, particularly with Te-rich alloys which are known to be prone to massive undercooling.

The ternary system is divided into three partial systems by the pseudobinary sections between $AuTe_2$–PbTe and Au–PbTe. The partial ternary systems $AuTe_2$–Te–PbTe and Au–$AuTe_2$–PbTe are simple eutectic systems. The partial ternary system Au–PbTe–Pb has not been experimentally determined but the reaction sequence proposed by [72 Leg] is accepted as plausible, with the proviso that the eutectoidal breakdown of Au_2Pb at 247°C in the Au–Pb binary system is not accepted.

There is a need for further experimental work to establish the composition of the ternary eutectic in the $AuTe_2$–Te–PbTe partial system and to confirm the proposed equilibria in the Au–PbTe–Pb partial system.

BINARY SYSTEMS

The Au–Pb diagram determined by [83 Eva] is accepted. The compounds Au_2Pb, $AuPb_2$ and $AuPb_3$ all form peritectically at temperatures of 434, 253, and 221·5°C respectively. The compositions of the liquid phase at the peritectic horizontals are 42·6, 73·8, and 82·2 at.-% Pb respectively. $AuPb_3$ and (Pb) form a eutectic at 212°C and 84·8 at.-% Pb. Au_2Pb was not found to decompose at 247°C in contradiction to [71 Leg]. The Au–Te system evaluated by [84 Oka] and the Pb–Te system evaluated by [86 Lin] are accepted.

SOLID PHASES

No ternary compounds were found by [68 Mot, 72 Leg]. Table 99 summarizes crystal structure data for the binary compounds.

Table 99. Crystal structures

Solid phase	*Prototype*	*Lattice designation*
(Au)	Cu	cF4
(Pb)	Cu	cF4
(Te)	γSe	hP3
Au_2Pb	Cu_2Mg	cF24
$AuPb_2$	$CuAl_2$	tI12
$AuPb_3$	α–V_3S	tI32
$AuTe_2$	$AuTe_2$	mC6
PbTe	NaCl	cF8

PSEUDOBINARY SYSTEMS

Both the Au–PbTe [68 Mot, 72 Leg] and $AuTe_2$–PbTe [72 Leg] sections are pseudobinary eutectic systems. The

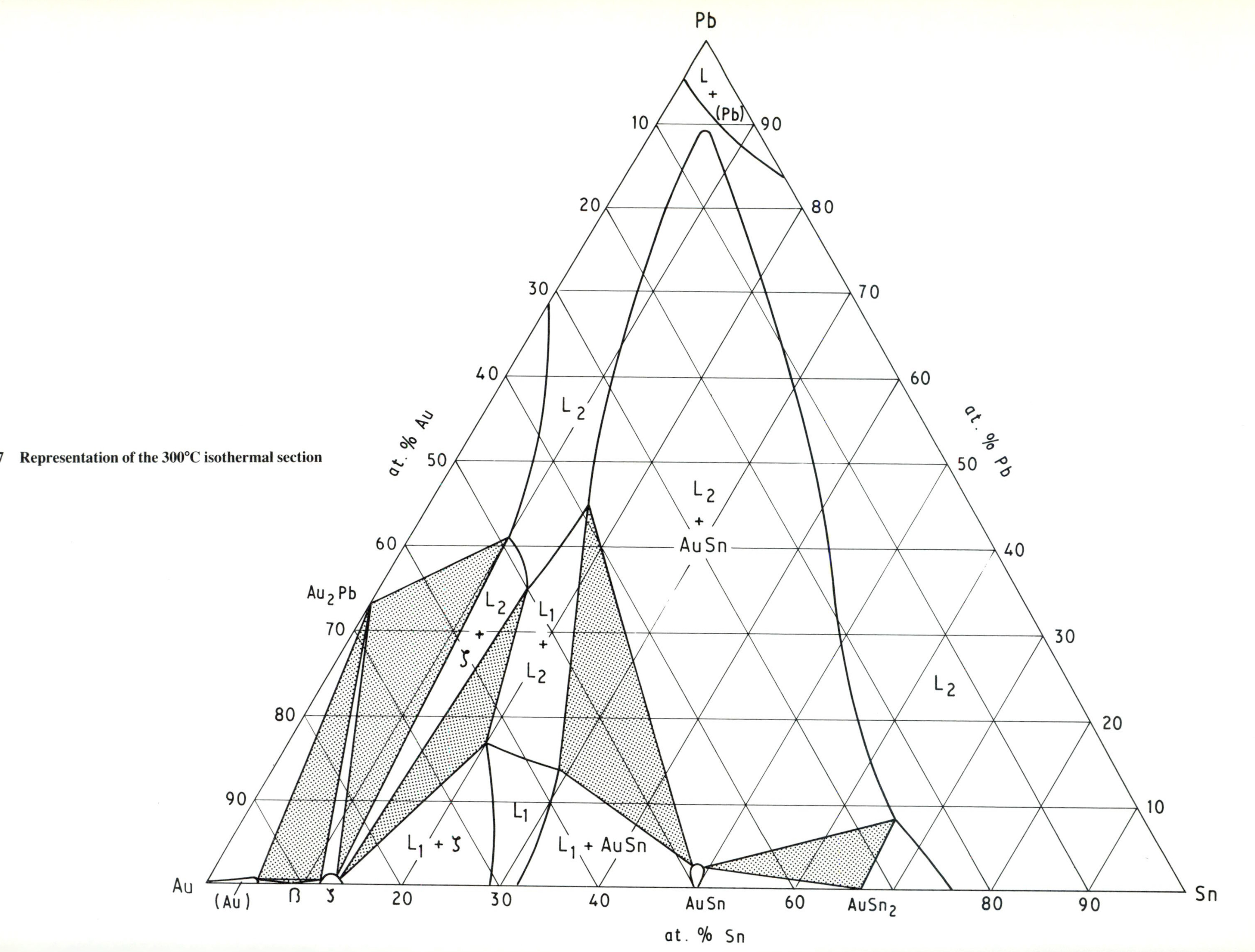

307 Representation of the 300°C isothermal section

liquidus data of [68 Mot] for the Au–PbTe section are some 20–30°C lower than those quoted by [72 Leg]. On the other hand two of the liquidus points plotted by [72 Leg] are well below (19 and 54°C) the liquidus given in Fig. 308. The eutectic temperature and compositions are 742 ± 5°C at 42 at.-% Au [68 Mot], 748°C and 40 at.-% Au [72 Leg]. Values of 746 ± 5°C and 41 at.-% Au are selected. The liquidus data of [72 Leg] are considered more reliable on the basis that they imply less undercooling associated with the primary phase separations.

The $AuTe_2$–PbTe section [72 Leg], Fig. 309, shows a eutectic at 426°C and 8·1 at.-% Pb. The eutectic composition is consistent with the extrapolated liquidus for the separation of primary PbTe. The eutectic temperature is probably accurate to ±2°C. In neither pseudobinary section was any change in lattice parameter of the phases found as the alloy composition varied across the sections. [72 Leg] regard this fact as evidence for no solid solution formation of Au in PbTe and vice versa, or of $AuTe_2$ in PbTe and vice versa. It should be noted that some solid solution formation may have been detected had the authors studied alloys closer in composition to each component in both sections.

INVARIANT EQUILIBRIA [72 Leg]

Table 100 and Fig. 310 give details of the invariant equilibria. Only the details of ternary eutectic reactions E_1 and E_2 were determined. The composition of the eutectic liquid given by [72 Leg] for reaction E_1 is accepted. The eutectic liquid composition for reaction E_2 quoted by [72 Leg] is considered incorrect. The method used by [72 Leg] to determine the eutectic point E_2 is illustrated in Fig. 311. In the absence of solubility of the components in each other the ternary eutectic point may be joined to each component. Such tie lines lie on the eutectic plane and separate the adjoining surfaces of secondary separation. A vertical section such as $AuTe_2$–6.67 Pb 93·33 Te will locate the point a which lies on the tie line joining Te to the ternary eutectic E_2. A vertical section $AuTe_2$–33.33 Pb 66·67 Te will locate the point b which lies on the tie line joining PbTe to E_2. Equations of the tie lines Te–E_2 and PbTe–E_2 can be derived and from these the composition of E_2 calculated. A rough check of the calculated composition for E_2 can also be made using the point of intersection of the vertical sections with the monovariant curves starting at the binary eutectic points e_5 (point c) and e_4 (point d). Figure 311 illustrates the selection of E_2 at a composition 17·6 at.-% Au, 7·8 at.-% Pb. [72 Leg] establish E_2 at 12.7 at.-% Au, 5·6 at.-% Pb. This value is not consistent with point c. Examination of the section $AuTe_2$–33·33 Pb 66·67 Te (Fig. 312) throws doubt on the validity of the composition of point b (26·67 at.-% Pb [72 Leg]). The data have been replotted in Fig. 311 and indicate a composition for point b of 21·5 at.-% Pb. Taking this value Fig. 312 was constructed to give E_2 at 12·7 Au, 5·6 Pb. This value is now consistent with the directions of the monovariant curves e_3–c and e_4–d. Indeed, by coincidence, these curves appear to be straight and if so treated produce virtually the same composition of E_2 as was calculated from the tie lines Te–a–E_2 and PbTe–b–E_2. [72 Leg] state that an alloy of their proposed ternary eutectic composition showed only one arrest on DTA. It should be noted that the ternary eutectic composition proposed by [72 Leg] would involve primary separation of Te followed by the eutectic reaction at 388°C if the amended composition for E_2 is accepted. It is suggested that the experimental conditions (see introduction) used by [72 Leg] would allow such undercooling of the primary Te separation that only the eutectic reaction would be detected. Further work is needed, incorporating metallographic corroboration, to establish the composition of E_2.

The invariant equilibria in the partial ternary system Au–PbTe–Pb as proposed [72 Leg] are plausible. Difficulty was found in establishing the equilibria since all ternary reactions U_1, U_2, and U_3 occurred at temperatures within 2°C of the Au–Pb binary peritectic reaction temperatures and the ternary eutectic at E_3 coincided in temperature with the binary Au–Pb eutectic temperature. Further work is needed to establish the ternary equilibria.

Table 100. Invariant equilibria

Reaction	*Temperature*, °C	*Phase*	*Composition*, at.-% *Au*	*Pb*
E_1	402	L	35·7	10·7
E_2	388	L	17·6	7·8

LIQUIDUS SURFACE

The liquidus surface presented in Fig. 313 differs markedly from that published [72 Leg] with respect to the isotherms associated with the field of primary PbTe separation. The liquidus, Fig. 313, was constructed in the PbTe primary field from a consideration of the data presented by [72 Leg] for the sections $AuTe_2$–33·33 Pb, 66·67 Te, $AuTe_2$–PbTe, PbTe–50 Au, 50 Te and Au–PbTe, together with the binary data for the Pb–Te system. Where liquidus isotherms are regarded as indefinite they are shown as broken curves.

ISOTHERMAL SECTIONS

Only vertical sections were studied [68 Mot, 72 Leg]. As the two partial ternary systems for which the equilibria have been established are both simple ternary eutectic systems, no additional information is conveyed by producing isothermal sections over what can be deduced from the vertical sections and the liquidus isotherms.

REFERENCES

68 Mot: D. J. Mottern and F. Wald: *Trans. Met. Soc. AIME*, 1968, **242**(1), 150–151

71 Leg: B. Legendre, S. Jaulmes and C. Souleau: *Compt. rend.*, 1971, **273C**(4), 356–358

72 Leg: B. Legendre and C. Souleau: *Bull. Soc. Chim. France*, 1972, (2), 473–479

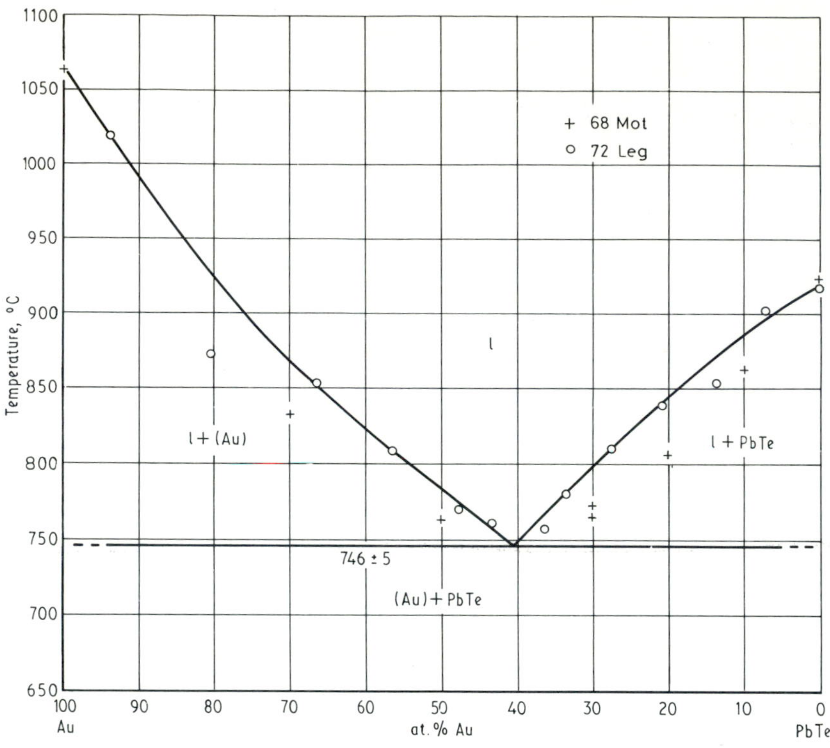

308 The Au–PbTe section

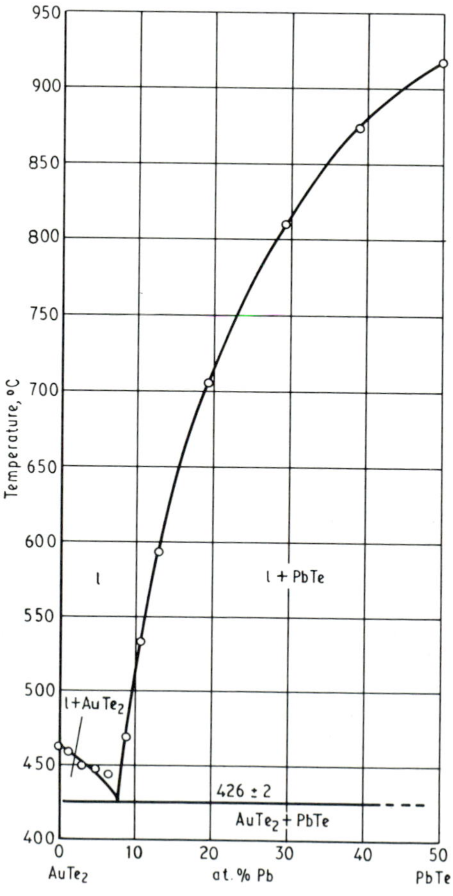

309 The $AuTe_2$–PbTe section

310 Reaction scheme for the Au–Pb–Te system

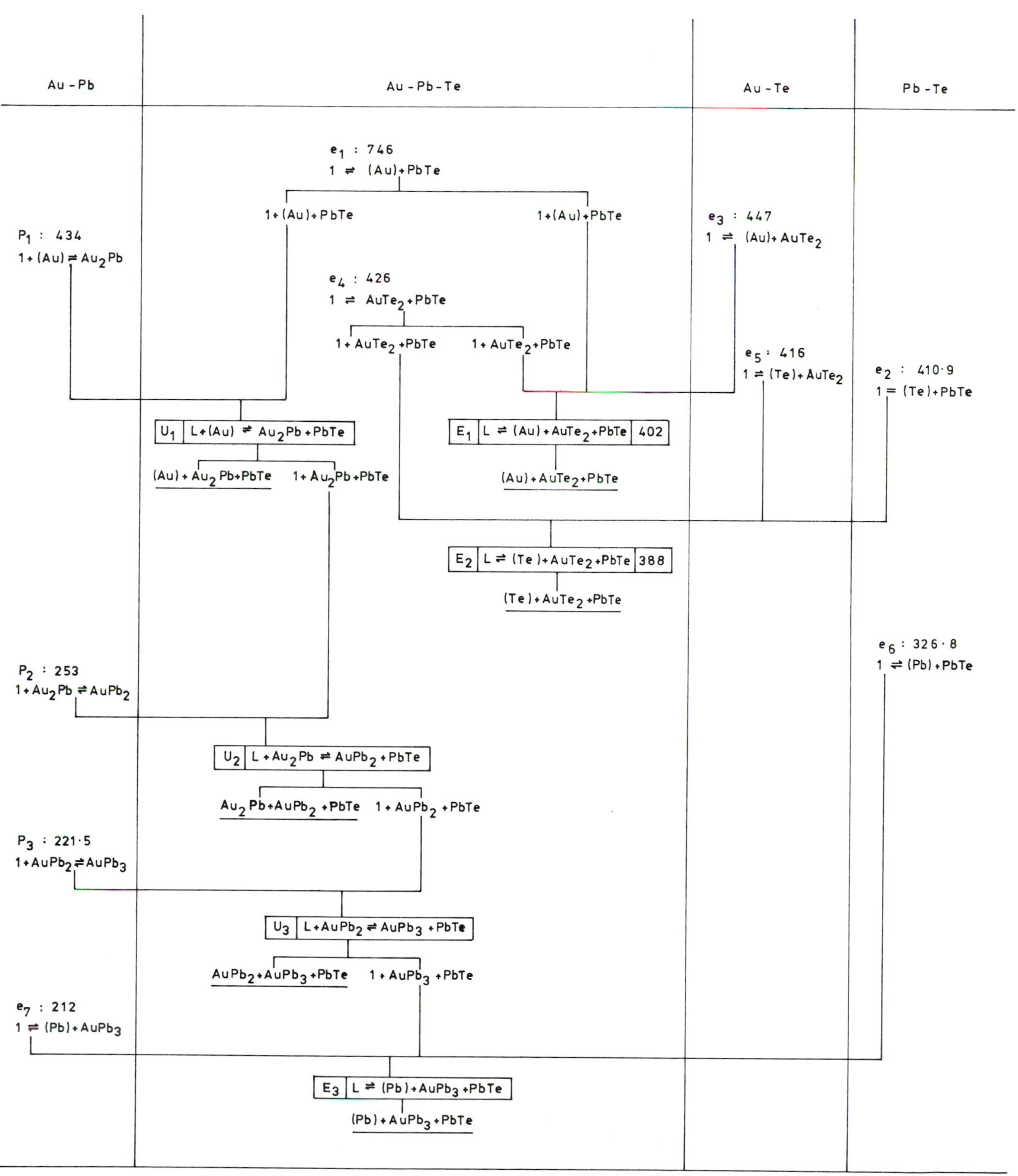

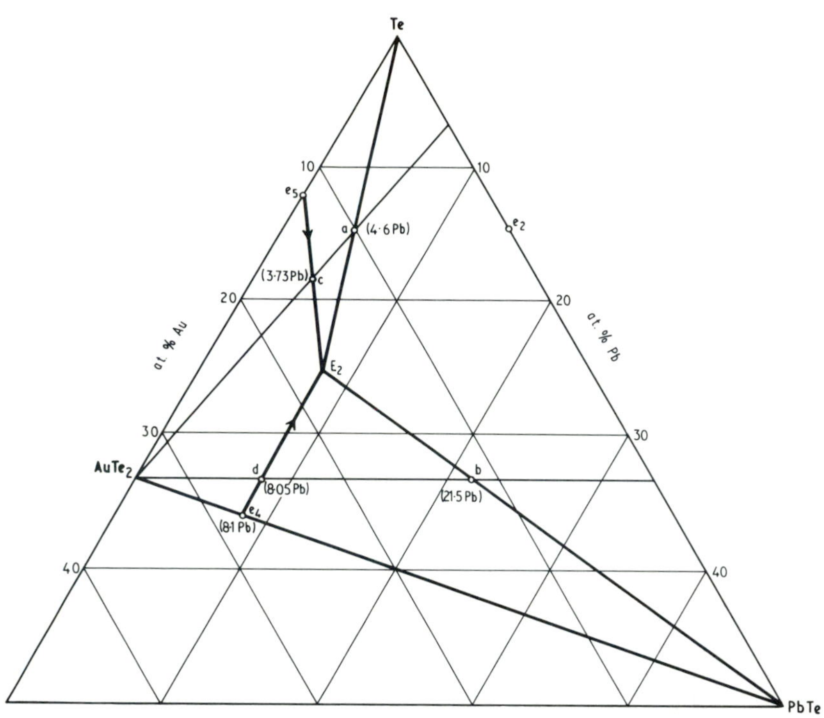

311 Determination of the ternary eutectic composition E_2

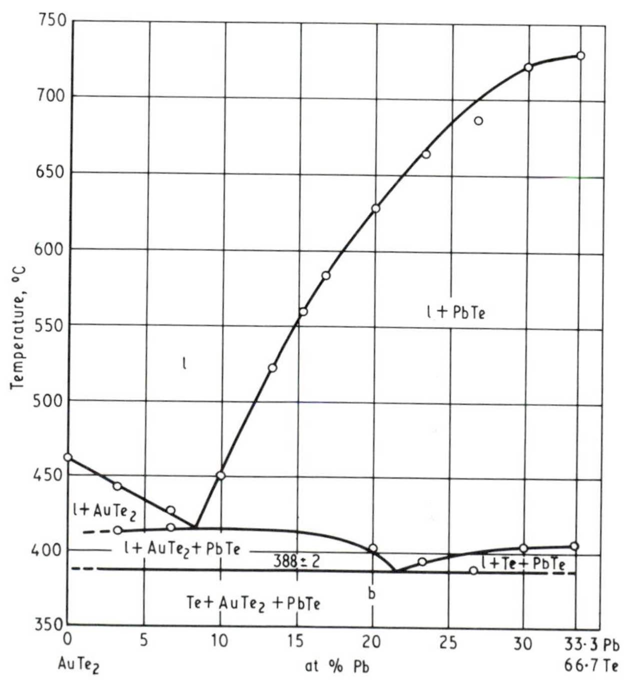

312 The $AuTe_2$–33·3 at.-% Pb, 66·7 at.-% Te section

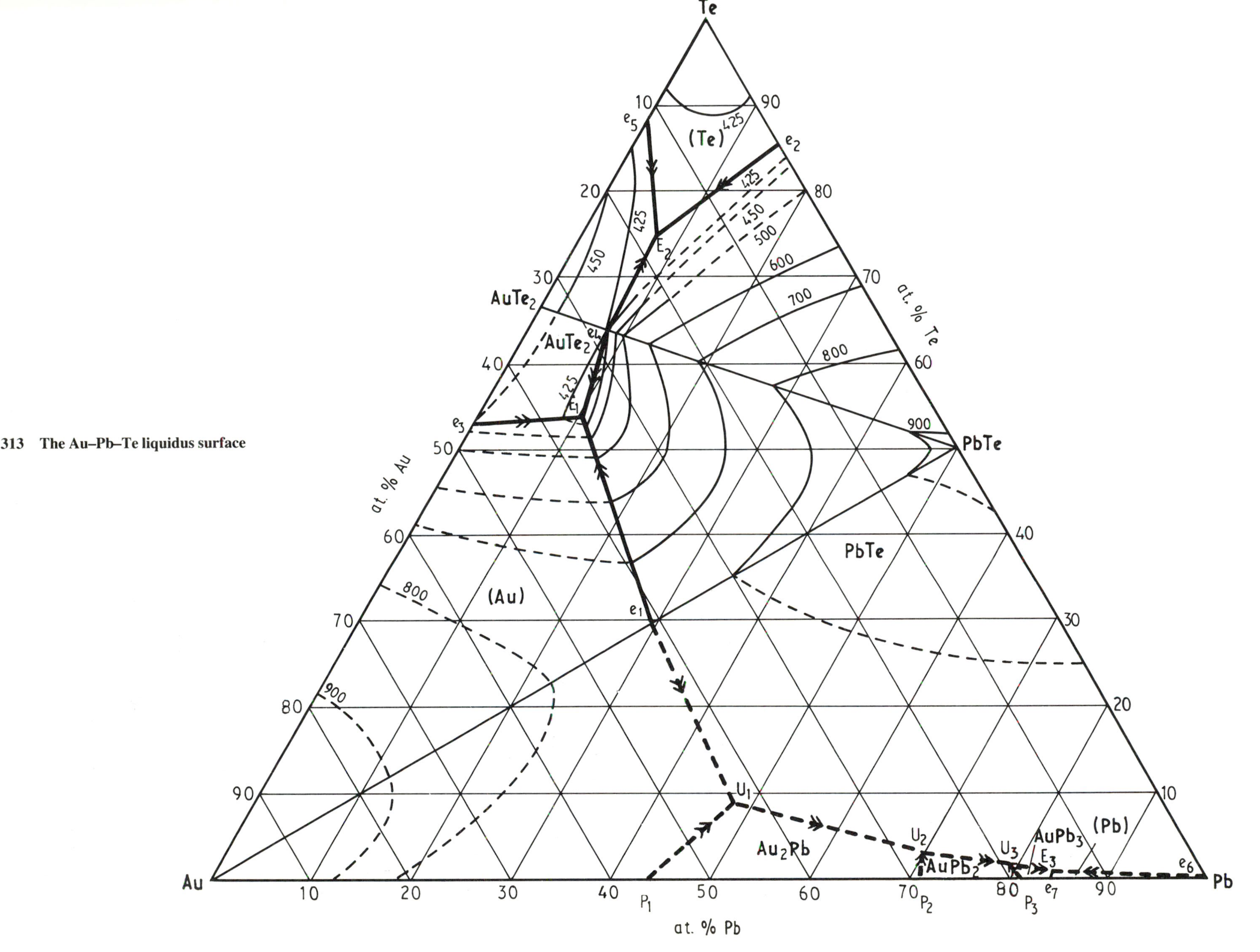

313 The Au–Pb–Te liquidus surface

83 Eva: D. S. Evans and A. Prince: *Mat. Res. Soc. Symp. Proc.*, 1983, **19**, 383–388

84 Oka: H. Okamoto and T. B. Massalski: *Bull. Alloy Phase Diagrams*, 1984, **5**, 172–177

86 Lin: J. C. Lin, K. C. Hsieh, R. C. Sharma and Y. A. Chang: 'Binary Alloy Phase Diagrams' (ed. T. B. Massalski), 1850–1851; 1986, ASM International

Au–Pb–Zn

[24 Tam] studied the partitioning of Au between immiscible liquid layers of Pb and Zn at 500 ± 5°C. Chemical analysis of the two layers gave a Au content of 4·08 wt-% in the Zn-rich layer and 0·06 wt-% in the Pb-rich layer. On the assumption that the Pb–Zn liquid immiscibility gap in the ternary is parallel to the Au–Pb and Au–Zn binary edges in the initial stages at 500°C, the Zn-rich layer will have a composition 1·4 Au, 0·3 Pb, 98·3 at.-% Zn and the Pb-rich layer will contain 0·06 Au, 93·9 Pb, 6·04 at.-% Zn. The tie line joining the two liquid compositions is skewed towards the Au–Pb binary edge. The degree to which the liquid immiscibility gap extends into the ternary system is unknown and study of the section AuZn–Pb would give a useful guide to the ternary equilibria.

[49 Jol] stated that the monovariant eutectic curve associated with the binary Pb–Zn eutectic rection occurs virtually at 325°C and follows the solubility product equation: log Au + 9 log Zn = −33·21, where Au and Zn are expressed in atom fractions. This relation implies equilibrium of the Pb-rich liquid with the epsilon phase of the Au–Zn binary system over the Au concentration range studied by [49 Jol]. The Au concentrations were minute, measured in gm per tonne. Conversion to at.-% gives compositions on the monovariant eutectic curve at 10^{-5} at.-% Au, 0·127 at.-% Zn; 10^{-4} Au, 0·095 Zn and $1{\cdot}6 \times 10^{-4}$ Au, 0·090 Zn.

[67 Peh] used a galvanic cell with a liquid chloride electrolyte to measure the EMF generated between pure liquid Zn and Zn in Pb–Au alloys. Discontinuities in the linear EMF-temperature plots gave an indication of liquidus temperatures. [67 Peh] studied Pb–Au alloys containing 1·2, 2·4, 3·6 at.-% Au with 1·2, 1·6, 2·0, 2·4 at.-% Zn added to each Au content. The tabulated liquidus temperatures are quoted as 640–680°C for all the alloys except the 3·6 Au, 2·0 Zn and 3·6 Au, 2·4 Zn compositions, which have liquidus temperatures of 680–720°C. A graphical plot of EMF-temperature data for alloys containing 2·0 at.-% Zn indicates discontinuities at 675°C for an alloy with 3·6 at.-% Au and 665°C for an alloy with 2·4 at.-% Au. The liquidus surface in the Pb-rich corner rises steeply with the addition of Zn to Pb–Au alloys. [67 Peh] determined the ternary interaction parameter, ϵ_{Zn}^{Au}, to be large, negative and strongly temperature dependent. The relationship is $\epsilon_{Zn}^{Au} = -\dfrac{27\,100}{T} + 21{\cdot}6$, where T = temperature Kelvin.

REFERENCES

24 Tam: G. Tammann and P. Schafmeister, *Z.anorg. Chem.*, 1924, **138**, 220–232

49 Jol: L. Jollivet, *Compt. rend.*, 1949, **228**, 1128–1130

67 Peh: R. D. Pehlke and K. Okajima, *Trans. Met. Soc. AIME*, 1967, **239**, 1351–1357

Au–Pd–Pt
(G. V. Raynor)

INTRODUCTION

Following earlier reviews of the alloys of palladium and platinum and their uses [43 Nem 1, 2], and limited experimental work by [47 Nem], the system was carefully examined by [55 Rau], using X-ray metallographic methods and relying chiefly upon the determination of the lattice spacings of alloys quenched from various temperatures after annealing for periods sufficient to give rise to sharp diffraction lines. The results indicate a miscibility gap in the solid state, arising from that which exists in the Au–Pt system, and that at temperatures exceeding the upper limit of the binary miscibility gap, the three components Au, Pd, and Pt form an uninterrupted series of solid solutions.

Subsequent work [71 Hay, 71 Kub] dealing with the thermodynamic quantities appropriate to the ternary system does not conflict with the interpretation in [55 Rau].

BINARY SYSTEMS

The Au–Pt equilibrium diagram evaluated by [85 Oka 1] is accepted, as is the Au–Pd system evaluated by [85 Oka 2]. No equilibrium diagrams appear to exist for the Pd–Pt alloys, though the system is reported [58 Han] as exhibiting complete solid solubility. [61 San], however, refers to empirical work by [59 Rau] which suggests that a miscibility gap might exist, with a critical temperature at ~800°C. No evidence for this was obtained in the work on the ternary system [55 Rau], in the course of which the lattice spacing/composition curve for the complete range of compositions from Pd to Pt was established. The presence of a miscibility gap is thus not proven, and is therefore not accepted.

SOLID PHASES

No ternary compounds have been reported. Table 101 summarizes crystal structure data for the binary systems. Since the only solid phase is the continuous solid solution (α) of the three components at temperatures close to the solidus, the system contains no pseudobinary sections or invariant reactions. The binary Au–Pt reaction at which α ⇌ α_1 (Au-rich) + α_2 (Pt-rich) projects into the ternary system over a range of compositions.

Table 101. Crystal structures

Solid phase	*Prototype*	*Lattice designation*
(Au)	Cu	cF4
(Pd)	Cu	cF4
(Pt)	Cu	cF4
(Au,Pd,Pt) solid solution	Cu	cF4

LIQUIDUS SURFACE

No definitive determination of the liquidus surface is available, but since the components are miscible in all proportions above ~1 250°C it must take the form of a continuous smooth surface, connecting the three melting points 1 769°C (Pt), 1 552°C (Pd), and 1 064°C (Au).

THE SOLID STATE

Raub and Wörwag [55 Rau] carried out careful X-ray lattice spacing determinations using four series of alloys, as follows:

Series 1: from 34·8 at.-% Au, 65·2 at.-% Pt to 49·6 at.-% Pd, 50·4 at.-% Pt

Series 2: from 40·8 at.-% Au, 59·2 at.-% Pt to the ternary composition 25·5 at.-% Au, 38·0 at.-% Pd, and 36·5 at.-% Pt

Series 3: from 49·7 at.-% Au, 50·3 at.-% Pt to 35·1 at.-% Au, 64·9 at.-% Pd

Series 4: from 49·7 at.-% Au, 50·3 at.-% Pt to the ternary composition 58·7 at.-% Au, 28·2 at.-% Pd, and 13·1 at.-% Pt.

The total number of compositions studied was 37, at 13 temperatures between 1 225 and 750°C. It may be noted that Series 2 runs from a composition approximating to the maximum temperature of the miscibility gap in the binary Au–Pt system in the direction of a composition close to 100% Pd.

The experimental method adopted was to measure the lattice spacings of the two fcc phases α_1 and α_2, contained in alloys whose compositions fell within the miscibility gap, as a function of quenching temperature. The temperature at which the curves for the α_1 and α_2 phases intersected for a given alloy composition was taken as that above which an alloy of the same composition was single phase. The temperature intervals were normally 50°C and the time of annealing before quenching was adjusted to give sharp X-ray diffraction lines. In this way a series of curves was obtained defining the boundary of the two-phase region for each of the sections mentioned above. Figure 314 illustrates the result for Series 2, and it will be seen that the maximum temperature of immiscibility for the binary Au–Pt alloys (determined as 1 257°C in this work) is rapidly decreased by the solution of Pd in the binary alloy containing 59·2 at.-% Pt. Thus with 40 at.-% Pd present, the critical temperature has fallen to 800°C. The results of this work, taken together with data on the binary Au–Pt alloys, allow constant temperature contours to be drawn for the ternary alloys, defining the extent of the miscibility gap at temperatures between 1 200 and 800°C inclusive. This is illustrated in Fig. 315, from which it is clear that at 800°C the binary Au–Pt miscibility gap projects into the ternary model up to a composition of approximately 41·5 at.-% Pd; at this temperature, therefore, two-phase alloys persist over a considerable range of compositions. As the temperature is increased the composition range corresponding to two-phase alloys shrinks and eventually disappears at 1 257°C on the binary Au–Pt face of the model (see Fig. 314). The regions of the ternary model corresponding to single and two-phase alloys are therefore separated by a surface approximating to half a dome, divided vertically. The contours in Fig. 315 are isothermal (horizontal) sections of this surface.

Thermodynamic quantities including heats of formation for the ternary alloys have been determined [71 Hay, 71 Kub].

REFERENCES

43 Nem 1: V. A. Nemilov: *Izvest. Sekt. Fiz. — Khim. Anal.*, 1943, **16**(1), 167–183

43 Nem 2: V. A. Nemilov: *Izvest. Sekt. Platiny.*, 1943, **19**, 21–14

47 Nem: V. A. Nemilov, T. A. Vidusova, A. A. Rudnitsky and M. M. Putsykina: *Invest., Sekt. Platiny.*, 1947, **20**, 176–224

55 Rau: E. Raub and G. Wörwag: *Z. Metall.*, 1955, **46**, 513–515

58 Han: M. Hansen and K. Anderko: 'Constitution of Binary Alloys'; 1958, New York, McGraw-Hill

59 Rau: E. Raub: *J. Less-Common Metals*, 1959, **1**, 3–18

61 San: V. V. Sanadze: *Dokl. Akad. Nauk. SSSR*, 1961, **140**, 133–136 (*Proc. Acad. Sci. USSR, Chem. Sect.*, 1961, **140**, 889–892)

71 Hay: F. H. Hayes and O. Kubaschewski: *Metal Sci.*, 1971, **5**, 35–40

71 Kub: O. Kubaschewski and J. F. Counsell: *Monatsh. Chem.*, 1971, **102**, 1724–1728

85 Oka 1: H. Okamoto and T. B. Massalski: *Bull. Alloy Phase Diagrams*, 1985, **6**, 46–56

85 Oka 2: H. Okamoto and T. B. Massalski: *Bull. Alloy Phase Diagrams*, 1985, **6**, 229–235

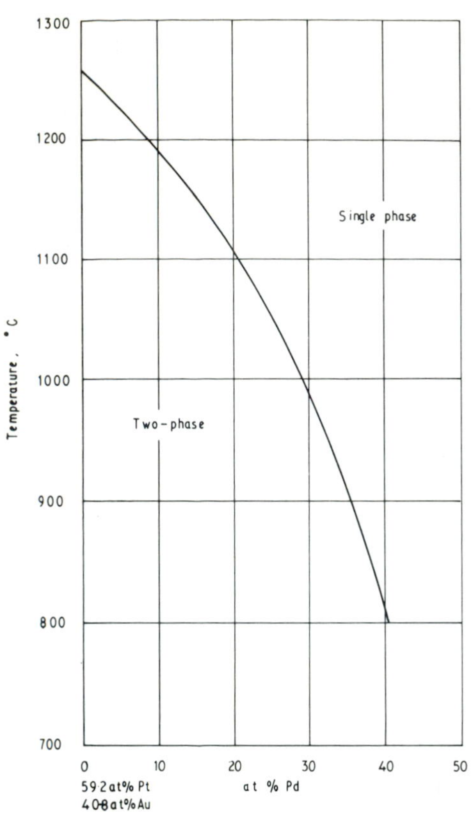

314 System Au–Pd–Pt: vertical section of ternary model for compositions corresponding to series 2 alloys, showing boundary between single and two-phase alloys, after [55 Rau]

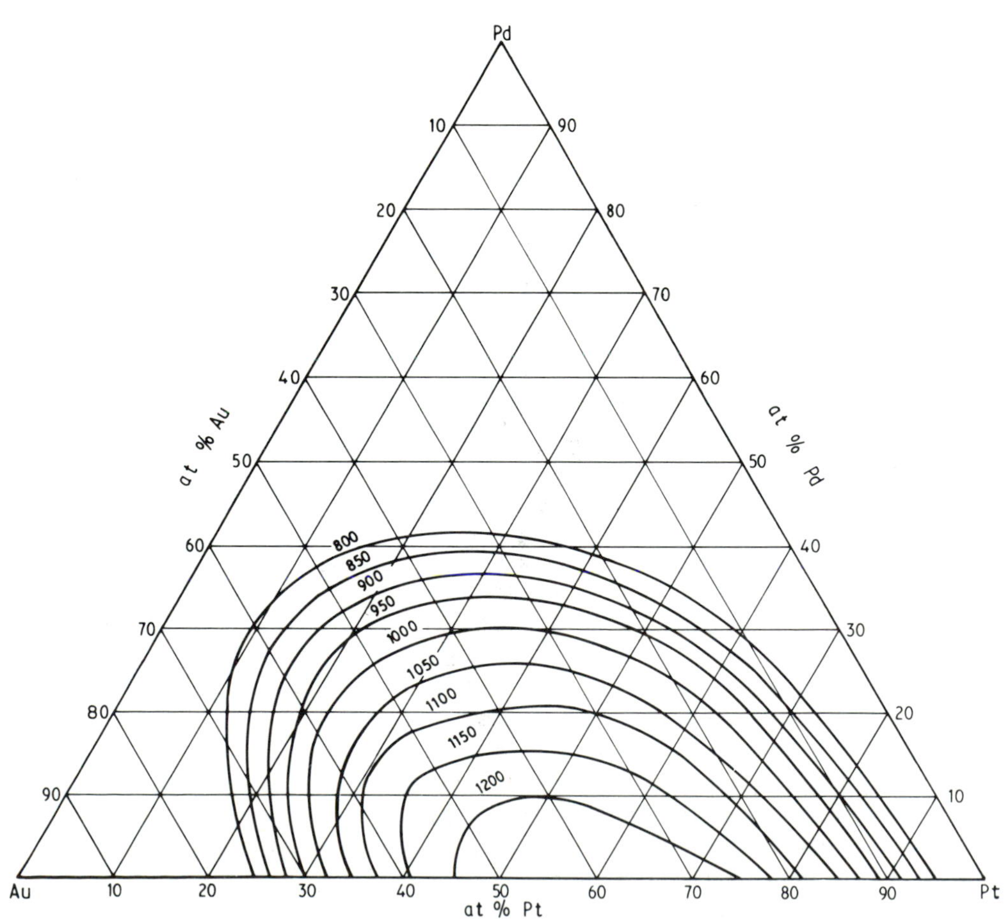

315 System Au–Pd–Pt: isotherms defining miscibility gap in ternary system, after [55 Rau]

Au–Pd–Rh
(G. V. Raynor)

INTRODUCTION

Alloys in the Au–Pd–Rh system have potential uses as electrical contacts and may lead to the partial replacement of platinum alloys for this purpose. Only one systematic examination of the system has been made in which metallography, mechanical properties, electrical resistivity, and thermoelectric effects were studied [60 Rud]. Because the alloys were slow-cooled and show liquid immiscibility, the phase diagrams represent the conditions described rather than an equilibrium form. A brief note [84 Zhm] describes the 1 000°C isothermal section in which the two-phase region (Rh) + (Au–Pd) extends to 68 at.-% Pd in the ternary system. [85 Sok] presented tie line data for the 1 000°C isothermal section.

BINARY SYSTEMS

The equilibrium diagram for the Au–Pd system evaluated by [84 Oka] is accepted. The Au–Rh system has been evaluated by [84 Oka]. There is considerable uncertainty as to the temperature and extent of the monotectic reaction. [84 Oka] gives two versions, one indicating liquid immiscibility from 17·8 to 90 at.-% Rh at 1 885°C, and the other liquid immiscibility from 1·4 to 70·5 at.-% Rh at 1 320°C.

Pd and Rh are completely miscible in the solid state above 845°C [65 Ell]. Below this temperature, the solid solution splits up into Pd-rich and Rh-rich solid solutions. The liquidus and solidus curves have not been accurately established.

SOLID PHASES

No ternary compounds have been reported. Clearly the possibility exists of complete solid solution formation between Au and Pd and between Pd and Rh, but not directly between Au and Rh.

TERNARY SYSTEM

[60 Rud] examined 59 ternary alloys made from Au prepared by reduction with ethanol, 99.9% pure Pd, and commercially refined Rh. 10 g samples were induction-melted in alumina crucibles using borax as the flux. The Rh and Pd were melted first and the Au then added. Most alloys were remelted twice. For metallographic examination, small samples were annealed for 120 h at 900–1200°C, depending on composition, and furnace cooled at 100°C per 24 h. The results for each alloy therefore refer to an unknown temperature at which diffusion became negligible. The extent of the immiscibility was determined by both macro- and micro-examination of cast specimens.

As a result of the work on cast samples, [60 Rud] produced a diagram showing the extent of penetration into the ternary system of the liquid immiscibility zone. According to this diagram, liquid immiscibility is not observed at Pd contents exceeding 47 at.-%. In Fig. 316, which shows the region in which two liquids may be observed in cast alloys, the results of [60 Rud] have been accepted except for the Au-rich alloys where the boundary has been very slightly modified to conform to the binary Au-Rh phase diagram showing liquid immiscibility from 1·4 to 70·5 at.-% Rh [84 Oka].

The constitution of alloys after slow cooling is given in Fig. 317. Most of the diagram corresponds to a two-phase (Rh) + (Au + Pd) region. The (Au–Pd) solid solution is confined to a narrow strip close to the Au–Pd axis for compositions <50 at.-% Pd, but widens considerably above this composition. The results may reflect the lack of equilibrium in slowly cooled alloys. Though the form of Fig. 317 establishes the nature of the low-temperature constitution, the shape of the (Au–Pd)-rich boundary must be treated with reserve.

The experiments carried out by [60 Rud] involving mechanical and electrical properties are consistent with the constitution shown in Figs 316 and 317 but do not serve to establish the phase boundaries more accurately.

No critical information exists with regard to the liquidus surfaces, although the form of Fig. 316 indicates that a dome-shaped surface corresponding with the separation into two liquids will exist, superimposed on a relatively smooth liquidus surface corresponding to the separation of the solid solution containing all these components. The publication by Zhmurko *et al.* [84 Zhm] is a short abstract describing the preparation of an unspecified number of alloys rich in Pd and Rh, by arc-melting on a water-cooled Cu hearth under an atmosphere of high-purity argon. The alloys were equilibrated by annealing for 100 h at 1 200°C, then 1 000 h at 1 000°C, and quenching in iced water. The two-phase region of the Au–Rh system penetrates into the ternary system up to 68 at.-% Pd. The tie lines are inclined at an angle to the Au–Rh binary edge, indicating a displacement of the critical point of the (Pd–Rh) solid solution towards the Pd–Rh binary. This is consistent with the (Rh) + (Au–Pd) phase region reaching the Pd–Rh binary at 845°C, corresponding to the critical temperature for the solid state miscibility gap in the Pd–Rh system. [85 Sok] gives tie lines for 20 at.-% Au alloys containing 35, 50, and 70 at.-% Rh annealed for 100 h at 1 000°C (Fig. 318). Phase compositions were determined by electron probe microanalysis. The data of [85 Sok] agrees with the descriptive abstract of [84 Zhm].

REFERENCES

60 Rud: A. A. Rudnitskii and A. N. Khotinskaya: *Zhur. Neorg. Khim.*, 1960, **5**, 2781–2790 (*Russian Inorg. Chem.*, 1960, **5**(12), 1341–1349)

65 Ell: R. P. Elliott: 'Constitution of Binary Alloys', 1965, New York, McGraw-Hill

84 Oka: H. Okamoto and T. B. Massalski: *Bull. Alloy Phase Diagrams*, 1984, **5**, 384–387

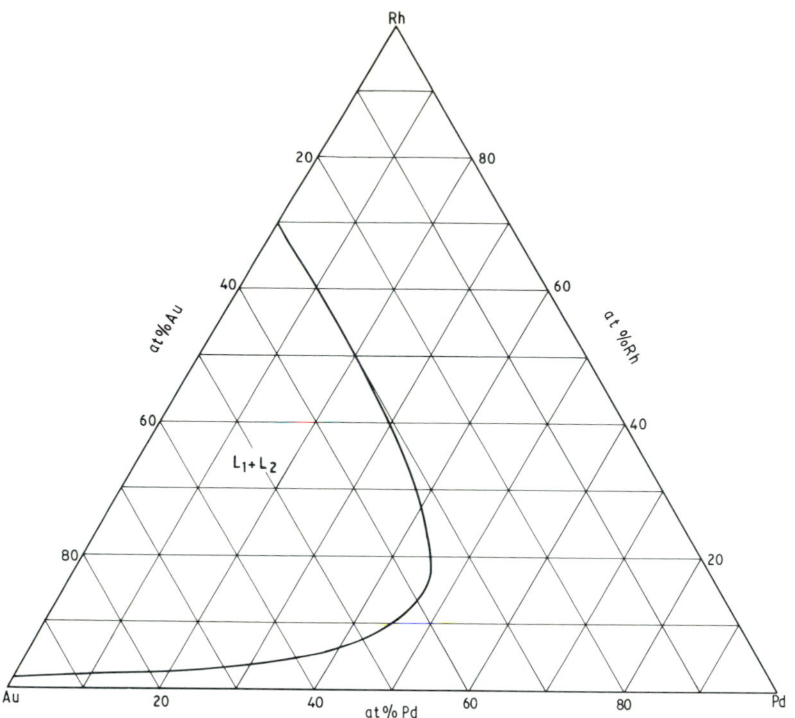

316 The region of two liquids in cast ternary alloys

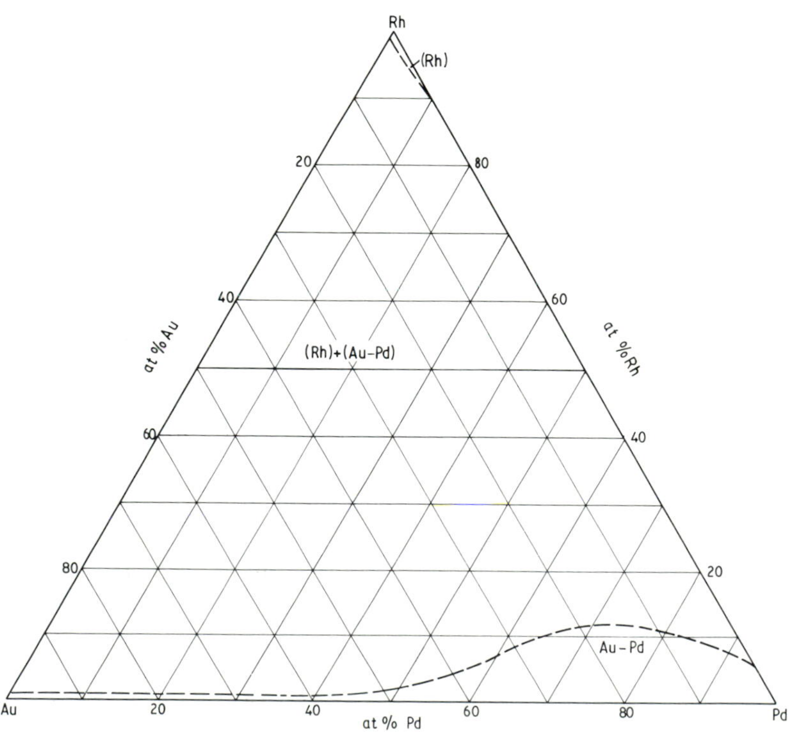

317 The constitution of ternary alloys after slow cooling

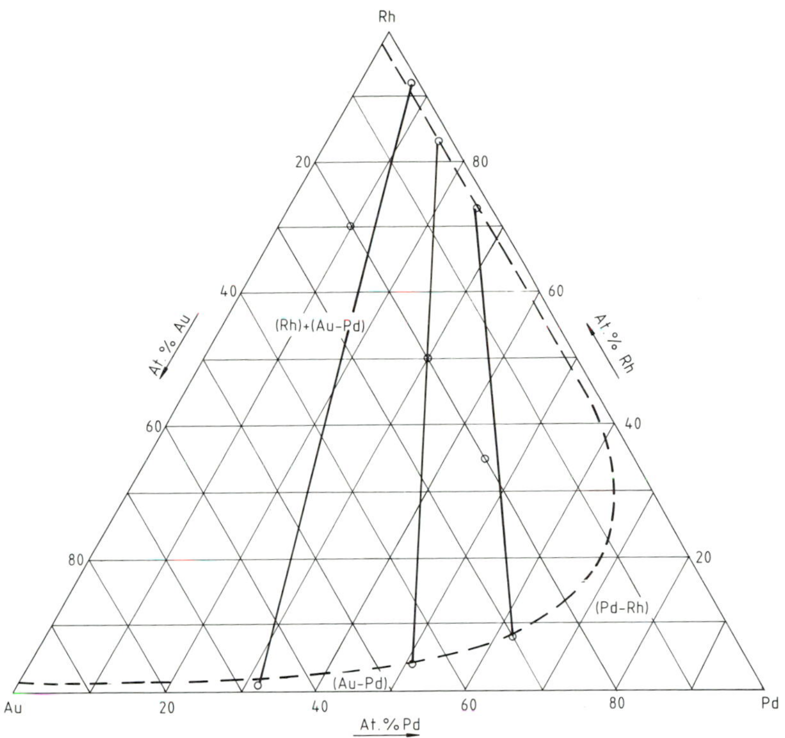

318 Isothermal section of the Au–Pd–Rh ternary system at 1 000°C [85 Sok]

84 Zhm: G. P. Zhmurko, E. M. Sokolovskaya, I. G. Sokolova, Yu. A. Brauer and E. I. Dobrodomova: *Vestn. Mosk. Univ., Ser. Khim.*, 1984, **25**(2), 207

85 Sok: E. M. Sokolovskaya, M. V. Raevskaya, G. P. Zhmurko, and V. V. Vasekin: 'Stable and metastable phase equilibria in metallic systems', (ed. M. E. Drits), 138–142; 1985, Moscow, Nauka

Au–Pd–Ru
(G. V. Raynor)

INTRODUCTION

The gold–palladium–ruthenium alloys are of importance in relation to the protection of ruthenium from oxidation at high temperatures by a coating of gold–palladium alloy, palladium having been added to raise the melting point of gold. Relatively little work, however, has been done on the system. Novikova [62 Nov] examined the effect of Ru on certain physical and mechanical properties of the Au–Pd alloys, and suggested that an extensive liquid immiscibility field existed in the ternary alloys. Since, however, the boundary drawn did not conform to the Pd–Ru phase diagram as now understood, it is not thought sufficiently established for further consideration. More recently isothermal sections at 1 200°C and 1 450°C were derived by Jones *et al.* [72 Jon]. In this work the phase boundaries were derived from conventional metallographic examination of samples homogenized at the relevant temperatures for not less than 65 h. and water-quenched. The initial preparation of specimens was by melting together of components of unspecified purity in an induction furnace under argon or, for ruthenium-rich alloys, in an argon-arc furnace. Two tie lines were established at each of the relevant temperatures. The phase compositions existing in quenched samples of two-phase alloys were measured by X-ray determination of lattice spacings. Further work would be helpful, to extend the range of temperatures within which information is available.

BINARY SYSTEMS

According to [85 Oka] the Au–Pd system consists of a continuous series of solid solutions, with liquidus and solidus curves rising smoothly from the melting point of Au to that of Pd. Information on the Au–Ru system is presented by [84 Oka] and is somewhat limited. The Ru-rich liquidus curve falls from the melting point of ruthenium to a monotectic point at ~2 012°C and approximately 73 at.-% Ru. At this temperature liquid of ~73 at.-% Ru is in equilibrium with a second liquid containing ~27 at.-% Ru and almost pure Ru. Between 2 012 and 1 066°C the diagram shows equilibrium between the gold-rich liquids and almost pure Ru. At 1 066°C liquid and (Ru) react together to give an extremely limited solid solution of Ru in Au. In the solid state alloys consist of mixtures of (Au) and (Ru) containing respectively very small amounts of the second component. The Pd–Ru system is, however, well established [65 Ell]; the liquidus falls from the melting point of Ru to a composition of about 99 at.-% Pd at a temperature variously reported as 1 579 and 1 593°C. At this temperature liquid reacts peritectically with the Ru-rich solid solution containing 17·5 at.-% Pd to form the Pd-rich solid solution containing 17% Ru. Though the existence of an intermediate phase at approximately 14·5–19 at.-% Ru, stable from 1 575 to 724°C has been suggested, careful work by Darling and Yorke [60 Dar] failed to confirm its presence, which is therefore not accepted in the present assessment. The solid solubilities of Ru in Pd and of Pd in Ru both decrease as the temperature is lowered.

SOLID PHASES

No ternary compounds have been reported. Table 102 summarizes the crystal structures known to be involved.

THE TERNARY SYSTEM

Fig. 319 shows the (Ru) + (Au–Pd) boundary as determined by [72 Jon] at 1 200°C, modified slightly close to the Pd–Ru axis to conform with the solubility limit of Ru in Pd. The symbol (Au–Pd) represents the wide mutual solution of Au and Pd.

Since binary Au–Pd alloys containing more than 91 at.-% Au are in the liquid state at 1 200°C, a hypothetical L + (Ru) + (Au–Pd) field has been included in Fig. 319. The L_1 + (Ru) field completes the description of the equilibria. The tie lines determined by [72 Jon] are also plotted in Fig. 319. Though the Pd-rich terminal tie line compositions do not lie exactly on the determined boundary, the direction of the tie lines is clearly established, and it is noteworthy that they diverge from compositions very close to the Ru corner of the diagram.

The equilibria for 1 450°C summarized in Fig. 320 are the same as those already described. The experimental (Ru) + (Au–Pd)/(Au–Pd) boundary due to [72 Jon] is shown, together with the tie lines reported by [72 Jon]; again they run from compositions close to the experimentally determined boundary. The work available is sufficient to establish the general form of the constitution at 1 200 and 1 450°C.

Table 102. Crystal structures

Solid phase	*Prototype*	*Lattice designation*
(Au)	Cu	cF4
(Pd)	Cu	cF4
(Ru)	Mg	hP2

REFERENCES

60 Dar: A. S. Darling and M. J. Yorke: *Platinum Metals Rev.*, 1960, **4**, 104–110

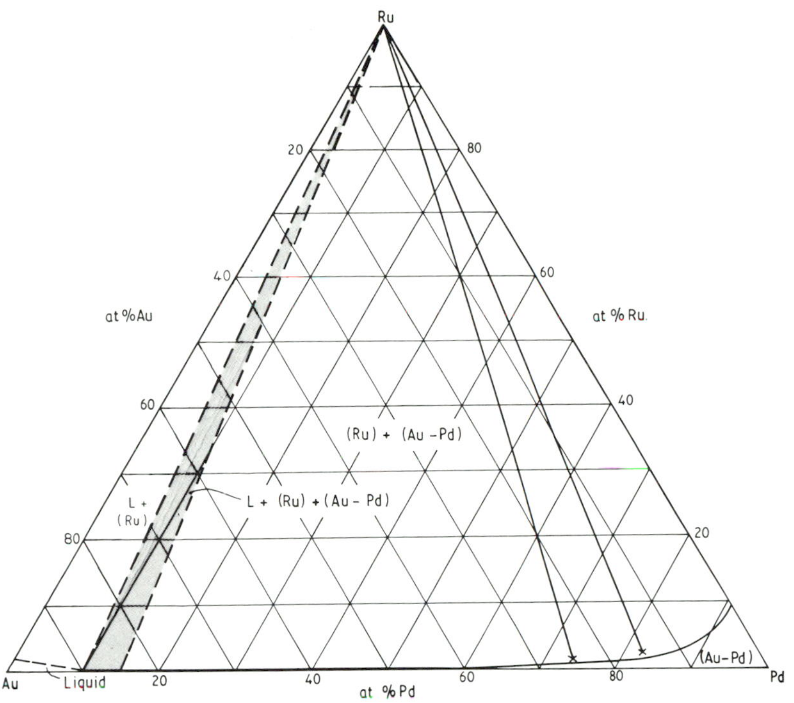

319 **System Au–Pd–Ru: constitution at 1 200°C, with experimentally determined tie lines, based on [72 Jon]**

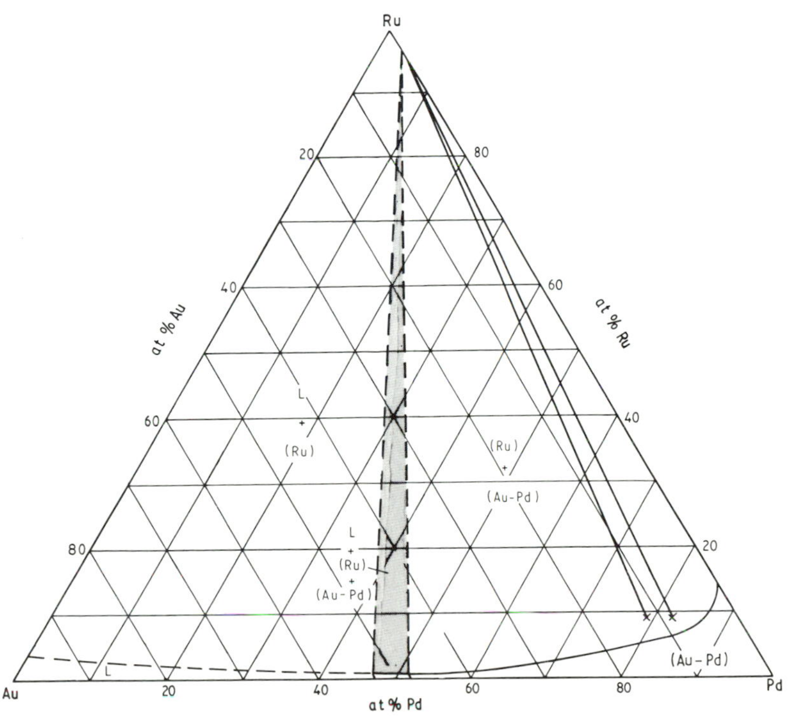

320 **System Au–Pd–Ru: constitution at 1 450°C, with experimentally determined tie lines, based on [72 Jon]**

62 Nov: O. A. Novikova: *Trudy Inst. Met. im A. A. Baikova, Akad. Nauk. SSSR*, 1962, (11), 148–154

65 Ell: R. P. Elliott: 'Constitution of Binary Alloys' (1st Suppl.); 1965, New York, McGraw-Hill

72 Jon: B. Jones, M. W. Jones and D. W. Rhys: *J. Inst. Metals*, 1972, **100**, 136–141

84 Oka: H. Okamoto and T. B. Massalski: *Bull. Alloy Phase Diagrams*, 1984, **5**, 388–390

85 Oka: H. Okamoto and T. B. Massalski: *Bull. Alloy Phase Diagrams*, 1985, **6**, 229–235

Au–Pd–Sb

Furuseth *et al.* [67 Fur] prepared eight ternary alloys on the $AuSb_2$–$PdSb_2$ section of the Au–Pd–Sb system. The alloys were prepared from 99·999% Pd, Sb and 99·995% Au by heating the elements in Al_2O_3 crucibles contained in evacuated and sealed silica tubes. Several ternary alloys were made by melting together $AuSb_2$ and $PdSb_2$. The alloys were either heated to 600°C for 14 days and then slow cooled, or heated to 1 000°C for 1 day and water quenched. In the latter case one would anticipate that all the alloys on the $AuSb_2$–$PdSb_2$ section would be molten before quenching, but no comment is made on the results obtained from the 1 000°C heat treatment. Great difficulty was experienced in achieving equilibrium; only reliable results were presented. X-ray powder diffraction analysis showed that Au substitutes for Pd in $PdSb_2$ up to the composition $Au_{0·25}Pd_{0·75}Sb_2$ (8·33 Au, 25 Pd, 66·67 Sb). This solid solution is followed at higher Au contents by a two-phase region in which $Au_{0·25}Pd_{0·75}Sb_2$ is in equilibrium with $AuSb_2$ at 600°C. The diffraction patterns for two-phase alloys between 10 Au, 23·33 Pd, 66·67 Sb and 30 Au, 3·33 Pd, 66·67 Sb ($Au_{0·3}Pd_{0·7}Sb_2$ and $Au_{0·9}Pd_{0·1}Sb_2$) showed on the X-ray spectra regions of distinct blackening between the sharp reflections corresponding to the $PdSb_2$ solid solution ($Au_{0·25}Pd_{0·75}Sb_2$) and $AuSb_2$. This could indicate that protracted annealing at ≤600°C may produce complete solid solubility between $AuSb_2$ and $PdSb_2$. Both these binary compounds have the FeS_2-type structure, cP12.

REFERENCES

67 Fur: S. Furuseth, K. Selte and A. Kjekshus: *Acta Chem. Scand.*, 1967, **21**, 527–536

Au–Pd–Te

[71 El-B] reported that an alloy containing 7 Au, 70 Pd, 23 at.-% Te after annealing for 1 day at 480°C contained Pd_3Te + (Pd). The (Pd) solid solution contained an unspecified Au content.

REFERENCE

71 El-B: M. El-Boragy and K. Schubert: *Z. Metall.*, 1971, **62**, 314–323

Au–Pd–V

Köster *et al.* [62 Kös] studied the composition dependence along the 9 at.-% V section of the electrical resistivity, Hall coefficient, and thermoelectric force for alloys quenched from 800°C, slow cooled from 800°C, and cold worked up to 70%. The maximum change in properties occurred for an alloy containing 45·5 at.-% Pd, 9 at.-% V. A similar behaviour was found at 45 at.-% Pd in the binary Au–Pd system. The property changes appear to be associated with changes in the degree of short-range order in the Au–Pd solid solution.

REFERENCE

62 Kös: W. Köster, H.-P. Rave and Y. Takeuchi: *Z. Metall.*, 1962, **53**, 749–753

Au–Pd–W
(G. V. Raynor)

INTRODUCTION

The gold–palladium–tungsten system was examined by [72 Jon] in the course of work to test the feasibility of protecting tungsten from oxidation by a coating of gold–palladium alloy. The effect of tungsten on certain physical and mechanical properties of gold–palladium alloys was also investigated by Pravoverov and Begatova [69 Pra], who also constructed a phase diagram relating to an unspecified temperature described as near to room temperature.

THE BINARY ALLOYS

The equilibrium diagram given by [85 Oka 1] for the Au–Pd system is accepted; the liquidus and solidus curves fall smoothly from the melting point of Pd to that of Au, and in the solid state a complete series of solid solutions is formed. No equilibrium diagram exists for the system Au–W. According to [85 Oka 2] a eutectic may be formed at ~0·1 at.-% W and 1 063°C. The phase diagram for the system Pd–W given in [69 Shu] may be accepted. The system contains no intermediate phases, and the major feature is the peritectic reaction L + (W) $\rightleftharpoons$ (Pd) at 1 815°C. The liquidus curve, which falls from the melting point of W to that of Pd, is not accurately established, while the solidus curve relating to the (W) solid solution is also uncertain. The solid solubility of Pd in W is approximately 5 at.-% at 1 815°C, decreasing to 1·5 at.-% Pd at 1 100°C. The solubility of W in Pd reaches a maximum of

22 at.-% at 1 815°C, and, according to the work described in [69 Shu], is almost independent of temperature, decreasing to 21·5 at.-% W at 1 100°C.

THE SOLID PHASES

No distinct ternary intermediate phases have been reported, so that in the ternary system the only solid phases are the (Au–Pd) solid solution (fcc) and the restricted solid solution of Pd in W (bcc). The structures are summarized in Table 103.

Table 103. Crystal structures

Solid phase	*Prototype*	*Lattice designation*
(W)	W	cl2
(Au–Pd) solid solution	Cu	cF4

THE TERNARY SYSTEM

The ternary alloys for the determination of equilibria were prepared by induction-melting under argon. Specimens were homogenized at 1 200 or 1 450°C for not less than 65 h and quenched in water before examination. The lattice spacings of the phases present in selected two-phase alloys containing appreciable quantities of both constituents were then determined by the use of the Debye–Scherrer X-ray diffraction technique. The derivation of phase-compositions from measured lattice spacings requires a detailed knowledge of the Au–Pd lattice spacings and the effect on them of the solution of W. The lattice spacing/composition curve for the solid solution of Pd in W is also required. These lattice spacing relationships are not explicitly described in [72 Jon], but results are given for the compositions of the conjugate phases in two duplex alloys at 1 200°C and also at 1 450°C. The tie lines at 1 450°C are shown in Fig. 321, and it is clear that the solubility of W in the (Au–Pd) solid solution is limited (ranging from 1 at.-% W at 40 at.-% Au to 3·5 at.-% W at 10 at.-% Au); the solubility of W in Pd is taken from the accepted binary diagram. At this temperature Au–Pd alloys with compositions between 50 and 55 at.-% Pd consist of both liquid and solid phases, requiring a L + (W) + (Au–Pd) three-phase triangle, as shown in Fig. 321. Owing to the small solid solubility of W in the (Au–Pd) solid solution, the position of this three-phase triangle cannot vary significantly from that shown. The isothermal section at 1 200°C is shown in Fig. 322. The tie line results are very similar to those at 1 450°C and again the solubility value for W in Pd is taken from the binary data. The semi-liquid region in the Au–Pd system now occurs at a much more Au-rich composition, with consequent change in the position of the L + (W) + (Au–Pd) three-phase triangle. At temperatures below the melting point of Au the alloys are entirely solid, and the constitution would then be represented by an isothermal of the same general form as that of Fig. 322, but with the (W) + (Au–Pd) region extending to the Au–W axis. A phase diagram of this type was proposed, for an unspecified temperature considered to be near to 'normal' or room temperature, in the work of [69 Pra], in which the main aim was to investigate the effect of W on the physical and mechanical properties of Au–Pd alloys. The diagram, based on metallographic examination of alloys annealed for 100h at 1 000°C and slowly cooled, shows equilibrium between (W) and the (Au–Pd) solid solution containing W, with the (Au–Pd) solvus curve joining the Au corner of the phase diagram to the point on the (Pd–W) axis representing the solid solubility of W in Pd. This solubility is, however, given as about 14·8 at.-% W, which is considerably lower than that given by [69 Shu]. The convexity of the ternary boundary towards the Pd corner of the diagram is also less pronounced than in Figs 321 and 322. The diagram is not reproduced since the temperature to which it refers is not known; it does, however, confirm the general nature of the equilibria.

REFERENCES

69 Pra: N. L. Pravoverov and I. S. Begatova: *Izvest. Akad. Nauk SSSR., Metally.*, 1969, (6), 177–180

69 Shu: F. A. Shunk: 'Constitution of Binary Alloys' (2nd Suppl.); 1969, New York, McGraw-Hill

72 Jon: B. Jones, M. W. Jones and D. W. Rhys: *J. Inst. Metals*, 1972, **100**, 136–141

85 Oka 1: H. Okamoto and T. B. Massalski: *Bull. Alloy Phase Diagrams*, 1985, **6**, 136–137

85 Oka 2: H. Okamoto and T. B. Massalski: *Bull. Alloy Phase Diagrams*, 1985, **6**, 229–235

Au–Pd–Zn
(G. V. Raynor)

INTRODUCTION

Although the constitution of the system Au–Pd is simple, those of the systems Au–Zn and Pd–Zn are relatively complex, and no extensive experimental work on the system Au–Pd–Zn has been carried out. The relationship between the ordered phases which occur at the equiatomic compositions in the Au–Zn and Pd–Zn systems has, however, been examined by Krompholtz and Weiss [76 Kro] in terms of the section AuZn–PdZn. Limited work [57 Wil] has been done on the extent of the Au_4Zn phase region at 200°C. A non-isothermal section by [87 Dob] disagrees with the conclusions of [76 Kro].

BINARY SYSTEMS

The equilibrium diagram given by [85 Oka] for the Au–Pd system is accepted: the liquidus and solidus curves fall smoothly from the melting point of Pd to that of Au. In the solid state a complete series of solid solutions is formed.

The Au–Zn system is characterized by the existence of

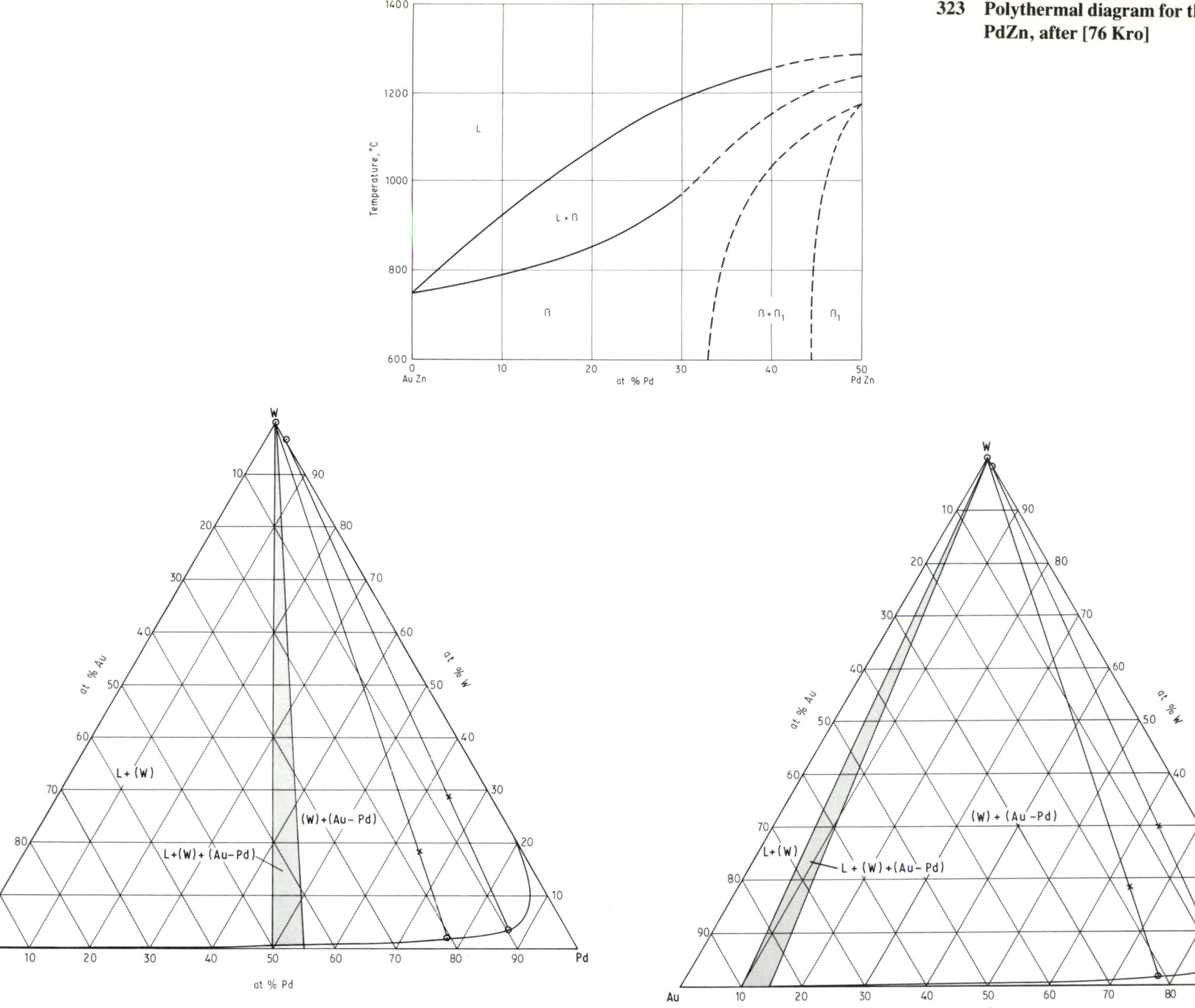

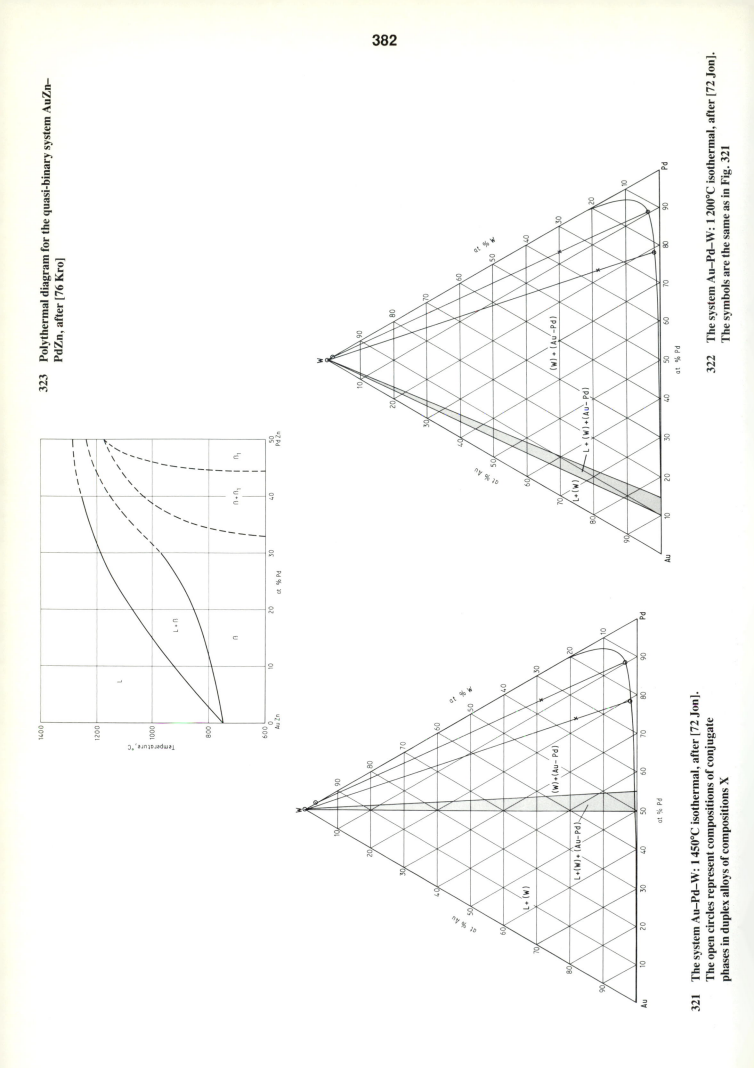

323 Polythermal diagram for the quasi-binary system AuZn–PdZn, after [76 Kro]

321 The system Au–Pd–W: 1 450°C isothermal, after [72 Jon]. The open circles represent compositions of conjugate phases in duplex alloys of compositions X

322 The system Au–Pd–W: 1 200°C isothermal, after [72 Jon]. The symbols are the same as in Fig. 321

at least nine intermediate phases [87 Oka]. There is a considerable solid solubility of Zn in Au, reaching a maximum of 31% Zn. The liquidus and solidus curves for this solid solution fall from the melting point of Au to a eutectic at 683°C (34% Zn), at which L $\rightleftharpoons$ (Au) + β, where (Au) represents the Au-rich solid solution. The β phase melts at a maximum of the liquidus of 751°C, which also corresponds with the composition AuZn. The β phase has a wide range of homogeneity and the crystal structure is of the CsCl (cP2) type. At 654°C and 63% Zn, the reaction L $\rightleftharpoons$ β + γ occurs: the γ phase has the typical γ-brass crystal structure. With further increase in Zn content, the phases γ_3 and ϵ are formed peritectically, and finally the Zn-rich solid solution is formed peritectically from liquid and ϵ. A phase γ_2 is formed from γ and γ_3 in the solid state at ~520°C. The phases γ_2 and γ_3, neither of which appears to have the ideal γ-brass structure, and the ϵ phase are not particularly relevant to the interpretation of the limited amount of work on the Au–Pd–Zn system. A series of gold-rich phases, one of which involves equilibrium with β, also exists below ~420°C and 31% Zn. The relevant phase boundaries are not known with certainty, and are only involved to a limited extent in the work so far available on the ternary system.

The system Pd–Zn is also somewhat complex, and in [58 Han] two variants are given. In relation to the discussion of the ternary system, the important feature is the existence of ordered body-centred cubic phases, over a composition range which reaches a maximum of ~33 to 66·5% Zn at 845°C. A bcc phase with the CsCl (cP2) structure, here denoted β, crystallizes from the liquid at ~1 375°C and a composition of 33% Zn. It enters into equilibrium with the palladium-rich solid solution (Pd) as a result of the eutectic reaction: L = (Pd) + β. At 845°C and 75% Zn occurs the reaction: L = β + a phase of the γ-brass type. It is uncertain whether this phase is of the ideal γ-brass structure, transforming in the solid state to a more complex orthorhombic variant (γ′), or whether the reverse is the case, with γ′ crystallizing from the melt and transforming with decrease in temperature to γ_1. It is considered that the former possibility is the more likely. The important feature of the β phase is that it transforms in the solid state, at a temperature depending on composition, into β_1 with a structure of the ordered AuCu I type (tP4). The maximum transformation temperature is 1 170°C at the composition PdZn. According to the phase diagram in [58 Han], the CsCl and AuCu I-type structures are separated by a two-phase region. The diagram also contains a δ-phase (Pd_2Zn) formed in the solid state at ~700°C. While the main features of the constitution of Pd–Zn alloys are clear, further detailed work is needed.

SOLID PHASES

No distinct ternary phases have been reported in the system Au–Pd–Zn. The structures of the phases referred to above, and which are relevant to the discussion of the limited amount of work available for the ternary system, are summarized in Table 104.

Table 104. Crystal structures of relevant phases

Solid phase	*Prototype*	*Lattice designation*
Au–Pd solid solution	Cu	cF4
Au_4Zn	Doubly-faulted Cu_3Au	Doubly-faulted cP4
Au_3Zn	Faulted Cu_3Au	Faulted cP4
Au–Zn β	CsCl	cP2
Au–Zn γ	γ-brass (Cu_5Zn_8)	cI52
Au–Zn γ_2	—	Cubic, 8 atoms/cell
Au–Zn γ_3	—	Hexagonal
Au–Zn ϵ	Mg	hP2
Pd–Zn δ	$PbCl_2$	Orthorhombic oP12
Pd–Zn β, β′, β″	CsCl	cP2
Pd–Zn β_1	AuCuI	tP4
Pd–Zn γ_1	γ-brass (Cu_5Zn_8)	cI52
Pd–Zn γ′	—	Orthorhombic
Zn	Mg	hP2

THE TERNARY SYSTEM

The study of the quasi-binary section AuZn–PdZn reported in [76 Kro] was carried out using metallographic and X-ray powder diffraction methods, thermal analysis and differential thermal analysis. Fifteen alloys of the series $Au_{1-x}Pd_xZn$ were prepared by melting together the appropriate quantities of 99.99% pure Au, Pd, and Zn, under argon in sealed quartz tubes. Synthetic compositions were accepted, following determinations of the weight loss in preparation. For the X-ray diffraction work, powdered samples were annealed at 530°C for a period in excess of 100 h. Figure 323 summarizes the results obtained; the solid lines indicate the differential thermal analysis results from heating curves, while the dashed lines are, for the liquidus and solidus, reasonable extrapolations to binary results and also, in the solid state, summarize the indications of the X-ray results. The diagram shows that there is a considerable liquidus–solidus separation, and that, at the higher temperatures in the solid state, a complete series of solid solutions is formed between the Au–Zn and Pd–Zn β-phases which have the CsCl structure. There is also a miscibility gap in which the CsCl (β) and AuCu I-type (β_1) structures are in equilibrium. The temperature dependence of the boundaries of this gap was not experimentally determined, but the extent of the two-phase area at 530°C was established by lattice spacing measurements. From the results of this work the solid solubility limit of β_1 PdZn in the CsCl-type β AuZn is at approximately 32·5% Pd. The AuCu I-type structure is stable, at 530°C, from 44·5 to 50% Pd. Alloys within the miscibility gap were characterized as two-phase both metallographically and by X-ray phase recognition.

Lattice spacing experiments were also carried out on alloys of the series $AuPd_xZn_{1-x}$ containing 50% Au, and the homogeneous AuZn phase was found to persist only as far as $0{\cdot}1 \leqslant x \leqslant 0{\cdot}2$. Thus the alloy containing 5% Pd and 45% Zn was homogeneous; those with greater amounts of Pd contained additional phases.

[87 Dob] examined the structure of 46 ternary alloys by both metallography and X-ray powder diffraction analysis, although they only tabulated the phase constitution of 12 ternary alloys. Alloys were prepared from 99·99% Au, 99·95% Pd and 99·9% Zn by melting in evacuated quartz

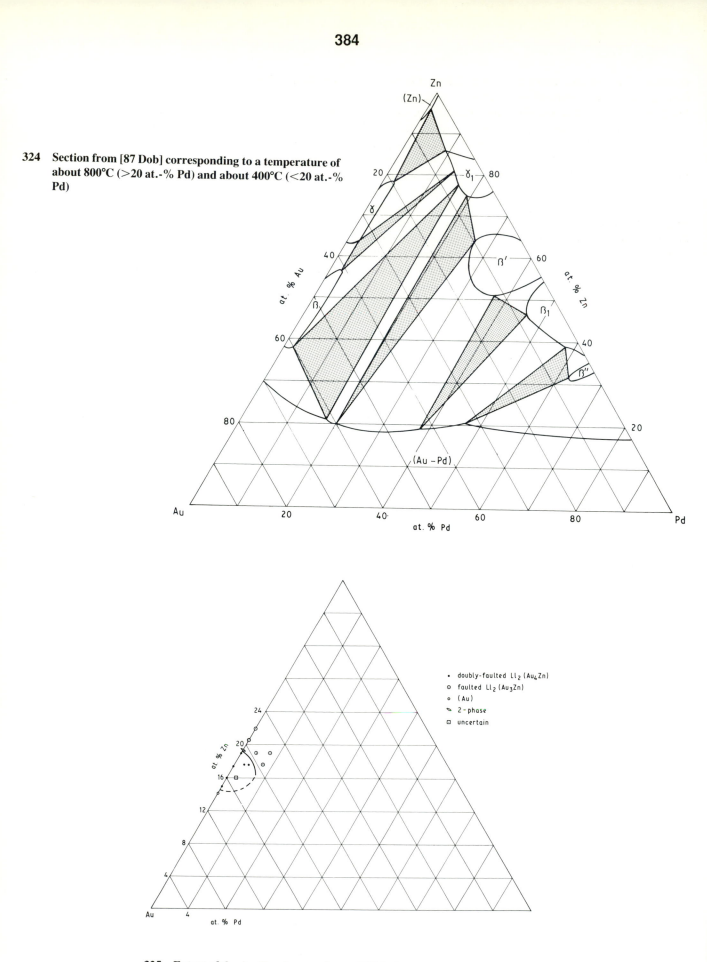

324 **Section from [87 Dob] corresponding to a temperature of about 800°C (>20 at.-% Pd) and about 400°C (<20 at.-% Pd)**

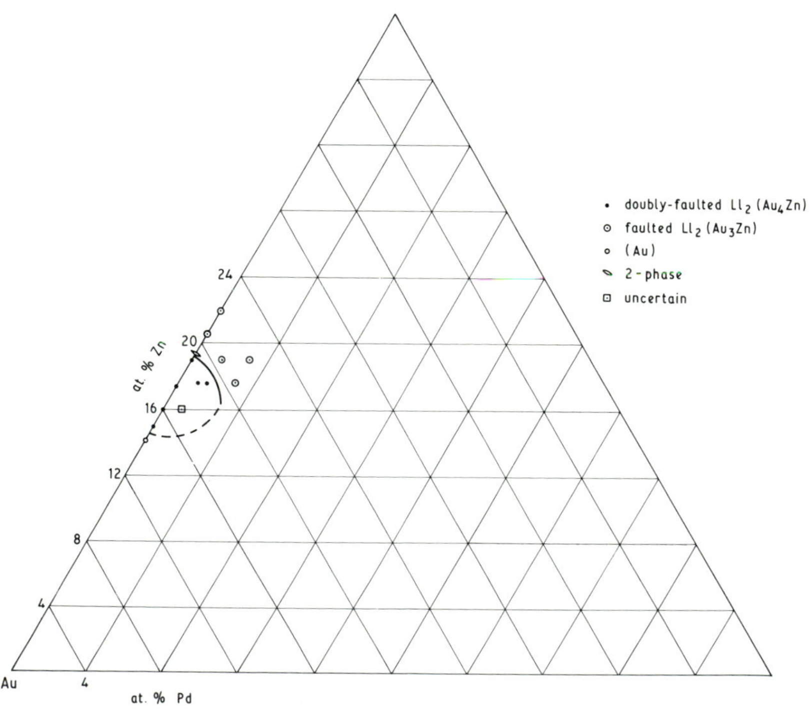

325 **Extent of the Au_4Zn phase region at 200°C [57 Wil]**

crucibles. They were subsequently annealed at temperatures ranging from 400 to 850°C. The variable annealing temperatures do not allow the representation of the data by an isothermal section. Figure 324 is a section [87 Dob] which corresponds to a temperature of about 800°C for alloys containing more than 20 at.-% Pd and to a temperature of about 400°C for alloys with less than 20 at.-% Pd. Only the Au–Zn γ phase was detected by [87 Dob]; no comment is made on the absence of the γ_2, γ_3 and ϵ phases, Table 104. The section in Fig. 324 does not confirm the results of [76 Kro], Fig. 323, for the AuZn–PdZn vertical section. According to [87 Dob] the (Au–Pd) solid solution forms three-phase equilibria with $\beta'' + \beta_1$, $\beta_1 + \beta'$, $\beta' + \gamma_1$ and $\beta + \gamma_1$. The γ_1 phase forms three-phase equilibria with β + (Au–Pd), $\beta + \gamma$ and γ + (Zn). There is a need for the determination of true isothermal sections for the Au–Pd–Zn system to resolve the discrepancy between [76 Kro] and [87 Dob].

No examination has been made of the liquidus and solidus surfaces, and the role of the gold-rich ordered phases which occur in the Au–Zn system below 420°C and 30% Zn in the ternary system are uncertain. [57 Wil] used the X-ray powder diffraction method on six ternary alloys, annealed for 8 days at 200°C, to delineate the phase region for the doubly-faulted $L1_2$-type structure based on Au_4Zn (Fig. 325).

REFERENCES

57 Wil: M. Wilkens and K. Schubert, *Z. Metallkunde*, 1957, **48**, 550–557

58 Han: M. Hansen and K. Anderko, *Constitution of Binary Alloys*, 1958, New York, McGraw-Hill

76 Kro: K. Krompholtz and A. Weiss, *J. Less-Common Metals*, 1976, **50**, 213–222

85 Oka: H. Okamoto and T. B. Massalski, *Bull. Alloy Phase Diagrams*, 1985, **6**, 229–235

87 Oka: H. Okamoto and T. B. Massalski, 'Phase Diagrams of Binary Gold Alloys', 1987, pp. 331–340. ASM INTERNATIONAL

87 Dob: M. Dobersek and I. Kosovinc, 'Precious Metals 1987', Proc. 11th Internat. Conf., Brussels, June 1987. Ed: G. Vermeylen and R. Verbeeck. pp. 325–334. Internat. Precious Metals Inst., Allentown, Pa, 1987

Au–Pd–Zr

Wang *et al.* [67 Wan] studied alloys on the 25 at.-% Zr section by X-ray diffraction techniques. They report the presence of three ordered close-packed phases on this section, but no further details were given. Au_3Zr has a Cu_3Ti-type structure (oP8) and Pd_3Zr a Ni_3Ti-type structure (hP16).

REFERENCE

67 Wan: R. Wang, B. C. Giessen and N. J. Grant: *Abst. Bull. Inst. Metals Div., Met. Soc. AIME*, 1967, **2**(2), 120

Au–Pt–Re

Au and Pt are completely miscible in the liquid state and in the solid state above 1 260°C. A solid state miscibility gap forms at 1 260°C and 61 at.-% Pt to give equilibrium of (Au) and (Pt) solid solutions [85 Oka]. The Pt–Re system is a simple peritectic system; the Au–Re system [84 Oka] shows virtual immiscibility in both the liquid and solid state. [66 Fla 1] reported the only work on the Au–Pt–Re ternary system. It includes data from [60 Rex] and [66 Fla 2]. Using 99·99%Au, 99·9% Pt, 99·8% Re [60 Rex] induction melted Au–19·1 at.-% Pt and Au–97·7 at.-% Pt and studied the diffusion of Re into the binary Au–Pt alloys after 256 h at 600°C. A diffusion zone was detected between Re and the Au–Pt alloys. The X-ray parameter of the Au–19·1 at.-% Pt alloy increased by 0·0006 nm and that of the Au–97.7 at.-% Pt alloy by 0·0005 nm. [66 Fla 2] prepared powder compacts of Au–20·1 at.-% Pt and Au–20·1 at.-%Pt–2·1 at.-% Re and heat treated the compacts at 900°C for 140 and 284 h. After 140 h at 900°C the Au–20·1 at.-% Pt alloy was two-phase, containing an fcc Au-rich phase and an fcc Pt-rich phase whose lattice parameters corresponded to Au–27 at.-% Pt and Au–92 at.-% Pt; it remained two-phase after 284 h at 900°C. It is difficult to accept these results since it is not possible for an alloy with an overall composition of 20·1 at.-% Pt to generate two phases containing 27 and 92 at.-% Pt respectively. The ternary alloy was two-phase after 140 h at 900°C and after 284 h at 900°C. A sample of the ternary alloy prepared some 45 months earlier was single phase after 284 h at 900°C. Although somewhat inconclusive, the results indicate that Re has a small solubility in the Au-rich phase of the Au–Pt binary. The results of [60 Rex] suggest a small Re solubility in the Pt-rich phase. In the Pt–Re system at 900°C, 40 at.-% Re is soluble in (Pt). One would anticipate the 900°C isothermal section of the Au–Pt–Re system to have the equilibria suggested in Fig. 326.

REFERENCES

60 Rex: J. Rexer: Thesis, Univ. Halle, 1960

66 Fla 1: K. Flammiger and F. Sauerwald: *Neue Hütte*, 1966, **11**, 218–221

66 Fla 2: K. Flammiger: Thesis, Univ. Halle, 1966

84 Oka: H. Okamoto and T. B. Massalski: *Bull. Alloy Phase Diagrams*, 1984, **5**, 383

85 Oka: H. Okamoto and T. B. Massalski: *Bull. Alloy Phase Diagrams*, 1985, **6**, 46–56

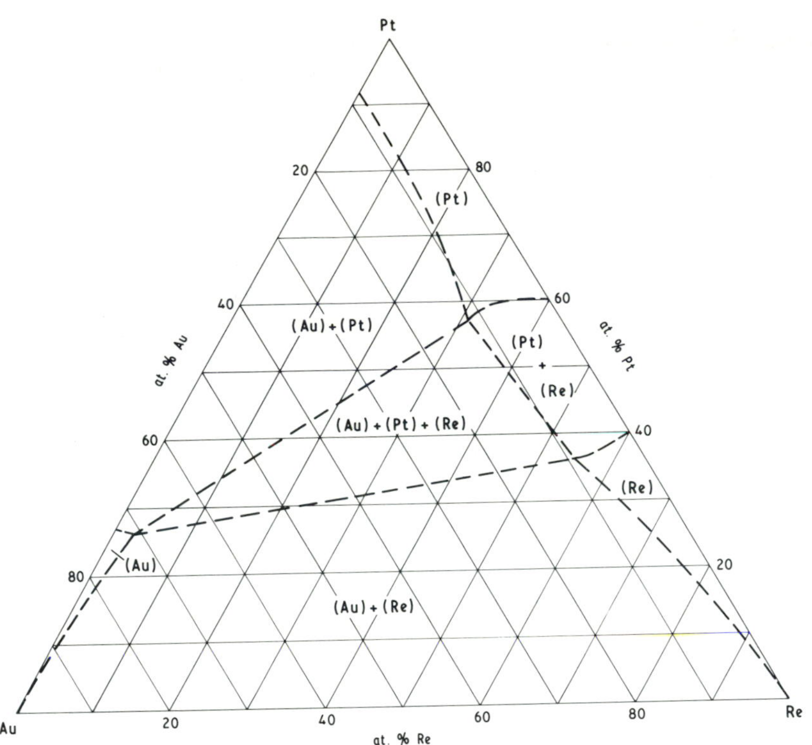

326 Suggested isothermal section at 900°C in the Au–Pt–Re ternary system

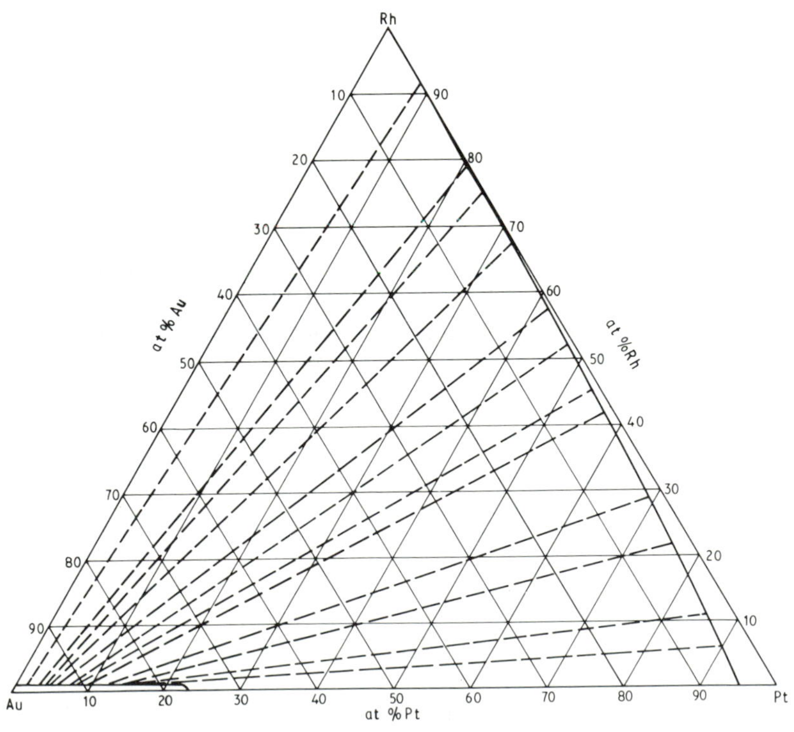

327 The system Au–Pt–Rh: 800°C isothermal section [64 Rau]

Au–Pt–Rh
(G. V. Raynor and A. Prince)

INTRODUCTION

Alloys belonging to the system gold–platinum–rhodium have found application as materials for the bulk handling of molten glass in the glass-making industry, and have been used also in the manufacture of spinnerets for the artificial fibres industry. Relatively little constitutional information, however, is available. Following earlier work on the general properties of Au–Pt alloys with small additions (up to 1·0 wt-%) of Rh [58 Sch], detailed X-ray diffraction studies were carried out by Raub and Falkenburg [64 Rau], interpretation of the results of which enabled establishment of the constitution of the system at 800°C. In addition, later work by [66 Sel] gives an indication of the constitution as a function of temperature and composition for a constant Pt content of 85 wt-%. The most recent work [85 Sok] presents vertical sections at 5, 10, and 20 at.-% Au and 10 and 20 at.-% Rh as well as a 1 000°C isothermal section of the ternary system.

BINARY SYSTEMS

The Au–Pt equilibrium diagram is well established [85 Oka]. It is characterized by a wide liquidus–solidus separation over the greater part of the composition range, and the establishment of a continuous series of solid solutions above a temperature of 1 260°C. Below this temperature a miscibility gap exists, separating fcc alloys of different compositions and lattice spacings. The maximum on the curve enclosing the two-phase region is at 61 at.-% Pt, and the region widens considerably as the temperature falls. The Au–Rh equilibrium diagram given in [84 Oka] is accepted as a basis for interpreting the ternary system. An important characteristic is the existence of a region of liquid immiscibility extending from 1 885°C to an unspecified high temperature. At 1 885°C, $L_1 \rightleftharpoons L_2 + (Rh)$, where the composition of L_1 is ~90 at.-% Rh, and (Rh) represents a solid solution at this temperature of 2·9 at.-% of Au in Rh. The composition of L_2, the gold-rich liquid, is ~18 at.-% Rh. A peritectic reaction at 1 068°C produces (Au) with ~1·6 at.-% Rh in solution. There is considerable uncertainty as to the temperature and the extent of the monotectic reaction. [84 Oka] give two versions, the one described above and a second version involving liquid immiscibility from 1·4 to 70·5 at.-% Rh at 1 320°C. In the case of Pt–Rh alloys a complete series of solid solutions is formed between the components [58 Han]; the liquidus curve falls smoothly from the melting point of Rh to that of Pt, and the solidus curve, though not accurately established, is shown as lying close to the liquidus. Various authors refer to the possibility of a miscibility gap in the solid state, with a relatively low critical temperature of 760°C, but such a gap has not been observed experimentally. In examination of this point [64 Rau] subjected binary Pt–Rh alloys to heat-treatments at 600 and 800°C lasting for more than four years. No duplex structures were observed. The lattice spacings of the one fcc phase present were found to be linear with composition between the spacings of Pt and Rh. The presence of a solid state miscibility gap in the system Pt–Rh is therefore not accepted.

SOLID PHASES

In the solid state the only crystal structure present in the ternary alloys is the fcc structure characteristic of the components. However, the wide two-phase region between solid (Au) and solid (Rh), together with the miscibility gap in the Au–Pt system below ~1 260°C, suggest that the various composition regions of the ternary solid solution may be designated (Au) and (Pt–Rh), all with the Cu, cF4, structure-type.

THE TERNARY SYSTEM

Alloys studied in the work of [64 Rau], by the method of X-ray diffraction and by limited thermal analysis and microscopy, lay along various lines in the ternary composition diagram. Three of these lines radiated from the Rh corner to the compositions 90 at.-% Au, 10 at.-% Pt (section 1), 65 at.-% Au, 35 at.-% Pt (section 2), and 10 at.-% Au, 90 at.-% Pt (section 3). In addition, two series of alloys with constant Rh atomic percentages of 4·6 (section 4) and 9·1 (section 5), and a series with a constant atomic percentage of 1·6 at.-% Au (section 6), were also examined. In all some 37 alloys were used, prepared from the 'purest available' materials in a tungsten-arc furnace. For X-ray work specimens were filed if ductile, or pulverized if brittle; for the micrographic work alloys were deformed prior to annealing. All annealing treatments were given *in vacuo* and terminated by quenching.

Inspection of the three binary diagrams relevant to the ternary system indicates that at 800°C, the highest temperature at which the ternary alloys were studied by [64 Rau], the constitution of the ternary system should consist of simple equilibrium between gold-rich Au–Pt alloys containing a little Rh, and alloys derived from the Pt–Rh system by the solution of Au in (Pt–Rh). The work of [64 Rau] confirms this expectation, and contains graphs of lattice spacings as functions of composition, after quenching from 600 and 800°C, for the two fcc phases present in the alloys lying along the six composition lines quoted above. Thus for the series lying on the composition line connecting Rh with the composition 65 at.-% Au, 35 at.-% Pt (section 2), lattice spacings are shown as functions of Rh content. At 0% Rh the two spacings are those of the (Au) and (Pt) solid solutions in the binary Au–Pt system at the relevant temperature. At other Rh percentages (x at.-%) the spacings are those of the Au-rich alloys and the alloys rich in Pt and Rh present in equilibrium, and thus forming the ends of tie lines passing through the appropriate two-phase alloy of composition $0{\cdot}65\,(100-x)$ at.-% Au, $0{\cdot}35\,(100 - x)$ at.-% Pt. If the variation of lattice spacing with composition along the 800°C boundary of the Au-rich solid solution (containing Pt and a very small amount of Rh) and of the 800°C boundary of the

(Pt–Rh) solid solution (containing varying quantities of Au) is known, the compositions corresponding to the observed spacings may be derived, and the tie line thus established. In the work of [64 Rau] lattice spacing/composition relationships were established for the binary Pt–Rh alloys, for the binary Au–Pt alloys, and also for ternary alloys containing ~1·6 at.-% Au, and from 36 to 74 at.-% Rh. Some alloys of the latter series must lie within the ternary Pt-rich (Pt–Rh) solid solution area at 800°C; others are likely to lie in the two-phase area (Au) + (Pt–Rh) at the higher Rh contents. The curve of lattice spacing against Rh content lies slightly but significantly above the curve for binary Pt–Rh alloys by about 0·004 Å at a given Rh composition. It would appear, therefore, that the actual spacing variation along the two-phase boundaries was not explicitly carried out. Raub and Falkenburg [64 Rau] show further diagrams which summarize the variation of the compositions of the two equilibrated phases with the rhodium contents of two-phase alloys lying along certain of the composition lines quoted above, as deduced from the measured lattice spacings. The report, however, does not explicitly discuss the manner in which the compositions are derived, and it is not clear what allowance was made for the Au content of the (Pt–Rh) solid solution, or for the Rh content of the Au-rich solid solution. Thus for the (Pt–Rh) terminal compositions of tie lines for two-phase alloys of sections 2, 3, and 4 the results plotted in [64 Rau] may be approximated to very closely by assuming that the binary Pt–Rh lattice spacing/composition curve was used to assess the Pt content relevant to the measured spacing of the (Pt–Rh) phase in the two-phase alloys. On the other hand the results plotted for section 5 cannot be accurately reproduced in this way, and the effect of the Au content of the (Pt–Rh) solid solution appears to have been taken fully into account. These results appear more accurate than those plotted in relation to the other sections. Lack of allowance for the third metal would not be serious for the Au-rich solid solution, owing to the very low solubility of Rh in Au at the temperature involved. For the Pt-rich (Pt–Rh) solid solution, however, neglect of the Au content would introduce errors. At high Rh contents such errors would be small, owing to the very limited solubility of Au in this composition region. For the more Pt-rich range of composition neglect of the Au content would lead to an overestimate of the Pt content corresponding to a given spacing, which would increase as the Au–Pt axis was approached to an amount of approximately 2 at.-% Pt.

The results, however, may be relied upon to show the general direction of the tie lines. The 800°C isothermal derived by [64 Rau] is shown in Fig. 327, and the tie lines derived from the X-ray work fan out regularly from the Au-rich corner to the Pt–Rh side of the diagram. The corresponding 600°C isothermal is of the same form and the boundaries are closely similar. The major features are the considerable widening of the Au–Pt miscibility gap by the addition of Rh, and the confinement of the (Pt–Rh) solid solution area to a narrow strip parallel to the Pt–Rh axis at Rh contents exceeding about 60 at.-%. These features are maintained in the 1 000 and 1 200°C isothermals presented by [85 Sok] (Fig. 328). The two isothermals were plotted by taking measurements from the diagrams given by [85 Sok]. The assessed isothermals lie at lower Au contents than those claimed by [85 Sok]; this point is discussed later when consideration is given to both the vertical sections and the isothermals presented by [85 Sok].

Further experimental evidence is available from the work of [66 Sel], which extended investigations to higher temperatures for Pt-rich alloys containing 85 wt-% Pt. Alloys were prepared from Pt and Rh sponge and Au grain, all of the highest available purity, by induction-melting in air using Al_2O_3 crucibles, and were chill-cast in thick copper moulds. Ingots were homogenized at 1 250°C for periods of up to 64 h, slowly cooled to 300°C, and held there for 16 h, before quenching into water. The specimens could then be cold-worked into sheet or wire. The total base metal impurity content of the samples was approximately 0·005%. The solidus temperatures for the 85 wt-% Pt section was studied by the metallographic method involving microexamination of samples quenched from successively higher temperatures until signs of melting were observed. The use of metallography and X-ray diffraction to locate accurately the steeply sloping boundary between homogeneous and two-phase alloys for the section under consideration proved unsatisfactory, and the boundary was therefore examined by electrical resistometric methods using a differential resistance bridge. Experiments showed that the miscibility gap between fcc structures of different compositions closes at approximately 1 182°C and a composition of 9 at.-% Au in this section, which extends from 74·9 at.-% Pt, 25·1 at.-% Rh to 85·1 at.-% Pt, 14·9 at.-% Au (Fig. 329). It may be noted that at 800°C the boundary of the two-phase region is in good agreement with the work of [64 Rau] shown in Fig. 327.

The most recent work on the ternary system emanates from the USSR [83 Sok, 85 Sok, 84 Zhm]. The work of [83 Sok] is presented as vertical sections at 5 and 10 at.-% Au, and 5 and 10 at.-% Rh. It appears to be preliminary work which was revised in the later publications [84 Zhm, 85 Sok]; it is not necessary to consider it further. [84 Zhm] published the complete 1 000°C isothermal section and the 1 200°C isotherm in the Pt-rich corner only. As this data is also included in the latest publication [85 Sok] it is only necessary to consider the work reported in [85 Sok], extracting from [83 Sok, 84 Zhm] details of experimental technique not included by [85 Sok]. Alloys were arc-melted and cast in pure He. The extent of the liquid immiscibility region, originating with the Au–Rh binary system, was determined by metallographic study of cast alloys. The ternary liquid immiscibility gap extends to about 35 at.-% Pt. Alloys were annealed for 1 000 h at 1 000°C and 40 h at 1 200°C for determination of the 1 000 and 1 200°C isotherms [84 Zhm]. Thermal effects on vertical sections were determined by DTA using the very high heating rates of 40°C min^{-1} and 80°C min^{-1} [83 Sok].

The isothermal sections at 1 000 and 1 200°C (Fig. 328) are drawn in accordance with [85 Sok] and in an amended

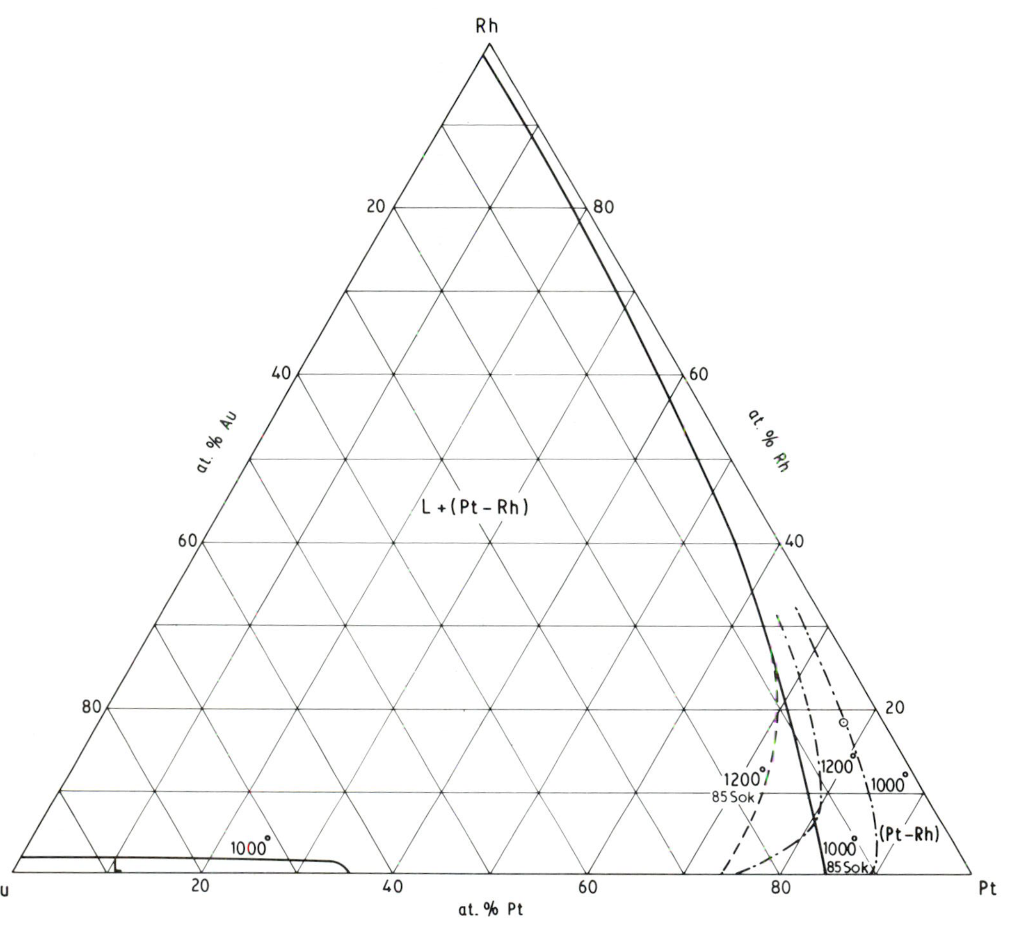

328 The system Au–Pt–Rh: 1 000°C and 1 200°C isothermals [85 Sok] and assessed isothermals

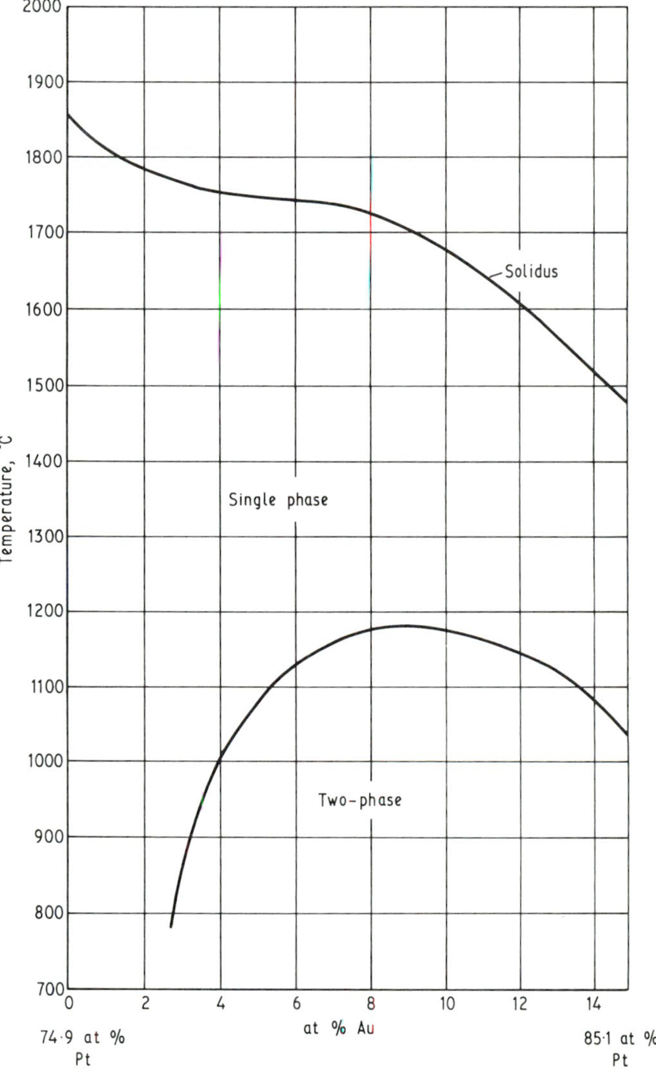

329 The system Au–Pt–Rh: vertical section for 85 wt-% Pt [66 Sel] (74·9 at.-% Pt, 25·1 at.-% Rh to 85·1 at.-% Pt, 14·9 at.-% Au)

form derived from an assessment of the vertical section presented by [85 Sok]. One data point on the 1 000°C amended isotherm is derived from [66 Sel] who determined the phase boundary by electrical resistance methods. Reliance is placed on the data of [66 Sel] (Fig. 329), which implies that the 1 000°C isotherm published by [85 Sok] should be redrawn at lower Au contents, as indicated in Fig. 328. The two-phase region (Fig. 329) does not quite reach 1 200°C in the 85 wt-% Pt section. The 1 200°C isotherm must follow a course that brings it within a few percent of Pt of the section studied by [66 Sel]. The suggested amendment to the 1 200°C data of [85 Sok] is shown in Fig. 328.

The vertical sections (Fig. 330*a–e*) are taken from [85 Sok] with the original data points included. To provide a set of self-consistent data the boundaries between phase regions have been amended from those drawn by [85 Sok]. In terms of the intersection of the sections with the Au–Pt and Au–Rh binary systems the liquidus–solidus gap for the 5 and 10 at.-% Au sections have been reduced to conform with [66 Sel]. Accepting that the Au–Rh monotectic occurs at 1 885°C has meant amending the 10 and 20 at.-% Au and the 20 at.-% Rh sections so that the $L_1 + L_2$ + (Pt–Rh) phase regions originates at 1 885°C, as against 1 850, 1 810, and 1 760°C respectively [85 Sok]. It is difficult to interpret the data for the liquid immiscibility region. [85 Sok] indicate a minimum in the $L_1 + L_2$ + (Pt–Rh) phase region for the 10 and 20 at.-% Au and the 20 at.-% Rh sections. Unfortunately the minimum occurs at 1 795, 1 763, and 1 661°C respectively and not at a constant temperature corresponding to a critical tie line joining the L_1, L_2, and (Pt–Rh) phases. No attempt has been made to draw the $L_1 + L_2$ + (Pt–Rh) phase region in Fig. 330*b,c,e*, in view of the discrepancies in the original data. The three-phase region L + (Au) + (Pt–Rh), present in Fig. 330*b–e*, must end at the peritectic horizontal on the Au–Rh binary system. A value of 1 068°C is accepted for the binary peritectic temperature. As plotted by [85 Sok] intersection with the Au–Rh binary occurred at variable temperatures; 1 066°C (Fig. 330*b*), 1 035°C (Fig. 330*c*), 1 047°C (Fig. 330*d*), 1 090°C (Fig. 330*e*). The L + (Au) + (Pt–Rh) phase region in Fig. 330*b–e* has been amended so that it is constrained to meet the Au–Rh binary at 1 068°C.

The 10 at.-% Au section (Fig. 330*b*) has been modified so as to recognize that the 10 at.-% Au, 90 at.-% Pt binary alloy is not single phase from 900°C to the solidus [85 Sok] but is two-phase (Au) + (Pt) up to 970°C and single phase from 970°C to the solidus. The data point of [66 Sel] also locates on the modified boundary between the (Au) + (Pt–Rh) and (Pt–Rh) phase regions. The 10 at.-% Rh section (Fig. 330*d*) has also been modified. The boundary between the (Au) + (Pt–Rh) and the (Pt–Rh) phase regions was originally placed below 80 at.-% Pt at 900°C [85 Sok]. The data point from [66 Sel] indicates that the boundary should be extended to ~85 at.-% Pt at 900°C. The liquidus is nearly horizontal but plunges to 1 090°C as the Pt is reduced from 30 at.-% to zero [85 Sok]. The uncertainties of the Au–Rh phase diagram make assessment of the liquidus of the 10 at.-% Rh section speculative, but it is considered that the liquidus of 10 at.-% Rh, 90 at.-% Au alloy will be of the order of 1 670°C rather than 1 090°C; this reflected in the amended section (Fig. 330*d*).

Of the sections examined three alloys lie on two sections; alloy 10 at.-% Au, 80 at.-% Pt, 10 at.-% Rh lies on both the 10 at.-% Au and the 10 at.-% Rh sections, alloy 20 at.-% Au, 60 at.-% Pt, 20 at.-% Rh lies on both the 20 at.-% Au and 20 at.-% Rh sections, alloy 20 at.-% Au, 70 at.-% Pt, 10 at.-% Rh lies on both the 20 at.-% Au and 10 at.-% Rh sections. The vertical lines on Fig. 330 *b–e* indicate the different temperatures scaled from the original diagrams [85 Sok]. Vertical sections at 65, 70, and 75 at.-% Pt have been generated from Fig. 330*a–e*. They are reproduced in Fig. 331*a–c*.

Isothermal sections can also be generated from the vertical sections of Figs 330 and 331. Sections at 1 700, 1 600, 1 280, 1 260, 1 200. and 1 100°C are given in Fig. 332*a–f* respectively. These isothermal sections are consistent with the amended vertical sections taken from [85 Sok]. In terms of the solidification behaviour of ternary Au–Pt–Rh alloys it can be assumed that the liquid miscibility gap, originating at the Au–Rh monotectic reaction, closes in a critical tie line joining the liquid phase to a (Pt–Rh) phase. As noted previously, the data of [85 Sok] indicates a minimum in the $L_1 + L_2$ + (Pt–Rh) phase region, but this data requires confirmation. At lower temperatures, i.e. 1 700°C and below, the ternary system is dominated by the L + (Pt–Rh) phase region Fig. 332*a,b*. The (Pt–Rh) phase region develops a pronounced kink with falling temperature, probably leading to a critical tie line between a liquid phase of composition 82 at.-% Au, 16 at.-% Pt, 2 at.-% Rh and a (Pt–Rh) solid solution of composition 20 at.-% Au, 75 at.-% Pt, 5 at.-% Rh (Fig. 333) at an estimated temperature of 1 290°C. This critical tie line is suggested to be the origin of the three-phase region L + (Au) + (Pt–Rh) which descends from 1 290°C to end at the peritectic reaction in the Au–Rh system at 1 068°C. At 1 280°C (Fig. 332*c*) a two-phase region (Au) + (Pt–Rh) exists in the ternary. This region expands with fall in temperature and touches the Au–Pt binary at 1 260°C, corresponding to the critical point in the Au–Pt solid state miscibility (Fig. 332*d*). The L + (Au) + (Pt–Rh) three-phase triangle swings round towards the Au–Rh binary with further fall in temperture (Fig. 332*e,f*). On completion of solidification at 1 068°C the equilibrium typified by the 800°C isothermal section [64 Rau] (Fig. 327), will be established. The path of the liquid phase, the (Au), and (Pt–Rh) phases from 1 290 to 1 068°C are plotted in Fig. 333.

The steep rise in the solidus for (Pt–Rh) solid solutions is evident from Figs 330*b,d,e* and 331*a,b*.

REFERENCES

58 Sch: H. Schmid: *Metall.*, 1958, **12**, 612–619

58 Han: M. Hansen and K. Anderko: 'Constitution of Binary Alloys'; 1958, New York, McGraw-Hill

64 Rau: E. Raub and G. Falkenburg: *Z. Metall.*, 1964, **55**, 392–397

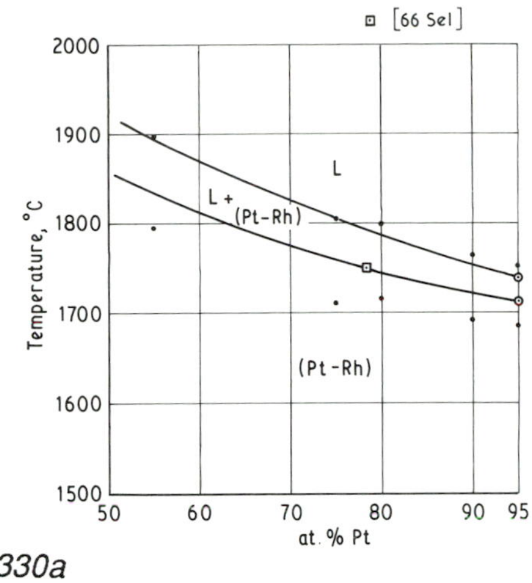

330a

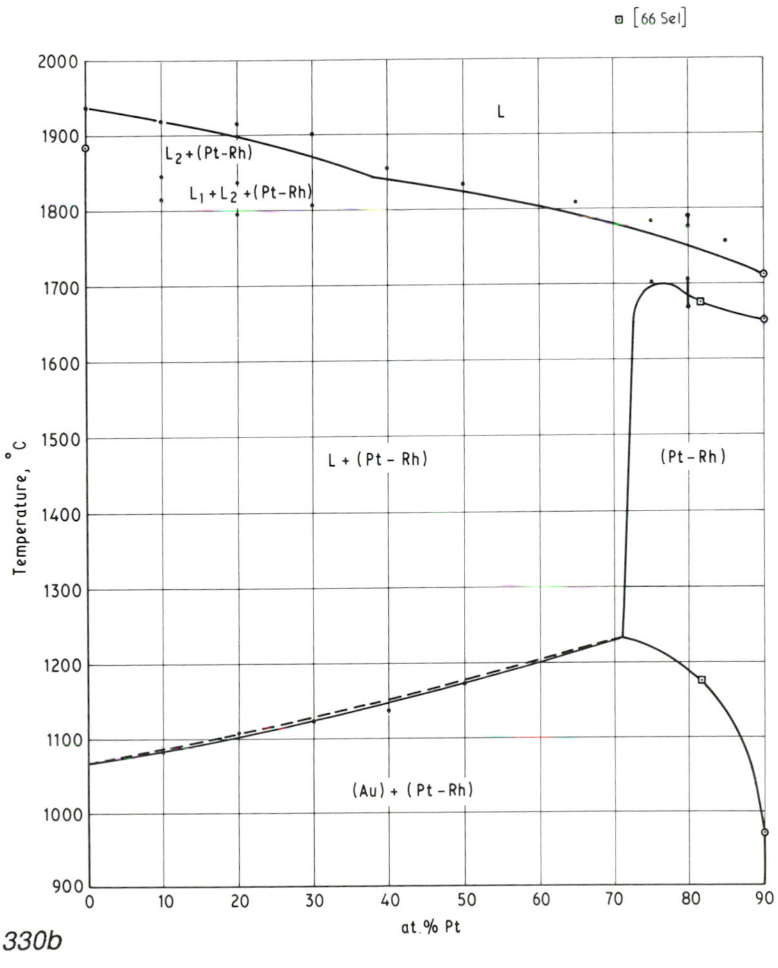

330b

330 The system Au–Pt–Rh: vertical sections *a* 5 at.-% Au; *b* 10 at.-% Au; *c* 20 at.-% Au; *d* 10 at.-% Rh; *e* 20 at.-% Rh

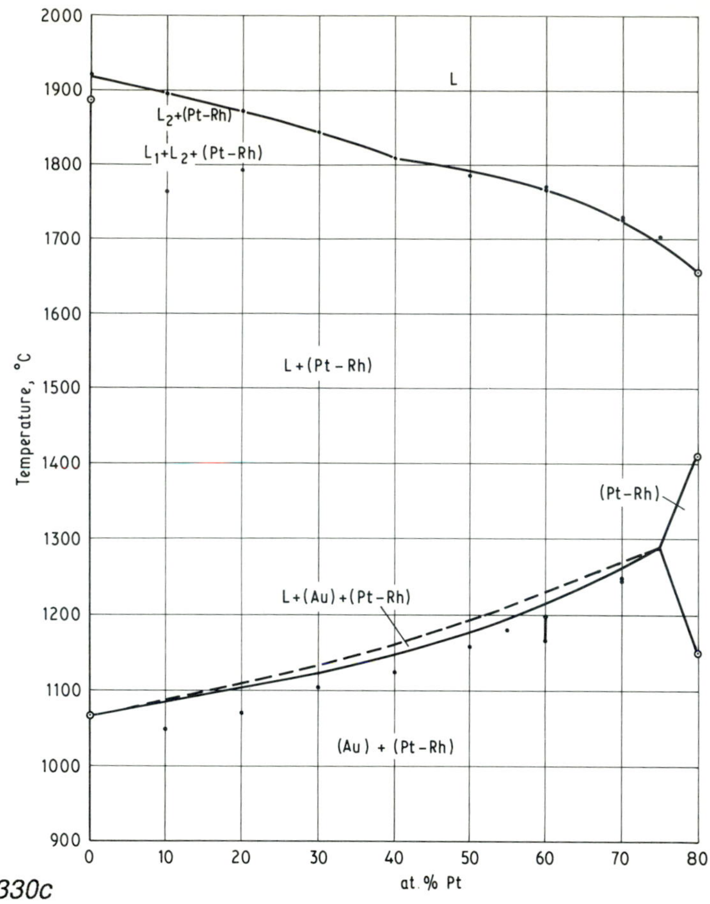
L
L2+(Pt-Rh)
L1+L2+(Pt-Rh)
L+(Pt-Rh)
(Pt-Rh)
L+(Au)+(Pt-Rh)
(Au) + (Pt-Rh)
Temperature, °C
at.% Pt

330c

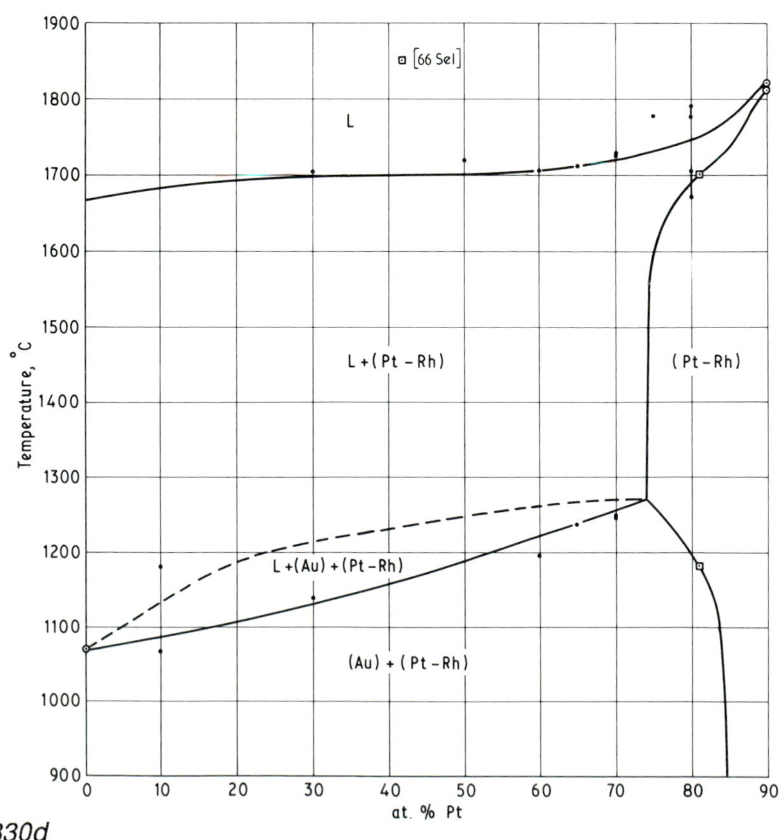
[66 Sel]
L
L+(Pt-Rh)
(Pt-Rh)
L+(Au)+(Pt-Rh)
(Au) + (Pt-Rh)
Temperature, °C
at.% Pt

330d

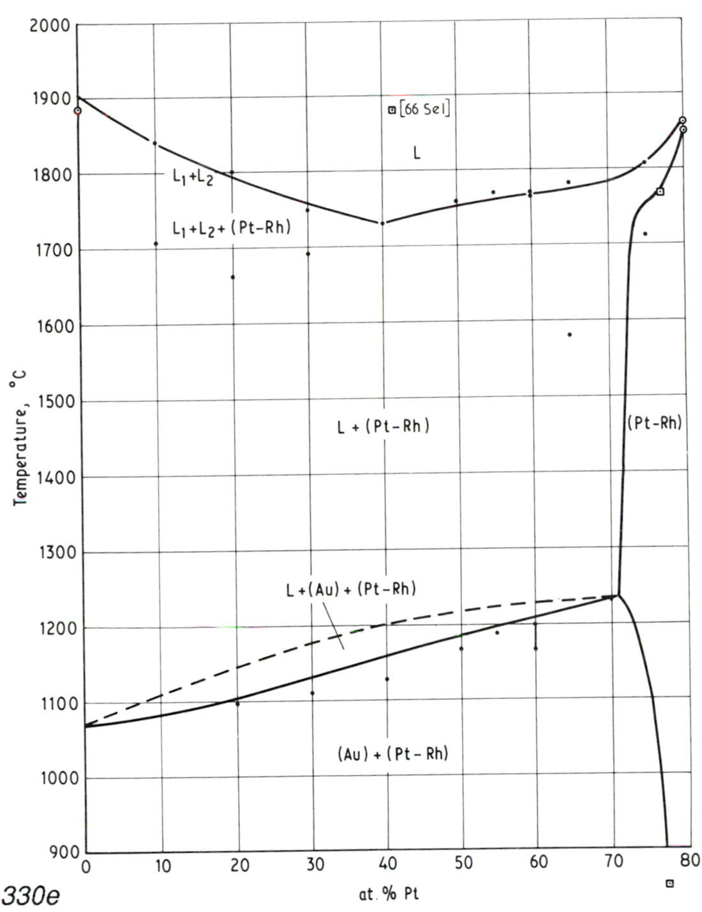
Temperature, °C
at. % Pt
[66 Sel]
L
L1+L2
L1+L2+(Pt-Rh)
L+(Pt-Rh)
(Pt-Rh)
L+(Au)+(Pt-Rh)
(Au)+(Pt-Rh)

330e

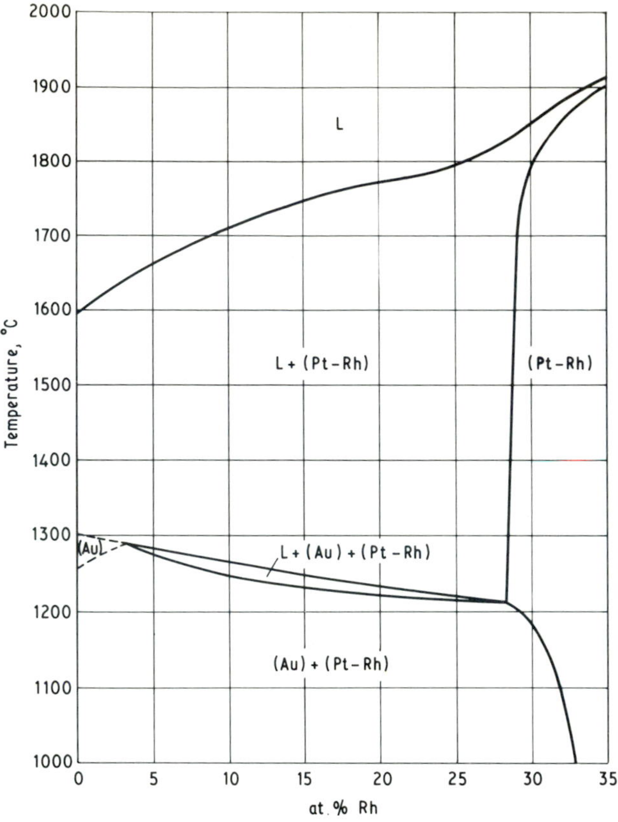

331a

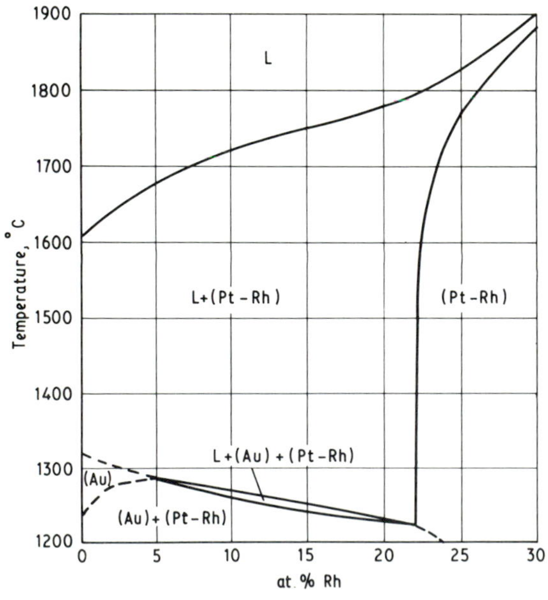

331b

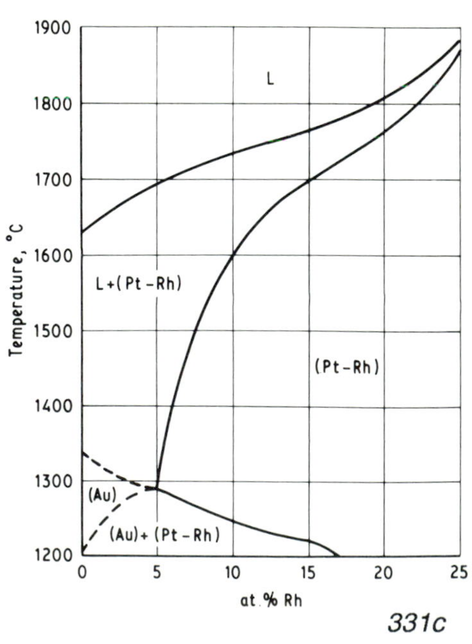

331c

331 The system Au–Pt–Rh: vertical sections *a* 65 at.-% Pt; *b* 70 at.-% Pt; *c* 75 at.-% Pt

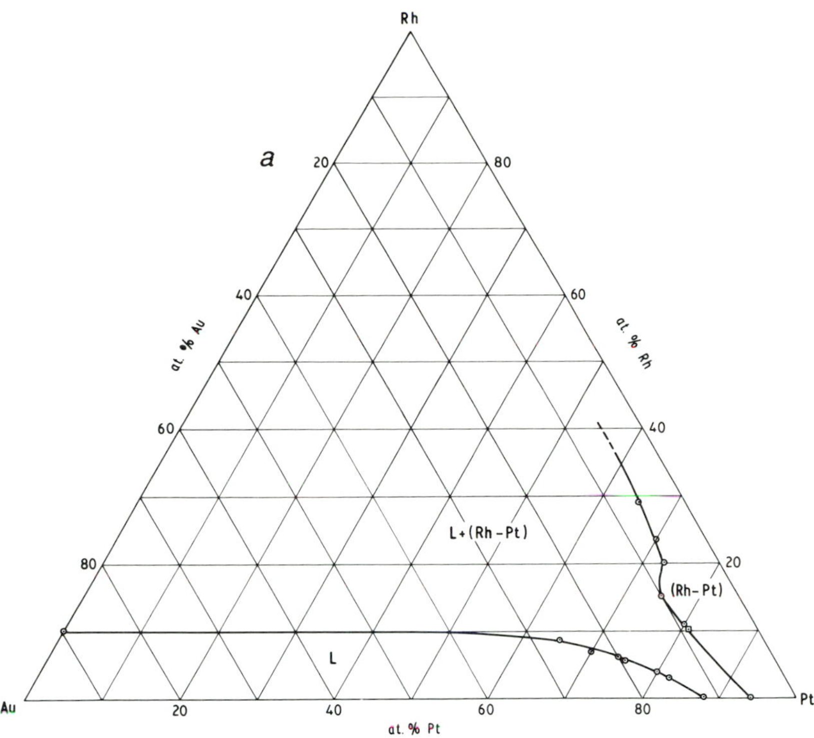

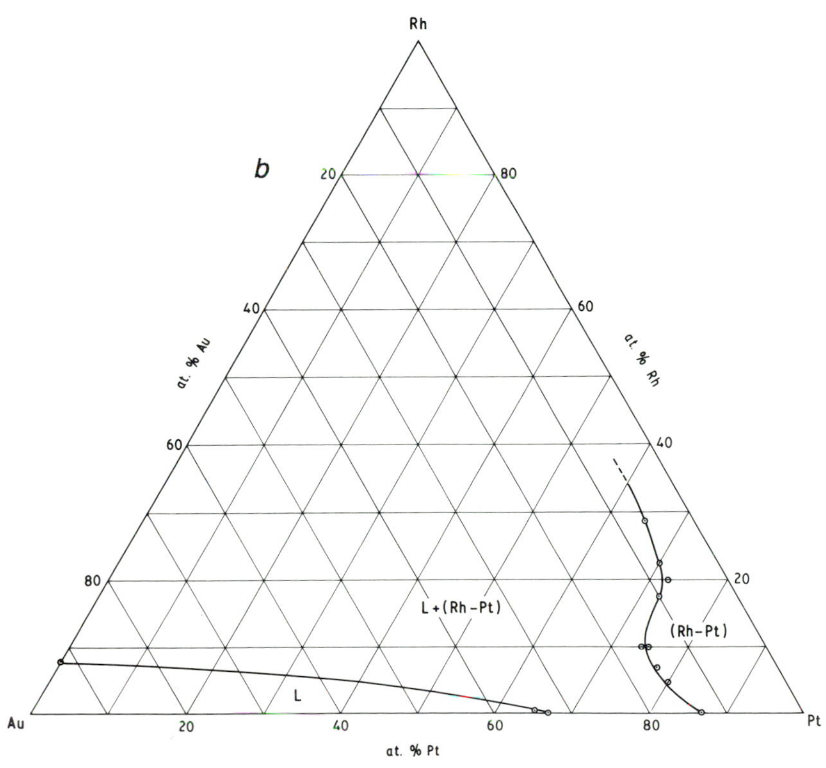

332 The system Au–Pt–Rh: isothermal sections derived from the vertical sections *a* 1 700°C; *b* 1 600°C; *c* 1 280°C; *d* 1 260°C; *e* 1 200°C; *f* 1 100°C

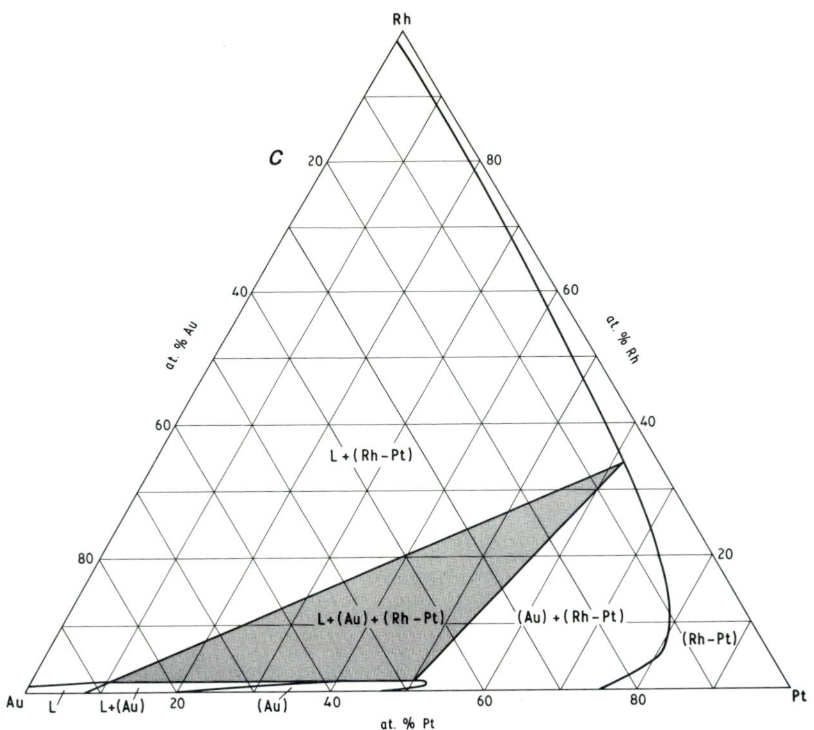
Rh
c
20
80
40
60
at. % Au
at. % Rh
60
40
L + (Rh – Pt)
80
20
L+(Au)+(Rh – Pt)
(Au) + (Rh – Pt)
(Rh – Pt)
Au
L
L+(Au)
20
(Au)
40
60
80
Pt
at. % Pt

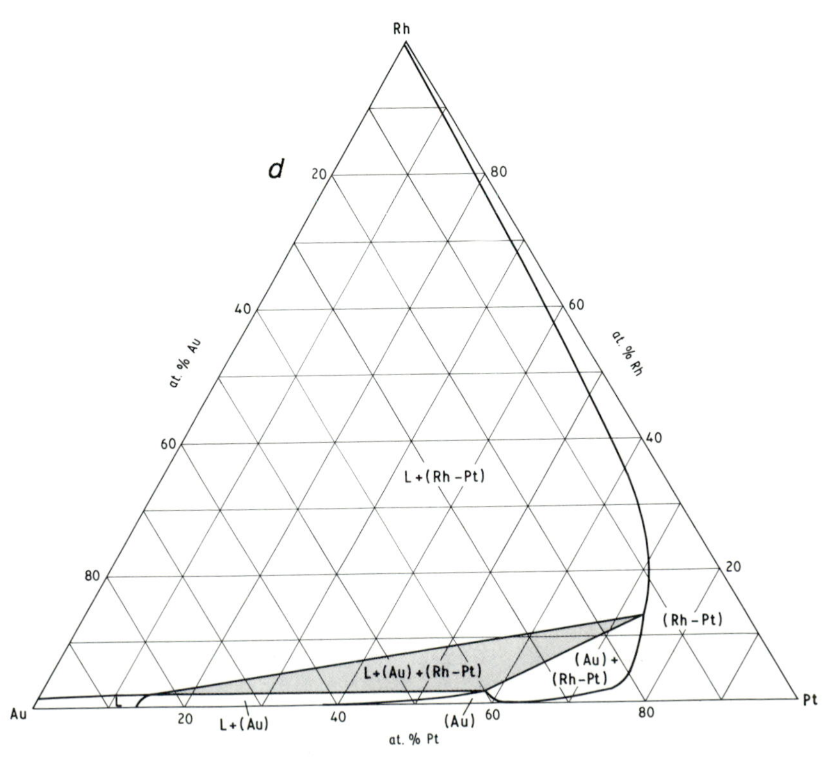
Rh
d
20
80
40
60
at. % Au
at. % Rh
60
40
L + (Rh – Pt)
80
20
(Rh – Pt)
L+(Au)+(Rh – Pt)
(Au) +
(Rh – Pt)
Au
20
L + (Au)
40
(Au)
60
80
Pt
at. % Pt

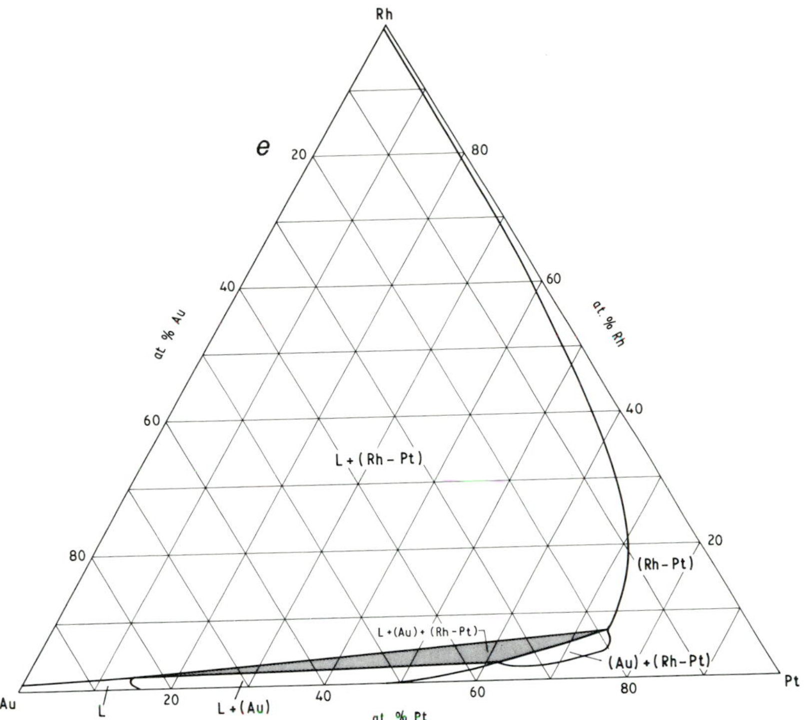

e
Rh
20
40
60
80
at % Au
80
60
40
20
at. % Rh
L + (Rh – Pt)
(Rh – Pt)
L + (Au) + (Rh – Pt)
(Au) + (Rh – Pt)
Au
L
20
L + (Au)
40
60
80
Pt
at. % Pt

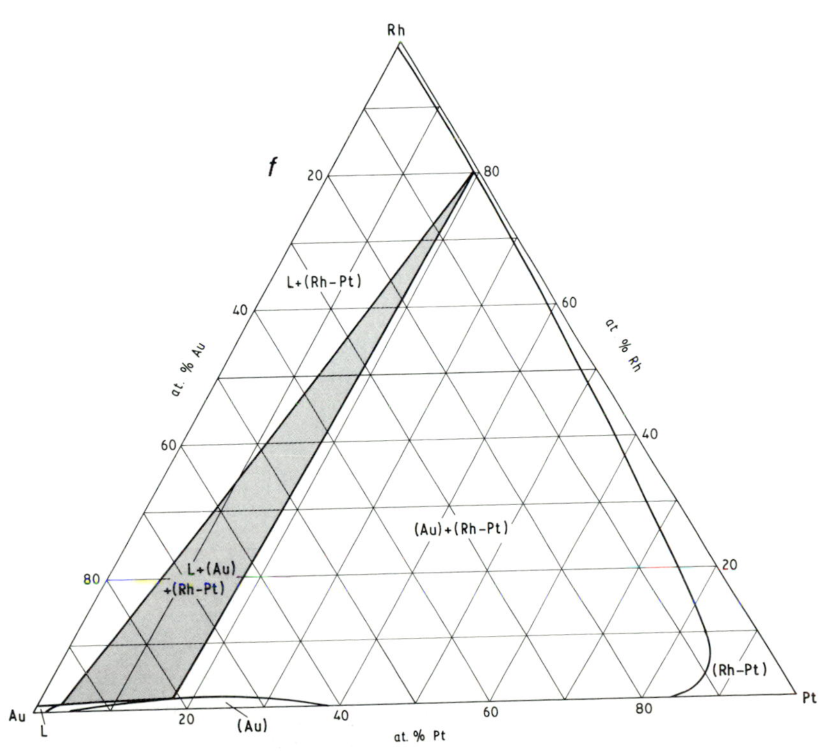

f
Rh
20
40
60
80
at. % Au
80
60
40
20
at. % Rh
L+(Rh–Pt)
(Au)+(Rh–Pt)
L+(Au)
+(Rh–Pt)
(Rh–Pt)
Au
L
20
(Au)
40
60
80
Pt
at. % Pt

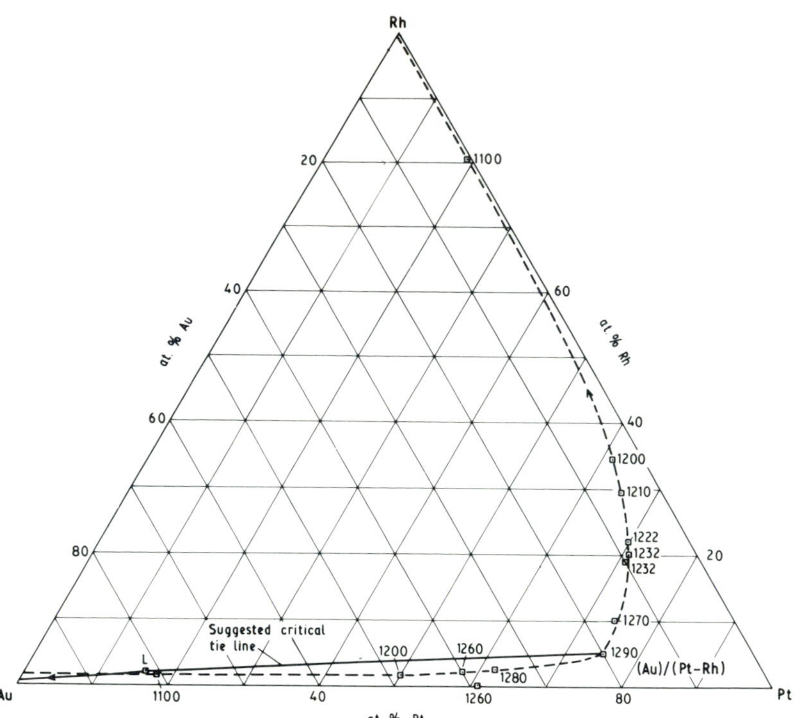

333 The system Au–Pt–Rh: suggested critical tie line at 1 290°C and path of monovariant curves associated with L, (Au), and (Pt–Rh) phases

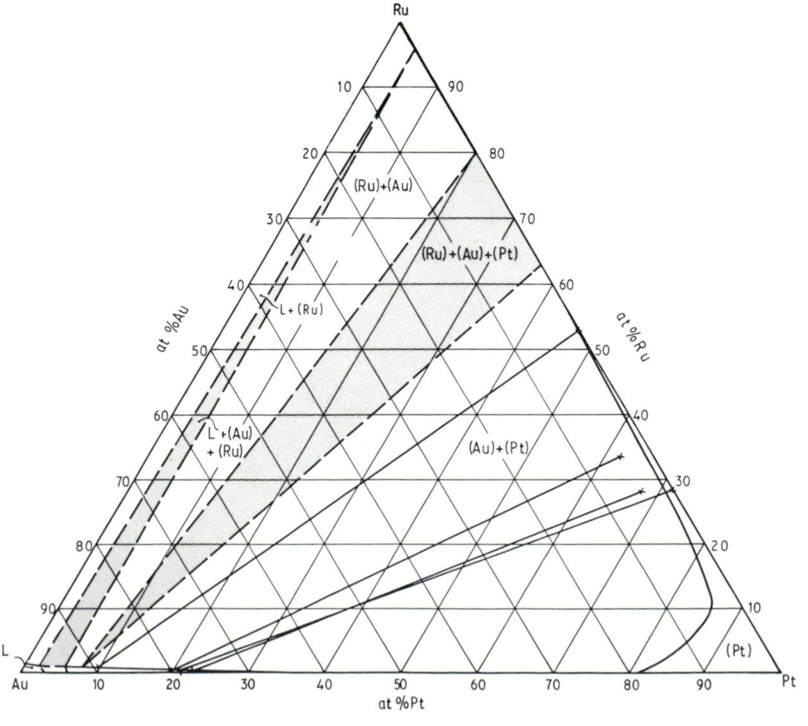

334 The system Au–Pt–Ru: the 1 100°C isothermal, showing the experimentally determined tie lines [72 Jon]

66 Sel: G. L. Selman, M. R. Spender and A. S. Darling: *Platinum Metals Rev.*, 1966, **10**, 54–59

83 Sok: E. M. Sokolovskaya, G. P. Zhmurko and Yu. D. Seropegin: *Tsvetn. Metall. Nauchno-Tekh. SB*, 1983, (22), 33–34

84 Zhm: G. P. Zhmurko, E. M. Sokolovskaya, M. V. Raevskaya, Yu. A. Brauer and V. V. Pristavko: *Vestn. Moskov Univ., Ser. 2, Khim.*, 1984, **25**(1), 107–108

84 Oka: H. Okamoto and T. B. Massalski: *Bull. Alloy Phase Diagrams*, 1984, **5**, 384–387

85 Oka: H. Okamoto and T. B. Massalski: *Bull. Alloy Phase Diagrams*, 1985, **6**, 46–56

85 Sok: E. M. Sokolovskaya, M. V. Raevskaya, G. P. Zhmurko, Yu. D. Seropegin and V. I. Rodionov: *J. Less-Common Metals,* 1985, **105**, 161–164

Au–Pt–Ru
(G. V. Raynor)

INTRODUCTION

The Au–Pt–Ru system is of importance with respect to the possible use of Au–Pt alloys to protect Ru from oxidation in high temperature applications; the function of Pt is to raise the melting point of Au. The system has been examined by metallographic, X-ray diffraction, and electron-probe microanalysis methods [72 Jon], and an 1 100°C isothermal is available.

THE BINARY SYSTEMS

The Au–Pt equilibrium diagram is well established [85 Oka]. It is characterized by a wide liquidus–solidus separation over the greater part of the composition range, and the establishment of a continuous series of solid solutions above a temperature of 1260°C. Below this temperature a miscibility gap exists, separating fcc alloys of different compositions and lattice spacings. The maximum on the curve enclosing the two-phase region is at 61 at.-% Pt, and the region widens considerably as the temperature falls.

The Au–Ru diagram given in [84 Oka] is based on somewhat limited information. The Ru-rich liquidus curve falls from the melting point of Ru to a monotectic point at ~2 012°C and approximately 73 at.-% Ru. At this temperature liquid of 73 at.-% Ru is in equilibrium with a second liquid containing ~27 at.-% Ru and almost pure Ru. Between 2 012 and 1 066°C the diagram shows equilibrium between the Au-rich liquid and almost pure Ru. At 1 066°C liquid and (Ru) react together to give an extremely limited solid solution of Ru in Au. In the solid state alloys consist of mixtures of (Au) and (Ru) containing respectively very small amounts of the second component. A tentative equilibrium diagram for the Pt–Ru system is presented by [72 Hut]. Though the liquidus and solidus curves are not accurately known, it appears that the liquidus falls smoothly from the melting point of Ru to that of Pt. At approximately 2 100°C a peritectic reaction occurs, at which liquid of composition ~55 at.-% Ru reacts with the Ru-rich solid solution containing ~20 at.-% Pt to form a solid solution of about 76 at.-% Ru in Pt. It will be convenient to refer to these solid solutions as (Ru) and (Pt) respectively. The two phase region (Pt) + (Ru) widens as the temperature falls. At 1 900°C it extends from 70 at.-% to 80 at.-% Ru, while at 1 000°C the range is 62 at.-% to 80 at.-% Ru. No accurate information appears to exist below 1 000°C, but between 2 000 and 1 000°C the phase boundaries appear well established.

THE SOLID ALLOYS

In the solid state the solid solutions based upon Au and Pt are fcc in crystal structure, while that based on Ru is close packed hexagonal. No ternary compounds have been observed, so that equilibrium in the ternary system involves (Au), (Pt), and (Ru) at temperatures corresponding to the miscibility gap in the Au–Pt system, and to (Au–Pt) and (Ru) above the critical temperature in this system; (Au–Pt) denotes the continuous high-temperature solid solution between Au and Pt. Table 105 summarizes the crystal structures in the solid alloys.

Table 105. Crystal structures

Solid phase	*Prototype*	*Lattice designation*
(Au)	Cu	cF4
(Pt)	Cu	cF4
(Ru)	Mg	hP2

THE TERNARY SYSTEM

The ternary alloys used for the determination of equilibria were prepared by melting the components together. Alloys rich in Ru were melted in an argon-arc furnace, and others by induction-melting under argon. Alloys were homogenized for not less than 65 h at 1 100°C and water-quenched before examination. The boundary between the single phase area and the two-phase area in the ternary diagram was determined metallographically. Tie lines were determined by electron probe microanalysis of the phases present in equilibrated two-phase specimens, or by X-ray diffraction methods. In the latter case, the lattice spacings of the phases present were determined and the compositions estimated, apparently from the known variation of lattice spacing with composition in the Au–Pt and Pt–Ru systems. In the account of this work [72 Jon] it is stated that the X-ray method was in general more satisfactory than the electron probe microanalysis. The tie lines [72 Jon] are plotted in Fig. 334, which also presents the phase diagram proposed as a result of the metallographic work. The tie lines run from the Au-rich region to points close to the Pt–Ru axis, and though the determined phase compositions do not lie exactly upon the proposed boundary and nor should tie lines intersect on an isothermal section, the general nature of the equilibria is established.

At 1 100°C, miscibility gaps exist in the systems Au–Pt and Pt–Ru so that the solid phases present are the Ru-rich Pt–Ru solid solution containing very small amounts of Au, the Pt-rich solid solution containing appreciable amounts of Ru and Au, and the Au-rich solid solution containing appreciable quantities of Pt but little Ru. These phases enter into equilibrium with each other, and a three-phase ((Au) + (Pt) + (Ru)) triangle exists. The position of this, shown in broken lines, is not accurately established, but the tie lines indicate that its position cannot vary greatly from that shown. Since 1 100°C is above the melting point of Au, a small area of liquid exists in the extreme Au-rich corner, so that the diagram must include a $Liquid_1$ + (Au) area, a $Liquid_1$ + (Ru) area, and the corresponding $Liquid_1$ + (Au) + (Ru) region. There is no further experimental information. In particular, no investigation has been made of the liquidus and solidus surfaces.

REFERENCES

72 Jon: B. Jones, M. W. Jones and D. W. Rhys: *J. Inst. Metals*, 1972, **100**, 136–141

72 Hut: J. M. Hutchinson: *Pt Metals Rev.*, 1972, **16**, 88–90

84 Oka: H. Okamoto and T. B. Massalski: *Bull. Alloy Phase Diagrams*, 1984, **5**, 388–390

85 Oka: H. Okamoto and T. B. Massalski: *Bull. Alloy Phase Diagrams*, 1985, **6**, 46–56

Au–Pt–Sb

Furuseth *et al.* [67 Fur] prepared an unspecified number of ternary alloys on the $AuSb_2$–$PtSb_2$ section by heating either the elements (99·999% Pt, Sb and 99·995% Au) or the binary compounds for 60 days at 600°C followed by slow cooling, or 1 day at 1 000°C followed by slow cooling or quenching, or 1 day at 1 200°C followed by quenching. X-ray powder diffraction analysis showed that the ternary alloys were mixtures of the binary compounds. No solid solubility was detected in the $AuSb_2$–$PtSb_2$ section, in contrast with the 8·33 at.-% Au that is soluble in the $PdSb_2$ phase. Both $AuSb_2$ and $PtSb_2$ have the FeS_2-type structure, cP12.

REFERENCE

67 Fur: S. Furuseth, K. Selte and A. Kjekshus: *Acta Chem. Scand.*, 1967, **21**, 527–536

Au–Pt–Ti

The only published work on the Au–Pt–Ti system is that of [65 Pie] and [68 Pie] concerning the $AuTi_3$–$PtTi_3$ section. Both $AuTi_3$ and $PtTi_3$ have the W_3O-type structure (cP8) and, metallographically, as-cast ternary alloys on the $AuTi_3$–$PtTi_3$ section were single phase. Analysis of data from nine ternary alloys using X-ray single crystal and powder diffraction techniques indicated that alloys heat treated for 50 days at 800°C and subsequently re-heat treated for 62 days at 500°C contained two coherent cubic phases, each of which possesses a primitive cubic space lattice (Table 106). One lattice gave sharp diffraction maxima and the other space lattice gave relatively diffuse maxima at slightly larger interplanar spacings. Some of the Pt-rich ternary alloys gave sharp maxima for both cubic lattices after annealing for 50 days at 800°C and 62 days at 500°C. There is undoubted difficulty in achieving equilibrium conditions at these temperatures. Microhardness measurements on bulk samples that had undergone the 800°C + 500°C anneal show a pronounced softening for the equiatomic alloy with 12·5 Au, 12·5 Pt, 75 Ti (Table 106). Pietrokowsky [68 Pie] speculates on the possibility of an ordered ternary phase of the W_3O-type occurring.

Table 106. Lattice parameters of alloys on the $AuTi_3$–$PtTi_3$ section [68 Pie]

Alloy composition, at.-%			*Lattice parameter*, nm				
			Annealed 50 days/800°C		*Annealed 50 days/800°C + 62 days/500°C*		
Au	*Pt*	*Ti*	*Sharp maxima*	*Diffuse maxima*	*Sharp maxima*	*Diffuse maxima*	*Hardness (DPN)*
22·5	2·5	75	0·509 0(5)		0·508 9(6)		
20	5·0	75	0·508 3(3)	0·508 8(9)	0·508 2(0)	0·508 8(9)	1 042
			0·508 3(0)	0·508 8(4)			
			0·508 4(0)	0·509 0(4)			
17·5	7·5	75	0·507 2(7)	0·507 9(5)	0·507 2(5)	0·507 9(4)	1 045
			0·507 3(3)	0·508 0(2)			
15	10	75	0·506 8(7)	0·507 7(7)	0·506 7(7)	0·507 5(7)	1 049
			0·506 6(5)	0·507 5(2)			
12·5	12·5	75	0·506 2(1)	0·506 8(5)			918
			0·506 1(6)	0·506 9(3)			
			0·506 2(3)				
10	15	75	0·505 4(6)	0·505 9(5)	0·505 3(0)	0·505 9(5)	1 066
7·5	17·5	75	0·504 8(2)	0·505 3(1)	0·504 4(6)	0·505 0(7)	—
5	20	75	0·504 1(9)	0·504 6(6)	0·503 9(8)	0·504 5(7)	1 080
			0·504 1(6)				
2·5	22·5	75	0·503 6(8)		0·503 6(5)		1 073

REFERENCES

65 Pie: P. Pietrokowsky: *Appl. Phys. Lett.*, 1965, **6**, 132–134

68 Pie: P. Pietrokowsky: *Scripta Metall.*, 1968, **2**, 379–383

Au–Pt–W
(G. V. Raynor)

INTRODUCTION

The very limited amount of information available on the system gold–platinum–tungsten was acquired by Jones *et al.* [72 Jon] in the course of an examination of the possibility of protecting refractory metals from oxidation by means of a coating of a gold–platinum alloy. The work involved the determination, by X-ray methods, of two tie lines at 1 100°C.

THE BINARY SYSTEMS

The Au–Pt equilibrium diagram is well established [85 Oka 1]. It is characterized by a wide liquidus–solidus separation over the greater part of the composition range, and the establishment of a continuous series of solid solutions above a temperature of 1 260°C. Below this temperature a miscibility gap exists, separating fcc alloys of different compositions and lattice spacings. The maximum on the curve enclosing the two-phase region is at 61 at.-% Pt, and the region widens considerably as the temperature falls.

No equilibrium diagram exists for the system Au–W. According to [85 Oka 2] a eutectic is formed at ~0·1 at.-% W and 1 063°C.

Information on the constitution of the Pt–W system is also sparse and partially contradictory. The phase diagram given by [58 Han] is tentative, and shows a simple peritectic system without intermediate phases. The suggested liquidus curve falls, on adding Pt, from the melting point of W to the peritectic reaction L + (W) $\rightleftharpoons$ (Pt) at ~2 460°C, and, with further increase in Pt content, to the melting point of Pt. The brackets denote solid solutions in W and Pt. The maximum solid solubilities at ~2 460°C are given as ~ 5 at.-% Pt in W, and ~63·5 at.-% W in Pt. The latter value, however, is suspect, since [80 Kna] indicates the existence of two intermediate phases, γ and ϵ, at 1 400°C. ϵ (45–52 at.-% W) is close-packed hexagonal in crystal structure; γ (33–37 at.-% W) has a tetragonally distorted fcc structure, and is shown as coming into equilibrium with a (Pt) solid solution containing only ~25–29 at.-%. The presence of γ and ϵ is accepted, and it would appear that the extent of the (Pt) solid solution is much smaller than that suggested in [58 Han], unless γ and ϵ are formed from this solid solution in the solid state. No information exists on this point, nor on the constitution at lower temperatures.

THE SOLID PHASES

The structures of the solid phases known to exist in the ternary system are summarized in Table 107. No discrete ternary phases have been reported.

Table 107. Crystal structures

Solid phase	*Prototype*	*Lattice designation*
(W)	W	cI2
(Au), (Pt) and (Au–Pt)	Cu	cF4
γ(system W–Pt)	—	Tetragonally distorted fcc
ϵ(system W–Pt)	Mg	hP2

THE TERNARY SYSTEM

The ternary alloys for the examination of equilibria were prepared by melting together the appropriate amounts of the components under argon in an induction furnace. Specimens were homogenized for not less than 65 h at 1 100°C and water-quenched. The compositions of the phases coexisting in two-phase alloys were determined by the measurement of lattice spacings, using powders prepared from the quenched samples, stress-relieved at the quenching temperature, and again quenched. The method in principle involves a knowledge of the variation of lattice spacings with composition in the ternary system, but since the compositions of the coexisting phases are given in terms of two of the components only, it would appear that reliance was placed on the lattice spacing variations only in the Au–Pt and Pt–W systems. There is no explicit reference to this point in [72 Jon], and no isothermal phase diagram is given.

Figure 335 shows the two tie lines determined for 1 100°C, and these run from Au-rich compositions on the Au–Pt axis to Pt-rich compositions on the Pt–W axis. At this temperature, alloys with compositions at the extreme Au-rich region of the Au–Pt phase diagram are in the liquid state; accordingly Fig. 335 contains L + (W) and L + (W) + (Au) areas, and the position of the three-phase triangle cannot vary significantly from that shown. Also indicated in Fig. 335 is constitutional information taken from the binary systems, assuming that the γ and ϵ phases detected at 1 400°C remain stable at 1 100°C without significant change of composition. In the Au–Pt system the (Au) + (Pt) miscibility gap has been established, and the tie lines clearly involve the (Au) phase, which is stable at 1 100°C up to ~38 at.-% Pt. The virtually Au-free compositions of the conjugate phase are, however, more difficult to reconcile with the constitution of the ternary system which would be expected at 1 100°C.

DISCUSSION

Nothing is known about the manner in which the γ and ϵ phases are involved in the ternary equilibria. If the tie lines determined by [72 Jon] are accepted, it is difficult to escape the conclusion that the equilibria entered into by the γ-phase are γ + (Au) and γ + (Pt). A schematic isothermal diagram for 1 100°C, assuming the stability of the γ and ϵ-phases at this temperature, is sketched in Fig.

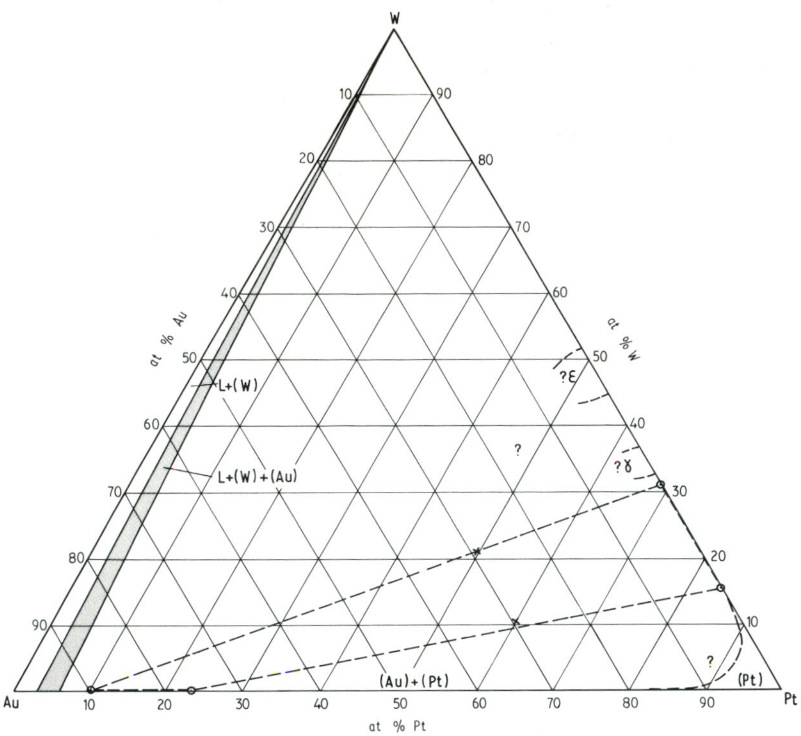

335 The system Au–Pt–W: tie lines established at 1 100°C [72 Jon]. The diagram also contains features derived from the binary diagrams

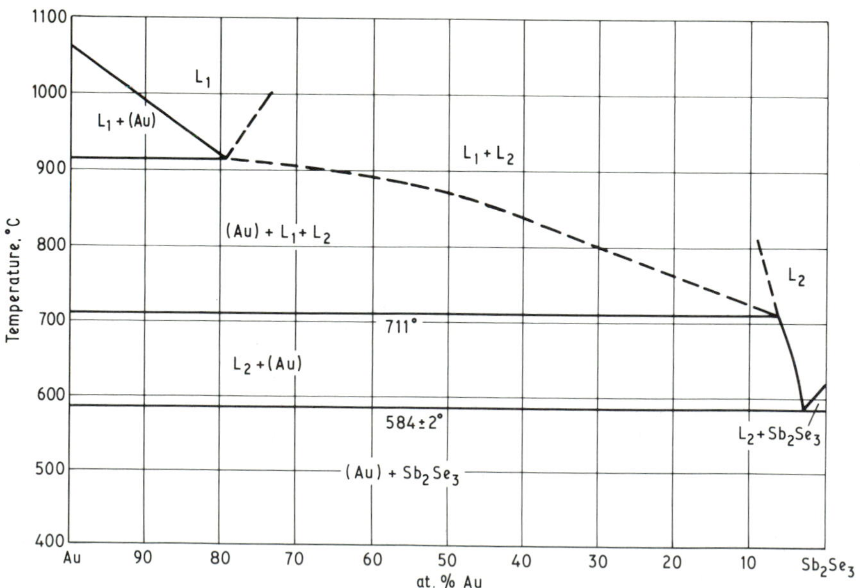

337 The Au–Sb_2Se_3 section

336, which is entirely hypothetical. Comparison of Figs. 335 and 336 shows that the composition of the more W-rich two-phase alloy examined by [72 Jon] falls within the hypothetical γ + (Au) field, and it is possible that the compositions of the co-existing phases really represent (Au) and γ, remembering that the γ-phase has a basically, though slightly tetragonally deformed, fcc structure. The other two-phase alloy, however, must fall in the (Au) + (Pt) field. The existence of this two-phase alloy containing a virtually Au-free (Pt) phase indicates that the γ + (Pt) field must be confined to a narrow region close to the Pt–W axis of the ternary diagram.

REFERENCES

58 Han: M. Hansen and K. Anderko: 'Constitution of Binary Alloys': 1958, New York, McGraw-Hill

72 Jon: B. Jones, M. W. Jones and D. W. Rhys: *J. Inst. Metals*, 1972, **100**, 136–141

80 Kna: A. G. Knapton: *Pt Metals Rev.*, 1980, **24**, 64–69

85 Oka 1: H. Okamoto and T. B. Massalski: *Bull. Alloy Phase Diagrams*, 1985, **6**, 46–56

85 Oka 2: H. Okamoto and T. B. Massalski: *Bull. Alloy Phase Diagrams*, 1985, **6**, 136–137

Au–Pt–Zr

Wang *et al.* [67 Wan] studied alloys on the 25 at.-% Zr section by X-ray diffraction techniques. They report the presence of four ordered close-packed phases on this section, but no further details were given. Au_3Zr has a Cu_3Ti-type structure (oP8) and Pt_3Zr a Ni_3Ti-type structure (hP16).

REFERENCE

67 Wan: R. Wang, B. C. Giessen and N. J. Grant: *Abst. Bull. Inst. Metals Div., Met. Soc. AIME*, 1967, **2**(2), 120

Au–Rb–Sn

Sinnen and Schuster [81 Sin] prepared the ternary compound $Rb_4Au_7Sn_2$ from the elements by melting in a Ta crucible enclosed in a steel ampoule under argon. The alloy was subsequently annealed for 10 days at 550°C. Its crystal structure was determined by single crystal X-ray diffraction analysis. $Rb_4Au_7Sn_2$ crystallizes in the hexagonal (rhombohedral) space group R3m with lattice parameters a = 0·680 1(3), c = 2·909 0(7) nm, c:a = 4·277. There are three formula units in the unit cell. The structure has similarities to that of the Laves phases of the $MgCu_2$-type.

REFERENCE

81 Sin: H.-D. Sinnen and H.-U. Schuster: *Z. Naturf.*, 1981, **36B**, 833–836

Au–Rh–Zr

Wang *et al.* [67 Wan] studied alloys on the 25 at.-% Zr section by X-ray diffraction techniques. They report the presence of three ordered close-packed phases on this section but no further details were given. Au_3Zr has a Cu_3Ti-type structure (oP8) and Rh_3Zr a Cu_3Au-type structure (cP4).

REFERENCE

67 Wan: R. Wang, B. C. Giessen and N. J. Grant: *Abst. Bull. Inst. Metals Div., Met. Soc. AIME*, 1967, **2**(2), 120

Au–Sb–Se

The only data is unpublished work [76 Gat] on the Au–Sb_2Se_3 section (Fig. 337). Below 711°C the system has pseudobinary characteristics. A schematic representation of the ternary equilibria, adapted from [76 Gat], is presented in Fig. 338 and the corresponding reaction scheme in Fig. 339. It is likely that the major part of the ternary system is occupied by two regions of liquid immiscibility, one stretching from the 963°C Au–Se monotectic to the 580°C Sb–Se monotectic, and the other invading the ternary system from the 760°C Au–Se monotectic. The postulated invariant reactions are given in Fig. 339. There is a need to extend the work of [76 Gat] to the complete ternary system.

REFERENCE

76 Gat: B. Gather: Thesis, Technischen Universität Clausthal, 1976

Au–Sb–Si

INTRODUCTION

[71 Gub] determined the liquidus curves in the region of primary solidification of Si for three sections close to the Au–Si binary edge of the ternary system. Measurements were made at 550, 650, 700, 750 and 800°C on the sections Au, 0·16 at.-% Sb–Si, Au, 1·13 at.-% Sb–Si and Au, 1·61 at.-% Sb–Si. The solubility of Si in the liquid phase was found to increase with the addition of Sb to the Au. The major work on the Au–Sb–Si ternary phase equilibria [86 Leg], involved the study of 76 alloys prepared from

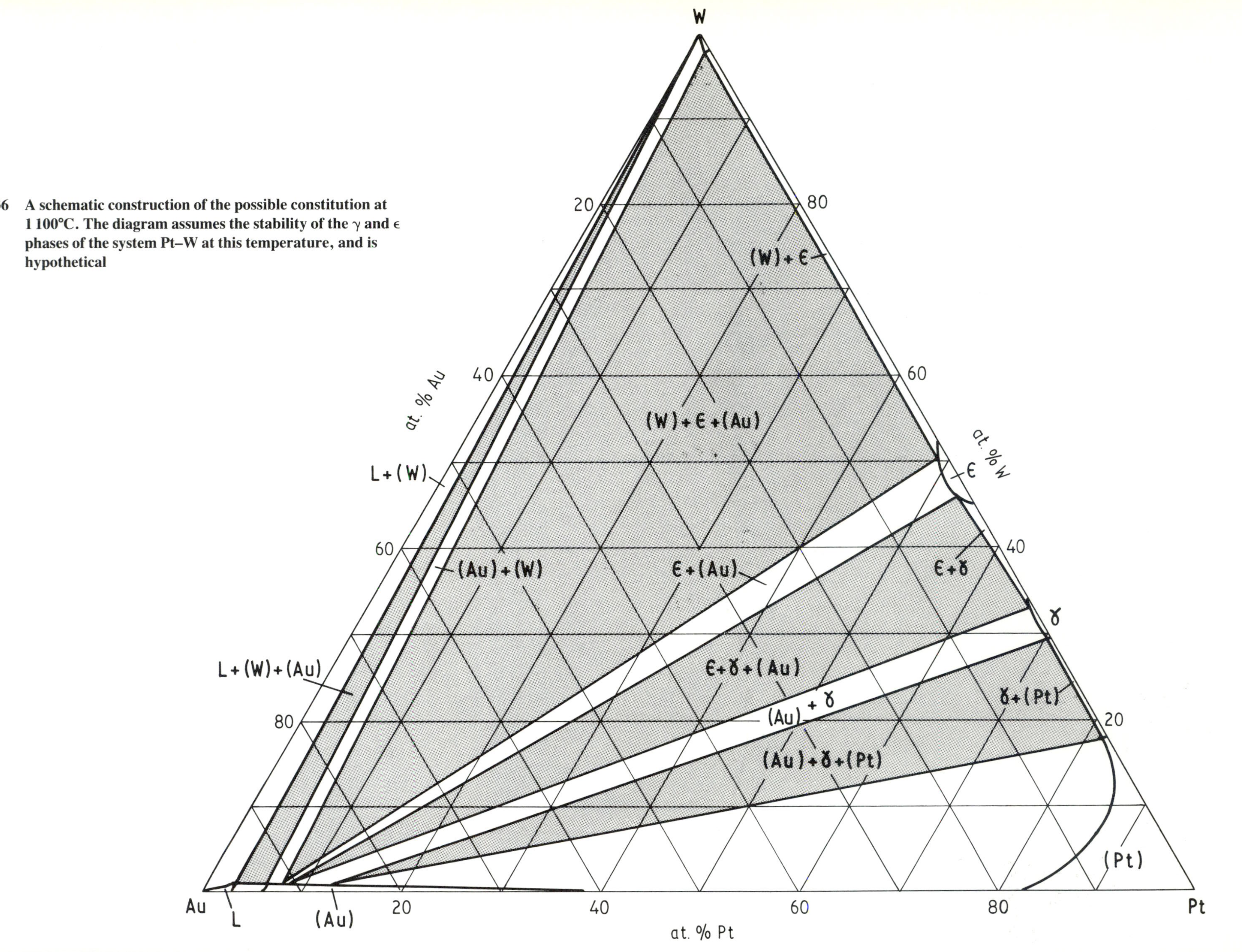

336 A schematic construction of the possible constitution at 1 100°C. The diagram assumes the stability of the γ and ϵ phases of the system Pt–W at this temperature, and is hypothetical

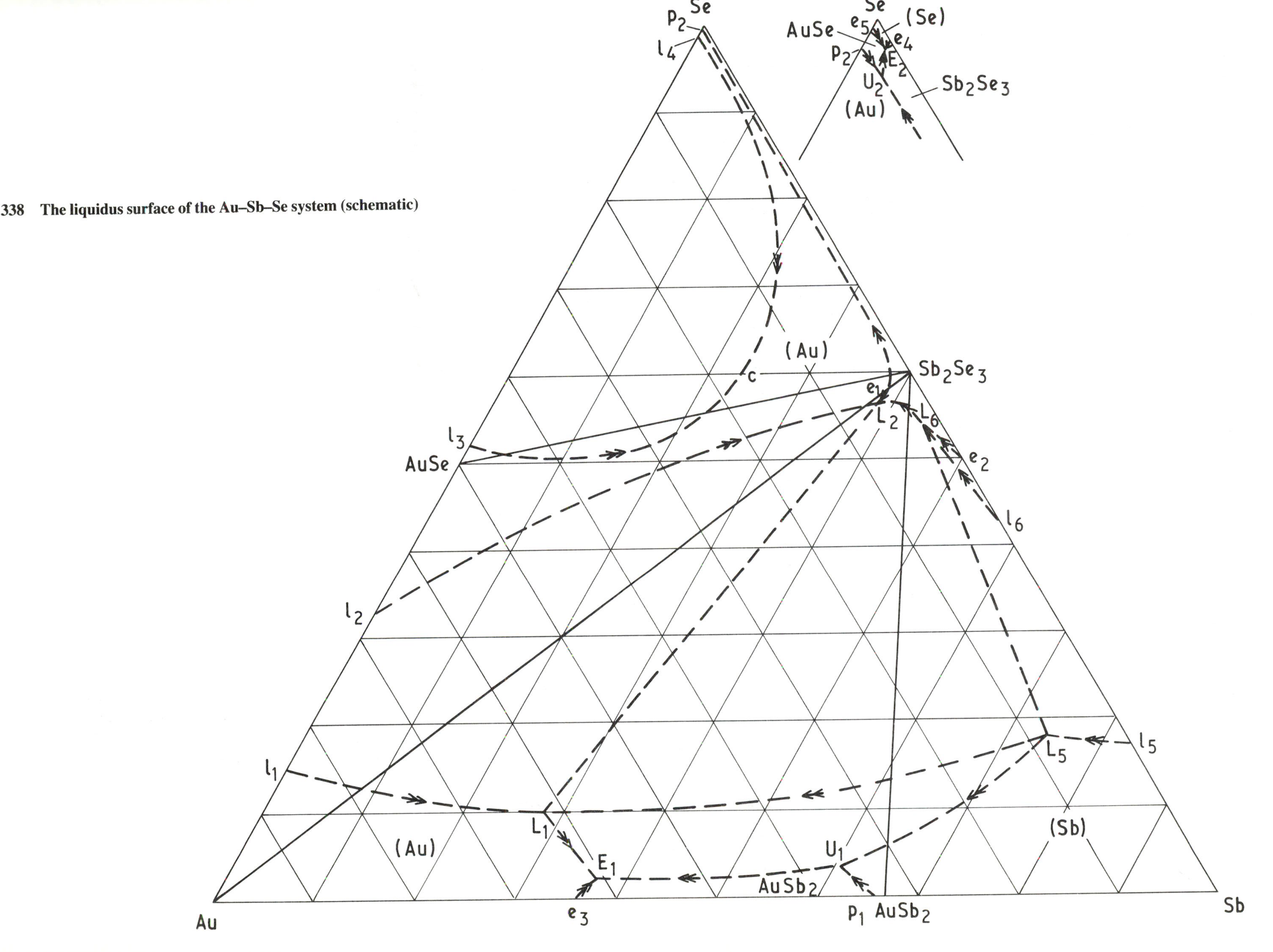

338 The liquidus surface of the Au–Sb–Se system (schematic)

99·99% Au and 99·999 9% Sb and Si. Differential thermal analysis at heating rates of 2 and 5°C min^{-1} and differential scanning calorimetry at heating rates of 0·1 and 2°C min^{-1} were used to characterize the equilibria. Two invariant reactions were detected. A transition reaction, L + (Sb) ⇌ $AuSb_2$ + (Si), occurs at 457·2 ± 0·5°C (Fig. 340). The composition of the liquid (U) is close to the composition of the binary peritectic liquid (p) as would be expected from the small difference in reaction temperatures. A ternary eutectic reaction, L ⇌ (Au) + $AuSb_2$ + (Si), occurs at 332·3 ± 0·5°C. The composition of the ternary eutectic liquid (E) is 20 ± 0·5 Sb, 10 ± 0·5 at.-% Si. A reaction scheme is given in Fig. 341. It should be noted that the liquidus isotherms determined by [86 Leg] are concave to the Au–Sb binary edge in the region of primary Si solidification. This is the reverse of the data reported by [71 Gub]. The results of the more extensive study of [86 Leg] are accepted.

BINARY SYSTEMS

The Au–Si phase diagram evaluated by [83 Oka] and the Sb–Si phase diagram evaluated by [85 Ole] are accepted. The Au–Sb phase diagram evaluated by [84 Oka] shows a peritectic liquid containing 66·6 at.-% Sb, only slightly differing from the composition of the compound $AuSb_2$. The work of [84 Eva] is preferred. They reported the peritectic liquid composition at 65·6 at.-% Sb with a peritectic temperature of 458 ± 1°C. The eutectic temperature is 356 ± 1°C at a eutectic composition of 36·1 at.-% Sb.

SOLID PHASES

Table 108 summarizes crystal structure data for the binary phases. No ternary phases have been detected.

Table 108. Crystal structures

Solid phase	*Prototype*	*Lattice designation*
(Au)	Cu	cF4
(Sb)	αAs	hR2
(Si)	C(diamond)	cF8
$AuSb_2$	FeS_2	cP12

INVARIANT EQUILIBRIA

[86 Leg] studied five vertical sections, at 10 at.-% Si, 10, 30 and 50 at.-% Sb and at 80 at.-% Au. The sections, Figs. 342–346 respectively, have been amended from the original drawings to give a consistent set of data. They clearly indicate the presence of a ternary transition reaction at 457·2°C and a ternary eutectic reaction at 332·3°C. The 30 at.-% Sb section (Fig. 344) was shown by [86 Leg] as containing a primary $AuSb_2$ phase region whose liquidus trace had a temperature maximum. However, the primary and secondary separations for an alloy containing 30 Sb, 10 at.-% Si were plotted at substantially different temperatures on the 30 at.-% Sb and 10 at.-% Si sections. On the 30 at.-% Sb section primary (Si) separated at 428°C and secondary (Si) + $AuSb_2$ at 353°C; on the 10 at.-% Si section the same alloy composition was shown as separating primary (Si) at 485°C and secondary (Si) + $AuSb_2$ at 390°C. The data given for the 10 at.-% Si section (Fig. 342) have been accepted on the basis that more alloy compositions were studied on this section and the data is consistent with the liquidus projection (Fig. 340). The 30 at.-% Sb section (Fig. 344), has been redrawn to give consistency with the 10 at.-% Si sections.

The composition of the ternary eutectic lies on or very near to the 10 at.-% Si section. [86 Leg] quote a composition 20 ± 0·5 Sb, 10 ± 0·5 at.-% Si. The transition reaction occurs at a temperature 1°C below that of the binary peritectic temperature in the Au–Sb system. The composition of the liquid U (Fig. 340) is close to the Au–Sb binary edge. The monovariant curve e_1U runs close to the Au–Sb binary edge for its whole length.

LIQUIDUS SURFACE

The liquidus surface, shown in Fig. 340, is dominated by the region of separation of primary (Si). The possibility of producing amorphous or microcrystalline alloys for compositions near to the ternary eutectic composition was noted by [86 Leg].

[71 Gub] reported liquidus traces for alloys on the sections Au, 0·16 Sb − Si, Au, 1·13 Sb − Si and Au, 1·61 Sb − Si that disagree with those given in Fig. 340. A disc of the Au–Sb alloy was placed on a (111) single-crystal Si slice, heated under vacuum to 550, 650, 700, 750 and 800°C, held for 20 min to allow dissolution of Si in the Au–Sb discs, and finally cooled to room temperature at 3°C min^{-1}. The liquid phase that forms is assumed to be saturated with Si to a composition on the ternary liquidus surface at each temperature. On cooling the primary (Si) deposits epitaxially on the Si slice. The weight of Si dissolved in the liquid phase was determined by measurement of the penetration depth (the depth of the epitaxially deposited Si layer plus the increased depth of the Au–Sb layer caused by formation of a eutectic with Si) and by selectively etching the Au–Sb reaction layer away and determining the loss of Si to the eutectic by weighing. As Table 109 indicates, the liquidus compositions reported by [71 Gub] for binary Au–Si alloys do not agree with the assessed values from [83 Oka] nor with values found by [67 Ger] using the penetration depth method. It is not possible to reconcile the liquidus data from [71 Gub] with the liquidus surface proposed by [86 Leg].

REFERENCES

67 Ger: W. Gerlach and B. Goel, *Solid-State Electron.*, 1967, **10**, 589–592

71 Gub: A. Ya. Gubenko and I. K. Kiparisova: *Izvest. Akad. Nauk SSSR., Neorgan. Materialy*, 1971, **7**, 1149–1152

83 Oka: H. Okamoto and T. B. Massalski: *Bull. Alloy Phase Diagrams*, 1983, **4**, 190–198

84 Oka: H. Okamoto and T. B. Massalski: *Bull. Alloy Phase Diagrams*, 1984, **5**, 166–171

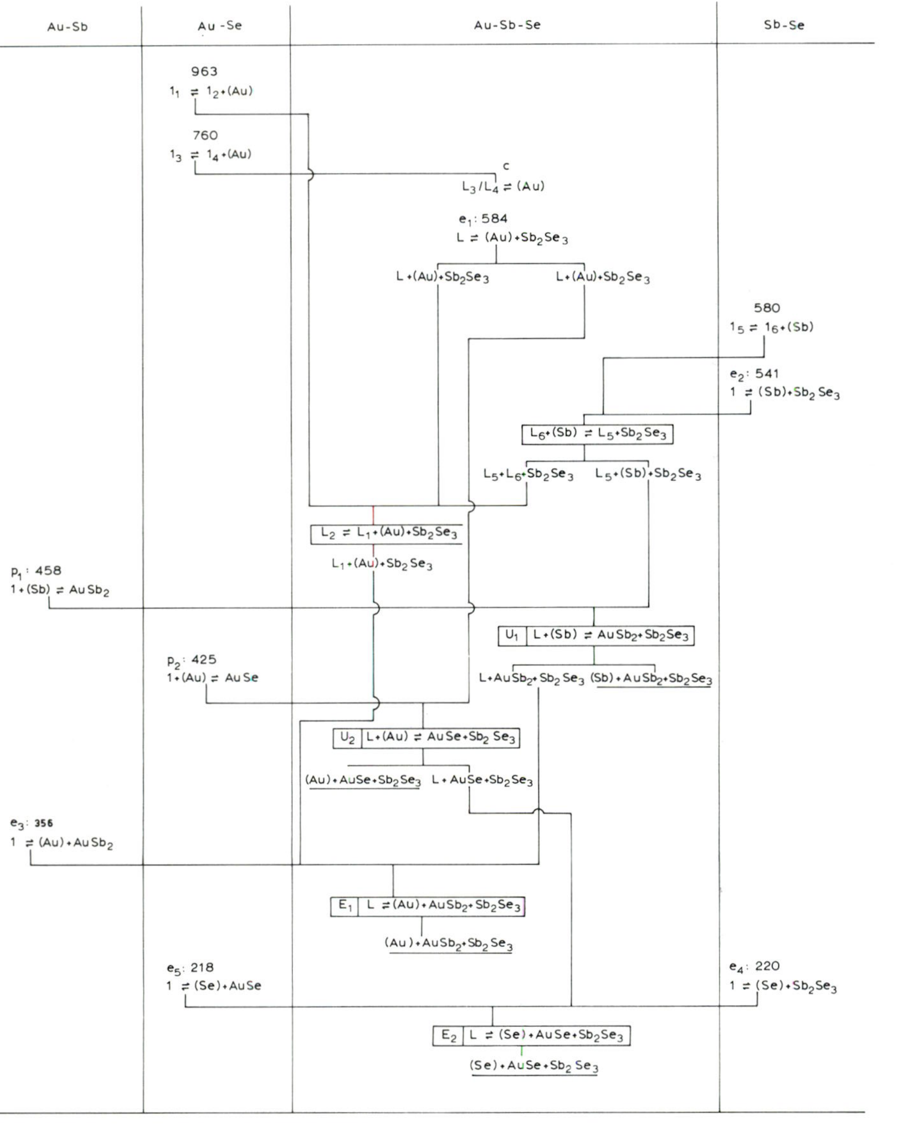

339 The reaction scheme for the Au–Sb–Se system (partially hypothetical)

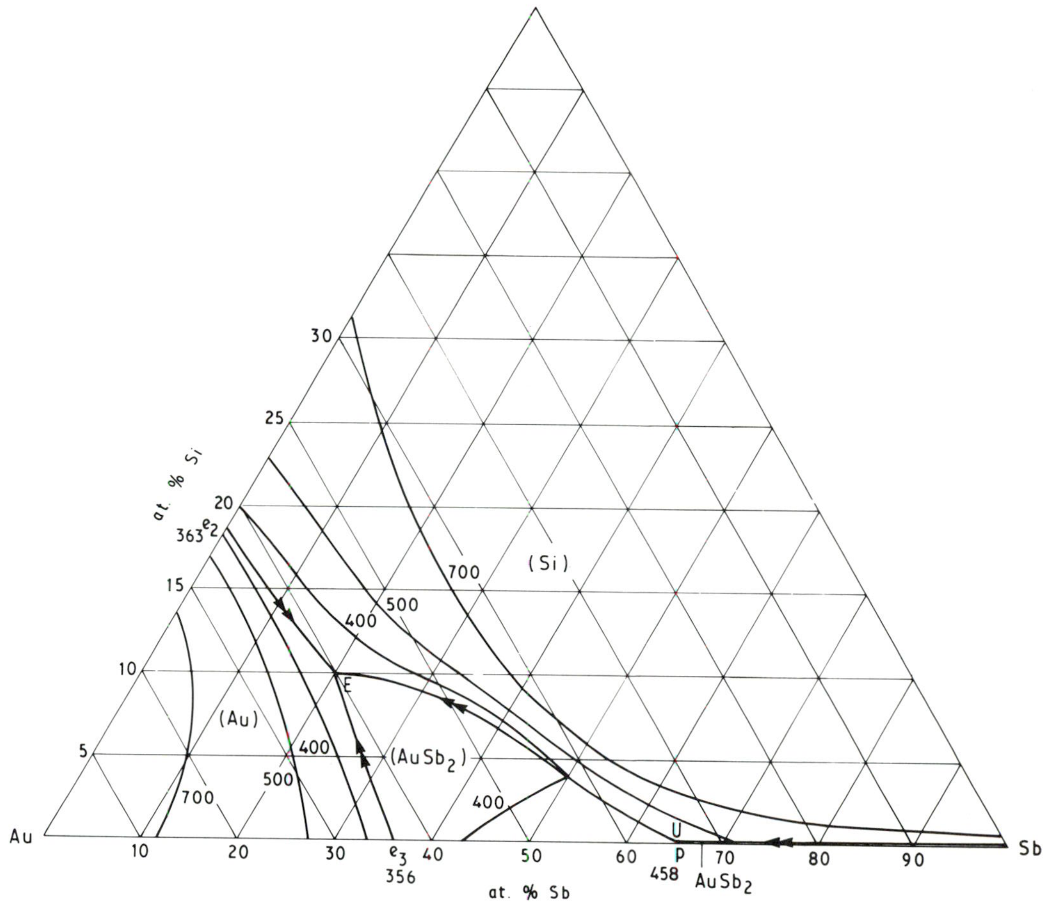

340 Liquidus projection of the Au–Sb–Si ternary system

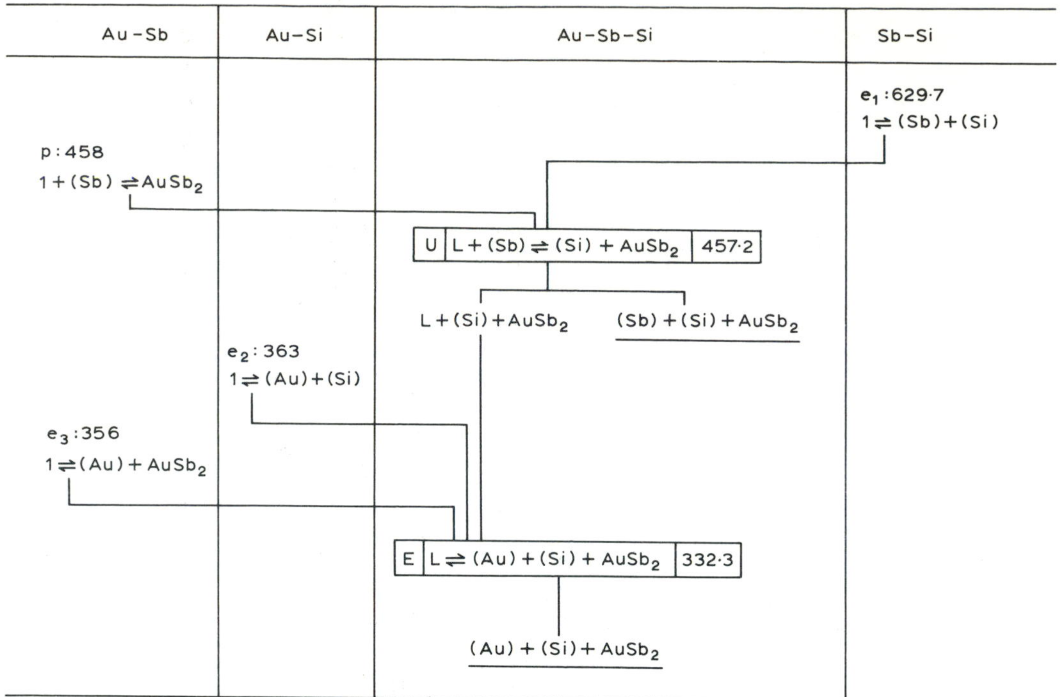

341 Reaction scheme for the Au–Sb–Si ternary system

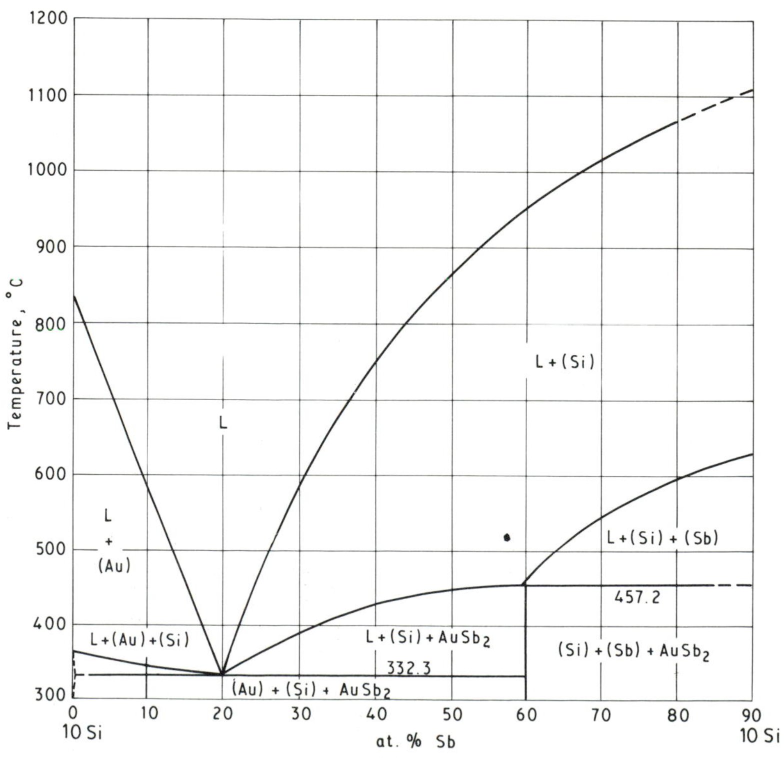

342 Vertical section at 10 at.-% Si

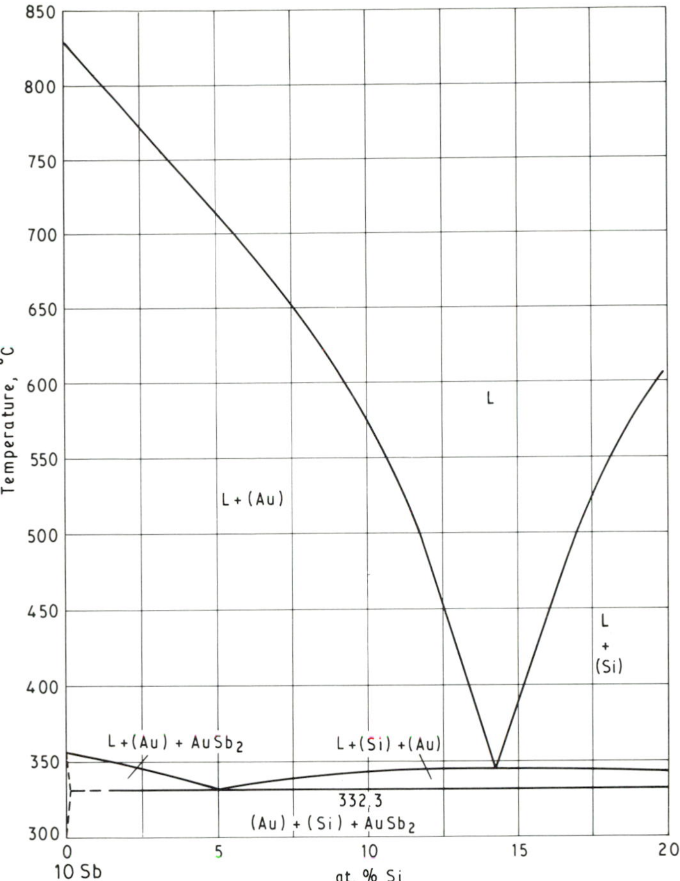

343 Vertical section at 10 at.-% Sb

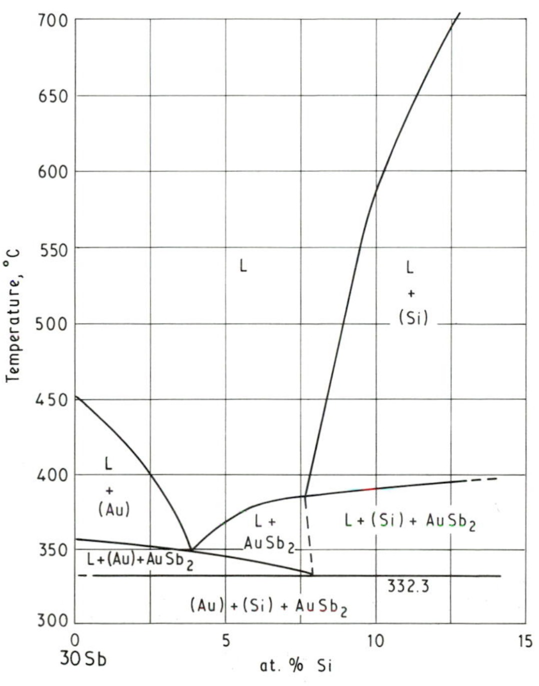

344 Vertical section at 30 at.-% Sb

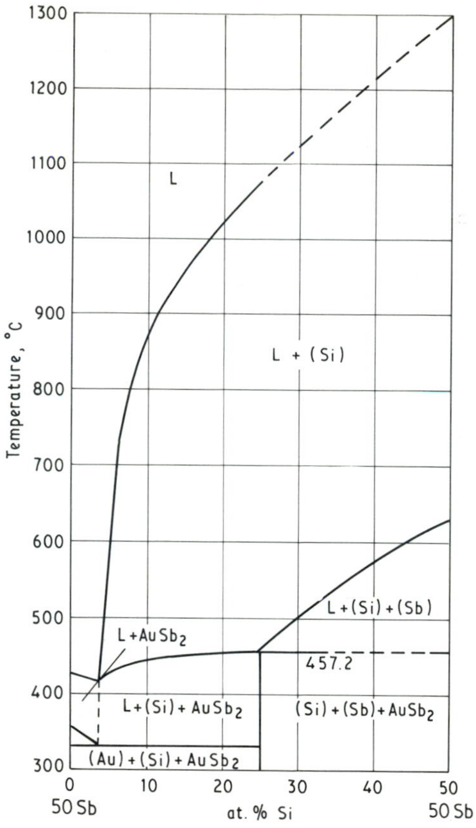

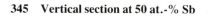

345 Vertical section at 50 at.-% Sb

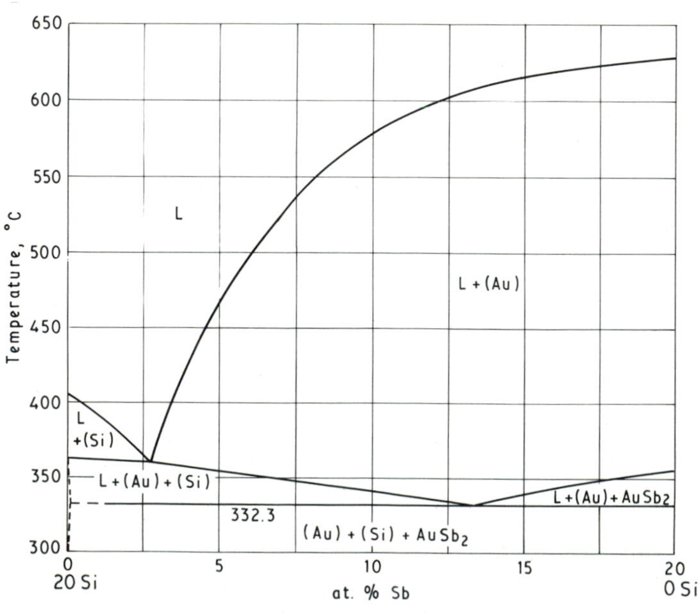

346 Vertical section at 80 at.-% Au

Table 109. Liquidus temperatures reported by [71Gub]

Temperature, °C	Liquid composition, at.-% Si					
	Au–Si			*Au,0·16Sb–Si*	*Au,1·13Sb–Si*	*Au,1·61Sb–Si*
	[83Oka]	[67Ger]	[71Gub]	[71Gub]	[71Gub]	[71Gub]
550	25·2	—	26·5	28·9	32·3	34·6
650	29·2	28·7	30·6	33·0	36·4	39·2
700	31·6	31·0	33·6	36·2	39·9	42·6
750	34·0	34·6	36·8	39·5	44·0	47·3
800	36·7	36·2	39·5	43·3	49·2	53·2

84 Eva: D. S. Evans and A. Prince, CALPHAD XIII, Villard de Lans, 1984

85 Ole: R. W. Olesinski and G. J. Abbaschian, *Bull. Alloy Phase Diagrams*, 1985, **6**, 445–448

86 Leg: B. Legendre and Chhay Hancheng, *Bull. Soc. Chim. France I*, 1986, (2), 138–144

Au–Sb–Sn

Schubert *et al.* [59 Sch] published a 250°C isothermal section based on the heat treatment of alloys initially melted in evacuated silica tubes. The solidified alloys were vacuum heat-treated for times ranging from 0 to 15 days at 250°C. Powder samples were also prepared and similarly heat treated. Examination of the cast and heat treated alloys was by metallography; the powder samples were studied by X-ray diffraction analysis. Figure 347 presents the 250°C section in an amended form to take into account the composition limits of the binary phases at 250°C. At 250°C Au–Sn alloys have an L + $AuSn_4$ region stretching from 80 to 89 at.-% Sn, followed by a liquid region to Sn. Sb–Sn alloys are liquid from Sn to 9·2 at.-% Sb and enter a L + βSbSn region from 9·2 to 41·5 at.-% Sb, according to [58 Han]. Figure 347 is therefore a guide to the 250°C isothermal section for alloys whose compositions are on the Au-rich side of a line joining βSbSn and $AuSn_2$. For alloys on the Au-poor side of this line some liquid phase will be present at 250°C; Fig. 347 relates to the equilibria anticipated at temperatures of 200°C and lower in this portion of the ternary system. Although the phases present in 24 alloys were included in the isothermal section a table of experimental data refers to 14 alloys only.

More recent work [71 Pre] on the Sb–Sn phase diagram shows the presence of an intermetallic compound Sb_2Sn_3 in addition to SbSn. Sb_2Sn_3 forms peritectically at 324°C and transforms by eutectoidal reaction at 242°C into (αSn) + SbSn. The isothermal section [59 Sch] (Fig. 347) at 250°C does not include Sb_2Sn_3 in the equilibria, and this section can only be reconciled with the Sb–Sn phase diagram of [71 Pre] if it is assumed that the temperature is below 242°C, rather than 250°C as stated by [59 Sch].

Humpston [85 Hum] has confirmed that the AuSn–Sb section is not a pseudobinary, but that the section $AuSb_2$–AuSn is a pseudobinary eutectic, except in the vicinity of the peritectically formed $AuSb_2$ phase. The eutectic temperature is 280°C and the eutectic composition is provisionally placed at 57 at.-% Sb. Accepting the Sb–Sn phase diagram of [71 Pre], a *speculative* reaction scheme is given in Fig. 348. The temperatures for the proposed invariant reactions U_2, U_5, U_8 and E are provisional data [85 Hum]. The phases associated with the invariant reactions have not yet been confirmed.

REFERENCES

58 Han: M. Hansen and K. Anderko: 'Constitution of Binary Alloys', 1958, New York, McGraw-Hill

59 Sch: K. Schubert, H. Breimer and R. Gohle: *Z. Metall.*, 1959, **50**, 146–153

71 Pre: B. Predel and W. Schwermann: *J. Inst. Metals*, 1971, **99**, 169–173

85 Hum: G. Humpston: PhD Thesis, Brunel University, 1985

Au–Sb–Ta

In studying solid solution formation between Cr_3Si-type phases (cP8), von Philipsborn [70 Phi] prepared a $Ta_3Au_{0\cdot5}Sb_{0\cdot5}$ alloy from the elements (Ta > 99·8%; Au, Sb > 99·9% purity) by arc-melting on a water-cooled Cu hearth under argon. Samples were annealed for 45 days at 550°C, 30 days at 800°C, and 9 days at 1 050°C. The phases present were characterized by the X-ray powder diffraction technique. It should be noted that Ta_3Sb has the Cr_3Si-type structure, but the Au–Ta phase with a Cr_3Si-type structure has been assigned various compositions (Au_3Ta, Au_4Ta, Au_5Ta).

Only the 550 and 1 050°C annealed samples contained a Cr_3Si-type phase with lattice parameter $a = 0{\cdot}526 \pm 0{\cdot}001$ nm (Ta_3Sb has an almost identical parameter). This sample also contained a W-type phase with $a = 0{\cdot}3310 \pm 0{\cdot}0003$ nm and an unidentified phase. Surprisingly the 800°C annealed sample contained no Cr_3Si-type phase; it showed two W-type phases with $a = 0{\cdot}3310$ and $0{\cdot}3166 \pm 0{\cdot}0003$ nm. Without further data it is not possible to interpret these results.

REFERENCE

70 Phi: H. von Philipsborn: *Z. Kristallogr.*, 1970, **131**, 73–87

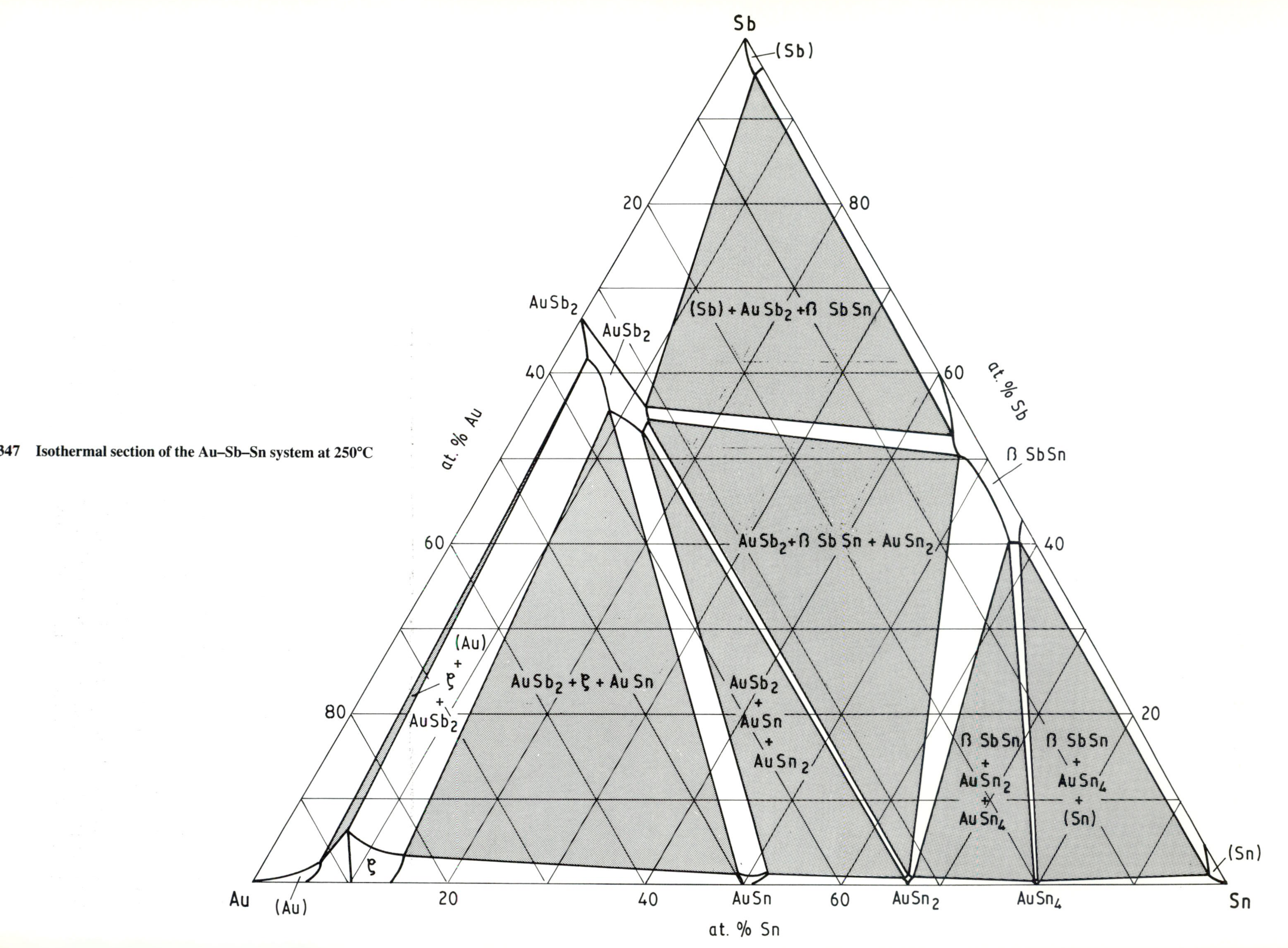

347 Isothermal section of the Au–Sb–Sn system at 250°C

348 Speculative reaction scheme

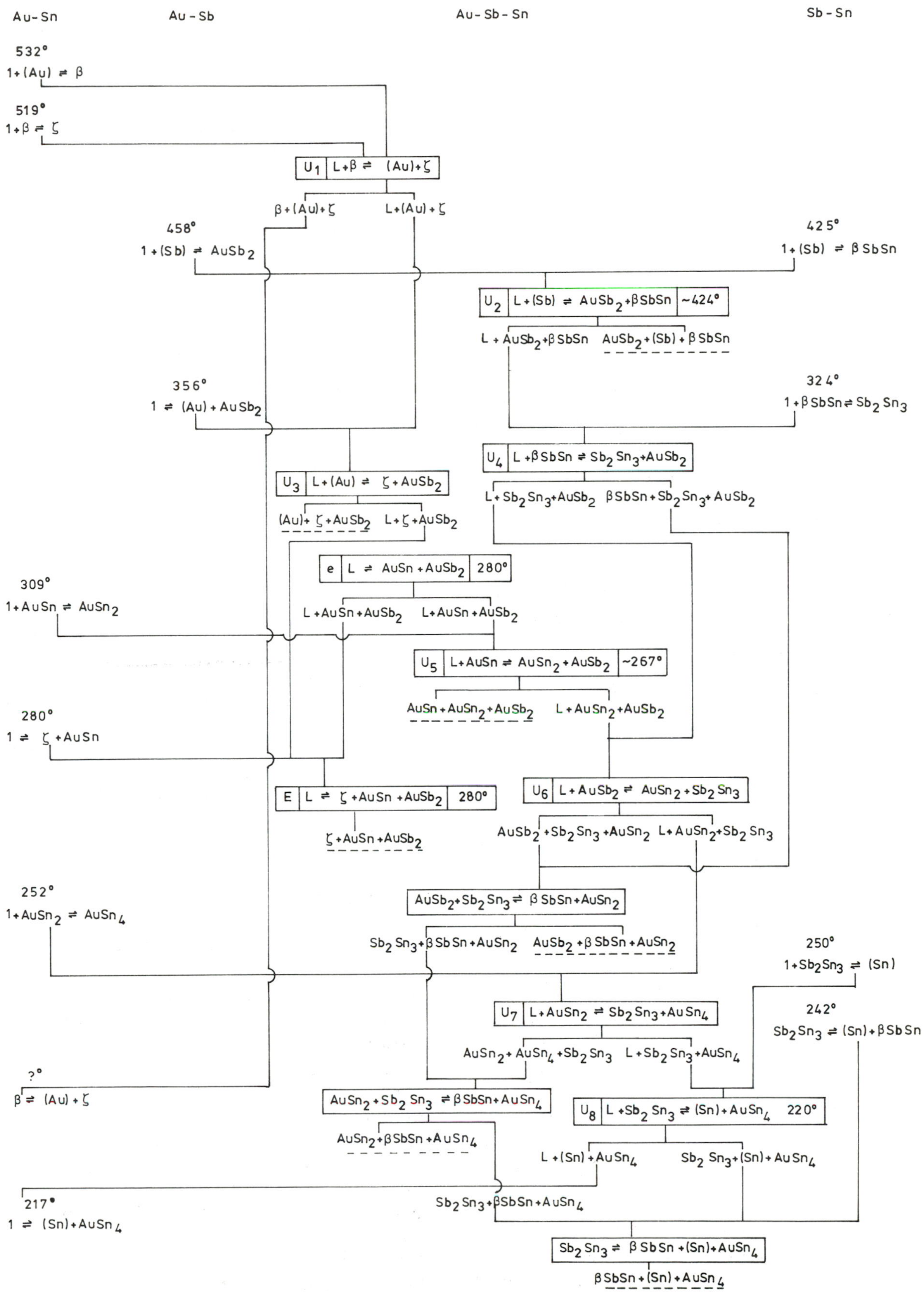

Au–Sb–Te

INTRODUCTION

The major work is a survey [76 Gat] of an unspecified number of alloys made by melting 99·995% purity elements in evacuated quartz containers and studying their constitution, after tempering all alloys for 60 days at 300°C, by DTA, X-ray diffraction, and metallographic techniques. The ternary system is dominated by a large, steeply-sloping field of primary solidification of Au and by a similarly large but relatively flat field of primary solidification of Sb_2Te_3. Pseudobinary eutectic systems Au–Sb_2Te_3 and $AuSb_2$–Sb_2Te_3 (essentially pseudobinary) divide the ternary system into three partial ternary systems. In the partial ternary system Au–Te–Sb_2Te_3 [76 Gat] located a ternary compound τ_1 of composition $AuSb_{0\cdot076}Te_{1\cdot955}$ which forms at 444°C by the reaction L_{u_5} + $AuTe_2 \rightleftharpoons \tau_1$ (Fig. 349). The section τ_1–Sb_2Te_3 contains a eutectic saddle point, e_4, at 430°C, but the section is not a pseudobinary since the monovariant curve u_5U_6 cuts through it. Transition reactions at U_4 and U_6 precede ternary eutectic reactions at E_2 (423°C) and E_3 (396°C) that complete solidification in the regions Au–τ_1–Sb_2Te_3 and Te–$AuTe_2$–Sb_2Te_3 respectively. Table 110 summarizes the liquid compositions associated with the ternary invariant reactions. Amendments to the compositions quoted [76 Gat] are noted in Table 110 and discussed in the section dealing with invariant equilibria. [76 Bla] give further information on τ_1, at variance with the finding of [72 Bac] that a ternary compound with the crystal structure of the mineral montbrayite was found at the composition $AuSb_{0\cdot122}Te_{1\cdot524}$ (37·8 at.-% Au, 4·6 at.-% Sb, 57·6 at.-% Te). [76 Gat, 76 Bla] quote τ_1 at 33 at.-% Au, 2·5 at.-% Sb, 64·5 at.-%Te.

Table 110. Invariant equilibria

Reaction	*Temperature*, °C	*Phase*	*Composition**, at.-%	
			Sb	*Te*
e_1	454	L	24·0	36·0
e_2	450	L	62·2 (62·75)	10·0 (8·75)
e_4	430	L	11·5 (11·5)	63·4 (63·5)
u_5	~444	L	10·0 (9·0)	58·0 (58·0)
u_6	430	L	33·7 (35·0)	7·0 (7·0)
U_1	454	L	67·0	4·5
U_2	~445	L	64·5	8·5
U_3	~443	L	66·0	7·0
U_4	428	L	9·5	51·0
U_5	427	L	33·2 (34·5)	7·0 (7·0)
U_6	425	L	11·5	65·5
U_7	395	L	39·0	3·5
E_1	~440	L	67·0	6·0
E_2	423	L	13·0	49·5
E_3	396	L	7·0	83·0
E_4	356	L	~34·0 (24·0)	~ 1·5 (36·0)

* Assessed composition; compositions in brackets from [76 Gat]

In the partial ternary system Au–Sb_2Te_3–$AuSb_2$ a ternary compound τ_2 is formed by the incongruent reaction, L_{u_6} + $Sb_2Te_3 \rightleftharpoons \tau_2$ (430°C). The composition is 33 at.-% Au, 36·5 at.-% Sb, 30·5 at.-% Te [76 Gat], corresponding to $AuSb_{1\cdot106}Te_{0\cdot924}$. The point u_6 is a maximum on the monovariant curve $U_5u_6U_7$. At U_5 the reaction is L_{U_5} + $Sb_2Te_3 \rightleftharpoons$ (Au) + τ_2: at U_7 the reaction is L_{U_7} + $Sb_2Te_3 \rightleftharpoons$ $AuSb_2$ + τ_2. Solidification ends in the partial ternary system Au–Sb_2Te_3–$AuSb_2$ at the ternary eutectic E_4. Gather and Blachnik [76 Gat] misquote the composition of the liquid at E_4. The composition given in Table 110 is an estimate from the liquidus surface presented by [76 Gat]. Structural data on τ_2 are given in [76 Bla] and [72 Bac]. The region of primary solidification of τ_2 is very small ($U_5u_6U_7E_4$).

The partial ternary system Sb–$AuSb_2$–Sb_2Te_3 contains three transition reactions, U_{1-3}, and a ternary eutectic E_1 [76 Gat]. To accommodate a ternary eutectic reaction at E_1 [76 Gat] assume that the β phase in the Sb–Te system exists as β′ and β″, where β′ is the composition of β at the Sb-rich boundary of the wide β homogeneity region and β″ is the composition at the Te-rich boundary of the β region. [76 Gat] propose a reaction $L_{E_1} \rightleftharpoons \beta' + \beta'' +$ $AuSb_2$. This is regarded as unlikely. It implies the appearance of a three-phase region 1 + β′ + β″, originating at a critical tie line on the β liquidus where 1 $\rightleftharpoons$ β, and descending to E_1. No experimental evidence is given [76 Gat] to justify the breakdown of the β phase into two distinct compositions, β′ and β″. It is thought more probable that the 1 + β + $AuSb_2$ phase regions descending from U_1 and U_3 meet at a minimum critical tie line at the temperature of E_1.

BINARY SYSTEMS

The Au–Te system evaluated by [84 Oka 2] is accepted. The Au–Sb system evaluated by [84 Oka 1] indicates a peritectic liquid containing 66·6 at.-% Sb, only slightly differing from the composition of the compound $AuSb_2$. [84 Eva] report the peritectic liquid at 65·6 at.-% Sb at 458 ± 1°C. The eutectic temperature is 356 ± 1°C at 36·1 at.-% Sb. The Sb–Te diagram [65 Ell], as updated by [86 Bor], is accepted. Gather and Blachnik [76 Gat] state that their X-ray investigation supports the equilibria given by Elliott.

SOLID PHASES

Table 111 summarizes crystal structure data for binary compounds. No structural data has been located for the β and γ phases in the Sb–Te system. No conclusions can be drawn from [76 Bla] on the crystal structure of τ_2. The ternary compound τ_1 is triclinic [76 Bla, 72 Bac]. Mineralogical literature refers to the mineral montbrayite as being based on Au_2Te_3. This composition is not a stable phase in synthetic alloys of the Au–Te system. Bachechi [72 Bac] shows that Au_2Te_3 is not a stable compound in natural minerals. It requires the presence of Sb to provide a stable compound which is equated to the τ_1 ternary compound found by Gather and Blachnik [76 Gat, 76 Bla].

PSEUDOBINARY SYSTEMS

The Au–Sb_2Te_3 section is a pseudobinary eutectic system

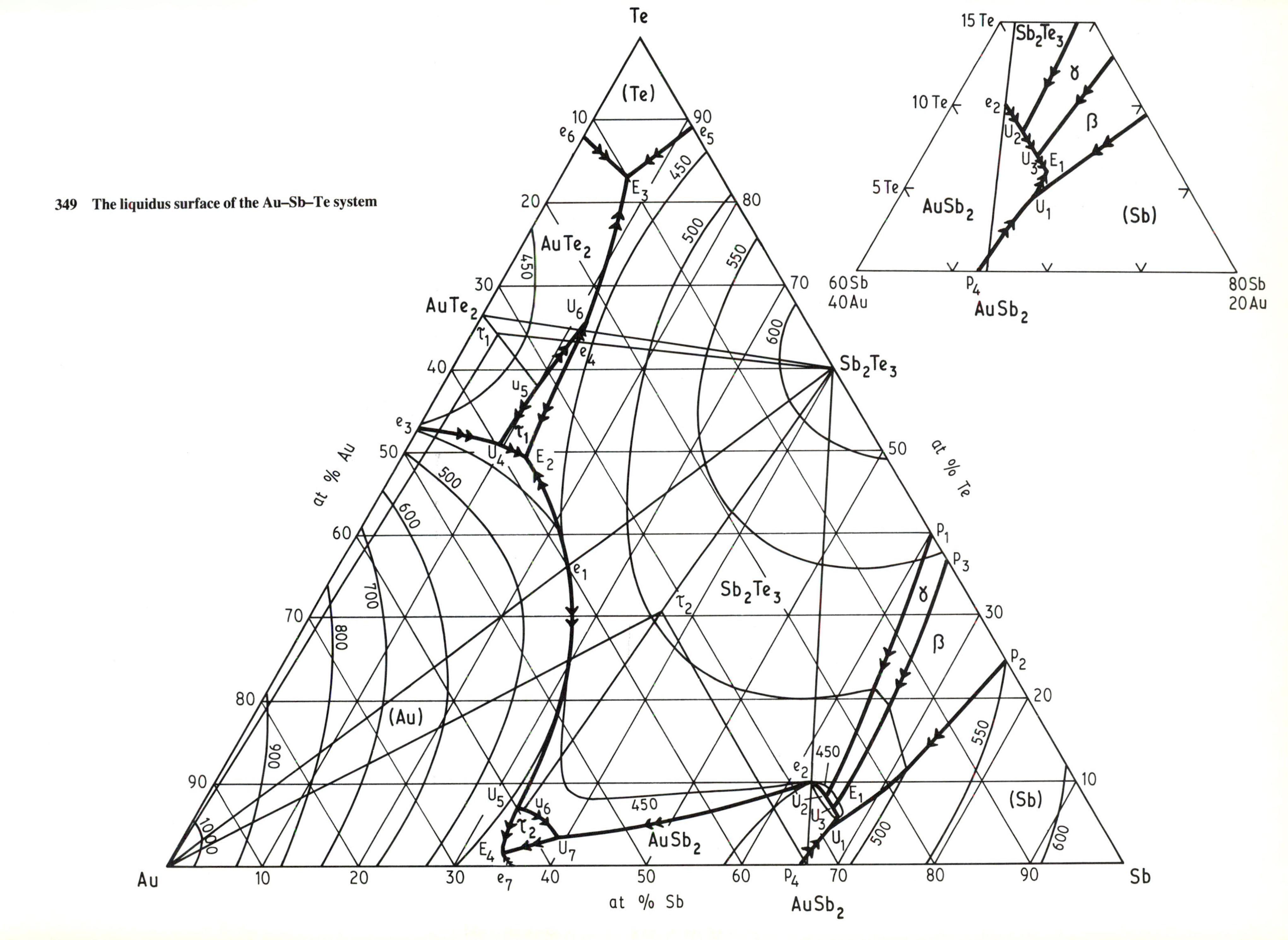

349 The liquidus surface of the Au–Sb–Te system

Table 111. Crystal structures

Solid phase	*Prototype*	*Lattice designation*
(Au)	Cu	cF4
(Sb)	As	hR2
(Te)	γSe	hP3
$AuTe_2$	$AuTe_2$	mC6
$AuSb_2$	FeS_2	cP12
Sb_2Te_3	Bi_2STe_2	hR5
β(Sb–Te)	—	—
γ(Sb–Te)	—	—
τ_1	—	—
τ_2	—	—

(Fig. 350). The eutectic temperature is given as 454°C and the eutectic composition, e_1, is 40 at.-% Au, 24 at.-% Sb, 36 at.-% Te [76 Gat]. The eutectic temperature is probably reliable to ±2°C and the composition to ±1 at.-% Au. The $AuSb_2$–Sb_2Te_3 section is pseudobinary (Fig. 351), except for alloy compositions close to $AuSb_2$. The eutectic temperature is 450°C [76 Gat] with a reliability of ±2°C and the eutectic composition e_2 (Table 110) is assessed as 27·8 at.-% Au, 62·2 at.-% Sb, 10·0 at.-% Te. The composition quoted by [76 Gat] was 28·5 at.-% Au, 62·75 at.-% Sb, 8·75 at.-% Te, but redrawing the data points scaled from the original publication indicates a Te content of 10 at.-%. This is regarded as accurate to ±1 at.-% Te. The section τ_1–Sb_2Te_3 contains a eutectic saddle point, e_4, at 430°C and a composition 25·1 at.-% Au, 11·5 at.-% Sb, 63·4 at.-% Te. This composition differs slightly from that given by [76 Gat]. The amended composition for e_4 lies on the section τ_1–Sb_2Te_3, whereas that given [76 Gat] lies slightly off the section. The presence of the monovariant curve originating at u_5 and descending to U_6 destroys the true pseudobinary nature of the τ_1–Sb_2Te_3 section. This section will appear similar to the $AuSb_2$–Sb_2Te_3 section (Fig. 351), in that it is pseudobinary for Sb contents greater than the point of intersection of u_5U_6 with the τ_1–Sb_2Te_3 section. [76 Gat] did not publish the τ_1–Sb_2Te_3 section. The formation of the ternary compounds τ_1 and τ_2 by reaction of melt with $AuTe_2$ and Sb_2Te_3 respectively means that the partial sections u_5–$AuTe_2$ and u_6–Sb_2Te_3 have a partially pseudobinary character. These sections will be discussed in the section devoted to invariant equilibria.

INVARIANT EQUILIBRIA

Table 110 and Fig. 352 give details of the invariant equilibria [76 Gat]. The Au–Sb_2Te_3 pseudobinary eutectic section divides the ternary system into the partial ternary system Au–Te–Sb_2Te_3 and Au–Sb–Sb_2Te_3. The latter partial ternary system can be regarded as split into two partial systems by the essentially pseudobinary $AuSb_2$–Sb_2Te_3 section. The equilibria in the partial ternary system Au–Te–Sb_2Te_3 are conditioned by the occurrence of the ternary compound τ_1 of composition $AuSb_{0\cdot076}Te_{1\cdot955}$. This compound forms by reaction of liquid u_5 with $AuTe_2$ at 444°C. All temperatures of invariant reactions are assigned a precision of ±2°C. From u_5 a three-phase equilibrium between $l + \tau_1 + AuTe_2$ descends to both U_4 at 428°C and U_6 at 425°C. The transition reaction $L_{U_4} + AuTe_2 \rightleftharpoons (Au) + \tau_1$ completes the solidification of alloys in the $(Au) + \tau_1 + AuTe_2$ region. The composition of u_5 (Table 110) differs from that published [76 Gat]. It is necessary for u_5 to have a composition on the line extending from $AuTe_2$–τ_1. Accepting the composition quoted by [76 Gat] for τ_1 the equation of the line $AuTe_2$–τ_1 is Te = 6·576 Au − 152·5. [76 Gat] quotes a composition for u_5 of 33 at.-% Au, 9 at.-% Sb, 58 at.-% Te. To satisfy the equation by substituting Au = 33 at.-% necessarily gives the composition of u_5 as identical to that of τ_1. Substituting Te = 58 at.-% gives a composition for u_5 of 32 at.-% Au, 10 at.-% Sb, 58 at.-% Te. The latter composition has been accepted for the liquid at u_5. The section τ_1–Sb_2Te_3 has a eutectic saddle point e_4 at 430°C with a slightly amended composition for e_4 (Table 110) compared with [76 Gat]. From e_4, three-phase equilibria between $1 + \tau_1 + Sb_2Te_3$ descend to both U_6 and E_2. Solidification of alloys in the $(Au) + \tau_1$ Sb_2Te_3 region ends with the reaction $L_{E_2} \rightleftharpoons (Au) + \tau_1 + Sb_2Te_3$ at 423°C. Similarly solidification of alloys in the $AuTe_2 + \tau_1 + Sb_2Te_3$ region ends at the transition reaction $L_{u_6} + \tau_1 \rightleftharpoons AuTe_2 + Sb_2Te_3$ at 425°C. [76 Gat] published details of the section $AuTe_2$–Sb_2Te_3 (Fig. 353). As can be noted (Fig. 349), the monovariant curves u_5U_6 and e_4U_6 meet just on the Te-rich side of the section $AuTe_2$–Sb_2Te_3. This section will possess a very small region of primary solidification of τ_1. The $l + \tau_1$ phase region will close at the intersection of the tie lines U_6–τ_1 and $AuTe_2$–Sb_2Te_3. This is calculated to occur at a composition 26 at.-% Au, 8·8 at.-% Sb, 65·2 at.-% Te, as shown in the enlarged insert (Fig. 353). The detection of a reaction at 396°C in the $AuTe_2$–Sb_2Te_3 section is a reflection of the normal tendency for incomplete reaction of the liquid phase at the transition reaction U_6. The non-equilibrium liquid finally solidifies at E_3. At E_3 solidification ends for alloys in the region $AuTe_2$–Te–Sb_2Te_3 with the ternary eutectic reaction $L_{E_3} \rightleftharpoons (Te) + AuTe_2 + Sb_2Te_3$.

The Au–Sb–Sb_2Te_3 partial ternary system can be regarded as divided by the essentially pseudobinary eutectic system $AuSb_2$–Sb_2Te_3 into the partial systems Au–$AuSb_2$–Sb_2Te_3 and Sb–$AuSb_2$–Sb_2Te_3. In the Au–$AuSb_2$–Sb_2Te_3 partial system the occurrence of the ternary compound τ_2 conditions the ternary equilibria. τ_2 forms by reaction of liquid u_6 with Sb_2Te_3 at 430°C. From u_6 the three-phase equilibria involving $l + \tau_2 + Sb_2Te_3$ descend to both U_5 at 427°C and U_7 at 395°C. Transition reactions associated with U_5 and U_7 are $L_{U_5} + Sb_2Te_3 \rightleftharpoons (Au) + \tau_2$ and $L_{U_7} + Sb_2Te_3 \rightleftharpoons AuSb_2 + \tau_2$. Solidification of alloys in the $(Au) + \tau_2 + Sb_2Te_3$ region ends at 427°C and in the $AuSb_2 + \tau_2 + Sb_2Te_3$ region at 395°C. Solidification of alloys in the $(Au) + \tau_2 + AuSb_2$ region ends with the ternary eutectic reaction at E_4, $L_{E_4} \rightleftharpoons (Au) + \tau_2 + AuSb_2$. The composition of τ_2 was given [76 Gat] as 33 at.-% Au, 36·5 at.-% Sb, 30·5 at.-% Te, corresponding to $AuSb_{1\cdot106}Te_{0\cdot924}$. Gather and Blachnik [76 Gat] quote an approximate formula for τ_2 of AuSbTe which is acceptable but, in a later paper [76 Bla] they quote AuSbTe for τ_2 and give the composition 33·3 at.-% Au, 33·3 at.-% Sb, 33·3 at.-%Te. The original composition [76 Gat] is accepted. The composition of u_6 was

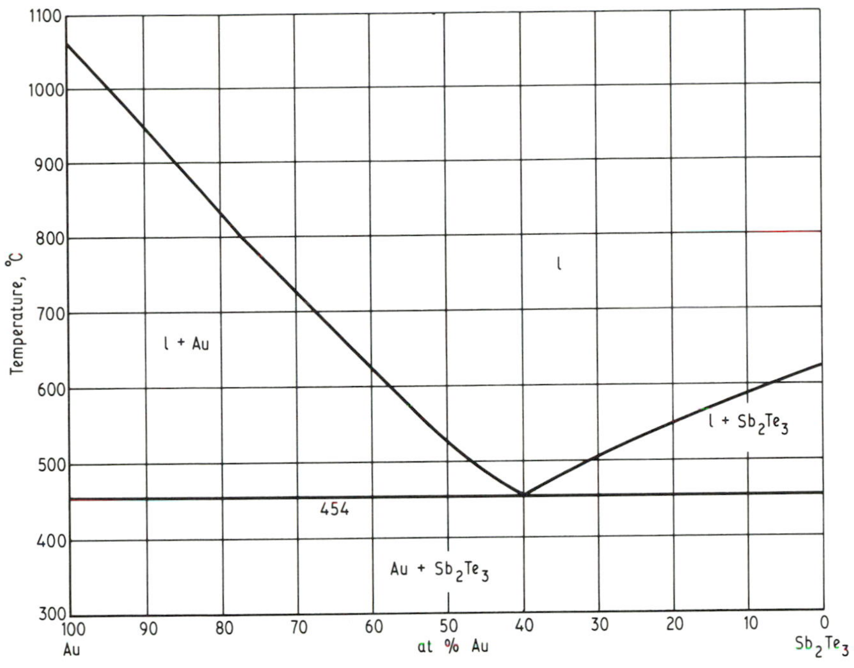

350 The Au–Sb_2Te_3 section

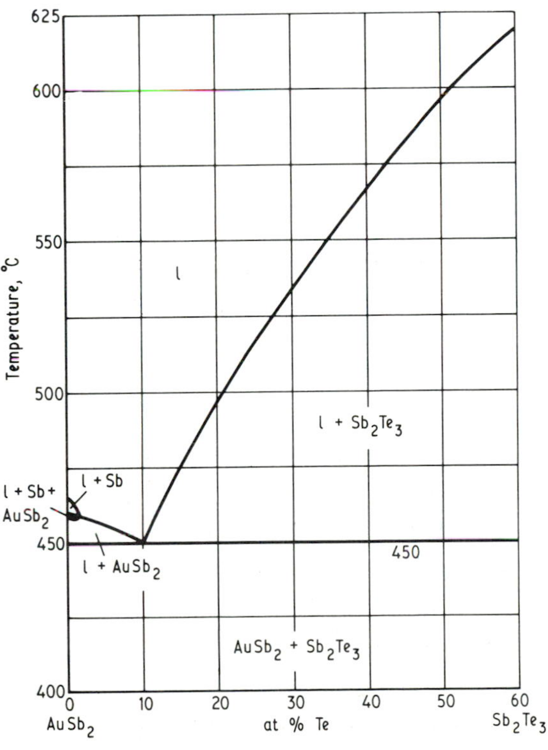

351 The $AuSb_2$–Sb_2Te_3 section

352 Reaction scheme for the Au–Sb–Te system

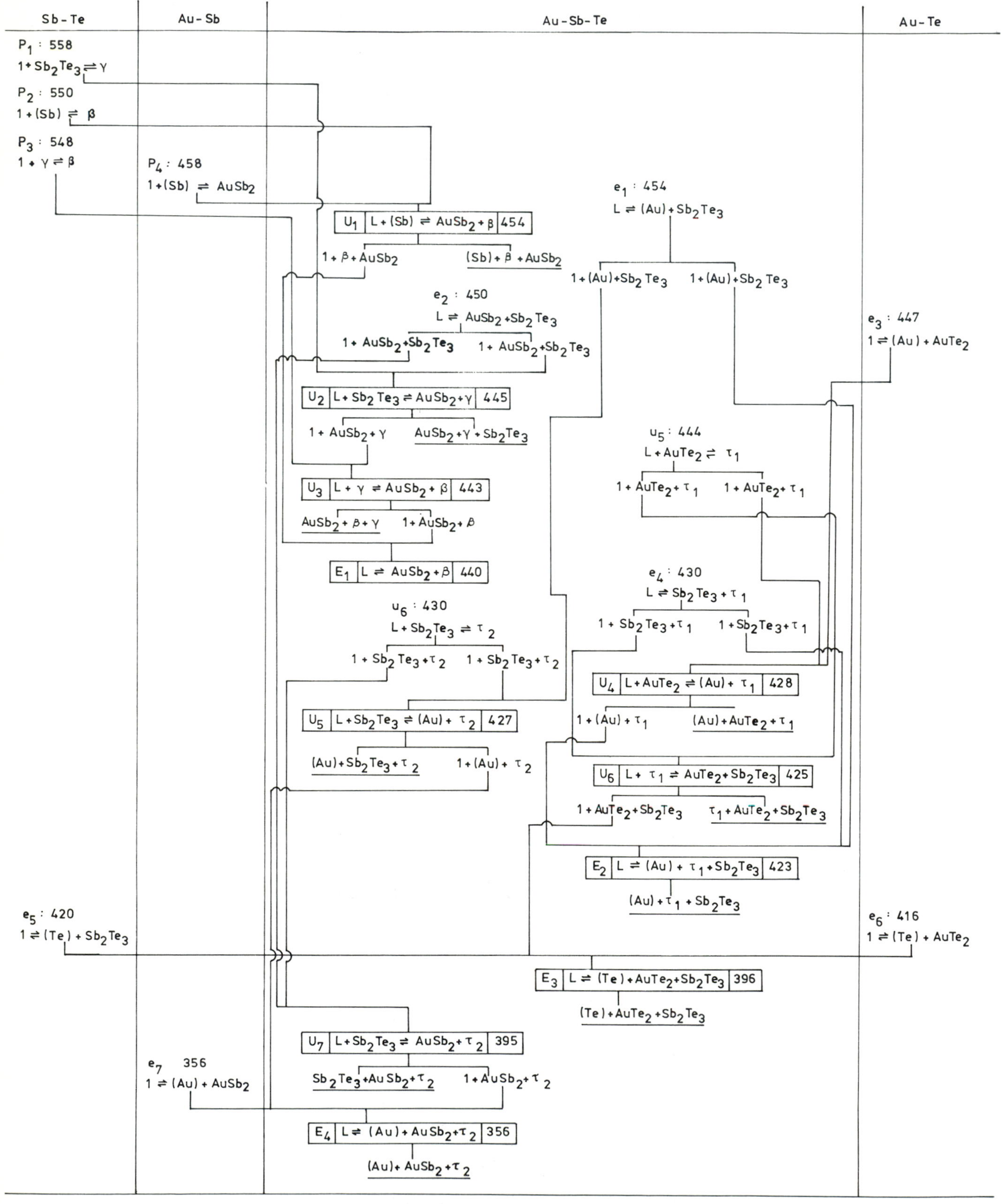

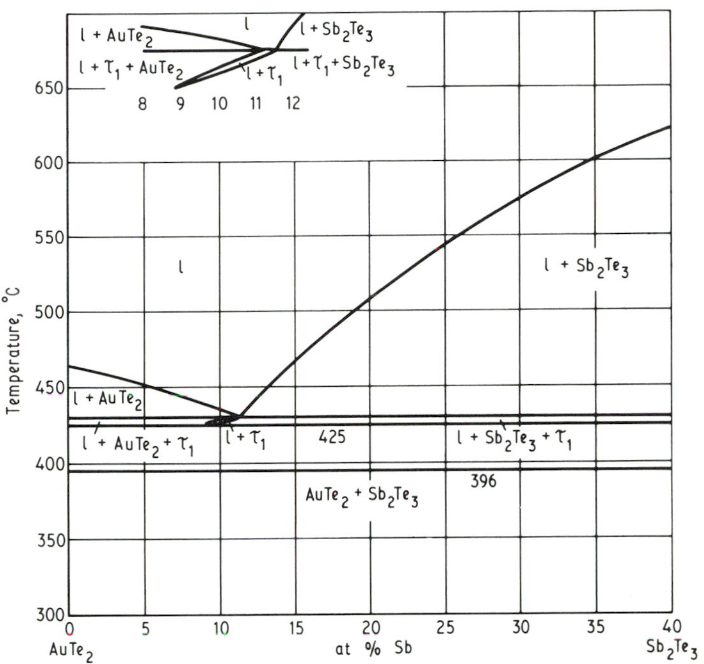

353 The $AuTe_2$–Sb_2Te_3 section

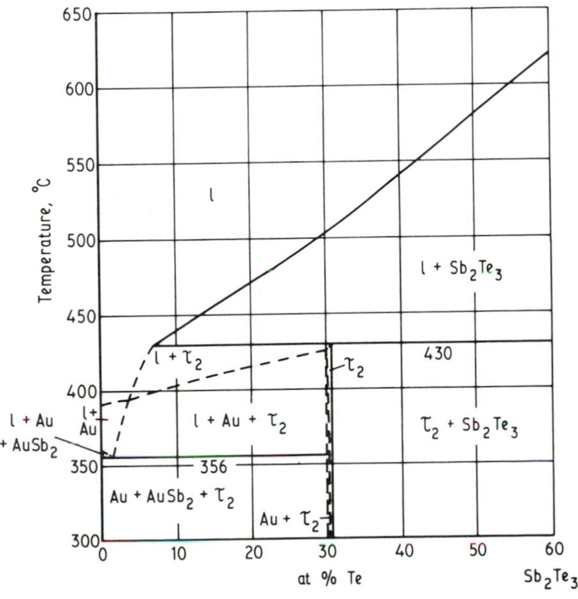

354 The 32·9 at.-% Sb, 67·1 at.-% Au–Sb_2Te_3 section

tabulated [76 Gat] as 58 at.-% Au, 35 at.-% Sb, 7 at.-% Te. This has been amended (Table 110) to 59·3 at.-% Au, 33·7 at.-% Sb, 7 at.-% Te on the basis that the composition quoted by [76 Gat] does not lie on the extension of the τ_2–Sb_2Te_3 tie line as u_6 must. As U_5 lies at the same Te content but contains 0·5 at.-% Sb less than u_6, the composition of U_5 [76 Gat] has been amended to show the same Te content and 0·5 at.-% Sb less than the amended composition for u_6 (Table 110). In Fig. 349 only the monovariant curve descending from u_6 to U_7 is designated by arrows; the proximity in compositions of u_6 and U_5 does not allow arrows to be drawn on the monovariant curve u_6 U_5. The field of primary solidification of τ_2 is quite small, U_5 u_6 U_7 E_4. As noted in the introduction, [76 Gat] quote the same compositions for e_1 and E_4. The composition of E_4 (Table 110) is an estimate of the composition taken from the liquidus surface presented by [76 Gat]. The sections $AuTe_2$–u_5 and Sb_2Te_3–u_6 are both pseudobinary in part. [76 Gat] did not present a diagram of either section. Figure 354 is a construction of the Sb_2Te_3–u_6 section extended to its intersection with the Au–Sb binary edge at 32·9 at.-% Sb, 67·1 at.-% Au.

The partial ternary system Sb–$AuSb_2$–Sb_2Te_3 contains the transition reactions U_2 and U_3 corresponding to L_{U_2} + $Sb_2Te_3 \rightleftharpoons AuSb_2 + \gamma$ at 445°C and $L_{U_3} + \gamma \rightleftharpoons AuSb_2 + \beta$ at 443°C. A further transition reaction was found at U_1 corresponding to L_{U_1} + (Sb) $\rightleftharpoons AuSb_2 + \beta$ at 454°C. The same three-phase equilibria, l + $AuSb_2$ + β, fall from both U_3 and U_1. As noted in the introduction [76 Gat] interpret final solidification at E_1 (440°C) as involving a ternary eutectic reaction $L_{E_1} \rightleftharpoons \beta' + \beta'' + AuSb_2$. The β phase is considered to show a miscibility gap in the ternary system such that, below a critical temperature where l $\rightleftharpoons$ β, a three-phase equilibrium is established between l + β′ + β″. At the temperature of the critical tie line β′ + β″ → β. No experimental evidence is presented for this solidification sequence, and it is more likely that the two three-phase equilibria originating from U_3 and U_2 fall to a critical tie line at 440°C. This implies that the compositions of the liquid, $AuSb_2$, and β phases lie on a tie line at 440°C which is at a minimum in the monovariant curve U_1U_3 associated with the liquid and with similar curves associated with $AuSb_2$ and β. It should be noted that the β phase is associated with a liquidus minimum in the binary Sb–Te system. [76 Gat] also published vertical sections at 20 at.-% Sb (Fig. 355) and 50 at.-% Au (Fig. 356). The phase region below 396°C in the 20 at.-% Sb section was incorrectly designated as τ_1 + $AuTe_2$ + Sb_2Te_3 by [76 Gat]. It has been corrected to Te + $AuTe_2$ + Sb_2Te_3 in Fig. 355.

LIQUIDUS SURFACE

The liquidus surface (Fig. 349) is substantially as published [76 Gat] with amendments to invariant liquid compositions noted in Table 110.

ISOTHERMAL SECTIONS

A room temperature isothermal section (Fig. 357) is based on the assumption of solubility of $AuSb_2$ in the β and γ phases of the Sb–Te system.

Table 112. Equivalence of symbols for invariant reactions

Reaction	*Assessment*	[76 Gat]
$L \rightleftharpoons AuSb_2 + \beta$	E_1	E_4
$L \rightleftharpoons (Au) + \tau_1 + Sb_2Te_3$	E_2	E_1
$L \rightleftharpoons (Te) + AuTe_2 + Sb_2Te_3$	E_3	E_2
$L \rightleftharpoons (Au) + AuSb_2 + \tau_2$	E_4	E_3
$L + (Sb) \rightleftharpoons AuSb_2 + \beta$	U_1	U_5
$L + Sb_2Te_3 \rightleftharpoons AuSb_2 + \gamma$	U_2	U_6
$L + \gamma \rightleftharpoons AuSb_2 + \beta$	U_3	U_7
$L + AuTe_2 \rightleftharpoons (Au) + \tau_1$	U_4	U_1
$L + Sb_2Te_3 \rightleftharpoons (Au) + \tau_2$	U_5	U_3
$L + \tau_1 \rightleftharpoons AuTe_2 + Sb_2Te_3$	U_6	U_2
$L + Sb_2Te_3 \rightleftharpoons AuSb_2 + \tau_2$	U_7	U_4
$L \rightleftharpoons (Au) + Sb_2Te_3$	e_1	sk_1
$L \rightleftharpoons AuSb_2 + Sb_2Te_3$	e_2	sk_3
$l \rightleftharpoons (Au) + AuTe_2$	e_3	e_1
$L \rightleftharpoons Sb_2Te_3 + \tau_1$	e_4	sk_2
$l \rightleftharpoons (Te) + Sb_2Te_3$	e_5	e_3
$l \rightleftharpoons (Te) + AuTe_2$	e_6	e_2
$l \rightleftharpoons (Au) + AuSb_2$	e_7	e_4
$L + AuTe_2 \rightleftharpoons \tau_1$	u_5	uk_1
$L + Sb_2Te_3 \rightleftharpoons \tau_2$	u_6	uk_2
$l + Sb_2Te_3 \rightleftharpoons \gamma$	p_1	u_4
$l + (Sb) \rightleftharpoons \beta$	p_2	u_2
$l + \gamma \rightleftharpoons \beta$	p_3	u_3
$l + (Sb) \rightleftharpoons AuSb_2$	p_4	u_1

MISCELLANEOUS

For the convenience of those wishing to consult the original work [76 Gat], Table 112 provides a list of the equivalence of the symbols used in this assessment and by [76 Gat] for the binary invariant reactions. Shanov *et al.* [84 Sha] have reported the synthesis of a primitive cubic phase, $a = 0{\cdot}301\,8 \pm 0{\cdot}000\,2$ nm, of composition $AuSb_6Te$ by heating to 1000°C under 40 kbar. The melt was quenched under pressure to room temperature and the crystalline phase characterized by X-ray diffraction analysis. The high-pressure phase extends from 68·75 to 81·25 at.-% Sb along the 12·5 at.-% Au section.

REFERENCES

65 Ell: R. P. Elliott: 'Constitution of Binary Alloys' (1st Suppl.); 1965, New York, McGraw-Hill

72 Bac: F. Bachechi: *Amer. Mineral.*, 1972, **57**, 146–154

76 Gat: B. Gather and R. Blachnik: *Z. Metall.*, 1976, **67**, 395–399

76 Bla: R. Blachnik and B. Gather: *Z. Naturf.*, 1976, **31B**, 526–527

84 Sha: V. N. Shanov, K. J. Range and F. Rau: 'High Pressure in Science and Technology', Part I (Ed. C. Homan, R. K. MacCrone and E. Whalley), 351–353; 1984. Materials Res. Soc. Symp. Proc., Vol. 22

84 Oka 1: H. Okamoto and T. B. Massalski: *Bull. Alloy Phase Diagrams*, 1984, **5**, 166–171

84 Oka 2: H. Okamoto and T. B. Massalski: *Bull. Alloy Phase Diagrams*, 1984, **5**, 172–177

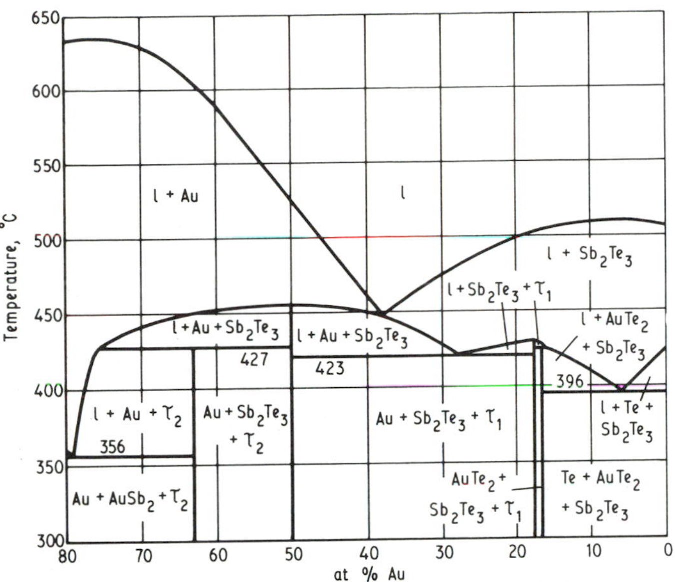

355 **The 20 at.-% Sb section**

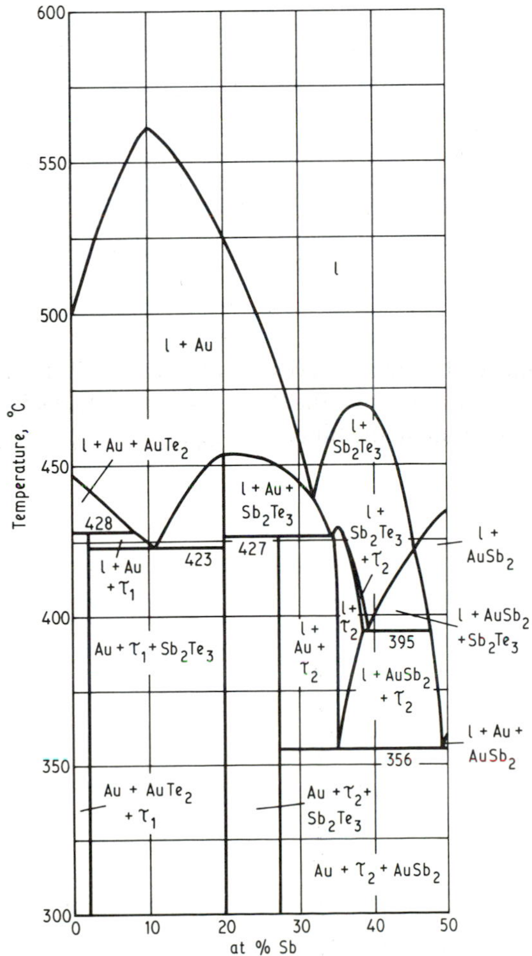

356 **The 50 at.-% Au section**

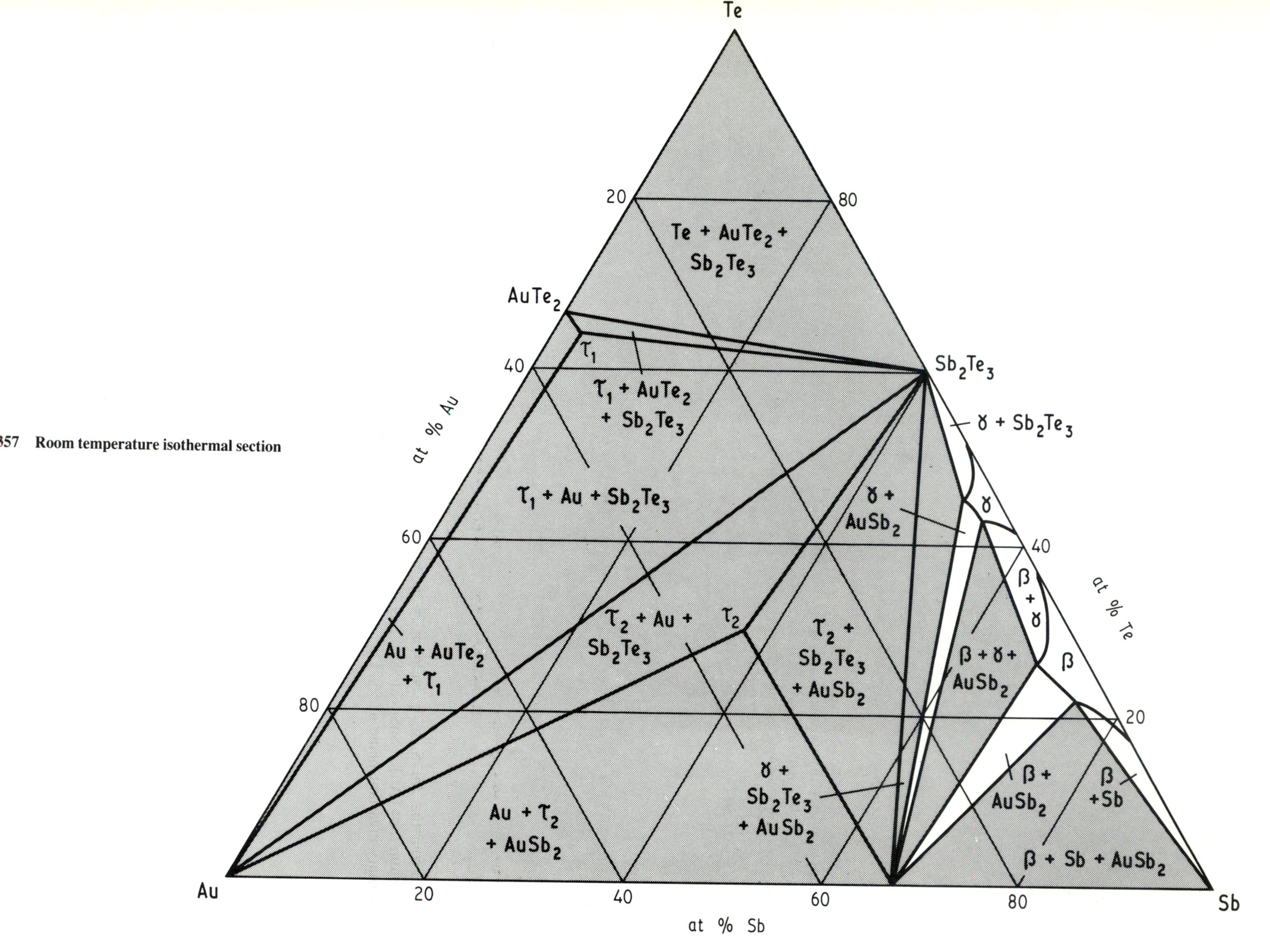

357 Room temperature isothermal section

84 Eva: D. S. Evans and A. Prince: CALPHAD XIII, Villard de Lans, 1984

86 Bor: S. Bordas, M. T. Clavaguera-Mora, B. Legendre and Chhay Hancheng: *Thermochim. Acta*, 1986, **107**, 239–265

Au–Se–Sn
(B. Gather and A. Prince)

INTRODUCTION

There are two publications on the Au–Se–Sn system. Rouland *et al.* [75 Rou] found the sections Au–$SnSe_2$ and AuSn–SnSe to be pseudobinary. [76 Rou] investigated 157 alloys made from 99·999% pure elements in a survey of the whole ternary system. The pseudobinary sections from [75 Rou] were included, with the addition of a polymorphic transformation in SnSe at 530°C. Vertical sections were examined using differential thermal analysis with supporting X-ray diffraction and metallographic studies. Not all the sections studied by [76 Rou] are included in their published paper. The vertical sections SnSe–$AuSn_2$ and SnSe–$AuSn_4$, as well as the 10% Se section from 50–90% Sn, were not given by [76 Rou].

The ternary system is divided into three partial systems by the pseudobinary systems Au–$SnSe_2$ and AuSn–SnSe (Fig. 358). The partial ternary system Au–$SnSe_2$–Se is substantially covered by two liquid miscibility gaps which lie on the primary crystallization surface of Au. In the Au-rich region the binary monotectic reaction at 963°C, $l_1 \rightleftharpoons l_2$ + (Au) sweeps down to a ternary invariant reaction at 730°C, $L_1 + L_2 \rightleftharpoons$ (Au) + SnSe. In doing so it crosses the Au–$SnSe_2$ pseudobinary section. In the Se-rich region the binary monotectic reaction at 760°C, $l_5 \rightleftharpoons l_6$ + (Au) descends to a ternary invariant reaction at 558°C, $L_3 \rightleftharpoons L_4$ + (Au) + $SnSe_2$. A three-phase equilibrium originates at a critical tie line L $\rightleftharpoons$ $SnSe_2$ which is established at about 630°C and ends as the $L_2 + L_4 + SnSe_2$ tie-triangle at 558°C. From the Au–$SnSe_2$ pseudobinary the eutectic equilibrium L $\rightleftharpoons$ (Au) + $SnSe_2$ at 580°C also descends to the ternary invariant plane at 558°C. Ternary reactions also occur at 380 and 214°C in this partial ternary system; they are considered in the section devoted to invariant reactions.

The partial ternary system Au–AuSn–SnSe–$SnSe_2$ is nearly totally covered by the liquid miscibility gap starting from the Au–Se binary and crossing the whole ternary until it reaches the Sn–Se binary. This partial ternary system is characterized by invariant reactions at 730, 564, and 276°C. A further invariant reaction at 530°C [76 Rou] is discussed below. The reaction $L_1 + L_2 \rightleftharpoons$ (Au) + SnSe at 730°C changes the liquid miscibility gap from the primary crystallization surface of (Au) to that of SnSe. In the Au–SnSe–$SnSe_2$ triangle the crystallization ends in the ternary eutectic point E_1 with the reaction $L_{E_1} \rightleftharpoons$ (Au) + SnSe + $SnSe_2$ at 564°C. In the Au–AuSn–SnSe triangle [76 Rou] propose a transition reaction L + (Au) $\rightleftharpoons \zeta$ + SnSe at 530°C. The L_1 + (Au) + SnSe equilibrium originating from the 730°C invariant reaction ($L_1 + L_2 \rightleftharpoons$ (Au) + SnSe) meets a L + (Au) + ζ equilibrium originating from the l + (Au) $\rightleftharpoons \zeta$ reaction at 530°C in the Au–Sn binary. This binary Au–Sn reaction has subsequently been shown [87 Leg] to consist of two peritectic reactions, l + (Au) $\rightleftharpoons \beta$ at 532°C, and l + $\beta \rightleftharpoons \zeta$ at 519°C. It is not possible to accept a transition reaction, L + (Au) $\rightleftharpoons \zeta$ + SnSe at a temperature of 530°C. It should be noted that [76 Rou] found the αSnSe $\rightleftharpoons$ βSnSe transition at 530°C. It is suggested that the recent data [87 Leg] on the Au–Sn equilibria lead to a reaction scheme (Fig. 359) incorporating a transition reaction U, L + $\beta \rightleftharpoons$ (Au) + ζ at a temperature just below 519°C, followed by the transition reaction U′, L + Au $\rightleftharpoons \zeta$ + SnSe [76 Rou] at a temperature of 510–515°C. Solidification ends in the ternary eutectic reaction $L_{E_2} \rightleftharpoons$ AuSn + ζ + SnSe at 226°C. In the partial ternary system AuSn–Sn–SnSe three invariant reactions are observed at 309, 252, and 217°C. While the first two were of ternary transition types L_{U_2} + AuSn $\rightleftharpoons$ $AuSn_2$ + SnSe at 309°C and L_{U_3} + $AuSn_2 \rightleftharpoons AuSn_4$ + SnSe at 252°C respectively, the third, at 217°C, is a ternary eutectic reaction, $L_{E_3} \rightleftharpoons AuSn_4$ + Sn + SnSe.

This partial ternary is nearly totally covered by the liquid miscibility gap which gently descends from the AuSn–SnSe pseudobinary section (858°C) to the SnSe binary (830°C). In general the interpretation of the experimental data in the different sections fit very well the positions of the invariant ternary points as well as the monovariant curves. Slight modifications are made to some boundaries of the miscibility gaps which are mentioned in the section on invariant equilibria. The same comments are valid for tie lines in the miscibility gaps, as well as for the interpretation of the experimental data for the second phase separation.

It must be mentioned that all concentration scales in [75 Rou, 76 Rou] which are said to be in mole percent are actually plotted in atomic percent. Liquidus isotherms have been constructed from the data [76 Rou] and they are included on the liquidus surface projection (Fig. 358).

BINARY SYSTEMS

The Au–Se system evaluated by [86 Oka] and the Se–Sn system evaluated by [86 Sha] are accepted. The Au–Sn system has been evaluated by [84 Oka] but considerable uncertainties exist concerning equilibrium reactions for Au-rich alloys. [87 Leg] have shown the existence of two peritectic reactions, l + (Au) $\rightleftharpoons \beta$ at 532°C and l + $\beta \rightleftharpoons \zeta$ at 519°C. The solid-state transformations require further study.

SOLID PHASES

Table 113 summarizes crystal structure data for the binary compounds. No ternary compounds were observed.

PSEUDOBINARY SYSTEMS

The Au–$SnSe_2$ and AuSn–SnSe sections are pseudobinary, as shown by Figs 360 and 361 respectively.

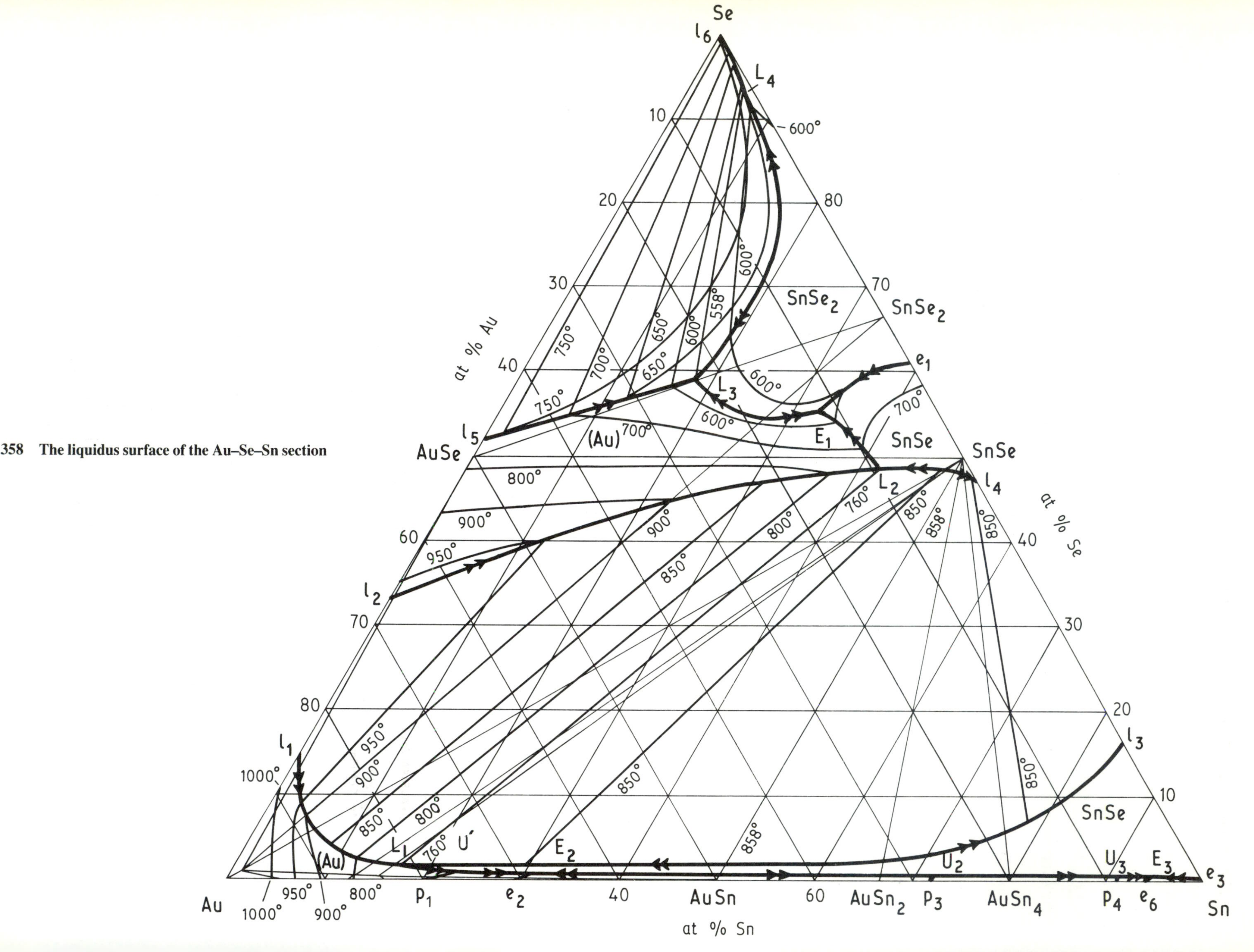

358 The liquidus surface of the Au–Se–Sn section

359 Reaction scheme for the Au–Se–Sn system

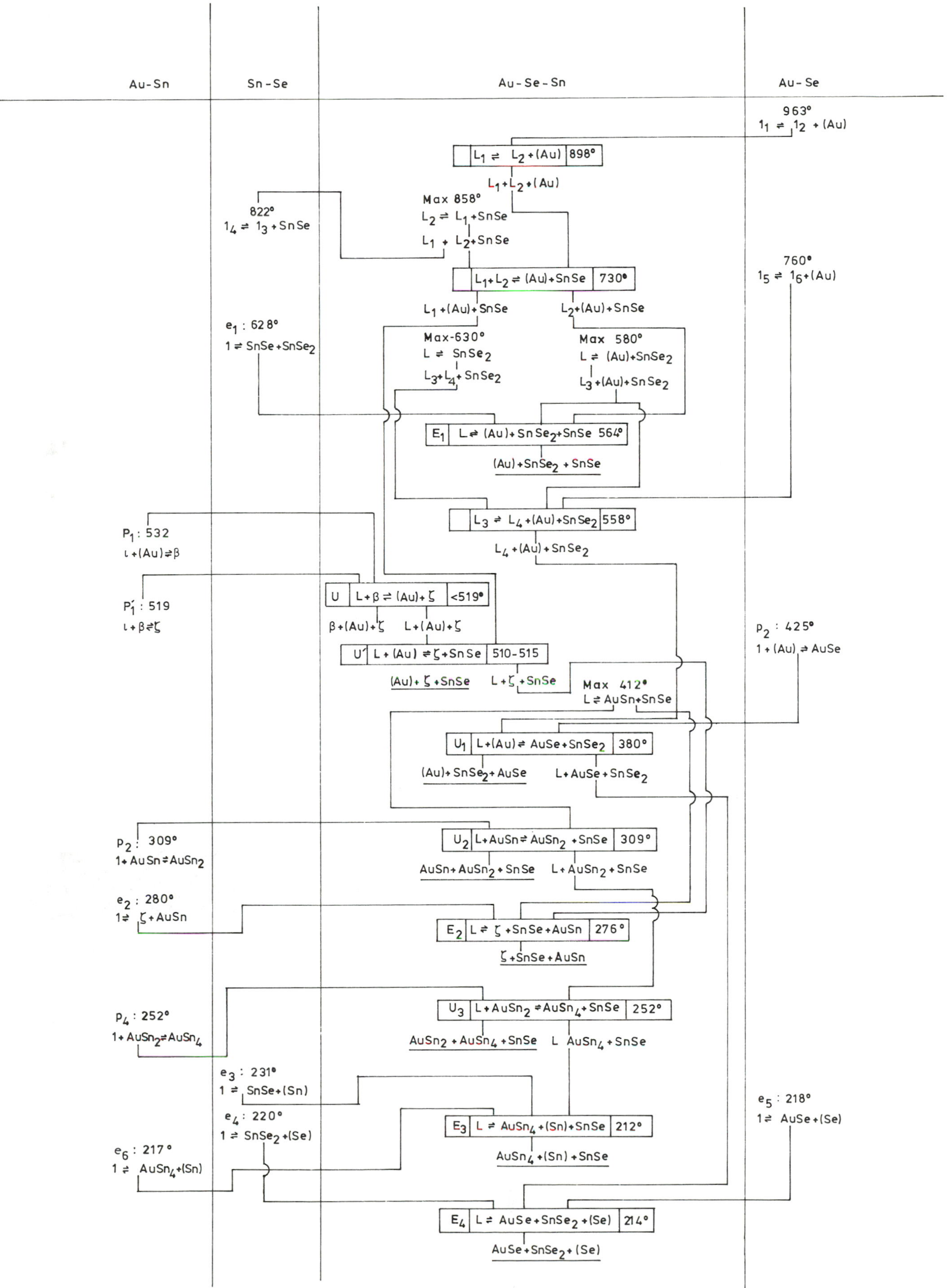

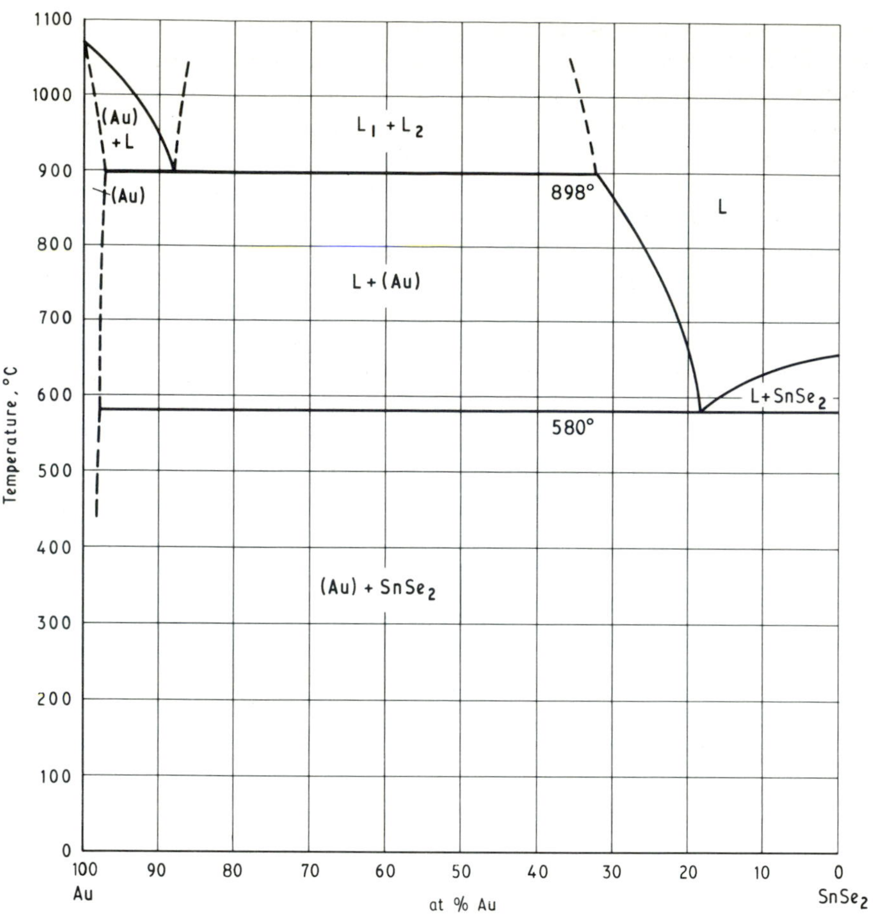

360 The Au–$SnSe_2$ section

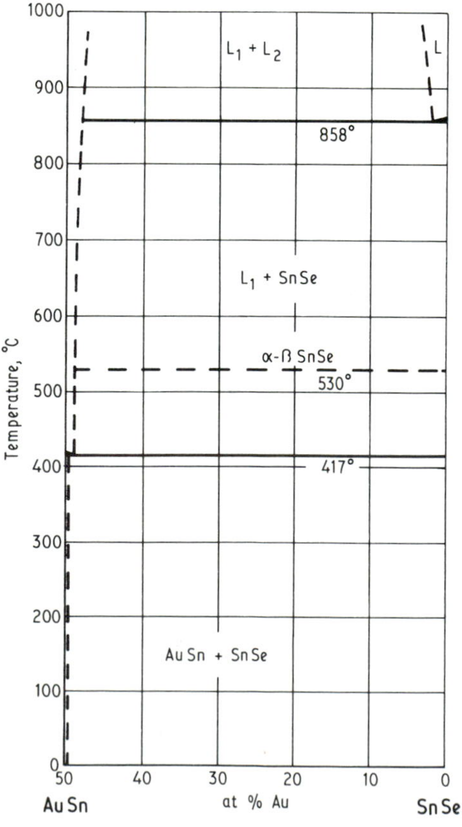

361 The AuSnSe section

The monotectic temperature of the Au–$SnSe_2$ section is given as 898°C, the eutectic temperature as 580°C. All temperatures are considered reliable to within ±2°C. The eutectic composition was scaled from the original diagram [75 Rou, 76 Rou] as 17·6 Au–27·5 Sn–54·9% Se and slightly modified in Fig. 360 to 18·2 Au–27·3 Sn–54·5% Se. The liquid compositions at the monotectic scaled from [75 Rou, 76 Rou] are 90·6 Au–3·1 Sn–6·3% Se and 33·4 Au–22·2 Sn–44·4% Se respectively. They are also modified to 88·0 Au–4·0 Sn–8·0% Se and 32·2 Au–22·6 Sn–45·2% Se to provide a better fit to the ternary data. A solid solution of 1% Se and 2% Sn in gold at the 998°C monotectic reaction is suspected.

Table 113. Crystal structures

Solid phase	*Prototype*	*Lattice designation*
(Au)	Cu	cF4
(Sn)	Sn	tI4
(Se)	γSe	hP3
AuSe	—	mC24
AuSn	NiAs	hP4
β	Ni_3Ti	hP16
ζ	Mg	hP2
$AuSn_2$	—	oP24
$AuSn_4$	$PtSn_4$	oC20
SnSe	GeS	oP8
$SnSe_2$	CdI_2	hP3

The monotectic temperature of the AuSn–SnSe section (Fig. 361) is given as 858°C, and the eutectic temperature as 417°C. The liquid compositions at the monotectic are scaled from [75 Rou, 76 Rou] and show 2 Au–50 Sn–48% Se and 48 Au–50 Sn–2% Se respectively. The eutectic composition, scaled from the original drawings to 49·5 Au–50 Sn–0·5 Se, is an estimated composition. However, the small difference between the eutectic temperature (412°C) and the melting point of AuSn (419°C) implies that the composition of the eutectic will be within 1% Se.

In addition to [75 Rou] a later paper [76 Rou] shows a thermal effect at 530°C which is regarded as a second order transition of SnSe. In all ternary sections this effect is drawn with broken lines because of uncertainty in its interpretation.

INVARIANT REACTIONS

In the partial ternary system Au–$SnSe_2$–Se there are three invariant reactions. At 558°C, $L_3 \rightleftharpoons L_4$ + (Au) + $SnSe_2$. The compositions of the liquid phases are given in Table 114. The tie line L_3–L_4 is substantiated by the reactions at 66·7% Se (Fig. 362), 76·7% Se (Fig. 363) and the section AuSn–Se (Fig. 364). Figure 364 also provides a good indication of the position of the L_3 composition as does the section at 56·7% Se and the sections at 20 and 30% Se (Figs 365–367). The composition of L_3 is considered to be established within ±0·3% of the elements.

The L_4 composition is fixed in terms of the Sn to Se ratio by the line L_3–L_4, but the Au content is uncertain. A value of 0·7% Au is assumed. The second invariant reaction (U_1) is of a transition type, L_4 + (Au) $\rightleftharpoons$ AuSe + $SnSe_2$, at 380°C (Fig. 368). A thermal effect at approximately 380°C was noted for one or two alloy compositions within 10 at.-% Sn of the Au–Se binary edge for sections containing 30, 56·67, 66·67, and 76·67% Se (Figs 367, 365, 362 and 363 respectively). The paucity of data related to this invariant reaction led [76 Rou] to represent it by a dashed line on the vertical sections. This practice has been followed and the interpretation of the 380°C reaction as a transition reaction is accepted. The composition of U_1 will lie close to the Se corner; Fig. 368 gives a schematic drawing of the reactions in Se-rich alloys. Solidification ends in the Au–Se–$SnSe_2$ partial ternary system at the ternary eutectic point E_4 with the reaction $L_{E_4} \rightleftharpoons$ (Se) + AuSe + $SnSe_2$ at 214°C.

The phase boundary separating the region of liquid immiscibility in Se-rich alloys from the region for primary crystallization of $SnSe_2$ is well defined by sections at 76·67, 66·67, 56·67% Se and by the AuSn–Se section (Figs 363, 362, 365 and 364 respectively). Amendments have been made to the original sections [76 Rou] to achieve consistent results. In particular the 66·67% Se section (Fig. 362) has been changed to show the intersection of the L_3 + L_4 + $SnSe_2$ phase regions at a lower Sn content than [76 Rou]. This change was made since the alloy common to the 66·67% Se section and the AuSn–Se section was given different temperatures for primary and secondary separations from the melt. The data presented in this assessment are now self-consistent. As a result the liquid miscibility gap on the Sn-rich side of the tie line L_3–L_4 is drawn at 2% Sn lower than [76 Rou].

Tie lines have been constructed on the L_3–L_4 surface of the (Au) + L_3 + L_4 and $SnSe_2$ + L_3 + L_4 phase regions. Liquidus isotherms have also been inserted from 600 to 1 000°C (Fig. 358). These data have been derived from the sections presented but were not plotted in the original [76 Rou].

The partial ternary system Au–AuSn–SnSe–$SnSe_2$ shows four ternary invariant reactions. The four phase reaction L_1 + $L_2 \rightleftharpoons$ (Au) + SnSe at 730°C is very well defined by the sections AuSn–Se, AuSn–$SnSe_2$ and the sections at 20%, Se, 30% Se, 6·7% Au and 16·7% Au. In all cases the sections give a very precise description of the direction of the tie line L_1–L_2 and of the composition of the invariant points L_1 and L_2 which [76 Rou] gives as L_1 = 82·0 Au–16·6 Sn–1·3% Se and L_2 = 9·0 Au–42·3 Sn–48·6% Se. The compositions of these points are anticipated to be correct to within ±0·5% of each element.

This four phase reaction divides the partial ternary described into two sub-ternary systems, Au–SnSe–$SnSe_2$ and Au–AuSn–SnSe. The Au–SnSe–$SnSe_2$ part behaves after the 730°C reaction as a simple ternary eutectic system with the eutectic reaction $L_{E_1} \rightleftharpoons$ (Au) + SnSe + $SnSe_2$ at 564°C. The eutectic composition given by [76 Rou] is 12 Au–32·6 Sn–55·3% Se. The ternary composition is well defined by the sections shown in Figs 364, 365, 369 and 370. The latter represent sections between the Au–$SnSe_2$ and the AuSn–SnSe pseudobinary systems at 6·67% Au and 16·67% Au respectively. The ternary eutectic composition is judged to be correct to within 0·5% of each element.

In the Au–AuSn–SnSe part of the partial ternary Au–AuSn–SnSe–$SnSe_2$ liquid of the composition L_1 separates (Au) + SnSe from 730°C until it reaches the in-

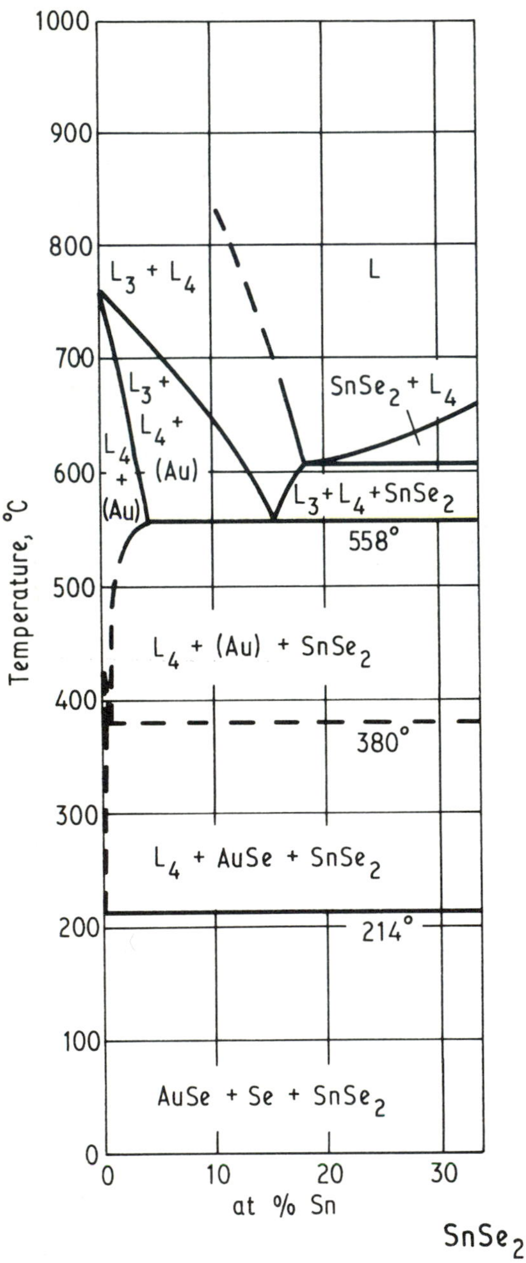

362 The 66·67% Se section

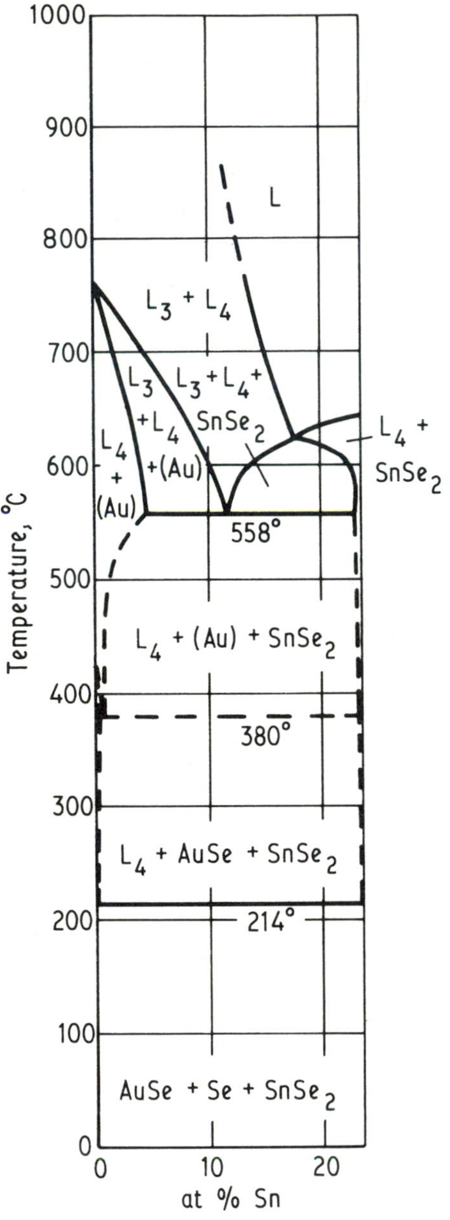

363 The 76·67% Se section

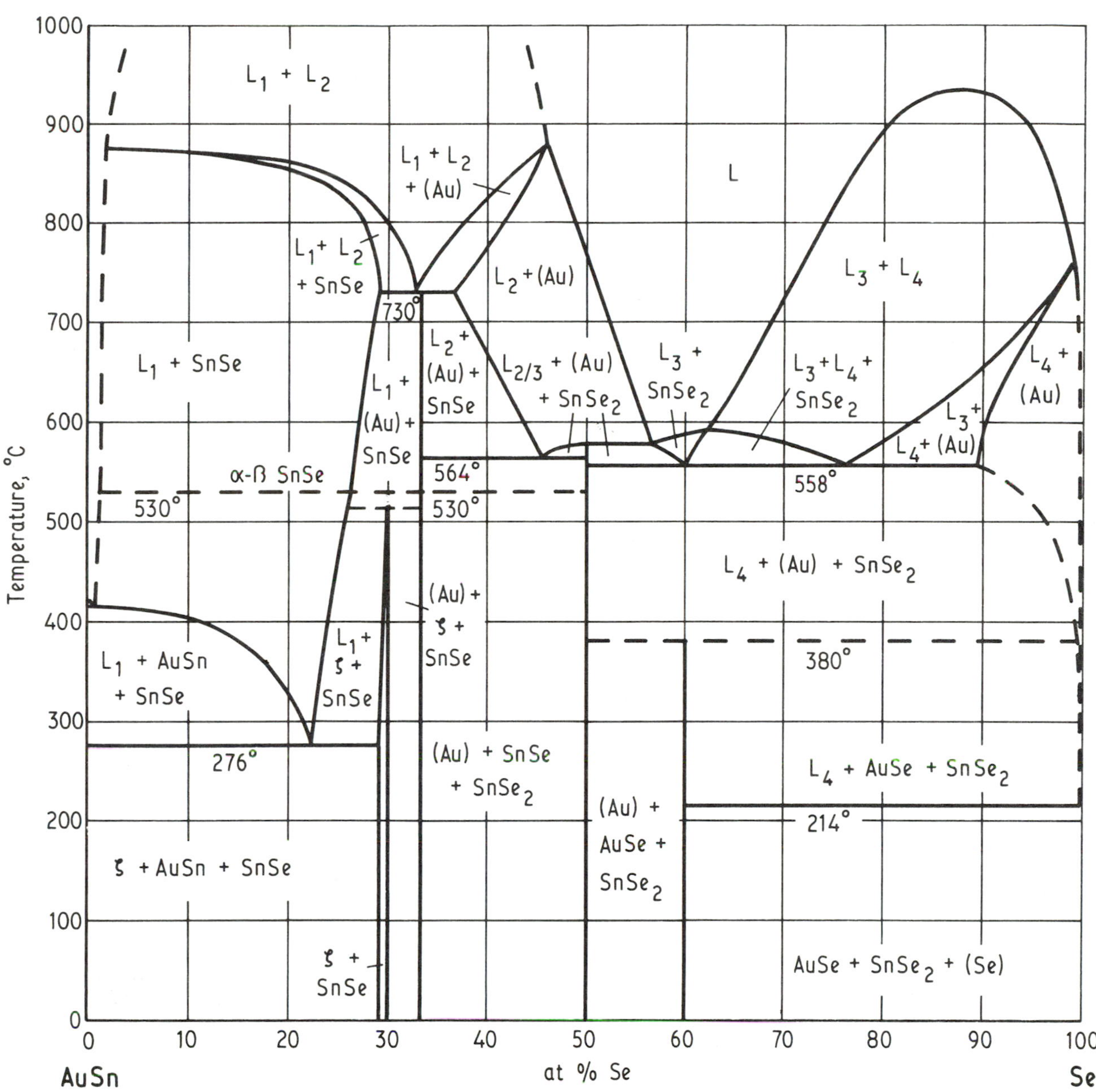

364 The AuSn–Se section

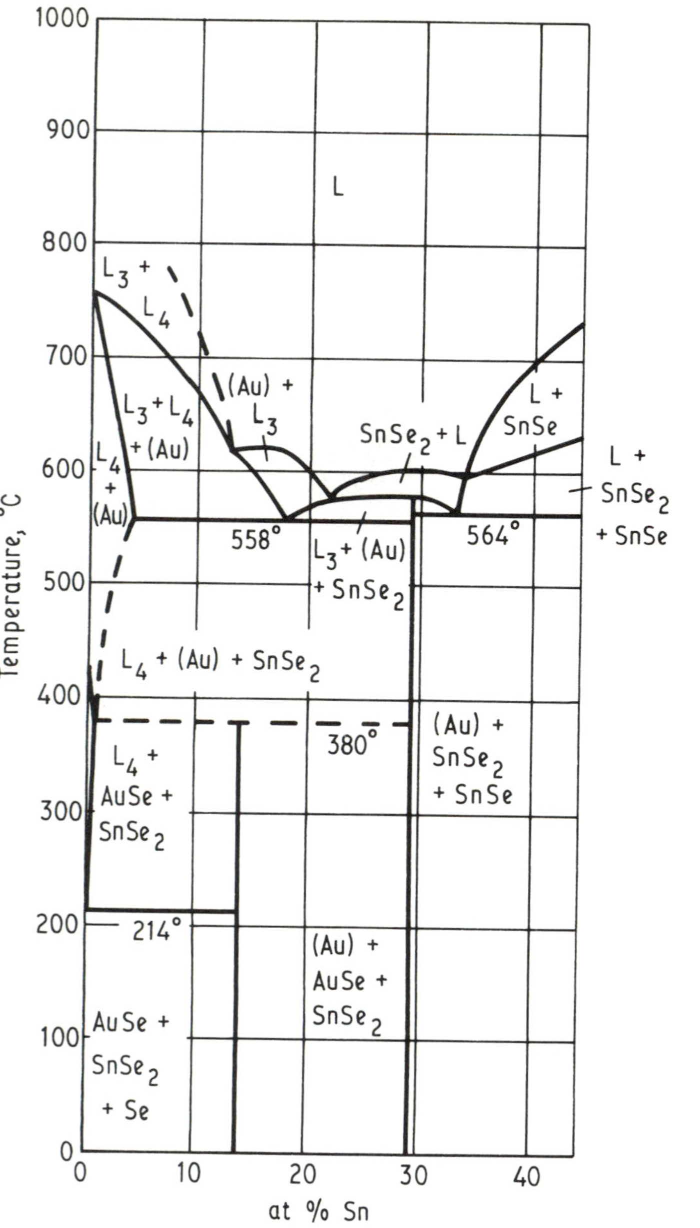

365 The 56·67% Se section

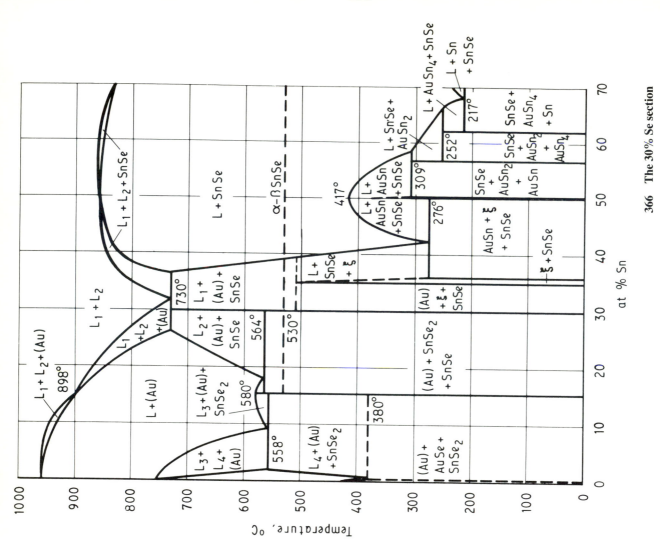

366 The 30% Se section

1000
900
800
700
600
500
400
300
200
100
0
Temperature, °C
0 10 20 30 40 50 60 70 80
at % Sn
$L_1 + L_2 + (Au)$
898°
$L_2 + (Au)$
$L_1 + L_2 + (Au)$
$L_1 + L_2$
858°
$L_1 + L_2 + SnSe$
730°
$L_1 + (Au) + SnSe$
$L_3 + L_4 + (Au)$
$L_3 + (Au) + SnSe_2$
$L_2 + (Au) + SnSe$
580°
564°
$L_4 + (Au)$
558°
L + SnSe
α-β SnSe
530°
530°
$L_4 + (Au) + SnSe_2$
380°
L + SnSe + ξ
417°
L + AuSn + SnSe
L + AuSn + SnSe
$L + SnSe + AuSn_2$
$L + AuSn_4 + SnSe$
309°
276°
252°
217°
L + Sn + SnSe
$(Au) + SnSe_2 + SnSe$
(Au) + ξ + SnSe
$(Au) + AuSe + SnSe_2$
AuSn + ξ + SnSe
$SnSe + AuSn_2 + AuSn$
$SnSe + AuSn_2 + AuSn_4$
$SnSe + AuSn_4 + Sn$
ξ + SnSe

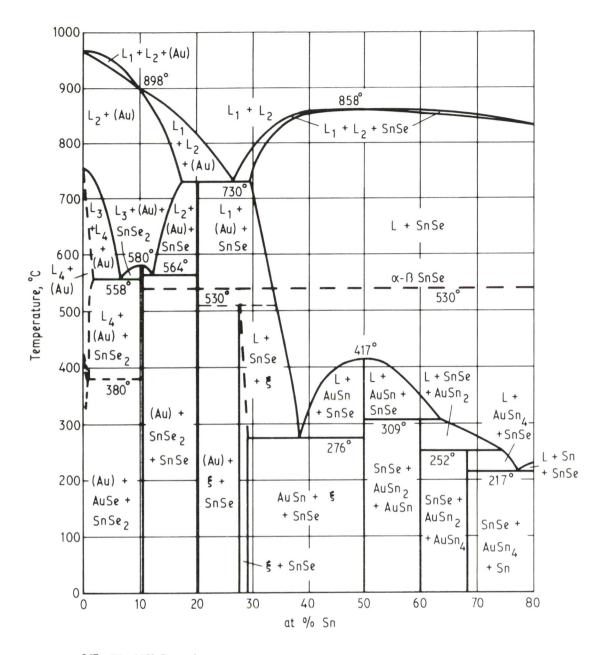

367 The 20% Se section

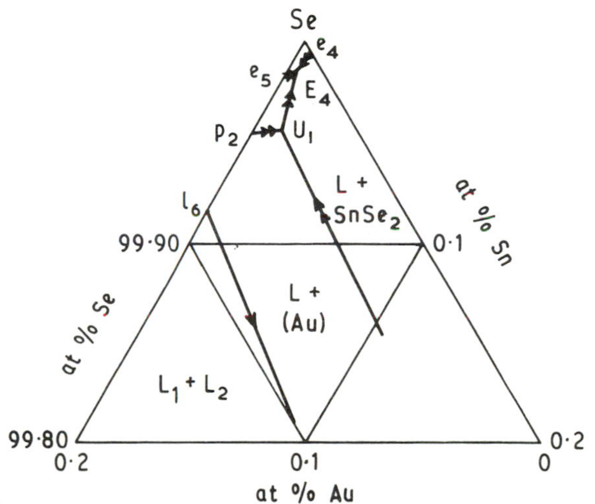

368 Schematic drawing of the reactions in the Se-rich region

variant point U′. The temperature is given as 530°C by [76 Rou]. As discussed in the introduction the recent confirmation [87 Leg] of two peritectic reactions in the Au-rich part of the Au–Sn system vitiates a ternary invariant reaction L + (Au) ⇌ ζ + SnSe as occurring at 530°C. It is considered that a thermal effect at 530°C is due to the transformation of αSnSe to βSnSe. Figure 359 contains a suggested reaction scheme incorporating a transition reaction U at a temperature just below 519°C and a transition reaction U′, L + (Au) ⇌ ζ + SnSe at 510–515°C. The estimated composition of U′ is 75·2 Au–24 Sn–0·8 at.-% Se. The compositions of U and U′ will be very close and only U′ is shown in Fig. 358.

The fourth invariant reaction takes place at 276°C and is of ternary eutectic type, L_{E_2} ⇌ ζ + AuSn + SnSe. The composition in terms of the Au–Sn ratio is well defined by the sections given in Figs 364, 366, 367, 369, 370 and 371. The composition of E_2 was scaled from the original drawing, which is very uncertain in this region, to give a value of 68·9 Au–30·3 Sn–0·8% Se. From the sections a composition of 68·7 Au–30·5 Sn–0·8% Se is estimated.

The partial ternary AuSn–Sn–SnSe is characterized by three invariant reactions, each associated with the AuSn–Sn binary reactions. A transition reaction (U_2) occurs at 309°C, L_{U_2} + AuSn ⇌ SnSe + $AuSn_2$. The temperature and the Au–Sn ratio of this invariant is well defined by the sections given in Figs 366 and 367. The Se content is expected to be very small, so U_2 will be very close to the Au–Sn binary edge. The composition of U_2 was scaled from the original drawing to 28·0 Au–71·3 Sn–0·7% Se. The second invariant reaction (U_3) takes place at 252°C, L_{U_3} + $AuSn_2$ ⇌ SnSe + $AuSn_4$. The composition of U_3 is scaled from the original [76 Rou] as 11·3 Au–88 Sn–0·7% Se. The evaluation of the sections given in Figs 367 and 366 leads to a composition of 8·7 Au–90·6 Sn–0·7% Se, which is 2·6% higher in Sn than [76 Rou] but better verified by the experimental data. The third invariant reaction is of ternary eutectic type (E_3) at 217°C, L_{E_3} ⇌ $AuSn_4$ + Sn + SnSe. The ternary eutectic composition was scaled from [76 Rou] as 6·4 Au–92·8 Sn–0·7% Se. The ternary eutectic composition was placed close to the Sn–$AuSn_4$ binary eutectic composition, e_6, on the basis that there is little difference in the binary and ternary eutectic temperatures. However [76 Rou] placed the Sn–$AuSn_4$ eutectic composition at 7·5% Au compared with the accepted value for e_6 of 6·3 Au. The composition of the ternary eutectic, E_3, is amended to an approximate composition 4·7 Au–94.6 Sn–0·7% Se.

The sections [76 Rou] associated with the Au–Sn–SnSe–AuSe region were evaluated in order to derive L_1–L_2 tie lines in the large area of liquid immiscibility. The sections presented are amendments of the original data [76 Rou], drawn to give self-consistent results for the L_1–L_2 tie lines. Compared with the original data the sections AuSn–Se, 20% Se, 30% Se, 6·67% Au, 16·67% Au, and AuSn–$SnSe_2$ (Figs 364, 367, 366, 369, 370, 371 respectively) have been amended in terms of the L_1 + L_2 + SnSe phase region to give a much narrower phase region. This reflects the fact that points on the L_1 + L_2/L_1 + L_2 + SnSe boundary lie on the L_1 + L_2 tie lines of isothermal sections, and points on the L_1 + L_2 + SnSe/L + SnSe boundary must necessarily lie on the L_1 + SnSe tie line of isothermal sections. The liquidus surface for primary crystallization of (Au) (Fig. 370) has been increased from 750 to 820°C at 35% Sn to give consistency with other sections. The surface of secondary separation between the L_1 + L_2 + (Au) phase regions (Fig. 372) has been amended to comply with the vertical sections at 20 and 30% Se (Figs 366 and 367). Additionally the composition of liquid L_1 at the intersection with the L + (Au) and L_1 + L_2 phase regions is placed at a lower Au content in conformity with the data presented for the Au–$SnSe_2$ section (Fig. 360). Table 114 summarizes the compositions of the liquid phase associated with the invariant reactions and Fig. 359 is a summary of the reaction scheme for the ternary system.

Table 114. Invariant equilibria

Reaction	*Temperature*, °C	*Phase*	*Composition*, at.-% *Au*	*Se*	*Sn*
E_1	564	L	12·0	55·3	32·6
E_2	276	L	68·7*	0·8*	30·5*
			6·4	0·7	92·8
E_3	217	L	4·7*	0·7*	94·6*
E_4	214	L		>99·9	
L_1/L_2	730	L_1	82·0	1·3	16·6
		L_2	9·0	48·6	42·3
L_3/L_4	558	L_3	23	59	18
		L_4	0·7	94	5·3
U′	510–515	L	75·2	0·8	24·0
U_1	380	L		>99·9	
U_2	309	L	28·0	0·7	71·3
U_3	252	L	11·3	0·7	88·0
			8·7*	0·7*	90·6*

* Suggested amendments to the compositions: see the discussion of invariant equilibria

LIQUIDUS SURFACE

The liquidus surface given in Fig. 358 agrees with that published in [76 Rou], except for minor modification of the extent of the liquid miscibility gap in the Au–$SnSe_2$ section and the limit of the miscibility gap on the primary crystallization surface of $SnSe_2$ in the Se-rich region.

ISOTHERMAL SECTIONS

Isothermal sections at temperatures of 800, 700, and 600°C (Figs 374, 375, and 376 respectively) have been generated from the vertical sections.

REFERENCES

75 Rou: J.-C. Rouland, C. Souleau and B. Legendre: *Compt. rend.*, 1975, **281C**, 719–722

76 Rou: J. C. Rouland, B. Legendre and C. Souleau: *Bull. Soc. Chim. France*, 1976, (11–12), 1614–1624

84 Oka: H. Okamoto and T. B. Massalski: *Bull. Alloy Phase Diagrams*, 1984, **5**, 492–503

86 Sha: R. C. Sharma and Y. A. Chang: *Bull. Alloy Phase Diagrams*, 1986, **7**, 68–72

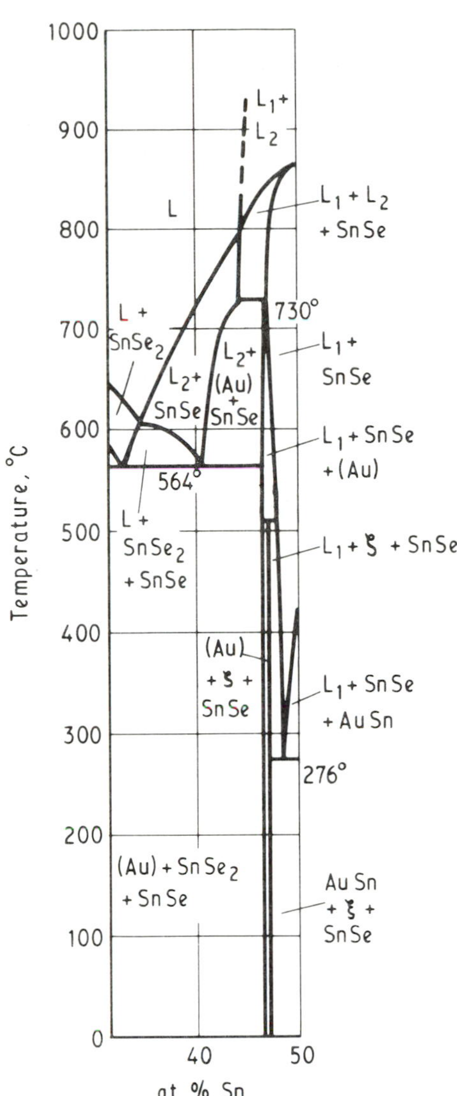

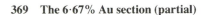

369 The 6·67% Au section (partial)

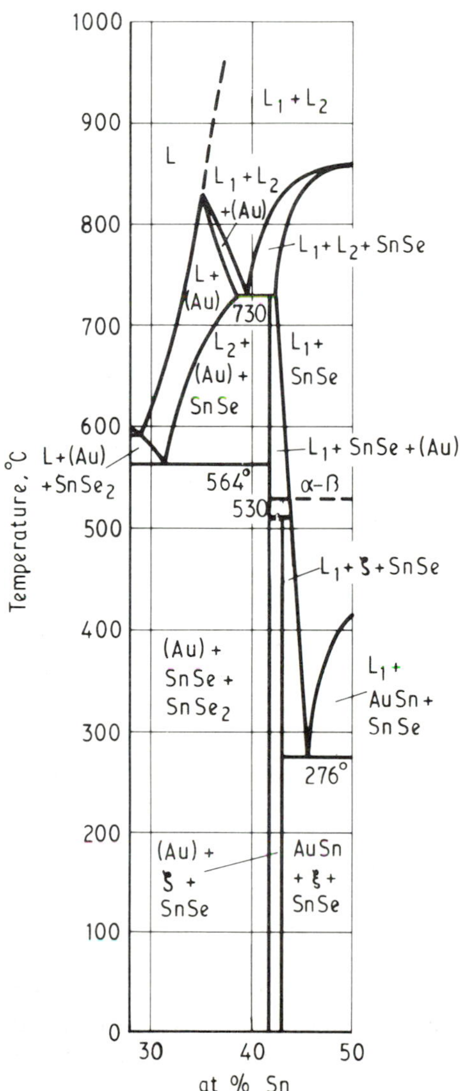

370 The 16·67% Au section (partial)

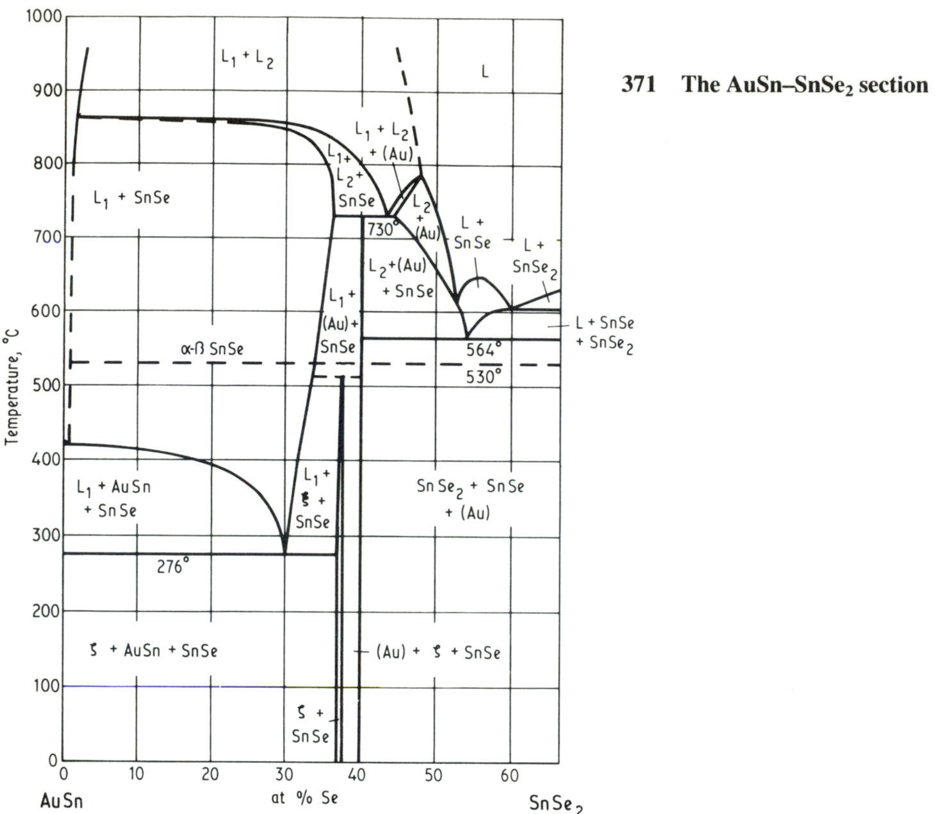

371 The AuSn–SnSe₂ section

372 The Au–SnSe section

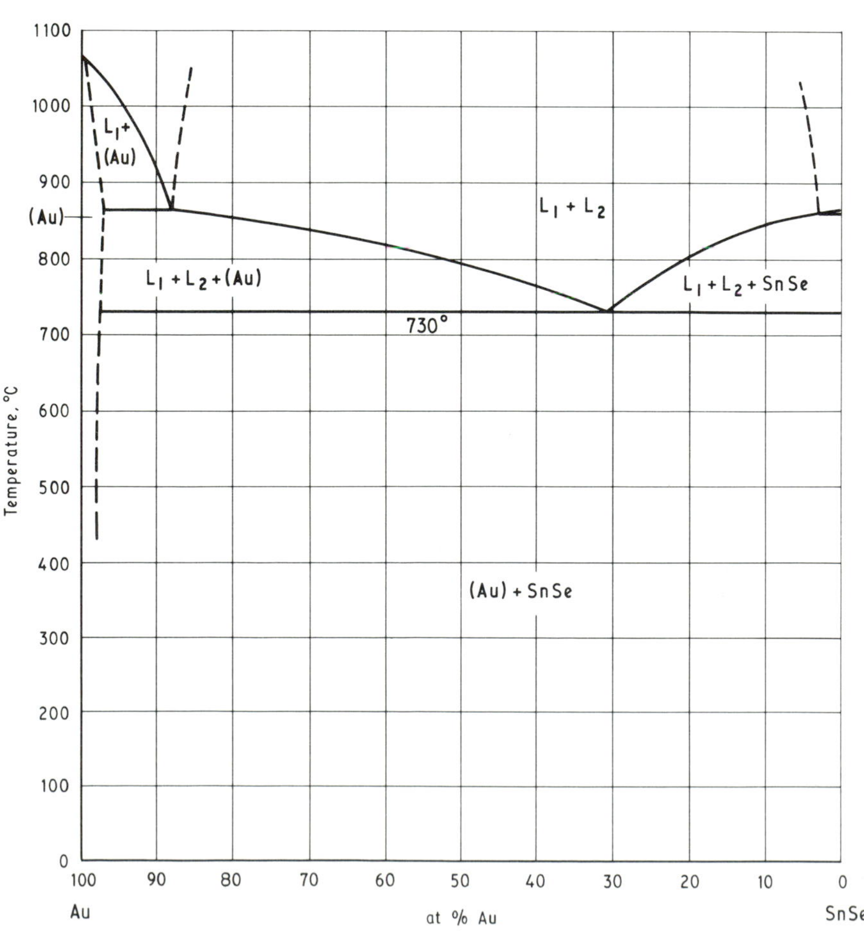

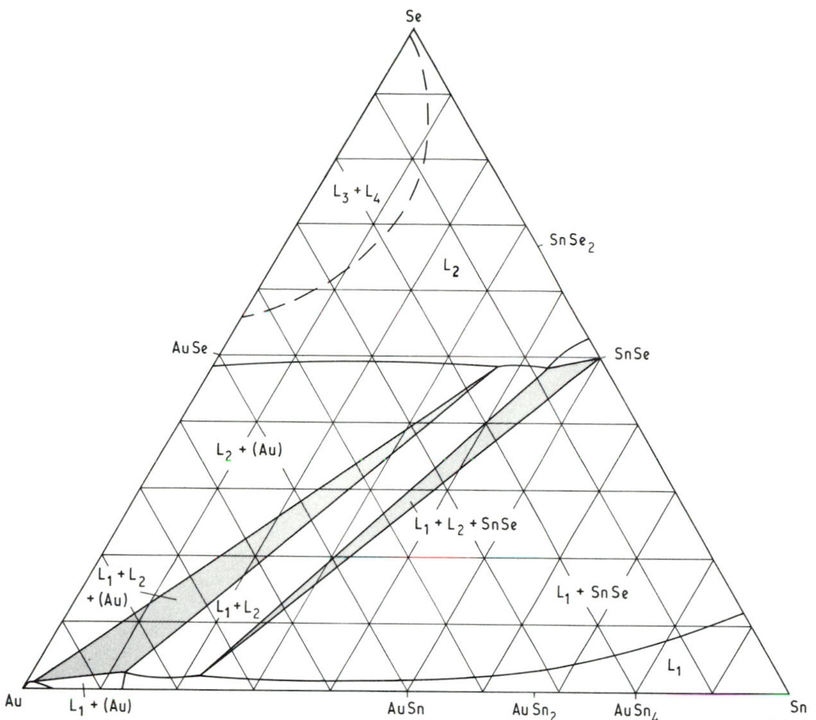

373 Isothermal section at 800°C

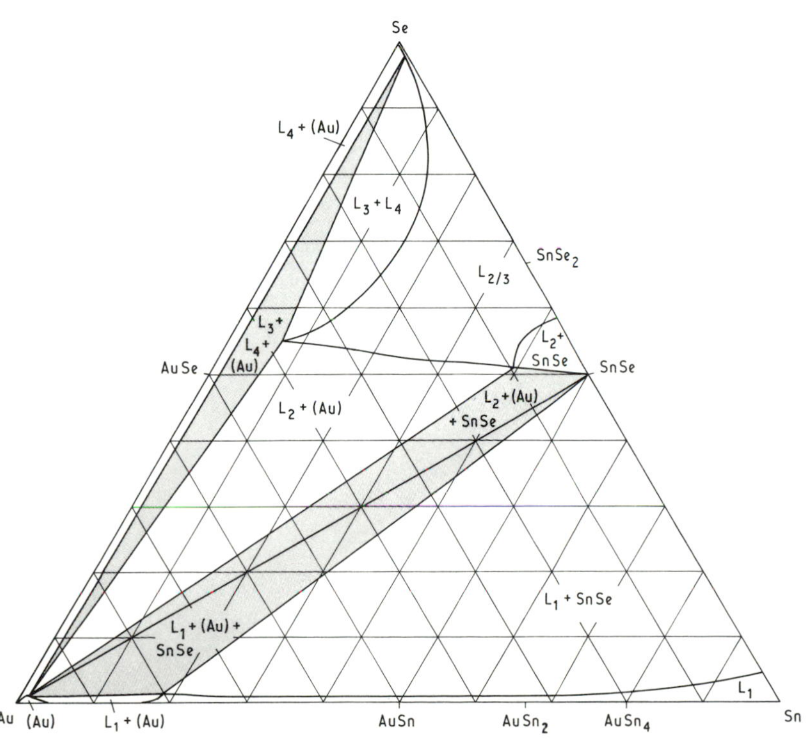

374 Isothermal section at 700°C

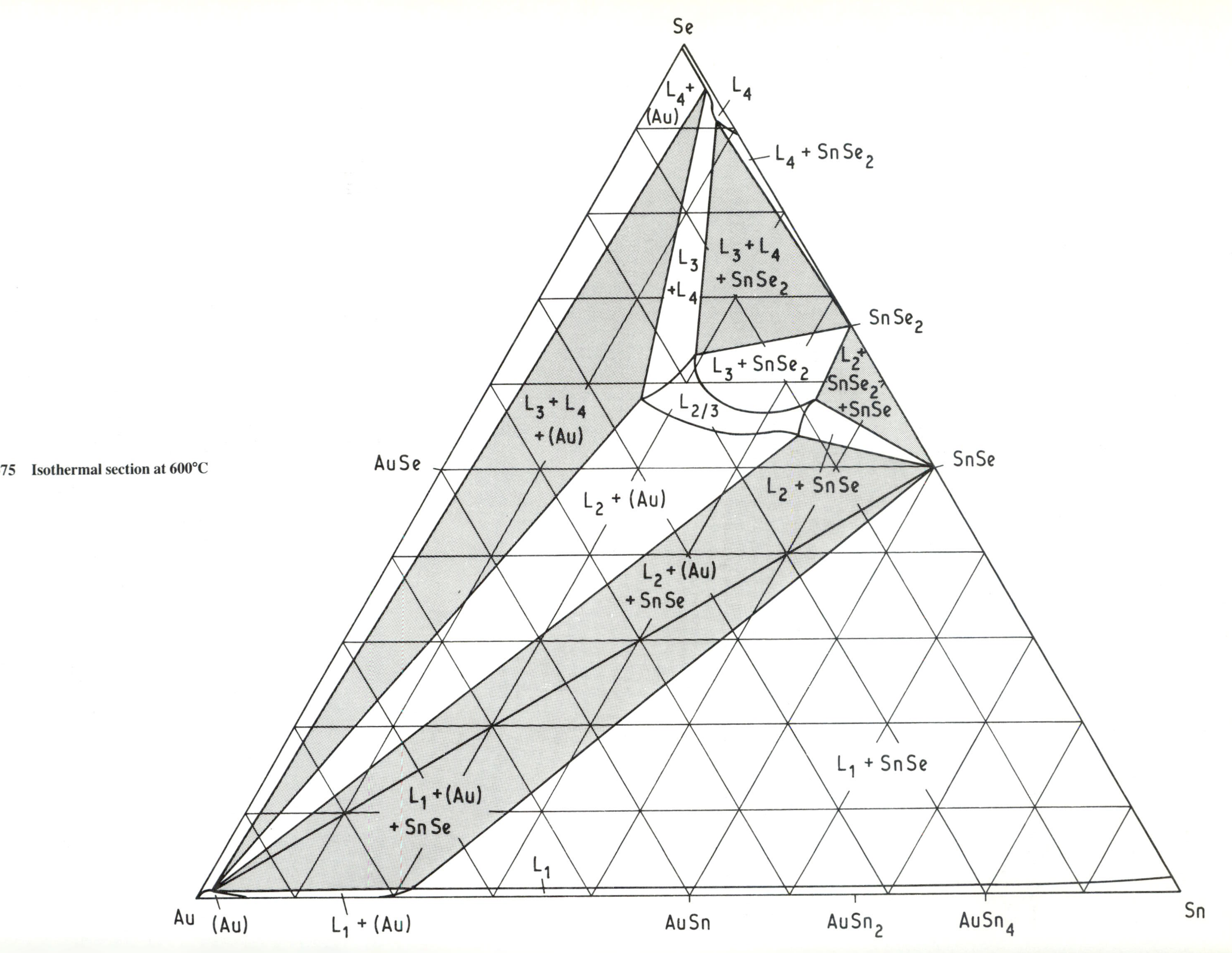

375 Isothermal section at 600°C

86 Oka: H. Okamoto and T. B. Massalski: *Binary Alloy Phase Diagrams* (ed. T. B. Massalski), Vol. 1, 309–312; 1986, ASM International

87 Leg: B. Legendre, Chhay Hancheng, F. Hayes, C.A. Maxwell, D. S. Evans and A. Prince: *Mater. Sci. Technol.*, 1987, **3**, 875–876

Au–Se–Te
(B. Gather and A. Prince)

The only published data is an investigation of an unstated number of samples in series of $AuTe_xSe_{1-x}$ and $AuSe_xTe_{2-x}$ compositions using X-ray powder diffraction analysis and DTA [68 Cra]. The samples were prepared by reaction of 99·999% pure gold and 99·999 9% pure Se and Te in sealed and evacuated Vycor tubes. The $AuSe_xTe_{2-x}$ alloys were annealed at 400–425°C prior to quenching for X-ray study. Examination of $AuTe_xSe_{1-x}$ alloys indicated no significant substitution of Se by Te in AuSe. A ternary compound, Au_2SeTe, melting at 471°C was identified, but its range of homogeneity was not determined. At the composition Au_2SeTe a primitive orthorhombic structure, with lattice parameters $a = 0{\cdot}892\,4 \pm 0{\cdot}000\,6$, $b = 0{\cdot}756 \pm 0{\cdot}001$, $c = 0{\cdot}574\,1 \pm 0{\cdot}000\,8$ nm and four formula units to the cell, was quoted. Examination of $AuSe_xTe_{2-x}$ alloys by X-ray powder diffraction indicated that Se replaces Te in monoclinic (pseudo-orthorhombic) $AuTe_2$ for $0 < x \leqslant 0{\cdot}30$ (Table 115) for alloys quenched from 400–425°C. In the composition range $0{\cdot}30 \leqslant x \leqslant 0{\cdot}35$ an unidentified second phase appeared. The 'melting points', probably solidus temperatures, fall from $AuTe_2$ as Se substitutes for Te (Table 115).

REFERENCE

68 Cra: G. E. Cranton and R. D. Heyding: *Can. J. Chem.*, 1968, **46**, 2632–2640

Au–Si–Sn

INTRODUCTION

An extensive study of the Au–Si–Sn ternary system [85 Leg] showed that the section AuSn–Si is a pseudobinary eutectic, with a eutectic temperature of 418·6°C and a composition of 0·3 at.-% Si. In the partial ternary system AuSn–Si–Sn the three invariant reactions occur at temperatures indistinguishable from those in the binary Au–Sn system. This would be expected from the presence of the pseudobinary eutectic at 0·3 at.-% Si and the degenerate form of the eutectic in the Si–Sn binary system. In the partial ternary system AuSn–Si–Au three invariant reactions involving the liquid phase were detected. At U_1 (Fig. 376) L + β ⇌ (Au) + ζ (490°C), at U_2 L + (Au) ⇌ (Si) + ζ (356·7°C), and at E_1 L ⇌ (Si) + ζ + AuSn (273·5°C). A thermal effect at approximately 175°C reflects the presence of a solid-state invariant reaction. The reaction scheme (Fig. 377) should be regarded as speculative in terms of the reaction proposed at 175°C. There is uncertainty concerning the solid-state equilibria in the binary Au–Sn system in the Au-rich region, and this uncertainty is reflected in the ternary solid-state equilibria for all Au-rich Au–Sn–X systems.

BINARY SYSTEMS

The Au–Si diagram evaluated by [83 Oka] is accepted, as is the Si–Sn diagram assessed by [84 Ole]. The Au–Sn phase diagram has been assessed by [84 Oka] but considerable uncertainties exist concerning equilibrium reactions for Au-rich alloys. The Au–Sn diagram accepted in this assessment is due to [85 Leg, 87 Leg] (Fig. 378). It differs in one major respect from the diagram proposed by [84 Oka]. Two peritectic reactions were identified, corresponding to the reactions l + (Au) ⇌ β at 532°C and l + β ⇌ ζ at 519°C [87 Leg]. The mechanism for the formation of γ and the low-temperature transformation of ζ to (Au) + γ require further study.

SOLID PHASES

Table 116 summarizes crystal structure data for the binary phases. No ternary compounds were found [85 Leg].

Table 116. Crystal structures

Solid phase	*Prototype*	*Lattice designation*
(Au)	Cu	cF4
(Si)	C	cF8
(Sn)	Sn	tI4
$AuSn_4$	$PtSn_4$	oC20
$AuSn_2$	—	Orthorhombic
AuSn	NiAs	hP4
γ	—	Hexagonal
ζ	Mg	hP2
β	Ni_3Ti	hP16

PSEUDOBINARY SYSTEMS

The AuSn–Si section is a pseudobinary eutectic system with a eutectic temperature of 418·6°C (Fig. 379). This temperature is only just below the melting point of AuSn,

Table 115. The effect of Se on $AuTe_2$ melting points

$AuSe_xTe_{2-x}$	at.-%			*Lattice parameter*, nm			*Melting point*, °C
	Au	*Se*	*Te*	*a*	*b*	*c*	
x = 0·1	33·33	3·33	63·33	0·718	0·439 4	0·505 7	450
x = 0·2	33·33	6·67	60·00	0·717	0·438 9	0·502 5	440
x = 0·3	33·33	10·0	56·67	0·717	0·438 5	0·501 4	434
				(±0·001)	(±0·000 4)	(±0·000 5)	

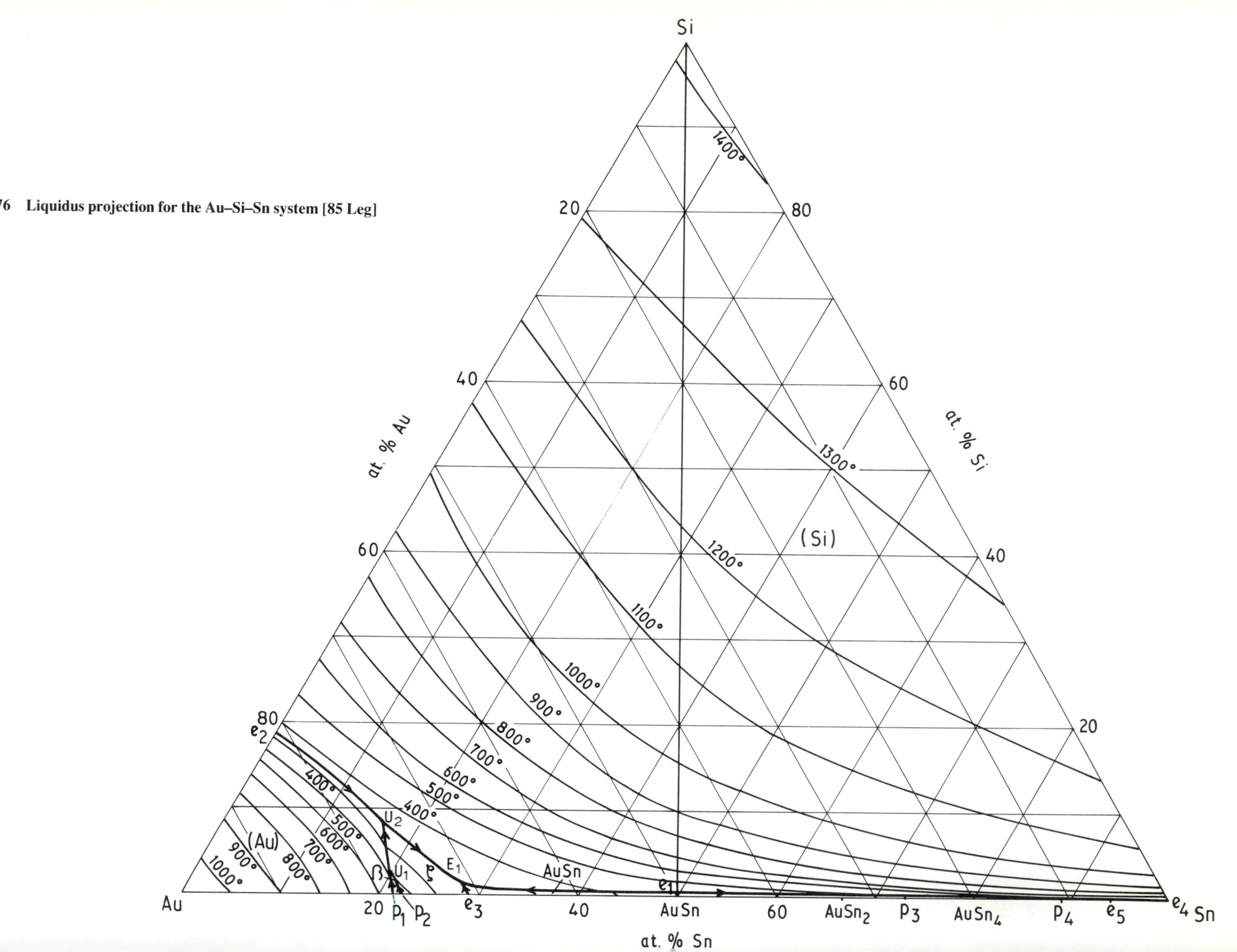

376 Liquidus projection for the Au–Si–Sn system [85 Leg]

377 Reaction scheme for the Au–Si–Sn system [85 Leg]

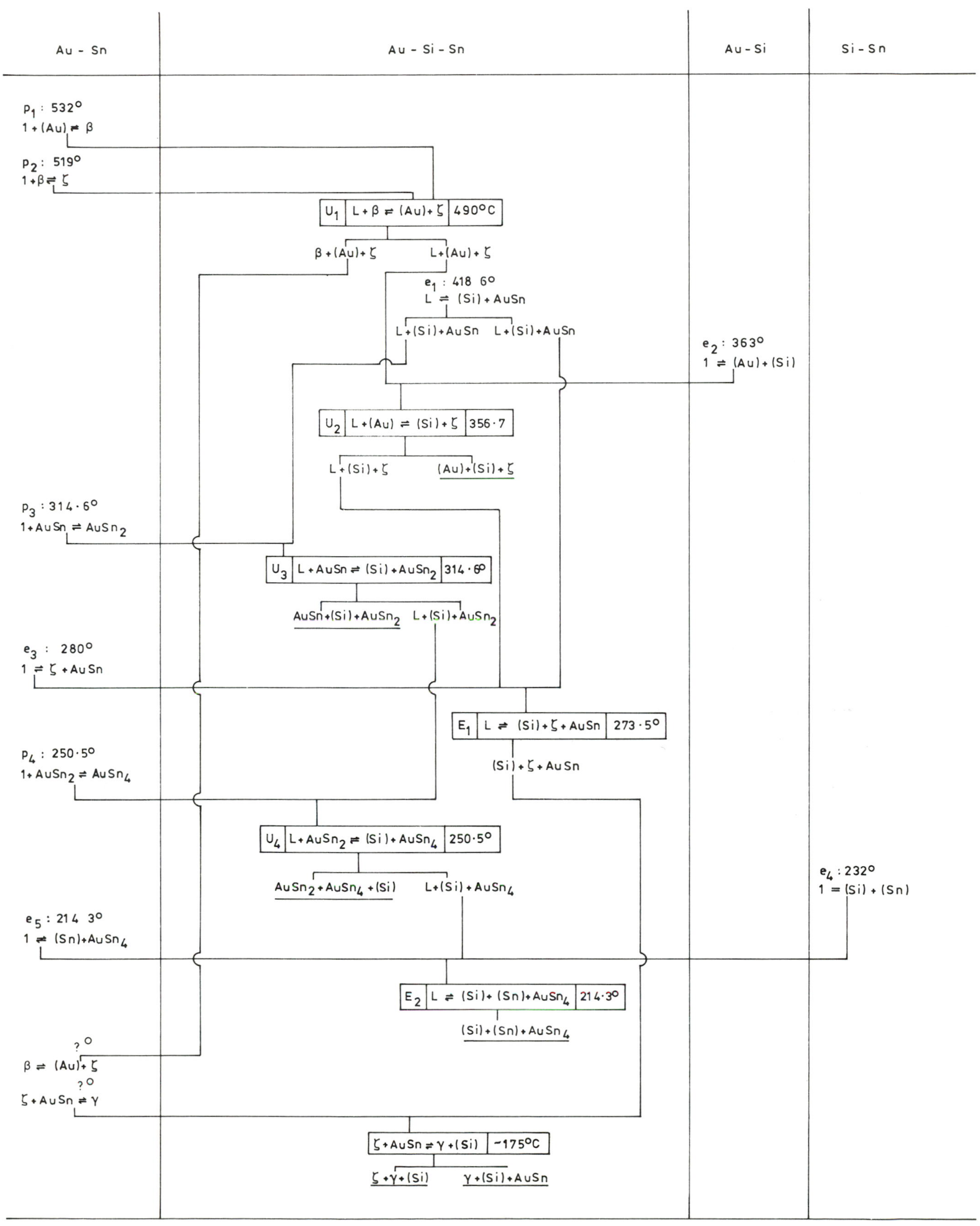

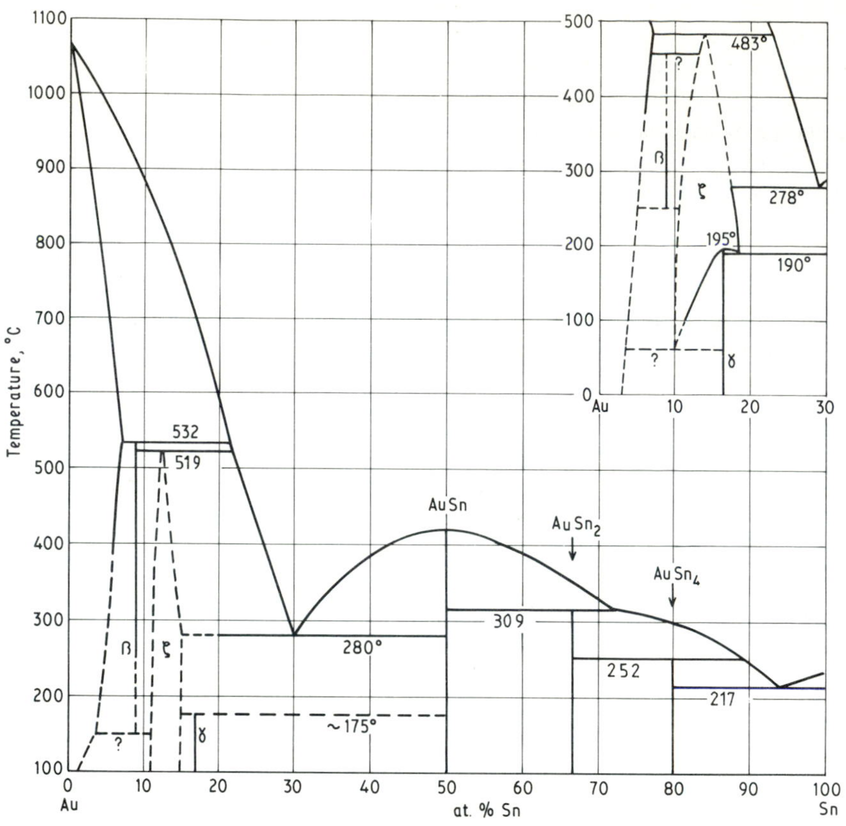

378 The Au–Sn phase diagram [85 Leg, 87 Leg] with inset of Au-rich region according to [84 Oka]

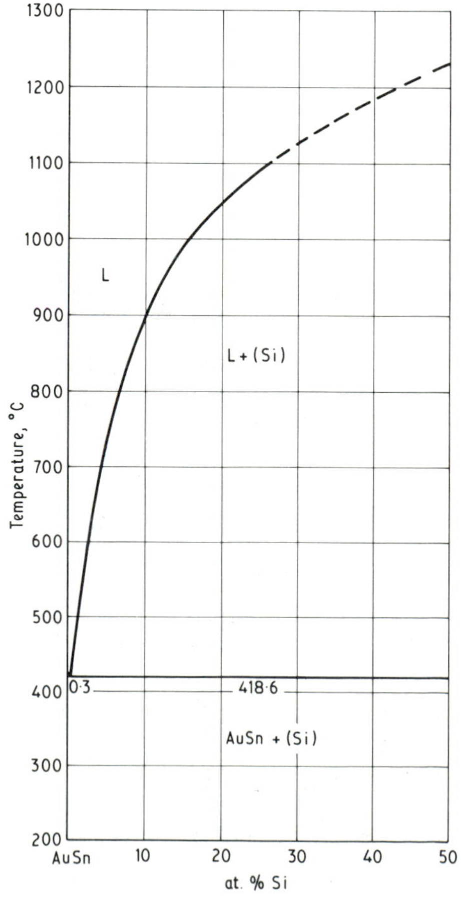

379 The AuSn–Si pseudobinary system

for which the preferred value is 419·8 ± 0·5°C [84 Hum]. It would be anticipated that the eutectic composition would lie close to AuSn. Assuming that the solid solubility of AuSn in Si and of Si in AuSn is negligible, the eutectic composition can be calculated from the Clausius–Clapeyron relation to be at 0·3 at.-% Si. The heat of fusion of AuSn is taken as 10210 J $mole^{-1}$ [85 Leg].

INVARIANT EQUILIBRIA

[85 Leg] surveyed the ternary equilibria using 120 alloys prepared from 99·99% Au, 99·9999% Sn, and Si. Alloys were melted in evacuated silica ampoules and then quenched into water. They were homogenized by heat treatment for 3 to 7 days at temperatures varying with composition. Alloys were studied on both heating and cooling by differential thermal analysis and differential scanning calorimetry, using heating/cooling rates of 2 and 5°C mn^{-1}. Seven sections were examined, at 10 at.-% Sn (Fig. 380), 16·67 at.-% Sn (Fig. 381), 22 at.-% Sn (Fig. 382), 2 at.-% Si (Fig. 383), 10 at.-% Si (Fig. 384) and 23·33 at.-% Si (Fig. 385).

The AuSn–Si pseudobinary divides the Au–Si–Sn system into two partial ternary systems. The AuSn–Si–Sn partial system contains three invariant ternary reactions denoted U_3, U_4, and E_2 in Fig. 377. These reactions are the ternary equivalents of the binary reactions of p_3, p_4, and e_5, and represent the equilibria L + AuSn ⇌ (Si) + $AuSn_2$ at U_3, L + $AuSn_2$ ⇌ (Si) + $AuSn_4$ at U_4, and L ⇌ (Si) + (Sn) + $AuSn_4$ at E_2 respectively. The ternary reactions occur at temperatures indistinguishable from the corresponding binary reactions, as would be anticipated from a consideration of the course of the monovariant curve from e_1 at 0·3 at.-% Si on the AuSn–Si section which ultimately links with the degenerate Si–Sn eutectic at the Sn corner of the ternary system. It should be noted (Fig. 377) that the binary reaction temperatures p_3, p_4, and e_5 do not agree with [84 Oka]. It is likely that the ternary invariant reactions will occur at 309°C (U_3), 252°C (U_4), and 217°C (E_2) to give correspondence with the temperatures associated with p_3, p_4, and e_5. Virtually the whole of the AuSn–Si–Sn partial ternary system is occupied by the region of primary crystallization of Si (Fig. 376).

The Au–Si–AuSn partial ternary system also contains three invariant reactions associated with the liquid phase — U_1, U_2, and E_1 (Figs 376 and 377). The formation of β by peritectic reaction in the Au–Sn system [87 Leg] implies that β is involved in a ternary invariant reaction. Evidence for a reaction at 490°C was found in the 10 at.-% Sn, 16·67 at.-% Sn, and 2 at.-% Si sections (Figs 380, 381 and 383 respectively). The reaction is interpreted to be L + β ⇌ (Au) + ζ, U_1. This is followed by the reaction L + (Au) ⇌ (Si) + ζ at U_2 with completion of solidification at E_1, L ⇌ (Si) + ζ + AuSn. The compositions of the phases taking part in the three invariant reactions are given in Table 117. The liquid phase compositions are considered accurate to ±0·5 at.-%. The compositions of the solid phases are enclosed in brackets to indicate that they are estimated, not experimentally determined. The solid-state reactions in the Au–Si–AuSn partial ternary system have not been elucidated. A reaction occurs at approximately 175°C, but its interpretation is dependent on a resolution of the solid-state equilibria in the binary Au–Sn system. The reaction scheme (Fig. 377) is speculative in terms of the ternary reaction at 175°C, ζ + AuSn ⇌ γ + (Si). If ζ transforms to (Au) + γ as [84 Oka] propose, there may well be a reaction (Au) + ζ ⇌ (Si) + γ, followed by eutectoidal breakdown of ζ, ζ ⇌ γ + (Si) + AuSn.

Table 117. Invariant equilibria

Reaction	*Temperature*, °C	*Phase*	*Composition*, at.-%		
			Au	*Si*	*Sn*
U_1	490	L	77·7	2·7	19·6
		(Au)	(94)	(1)	(5)
		β	(91)	—	(9)
		ζ	(86·5)	(0·5)	(13)
U_2	356·7	L	75·6	8·2	16·2
		(Au)	(96)	(1)	(3)
		ζ	(86)	(1)	(13)
		(Si)	—	(100)	—
E_1	273·5	L	71	1·7	27·3
		ζ	(86)	(1)	(13)
		AuSn	(50)	—	(50)
		(Si)	—	(100)	—

LIQUIDUS SURFACE

Apart from a small area of primary crystallization of (Au) and much smaller areas of primary crystallization of the Au–Sn binary compounds, some 95% of the ternary system corresponds to primary crystallization of Si (Fig. 376).

REFERENCES

83 Oka: H. Okamoto and T. B. Massalski: *Bull. Alloy Phase Diagrams*, 1983, **4**(2), 190–198, 362

84 Ole: R. W. Olesinski and G. J. Abbaschian: *Bull. Alloy Phase Diagrams*, 1984, **5**(3), 273–276

84 Oka: H. Okamoto and T. B. Massalski: *Bull. Alloy Phase Diagrams*, 1984, **5**(5), 492–503

84 Hum: G. Humpston and B. L. Davies: *Metal Sci.*, 1984, **18**, 329–331

85 Leg: B. Legendre, Chhay Hancheng and A. Prince: *Bull. Soc. Chim. France*, 1985, I-50–I-57

87 Leg: B. Legendre, Chhay Hancheng, F. Hayes, C. A. Maxwell, D. S. Evans and A. Prince: *Mater. Sci. Technol.*, 1987, **3**, 875–876

Au–Si–Te

INTRODUCTION

The only published data is a survey of 240 alloys made from 99·999% pure elements [78 Leg]. Vertical sections were examined using differential thermal analysis with supporting X-ray diffraction, metallography and micro-

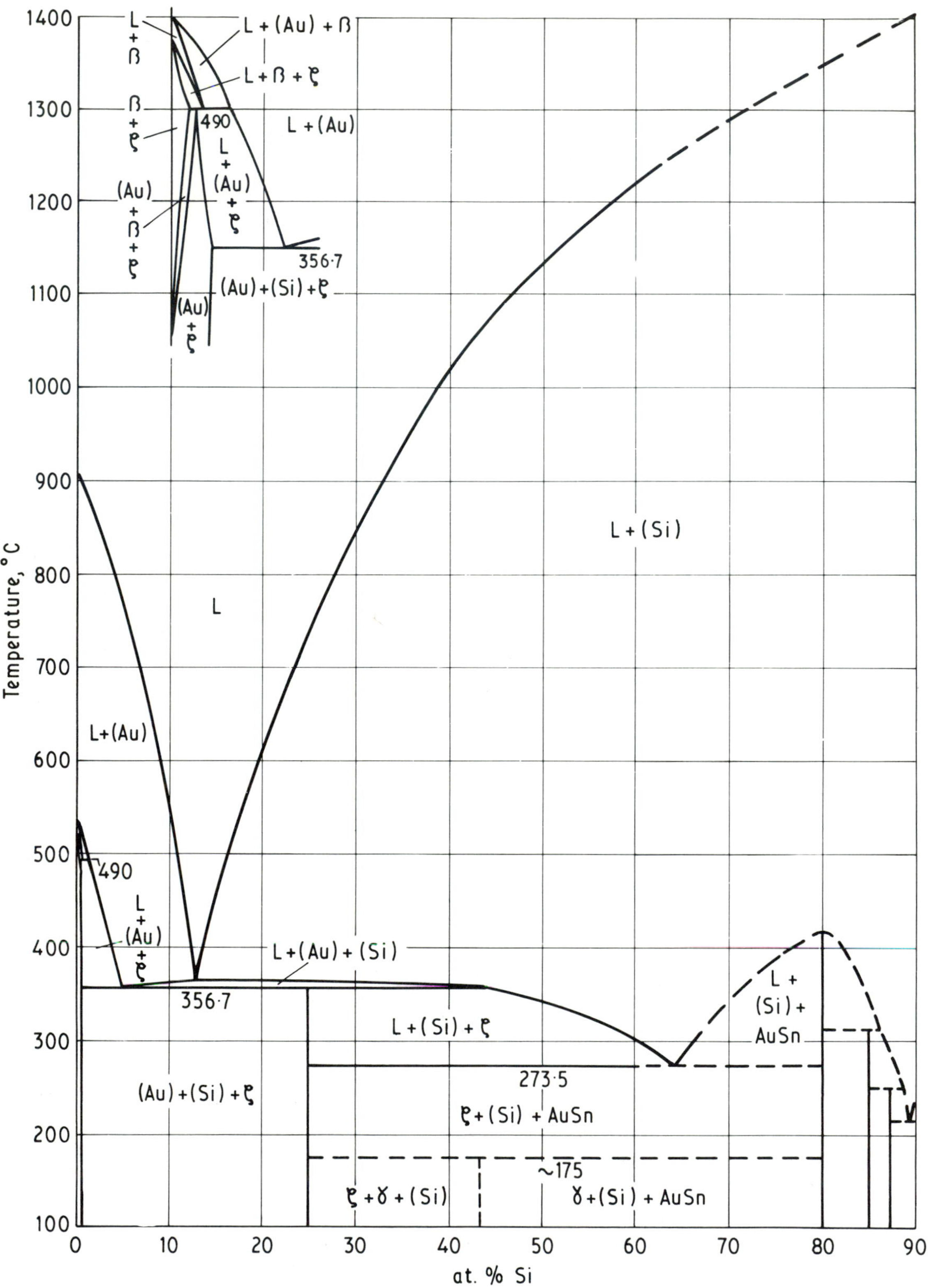

380 The 10 at.-% Sn section

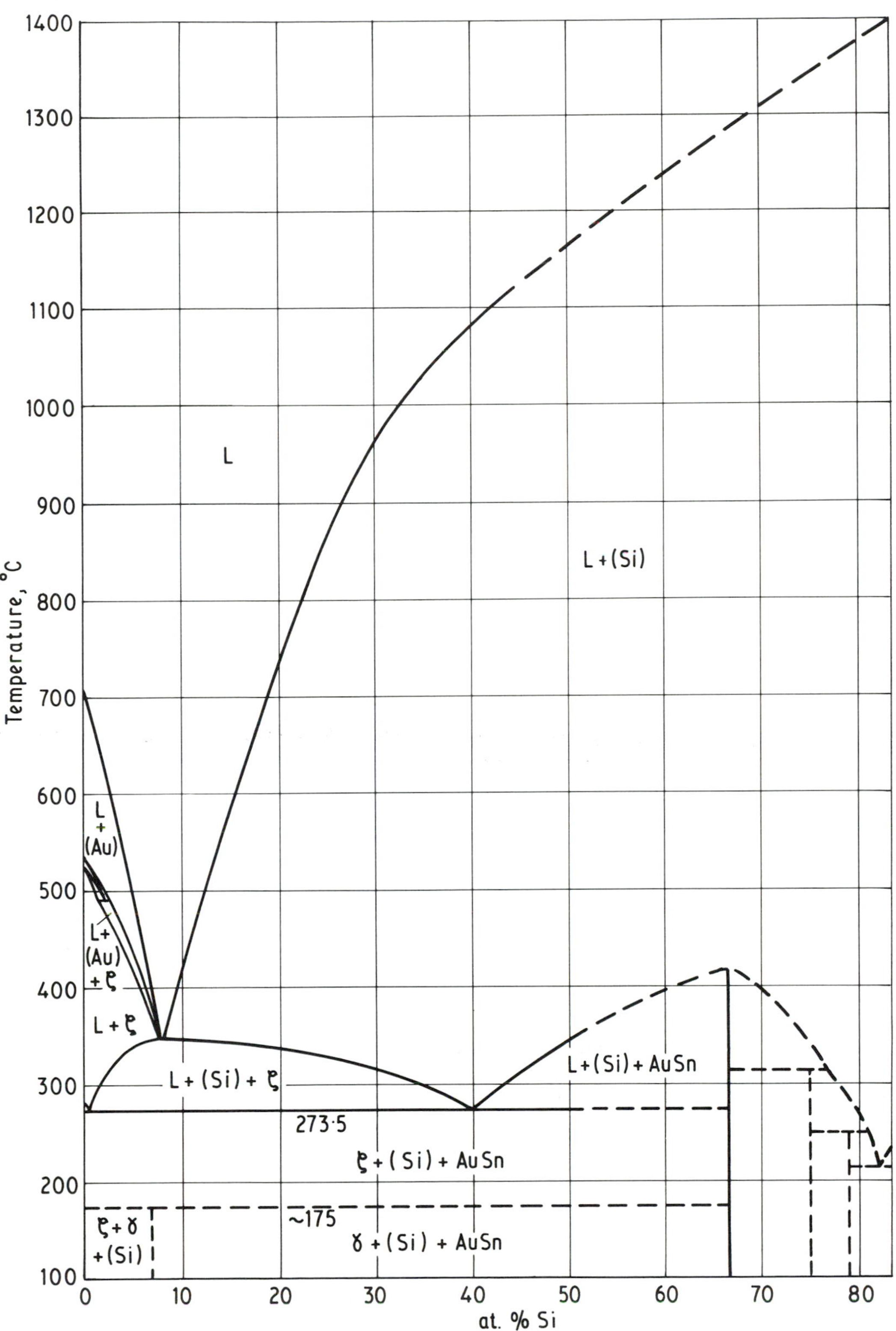

381 The 16·67 at.-% Sn section

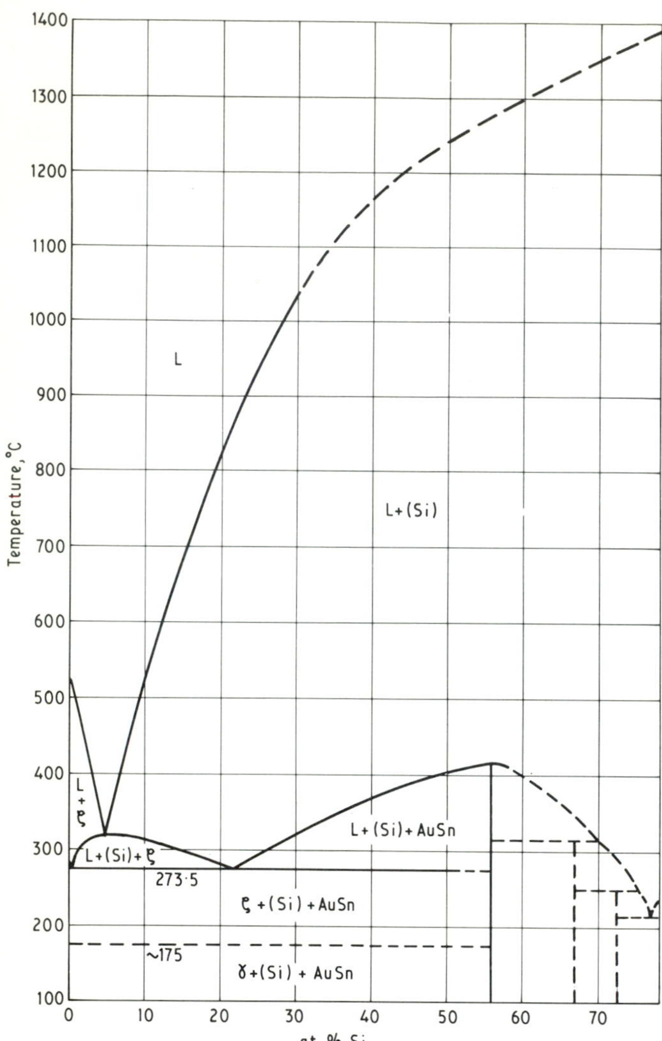

382 The 22 at.-% Sn section

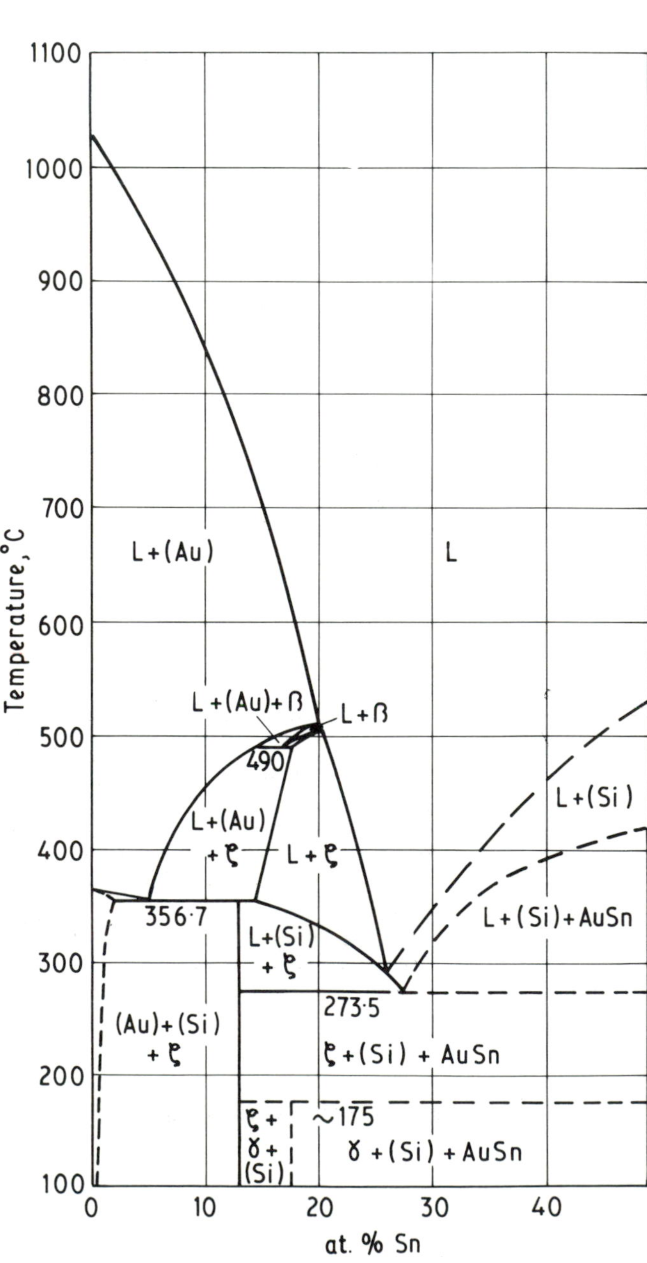

383 The 2 at.-% Si section

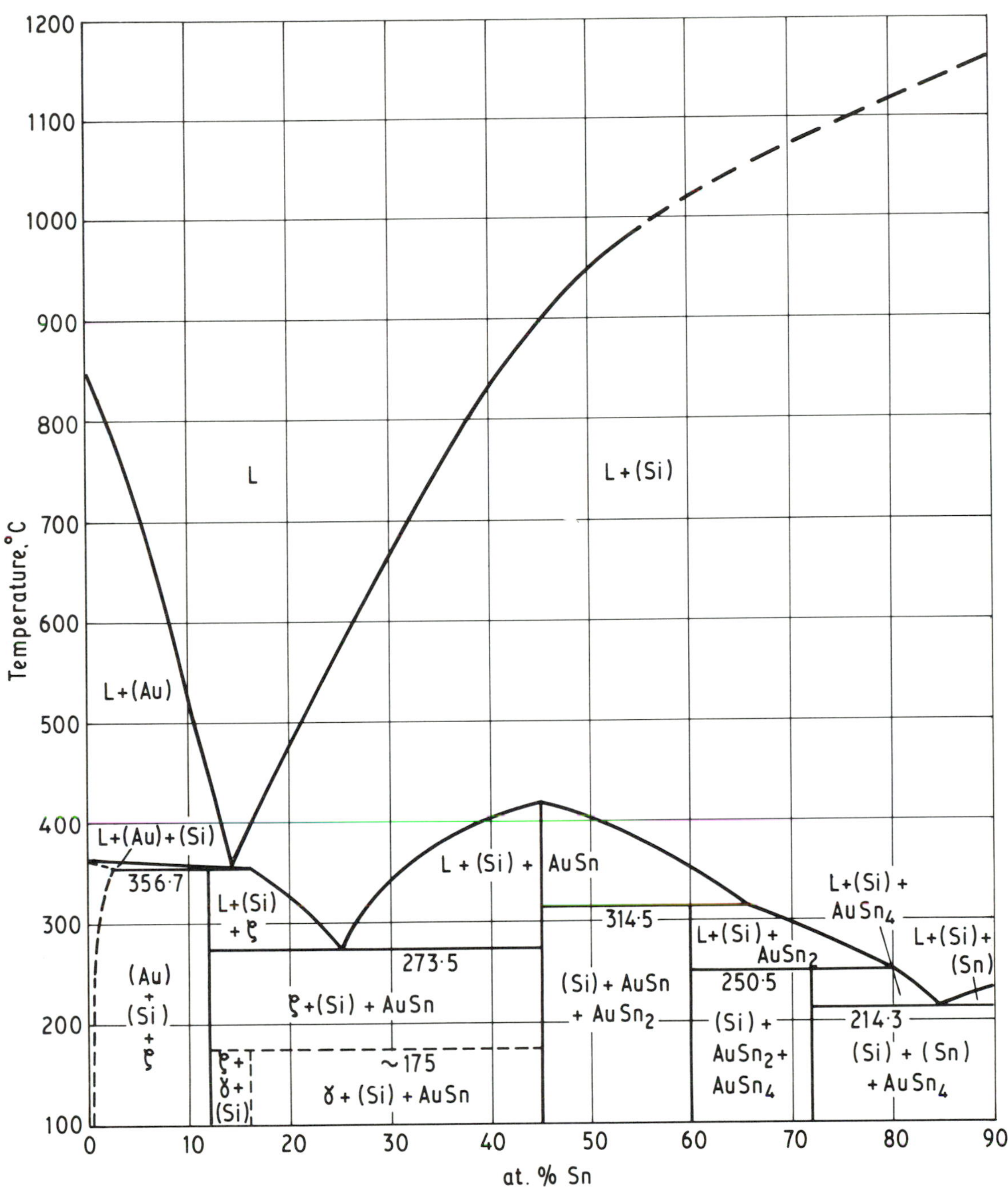

384 The 10 at.-% Si section

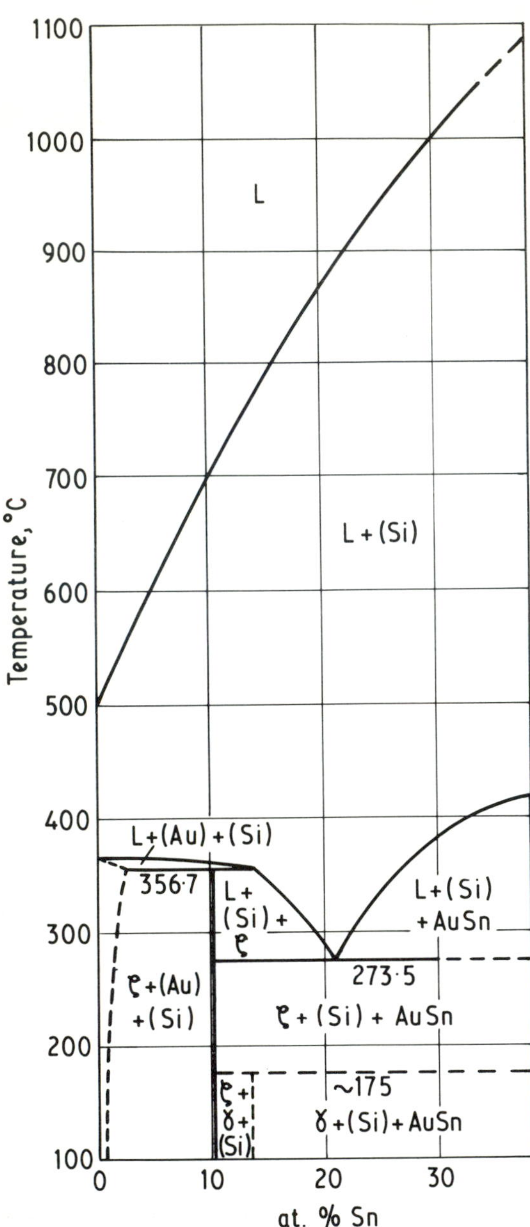

385 The 23·33 at.-% Si section

386 The section $AuTe_2$–Si_2Te_3

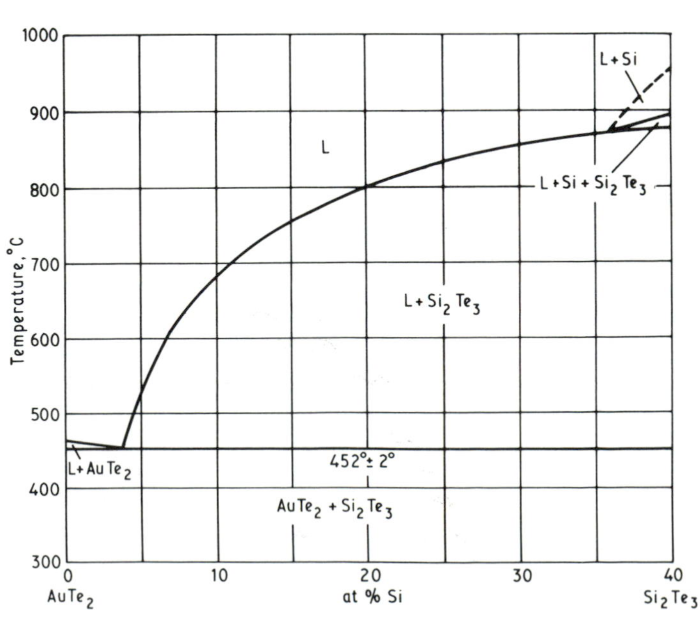

hardness testing. The ternary system is divided into two partial systems by the $AuTe_2$–Si_2Te_3 section. Although not a true pseudobinary because of the peritectic nature of the formation of Si_2Te_3, the $AuTe_2$–Si_2Te_3 section nevertheless can be considered as such. The partial ternary system $AuTe_2$–Si_2Te_3–Te is a simple ternary eutectic system. The partial ternary system Au–$AuTe_2$–Si_2Te_3–Si is rather complex, exhibiting five invariant reactions at 362, 437, 540, 628, and 776°C respectively. The compositions of the liquid phase associated with the 776 and 628°C invariant reactions are very uncertain. The reaction at 776°C is a true ternary peritectic, L_{U_1} + (Si) + $Si_2Te_3 \rightleftharpoons \tau$, where τ is a ternary compound whose composition was not established. This compound transforms at 628°C by the reverse reaction, $\tau \rightleftharpoons L_{U_2}$ + (Si) + Si_2Te_3. It was observed by X-ray diffraction and metallographic methods. The 540°C reaction, also referred to as a 530°C reaction by [78 Leg], involves ternary liquid immiscibility in the Au-rich region: $L_1 + L_2 \rightleftharpoons$ (Au) + Si_2Te_3. The compositions of L_1 and L_2 and the extent of the $L_1 + L_2$ phase region are uncertain. Reactions at 362°C and 437°C are ternary eutectic reactions: $L_{E_3} \rightleftharpoons$ (Au) + (Si) + Si_2Te_3 and $L_{E_1} \rightleftharpoons AuTe_2 + Si_2Te_3$ + (Au) respectively.

It is not possible to accept all the interpretations placed on their data by Legendre *et al.* [78 Leg]. There are many inconsistencies in the data themselves and these are discussed in the section on invariant equilibria. The partial ternary system Au–$AuTe_2$–Si_2Te_3–Si needs further experimental work to provide a definitive version of the equilibria.

BINARY SYSTEMS

The Au–Si system has been assessed by [83 Oka]. The system was confirmed as a simple eutectic with the eutectic composition being 18·6 ± 0·5 Si. The eutectic temperature was quoted as 363 ± 3°C: however, it is suggested that error limits of ±1°C are more appropriate.

The Au–Te system has been assessed by [84 Oka]. The compound $AuTe_2$ melts congruently at 464°C and forms a eutectic with Au at 54% Te and 447°C, and a second with Te at 88% Te and 416°C.

The Si–Te system shows peritectic formation of Si_2Te_3, at 895°C, L_{p_1} + (Si) $\rightleftharpoons Si_2Te_3$, followed by a eutectic reaction, $L_{e_4} \rightleftharpoons Si_2Te_3$ + (Te), at 406°C and 82·5% Te [78 Leg]. The composition of the peritectic liquid is given by [78 Leg] as 63·5% Te but a value of 61·8% Te is obtained by scaling from the phase diagram given by [78 Leg]. [66 Bai] has determined the peritectic at 892°C with a peritectic liquid composition of Si–62% Te and the eutectic at ~85% Te and 409°C, in reasonable agreement with [78 Leg]. There is conflicting evidence on whether Si_2Te_3 forms congruently or incongruently. In contrast to [78 Leg, 66 Bai, 73 Pet], congruent melting was claimed by Ploog *et al.* [76 Plo]. Incongruent melting is accepted on the weight of the experimental evidence to date. Si_2Te_3 is hexagonal with a superstructure with lattice parameters a = 0·742 2 nm, c = 1·346 4 nm [78 Leg] or a = 0·743 0 nm, c = 1·348 2 nm [76 Plo].

SOLID PHASES

The ternary compound τ was not characterized by [78 Leg]. Table 118 summarizes crystal structure data for the binary compounds.

Table 118. Crystal structures

Solid phase	*Prototype*	*Lattice designation*
(Au)	Cu	cF4
(Si)	C	cF8
(Te)	γSe	hF3
$AuTe_2$	$AuTe_2$	mC6
Si_2Te_3	CdI_2	hP3
τ	—	—

PSEUDOBINARY SYSTEMS

The $AuTe_2$–Si_2Te_3 section can be regarded as virtually pseudobinary, as shown by Fig. 386. The eutectic temperature is given as 452°C and the eutectic composition e_1 is 30·3 Au–3·6 Si–66% Te. The eutectic temperature is considered reliable to within ±2°C, but the eutectic composition can only be quoted to ±1% Si and ±0·2% Te since the value was not quoted by [78 Leg], and had to be scaled from the original small published diagram.

INVARIANT EQUILIBRIA

Table 119 and Fig. 387 give details of the invariant equilibria. The partial ternary system $AuTe_2$–Te–Si_2Te_3 is a simple ternary eutectic system with a ternary eutectic temperature at 397 ± 2°C and a composition of the ternary eutectic E_2 given as 4·33 Au–14·2 Si–81·4% Te (total 99·93%). The two sections of the ternary system presented by [78 Leg], at 66·67% Te and 86·67% Te, are reproduced to a scale that makes it difficult to check the validity of the quoted composition for E_2.

Table 119. Invariant equilibria

Reaction	*Temperature,* °C	*Phase*	*Composition,* at.-% *Au*	*Si*	*Te*
E_1	437	L	36·67	3·33	60
E_2	397	L	4·4	14·2	81·4
E_3	362	L	79·0	19·3	1·7
U_1	776	L	70	20	10
		L	55*	25*	20*
U_2	628	L	73	20	7
L_1/L_2	540	L_1	53·4	6·6	40
			49·4*	10·6*	40*
L_1/L_2	540	L_2	63·4	16·6	20
			70·0*	16*	14*

* Suggested amendments to the compositions: see the discussion of invariant equilibria

There are five invariant reactions in the partial ternary system Au–$AuTe_2$–Si_2Te_3–Si. A ternary eutectic reaction, $L_{E_3} \rightleftharpoons$ (Au) + (Si) + Si_2Te_3, was found 1°C below the binary Au–Si eutectic temperature with the composition of E_3 close to the Au–Si binary edge. [78 Leg] quote E_3 as containing 79 Au–19·3 Si–0·7% Te (total 99·0%). The tie line E_3–Si_2Te_3 is intersected by six of the sections presented. If this composition of E_3 given by [78 Leg] is modified by increasing the Te content to 1·7% Te to make the constituents total 100%, then intersection with the

387 The reaction scheme for the Au–Si–Te system

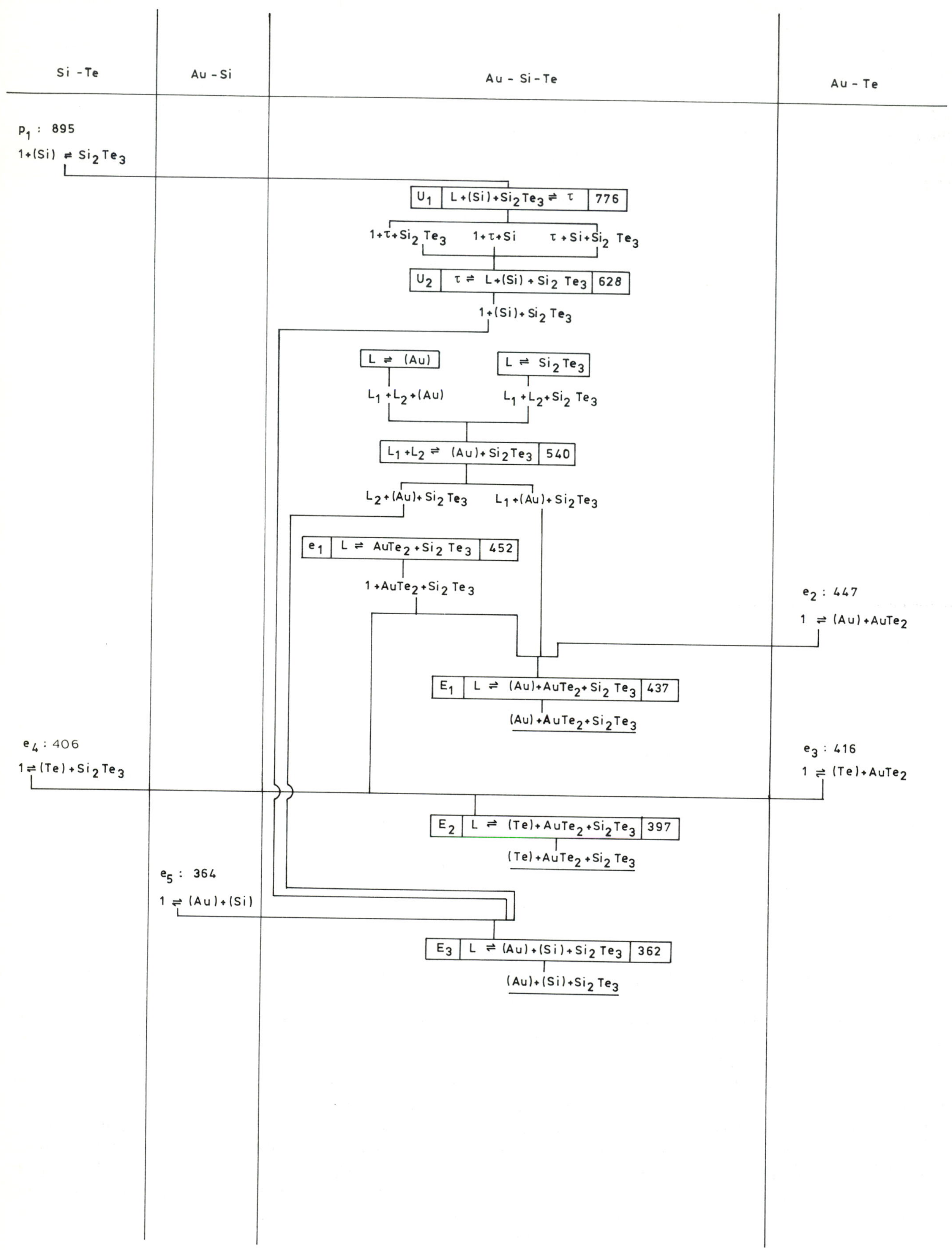

20% Au, $AuTe_2$–Si, 40% Te, 40% Au, 20% Te, and 60% Au sections should occur at 35·0, 34·4, 33·0, 29·6, 26·0, and 24·6% Si respectively. Scaled from the published sections, intersection occurs at 35·0, 34·7, 37·1, 31·6, 29·0, and 24·6% Si respectively. It should be noted that in the three cases where intersection takes place at a Si content in agreement with the proposed composition for E_3, namely the 20% Au, $AuTe_2$–Si, and 60% Au sections, [78 Leg] provided no experimental data to allow the estimation of the intersection points. Most reliance for the composition of E_3 is placed on the intersection with the 20% Te section at the experimental value of 29% Si (Fig. 388) compared with a value of 26% Si. It is apparent that the composition of the ternary eutectic E_3 is probably at a higher Si content than [78 Leg] suggest, although the 1°C difference in temperature from the binary Au-Si eutectic temperature implies a ternary eutectic E_3 reasonably near to the Si content of the binary eutectic liquid.

An alternative interpretation, for which there is no experimental evidence, is that the tie line E_3–Si_2Te_3 does not extrapolate to the binary compound Si_2Te_3 but to the Si-rich side of the composition Si_2Te_3. This would imply some solubility of Au in Si_2Te_3, the binary line compound broadening into a phase region in the ternary. This suggestion would produce a composition for E_3 of 78·3 Au–20 Si–1·7% Te in reasonable agreement with [78 Leg]. The ternary eutectic reaction, $L_{E_1} \rightleftharpoons (Au) + AuTe_2 + Si_2Te_3$, was found at 437°C, E_1 having the composition 36·67 Au–3·33 Si–60% Te [78 Leg]. The intersection points for the tie line Au–E_1 are all at low Si contents, and this makes it difficult to check the composition of E_1 from the original figures. With the exception of the $AuTe_2$–Si and the 20% Au sections, all sections intersect the tie line Au–E_1 on the 437°C ternary eutectic plane. The $AuTe_2$–Si and the 20% Au sections intersect the tie line Si_2Te_3–E_1 and, according to [78 Leg], this tie line stretches horizontally across the diagram at 60% Te. However, measurement of the intersections from the published sections gives a value of 8·3% Si against the required 10% Si for the $AuTe_2$–Si section and a value of 21·3% Si against the required 20% Si for the 20% Au section. There is some uncertainty with regard to the composition of E_1 with the possibility that it is more Te-rich than 60% Te. It should be noted that the tie line obtained from intersection with the $AuTe_2$–Si and the 20% Au sections extrapolate back to a composition on the Si-rich side of Si_2Te_3.

Major difficulty occurs with any attempt to locate the liquid compositions for the remaining three invariant reactions. At 776°C, Legendre *et al.* [78 Leg] postulate the peritectic formation of a ternary compound by the reaction $L_{U_1} + (Si) + Si_2Te_3 \rightleftharpoons \tau$. A peritectic fold originating at p_1 (895°C) on the Si–Te binary edge, crosses the section $AuTe_2$–Si_2Te_3 and terminates at U_1 (776°C), the composition of which is given as 70 Au–20 Si–10% Te. [78 Leg] state that this composition is indicative only. The experimental data allow the course of the peritectic fold p_1U_1 to be traced in a reasonably consistent manner to the 40% Au section. Beyond this, its course is a matter for conjecture. The monovariant curve is intersected by the Au–Si_2Te_3 section at about 17% Au, as shown by Fig. 389. The 20% Au section given in Fig. 390 differs from that presented by [78 Leg] in that the 40% Si alloy is plotted by [78 Leg] as on the Si_2Te_3 liquidus surface, whereas the same alloy is shown to separate primary Si in the 40% Te section. Accepting the latter data implies that the curve p_1U_1 intersects the 20% Au section at a much lower Si content. Conversely, [78 Leg] ascribe too low a Si content for the intersection of p_1U_1 with the $AuTe_2$–Si section. An alloy with just over 30% Si is placed as undergoing primary Si separation followed by the crystallization of Si + Si_2Te_3, whereas this alloy should be placed, as shown by Fig. 391, on the Si_2Te_3 liquidus. Curve p_1U_1 intersects the 40% Te section at 30% Si [78 Leg], as is shown by Fig. 392. By amending the data of [78 Leg] to achieve a degree of self-consistency in terms of the sections represented by Figs 390–392, the course of the monovariant curve p_1U_1 can be traced to the 30% Au section.

The 40% Au section presents problems. The Si_2Te_3 liquidus, indicated as a dashed line in Fig. 393, appears consistent, but the temperature of the intersection with the Si liquidus, 900°C, is far too high: it follows that the corresponding Si content is also too high. The 20% Si alloy in the 40% Te section (Fig. 392) and the 40% Au alloy in the Au–Si_2Te_3 section (Fig. 389) have liquidus temperatures, marked as a and b respectively, some 60°C below the reported liquidus temperatures in Fig. 393. The liquidus curve has been redrawn on the basis of accepting these two temperatures and rejecting the liquidus, shown as a dashed line in Fig. 393. This construction places the intersection of p_1U_1 with the 40% Au section at 810°C and about 28% Si.

It is not possible to trace p_1U_1 further with any degree of certainty. [78 Leg] suggest an indicative composition for U_1 of 70 Au–20 Si–10% Te. By contrast, their 20% Te section shows p_1U_1 intersecting this section at 776°C, the temperature of the invariant reaction at U_1. The projection of the liquidus [78 Leg] also shows U_1 as placed between the 600 and 700°C isotherms, instead of the requirement to place it at a temperature of 776°C. The available evidence suggests a composition for U_1 nearer 20% Te than 10% Te. On this basis, the 20% Te section proposed by [78 Leg] has been amended to conform with the phase rule, and is given in Fig. 388.

From U_1, two monovariant valleys descend to U_2 at 628°C. The reaction at U_2 is the reverse of that at U_1, namely $\tau \rightleftharpoons L_U + (Si) + Si_2Te_3$. [78 Leg] give an indicative composition for U_2 as 73 Au–20 Si–7% Te: it is not possible to comment critically on this due to insufficient data. If the above composition for U_2 is accepted then the tie line U_2–Si_2Te_3 lies vertically underneath the tie line U_1–Si_2Te_3 (with the amended composition for U_1 (Fig. 388) of 55 Au–25 Si–20% Te). This produces acceptable sections, except for the 20% Te section where [78 Leg] plot data points on the 628°C horizontal extending below 25% Si. The reactions at U_1 and U_2 are unusual in that alloys pass through a L + (Si) + Si_2Te_3 phase region, undergo peritectic reaction at U_1, suffer the reverse transformation at U_2, and then pass again into a L + (Si) + Si_2Te_3 phase region. To prove this sequence of formation and decomposition reactions of the compound τ, [78 Leg] prepared an unspecified alloy in the U_2–Si–Si_2Te_3 tri-

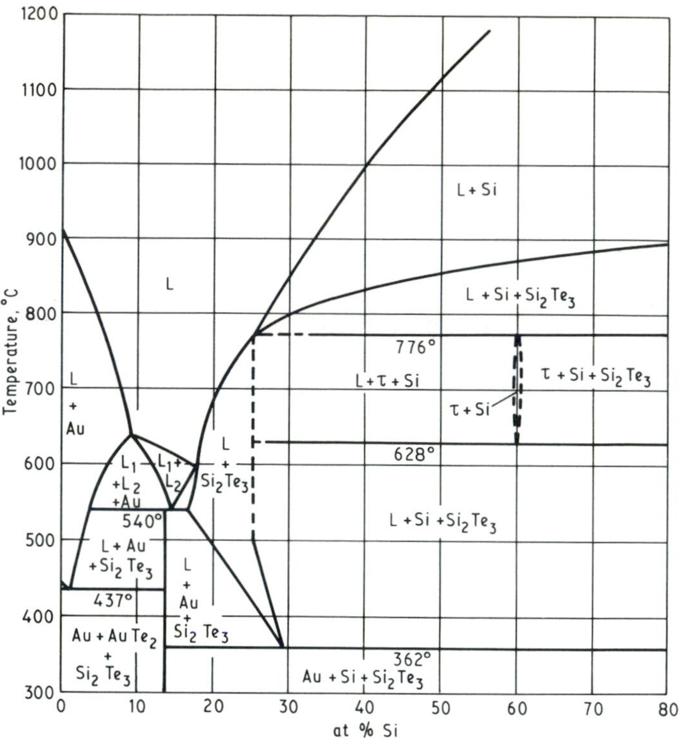

388 **The 20 at.-% Te section**

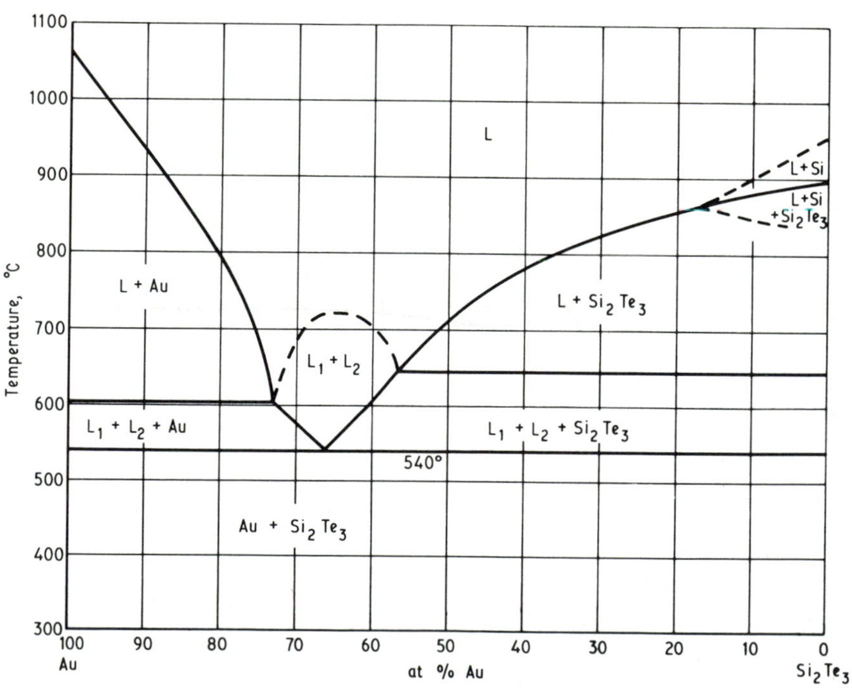

389 **The section Au–Si_2Te_3**

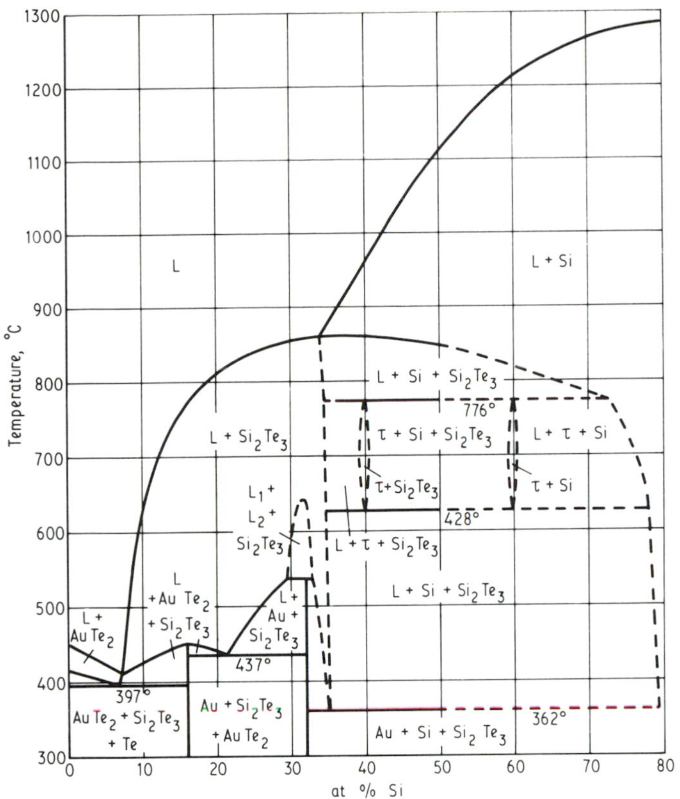

390 The 20 at.-% Au section

391 The section $AuTe_2$–Si

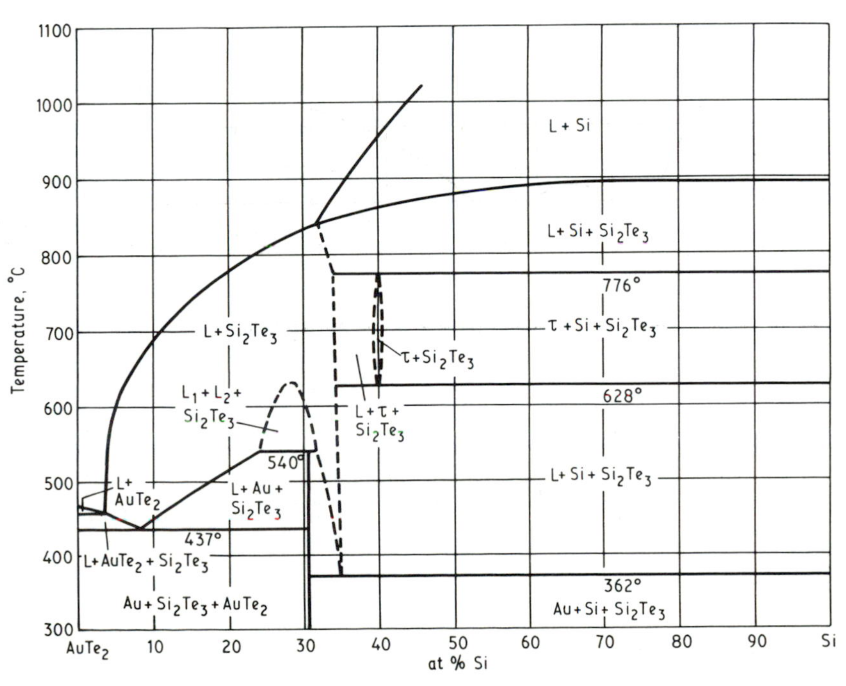

angle and, after differential thermal analysis, observed by X-ray diffraction τ, Si, Au, and Si_2Te_3. The alloy was then heated to above 776°C (i.e. above U_1) and quenched. Only Si, Si_2Te_3, and Au were observed. Finally the alloy was reheated to 650°C and cooled slowly, when again all four phases were observed. [78 Leg] did not do a long term heat treatment at 600°C (i.e. below U_2) to prove that no ternary compound was then observed. Further work needs to be done on the formation of the compound τ and the subsequent reactions in which it takes part.

The final invariant reaction 540°C is connected with a ternary liquid immiscibility region. The reaction at 540°C is: $L_1 + L_2 \rightleftharpoons$ (Au) + Si_2Te_3. The authors quote the compositions of L_1 and L_2 as being known to medium accuracy. L_1 lies on the 40% Te section at 6·6% Si and L_2 on the 20% Te section at 16·6% Si [78 Leg]. The 40% Te section shown in Fig. 392 indicates that the 540°C invariant reaction should be placed at 10·6% Si and not 6·6% Si. The published diagram for the 20% Te section is incorrectly interpreted if the composition of L_2 lies on this section. The original diagram of [78 Leg] includes a L_1 + L_2 phase region which could not appear if L_2 has the composition 63·4 At–16·6 Si–20% Te. However, the Au–Si_2Te_3 section (Fig. 389) indicates a $L_1 + L_2$ phase region stretching from 57% Au to 73% Au. If L_2 lies on the 20% Te section, the $L_1 + L_2$ phase region should not extend as far as 73% Au on the Au–Si_2Te_3 section. Acceptance of the Au–Si_2Te_3 section means that L_2 must lie below 20% Te. Reasonable consistency between the data is obtained if L_2 is assigned a composition 70 Au–16 Si–14% Te. The 20% Te section shown in Fig. 388 has been re-interpreted from the original on this assumption. The $L_1 + L_2$ phase region is shown as extending in temperature from 640 to 600°C, compared with 845 to 620°C [78 Leg]. The value of 845°C is not in accord with the $L_1 + L_2$ phase region plotted on the projection of the liquidus surface by [78 Leg].

Table 119 summarizes the invariant equilibrium reactions, and includes the recommended compositions of the liquid phase associated with each reaction according to [78 Leg], and suggested amendments.

LIQUIDUS SURFACE

The liquidus surface presented in Fig. 394 differs from that published [78 Leg] with respect to the location of U_1, L_1, and L_2, and the positioning of the isotherms in the Au-rich region to recognize the temperatures at which reactions U_1 and U_2 occur. Broken lines are indicative of uncertainty in the data.

ISOTHERMAL SECTIONS

The data does not warrant the conversion of the results from vertical sections to isothermal sections.

REFERENCES

66 Bai: L. G. Bailey: *J. Phys. Chem. Solids*, 1966, **27**, 1593–1598

73 Pet: K. E. Petersen, U. Birkholz and D. Adler: *Phys. Rev. B*, 1973, **8**, 1453–1461

76 Plo: K. Ploog, W. Stetter, A. Nowitzki and E. Schönherr: *Mater. Res. Bull.*, 1976, **11**, 1147–1154

78 Leg: B. Legendre, C. Souleau, Chhay-Hancheng and N. Rodier: *J. Chem. Res. (S)*, 1978, (5), 2139–2168

83 Oka: H. Okamoto and T. B. Massalski: *Bull. Alloy Phase Diagrams*, 1983, **4**, 190–198

84 Oka: H. Okamoto and T. B. Massalski: *Bull. Alloy Phase Diagrams*, 1984, **5**, 172–177

Au–Si–Th

Marazza *et al.* [77 Mar] prepared the ternary compound $ThAu_2Si_2$ by induction melting >99·99% Si and >99.9% Au,Th under argon. The alloy was subsequently annealed in a silica capsule under an inert atmosphere for 1 week at 500°C. Metallographically the alloy was stated to be nearly homogeneous. X-ray powder diffraction analysis established that $ThAu_2Si_2$ has a tetragonal $ThCr_2Si_2$-type structure, tI10, with lattice parameters a = 0·4295 ± 0·0002, c = 1·0347 ± 0·0005 nm and c/a = 2·409. The analogous compounds $ThAu_2Ge_2$ and UAu_2Si_2 have the same crystal structure but [77 Mar] did not succeed in preparing homogeneous samples of UAu_2Ge_2 by the techniques quoted.

REFERENCE

77 Mar: R. Marazza, R. Ferro, G. Rambaldi and G. Zanicchi: *J. Less-Common Metals*, 1977, **53**, 193–197

Au–Si–U

Marazza *et al.* [77 Mar] prepared the ternary compound UAu_2Si_2 by induction-melting >99·99% Si and >99·9% Au,U under argon. The alloy was subsequently annealed in a silica capsule under an inert atmosphere for 1 week at 500°C. Metallographically the alloy was stated to be nearly homogeneous. X-ray powder diffraction analysis established that UAu_2Si_2 has a tetragonal $ThCr_2Si_2$-type structure, tI10, with lattice parameters a = 0·4229 ± 0·0002, c = 1·0250 ± 0·0005 nm and c/a = 2·424. The analogous compounds $ThAu_2Ge_2$ and $ThAu_2Si_2$ have the same crystal structure but [77 Mar] did not succeed in preparing homogeneous samples of UAu_2Ge_2 by the techniques quoted. Buschow and Mooij [86 Bus] prepared UAu_2Si_2 from >99·9% pure elements by arc-melting. The alloy was subsequently wrapped in Ta foil and vacuum annealed in a silica tube for 3 weeks at 800°C. X-ray powder diffraction study confirmed the body-centred tetragonal structure of the $ThCr_2Si_2$ type with lattice parameters a = 0·42141, c = 1·02741 nm, and c/a = 2·438.

REFERENCES

77 Mar: R. Marazza, R. Ferro, G. Rambaldi and G. Zanicchi: *J. Less-Common Metals*, 1977, **53**, 193–197

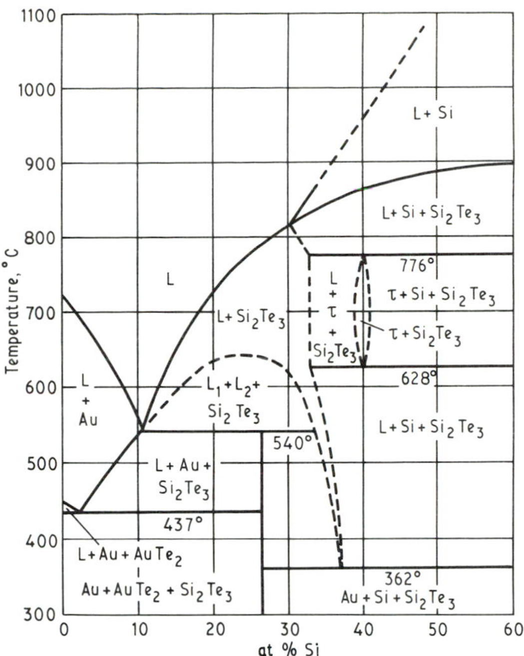

392 The 40 at.-% Te section

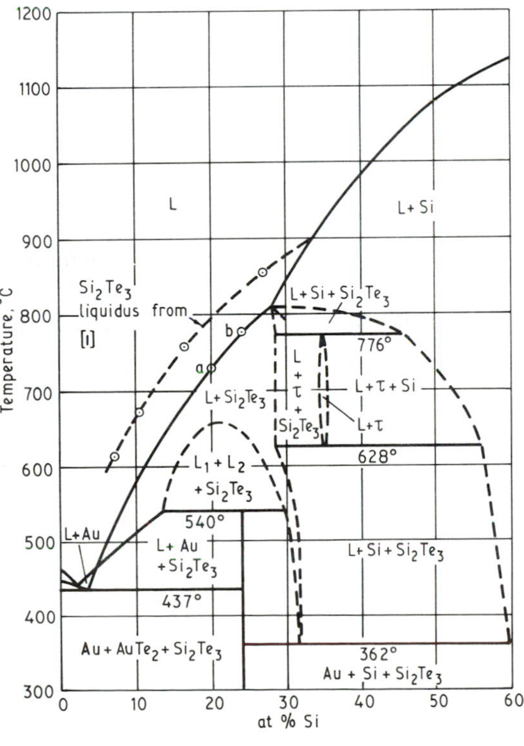

393 The 40 at.-% Au section

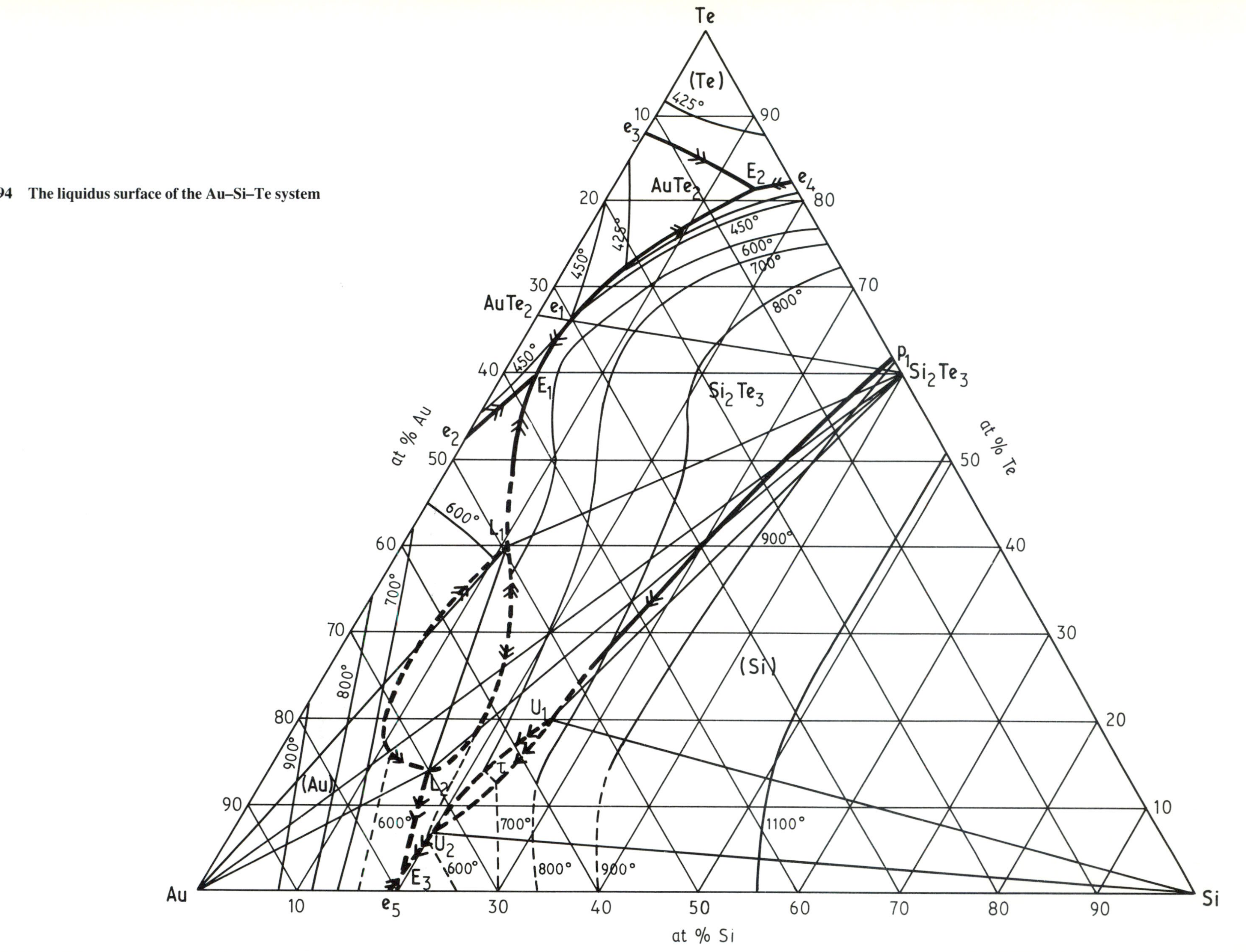

394 The liquidus surface of the Au–Si–Te system

395 Reaction scheme for the Au–Sn–Te system

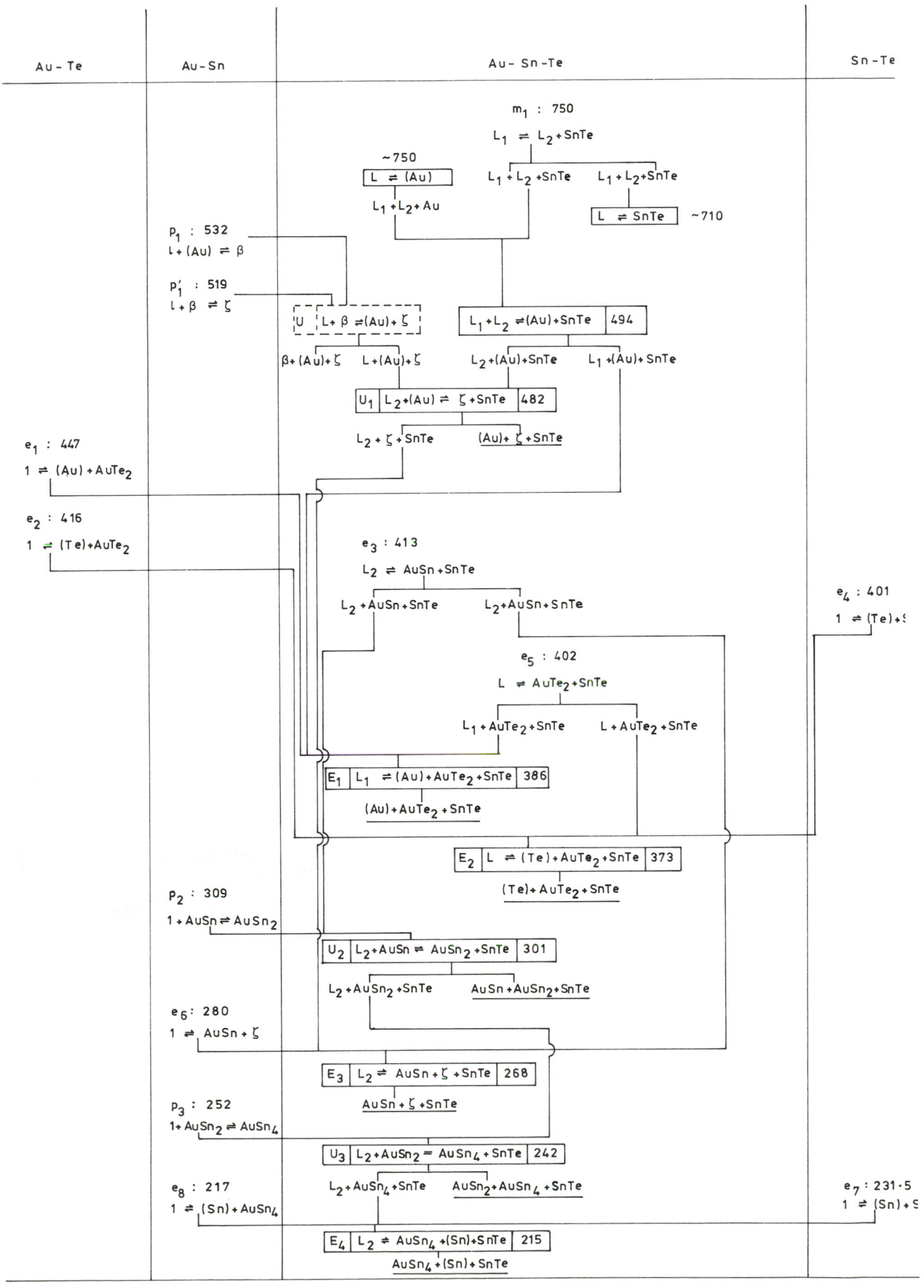

86 Bus: K. H. J. Buschow and D. B. De Mooij: *Philips J. Res.*, 1986, **41**, 55–76

Au–Si–V

In studying solid solutions formed between Cr_3Si-type phases (cP8) von Philipsborn [70 Phi] prepared a $V_3Au_{0\cdot5}Si_{0\cdot5}$ alloy by arc-melting V_3Au and V_3Si on a water-cooled Cu hearth under argon. Samples were annealed 45 days at 550°C, 4 days at 850°C, and 20 days at 1 050°C in evacuated quartz ampoules. The phases present were characterized by the X-ray powder diffraction technique.

The as-cast, 550 and 850°C annealed samples contained three Cr_3Si-type phases with lattice spacings a = 0·478 0, 0·481 9, 0·487 0 ± 0·000 3 nm. In addition a V_3AuO phase with a Cu_3Au-type structure was found. After the 1 050°C anneal more of the V_3AuO phase was found, none of the Cr_3Si-type phases, and two unidentified phases. Without further data it is not possible to interpret these results. The absence of binary compounds in the Au–Si system means that V_3Au and V_3Si should form a stable tie line. The absence of a Cr_3Si-type phase at 1 050°C could indicate the presence of ternary silicide intermetallic compounds that prevent equilibrium of V_3Au and V_3Si at 1 050°C. It should be noted that two of the Cr_3Si-type phases reported in as-cast, 550, and 850°C anneals have lattice parameters similar to V_3Au (0·488 0) and V_3Si (0·472 5).

REFERENCE

70 Phi: H. von Philipsborn: *Z. Kristallogr.*, 1970, **131**, 73–87

Au–Si–Zn

Krompholz [76 Kro] noted that an alloy with the composition AuSiZn was not single phase. No further details were given.

REFERENCE

76 Kro: K. Krompholz and A. Weiss: *Z. Metall.*, 1976, **67**, 400–403

Au–Sn–Te

INTRODUCTION

The sections $AuTe_2$–SnTe and AuSn–SnTe were found to be pseudobinaries [72 Leg] on the basis of examination of 15 alloys on each section by differential thermal analysis with supporting X-ray diffraction and metallographic examination. The $AuTe_2$–SnTe section is a simple eutectic and the AuSn–SnTe section incorporates a liquid miscibility gap associated with a monotectic reaction followed by a eutectic reaction at a lower temperature. The ternary system is therefore divided into three partial systems, the Te–$AuTe_2$–SnTe system, the Au–$AuTe_2$–SnTe–AuSn system, and the Sn–SnTe–AuSn system. As would be anticipated the Te–$AuTe_2$–SnTe partial ternary system is a simple ternary eutectic system [77 Leg] with the ternary eutectic temperature at 373 ± 1°C. This conclusion was based on the study of 40 alloys on three sections through the partial ternary systems using the same techniques as [72 Leg]. The Au–$AuTe_2$–SnTe–AuSn system is complicated by the liquid immiscibility that originates from the AuSn–SnTe pseudobinary section. This partial ternary system was studied by preparing 100 alloys on eight sections, relying on differential thermal analysis as the main experimental technique [75 Leg]. Four invariant reactions were found at temperatures of 503, 496, 398, and 278°C [75 Leg]. Scaling data points from the published sections gives temperatures of 494 ± 8, 482 ± 4, 386 ± 7, and 268 ± 7°C with calculated standard deviation of the data from the quoted means. The compositions of the liquid phase(s) involved in the four invariant reactions are only quoted [75 Leg] for the ternary eutectic reaction $L_{E_1} \rightleftharpoons$ (Au) + $AuTe_2$ + SnTe at 386°C. The original data has been reinterpreted to give an amended composition for E_1 and to produce recommended compositions for the liquid phase(s) entering the other three invariant reactions, i.e. $L_1 + L_2 \rightleftharpoons$ (Au) + SnTe at 494°C, L_{U_1} + (Au) $\rightleftharpoons \zeta$ + SnTe at 482°C, and $L_{E_3} \rightleftharpoons \zeta$ + AuSn + SnTe at 268°C. A discussion of the invariant reactions is contained in the section on invariant equilibria. The Sn–SnTe–AuSn partial ternary system was covered by the study of 20 alloys only [75 Leg]. The liquid immiscibility gap originating from the AuSn–SnTe pseudobinary section closes in a critical tie line, $L_1/L_2 \rightleftharpoons$ SnTe, within this partial ternary system. The three invariant reactions that reflect the Au–Sn binary invariant reactions (l + AuSn $\rightleftharpoons$ $AuSn_2$, l + $AuSn_2 \rightleftharpoons AuSn_4$, and l $\rightleftharpoons$ (Sn) + $AuSn_4$) lie close to the AuSn–Sn binary edge and no definite conclusions can be reached on the liquid compositions associated with the three ternary invariant reactions. Their reaction temperatures are quoted [75 Leg] as L_{U_2} + AuSn $\rightleftharpoons$ $AuSn_2$ + SnTe at 309°C (scaled from the data as 301 ± 7°C), L_{U_3} + $AuSn_2 \rightleftharpoons AuSn_4$ + SnTe at 252°C (scaled as 242 ± 13°C), and $L_{E_4} \rightleftharpoons$ (Sn) + $AuSn_4$ + SnTe at 217°C (scaled from the two data points as 215°C).

The major features of the ternary equilibria are the presence of an extensive region of liquid immiscibility for alloys containing from 2 to 45 at.-% Te and the dominance of SnTe as the primary phase to separate from the liquid alloys in a large area of the ternary system. There is a need for re-examination of the partial ternary system Au–$AuTe_2$–SnTe–AuSn to provide a more definitive version of the equilibria.

BINARY SYSTEMS

The Au–Te system evaluated by [84 Oka 1] is accepted. The Au–Sn system has been evaluated by [84 Oka 2] but considerable uncertainties exist concerning equilibrium reactions for Au-rich alloys. The Au–Sn diagram accepted in this assessment is due to [85 Leg, 87 Leg]. It differs in one major respect from the diagram proposed by [84 Oka 2]. Two peritectic reactions were identified, corresponding to the reactions l + (Au) $\rightleftharpoons$ β at 532°C and l + β $\rightleftharpoons$ ζ at 519°C [87 Leg]. The mechanism for formation of a γ phase with 17 at.-% Sn and the low-temperature solid-state transformations require further study. The Sn–Te system evaluated by [86 Sha] is accepted.

SOLID PHASES

No ternary compounds were observed. Table 120 summarizes crystal structure data for the binary compounds.

Table 120. Crystal structures

Solid phase	*Prototype*	*Lattice designation*
(Au)	Cu	cF4
(Sn)	Sn	tI4
(Te)	γSe	hP3
$AuTe_2$	$AuTe_2$	mC6
SnTe	NaCl	cF8
ζ	Mg	hP2
AuSn	NiAs	hP4
$AuSn_2$	—	oP24
$AuSn_4$	$PtSn_4$	oC20

PSEUDOBINARY SYSTEMS

[72 Leg] found the sections $AuTe_2$–SnTe and AuSn–SnTe to be pseudobinaries. Fifteen alloys were prepared on each section from 99·999% pure elements. All alloys were encapsulated in evacuated silica tubes and heated for 48 h at 250°C, 48 h at 500°C, and 3 h at 1 100°C before cooling to room temperature at 2°C min^{-1}. They were studied by DTA using heating and cooling rates of 2°C min^{-1}, X-ray diffraction, and metallography. The $AuTe_2$–SnTe section (Fig. 396) is a pseudobinary eutectic system with a eutectic composition at 12·1 at.-% Sn and a eutectic temperature of 402 ± 3°C. The AuSn–SnTe section (Fig. 397) exhibits a liquid immiscibility gap associated with the monotectic reaction $L_1 \rightleftharpoons L_2$ + SnTe at 750 ± 4°C. L_1 and L_2 have compositions approximating to 41·5 at.-% Te and 8·2 at.-% Te respectively. The temperature over which liquid immiscibility occurs is uncertain. A eutectic reaction was located at 413 ± 3°C, the eutectic composition being quoted [72 Leg] as 0·66 at.-% Te, but this value is uncertain.

INVARIANT EQUILIBRIA

The Te–$AuTe_2$–SnTe partial ternary system was studied [77 Leg] by preparing 40 alloys on three sections (66·67 at.-% Te, 16·67 at.-% Au and $AuTe_2$–6·67 Sn, 93·33 Te). Alloys were examined by DTA, using heating and cooling rates of 2°C min^{-1}, X-ray diffraction, and metallography. This partial ternary system is a simple ternary eutectic system. The eutectic temperature is 373°C ± 3°C. It was previously quoted, without supporting data, as 378°C [75 Leg]. The eutectic composition was derived as the point of intersection of the tie lines SnTe–a, $AuTe_2$–b, and Te–c when extrapolated. Points a, b, and c (Figs 398, 399, and 400 respectively) are the intersection points of the three sections with the tie lines SnTe–E_2, AuTe–E_2, and Te–E_2 lying on the ternary eutectic plane at 373°C. Table 121 summarizes the quoted compositions [77 Leg] for points a, b, and c and for the points u, v, and w which represent the intersection of the liquidus surfaces in each of the three sections. Points u and v will lie on the monovariant curve e_5E_2 and point w will lie on the monovariant curve e_2E_2. Amended values for the compositions of points a, b, c, u, v, and w were produced by redrawing the data of [77 Leg]. The need to reinterpret the published data arose from inconsistencies in the data. Point b is quoted as having a composition on the 16·67 at.-% Au section of 10 at.-% Sn, but [77 Leg] produced this section with a composition for b at 6·67 at.-% Sn. Point c on the $AuTe_2$–6·67 at.-% Sn, 93·33 at.-% Te section is quoted [77 Leg] as 6·10 at.-% Sn and 89·8 at.-% Te. This composition does not lie on the section whose equation is (Te) = 4 (Sn) + 66·67. Reinterpretation of the data leads to a composition of the ternary eutectic E_2 at 9·1 at.-% Au, 10·9 at.-% Sn, 80 at.-% Te. Liquidus isotherms for the partial ternary system Te–$AuTe_2$–SnTe are given in Fig. 401. In compiling the isotherms for 390 and 400°C data was taken from the surfaces of secondary separation which are intersected by the three sections studied [77 Leg] along curves such as ua, vb, and wc. For example the composition corresponding to 390°C on the curve ua (Fig. 398), when joined to SnTe and extrapolated, will meet the monovariant curve e_5E_2 at a composition corresponding to a temperature of 390°C.

The remainder of the ternary system, Au–$AuTe_2$–SnTe–Sn, is divided into two partial ternary systems by the pseudobinary system AuSn–SnTe. A total of 120 alloys were prepared [75 Leg] from 99·999% pure elements in addition to the 30 alloys used to define the pseudobinary systems $AuTe_2$–SnTe and AuSn–SnTe. Alloys were examined by DTA using heating and cooling rates of 2 and 5°C min^{-1}, X-ray diffraction, and metallography. Invariant temperatures are quoted [75 Leg] to a precision of ±3°C; temperatures for primary and secondary separation from the melt are not known with good

Table 121. Compositions of the intersection points of vertical sections in the partial ternary system Te–$AuTe_2$–SnTe

Section	*Point*	*Composition*					
		According to [77 Leg]			*Assessed*		
		Au	*Sn*	*Te*	*Au*	*Sn*	*Te*
66·67 Te	a	5	28·33	66·67	5	28·33	66·67
	u	20	13·33	66·67	20·33	13	66·67
16·67 Au	b	16·67	10	73·33	16·67	7·5	75·83
	v	16·67	13·33	70	16·67	12·8	70·53
$AuTe_2$–6·67 Sn, 93·33 Te	c	4	6·1	89·8	4·8	5·7	89·5
	w	11·66	4·4	84	10·8	4·5	84·7
—	E_2	9·1	12·3	78·6	9·1	10·9	80

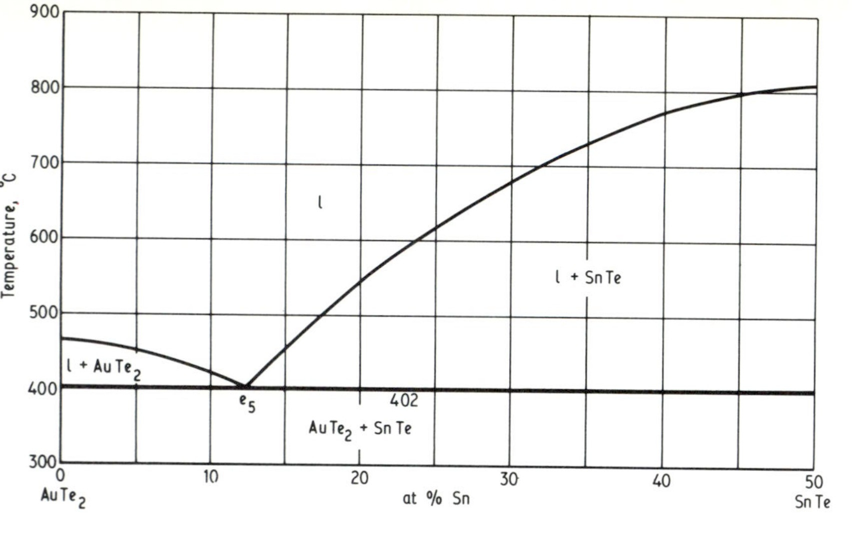

396 The $AuTe_2$–SnTe section

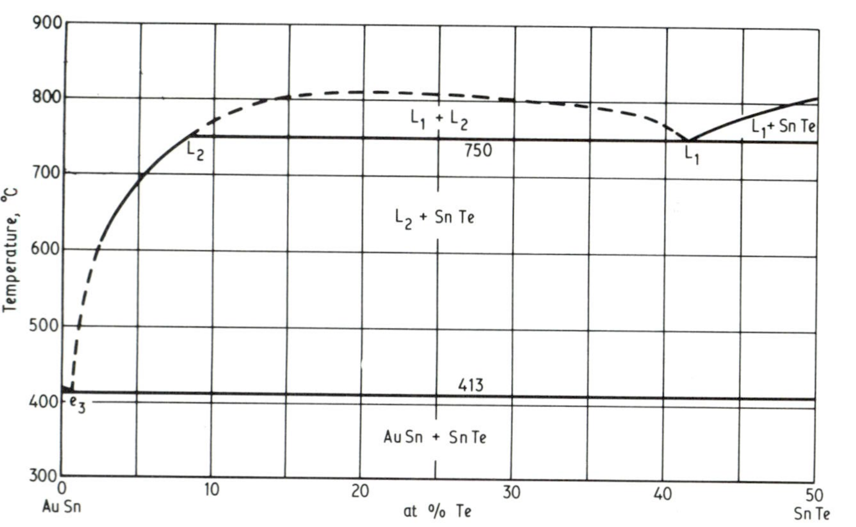

397 The AuSn–SnTe section

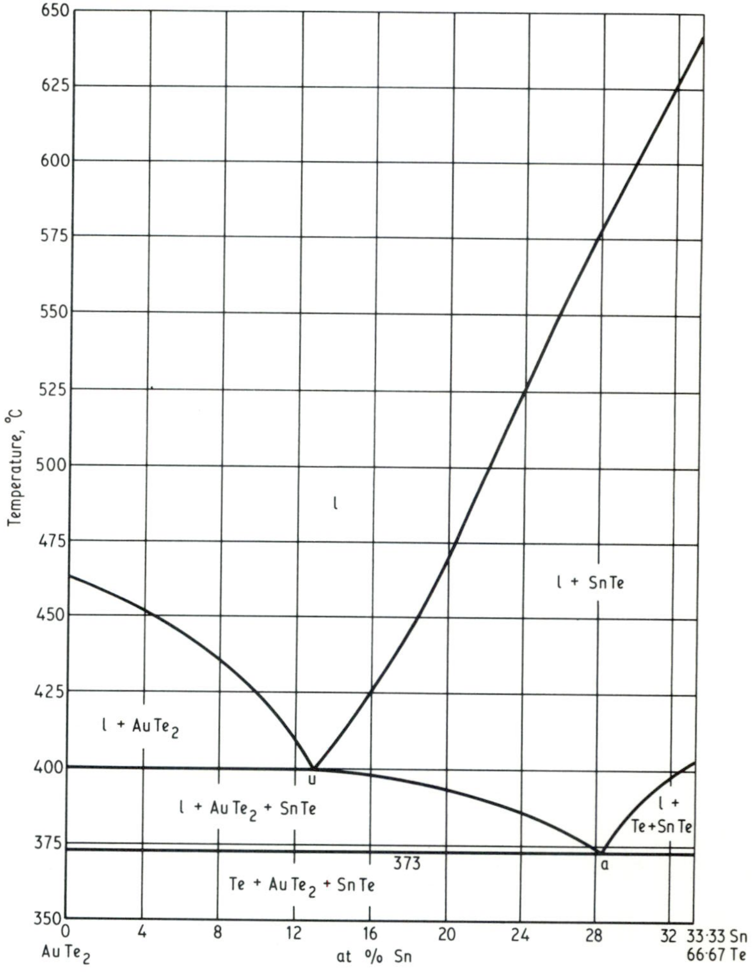

398 The 66·67 at.-% Te section

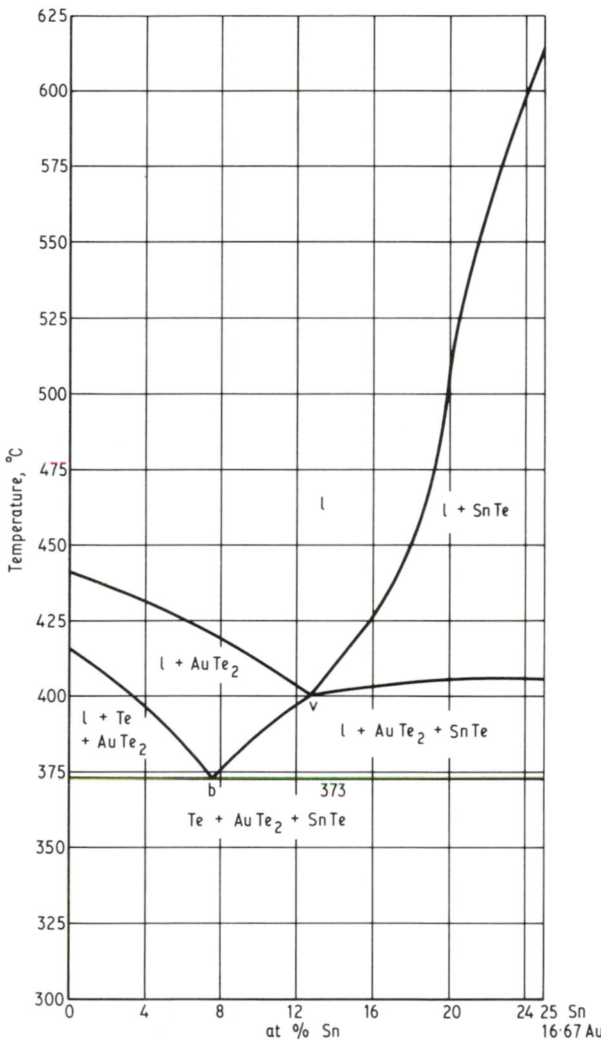

399 The 16·67 at.-% Au section

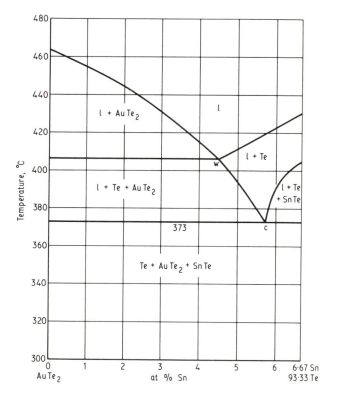

400 The $AuTe_2$–6·67 at.-% Sn, 93·33 at.-% Te section

401 Liquidus surface of the partial ternary system Te–AuTe$_2$–SnTe

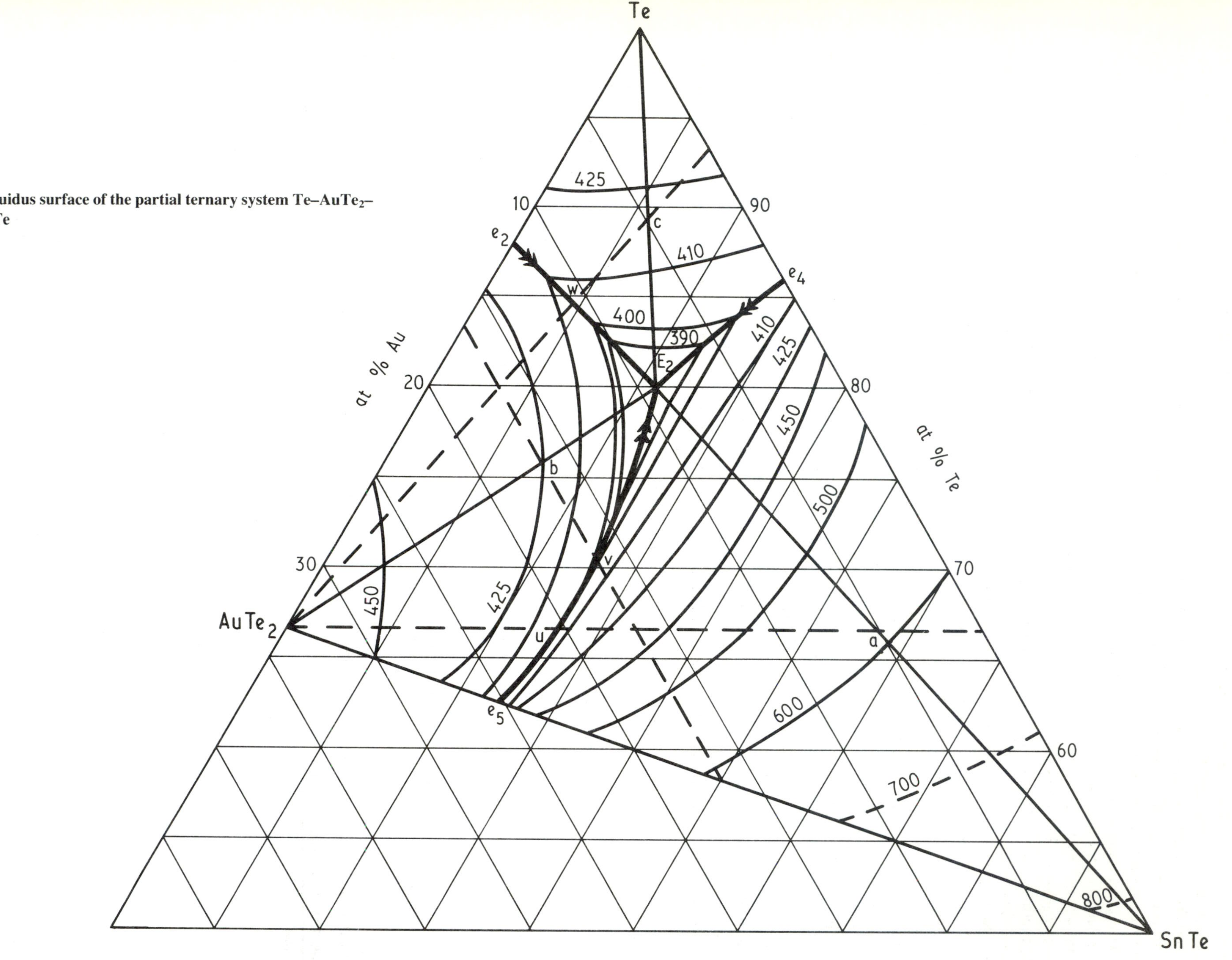

precision because of the weak thermal effects associated with separation of solid(s) from the melt. Eight polythermal sections are presented [75 Leg] at 10, 20, 33·33, 50, 70, and 80 at.-% Au, 40 at.-% Te, and along Au–SnTe.

The Au–$AuTe_2$–SnTe–AuSn partial ternary system contains a ternary liquid miscibility gap which originates from the pseudobinary monotectic reaction on the AuSn–SnTe section and from a critical tie line in the Au-rich region where L $\rightleftharpoons$ (Au) becomes a $L_1 + L_2$ + (Au) three-phase region as the temperature falls from the critical temperature. An invariant ternary reaction $L_1 + L_2 \rightleftharpoons$ (Au) + SnTe occurs at 503 ± 3°C [75 Leg], but it is scaled from the authors' data as 494 ± 8°C (Fig. 402). In the triangle Au–$AuTe_2$–SnTe a monivariant curve L_1E_1 falls to the ternary eutectic at E_1 where the reaction is $L_{E_1} \rightleftharpoons$ (Au) + AuTe + SnTe. Legendre *et al.* [75 Leg] quote a 398 ± 3°C eutectic temperature, but this does not comply with data plotted on six of eight vertical sections. Analysis of the data suggests a eutectic temperature for E_1 of 386 ± 7°C. The eutectic composition at E_1 is quoted as 33·33 at.-% Au, 16·67 at.-% Sn, 50 at.-% Te [75 Leg]. The tie line $AuTe_2$–E_1 would lie on the 33·33 at.-% Au section and necessarily the liquidus curve for primary separation of $AuTe_2$ would descend from the melting point of $AuTe_2$ to E_1 without any secondary separation from the melt. [75 Leg] indicate the $AuTe_2$ liquidus curve as being intersected by the Au liquidus curve before the eutectic E_1 is reached, and they also plot points for the secondary separation of Au + $AuTe_2$ from the liquid. As far as can be judged from the data, E_1 lies at a composition 36 at.-% Au, 16 at.-% Sn, 48 at.-% Te, leading to amendment of the 33·33 at.-% Au section (Fig. 403).

The locations of L_1 and L_2 are uncertain. They are not quoted by [75 Leg]. They should be determined by plotting the compositions on the tie lines Au–L_1, L_1–SnTe, SnTe–L_2, and L_2–Au which correspond to the extremeties of the 494°C invariant horizontal in the sections studied. A further check on the compositions of L_1 and L_2 can be obtained by plotting the intersection point of the surfaces of secondary separation ($L_1 + L_2$ + Au and $L_1 + L_2$ + SnTe) since such intersection points will lie on the tie line L_1L_2. Study of the appropriate sections [75 Leg] does not allow any definite conclusions as to the compositions of L_1 and L_2. It is suggested that L_1 has a composition 28 at.-% Au, 27 at.-% Sn, 45 at.-% Te, and L_2 a composition 80 at.-% Au, 18 at.-% Sn, 2 at.-% Te. These compositions are based to a large extent on an estimate of the tie line L_1–L_2 and, for L_1, the intersection of L_1L_2 with the tie line L_1–SnTe. L_2 is considered to lie on the 80 at.-% Au section and extrapolation of the tie line from L_1 to the 80 at.-% Au section provides an estimate of the composition of L_2. Table 122 summarizes the original tabulated data [75 Leg] and the recommended amendments for the compositions of the intersection points of vertical sections with the 494°C invariant plane. The amended data has been used to construct modified vertical sections. These have then been used to construct liquidus isotherms for the partial ternary system Au–$AuTe_2$–SnTe–AuSn (Fig. 402). The inclusion of the relevant three-phase triangles in Fig. 402 allows the 600 and 700°C isothermal sections to be deduced from the amended data for the vertical sections. The reinterpretation placed on the original data [75 Leg] necessarily implies ignoring some of the original thermal effects. Figures 404–410 are vertical sections that are consistent between themselves and consistent with the liquidus surface presented in Fig. 402. In Figs 404–410 data points from [75 Leg] that do not correspond with the recommended reinterpretation of the original data are circled. The 70 at.-% Au section (Fig. 405) was originally drawn [75 Leg] with the boundaries of the $L_1 + L_2$ + Au phase region displaced to lower Sn contents. This interpretation is not consistent with the data points for an alloy with 70 at.-% Au, 15 at.-% Sn on the Au–SnTe section (Fig. 409). Nor is it consistent with the alloy compositions over which liquid immiscibility was found. [75 Leg] indicate the $L_1 + L_2$ phase region stretching to within 2·5% of the Au–Te binary edge for the 70 at.-% Au section. In no other section was liquid immiscibility found so close to the Au–Te binary edge. The 50 at.-% Au section (Fig. 406) included thermal effects on the Au-rich side of the Au–L_1 tie line which were ignored by [75 Leg] in their interpretation of the equilibria. The circled points on the Sn-rich side, corresponding to intersection of the section with the L_2–SnTe tie line (Table 122), are considered to be non-

Table 122. Compositions of the intersection points of vertical sections with the 494°C invariant plane

Section	*Tie-lines intersected by 494°C invariant plane*	*Composition of intersection points*					
		According to [75 Leg]			*Assessed*		
		Au	*Sn*	*Te*	*Au*	*Sn*	*Te*
80 Au	Au–L_1	80	8·3	11·6	80	7·5	12·5
	Au–L_2	80	18·3	1·6	80	18	2·0
70 Au	Au–L_1	70	10·8	19·2	70	11·25	18·75
	L_2–SnTe	70	25·3	4·6	70	22	8
50 Au	Au–L_1	50	18·6	31·3	50	18·75	31·25
	L_2–SnTe	50	36·6	13·3	50	30	20
33·33 Au	Au–L_1	33·33	23·33	43·33	33·33	25	41·67
	L_2–SnTe	33·33	43·33	23·33	33·33	36·67	30
20 Au	L_1–SnTe	20	33·33	46·66	20	33·57	46·43
	L_2–SnTe	20	41·66	38·33	20	42	38
10 Au	L_1–SnTe	10	43·33	46·66	10	41·8	48·2
	L_2–SnTe				10	46	44
40 Te	Au–L_1	40	20	40	36	24	40
	L_2–SnTe	16·66	43·33	40	16·67	43·33	40

402 Liquidus surface of the partial ternary system Au–AuTe$_2$–SnTe–Sn

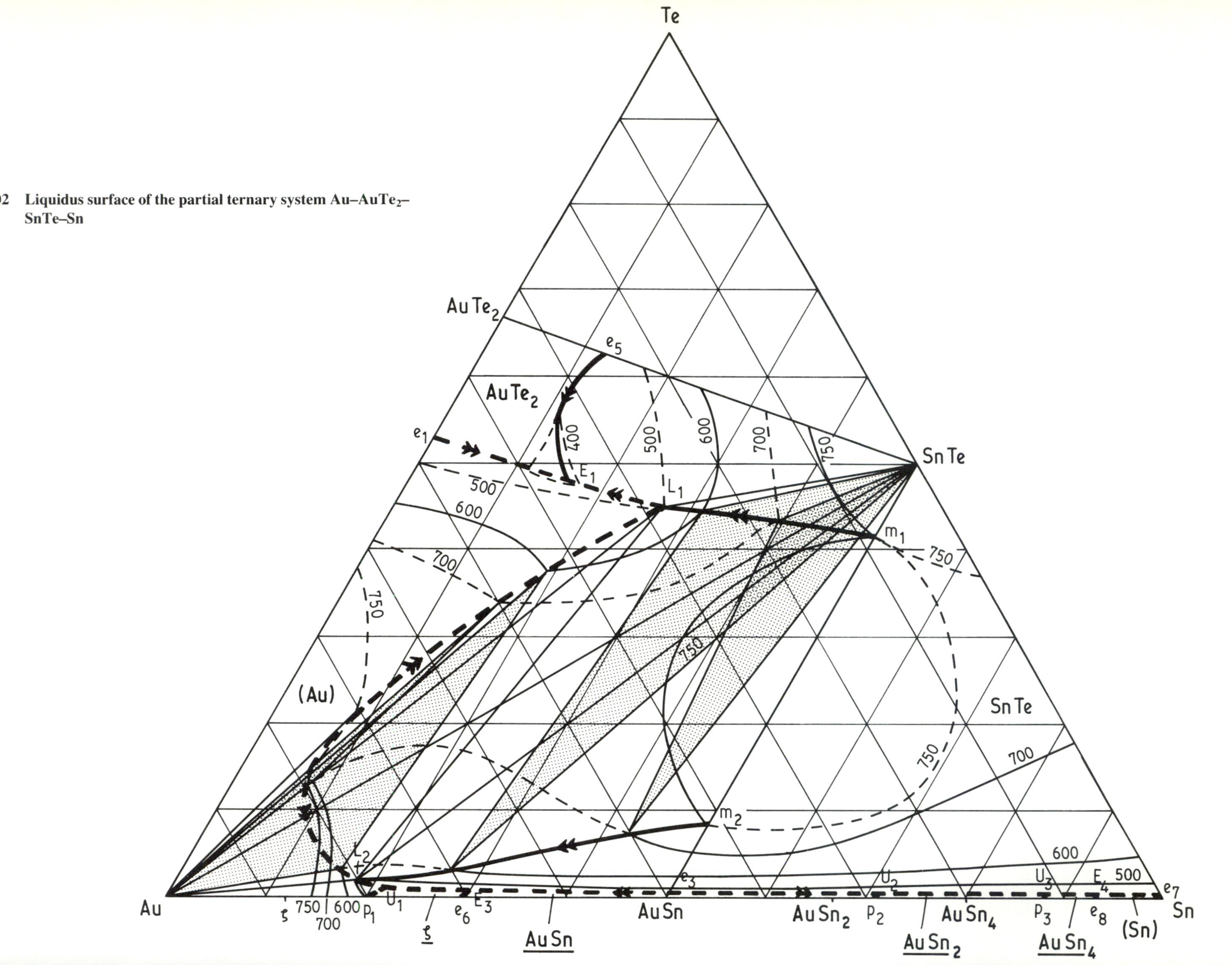

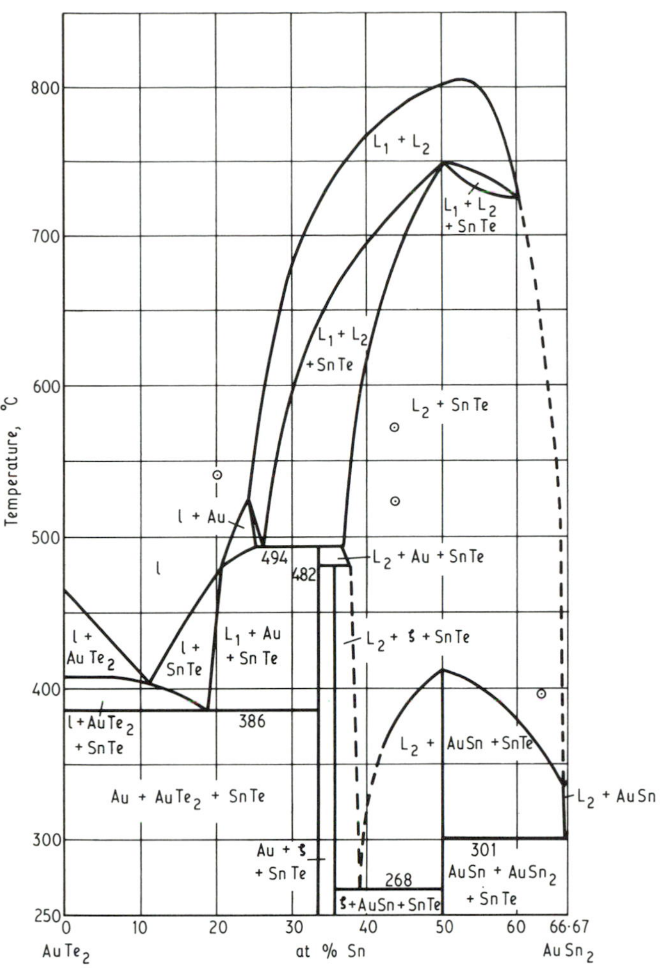

403 The 33·33 at.-% Au section

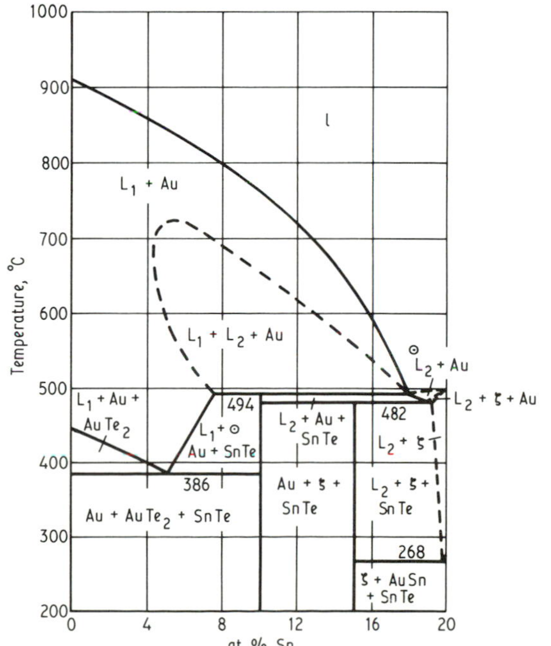

404 The 80 at.-% Au section

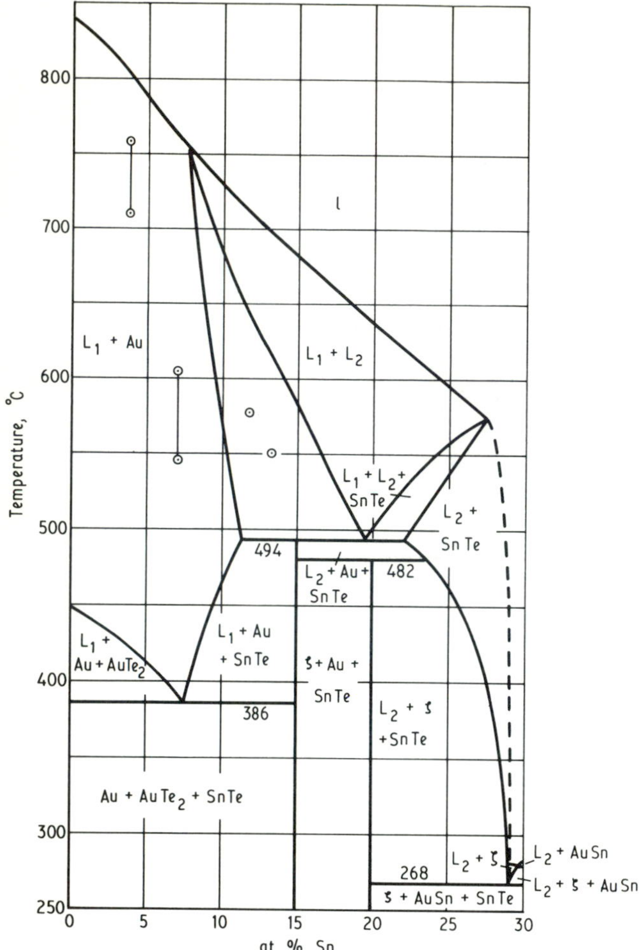

405 The 70 at.-% Au section

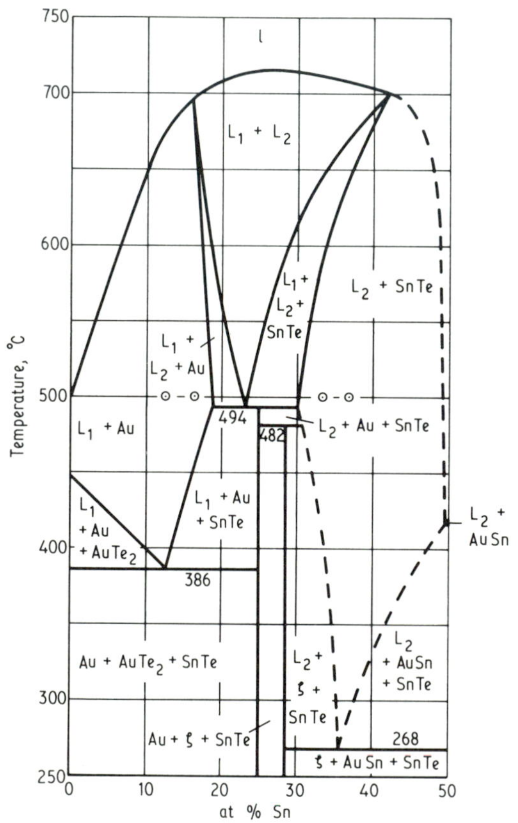

406 The 50 at.-% Au section

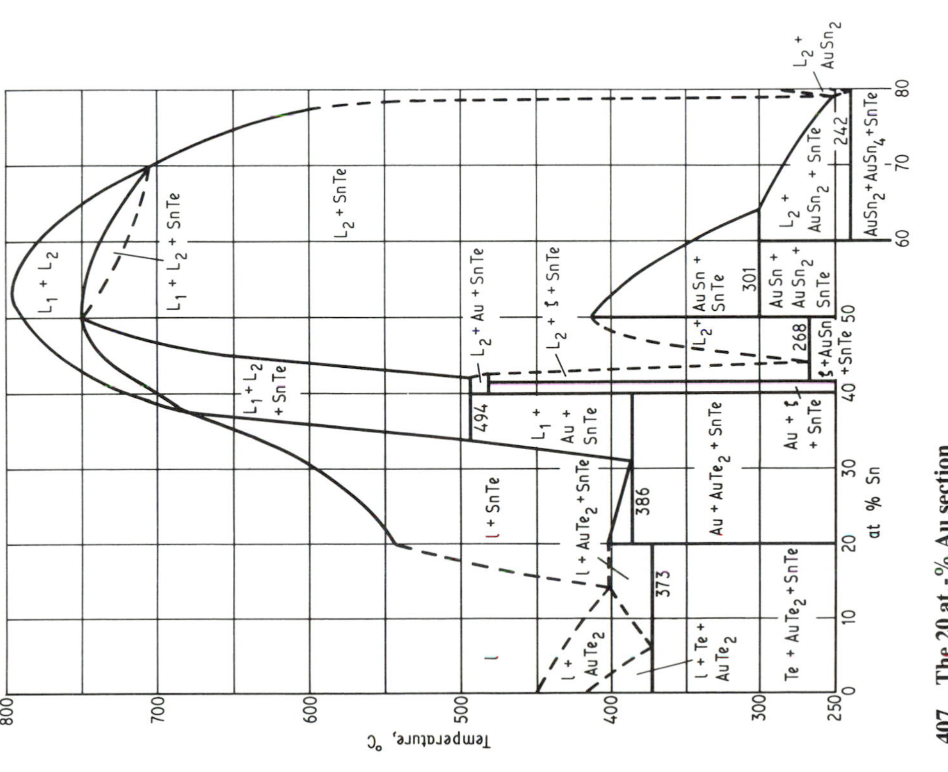

407 The 20 at.-% Au section

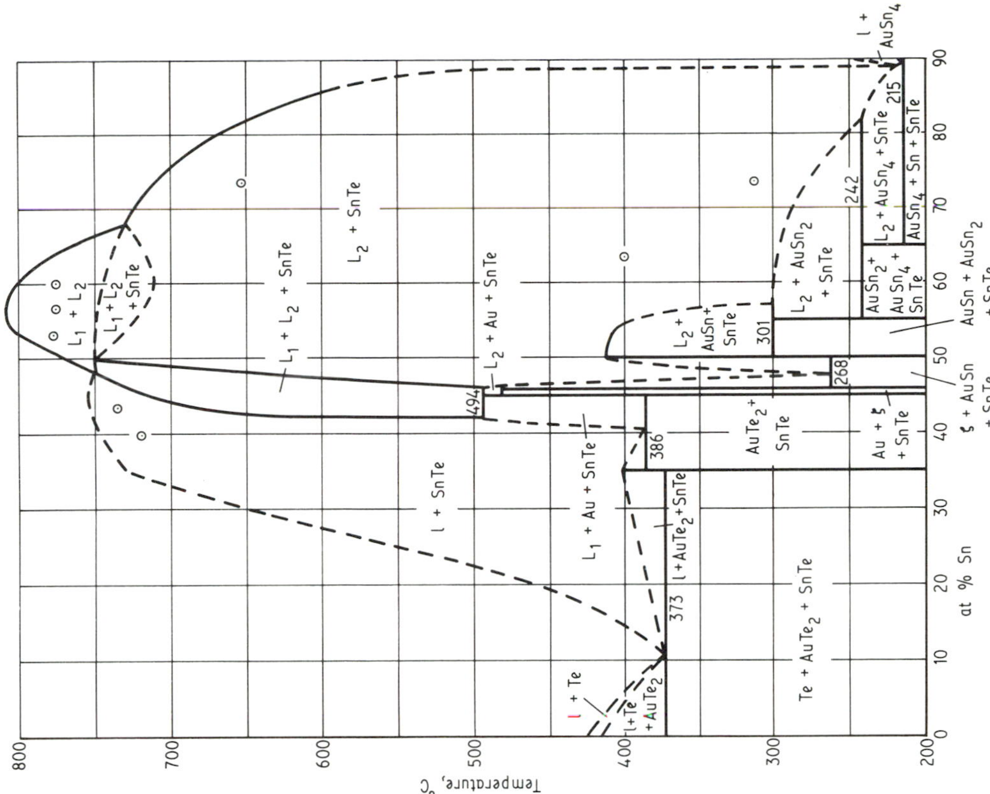

408 The 10 at.-% Au section

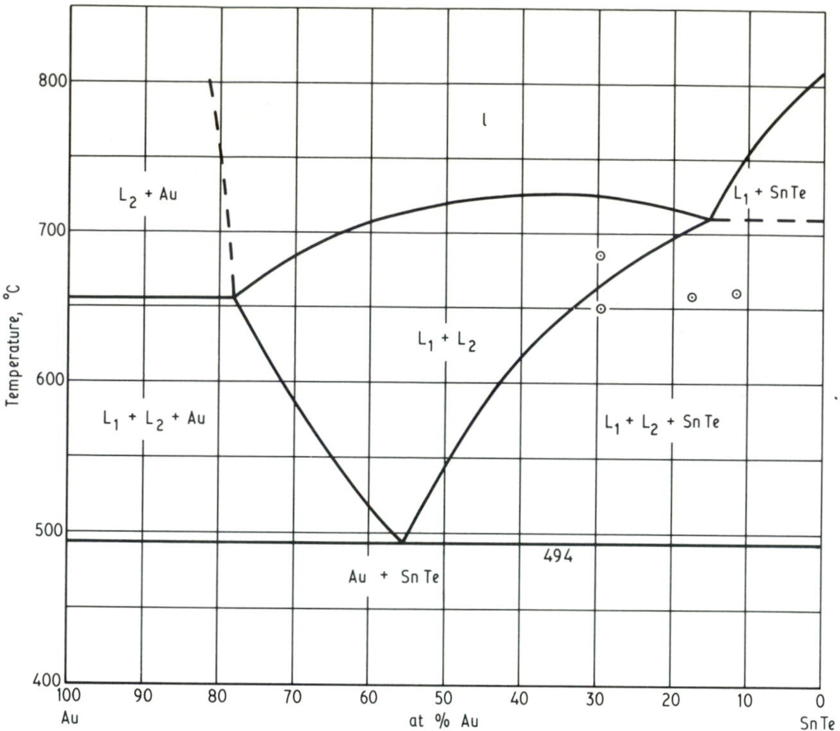

409 The Au–SnTe section

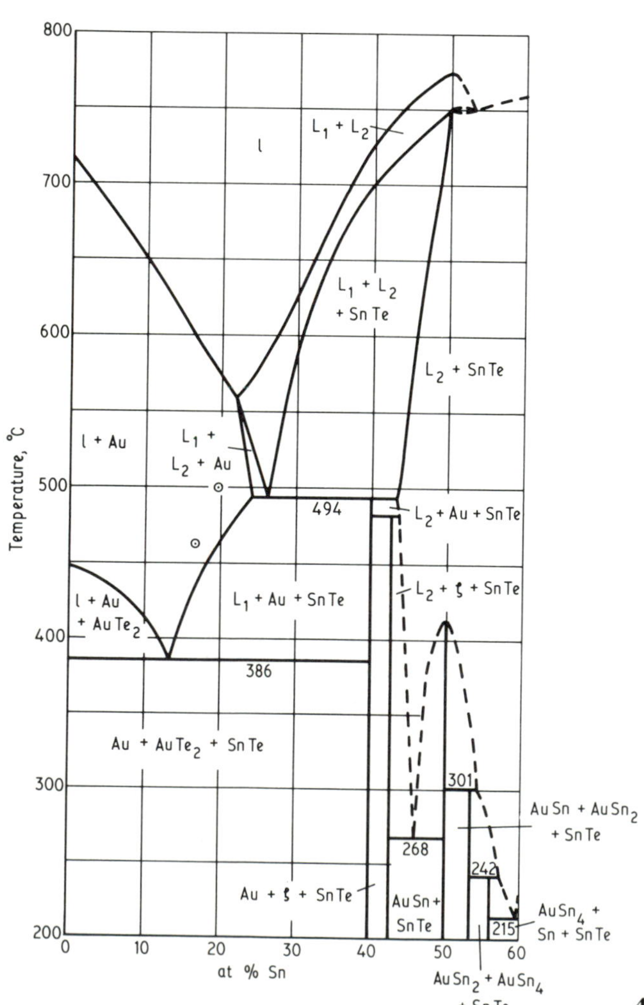

410 The 40 at.-% Te section

equilibrium thermal effects, although they were treated as equilibrium arrests by [75 Leg]. The 33·33 at.-% Au section (Fig. 403) presents problems in interpretation. [75 Leg] regarded L_1 as lying on this section. As the section presented by [75 Leg] vitiates this conclusion, it is to be anticipated that any reinterpretation of the original data will entail considerable departure from the original section. Figure 403 indicates, by the circled data points, the thermal effects not accounted for in the amended section. The two thermal effects in an alloy containing 43·5 at.-% Sn were considered [75 Leg] to embrace a phase field between $L_1 + L_2 +$ SnTe and $L_2 +$ SnTe. No such phase field can exist. The thermal effect in the 63 at.-% Sn alloy was regarded [75 Leg] as a liquidus point on the AuSn phase region. This cannot be so since the eutectic on the AuSn–SnTe pseudobinary section contains ~0·66 at.-% Te and the ternary reactions U_2, U_3, and E_4 (Fig. 402) will all have liquid compositions close to the Au–Sn binary edge. The 20 at.-% Au section (Fig. 407) is similar to that proposed by [75 Leg]. The 10 at.-% Au section (Fig. 408) differs in several respects from the original. The Sn-rich part has been considerably amended in so far as [75 Leg] indicated that the section intersected the liquidus surfaces for the separation of SnTe, AuSn, $AuSn_2$, and $AuSn_4$. If this were so the monovariant curves e_3U_2, p_2U_2, and p_2U_3 would necessarily intersect the 10 at.-% Au section at Sn contents that would place U_2 and U_3 close to the Sn–Te binary edge. The 301°C invariant reaction at U_2 is shown extending to 64 at.-% Sn in the 20 at.-% Au section (Fig. 407). Accepting this value implies that the 301°C invariant plane extends to 57 at.-% Sn in the 10 at.-% Au section (Fig. 408) and not to about 85 at.-% Sn as drawn by [75 Leg]. The thermal effects at 400°C for the 63·5 at.-% Sn alloy and at 312°C for the 73·5 at.-% Sn alloy are considered to be spurious (Fig. 408). The liquidus point at 652°C for the 73·5 at.-% Sn alloy is also well below the true liquidus temperature. The three thermal effects at 775°C are spurious. They were considered by [75 Leg] to lie on the $L_1 + L_2 +$ SnTe surface of secondary separation. As the 750°C monotectic reaction along the AuSn–SnTe section is depressed in temperature in the AuSn–SnTe–Sn partial ternary system, the $L_1 + L_2 +$ SnTe phase field must lie below 750°C. It is shown (Fig. 408) as having a minimum at about 710°C. This is considered to be an approximation to the temperature of the critical tie line connecting L_1/L_2 with SnTe in the AuSn–SnTe–Sn partial ternary system. Figure 409 represents the amended Au–SnTe section. In the original [75 Leg] the L_1 + SnTe phase region was shown as extending to about 27 at.-% Au. This is not consistent with the course of the monovariant curve m_1L_1, nor with the 40 at.-% Te section (Fig. 410), where the alloy with 40 at.-% Te, 20 at.-% Au is shown to separate $L_1 + L_2$ from the liquid on cooling, whereas [75 Leg] indicate this alloy as separating SnTe on the Au–SnTe section. The circled points in Fig. 409 are not consistent with the other data, e.g. the primary and secondary arrest points for the 30·5 at.-% Au alloy are not consistent with arrest temperatures for the 33·33 at.-% Au alloy as read from Fig. 403. The 40 at.-% Te section (Fig. 410) is similar to that presented by [75 Leg] with the exception of the extent of the 494°C invariant plane on the Au-rich side (see also Table 120).

Figure 395 summarizes the invariant equilibria in the form of a reaction scheme. Figure 402 indicates the reaction originating from p_1 in the Au–Sn binary. To avoid a confusion of lines, the curve originating from p_1' and running to U has been omitted and only the curve p_1U_1 drawn.

LIQUIDUS SURFACE

The liquidus surface of the Te–$AuTe_2$–SnTe partial ternary system (Fig. 401) is reasonably characterized. The ternary eutectic point E_2 is considered to lie at the same Au content as proposed by [77 Leg], but after a lower Sn content. The liquidus surface of the Au–$AuTe_2$–SnTe–Sn, partial ternary system (Fig. 402) is not well characterized and the isotherms drawn with broken lines are indicative of uncertainty in the data. The composition of the ternary eutectic point E_1 was amended to 36 at.-% Au, 16 at.-% Sn, 48 at.-% Te, but this is not to be regarded as definitive. The compositions of the invariant points L_1, L_2, U_1, E_3, U_2, U_3, and E_4 were not specified by [75 Leg], but the original data has been assessed to provide a reasonably consistent set of data. The liquidus surface derived from the reinterpretation of the data (Fig. 402) requires verification by further experiment.

ISOTHERMAL SECTIONS

The 600 and 700°C isothermal sections can be derived from the three-phase tie triangle drawn on the liquidus surface (Fig. 402).

REFERENCES

72 Leg: B. Legendre, C. Souleau and J.-C. Rouland: *Compt. rend.*, 1972, **275C**, 805–808

75 Leg: B. Legendre, R. Ceolin and C. Souleau: *Bull. Soc. Chim. France*, 1975, (11–12), 2475–2480

77 Leg: B. Legendre and C. Souleau: *Compt. rend.*, 1977, **284C**, 739–742

84 Oka 1: H. Okamoto and T. B. Massalski: *Bull. Alloy Phase Diagrams*, 1984, **5**, 172–177

84 Oka 2: H. Okamoto and T. B. Massalski: *Bull. Alloy Phase Diagrams*, 1984, **5**, 492–503

85 Leg: B. Legendre, Chhay Hancheng and A. Prince: *Bull. Soc. Chim. France*, 1985, I-50–I-57

86 Sha: R. C. Sharma and Y. A. Chang: *Bull. Alloy Phase Diagrams*, 1986, **7**, 72–80

87 Leg: B. Legendre, Chhay Hancheng, F. Hayes, C. A. Maxwell, D. S. Evans and A. Prince: *Mater. Sci. Technol.*, 1987, **3**, 875–876

Au–Sn–V

In studying solid solution formation between Cr_3Si-type phases (cP8) von Philipsborn [70 Phi] prepared a $V_3Au_{0.5}Sn_{0.5}$ alloy by arc-melting V_3Au and V_3Sn on a water-cooled Cu hearth under argon. Samples were annealed for 45 days at 550°C, 4 days at 850°C, and 20 days at 1 050°C in evacuated quartz ampoules. The phases present were characterized by the X-ray powder diffraction technique.

The as-cast sample contained a Cr_3Si-type phase with lattice parameter $a = 0{\cdot}493 \pm 0{\cdot}001$ nm and an unidentified phase. The samples annealed at 550 and 850°C had the same phases as the as-cast sample, plus a V_3AuO phase with a Cu_3Au-type structure. The 1 050°C annealed sample contained more of the V_3AuO phase, a Cr_3Si-type phase with $a = 0{\cdot}497 \pm 0{\cdot}001$ nm, and an unidentified phase. Without further data it is not possible to interpret these results.

REFERENCE

70 Phi: H. von Philipsborn: *Z. Kristallogr.*, 1970, **131**, 73–87

Au–Sn–Zn

Wilkens and Schubert [57 Wil] used the X-ray powder diffraction method on two ternary alloys, annealed for 8 days at 200°C, and showed that both alloys were in the phase region for the doubly-faulted $L1_2$-type structure based on Au_4Zn (Fig. 411).

Krompholz and Weiss [76 Kro] noted that an alloy with the composition AuSnZn was not single phase; no further details were given.

REFERENCES

57 Wil: M. Wilkens and K. Schubert: *Z. Metall.*, 1957, **48**, 550–557

76 Kro: K. Krompholz and A. Weiss: *Z. Metall.*, 1976, **67**, 400–403

Au–Sn–Zr

In studying solid solution formation between Cr_3Si-type phases (cP8) von Philipsborn [70 Phi] prepared a $Zr_3Au_{0.5}Sn_{0.5}$ alloy from the elements (Zr > 99·8%; Au, Sn > 99·9% purity) by arc-melting on a water-cooled Cu hearth under argon. Samples were annealed for 4–5 days at 550°C, 30 days at 800°C, and 9 days at 1 050°C. The phases present were characterized by the X-ray powder diffraction technique.

The as-cast, 550, 800, and 1 050°C annealed samples all showed the same structure; a Cr_3Si-type phase with lattice parameter $a = 0{\cdot}556 \pm 0{\cdot}001$ nm and an unidentified phase.

REFERENCE

70 Phi: H. von Philipsborn: *Z. Kristallogr.*, 1970, **131**, 73–87

Au–Te–Tl

Wegener [65 Weg] published the only study of Au–Te–Tl alloys, and it was limited to the preparation and characterization of the ternary compound AuTeTl. This ternary compound forms by peritectic reaction at a temperature of the order of 450°C, and it is postulated [65 Weg] that the liquid reacts with (Au) and a Te–Tl compound to form AuTeTl. The crystal structure is orthorhombic with $a = 0{\cdot}901\,5 \pm 0{\cdot}000\,09$, $b = 0{\cdot}891\,2 \pm 0{\cdot}000\,9$, $c = 0{\cdot}786\,8 \pm 0{\cdot}000\,8$ nm and eight formula units per unit cell.

REFERENCES

65 Weg: H. A. R. Wegener: Thesis, Polytechnic Institute of Brooklyn, USA, 1965

Au–Ti–V

Toth *et al.* state, without elaboration, that the ordered Au_4X phase is stable on the section $Au_4(Ti_{1-x}V_x)$ with $0 \leqslant x \leqslant 1$. This would be expected since Au_4Ti and Au_4V both have the ordered Ni_4Mo-type structure, tI10, with disordering temperatures of 1 172 and 565°C respectively.

REFERENCE

60 Tot: R. S. Toth, A. Arrott, S. S. Shinozaki, S. A. Werner and H. Sato: *J. Appl. Phys.*, 1969, **40**, 1373–1375

Au–Tl–Zn

[24 Tam] studied the partitioning of Au between immiscible liquid layers of Tl and Zn at 445 ± 5°C. Chemical analysis of the two layers gave a Au content of 4·57 wt-% in the Zn-rich layer and 0·05 wt-% in the Tl-rich layer. On the assumption that the Tl–Zn immiscibility gap in the ternary is parallel to the Au–Tl and Au–Zn binary edges in the initial stages at 445°C, the Zn-rich layer will have a composition 1·6 Au, 1·1 Tl, 97·3 at.-% Zn and the Tl-rich layer will contain 0·05 Au, 84·80 Tl, 15·15 at.-% Zn. The tie line joining the two liquid compositions is skewed towards the Au–Tl binary edge.

REFERENCE

24 Tam: G. Tammann and P. Schafmeister, *Z. anorg. Chem.*, 1924, **138**, 220–232

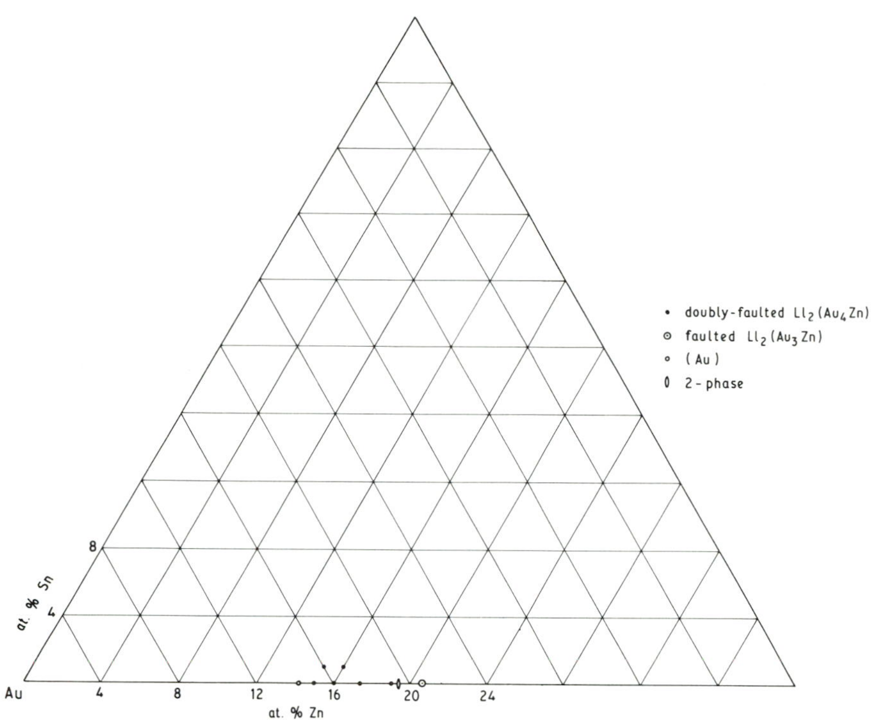

411 The structure of alloys examined by Wilkens and Schubert [57 Wil]

Relevant Binary Systems

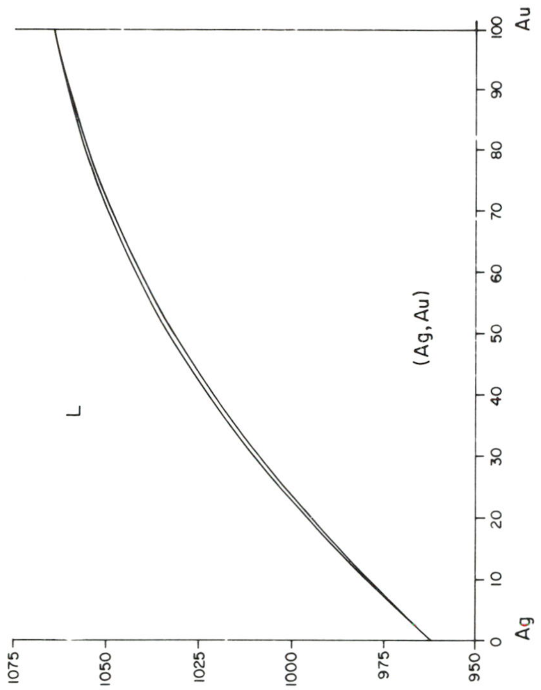
L
(Ag)
β
δ
μ
(Al)
Ag
Al
1000
900
800
700
600
500
400
300
200
0
10
20
30
40
50
60
70
80
90
100

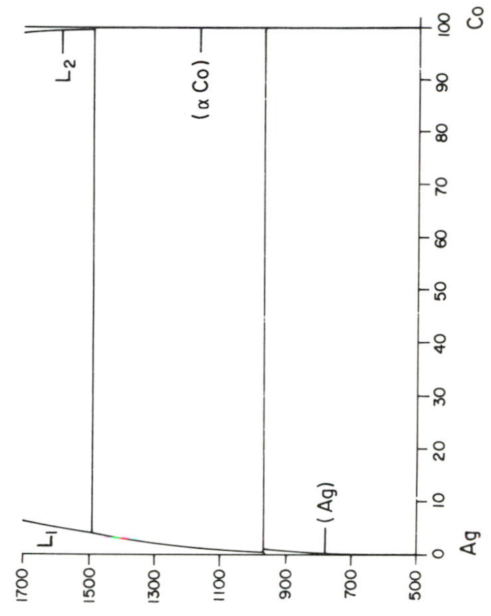
L
(Ag, Au)
Ag
Au
1075
1050
1025
1000
975
950
0
10
20
30
40
50
60
70
80
90
100

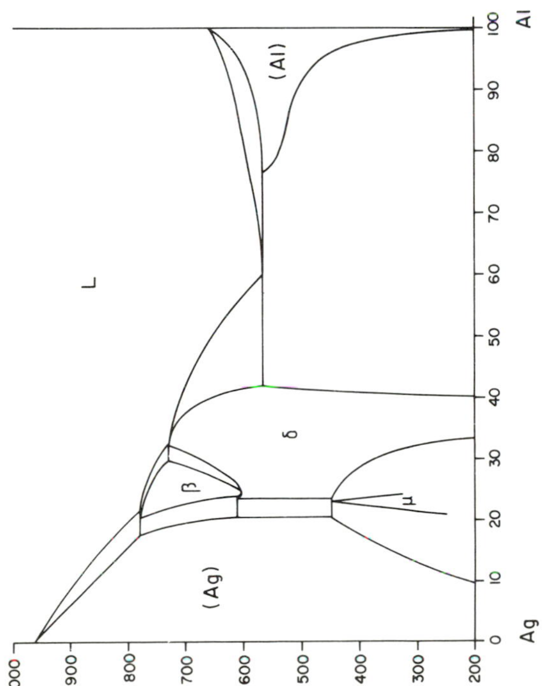
L
(Ag)
β
β'
γ
γ'
ζ
ε
(Cd)
Ag
Cd
1000
900
800
700
600
500
400
300
200
100
0
10
20
30
40
50
60
70
80
90
100

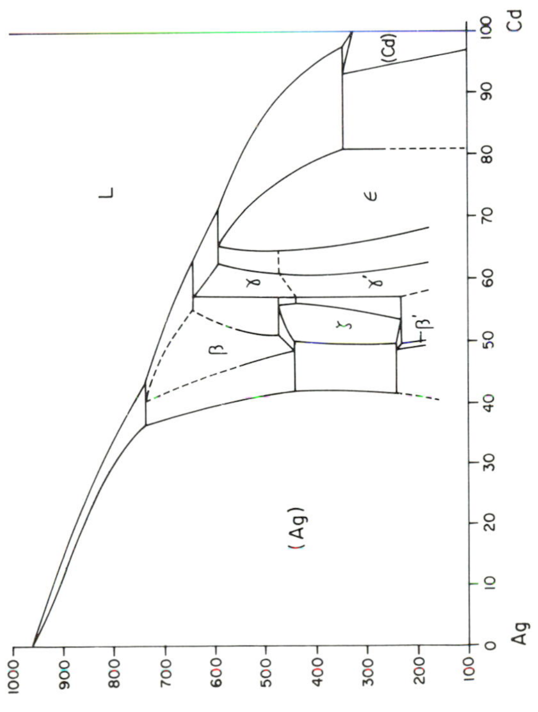
L1
L2
(αCo)
(Ag)
Ag
Co
1700
1500
1300
1100
900
700
500
0
10
20
30
40
50
60
70
80
90
100

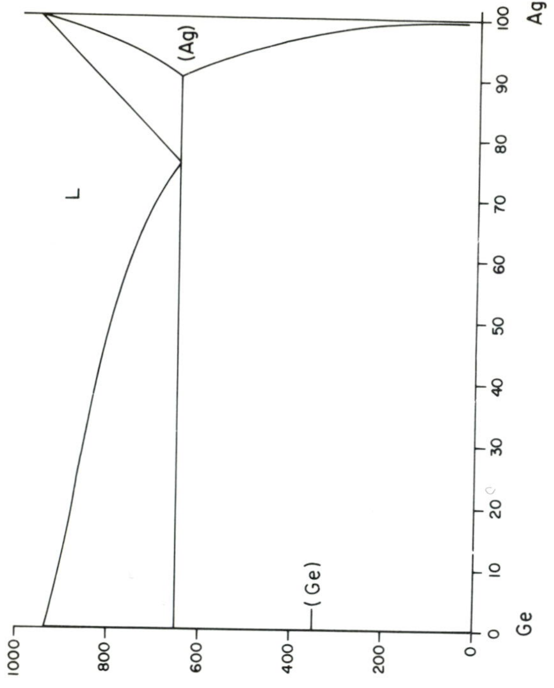
L
(Ag)
(Cu)
Ag
Cu

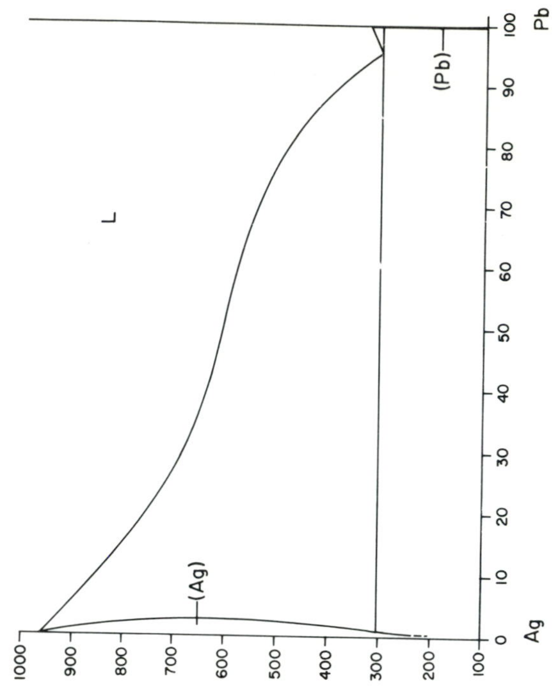
L
(Ag)
(Ge)
Ge
Ag

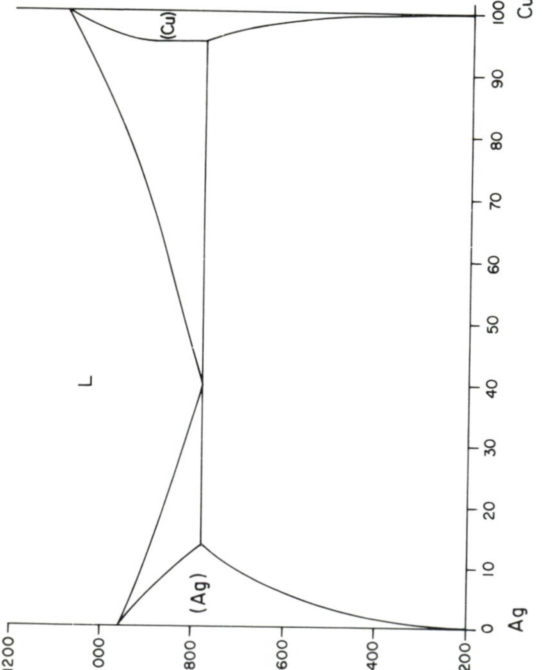
$L_1 + L_2$
L_1
L_2
(Ni)
(Ag)
Ni
Ag

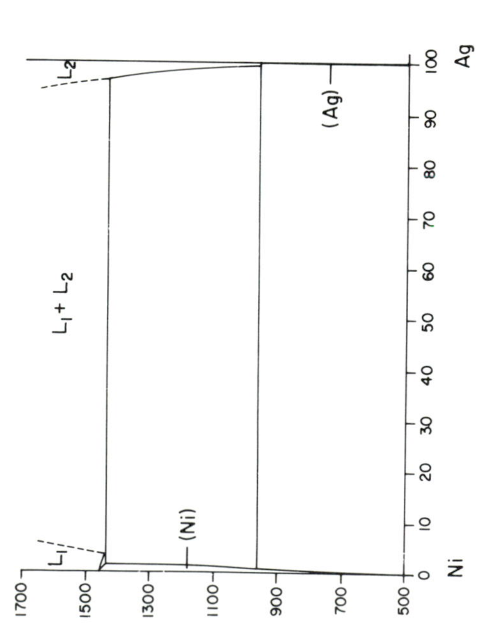
L
(Ag)
(Pb)
Ag
Pb

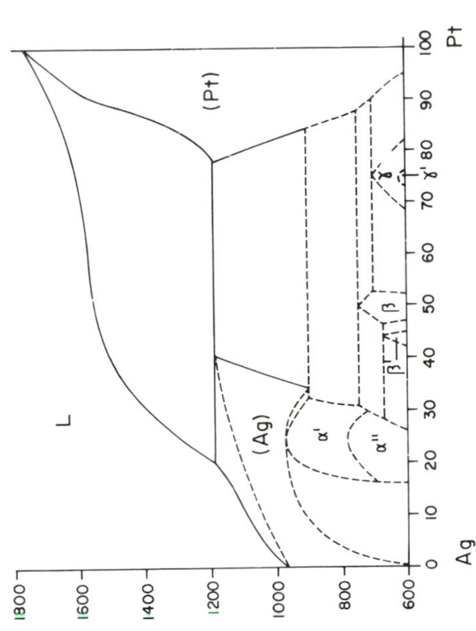
L
(Pt)
(Ag)
α'
α''
β
β'
γ
γ'
Ag
Pt
0
10
20
30
40
50
60
70
80
90
100
600
800
1000
1200
1400
1600
1800

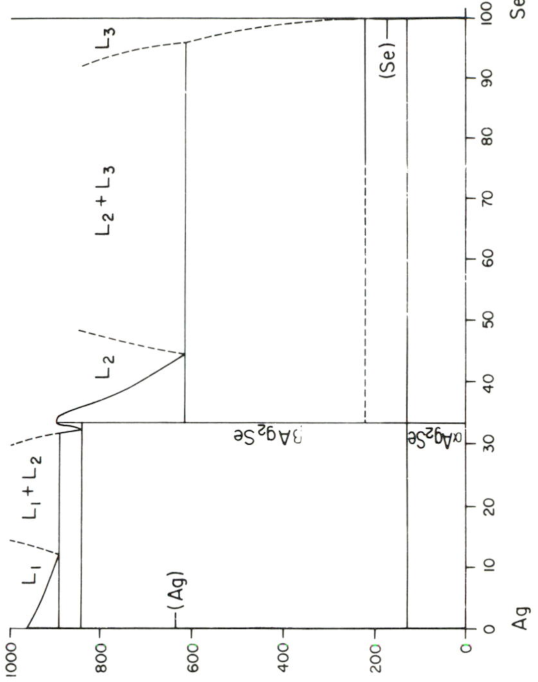
L₁
L₁ + L₂
L₂
L₂ + L₃
L₃
(Ag)
βAg₂Se
αAg₂Se
(Se)
Ag
Se
0
10
20
30
40
50
60
70
80
90
100
0
200
400
600
800
1000

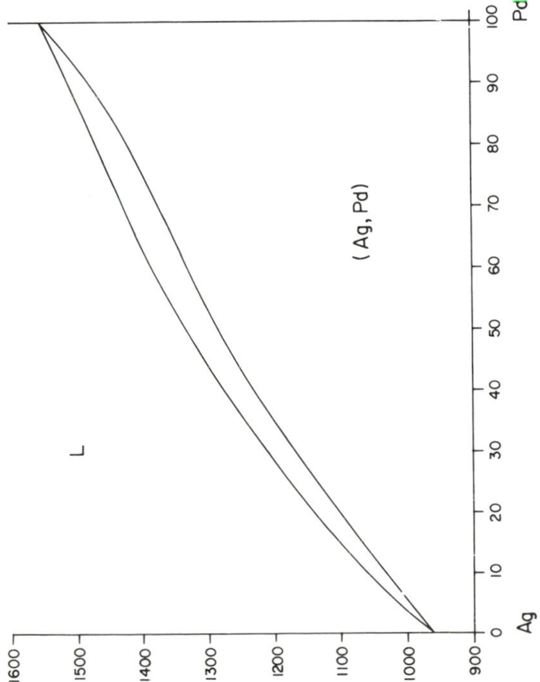
L
(Ag, Pd)
Ag
Pd
0
10
20
30
40
50
60
70
80
90
100
900
1000
1100
1200
1300
1400
1500
1600

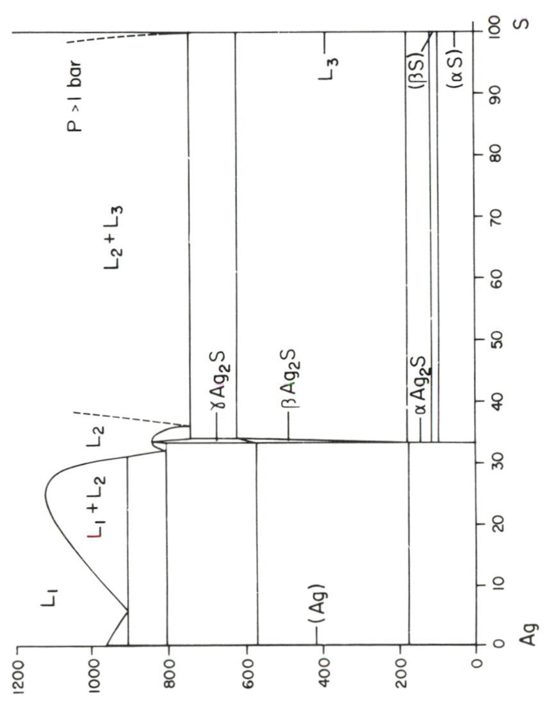
L₁
L₁ + L₂
L₂
L₂ + L₃
L₃
P >1 bar
γAg₂S
βAg₂S
αAg₂S
(Ag)
(βS)
(αS)
Ag
S
0
10
20
30
40
50
60
70
80
90
100
0
200
400
600
800
1000
1200

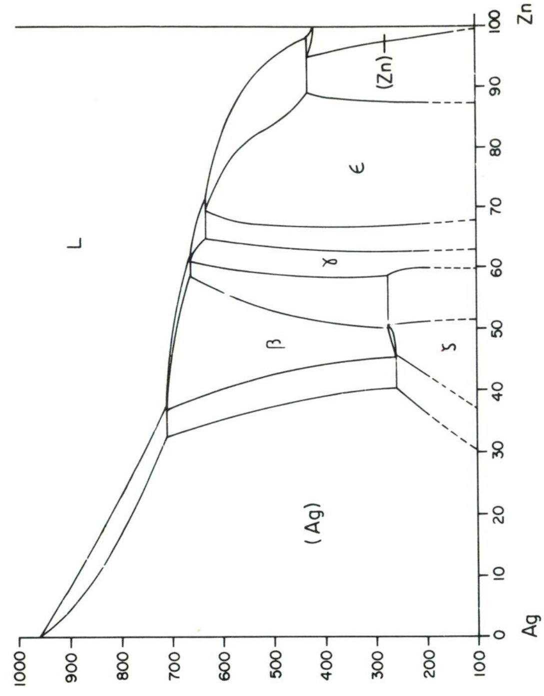
L
β
γ
ε
ζ
(Ag)
(Zn)
Ag
Zn
1000
900
800
700
600
500
400
300
200
100
0
10
20
30
40
50
60
70
80
90
100

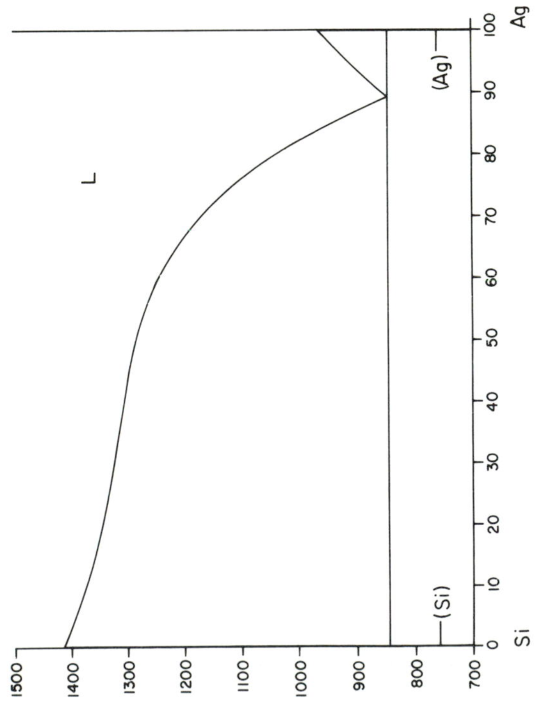
L
(Si)
(Ag)
Si
Ag
1500
1400
1300
1200
1100
1000
900
800
700
0
10
20
30
40
50
60
70
80
90
100

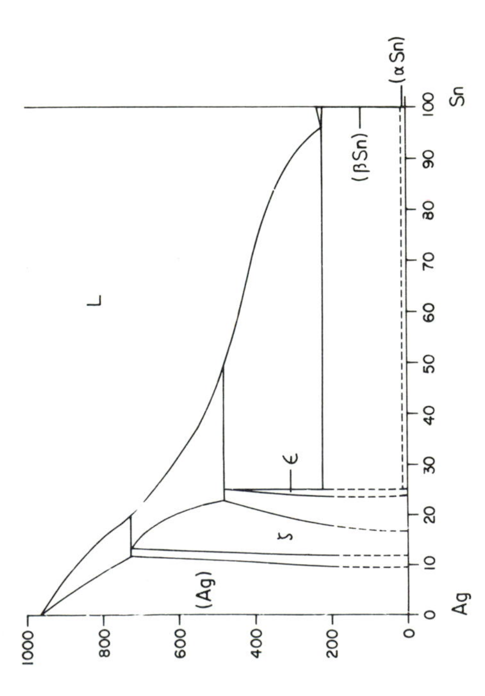
L
(Ag)
ζ
ε
(βSn)
(αSn)
Ag
Sn
1000
800
600
400
200
0
10
20
30
40
50
60
70
80
90
100

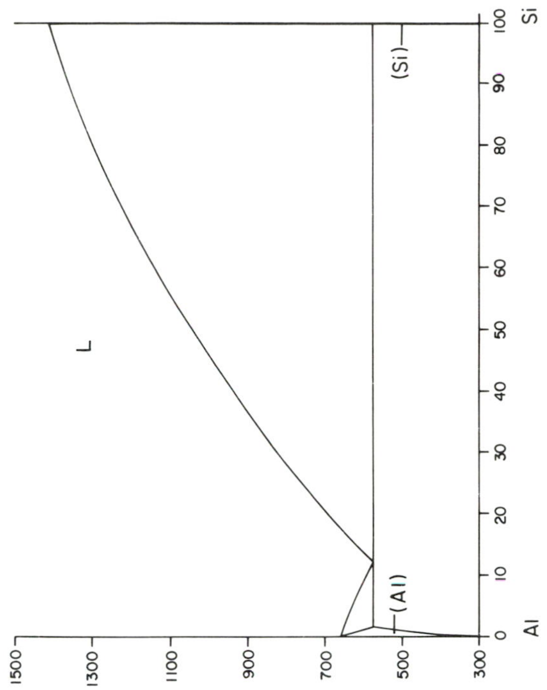

1100 1000 900 800 700 600 500 400
L
Al_2Au
(Au)
$\gamma AlAu_2$
β
$AlAu_4$
Al_2Au_5
$\alpha AlAu_2$
$\beta AlAu_2$
AlAu
(Au)
0 10 20 30 40 50 60 70 80 90 100
Al
Au

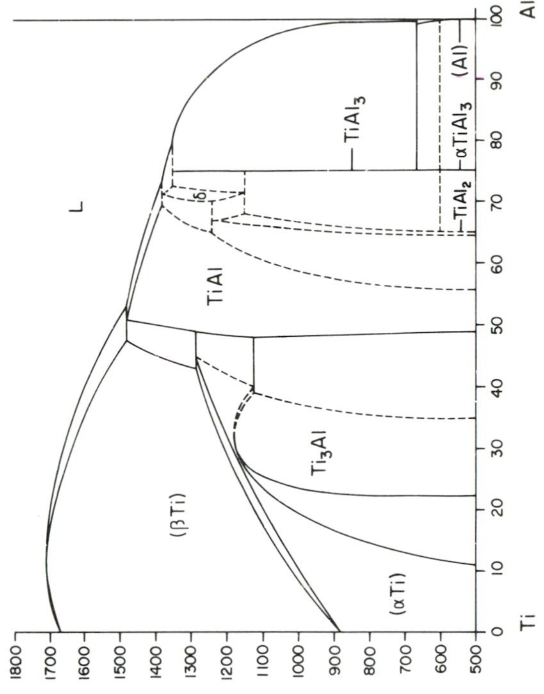

1500 1300 1100 900 700 500 300
L
(Al)
(Si)
0 10 20 30 40 50 60 70 80 90 100
Al
Si

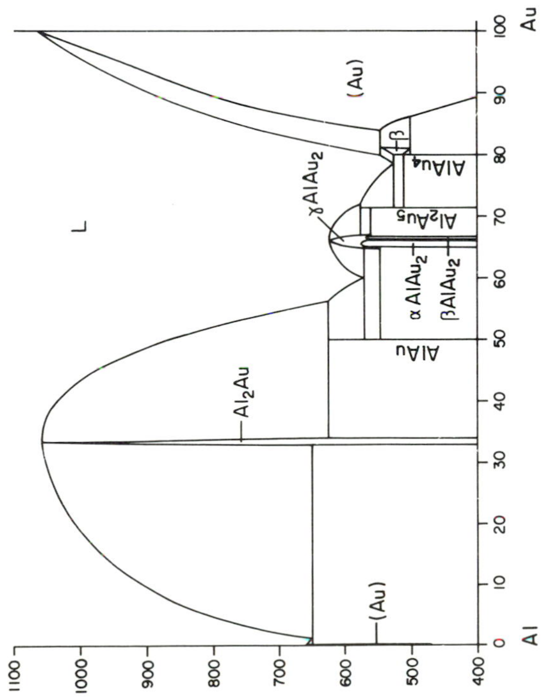

700 600 500 400 300 200 100
L
(Al)
(βSn)
0 10 20 30 40 50 60 70 80 90 100
Al
Sn

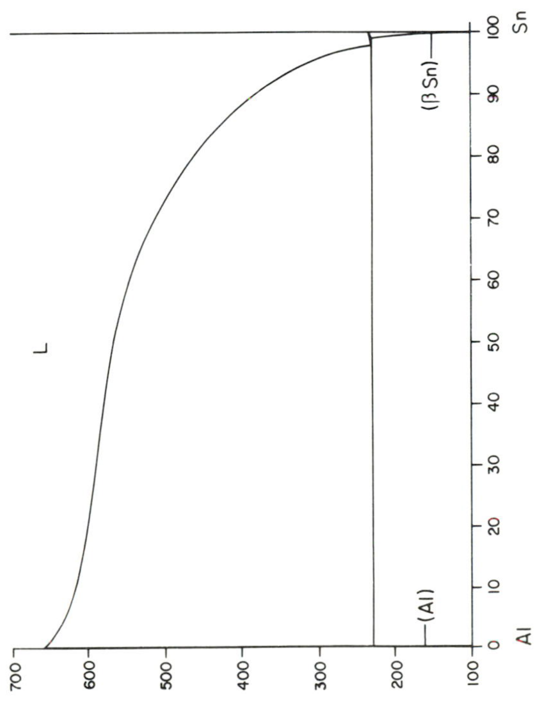

1800 1700 1600 1500 1400 1300 1200 1100 1000 900 800 700 600 500
L
(βTi)
(αTi)
Ti_3Al
TiAl
δ
$TiAl_3$
$TiAl_2$
$\alpha TiAl_3$
(Al)
0 10 20 30 40 50 60 70 80 90 100
Ti
Al

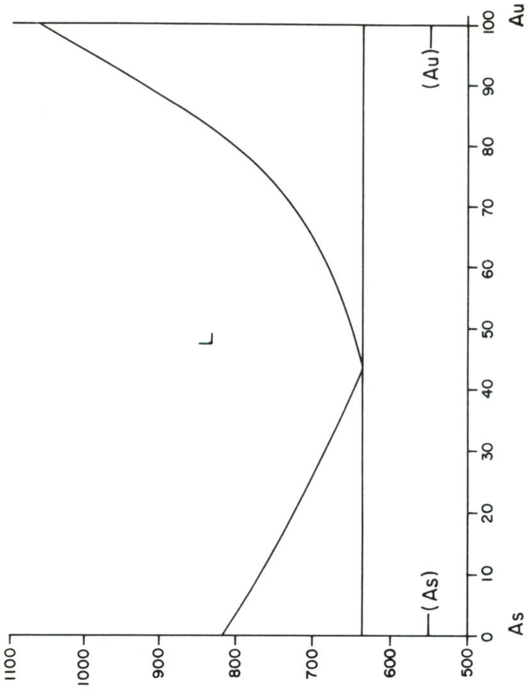
L
(As)
(Au)
As
Au
0
10
20
30
40
50
60
70
80
90
100
500
600
700
800
900
1000
1100

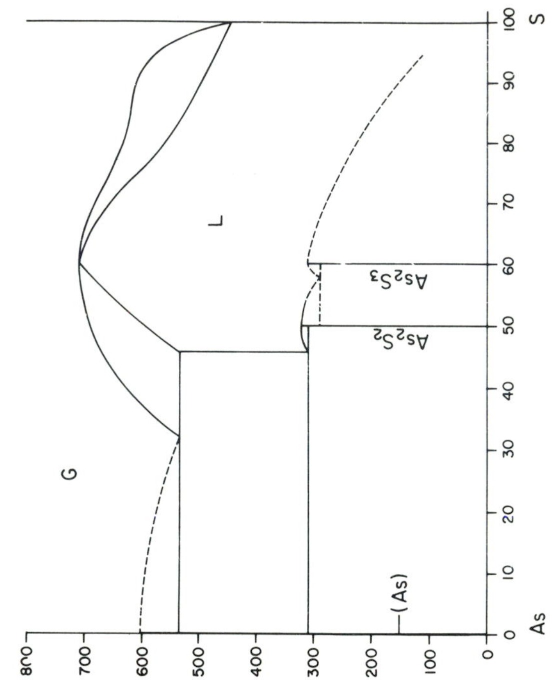
G
L
(As)
As_2S_2
As_2S_3
As
S
0
10
20
30
40
50
60
70
80
90
100
0
100
200
300
400
500
600
700
800

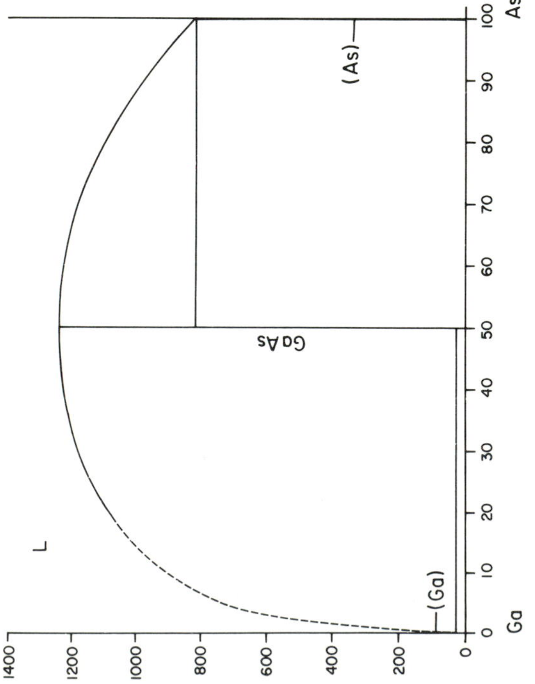
L
(Ga)
(As)
GaAs
Ga
As
0
10
20
30
40
50
60
70
80
90
100
0
200
400
600
800
1000
1200
1400

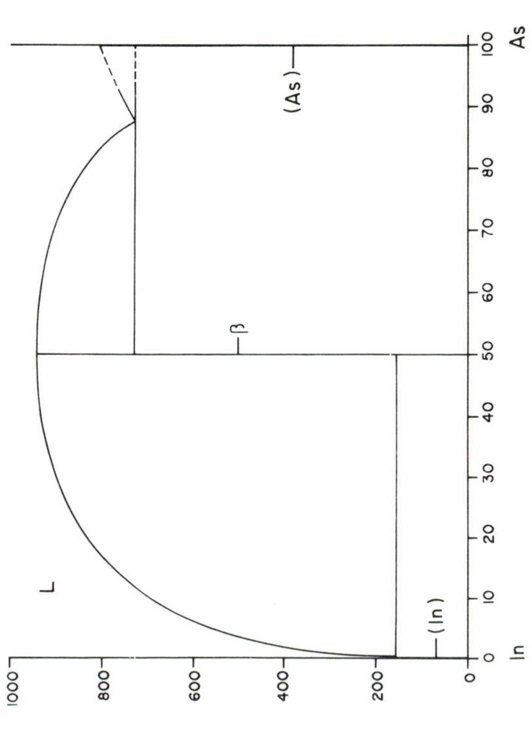
L
(In)
β
(As)
In
As
0
10
20
30
40
50
60
70
80
90
100
0
200
400
600
800
1000

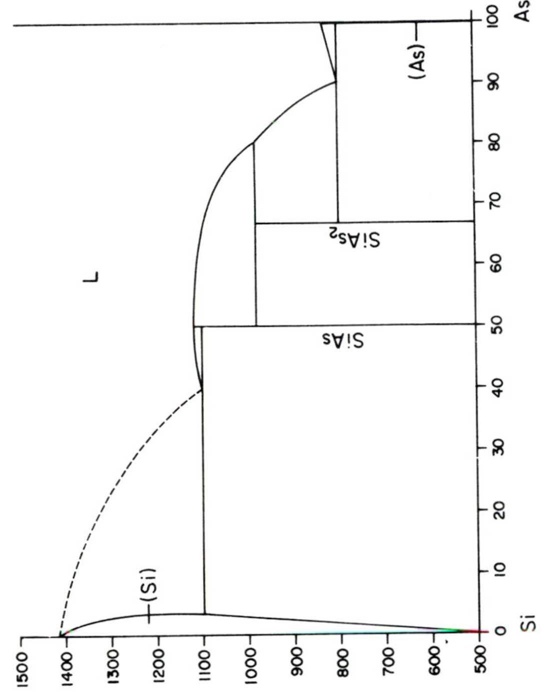
L
(Si)
SiAs
$SiAs_2$
(As)
Si
As
1500
1400
1300
1200
1100
1000
900
800
700
600
500
0
10
20
30
40
50
60
70
80
90
100

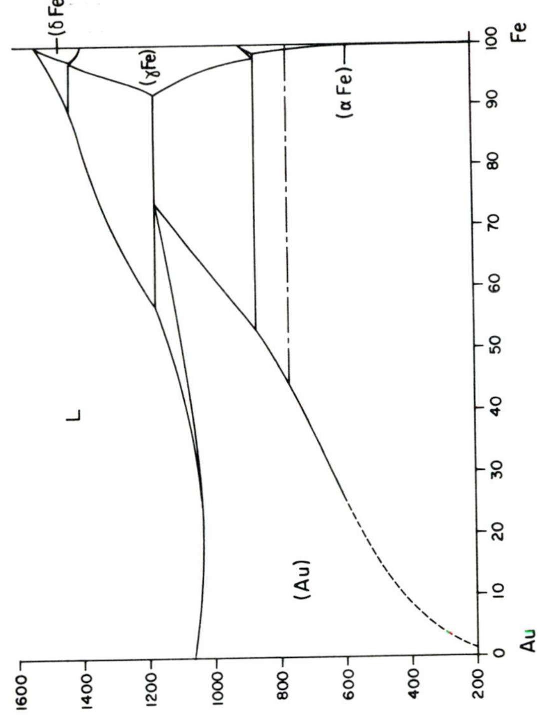
L
(δFe)
(γFe)
(αFe)
(Au)
Au
Fe
1600
1400
1200
1000
800
600
400
200
0
10
20
30
40
50
60
70
80
90
100

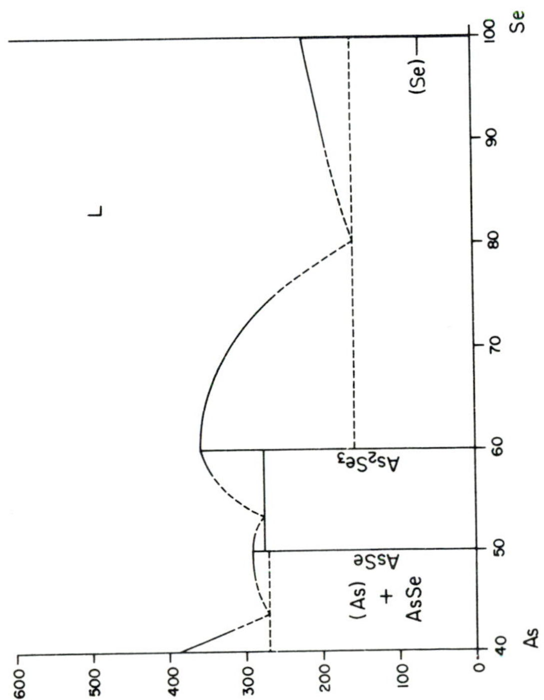
L
(As) + AsSe
AsSe
As_2Se_3
(Se)
As
Se
600
500
400
300
200
100
0
40
50
60
70
80
90
100

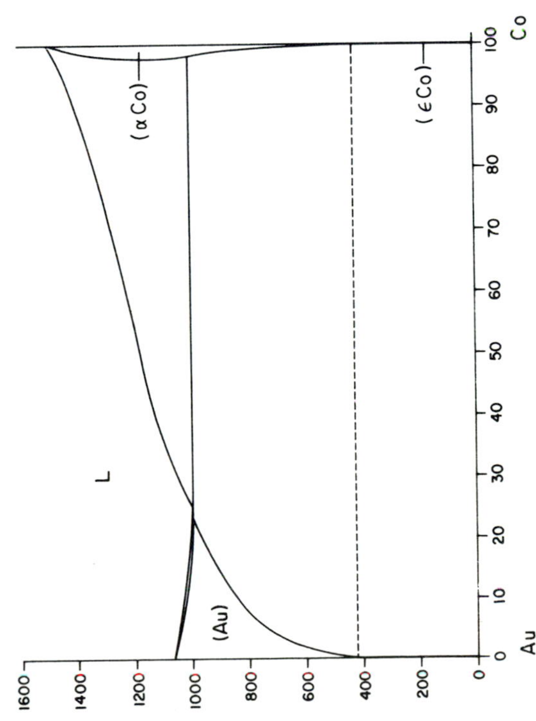
L
(αCo)
(εCo)
(Au)
Au
Co
1600
1400
1200
1000
800
600
400
200
0
10
20
30
40
50
60
70
80
90
100

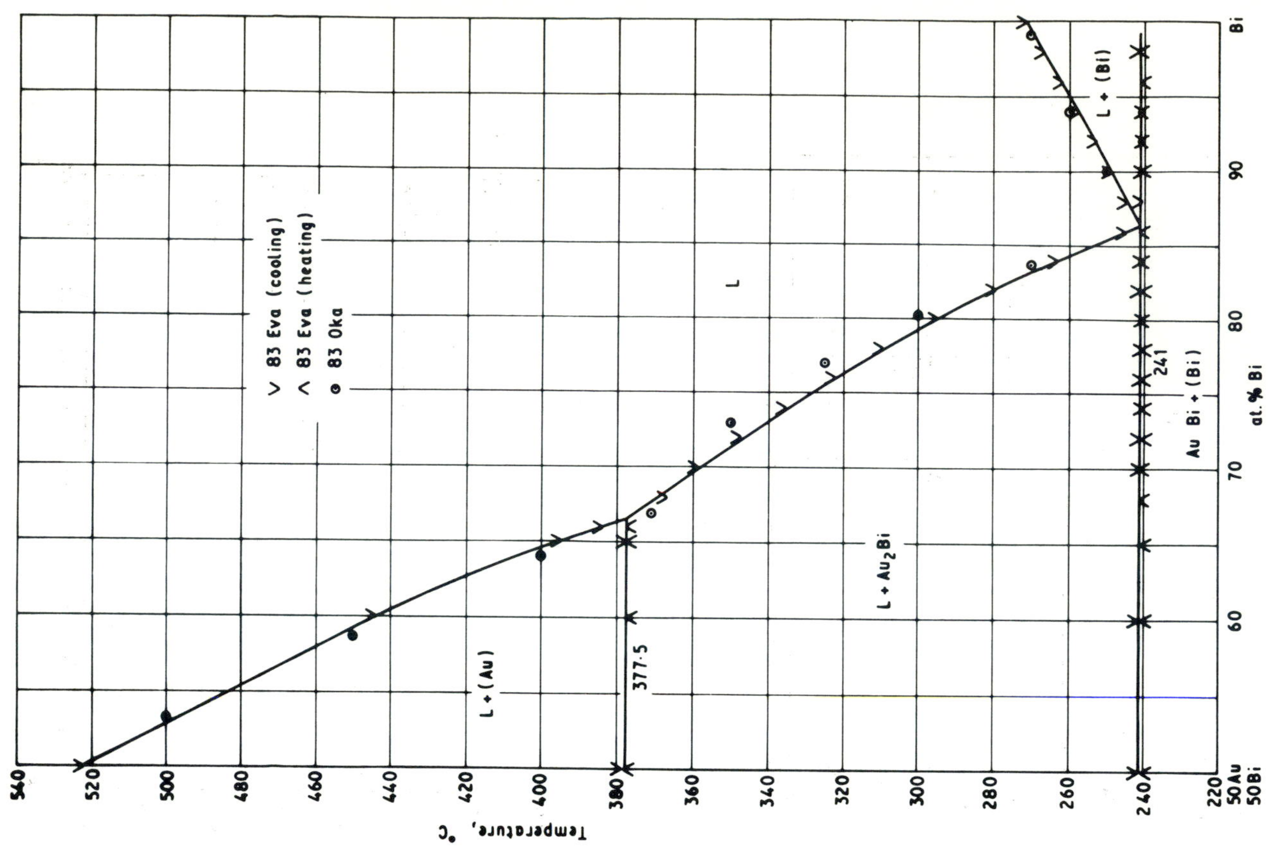
V 83 Eva (cooling)
Λ 83 Eva (heating)
⊙ 83 Oka
L
L + (Au)
377.5
L + Au$_2$Bi
241
Au Bi + (Bi)
L + (Bi)
Temperature, °C
at.% Bi
50Au
50Bi
Bi

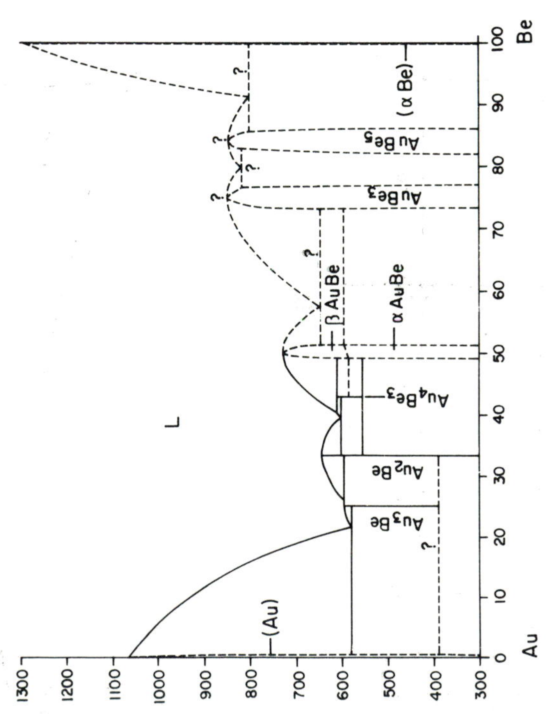
L
(Au)
Au$_3$Be
Au$_2$Be
Au$_4$Be$_3$
β AuBe
α AuBe
AuBe$_3$
AuBe$_5$
(α Be)
Au
Be

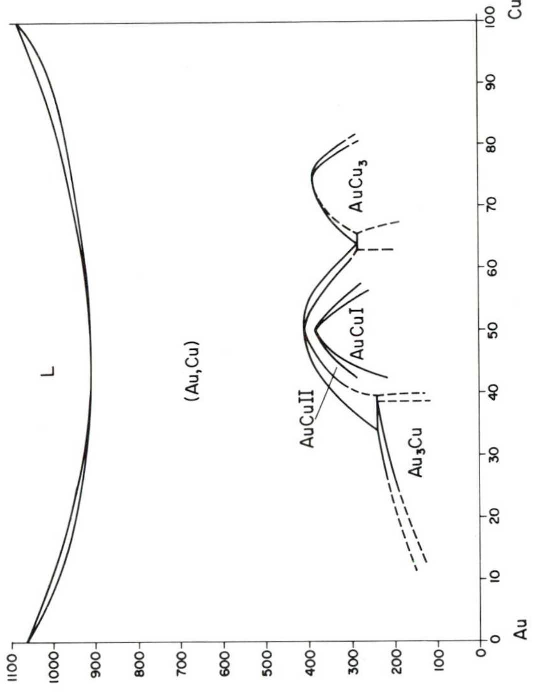
L
(Au,Cu)
AuCuII
AuCuI
AuCu$_3$
Au$_3$Cu
Au
Cu

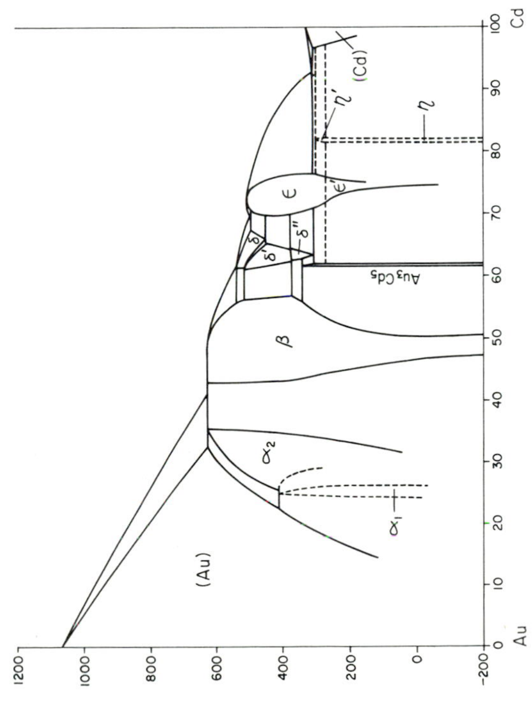
(Au)
α$_2$
α$_1$
β
Au$_3$Cd$_5$
δ'
δ
δ''
ε
ε'
η'
η
(Cd)
Au
Cd

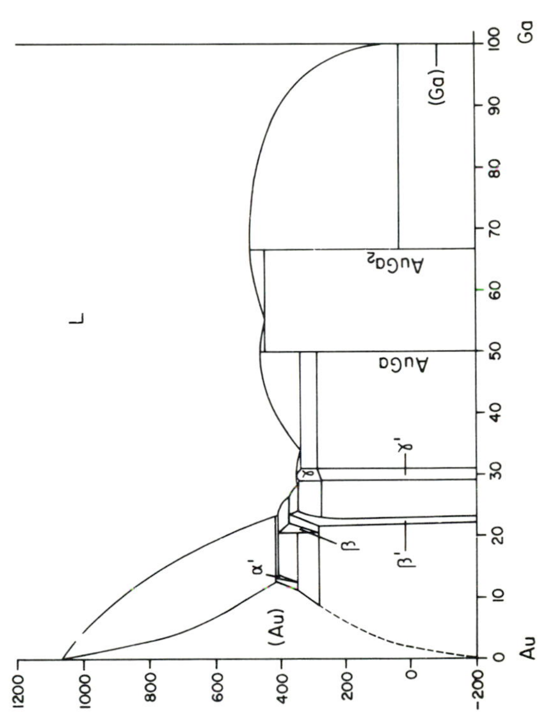
L
(Au)
α'
β
β'
γ
γ'
AuGa
AuGa$_2$
(Ga)
Au
Ga

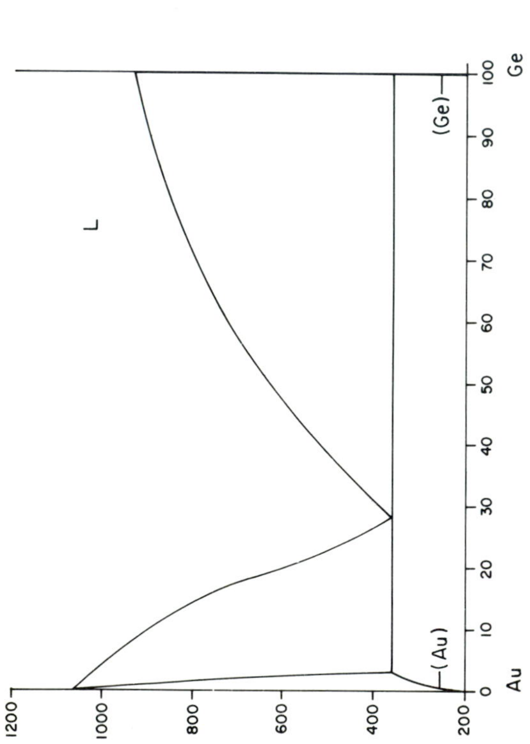
L
(Au)
(Ge)
Au
Ge

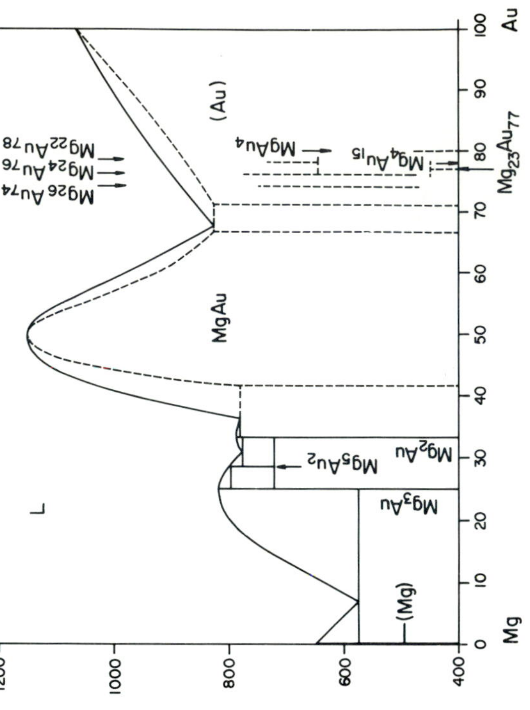
L
(Au)
MgAu
(Mg)
Mg_3Au
Mg_2Au
Mg_5Au_2
$MgAu_4$
Mg_4Au_{15}
$Mg_{23}Au_{77}$
$Mg_{26}Au_{74}$
$Mg_{24}Au_{76}$
$Mg_{22}Au_{78}$
Mg
Au

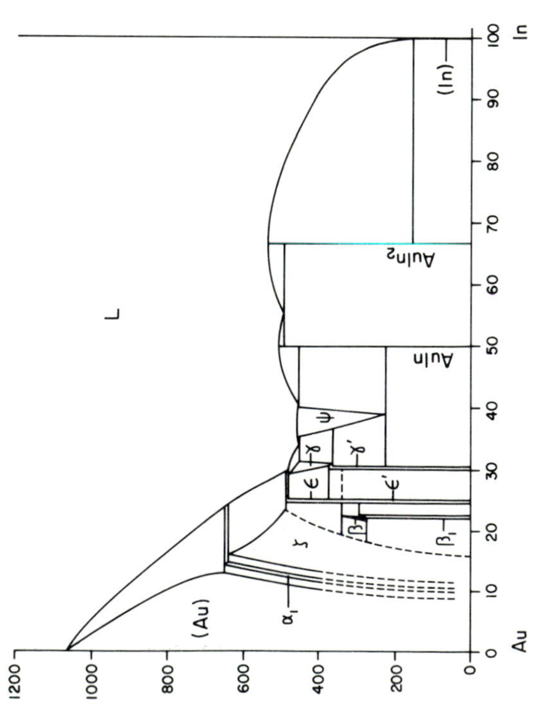
L
(Au)
(In)
AuIn
$AuIn_2$
Au
In

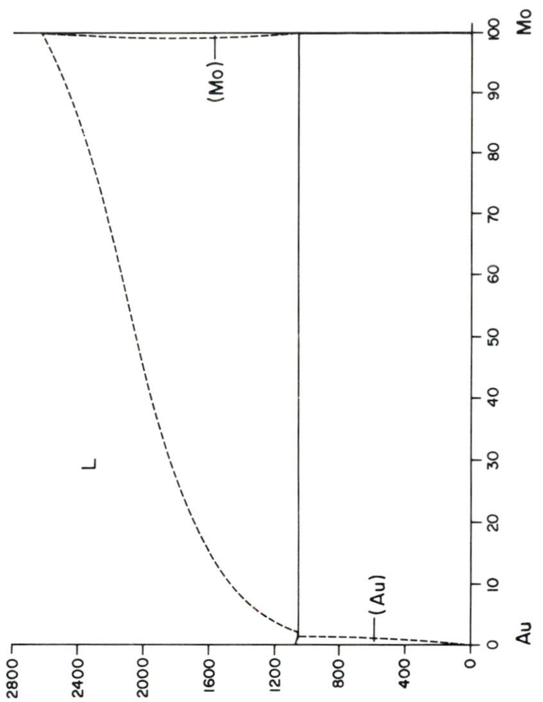
L
(Au)
β
(γMn)
(αMn)
(δMn)
(βMn)
$AuMn_2$
Au
Mn

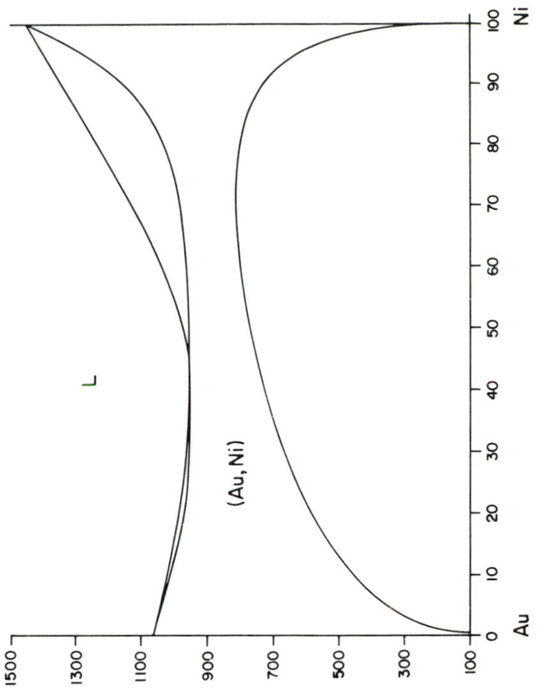
L
(Au)
(Mo)
Au
Mo

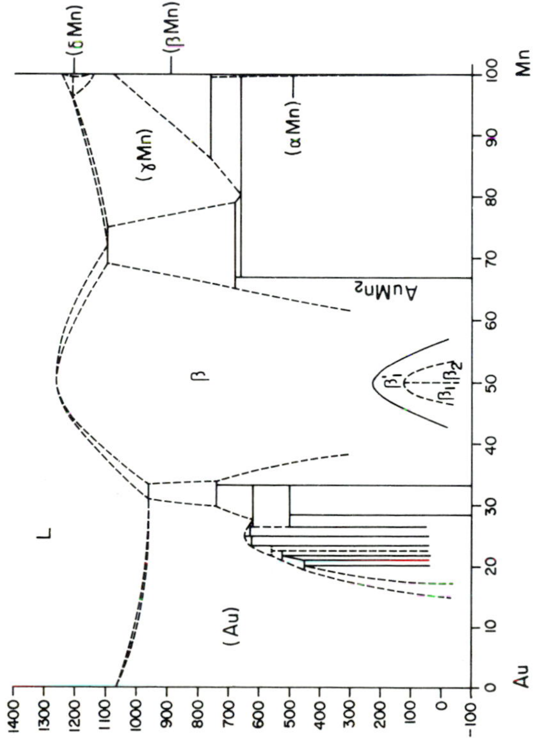
L
(Au)
(Nb)
Au_2Nb
Au_2Nb_3
$AuNb_3$
Au
Nb

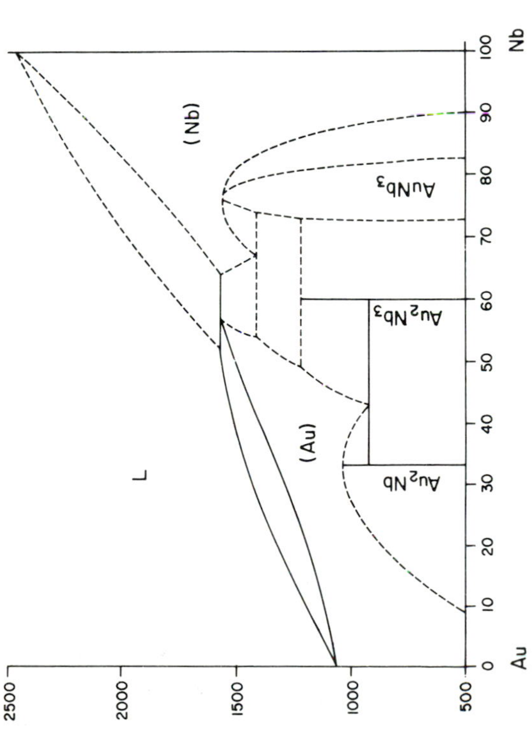
L
(Au, Ni)
Au
Ni

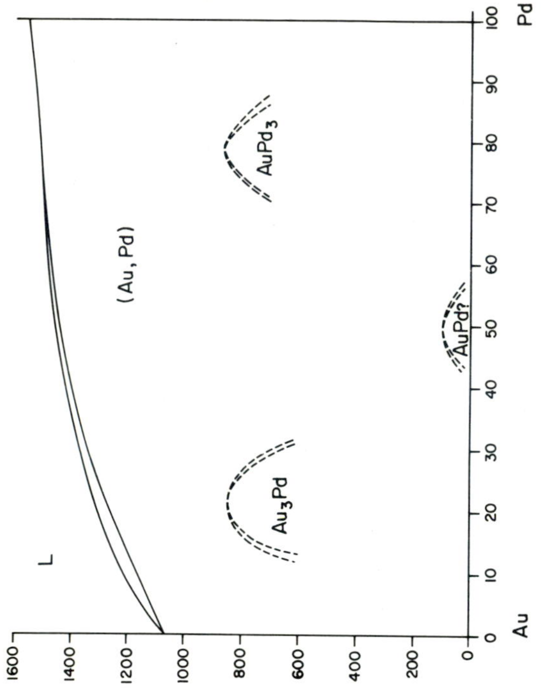
L
G
(Au)
G
L
(α P)
Au
P

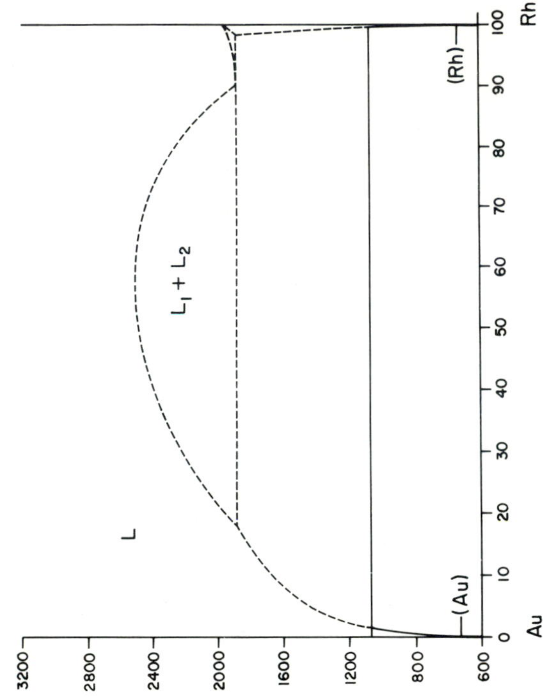
L
(Au, Pd)
Au_3Pd
AuPd?
$AuPd_3$
Au
Pd

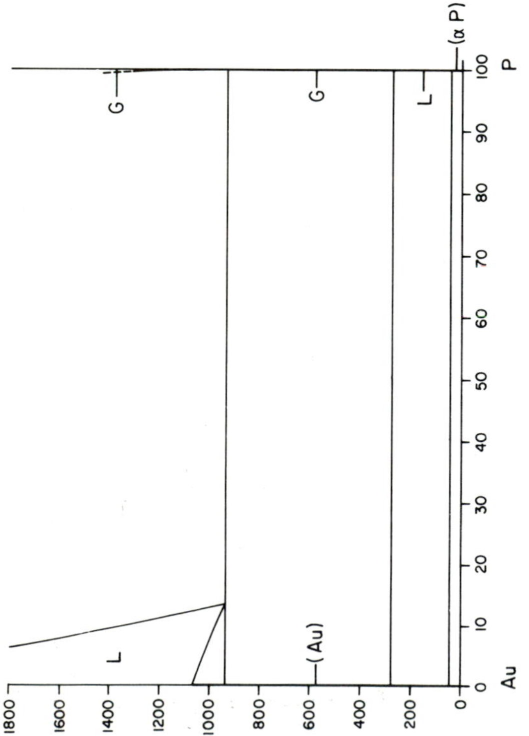
L
(Au,Pt)
(Au) + (Pt)
Au
Pt

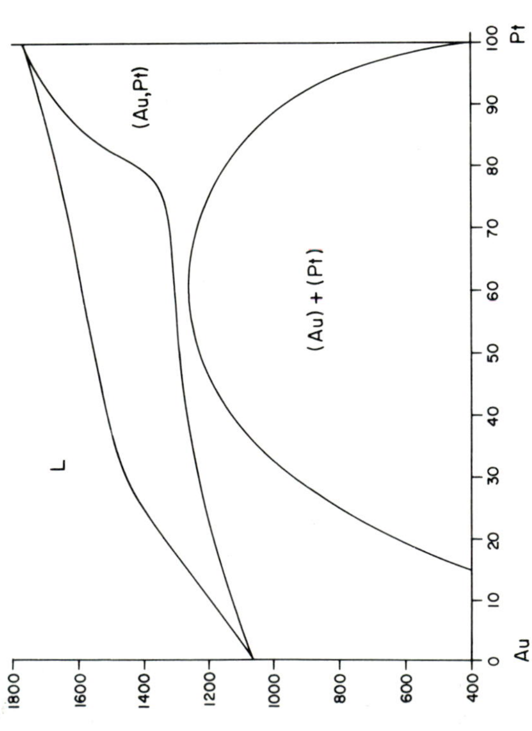
L
$L_1 + L_2$
(Au)
(Rh)
Au
Rh

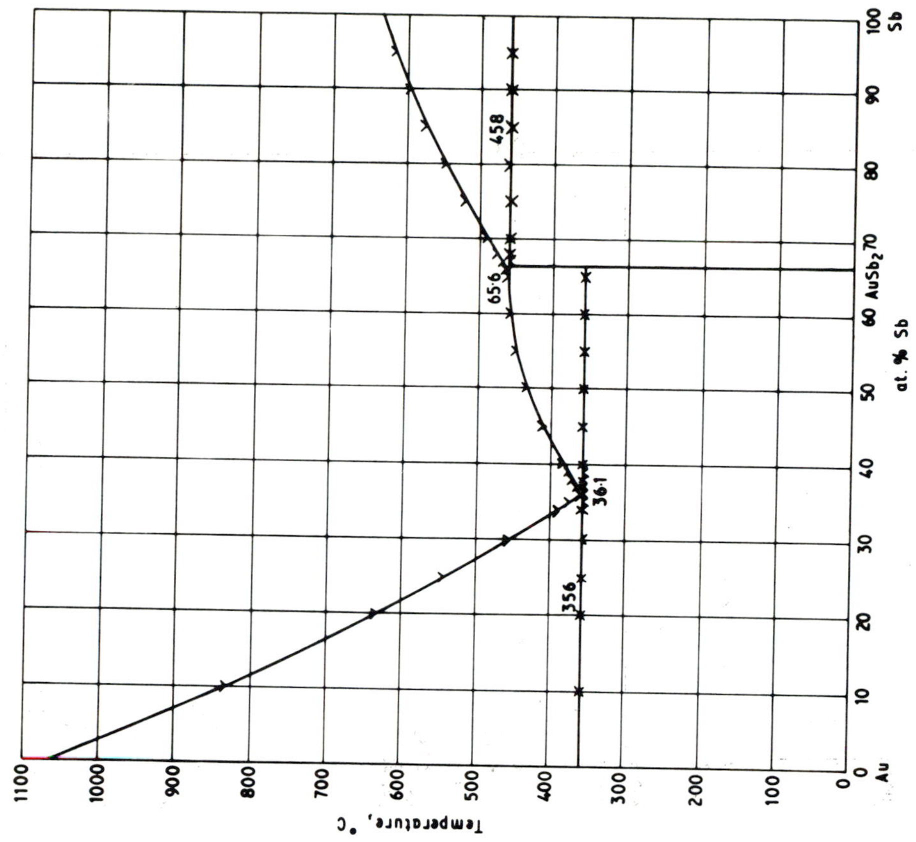

Temperature, °C
at. % Sb
Au
Sb
$AuSb_2$
65·6
458
36·1
356

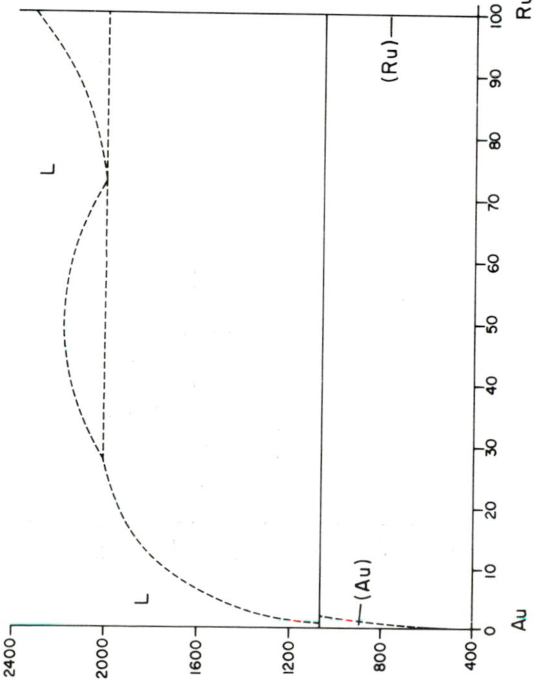

L
(Au)
(Ru)
Au
Ru

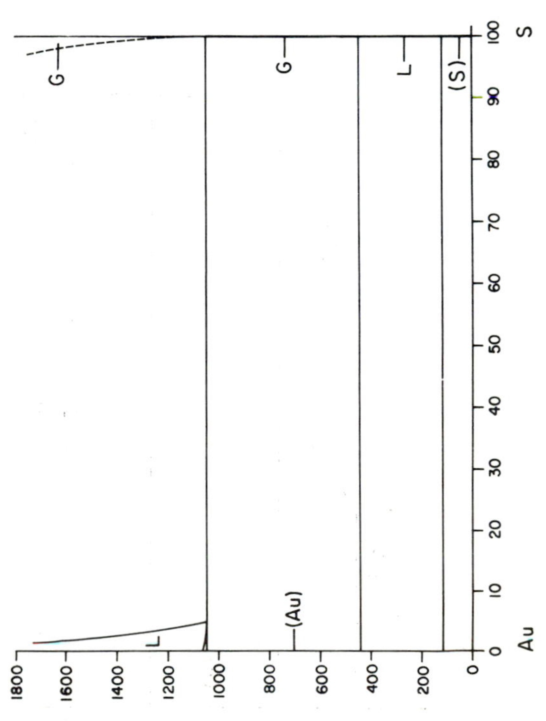

G
L
(Au)
(S)
Au
S

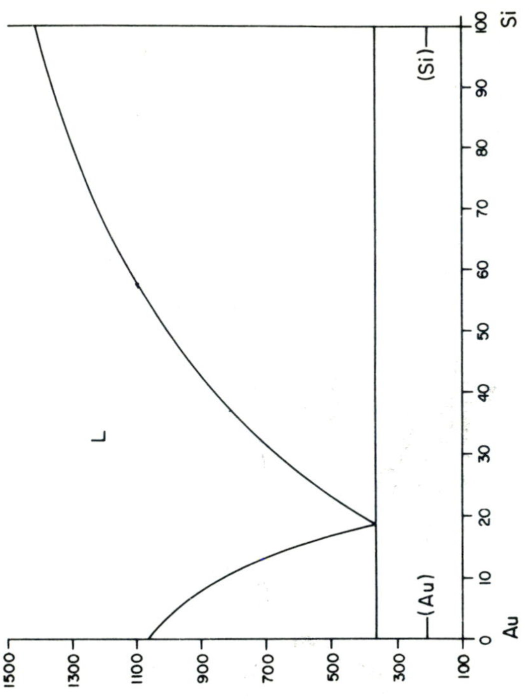
L1
L1+L2
L2
L2+L3
L3
(Au)
αAuSe
(Se)
Au
Se

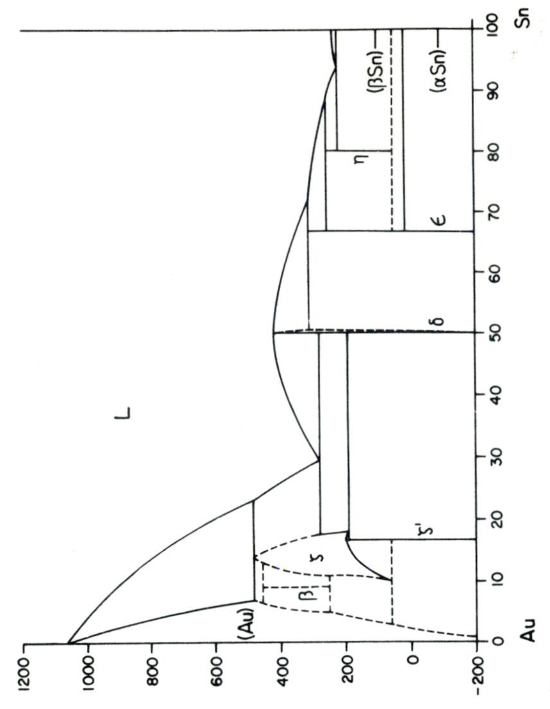
L
(Au)
(Si)
Au
Si

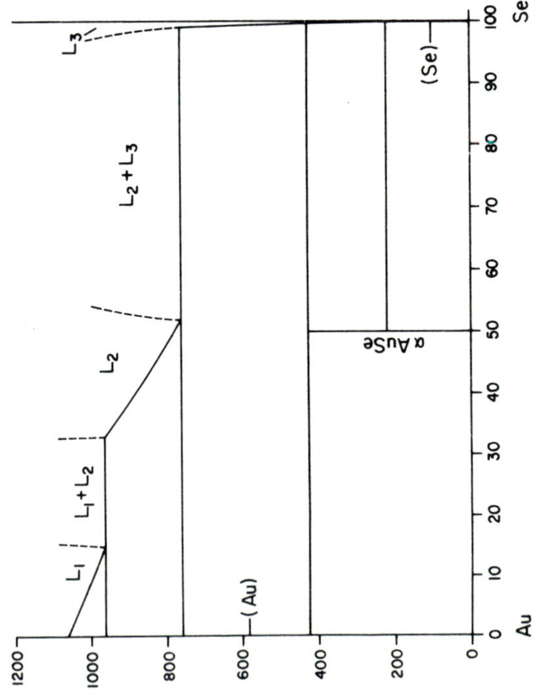
L
L + β
L + ζ
β
(Au)
Au
Sn

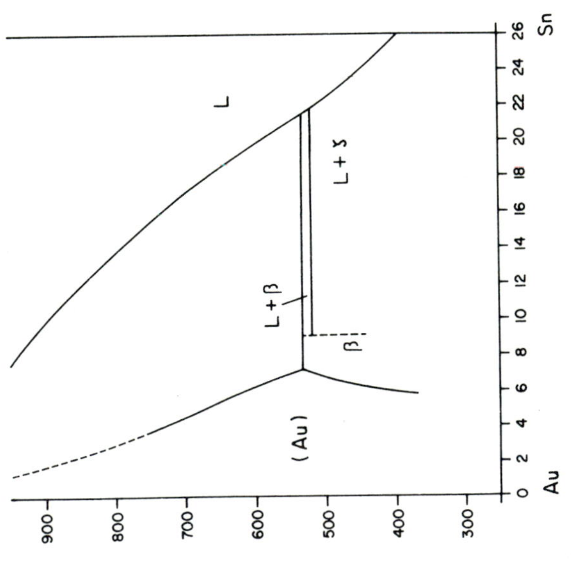
L
(Au)
ζ
β
ζ'
δ
ε
η
(βSn)
(αSn)
Au
Sn

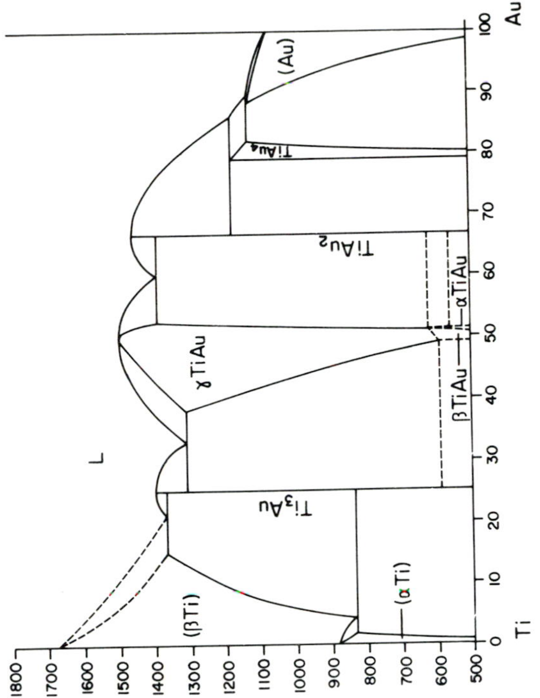
L
(βTi)
Ti$_3$Au
γTiAu
βTiAu
αTiAu
TiAu$_2$
TiAu$_4$
(Au)
(αTi)
Ti
Au

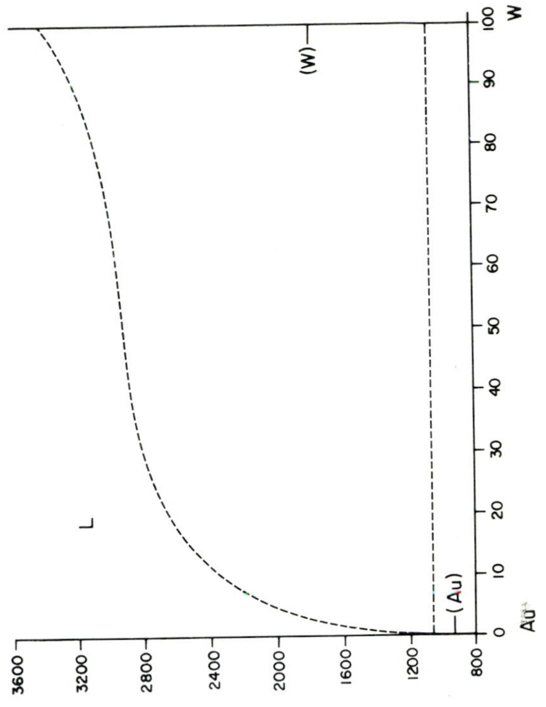
L
(W)
(Au)
Au
W

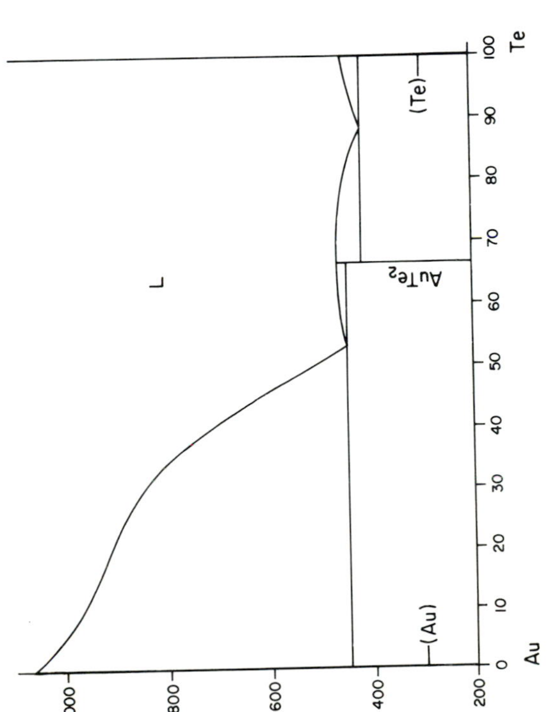
L
AuTe$_2$
(Te)
(Au)
Au
Te

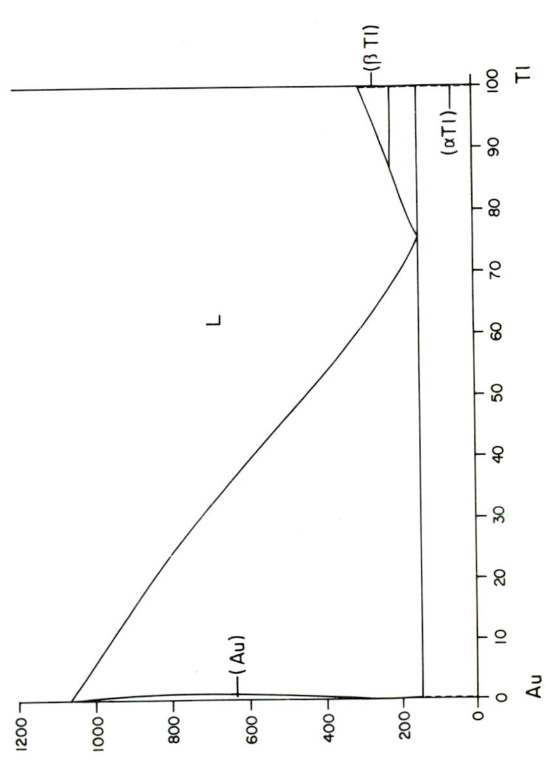
L
(Au)
(βTl)
(αTl)
Au
Tl

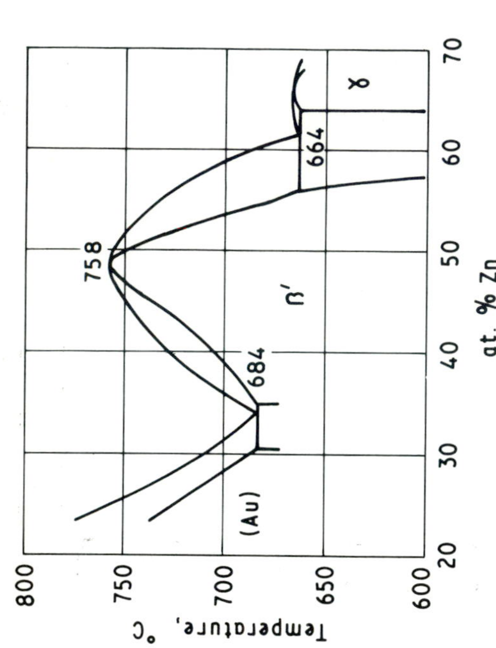
Temperature, °C
1400
1300
1200
1100
1000
900
800
700
600
500
Au
10
20
30
40
50
60
70
80
90
U
at. % U
$Au_{51}U_{14}$
Au_2U
1275
840
1185
1075
745
660
(γU)
(βU)
(αU)

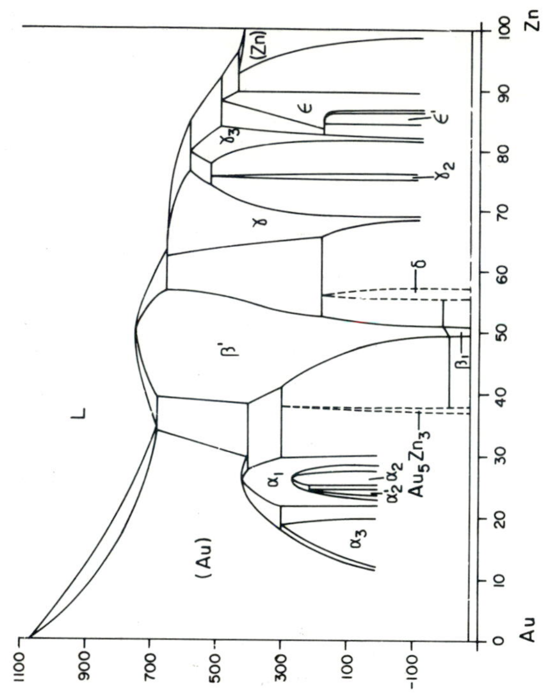
Temperature, °C
800
750
700
650
600
20
30
40
50
60
70
at. % Zn
758
684
664
(Au)
β'
γ

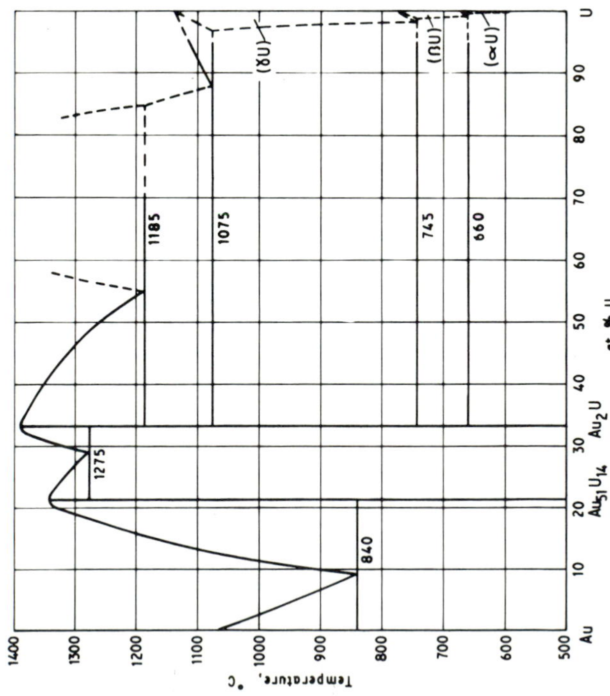
1100
900
700
500
300
100
-100
Au
0
10
20
30
40
50
60
70
80
90
100
Zn
L
(Au)
β'
γ
ε
ε'
(Zn)
δ
α_1
α_3
α_2' α_2
Au_5Zn_3
β_1
γ_2
γ_3

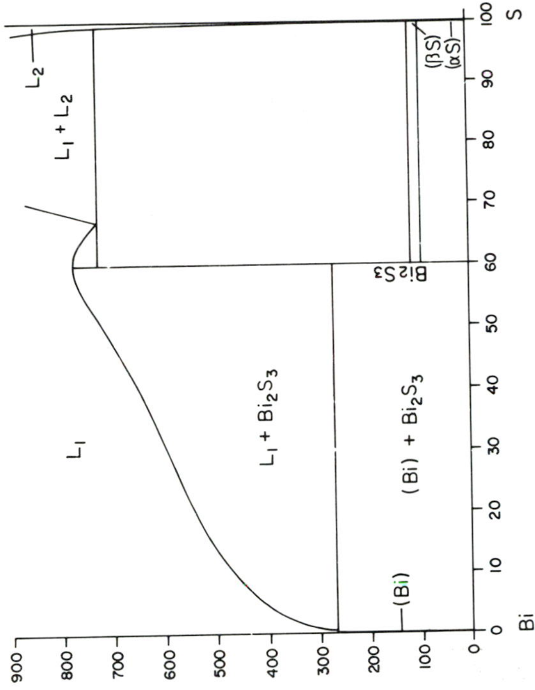
L$_2$
L$_1$ + L$_2$
L$_1$
L$_1$ + Bi$_2$S$_3$
(Bi) + Bi$_2$S$_3$
(Bi)
Bi$_2$S$_3$
(βS)
(αS)
Bi
S

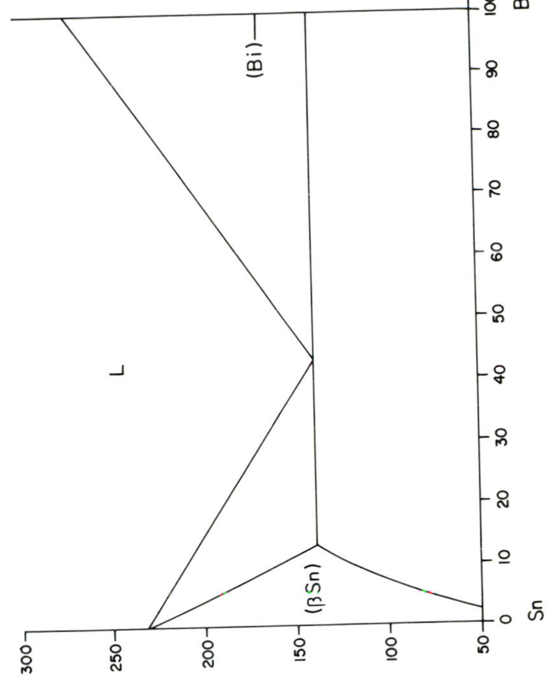
L
(βSn)
(Bi)
Sn
Bi

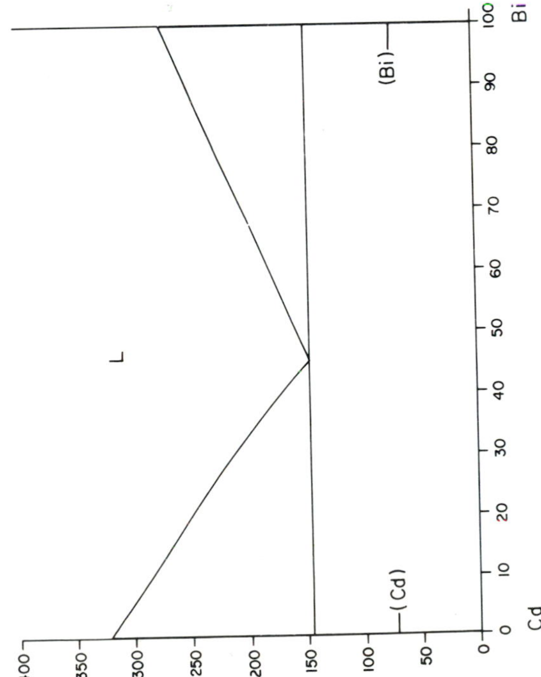
L
(Cd)
(Bi)
Cd
Bi

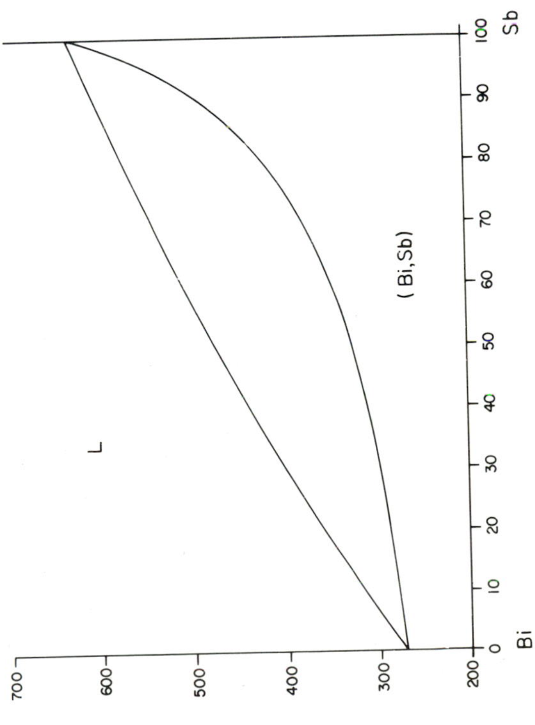
L
(Bi,Sb)
Bi
Sb

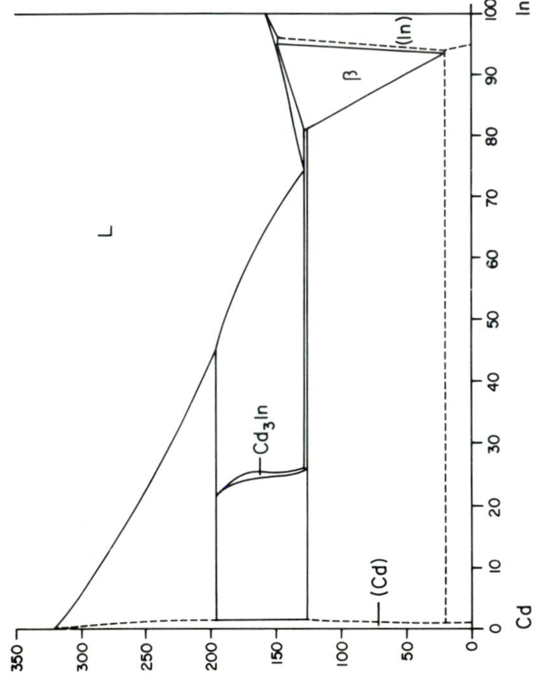
L
(Bi)
$Bi_{14}Te_6$
Bi_2Te
BiTe
Bi_2Te_3
(Te)
Bi
Te

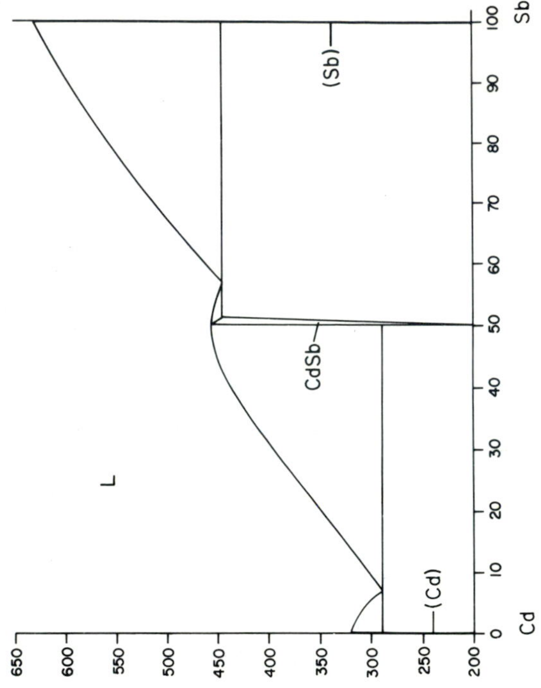
L
Cd_3In
(Cd)
β
(In)
Cd
In

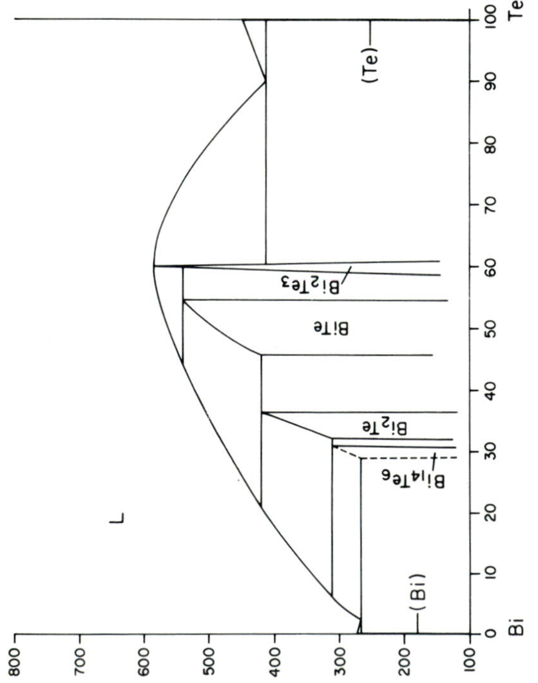
L
(Cd)
(Pb)
Cd
Pb

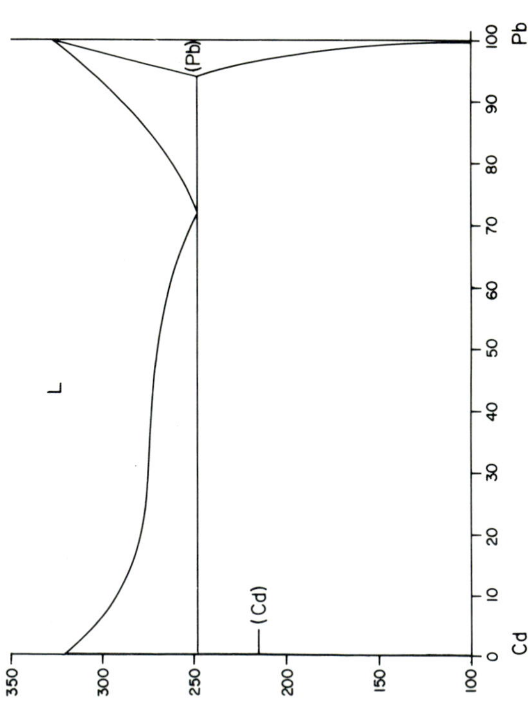
L
CdSb
(Cd)
(Sb)
Cd
Sb

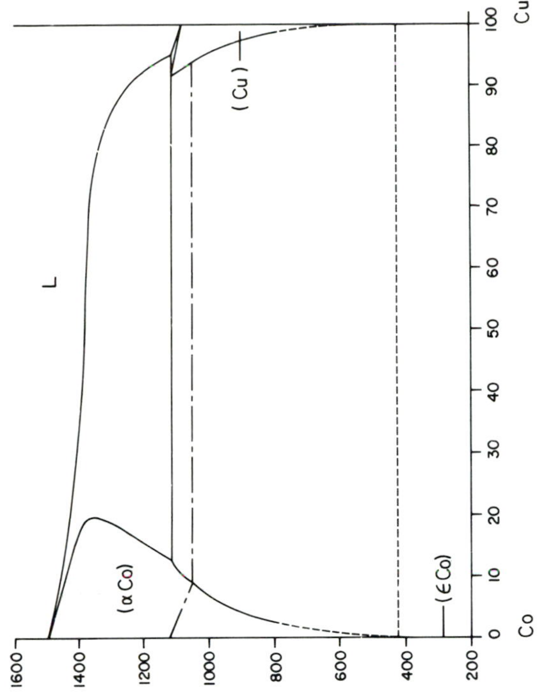
1600
1400
1200
1000
800
600
400
200
0 10 20 30 40 50 60 70 80 90 100
Co
Cu
L
(αCo)
(εCo)
(Cu)

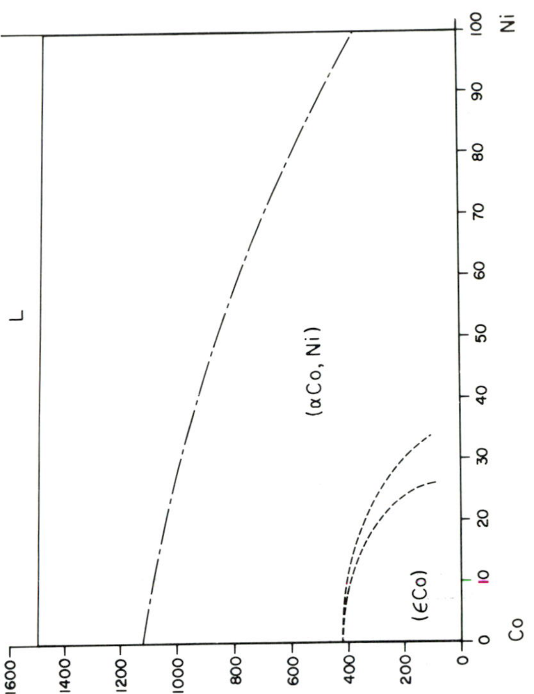
1600
1400
1200
1000
800
600
400
200
0
0 10 20 30 40 50 60 70 80 90 100
Co
Ni
L
(αCo, Ni)
(εCo)

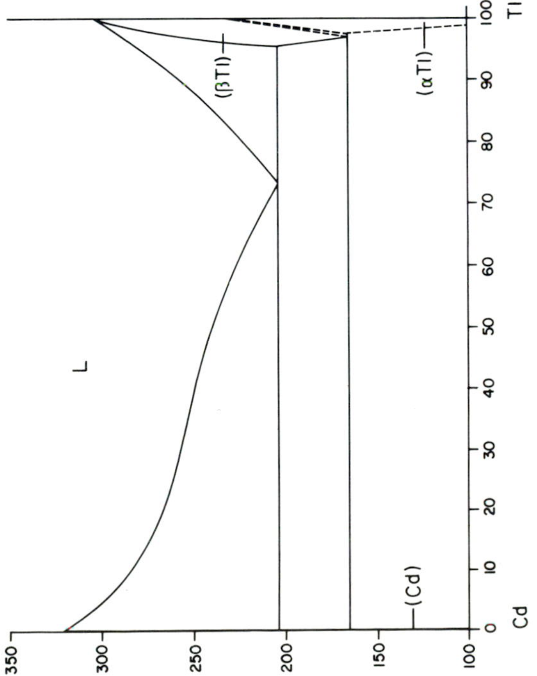
350
300
250
200
150
100
0 10 20 30 40 50 60 70 80 90 100
Cd
Tl
L
(βTl)
(αTl)
(Cd)

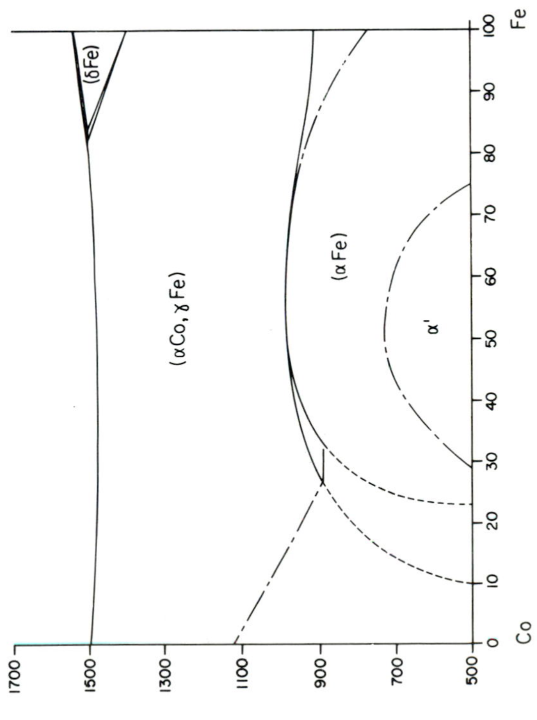
1700
1500
1300
1100
900
700
500
0 10 20 30 40 50 60 70 80 90 100
Co
Fe
(δFe)
(αCo, γFe)
(αFe)
α'

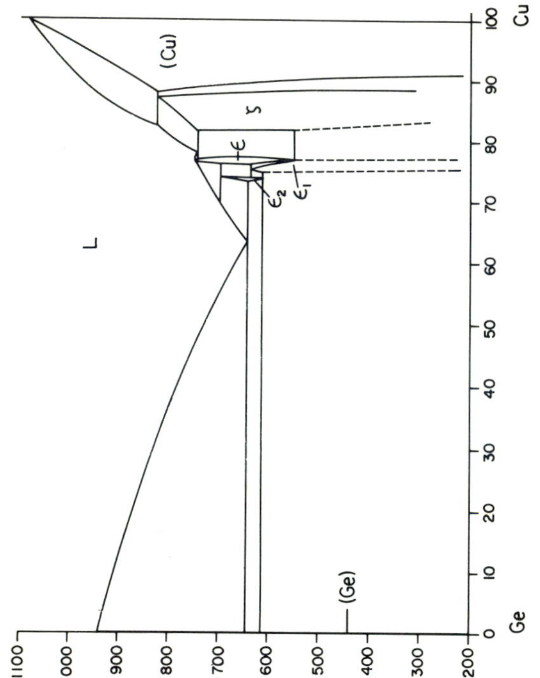
L
(Cu)
ζ
ε
ε2
ε1
(Ge)
Ge
Cu

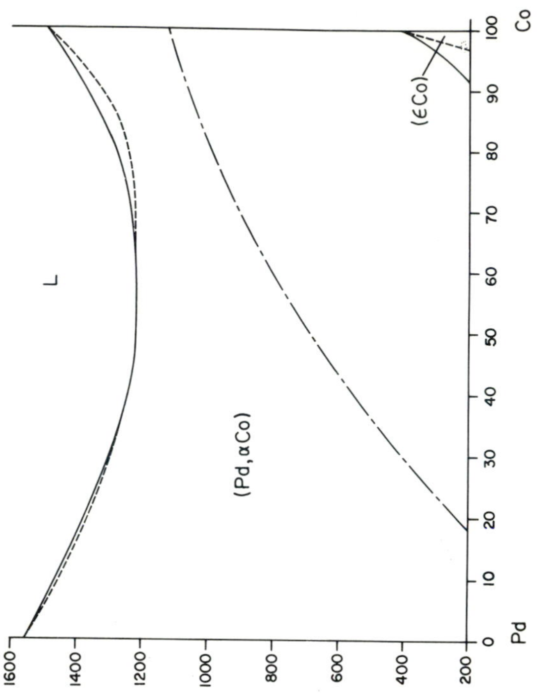
L
(Pd, αCo)
(εCo)
Pd
Co

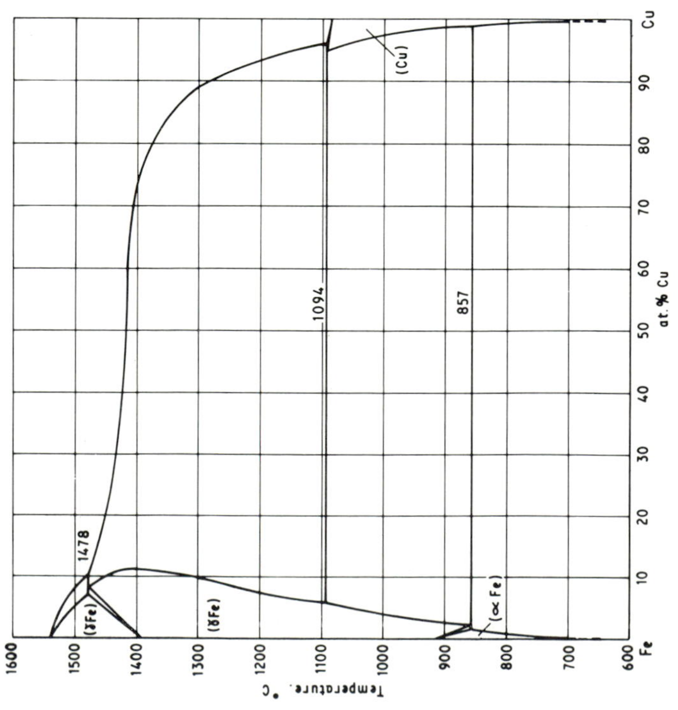
Temperature, °C
at.% Cu
1478
1094
857
(γFe)
(δFe)
(αFe)
(Cu)
Fe
Cu

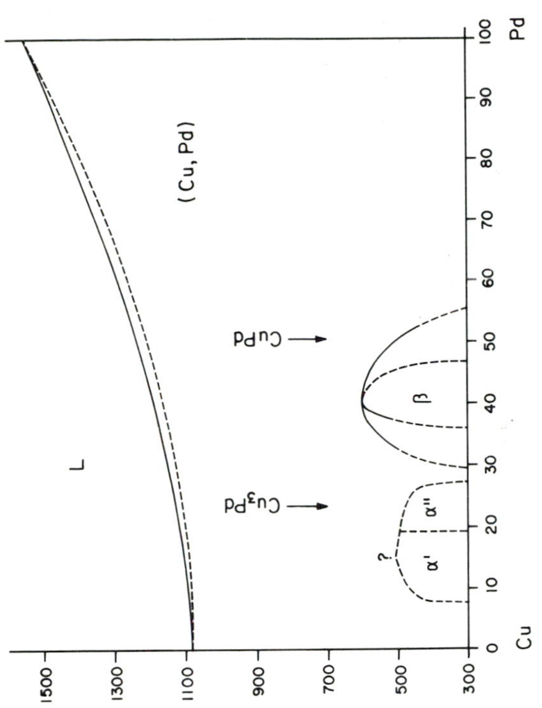
L
(Cu, Pd)
Cu_3Pd
CuPd
α'
α''
?
β
Cu
Pd

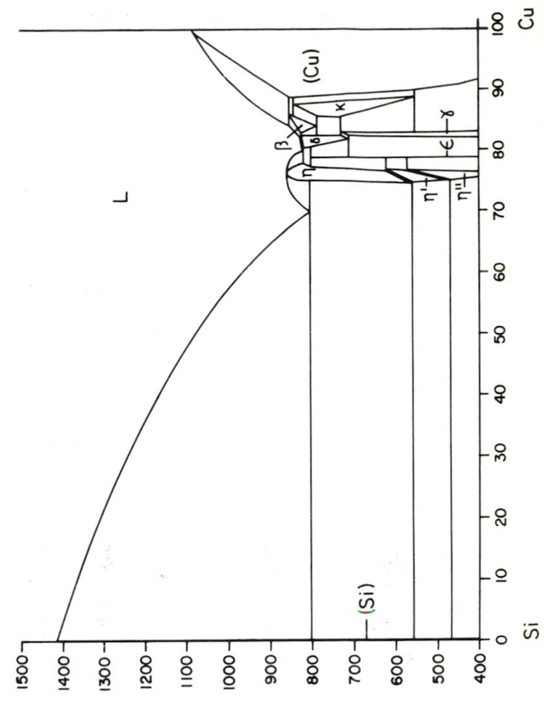
L
(Cu)
(Si)
β
δ
κ
η
η'
η''
ε
γ
Si
Cu

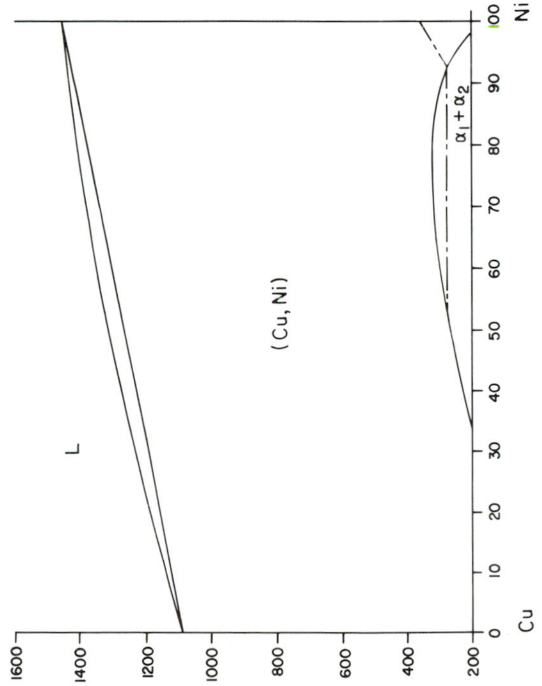
L
(Cu, Ni)
$α_1 + α_2$
Cu
Ni

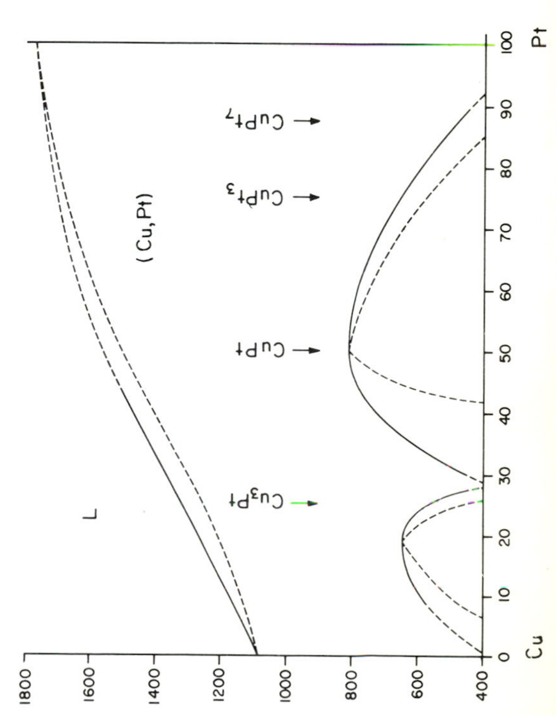
L
(Cu, Pt)
Cu_3Pt
CuPt
$CuPt_3$
$CuPt_7$
Cu
Pt

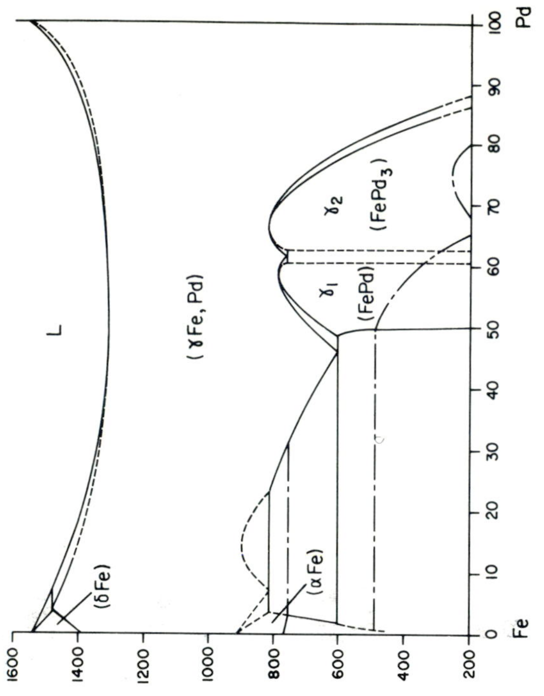
L
γ
δ
α
γ1
γ2
γ'
Fe
Ni

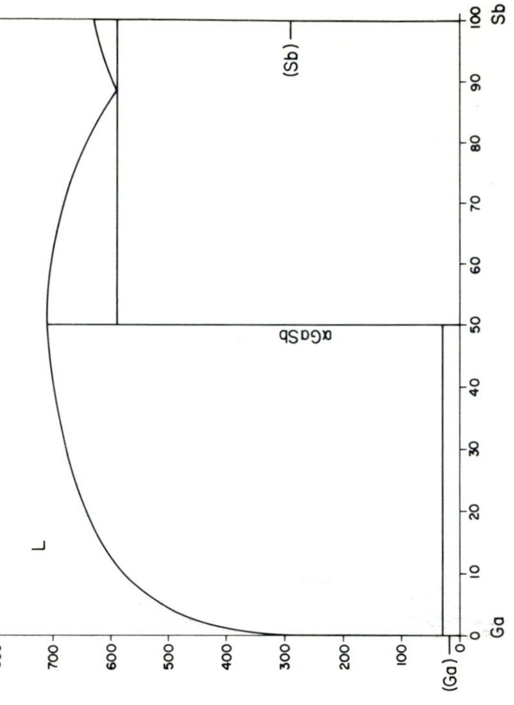
L
(γFe, Pd)
(δFe)
(αFe)
γ1
(FePd)
γ2
(FePd3)
Fe
Pd

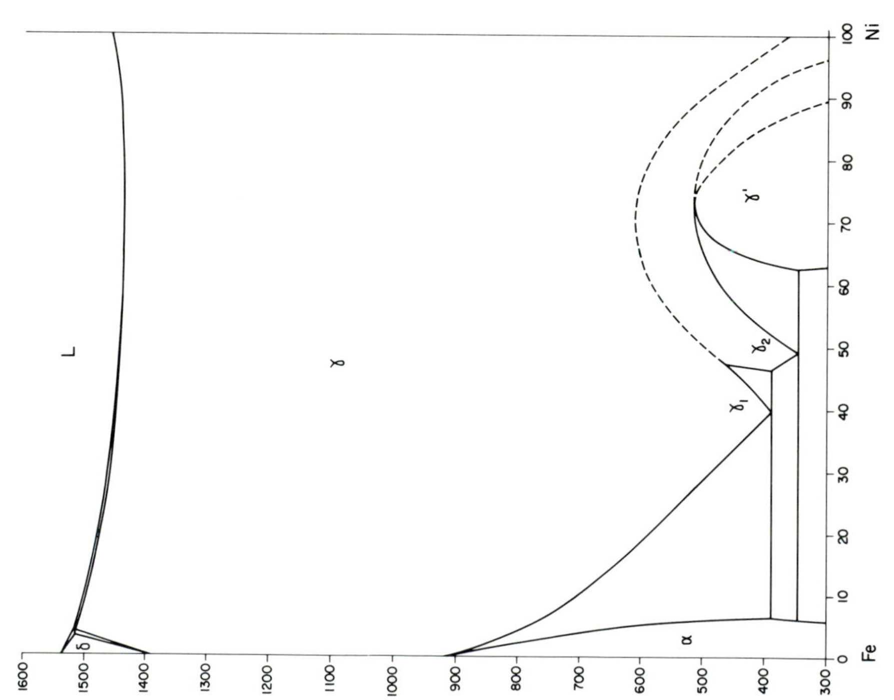
L
αGaSb
(Sb)
(Ga)
Ga
Sb

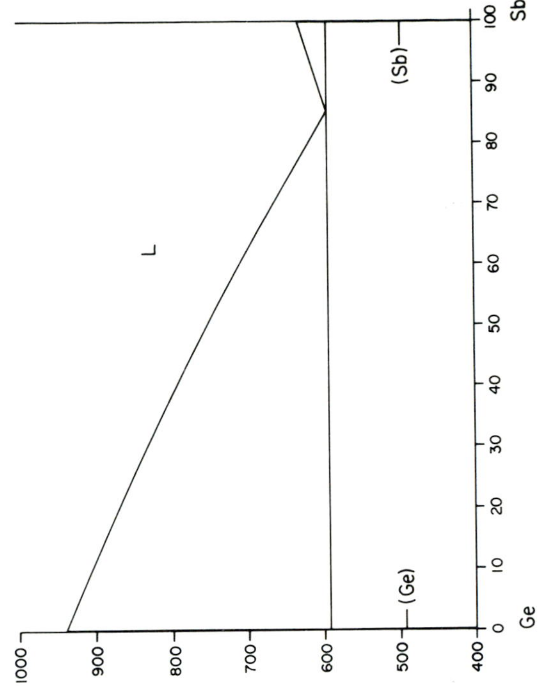
L
(Sb)
(Ge)
Ge
Sb
0
10
20
30
40
50
60
70
80
90
100
400
500
600
700
800
900
1000

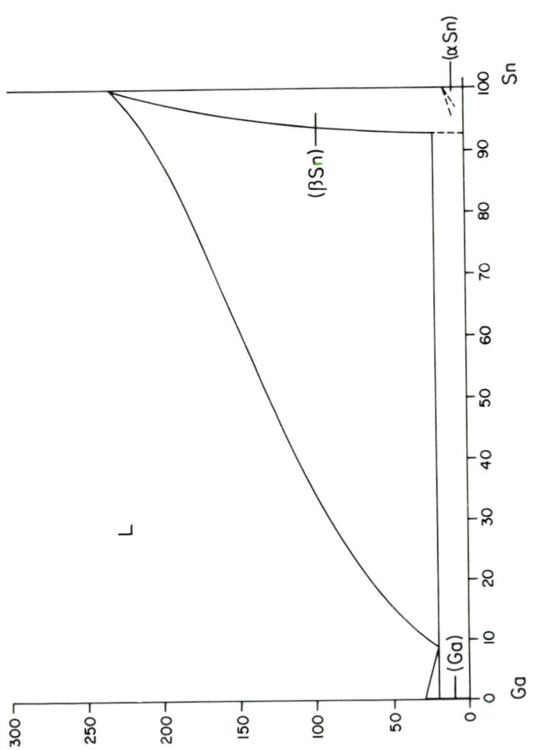
L
(βSn)
(αSn)
(Ga)
Ga
Sn
0
10
20
30
40
50
60
70
80
90
100
0
50
100
150
200
250
300

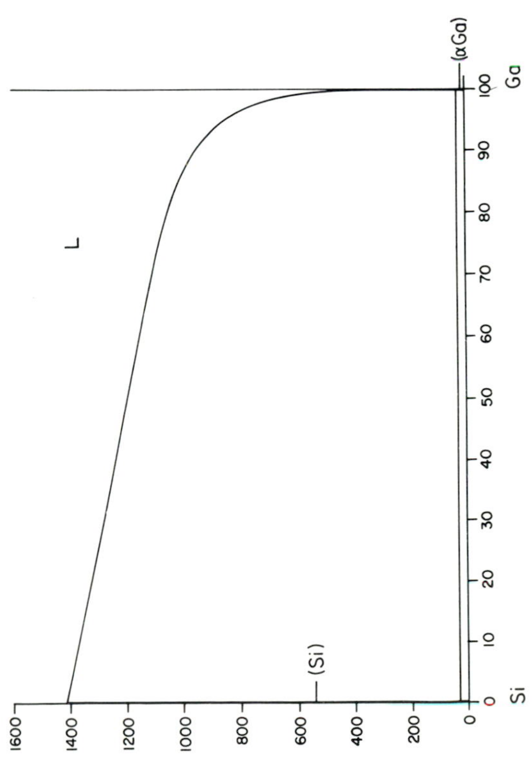
L
(Si)
(αGa)
Si
Ga
0
10
20
30
40
50
60
70
80
90
100
0
200
400
600
800
1000
1200
1400
1600

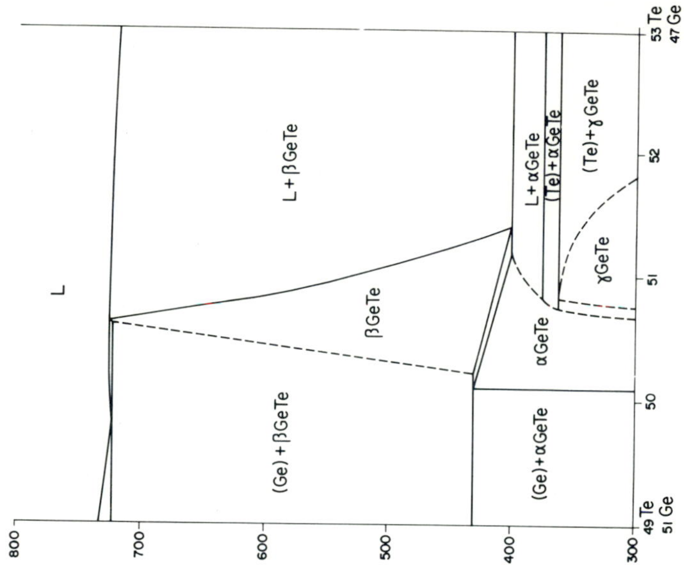
L
L + βGeTe
βGeTe
(Ge) + βGeTe
(Ge) + αGeTe
αGeTe
L + αGeTe
(Te) + αGeTe
γGeTe
(Te) + γGeTe
800
700
600
500
400
300
49 Te
51 Ge
50
51
52
53 Te
47 Ge

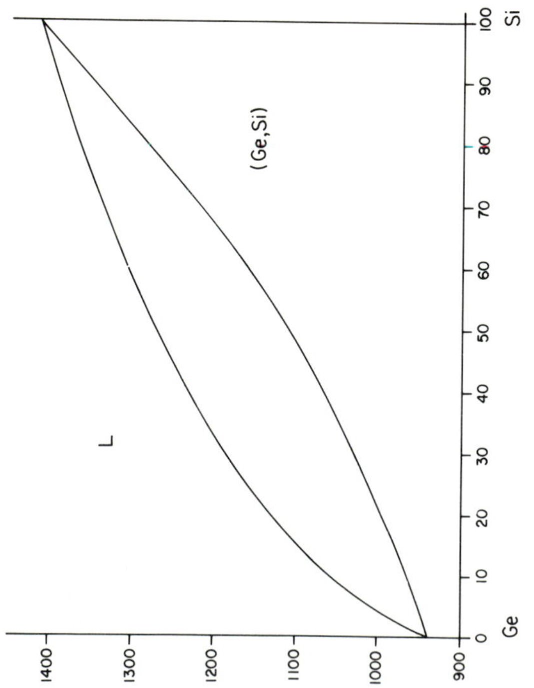
L
(Ge,Si)
1400
1300
1200
1100
1000
900
0
10
20
30
40
50
60
70
80
90
100
Ge
Si

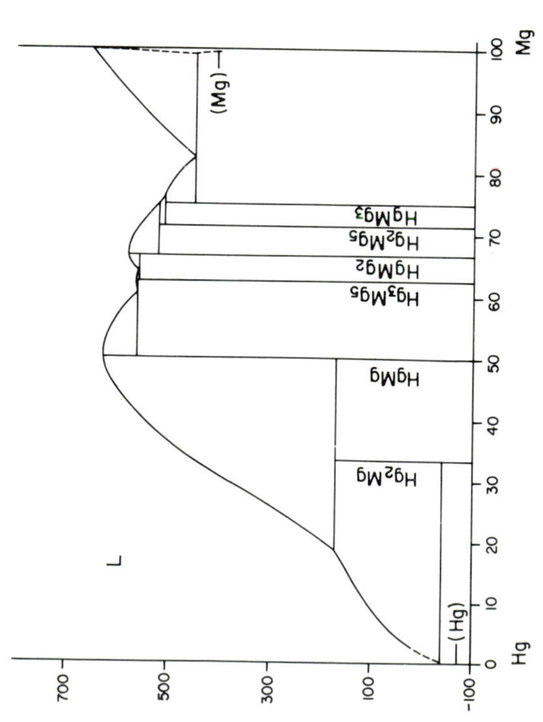
L
(Mg)
(Hg)
HgMg
Hg_2Mg
Hg_3Mg_5
$HgMg_2$
Hg_2Mg_5
$HgMg_3$
700
500
300
100
-100
0
10
20
30
40
50
60
70
80
90
100
Hg
Mg

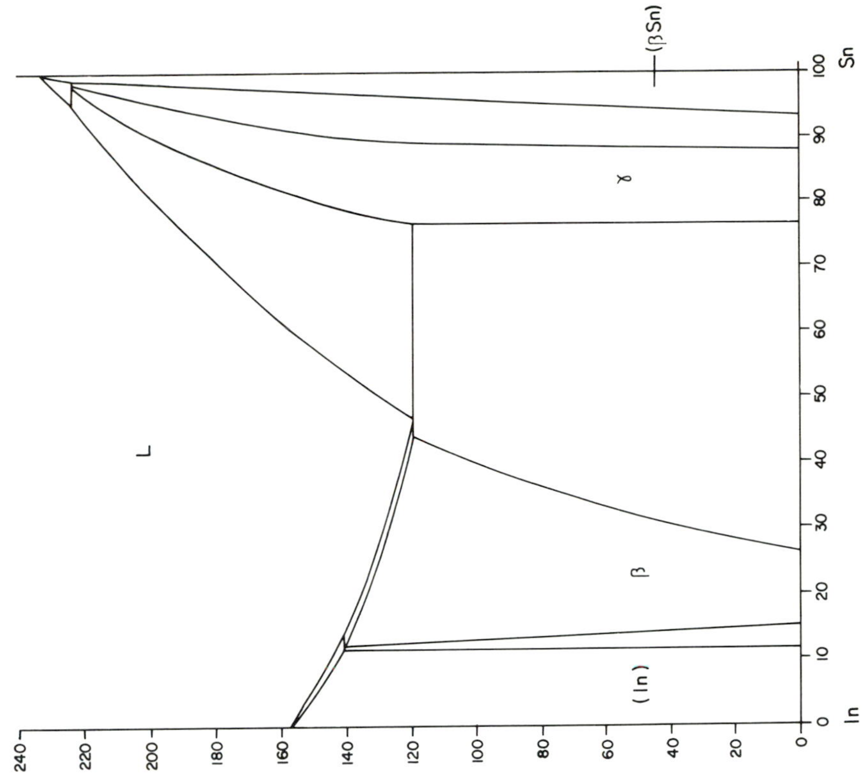

L
β
γ
(In)
(βSn)
In
Sn
0 10 20 30 40 50 60 70 80 90 100
0 20 40 60 80 100 120 140 160 180 200 220 240

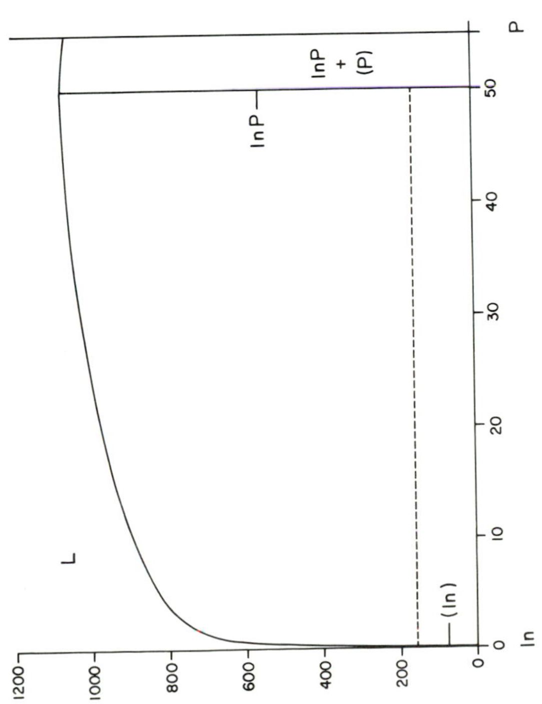

L
InP
InP + (P)
(In)
In
P
0 10 20 30 40 50
0 200 400 600 800 1000 1200

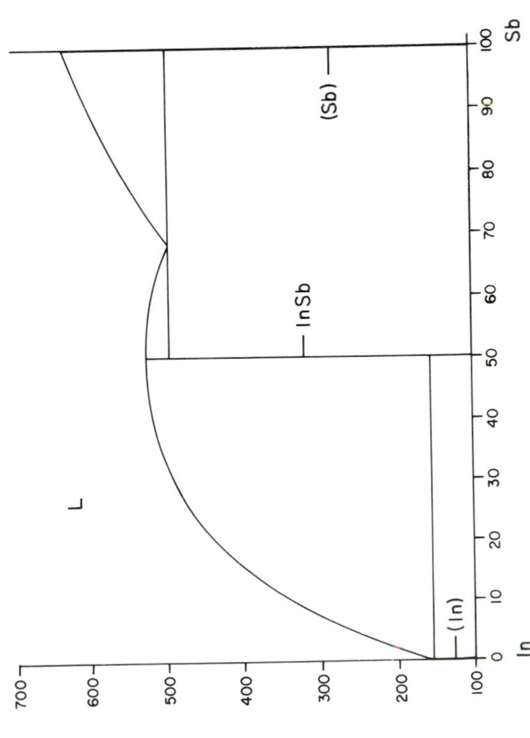

L
InSb
(Sb)
(In)
In
Sb
0 10 20 30 40 50 60 70 80 90 100
100 200 300 400 500 600 700

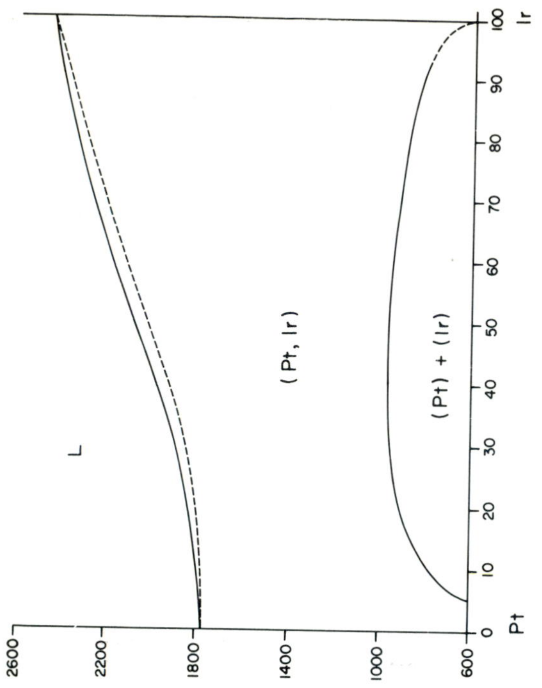
L
(Pd, Ir)
(Pd) + (Ir)
1800
1600
1400
1200
1000
800
600
0
10
20
30
40
50
60
70
80
90
100
Pd
Ir

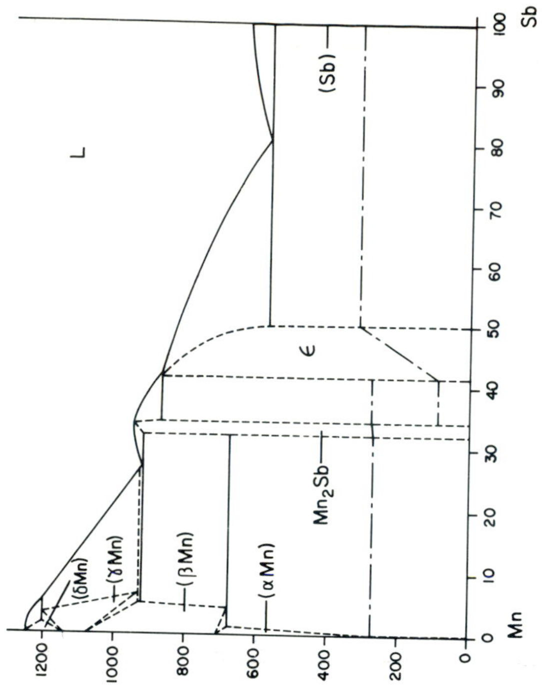
L
(Pt, Ir)
(Pt) + (Ir)
2600
2200
1800
1400
1000
600
0
10
20
30
40
50
60
70
80
90
100
Pt
Ir

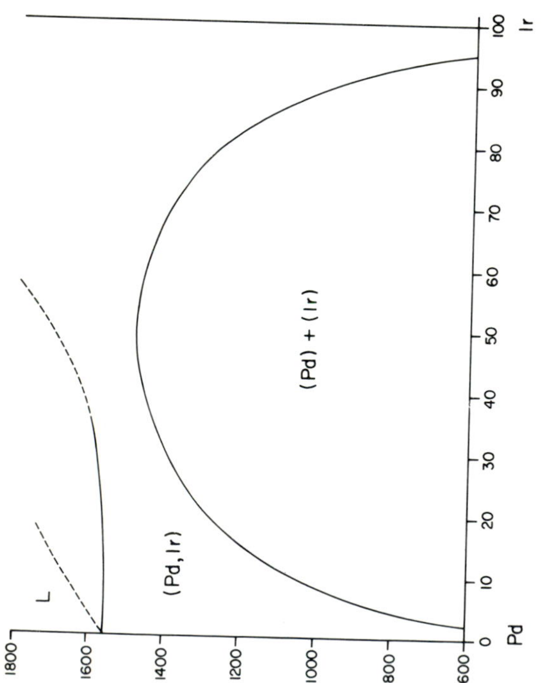
L
Mg2Si
(Mg)
(Si)
1400
1300
1200
1100
1000
900
800
700
600
0
10
20
30
40
50
60
70
80
90
100
Mg
Si

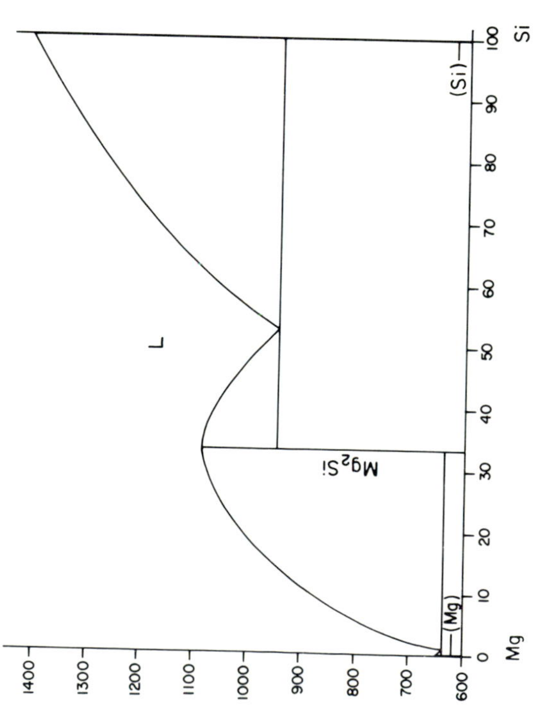
L
(δMn)
(γMn)
(βMn)
(αMn)
Mn2Sb
ε
(Sb)
1200
1000
800
600
400
200
0
0
10
20
30
40
50
60
70
80
90
100
Mn
Sb

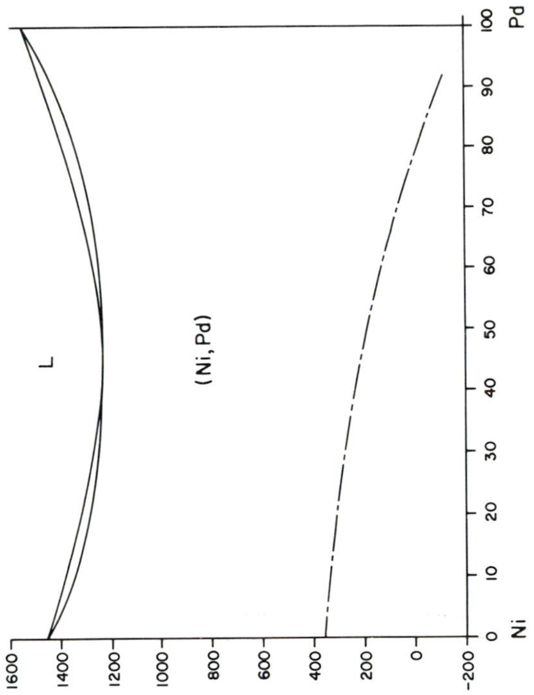
L
(Ni, Pd)
1600
1400
1200
1000
800
600
400
200
0
-200
0 10 20 30 40 50 60 70 80 90 100
Ni
Pd

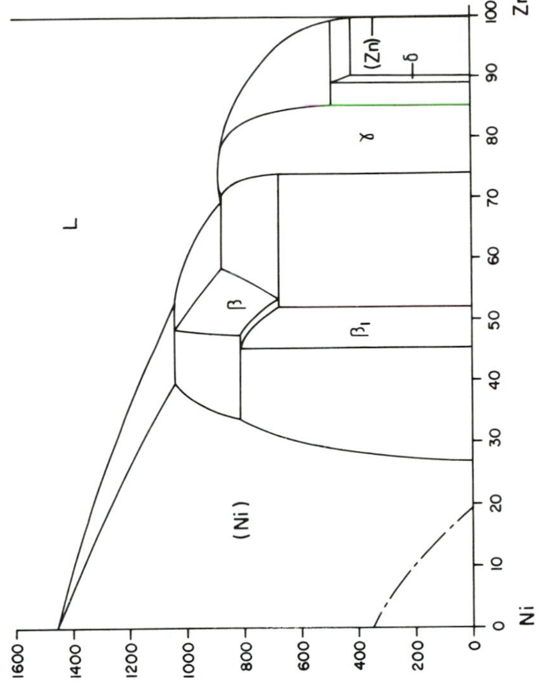
L
(Ni)
β
$β_1$
γ
(Zn)
δ
1600
1400
1200
1000
800
600
400
200
0
0 10 20 30 40 50 60 70 80 90 100
Ni
Zn

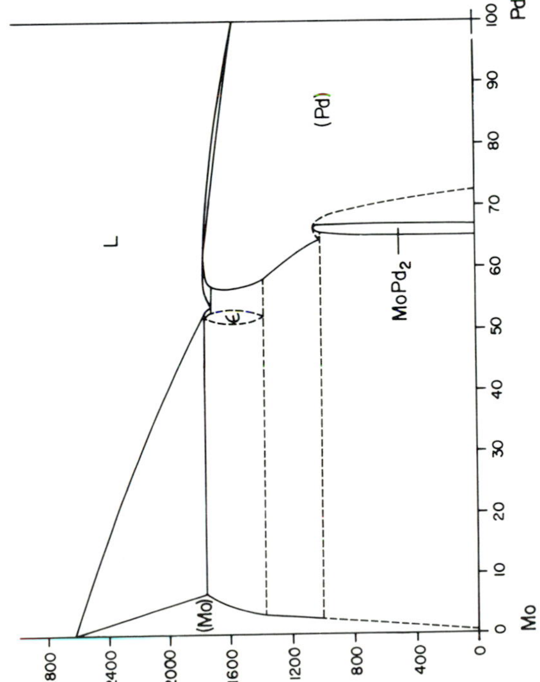
L
(Mo)
ε
$MoPd_2$
(Pd)
2800
2400
2000
1600
1200
800
400
0
0 10 20 30 40 50 60 70 80 90 100
Mo
Pd

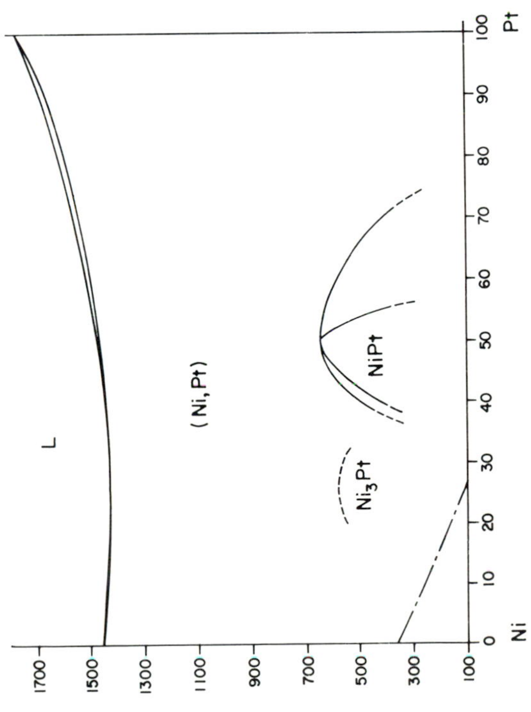
L
(Ni, Pt)
Ni_3Pt
NiPt
1700
1500
1300
1100
900
700
500
300
100
0 10 20 30 40 50 60 70 80 90 100
Ni
Pt

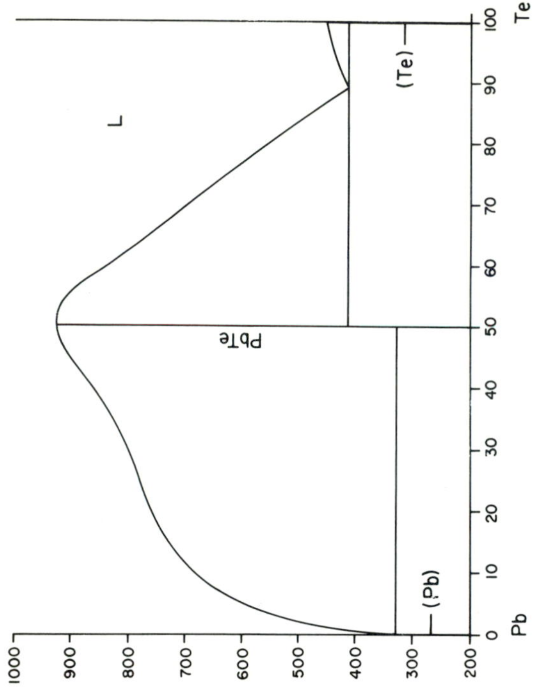
1000
900
800
700
600
500
400
300
200
0 10 20 30 40 50 60 70 80 90 100
Pb
Te
L
PbTe
(Pb)
(Te)

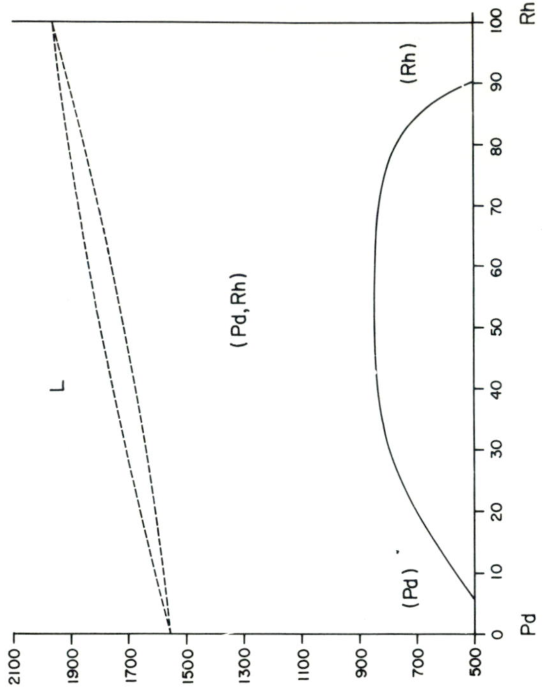
2100
1900
1700
1500
1300
1100
900
700
500
0 10 20 30 40 50 60 70 80 90 100
Pd
Rh
L
(Pd,Rh)
(Pd)
(Rh)

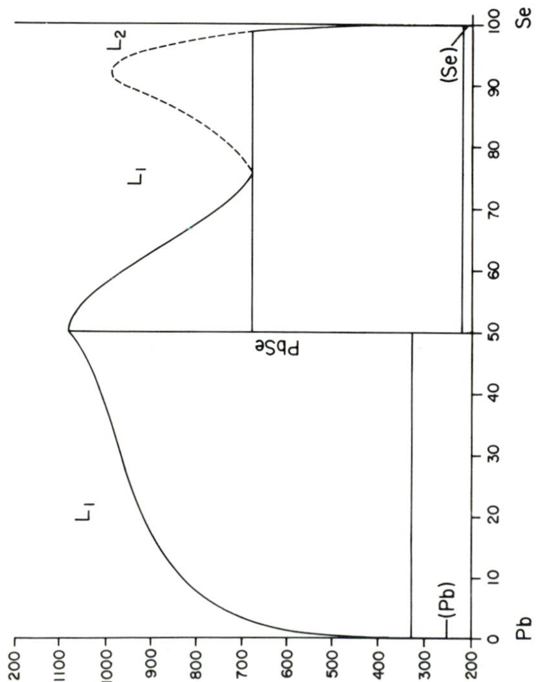
1200
1100
1000
900
800
700
600
500
400
300
200
0 10 20 30 40 50 60 70 80 90 100
Pb
Se
L_1
L_2
PbSe
(Pb)
(Se)

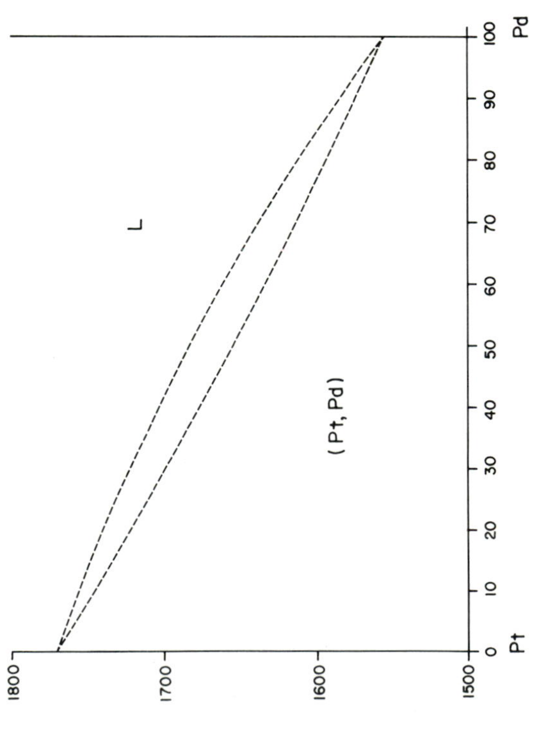
1800
1700
1600
1500
0 10 20 30 40 50 60 70 80 90 100
Pt
Pd
L
(Pt, Pd)

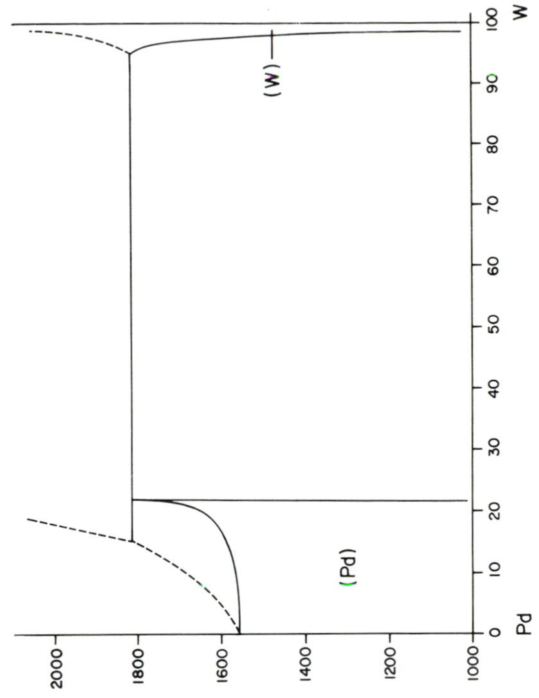
L
(Ru)
(Pd)
Pd
Ru
2400
2200
2000
1800
1600
1400
1200
1000
800
600
0
10
20
30
40
50
60
70
80
90
100

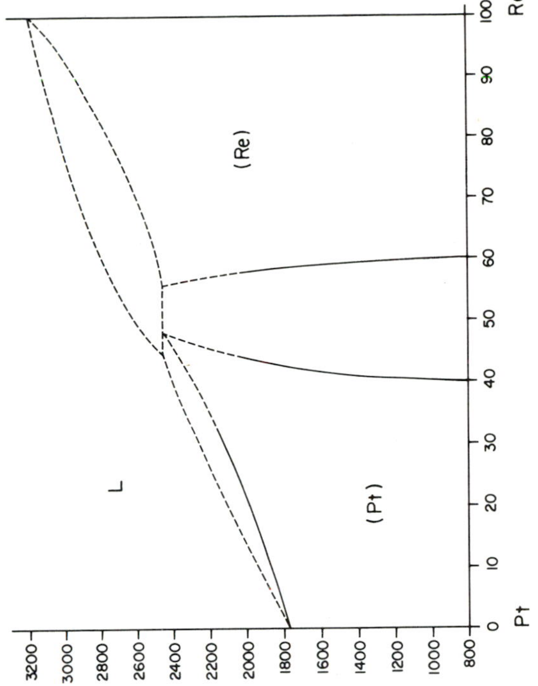
(W)
(Pd)
Pd
W
2000
1800
1600
1400
1200
1000
0
10
20
30
40
50
60
70
80
90
100

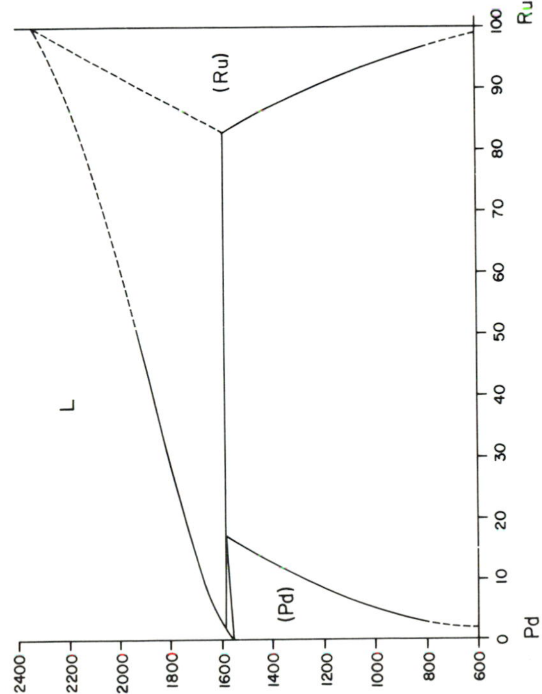
L
(Pd)
β''
δ
β_1
β'
γ'
γ
(Zn)
Pd
Zn
1500
1300
1100
900
700
500
300
0
10
20
30
40
50
60
70
80
90
100

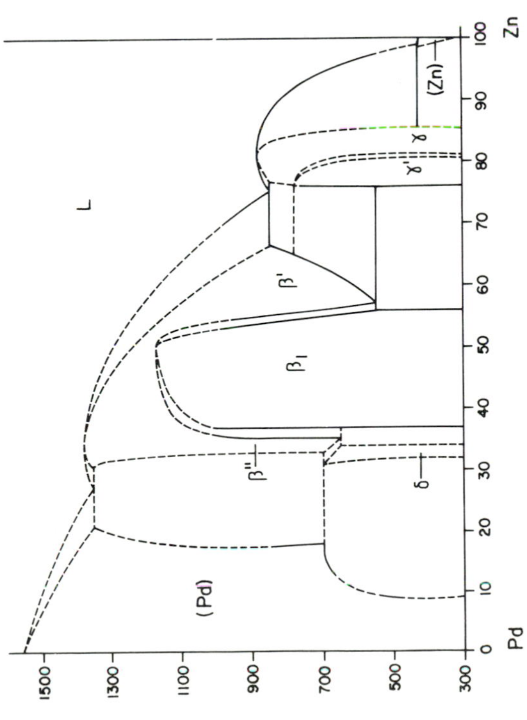
L
(Pt)
(Re)
Pt
Re
3200
3000
2800
2600
2400
2200
2000
1800
1600
1400
1200
1000
800
0
10
20
30
40
50
60
70
80
90
100

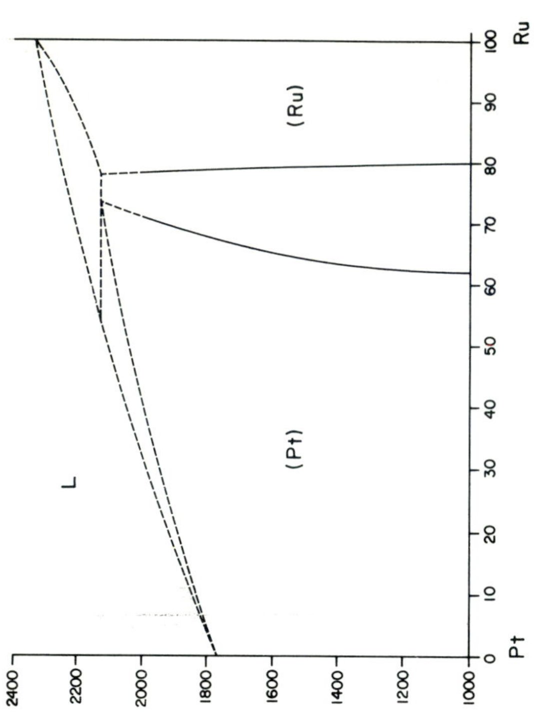
2100
2000
1900
1800
1700
0 10 20 30 40 50 60 70 80 90 100
Pt
Rh
L
(Pt, Rh)

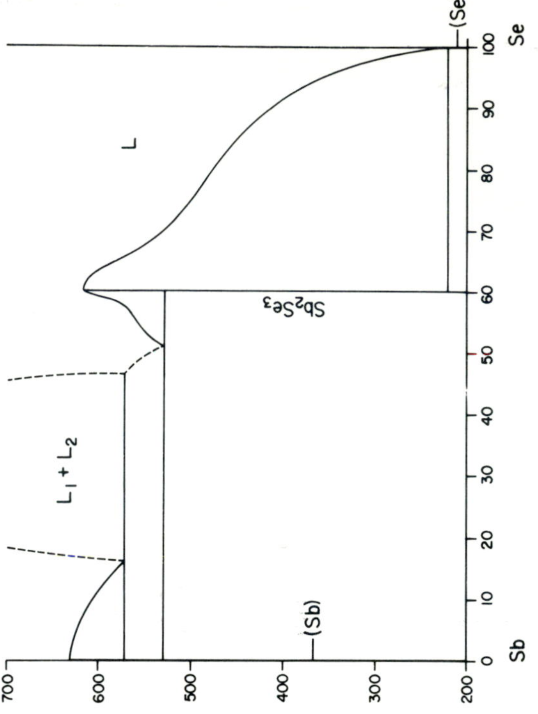
2400
2200
2000
1800
1600
1400
1200
1000
0 10 20 30 40 50 60 70 80 90 100
Pt
Ru
L
(Pt)
(Ru)

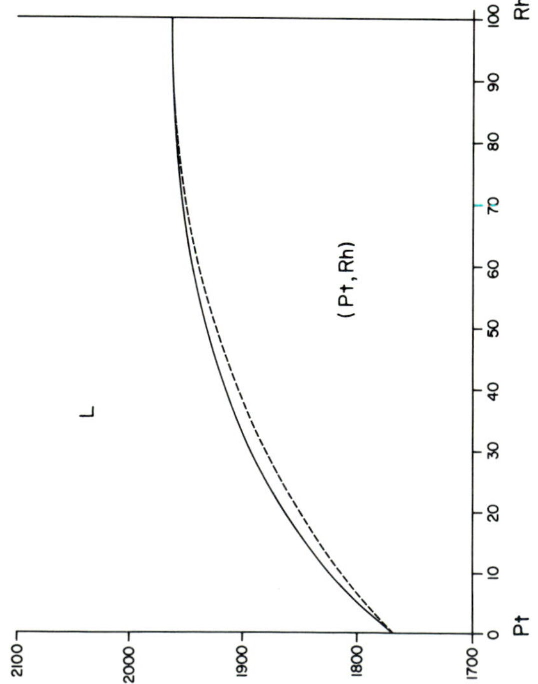
3600
3200
2800
2400
2000
1600
0 10 20 30 40 50 60 70 80 90 100
Pt
W
L
(Pt)
(W)

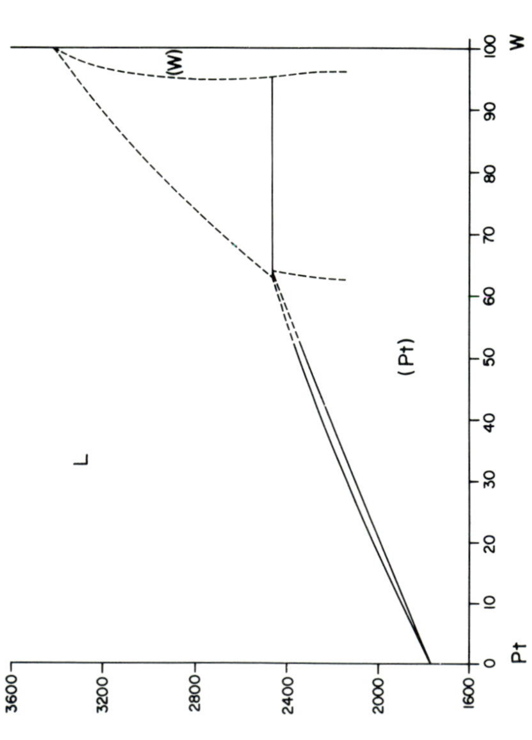
700
600
500
400
300
200
0 10 20 30 40 50 60 70 80 90 100
Sb
Se
L
L1 + L2
Sb2Se3
(Sb)
(Se)

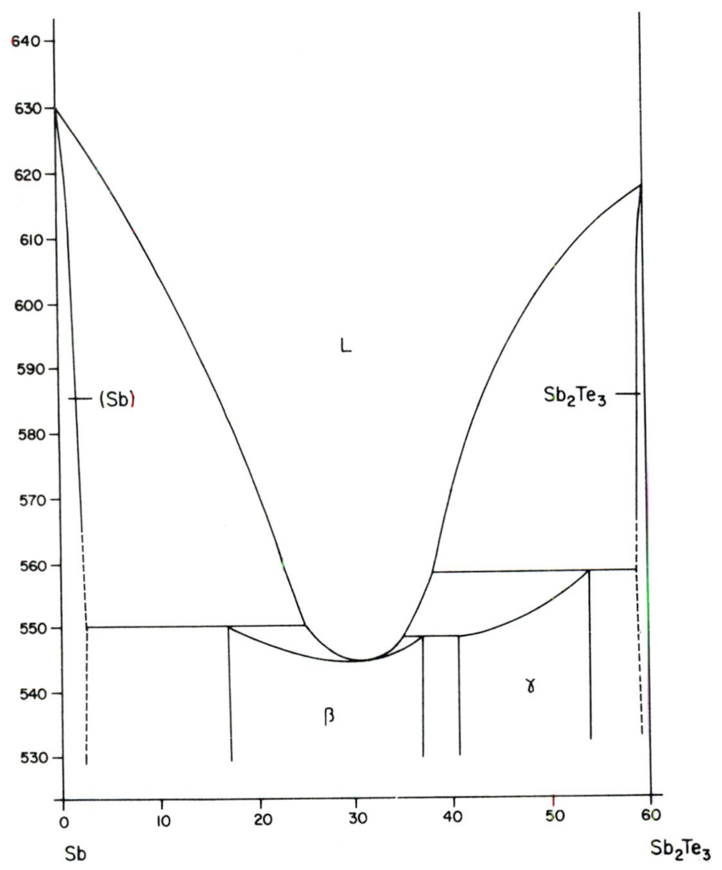

640
630
620
610
600
590
580
570
560
550
540
530
L
(Sb)
Sb_2Te_3
β
γ
0
10
20
30
40
50
60
Sb
Sb_2Te_3

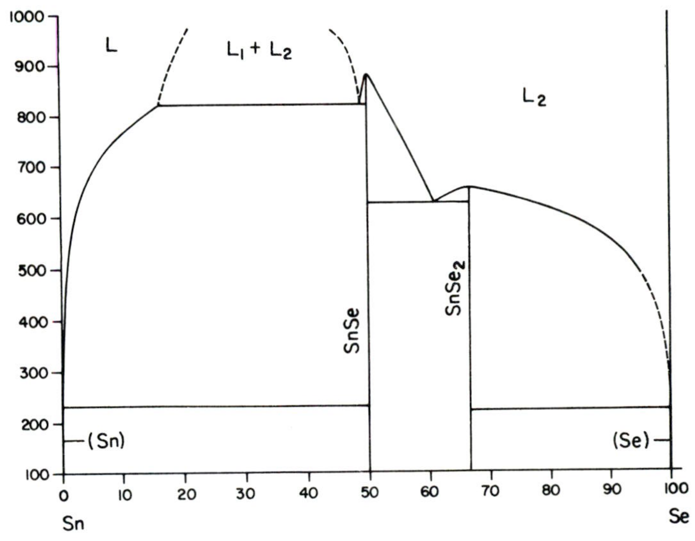

1000
900
800
700
600
500
400
300
200
100
L
$L_1 + L_2$
L_2
SnSe
$SnSe_2$
(Sn)
(Se)
0
10
20
30
40
50
60
70
80
90
100
Sn
Se

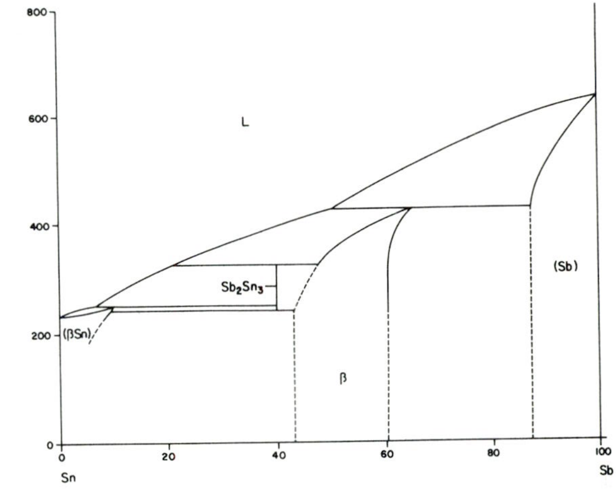

800
600
400
200
0
L
Sb_2Sn_3
(Sb)
(βSn)
β
0
20
40
60
80
100
Sn
Sb

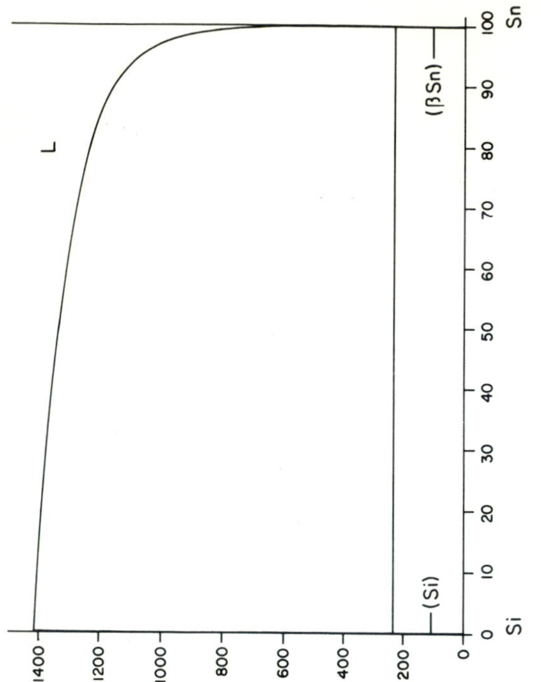

L
(Si)
(βSn)
Si
Sn
0
200
400
600
800
1000
1200
1400
0
10
20
30
40
50
60
70
80
90
100

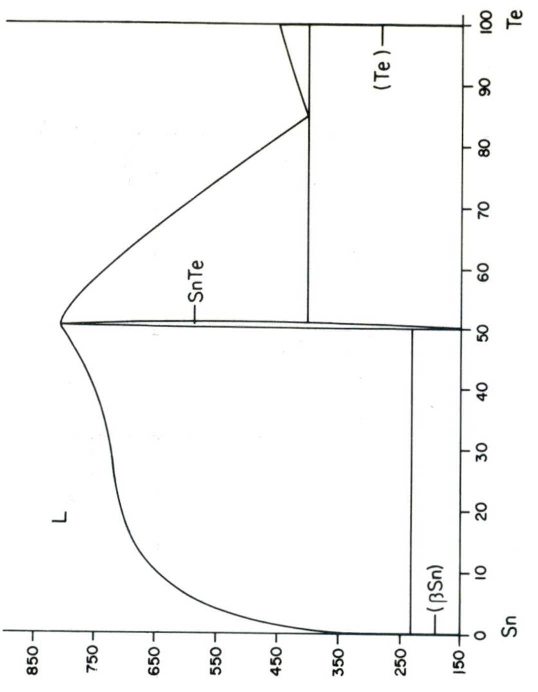

L
SnTe
(βSn)
(Te)
Sn
Te
150
250
350
450
550
650
750
850
0
10
20
30
40
50
60
70
80
90
100

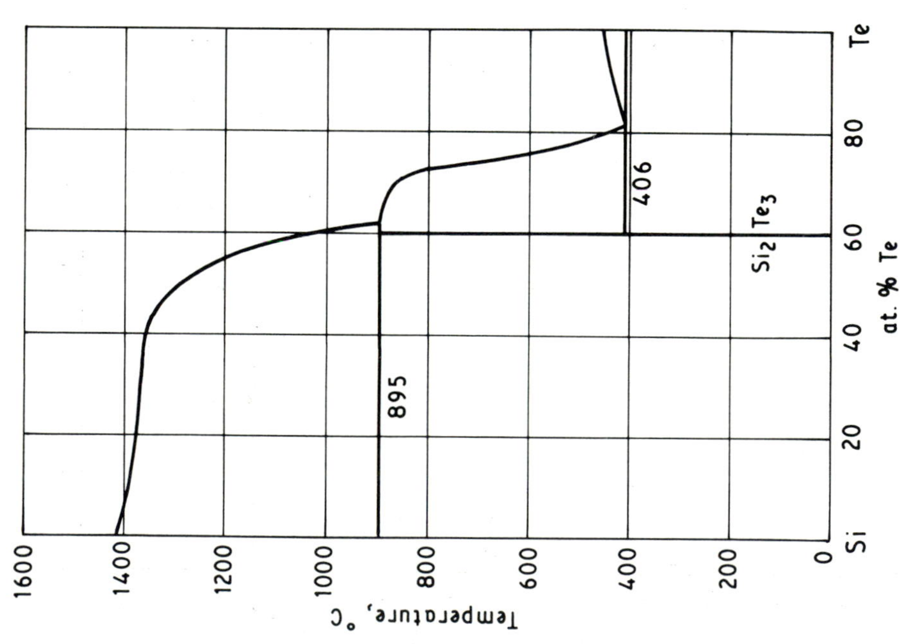

Temperature, °C
at. % Te
Si
Te
895
406
Si_2Te_3
0
200
400
600
800
1000
1200
1400
1600
20
40
60
80

Appendix I

Pure Component Transition Data

(compiled by A. T. Dinsdale, National Physical Laboratory, UK, 4 November 1988)

Element	*Transformation*	*Temperature, °C*	*Temperature, K*
Ag	L ⇌ S	961·93	1 235·08
Al	L ⇌ S	660·4	933·6
As	L ⇌ S	817	1 090
Au	L ⇌ S	1 064·43	1 337·58
B	L ⇌ S	2 075	2 348
Ba	L ⇌ S	727	1 000
Be	L ⇌ β	1 287	1 560
	β ⇌ α	1 254	1 527
Bi	L ⇌ S	271·37	544·52
C	L ⇌ S	4 492·1	4 765·3
Ca	L ⇌ β	842	1 115
	β ⇌ α	443	716
Cd	L ⇌ S	321·11	594·26
Ce	L ⇌ δ	798	1 071
	δ ⇌ γ	726	999
Co	L ⇌ α	1 495	1 768
	α ⇌ ε	421·84	694·99
Cr	L ⇌ S	1 906·4	2 179·6
Cs	L ⇌ S	28·44	301·59
Cu	L ⇌ S	1 084·87	1 358·02
Dy	L ⇌ β	1 411	1 684
	β ⇌ α	1 386	1 659
Er	L ⇌ S	1 529	1 802
Eu	L ⇌ S	822	1 095
Fe	L ⇌ δ	1 538	1 811
	δ ⇌ γ	1 394·3	1 667·5
	γ ⇌ α	911·6	1 184·8
Ga	L ⇌ S	29·77	302·92
Gd	L ⇌ β	1 314	1 587
	β ⇌ α	1 262	1 535
Ge	L ⇌ S	938·3	1 211·5
Hf	L ⇌ β	2 231	2 504
	β ⇌ α	1 743	2 016
Hg	L ⇌ S	−38·84	234·31
Ho	L ⇌ β	1 472	1 745
	β ⇌ α	1 430	1 703
In	L ⇌ S	156·63	429.78
Ir	L ⇌ S	2 447	2 720
K	L ⇌ S	63·71	336·86
La	L ⇌ γ	920	1 193
	γ ⇌ β	861	1 134
	β ⇌ α	277	550
Li	L ⇌ S	180·54	453·69
Lu	L ⇌ S	1 663	1 936
Mg	L ⇌ S	650	923
Mn	L ⇌ δ	1 243·85	1 517·01
	δ ⇌ γ	1 137·29	1 410·44
	γ ⇌ β	1 086·93	1 360·08
	β ⇌ α	707·07	980·22
Mo	L ⇌ S	2 623	2 896
Na	L ⇌ S	97·85	371·00
Nb	L ⇌ S	2 477	2 750
Nd	L ⇌ β	1 016	1 289
	β ⇌ α	855	1 128
Ni	L ⇌ S	1 455	1 728
Os	L ⇌ S	3 033	3 306
P	L ⇌ α	44·1	317·3
Pb	L ⇌ S	327·50	600·65
Pd	L ⇌ S	1 555·3	1 828·5
Pr	L ⇌ β	931	1 204
	β ⇌ α	795	1 068
Pt	L ⇌ S	1 768·9	2 042·1
Pu	L ⇌ ε	640	913
	ε ⇌ δ′	482·5	755·7
	δ′ ⇌ δ	463	736
	δ ⇌ γ	319·9	593·1
	γ ⇌ β	214·7	487·9
	β ⇌ α	124·4	397·6
Rb	L ⇌ S	39·32	312·47
Re	L ⇌ S	3 186	3 459
Rh	L ⇌ S	1 964	2 237
Ru	L ⇌ S	2 334	2 607
S	L ⇌ γ	115·21	388·36
	γ ⇌ β	101	374
	β ⇌ α	95·39	368·54
Sb	L ⇌ S	630·75	903·90
Sc	L ⇌ β	1 541	1 814
	β ⇌ α	1 335	1 608
Se	L ⇌ S	221	494
Si	L ⇌ S	1 414	1 687
Sm	L ⇌ β	1 072	1 345
	β ⇌ α	917	1 190
Sn	L ⇌ S	231·97	505·12
Sr	L ⇌ β	768	1 041
	β ⇌ α	555	828
Ta	L ⇌ S	2 985	3 258
Tb	L ⇌ β	1 359	1 632
	β ⇌ α	1 289	1 562
Te	L ⇌ S	449·57	722·72
Th	L ⇌ β	1 750	2 023
	β ⇌ α	1 360	1 633
Ti	L ⇌ β	1 670	1 943
	β ⇌ α	882	1 155
Tl	L ⇌ β	304	577
	β ⇌ α	234	507
Tm	L ⇌ S	1 545	1 818
U	L ⇌ γ	1 135	1 408
	γ ⇌ β	776	1 049
	β ⇌ α	669	942
V	L ⇌ S	1 910	2 183
W	L ⇌ S	3 421·7	3 694·9
Y	L ⇌ β	1 526	1 799
	β ⇌ α	1 479	1 752
Yb	L ⇌ β	824	1 097
	β ⇌ α	760	1 033
Zn	L ⇌ S	419·58	692·73
Zr	L ⇌ β	1 854·71	2 127·86
	β ⇌ α	866·30	1 139·45

Appendix II

Atomic Weights of the Elements

Ag	107·868 2	Nb	92·906 4
Al	26·981 54	Nd	144·24
As	74·922	Ni	58·69
Au	196·966 5	O	15·999 4
B	10·81	Os	190·2
Ba	137·33	P	30·973 76
Be	9·012 18	Pb	207·2
Bi	208·980 4	Pd	106·42
C	12·011	Pr	140·907 7
Ca	40·08	Pt	195·08
Cd	112·41	Pu	244
Ce	140·12	Rb	85·467 8
Co	58·933 2	Re	186·207
Cr	51·996	Rh	102·905 5
Cs	132·905 4	Ru	101·07
Cu	63·546	S	32·06
Dy	162·50	Sb	121·75
Er	167·26	Sc	44·955 9
Eu	151·96	Se	78·96
Fe	55·847	Si	28·085 5
Ga	69·72	Sm	150·36
Gd	157·25	Sn	118·69
Ge	72·59	Sr	87·62
H	1·007 94	Ta	180·947 9
Hf	178·49	Tb	158·925 4
Hg	200·59	Te	127·60
Ho	164·930 4	Th	232·038 1
In	114·82	Ti	47·88
Ir	192·22	Tl	204·383
K	39·098 3	Tm	168·934 2
La	138·905 5	U	238·028 9
Li	6·941	V	50·941 5
Lu	174·967	W	183·85
Mg	24·305	Y	88·905 9
Mn	54·938 0	Yb	173·04
Mo	95·94	Zn	65·38
N	14·006 7	Zr	91·22
Na	22·989 77		

Appendix III

Conversion of atomic percent to weight percent (mass percent) and vice versa

Alloy compositions in this monograph are presented in terms of atomic percentages. These compositions can be converted into weight (mass) percentages by using the atomic weights of the elements given in Appendix II.

Let a ternary alloy contain (x) atomic % of component A, (y) atomic % of component B and (z) atomic % of component C.

The atomic weights of components A, B and C are denoted by (at.wt.A), (at.wt.B) and (at.wt.C) respectively.

The weight (mass) of component A in the ternary alloy A–B–C = (x) (at.wt.A)

Similarly the weight of component B in the ternary alloy A–B–C = (y) (at.wt.B)

and the weight of component C in the ternary alloy A–B–C = (z) (at.wt.C)

The weight % of component A in the ternary alloy A–B–C

$$= \frac{(x)\,(\text{at.wt.A})\,(100)}{(x)\,(\text{at.wt.A}) + (y)\,(\text{at.wt.B}) + (z)\,(\text{at.wt.C})}$$

The weight % of component B in the ternary alloy A–B–C

$$= \frac{(y)\,(\text{at.wt.B})\,(100)}{(x)\,(\text{at.wt.A}) + (y)\,(\text{at.wt.B}) + (z)\,(\text{at.wt.C})}$$

The weight % of component C in the ternary alloy A–B–C

$$= \frac{(z)\,(\text{at.wt.C})\,(100)}{(x)\,(\text{at.wt.A}) + (y)\,(\text{at.wt.B}) + (z)\,(\text{at.wt.C})}$$

Example

Conversion of the composition of a Ag–Au–Cu alloy from atomic % to weight % (mass %). The alloy contains 70 atomic % Ag, 20 atomic % Au, 10 atomic % Cu. The atomic weights of Ag, Au and Cu are taken as 107·87, 196·97 and 63·55 respectively.

The weight (mass) of Ag in the Ag–Au–Cu alloy $= (70)(107{\cdot}87)$
The weight of Au $= (20)(196{\cdot}97)$
The weight of Cu $= (10)(63{\cdot}55)$

The weight % (mass %) of Ag in the Ag–Au–Cu alloy

$$= \frac{(70)(107{\cdot}87)(100)}{(70)(107{\cdot}87) + (20)(196{\cdot}97) + (10)(63{\cdot}55)}$$

$$= \frac{755\,090}{12\,125{\cdot}8}$$

$= 62{\cdot}27$ weight % Ag.

The weight % of Au in the Ag–Au–Cu alloy

$$= \frac{(20)(196{\cdot}97)(100)}{12\,125{\cdot}8}$$

$= 32{\cdot}49$ weight % Au.

The weight % of Cu in the Ag–Au–Cu alloy

$$= \frac{(10)(63{\cdot}55)(100)}{12\,125{\cdot}8}$$

$= 5{\cdot}24$ weight % Cu.

The alloy contains 62·27 weight % Ag, 32·49 weight % Au, 5·24 weight % Cu.

Conversion of weight percent (mass percent) to atomic percent

Let a ternary alloy contain (a) weight % of component A, (b) weight % of component B and (c) weight % of component C. The atomic weights of components A, B and C are denoted by (at.wt.A), (at.wt.B) and (at.wt.C) respectively.

The number of atoms of component A in the ternary alloy A–B–C

$$= \frac{(a)}{(at.wt.A)}$$

Similarly the number of atoms of components B and C in the ternary alloy A–B–C equals

$$\frac{(b)}{(at.wt.B)} \quad \text{and} \quad \frac{(c)}{(at.wt.C)} \quad \text{respectively.}$$

The number of atoms of component A in the ternary alloy A–B–C

$$= \frac{\dfrac{(100)\,(a)}{(at.wt.A)}}{\dfrac{(a)}{(at.wt.A)} + \dfrac{(b)}{(at.wt.B)} + \dfrac{(c)}{(at.wt.C)}}$$

Similar expressions apply for the atomic % of components B and C.

Example

Conversion of the composition of a 62·27 weight % Ag, 32·49 weight % Au, 5·24 weight % Cu alloy to atomic %.

The number of atoms of Ag $= \dfrac{62{\cdot}27}{107{\cdot}87} = 0{\cdot}577\,3$

The number of atoms of Au $= \dfrac{32{\cdot}49}{196{\cdot}97} = 0{\cdot}164\,9$

The number of atoms of Cu $= \dfrac{5{\cdot}24}{63{\cdot}55} = 0{\cdot}082\,5$

The atomic % of Ag $= \dfrac{(0{\cdot}577\,3)(100)}{0{\cdot}577\,3 + 0{\cdot}164\,9 + 0{\cdot}082\,5}$

$$= \frac{57{\cdot}73}{0{\cdot}824\,7}$$

$= 70{\cdot}0$

The atomic % of Au $= \dfrac{(0{\cdot}164\,9)(100)}{0{\cdot}824\,7} = 20{\cdot}0$

The atomic % of Cu $= \dfrac{(0{\cdot}082\,5)(100)}{0{\cdot}824\,7} = 10{\cdot}0$

The alloy contains 70 atomic % Ag, 20 atomic % Au, 10 atomic % Cu.

Appendix IV

Conversion of mole fractions to atomic percent

Vertical sections through ternary systems sometimes have a composition scale reported in mole fractions or mole percentages. If the section is constructed between a compound AB_2 and a compound AC and a eutectic composition is shown at 0·5 mole fraction of compound AC, conversion of the eutectic composition to atomic percentages of components A, B and C may be done as follows.

At the eutectic composition there are 0·5 mole of component A, 0·5 mole of component C (compound AC), 0·5 mole of component A and 0·5 (2) mole of component B (compound AB_2) = 2·5 mole in total.

The atomic fraction of A $= \dfrac{0{\cdot}5 + 0{\cdot}5}{2{\cdot}5} = 0{\cdot}4$

The atomic fraction of B $= \dfrac{1{\cdot}0}{2{\cdot}5} = 0{\cdot}4$

The atomic fraction of C $= \dfrac{0{\cdot}5}{2{\cdot}5} = 0{\cdot}2$

The eutectic composition at 0·5 mole fraction AC, 0·5 mole fraction AB_2 converts to a composition containing 40 atomic % A, 40 atomic % B, 20 atomic % C.